T0348731

FREE ELECTRON LASERS 2003

FREE ELECTRON LASERS 2003

Proceedings of the 25th International Free Electron Laser
Conference, and the 10th FEL Users Workshop,
Tsukuba, Ibaraki, Japan, September 8–12, 2003

Editors

E.J. Minehara
R. Hajima
M. Sawamura

Japan Atomic Energy Research Institute, Tokai, Ibaraki, Japan

2004

ELSEVIER

Amsterdam – Boston – Heidelberg – London – New York – Oxford
Paris – San Diego – San Francisco – Singapore – Sydney – Tokyo

ELSEVIER B.V.
Sara Burgerhartstraat 25
P.O. Box 211, 1000 AE Amsterdam
The Netherlands

ELSEVIER Inc.
525 B Street, Suite 1900
San Diego, CA 92101-4495
USA

ELSEVIER Ltd
The Boulevard, Langford Lane
Kidlington, Oxford OX5 1GB
UK

ELSEVIER Ltd
84 Theobalds Road
London WC1X 8RR
UK

First edition 2004

ISBN: 0-444-51727-8

Part I is reprinted from:
NUCLEAR INSTRUMENTS AND METHODS IN PHYSICS RESEARCH
Section A: Accelerators, Spectrometers, Detectors and Associated Equipment, Volume 528/1-2

Transferred to digital print 2007
Printed and bound by CPI Antony Rowe, Eastbourne

Preface

The 25th International Free-Electron Laser (FEL) Conference and 10th Free-Electron Laser Users Workshop was held on September 8–12th at Epochal Tsukuba, Tsukuba, Ibaraki, Japan. The Conference and the Workshop, which is abbreviated as FEL2003, was organized by FEL laboratory at Tokai, Advanced Photon Research Center, Kansai Research Establishment, Japan Atomic Energy Research Institute JAERI. A total of 221 participants from 16 different countries attended the conference, presented 51 talks and 204 posters, and submitted 180 papers, which are published in the present volume of the conference proceedings. These countries included Armenia, China, France, Germany, Israel, Italy, Japan, Korea, Mexico, Russia, Sweden, Taiwan, The Netherlands, Ukraine, United Kingdom, and United States of America.

The scope of the conference was set as usual to include FEL Prize talks, First lasings, High-Power Long Wavelength FELs, FEL Technologies, Storage Ring FELs, High Brightness Electron Beams, High-Gain Short Wavelength FELs, FEL Users Workshop, Joint Session for FEL developers and users, and New Concepts and Proposals.

Since the re-invention of the FEL of 1977, FELs have made remarkable expansion and progresses in average and peak power, and wavelength regions, and in many scientific and industrial applications. Currently, many experiments are focusing on high-gain FELs in the X-ray region as a means to achieve a next-generation light source, as well as high average power long-wavelength FELs for many scientific and industrial applications. The program of FEL2003 clearly shows us such a written milestone that FEL research is still very exciting and expanding. We thank the members of the International Program Committee and the Session Chairs for the scientific success of the conference.

The conference included an excursion to Nikko Mountainous area, World-Heritage of historical Nikko Shrines and Temples, a banquet and a beautiful performance of classical music by four young ladies of Flute quartet after the excursion. Following the tradition, the winner of the 2003 FEL prize was recognized and announced during the banquet. This year's winner was Dr. Li Hua Yu who from Brookhaven National Laboratory for his contributions to high gain SASE FELs and for proposing and demonstrating the concept of High Gain Harmonic Generation (HGHG).

It was unfortunate that about several scientists from China, Russia, and other countries were not able to receive visas or financial supports to enter Japan and to attend the conference due to the heightened security measures in place after Iraq War and SARS disease outbroke in the first and second quarters of the 2003.

On behalf of the International Executive Committee and the Local Organizing Committee, we would like to express our sincere appreciation and many thanks to all of the Industrial Exhibition Participants listed in the proceedings, and especially to the two sponsors of the 2003 FEL Prize: ACCEL Instruments GmbH, and Elsevier Science B.V., and the financial supporter for the FEL 2003: Tsuchiura Tsukuba Convention Bureau/The Science and Technology Promotion Foundation of Ibaraki Prefecture, Ibaraki, Japan.

Another distinguishing feature of the conference was the Industrial Exhibition in large scale, which attracted 63 leading scientific instrument makers, trading companies, publishers and academic institutions. Their performed exhibition was planned to show currently available industrial technologies to construct operating FELs inside and outside Japan.

0168-9002/$ - see front matter © 2004 Elsevier B.V. All rights reserved.
doi:10.1016/j.nima.2004.04.004

The fee from the Industrial Exhibition Participants helped to balance the conference budget very much.

We would like to express our gratitude to the members of the Local Organizing Committee for their hard work before and after the conference. Special thanks are due to the Conference Coordinator, Toyoaki Kimura, who took care of many administrative tasks small and large, including the conference auditing works, and tough negotiation with JAERI and Japanese Government administrative officers. The proceedings publication follows the policy established up to now. Only those papers presented by the authors during the conference have been accepted. The first part of the proceedings contains papers that have been refereed by members of the International Program Committee, while the second part contains extended abstracts. We hope that the volume will become useful to the readers in learning the current status and in seeking the future directions of FEL research.

Eisuke J. Minehara (Conference Chair)
E-mail: minehara@popsvr.tokai.jaeri.go.jp
Ryoichi Hajima (Program Chair)
Masaru Sawamura (Guest Editor)

Conference Committees

Conference Chairman
Eisuke J. Minehara (JAERI)
Users Workshop Chairman
Kunio Awazu (Osaka University)
Local Organizing Committee Chairman
Koichi Shimoda (University of Tokyo)

Program Committee Chairman
Ryoichi Hajima (JAERI)
Conference Coordinator
Toyoaki Kimura (JAERI)

Local Organizing Committee
K. Awazu (Osaka University)
H. Hama (Tohoku University)
T. Harami (JAERI)
K. Imasaki (Osaka University)
H. Kimura (JAERI)
T. Kimura (JAERI)
H. Kuroda (FEL-SUT)
K. Shimoda (University of Tokyo)
T. Tomimasu (Saga L.S.)
T. Tomita (JAERI)
K. Yamada (AIST)
T. Yamazaki (Kyoto University)

International Executive Committee
I. Ben-Zvi (BNL)
W.B. Colson (NPGS)
M.-E. Couprie (LURE)
A. Gover (Tel Aviv University)
H. Hama (Tohoku University)
K.-J. Kim (ANL & University of Chicago)
Y. Li (IHEP)
V.N. Litvinenko (Duke University)
E.J. Minehara (JAERI)
G.R. Neil (TJNAF)
C. Pellegrini (UCLA)
M.W. Poole (CLRC)
A. Renieri (ENEA)
C.W. Roberson (ONR)
J. Rossbach (DESY)
T. Smith (Stanford University)
A.F.G. van der Meer (FOM)
N.A. Vinokurov (BINP)
R. Walker (CLRC)

International Program Committee
K. Awazu (Osaka University)
R. Bakker (ELETTRA)
S. Benson (TJNAF)
C. Brau (Vanderbilt University)
M.-E. Couprie (LURE)
J. Dai (BFEL)
G. Dattoli (ENEA)
B. Faatz (DESY)
H.P. Freund (SAIC)
J.N. Galayda (SLAC)
N. Ginzburg (IAPNN)
F. Glotin (CLIO)
R. Hajima (JAERI)
H. Hama (Tohoku University)
S. Hiramatsu (KEK)
R. Kato (Osaka University)
S. Krinsky (BNL)
B.C. Lee (KAERI)
A.H. Lumpkin (ANL)
D. Nguyen (LANL)
H. Ohgaki (Kyoto University)
J. Pflüger (DESY)
S. Reiche (UCLA)
Y. Shimizu (AIST)
M. Shinn (TJNAF)
T. Shintake (RIKEN)
X. Shu (IAPCM)
T. Smith (Stanford University)
N.A. Vinokurov (BINP)
M. Xie (LBL)
K. Yamada (AIST)

doi:10.1016/j.nima.2004.04.002

Exhibitors

FEL2003 Exhibitors

Udo Klein
Peter om Stein
ACCEL Instruments GmbH
Friedrich-Ebert-Strasse 1
Bergisch, Gladbach 51429
Germany
Tel.: 49.2204.84.2288
Fax: 49.2204.84.2599
E-mail: klein@accel.de
E-mail: Stein@accel.de

Machiel Kleemans
Elsevier Science
P.O. Box 103, 1000 AC
The Netherlands
Tel.: +31 20 485 2524
Fax: +31 20 485 2580
E-mail: M.Kleemans@elsevier.nl
URL: http://www.elsevier.nl

Eiji Tanabe
Koichi Kanno
AET Japan, Inc.
URSIS Bldg. 9th Floor
1-2-3 Manpukuji, Kawasaki-City
Kanagawa 215-0004
Japan
Tel.: +81-44-966-9981
Fax: +81-44-951-1572
E-mail: etanabe@aetjapan.com
E-mail: kanno@aetjapan.com
URL: http://www.aetjapan.com

Tusyoshi kobiyama
Akiko Sukigara
EPSON Sales, Japan Corporation.
3-4-14 Minami-machi, Mito 310-0021
Japan
Tel.: +81-29-228-2532
Fax: +81-29-228-2533
E-mail: KOBIYAMA@
exc.ehb.epson.co.jp
E-mail: sukigara.akiko@
exc.ehb.epson.co.jp
URL: http://www.epson.co.jp

Shunsuke Takahashi
Hiroshi Takagi
HAKUTO Co., Ltd.
1-1-13 Shinjuku, Shinjukuku
Tokyo 160-8910
Japan
Tel.: +81-3-3225-8051
Fax: +81-3-3225-9011
E-mail: takahashi-shunsuke@
hakuto.co.jp
E-mail: takagi-h@hakuto.co.jp

Shinya Yanase
Takeshi Saito
HAKUTO Co., Ltd.
1-13 Shinjuku, 1-Chome
Shinjukuku, Tokyo 160-8910
Japan
Tel.: +81-3-3355-7620
Fax: +81-3-3355-7681
E-mail: yanase-s@hakuto.co.jp
E-mail: saito-t@hakuto.co.jp
URL: http://www.hakuto.co.jp

Toshitaka Nagayama
Kaneichi Momotsuka
Hitachi Haramachi electronic Ltd.
3-10-2 Bentencho, Hitachi-shi
Ibaraki-ken 317-0072
Japan
Tel.: +81-294-55-7425
Fax: +81-294-55-9648
E-mail: nagayama@
haraden.hitachi.co.jp
E-mail: momozuka@
haraden.hitachi.co.jp
URL: http://www.haraden.co.jp

Tetsuo Mori
Masaharu Fukumuro
INFRARED LIMITED
6-17-32 Shakujii-Machi, Nerima-Ku
Tokyo 177-0041
Japan
Tel.: +81-3-5372-7575
Fax: +81-3-5372-7577
E-mail: mori@infrared.co.jp
E-mail: fukumuro@infrared.co.jp
URL: http://infrared.co.jp

Koichi Katayama
Ryo Hamanaka
Ishikawajima-Harima Heavy
Industries Co., Ltd.
isogo-ku, sin-nakahara-cho
yokohama, Kanagawa-ken
235-8501
Japan
1, sin-nakahara-cho, isogo-ku
yokohama, Kanagawa-ken
235-8501
Japan
Tel.: +81-45-759-2061
Fax: +81-45-759-2053
E-mail:
kouichi_katayama@ihi.co.jp
E-mail: ryo_hamanaka@ihi.co.jp
URL: http://www.ihi.co.jp

Junya Kamino
Shinji Tanaka
Kawasaki Heavy Industries, Ltd.
2-4-1 Hamamatsu-cho,
Minato-ku
Tokyo 105-6116
Japan
Tel.: +81-3-3435-2959
Fax: +81-3-3578-1573
E-mail: kamino_j@khi.co.jp
E-mail: tanaka_shi@khi.co.jp
URL: http://www.khi.co.jp

Seigo Uehara
Takeshi Nakamura
Nichizo Denshi Seigyo Co., Ltd.
Palace-side Bldg 7F

1-1 Hitotsubashi, 1-chome
Chiyoda-ku, Tokyo 100-8121
Japan
Tel.: +81-3-3217-8499
Fax: +81-3-3201-0522
E-mail: uehara@ndssf.co.jp
3-28 Nishikujo, 5-chome
Konohanaku, Osaka 554-0012
Japan
Tel.: +81-6-6468-9753
Fax: +81-6-6468-9744
E-mail: nakamura_ta@ndssf.co.jp
URL: http://www.ndssf.co.jp

Mitsuo Niki
Asao Suzuki
NIKI Glass Co., Ltd.
3-9-7 Mita Minato-Ku
Tokyo 108-0073
Japan
Tel.: +81-3-3456-4700
Fax: +81-3-3456-3423
E-mail: asachan@nikiglass.com
URL: http://nikiglass.com

Takeshi Kuramochi
Mitsugu Matsuzaki
Pulse Electronic Engineering Co., Ltd.
274 Michisita Futatsuzuka, Noda
chiba, 278-0016
Japan
Tel.: +81-4-7123-0611
Fax: +81-4-7123-0620
E-mail: peec5@pE-mail.ne.jp
E-mail: vz6m-yur@asahi-net.or.jp
URL: http://mmjp.or.jp/peec/

Aritaka Koizumi
Masashi Kamaishi
Sharan Instruments Corporation
13 Fukusawakubo, Same-machi
Hachinohe, Aomori-ken 031-0841
Japan
Tel.: +81-178-34-5011
Fax: +81-178-31-2711
E-mail: rao@sharan.co.jp
URL: http://www.sharan.co.jp

Kazumasa Konishi
Toyomi Ishiii
Sumitomo Electric Hardmetal
3-12 Moto-akasaka, 1-chome
Minato-ku, Tokyo 107-8468
Japan
Tel.: +81-3-3423-5611
Fax: +81-3-3423-5610
E-mail: konishi-kazuhisa@sei.co.jp
E-mail: ishii-toyomi@gr.sei.co.jp
URL: http://www.hm.sei.co.jp

Chie Okamoto
Sumitomo Heavy Industries, Ltd.
5-9-11 Kitashinagawa
Shinagawa-ku, Tokyo 141-8686
Japan
Tel.: +81-3-5488-8412
Fax: +81-3-5488-8302
E-mail: Che_Okamoto@shi.co.jp
URL: http://www.shi.co.jp/cryopage

doi:10.1016/j.nima.2004.04.005

Nobuaki Doi
Yasuo Itoh
Thamway Co., Ltd.
3-9-2 Imaizumi, Fuji, Shizuoka
417-0001
Japan
Tel.: +81-545-53-8965
Fax: +81-545-8978
E-mail: doi@thamway.co.jp
E-mail: itoh@thamway.co.jp
URL: http://www.thamway.co.jp/

Ichiji Kojima
Shigeo Kowada
IDX Corporation
2452-15 Ishigamitojuku
Tokai-mura, Naka-gun
Ibaraki-ken 319-1101
Japan
Tel.: +81-29-306-3520
Fax: +81-29-306-3525
E-mail: i.kojima@idx-net.co.jp
8-12 Higashiarai, Tsukuba-shi
Ibaraki-ken 305-0033
Japan
Tel.: +81-29-860-2217
Fax: +81-29-860-2218
E-mail: kowada@sano.idx-net.co.jp
URL: http://www.idx-net.co.jp/

Norio Kase
Toshihide Takeuchi
TOYO Corporation
1-6 Yaesu, 1-chome, Chuo-ku
Tokyo 103-8284
Japan
Tel.: +81-3-3279-0771
Fax: +81-3-3246-0645
E-mail: kase@toyo.co.jp
E-mail: takeuchit@toyo.co.jp
URL: http://www.toyo.co.jp

Osamu Yushiro
Kenichi Hayashi
Toshiba Electron Tubes &
Deviced Co., Ltd (TETD)
682-2 Nishiki-cho, Omiya-ku
saitama-city, 331-0851
Japan
Tel.: +81-48-640-1224
Fax: +81-48-640-1193
E-mail: osamu.yushiro@
toshiba.co.jp
1385 Shimo-ishigami, Otawara
Tochigi 243-8550
Japan
Tel.: +81-287-26-6578
Fax: +81-287-26-6061
E-mail: kenichi1.hayashi@
toshiba.co.jp

Anthony J. Favale
Alan Todd
Advanced Energy Systems, Inc.
P.O. Box 7455
501 Forrestal Road
Suite 316 Princeton
NJ 08543-7455
Tel.: +1-609-514-0316
Fax: +1-609-514-0318
Cell: +1-609- 841-5607
E-mail: tony_favale@mail.aesys.net
E-mail: alan_todd@mail.aesys.net

Yukimori Akutsu
ARAKI ELETEC CORP
2-11-6 Ebisu, Sibuya-ku
Tokyo 150-0013
Japan
Tel.: +81-3-3440-1001
Fax: +81-3-3443-0163
E-mail: akutsu@araki-eletec.co.jp

Yoshio Ogura
Taiki Iida
LeCroy Japan Cop.
2-1-6 Sasazuka, Sibuya-ku
Tokyo 151-0073
Japan
Tel.: +81-3-3376-9400
Fax: +81-3-3376-9587
E-mail: yoshio.ogura@lecroy.com
E-mail: taiki_iida@lecroy.com

Kazuo Matsuhashi
CHUO PRECISION INDUSTRIAL
CO., LTD
1-9 Kanda-Awaji-Cho
Chiyoda-Ku, Tokyo 101-0063
Japan
Tel.: +81-3-3257-1911
Fax: +81-3-3257-1915
E-mail: k-matsu@chuo.co.jp
URL: http://www.chuo.co.jp

Marv Eberhardt
CPI EIMAC
301 Industrial Way, San Carlos
CA 94070
USA
Tel.: +1-650-592-1221
Fax: +1-650-592-9988
URL: http://www.cpii.com/eimac/
main.html

Hideki Kawamura
Shinji Watanabe
EIKO ENGINEERING
50 Yamazaki, Hitachinaka
Ibaraki 311-1251
Japan
Tel.: +81-29-265-7401
Fax: +81-29-265-7406
E-mail: eiko1974@ap.wakwak.com
URL: http://www.kagaku.com/eiko/

Yasuhiro Kakimi
Daisuke Sanjou
AIR WATER INC.
1-20-16 higashishinnsaibashi, Chuo-
Ku, Osaka, Osaka 5420083
Japan
Tel.: +81-6-6252-5411
Fax: +81-6-6252-3965
E-mail: sanjou-dai@awi.co.jp
URL: http://www.awi.co.jp/

Akihiro Sawaki
Haruhito Iio
HAMAMATSU PHOTONICS K.K.
325-6 sunayama-cho
Hamamatsu City
Shizuoka 430-8587
Japan
Tel.: +81-53-452-2148
Fax: +81-53-452-2139
E-mail: salesi@sys.hpk.co.jp
5th Floor, Toranomon
33 Mori Building
3-8-21 Toranomon, Minato-ku
Tokyo 105-0001
Japan
Tel.: +81-3-3436-0491
Fax: +81-3-3433-6997
URL: http://www.hamamatsu.com

Kenichi Kosaka
Japan Superconductor Technology
Inc.
5-9-12 Kitashinagawa
Shinagawa-ku, Tokyo 141-8688
Japan
Tel.: +81-298-56-6300
Fax: +81-298-56-6301
Tel.: +81-3-5739-5210
Fax: +81-3-5739-5211
E-mail: kosaka-jastec@kobelco.jp
URL: http://www.JASTEC.org

Morio Kawasaki
Kawasaki Science Co., Ltd.
2-23-29 Umezono Tsukuba
Ibaraki 305-0045
Japan
Tel.: +81-298-56-6300
Fax: +81-298-56-6301
E-mail: vac@k-science-inc.com
URL: http://www.k-science-inc.com

Yoshimasa Aihara
KEYENCE Corporation
1-3-14, higashi-nakajima
higashi-yodogawa-Ku
Osaka 533-8555
JAPAN
Tel.: +81-6-6379-2211
Fax: +81-6-6379-2131
E-mail: aiharay@sales.
keyence.co.jp
URL: http://www.keyence.co.jp/

Yoshi Mazda
MARUBUN CORPORATION
8-1 Nihonbashi, Odenmacho
Chuo-ku, Tokyo 103-8577
Japan
Tel.: +81-3-3639-9814
Fax: +81-3-3661-7473
E-mail: ymazda@marubun.co.jp

Osamu Asada
Shinsuke Morohuji
Mitubishi Electric Corporation
2-1-12 Funaishikawaekinishi
Tokai-Mura, Naka-gun
Ibaraki-ken 319-1116
Japan
Tel.: +81-29-284-0512
Fax: +81-29-284-0516
E-mail: Osamu.Asada@hq.melco.
co.jp
URL: http://www.mitsubishielectric.
co.jp

Koichi Ohkubo
Akira Zakou
Mitsubishi Heavy Industries, Ltd.
Kobe Shipyard & Machinery Works
1-1 Wadasaki-cho, 1-chone
Hyogo-ku, Kobe, Hyogo 652-8585
Japan
Tel.: +81-78-672-2910
Tel.: +81-78-672-2908
Fax: +81-78-672-2920
E-mail: koichi_okubo@mhi.co.jp
E-mail: akira_zakou@mhi.co.jp

Takashi Namae
Kenichi Kon
NEC TOKIN Corporation
5-8 Kita-Aoyama, 2-chome
Minato-ku, Tokyo 107-8620
Japan
Tel.: +81-3-3402-3901
Fax: +81-3-3402-9892
E-mail: namae@nec-tokin.com
E-mail: kon@nec-tokin.com
URL: http://www.nec-tokin.com

Shoji Ueda
TEKTRONIX JAPAN, LTD.
5-9-31 Kitashinagawa
Shinagawa-ku, Tokyo 141-0001
Japan
Tel.: +81-3-3448-3008
Fax: +81-3-3448-3672
E-mail: ueda.shoji@tektronix.com
URL: http://www.tektronix.co.jp/

Eiji Oshita
Yoshinori Noguchi
NISSIN Electric Co., Ltd.
1 Knada-Izumi-cho, Chiyoda-ku
Tokyo 101-0024
Japan
Tel.: +81-3-5821-5909
Fax: +81-3-5821-0380
E-mail: oshita@nhv.nissin.co.jp
E-mail: noguchi@nhv.nissin.co.jp
URL: http://www.nissin.co.jp/,
http://www.nhv.jp/

Koujirou Oka
Kiyoshi Ishiwata
NIPPON SANSO CORPORATION
6-2 Kojima-Cho, Kawasaki-Ku
Kawasaki-City, Kanagawa-Ken
210-0861
Japan
Tel.: +81-44-288-6963
Fax: +81-44-288-2727
E-mail: Koujirou_Oka@sanso.co.jp
E-mail: Kiyoshi_Ishiwata@
sanso.co.jp
URL: http://www.sanso.co.jp

Toshio Ando
Yosiaki Asano
NIKOHA Co., Ltd.
1119 Nakayama, Midori-ku
yokohama, Kanagawa 226-0011
Japan
Tel.: +81-45-939-6961
Fax: +81-45-932-1900
E-mail: ando@nikoha.co.jp
URL: http://www.nikoha.co.jp

Toshio Miyake
Kei Isokawa
Ohyo Koken Kogyo Co., Ltd.
1642-26 Kumagawa, Fussa-City
Tokyo 197-0003
Japan
Tel.: +81-42-552-4511
Fax: +81-42-552-5750
E-mail: miyake@oken.co.jp
E-mail: isokawa@oken.co.jp
URL: http://www.oken.co.jp

Mitsuru Kawamoto
Junya Tasaka
Pechiney Japon
Shinjuku-Mitsui Bldg 29F
2-1-1 Nishishinjuku, Shinjuku-ku
Tokyo 163-0429
Japan
Tel.: +81-3-3349-6671
Fax: +81-3-3349-6770
E-mail: mitsuru.kawamoto@
pechiney.com
E-mail: junya.tasaka@
pechiney.com
URL: http://www.pechineyjapon.com

Yasuo Matsumoto
Sunao Takahashi
Light Stone Corp.
3-4-16 HigashiShinkoiwa
Katsushika-ku, Tokyo 124-0023
Japan
Tel.: +81-3-5670-0301
Fax: +81-3-5670-0311
E-mail: matsumoto@
lightstone.co.jp
E-mail: sunao@lightstone.co.jp

Yoshiyuki Nonaka
Laboratory Equipment Corporation
1-7-3 Minato-Machi, Tutiura
Ibaraki 300-0034
Japan
Tel.: +81-29-821-6051
Fax: +81-29-281-6054
E-mail: nonaka@labo-eq.co.jp

Shokichi Hirasawa
Mikinori Ochiai
REPIC Corporation
28-3 Kita-Otsuka, Toshima-ku
Tokyo 170-0004
Japan
Tel.: +81-3-3918-5326
Fax: +81-3-3918-5712
E-mail: hirasawa@repic.co.jp
E-mail: ochiai@repic.co.jp
URL: http://www.repic.co.jp

Tamaki Mori
Masahiro Watanabe
SAES Getters Japan Co., Ltd.
5-23-1 2nd Gotanda Fujikoshi Bldg
Higashigotanda, Shinagawa
Tokyo 141-0022
Japan
Tel.: +81-3-5420-0542
Tel.: +81-3-5420-0434
Fax: +81-3-5420-0438
E-mail: tamaki_mori@
saes-group.com
E-mail: masahiro_watanabe@
saes-group.com
URL: http://www.saesgetters.com

Takeshi Morita
Shigemi Komatsu
SEIKO EG&G Co., LTD
11-1, Nihonbashi, Tomizawa-cho
Chuo-ku, Tokyo 103-0006
Japan
Tel.: +81-3-5645-1779
Fax: +81-3-5645-1770
E-mail: takeshi.morita@sii.co.jp
E-mail: shigemi.komatsu@sii.co.jp
URL: http://www.sii.co.jp/segg/

Tsuyoshi Watanabe
SETEC CORPORATION
5-2-3 Asakusabashi, Taitouku
Tokyo 111-0953
Japan
Tel.: +81-3-3864-6031
Fax: +81-3-3864-6089
E-mail: t-watanabe@setec-k.com

Hideki Kobayashi
Shin-Etsu Chemical Co., Ltd.
2-1-5 Kitago, Takefu
Fukui 915-8515
Japan
Tel.: +81-778-21-8142
Fax: +81-778-23-8538
E-mail: h_kobayashi@shinetsu.jp

Y. Hakoishi
SIGMA KOKI
5F, SIGMA KOKI Tokyo Head office

19-9, Midori 1 chome, Sumida-ku
Tokyo, 130-0021
JAPAN
Tel.: +81-3-5638-8228
Fax: +81-3-5638-6550
E-mail: y.hakoishi@sigma-koki.com
E-mail: international@
sigma-koki.com
URL: http://www.sigma-koki.
com/english

Hiroyuki Tsukada
Tomoki Ishikawa
Spectra Physics K. K.
4-6-1 Nakameguro, Meguro-ku
Tokyo 153-0061

Japan
Tel.: +81-3-3794-5511
Fax: +81-3-3794-5510
E-mail: tishikawa@splasers.co.jp
URL: http://www.spectra-physics.jp

Take-Off, Co., Ltd.
2-31-6 Oshima Koutou-ku, Tokyo
Japan
Tel.: +81-3-5628-9775
Fax: +81-3-5628-9618
E-mail: science@takeoff-ltd.co.jp

JAERI
Japan Atomic Energy Research
Institute, Tokai Research

Establishment
2-4 Shirakata shirane, Tokai, Naka
Ibaraki 319-1195
JAPAN
Tel.: +81-29-282-5464
Tel.: +81-29-282-6752
Fax: +81-29-282-6057
c/o Esuke J. Minehara
E-mail: eisuke@jfel.tokai.jaeri.go.jp

Takenori Kato
Naotaka Hokkyou
SOKKIA TOKYO CO., LTD

5-26-10 kamiyouga, setagaya
Tokyo 158-0098
Japan
Tel.: +81-3-3708-4911
Fax: +81-3-3708-4910
E-mail: t.kato@sokkia.co.jp
E-mail: n.hokkyou@sokkia.co.jp
URL: http://sokkia.co.jp

Yukinari Tsutsumi
Sumie Ariga
Suruga Seiki Co., Ltd.
5F, Sudacho, Towa Bldg, 2-2-4
Kanda-Sudacho, Chiyoda-Ku

Tokyo 101-0041
Japan
Tel.: +81-3-5156-9911
Fax: +81-3-5156-9913
E-mail: sariga@suruga-g.co.jp
URL: http://www.surugaost.jp/

Keisuke Endo
Hiroyuki Katayama
TOYAMA Co., Ltd.
4-13-16 Hibarigaoka, Zama-shi
Kanagawa 228-0003
Japan
Tel.: +81-46-253-1411
Fax: +81-46-253-1412
E-mail: joe-endo@toyama-jp.com
E-mail: katayama@toyama-jp.com
URL: http://www.toyama-jp.com

BINP
Siberian Branch of Russian
Academy of Science

The Budker Institute for Nuclear
Physics (BINP)
630090 Novosibirsk
Russia
Tel.: +7-3832-39-4000
Fax: +7-3832-34-2183
c/o Dr. Nikolai A. Vinokurov
E-mail: N.A.Vinokurov@inp.nsk.su

Yutaka Touchi
Yuji Matsubara
SUMITOMO HEAVY
INDUSTRIES, LTD
5-33 Kitahama, 4-Chome Chuo-Ku
Osaka, Osakafu 541-7169
Japan
Tel.: +81-6-6223-7169
Fax: +81-6-6223-7234
E-mail: Ytk_Touchi@shi.co.jp
9-11 Kitasinagawa, 5-Chome
Sinagawaku, Tokyo 141-8686
Japan
Tel.: +81-3-5488-8327
Fax: +81-3-5488-8321
E-mail: Yji_Matsubara@shi.co.jp
URL: http://www.shi.co.jp/

Yoshihiro Shibanuma
Haruyoshi Watanabe
Taiyoukeisoku Corporation
2-1-6 Ekinishi, Funaishikawa
Tokai-Mura, Naka-Gun

Ibaraki-Ken 319-1116
Japan
Tel.: +81-29-287-2151
Fax: +81-29-287-2156
E-mail: shibanuma@taiyo.co.jp
URL: http://www.taiyo.co.jp
2074 Uenomuro,Tsukuba-City
Ibaraki-Ken 305-0023
Japan
Tel.: +81-29-857-2452
Fax: +81-29-857-4629
E-mail: watanabe@taiyo.co.jp
URL: http://www.yokogawa.co.jp

Nobuyuki Tsuji
Tsuji Electronics Co., Ltd
3739 Kandatsu-Machi
Tsuchiura-City
Ibaraki-Ken 300-0013
Japan
Tel.: +81-29-832-3031
Fax: +81-29-832-2662
E-mail: tsuji@tsuji-denshi.co.jp
URL: http://www.tsujicon.jp

Tsuchiura Tsukuba Convention
Bureau
Tsukuba Minami 1 Park
Building 1F
1-4-1,Takezono,Tsukuba
Ibaraki 305-0032
Japan
Tel.: +81-298-61-7171
Fax: +81-298-61-7161
E-mail: ttcb@intio.or.jp
URL: http://www.intio.or.jp/ttcb/
english/index.html

Yoshio Kobayashi
Nobuko Kawaguchi
Sumitomo Special Metals Co., LTD

4-7-19 Kitahama Chuou-Ku
Osakafu 541-0041
Japan
Tel.: +81-6-6220-8807
Fax: +81-6-6226-0089
E-mail: KOBAYASHI.Y@ssmc.co.jp
E-mail: KAWAGUCHI.N@
ssmc.co.jp
URL: http://www.ssmc.co.jp

Kiyoshi Miki
Noboru Endoh
Kotoyo Ikeda
Thales International Japan K.K.
TBR-Bldg. 5-7 Koji-machi,
Chiyoda-ku, Tokyo 102-0083
Japan
Tel.: +81-3-3264-6346
Fax: +81-3-3264-6696
E-mail: info@ttej.ne.jp
URL: http://www.thalesgroup.com/
home/home/index.html

Yuuichi Shimodaira
YAMATO Scientific Co., Ltd.
2-10-4 Higashi, Tsukuba-Shi
Ibaraki-Ken 305-0046
Japan
Tel.: +81-29-852-3411
Fax: +81-29-852-1691
E-mail: shimodairay@
yamato-net.co.jp

Sponsors

The Organizers of FEL 2003 would like to thank the following sponsors for their important support that they have provided.

Atomic Energy Society of Japan
The Physical Society of Japan
Japan Society of Applied Physics
The Laser Society of Japan
The Institute of Electrical Engineers of Japan
The Japanese Society of Synchrotron Radiation Research
The Japan Society of Infrared Science and Technology
Japan Society for Laser Surgery and Medicine
High Energy Accelerator Research Organization
Japan Synchrotron Research Institute

doi:10.1016/j.nima.2004.04.003

2003 International Free Electron Laser Prize Winner:
Li Hua Yu

This year the FEL Prize Committee awarded the FEL prize to Li Hua Yu from Brookhaven National Laboratory for his contributions to high gain SASE FELs and for proposing and demonstrating the concept of High Gain Harmonic Generation (HGHG).

Dr. Yu has made seminal contributions to the theoretical development of the high gain SASE FELs. In collaboration with Dr. J.M. Wang and Sam Krinsky he developed both the 1-D and 3-D formulations of the start-up from noise in SASE devices. This formulation is quite general and has been quite useful in analyzing experimental results in SASE FELs. With Krinsky and R.L. Gluckstern he developed a universal scaling function for the gain of a SASE FEL that allows one to quickly calculate the gain length of an FEL and easily optimize the design. The FEL gain was expressed in a form of a universal scaling function with only four independent scaling parameters.

0168-9002/$ - see front matter
doi:10.1016/j.nima.2004.04.139

He then turned his attention to the allowable errors in the long wigglers required for SASE devices. He found that the allowable wiggler field amplitude errors were much larger than had been thought, varying as the square root of the Pierce parameter rather than linearly. The important design parameter was the electron beam trajectory straightness in the wiggler, which could be achieved using steering coils.

Driven by his desire to improve the coherence and stability of SASE FELs, Dr. Yu developed a method to seed a SASE device using a sub-harmonic of the desired wavelength. This allows one to greatly reduce the intensity and wavelength fluctuations in a high-gain FEL and provide nearly transform-limited pulses to users. He also derived variations on the original idea by cascading harmonic generation so that one might reach the hard X-ray using a conventional laser seed and proposed a fresh-bunch technique allowing one to have even shorter wigglers and more coherence in the shorter wavelength devices.

Not content to just develop the theory of HGHG FELs Dr. Yu led a team that convincingly demonstrated the concept, first with second harmonic generation in the infrared at 5.3 μm, then with third harmonic generation in the ultra-violet at 266 nm. These pioneering experiments showed very high stability and near Fourier transform limited operation. The third harmonic generated in the UV HGHG device at 89 nm is sufficiently high energy that it is now being used for chemistry experiments.

The idea of cascaded HGHG has now been included in the designs of several proposed X-ray facilities world-wide, indicating how desirable the concept has become. Clearly Dr. Yu has contributed greatly to the field of FELs in general and high gain FELs in particular. He continues to come up with clever new ways to improve on the performance of both SASE and HGHG devices.

Stephen Benson
Chairman FEL-Prize Committee

Conference photos

0168-9002/$ - see front matter
doi:10.1016/j.nima.2004.04.006

Conference photos

Conference photos

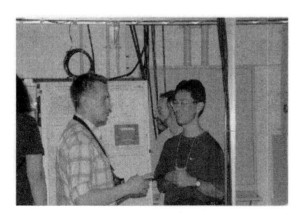

Available online at www.sciencedirect.com

SCIENCE DIRECT®

ELSEVIER   Nuclear Instruments and Methods in Physics Research A 528 (2004) xxi–xxxii

**NUCLEAR
INSTRUMENTS
& METHODS
IN PHYSICS
RESEARCH**
Section A

www.elsevier.com/locate/nima

Contents

FEL 2003
Proceedings of the 25th Free Electron Laser Conference
and 10th FEL Users Workshop
Tsukuba, Ibarabi, Japan, September 8–12, 2003

Editors: E.J. Minehara, R. Hajima, M. Sawamura

doi:10.1016/S0168-9002(04)01370-1

Section 2. FEL Theory

Section 3. High-Power, Long Wavelength FELs

Section 4. FEL Technologies

Section 5. Storage Ring FELs

Section 6. High-Brightness Electron Beams

Section 7. High-Gain, Short-Wavelength FELs

Section 8. New Concepts and Proposals

Section 9. Applications of FELs

**Part II (Extended Abstracts: The text of these papers is included only in the book edition of the proceedings,
ISBN 0-444-51727-8)**

Available online at www.sciencedirect.com

SCIENCE  DIRECT°

Nuclear Instruments and Methods in Physics Research A 528 (2004) 1–7

NUCLEAR INSTRUMENTS & METHODS IN PHYSICS RESEARCH
Section A

www.elsevier.com/locate/nima

Viability of infrared FEL facilities

H. Alan Schwettman*

Department of Physics and W. W. Hansen Experimental Physics Laboratory, Stanford University, Stanford, CA 94305, USA

Abstract

Infrared FELs have broken important ground in optical science in the past decade. The rapid development of optical parametric amplifiers and oscillators, and THz sources, however, has changed the competitive landscape and compelled FEL facilities to identify and exploit their unique advantages. The viability of infrared FEL facilities depends on targeting unique world-class science and providing adequate experimental beam time at competitive costs.
© 2004 Elsevier B.V. All rights reserved.

PACS: 41.60.Cr; 07.57.Hm

Keywords: Superconducting linac; FEL facility; Optical science

1. Introduction

Infrared FELs have broken important ground in optical science in the past decade, opening new vistas in physics, chemistry, biology, and medicine [1]. There are now a number of FEL facilities throughout the world committed to infrared (IR) optical science; and the capabilities of these facilities are remarkably different, one from the other. To cite an example, linac-driven IR FELs have many attractive features, most notably wavelength tunability, control of spectral and temporal pulse width, and exceptional beam quality and stability. However, the normal conducting version is capable of delivering a high average power optical pulse train for a few

microseconds, while the superconducting version is capable of delivering a modest average power optical pulse train that continues indefinitely. These differences, in fact, represent important capabilities that individual FEL facilities can exploit in scientific pursuits.

One decade ago the most important viability issue for IR FEL facilities was FEL reliability. As the facilities have matured, this issue has thankfully passed and in its stead are the central issues of providing sufficient experimental time at competitive costs and providing world-class scientific opportunities. These issues have everything to do with attracting outstanding scientific talent to the facility and with attracting adequate financial support that is stable over time. In this paper, I will use the superconducting linac driven FEL to illustrate viability issues and will derive most of my observations from our experience at Stanford.

*Tel.: +1-650-723-0305; fax: +1-650-725-8311.
E-mail address: has@stanford.edu (H.A. Schwettman).

0168-9002/$ - see front matter © 2004 Elsevier B.V. All rights reserved.
doi:10.1016/j.nima.2004.04.007

2. Superconducting linacs for infrared FELs

The very first superconducting linac was built at Stanford in the early 1970s, now more than 30 years ago [2]. The landmark FEL oscillator experiment [3] depended critically on the unique capabilities of that linac, and that same linac has served well in support of our FEL Optical Science Center. Our linac, is located in a tunnel some 10 m below the main floor of the laboratory. In the tunnel there are two undulators, one to provide mid-IR radiation and the other to provide far-IR radiation. On the main floor of the laboratory there are 10 experimental rooms for optical science fed by the two undulators. In the early history of our Center, we were forcefully reminded by our colleagues that a laser is not an experiment! In fact, the 10 experimental rooms are filled with a vast array of optics, detectors, electronics, superconducting magnets, lasers, and other sophisticated optical instrumentation. The total investment in these probably exceeds that in the FEL itself.

In the past decade superconducting technology has progressed enormously and thus we are now replacing the historic superconducting structures with state-of-the-art structures, following the recent DESY design [4].[1] In the DESY design the helium vessel is an integral part of a 9-cell structure and the liquid helium capacity of this vessel is a mere 25 l. This is a significant departure from the original Stanford design and from the Jlab design. For continuous operation, as envisioned for the IR FEL, realistic energy gradients for the one meter long DESY structures are 10–15 MeV/m and at 2 K the refrigeration required is 10–20 W. The cryomodule requirements are dramatically different from those at DESY and thus a satisfactory design has been generated in a collaboration between Stanford and FZR Rossendorf. The cryomodule holds two structures and a cutaway view of one half of a cryomodule is shown in Fig. 1.

The new superconducting structures will have a dramatic impact on FEL operations at Stanford. The peak and average FEL power will increase by a factor of 5, and we will once again be able to deliver a continuous optical pulse train. But perhaps most important, liquifying a mere 150 l will be sufficient to bring the FEL into operation, rather than the few thousand liters required now, thus providing great flexibility in scheduling experiments. One of the new cryomodules is already installed on the linac and all cryomodules will be replaced this year.

3. Viability issue: providing experimental time

Providing adequate experimental time at an IR FEL facility is not as simple as it sounds. Existing facilities typically provide 3000 h of experimental time per year and typically 25% of this time is allocated to a core experimental group. At Stanford in the mid-1990s a broad collaboration built around Professor Fayer's group carried out the first comprehensive vibrational dynamics experiments on glass-forming liquids and proteins. For these pioneering experiments Fayer was awarded the Earl K. Plyler Prize for Molecular Spectroscopy in 2000. Unfortunately, the scientific interest in such experiments far exceeded the 750 h of time that could be allocated, and thus as optical parametric amplifiers and oscillators improved, Fayer sought and gained funding to build three mid-IR laser systems which now provide for his group approximately 7500 h of experimental time per year. It is clear that a scientifically successful FEL facility providing 10,000 h per year is still not providing enough experimental time.

At a synchrotron radiation facility tens of experimental stations are located around the perimeter of the synchrotron and experiments can be performed at each of these simultaneously. The radiation produced by a synchrotron covers a broad spectral range and each User can select an appropriate portion of the spectrum. The crux of the problem here is that the FEL by its nature provides a narrow spectral line, and each individual User will want to control the wavelength of that line. There are means, however, for an IR

---

[1] The author would particularly like to thank Dercy Pooch and his group at DESY for advise and assistance in producing the Stanford structures.

Gaseous He Output

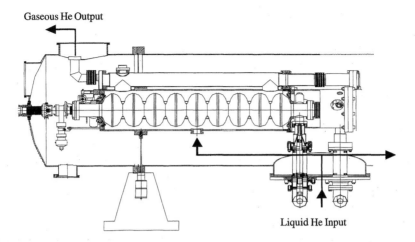

Liquid He Input

Fig. 1. Cutaway view of half a cryomodule indicating one of the 9-cell superconducting structures with integral helium vessel. Liquid helium is fed into the bottom of the vessel. Cold helium gas is collected in the tube located directly above the vessel and then transported to the refrigerator. RF power is fed to the structures from below at the center of the cryomodule.

FEL to effectively serve multiple experiments, for instance by switching from one experiment to another on a macropulse basis. In an FEL driven by a superconducting linac, the linac parameters, and therefore the e-beam and optical beam parameters can be changed in a time the order of 10 ms. Thus it is possible to switch between two linac states, ultimately providing a 40 ms optical beam to each of two experiments at a 10 Hz rate, and each can have independent control of its own beam. At Stanford we implemented such macropulse switching between two undulators a few years ago. One might imagine that there is no need to switch linac parameters if there are two undulators, but that proved to be naïve. If the two experiments require widely different wavelengths or widely different pulse lengths, macropulse control of e-beam parameters is essential. The Stanford system [5] provides macropulse-to-macropulse control of the gun current, the phase of the sub-harmonic buncher, the amplitude and phase of the accelerating structures and the electron beam steering and focussing. The switching time in the Stanford system is approximately 40 ms, limited by switching of the quadrupole settings. Careful design of the quadrupoles would permit switching in the 10 ms time referred to above.

It is also possible to macropulse switch in a single undulator. The interleaved optical macropulses must then be separated and delivered to the proper experiment. At Stanford in the past month, we have demonstrated an optical switching system [6] that time-sorts the interleaved macropulses and transports them to separate experimental rooms. In principle there is no reason an FEL could not switch between four different states, providing an independently controllable 15 ms optical beam to each of four different experiments. To accomplish this, electron beam macropulses in two different states must be delivered to each of two different undulators. The nominal 3000 h of experimental time could then be expanded to 12,000 h. Increasing experimental time in this way would diminish the availability issue, and would significantly reduce the cost. At Stanford the operations cost for experimental time could decrease from $600 per hour to $150 per hour.

4. Viability issue: scientific opportunities

In the early 1990s FELs were virtually alone in providing high quality picosecond (ps) optical beams in the mid- and far-IR. At FEL facilities, experiments have been performed exploring vibrational dynamics, timescales and pathways of energy flow, IR near-field microscopy, THz spectroscopy of semiconducting nanostructures, and many other topics. The rapid development of optical parametric amplifiers and oscillators, and THz sources, however, has changed the landscape and has encouraged Users to identify and exploit the special advantages of the FEL.

4.1. Vibrational dynamics experiments

Superconducting linac-driven FELs, as an example, have a special advantage that can be exploited in vibrational dynamics experiments. They provide an optical pulse train at 10–20 MHz that continues indefinitely, and from this pulse train it is possible to select optical pulses at a repetition rate that matches the experiment at hand. To illustrate the experimental problem consider two extreme situations. The first is a pump–probe experiment performed at the JLab FEL to study vibrational relaxation of hydrogen defects in silicon [7], a topic of great technological importance. There are many absorption features in proton-implanted crystalline silicon, corresponding to specific hydrogen defects, as illustrated in Fig. 2. Characterization of such defects requires a detailed understanding of timescales and pathways of energy flow from the defects, to local modes,

and then to the phonon bath of the host material. The lifetimes of the Si–H stretch modes, determined by the pump–probe experiments at JLab, represent a first step in this detailed understanding, and show that the lifetime is strongly dependent on local structure, ranging from ps for interstitial-like hydrogen to hundreds of ps for hydrogen-vacancy complexes.

Let us be certain we understand the nature of this vibrational relaxation measurement. In the mid-IR at room temperature the bound hydrogen impurities are in the vibrational ground state. The pump beam excites some of these to the first excited state and they subsequently relax back to the ground state by thermalization processes. That vibrational relaxation process can be monitored by the probe beam. The attenuation of the probe beam as it passes through the sample is proportional to the population difference between the ground state and the first excited state. Thus by measuring the probe beam attenuation as a function of the delay between pump and probe, one can determine the vibrational relaxation time constant, T_1.

The problem in this pump–probe experiment is that the concentration of hydrogen impurities in the experiment is small and the cross-section of vibrational transitions is small as well. The saving grace is that in the wavelength region of interest, silicon is nearly transparent. Thus it is possible to utilize the full 20 MHz rep rate of the FEL, delivering an average power of 10 W to the experiment. The high rep rate, high average power and exceptional beam stability of the FEL transformed an experiment that was marginal using an optical parametric amplifier at 1 KHz and 5 mW average power to an eminently viable experiment.

At the other extreme, we have the situation encountered in the vibrational dynamics experiments on glass-forming liquids and proteins performed at Stanford. Here the sample concentrations are much higher, but the solvent is rather strongly absorbing. In vibrational dynamics one is interested in the pure dephasing time which is a measure of ps fluctuations of the local environment. This measurement is a bit more complicated than the simple relaxation measurement discussed

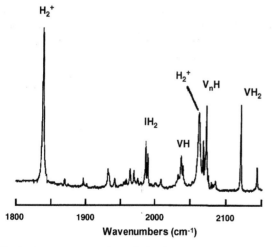

Fig. 2. FTIR spectrum of protein-implanted crystalline silicon, illustrating absorption features corresponding to specific hydrogen defects.

above. To determine the pure dephasing time one must perform a photon echo experiment, the optical analogue of the spin echo, in addition to the vibrational relaxation experiment. At Stanford pure dephasing rates as a function of temperature have been measured for a number of systems. Results are shown in Fig. 3 for myoglobin, a small protein that has the biological function of reversible binding and transport of oxygen in muscle tissue [8]. The observed temperature dependence below 200 K is reminiscent of the two level system dynamics of conventional low-temperature glasses, and suggests that the observed behavior arises from tunneling dynamics of an ensemble of protein two-level systems. Above 200 K, however, the data indicate the emergence of an exponentially activated process.

But what is an appropriate optical pulse rep rate and single pulse energy for such an experiment? Among other things that depends on the choice of solvent, but let us consider water, the solvent of preference for biological systems. If the sample is confined in a chamber, let us say with a CaF_2 window at both the front and back surface, and if you assume the heat conduction path is through the CaF_2 to a heat sink, one can calculate the temperature distribution in the water, given the

pulse rep rate, the single pulse energy, and the thermal conductivity, heat capacity and absorption coefficient of water at the wavelength of interest. At $6.45\,\mu m$, the wavelength of the Amide II mode in proteins, the absorption coefficient of water is $825\,cm^{-1}$ corresponding to a penetration depth of approximately $12\,\mu m$. For a water sample $20\,\mu m$ thick and a train of $1\,\mu J$ pulses repeating at $10\,kHz$ ($10\,mW$ average power), the steady state temperature at the water/CaF_2 interface rises by $2.5\,K$ and the maximum temperature in the sample rises by $5.2\,K$. The calculated thermal relaxation time of the system is approximately $200\,\mu s$ and thus it is clear that the assumed repetition rate is not quite fast enough to smooth the sample temperature in time. In fact, a single $1\,\mu J$ optical pulse elevates the water temperature at the surface by $1.4\,K$. The incremental temperatures calculated above are acceptable for biological experiments, but increasing either the single pulse energy or the average power by an order of magnitude will lead to unacceptable conditions. The Stanford experiments on proteins, in fact, utilized a rep rate 2.5 times larger and a single pulse energy 2.5 times smaller than assumed in the calculation.

4.2. Gas phase experiments

At the time we wrote the proposal for our optical science center at Stanford gas phase experiments with our FEL seemed marginal at best. My interest in gas phase experiments was rekindled, however, in the early 1990s when I served as chair of a DOE committee to review a proposed $600\,W$ IR FEL to be constructed at Berkeley as part of the Chemical Dynamics Research Laboratory. I suggested to Professor Yuan Lee, the principal scientist behind the proposed facility, that if the project were approved it would be interesting to consider doing these experiments either as FEL intracavity experiments or as synchronously pumped external cavity experiments. Either would greatly reduce the power required in the FEL. These suggested possibilities, illustrated in Fig. 4, were motivated in part by two projects underway at Stanford. At that time a graduate student, Ken Berryman, was designing our FIR FEL system [9] and we were

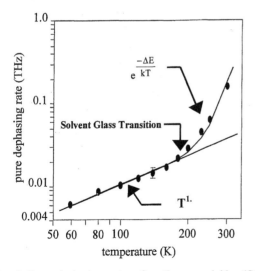

Fig. 3. Pure dephasing rate of native myoglobin (CO), indicating transition from region governed by the tunneling dynamics of protein two level systems to region governed by an exponentially activated process.

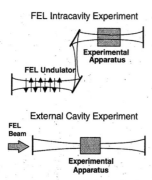

Fig. 4. Possible intracavity and external cavity configurations that could accommodate gas-phase optical experiments. Each of these would dramatically increase the optical fluence available for experiments.

already committed to building a three mirror FEL cavity that reached out of the shielded area into an experimental room. It did not seem out of the question to make a four mirror FEL cavity that could then accommodate a gas-phase experiment. Another graduate student, Paul Haar, was beginning experiments, to synchronously pump an external cavity [10], and he soon achieved pulse energies in the external cavity that were nearly 100 times the energy of the individual incident pulses, demonstrating remarkable coherence in the optical pulse train.

Unfortunately, the FEL plans at Berkeley never materialized. At Stanford no intracavity experiment was attempted, however, the external cavity was reconfigured and used to demonstrate the feasibility of synchronously pumped cavity ringdown spectroscopy. Cavity ringdown has often been used to measure absorption in gaseous systems that are very dilute or very weakly absorbing. In this technique, light from a laser source is coupled into a high Q optical cavity that encloses the gas of interest. When the light source is interrupted, optical energy stored in the cavity decays exponentially due to cavity losses and absorption by the gaseous medium. The absorption spectrum of the gas is obtained by measuring the decay rate with medium and subtracting the decay rate of the empty cavity. High sensitivity is achieved since the method provides a very long effective pathlength for absorption by the gas and

is insensitive to amplitude fluctuations of the laser source.

The high pulse repetition rate of the FEL pulse train has a dramatic impact on cavity ringdown spectroscopy (CRDS). To achieve high sensitivity, the mirror transmission must be small, let us assume 10^{-3}. Thus a single pulse, coupled through the front mirror into the cavity will be smaller than the incident pulse by this factor. The exponential decay of this pulse will be monitored by the energy coupled out of the cavity through the back mirror and this is smaller than the incident pulse by the factor 10^{-6}. Synchronous pumping addresses this problem. The pulse energy in the cavity builds to 10^{+3} times the incident pulse energy and the energy coupled out of the cavity through the back mirror initially is equal to the incident pulse energy and then decays exponentially in time. To demonstrate the feasibility of the technique we measured the (8,2,6)–(9,3,7) transition of H_2O diluted in helium [11]. The measured transition spectrum is compared to HITRAN96 in Fig. 5.

To proceed quickly with a demonstration of this CRDS technique, a number of compromises were made. A single detector was used at the output of the spectrometer instead of a commercially available 30 element array. Furthermore, our 30 year old FEL delivered optical pulses at 10 Hz, and without cavity length stabilization, synchronized pumping actually occurred at less than 1 Hz.

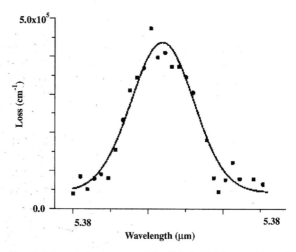

Fig. 5. Measured spectrum of the (8,2,6)–(9,3,7) transition of H_2O diluted in helium compared to HITRAN96.

Despite these compromises, the sensitivity demonstrated was comparable to the best achieved in the literature. Now, however, with the new linac structures we expect a pumping rate of 100 Hz. This development and the array detector will dramatically improve the data collection rate and the sensitivity and thus open new opportunities in gas-phase spectroscopy. But is it realistic to synchronously pump an external cavity at 100 Hz? With the new linac we will once again be able to operate the FEL continuously. If we modulate the cavity length at 50 Hz, we will pass through the synchronous pumping resonance condition at a 100 Hz rate. And each time we pass through resonance, we will have acquired both a measured decay and a measurement of drift in the cavity length which can be used to stabilize operation. The remaining question is whether passage through resonance is adiabatic. For reasonable cavity parameters, the resonance width (change in cavity length) is 4 nm and the cavity decay time is 2 μs. With a modulation amplitude of 300 nm and a frequency of 50 Hz the velocity is 10^{-4} m/s which satisfies the adiabatic condition by an order of magnitude.

The scheme for modulating the length of an external cavity to achieve periodic synchronous pumping might also be useful in other gas phase spectroscopic techniques such as FT Ion Cyclotron Resonance Mass Spectroscopy. With the new linac installed the single optical pulse energy will increase from 1 to 5 μJ and the pulse circulating in the cavity will be 500 μJ. For the modulation conditions described above the FWHM time on resonance is approximately 40 μs and the total energy passing through the interaction region is more than 1 J in that time.

5. Conclusions

With the emergence of optical parametric amplifiers and oscillators, and THz sources, the domain of infrared FELs is less that it had been one decade ago. None-the-less, FELs have unique advantages that can be exploited in important classes of optical experiments. The task at hand for IR FEL facilities is to identify those opportunities and provide adequate experimental time at competitive costs.

Acknowledgements

The author would like to acknowledge the remarkable contributions of the many individuals that have been part of the FEL program at Stanford over the past 30 years, but most particularly, his long term colleague Todd Smith. This work was supported in part by the Air Force Office of Scientific Research, grant number F49620-00-1-0349.

References

[1] W.B. Colson, E.D. Johnson, M.L. Kelley, H.A. Schwettman, Phys. Today (2003) 35.
[2] M.S. McAshan, K. Mittag, H.A. Schwettman, L.R. Suelzle, J.P. Turneaure, Appl. Phys. Lett. 22 (11) (1973) 605.
[3] D.A.G. Deacon, L.R. Elias, J.M.J. Madey, G.J. Ramian, H.A. Schwettman, T.I. Smith, Phys. Rev. Lett. 38 (16) (1977) 892.
[4] TESLA reports describing the DESY structure are available at http://tesla.desy.de.
[5] E.R. Crosson, G.E. James, H.A. Schwettman, T.I. Smith, R.L. Swent, Multi-user operation at an FEL facility, Nucl. Instr. and Meth. B 144 (1998) 25.
[6] Private communications from Doug King, George Marcus, Richard Swent.
[7] G. Luepke, N.H. Tolk, L.C. Feldman, J. Appl. Phys. 93 (5) (2003) 2317.
[8] C.W. Rella, A. Kwok, K. Rector, J.R. Hill, H.A. Schwettman, D.D. Dlott, M.D. Fayer, Phys. Rev. Lett. 77 (8) (1996) 1648.
[9] K.W. Berryman, Design operation, and applications of a far-infrared free electron laser, Ph.D. Dissertation, Stanford University, CA, USA, 1995.
[10] P. Haar, Pulse stacking in the Stanford external cavity and photo-induced reflectivity in the infrared, Ph.D. Dissertation, Stanford University, CA, USA, 1997.
[11] E.R. Crosson, P. Haar, G.A. Marcus, H.A. Schwettman, B.A. Paldus, T.G. Spence, R.N. Zare, Rev. Sci. Instrum. 70 (1) (1999) 4.

Available online at www.sciencedirect.com

**NUCLEAR
INSTRUMENTS
& METHODS
IN PHYSICS
RESEARCH**
Section A

ELSEVIER   Nuclear Instruments and Methods in Physics Research A 528 (2004) 8–14

www.elsevier.com/locate/nima

FELs, nice toys or efficient tools?

A.F.G. van der Meer

FOM Institute for Plasma Physics 'Rijnhuizen', Edisonbaan 14, 3439 MN Nieuwegein, The Netherlands

Abstract

An FEL is an intrinsically interesting device and pushing its performance presents a natural challenge to a physicist. Nonetheless, the main justification for doing FEL research is of course its potential as a unique, versatile source of radiation to be employed for something useful. After 25 years of FEL research, one may wonder how efficient these tools have become. In this paper, I will reflect on this issue from the perspective of 10 years of operation of FELIX as a user facility.
© 2004 Elsevier B.V. All rights reserved.

PACS: 41.60Cr; 42.62Fi; 82.80Gk; 82.50Bc

Keywords: Free-electron lasers; Infrared; Spectroscopy

1. Introduction

Whereas the (technological) challenges of a free electron laser probably present the main personal motivation for the people active in the field, its potential as a unique and versatile source of radiation that could be used for a variety of purposes, has always been the main justification for the research on FELs. More than 25 years after the 'invention' and first demonstration of operation of an FEL [1], it seems appropriate to ask ourselves how far we have come on the way to realizing this potential. In this paper, I will address this question with the experience gained in 10 years of operation of the IR User Facility FELIX. I will therefore start with an evaluation of the efficiency

E-mail address: a.f.g.vandermeer@rijnh.nl
(A.F.G. van der Meer).

of FELIX as a tool for scientific research, before attempting to generalize to other (types of) FELs and applications.

2. The IR user facility 'FELIX'

FELIX consists of a normal conducting, 12–45 MeV linear accelerator that alternatively drives a far-IR FEL with partial waveguide that covers the spectral range from 25 to 250 μm, or a mid-IR FEL with a spectral range from 5 μm (3 μm on 3rd harmonic) to 40 μm. The general layout of FELIX is shown in Fig. 1. As usual for this type of linear accelerator, the output consists of bursts (macropulses) of micropulses. The spacing between the micropulses is either 1 ns (1 GHz-mode) or 40 ns (25 MHz-mode), the latter corresponding to the roundtrip time of the 6 m

0168-9002/$ - see front matter © 2004 Elsevier B.V. All rights reserved.
doi:10.1016/j.nima.2004.04.008

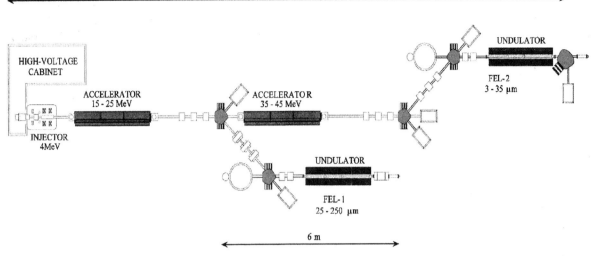

Fig. 1. General layout of FELIX showing the two beamlines for far- and mid-infrared generation.

Table 1
Characteristic parameters of FELIX

Tuning range	3–250 μm (3300–40 cm^{-1})
Rapid tuning	> 1 octave in a minute
Micropulse rep. rate	1 ns, 40 ns, single pulse
Macropulse rep. rate	up to 10 Hz
Micropulse energy	1–50 μJ
Micropulse power	up to 100 MW
Macropulse duration	4–8 μs
Bandwidth (adjustable)	0.4–6% (transform limited)
Polarization	> 99%
Beam quality	near diffraction limit
Beam hours	> 3000
Unscheduled downtime	< 5%

cavity. Using a transient optical switch, it is possible to slice a single micropulse out of the pulse train with an efficiency of more than 50%. The main characteristics of the output are listed in Table 1. The facility is operated in two-shift mode, 5 days a week, providing more than 3000 h of beam time for user experiments.

3. User experiments at FELIX

Presently, these experiments fall predominantly in one of two classes: relaxation phenomena in condensed matter or spectroscopy of gas phase species, (bio)molecules and clusters, either neutral or ionized. Experiments in the first class will use the 25 MHz or single-micropulse mode in view of the relaxation times involved, not only of the primary process but also of the temperature transient due to the energy absorbed. Especially in the case of biological samples the latter is usually more limiting. The second class is characterized by (very) low absorption cross-sections and, because detection is based on dissociation or ionization of the species, typically a strongly nonlinear dependence of the signal on the laser fluence, implying the use of the 1 GHz-mode.

For illustration, two examples of either class will be discussed briefly. The first example of the first class concerns the relaxation of the stretch vibration of hydrogen and deuterium in amorphous silicon. Hydrogen is often used to passivate the dangling bonds, thereby enhancing the characteristics of this technologically important material, but the beneficial effect strongly reduces with time. Recently, it was found that this aging effect is much smaller when deuterium is used instead. The experimental result for hydrogen is given in Fig. 2 [2] and for deuterium in Fig. 3 [3]. Whereas the decay for deuterium is clearly single-exponential, it is not for hydrogen. This observation can be related to the striking difference in energy decay

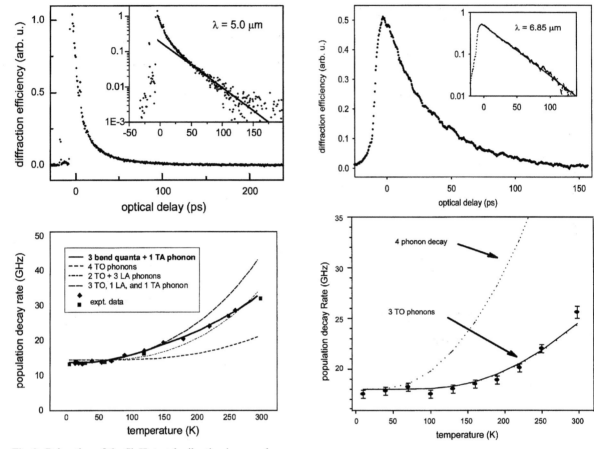

Fig. 2. Relaxation of the Si–H stretch vibration in amorphous Si: time dependence (upper panel) and temperature-dependent rate (lower panel) [2].

Fig. 3. Relaxation of the Si–D stretch vibration in amorphous Si: time dependence (upper panel) and temperature-dependent rate (lower panel) [3].

channel for both cases: whereas the relaxation for *H* occurs mainly via an almost resonant energy transfer to the bending mode at the same site and is therefore a local event, the relaxation of D primarily involves phonons, so modes of the bulk. This non-localized energy release is now believed to be the main reason for the strongly reduced 'aging' effect.

In the second example, far-IR radiation of FELIX was used to probe the time-dependent exciton density in a GaAs quantum well after excitation across the bandgap with a synchronized pump laser. The result was compared to a conventional measurement, i.e. by monitoring the luminescence of the sample in the spectral interval associated with the presence of excitons (Fig. 4) [4].

Whereas there is a prompt absorption signal, resulting from 1s–2p transitions of the bound electron–hole pairs, when the sample is excited very close to the bandgap, there is no signal if the excitation is well above the bandgap. In the latter case, the conventional method does show a rapid signal though. This result seems to provide experimental evidence for the possibility of exciton-like emission without excitons being present. Recently, this behaviour was predicted theoretically and is attributed to a correlation of the unbound electron and hole population. The rise time of the photo-luminescence reflects the time for the electrons and holes to relax into the low-lying k-states at the bottom of the band.

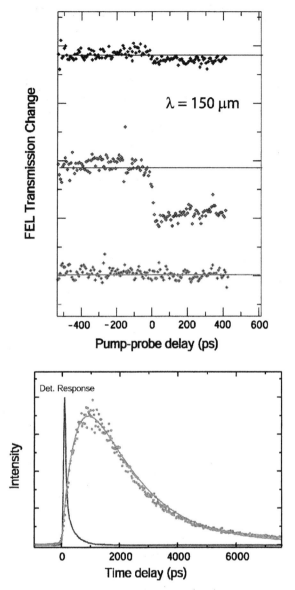

Fig. 4. The transient transmission change of the FIR probe pulse is shown in the upper panel for three cases: exciting at resonance (middle), 36 meV below resonance (bottom) and 62 meV above resonance (top). In the lower panel the transient photoluminescence at the 'exciton peak' is plotted while exciting 62 meV above resonance [4].

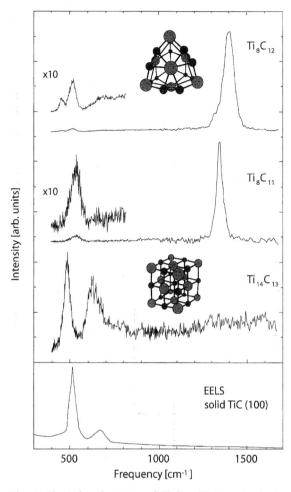

Fig. 5. The infrared spectra of Ti_8C_{12}, Ti_8C_{11} and $Ti_{13}C_{14}$ clusters measured using IR-REMPI as a function of the FELIX wavelength (three upper traces) [5]. The insets show the previously reported structures. As a comparison the lowest trace shows the EELS spectrum of bulk TiC(100).

As a first example of an experiment of the second class, a measurement of the IR spectrum of titanium-carbide clusters is shown in Fig. 5. A YAG-laser is used to ablate material from a titanium rod and subsequently a gas-puff, in this case methane, is applied, resulting in a molecular beam containing titanium-carbide clusters.

Previous work, using UV-lasers to ionize these clusters, showed that clusters consisting of 8 titanium and 12 carbon atoms were particularly stable, as well as clusters with, respectively, 14 and 13 atoms. Due its high fluence, it proved to be possible to ionize the clusters with the IR beam of FELIX and record an infrared spectrum [5]. A very strong signal around 8 μm, characteristic for a C–C stretch vibration, is present for the Ti_8C_{12} cluster and absent for the $Ti_{14}C_{13}$ cluster, in

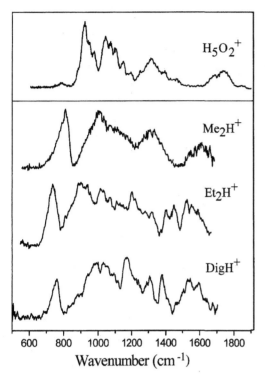

Fig. 6. IR-spectra of protonated water dimers [7] and p-methyl ether dimer, p-ethyl ether dimer and p-diglyme, respectively [8].

accordance with the ordering of the C-atoms in the structures that had been proposed (see insets in Fig. 5). The signal around $20 \mu m$ for the $Ti_{14}C_{13}$ cluster shows a great similarity with a spectrum recorded with EELS on bulk TiC, suggesting that it is indeed a nano-crystal. Moreover, this spectral feature is now believed to be the source of a hitherto unexplained strong emission around post asymptotic giant-branch stars [6]. In a similar manner, by looking for dissociation rather than ionization, the IR spectrum of ions can be recorded. Usually the ions are confined in a trap in order to increase the density. Recently, the IR spectrum of two water molecules bridged by a proton was measured using a tandem ion trap separated by a mass selector (Fig. 6) [7]. Using an FTICR high-resolution mass spectrometer, proton bridging of larger molecules was investigated [8]. The similarity of the main features of the spectra shows that these are really characteristic for proton bridging.

4. Performance assessment

From the above, I hope it is obvious that FELIX can be used for high-quality research, but what about productivity? A typical 3rd generation synchrotron has some 40 beam lines and an annual output of some 500 papers, whereas the number of user papers for the FELIX facility is only 20–25 per year. Given the difference in investment and running costs, typically a factor of 15, the relative output of FELIX however comes close to that of a synchrotron. So it seems justified to conclude that an FEL can be an efficient tool for scientific research. On the other hand, the total number of user papers that have been produced at FEL facilities is substantially less than twice the annual production of a single synchrotron! So, on the whole, the impact of FELs on scientific research has until now been almost negligible and does not yet justify the efforts put into their development.

5. Outlook

May we expect this balance to improve in the future, either by more and high-impact scientific research applications or applications in different areas, for example industrial processing? Following the successful demonstration of a 1 kW average power IR FEL, based on superconducting accelerator technology combined with energy recovery at Jefferson labs, very high average power (> 100 kW) units are now under study. At these power levels, military applications once again appear at the horizon, but history-based skepticism still seems justified. High average power would in principle also greatly increase the number of applications in industrial processing, and a successful application in a billion dollar market could already affect the balance. An industrial application that is clearly economically sound still needs to be demonstrated though. Returning to the use of FELs for scientific research, it should be noted that to my opinion the future of IR FELs is not very bright. Primarily because of the rapid development of alternative sources over the past 10 years, especially those based on today's work horse, the Ti:Sapphire laser. Table-top sources

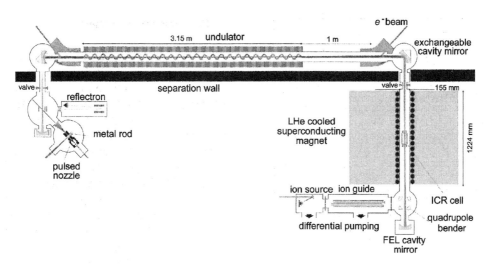

Fig. 7. General layout of the FELICE cavity including two intra-cavity setups: for IR-REMPI and IR-MPD experiments a high-resolution FTICR mass spectrometer and a molecular beam machine.

based on parametric generation are very competitive in the mid-IR, especially in the field of relaxation studies for which the pulse structure is often better suited for the time scales involved. In the far-IR, the micropulse energies available from FELs are still unchallenged and applications requiring high peak powers to pump the system far from equilibrium will also in the foreseeable future require the availability of an FEL. If the far-IR radiation is just used for probing, it is however not power that counts but sensitivity. Broad-band, 'single-cycle' THz pulses can be generated by focussing a short-pulse laser, again usually a Ti:Sapphire laser, onto a semiconductor such as GaAs. Typically these pulses have rather low energy, at the pJ level, but by using detection schemes that are based on (e.g. electro-optic) sampling with part of the pulse used for the generation, a very high sensitivity has been obtained: about 1 part in 10^8 for a 1 s integration time [9]. By varying the delay between the THz pulse and the sampling beam, very high time resolution can also be obtained. As the detection is often sensitive to the electric field of the THz pulse rather than its intensity, it has the added advantage that also phase information of the different frequency components within the bandwidth is obtained. As a matter of fact, the use of an FEL for the experiments given above as examples of the

first class can no longer be justified. FELs will most likely continue to have an advantage for experiments of the second class. This field of research is rapidly expanding at our facility and recently funding was obtained for the construction of a third beam line [10] that will allow this class of experiments to benefit from the much higher powers present within the FEL cavity. A schematic of FELICE, the Free Electron Laser for Intra-Cavity Experiments, that should cover the wavelength range from 3 to 100 μm, is shown in Fig. 7. Nonetheless, the niche for IR FELs has decreased quite significantly compared to when FELIX started operation and even though there will certainly be a need for a number of FEL-based IR facilities in the future, it is not realistic to expect an increase of the impact of FELs in the IR. Also in this respect the recent progress made towards very short-wavelength operation of FELs is of course very encouraging for the FEL community. But based on the experience in the infrared it should be realized that the success of X-FELs is not necessarily guaranteed merely by their realization.

References

[1] L.R. Elias, et al., Phys. Rev. Lett. 36 (1976) 717.

[2] M. van der Voort, et al., Phys. Rev. Lett. 84 (2000) 1236.

[3] J-P.R. Wells, et al., Phys. Rev. Lett. 89 (2002) 125504.

[4] R. Chari, et al., Phys. Rev. Lett., submitted for publication.

[5] D. van Heijnsbergen, et al., Phys. Rev. Lett. 83 (1999) 4983.

[6] G. von Helden, et al., Science 288 (2000) 313–316.

[7] K.R. Asmis, et al., Science 299 (2003) 1375.

[8] D.T. Moore, et al., Chem. Phys. Chem., (2004) in press.

[9] A. Leitenstorfer, et al., Physica B 314 (2002) 248.

[10] B.L. Militsyn, et al., Nucl. Instr. and Meth. A 507 (2003) 494.

Available online at www.sciencedirect.com

Nuclear Instruments and Methods in Physics Research A 528 (2004) 15–18

NUCLEAR
INSTRUMENTS
& METHODS
IN PHYSICS
RESEARCH
Section A

www.elsevier.com/locate/nima

First lasing at the high-power free electron laser at Siberian center for photochemistry research

E.A. Antokhin, R.R. Akberdin, V.S. Arbuzov, M.A. Bokov, V.P. Bolotin, D.B. Burenkov, A.A. Bushuev, V.F. Veremeenko, N.A. Vinokurov*, P.D. Vobly, N.G. Gavrilov, E.I. Gorniker, K.M. Gorchakov, V.N. Grigoryev, B.A. Gudkov, A.V. Davydov, O.I. Deichuli, E.N. Dementyev, B.A. Dovzhenko, A.N. Dubrovin, Yu.A. Evtushenko, E.I. Zagorodnikov, N.S. Zaigraeva, E.M. Zakutov, A.I. Erokhin, D.A. Kayran, O.B. Kiselev, B.A. Knyazev, V.R. Kozak, V.V. Kolmogorov, E.I. Kolobanov, A.A. Kondakov, N.L. Kondakova, S.A. Krutikhin, A.M. Kryuchkov, V.V. Kubarev, G.N. Kulipanov, E.A. Kuper, I.V. Kuptsov, G.Ya. Kurkin, E.A. Labutskaya, L.G. Leontyevskaya, V.Yu. Loskutov, A.N. Matveenko, L.E. Medvedev, A.S. Medvedko, S.V. Miginsky, L.A. Mironenko, S.V. Motygin, A.D. Oreshkov, V.K. Ovchar, V.N. Osipov, B.Z. Persov, S.P. Petrov, V.M. Petrov, A.M. Pilan, I.V. Poletaev, A.V. Polyanskiy, V.M. Popik, A.M. Popov, E.A. Rotov, T.V. Salikova, I.K. Sedliarov, P.A. Selivanov, S.S. Serednyakov, A.N. Skrinsky, S.V. Tararyshkin, L.A. Timoshina, A.G. Tribendis, M.A. Kholopov, V.P. Cherepanov, O.A. Shevchenko, A.R. Shteinke, E.I. Shubin, M.A. Scheglov

Budker Institute of Nuclear Physics, Acad. Lavrentyev Prospect 11, 630090 Novosibirsk, Russia

Abstract

The first lasing near wavelength 140 μm was achieved in April 2003 on a high-power free electron laser (FEL) constructed at the Siberian Center for Photochemical Research. In this paper, we briefly describe the design of FEL driven by an accelerator–recuperator. Characteristics of the electron beam and terahertz laser radiation, obtained at the first experiments, are also presented in the paper.
© 2004 Elsevier B.V. All rights reserved.

PACS: 41.60.Cr

Keywords: Free electron laser; Energy recovery linac

---

*Corresponding author. Tel.: +7-3832-394003; fax: +7-3832-342163.
E-mail address: vinokurov@inp.nsk.su (N.A. Vinokurov).

0168-9002/$ - see front matter © 2004 Elsevier B.V. All rights reserved.
doi:10.1016/j.nima.2004.04.009

1. Introduction

A new source of terahertz radiation was commissioned recently in Novosibirsk. It is CW FEL based on an accelerator–recuperator, or an energy recovery linac (ERL). It differs from the earlier ERL-based FELs [1,2] in the low-frequency non-superconducting RF cavities and longer wavelength operation range. The terahertz FEL is the first stage of a bigger installation, which will be built in 3 years and will provide shorter wavelengths and higher power. The facility will be available for users in 2004. The first radiation study results are discussed in this paper.

2. Accelerator–recuperator

Full-scale Novosibirsk free electron laser is to be based on multi-orbit 50 MeV electron accelerator–recuperator. It is to generate radiation in the range 3 µm–0.3 mm [3,4]. The first stage of the machine contains a full-scale RF system, but has only one orbit. Layout of the accelerator–recuperator is shown in Fig. 1. The 2 MeV electron beam from an injector passes through the accelerating structure, acquiring 12 MeV energy, and comes to the FEL, installed in the straight section. After interaction with radiation in the FEL the beam passes once more through the accelerating structure, returning the power, and comes to the beam dump at the injection energy. Main parameters of the accelerator are listed in Table 1.

The FEL is installed in a long straight section of a single-orbit accelerator–recuperator. It consists of two undulators, a magnetic buncher, two mirrors of the optical resonator, and an outcoupling system. Both electromagnetic planar undulators are identical. The length of an un-

dulator is 4 m, period is 120 mm, the gap is 80 mm, and deflection parameter K is up to 1.2. One can use one or both undulators with or without a magnetic buncher. The buncher is simply a three-pole electromagnetic wiggler. It is necessary to optimize the relative phasing of undulators and is used now at low longitudinal dispersion $N_d < 1$. Both laser resonator mirrors are identical, spherical, 15 m curvature radius, made of gold plated copper, and water cooled. In the center of each mirror there is a 3.5 mm diameter hole. It serves for mirror alignment (using He–Ne laser beam) and output of small amount of radiation. The distance between mirrors is 26.6 m. The outcoupling system contains four adjustable planar 45 copper mirrors (scrapers). These mirrors cut the tails of Gaussian eigenmode of the optical resonator and redirect radiation to calorimeters. This scheme preserves the main mode of optical resonator well and reduces amplification of higher modes effectively.

Table 1
Accelerator parameters (the first stage)

RF wavelength (m)	1.66
Number of RF cavities	16
Amplitude of accelerating voltage at one cavity (MV)	0.7
Injection energy (MeV)	2
Final electron energy (MeV)	12
Bunch repetition rate (MHz)	1.4–22.5
Average current (mA)	2–40
Beam emittance (mm mrad)	1
Final electron energy spread (%)	1
Final electron bunch length (ns)	0.02–0.1
Final peak electron current (A)	40–10

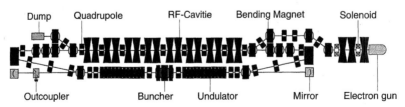

Fig. 1. Scheme of the first stage of Novosibirsk high-power free electron laser.

3. FEL commissioning

For FEL commissioning, we used both undulators. Beam average current was typically 5 mA at repetition rate 5.6 MHz, which is the round-trip frequency of the optical resonator and 32th subharmonics of the RF frequency $f \approx 180$ MHz. Most of measurements were performed without scrapers recording radiation flux from one of the mirror apertures. Instead of fine tuning of the optical resonator length we tuned the RF frequency. The tuning curve is shown in Fig. 2.

Typical results of spectrum measurement with rotating Fabri–Perot interferometer are shown in Fig. 3. They were used to find both wavelength and line width of radiation. Radiation wavelengths were in the range 120–180 μm depending on the undulator field amplitude. The shortest wavelength is limited by the gain decrease at a low undulator field, and the longest one—by the optical resonator diffraction loss increase. Relative line width (FWHM) was near $3 \cdot 10^{-3}$. The corresponding coherence length $\lambda^2/2\Delta\lambda = 2$ cm is close to the electron bunch length, therefore we, probably, achieved the Fourier-transform limit.

The loss of the optical resonator was measured with a fast Schottky diode detector [5]. Its typical output is the pulse sequence with 5.6 MHz repetition rate. Switching off the electron beam, we measured the decay time (see Fig. 4). The typical round-trip loss values were from 5% to 8%.

The FEL oscillation was obtained not only at $f_0 = 5.6$ MHz bunch repetition rate, but at $f_0/2$, $f_0/3$, $f_0/4$ and $2f_0/3$. The time dependence of intensity at bunch repetition rate $f_0/4$ is shown in Fig. 5. Radiation decay time (and therefore resonator loss) can also be measured from this dependence. The dependence of power on loss is shown in Fig. 6. For example, operation at bunch repetition rate $f_0/4$ corresponds to four time more loss per one amplification. It indicates that our maximum gain is about 30%.

The absolute power measurements were performed in two ways. First we measured the power coming through the hole in the mirror without scrapers. Output coupling is very weak in this case, so the power was about 10 W. It corresponds to intra-cavity average power near 2 kW. Other measurements were performed with two (right

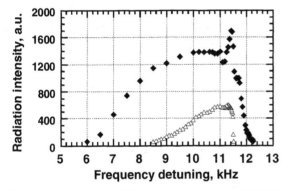

Fig. 2. Laser intensity vs RF frequency detuning f-180400 kHz (diamonds at repetition rate 5.6 MHz, triangles at repetition rate 2.8 MHz).

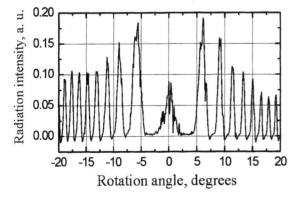

Fig. 3. Results of the Fabri–Perot interferometer rotation angle scanning (laser wavelength $\lambda = 136$ μm).

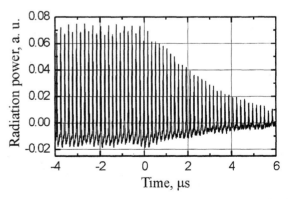

Fig. 4. Time dependence of the output radiation power after switching the electron beam off.

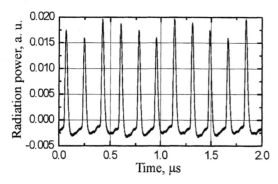

Fig. 5. The output radiation time dependence. Electron bunch repetition rate 1.4 MHz is four time less then the optical resonator round-trip frequency 5.6 MHz.

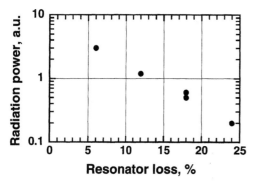

Fig. 6. Average intra-cavity power vs loss per one amplification.

Table 2
Expected radiation parameters for users

Wavelength (mm)	0.11–0.18
Pulse length (ns)	0.1
Peak power (MW)	0.1
Maximum repetition rate (MHz)	5.6–22.5
Average power (W)	100

undulator and high electron energy spread. Attempts to get oscillation with one undulator switched off are in progress. Possible way for decreasing of the energy spread—the installation of a 3rd harmonic (540 MHz) cavity—is under examination.

4. Further development

A beamline for transport radiation out of the accelerator hall to the user station rooms is under construction. The first experimental station is designed. The facility is to start operation for users in 2004. Expected radiation parameters for users are shown in Table 2.

and left) scrapers inserted. The insertion depth was chosen to decrease intra-cavity power twice. The measured power in each calorimeter was 20 W. Taking into account other resonator loss one can estimate the total power loss as 100 W. The electron beam power was 50 kW. Therefore, an electron efficiency is about 0.2%. The possible explanation of such a low value is too long

References

[1] G.R. Neil, et al., Phys. Rev. Lett. 84 (2000) 662.
[2] E.J. Minehara, Nucl. Instr. and Meth. A 483 (2002) 8.
[3] N.G. Gavrilov, et al., IEEE J. Quantum Electron. QE-27 (1991) 2626.
[4] V.P. Bolotin, et al., Proceedings of FEL-2000, Durham, USA, 2000, p. II-37.
[5] V.V. Kubarev, G.M. Kazakevich, Y.V. Jeong, B.C. Lee, Nucl. Instr. and Meth. A 507 (2003) 523.

Available online at www.sciencedirect.com

SCIENCE  DIRECT°

ELSEVIER   Nuclear Instruments and Methods in Physics Research A 528 (2004) 19–22

**NUCLEAR
INSTRUMENTS
& METHODS
IN PHYSICS
RESEARCH**
Section A

www.elsevier.com/locate/nima

First lasing of the IR upgrade FEL at Jefferson lab

C. Behre[a], S. Benson[a],*, G. Biallas[a], J. Boyce[a], C. Curtis[a], D. Douglas[a],
H.F. Dylla[a], L. Dillon-Townes[a], R. Evans[a], A. Grippo[a], J. Gubeli[a], D. Hardy[a],
J. Heckman[a], C. Hernandez-Garcia[a], T. Hiatt[a], K. Jordan[a], L. Merminga[a],
G. Neil[a], J. Preble[a], H. Rutt[b], M. Shinn[a], T. Siggins[a], H. Toyokawa[c],
D.W. Waldman[a], R. Walker[a], N. Wilson[a], B. Yunn[a], S. Zhang[a]

[a] Jefferson Lab, MS 6A, TJNAF, Newport News, 12000 Jefferson Avenue, VA 23606, USA
[b] Southampton University, Southampton, UK
[c] KEK, Tsukuba, Japan

Abstract

We report initial lasing results from the IR Upgrade FEL at Jefferson Lab (Proceedings: 2001 Particle Accelerator Conference, IEEE, Piscataway, NJ, 2001). The electron accelerator was operated with low average current beam at 80 MeV. The time structure of the beam was 120 pC bunches at 4.678 MHz with up to 750 μs pulses at 2 Hz. Lasing was established over the entire wavelength range of the mirrors (5.5–6.6 μm). The detuning curve length, turn-on time, and power were in agreement with modeling results assuming a 1 ps FWHM micropulse. The same model predicts over 10 kW of power output with 10 mA of beam and 10% output coupling, which is the ultimate design goal of the IR Upgrade FEL. The behavior of the laser while the dispersion section strength was varied was found to qualitatively match predictions. Initial CW lasing results also will be presented.
© 2004 Elsevier B.V. All rights reserved.

PACS: 29.20.c; 29.27.−a; 41.60.Cr; 42.60.−v

Keywords: High power; Undulators

1. Introduction

In previous work [1] we have described the design of the IR upgrade FEL. The layout is shown in Fig. 1. The injector is similar to the one used in the IR Demo FEL [2] but is capable of 10 mA of current. The linear accelerator will have three cryomodules capable of accelerating the injected 9.2 MeV beam to 150 MeV. Like the IR Demo, the beam transport lattice uses a Bates bend design but the magnets and vacuum chamber have been modified to accept 15% energy spread. The FEL gain medium is an optical klystron [3], operated at low dispersion, and

*Corresponding author. Tel.: 1-757-269-5026; fax: 1-757-269-5519.
E-mail address: felman@jlab.org (S. Benson).

0168-9002/$ - see front matter © 2004 Elsevier B.V. All rights reserved.
doi:10.1016/j.nima.2004.04.010

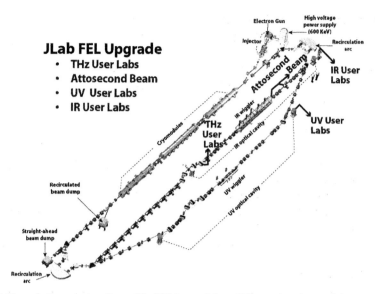

JLab FEL Upgrade
- **THz User Labs**
- **Attosecond Beam**
- **UV User Labs**
- **IR User Labs**

Fig. 1. Layout of the IR Upgrade Free-electron Laser. The UV line and the middle accelerating module are not yet installed. The first arc 180° dipole was temporarily replaced with one from the IR Demo.

the optical cavity is 32 m in length with a deformable high reflector to keep the Rayleigh range constant as the output coupler distorts due to absorbed power.

This paper reports on initial lasing results in two configurations. In the "first-light" configuration only two of the three cryomodules were installed, allowing an electron beam energy of 80 MeV. The first 180° bend magnet was not available and was replaced with a dipole magnet from the IR Demo capable of accepting an 80 MeV beam. The second arc was not complete. The electron beam downstream of the optical klystron was deflected to an air-cooled beam dump.

The "high-power" configuration added the second arc and an energy recovery beam dump, allowing energy recovery and full continuous current operation to 10 mA at 80 MeV. The only items in the final machine design not present are the first 180° bend magnet, the second cryomodule in the linac, the final sextupole magnets capable of operating at 210 MeV, and the octupoles necessary for high extraction efficiency operation at 150 MeV. None of these are required for 10 kW operation.

2. First lasing results

Installation of all systems necessary for the first-light mode of operation was complete by the time of first electron emission from the gun on May 7, 2003. After 3 weeks dedicated to processing the injector SRF cavities up to full gradient, the beam was accelerated to full energy and quickly steered to the temporary low-power dump. A week spent in careful setup of the machine allowed the start of the search for the proper cavity length, however no lasing was seen. It was discovered that the vault floor had expanded since the cavity length was initially surveyed and the proper length was out of the scan range of the cavity. When this was corrected the system lased easily on June 17, just 6 weeks after first beam.

The optical cavity for first light had a deformable 6 μm high reflector and an uncooled output coupler with 98% reflectivity. Since the air-cooled dump could not handle high power, the accelerator was limited to 600 μA average current during a 750 μs macropulse at 2 Hz. The bunch charge was approximately 120 pC.

A Joulemeter was used to measure the macropulse energy and a fast quantum detector was used to measure the macropulse shape. From these measurements we determined that the power extracted from the FEL during the macropulse was 280 W. The cavity length detuning curve was 8–10 μm in length and the turn-on time was 40 μs. The laser tuned easily from 5.5 to 6.6 μm (the reflectivity range of the mirrors). The FEL also lased at the third harmonic when the fundamental wavelength was set to 18 μm.

The turn-on time and detuning curve length indicate that the small signal gain is approximately 18%. The threshold of 16 pC, the power output, and the gain at 120 pC are consistent with having a charge distribution consisting of 80% of the charge in a 1 ps FWHM bunch and the remainder in low current tails.

In agreement with Ref. [3], the performance of the optical klystron is sensitive to the exact value of the dispersion when operated in the small dispersion regime. Gain peaks shift within the overall gain envelope as the dispersion is varied. When a gain peak is centered on the envelope, the gain is highest. When two gain peaks are present and located symmetric about the center of the overall gain envelope, the gain is lowest. The gain varies in a cycloidal fashion as the dispersion is varied. There is a stair step behavior as well due to interference between the wiggler ends. This leads to the gain being slightly higher for the even values of dispersion than for odd values.

The power was measured in the IR upgrade as the dispersion was varied. The results are shown in

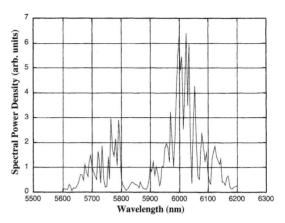

Fig. 3. Bistable lasing spectrum. The curve shown is the long-term average behavior. During each macropulse, the laser is lasing in one or the other line.

Fig. 2. The power shows a variation similar to the expected gain dependence including the stair step and the cycloidal dependence.

When the gain-to-loss ratio of the two gain peaks in the optical klystron is the same, the laser wavelength is very sensitive to initial conditions and can jump from one peak to the other on different macropulses. This behavior was seen in the IR upgrade. The spectrum in Fig. 3 was taken when lasing in either wavelength was equally likely. The spectrum is noisy because the wavelength is jumping back and forth between the two lines. The wavelength was constant for each macropulse and was centered on one spectral line or the other. The group at Stanford University has documented similar behavior using their Firefly wiggler operated in optical klystron mode [4]. When the dispersion was adjusted for stable, single spectral line operation, the wavelength shifted uniformly as the dispersion changed.

3. Present status

On June 25, 2003 the accelerator was shut down so that the second arc string and the energy recovery dump could be completed. A set of high-power optics centered at 10 μm was also installed. Operations recommenced one month later. In 2 weeks, we set up the beam through the entire machine and delivered beam to the energy

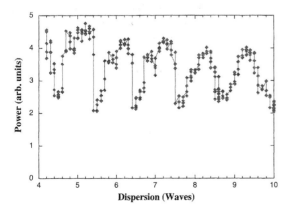

Fig. 2. Power vs. optical klystron dispersion. The dispersion is the net optical phase advance between the wigglers.

recovery dump. After commissioning the machine protection system, CW beam up to 2.4 mA was run. On August 19, we achieved over 300 W of CW laser power using the first light optics at 6 μm. Since the output coupler was uncooled we did not operate at higher power. We also achieved 30 W with a 6% duty cycle, which corresponds to 500 W during the macropulse.

After achieving CW lasing at 6 μm we attempted to lase CW with the 10 μm mirrors. Though we managed to obtain 10 μm lasing, the extraction efficiency was quite poor. We think this was due to a mirror figure error caused by improper mounting. We therefore replaced the 10 μm output couplers and installed a high power water-cooled 6 μm output coupler. We could not get the 6 μm coupler to lase, again due to a mounting problem. We have managed to get two different 10 μm mirror sets to lase but the absorption of these mirrors is much higher than expected. We could obtain 250 W from this mirror set and 600 W pulsed but the mirror distortion kept the power from going higher. We demonstrated that the high reflector could be used to keep the Rayleigh range constant as the output coupler distorted. On-line mirror diagnostics allowed us to monitor the change in the radius of curvature vs. power and estimate the absorbed power.

4. Conclusions

To date, commissioning has been quite rapid and smooth, demonstrating the validity of the design model. The IR Upgrade accelerator has performed in good agreement with design expectations. The one area where performance falls below specifications is the bunch length at the FEL. The bunch length appears to be at least twice as large as the design. We are working on improving the diagnostics to allow us to set up the longitudinal match to get better bunching at the FEL.

The optical klystron behaves as expected. The high reflectors show no sign of absorption with 2500 W circulating in the cavity as expected. There was also no sign of absorption in the first light output coupler. We are working to determine the source of the high absorption in the 10 μm output couplers. We should be able to correct the problem with the figure error in the 6 μm output coupler shortly and attempt to get high power with that mirror set.

We plan to install the third cryomodule this fall along with the second 210 MeV 180° dipole magnet and the sextupoles. This should allow operation at much shorter wavelengths and higher power levels.

Acknowledgements

This work supported by the Office of Naval Research, the Joint Technology Office, the Commonwealth of Virginia, the Air Force Research Laboratory, and by DOE Contract DE-AC05-84ER40150.

References

[1] S.V. Benson, G. Biallas, J. Boyce, D. Douglas, H.F. Dylla, R. Evans, A. Grippo, J. Gubeli, K. Jordan, G. Krafft, R. Li, J. Mammosser, L. Merminga, G.R. Neil, L. Phillips, J. Preble, M. Shinn, T. Siggins, R. Walker, B. Yunn, A 10 kW IRFEL Design For Jefferson Lab, in: P.W. Lucas, S. Webber (Eds.), Proceedings: 2001 Particle Accelerator Conference, IEEE, Piscataway, NJ, 2001.

[2] G.R. Neil, S. Benson, G. Biallas, C.L. Bohn, D. Douglas, H.F. Dylla, R. Evans, J. Fugitt, A. Grippo, J. Gubeli, R. Hill, K. Jordan, R. Li, L. Merminga, P. Piot, J. Preble, M. Shinn, T. Siggins, R. Walker, B. Yunn, Phys. Rev. Lett. 84 (2000) 662.

[3] S. Benson, Nucl. Instr. and Meth. A 507 (2003) 40.

[4] K.W. Berryman, T.I. Smith, Nucl. Instr. and Meth. A 375 (1996) 539.

Available online at www.sciencedirect.com

NUCLEAR INSTRUMENTS & METHODS IN PHYSICS RESEARCH
Section A

ELSEVIER Nuclear Instruments and Methods in Physics Research A 528 (2004) 23–27

www.elsevier.com/locate/nima

Radiation measurements in the new tandem accelerator FEL

A. Gover[a,*], A. Faingersh[a], A. Eliran[a], M. Volshonok[a], H. Kleinman[a],
S. Wolowelsky[a], Y. Yakover[a], B. Kapilevich[b], Y. Lasser[b], Z. Seidov[b], M. Kanter[b],
A. Zinigrad[b], M. Einat[b], Y. Lurie[b], A. Abramovich[b], A. Yahalom[b], Y. Pinhasi[b],
E. Weisman[c], J. Shiloh[c]

[a] *Department of Physical Electronics, Faculty of Engineering, Tel Aviv University, Tel Aviv, Israel*
[b] *Department of Electrical and Electronic Engineering, The College of Judea and Samaria, Ariel, Israel*
[c] *Rafael, Haifa, 31021, Israel*

Abstract

The Israeli tandem electrostatic accelerator FEL (EA-FEL), which is based on an electrostatic Van der Graaff accelerator was relocated to Ariel 3 years ago, and has now returned to operation under a new configuration. In the present FEL, the millimeter-wave radiation generated in the resonator is separated from the electron beam by means of a perforated Talbot effect reflector. A quasi-optic delivery system transmits the out-coupled power through a window in the pressurized gas accelerator tank into the measurement room (in the previous configuration, radiation was transmitted through the accelerator tubes with 40 dB attenuation). This makes it possible to transmit useful power out of the accelerator and into the user laboratories.

After re-configuring the FEL electron gun and the e-beam transport optics and installing a two stage depressed collector, the e-beam current was raised to 2 A. This recently enabled us to measure both spontaneous and stimulated emissions of radiation in the newly configured FEL for the first time. The radiation at the W-band was measured and characterized. The results match the predictions of our earlier theoretical modeling and calculations.
© 2004 Elsevier B.V. All rights reserved.

PACS: 41.60.−m; 41.60.Cr

Keywords: FEL; Linear accelerators

1. Introduction

The Israeli electrostatic accelerator FEL (EA-FEL) is based on a 6 MeV EN-Tandem Van de Graaff accelerator (shown in Fig. 1), which was originally used as an ion accelerator for nuclear

physics experiments [1]. The scheme employs straight geometry for the electron beam transport, where the electron gun and the collector are installed outside the accelerator region, as illustrated in Fig. 2. Lasing was reported in a previous configuration, where radiation was transmitted through the accelerator tubes with 40 dB attenuation [2,3].

In the present version of the FEL, which was relocated to Ariel, the millimeter-wave radiation

---

*Corresponding author.
E-mail address: gover@eng.tau.ac.il (A. Gover).

0168-9002/$ - see front matter © 2004 Elsevier B.V. All rights reserved.
doi:10.1016/j.nima.2004.04.011

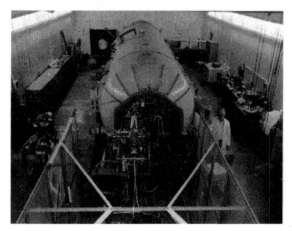

Fig. 1. The EN-Tandem electrostatic accelerator of the Israeli FEL.

Table 1
Parameters of the tandem electrostatic accelerator FEL

Accelerator	
Electron beam energy:	$E_k = 1-3\,\text{MeV}$
Beam current:	$I_0 = 1-2\,\text{A}$
Undulator	
Type:	Magneto-static planar wiggler
Magnetic induction:	$B_W = 2\,\text{kG}$
Period length:	$\lambda_W = 4.444\,\text{cm}$
Number of periods:	$N_W = 20$
Resonator	
Waveguide:	Curved-parallel plates
Transverse mode:	TE_{01}
Round-trip length:	$L_C = 2.62\,\text{m}$
Out-coupling coefficient:	$T = 7\%$
Total round-trip reflectivity:	$R = 65\%$

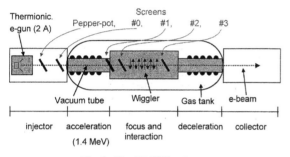

Fig. 2. The EA-FEL scheme.

generated in the resonator is separated from the electron beam by means of a perforated Talbot effect reflector [4,5]. A quasi-optic delivery system transmits the out-coupled power through a window in the pressurized gas accelerator tank. The basic parameters of the FEL are summarized in Table 1. The acceleration voltage is set to be $E_k = 1.4\,\text{MeV}$ in order to tune the frequency of the FEL radiation to the W-band near 100 GHz.

In the following sections, we present an analysis and the results of spontaneous and stimulated emissions measurements carried out recently.

2. Spontaneous emission in a resonator

Random electron distribution in the e-beam causes fluctuations in current density, identified as shot noise in the beam current. Electrons passing through a magnetic undulator emit a partially coherent radiation, which is called *undulator synchrotron radiation*. The electromagnetic fields excited by each electron add incoherently, resulting in a *spontaneous emission* with generated power spectral density [6]:

$$\frac{dP_{sp}(L_W)}{df} = \tau_{sp} P_{sp}(L_W) \text{sinc}^2\left(\frac{1}{2}\theta L_W\right) \qquad (1)$$

where $P_{sp}(L_W)$ is the *expected value* of the spontaneous emission power, $\tau_{sp} = |(L_W/V_{z0}) - (L_W/V_g)|$ is the slippage time and $\theta = (2\pi f/V_{z0}) - (k_z + k_w)$ is the detuning parameter (V_{z0} is the axial velocity of the accelerated electrons and V_g is the group velocity of the generated radiation). The spontaneous emission null-to-null bandwidth is approximately $2/\tau_{sp} \approx 2(f_0/N_W)$. In a FEL, utilizing a magneto-static planar wiggler, the total power of the spontaneous emission is given by:

$$P_{sp}(L_W) = \frac{1}{8}\frac{eI_0}{\tau_{sp}}\left(\frac{a_w}{\gamma\beta_{z0}}\right)^2 \frac{Z}{A_{em}}L_W^2 \qquad (2)$$

where $Z \approx 2\pi f \mu_0/k_z$ is the mode impedance, and I_0 is the DC beam current. The expected value of the total spontaneous emission power generated inside the cavity is about $P_{sp}(L_W)/I_0 = 60\,\mu\text{W A}^{-1}$. The calculated spectrum of the spontaneous emission power of the Israeli EA-FEL, has a null-to-null bandwidth of 18 GHz.

At the resonator output, the spontaneous emission spectrum generated inside the resonator is modified by a Fabry–Perot spectral transfer-function [7]:

$$\frac{dP_{out}}{df} = \frac{T}{(1 - \sqrt{R})^2 + 4\sqrt{R}\sin^2(\frac{1}{2}k_z L_c)} \cdot \frac{dP_{sp}(L_W)}{df} \tag{3}$$

where L_c is the resonator (round-trip) length, R is the total power reflectivity of the cavity, T is the power transmission of the out-coupler and $k_z(f)$ is the axial wavenumber of the waveguide mode. The maxima of the resonator transfer function factor occur when $k_z(f_m) \cdot L_c = 2m\pi$ (where m is an integer), which defines resonant frequencies f_m of the longitudinal modes. The *free-spectral range* (FSR) (the inter-mode frequency separation) is given by $FSR = v_g/L_c = 113$ MHz. The transmission peak is $T/(1 - \sqrt{R})^2 = 1.6$ with *full-width half-maximum* (FWHM) bandwidth of $FWHM = FSR/F = 7.76$ MHz, where $F = \pi\sqrt[4]{R}/(1 - \sqrt{R}) = 14.56$ is the *Finesse* of the resonator. The spectral line-shape of the spontaneous emission power obtained at the resonator output of the EA-FEL is shown in Fig. 3.

The *noise equivalent bandwidth* is defined as the bandwidth of an ideal band-pass filter producing the same noise power at its output. The noise equivalent bandwidth of a single resonant longitudinal mode is $B = (\pi/2)$ FWHM = 12.2 MHz. Consequently, the spontaneous emission power of mode m is given by

$$P_{sp}^{out}(m) = \frac{T}{(1 - \sqrt{R})^2} \frac{dP_{sp}(L_W)}{df}\bigg|_{f_m} B. \tag{4}$$

The typical bandwidth of the generated spontaneous emission power spectrum (1) is $1/\tau_{sp} = 9$ GHz. The number of longitudinal modes within the spontaneous emission bandwidth is then $N_{modes} = (1/\tau_{sp}) \cdot (1/FSR) \cong 80$. Thus the total spontaneous emission power measured at the output of the resonator is given as follows:

$$P_{sp}^{out} = N_{modes} P_{sp}^{out} m \cong \frac{T}{(1 - R)^2} \cdot P_{sp}(L_W). \tag{5}$$

Using Eq. (2), we expect up to $P_{sp}(L_W) \cong 120\,\mu W$ spontaneous emission power to be generated inside the resonator. From (5), the power emitted from the resonator out-coupler is reduced to $P_{sp}^{out} = 24\,\mu W$. The attenuation of the wave-guiding system, which delivers the power

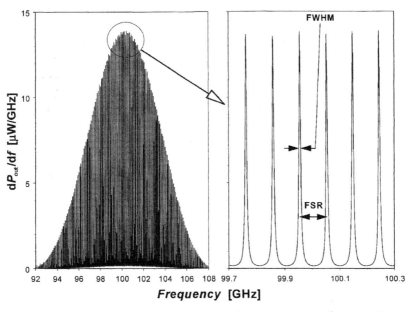

Fig. 3. Spontaneous emission power spectrum at resonator output (for $I_0 = 1$ A).

from the resonator, located inside the high-voltage terminal, to the measurement apparatus is 10 dB. Consequently, the spontaneous emission power expected at the detector sight is 2.4 μW. The traces shown in Fig. 4 describe the electron beam current pulse and the signal obtained at the detector video output correspond to the measured spontaneous emission RF power.

3. Stimulated emission

In the present operation regime of the FEL, the efficiency of energy extraction from the electron beam is given in terms of the number of wiggler's periods N_W by the approximate formula $\eta_{ext} \cong 1/2N_W = 2.5\%$. The stimulated radiation power generated inside the resonator at steady state is given as follows:

$$\Delta P = \eta_{ext} E_k I_0 \tag{6}$$

where $\Delta P \cong 35$ kW for a beam current of $I_0 = 1$ A. The resulting power obtained from the out-coupler is given as follows:

$$P_{out} = \frac{T}{1-R}\Delta P \tag{7}$$

and evaluated to be $P_{out} = 7$ kW. Considering the attenuation of the transmission system, 700 W is

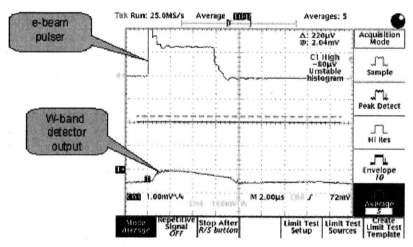

Fig. 4. Spontaneous emission power measurement.

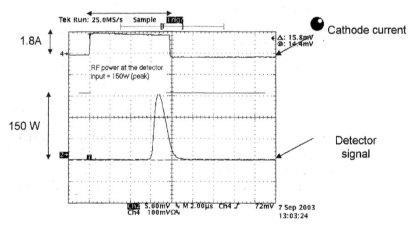

Fig. 5. Stimulated emission (lasing) power measurement.

expected at the detector. Fig. 5 shows the recent measurement of 150 W radiation power at the end of the optical transmission line in the measurement room. We note that in the present preliminary experiments, only a fraction of the cathode current was transported through the wiggler, and no beam circulation (transport up to the collector) was achieved. The charging of the terminal caused a voltage drop of the terminal of 125 kV during the pulse duration. Evidently, the FEL had not yet reached saturation because the radiation mode built inside the resonator went out of synchronism with the beam before reaching saturation.

Acknowledgements

This work was carried out at the Israeli FEL National Knowledge Center supported by the Ministry of Science, Israel, and was supported in part by the Ministry of Infrastructure.

References

[1] A. Gover, E. Jerby, H. Kleinman, I. Ben-Zvi, B.V. Elkonin, A. Fruchtman, J.S. Sokolowski, B. Mandelbaum, A. Rosenberg, J. Shiloh, G. Hazak, O. Shahal, Nucl. Instr. and Meth. A 296 (1990) 720.

[2] A. Abramovich, M. Canter, A. Gover, J.S. Sokolovski, Y.M. Yakover, Y. Pinhasi, I. Schnitzer, J. Shiloh, Phys. Rev. Lett. 82 (1999) 6774.

[3] A. Abramovich, et al., Appl. Phys. Lett. 71 (1997) 3776.

[4] I. Yakover, Y. Pinhasi, A. Gover, Nucl. Instr. and Meth. Phys. Res. A 445 (1996) 260.

[5] B. Kapilevich, A. Faingersh, A. Gover, Microwave Opt. Technol. Lett. 36 (2003) 303.

[6] Y. Pinhasi, Yu. Lurie, Phys. Rev. E 65 (2002) 026501.

[7] Y. Pinhasi, Yu. Lurie, A. Yahalom, Nucl. Instr. and Meth. A (2003), these Proceedings.

Available online at www.sciencedirect.com

SCIENCE  DIRECT°

ELSEVIER   Nuclear Instruments and Methods in Physics Research A 528 (2004) 28–33

NUCLEAR
INSTRUMENTS
& METHODS
IN PHYSICS
RESEARCH
Section A

www.elsevier.com/locate/nima

Femtosecond X-ray pulses from a frequency-chirped SASE FEL

Z. Huang[a],*, S. Krinsky[b]

[a] *Stanford Linear Accelerator Center, Stanford, CA 94309, USA*
[b] *Brookhaven National Laboratory, Upton, NY 11973, USA*

Abstract

We discuss the temporal and spectral properties of self-amplified spontaneous emission (SASE) utilizing an energy-chirped electron beam. A short temporal pulse is generated by using a monochromator to select a narrow radiation bandwidth from the frequency-chirped SASE. For the filtered radiation, the minimum pulse length is limited by the intrinsic SASE bandwidth, while the number of modes and the energy fluctuation can be controlled through the monochromator bandwidth. Two cases are considered: (1) placing the monochromator at the end of a single long undulator; (2) placing the monochromator after an initial undulator and amplifying the short-duration output in a second undulator. We analyze these cases and show that tens of femtosecond X-ray pulses may be generated for the linac coherent light source.
© 2004 Elsevier B.V. All rights reserved.

PACS: 41.60.Cr; 42.25.Kb

Keywords: Femtosecond X-ray; Frequency-chirped self-amplified spontaneous emission

1. Introduction

The generation of femtosecond X-ray pulses is critical to exploring the ultra-fast science at an X-ray free-electron laser (FEL) facility based on self-amplified spontaneous emission (SASE) [1,2]. Since the pulse length of typical electron bunches that drive the SASE FEL is on the order of 100 fs, many schemes have been proposed to reduce the pulse length of X-ray pulses generated from the

electron bunches that drive the SASE FEL [3–11]. Among them, a number of optical schemes are based on X-ray manipulations (such as compressing and slicing) of a frequency-chirped SASE [3,4,7]. In this paper, we study the temporal and spectral properties of a frequency-chirped SASE and show that the minimum X-ray pulse length in these optical schemes is determined by the SASE bandwidth and the amount of energy chirp allowed by the accelerator and the FEL performance. In particular, we consider two approaches that use a monochromator to control the radiation pulse for the linac coherent light source (LCLS)

---
*Corresponding author.
E-mail address: zrh@slac.stanford.edu (Z. Huang).

0168-9002/$ - see front matter © 2004 Elsevier B.V. All rights reserved.
doi:10.1016/j.nima.2004.04.012

and discuss the expected pulse length, the energy fluctuation, and the number of X-ray photons.

2. Statistical properties of a frequency-chirped SASE

In this section, we briefly discuss the statistical properties of a frequency-chirped SASE, based on the one-dimensional FEL analysis of Ref. [12]. Consider an energy-chirped electron bunch passing through a planar undulator having period $\lambda_w = 2\pi/k_w$ and rms undulator strength parameter a_w. The jth electron has energy γ_j (in units of its rest mass) and arrives at the undulator entrance at time t_j. Suppose that the electron energy chirp is linear and is specified by

$$\frac{\gamma_j - \gamma_0}{\gamma_0} = \alpha \frac{t_j}{T_b} \tag{1}$$

where $\gamma_0 mc^2$ is the central energy of the electron bunch, and T_b is the full pulse length of a flat-top electron bunch. Due to the resonance condition, the undulator radiation has a linear frequency chirp given by

$$\omega_j = k_j c = \frac{2\gamma_j^2 k_w c}{1 + a_w^2} \approx \omega_0 + u t_j \tag{2}$$

where $u = 2\alpha\omega_0/T_b$, and $\omega_0 = k_0 c = 2\gamma_0^2 k_w c/(1 + a_w^2)$.

In the exponential growth regime before saturation, the electric field of the frequency-chirped SASE at the undulator distance z is

$$E(z,t) \propto \sum_j e^{ik_j z - i\omega_j(t - t_j)} g(z, t - t_j, u). \tag{3}$$

To the first order in u, the Green's function can be approximated by

$$g(z, t - t_j, u) \approx \exp\left[(\sqrt{3} + i)\rho k_w z - \frac{3}{4}\left(1 + \frac{i}{\sqrt{3}}\right)\right.$$
$$\times \sigma_\omega^2 \left(t - t_j - \frac{z}{v_g}\right)^2\right]$$
$$\times \exp\left[-\frac{iu}{2}\left(t - t_j - \frac{z}{v_0}\right)\right.$$
$$\times \left.\left(t - t_j - \frac{z}{c}\right)\right] \tag{4}$$

where ρ is the FEL scaling parameter [13], σ_ω is the SASE bandwidth [14,15],

$$\frac{\sigma_\omega}{\omega_0} = \left(\frac{3\sqrt{3}}{k_w z}\rho\right)^{1/2} \tag{5}$$

$v_g = \omega_0/(k_0 + 2k_w/3)$ is the group velocity of the radiation wave packet, and $v_0 = \omega_0/(k_0 + k_w)$ is the average electron velocity. We note that in addition to the dependence of frequency on arrival time, $\omega_j \approx \omega_0 + ut_j$, the frequency chirp also appears in the Green's function since the gain process depends on the electrons within one radiation slippage length.

Eq. (3) determines the temporal and the spectral properties of a frequency-chirped SASE. Treating the arrival time t_j as a random variable, we can average over the stochastic ensemble to obtain the coherence time [12,16,17]

$$\tau_{coh} = \int d\tau \left|\frac{\langle E(z, t - \tau/2)E^*(z, t + \tau/2)\rangle}{\langle |E(z,t)|^2\rangle}\right|^2$$
$$= \frac{\sqrt{\pi}}{\sigma_\omega} \tag{6}$$

which is independent of the electron energy chirp. Fourier transforming Eq. (3), we obtain the radiation field in frequency domain $\tilde{E}(z, \omega)$. The range of spectral coherence is given by [12]

$$\Omega_{coh} = \int d\tau \left|\frac{\langle E(z, \omega - \Omega/2)E^*(z, \omega + \Omega/2)\rangle}{\langle |E(z,\omega)|^2\rangle}\right|^2$$
$$= |u|\frac{\sqrt{\pi}}{\sigma_\omega} = |u|\tau_{coh} \tag{7}$$

where we have assumed $u\tau_{coh} \gg 2\pi/T_b$ for a long electron pulse. Because of the noisy start-up, the frequency-chirped SASE is organized into M independent longitudinal modes in both time and frequency domains, where

$$M = \frac{T_b}{\tau_{coh}} = \frac{|u|T_b}{\Omega_{coh}}. \tag{8}$$

The relative radiation energy fluctuation is given by $M^{-1/2}$.

3. Short X-ray pulse generation

A monochromator can be used to select a short portion of the frequency-chirped SASE pulse. Let us assume that the monochromator is centered at ω_m with an rms bandwidth σ_m, the filtered electric field after the monochromator is

$$E_F(z, t) = \int \frac{d\omega}{2\pi} e^{-i\omega t} \tilde{E}(z, \omega) \exp\left[-\frac{(\omega - \omega_m)^2}{4\sigma_m^2}\right]. \quad (9)$$

The intensity of the sliced X-ray pulse can be computed as [12]

$$I_F \propto \langle |E_F(z, t)|^2 \rangle \propto e^{2\sqrt{3}\rho k_w z} \exp\left[-\frac{(t - t_m)^2}{2\sigma_t^2}\right] \quad (10)$$

where the pulse arrival time is

$$t_m(z) = \frac{\omega_m - \omega_0}{u} + \frac{1}{2}\left(\frac{z}{v_0} + \frac{z}{c}\right) \quad (11)$$

and the rms pulse duration σ_t is given by

$$\sigma_t^2(z) = \frac{\sigma_\omega^2(z) + \sigma_m^2}{u^2} + \frac{1}{4\sigma_m^2}. \quad (12)$$

Eq. (12) shows that the pulse duration cannot be made smaller than σ_ω/u, as illustrated in the phase space geometry of Fig. 1. This limitation to the attainable pulse length is applicable to both pulse slicing and compression. Another limitation of these optical schemes is the bandwidth of optical elements. In the case of a monochromator, an extremely narrow bandwidth will stretch the pulse due to Fourier transform limit, as expressed by the last term in Eq. (12). The optimal monochromator bandwidth that yields the minimum

pulse duration is

$$\frac{(\sigma_m)_{opt}}{\omega_0} = \sqrt{\frac{|u|}{2\omega_0^2}}. \quad (13)$$

The corresponding minimum rms pulse duration is

$$(\sigma_t)_{min} = \sqrt{\frac{\sigma_\omega^2 + |u|}{u^2}}. \quad (14)$$

In principle, for a given SASE bandwidth, one can increase the energy chirp α and hence the frequency chirp u to select a shorter pulse. The maximum chirp without significantly degrading the FEL gain is $u \sim \sigma_\omega^2$ [7]. From Eq. (14), we see that $(\sigma_t)_{min} \sim \sqrt{2}/\sigma_\omega \sim \tau_{coh}$, i.e., a single coherent spike may be selected. However, as shown in the following section, the required energy chirp for a single spike selection is typically too large for the normal accelerator operations.

Finally, the fractional shot-to-shot energy fluctuation σ_W/W after the monochromator can be expressed as [12]

$$\frac{\sigma_W^2}{W^2} = \frac{|u|}{\sqrt{u^2 + 4\sigma_m^2\sigma_\omega^2}} \equiv \frac{1}{M_F} \quad (15)$$

where

$$M_F = \sqrt{\frac{4\sigma_m^2\sigma_\omega^2}{u^2} + 1} \approx \frac{2\sigma_m}{|u|/\sigma_\omega} = \frac{2\sqrt{\pi}\sigma_m}{\Omega_{coh}}, \quad (16)$$

when $2\sqrt{\pi}\sigma_m \gg \Omega_{coh}$. Thus, M_F is the number of spectral modes that pass the monochromator with a "full" bandwidth $2\sqrt{\pi}\sigma_m$. The coherence time of the filtered radiation can be calculated similar to Eq. (6) as

$$(\tau_{coh})_F = \frac{2\sqrt{\pi}\sigma_t}{M_F}. \quad (17)$$

4. Application to the LCLS

We now apply these results to pulse slicing schemes proposed for the LCLS operated at the fundamental wavelength 1.5 Å (i.e., $\omega_0 = 1.2 \times 10^{19}$ s^{-1}) [1]. The norminal LCLS design has a nearly flat-top electron bunch with a FWHM duration $T_b = 230$ fs. Since the SASE intrinsic

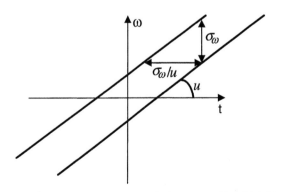

Fig. 1. Longitudinal phase space of a frequency-chirped SASE.

bandwidth is an important limiting factor to the attainable X-ray pulse length, we use the time-dependent FEL simulation code GINGER [18] to determine it along the LCLS undulator. Fig. 2 shows the normalized rms bandwidth $\sigma_\omega/(\rho\omega_0)$ and the power in the normal SASE operations of the LCLS. Near saturation, the minimum rms bandwidth is $(\sigma_\omega)_{\min} \approx \rho = 5 \times 10^{-4}$, as expected from the 1D formula (i.e., Eq. (5)). However, in the early exponential growth regime, the SASE bandwidth is substantially larger than the 1D prediction ($\propto z^{-1/2}$) due to the existence of higher-order transverse modes. Note that the coherence time near saturation is $\tau_{\mathrm{coh}} = \sqrt{\pi}/(\sigma_\omega)_{\min} \approx 0.3$ fs and the number of longitudinal modes is $M \approx 766$.

The maximum allowable chirp without significantly degrading the FEL gain is $u \sim \sigma_\omega^2$, corresponding to a linear energy chirp $\alpha \sim 35\%$, which is obviously too large for the normal acceleration operations. Tracking studies suggest that a maximum energy chirp of 2% across the whole bunch is possible for the LCLS accelerator, although a 1% energy chirp is more accessible [1]. Thus, we take $\alpha = 0.01$, $u = 7 \times 10^{-9}\omega_0^2 \ll \sigma_\omega^2$ and consider two cases: (1) placing the monochromator at the end of a single long undulator in the one-stage approach; (2) placing the monochromator after an initial undulator and amplifying the short-duration output in a second undulator in the two-stage approach.

4.1. One-stage approach

In the one-stage approach, the monochromator is placed at the end of the undulator where the frequency-chirped SASE reaches power saturation [3]. At this point, the SASE bandwidth also reaches the minimum. We can further minimize the sliced pulse length by choosing the optimal monochromator bandwidth given by Eq. (13), i.e.,

$$\frac{(\sigma_{\mathrm{m}})_{\mathrm{opt}}}{\omega_0} = \sqrt{\frac{|u|}{2\omega_0^2}} = 6 \times 10^{-5}. \tag{18}$$

Since $|u| \ll \sigma_\omega^2$, the minimum rms pulse duration from Eq. (14) is

$$(\sigma_t)_{\min} \approx \frac{(\sigma_\omega)_{\min}}{u} = 6 \text{ fs} \tag{19}$$

and the minimum FWHM pulse duration is about 15 fs. At this minimum, the number of longitudinal modes is $M_F \approx 2\sigma_{\mathrm{m}}\sigma_\omega/u = 9$, corresponding to a fractional energy fluctuation $M_F^{-1/2} = 33\%$.

In Fig. 3, we plot the sliced rms pulse duration σ_t and the number of modes M_F as a function of the monochromator bandwidth, given by Eqs. (12) and (16), respectively. The minimum of σ_t is broad as it is dominated by the relatively large SASE bandwidth. Since M_F depends nearly linearly on the monochromator bandwidth, one can change σ_{m} over the range $(\sigma_{\mathrm{m}})_{\mathrm{opt}} < \sigma_{\mathrm{m}} < \sigma_\omega$ to reduce the energy fluctuation with little increase in the pulse duration. For instance, at a much larger

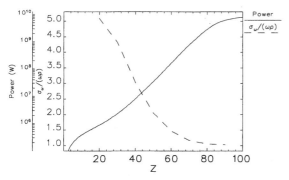

Fig. 2. GINGER simulation of SASE power (solid) and bandwidth (dashed) as a function of the undulator distance z in the LCLS undulator.

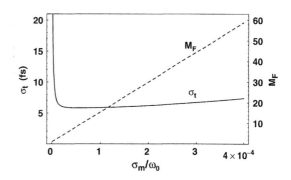

Fig. 3. RMS pulse duration σ_t (solid) and the number of modes M_F (dashed) as a function of the rms bandwidth σ_{m} of the monochromator placed at the end of the LCLS undulator ($\sigma_\omega/\omega \approx \rho = 5 \times 10^{-4}$) for a (full) energy chirp of 1%.

monochromator bandwidth $\sigma_m = 4 \times 10^{-4}$, Fig. 3 shows that the rms pulse duration is slightly increased to 7.5 fs, but the fractional energy fluctuation can be reduced to $1/\sqrt{60} \approx 13\%$. Since the number of X-ray photons passing through the monochromator is proportional to $\sigma_m \sigma_t \approx \sigma_m \sigma_\omega / u$ near the minimum pulse duration, a larger monochromator bandwidth also increases the available X-ray photons for the experimental stations. On the other hand, choosing a monochromator bandwidth $\sigma_m \approx 10^{-5}$ will produce a nearly Fourier transform-limited pulse with $M_F \approx 2$ and a similar pulse duration.

4.2. Two-stage approach

In the two-stage approach, the monochromotor is placed after the first undulator to select a short pulse from a frequency-chirped SASE in the exponential growth regime. The short-duration radiation is then amplified to saturation in the second undulator by the same energy-chirped electron bunch [7]. Since the peak power after the first undulator is much less than the saturation power, the damage to optical elements of the monochromator is reduced. However, the SASE bandwidth is also larger than the minimum bandwidth at saturation. For example, Fig. 2 shows that the relative rms SASE bandwidth is about $2\rho = 1 \times 10^{-3}$ at $z = 50\,\text{m}$ in the LCLS undulator, twice as large as the minimum bandwidth near saturation. From the analysis of previous sections, we see that the minimum rms pulse duration sliced at this undulator location is 12 fs (30 fs FWHW), two times larger than that given in Eq. (19). After its microbunching is destroyed in a bypass chicane, the chirped electron bunch is brought into the second undulator to amplify the short radiation pulse until it reaches the full saturation power. Note that for effective FEL interaction in the second undulator, the radiation pulse after the monochromator should overlap with the resonant part of the chirped electron bunch having the appropriate energy. Thus, the relative time delay between the optical and electron paths at the entrance of the second undulator can not be much longer than 10 fs.

5. Conclusions

In this paper, we present the statistical properties of a frequency-chirped SASE FEL, which form the basis for any optical manipulation of such a radiation pulse. We apply this analysis to the pulse slicing schemes using a monochromator and determine the minimum pulse duration, the number of modes and the energy fluctuation. Two pulse slicing configurations for the LCLS are discussed. In the one-stage approach, a shortest X-ray pulse (~ 10 fs) can be selected at the FEL saturation because the SASE bandwidth reaches the minimum. Increasing the monochromator bandwidth just below the SASE bandwidth only slightly lengthens the selected pulse, but significantly reduces the energy fluctuation and increases the X-ray photons for experiments. In the two-stage approach, the selected pulse is somewhat longer than that obtained from the one-stage approach because the slicing occurs at a larger SASE bandwidth before saturation, but it can be amplified in the second undulator to reach the same peak power as a saturated FEL.

Acknowledgements

We thank J. Arthur, M. Cornacchia, P. Emma, J. Galayda, J. Hastings, C. Pellegrini, S. Reiche, E. Saldin and C. Schroeder for useful discussions. This work was supported by the Department of Energy Contracts DE-AC03-76SF00515 and DE-AC02-76CH00016.

References

[1] Linac Coherent Light Source Conceptual Design Report, SLAC-R-593, 2002.
[2] TESLA Technical Design Report, DESY TESLA-FEL 2001-05, 2001.
[3] V. Bharadwaj, et al., LCLS-TN-00-8, 2000.
[4] C. Pellegrini, Nucl. Instr. and Meth. A 445 (2000) 124.
[5] W. Brefeld, et al., DESY 01-063, 2001.
[6] W. Brefeld, et al., Nucl. Instr. and Meth. A 483 (2002) 75.
[7] C.B. Schroeder, et al., Nucl. Instr. and Meth. A 483 (2002) 89;
C.B. Schroeder, et al., J. Opt. Soc. Am. B 19 (2002) 1782.

[8] E.L. Saldin, E.A. Schneidmiller, M.V. Yurkov, Opt. Commun. 212 (2002) 377.

[9] S. Reiche, P. Emma, C. Pellegrini, Nucl. Instr. and Meth. A 507 (2003) 426.

[10] P. Emma, Z. Huang, Nucl. Instr. and Meth. A (2003), these Proceedings.

[11] P. Emma, et al., SLAC-PUB-10002, Phys. Rev. Lett. 92, 074801 (2004).

[12] S. Krinsky, Z. Huang, Phys. Rev. ST Accel. Beams 6 050702 (2003).

[13] R. Bonifacio, C. Pellegrini, L.M. Narducci, Opt. Commun. 50 (1984) 373.

[14] K.-J. Kim, Nucl. Instr. and Meth. A 250 (1996) 396.

[15] J.-M. Wang, L.-H. Yu, Nucl. Instr. and Meth. A 250 (1986) 484.

[16] J.W. Goodman, Statistical Optics, Wiley, New York, 1985.

[17] E.L. Saldin, E.A. Schneidmiller, M.V. Yurkov, The Physics of Free Electron Lasers, Springer, Berlin, 2000.

[18] W. Fawley, Report LBNL-49625, 2002.

Available online at www.sciencedirect.com

SCIENCE DIRECT®

ELSEVIER   Nuclear Instruments and Methods in Physics Research A 528 (2004) 34–38

NUCLEAR
INSTRUMENTS
& METHODS
IN PHYSICS
RESEARCH
Section A

www.elsevier.com/locate/nima

Analysis of an FEL oscillator at zero detuning length of an optical cavity

Nobuyuki Nishimori*

Free Electron Laser Laboratory, Japan Atomic Energy Research Institute, Advanced Photon Research Center, 2-4 Shirakata-Shirane, Tokai Naka, Tokai-mura, Ibaraki 319-1195, Japan

Abstract

The field of an FEL oscillator at zero detuning length of an optical cavity ($\delta L = 0$) is characteristic of self-amplified spontaneous emission (SASE) FEL. For both cases, the field is composed of the precedent incoherent and subsequent coherent states. The evolution and saturation of the coherent field are studied analytically. The field during evolution is found to scale with FEL parameter ρ and the round-trip number, and that at saturation with ρ and an optical cavity loss. The field at $\delta L = 0$ is similar to that of the leading edge of SASE FEL with high electron beam density.
© 2004 Elsevier B.V. All rights reserved.

PACS: 41.60.Cr

Keywords: FEL oscillator; SASE; Zero detuning length; FEL parameter; Round-trip number; Optical cavity loss

1. Introduction

A recent experiment in the Japan Atomic Energy Research Institute (JAERI) has clearly shown that efficiency in an FEL oscillator can reach maximum at zero detuning length of an optical cavity ($\delta L = 0$) despite the lethargy effect [1]. The efficiency detuning curve obtained has been already reproduced by a time-dependent simulation code including shot-noise effects [2], but the physics responsible for the FEL at $\delta L = 0$ has not been clearly explained yet. A few

_____
*Tel.: +81-29-282-6315; fax: +81-29-282-6057.

E-mail address: nisi@milford.tokai.jaeri.go.jp
(N. Nishimori).

theoretical studies have followed the experiment and attributed the sideband instability [3] or the superradiance in short-pulse FELs [4] to the lasing at $\delta L = 0$. However, these studies are still based on numerical simulations.

The main difference of FELs between $\delta L = 0$ and $\delta L < 0$ is whether incident electrons interact with the field characterized by the steep intensity gradient on the leading edge similar to self-amplified spontaneous emission (SASE) [5]. The present study analyzes the interaction between a SASE field and electrons, and shows that intense few-cycle FELs are generated as a result of intensive energy transfer from electrons to the field at the peak at $\delta L = 0$. It is also found that the field during evolution scales with the FEL

0168-9002/$ - see front matter © 2004 Elsevier B.V. All rights reserved.
doi:10.1016/j.nima.2004.04.013

parameter ρ [6] and the round-trip number n, and that at saturation scales with ρ and an optical cavity loss α. The present analysis is based on the basic FEL process, that is energy modulation of electrons, the subsequent bunching and final radiation.

2. Basic equations

The slowly varying envelope approximation (SVEA) [7] is assumed in the present study. The initial electron energy $\gamma_0 mc^2$ is assumed to be resonant for radiation wavelength $\lambda = \lambda_w(1 + a_w^2)/(2\gamma_0^2)$, where $\lambda_w = 2\pi/k_w$ and a_w are undulator period and parameter, respectively. In order to deal with few-cycle fields, I choose unity for the number of undulator periods, through which electrons pass in a scaled time, instead of $1/(4\pi\rho)$ [6] or the total number of undulator periods N_w [8]. The fundamental FEL parameter ρ [6] is defined by

$$\rho = [ea_w F\sqrt{n_e/(\varepsilon_0 m)}/(4ck_w)]^{2/3}/\gamma_0 \qquad (1)$$

in MKSA units, where n_e is the electron beam density and F is unity for a helical undulator or Bessel function [JJ] for that of planar type [6]. I define dimensionless time by $\tau = ct/\lambda_w$, dimensionless optical field by

$$a(\zeta,\tau) = \frac{2\pi e a_w \lambda_w F}{\gamma_0^2 mc^2} E(\zeta,\tau)\exp[i\phi(\zeta,\tau)] \qquad (2)$$

with phase $\phi(\zeta,\tau)$, and dimensionless beam current by $[4\pi\rho(\zeta,\tau)]^3$ as similar to Ref. [8]. Here $E(\zeta,\tau)$ is the rms optical field strength. The longitudinal position, dimensionless energy and phase of the ith electron are, respectively, defined by $\zeta_i(\tau) = [z_i(t) - ct]/\lambda$, $\mu_i(\tau) = 4\pi[\gamma_i(t) - \gamma_0]/\gamma_0$ and $\psi_i(\tau) = (k_w + k)z_i(t) - \omega t$. In the present definition, the electron dynamics is represented by the following pendulum equations [8]:

$$\frac{d\mu_i(\tau)}{d\tau} = 2|a[\zeta_i(\tau),\tau]|\cos\{\psi_i(\tau) + \phi[\zeta_i(\tau),\tau]\} \qquad (3)$$

$$\frac{d\psi_i(\tau)}{d\tau} = \mu_i(\tau). \qquad (4)$$

The evolutions of FEL phase and amplitude are given by [8]

$$\frac{\partial\phi(\zeta,\tau)}{\partial\tau}$$
$$= \frac{[4\pi\rho(\zeta,\tau)]^3}{|a(\zeta,\tau)|}\langle\sin\{\psi_i(\tau) + \phi[\zeta_i(\tau),\tau]\}\rangle_{\zeta_i=\zeta} \qquad (5)$$

$$\frac{|a(\zeta,\tau)|}{\partial\tau}$$
$$= -[4\pi\rho(\zeta,\tau)]^3\langle\cos\{\psi_i(\tau) + \phi[\zeta_i(\tau),\tau]\}\rangle_{\zeta_i=\zeta}. \qquad (6)$$

The angular bracket shows an average over electrons around ζ within λ along the propagating direction [7].

3. Phase modulation of electrons at $\delta L = 0$

At $\delta L = 0$, the head of a round-trip FEL coincides with that of an incident electron pulse at the entrance to an undulator. Fig. 1 shows a semi-log plot of an FEL amplitude at saturation as a function of longitudinal position in units of resonant wavelength λ. This is obtained in a time-dependent simulation. The right-hand side shows the front edge of the FEL field. The zero is the

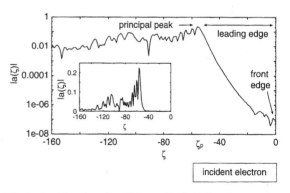

Fig. 1. Semi-log plot of an FEL amplitude $|a(\zeta)|$ at $\delta L = 0$ after saturation with respect to the longitudinal position ζ (solid line) together with an electron pulse at the entrance to an undulator. The positions of the front edge and the principal peak are 0 and ζ_p, respectively. The inset is a linear plot of $|a(\zeta)|$. The electron bunch length and slippage length are 100 λ and 60 λ, respectively. The FEL parameter is $\rho = 0.0045$ and an optical cavity loss is $\alpha = 0.05$.

position of the head of incident electrons at the entrance to an undulator. The position of the principal peak of the field is represented as ζ_p. The region from 0 to ζ_p is called the leading edge in this paper. The inset shows the linear plot of the amplitude.

The exponential increase on the leading edge can be attributed to the lethargy, because the trailing part of the field is amplified due to the lethargy [9]. In fact, the field at the first round trip in an oscillator is identical to that of SASE. It is therefore reasonable to assume that the input field on the leading edge at the round-trip number n is the same as the SASE with FEL parameter ρ_n. The input field on the leading edge is given by [8]

$$|a_n(\zeta)| = [|a(0)|/3] \exp[-2\pi\sqrt{3}\rho_n\zeta]$$
$$\phi_n(\zeta) = \phi(0) - 2\pi\rho_n\zeta. \tag{7}$$

Here $\rho_2 = \rho$. The interaction between the field similar to SASE and electrons is unique to FELs at $\delta L = 0$. An optical field at $\delta L < 0$ is pushed forward with round trips, and electrons interact with a field much stronger than that on the leading edge at $\delta L = 0$. The phase of the ith electron remains almost unchanged during the interaction with the leading edge field at $\delta L = 0$: $\psi_i(\tau) \approx \psi_i(0)$.

The energy modulation given to the ith electron at τ' is represented by $\Delta\mu_i(\tau') = 2|a_n[\zeta_i(\tau')]| \cos\{\psi_i(0) + \phi_n[\zeta_i(\tau')]\}\Delta\tau'$ from Eq. (3). The phase modulation at τ due to $\Delta\mu_i(\tau')$ is given by $\Delta\mu_i(\tau')(\tau - \tau')$ from Eq. (4). The phase of the ith electron is derived from the sum of those phase modulations during τ

$$\psi_i(\tau) = \psi_i(0) + 2|a_n[\zeta_i(0)]| \int_0^\tau e^{2\pi\sqrt{3}\rho_n\tau'} \cos\{\psi_i(0)$$
$$+ \phi_n[\zeta_i(\tau')]\}(\tau - \tau') \, d\tau'. \tag{8}$$

Substitution of Eq. (7) into Eq. (8) yields

$$\psi_i(\tau) = \psi_i(0)$$
$$+ \frac{|a_n[\zeta_i(\tau)]|}{(8\pi^2\rho_n^2)}(\cos\{\psi_i(0) + \phi_n[\zeta_i(\tau)] - \pi/3\}$$
$$- e^{-2\pi\sqrt{3}\rho_n\tau} \cos\{\psi_i(0) + \phi_n[\zeta_i(0)] - \pi/3\}$$
$$- 4\pi\rho_n\tau e^{-2\pi\sqrt{3}\rho_n\tau} \cos\{\psi_i(0)$$
$$+ \phi_n[\zeta_i(0)] - \pi/6\}). \tag{9}$$

4. Phase shift and gain of FELs at $\delta L = 0$

The phase shift and gain of $a_n(\zeta, \tau)$ due to an electron micro-bunch in units of λ, whose initial positions are around $\zeta + \tau$, are obtained by substituting Eq. (9) into Eqs. (5) and (6), respectively, as follows:

$$\frac{\partial\phi_n(\zeta, \tau)}{\partial\tau}$$
$$= 4\pi(\rho^3/\rho_n^2)\{1/2 - e^{-2\pi\sqrt{3}\rho_n\tau}[\cos(2\pi\rho_n\tau + \pi/3)$$
$$+ 4\pi\rho_n\tau\cos(2\pi\rho_n\tau + \pi/6)]\} \tag{10}$$

$$\frac{1}{|a_n(\zeta, \tau)|}\frac{\partial|a_n(\zeta, \tau)|}{\partial\tau}$$
$$= 4\pi(\rho^3/\rho_n^2)\{\sqrt{3}/2$$
$$- e^{-2\pi\sqrt{3}\rho_n\tau}[\sin(2\pi\rho_n\tau + \pi/3)$$
$$+ 4\pi\rho_n\tau\sin(2\pi\rho_n\tau + \pi/6)]\} \tag{11}$$

when $|a_n(\zeta)|/(8\pi^2\rho_n^2) \ll 1$ is satisfied. Eq. (11) is asymptotically equal to $2\pi\sqrt{3}\rho^3/\rho_n^2$ for $\rho_n\tau > 0.4$. This means that incident electrons lose almost the same amount of energy at ζ regardless of their initial positions, when the electrons have slipped back over the distance longer than $0.4/\rho_n$ on the leading edge. This results in an intensive energy transfer from electrons to the field at ζ_p. The intensive energy transfer is also reproduced in a time-dependent simulation. In order to understand why electrons lose their energy at ζ_p intensively, let us suppose that we stand at the position ζ_p and watch the electrons slipped back toward us. The electrons slipped back to us are largely modulated in energy at first, but not bunched so considerably. This is because the drift space is not long enough for bunching. However, at time greater than $0.4/\rho_n$ we see the electrons bunched in a similar way independent of their initial positions. This is because the effect of the precedent weak fields is negligibly small. Namely, the electrons are mainly bunched in strong fields in front of us.

The phase of the ith electron is asymptotically equal to

$$\psi_i(\tau) = \psi_i(0) + \{|a_n[\zeta_i(\tau)]|/(8\pi^2\rho_n^2)\}$$
$$\times \cos\{\psi_i(0) + \phi_n[\zeta_i(\tau)] - \pi/3\} \tag{12}$$

for $\rho_n\tau > 0.4$ from Eq. (9). The energy transfer from those electrons to the field is proportional to

$$-\left\langle \cos\left(\psi_i(0) + \phi_n(\zeta) + \frac{|a_n[\zeta_i(\tau)]|}{(8\pi^2\rho_n^2)} \cos\{\psi_i(0) + \phi_n[\zeta_i(\tau)] - \pi/3\}\right)\right\rangle_{\zeta_i(\tau)=\zeta} \tag{13}$$

from Eq. (6) together with Eq. (12). The calculation of the bracket as a function of $|a_n[\zeta_i(\tau)]|/(8\pi^2\rho_n^2)$ shows that $|a_n[\zeta_i(\tau)]|/(8\pi^2\rho_n^2) = 1.84$ at the maximum gain. In the calculation, a value of $\psi_i(0) + \phi_n[\zeta_i(\tau)]$ is distributed uniformly over 2π. Hence, the peak amplitude is given by

$$|a(0)|e^{-2\pi\sqrt{3\rho_n}\zeta_p}/3 = 1.84 \times 8\pi^2\rho_n^2 \tag{14}$$

where both an electron pulse L_b and a slippage length N_w are assumed to be longer than $|\zeta_p|$. From Eq. (14), the peak intensity is $|a(\zeta_p)|^2 = 2.1 \times 10^4\rho_n^4$, which is 60% of $1.4(4\pi\rho_n)^4$—peak intensity in the steady-state regime of a SASE with FEL parameter ρ_n [6]. The number of cycles, during which the intensity rises from 10% to 90% of the peak, is given by

$$(\ln 9)/(4\pi\sqrt{3}\rho_n) = 0.10/\rho_n. \tag{15}$$

The synchrotron oscillation of electrons leads to both steep decrease in intensity in rear of the principal peak and formation of subsequent peaks in $\zeta < \zeta_p$. The intensity of such secondary peaks are lower than the half of the principal peak (see Fig. 1). Thus the FWHM pulse width is nearly equal to Eq. (15).

The field gain per round trip due to electrons $[\partial a_n(\zeta)/\partial n]/a_n(\zeta) = [\partial|a_n(\zeta)|/\partial n]/|a_n(\zeta)| + i[\partial\phi_n(\zeta)/\partial n]$ is derived from integrations of Eqs. (10) and (11) from 0 to $-\zeta$ as follows:

$$\frac{1}{a_n(\zeta)}\frac{\partial a_n(\zeta)}{\partial n} = \left(\frac{\rho}{\rho_n}\right)^3[-4\pi\rho_n\zeta e^{i\pi/6}(1 + e^{4\pi\rho_n\zeta e^{i\pi/6}}) - 2(1 - e^{4\pi\rho_n\zeta e^{i\pi/6}})]. \tag{16}$$

Eq. (16) is asymptotically equal to

$$[\partial a_n(\zeta)/\partial n]/a_n(\zeta) = (\rho/\rho_n)^3(-4\pi\rho_n\zeta e^{i\pi/6} - 2) \tag{17}$$

for $\rho_n\zeta < -0.2$. The differentiation of Eq. (7) by n gives the field evolution per round trip

$$[\partial a_n(\zeta)/\partial n]/a_n(\zeta) = -4\pi e^{i\pi/6}(\partial\rho_n/\partial n)\zeta. \tag{18}$$

When the gain from electrons represented by Eq. (17) is much higher than $\alpha/2$, the gain is equal to the field evolution given by Eq. (18). This leads to the evolution of ρ_n:

$$\rho_n \approx \rho(3n - 5)^{1/3}. \tag{19}$$

Eq. (19) confirms the validity of the assumption that the field on the leading edge at $\delta L = 0$ is represented as SASE with FEL parameter ρ_n. Although the second term in the right-hand side of Eq. (17) is neglected for derivation of Eq. (19), a numerical calculation using Eqs. (16) and (18) confirms that the ρ_n increases with n in a similar way at $\rho\zeta < -0.1$, where the field of the first round-trip can be given by Eq. (7).

The gain at ζ in the nth round trip is approximately given by

$$[\partial|a_n(\zeta)|/\partial n]/|a_n(\zeta)| \approx -2\pi\sqrt{3}\rho\zeta(3n - 5)^{-2/3} \tag{20}$$

from Eqs. (17) and (19). The gain decreases with n and becomes equal to $\alpha/2$ at saturation. Since the gain decreases almost linearly with the decrease of $|\zeta|$, the evolution of ρ_n at smaller $|\zeta|$ stops at the smaller number of round trips. Therefore, ρ_n at saturation (ρ_s) depends on ζ and increases with the increase of $|\zeta|$.

A procedure to obtain ρ_s as functions of ρ and α is mentioned here. Since it seems difficult to analytically obtain an exact form of $\rho_s(\zeta)$ which includes the dependence on ζ, I make assumptions that $\rho_s(\zeta)$ is constant independent of ζ and that the gain at the peak is equal to $\alpha/2$. This results in $(\rho/\rho_s)^3(-2\pi\sqrt{3}\rho_s\zeta_p - 2) = \alpha/2$, where $\rho_s = \rho_s(\zeta_p)$. Substitution of the above equation into Eq. (14) yields

$$[|a_s(0)|/3]e^{\rho_s^3\alpha/(2\rho^3)+2} = 1.84 \times 8\pi^2\rho_s^2. \tag{21}$$

The amplitude $|a_s(0)|$ at the front edge is random superposition of incoherent radiation and is proportional to $\sqrt{\rho^3/\alpha}$ at the round-trip number greater than $1/\alpha$. Substitution of $|a_s(0)| \propto \sqrt{\rho^3/\alpha}$ into Eq. (21) results in $\rho_s = 3\rho\alpha^{-1/3}$ at ζ_p [10]. The peak amplitude and pulse width of an intense few-cycle field at $\delta L = 0$, which are given by Eqs. (14)

and (15), respectively, are found to scale with $(\rho^3\alpha^{-1})^{2/3}$ and $(\rho^3\alpha^{-1})^{-1/3}$, respectively, at saturation. The experimental results at JAERI-FEL [1] agree well with the scaling. The details will be described elsewhere [10].

5. Conclusion

An FEL interaction between incident electrons and a field with a steep intensity gradient on the leading edge is studied analytically. It is found that an intensive energy transfer, which is unique to the FEL at $\delta L = 0$, occurs at the principal peak on the leading edge due to the interaction. The intensive energy transfer accounts for generation of an intense few-cycle field at $\delta L = 0$. It is also found that the field during evolution scales with the FEL parameter ρ and the round-trip number, and that at saturation with ρ and a cavity loss.

References

[1] N. Nishimori, R. Hajima, R. Nagai, E.J. Minehara, Phys. Rev. Lett. 86 (2001) 5707;
    N. Nishimori, R. Hajima, R. Nagai, E.J. Minehara, Nucl. Instr. and Meth. A 483 (2002) 134.
[2] R. Hajima, N. Nishimori, R. Nagai, E.J. Minehara, Nucl. Instr. and Meth. A 475 (2001) 270.
[3] Z.-W. Dong, et al., Nucl. Instr. and Meth. A 483 (2002) 553.
[4] R. Hajima, R. Nagai, Phys. Rev. Lett. 91 (2003) 024801.
[5] K.-J. Kim, Phys. Rev. Lett. 57 (1986) 1871;
    K.-J. Kim, Nucl. Instr. and Meth. A 250 (1986) 396.
[6] R. Bonifacio, et al., Riv. Nuovo Cimento 13 (1990) 9.
[7] W.B. Colson, S.K. Ride, Phys. Lett. 76A (1980) 379.
[8] W.B. Colson, in: W.B. Colson, C. Pellegrini, A. Renieri (Eds.), Laser Handbook, Vol. 6, North Holland, Amsterdam, 1990, pp. 115–193.
[9] G. Dattoli, A. Renieri, in: M.L. Stitch, M. Bass (Eds.), Laser Handbook, Vol. 4, North Holland, Amsterdam, 1985, p. 75.
[10] N. Nishimori, unpublished.

Available online at www.sciencedirect.com

Nuclear Instruments and Methods in Physics Research A 528 (2004) 39–43

ELSEVIER

**NUCLEAR
INSTRUMENTS
& METHODS
IN PHYSICS
RESEARCH**
Section A

www.elsevier.com/locate/nima

A deeper analytical insight into the longitudinal dynamics of a storage-ring free-electron laser

G. De Ninno[a],*, D. Fanelli[b]

[a] Sincrotrone Trieste, Area Science Park, 34012 Basovizza, Trieste, Italy
[b] Department of Cell and Molecular Biology, Karolinska Insitute, SE-171 77 Stockholm, Sweden

Abstract

In this paper, a deep characterization is provided of the longitudinal dynamics of a storage-ring free-electron laser. Closed analytical expressions are derived for the main statistical parameters of the system (i.e. beam energy spread, intensity, centroid position and rms value of the laser distribution) as a function of the light-electron beam detuning at each pass inside the interaction region. Moreover, the transition between the stable "cw" regime and the unstable pulsed behaviour is shown to be a Hopf bifurcation. Finally, a feedback procedure is introduced which suppresses the bifurcation and significantly improves the system stability.
© 2004 Published by Elsevier B.V.

PACS: 29.20. Dh; 41.60.Cr

Keywords: Storage ring; Free electron laser; Nonlinear dynamics; Hopf bifurcation

1. The initial model

The starting point of our analysis is the well-known theoretical model, introduced in Ref. [1] and further improved in Refs. [2,3], which allows to capture the main features of the longitudinal dynamics of a Storage-Ring Free-Electron Laser (SRFEL). Such a model describes the coupled evolution of the laser temporal profile and laser-induced beam energy spread. The laser profile, y_n, is updated after each interaction according to

$$y_{n+1}(z) = R^2 y_n(z - \varepsilon)[1 + g_n(z)] + i_s(z) \tag{1}$$

where z is the temporal position of the electron bunch distribution with respect to the centroid and $R = \sqrt{1 - P}$ (where P stands for the cavity losses) is the mirror reflectivity, the detuning parameter ε represents the difference between the electrons revolution period (divided by the number of bunches) and the period of the photons inside the cavity. The term g_n stands for the optical gain (assumed to have the same Gaussian profile of the electron bunch) and $i_s(\tau)$ accounts for the profile of the spontaneous emission of the optical klystron [4].

The cumulative laser-electron beam delay induced by a finite ε value is responsible for the behaviour of the laser intensity, experimentally observed on the Super-ACO and UVSOR FELs: a

*Corresponding author.
E-mail address: giovanni.deninno@elettra.trieste.it (G. De Ninno).

0168-9002/$ - see front matter © 2004 Published by Elsevier B.V.
doi:10.1016/j.nima.2004.04.014

"cw" regime[1] when ε is zero or close to zero, and a stable pulsed regime when ε exceeds a given threshold.

2. An explicit, simplified formulation

Eq. (1) contains the evolution of the main statistical parameters of the laser distribution, namely the intensity I_n (zero-order moment), the centroid position with respect to the centre of the electron bunch τ_n (first-order moment) and the rms value $\sigma_{l,n}$ (second-order moment). In order to gain analytical insight, we have assumed the laser to keep a Gaussian profile and we have calculated the first three moments of the distribution. The details of the calculation are given elsewhere [5]. As a result, by approximating finite differences with differentials, the explicit coupled evolution of the FEL statistical parameters can be cast in the form

$$\frac{d\sigma}{dt} = \frac{\alpha_1}{\Delta T} \frac{1}{2\sigma} [\alpha_2 I + 1 - \sigma^2]$$

$$\frac{dI}{dt} = \frac{R^2 I}{\Delta T}\left[-\frac{P}{R^2} + \frac{g_i \alpha_3}{2\sigma^3}\alpha_4^{(\sigma^2-1)/\alpha_2}\left(\frac{2\sigma^2}{\alpha_3} - \sigma_l^2 - \hat{\tau}^2\right)\right]$$
$$+ \frac{I_s}{\Delta T}$$

$$\frac{d\tau}{dt} = -\frac{\tau}{\Delta T} + \frac{\hat{\tau}}{\Delta T}\left[1 - \frac{g_i}{\sigma}\alpha_3\alpha_4^{(\sigma^2-1)/\alpha_2}\frac{\sigma_l^2}{\sigma^2}\right]$$

$$\frac{d\sigma_l}{dt} = -\frac{1}{\Delta T}\frac{g_i}{2}\alpha_3\alpha_4^{(\sigma^2-1)/\alpha_2}\frac{\sigma_l^3}{\sigma^3} + \frac{1}{\Delta T}\frac{I_s}{I}\frac{1}{2\sigma_l}$$
$$\times \left(\frac{\sigma^2}{\alpha_3} + \tau^2\right) \tag{2}$$

where $\hat{\tau} = \tau + \varepsilon$ and

$$\alpha_1 = \frac{2\Delta T}{\tau_s}, \quad \alpha_2 = \frac{\sigma_e^2 - \sigma_0^2}{\sigma_0^2} \tag{3}$$

$$\alpha_3 = \left(\frac{\Omega}{\sigma_0\alpha}\right)^2, \quad \alpha_4 = \frac{P\sigma_e}{g_i\sigma_0}. \tag{4}$$

Here σ represents the laser-induced energy spread normalized to laser-off value σ_0, ΔT is the

---

[1] The FEL dynamics is naturally pulsed on the temporal scale of the inter-bunch period, while it appears as "continuous-wave" on a larger, millisecond, temporal scale.

bouncing period of the laser pulse inside the optical cavity, g_i stands for the small signal gain, Ω is the synchrotron frequency, τ_s the synchrotron damping time and α the momentum compaction, and σ_e represents the equilibrium value (i.e. that reached at the laser saturation) of the beam energy spread at the perfect tuning (i.e. $\varepsilon = 0$).

3. Equilibrium statistical parameters

Formulation (2) opens up the perspective of a deep analytical study of the SRFEL dynamics. The analysis of the fixed points of Eq. (2) allows, in fact, to characterize the functional dependence of the statistical parameters versus the light electron beam detuning.

The fixed points $(\bar{I}, \bar{\sigma}, \bar{\tau}, \overline{\sigma_l})$ are found by imposing $dI/dt = d\sigma/dt = d\tau/dt = d\sigma_l/dt = 0$, and solving the corresponding system. Assume hereon $\varepsilon > 0$, the scenario for $\varepsilon < 0$ completely being equivalent. After some algebraic calculations, one can express the equilibrium values $\bar{I}, \bar{\tau}, \overline{\sigma_l}$ as function of $\bar{\sigma}$:

$$\bar{I} = \frac{\bar{\sigma}^2 - 1}{\alpha_2} \tag{5}$$

$$\bar{\tau} = \left\{\frac{1}{2}\left[-\frac{\bar{\sigma}^2}{\alpha_3} + \sqrt{\left(\frac{\bar{\sigma}^2}{\alpha_3}\right)^2 + 4\varepsilon^2 A}\right]\right\}^{1/2} \tag{6}$$

$$\overline{\sigma_l} = \left\{\frac{I_s}{2g_i\alpha_3}\alpha_4^{(1-\bar{\sigma}^2)/\alpha_2}\alpha_2\frac{\bar{\sigma}^2}{\bar{\sigma}^2 - 1}\right.$$
$$\left.\times \left[\frac{\bar{\sigma}^2}{\alpha_3} + \sqrt{\left(\frac{\bar{\sigma}^2}{\alpha_3}\right)^2 + 4\varepsilon^2 A}\right]\right\}^{1/4} \tag{7}$$

where

$$A = \frac{\bar{\sigma}^3(\bar{\sigma}^2 - 1)}{\alpha_2 I_s}\frac{\alpha_4^{(1-\bar{\sigma}^2)/\alpha_2}}{g_i\alpha_3}. \tag{8}$$

The equilibrium value of the energy spread $\bar{\sigma}$ can be found numerically by solving the second of Eq. (2), after imposing $dI/dt = 0$, and making explicit use of Eqs. (5–7)(The results are represented with a solid line in Fig. 1 and display an excellent agreement with simulations based on Eq. (2) (symbols in Fig. 1). Alternatively, it is

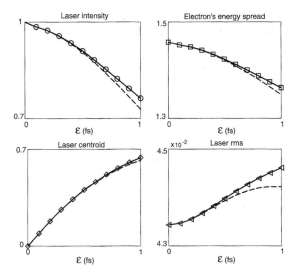

Fig. 1. The fixed points are plotted as function of the detuning parameter ε. Top left panel: normalized laser intensity. Top right panel: normalized electron-beam energy spread. Bottom left panel: Laser centroid. Bottom right panel: rms value of the laser distribution. The symbols refer to the simulations performed using Eqs. (2), the solid line stands for the analytic approach based on the numerical solution of the second of Eqs. (2) (after imposing $dI/dt = 0$), while the long-dashed lines represent the result obtained using the closed analytical expressions (9)–(13).

possible to derive a closed analytical expression for $\bar{\sigma}$ by performing a perturbative calculation. A quite cumbersome calculation leads to

$$\bar{\sigma} = \left(\frac{\sigma_e^2}{\sigma_0^2} + \delta \right)^{1/2} \tag{9}$$

where

$$\delta = \frac{2\left(\frac{\sigma_e P}{\alpha_4 g_i \sigma_0 R^2} - 1 + \frac{1}{2}\alpha_3 \frac{\sigma_0^2}{\sigma_e^2} \Gamma_1 \right)}{\left(\frac{2\log\alpha_4}{\alpha_2} - \frac{\sigma_0^2}{\sigma_e^2} \Gamma_1 \right)\left(2 - \alpha_3 \frac{\sigma_0^2}{\sigma_e^2} \Gamma_1 \right) - \alpha_3 \frac{\sigma_0^2}{\sigma_e^2}\left(\Gamma_2 - \frac{\sigma_0^2}{\sigma_e^2} \Gamma_1 \right)} \tag{10}$$

with

$$\Gamma_1 = \frac{1}{2}\left(-\frac{1}{\alpha_3}\frac{\sigma_e^2}{\sigma_0^2} + \sqrt{c} \right) \tag{11}$$

$$\Gamma_2 = \frac{1}{2\sqrt{c}}\left(-\frac{\sqrt{c}}{\alpha_3} + \frac{1}{\alpha_3^2}\frac{\sigma_e^2}{\sigma_0^2} + 2\varepsilon^2 b \right) \tag{12}$$

and

$$b = \frac{\sigma_e/\sigma_0}{g_i I_s \alpha_2 \alpha_3 \alpha_4}\left[\frac{3}{2}\left(\frac{\sigma_e^2}{\sigma_0^2} - 1 \right) \right.$$
$$\left. + \frac{\sigma_e^2}{\sigma_0^2} - \frac{\log\alpha_4}{\alpha_2}\frac{\sigma_e^2}{\sigma_0^2}\left(\frac{\sigma_e^2}{\sigma_0^2} - 1 \right) \right] \tag{13}$$

$$c = \frac{1}{\alpha_3^2}\frac{\sigma_e^2}{\sigma_0^2} + 4\frac{\sigma_e^2 \sigma_0^2}{g_i I_s \alpha_2 \alpha_3 \alpha_4}\left[\left(\frac{\sigma_e^2}{\sigma_0^2} - 1 \right)\frac{\sigma_e^2}{\sigma_0^2} \right]\varepsilon^2. \tag{14}$$

The above solution is represented in Fig. 1 with a long-dashed line, displaying satisfactory agreement with simulations. Calculations have been performed using the case of the Super-ACO FEL as reference. The values of the relevant parameters are: $\Delta T = 120$ ns, $\tau_s = 8.5$ ms, $\sigma_0 = 5 \times 10^{-4}$, $\sigma_e/\sigma_0 = 1.5$, $\Omega = 14$ kHz, $g_i = 2\%$, $P = 0.8\%$, $I_s = 1.4 \times 10^{-8}$.

To our knowledge, this study represents the first attempt to characterize the analytic dependence of the equilibrium statistical parameters of the SRFEL versus the temporal detuning ε, over the whole region of "cw" behaviour. As a straightforward application of the above derivation, let us focus on the expression for $\bar{\sigma}_l$ at perfect tuning, i.e. $\varepsilon = 0$. We obtain[2]

$$\bar{\sigma}_l \simeq \left(\frac{I_s}{P} \right)^{1/4}\left[1 + \frac{1}{2}\log\left(\frac{g_i}{P} \right) \right]\sigma_{\tau,0} \tag{15}$$

where we made use of the relation $\sigma_e/\sigma_0 \simeq 1 + 0.5\log(g_i/P)$ derived in Ref. [6]. Relation (15) was applied to the case of Super ACO and was shown to reproduce quantitatively the experimental value [5]. Further, a comparison between the estimate of σ_l derived in the context of super-modes theory [7] and the prediction based on Eq. (15) was drawn, the latter resulting systematically in values closer to the ones measured experimentally [5].

The stability of the fixed point $[\bar{I}(\varepsilon), \bar{\sigma}(\varepsilon), \bar{\tau}(\varepsilon), \overline{\sigma}_l(\varepsilon)]$ can be determined by studying the eigenvalues of the Jacobian matrix associated to system (2). The real part of the eigenvalues are reported in Fig. 2. The system is by definition

---

[2] It should be remarked that when $\varepsilon = 0$, $\delta = 1/R^2 - 1$. Thus, formally, $\bar{\sigma}(\varepsilon = 0) \neq \sigma_e/\sigma_0$, which is in apparent contradiction with the assumption of the model. However, since $R^2 = 1 - P \sim 1$, $\delta(\varepsilon = 0) \sim 0$, the residual small discrepancy being related to the approximations involved in the calculation

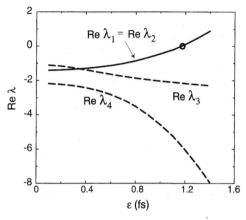

Fig. 2. Behaviour of the real parts of the eigenvalues of the Jacobian matrix as a function of the detuning amount. The calculation has been done for the case of the Super-ACO FEL. The transition occurs around 1.3 fs, a value which is in good agreement with experiments.

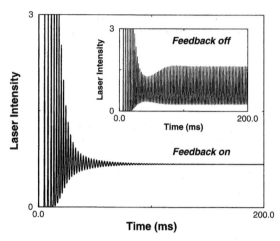

Fig. 3. Behaviour of the FEL intensity in absence (see inset) and in presence of the closed-loop feedback. The parameters utilized for the simulations are those of Super ACO.

stable when all the real parts of the eigenvalues are negative. The transition to an unstable regime occurs when at least one of those becomes positive. In general, the loss of stability takes place according to different modalities. Consider, for instance, a Jacobian matrix with a pair of complex conjugate eigenvalues and assume the real parts of all the eigenvalues to be negative. A Hopf bifurcation occurs when the real part of the two complex eigenvalues become positive, provided the others keep their sign unchanged [8]. This situation is clearly displayed in Fig. 2, thus allowing to conclude that the transition between the "cw" and the pulsed regime in a SRFEL is a Hopf bifurcation.

4. Stabilization of the pulsed regime

Having characterized the transition from the stable to the unstable steady state in terms of Hopf bifurcation opens up interesting perspectives for the improvement of the system performance. In fact, as it has been shown by several authors [9,10], the chaotic behaviour induced in conventional lasers can be stabilized by using a self-controlled (closed-loop) procedure. In particular, in Ref. [11] the dynamics of a conventional laser was stabilized

around the unstable steady state, arising from a Hopf bifurcation. By virtue of the results of the previous sections, this approach can be extended to the case of a SRFEL. For this purpose the constant detuning ε is replaced with the time-dependent quantity

$$\varepsilon = \varepsilon_0 + \beta \Delta T \frac{dI}{dt} \qquad (16)$$

which is added to system (2).

Here ε_0 is assumed to be larger than ε_c. As it is shown in Fig. 3, when the control is switched off, i.e. $\beta = 0$, the laser is unstable and displays periodic oscillations. For β larger than a certain threshold, β_c, the oscillations are damped and the laser behaves as if it was in the "cw" region. Preliminary experimental results have been already obtained at Super ACO [12].

5. Conclusions

In this paper, we have shown that making explicit the evolution of the statistical parameters of a SRFEL allows to gain a deep insight into the longitudinal dynamics of the system. Exploiting this analytical characterization, we have introduced a suitable feedback procedure to enlarge the region of stable signal. This result opens up the perspective of improving the performance of last generation SRFELs.

Acknowledgements

The work of GDN has been partially supported by EUFELE, a Project funded by the European Commission under FP5 Contract No. HPRI-CT-2001-50025. DF thanks U. Skoglund for the useful discussions.

References

[1] M. Billardon, et al., Phys. Rev. Lett. 69 (1992) 2368.

[2] G. De Ninno, D. Fanelli, M.E. Couprie, Nucl. Instr. and Meth. A 483 (2002) 177.
[3] G. De Ninno, et al., Europhys. J. D 22 (2003) 269.
[4] N.A. Vinokurov, et al., preprint INP77.59 Novosibirsk, unpublished, 1977.
[5] G. De Ninno, D. Fanelli, ELETTRA Internal Note ST/SL-03/03, 2003.
[6] G. De Ninno, D. Fanelli, Phys. Rev. Lett. 92 (2004) 094801.
[7] G. Dattoli, private communication.
[8] N. Berglund, Nonlinearity 13 (2000) 225.
[9] V. Petrov, et al., J. Chem. Phys. 96 (1992) 7503.
[10] S. Bielawski, et al., Phys. Rev. E 49 (1994) 971.
[11] S. Bielawski, et al., Phys. Rev. A 47 (1993) 3276.
[12] M.E. Couprie, Nucl. Instr. and Meth. A, (2004) these proceedings.

Available online at www.sciencedirect.com

SCIENCE  DIRECT®

ELSEVIER   Nuclear Instruments and Methods in Physics Research A 528 (2004) 44–47

NUCLEAR
INSTRUMENTS
& METHODS
IN PHYSICS
RESEARCH
Section A

www.elsevier.com/locate/nima

Nonlinear harmonic generation in free-electron laser oscillators

H.P. Freund[a,*], P.G. O'Shea[b], S.G. Biedron[c]

[a] Science Applications International Corporation, 1710 Goodridge Drive, McLean, VA 22102, USA
[b] Institute of Research in Electronics and Applied Physics, University of Maryland, College Park, MD 20742, USA
[c] MAX-Laboratory, University of Lund, SE-22100 Lund, Sweden

Abstract

Nonlinear harmonic generation has been described theoretically, and observed, in free-electron laser (FEL) amplifiers. The mechanism relies on the nonlinear, higher-order bunching when the fundamental reaches high intensities, and can also be expected in high-power FEL oscillators. We use the 3D polychromatic simulation code MEDUSA to study the nonlinear harmonic generation in FEL oscillators. Strong harmonic growth is, indeed, found when an oscillator is near saturation, and the third (fifth) harmonic can reach power levels approaching 1% (0.1%) that of the fundamental.
© 2004 Elsevier B.V. All rights reserved.

PACS: 41.60.Cr; 52.59.Rz

Keywords: Free-electron lasers; Nonlinear harmonic generation

1. Introduction

Nonlinear harmonic generation has been described theoretically [1] in free-electron laser (FEL) amplifiers, and has been observed in FELs operating by Self-Amplified Spontaneous Emission (SASE) at Argonne National Laboratory [2] and Brookhaven National Laboratory [3] as well as in the high-gain harmonic generation experiment at Brookhaven National Laboratory [4]. This mechanism relies on nonlinear, higher-order

bunching when the fundamental reaches high power levels. As such, the nonlinear harmonic generation can also be expected in high-power oscillators.

We use the three-dimensional, polychromatic simulation code MEDUSA [1] to study the nonlinear harmonic generation in FEL oscillators. MEDUSA can model both planar and helical wigglers and treats the optical field as a superposition of either Gauss–Hermite or Gauss–Laguerre modes. The field equations are integrated simultaneously with the Lorentz force equations for an ensemble of electrons. No wiggler-average orbit approximation is used, and MEDUSA propagates the electron beam through a complex wiggler/transport line including multiple wiggler

---

*Corresponding author. Tel.: +1-202-767-0034; fax: +1-202-734-1280.

E-mail addresses: henry.p.freund@saic.com, freund@mmace.nrl.navy.mil (H.P. Freund).

0168-9002/$ - see front matter © 2004 Elsevier B.V. All rights reserved.
doi:10.1016/j.nima.2004.04.015

sections, quadrupole and dipole corrector magnets, FODO lattices, and magnetic chicanes. Since it is polychromatic, MEDUSA can treat both sidebands and harmonic radiation simultaneously with the fundamental.

In an amplifier or SASE FEL, the fundamentals and harmonics are amplified exponentially in a single pass through the undulator. This amplification is a consequence of electron beam bunching at radiation wavelengths. This bunching, and the associated amplification, is cumulative and the particles ultimately become trapped in the troughs of the wave. Saturation occurs after the beam executes approximately one-half of a trapped oscillation at which point there are as many electrons extracting energy from the wave as there are electrons gaining energy from the wave. Harmonics are driven by nonlinearities in the bunching that are proportional to powers of the electric field of the fundamental. Thus, if the electric field varies as $E_0 \exp(ikz - i\omega t + \Gamma z)$, where ω and k denote the angular frequency and wavenumber and Γ is the spatial growth rate, then the hth harmonic amplification and bunching will be driven by a source current which varies as $E_0^h \exp(ihkz - ih\omega t + h\Gamma z)$. Nonlinear harmonic generation only becomes important when fundamental intensities reach high levels but, once this happens, the harmonics grow rapidly and the growth rates scale as the product of the harmonic number and the fundamental growth rate ($h\Gamma$). Further, saturation of the harmonics generally occurs slightly prior to that of the fundamental.

In contrast, the radiation in oscillators generally does not grow exponentially in a single pass through the undulator; instead, the radiation grows incrementally over many passes. In oscillators driven by radio frequency linear accelerators, the electron beam consists of a sequence of pulses, and the length of the optical resonator must be chosen so that the radiation pulse after one pass is re-injected into the undulator for the next pass in synchronism with the next electron pulse. Each pass involves the interaction of a "fresh" electron bunch with the light. Saturation occurs through the same electron trapping mechanism as in amplifiers, but in oscillators there is no appreciable bunching or trapping of the electrons until the radiation has built up to high intensities. Once this occurs, the harmonics grow rapidly.

2. Nonlinear simulation

MEDUSA is a steady-state simulation code in the frequency—as opposed to the time-domain—and it can model the evolution (growth or damping) of the radiation (fundamental and harmonics) during a single pass through the undulator. As a result, the formulation is valid as long as the slippage of the light relative to the electrons through the undulator is much less than the length of the electron bunch. This condition is well satisfied in the examples considered in this paper. In order to simulate oscillators, therefore, we generate a drive curve describing the output power from the undulator (at the fundamental and harmonics) as a function of the input (or drive) power in the fundamental. This allows us to obtain the single pass gain so that the power after the $(n+1)$th pass can be determined based on the power on the nth pass (P_n). This takes the form $P_{n+1} = G[(1 - \varepsilon)P_n]$, where ε is the fractional power which is either lost or out-coupled, and G is the gain function. Saturation occurs when the single pass gain for a given drive power has fallen to the level of the resonator losses, including the fraction of power which is out-coupled from the resonator. Oscillator codes exist which can simulate many passes through the undulator in complex resonator structures, such as the FELEX code developed at Los Alamos National Laboratory [5], but these codes cannot treat the simultaneous growth of harmonics.

The case we consider is that of a short Rayleigh range oscillator, where "short" refers to a Rayleigh range shorter than the undulator. Short Rayleigh ranges have been proposed for high-power FEL oscillators [6] so that the optical beam at the resonator mirrors is large enough that the mirror loading is reduced below that which causes optical damage. Of course, optical mode expansion within the undulator in these designs is large and the peak current must be high to ensure a sufficiently high enough that single pass gain can overcome resonator losses. The practical importance of harmonic growth in oscillators arises from the possibility of mirror damage due to harmonics in the ultraviolet.

The wiggler model we use describes a parabolic-pole-face design for focusing in both transverse dimensions. We assume an amplitude of 10.1 kG, a period of 3.0 cm, and a length of 26 wiggler periods. MEDUSA models both the injection and ejection of the beam from the wiggler, and we also assume that these transition regions are 3 wiggler periods in length, so that the uniform field region of the wiggler is 20 periods long. The electron beam energy is 140 MeV and the peak current is 800 A with a normalized emittance of 1.9 mm mrad and an energy spread of 0.1%. The matched beam radius is approximately 80 µm. These parameters yield an optimal resonance at a wavelength of 1.04 µm and we assume that the optical mode waist is centered in the wiggler with a radius of 80 µm corresponding to the matched electron beam radius.

The Rayleigh range in this case is 1.93 cm, which is shorter than the wiggler period. Therefore, the area (radius) of the optical mode at the ends of the uniform wiggler region is a factor of about 246 (15.5) times that at the center of the wiggler. Normally, this kind of extreme mode expansion would result in a sharp reduction in the coupling between the electrons and the radiation with a resulting decline in the gain. However, the high peak current compensates for this, and the linear single-pass gain is close to 100%. This is shown in Fig. 1 where the single pass gain of the fundamental and the powers in the third and fifth harmonics are plotted versus the input power in the fundamental. As shown, the linear regime extends up to input powers of about 10 MW where the single pass gain is about 97%. Nonlinear bunching causes the gain to decline rapidly as the power increases beyond this point and falls to near zero at an input power of about 30 GW. The saturation point depends upon resonator losses. If this loss results predominantly from the out-coupling of the radiation, then the oscillator would saturate at an intra-cavity power of approximately 4.1 GW for an out-coupling fraction of 25%. Of course this is a peak power estimate, and the average power of the

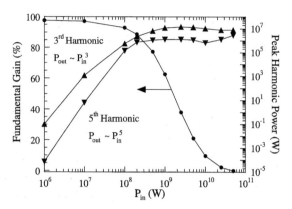

Fig. 1. The gain of the fundamental and powers in the third and fifth harmonics as a function of drive power at the fundamental.

oscillator will depend upon the duty factor of the accelerator.

The growth of the power at the third and fifth harmonics is shown in Fig. 1 as functions of drive power at the fundamental. It is clear that the harmonic powers are small as long as the fundamental power is less than about 1–10 MW, but grow rapidly as the fundamental power increases further. As expected, the third harmonic power grows as the power in the fundamental raised to the third power and the fifth harmonic grows as the fifth power of the fundamental power in the linear regime, but saturate at fundamental drive powers of about 1 GW. If we assume the same out-coupling fraction of 25% as above, then the third harmonic will reach a power of about 14.8 MW and the fifth harmonic will reach a power of 1.16 MW. Thus, the third (fifth) harmonic power reaches about 0.36% (0.028%) that of the fundamental.

Note that the harmonic power levels are relatively flat for fundamental drive powers above 1 GW. This means that the relative power ratios between the harmonics and the fundamental are sensitive to the out-coupling fraction. For an out-coupling fraction of 50%, the fundamental reaches a power level of about 1.56 GW and the third (fifth) harmonic reaches powers of 20.3 MW (1.36 MW) for a power ratio between the third (fifth) harmonic and the fundamental of 1.3% (0.087%). Of course, these are peak powers. Of

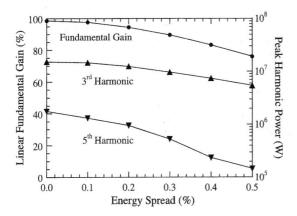

Fig. 2. The variation in the gain of the fundamental (left axis) and the harmonic powers (right axis) versus energy spread.

particular concern, however, is the fifth harmonic at a wavelength of about 200 nm because of the possibility of damage to mirror coatings.

The sensitivity of the nonlinear harmonic generation mechanism to wiggler imperfections and beam energy spread and emittance has been studied for amplifier and SASE FELs [7,8]. The earlier findings suggest that the sensitivity to these factors of the nonlinear harmonic generation mechanism in amplifier and SASE FELs are similar to the sensitivity of the fundamental interaction. While wiggler imperfections may be a serious consideration in long amplifier and SASE FELS, it is not important in oscillators where the wiggler is short; hence, we focus here on the effect of electron beam quality on the nonlinear harmonic generation in oscillators. This issue is addressed in Fig. 2 where the fundamental gain in the linear regime (right axis) and the maximum third and fifth harmonic powers are plotted versus beam energy spread. It is clear that as the energy spread increases to 0.5%, the fundamental gain decreases by about 20%. While the fundamental gain drops, the saturated power at the fundamental remains relatively flat, and for an out-coupling of 25%, the fundamental saturates at a power of approximately 3.5 GW for an energy spread of 0.5% which is down from 4.1 GW for an energy spread of 0.1%. The maximum power at the third harmonic drops from about 15 MW to about 5.4 MW, and the fifth harmonic power decreases

from about 1.76 MW to about 146 kW. Therefore, while the harmonic powers drop off somewhat faster than the fundamental, they are still at appreciable power levels for energy spreads of 0.5%.

3. Summary and conclusions

In this paper, we have extended previous analyses of nonlinear harmonic generation in FEL amplifiers and SASE FELs to study nonlinear harmonic generation in FEL oscillators. The results indicate that nonlinear harmonic generation in FEL oscillators display characteristics similar to what has been found in amplifiers. Specifically, that harmonic powers can reach substantial power levels and that the nonlinear harmonic generation mechanism is much less sensitive to beam quality than the linear harmonic instability. In conclusion, while nonlinear harmonic generation can provide a means of reaching short wavelengths with lower beam energies in SASE FELs, they can be a serious problem for high-power FEL oscillators.

Acknowledgements

This work was supported by NAVSEA and the JTO.

References

[1] H.P. Freund, et al., IEEE J. Quantum Electron. 36 (2000) 275.
[2] S.G. Biedron, et al., Nucl. Instr. and Meth. A 483 (2002) 94.
[3] A. Tremaine, et al., Nucl. Instr. and Meth. A 507 (2003) 445.
[4] A. Doyuran, et al., Phys. Rev. Lett. 86 (2001) 5902.
[5] B.D. McVey, Nucl. Instr. and Meth. A 250 (1986) 449.
[6] W.B. Colson, et al., Nucl. Instr. and Meth. A 507 (2003) 48.
[7] H.P. Freund, et al., IEEE J. Quantum Electron. 37 (2001) 790.
[8] S.G. Biedron, et al., Nucl. Instr. and Meth. A 483 (2002) 101.

Available online at www.sciencedirect.com

SCIENCE  DIRECT®

ELSEVIER   Nuclear Instruments and Methods in Physics Research A 528 (2004) 48–51

**NUCLEAR
INSTRUMENTS
& METHODS
IN PHYSICS
RESEARCH**
Section A

www.elsevier.com/locate/nima

Fluctuation-induced linewidth in oscillator FEL

Oleg A. Shevchenko*, Nikolay A. Vinokurov

Budker Institute of Nuclear Physics, 11 Acad. Lavrentyev Prosp., 630090 Novosibirsk, Russia

Abstract

Due to relatively small number of microscopic radiators, the noise (spontaneous emission contribution to the radiation field) level in free electron lasers (FELs) is much more than in other types of lasers. The influence of this noise on the FEL linewidth is considered. A low gain optical klystron model is used. It is shown that in the most of practically interesting cases the noise-induced linewidth is negligible.
© 2004 Elsevier B.V. All rights reserved.

PACS: 41.60.Cr

Keywords: Free electron laser

1. Introduction

Linewidths of contemporary free electron lasers (FELs) is typically limited by the Fourier-transform limit due to short length of electron bunches. Sideband instability causes the line widening frequently, especially for the high-gain regime. But in some cases electron bunches may be relatively long (one can try to do it just in order to obtain narrow line), or electron beam is not bunched at all, as it takes place in electrostatic accelerators.

On the other hand, one can compare FELs with other types of lasers. As the number of initially independent radiators (radiating electrons) is much less for FEL, the statistical noise (sponta-

neous emission) is much more intensive for FELs. Some linewidth estimations were done earlier by different authors [1,2]. In the present paper, we consider a general method to obtain the steady-state linewidth. It is based on estimation of the optical phase diffusion coefficient (see, e.g. textbooks [3,4]). To simplify equations a low-gain optical klystron (OK) is considered. For FELs with other magnetic configuration the result, expressed through the spontaneous emission linewidth, is obviously the same.

2. Basic equations

We suppose, that our OK has an optical resonator, and will be interested in the radiation field projection on the transverse fundamental eigenmode (Gaussian beam) of this resonator. Consider the particle entering the first undulator

---

*Corresponding author. Tel.: +7-3832-394859; fax: +7-3832-342163.

E-mail address: o.a.shevchenko@inp.nsk.su (O.A. Shevchenko).

of OK at time t_0. Then the particle radiation field $E(t)$ has the form

$$E(t) = E_1(t - t_0) + E_2[t - t_0 - \Delta(1 - 2\eta)], \qquad (1)$$

where E_1 and E_2 are the wave trains from first and second undulators of OK, respectively, η is relative electron energy deviation (from the average electron energy γmc^2 in the beam), and Δ is the radiation delay for the average energy. It is more convenient to represent radiation field using a analytical signal, i.e., $E(t) = \mathrm{Re}\left[b(t)\mathrm{e}^{-\mathrm{i}\omega t}\right]$, with slowly varying complex amplitude $b(t)$

$$
\begin{aligned}
b(t) = {}& b_1(t - t_0)\mathrm{e}^{\mathrm{i}\omega t_0} + b_2[t - t_0 - \Delta(1 - 2\eta)] \\
& \times \mathrm{e}^{\mathrm{i}\omega[t_0 + \Delta(1 - 2\eta)]}.
\end{aligned}
\qquad (2)
$$

Carrier frequency ω is close to the central frequency of spontaneous emission spectrum. To simplify consideration we suppose the number of periods in undulator N to be small enough, i.e., $\eta \ll 1/(4\pi N)$, to neglect the dependence of amplitudes b_1, b_2 on energy deviation η.

When an external wave $E_{\mathrm{in}}(t)$ is propagating together with electrons, the energy at the exit of the first undulator will be changed

$$
\begin{aligned}
\eta = {}& \eta_0 - \frac{S}{2\pi\gamma mc} \int_{-\infty}^{\infty} E_1(t - t_0)E_{\mathrm{in}}(t)\,\mathrm{d}t \\
= {}& \eta_0 - \frac{S}{4\pi\gamma mc} \mathrm{Re}\left[\mathrm{e}^{\mathrm{i}\omega t_0} \int_{-\infty}^{\infty} b_1(t - t_0)b_{\mathrm{in}}^*(t)\,\mathrm{d}t\right],
\end{aligned}
\qquad (3)
$$

where S is the eigenmode cross-section. For small amplitude of the external wave b_{in} the particle radiation Eq. (2) may be expanded over it

$$
\begin{aligned}
b(t) \approx {}& b_1(t - t_0)\mathrm{e}^{\mathrm{i}\omega t_0} + b_2[t - t_0 \\
& - \Delta(1 - 2\eta_0)]\mathrm{e}^{\mathrm{i}\omega[t_0 + \Delta(1 - 2\eta_0)]} \\
& + \frac{\mathrm{i}S\omega\Delta}{4\pi\gamma mc} b_2[t - t_0 - \Delta(1 - 2\eta_0)]\mathrm{e}^{\mathrm{i}\omega\Delta(1 - 2\eta_0)} \\
& \times \int_{-\infty}^{\infty} b_1^*(t' - t_0)b_{\mathrm{in}}(t')\,\mathrm{d}t' \\
& - \frac{\mathrm{i}}{2} b_2[t - t_0 - \Delta(1 - 2\eta_0)]\mathrm{e}^{\mathrm{i}\omega\Delta(1 - 2\eta_0)} \\
& \times \left(\frac{S\omega\Delta}{4\pi\gamma mc}\right)^3 \int_{-\infty}^{\infty} b_1^*(t' - t_0)b_{\mathrm{in}}(t')\,\mathrm{d}t' \\
& \times \left|\int_{-\infty}^{\infty} b_1^*(t' - t_0)b_{\mathrm{in}}(t')\,\mathrm{d}t'\right|^2 + \cdots
\end{aligned}
\qquad (4)
$$

Here we keep only high-order terms, which do not contain oscillating non-zero powers of $\mathrm{e}^{\mathrm{i}\omega t_0}$. First two terms in Eq. (4) are referred to a spontaneous emission (into the resonator eigenmode) field. They are phased to radiating particle and therefore are proportional to $\mathrm{e}^{\mathrm{i}\omega t_0}$. The terms, which do not contain oscillating non-zero powers of $\mathrm{e}^{\mathrm{i}\omega t_0}$, are referred to stimulated emission field.[1] The last term of Eq. (4) describes saturation.

To calculate the field transformation in the OK one has to sum the radiation fields of particles with different t_0 and η_0. In the OK approximation one can suppose both undulators to be short, i.e., b_1 and b_2 can be treated as δ functions in the integrals. Then the field transformation has form

$$
\begin{aligned}
a_{\mathrm{out}}(t) = {}& (1 - \Gamma)a_{\mathrm{in}}(t) + \mathrm{i}\mathrm{e}^{\mathrm{i}\omega\Delta}B(t - \Delta)a_{\mathrm{in}}(t - \Delta) \\
& \times \left\lfloor 1 - |a_{\mathrm{in}}(t - \Delta)|^2 \right\rfloor + a_{\mathrm{sp}}(t),
\end{aligned}
\qquad (5)
$$

where Γ represents round-trip loss in the optical resonator ($\Gamma \ll 1$),

$$a(t) = b(t)\frac{S\omega\Delta}{4\sqrt{2}\pi\gamma mc} \int_{-\infty}^{\infty} b_1^*(t)\,\mathrm{d}t, \qquad (6)$$

a_{sp} is the normalized amplitude of the spontaneous radiation field, and A is proportional to time-dependent small signal gain G. To simplify Eq. (5) further one can choose the carrier frequency ω at the maximum amplification $\mathrm{i}\mathrm{e}^{\mathrm{i}\omega\Delta}B = |B| \equiv A$

$$
\begin{aligned}
a_{\mathrm{out}}(t) = {}& (1 - \Gamma)a_{\mathrm{in}}(t) + A(t - \Delta)a_{\mathrm{in}}(t - \Delta) \\
& \times \left\lfloor 1 - |a_{\mathrm{in}}(t - \Delta)|^2 \right\rfloor + a_{\mathrm{sp}}(t),
\end{aligned}
\qquad (7)
$$

For long enough electron bunches $A = G/2$. In this paper, we consider the simplest case of unbunched electron beam, when A is constant.

3. Explicit solution for the case of unbunched electron beam

Let us consider the transformation

$$
\begin{aligned}
a_{n+1}(t) = {}& (1 - \Gamma)a_n(t) + Aa_n(t - \Delta) \\
& \times \left\lfloor 1 - |a_n(t - \Delta)|^2 \right\rfloor + a_{\mathrm{sp}}^n(t),
\end{aligned}
\qquad (8)
$$

---

[1] In classical physics one cannot separate stimulated emission and absorption of light, so we use expression "stimulated emission" to describe both these phenomena (sometimes people use term "reabsorption").

where n is radiation pass number and A is a constant which is supposed to be positive. In the absence of stochastic term $a_{sp}^n(t)$ the transformation (8) has a non-zero stationary points $a_n = $ const, $|a_n| = R_0 = \sqrt{1 - \Gamma/A}$. The lasing condition is $A > \Gamma$.

It is convenient to introduce "polar" coordinate system for the amplitude and phase of the field $a_n(t)$:

$$r_n(t) = |a_n(t)| = R_0 + \rho_n(t), \quad \varphi_n(t) = \arg(a_n), \quad (9)$$

$$\lambda_n(t) = |a_{sp}^n(t)|, \quad \alpha_n(t) = \arg(a_{sp}^n). \quad (10)$$

In most practically interesting cases $\lambda_n \ll R_0$ and it is natural to suppose that the amplitude will undergo only small deviations from the stationary state. As for the phase, it can change freely, but $|\varphi(t) - \varphi(t - \Delta)| \ll 1$ that follows from the solution below. Now we can write down the linearized transformation for the amplitude deviation $\rho_n(t)$ and phase $\varphi_n(t)$:

$$\rho_{n+1}(t) = (1 - \Gamma)\rho_n(t) + (3\Gamma - 2A)\rho_n(t - \Delta) + \lambda_n(t)\cos[\alpha_n(t) - \varphi_n(t)], \quad (11)$$

$$\varphi_{n+1}(t) = (1 - \Gamma)\varphi_n(t) + \Gamma\varphi_n(t - \Delta) + \frac{\lambda_n(t)}{R_0}\sin[\alpha_n(t) - \varphi_n(t)]. \quad (12)$$

One can easily find that the stability criterion is $\Gamma < A < 2\Gamma$. The dependence of the last term of Eq. (12) on $\varphi_n(t - \Delta)$ does not influence on its statistical properties, so we can replace it by stationary random process $\xi_n(t) = \text{Im}((a_{sp}^n(t))/R_0$. Its power spectral density near the zero tune Ω is (see Eq. (6)).

$$\langle|\xi_n(\Omega)|^2\rangle \approx \langle|\xi_n(0)|^2\rangle$$
$$= \frac{I}{e}\left(\frac{S\omega\Delta}{4\sqrt{2}\pi\gamma mc R_0}\right)^2\left|\int_{-\infty}^{\infty} b_1(t)\,dt\right|^4. \quad (13)$$

The factor $2I/e$ is the spectral density of the beam shot noise. It appears in Eq. (13) due to the absence of correlations between wave trains radiated by different electrons.

To simplify Eq. (13) we considered identical undulators, i.e., $b_1 = b_2$. It is worth noting that as the spectral power density Eq. (13) does not depend on frequency Ω, $\xi_n(t)$ is delta correlated.

The phase spectral density evolution may be obtained from Eq. (12).

$$\langle|\varphi_{n+1}(\Omega)|^2\rangle = |1 - \Gamma + \Gamma e^{i\Omega\Delta}|^2\langle|\varphi_{n+1}(\Omega)|^2\rangle + \langle|\xi_n(\Omega)|^2\rangle. \quad (14)$$

The stationary solution of Eq. (14) is

$$\langle|\varphi(\Omega)|^2\rangle = \frac{\langle|\xi_n(\Omega)|^2\rangle}{4\Gamma(1 - \Gamma)\sin^2((\Omega\Delta)/2)}. \quad (15)$$

It has singularity $\langle|\varphi(\Omega)|^2\rangle \approx 2D/\Omega^2$ at $\Omega = 0$, which corresponds to the phase diffusion with the diffusion coefficient D. Then the relative line width is

$$\frac{\delta\omega}{\omega} = \frac{D}{\omega} \approx \frac{\langle|\xi_n(0)|^2\rangle}{2\Gamma\omega\Delta^2}$$
$$= \frac{I\omega}{e\Gamma}\left(\frac{S}{8\pi\gamma mc R_0}\right)^2\left|\int_{-\infty}^{\infty} b_1(t)\,dt\right|^4. \quad (16)$$

For planar undulator with length L and undulator deflection parameter K

$$\left|\int_{-\infty}^{\infty} b_1(t)\,dt\right| = \frac{2\pi e L K}{S\gamma c}\left[J_0\left(\frac{K^2}{4 + 2K^2}\right) - J_1\left(\frac{K^2}{4 + 2K^2}\right)\right]. \quad (17)$$

Then

$$\frac{\delta\omega}{\omega} = \frac{eI\omega}{\Gamma}\left(\frac{\pi K^2 L^2 e}{2\gamma^3 mc^3 R_0 S}\right)^2 (JJ)^4$$
$$= \frac{eIcN^2}{\lambda_0 I_0^2\Gamma} 2\pi^3\left[\frac{K^2(JJ)^2}{(1 + K^2/2)^{3/2}} R_0^{-2}\frac{\lambda L}{S}\right]^2, \quad (18)$$

where λ_0 is the undulator period, $N = L/\lambda_0$, $I_0 = mc^3/e \approx 17\,\text{kA}$, and $\lambda = 2\pi c/\omega$. The order of magnitude estimation for Eq. (18) is

$$\frac{\delta\omega}{\omega} = \frac{eIcN^2}{\lambda_0 I_0^2\Gamma}. \quad (19)$$

The value, given by Eq. (19) for typical FEL parameters, is less than 10^{-10}.

It is also interesting to represent Eq. (16) in the form, similar to the Schawlow–Townes formula (see, e.g. Ref. [3]). The output power P is

$$P = \Gamma\frac{ScR_0^2}{4\pi}\left(\frac{\omega\Delta e L K(JJ)}{2\sqrt{2}\gamma^2 mc^2}\right)^{-2}. \quad (20)$$

Then

$$\delta\omega = \frac{\pi e I}{8 P \Delta^2 S c}\left[\frac{K(JJ)L}{\gamma}\right]^2. \tag{21}$$

As the spontaneous emission power spectral density $P_{\mathrm{sp}}(\omega)$ is

$$P_{\mathrm{sp}}(\omega) = \frac{IcS}{e8\pi}\left|\int_{-\infty}^{\infty} b_1(t)\,\mathrm{d}t\right| = \frac{\pi e I}{2Sc}\left[\frac{K(JJ)L}{\gamma}\right]^2, \tag{22}$$

$$\delta\omega = \frac{P_{\mathrm{sp}}(\omega)}{4P\Delta^2}. \tag{23}$$

4. Conclusion

The results of our paper show that the spontaneous emission contribution to the oscilla-tor FEL linewidth is negligible in practically interesting cases (probably, excluding some electrostatic accelerator based FELs). The single-mode case, where one cannot use Eq. (8), requires additional consideration.

References

[1] K.-J. Kim, Nucl. Instr. and Meth. A 304 (1991) 458.
[2] B. Levush, T.M. Antonsen, Nucl. Instr. and Meth. A 285 (1989) 136.
[3] A. Yariv, Quantum Electronics, Wiley, New York, 1975.
[4] Yu.L. Klimontovich, Statistical Physics, Nauka, Moscow, 1982.

Available online at www.sciencedirect.com

SCIENCE @ DIRECT°

Nuclear Instruments and Methods in Physics Research A 528 (2004) 52–55

NUCLEAR
INSTRUMENTS
& METHODS
IN PHYSICS
RESEARCH
Section A

www.elsevier.com/locate/nima

Saturation of a high-gain single-pass FEL

S. Krinsky*

Brookhaven National Laboratory, Bldg. 725B, National Synchrotron Light Source, Upton, NY 11973, USA

Abstract

We study a perturbation expansion for the solution of the nonlinear one-dimensional FEL equations. We show that in the case of a monochromatic wave, the radiated intensity satisfies a scaling relation that implies, for large distance z traveled along the undulator, a change in initial value of the radiation field corresponds to a translation in z (lethargy). Analytic continuation using Padé approximates yields accurate results for the radiation field early in saturation.
© 2004 Elsevier B.V. All rights reserved.

PACS: 41.60.Cr

Keywords: FEL saturation; Perturbation expansion; Padé approximate

1. Introduction

Free-Electron Laser (FEL) amplifiers in the exponential growth regime are accurately described by linear equations that are very well understood [1]. On the other hand, although there has been interesting work [2–7] on the theory of the saturation of the gain process, the description of the nonlinear phenomena involved is in a less-advanced state. At present, most studies of saturation are based upon computer simulation [1]. In this paper, we use a perturbation expansion to treat the nonlinearity in the one-dimensional (1-D) free-electron laser equations. For a monochromatic wave, the resulting Taylor series for the radiation field has a finite radius of convergence. We find that analytic continuation using Padé

approximates [8] yields results in agreement with numerical integration of the 1-D FEL equations, well into saturation.

A more detailed exposition of the work reported in this paper as well as discussion of a simplified model for SASE statistics in saturation can be found in Ref. [9].

2. Perturbation expansion

The scaled equations [2] for the evolution of a 1-D electron distribution and a monochromatic radiation field are

$$d\theta_j/dZ = p_j \tag{1}$$

$$dp_j/dZ = -Ae^{i\theta_j} - A^*e^{-i\theta_j} \tag{2}$$

$$dA/dZ = \langle e^{-i\theta_j} \rangle. \tag{3}$$

*Tel.: +1-631-344-4740; fax: +1-631-344-4745.
E-mail address:* krinsky@bnl.gov (S. Krinsky).

0168-9002/$ - see front matter © 2004 Elsevier B.V. All rights reserved.
doi:10.1016/j.nima.2004.04.017

$\theta_j = (k_s + k_w)z - \omega_s t_j(z)$ is the phase of the jth electron relative to the radiation and $p_j = (\gamma - \gamma_0)/\rho\gamma_0$ is its (scaled) energy deviation. We define: γ the relativistic parameter; $Z = 2\rho k_w z$ the scaled distance along the undulator axis; $2\pi/k_w$ the undulator period; $2\pi/k_s$ the radiation wavelength; and $t_j(z)$ the arrival time of the jth electron at position z. The radiated electric field has the form $E \exp[ik_s(z - ct)]$ and the scaled amplitude $A \equiv E/\sqrt{\rho n_0 \gamma_0 mc^2/\varepsilon_0}$ (mks units), where ρ is the Pierce parameter and n_0 the electron density. The bracket $\langle \ \rangle$ indicates the average over the initial electron distribution.

We develop the solution of Eqs. (1)–(3) as a perturbation expansion in the small parameter ε, which we take to be the initial value of the radiation amplitude, $A(0) = \varepsilon$. Without loss of generality we consider $\varepsilon \ll 1$ to be real. Expanding in powers of ε, we write

$$\theta(Z,\theta_0,p_0) = \theta_0 + p_0 Z + \varepsilon\theta_1(Z,\theta_0,p_0)$$
$$+ \varepsilon^2\theta_2(Z,\theta_0,p_0) + \cdots \quad (4)$$

$$A(Z) = \varepsilon A_1(Z) + \varepsilon^3 A_3(Z) + \varepsilon^5 A_5(Z) + \cdots . \quad (5)$$

The constraints: $\theta_n(0) = \theta'_n(0) = 0 \ (n \geqslant 1)$, and $A_1(0) = 1$, $A_n(0) = 0 \ (n \geqslant 3)$ assure that $\theta(0) = \theta_0$, $\theta'(0) = p_0$, and $A(0) = \varepsilon$. For an initially uniform, monoenergetic $(p_0 = 0)$ electron beam, and a monochromatic electromagnetic wave, the system is periodic so we can restrict our attention to the interval $0 \leqslant \theta_0 \leqslant 2\pi$. $\langle e^{-im\theta_0} \rangle = \delta_{m,0}$, where $\delta_{m,0}$ is the Kronecker delta which equals unity for $m = 0$ and vanishes for all $m \neq 0$.

Eqs. (1)–(3) imply

$$\theta'' = -Ae^{i\theta} - A^* e^{-i\theta} \quad (6)$$

$$A''' - iA = iA^*\langle e^{-2i\theta} \rangle - \langle \theta'^2 e^{-i\theta} \rangle . \quad (7)$$

The prime denotes derivative with respect to Z. We insert the expansions of Eqs. (4) and (5) into Eqs. (6) and (7), and equate terms having equal powers of ε. The first-order amplitude has the well-known solution [1], $A_1(Z) = (e^{sZ} + e^{-s*Z} + e^{-iZ})/3$ where $s = (\sqrt{3} + i)/2$. There are three modes: growing, decaying and oscillating. For $Z \gg 1$, the exponentially growing mode dominates, $\varepsilon A_1(Z) \approx A_L(Z) \equiv (\varepsilon/3)\exp(sZ)$, and the perturba-

tion coefficients θ_n and A_n have the form

$$\varepsilon^n \theta_n(Z,\theta_0) = \sum_{k=0}^{n} b(n, n-2k) A_L^{n-k}(Z)$$
$$\times A_L^{k\,*}(Z)e^{i(n-2k)\theta_0} \quad (n \geqslant 1) \quad (8)$$

and

$$\varepsilon^{2m+1} A_{2m+1}(Z) = a(m) A_L(Z)|A_L(Z)|^{2m} \quad (m \geqslant 0). \quad (9)$$

$b(n, n-2k)$ and $a(m)$ are complex constants independent of Z, determined recursively from Eqs. (6) and (7). We know that $a(0) = 1$ and find θ_1, θ_2 from Eq. (6) and then A_3 from Eq. (7). Next, θ_3, θ_4 are determined from Eq. (6). Once this is accomplished, A_5 is found from Eq. (7). In general, suppose we know $\theta_1, \theta_2, ..., \theta_{2m}$ and $A_1, A_3, ..., A_{2m+1}$, then θ_{2m+1} and θ_{2m+2} can be determined from Eq. (6), and then A_{2m+3} can be found from Eq. (7).

It is seen from Eqs. (5) and (9) that the radiation amplitude can be expressed in terms of the linear solution, $A_L(Z) = (\varepsilon/3)\exp(sZ)$, as

$$A(Z;\varepsilon) \cong A_L(Z)h(|A_L(Z)|^2) \quad (Z \gg 1) \quad (10)$$

with

$$h(\xi) = \sum_{m=0}^{\infty} a(m)\xi^m. \quad (11)$$

Using Mathematica we have computed the coefficients $a(1), ..., a(12)$ of the power series in Eq. (11). In Table 1, columns 2 and 3, we present

Table 1
Ratios of coefficients in expansion for $h(\xi)$ [Eq. (11)]

n	$|a(n)/a(n-1)|$	$\mathrm{Arg}[a(n)/a(n-1)]$	$|a(n)/a(n-1)|\dfrac{n}{n-1/2}$
1	0.216951	2.55393	0.433903
2	0.272966	2.43870	0.363955
3	0.298157	2.42034	0.357788
4	0.310309	2.40888	0.354639
5	0.318838	2.40122	0.354264
6	0.325133	2.39864	0.354690
7	0.329361	2.39838	0.354696
8	0.332254	2.39793	0.354404
9	0.334581	2.39709	0.354262
10	0.336581	2.39662	0.354296
11	0.338190	2.39659	0.354294
12	0.339444	2.39654	0.354203

the magnitude and argument of the complex ratios, $a(n)/a(n-1)$. We see that after the first few values of n, the argument of this ratio remains close to 2.397 rad. The magnitude of the ratio also varies slowly. The variation is further reduced if we multiply by $n/(n-\frac{1}{2})$. These results suggest that there exists an inverse square root branch point at $\xi_0 \cong \exp(-i2.397)/0.354$. This singularity limits the radius of convergence of the power series in Eq. (11). Therefore, in order to use it to study saturation, we need to carry out an analytic continuation. A Taylor series can be analytically continued by the use of Padé approximates [8]. One constructs a sequence of rational functions to approximate the unknown function such that when the rational functions are expanded, the coefficients match the original series expansion as well as possible.

For $Z \gg 1$, Eq. (10) implies that the radiation intensity has the form

$$|A(\varepsilon, Z)|^2 \cong I(\xi) \equiv \xi \left| \sum_{m=0}^{\infty} a(m)\xi^m \right|^2 \qquad (12)$$

where the coefficients $a(m)$ are complex and the scaling variable,

$$\xi \equiv \tfrac{1}{9}\varepsilon^2 e^{\sqrt{3}Z} \qquad (13)$$

is real. Eq. (12) shows that for large Z, the intensity does not depend on ε and Z independently, but only in the combination specified in Eq. (13). Therefore, a change in the initial value of the radiation field, ε, corresponds to a translation in Z. This is a mathematical expression of the intuitive idea that in a process with exponential growth the initial conditions are "forgotten". In FEL physics this property is sometimes referred to as "lethargy".

The singularity limits the radius of convergence of the power series in Eq. (11). Therefore, in order to use it to study the saturation of the FEL, we need to carry out an analytic continuation. One approach to the analytic continuation of a Taylor series is the use of Padé approximates [8]. In this approach, one constructs a sequence of rational functions to approximate the unknown function. The rational functions are chosen such that when they are expanded, the coefficients match the

original series expansion as well as possible. As an example [8], let us consider the function

$$f(x) = \left(\frac{1+2x}{1+x} \right)^{1/2}$$
$$= 1 + \frac{1}{2}x - \frac{5}{8}x^2 + \frac{13}{16}x^3 - \frac{141}{128}x^4 + \cdots . \qquad (14)$$

Clearly, the Taylor series fails to converge for any value of $x > \frac{1}{2}$. The first Padé approximate is

$$\frac{1+(7/4)x}{1+(5/4)x} = 1 + \frac{1}{2}x - \frac{5}{8}x^2 + \frac{25}{32}x^3$$
$$- \frac{125}{128}x^4 + \cdots . \qquad (15)$$

This simple approximation has the value 1.4 at $x = \infty$ which should be compared to the exact value, $\sqrt{2}$. The next approximation is

$$\frac{1+(13/4)x+(41/16)x^2}{1+(11/4)x+(29/16)x^2} \qquad (16)$$

whose value is 1.4138 at $x = \infty$.

We expand the right-hand side of Eq. (12) in powers of ξ and analytically continue using Padé approximates. We denote by $[M, N]$, the Padé approximate in which the numerator is a polynomial of degree M and the denominator is a polynomial of degree N. In Fig. 1, we plot the intensity $|A(Z)|^2$ versus Z for the $[N, N]$ approximates, with $N = 1, \ldots, 6$. It is seen that convergence out to about $Z = 10$ has been achieved for the [5,5] and [6,6] approximates. The

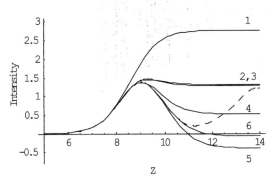

Fig. 1. The dimensionless intensity $|A(Z)|^2$ as derived from the $[N, N]$ ($N = 1, \ldots, 6$) Padé approximates (for $\varepsilon = 0.003$) versus dimensionless distance Z travelled along the undulator. The dashed curve shows the result of a numerical integration of Eqs. (1)–(3).

[6,6] approximate agrees very accurately with the result of direct numerical integration of Eqs. (1)–(3) (dashed curve) out to $Z = 11$. We have also used Eqs. (10) and (11) to calculate a power series expansion for the phase of the radiation field. We found that the [6,6] Padé approximation agrees very accurately with the numerical solution for the phase out to $Z = 10$ [9].

Acknowledgements

I wish to thank Dr. R.L Gluckstern for stimulating discussions and collaboration in analyzing the series coefficients presented in Table 1, and Dr. Z. Huang for enlightening comments and discussion of results from his time-dependent FEL code. I also wish to thank the Stanford Linear Accelerator Center for its hospitality during the course of this work. This work was supported by Department of Energy Contracts DE-AC03-76SF00515 and DE-AC02-98CH10886.

References

[1] E.L. Saldin, E.A. Schneidmiller, M.V. Yurkov, The Physics of Free Electron Lasers, Springer, Berlin, 2000.

[2] R. Bonifacio, F. Casagrande, L. De Salvo Souza, Phys. Rev. A 33 (1986) 2836.

[3] C. Maroli, N. Sterpi, M. Vasconi, R. Bonifacio, Phys. Rev. A 44 (1991) 5206.

[4] R.L. Gluckstern, S. Krinsky, H. Okamoto, Phys. Rev. E 47 (1993) 4412.

[5] Z. Huang, K.J. Kim, Nucl. Instr. and Meth. A 483 (2002) 504.

[6] N.A. Vinokurov, Z. Huang, O.A. Shevenko, K.J. Kim, Nucl. Instr. and Meth. A 475 (2001) 74.

[7] G. Dattoli, P.L. Ottaviani, Opt. Commun. 204 (2002) 283.

[8] G.A. Baker, Essentials of Padé Approximates, Academic Press, New York, 1975.

[9] S. Krinsky, SLAC-PUB-9619.

Available online at www.sciencedirect.com

SCIENCE @ DIRECT°

ELSEVIER

Nuclear Instruments and Methods in Physics Research A 528 (2004) 56–61

NUCLEAR
INSTRUMENTS
& METHODS
IN PHYSICS
RESEARCH
Section A

www.elsevier.com/locate/nima

The stability of electron orbits due to the wiggler field amplitude in free-electron laser

S.-K. Nam*, K.-B. Kim, E.-K. Paik

Department of Physics, Kangwon National University, Chunchon 200-701, Republic of Korea

Abstract

The relativistic electron orbits have been studied in the helical–wiggler field, which included self-field effect in free-electron laser. The equations of motion are derived from the Hamiltonian which includes self-field, and the steady-state orbit solutions were found from the equations of motion. We have analyzed the stability of electron orbits due to the wiggler field amplitude by numerical simulations. We have estimated the maximal Lyapunov exponents, which required estimation of the threshold value of the wiggler field amplitude for the onset of chaos in electron orbits.
© 2004 Elsevier B.V. All rights reserved.

PACS: 41.60.Cr

Keywords: Free-electron lasers; Self-field; Chaotic motion

1. Introduction

The free-electron laser (FEL) has several remarkable properties, such as frequency tunability, high power, and high efficiency. When the FEL operates at a high-current regime and an intense wiggler field regime to obtain a sufficiently large gain, the electron motion can be altered by a self-field.

We studied the chaotic behavior of the relativistic electron orbits in a magnetic field, which consists of a helical–wiggler field and a self-field. The equations of motion are derived from the Hamiltonian which includes self-field, and the steady-state orbit solutions were obtained from the equations of motion. We analyzed the stability of electron orbits, induced by the wiggler field amplitude, through the numerical simulation. Poincarè maps have been generated to demonstrate the chaos phenomena in the vicinity of the steady-state orbit. The results were confirmed by estimating the maximal Lyapunov exponents required to find the threshold value of the wiggler field amplitude for the onset of chaos in electron orbits.

2. Theoretical model

The helical–wiggler field can be defined as

$$\mathbf{B}_w = -2\frac{B_w}{k_w} \vec{\nabla}[I_1(k_w r)\cos(k_w z - \theta)] \tag{1}$$

*Corresponding author. Tel.: +82-33-250-8463; fax: +82-33-257-9689.

E-mail address: snam@kangwon.ac.kr (S.-K. Nam).

0168-9002/$ - see front matter © 2004 Elsevier B.V. All rights reserved.
doi:10.1016/j.nima.2004.04.018

and the vector potential of the helical–wiggler field in cylindrical coordinate system is given by

$$\mathbf{B}_w = \nabla \times \mathbf{A}_w$$

where

$$
\mathbf{A}_w = \frac{mc^2 a_w}{e}\{[I_0(k_w r)\cos(k_w z - \theta) - I_2(k_w r) \\
\times \cos(k_w z - \theta)]\hat{e}_r + [I_0(k_w r)\sin(k_w z - \theta) \\
- I_2(k_w r)\sin(k_w z - \theta)]\hat{e}_\theta\}. \tag{2}
$$

Assuming that $k_w^2 r^2 < 1$, the vector potential can be expanded as $\mathbf{A}_w = A_w^{(0)} + A_w^{(2)} + O(k_w^4 r^4)$

$$
\mathbf{A}_w \cong \frac{mc^2 a_w}{e}\left[\left(1 + \frac{k_w^2 r^2}{8}\right)\right]\cos(k_w z - \theta)\hat{e}_r \\
+ \left(1 + \frac{3k_w^2 r^2}{8}\right)\sin(k_w z - \theta)\hat{e}_\theta\Bigg], \tag{3}
$$

where wiggler period $\lambda_w = 2\pi/k_w$, $\theta = \tan^{-1}(y/x)$, and $r = (x^2 + y^2)^{1/2}$.

Considering a relativistic electron beam in the externally applied magnetic field; the electron beam can be assumed to have a uniform density.

$$
n_b^0 = \begin{cases} n_b = \text{const} & \text{for } 0 \leqslant r < r_b, \\ 0 & \text{for } r > r_b. \end{cases} \tag{4}
$$

This assumption is valid if the strength of self-field is not considered. However, the results of simulation yield identical patterns for uniform density and Gaussian density distribution even when the self-field was included. It is readily shown from the steady-state Maxwell equations that the beam space charge and current generate both self-electric and self-magnetic fields. When the equilibrium field can be conveniently represented as a scalar potential $\mathbf{E}_s = -\nabla \Phi_s$, the electrostatic potential becomes $\Phi_s = (m\omega_{pb}^2 r^2)/(4e)$ for $0 \leqslant r < r_b$. In such a case, the vector potential of the self-magnetic field [1,2] is given by

$$\mathbf{B}_s = \nabla \times \mathbf{A}_s$$

where

$$
A_s = \frac{m\omega_p^2 \beta_b r^2}{4e\hat{e}_z} = \beta_b \Phi_s \hat{e}_z. \tag{5}
$$

In Eq. (5), $\omega_p = (4\pi n_b e^2/m_e)^{1/2}$ is the plasma frequency of the electron beam, m is the electron rest mass, and e is the electron charge.

It is useful to introduce the dimensionless parameter $\kappa_s = \omega_p^2/(c^2 k_w^2)$, which is the strength of the self-field. For uniform density profile and uniform energy, the total beam current is estimated by

$$
I = 2\pi e \int_0^{r_b} r n_b v_z \, dr = \pi c e n_b r_b^2 \beta_z = I_A \left(\frac{r_b}{2\delta}\right)^2, \tag{6}
$$

where $\delta = \sqrt{(\gamma_b c)/\omega_p}$ is the collisionless skin depth, and the Alfven current is estimated by

$$
I_A = \frac{\beta_b \gamma_b mc^3}{e}.
$$

The total vector potential and the canonical momentum can be expressed as $\mathbf{A} = \mathbf{A}_w + \mathbf{A}_s$ and $\mathbf{P} = \mathbf{P}_r \hat{e}_r + (\mathbf{P}_\theta/(k_w r \hat{e}_\theta)) + \mathbf{P}_z \hat{e}_z$, respectively. The equations of motion for a test electron within the beam $(0 < r < r_b)$ can be derived from the Hamiltonian

$$
\mathbf{H} = \sqrt{(c\mathbf{P} + e\mathbf{A})^2 + m^2 c^4} - e\Phi_s \equiv \gamma mc^2 - e\Phi_s. \tag{7}
$$

The generalized Hamiltonian with the self-field effects and the expanded vector potential can be expressed as

$$
\bar{\mathbf{H}} = \sqrt{1 + (\bar{\mathbf{P}} + \bar{\mathbf{A}})^2} - \bar{\Phi}_s \\
= [\{\bar{P}_r + a_w(1 + \bar{r}^2/8)\cos(\theta - \bar{z})\}^2 \\
+ \{\bar{P}_\theta/\bar{r} - a_w(1 + \bar{r}^2/8)\sin(\theta - \bar{z})\}^2 \\
+ \{\bar{P}_z + \beta_b \bar{\Phi}_s\}^2]^{1/2} + \kappa_s \bar{r}^2/4, \tag{8}
$$

where $\bar{\mathbf{H}} = \mathbf{H}/mc^2$, $\bar{\Phi}_s = ((e\Phi_s)/(mc^2))$, $\bar{P} = (P/(mc))$, $\bar{r} = k_w r$, and $\bar{z} = k_w z$.

$a_w = ((eB_w)/(mc^2 k_w))$ is the usual dimensionless measure of the wiggler field amplitude.

3. Steady-state orbits analysis

It is useful to perform the canonical transformation due to the combination of $\sin(\theta - \bar{z})$ and $\cos(\theta - \bar{z})$ terms in Eq. (8). The new variables $(\bar{r}, \psi, \bar{z}', \bar{P}_r, \bar{P}_\psi, \bar{P}_{z'})$ are obtained from the generating function $\mathbf{F}_2(\bar{P}_\psi, \bar{P}_{z'}; \psi, \bar{z}) = (\theta - \bar{z})\bar{P}_\psi + \bar{z}\bar{P}_{z'}$

$$
\psi = \frac{\partial \mathbf{F}_2}{\partial \bar{P}_\psi} = \theta - \bar{z}, \quad \bar{z}' = \frac{\partial \mathbf{F}_2}{\partial \bar{P}_{z'}} = \bar{z},
$$

$$
\bar{P}_\theta \equiv \frac{\partial \mathbf{F}_2}{\partial \theta} = \bar{P}_\psi, \quad \bar{P}_z \equiv \frac{\partial \mathbf{F}_2}{\partial \bar{z}} = \bar{P}_{z'} - \bar{P}_\psi. \tag{9}
$$

The equations of motion are derived from the Hamiltonian. If the equations of motion are independent of time, i.e.,

$$\frac{d\bar{r}}{d\tau} = \frac{d\psi}{d\tau} = \frac{d\bar{P}_r}{d\tau} = \frac{d\bar{P}_\psi}{d\tau} = 0,$$

we could find a steady-state solution as shown in Eq. (10).

$$\bar{P}_{r0} = 0, \quad \psi_0 = \frac{3\pi}{2},$$

$$\bar{P}_{\psi 0} = \frac{\bar{r}_0[4\beta_b\kappa_s\bar{r}_0 - 8a_w + 3a_w\bar{r}_0^2 + \zeta]}{16},$$

$$\bar{P}_{z'} = \frac{[4\beta_b\kappa_s\bar{r}_0 + a_w(8 + \bar{r}_0^2 + 3\bar{r}_0^4) + \zeta(1 + \bar{r}_0^2)]}{16\bar{r}_0}, \quad (10)$$

where

$$\zeta^2 = 16\kappa_s(\beta_b^2\kappa_s - 8\gamma)\bar{r}_0^2 + 8a_w\beta_b\kappa_s\bar{r}_0(8 + 9\bar{r}_0^2) + a_w^2(8 + 9\bar{r}_0^2)^2.$$

Substituting the steady-state solution obtained from Eq. (10) into new Hamiltonian, we obtained

$$\gamma = \bar{H} + \bar{\Phi}_s = \frac{1}{16}\sqrt{256 + \chi^2 + (\chi/\bar{r})^2}, \quad (11)$$

where

$$\chi = 8a_w + \zeta + 4\beta_{zb}\kappa_s\bar{r}_0 + 9a_w\bar{r}_0^2.$$

The wiggler parameter can be deduced from Eq. (11)

$$a_w = \frac{4\bar{r}_0}{8 + 9\bar{r}_0^2}\left[\frac{\gamma_0(\kappa_s + 2\gamma_0 + \kappa_s\bar{r}_0^2) - 2}{\sqrt{(1 + \bar{r}_0^2)(1 - \gamma_0^2)}} - \beta_{zb}\kappa_s\right]. \quad (12)$$

If the self-field is negligibly small, the wiggler parameter has a maximum $a_w^c \approx 0.37\sqrt{\gamma^2 - 1}$ at $\bar{r}_0 \approx 0.68$. This result agrees with that obtained by Chen [3].

4. Numerical analysis

It is useful to introduce the dimensionless parameter, as it helps us analyze

$$\eta \equiv a_w / \sqrt{2(\gamma^2 - 1 - a_w^2)} \quad (13)$$

the gyro-radius $\bar{r}_{gy}$ of the helical orbit in the steady state. Eq. (13) can have two real solutions when the

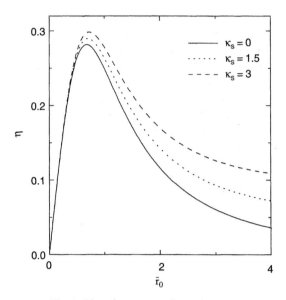

Fig. 1. Plot of η versus $\bar{r}_0$ for various κ_s.

value of η is in the range $0 \leqslant \eta < \eta^c$. Fig. 1 shows the dimensionless parameter η versus initial electron radius $\bar{r}_0$, and η^c which is the critical value of η gets altered by the self-field. The η^c and $\bar{r}_0^{peak}$ are increased monotonically by increasing the self-field parameter by $\eta^c \approx 0.28 + 0.006\kappa_s$ at $\bar{r}_0^{peak} \approx 0.68 + 0.017\kappa_s$. Poincarè maps have been generated to demonstrate the chaoticity in the vicinity of the steady-state orbit with $\bar{r}_0 = \bar{r}^<$, where $\bar{r}^<$ is the smaller of the two solutions obtained from Eq. (13). The equation of motion derived from Hamiltonian is numerically solved with the initial conditions of Eq. (10) and $\bar{r}_0 = \bar{r}_{gy0} + \bar{r}_{gc0}$. The $\bar{r}_{gy0}(\bar{r}_{gc0})$ is an initial value of the normailzed gyro-radius (guiding-center radius) of the steady state orbits. Fig. 2 shows the Poincarè maps in the $(\psi, \bar{P}_\psi)$ plane without self-field for different values of η. The contour size increased on increasing η, which suggests that the amplitude of the betatron oscillation increases, and the coupling between the helical motion and the betatron oscillation also becomes stronger, leading to chaos. Moreover, as η increases, the electron orbits become easily unstable on increasing the guiding-center radius $\bar{r}_{gc}$. In our simulation, we choose the parameters $\lambda_w = 4$ cm, $r_b = 4$ mm, $\gamma = 10$, and observed that the threshold value of η was independent of the electron energy up to $\gamma = 100$. The dependency of self-field strength κ_s is shown in Fig. 3. We found that the contour size decreased and stable region increased on increasing the self-field parameter κ_s.

We estimated the maximal Lyapunov exponent from the time series data [4,5], which are based on the method of delays [6]. Fig. 4 shows the maximal Lyapunov exponent for various values of κ_s at $\eta = 0.22$. This result shows that the critical guiding-center radius, which represents the onset of chaos, is increased by increasing self-field parameter κ_s.

The Fourier transformation without self-field is shown in Fig. 5. The periodic motions

disappeared at the onset of chaos as shown in Fig. 4(a).

Fig. 6 shows the onset of chaos in the electron orbit for various strengths of self-field κ_s. The regular regime of electron orbits broadened on increasing the self-field parameter. For $\bar{r}_{gc} < 0.2$ regime, the critical parameter $\eta^c \sim 0.281$ corresponds to the critical wiggler parameter, $a_w^c = 3.59$ and the critical wiggler field amplitude, $B_w^c = 0.962$ T in the case of without self-field ($\kappa_s = 0$). On the other hand, $\kappa_s = 1$ (2) corresponds to the beam current $I = 1.67$ kA (3.34 kA) and $\eta^c \sim 0.285$

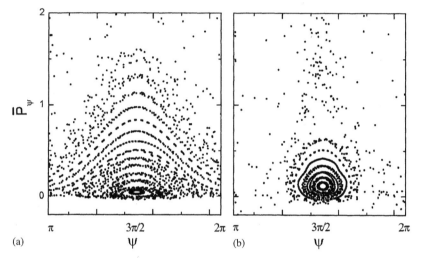

Fig. 2. Poincarè maps without self-field ($\kappa_s = 0$) for (a) $\eta = 0.18$ and (b) $\eta = 0.22$.

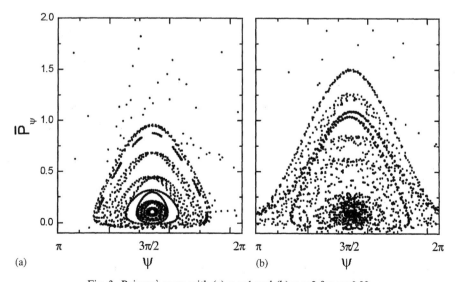

Fig. 3. Poincarè maps with (a) $\kappa_s = 1$ and (b) $\kappa_s = 2$ for $\eta = 0.22$.

(0.292) corresponds to $a_w^c = 3.63$ (3.71) and $B_w^c = 0.974\,T$ (0.994 T) in Fig. 6. The value of 285 (0.292) corresponds to $a_w^c = 3.63$ (3.71) and $B_w^c = 0.974\,T$ (0.994 T) in Fig. 6. A stable region was observed up to magnetic field of 0.962 T and guiding-center radius $r_{gc} = 0.127$ cm, which suggests that this region was the maximum stable region of electron orbits. However, an unstable region appeared when the gyro-radius was greater than the critical guiding-center radius of $r_{gc} = 0.127$ cm.

5. Conclusion

We have investigated the effects of the self-field of electron orbits in a helical–wiggler. In order to investigate the stability of a dynamical system, we

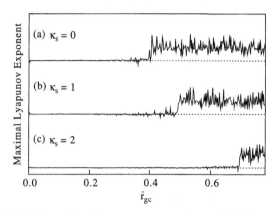

Fig. 4. The maximal Lyapunov exponent for $\kappa_s = 0,1,2$. Dotted lines indicate that the maximal Lyapunov exponent is zero.

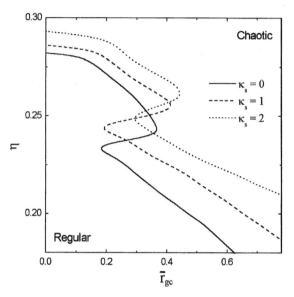

Fig. 6. The onset of chaos in electron orbits for $\kappa_s = 0$ (solid line), 1 (dashed line), and 2 (dotted line).

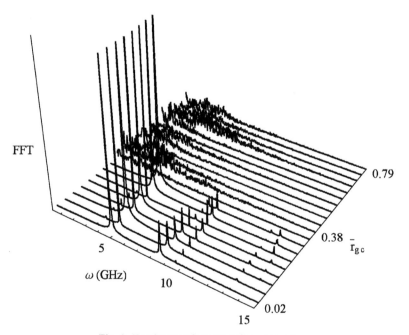

Fig. 5. Fourier transformation for various $\bar{r}_{gc}$.

used Poincarè maps. We estimated the maximal Lyapunov exponents in order to find the critical value of the wiggler field amplitude for the onset of chaos in the electron orbits. The regular regime of the electron orbits broadened on increasing self-field parameters.

Acknowledgements

This work was supported by a grant from the Korea Research Foundation.

References

[1] R.C. Davidson, Physics of Nonneutral Plasmas, Addison-Wesley, San Francisco, 1990.

[2] S.K. Nam, K.B. Kim, Nucl. Instr. and Meth. A 507 (2003) 69.

[3] C. Chen, R.C. Davidson, Phys. Rev. A 43 (1991) 5541.

[4] H. Kantz, T. Schreiber, Nonlinear Time Series Analysis, Cambridge University Press, UK, 1997.

[5] M.T. Rosenstein, J.J. Collins, C.J. De Luca, Physica D 65 (1993) 117.

[6] N.H. Packard, J.P. Crutchfield, J.D. Farmer, R.S. Shaw, Phys. Rev. Lett. A 45 (1980) 117.

Available online at www.sciencedirect.com

SCIENCE ⒟ DIRECT°

Nuclear Instruments and Methods in Physics Research A 528 (2004) 62–66

ELSEVIER

NUCLEAR
INSTRUMENTS
& METHODS
IN PHYSICS
RESEARCH
Section A

www.elsevier.com/locate/nima

Study of radiation spectrum in a free-electron laser oscillator from noise to saturation

Yosef Pinhasi*, Yuri Lurie, Asher Yahalom

Department of Electrical and Electronic Engineering, Faculty of Engineering, The College of Judea and Samaria, P.O. Box 3, Ariel 44837, Israel

Abstract

A three-dimensional, space–frequency model for analysis and simulation of radiation excitation in a FEL oscillator is developed. The total electromagnetic field is presented in the frequency domain as an expansion in terms of transverse eigenmodes of the (cold) resonator. Coupled-mode formulation is utilized in an analytical approach and in a numerical simulation to follow linear and nonlinear processes, taking place during radiation buildup, which lead to the establishment of single-mode lasing. The evolution of the radiation spectrum is investigated in the case of 'grazing', when the group velocity of the radiation is equal to the axial velocity of the electrons.
© 2004 Elsevier B.V. All rights reserved.

PACS: 41.60.Cr

Keywords: FEL; Grazing; Radiation build-up

1. Introduction

Theoretical studies of free-electron laser (FEL) oscillators have been carried out by several groups investigating the linear, nonlinear and saturation processes taking place in the radiation buildup [1–9]. These effects were shown to play an important role in the longitudinal mode competition, leading to the establishment of single-mode lasing at steady state. In these studies, carried out in the framework of a 1D model, the excitation of the electromagnetic radiation is described by a space–time partial differential equation derived from Maxwell's equations. The approximations taken in such time domain approaches, limit the use of the model for studies of wide-band interactions, in which dispersion effects emerge from the resonator and gain medium (electron beam). A difficulty is clearly recognized when the space–time formulation is applied for studying radiation excitation in FELs operating in the grazing incidence, where the group velocity v_g of the electromagnetic wave is equal to the axial velocity v_{z_0} of the electron beam. In that case of zero slippage, space–time approaches fail to describe the electromagnetic field excitation accurately.

Coupled-mode theory of the FEL interaction in the frequency domain was previously developed

---

*Corresponding author. Tel.: +972-3-906-6197; fax: +972-3-906-6238.

E-mail address: yosip@eng.tau-ac.il (Y. Pinhasi).

0168-9002/$ - see front matter © 2004 Elsevier B.V. All rights reserved.
doi:10.1016/j.nima.2004.04.019

[10]. The model, which utilizes expansion of the total excited field in terms of transverse eigen modes of the resonator, was employed to investigate 3D effects and 'supermode' establishment in an FEL oscillator at a single frequency [11,12]. In the work described in this paper, we extend the theory to deal with a continuum of frequencies and apply it in a numerical space–frequency simulation WB3D aimed at the simulation of wide-band interaction processes in FELs [13,14]. The theory is general, enabling the spectral study of radiation buildup in a free-electron maser (FEM) oscillator, also in the interesting case of grazing.

2. Space–frequency model for FEL oscillators

In laser oscillators, a part of the radiation excited in the gain medium is coupled-out, while the remainder is circulated by a feedback mechanism. Assuming a uniform cross-section resonator (usually a waveguide), the total electromagnetic field at every plane z, can be expressed in the frequency domain as a sum of a set of transverse (orthogonal) eigenfunctions with profiles $\widetilde{\mathscr{E}}_q(x,y)$ and related axial wave number $k_{z_q}(f)$ [11,12]. At the beginning of a round-trip n, each of the modes is assumed to have an initial amplitude $C_q^{(n)}(0,f)$ and the total field at $z = 0$ is given by

$$\widetilde{\mathbf{E}}^{(n)}(x,y,z=0;f) = \sum_q C_q^{(n)}(0,f)\widetilde{\mathscr{E}}_q(x,y). \tag{1}$$

The field obtained at the end of the interaction (wiggler) region $z = L_\mathrm{w}$ can be written as

$$\widetilde{\mathbf{E}}^{(n)}(x,y,z=L_\mathrm{w};f)$$
$$= \sum_q C_q^{(n)}(L_\mathrm{w},f)\widetilde{\mathscr{E}}_q(x,y)\mathrm{e}^{+jk_{z_q}(f)L_\mathrm{w}}. \tag{2}$$

Here the amplitude of the qth mode excited by a driving current density of the electron beam $\widetilde{\mathbf{J}}^{(n)}(x,y,z;f)$ is given by

$$C_q^{(n)}(L_\mathrm{w},f)$$
$$= C_q^{(n)}(0,f) - \frac{1}{2\mathscr{N}_q}\int_0^{L_\mathrm{w}}\int\int \widetilde{\mathbf{J}}^{(n)}(x,y,z;f)$$
$$\cdot \widetilde{\mathscr{E}}_q^*(x,y)\mathrm{e}^{-jk_{z_q}(f)z}\,\mathrm{d}x\,\mathrm{d}y\,\mathrm{d}z, \tag{3}$$

where the normalization of the mode amplitude is made via the complex Poynting vector power

$$\mathscr{N}_q = \int\int [\widetilde{\mathscr{E}}_{\perp_q}(x,y) \times \widetilde{\mathscr{H}}_{\perp_q}^*(x,y)]\cdot\hat{\mathbf{z}}\,\mathrm{d}x\,\mathrm{d}y. \tag{4}$$

The spectral density of the energy flow after the interaction with the electron beam at the nth round-trip is

$$\frac{\mathrm{d}\mathscr{W}^{(n)}(L_\mathrm{w})}{\mathrm{d}f} = \sum_q |C_q^{(n)}(L_\mathrm{w},f)|^2 \tfrac{1}{2}\mathfrak{R}\{\mathscr{N}_q\}. \tag{5}$$

After a round-trip in the resonator of length L_c, the field fed back into the entrance of the interaction region is

$$\widetilde{\mathbf{E}}^{(n+1)}(x,y,z=0;f) = \sum_{q'} C_{q'}^{(n+1)}(0,f)\widetilde{\mathscr{E}}_{q'}(x,y)$$
$$= \sum_{q'}\left[\sum_{q''}\rho_{q'q''}C_{q''}^{(n)}(L_\mathrm{w},f)\right]$$
$$\times \widetilde{\mathscr{E}}_{q'}(x,y)\mathrm{e}^{+jk_{z_{q'}}(f)L_\mathrm{c}}. \tag{6}$$

$\rho_{q'q''}$ is a complex coefficient, expressing the intermode field reflectivity of transverse mode q'' to mode q', due to scattering of the resonator mirrors or any other passive elements in the entire feedback loop. Scalar multiplication of both sides of Eq. (6) by $\widetilde{\mathscr{E}}_q^*(x,y)$, results in the initial mode amplitude

$$C_q^{(n+1)}(0,f) = \sum_{q''}\rho_{qq''}C_{q''}^{(n)}(L_\mathrm{w},f)\mathrm{e}^{+jk_{z_q}(f)L_\mathrm{c}} \tag{7}$$

which is required in Eq. (3) to solve the field excited in the consecutive round-trip. In the frequency domain, the total out-coupled radiation obtained at the oscillator output after N round-trips is composed of a summation of the circulated fields (2) inside the resonator:

$$\widetilde{\mathbf{E}}_\mathrm{out}(f) = \sum_q \Upsilon_q \sum_{n=0}^{N} C_q^{(n)}(L_\mathrm{w},f)\widetilde{\mathscr{E}}_q(x,y)\mathrm{e}^{+jk_{z_q}(f)L_\mathrm{w}},$$
$$\tag{8}$$

where Υ_q is qth mode field transmission of the out-coupler. The energy spectrum of the electromagnetic radiation obtained at the output after N

round-trips is given by

$$\frac{\mathrm{d}\mathscr{W}_{\mathrm{out}}(N)}{\mathrm{d}f} = \sum_{q} \mathscr{T}_q \left| \sum_{n=0}^{N} C_q^{(n)}(L_{\mathrm{w}}, f) \right|^2 \frac{1}{2} \Re\{\mathscr{N}_q\},$$

(9)

where $\mathscr{T}_q = |\varUpsilon_q|^2$ is the power transmission coefficient of mode q.

3. Characteristics of a waveguide resonator

To simplify the analysis, we assume a single transverse mode q. Using the formalism presented in the preceding section, we first show that it leads, in an absence of a gain medium (electron beam), to the analytical expression for the power transfer line shape of a 'cold' resonator. Without the electron beam, the driving current substituted in (3) is $\tilde{\mathbf{J}}^{(n)}(x, y, z; f) = 0$, resulting in $C_q^{(n)}(L_{\mathrm{w}}, f) = C_q^{(n)}(0, f)$. From Eq. (7) we derive the relation

$$C_q^{(n)}(0, f) = \rho^n \mathrm{e}^{jnk_{zq}(f)L_{\mathrm{c}}} C_q^{(0)}(0, f).$$

(10)

The total field obtained at the resonator output after $N \to \infty$ round-trips is then

$$\begin{aligned}
\tilde{\mathbf{E}}_{\mathrm{out}}(f) &= \varUpsilon \sum_{n=0}^{\infty} \rho^n \mathrm{e}^{jnk_{zq}(f)L_{\mathrm{c}}} C_q^{(0)}(0, f) \tilde{\mathscr{E}}_q(x, y) \mathrm{e}^{jk_{zq}(f)L_{\mathrm{w}}} \\
&= \frac{\varUpsilon}{1 - \rho \mathrm{e}^{jk_{zq}(f)L_{\mathrm{c}}}} C_q^{(0)}(0, f) \tilde{\mathscr{E}}_q(x, y) \mathrm{e}^{jk_{zq}(f)L_{\mathrm{w}}}.
\end{aligned}$$

(11)

The resulted energy spectrum at the resonator output is

$$\begin{aligned}
\frac{\mathrm{d}\mathscr{W}_{\mathrm{out}}}{\mathrm{d}f} &= \frac{\mathscr{T}}{(1 - \sqrt{\mathscr{R}})^2 + 4\sqrt{\mathscr{R}}\sin^2\left[\frac{1}{2}k_z(f)L_{\mathrm{c}}\right]} \\
&\quad \times |C_q^{(0)}(0, f)|^2 \frac{1}{2}\Re\{\mathscr{N}_q\},
\end{aligned}$$

(12)

where $\mathscr{R} = |\rho|^2$ is the total round-trip power reflectivity. Eq. (12) consists of the well-known expression describing the power transfer characteristics of the Fabry–Perot resonator [15]. The spectral line shape of the waveguide-based resonator utilized in the millimeter wave FEM of Table 1, is shown in Fig. 3. Maximum transmission occurs at resonant frequencies f_m, where $k_z(f_m)L_{\mathrm{c}} = 2m\pi$ (m is an integer). The *free-spectral*

Table 1
Operational parameters of millimeter wave free-electron maser

Accelerator	
Electron beam energy:	$E_k = 1.066$ MeV
Electron beam current:	$I_0 = 2$ A
Wiggler	
Magnetic induction:	$B_{\mathrm{w}} = 2000$ G
Period:	$\lambda_{\mathrm{w}} = 4.444$ cm
Number of periods:	$N_{\mathrm{w}} = 20$
Waveguide	
Rectangular waveguide:	10.1×9.005 mm
Mode:	TE_{01}
Resonator	
Total round-trip power reflectivity:	$\mathscr{R} = 80\%$
Power transmission coefficient:	$\mathscr{T} = 1\%$
Resonator length:	$L_{\mathrm{c}} = 1.70$ m

range is the intermode frequency separation $\mathrm{FSR} = f_{m+1} - f_m \cong 1/t_r = 163.5$ MHz, where $t_r = L_{\mathrm{c}}/v_{\mathrm{g}} = 6.12$ nS is the round-trip time. The *full-width half-maximum* of the transmission peaks is given by $\mathrm{FWHM} = \mathrm{FSR}/\mathscr{F} = 5.81$ MHz where $\mathscr{F} = \pi\sqrt[4]{\mathscr{R}}/(1 - \sqrt{\mathscr{R}}) = 28.1$ is the finesse of the resonator.

4. Radiation buildup in a FEL oscillator at grazing incidence

In the presence of electron beam, the radiation circulating in the resonator is re-amplified in the interaction region at each round-trip. At the first stage of the oscillator self-excitation, the synchrotron undulator radiation emitted by the individual electrons entering the interaction region at random, interfere and combine coherently with the circulating field in the resonator. If the single-pass gain is higher than the total resonator losses (transmission and internal losses), the radiation intensity inside the resonator increases and becomes more coherent. After several round-trips, the radiation power is built-up in the linear regime until the oscillator arrives to its non-linear stage of operation and to saturation.

The space–frequency numerical simulation WB3D was employed to demonstrate the development of the radiation, generated in a millimeter wave FEM oscillator, with the operational parameters given in Table 1. When the beam energy is set to approximately 1.066 MeV, a single synchronism frequency of about 44.3 GHz is obtained (grazing). The curves of energy spectral density drawn in Fig. 1, are for the first round-trips of the initial stage of radiation excitation due to spontaneous emission. The evolution of the spectrum during radiation round-trips is shown in Fig. 2. Note, that radiation growth is obtained at slightly lower frequency range than that of spontaneous emission, due to the detuning of the gain curve from synchronism. The total energy buildup is described in Fig. 3.

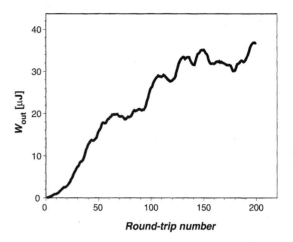

Fig. 3. Total energy buildup in the FEM oscillator.

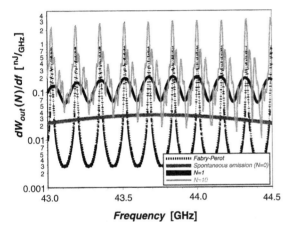

Fig. 1. Fabry–Perot line shape and power spectra at the initial stage of oscillator excitation.

5. Summary and conclusions

The presented coupled-mode theory, formulated in the frequency domain, enables development of three-dimensional models, which can accurately describe wide-band interactions between radiation and electron beam. Space–frequency solution of the electromagnetic equations inherently takes into account dispersive effects arising from the resonator and gain medium. Such effects play an role in the special case of grazing, and cannot be correctly considered in approximated space–time approaches. We note that our space–frequency model described here, also facilitates the consideration of statistical features of the electron beam and the excited radiation.

Acknowledgements

The research was supported by the Israel Science Foundation and the Israel Ministry of Science.

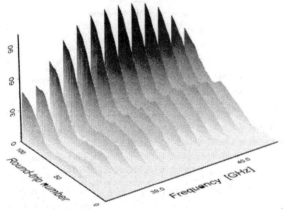

Fig. 2. Evolution of radiation spectrum.

References

[1] V.L. Bratman, N.S. Ginzburg, M.I. Petelin, Sov. Phys. JTEP 49 (1979) 469.
[2] V.L. Bratman, N.S. Ginzburg, M.I. Petelin, Opt. Comm. 30 (1979) 409.

[3] I. Kimmel, L. Elias, Phys. Rev. A 35 (1987) 3818.

[4] I. Kimmel, L. Elias, Phys. Rev. A 38 (1988) 2889.

[5] B. Levush, T.M. Antonsen, Nucl. Instr. and Meth. A 272 (1988) 375.

[6] B. Levush, T.M. Antonsen, Nucl. Instr. and Meth. A 285 (1988) 136.

[7] T.M. Antonsen, B. Levush, Phys. Fluids B 1 (1989) 1097.

[8] T.M. Antonsen, B. Levush, Phys. Rev. Lett. 62 (1989) 1488.

[9] B. Levush, T.M. Antonsen, IEEE Trans. Plasma Sci. 18 (1990) 260.

[10] Y. Pinhasi, A. Gover, Phys. Rev. E 51 (1995) 2472.

[11] Y. Pinhasi, A. Gover, Nucl. Instr. and Meth. A 375 (1996) 233.

[12] Y. Pinhasi, V. Shterngartz, A. Gover, Phys. Rev. E 54 (1996) 6774.

[13] Y. Pinhasi, Yu. Lurie, A. Yahalom, Nucl. Instr. and Meth. A 475 (2001) 147.

[14] Y. Pinhasi, Yu. Lurie, Phys. Rev. E 65 (2002) 026501.

[15] A. Yariv, Optical Electronics, Holt Rinehart and Winston, New York, 1991.

Available online at www.sciencedirect.com

Nuclear Instruments and Methods in Physics Research A 528 (2004) 67–70

NUCLEAR
INSTRUMENTS
& METHODS
IN PHYSICS
RESEARCH
Section A

www.elsevier.com/locate/nima

Regime of non-resonant trapping in a Bragg-cavity FEM oscillator

I.V. Bandurkin[a,*], N.Yu. Peskov[a], A.D.R. Phelps[b], A.V. Savilov[a]

[a] *Institute of Applied Physics, Russian Academy of Sciences, 46 Ulyanov Str., Nizhny Novgorod 603950, Russia*
[b] *Department of Physics, University of Strathclyde, Glasgow G4 0NG, UK*

Abstract

A method for realization of the regime of non-resonant trapping of electrons in a free electron maser (FEM)–oscillator is proposed and theoretically studied. A possibility to achieve a high (over 50%) electron efficiency and a very weak sensitivity to the electron-beam quality is demonstrated for a moderately relativistic mm-wavelength FEM.
© 2004 Elsevier B.V. All rights reserved.

PACS: 41.60.Cr

Keywords: Free-electron maser; Oscillator; Regime of trapping

1. Introduction

One of the main disadvantages of regimes of electron–wave interaction with a high Doppler up-conversion of the frequency of electron oscillations, which are traditionally used in free-electron masers (FEMs), is a strong sensitivity to the spread in electron velocity. This is caused by the fact that both the regime of inertial bunching [1] and the regime of trapping and adiabatic deceleration of electrons [2–4] require fulfilment of the electron–wave resonance condition,

$$\omega \approx hv_z + \Omega_e \qquad (1)$$

from the very beginning of the interaction region. Here Ω_e is the frequency of electron oscillation (for the FEM-ubitrons, Ω_e is the bounce frequency).

In order to get rid of this disadvantage, a different regime of electron–wave coupling (the regime of non-resonant trapping) was proposed [5,6] for FEM-amplifiers with profiled parameters of the interaction region. Its main idea is to realize a non-resonant (in a certain sense) regime of the electron–wave interaction. Namely, in this regime resonant condition (1) fulfils not at the beginning but at any arbitrary point inside the interaction region with a profiled resonant parameter. The trapping at this resonant point is possible due to "deepening" of the potential well, which is caused by an increase of the RF amplitude with the coordinate. In amplifiers, this regime not only diminishes the influence of the spread in electron

*Corresponding author. Tel.: +7-8312-160-637; fax: +7-8312-362-061.
E-mail address: iluy@appl.sci-nnov.ru (I.V. Bandurkin).

0168-9002/$ - see front matter © 2004 Elsevier B.V. All rights reserved.
doi:10.1016/j.nima.2004.04.020

velocity, but also provides a broad band of frequency tuning: since the location of the resonant point is not fixed, the resonant parameters of the system (the operating frequency and the electron velocity) are also not fixed.

In this work, we propose a method for realization of a similar regime in FEM-oscillators. Similar to work [6], a moderately relativistic mm-wavelength FEM with so-called "reversed" guiding magnetic field is examined. It is shown that very high (over 50%) efficiency and very weak sensitivity to the spread in electron velocity can be achieved.

2. Regime of non-resonant trapping

Fig. 1 illustrates the regime of non-resonant trapping on the phase plane (γ, θ), where γ is the electron Lorentz factor and θ is the electron phase with respect to the RF wave [2,4]. Both at the beginning and at the end of the operation region electrons are far from the resonance with the RF wave; the resonance condition (1) is fulfilled at an arbitrary point inside the interaction region. During the interaction process, the separatrix ("bucket"), which separates finite-phase trajectories of resonant particles and infinite ones of non-resonant particles, goes down due to decreasing of the resonant energy $\gamma_r(z)$ (the energy that corresponds to the exact resonance condition). At some resonant point in the middle of the interac-

tion region the "bucket" passes through the level $\gamma = \gamma_0$. If the RF-wave amplitude a, and, therefore, the "bucket" energy size, $U \propto \sqrt{|a|}$, increase significantly with the coordinate in the region close to this, then electrons are trapped by the "bucket". The further decrease of γ_r provides the decrease of the energy of the trapped electrons similar to the traditional 'resonant' scheme of the regime of trapping [2–4].

A principal requirement for realization of non-resonant trapping is a fast increase of the RF amplitude and, therefore, of the "bucket" energy size, in the region close to the resonant point. In amplifiers this is provided due to the RF-amplitude growth caused by the resonant electron–wave interaction; the RF-wave longitudinal inhomogeneity is significant if a high enough electron current is provided. As for oscillators with relatively high-Q operating cavities, the longitudinal RF structure is less inhomogeneous as it is fixed (at least partially) by the cavity.

In order to solve this problem, we propose to provide an "artificial" growth of the RF amplitude in the resonant point by using a proper microwave system. We propose to use an operating cavity, which is composed of a piece of waveguide terminated by two (input and output) mirrors (Fig. 2). It is important that the input mirror is a quite long Bragg-type reflector providing 100%

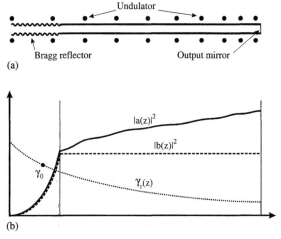

Fig. 2. Scheme of the regime of trapping in an oscillator. Distribution of fields of (a) forward- and (b) backward-propagating waves inside the cavity.

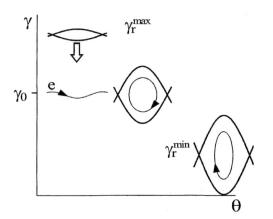

Fig. 1. Motion of the "bucket" through the level of the electron–wave resonance.

reflection (at the center of its band) of the backward feedback wave into the forward operating wave. This reflector represents the first section of the electron–wave interaction region; moreover, the resonant points of most of the velocity electron fractions are situated inside this reflector, where the amplitude of the operating wave monotonically increases with the coordinate (Fig. 2). This inhomogeneity is used to provide non-resonant trapping of electrons inside the reflector.

3. Simulations for a mm-wave FEM-oscillator

We study possibilities for realization of the "oscillator" version of the regime of the non-resonant trapping in the FEM with so-called "reversed" guiding magnetic field [7,8]. The rotating mode TE_{11} of a circular waveguide is excited by the helical axis-encircling electron beam, which moves in a combination of two magnetostatic fields: periodic field of a helical undulator, $\mathbf{B}_U$, and uniform guiding field, $\mathbf{B}_0$ (the "reversed" character of $\mathbf{B}_0$ means that the direction of electron rotation in the undulator field is contrary to the direction of imaginary electron gyro-rotation in the guiding field). The following spatio-temporal equations [8] for the electron motion and for the wave amplitudes are used

$$\frac{\partial \gamma}{\partial Z} = -\frac{p_\perp}{2p_\parallel} \, \mathrm{Im}(a e^{i\theta}), \tag{2}$$

$$\frac{\partial \theta}{\partial Z} = \frac{\gamma}{p_\parallel} - \frac{h}{k} - \mu + F, \tag{3}$$

$$\frac{\partial a}{\partial Z} = iG \left\langle \frac{p_\perp}{p_\parallel} e^{-i\theta} \right\rangle + i\sigma b, \tag{4}$$

$$\frac{\partial b}{\partial Z} - \left(\frac{1}{\beta_{ap}} + \frac{1}{\beta_{an}} \right) \frac{\partial b}{\partial \tau} = -i\sigma a. \tag{5}$$

Here $Z = kz$ is the normalized longitudinal coordinate, $\tau = \omega t - \int_0^Z dz/\beta_{an}$ is the normalized time, $p_{\perp,\parallel} = \gamma v_{\perp,\parallel}/c$ are the normalized rotational and translation components of the electron momentum, b is the amplitude of the feedback backward wave, σ is the waves' coupling factor ($\sigma = 0$ in the regular section),

Table 1
Parameters of simulations

Electron voltage	850 kV
Electron current	200 A
Operating wavelength	8 mm
Operating mode	TE_{11}
Operating waveguide radius	8.5 mm
Initial undulator velocity	0.35
Initial ratio between bounce-frequency and cyclotron frequency	2
Initial/final undulator period	5/1 cm
Input Bragg reflector length	15 cm
Input reflection factor	0.98
Regular waveguide section length	80 cm
Output reflection factor	0.64

$k = \omega/c$, h is the longitudinal wave number, $\mu = h_U/k$, F is the forcing term [8], $G = (((k^2 - h^2))/(Nhk))((eI)/(mc^3))$ is the excitation factor, I is the electron current, and N is the wave norm. The rotational momentum is given by the following formula:

$$p_\perp = a_U \frac{\Omega_e}{\Omega_e + \Omega_c}, \tag{6}$$

where a_u is the normalized vector-potential of the undulator field and $\Omega_c = eB_0/mc\gamma$ is the cyclotron frequency.

The parameters of the FEM-oscillator are chosen to be close to the FEM experiments based on the accelerator LINAC-LIU-3000 (JINR, Dubna) [7] (Table 1). As all the electrons pass the same accelerating voltage, we neglect the spread in electron energy. In order to take into account the spread in electron velocity, the transverse velocity is represented in the form $\beta_\perp^2 = \beta_U^2 + \beta_S^2$, where β_U is the rotational velocity, imposed by the undulator, whereas β_S is caused by "spurious" electron oscillations; the "spurious" velocity is supposed to be distributed uniformly over the interval $0 \leqslant \beta_S \leqslant \varepsilon$. Profiling of the resonant electron energy, $\gamma_r(z)$, is provided by linear profiling of the undulator period, $\lambda_U(z)$.

According to simulations, in the proposed system a stationary single-frequency operation with a high (over 50%) efficiency can be achieved. Fig. 3 illustrates the temporal dynamics of excita-

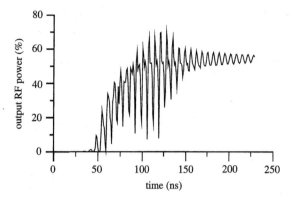

Fig. 3. Temporal dynamics of the oscillator in the case of an ideal beam.

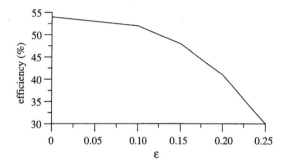

Fig. 4. Electronic efficiency versus value of velocity spread.

tion of the oscillator in the case of an ideal beam ($\varepsilon = 0$). Fig. 4 illustrates a very weak sensitivity of the proposed oscillator to the velocity spread; it proves to be weaker as compared to the similar scheme of the FEM amplifier [6]. One more advantage of the oscillator scheme is a weaker sensitivity to the value of the electron current; according to simulations, halving the current leads to as small a loss in efficiency as 10%. Actually, in the amplifier scheme of the regime of non-resonant trapping, the value of the electron current deter-

mines the increment of the spatial growth of the RF field and, therefore, the efficiency of electron trapping, whereas in the oscillator scheme the proper spatial RF structure is fixed by the microwave system.

Acknowledgements

The authors are grateful to Prof. V.L. Bratman for useful discussions. This work is supported by the Russian Foundation for Basic Research, Project 02-02-17205 and by the UK EPSRC and PPARC.

References

[1] V.L. Bratman, N.S. Ginzburg, M.I. Petelin, Opt. Commun. 30 (1979) 409.
[2] E.D. Belyavsky, Radiotekhnika i Electronica 16 (1971) 208;
    N.M. Kroll, P.L. Morton, M.N. Rosenbluth, IEEE J. Quantum Electron. EQ-17 (1981) 1436;
    P. Sprangle, C.-M. Tang, W.N. Manheimer, Phys. Rev. Lett. A 21 (1980) 302;
    N.S. Ginzburg, et al., in: A.V. Gapanov-Grekhov (Ed.), Relativistic HF Electronics, Institute of Applied Physics, Gorky, USSR, 1988, Issue 5, p. 37.
[3] T. Orzechowski, et al., Phys. Rev. Lett. 57 (1986) 2172.
[4] G.S. Nusinovich, Phys. Fluids B 4 (1992) 1989;
    V.L. Bratman, N.S. Ginzburg, A.V. Savilov, in: A.V. Gapanov-Grekhov (Ed.), Relativistic HF Electronics, Institute of Applied Physics, Gorky, USSR, 1992, Issue 7, p. 22.
[5] A.V. Savilov, Phys. Rev. E 64 (2001) 066501.
[6] A.V. Savilov, I.V. Bandurkin, N.Yu. Peskov, Nucl. Inst. and Meth. in Phys. Res. A 507 (2003) 158.
[7] N.S. Ginzburg, A.A. Kaminsky, A.K. Kaminsky, et al., Phys. Rev. Lett. 24 (2000) 3574.
[8] N.S. Ginzburg, N.Yu. Peskov, Zh. Tekh. Fiz. 58 (1988) 859;
    Ya.L. Bogomolov, N.S. Ginzburg, A.S. Sergeev, Radiotekhnika i Electronika 31 (1986) 102.

Available online at www.sciencedirect.com

Nuclear Instruments and Methods in Physics Research A 528 (2004) 71–77

NUCLEAR
INSTRUMENTS
& METHODS
IN PHYSICS
RESEARCH
Section A

www.elsevier.com/locate/nima

ELSEVIER

Overview of the 100 mA average-current RF photoinjector

D.C. Nguyen[a],*, P.L. Colestock[a], S.S. Kurennoy[a], D.E. Rees[a], A.H. Regan[a], S. Russell[a], D.L. Schrage[a], R.L. Wood[a], L.M. Young[a], T. Schultheiss[b], V. Christina[b], M. Cole[b], J. Rathke[b], J. Shaw[c], C. Eddy[c], R. Holm[c], R. Henry[c], J. Yater[c]

[a] Los Alamos National Laboratory, MS H851, Los Alamos, NM 87545, USA
[b] Advanced Energy Systems, Medford, NY 11763, USA
[c] Naval Research Laboratory, Washington, DC 11763, USA

Abstract

High-average-power FELs require high-current, low-emittance and low-energy-spread electron beams. These qualities have been achieved with RF photoinjectors operating at low-duty factors. To date, a high-average-current RF photoinjector operating continuously at 100% duty factor is yet to be demonstrated. The principal challenges of a high-duty-factor normal-conducting RF photoinjector are related to applying a high accelerating gradient continuously, thus generating large ohmic losses in the cavity walls, cooling the injector cavity walls and the high-power RF couplers, and finding a photocathode with reasonable Q.E. that can survive the poor vacuum of the RF photoinjector. We present the preliminary design of a normal-conducting 700 MHz photoinjector with solenoid magnetic fields for emittance compensation. The photoinjector is designed to produce 2.7 MeV electron beams at 3 nC bunch charge and 35 MHz repetition rate (100 mA average current). The photoinjector consists of a $2\frac{1}{2}$-cell, π-mode, RF cavity with on-axis electric coupling, and a non-resonant vacuum plenum. Heat removal in the resonant cells is achieved via dense arrays of internal cooling passages capable of handling high-velocity water flows. Megawatt RF power is coupled into the injector through two tapered ridge-loaded waveguides. PARMELA simulations show that the $2\frac{1}{2}$-cell injector can produce a 7 μm emittance directly. Transverse plasma oscillations necessitate additional acceleration and a second solenoid to realign the phase space envelopes of different axial slices at higher energy, resulting in a normalized rms emittance of 6.5 μm and 34 keV rms energy spread. We are developing a novel cesiated p-type GaN photocathode with 7% quantum efficiency at 350 nm and a cesium dispenser to replenish the cathode with cesium through a porous silicon carbide substrate. These performance parameters will be necessary for the design of the 100 kW FEL.
© 2004 Published by Elsevier B.V.

PACS: 41.60.Cr; 29.17.+w; 29.27.Bd; 41.75.Fr

Keywords: FEL; Photoinjector; High current; High brightness; cw; Photocathode

*Corresponding author. Tel.: +1-505-667-9385; fax: +1-505-667-8207.
E-mail address: dcnguyen@lanl.gov (D.C. Nguyen).

0168-9002/$ - see front matter © 2004 Published by Elsevier B.V.
doi:10.1016/j.nima.2004.04.021

1. Introduction

High-brightness electron photoinjectors (PI) have long been recognized as the enabling technology for short-wavelength vacuum ultraviolet and X-ray FEL, the so-called fourth-generation light sources [1]. PI are also the key component for high-power, high-duty-factor FEL at longer wavelengths, as the high-current, low-emittance and low-energy-spread electron beams lead to high gain, efficient and robust lasing, and reduce unwanted stray particles that can impact or even preclude high-duty operation.

State-of-the-art high-average-current PI come in three different designs: DC PI consisting of a DC electron gun followed by superconducting RF linac, superconducting RF PI with niobium cavities, and normal conducting RF PI with copper cavities. An example of the DC PI is the Jefferson Laboratory DC gun [2] which is the only true cw photoinjector. This DC PI has produced bunch charges of about 0.14 nC using a GaAs photocathode and a green drive laser. At a repetition rate of 75 MHz the DC PI produces an average current of 10 mA and normalized emittance of 7.5 μm. The superconducting RF PI is under development at FZ Rossendorf [3]. So far the SRF PI has produced low bunch charges.

The normal conducting RF PI, first invented at Los Alamos [4], is now employed in several FEL around the world. These PI typically operate at low-duty factor, very high gradients (20–100 MV/m) and nanocoulomb bunch charges. None of the normal conducting RF PI has operated at 100% duty factor, defined as the fraction of time RF is applied to the PI structure. The highest average current, 32 mA, has been achieved with the Boeing PI (in collaboration with Los Alamos) at 25% duty factor [5]. It uses a K_2CsSb photocathode and 433 MHz cavities to produce 3 nC, 7 μm beams at 5 MeV. The photocathode lifetime is 2–3 h at this duty.

Our objective is to increase the duty factor of normal conducting RF PI to full cw operation. Operating a normal conducting PI continuously presents three main challenges. First, a high accelerating gradient must be applied at the cathode to control the space charge induced expansion that would lead to space-charge and RF induced emittance growth. Second, the continuously applied RF field generates surface current that causes ohmic losses in the cavity walls. These ohmic losses, which scale with the square of the accelerating gradient, drive the RF power requirement to megawatt klystrons and present significant thermal engineering challenges. Third, with cw operation the vacuum in the PI cavities could degrade to a point where any photocathode with reasonable Q.E. can only survive for a short time as contaminants released by stray energetic electrons strike the cathode and poison it.

To address the above challenges, one must optimize the RF PI operating parameters such as RF frequency, accelerating gradient, photoemission radius, cavity cooling, and vacuum requirements for photocathode Q.E. and lifetime. An optimal design is a trade-off involving beam dynamics requirements, RF power, and feasibility of injector cavity cooling.

We present an overview of the cw normal-conducting 700 MHz PI capable of delivering more than 100 mA average current (e.g. 3 nC per bunch at 35 MHz repetition rate) at 2.7 MeV. The average current is scalable to higher values by increasing the bunch repetition rate. The RF PI is designed to operate at a cw gradient of 7 MV/m. The beam transverse normalized rms emittance is 6.5 μm and rms energy spread is 34 keV. We are considering the use of Cs_2Te, Cs:GaN and K_2CsSb photocathodes that require frequency quadrupled, tripled and doubled Nd drive laser, respectively.

2. Injector design

2.1. Cavity design

The cw RF PI is a $2\frac{1}{2}$-cell, π-mode, oxygen-free copper-on-Glidcop® 700 MHz structure with on-axis electric coupling and emittance compensation [6]. The design is a compromise between maximizing the gradient and minimizing the copper surface heat flux. Details of the cavity design can be found in Ref. [7].

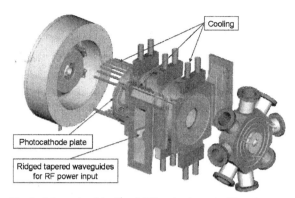

Fig. 1. Schematic of the $2\frac{1}{2}$-cell RF cavity (center) with emittance-compensating magnets (left) and vacuum plenum (right).

Table 1
Normal-conducting cw RF photoinjector parameters

Bunch charge	3 nC
Accelerating gradient	7 MV/m
Beam energy at injector end	2.7 MeV
Cavity ohmic losses	780–820 kW
Solenoid magnetic field, max	0.37 T
Normalized emittance, rms	7 μm
Bunch length, rms	9 ps
Energy spread, rms	34 keV

The RF PI model, including cooling and RF tapered ridge-loaded waveguides, is shown in Fig. 1. With aperture radius of 65 mm and septum thickness of 20 mm, the cell coupling is 0.03 and the π-mode is well separated from its nearest neighbors. The operating parameters of the cw RF PI are listed in Table 1. The choice of 7 MV/m gradient is dictated by space charge mitigation and cavity thermal management. The on-axis electric field in the π-mode PI has a flat distribution (Fig. 2). The field alternates from one cell to another, and reaches about 10 MV/m near the cathode. RF fields in the vacuum plenum are small since it is resonant at 650 MHz.

Wall power densities are calculated with Superfish and MicroWave Studio. The surface current distribution is shown in Fig. 3. The highest power density is on the cell septa and on the 1st half-cell end wall, where it reaches 105 W/cm^2 at 7 MV/m.

The RF feeds are in the 3rd cell of the $2\frac{1}{2}$-cell PI. Two symmetrically placed ridge-loaded tapered waveguides are connected to the cavity via "dog-bone" coupling irises—long narrow slots with two circular holes at their ends—in a 0.5″-thick copper wall. This design [8] is based on the LEDA RFQ and SNS high-power RF couplers. The required cavity-waveguide coupling is given by the coefficient $\beta_c = (P_w + P_b)/P_w \approx \frac{4}{3}$, where $P_w = 780$ kW is the cw ohmic losses, and $P_b = 270$ kW is the beam power of a 100 mA beam at the injector exit. The coupling is adjusted by changing the radius of the holes at the ends of the coupler slot.

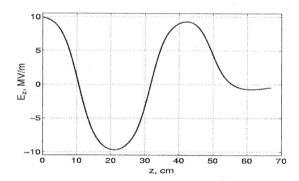

Fig. 2. Electric field distribution in the π-mode of $2\frac{1}{2}$-cell cavity.

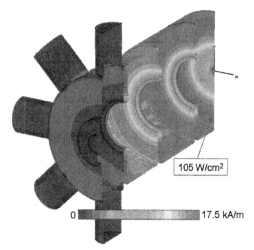

Fig. 3. Surface current distribution in the π-mode of $2\frac{1}{2}$-cell cavity.

2.2. Cathode radius optimization

Space-charge induced emittance scales inversely with the photoemission radius

$$\varepsilon_{n,sc} = \frac{I}{\gamma' I_A \left(3\sigma_r/\sigma_z + 5\right)} \tag{1}$$

where $\gamma' = eE/mc^2$ is the normalized gradient [9]. In the absence of a solenoid, the RF emittance term scales with the square of rms photoemission radius [9]

$$\varepsilon_{n,rf} = \gamma' k_{rf}^2 \sigma_z^2 \sigma_r^2. \tag{2}$$

Thermal emittance scales linearly with the photoemission radius

$$\varepsilon_{n,thermal} = \frac{r}{2}\sqrt{\frac{kT}{mc^2}}. \tag{3}$$

The normalized emittance for a thermalized beam is the quadrature sum of the above three terms.

$$\varepsilon_n = \sqrt{\varepsilon_{n,sc}^2 + \varepsilon_{n,rf}^2 + \varepsilon_{n,thermal}^2}. \tag{4}$$

Without a solenoid field, an optimum cathode radius exists where emittance is minimized. With solenoid magnetic focusing near the cathode, the RF emittance term, caused by transverse RF field at the apertures, is significantly reduced as the beam size decreases at the apertures. Thus, solenoid magnetic focusing leads to a smaller normalized emittance, approaching the thermal emittance limit.

2.3. Beam dynamics simulations

Beam dynamics simulations with PARMELA illustrate the effects of transverse plasma oscillations [10]. For the $2\frac{1}{2}$-cell injector and an initial double Gaussian temporal profile and top-hat spatial distribution, PARMELA output predicts an evolution of emittance that has two minima along the z-axis (Fig. 4). The first minimum occurs when different phase space ellipses line up as a result of the emittance compensation, and the second minimum occurs at a larger emittance when the phase-space ellipses realign. Adding booster linacs to increase the beam energy and a second solenoid to rotate the phase-space ellipses further reduce the normalized rms emit-

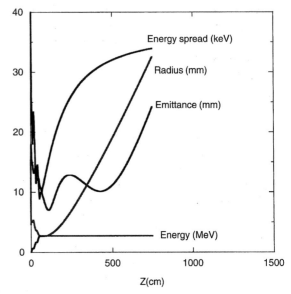

Fig. 4. Calculated normalized emittance, radius, energy and energy spread versus distance ($2\frac{1}{2}$-cell injector with a solenoid magnet).

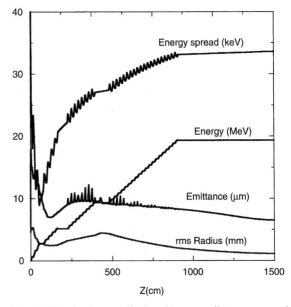

Fig. 5. Calculated normalized emittance, radius, energy and energy spread versus distance ($2\frac{1}{2}$-cell injector with a solenoid plus a 4-cell booster, two additional linacs and a second solenoid).

tance (Fig. 5). Transverse plasma oscillations cause different phase-space ellipses to rotate and realign at higher beam energy. This results in a second minimum further downstream with a lower emittance of 6.5 μm. The rms energy spread is 34 keV. At 19.3 MeV beam energy, the relative energy spread is 0.18%.

3. Thermal management

About 800 kW of RF power is deposited on the walls of the water-cooled PI $2\frac{1}{2}$-cell cavity. This requires a dense array of cooling channels close to the RF surface and a large flow of cooling water for heat removal [11]. Flow requirements based on RF heat loads at surface temperature are reasonable: 36 l/s at 5 m/s with inlet water temperature of 20°C and mean outlet temperature 26°C; see temperature distribution in Fig. 6.

An axisymmetric model of the cavity was developed using the finite element code ANSYS. RF loads were mapped on the surface from SF output and modified based on the surface temperature. Significant stresses generally occur between the heated RF surface and cooling channels. Glidcop® is used throughout this cavity. It has yield strength of 270 MPa and provides significant margin for this design. The resulting steady-state von Mises local stresses do not exceed 68 MPa (Fig. 7). The calculated surface current distribution near the RF coupler irises is shown in Fig. 8. The localized regions of high heat flux will be cooled with dedicated channels in the thick cavity walls.

4. Photocathode development

The choice of photocathode suitable for use in the high-current RF PI is determined by the

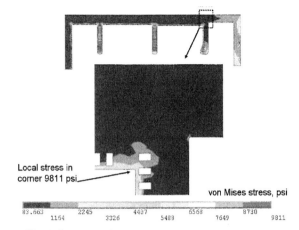

Fig. 7. Stress distribution in the RF photoinjector cavity.

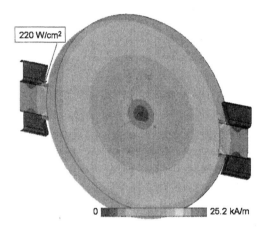

Fig. 8. Surface current distribution near the RF coupler irises.

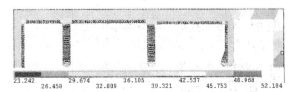

Fig. 6. Temperature distribution of the RF PI with realistic cooling.

cathode Q.E., the photon energy needed for photoemission, and the cathode ruggedness toward contamination. We consider three different photocathode materials: Cs_2Te, cesiated p-type GaN, and K_2CsSb. Table 2 summarizes the three photocathodes under consideration for use in the high-current RF PI.

Cesiated p-type GaN is a new photocathode material that has high Q.E. at the third harmonic of Nd lasers. Cesiated GaN is a compromise between the rugged but UV-driven Cs_2Te cathode and the short-lived, green-driven K_2CsSb cathode. We have studied the sensitivity of Cs:GaN to exposure to carbon monoxide. The Q.E. versus wavelength at different CO exposure durations are plotted in Fig. 9.

Table 2
Photocathodes being considered for use in the cw RF photoinjector

	Q.E. (%)	Wavelength (nm)	Lifetime (h)
Cs_2Te	3	263	40
Cs:p-GaN	7	350	20
K_2CsSb	8	530	3

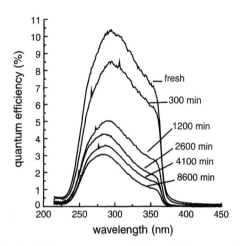

Fig. 9. Cs:GaN response curves at different CO exposure times.

A cathode that responds to lower energy photons is a better choice since it uses less laser power. At the same Q.E., K_2CsSb uses half as much laser power as Cs_2Te. With a 20% green-to-UV conversion efficiency, K_2CsSb uses ten times less IR power than Cs_2Te. Cs_2Te thus requires a larger and more complex drive laser than K_2CsSb.

However, K_2CsSb cathodes lose their Q.E. in the matter of a few hours. Their photoresponse can be rejuvenated with occasional re-cesiation but this process is time consuming and is not conducive to continuous operation. A solution to this problem is to use a cesium dispenser behind the photocathode to continuously replenish it with cesium. To this end, we use semi porous SiC as possible substrates for the above photocathodes. Preliminary studies show that at 400–600°C the released Cs diffuses through SiC and deposits on the surface at a rate of 1–3 Å s⁻¹.

5. Conclusions

We present an overview of the physics design of the 100 mA average-current RF photoinjector. This normal-conducting cw RF photoinjector is a $2\frac{1}{2}$-cell, π-mode, 700 MHz structure made out of oxygen-free copper on Glidcop® with on-axis electric coupling and emittance compensation. At an accelerating gradient of 7 MV/m, the output beam will have 2.7 MeV kinetic energy, 7 μm normalized rms emittance, 9 ps rms bunch length and 34 keV rms energy spread. The total ohmic losses in the injector are about 800 kW, with a maximum heat flux slightly above 100 W/cm². Thermal analysis shows that the heat due to ohmic losses can be effectively removed with water cooling through a dense array of cooling channels.

Cooling the regions around the coupling irises is a challenging thermal management issue because the surface heat flux exceeds 200 W/cm² at a few hot spots. However, these hot spots are localized around the ends of the dog-bones and can be cooled effectively with cooling channels placed properly around the slots in the thick cavity wall.

We are in the process of finalizing the engineering design and fabricating the cold cavity model. Our plan is to construct the full-power prototype and install it in the Low Energy Demonstration Accelerator facility at LANL to perform RF and thermal tests without beam in late 2004. Once the RF tests and thermal management with high heat flux have been successfully demonstrated, the system will be upgraded to operate at 100 mA average current with the addition of a photocathode, a drive laser, a second klystron, and beam diagnostics.

References

[1] D.C. Nguyen, et al., Phys. Rev. Lett. 81 (1998) 810.
[2] T. Siggins, et al., Nucl. Instr. and Meth. A 475 (2001) 549.
[3] E. Barhels, et al., Nucl. Instr. and Meth. A 445 (2000) 408.
[4] J.S. Fraser, R.L. Sheffield, E.R. Gray, Nucl. Instr. and Meth. A 250 (1986) 71.
[5] D.H. Dowell, et al., Appl. Phys. Lett. 63 (1993) 2035.
[6] B.E. Carlsten, Nucl. Instr. and Meth. A 285 (1989) 313.

[7] S.S. Kurennoy, et al.,'Development of Photoinjector RF Cavity for High Power FEL,' Nucl. Instr. and Meth. A (2003), these Proceedings.

[8] S.S. Kurennoy, L.M. Young, 'RF Coupler for High-Power CW FEL Photoinjector,' Proceedings of the 2003 Particle Accelerator Conference, p. 3515.

[9] K.J. Kim, Nucl. Instr. and Meth A 275 (1989) 201.

[10] L. Serafini, J.B. Rosenzweig, Phys. Rev. E 55 (1997) 7565.

[11] A.M.M. Todd, et al.,'High-power electron beam injectors for 100 kW free-electron lasers,' Proceedings of the 2003 Particle Accelerator Conference, p. 977.

Available online at www.sciencedirect.com

NUCLEAR
INSTRUMENTS
& METHODS
IN PHYSICS
RESEARCH
Section A

ELSEVIER   Nuclear Instruments and Methods in Physics Research A 528 (2004) 78–82

www.elsevier.com/locate/nima

On the mechanism of high selectivity of two-dimensional coaxial Bragg resonators

N.S. Ginzburg[a], N.Yu. Peskov[a,*], A.S. Sergeev[a], I.V. Konoplev[b], K. Ronald[b], A.D.R. Phelps[b], A.W. Cross[b]

[a] Institute of Applied Physics, Russian Academy of Sciences, 46 Uljanov St., Nizhny Novgorod 603950, Russia
[b] Department of Physics, University of Strathclyde, Glasgow G4 0NG, UK

Abstract

The analysis of electrodynamic properties of two-dimensional (2D) Bragg resonators of coaxial-geometry realizing 2D distributed feedback was carried out with diffraction effects taken into account. Such resonators represent coaxial waveguide sections with a shallow double-periodic corrugation, which provides mutual coupling of four partial waves. It is shown that the high selectivity of 2D Bragg resonators over the azimuthal mode index can be explained by different behavior of the normal symmetrical and non-symmetrical waves near the Bragg resonance in an unbound double-periodic corrugated waveguide.

© 2004 Published by Elsevier B.V.

PACS: 41.60.Cr; 84.40.Ik; 84.40.Fe

Keywords: Free-electron masers; Powerful microwave radiation; Bragg resonators

1. Introduction

The use of two-dimensional (2D) distributed feedback has been proposed [1] as a method of producing spatially coherent radiation from either sheet or annular high-current relativistic electron beams with the transverse size greatly exceeding the wavelength. The 2D distributed feedback can be realized in planar and coaxial 2D Bragg cavities with a double-periodic corrugation of the walls.

On this corrugation mutual scattering of the electromagnetic energy fluxes propagating in the forward, backward and transverse directions (relative to the direction of the electron beam propagation) takes place. These transverse waves can act to synchronize radiation from different parts of a large size electron beam. Experimental studies of FEMs exploiting this novel feedback mechanism have been developed currently [2,3]. In support of the FEM experiments detailed studies of the microwave properties of 2D Bragg resonators is required. The present paper is devoted to the analysis of 2D Bragg cavities of coaxial geometry with the diffraction effects taken into account.

*Corresponding author. Tel.: +7-8312-384575; fax: +7-8312-362061.

E-mail address: peskov@appl.sci-nnov.ru (N.Yu. Peskov).

0168-9002/$ - see front matter © 2004 Published by Elsevier B.V.
doi:10.1016/j.nima.2004.04.022

2. Model and basic equations

Let us consider a coaxial 2D Bragg structure (Fig. 1) consisting of two conductors of mean radius r_0 having a shallow double-periodical corrugation

$$a = a_1[\cos(\bar{h}_z z - \bar{M}\varphi) + \cos(\bar{h}_z z + \bar{M}\varphi)], \quad (1)$$

where $4a_1$ is the corrugation depth, $\bar{h}_z = 2\pi/d_z$, d_z is the period of the corrugation along the longitudinal coordinate z and $\bar{M}$ is the number of variations along the corrugation over the azimuthal coordinate φ. For simplicity, we suppose that $\bar{h}_z = \bar{M}/r_0 = \bar{h}$.

In the assumption of small curvature of the resonator surface it is appropriate to adopt a planar system with the transverse coordinate $x = r_0\varphi$. We present the electromagnetic field inside the structure in the form of four partial wave-beams:

$$A_+\vec{E}_1^0 e^{ih_z z} + A_-\vec{E}_1^0 e^{-ih_z z} + B_+\vec{E}_2^0 e^{ih_x x}$$
$$+ B_-\vec{E}_2^0 e^{-ih_x x}, \quad (2)$$

where h_z is the longitudinal and $h_x = M/r_0$ is the transverse (azimuthal) wavenumbers, and the structure $\vec{E}_{1,2}^0(r)$ coincide with one of the modes of a coaxial waveguide. This corresponds to the coupling of the following partial wave-beams: longitudinally propagating wave-beam $(A_\pm)$ consisting of a number of $TE_{M,0}$-type modes of a coaxial waveguide with low azimuthal indices $(M = 0, 1, ...)$ including the lowest TEM-mode and transversely propagating wave-beams $(B_\pm)$ consisting of $TE_{M,0}$-modes with high azimuthal indexes $(M \gg 1)$. The four partial waves undergo coupling on the double periodic structure if the Bragg resonance condition is satisfied: $h_z \approx h_x \approx \bar{h}$.

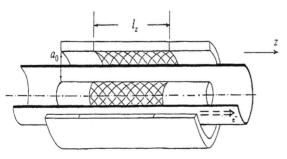

Fig. 1. Schematic diagram of coaxial 2D Bragg FEM.

Under the conditions described above the process of the partial waves scattering may be described by the set of coupled-wave equations [1,4], where for azimuthally propagating near cut-off waves $B_\pm$ the parabolic type equations including diffraction effects have been used:

$$\pm\frac{\partial A_\pm}{\partial z} + i\delta A_\pm + i\alpha(B_+ + B_-) = 0$$
$$\frac{i}{2\bar{h}}\frac{\partial^2 B_\pm}{\partial z^2} \pm \frac{\partial B_\pm}{\partial x} + i\delta B_\pm + i\alpha(A_+ + A_-) = 0. \quad (3)$$

Here $\delta = (\omega - \bar{\omega})/c$ is the frequency mismatch from the Bragg resonance $\bar{\omega} = \bar{h}c$ and $\alpha = a_1\bar{h}/a_0$ is the wave coupling coefficient.

Due to the coaxial geometry of the cavity the wave-beams should satisfy the cyclic boundary conditions: $A_\pm, B_\pm(x + l_x, z) = A_\pm, B_\pm(x, z)$, where $l_x = 2\pi r_0$ is the resonator perimeter. This allows a solution of Eqs. (3) to be presented as

$$A_\pm, B_\pm = \sum_{m=-\infty}^{\infty} A_\pm^m(z), B_\pm^m(z)e^{2\pi i m x/l_x}. \quad (4)$$

Each Fourier term in Eq. (4) with its own index m may be considered as a normal wave of the cavity which satisfies to equations

$$\pm\frac{dA_\pm^m}{dz} + i\delta A_\pm^m + i\alpha(B_+^m + B_-^m) = 0$$

$$\frac{i}{2\bar{h}}\frac{d^2 B_\pm^m}{dz^2} + \left(\delta \pm \frac{2\pi m}{l_x}\right)B_\pm^m + \alpha(A_+^m + A_-^m) = 0. \quad (5)$$

Note, that in correspondence with Eq. (2) for the partial waves $B_\pm$ the index m must be considered in addition to the azimuthal index $\bar{M}$. Thus, the normal wave with index $m = 0$, which is called a symmetrical wave consisting of both the azimuthally symmetrical partial waves $A_\pm$ and partial waves $B_\pm$ with high azimuthal index $\bar{M}$. In the general case, the normal wave with index m is a set of four coupled partial waves $A_\pm^m$ and $(B_+^m; B_-^m)$, which are characterized by azimuthal indexes m and $(m + \bar{M}; m - \bar{M})$, respectively.

3. Dispersion properties of the normal waves

Let us consider normal waves in an unbound coaxial waveguide having a double-periodic corrugation (Eq. (1)). Looking for solution of Eqs. (5) in the form $\sim e^{i\Gamma z}$ we get the dispersion equation

$$\left(2\bar{h}\delta - 2\bar{h}\frac{m}{r_0} - \Gamma^2\right)\left(2\bar{h}\delta + 2\bar{h}\frac{m}{r_0} - \Gamma^2\right)(\delta^2 - \Gamma^2)$$
$$= 8\alpha^2\bar{h}\delta(2\bar{h}\delta - \Gamma^2). \qquad (6)$$

At $\alpha = 0$ this equation reduces to four separate equations: two equations describing longitudinally propagating waves $A_\pm$, and two equations describing dispersion of the transversely propagating waves $B_\pm$:

$$\delta = \pm\Gamma, \quad 2\bar{h}\delta = \pm 2\bar{h}\frac{m}{r_0} + \Gamma^2. \qquad (7)$$

For $\alpha \neq 0$ the dispersion diagrams for normal waves with $m = 0$ and $m = \pm 1$ are shown in Fig. 2. From comparison with the asymptotes given by Eq. (7) and shown by thin lines one can conclude that the branches "1" and "2" originate from the partial waves $A_\pm$ while the branches "3" and "4"

originated from the quasi cut-off partial waves $B_\pm$. The main difference in the diagrams presented as compared to traditional 1D Bragg structures is the existence of a transparent band near the Bragg frequency $\omega \approx \bar{\omega}$ ($\delta \approx 0$). This peculiarity is obviously related with the participation in the scattering process of the quasi cut-off partial waves $B_\pm$.

Even more important is the fact that the dispersion diagrams have totally different behavior in the case of azimuthally symmetric ($m = 0$) and non-symmetric normal waves. When $m = 0$ for the partial waves $B_\pm$ the convergence of unperturbed dispersion curves (Fig. 2b) takes place and in the case of the coupling waves $\alpha \neq 0$ the dispersion equation is in the form

$$(2\bar{h}\delta - \Gamma^2)(\delta^2 - \Gamma^2) = 8\alpha^2\bar{h}\delta. \qquad (8)$$

It follows from Eq. (8) that near the Bragg frequency instead of the traditional parabolic dependence of frequency on the wave number (which is realized, in particular, near the Bragg reflection zone in the case of traditional 1D corrugation [5]) we get fourth order dependence [4] $\Gamma^4 = 8\alpha^2\bar{h}\delta$. Thus, near the Bragg frequency not only does the group velocity of the normal wave tend to zero but its first derivative as well. This

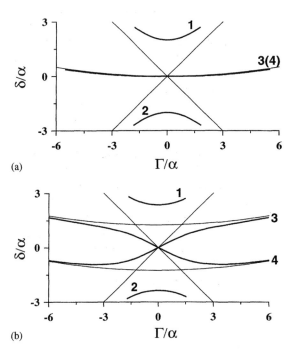

Fig. 2. Dispersion diagrams of the normal waves in the coaxial 2D Bragg structure unbound in the longitudinal direction when $\bar{h}/\alpha = 35$: (a) symmetrical waves $m = 0$ and (b) non-symmetrical waves $m = \pm 1$. Thin lines correspond to the dispersion curve of the partial waves.

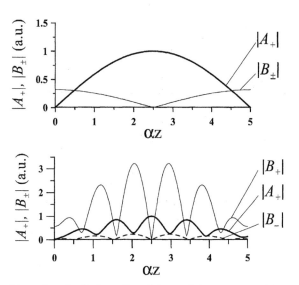

Fig. 3. Longitudinal profiles of the partial waves A_+ and $B_\pm$ for (a) symmetrical mode ($m = 0; n = 1$) with the frequency located near $\delta \approx 0$ and (b) non-symmetrical mode $m = 1$ having the maximum-Q factor: $\bar{h}/\alpha = 35, \alpha l_{x,z} = 5$.

peculiarity of the dispersion characteristic of the normal wave $m = 0$ near the Bragg frequency provides the conditions for the formation of an eigenmode with a Q-factor essentially exceeding the Q-factors of other modes.

4. Mode selection in 2D Bragg resonators

To find eigenmodes of the 2D Bragg cavity having a double-periodic corrugation of finite length l_z we assume an ideally matched system with zero reflections for all partial waves from the ends of the corrugations, that corresponds to the boundary conditions

$$A_{\pm}^{m}\left(x, z = \mp \frac{l_z}{2}\right) = 0$$

$$\left[\frac{\mathrm{d}B_{\pm}^{m}}{\mathrm{d}z} \mp \mathrm{i}\sqrt{2\bar{h}\left(\pm\frac{2\pi m}{l_x} + \delta\right)}B_{\pm}^{m}\right]\Bigg|_{z=\mp l_z/2} = 0. \quad (9)$$

In the case of the azimuthally symmetric modes $m = 0$ near the Bragg frequency the analytical solution of full characteristic equation based on dispersion equation (6) and boundary conditions (9) for the complex eigenfrequencies can be presented as

$$\delta_{0,n} = \frac{\pi^4 n^4}{8\bar{h}\alpha^2 l_z^4} + \mathrm{i}\frac{\pi^4 n^4}{4\bar{h}\alpha^3 l_z^5}, \quad (10)$$

where n is the longitudinal index. It follows from Eq. (10) that the highest Q-factor (lowest diffraction losses) is realized for the mode with one longitudinal variation $n = 1$ shown in Fig. 3a.

The analytical solution (10) coincides well with results of numerical simulation of the full characteristic equation. For non-symmetrical waves the maximum Q-factor is realized when the mode eigenfrequency was positioned near the minimums on the dispersion curves "4" (Fig. 2b), i.e. occurs near the point where the group velocity tends to zero. The profiles of the partial waves for the high-Q non-symmetrical $m = 1$ eigenmode are shown in Fig. 3b. However, we have a non-zero second derivative here and, as a result, this mode has significantly lower Q-factors as compared to the fundamental symmetrical mode.

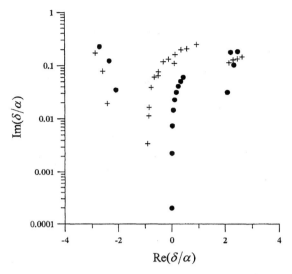

Fig. 4. The eigenmode spectrum for $\bar{h}/\alpha = 35$ and $\alpha l_{x,z} = 5$: symmetrical modes $m = 0$ are shown by the dots, non-symmetrical modes $m = 1$ by the crosses.

For the resonator geometry $\alpha l_x = \alpha l_z = 5$ the spectrum of the eigenmodes presented in Fig. 4 shows that the diffraction losses for the fundamental mode are more than an order of magnitude less than the losses for all other modes. Note also, that the high-Q non-symmetrical modes possess a large number of longitudinal variations of the field as well as a relatively low amplitude of the longitudinally propagating partial waves, i.e. $A_{+}^{\max}/B_{+}^{\max} \ll 1$ (compare Fig. 3a and b). Therefore, some additional mechanisms of so-called electronic selection exist for these modes when interacting with the electron beam.

Let us add in summary that the parameters of the resonator taken here for the modelling are close to the parameters of an experiment studying a coaxial FEM with 2D distributed feedback based on a high-current accelerator at the University of Strathclyde [3].

Acknowledgements

The authors acknowledge INTAS, Russian RFBR, QinetiQ and the UK EPSRC for support of this work.

References

[1] N.S. Ginzburg, N.Yu. Peskov, A.S. Sergeev, Opt. Commun. 96 (1993) 254; Opt. Commun. 112 (1994) 151.

[2] A.V. Arzhannikov, N.V. Agarin, V.B. Bobylev, et al., Nucl. Instr. and Meth. A 445 (2000) 222.

[3] A.W. Cross, I.V. Konoplev, K. Ronald, et al., Appl. Phys. Lett. 80 (2002) 1517.

[4] N.S. Ginzburg, A.S. Sergeev, I.V. Konoplev, Sov. Tech. Phys. 66 (1996) 108 (in Russian).

[5] V.L. Bratman, G.G. Denisov, N.S. Ginzburg, M.I. Petelin, IEEE J. Quantum Electron. QE-19 (1983) 282.

Available online at www.sciencedirect.com

SCIENCE @ DIRECT°

**NUCLEAR
INSTRUMENTS
& METHODS
IN PHYSICS
RESEARCH**
Section A

Nuclear Instruments and Methods in Physics Research A 528 (2004) 83–87

www.elsevier.com/locate/nima

New results of the 'CLIO' infrared FEL

R. Prazeres*, F. Glotin, J.M. Ortega

LURE, bat 209D, BP34, 91898 Orsay Cedex, France

Abstract

This paper shows the new results obtained with the CLIO infrared FEL. The spectral range of the CLIO FEL has recently been extended, and it is now spanning from $\lambda = 3$–$120\,\mu m$. We present here some features obtained with CLIO at large wavelengths. We show also measurements of electron energy spectrum after FEL interaction, and we make a comparison with the measured laser power.
© 2004 Elsevier B.V. All rights reserved.

PACS: 41.60.Cr; 42.72.Ai; 07.57.Hm

Keywords: Infrared; FEL; Harmonic

1. Introduction

The 'CLIO' Free Electron Laser (FEL) is an infrared tuneable laser source [1]. In the initial configuration of CLIO, in 1992, the wavelength range spanned from 3 to 18 μm. The spectral range is tuneable now from 5 to 120 μm. In order to do that, most parts of the FEL components have been modified: the optical cavity, the undulator and the linac. The maximum wavelength has been recently obtained using a hybrid cavity with a toroidal front mirror and a symmetric rear mirror. The best configuration for mirrors was obtained with a numerical code [2,3], which calculates the propagation of the wave $A(x, y)$ in the optical cavity. It gives the total cavity losses L and the 'extraction rate' T_x of hole coupling at the saturation regime. The amount of energy produced in FEL interac-

tion by each electron micro-bunch is $\Delta W_e = \eta W_e$, where η is the 'Electron efficiency', and $W_e = Q(\gamma mc^2/e)$ is the electron bunch energy, with Q the charge of electron bunch. At saturation, the energy of the laser micro-pulse, inside of the cavity and at undulator exit, is $W_i = \eta W_e/L$. The extracted laser energy $W_x = T_x W_i = \eta W_e T_x/L$ depends on the hole-coupling 'extraction rate' $T_x = W_x/W_i$. The ratio T_x/L is the 'extraction ratio' between the losses by hole coupling and the total losses. The extracted peak power (at macro-bunch scale) is: $P_x = f_\mu \eta W_e T_x/L$, where f_μ is the repetition rate of micro-pulses. And the average extracted power is:

$$\langle P_x \rangle \cong W_e \eta \Delta T_{sat} \frac{T_x}{L} f_\mu f_M \tag{1}$$

where f_M is the repetition rate of macro-pulses, and ΔT_{sat} is the time duration of saturation in the macro-pulse.

---

*Corresponding author.
E-mail address: rui.prazeres@lure.u-psud.fr (R. Prazeres).

0168-9002/$ - see front matter © 2004 Elsevier B.V. All rights reserved.
doi:10.1016/j.nima.2004.04.023

2. Measurements

2.1. Power measurements at large wavelength

The power measurement at 20 MeV, using a symmetrical cavity 'S' (see Table 1), is displayed in Fig. 1; and the curve at 14 MeV using a hybrid cavity 'H' is displayed in Fig. 2. Measurements are represented by dots and simulations by a line curve. Both figures exhibit a 'hole burning' at about $\lambda = 50\,\mu$m. It is not due to absorption by air, on mirrors or in the output diamond window. The numerical simulation gives the 'extraction ratio' T_x/L which is proportional to the FEL extracted power $\langle P_x \rangle$; see expression (1). In the first approximation, ΔT_{sat} and η are considered independent of wavelength. The simulation shows the same feature as measurements. The decrease in the measured power at short wavelength in Fig. 1 is not well fit by the simulation, because it is due to a misalignment of electron beam trajectory at large undulator gaps. The 'hole burning' at $\lambda = 50\,\mu$m is well explained by Fig. 3, which shows the amplitude profile $A(x,y)$ on the extraction front mirror, at 14 MeV. It is close to a TEM_{02}, with a minimum of energy through the extraction hole, and creating a large amount of cavity losses L. It shows that the power variations near a wavelength of 50 μm are due to modifications of the intracavity transverse mode of the laser.

2.2. Switching laser between fundamental and harmonics

The top of Fig. 4 shows the measured FEL detuning curve, at $\lambda = 80\,\mu$m and $\gamma mc^2 = 16$ MeV,

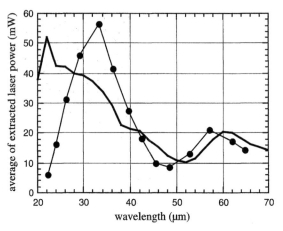

Fig. 1. Extracted power vs. wavelength, for 20 MeV electron beam energy and 'S' type cavity. The numerical simulation is represented by the thick line, and the measurements by dots.

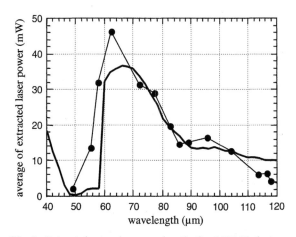

Fig. 2. Extracted power vs. wavelength, for 14 MeV electron beam energy and 'H'-type cavity. The numerical simulation is represented by the thick line and the measurements by dots.

Table 1
The various optical cavities for CLIO

Notation (in main text)	Type of cavity	Wavelength range (μm)	Radius of curvature of mirrors				
			Rear mirror (M1)		Front mirror (M2) (with hole coupling)		
			Rx_1 (m)	Ry_1 (m)	Rx_2 (m)	Ry_2 (m)	Φ hole (mm)
S	Symmetrical	$\lambda < 50$	3	3	3	3	2
T_inf	Toroidal	$\lambda \gg 100$	3	1.7	3	1.1	—
T	Toroidal	$\lambda > 100$	3	1.9	3	1.9	4
H	Hybrid	$50 < \lambda < 200$	3	3	3	1.7	3

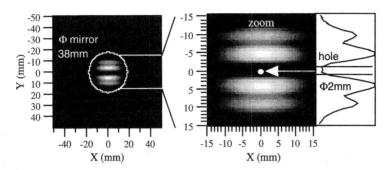

Fig. 3. Transverse profile of $A(x, y)$ on the extraction front mirror, calculated for $\lambda = 50\,\mu m$ and 14 MeV configuration. The right-hand side picture is an enlarged view.

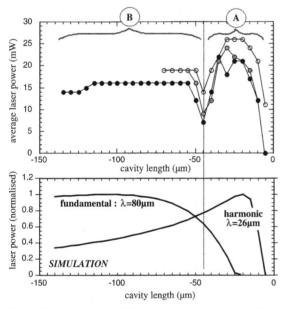

Fig. 4. Measurement of FEL detuning curve at $\lambda = 80\,\mu m$, represented by dots in the top of the figure: it shows the average extracted power $\langle P_x \rangle$ of the laser versus cavity length variation Δd. Three series of measurements have been done in succession. The detuning curves represented by line curves on the bottom of the figure are simulations.

using an 'S' type cavity. The detuning curve exhibits a 'hole burning' for the cavity length variation $\Delta d = -50\,\mu m$. We have checked the wavelength and shown that part B in Fig. 4 corresponds to lasing on fundamental at $\lambda = 80\,\mu m$, whereas part A corresponds to lasing on the third harmonic at $\lambda/3 = 27\,\mu m$. The bottom of Fig. 4 shows the theoretical [4] detuning curve

for $\lambda = 80$ and $27\,\mu m$. It shows that laser operation on fundamental is forbidden in region A, and there is then a maximum of power on the third harmonic. When working in part B, the laser becomes naturally favourable to the fundamental wavelength. In the intermediate region, the laser power is close to zero. We have verified that such effect is not due the shift of 'zero detuning', between fundamental and harmonic, induced by the variation of group velocity in the waveguide.

2.3. Electron efficiency for fundamental operation

An electron spectrometer [1] is installed at the exit of the undulator. It gives, in Fig. 5, the energy distribution of the electron beam after the FEL interaction. The horizontal axis is the time scale. The vertical axis is the variation, in %, of the electron beam energy $W_e(t)$ during the macropulse. The continuous white line is the average of energy $W_e(t)$. The spectra shown here are, respectively, for laser 'OFF', laser 'ON' working on fundamental and laser 'ON' working on harmonic. A comparison of the spectrum, with laser 'OFF' and laser 'ON', gives the 'Electron efficiency': $\eta(\tau) = \Delta W_e / W_e$. The relation between η and the extracted power $\langle P_x \rangle$ is displayed in expression (1). The losses L and the 'extraction rate' T_x are deduced from the numerical simulation. We have done a comparison with the measurement of the extracted power.

With an electron beam of 16 MeV, a laser wavelength (on fundamental) $\lambda = 80\,\mu m$ and a symmetrical cavity 'S', the numerical simulation

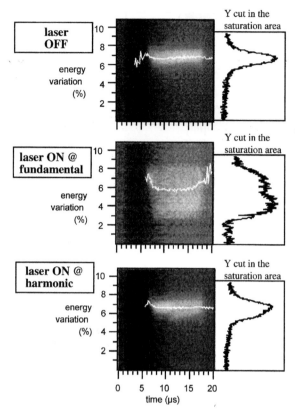

Fig. 5. Energy distribution (in %) of the electron beam, after the FEL interaction (arbitrary origin of %); respectively for laser OFF, laser ON when working at fundamental and laser ON at harmonic. The picture on right-hand side is a vertical cut within the FEL saturation time domain.

gives $L = 33\%$ and $T_x = 2.0\%$. For fundamental operation, we obtain $\langle P_x \rangle = 82 \, \text{mW}$. Now, the measurement on the detector gives $50 \, \text{mW}$. Taking into account the error that can be done in the calculation of W_e and in the measurement of η, we can consider that the values for the power are in good agreement. This means that the numerical simulation, giving L and T_x, is consistent with the real case. The same kind of analysis has been done for another wavelength $\lambda = 93 \, \mu\text{m}$. The numerical simulation gives: $L = 44\%$ and $T_x = 1.9\%$. The 'Electron efficiency measurement' gives $\langle P \rangle = 41 \, \text{mW}$ and the extracted power measurement gives $\langle P \rangle = 27 \, \text{mW}$. Taking into account the same sources of error, we can also consider that the values for the power are in good agreement.

2.4. Electron efficiency for harmonic operation

The same measurements have been done for harmonic operation of the FEL, at $\lambda = 27 \, \mu\text{m}$ (bottom of Fig. 5). The 'Electron efficiency' for harmonic operation is about $\eta = 0.2\%$, which is much lower than $\eta = 1\%$ for fundamental operation. The numerical simulation gives $L = 9.4\%$ and $T_x = 1.9\%$, corresponding to $T_x/L = 20\%$. This value is much larger here than for fundamental (see Section 2.3) because the cavity losses are smaller here. Using the measurement of η, we obtain the average extracted power $\langle P_x \rangle = 38 \, \text{mW}$. The measurement of laser extracted power gives $\langle P_x \rangle = 25 \, \text{mW}$. Taking into account the same sources of error as in Section 2.3, we can consider that the values for the power are in good agreement.

The 'Electron efficiency' is intrinsically smaller for harmonic operation than for fundamental operation, because the gain line width on the third harmonic is three times narrower that that of the fundamental. A theory of FEL saturation [5] gives an approximate expression for the 'Electron efficiency' η (at the fundamental wavelength): $\eta = 1.7/(2\pi N) \cong 1/4N$. For the nth harmonic lasing, the efficiency is $\eta = 1/4nN$. For $N = 38$ undulator periods and the third harmonic, it gives $\eta = 1/4nN = 0.22\%$. This value is in good agreement with the measurement of efficiency $\eta = 0.2\%$ from the electron energy spectrum for harmonic operation. For the first harmonic operation, this theory gives $1/4N = 0.66\%$ which is lower than the measured value $\eta = 1\%$. However, the approximation $\eta = 1/4N$ involves a monochromatic wave for the laser. It is not the case in real conditions, and especially at the maximum power of FEL. In addition, "Self-Amplified Spontaneous Emission" can enhance the efficiency η, especially for fundamental laser. Inspite of these effects, our analysis clearly explains the different FEL characteristics observed on the fundamental and harmonic.

3. Conclusion

Our experiments have allowed us to obtain lasing at long wavelengths with the CLIO FEL, originally designed for mid-infrared. The range from 3 to 120 µm is the largest spectral interval ever obtained with a single structure and a single electron beam line. The large cavity losses are due to the small size (38 mm) of the cavity mirrors. In the near future, a new optical cavity chamber will allow us to exchange different mirrors and to increase their size up to 50 mm. This will greatly enhance the available laser power at long wavelengths and increase even further the FEL longest possible wavelength up to about $\lambda = 180$ µm.

References

[1] R. Prazeres, F. Glotin, J.M. Ortega, C. Rippon, R. Andouart, J.M. Berset, E. Arnaud, R. Chaput, Nucl. Instr. and Meth. A 445 (2000) 204.

[2] R. Prazeres, M. Billardon, Nucl. Instr. and Meth. A 318 (1992) 889.

[3] R. Prazeres, Eur. Phys. J. AP 16 (2001) 209.

[4] G. Dattoli, P.L. Ottaviani, Opt. Commun. 204 (2002) 283.

[5] F. Ciocci, G. Dattoli, A. Torre, A. Renieri, Insertion devices for synchrotron radiation and free electron laser, ENEA INN-FIS, Frascati, Rome, Italy, ISBN 981-02-3832-0.

Available online at www.sciencedirect.com

NUCLEAR
INSTRUMENTS
& METHODS
IN PHYSICS
RESEARCH
Section A

ELSEVIER   Nuclear Instruments and Methods in Physics Research A 528 (2004) 88–91

www.elsevier.com/locate/nima

Laboratory-scale wide-band FIR user facility based on a compact FEL

Young Uk Jeong*, Grigori M. Kazakevitch, Byung Cheol Lee, Seong Hee Park, Hyuk Jin Cha

Laboratory for Quantum Optics, Korea Atomic Energy Research Institute, (KAERI), P.O. Box 105, Yusong, Taejon 305-600, Republic of Korea

Abstract

We have developed a laboratory-scale user facility with a compact far infrared (FIR) FEL. The FEL driven by a magnetron-based microtron operates in the wavelength range of 100–300 μm. The microtron accelerator has been upgraded to drive stable operation of the FEL by stabilizing current and bunch repetition rate of the electron beam. The FEL can be operated for several hours without further adjustment and the fluctuation in pulse energy during the period is less than 10% in rms value. The FIR radiation is divided by an FIR beam splitter (crystal quartz) and is detected by calibrated FIR detectors. The scheme facilitates high accuracy measurement of pulse energy with a fluctuation of $\sim 1\%$. When we used liquid-helium cooled detectors, the signal-to-noise ratio was more than 10^7, even in the case of a single pulse measurement. We could measure the transmission FIR spectrum of the vapour and liquid water as preliminary data of the molecular spectroscopy and imaging for living species. We are also preparing FIR imaging systems by 2-D scanning for high resolution and by the electro-optic method for single-pulse capturing. The area occupied by the compact FIR FEL is only 4 m^2.
© 2004 Elsevier B.V. All rights reserved.

PACS: 41.60.Cr; 07.57.Hm

Keywords: Far infrared; Free-electron laser; Microtron; Compact FEL

1. Introduction

There is an increasing demand for advanced infrared (IR) and far infrared (FIR) radiation sources for the applications in biomedical research, solid-state physics, gas spectroscopy, and so on [1–4]. The newly constructed synchrotron IR/FIR beamlines [5,6] and free-electron lasers [7,8] are also based on the demands. Table-top FIR sources generated by conventional lasers have been developed and used for these applications. Inexpensive and compact FIR FEL [8–10] can play an important role in encouraging FIR applications due to its higher power and spectral brightness compared to the table-top sources.

*Corresponding author. Tel.: +82-42-868-8342; fax: +82-42-861-8292.

E-mail address: yujung@kaeri.re.kr (Y.U. Jeong).

0168-9002/$ - see front matter © 2004 Elsevier B.V. All rights reserved.
doi:10.1016/j.nima.2004.04.024

We have developed a compact FIR FEL driven by a magnetron-based microtron [9]. We could extend the FEL wavelength range of 100–1000 μm with the variable-energy microtron and we could construct a user experimental stage for the wavelength of 100–300 μm [10,11]. The FEL and FIR experimental setups are placed in a laboratory with an area of 40 m². The microtron accelerator of the FEL has been stabilized in its current and RF frequency, which provides stable operation of the FEL without further adjustment for several hours. The FIR radiation at the experimental stage is well collimated and the diverging angle of the radiation is less than 1 mrad at the stage. We could measure the absorption and reflectance signal from samples with a fluctuation of ~1% by monitoring the pulse energy of the FEL beam. The S/N ratio of the single pulse measurement could be more than 10^7 if a liquid-helium cooled Ge:Ga detector is used. With the system, we could measure preliminary data of FIR imaging with a leaf and spectrum of liquid and vapour water.

2. Stabilization of the system

The microtron accelerator with an RF generator of magnetron is a simple and inexpensive driver for compact FIR FEL. However, the electron dynamics in the microtron is not very simple due to the strong coupling between the magnetron and an accelerating cavity with an electron gun inside. The main parameters of the gun, RF generator and cavity, cannot be independently controlled to obtain the optimal condition of the electron beam. We have investigated the accelerator by numerical simulation for the transient state of the coupled magnetron–microtron cavity system. The results show good agreement with the experiments and we could find optimal working parameters of the beam current and RF frequency for stable operation of the FEL. The main results are shown in Ref. [12].

Additionally, we could stabilize the pulse-to-pulse current of the electron beam by the feedback control of the emission current from the gun. With the stabilization, we could operate the accelerator for several hours without a breakdown inside the

cavity [12]. The fluctuation in macropulse current was less than 1% during our observation for several hours. The fluctuation of the macropulse energy of the FEL was measured to be less than 10% during the time.

The stability of the radiation measurement has been improved by monitoring the pulse energy with a beam splitter of crystal quartz. The 45° beam splitter from the FIR beam axis shows that 28% of the incident beam is reflected and 46% is transmitted for a wavelength of 110 μm. The loss of FIR radiation in the material is not very small, but for the optical alignment with a visible laser, the crystal quartz is more preferable than an opaque material. Polarization of the FEL beam at the measuring table is dominant in the vertical direction. The FIR measurement along with the monitoring of pulse energy allows low fluctuation of ~1% in rms. value. The S/N value of the FIR beam measured with a highly sensitive Ge:Ga detector, operating in liquid helium temperature, is more than 10^7 for a single pulse measurement and the value can be increased by suppressing the RF noise from the accelerator.

The FIR radiation is transported by relay optics in a vacuum channel and the focal length of the relay optics is approximately 10 m from a collimating lens near an output coupler mirror of the FEL resonator. The focal point is set between the experimental tables for applications. Fig. 1 shows measured spatial distribution of the FIR radiation

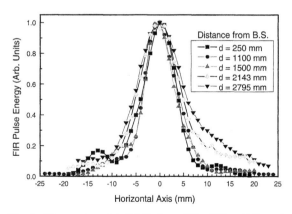

Fig. 1. Measured spatial distribution of the FIR radiation at the experimental tables of the facility. The distances noted in the figure are values between the measured points and the beam splitter for pulse energy monitoring.

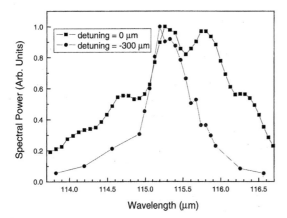

Fig. 2. Measured spectra of the FEL pulses for different detuning values. When the detuning is close to the resonance cavity length, the spectral width of the radiation is broadened to $2\,cm^{-1}$, and the pulse energy also increases depending on the spectral width.

at the experimental tables. The distances in the figure are values between the measured points and the beam splitter. We can see that the focal point of the radiation is located between 1.5 and 2 m from the beam splitter, and the minimum spot size is 7 mm in FWHM. We could obtain high accuracy in the FIR measurements by improving the driver accelerator and diagnostics.

The FEL shows different spectral width and pulse energy depending on cavity detuning [10,13]. Fig. 2 shows spectra of the FEL pulses for different detuning values. The spectral width of the radiation with $-300\,\mu m$ detuning length from the cavity length resonant to the electron-bunch repetition rate shows spectral width of $0.5\,cm^{-1}$. When the detuning length is close to the resonance cavity length, the spectral width of the radiation broadens to $2\,cm^{-1}$ and the pulse energy also increases depending on the spectral width. The result shows that the FEL can generate micropulses with short bunch length of several picoseconds.

3. Preliminary results of FIR applications

We have measured the preliminary data of spectral absorption on liquid and vapour water. The absorption coefficient of liquid water at a

wavelength near $110\,\mu m$ was measured to be approximately $400\,cm^{-1}$ and the result agrees well with the previous measurement [14]. Hereafter, we want to study the absorption characteristics of air, mainly the effect of water vapour, in the FIR wavelength range, when the main part of the experimental stage is exposed to air. Additionally, we also want to use the spectral data to determine the absolute value of the FIR wavelength with high accuracy. Fig. 3 shows a spectrum of air absorption. The spectral width of the FIR radiation was $0.2\,cm^{-1}$ using a monochromator. It can be observed that the spectrum is not well resolved as fine peaks. The calculated absorption lines of the water vapour near the wavelength range also show several lines or group of lines separated by a gap of approximately $0.5\,cm^{-1}$. To measure the spectrum in more detail, we have developed a Fabry–Perot interferometer with a spectral resolution of $0.05\,cm^{-1}$ by using two nickel meshes with a period of $20\,\mu m$ and in the near future we would again measure the spectrum with a better accuracy.

We have constructed an FIR imaging system by scanning with a small aperture. One-dimensional scanning with an aperture of 0.1 mm was tested with a bamboo leaf and the result was compared with a scanned intensity profile of the photograph, which is shown in the upper part of Fig. 4(b). The photograph of the leaf is shown in Fig. 4(a) with the same scale as the scanned profiles in Fig. 4(b). Both the profiles adequately represent the periodic

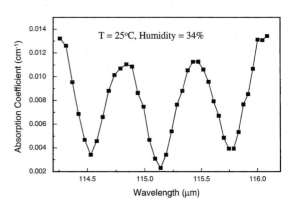

Fig. 3. Measured spectrum of air absorption using the FIR FEL. The spectral width of the FIR radiation was $0.2\,cm^{-1}$ measured with a monochromator.

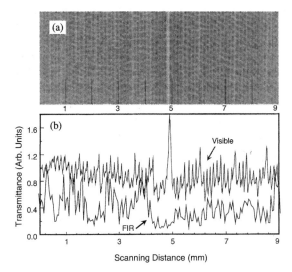

Fig. 4. One-dimensional FIR scanning with an aperture of 0.1 mm performed on a bamboo leaf. The result was compared with a scanned intensity profile of the photograph shown in the upper part of Fig. 4(b). The photograph of the leaf is shown in Fig. 4(a) with the same scale as the scanned profiles in Fig. 4(b).

structure of the leaf, which has a period of 0.1–0.2 mm. The profiles should be in relation of inverse image due to the extremely big difference in the absorption coefficients of liquid water for the wavelength ranges. However, we cannot observe a strong correlation between the profiles, which might show the usefulness of the FIR imaging.

We are constructing a fast-imaging system with a single micropulse of the FEL beam by using the electro-optic (EO) detection and switching method. The linearly polarized visible or IR laser beam is collinearly incident to the EO crystal with the FIR beam. The image of the FIR beam is transferred to the visible or IR laser beam and we will use an intensified CCD camera to capture the image.

4. Conclusion

We have developed a laboratory-scale users facility by using a compact FIR FEL. We hope that the system can be used for advanced experiments on FIR wavelength range. The efficiency of the FEL from the electron energy to the FIR radiation is less than 0.1% due to the low-energy spread of the electron beam and large slippage during the FEL oscillation. We are considering a tapered undulator to increase the efficiency. We will analyse the possibility with the theoretical calculation. The compact FEL with an efficiency of more than several percentages can generate the FIR radiation with an average power of 1 W.

References

[1] B.M. Fischer, et al., Phys. Med. Biol. 47 (2002) 3807.
[2] M. Walther, et al., Chem. Phys. Lett. 332 (2000) 389.
[3] T. Dekorsy, et al., Phys. Rev. Lett. 90 (2003) 055508.
[4] G. Lupke, et al., Phys. Rev. Lett. 88 (2002) 135501.
[5] M. Abo-Bakr, et al., Phys. Rev. Lett. 88 (2003) 254801.
[6] F. Sannibale, et al., CIRCE: a dedicated storage ring for Far-IR THz coherent synchrotron radiation, International Workshop on Infrared Microscopy and Spectroscopy with Accelerator-Based Sources, July 2003, Lake Tahoe, CA, USA.
[7] D. Douglas, et al., Proceedings of the 2001 Particle Accelerator Conference, Chicago, IL, USA, p. 247.
[8] A. Doria, et al., CATS: a compact free electron source in the THz region, International Workshop on Infrared Microscopy and Spectroscopy with Accelerator-Based Sources, July 2003, Lake Tahoe, CA, USA.
[9] Y.U. Jeong, et al., Nucl. Instr. and Meth. A 475 (2001) 47.
[10] Y.U. Jeong, et al., Nucl. Instr. and Meth. A 507 (2003) 125.
[11] G.M. Kazakevitch, et al., Nucl. Instr. and Meth. A 507 (2003) 146.
[12] G. Kazakevich, et al., Stabilization of the microtron-injector for a wide-band compact FIR FEL, Nucl. Instr. and Meth. A, (2004) these proceedings.
[13] N. Nishimori, et al., Phys. Rev. Lett. 86 (2001) 5707.
[14] J.K. Vij, F. Hufnagel, Chem. Phys. Lett. 155 (1989) 153.

Available online at www.sciencedirect.com

SCIENCE DIRECT®

ELSEVIER   Nuclear Instruments and Methods in Physics Research A 528 (2004) 92–95

NUCLEAR
INSTRUMENTS
& METHODS
IN PHYSICS
RESEARCH
Section A

www.elsevier.com/locate/nima

Recent research activities at BFEL laboratory

G. Wu*, J.P. Dai, M.K. Wang, J.Q. Xu, J.B. Zhu, X.P. Yang

Institute of High Energy Physics, Chinese Academy of Sciences, P.O. Box 2732, Beijing 100080, China

Abstract

Beijing free electron laser (BFEL) is a Compton-type mid-infrared FEL facility, with first lasing in 1993. In the past two years, two experimental stations were constructed, and the reliability and stability of BFEL improved much. The lasing time provided to the users was more than 1000 hours, and about 250 samples have been irradiated, many application items concerned. Next, BFEL will be upgraded to be user's facilities, and other three experimental stations will be furnished. This paper describes the operation, application and upgrade plan of BFEL in details. In addition, the development of several research activities are mentioned.
© 2004 Published by Elsevier B.V.

PACS: 41.60 Cr

Keywords: Free electron laser; Application; Facilities; Development

1. Introduction

In 1993, Beijing free electron laser (BFEL) realized its first lasing at a wavelength range $\lambda \sim 9$–$11\,\mu m$, and next year the output power reached its saturation within a more narrow range [1]. Lots of efforts have been made to upgrade the facilities since then [2], the power stability improved greatly, the wavelength range expanded to 6–17 μm, and one user station was built up. One thousand user hours for their experiments are provided each year, and some significant experiments have been accomplished.

Originally, this IR FEL was designed for the purpose of demonstration of the technical feasi-

bility and FEL Physics researches. The simplified project was adopted to minimize the costly investment. There is a long way for us to improve its levels to user's facilities, since too many factors for future development were ignored in its first step, which should at least include two aspects, the experimental conditions and the peripheral equipments.

2. Status of the facilities

The stability of the facilities was improved obviously in the recent years, and the fluctuation of the wavelength was reduced to $\pm 1\%$ from the original $> 5\%$. The adjustment of laser wavelength could be finished in several minutes as asked by

---
*Corresponding author.
E-mail address: wug@ihepa.ac.cn (G. Wu).

0168-9002/$ - see front matter © 2004 Published by Elsevier B.V.
doi:10.1016/j.nima.2004.04.025

users, which covered the limited range between 8 and 14 μm.

We have finished the model of dynamic analysis experimental station, and accomplished the designs of the optical transmission line and the related facilities' processing such as the motor sliding plate and some adjusting shelves. For the stabilized analysis experimental station (Fig. 1), we have prepared all kinds of optical adjusting shelves, optics devices, and installed the laser energy Radiometer of double probes and also fulfilled the manufacture of many detector chips.

We have rebuilt the optical system and realized the on-line diagnostics of laser characteristics, such as the wavelength, the power and the spectral width. This provides the reliable guarantee for the accelerator and the application experiments. Based on the application experiments, we assumed that we could use carbon dioxide laser to simulate the on-line applied system in FEL. This project is brought into practice step by step, and we think this measurement system has not only improved greatly the optical modulation but also played an important vole in build up of the experimental station. To build the spectral analysis experimental station, we are purchasing a Fourier transformation infrared spectrograph, which will be very helpful for spectral experimental analysis. Besides, a short period wiggler was made for short wavelength FEL applications.

The following characteristics of BFEL satisfied some requirements from application research in the field of material sciences and life sciences [3]:

adjustable wavelength in mid-IR waveband; coherent light and higher peak power; extra narrow pulse time structure and the diffraction limit light beam. Some obvious effects were found when FEL irradiation was used for the experiments research on photoelectric function materials, semi-conductor materials, giant magnetic materials, huge biological molecules, biochemistry samples, and so on.

3. Other research activities

3.1. X-ray generation

At BFEL, we conducted a series of experiments in recent years. On X-ray ranges by Compton Scattering, the method of which is to use FEL beam (wavelength 9.4 μm) backscattering a train of electron micro-bunches (energy 30 MeV) for generating X-rays, the experimental result with average X-ray flux 10^2 photons/s and X-ray energy 1.9 keV has been obtained [4].

We designed and fabricated a new beam transport line for the preliminary experiment for Parameteric X-ray Radiation (PXR) project, and the importance to restrain the radiation noise from the running background was educed. We will find a good shield for it on next step.

3.2. Study on photocathode RF gun

Low emittance and high-intensity electron beams are needed for the high gain, short wavelength FELs. To meet the requirements of the proposed DUV FEL facility, we designed a 1.5-cell, S-band photocathode RF gun using the emittance compensation techniques, which can produce a beam of $\varepsilon_n = 2.3\,\pi$ mm mrad, $Q = 2$ nC/bunch, $E_k = 4.8$ MeV at the gun exit with the field gradient of 100 MV/m on the cathode. Finally, the cavity mechanic design and manufacture were finished, and the cool-test of microwave give a good agreement with the simulation results. In addition, the design of the coils used for emittance compensation has also been accomplished, and they will be fabricated soon.

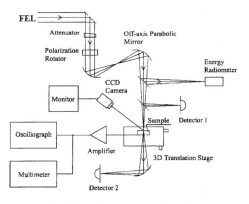

Fig. 1. Steady-state analysis station.

Simultaneously, we also succeeded to develop a Ti:Sapphire ultrafast pulse self-mode-locking laser oscillator, with a set of synchronization pulse phase-locking circuit. After system's regulation, a stable ultrafast laser pulse output was achieved with an average power of 180 mW, a wavelength of 780 ns, a pulse length of 70 fs, and a pulse repetition rate of 102 MHz. We will take on-line test in the future.

3.3. Study on bunch length measurement by CTR

Coherent transition radiation (CTR) is an advanced method to measure the bunch longitudinal length of less than 1 ps especially. We are developing bunch length diagnostic setup similar to Martin–Puplett interferometer [5]. The energy spectra and interferograms of RF electron bunches of BFEL are numerically calculated, and their bunch form factors and electron density distributions are deductively given. The software is developed on LabVIEW to control the step machine, for data acquisition, filter and analyzing of CTR signal, and to show the interference pattern, the radiation spectrum and bunch distribution curve. Till now, the technique to collimate and adjust the system has been mastered expertly. In order to test effectiveness of the setup and the software an off-line experiment is designed to measure the interferogram of long wave radiation from a 900°C blackbody radiator. On-line experiment will be performed immediately once this off-line experiment succeeds.

3.4. Thermionic cathode RF gun

Using a copper cavity mirror with a hole in the center, BFEL's output energy of the optical macro-pulse has been kept at 2–6 mJ, which is too low compared to the original design goal (Table 1), and we have not found any effective means to improve it. By deep research, we found that there is a great ununiformity in the electron beam energy spread at the exit of the accelerator (Fig. 2). It was observed that the electron energy in front of the current pulse rose so higher than the average that this portion of electron bunches lost in the bending segment and could not devote its

Table 1
Design parameters for BFEL user's facilities

Parameter	TC RF gun	PC RF gun
Wavelength	3–50 μm	2–50 μm
Spectrum bandwidth	0.5%	0.2%
Repetition rate	2856 MHz	102 MHz
Pulse length	1 ps	10 ps
Pulse energy	10 μJ	50–100 μJ

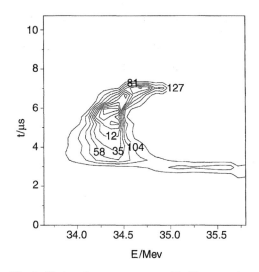

Fig. 2. Electron beam energy spread inside one pulse.

energy to laser gain. This disfigurement was considered to be caused by the difference of the building time of microwave field inside the RF gun and the accelerator. Normally, it will cost lesser time to build RF field inside a single resonance cavity than inside an accelerator tank, but in BFEL cases, this process accompanies with emitting of electrons from the cathode. With the loading effect, the time of building field in the RF gun is 0.5 μs longer than that in the accelerator.

In our experiments, we can verify that a 0.5 μs enlarging of macro-pulse of electron beam, can cause a twice rise in output energy of laser light. For the purpose of increasing the output power, we designed a cavity with low-quality factor of 4000 to shorten the difference from both. Simulations indicated that this new gun can be fed to its working level in 0.46 μs with electron beam loading. A more uniform electron train in energy,

energy spread and emittance inside one macro-pulse, are expected with this new gun used.

4. Future developments

As user's facilities, we must increase its output energy, expand wavelength range, build user's stations, and optimize their hardware and software, select the application items related to the BFEL superiority, develop new experiment method.

All of these works will be started by next year when the facilities are moved to the new place which our institute have arranged for us. We have designed better radiation shield for larger room to upgrade the facilities.

References

[1] Jialin Xie, et al., Nucl. Instr. and Meth. A 358 (1995) 256.
[2] Jialin Xie, Yonggui Li, et al., Nucl. Instr. and Meth. A 407 (1998) 146.
[3] Yonggui Li, et al., Nucl. Instr. and Meth. B 144 (1998) 140.
[4] Yu Zhao, Jiejia Zhuang, Nucl. Instr. and Meth. A 475 (2001) 445.
[5] Zhihui Li, High Energy Physics and Nuclear Physics, 2003.

Available online at www.sciencedirect.com

SCIENCE $\textit{d}$ DIRECT°

ELSEVIER   Nuclear Instruments and Methods in Physics Research A 528 (2004) 96–100

NUCLEAR
INSTRUMENTS
& METHODS
IN PHYSICS
RESEARCH
Section A

www.elsevier.com/locate/nima

Possible ways of improvement of FEM oscillator with Bragg resonator

A.V. Elzhov[a], N.S. Ginzburg[b], A.K. Kaminsky[a],*, S.V. Kuzikov[b], E.A. Perelstein[a], N.Yu. Peskov[b], S.N. Sedykh[a], A.P. Sergeev[a], A.S. Sergeev[b]

[a] Laboratory of Particle Physics, Joint Institute for Nuclear Research, 6 Joliot-Curie St., Dubna, Moscow Region 141980, Russia
[b] RAS Institute of Applied Physics, 46 Ulyanova St., Nizhny Novgorod 603600, Russia

Abstract

Two methods of efficiency enhancement of free-electron maser (FEM) oscillator with the Bragg resonator having a shift of corrugation phase were studied. The first one is a "starting mode" regime using an additional high-frequency mode located in the vicinity of the main mode. The Bragg resonator having a shift of corrugation phase seems to represent the most attractive oscillator configuration. Another possibility is a special variation of corrugation depth of Bragg reflectors (mirror profiling) providing the possibility of optimizing the longitudinal distribution of the RF field amplitude inside the interaction space and depressing of parasitic side modes of the resonator.

Optimization of RF power extraction from FEM, based on Talbot effect, has also been considered. Perspectives of future FEM development are discussed including schemes of pulse power compression and submillimeter wave generation.
© 2004 Elsevier B.V. All rights reserved.

PACS: 41.60.Cr; 84.40.Ik; 84.40.Fe

Keywords: Free-electron maser; Oscillator; Bragg resonator; High-power microwaves; Mode dynamics

1. Introduction

High-efficiency narrow-band free-electron maser (FEM) can be used as pulse microwave power source suitable for testing high-gradient accelerating structures of electron–positron linear colliders. An FEM oscillator with Bragg resonator having an operating frequency of 30 GHz (corresponding to CLIC collider [1]) has been developed and investigated by JINR-IAP collaboration.

2. Improvement of Bragg resonator for enhancement of oscillator efficiency

One possible way to increase the oscillator efficiency is employing the so-called starting-mode

*Corresponding author. Tel.: +7-09621-65-443; fax: +7-09621-65767.

E-mail address: alikk@sunse.jinr.ru (A.K. Kaminsky).

0168-9002/$ - see front matter © 2004 Elsevier B.V. All rights reserved.
doi:10.1016/j.nima.2004.04.026

regime. It is well known that FEM maximum efficiency corresponds to conditions in which electron initial energy E_{in} strongly exceeds the synchronous value. But in case of too large mismatch from synchronism, an oscillator cannot start up from the noise level because the self-excitation condition for operating mode is not fulfilled. However, such a condition can be fulfilled for another mode with higher frequency, which can be called starting mode. In the nonlinear regime, the starting mode initiates growth of the operating mode which suppresses the starting mode in final score due to nonlinear competition.

A similar scenario of generation was described by Bogomolov and Vlasov for gyrotrons [2]. For Bragg FEM, the first simulation of the above regime was reported in Ref. [3], for a simple model of Bragg resonator formed by a regularly corrugated waveguide section. This simulation demonstrated the possibility of obtaining a considerable gain in the oscillator efficiency.

In this paper, starting mode regime is studied for Bragg resonator with shift of corrugation phase $\Delta\varphi = \pi$ under larger area of parameters including coupling factor between forward and backward waves in the resonator and lengths of the Bragg reflectors. In such a resonator there are three eigenmodes: the frequencies of two modes are located at the edges of Bragg reflection zone—low-frequency (LF) and high-frequency (HF) modes—and one fundamental mode has a frequency close to the center of the above zone (CF). Therefore, two types of starting mode regimes are possible, with frequency shift from HF to CF and that from CF to LF. The areas of occurrence of the starting mode regimes are hatched in Fig. 1, where a particular distribution of FEM efficiency is presented. Unfortunately, the areas cover rather narrow ranges in energy, not greater than 0.2 in units of γ. The oscillator demonstrates the peak efficiencies only in the starting mode regimes, especially in the case of HF to CF frequency shift.

Another possibility to increase the oscillator efficiency with direct excitation of only the operating central mode is the optimization of Bragg structure profile. In order to have maximal beam deceleration at the very end of the interaction length, it is reasonable to reduce the length of

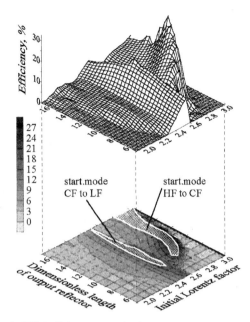

Fig. 1. FEM efficiency versus initial electron beam energy and the resonator length for Bragg resonator with shift of corrugation phase. The wave coupling factor $\alpha = 0.02$, guide magnetic field $b_g = -1.2$, wiggler field $b_w = 0.8$. The input reflector for second-type resonator is fixed at the length of $L_1 = 19$. All values are in dimensionless units [3].

the output Bragg reflector. However, a very short output reflector results in a low quality factor of the central mode, which is highly attractive for use as the operating mode. To maintain the quality factor of this mode, it is worth profiling Bragg resonator with the increasing of the coupling coefficient of the output reflector. However, parasitic excitation of other resonator eigenmodes (LF and HF modes) can prevent the generation at the operating central mode. To depress side modes, one can introduce a proper variation of the coupling along the Bragg reflector, thus smoothing the reflectivity spectral characteristic of the resonator.

A series of simulations were performed to optimize the corrugation profile to increase the efficiency of the FEM oscillator in single-mode regime (generation at central frequency without excitation of side modes). For example, an efficiency increase was obtained in a resonator with linear spatial distribution of the coupling

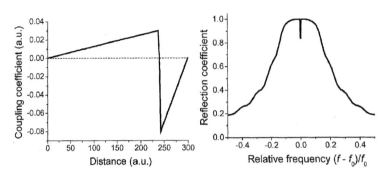

Fig. 2. Optimized distribution of the coupling coefficient for each Bragg reflector (left) and spectral distribution of the reflection coefficient of the resonator (right). Minimum in the center corresponds to the position of the operating mode.

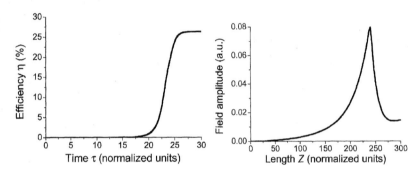

Fig. 3. Time dependencies of efficiency η and spatial distribution of electric field amplitude (in steady-state regime) for an FEM oscillator with profiled resonator. $L_1 = 24$, $L_2 = 6$, $b_g = -1.2$, $b_w = 0.7$, and $\gamma_0 = 2.7$.

coefficient at each Bragg reflector (Fig. 2 left). The corresponding spectral distribution of the reflection coefficient of the resonator is given on the right-hand side of Fig. 2.

Time dependence of the efficiency and field amplitude distribution in steady-state regime obtained in a simulation, including interaction with electron beam, are presented in Fig. 3. The electron efficiency reaches a maximum value of $\eta = 26.3\%$ instead of 16.8% for regularly corrugated mirrors.

3. Optimization of microwave power extraction system

An output system in any powerful RF device must provide the following: (1) electron beam collection, (2) measurement of electron beam current, (3) formation of high-quality radiation wavebeam, (4) eliminating the breakdown at the output window and (5) minimal wave reflection from the window. As a rule, these requirements are contradictory.

Employing the phenomenon of wave-front self-reproduction in oversized waveguide (Talbot effect) [4] yields the possibility to simultaneously optimize output system over several criteria. According to the simulation, the wavebeam profile is reproduced at a length of about 90 cm for the waveguide used in the experiment. Therefore, there is sufficient space for beam collection and accurate beam current measurements. Besides to eliminate the breakdown at the window, it is possible to choose a position for vacuum window where the electric field has relatively uniform distribution.

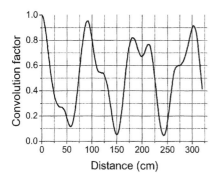

Fig. 4. Wavebeam transformation in the oversized waveguide.

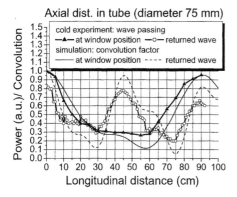

Fig. 5. Forward and reflected wave in FEM output waveguide: simulation and cold measurement.

The convolution factor, a numerical indicator of coincidence of transverse wavebeam distribution in a given cross-section with the initial one, has been studied by numerical simulation. The simulated spatial evolution under experimental conditions is presented in Fig. 4. The initial Gaussian beam is a superposition of several eigenmodes; therefore, the behavior of the convolution is rather complicated.

Employing the obtained distribution allows one to define the optimal length of the output waveguide to obtain a high-quality output wavebeam.

The convolution factor for reflected wave was also studied by numerical simulation. It corresponds to the fraction of the RF power reflected from the window and accepted by the output horn. Therefore, minimal reflection from the window back to the oscillator or minimal RF field amplitude along the window can be an optimization criterion for the longitudinal position of the window inside the waveguide.

Both simulated functions show reasonable agreement with the measured transmission and reflection coefficients (Fig. 5).

4. Perspectives of future FEM development

Several schemes for extending of the FEM facility to high peak power and submillimeter range are currently under study.

The powerful feedback wave of a millimeter-range FEM can be used as an electromagnetic wiggler for generation of short-wavelength (submillimeter) radiation using a moderately relativistic electron beam. Two possible variants of such a device are currently under study. The spatial separation of the regions generating millimeter and submillimeter waves is used in both schemes.

The first scheme is based on single pass amplification with preliminary bunching at a subharmonic frequency [5].

A feedback at HF is used in the second scheme. The preliminary bunching is not needed. Therefore, the HF section can be placed prior to the LF one, viewing along the beam propagation. The advantage of this scheme is that the beam is not perturbed by the high-power LF wave. The mirrors based on Talbot effect are used in the HF resonator.

One mirror splits the wavebeam into two identical partial beams and possesses constant reflectivity close to unity. The second mirror provides the variation in phase one of the two partial reflected waves and contains two supplemental waveguides. The reflection coefficient can be altered from 0 to 1 by a mechanical shift of the reflecting surface. It allows the optimization of the Q-factor of the resonator. The use of different waveguides for extraction of the HF and LF radiation is very convenient.

A similar resonator can be the base for active RF pulse compressor, which is an attractive way of sufficiently increasing the peak output power of the oscillator. Discharge tubes should be installed instead of movable reflector.

5. Conclusions

Nonlinear simulation shows the possibility of increasing the FEM oscillator efficiency by realization of the starting mode regime, the most attractive configuration being the resonator with phase shift. Proper variation of the Bragg mirror's coupling coefficient can be used for keeping the quality factor of central mode high enough, even if the output mirror is very short. It is an alternative way to increase the FEM oscillator efficiency. Variation of the coupling coefficient along the mirror can depress the parasitic side modes of the Bragg resonator with corrugation phase shift.

The FEM output optimization using Talbot effect can minimize power loss. The developed millimeter FEM facility possesses promising perspectives to enhance peak power and advance to submillimeter waves.

Acknowledgements

This work is supported by Grants ##03-02-16530, 02-02-17438 of Russian Foundation for Basic Research.

References

[1] J.P. Delahaye, et al., Proceedings of European Particle Accelerator Conference (EPAC'98), Stockholm, 1998, p. 58.
[2] N.S. Ginzburg, N.F. Kovalev, M.I. Petelin, Oscillators and Amplifiers Based on Relativistic Electron Beams, Moscow State University, Moscow 1987, p. 142 (in Russian).
[3] N.S. Ginzburg, A.V. Elzhov, A.K. Kaminsky, et al., Nucl. Instr. and Meth. A 483 (2002) 225.
[4] G.G. Denisov, S.V. Kuzikov, Strong Microwaves in Plasmas, Vol. 2, Nizhny Novgorod, IAP RAS (Institute of Applied Physics, Russian Academy of Sciences), 2000, p. 960.
[5] A.V. Savilov, N.Yu. Peskov, A.K. Kaminsky, Nucl. Instr. and Meth. A 507 (2003) 162.

Available online at www.sciencedirect.com

SCIENCE 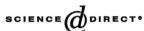 DIRECT°

ELSEVIER   Nuclear Instruments and Methods in Physics Research A 528 (2004) 101–105

**NUCLEAR
INSTRUMENTS
& METHODS
IN PHYSICS
RESEARCH**
Section A

www.elsevier.com/locate/nima

Experimental and computational studies of novel coaxial 2D Bragg structures for a high-power FEM

I.V. Konoplev[a],*, A.D.R. Phelps[a], A.W. Cross[a], K. Ronald[a], P. McGrane[a],
W. He[a], C.G. Whyte[a], N.S. Ginzburg[b], N.Yu. Peskov[b], A.S. Sergeev[b], M Thumm[c]

[a] *Department of Physics, University of Strathclyde, Glasgow, G4 0NG, UK*
[b] *Institute of Applied Physics, RAS, Nizhny Novgorod 603950, Russia*
[c] *Universität Karlsruhe, Institut für Hochfrequenz technik und Elektronik, D-76021 Karlsruhe, Germany*

Abstract

Two-dimensional (2D) coaxial Bragg structures have been suggested for use in high-power Free Electron Masers (FEM) to synchronize radiation from different parts of an oversized annular electron beam. In this paper, the simulations of field evolution using the three-dimensional code MAGIC are carried out and results are presented. An investigation of 2D Bragg structures obtained by corrugating the inner surface of the outer conductor of a coaxial waveguide or by lining the surface of a smooth waveguide with a dielectric material, which has a bi-periodic permittivity, has been conducted. Experimental studies of 2D Bragg structures were also undertaken and the good agreement between experimental measurements and theoretical predictions is demonstrated. Measurements of a 7 cm diameter annular electron beam produced by a high-current accelerator to be used to drive the FEM are presented and the experimental set-up discussed.
© 2004 Elsevier B.V. All rights reserved.

PACS: 41.60.Cr; 84.40.Fe; 84.40.Ik; 84.70. + p; 41.85.Ja

Keywords: FEM; 2D feedback; Microwaves; Annular e-beam

1. Introduction

Two-dimensional (2D) coaxial Bragg structures have recently been under intensive theoretical [1–5] and experimental [6–8] study. These structures have been proposed for application in high-power microwave electronics [1] to synchronize the radiation from different parts of an oversized electron beam which can be used to drive a high-power Free Electron Maser (FEM). An example of such a structure with periodic perturbation on the surface of the inner conductor is shown in Fig. 1. The studies of novel 2D Bragg structures for a high-power FEM driven by an annular electron beam have been conducted and results are presented. We also present measurements of the annular electron beam produced by a high-current accelerator (HCA), which will be used to power

---

*Corresponding author. Tel.: + 44-141-548-4272; fax: + 44-141-552-2891.

E-mail address: acp96115@strath.ac.uk (I.V. Konoplev).

0168-9002/$ - see front matter © 2004 Elsevier B.V. All rights reserved.
doi:10.1016/j.nima.2004.04.027

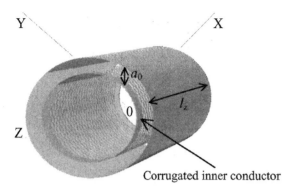

Fig. 1. The schematic diagram of the coaxial 2D Bragg structure with corrugated inner conductor.

the FEM experiment. In Section 2 of the paper the basic model is discussed, results of numerical simulations are presented and compared with experimental data. In Section 3, the field evolution inside the single section 2D Bragg cavity is considered. Experimental measurements of the annular electron beam, to be used to drive the high-power FEM are presented in Section 4. From these measurements the operating parameters of the FEM experiment are estimated in the Conclusion.

2. Basic model

The 2D Bragg corrugation of the waveguide on either inner or outer conductor surfaces can be presented as

$$r(z, \varphi) = R_{in,out} + a_1 \cos(\bar{k}_z z)\cos(\bar{m}\varphi) \qquad (1)$$

where $R_{in,out}$ are the radii of inner (in) and outer (out) conductors, a_1 is the amplitude of the corrugation, $\bar{k}_z = 2\pi/d_z$ and d_z is the period of the corrugation along the longitudinal coordinate z and $\bar{m}$ is the corrugation variation number along the azimuthal coordinate φ. In the Bragg structures studied, the amplitude of the perturbations a_1 is small in comparison with the operating wavelength λ_0 and the distance between the inner and outer conductors. The dispersion equation for the eigenwaves of the coaxial waveguide takes the

following form [9]:

$$k^2 = \omega^2/c^2 = k_z^2 + k_\perp^2 \qquad (2)$$

where ω is the wave frequency, k_z and $k_\perp$ are the longitudinal and transverse wavenumbers, respectively. The RF field inside the 2D Bragg structure can be presented as a superposition of four waves: $A_\pm$ propagating in $\pm z$ direction and $B_\pm$ are near cut-off waves as shown in the schematic diagram of Fig. 2a with the structure's eigenvectors represented as $\vec{k}_\pm$. The 2D periodic corrugation provides a wave scattering such that counter-propagating waves are coupled indirectly. The partial wave A_+ propagating in the $+z$ direction is scattered into near cut-off waves $B_\pm$, which scatter into waves $A_\pm$. This ensures that the following loop $A_+ \leftrightarrow B_\pm \leftrightarrow A_-$ is completed and allows the formation of a 2D feedback circle. Taking into account the case when the partial wave A_+ is represented by a TEM wave, one finds that to obtain an efficient coupling of the waves $A_\pm \leftrightarrow B_\pm$ the following resonance conditions should be satisfied:

$$k_z = k_z' \cong \bar{k}_z, \ |\bar{m}| = |M| \quad \text{and} \quad |k| \approx |k_\perp'| = |k_\perp| \quad (3)$$

where M is the azimuthal wavenumber and $k_\perp, k_\perp'$ are the full transverse wavenumbers of the partial waves $B_\pm$ and k is the full wavenumber of the incident wave A_+.

Using the three-dimensional PIC code MAGIC, the field scattering on the 2D Bragg corrugation has been investigated. The snapshot at the cross-section corresponding to the centre of the structure

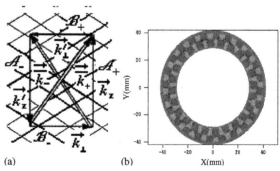

(a)   (b)

Fig. 2. (a) The schematic diagram of the 2D feedback circle; and (b) the snapshot of the longitudinal component of the magnetic field inside the 2D Bragg structure.

shows the profiles of the waves B$_+$ (Fig. 2b) due to scattering of the launched TEM wave of the coaxial waveguide. The TEM wave has no field variation along the radial and azimuthal coordinate, also the field component (H_z) observed in simulations (Fig. 2b) does not exist in the field of either incident or reflected TEM waves (A$_+$) and can only be attributed to the near cut-off wave TE$_{24,1}$ of the coaxial waveguide (partial waves B$_+$). The change of the polarity along the azimuthal and radial coordinates is obvious and (24,1) variations are clearly evident. This agrees well with conditions (3) and the waveguide dispersion relation. Considering dispersion relation (2) and condition (3), one notes that to excite a TE mode with zero radial variation for the same waveguide parameters and $\bar{k}_z$ the azimuthal variation index of the corrugation should be $\bar{m} = 28$ (Fig. 3).

In the recent studies [7,8], the existence of 2D scattering has been experimentally demonstrated and shown that the 2D Bragg structure can operate as a narrow frequency band reflector. Thus in Fig. 4, the transmission coefficient from the 2D Bragg structure with a corrugated inner conductor obtained using analytical theory developed in Refs. [4–8] is presented (bold line) and compared with experimental result (thin line) obtained using a Scalar Network Analyser (Fig. 4a). The results are also compared against data obtained from numerical simulations conducted using the code MAGIC (Fig. 4b) and the good agreement between the results obtained using the different methods is clearly evident.

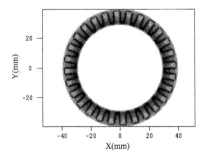

Fig. 3. The snapshot of the longitudinal component of the magnetic field inside the 2D Bragg structure.

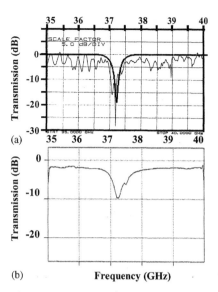

Fig. 4. The transmission coefficient from the 2D Bragg structure obtained (a) from analytical formula [4–8] (thick line) and experimental data (thin line); and (b) using the PIC code MAGIC.

3. Computational study of RF field evolution

Let us consider the 2D Bragg structure as a part of the interaction space of an FEM. The selective properties of the 2D Bragg structure as a single section cavity has been recently under intensive investigation [1–5,10]. It has been demonstrated that after some period of time only the fundamental mode with the highest Q-factor having one variation along the longitudinal coordinate can be found in the cavity. This has also been confirmed for the cavity based on a 2D Bragg structure considered in this paper. Analysing the time decay of the field inside the cavity excited by RF pulses [10] formed by TEM and TE$_{1,0}$ waves, respectively, it has been found that the time decay of the "non-symmetric" mode is $\tau \approx 10$ ns which corresponds to $Q \approx 1100$ while the time decay of the "symmetric" mode is $\tau \approx 15$ ns ($Q \approx 1600$). This may result in effective mode selection over the wave's azimuthal index allowing single mode operation of an FEM driven by an oversized beam. One notes that the mode's Q-factor tends to decrease with increase in the azimuthal index [2–5,7].

Considering a two-mirror cavity based on the 2D Bragg structures, an effective mode selection over the wave azimuthal index is also expected. One of the ways to provide the mode selection is to ensure that spurious azimuthally non-symmetric modes (for instance $TE_{1,0}$) excited inside the cavity lie outside the reflection band of either the input or output mirrors. This can be achieved if the input and output 2D Bragg mirrors have different azimuthal variation indices, i.e. $\bar{m}_1 \neq \bar{m}_2$. In the high-power Strathclyde FEM experiment, the use of input and output mirrors with $\bar{m}_1 = 24$, $\bar{m}_2 = 28$, respectively, is under consideration. Comparing Fig. 4b where $\bar{m}_1 = 24$ and Fig. 5 where $\bar{m}_1 = 28$, it is clear that the reflection bands for the azimuthally symmetric wave are located in the same frequency band. Assuming now that a spurious $TE_{1,0}$ wave is excited inside the cavity, one finds that the cut-off frequencies of the partial waves coupled on the corrugation and which define the location of the reflection zones, are different for input and output reflectors. This results in frequency shift of the reflection zones' centres and therefore to avoid any possibility of the excitation of such a mode, the overlap of reflection zones for the input and output reflectors should be avoided. This can be achieved by varying the coupling coefficient, whose value affects the effective width of the reflection zone [5–8]. In Fig. 3, the snapshot of the profile of the waves $B_{\pm}$ (H_z) is shown at a cross-section, in the centre of the structure. Taking into account the field structure of the $TE_{28,0}$ mode of a coaxial waveguide, one notes that to obtain an effective wave coupling the corrugation has to be machined on the surface of the outer conductor. We are also considering using a 2D Bragg structure which can be achieved by lining the surface of the outer conductor with a bi-periodic dielectric which is beneficial if a pulsed guide solenoid is used to confine the electron beam. Using the PIC code MAGIC field scattering on the outer conductor surface lined by a bi-periodic dielectric has been simulated and results similar to those presented in Fig. 4 have been obtained.

4. Annular electron beam formation

The HCA which is to be used to drive the FEM, is based on a magnetically insulated explosive emission carbon cathode able to produce a thin annular electron beam. In Fig. 6a, the photograph of the experiment is presented and the following components are indicated: (a) pulsed wiggler and capacitor bank power supply; (b) table containing ten $250\,\mu F$, $20\,kV$ capacitors used to generate up to $0.8\,T$ from a large diameter ($30\,cm$) pulsed solenoid of length $2.5\,m$; (c) X-ray shielded enclosure; (d) coaxial cavity, electron beam diagnostics and guide solenoid; (e) ignitron switches with solid-state trigger units; (f) diode tank containing electron gun and transmission line output spark gap. In Fig. 6b, a schematic diagram of the HCA shows: (I) the Marx pulsed power

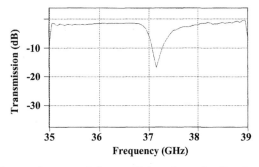

Fig. 5. The transmission coefficient from the 2D Bragg structure obtained using the PIC code MAGIC.

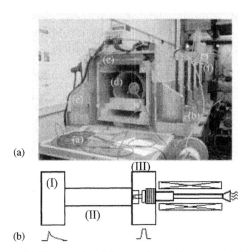

Fig. 6. (a) Photographs of the experimental set-up; and (b) the schematic diagram of the experimental set-up.

supply (a 15 stage, 1.5 MV Marx bank generator) connected via a transmission line (II) to the plasma flare emission electron gun (III). The power supply allows the intermediate electrode of the transmission line to be charged up to 1 MV and generates an output pulse of duration ~ 300 ns. The transmission line of length 2 m is a coaxial line which has an outer conductor diameter of 1 m, filled with high-purity de-ionized water (18 MΩ/cm), an impedance of 4.7 Ω and a total capacitance of 25 nF. The transmission line is switched into the load using a high-pressure (10 bar) nitrogen-filled spark-gap. At the input of the load a rectangular voltage pulse with a flat top of length up to 200 ns is formed. The high-voltage accelerator was connected to the electron gun and triggered to coincide with the firing of the pulsed solenoid. A high-current (1.6 kA), high-voltage (up to 500 kV) large-diameter (mean diameter is 7 cm) annular electron beam was generated. In Fig. 7a, a witness plate beam diagnostic positioned at the input to the coaxial interaction region, 15 cm from the

anode plate, was used to obtain the transverse position of the electron beam, with a beam current of 1.6 kA measured (Fig. 7b) using an in-line Rogowski coil beam diagnostic.

5. Conclusion

We have discussed the basic model and field scattering on the 2D Bragg corrugation. The formation of 2D feedback loop has been investigated. Good agreement between simulations and basic model predictions is demonstrated. The results of simulations have also demonstrated the possibility to obtain efficient azimuthal mode selection in an oversized cavity based on 2D Bragg structures. Taking into account the experimental measurements obtained from the HCA, the parameters of the FEM experiment may be estimated. An electron beam of power 750 MW has been measured. Assuming an FEM output efficiency of 15% which has been predicted in Refs. [1–7], the generation of ~ 100 MW of power at the operating frequency of 37.5 GHz is predicted.

Acknowledgements

The authors would like to thank QinetiQ, EPSRC, INTAS and the University of Strathclyde for support of this work.

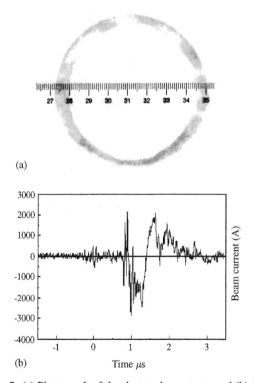

(a)

(b)   Time μs

Fig. 7. (a) Photograph of the electron beam trace; and (b) the electron beam current measured using a Rogowski coil.

References

[1] N.S. Ginzburg, et al., Opt. Commun. 96 (1993) 254.
[2] N.S. Ginzburg, et al., Tech. Phys. 41 (5) (1996) 465.
[3] N.S. Ginzburg, et al., IEEE Trans. Plasma Sci. 24 (1996) 770.
[4] I.V. Konoplev, et al., Nucl. Instr. and Meth. A 445 (2000) 236.
[5] N.S. Ginzburg, et al., J. Appl. Phys. 92 (2002) 1619.
[6] I.V. Konoplev, et al., Proceedings of IEE Symposium on Pulsed Power 2000, 2000, p. 412.
[7] A.W. Cross, et al., Nucl. Instr. and Meth. A 475 (2001) 164.
[8] A.W. Cross, et al., Appl. Phys. Lett. 80 (2002) 1517.
[9] A.F. Harvey, Microwave Engineering, Academic Press, London, 1963.
[10] N.S. Ginzburg, et al., Technical Physics, 48 (12) (2003) 54.

Available online at www.sciencedirect.com

Nuclear Instruments and Methods in Physics Research A 528 (2004) 106–109

ELSEVIER

NUCLEAR INSTRUMENTS & METHODS IN PHYSICS RESEARCH
Section A

www.elsevier.com/locate/nima

High-power infrared free electron laser driven by a 352 MHz superconducting accelerator with energy recovery

Byung Cheol Lee[a],*, Young Uk Jeong[a], Seong Hee Park[a], Young Gyeong Lim[a], Sergey Miginsky[b]

[a] *Korea Atomic Energy Research Institute (KAERI), P.O. Box 105, Yusong, Daejon, South Korea*
[b] *Budker Institute of Nuclear Physics, 630090 Novosibirsk, Russia*

Abstract

A high-average power infrared free electron laser driven by a 40 MeV energy recovery superconducting accelerator is being developed at the Korea Atomic Energy Research Institute. The free electron laser is composed of a 2-MeV injector, two superconducting acceleration cavities, a recirculation beamline, and an undulator. Each accelerator module contains two 352-MHz 4-cell superconducting cavities and can generate an acceleration gain of 20 MeV. The 2-MeV injector has been completed, and generates stably an average current of 10 mA. One of the 20 MeV superconducting accelerator modules has been installed, cooled down to 4.5 K, and RF-tested. An average power of 1 kW at the wavelengths of 3–20 μm is expected.
© 2004 Published by Elsevier B.V.

PACS: 41.60.Cr; 41.75.Fr; 29.27.Ac

Keywords: Free electron laser; FEL; Superconducting; Accelerator; Laser; Electron

1. Introduction

There are several potential applications of tunable high-average power free electron lasers (FELs) in nuclear industry. For example, it is possible to separate stable isotopes such as silicon (Si^{29}), carbon (C^{11}), oxygen (O^{18}), etc., by using selective multi-photon dissociation of molecules. Tunable ultraviolet FELs are useful in separating long-lifetime elements from nuclear spent fuels

using selective photochemical reactions. In these applications, few tens of kilowatts in average power of radiation at a reasonably low cost is needed. One of the best ways of generating high-average power FEL radiation at low cost is using superconducting accelerator with energy recovery.

After successful development of a millimeter-wave FEL driven by a 0.4-MeV electrostatic accelerator [1], and an infrared FEL driven by an 8-MeV microtron accelerator [2], and being encouraged by the pioneering demonstration of energy recovery FELs by the Jefferson Laboratory [3], and JAERI [4], the Korea Atomic Energy Research Institute (KAERI) started a project for

---

*Corresponding author. Tel +82-42-868-8378; fax +82-42-868-2969.
E-mail address: bclee4@kaeri.re.kr (B.C. Lee).

0168-9002/$ - see front matter © 2004 Published by Elsevier B.V.
doi:10.1016/j.nima.2004.04.028

the development of a high-power infrared FEL driven by a 40-MeV superconducting accelerator with energy recovery. This paper describes design concept of the FEL and the results of the first stage of the project.

2. Energy recovery superconducting linac

2.1. Schematic layout of the FEL

The KAERI infrared FEL is composed of a 300-keV electron gun, a 2-MeV pre-accelerator, two 20-MeV 352-MHz superconducting acceleration cavities, an undulator unit, and an energy recovery beamline. Schematic layout of the FEL is shown in Fig. 1, and design parameters of the FEL are listed in Table 1.

2.2. 2-MeV injector

The 2-MeV injector [5] is composed of a 300-keV electron gun, one RF bunching cavity, and two normal-conducting RF acceleration cavities. The kinetic energy of the electron beam is 1.5 MeV nominally, and 2 MeV at maximum. The duration of a pulse is 350 ps and its repetition rate is variable from 11 kHz to 22 MHz. The peak current is 6 A, and the average current at the maximum

repetition rate is 45 mA. The resonant frequency of the RF cavities is 176 MHz, which is a half of the resonance frequency (352 MHz) of the superconducting accelerator.

2.3. Superconducting cavities

Each main accelerator module from CERN contains two 352-MHz 4-cell superconducting cavities and can generate an acceleration gain of 20 MeV. The superconducting cavities are made of OFHC copper with thin Nb coating. Each cavity is equipped with one RF coupler and four

Table 1
Parameters of the KAERI infrared FEL

Accelerator	
Energy	20–40 MeV
Average current	10 mA
RF frequency	352 MHz
Undulator	
Period	35 mm
Magnetic field	1.7–3 kG
Gap	20 mm
FEL	
Wavelength	3–20 μm
Pulse width	~50 ps
Average power	1–10 kW

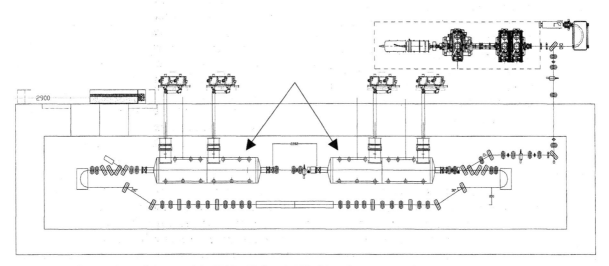

Fig. 1. Schematic layout of the KAERI infrared FEL driven by a superconducting energy-recovery linac.

higher-order mode (HOM) couplers. The RF coupler can deliver an average power of 150 kW to the SC cavity. The resonant frequency of the cavity can be tuned by controlling the cavity length, which in turn, is adjusted by changing the lengths of the nickel bars attached to the cavities. Table 2 shows the parameters of the super-conducting acceleration cavities, and Fig. 2 shows the schematic of the cavity.

Numerical and analytical analysis [6] shows that the wakefields generated by the bunched electrons passing through the superconducting cavities affect the beam trajectory negligibly at an average current of 50 mA. The HOM powers excited by the repetitive electrons will be about 140 W and 1.1 kW for the TE_{111} and the TM_{110} modes, respectively.

One superconducting cavity module has been installed successfully, and cooled down to 4.5 K. A 352-MHz RF generator for the superconducting cavity with average power of 100 kW has been installed and tested successfully. Fig. 3 shows the photograph of the superconducting cavity.

Table 2
Parameters of the superconducting cavities

Resonant frequency	352 MHz
Cavity material	Nb-Coated copper
Number of cell	4-Cell 1 cavity
Q_0	3.4×10^9 @ 6 MV/m
R/Q	500
Tuning range	50 kHz
Active length	1.7 m

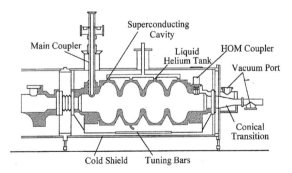

Fig. 2. Schematic of the 352-MHz superconducting acceleration cavity.

Fig. 3. Photograph of the 352-MHz superconducting.

2.4. Energy recovery beamline

Since the injection energy is rather low, there is a limitation in the pulse duration of the injected electron beam caused by the induced energy spread in the cavity. The maximum duration (if no "linearizer"[7]) is

$$\Delta t \approx \frac{1}{\omega}\sqrt{\frac{2\Delta E}{E}} \approx \pm 100 \text{ ps in our case.}$$

The energy recovery beamline is still under design. We must compromise between wide-tunability of the electron beam energy (which results in wide-tunability of the FEL radiation) and high-average power. One of the candidates is achromatic bend scheme as shown in Fig. 4. In this case, in contrast to the simplest ones, as in the first stage of BINP FEL [8], the dispersion can be controlled, and θ-function is less for the given dispersion (up to twice). The total beamline length spread can be compensated by the dispersion. Estimation shows that the permissible track length spread is ~10 mm, While tuning electron energy within 20–40 MeV the appropriate length varies within ± 6 mm, that is less than the permissible track length spread. Estimation of the effects of energy spread and injection pulse duration on the track length error show that by reducing the injection pulse duration from 100 to 50 ps will not only enhance FEL power but also enhance the flexibility in the design of beamline. Table 3 shows a result of the calculation of beam trajectory in the achromatic bend.

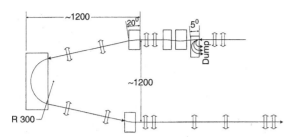

Fig. 4. Achromatic bend for energy recovery beamline.

Table 3
Results of beam trajectory in the achromatic bends

	β_x, m	α_x	β_y, m	α_y
Left bent				
Entrance	3–12	±1	3–12	±1
Exit	0.3–3	0	0.3–3	0
Right bent				
Entrance	0.3–3	0	0.3–3	0
Exit	3–12	±1	3–12	±1

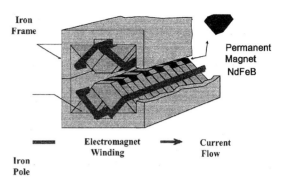

Fig. 5. Schematic of the electromagnet undulator.

3. Undulator

An upgraded version of the high-precision electromagnetic undulator assisted by permanent magnets [9] will be used in the high-power infrared FEL. This type of undulator is very compact in its structure, cost-effective, and generates high magnetic field with very high precision. The field strength changes depending on the electric current through main coil. Fig. 5 shows a schematic of the undulator.

4. Summary

A high-average power infrared free electron laser driven by a 40 MeV energy recovery superconducting accelerator is being developed at KAERI. The 2-MeV injector has been completed, and generates stably an average current of 10 mA. The 2-MeV electron beam has been transported successfully to the entrance of the superconducting cavity. One of the superconducting cavities from CERN has been installed and cooled down to its operating temperature, 4.5 K. Two 50-kW 352-MHz RF generators have been installed and tested successfully. Low-power RF test of the cavity has been finished, and high-power RF test will be performed soon, which will be follower by acceleration of electron beam up to 20 MeV.

Acknowledgements

The superconducting cavities used in this project has been donated by CERN. Authors thank to CERN for donation. Many parts of the accelerator has been fabricated on contract base by the Budker Institute of Nuclear Physics, Novosibirsk. The authors thank to the members for their contributions.

References

[1] B.C. Lee, et al., Nucl. Instr. and Meth. A 375 (1996) 28.
[2] Y.U. Jeong, et al., Nucl. Instr. and Meth. A 475 (2001) 47.
[3] S. Benson, et al., Nucl. Instr. and Meth. A 429 (1999) 27.
[4] R. Hajima, et al., Nucl. Instr. and Meth. A 507 (2003) 115.
[5] B.C. Lee, et al., Nucl. Instr. and Meth. A 429 (1999) 352.
[6] S.O. Cho, et al., J. Korean Phys. Soc. 41 (2002) 662.
[7] D.H. Dowell, et al., Nucl. Instr. and Meth. A 393 (1997) 184.
[8] N. Vinokurov, et al., Nucl. Instr. and Meth. A, (2004) these Proceedings.
[9] Y.U. Jeong, et al., Nucl. Instr. and Meth. A 483 (2002) 363.

Available online at www.sciencedirect.com

SCIENCE ⓓ DIRECT°

NUCLEAR
INSTRUMENTS
& METHODS
IN PHYSICS
RESEARCH
Section A

ELSEVIER   Nuclear Instruments and Methods in Physics Research A 528 (2004) 110–114

www.elsevier.com/locate/nima

Study of millimeter wave high-power gyrotron for long pulse operation

A. Kasugai*, K. Sakamoto, R. Minami, K. Takahashi, T. Imai

Naka Fusion Research Establishment, Japan Atomic Energy Research Institute (JAERI), 801-1 Mukoyama,
Naka-machi, Naka gun, Ibaraki-ken 311-0193, Japan

Abstract

The development of a high-power millimeter wave source, gyrotron, is under way for fusion application. A performance of ~10 s oscillation has been attained at 1 MW level output for 170 GHz frequency. Although the pulse extension was interrupted by a sudden outgassing in the gyrotron, the cause was confirmed to be the local heating of the internal component due to stray RF deposition. To suppress the heating, the inner surface of the component (bellows for RF beam steering mirror) was coated with copper, which reduced the Ohmic loss to $\frac{1}{10}$ of the original one. Moreover, forced water cooling for the bellows was incorporated. As a result, no sudden pressure increase was observed, and a quasi-steady-state oscillation of 100 s with 0.5 MW power level was demonstrated at 170 GHz. The temperature of the major components of the gyrotron stabilized, which indicates a prospect for a 1 MW-CW, 170 GHz gyrotron.
© 2004 Elsevier B.V. All rights reserved.

PACS: 52.59.Rz

Keywords: Gyrotron; Fusion; ITER; Millimeter wave; ECH/ECCD; Stray radiation

1. Introduction

A high-power RF source with frequencies of 100 GHz band is necessary for a fusion reactor. In International Thermonuclear Experimental Reactor (ITER), 170 GHz high-power gyrotron system with a total power of 24 MW is planned for electron cyclotron heating (ECH), current drive (ECCD), suppression of plasma instability, and start up of plasma [1]. The development of a 170 GHz gyrotron with 1 MW, CW (continuous wave) operation and 50% efficiency is targeted to satisfy the requirements of the ITER [2].

The Japan Atomic Energy Research Institute (JAERI) initiated the development of a high-power gyrotron with 100 GHz band in order to apply it to fusion devices, in 1980s [3]. Several breakthroughs with respect to high power, long pulse, and high efficiency operation of the gyrotron have been achieved by JAERI. These include a 1 MW oscillation of $TE_{31,8}$ high-order mode with an oversized open-ended cavity [4],

*Corresponding author. Tel.: +81-29-270-7562; fax: +81-29-270-7569.

E-mail address: kasugai@naka.jaeri.go.jp (A. Kasugai).

0168-9002/$ - see front matter © 2004 Elsevier B.V. All rights reserved.
doi:10.1016/j.nima.2004.04.029

energy recovery for 50% total efficiency by a depressed collector [5,6], and a synthetic diamond window that is capable of more than 1 MW-CW transmission [7–9]. In addition, the parasitic oscillation in the beam tunnel was suppressed by the installation of an RF absorber (silicon carbide) [10]. By integration of the above-mentioned breakthrough technologies, stable long pulse operations such as 0.9 MW/9.2 s, 0.5 MW/30 s, 0.3 MW/60 s, and 0.2 MW/133 s were demonstrated for a 170 GHz ITER gyrotron [11].

Further extension of the pulse duration was prevented by a sudden and rapid pressure increase in the gyrotron. This paper describes the study of sudden pressure increase and its countermeasures and the results of pulse extension to 100 s.

2. Structure and operational results of 170 GHz gyrotron

A schematic cross-sectional view of the gyrotron is shown in Fig. 1. The total height of the gyrotron is about 3 m, and its weight is about 800 kg. A gyrotron for high-power and long pulse operation is usually baked at a temperature condition of

450°C for a week during the manufacturing process to maintain a high vacuum of $< 10^{-8}$ Pa. The major components of the gyrotron are a magnetron injection gun (MIG), a cylindrical open-ended cavity, a built-in mode converter (RF radiator and mirrors), a collector, and a synthetic diamond output window.

A rotational electron beam with hollow distribution is generated by a triode-type MIG and is accelerated to about 75 keV by the highly stabilized applied field within $\pm 0.5\%$. The electron beam is guided to the cavity along a compressed magnetic field of ~ 6.7 T. A millimeter wave at 170 GHz is oscillated in the cavity through interaction between electron beam and electromagnetic wave by a cyclotron resonance maser (CRM). To reduce heat deposition on the inner wall of the cavity due to Ohmic loss, its radius is expanded to 17.9 mm using the higher-order mode of $TE_{31,8}$. The calculated peak heat load at the cavity is 2.2 kW/cm^2 for the 1.2 MW oscillation, which is acceptable for a CW operation.

The oscillation wave of $TE_{31,8}$ mode is transformed into a Gaussian beam (TEM$_{00}$) by a quasi-optical radiator and two-phase correction mirrors. The inner radius of the radiator is deformed by

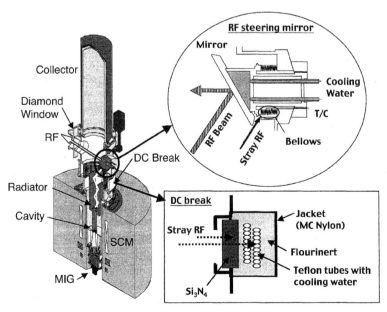

Fig. 1. Cross-sectional view of 170 GHz ITER gyrotron.

~0.1 mm periodically [12]. The RF beam is guided to an output window by the RF steering mirror. The window is made of artificial diamond, which enables 1 MW-CW output with the Gaussian beam at 100 GHz band [13] because of extremely high thermal conductivity (~2000 W/mK) and a low loss tangent (~2 × 10^{-5}).

After intensive efforts, the gyrotron had almost achieved the target of the ITER engineering design activities (1 MW/10 s/50% efficiency). However, the extension of the pulse duration was limited by a sudden pressure increase in the gyrotron. For example, outgassing occurred 133 s after the oscillation of 0.2 MW output power started, though the pressure was kept at a very low level of <1 × 10^{-6} Pa before the sudden increase. The pressure degradation always occurred at the same output energy of 15–20 MJ. It is considered that the rapid outgassing is caused by local heating with poor cooling of the inner component (stainless steel bellows), whose temperature exceeded the baking temperature. The source of the heating power is stray radiation resulting from diffraction loss in the quasi-optical built-in mode converter.

Although 93.8% of the oscillation power passes through the window, the diffraction loss power is 6.2% of the oscillation power in the calculation [11]. Most of the power escapes from the insulators (DC break, sub-window) to outside of the gyrotron, and a small part of the power is absorbed as Ohmic loss on the inner surface of the gyrotron. Table 1 shows an example of the measured power. The stray power that escaped through the DC break was estimated by the temperature increase in the cooling water that flowed in Teflon tubes covering the silicon nitride (Si$_3$N$_4$) DC break and by the temperature increase

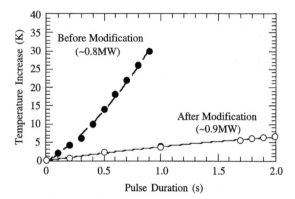

Fig. 2. Temperature increase of the bellows section before and after modification.

in flourinert (FX3300, 3 M) that cooled the surface of Si$_3$N$_4$. Backward power through a cavity is absorbed in a SiC block which is applied to the beam tunnel to suppress the parasitic oscillation. As a result, the measured stray radiation power is estimated to be about 9% of the total oscillation power.

The stainless steel bellows that control the RF beam direction to the output window was the component that was responsible for the large power deposition of stray radiation. The temperature on the bellows surface measured by a thermocoupler linearly increases with the pulse length as shown in Fig. 2 (closed circle). This result indicates that the temperature would exceed the baking temperature after about 15 s at 0.8 MW, which is consistent with the experimental results. Therefore, the stainless steel bellows were identified to be the source of the outgassing.

3. Improvement of long pulse performance

One of the countermeasures to reduce the power deposition on the bellows is to improve the bellows section design to reduce the Ohmic loss and to enable forced water cooling. In the old structure, the stray radiation power from the direction of the main RF beam was directly deposited with ease from the small gap. To avoid the RF deposition on the bellows, the configuration was modified to decrease the RF irradiation. In addition, the

Table 1
Typical power of output and stray radiation

		Power (kW)	Ratio (%)
Output		708	91
Stray radiation	DC break	50	
	Sub-window	16	9
	SiC block	3.7	
Total		778	100

welded bellows was replaced with molded bellows to reduce the surface area. The next modification was a copper coating with a thickness of $\sim 11\,\mu m$ on the inner surface of the molded bellows, made of stainless steel, to reduce the Ohmic loss by stray radiation. Finally, forced water cooling was employed for the CW operation.

As a result of the above improvements, the heating rate of the bellows was reduced to less than $\frac{1}{10}$ of the original rate as shown in Fig. 2 (open circle). Moreover, the time constant of thermal diffusion without cooling condition was reduced to about $\frac{1}{3}$ (from 270 to 90 s). In the operation of the modified gyrotron, sudden and rapid pressure increase was not observed, and the performance of 100 s operation with the output power of 0.5 MW level was demonstrated without significant conditioning. It was confirmed that the temperature of the major components reached steady state in the 100 s operation.

Fig. 3 shows an example of time evolution of the temperature changes at the collector surface (a), the center of the output window, which is made of a synthetic diamond disk and cooled by fluorinert at the edge (b), and flourinert for DC break cooling (c) at 100 s operation. Collector cooling is sufficient for 0.5 MW power (55 kV, 25 A) and the increase in the surface temperature stabilized at 120°C, which indicates the ΔT is acceptable for 1 MW operation. The edge of the diamond window with a loss tangent of $\sim 2 \times 10^{-5}$ is cooled by the flourinert. Since ΔT of the center of the disk measured by an IR camera stabilized at $\Delta T \sim 60°C$, flourinert cooling was sufficient for 0.5 MW power transmission. It could be possible to pass the 1 MW power by the enhancement of the flow speed of the coolant. A DC break ceramic insulator (Si_3N_4) for the depressed collector was also cooled by flourinert since it was required to isolate $\sim 30\,kV$. The temperature increase of the DC break was mainly due to the absorption of the stray radiation. The temperature of the flourinert flow stabilized at $\Delta T \sim 30°C$, which was again acceptable. Therefore, these results indicate that the 1 MW operation is possible.

Stable oscillation of 100 s with 0.5 MW output power level was achieved. Here, the power changed from ~ 0.6 to 0.46 MW because the

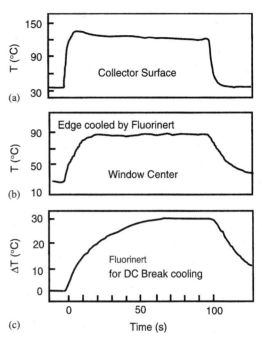

Fig. 3. Time evolution of temperature changes of major components in the gyrotron for a 100 s pulse operation with 0.5 MW level output power.

electron current slumped from 35 to 25 A for a 100 s operation. Although 1 MW operation has a large current decreasing rate of $\sim 1\,A/s$, it is also possible to extend the pulse duration by the enhancement of the cathode heater and tuning of magnetic field.

Suppression of power deposition on the inner surface of the gyrotron is the key to a 1 MW-CW operation. For further stable and reliable operation of the 1 MW-CW gyrotron, improvements for suppression of the diffraction loss from the mode converter will be done.

4. Conclusion

It was clarified that the sudden pressure increase in the gyrotron, which prevented the pulse extension, was caused by local and intensive heating of the bellows section due to power deposition by stray radiation. An improvement of the bellows to avoid power deposition by stray radiation and enhancement of the cooling capacity and quasi-steady-state oscillation of 170 GHz gyrotron was achieved with 0.5 MW level for 100 s. The temperatures of the major components stabilized in the 100 s operation, which indicates a prospect for the 1 MW-CW operation.

Acknowledgements

We thank Yu. Ikeda and H. Ouchi for their support in the gyrotron experiments. We also thank Drs. T. Kariya, Y. Mitsunaka, and K. Hayashi of Toshiba Corporation for the useful discussions as well as Drs. S. Seki, H. Takatsu, H. Tsuji, and M. Seki for their encouragement.

References

[1] Technical Basis for the ITER Final Design Report, 2001.
[2] T. Imai, et al., Fusion Eng. Des. 55 (2001) 281.
[3] A. Kasugai, et al., Fusion Eng. Des. 26 (1995) 281.
[4] K. Sakamoto, et al., J. Phys. Soc. Japan 65 (1966) 1888.
[5] K. Sakamoto, et al., Phys. Rev. Lett. 7 (1994) 3532.
[6] K. Sakamoto, et al., J. Plasma Fusion Res. 71 (1995) 1029.
[7] O. Braz, et al., Int. J. Infrared Millimeter Waves 18 (1997) 1495.
[8] A. Kasugai, et al., Rev. Sci. Instrum. 69 (1998) 2160.
[9] K. Sakamoto, et al., Rev. Sci. Instrum. 70 (1999) 208.
[10] H. Shoyama, et al., Jpn. J. Appl. Phys. 40 (2001) L906.
[11] K. Sakamoto, et al., Nucl. Fusion 43 (2003) 729.
[12] K. Sakamoto, et al., Int. J. Infrared Millimeter Waves 18 (1997) 1637.
[13] A. Kasugai, et al., Fusion Eng. Des. 53 (2001) 399.

Available online at www.sciencedirect.com

Nuclear Instruments and Methods in Physics Research A 528 (2004) 115–119

**NUCLEAR
INSTRUMENTS
& METHODS
IN PHYSICS
RESEARCH**
Section A

www.elsevier.com/locate/nima

Stabilization of the microtron-injector for a wide-band compact FIR FEL

Grigori M. Kazakevitch[a,*], Young Uk Jeong[a], Viatcheslav M. Pavlov[b],
Byung Cheol Lee[a]

[a] *Laboratory for Quantum Optics, Korea Atomic Energy Research Institute, P.O. Box 105, Yusong, Taejon 305-600, South Korea*
[b] *Budker Institute of Nuclear Physics RAS, Academician Lavrentyev 11, Novosibirsk 630090, Russia*

Abstract

To provide parameters of a simple and inexpensive magnetron-driven microtron-injector acceptable for a wide-band FIR FEL, the microtron has been improved through stabilization of the beam current and the magnetron frequency. The beam current was stabilized during the macro-pulse by increasing the magnetron anode current. The pulse stabilization of the emission current makes possible the microtron operation with the maximal accelerated current, without risk of break-downs in the cavity and keeps the instability of the accelerated current at approximately 1% during long-time experiments. The magnetron frequency was stabilized using the microtron accelerating cavity as a stabilizing external resonator in a simple scheme that involved the cavity loading of the magnetron through a ferrite insulator. The scheme provides stabilization of the magnetron frequency with a coefficient of 3.5. The stabilization of current and the frequency at the microtron FIR FEL-injector provides satisfactory intrapulse stability of the extracted lasing power of 40–50 W in the FIR macro-pulse having duration of 3–4 μs and a long-time pulse-to-pulse instability of the FIR pulse energy in the range of ⩽10%. Results of simulations of the stabilization based on the microtron operating parameters and measured results are presented and discussed.
© 2004 Elsevier B.V. All rights reserved.

PACS: 41.60.Cr; 07.57Hm; 29.20.−c

Keywords: Microtron; Magnetron; Stabilization; Frequency pulling; Free electron laser; Deviations

1. Introduction

An inexpensive, simple, reliable magnetron-driven, 12-turn classical microtron having an

---
*Corresponding author. Tel.: +82-42-868-8253; fax: +82-42-861-8292.
E-mail address: gkazakevitch@yahoo.com
(G.M. Kazakevitch).

internal cathode provides good bunching properties for the electron beam, and this makes it acceptable as an injector for a FIR FEL [1]. The microtron has been upgraded to increase the accelerated current up to 50–70 mA in the macro-pulse having duration of 6 μs for a wide-band FIR FEL. By such parameters of the accelerated beam the emission current of the microtron has an increase of 25–30% during the macro-pulse

0168-9002/$ - see front matter © 2004 Elsevier B.V. All rights reserved.
doi:10.1016/j.nima.2004.04.030

because of the back-streaming electrons over-heating the cathode. This phenomenon causes an abatement of the accelerated current. We stabilize the current by increasing the magnetron power, increasing the magnetron current during the macropulse [2]. Respective deviations of the magnetron frequency are stabilized with a simplified scheme using a frequency pulling in the magnetron through the wave reflected from the accelerating cavity. An additional stabilization of the pulse emission current in the microtron provides a pulse-to-pulse instability of the lasing energy in the range of ⩽10% during a long-time operation.

Ideas, methods and measured results permitting us to develop the inexpensive, stable, wide-band tunable FEL, having tens of Watts in the extracted FIR macro-pulse power, based on the cheap, simple and reliable microtron-injector are presented and discussed.

2. Intrapulse stabilization of the accelerated current

The effect of the incremental emission current on the shape of the accelerated current was investigated using a numerical simulation of the coupled magnetron–microtron cavity system considering the motion of the electrons in the median plane.

As a first step of the simulation we calculated the shape of the accelerated current for the measured time-dependent emission current by constant power of the magnetron during the macro-pulse. For the calculation was used as an approximation the abridged Equation [3] describing the transient state in the accelerating cavity:

$$\left\{ \frac{d}{d\tau} + \left[1 - i\frac{Q_0}{1+\beta_C}\left(\frac{\omega_0}{\omega} - \frac{\omega}{\omega_0}\right) \right] \right\} \tilde{V}_C$$
$$= \frac{2\beta_C}{1+\beta_C}\tilde{V}_{FC} - \frac{R_{sh}}{2(1+\beta_C)}\tilde{J}_C \cdot \exp(i\varphi_C). \quad (1)$$

Here, $\tau = t/\tau_{C0}$, τ_{C0} is the fill-time of the cavity, Q_0 is the wall quality factor, ω_0 is the circular eigen frequency of the cavity, β_C is the cavity coupling coefficient, $\tilde{V}_C$ and $\tilde{V}_{FC}$ are complex amplitudes of the oscillation in the cavity and in the forward wave, respectively, R_{Sh} is the shunt impedance of the cavity, $\tilde{J}_C$ is the complex

amplitude of the first-time harmonic of the loading current, and φ_C is the phase of the complex amplitude of $\tilde{V}_C(t)$. By simulation of the transient state of the microtron cavity, the cardinal problem is the calculation of the loading current $\tilde{J}_C$ [3]:

$$\tilde{J}_C = \frac{1}{w^*} \cdot \int_v \tilde{\vec{J}}_1(t,\vec{r})\,\vec{e}(\vec{r})\mathrm{d}V \quad (2)$$

where $\vec{e}(\vec{r})$ is the normalized cavity electric field distribution, $\tilde{\vec{J}}_1(t,\vec{r})$ is the first-time harmonic of the current density, which depends on the emission current $I_0(t)$ and coordinates and velocities of the electrons passing through the cavity, $w^* = \int_{-L/2}^{L/2} e_z(\vec{r}_\perp = 0, z)\,e^{-i\omega_0(z/v_0)}\,\mathrm{d}z$. Here L is the cavity length, and v_0 is the average velocity of the electron.

Note that the loading current $\tilde{J}_C$ expressed by Eq. (2) was calculated as a sum of the currents for all of the electrons simultaneously passing through the cavity from all of the orbits. The current on each of the orbits was calculated using 2-D tracking simulating the motion of the electrons. The Lorentz-force equation in the dimensionless variables [4] was used for the tracking. This method allows us to take into account the loading current of all the accelerated electrons, synchronous and non-synchronous as well.

The result of the calculation of the accelerated current for the 12th orbit is shown in Fig. 1, curve 2. As follows from the curve, the accelerated current has a decrease of up to half during the

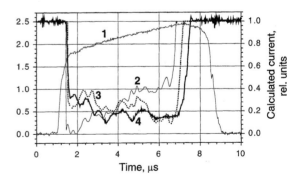

Fig. 1. 1—Measured macro-pulse emission current, vert. scale is 250 mA/div., 2—calculated shape of the accelerated current by a constant magnetron current during the macro-pulse, 3—calculated shape of the accelerated current by the incremental magnetron power, 4—measured accelerated current by the incremental magnetron power, vert. scale is 10 mA/div.

macro-pulse versus an increase of the emission current, Fig. 1, curve 1. The increase causes an additional loading in the accelerating cavity and as a result, the abatement of the accelerating field causing the abatement of the current. That makes a serious problem with the FEL operation in the full scale of the macro-pulse duration. To avoid that and to compensate for the effect of additional loading in the cavity because of the increase of the emission current, we stabilized the accelerated current by increasing the magnetron power during the macro-pulse. For that the modulator charging line was tuned to provide linear enhancement of the magnetron current for $\approx 10\%$ during the macro-pulse [2].

The shape of the accelerated current for measured time-dependent magnetron and emission currents, was calculated using the equation system including the abridged equation for the accelerating cavity (1) and a similar one, which described the transient state in the magnetron this equation differs from Eq. (1) with the noise term [5]. The loading current for the microtron cavity was calculated by the tracking in the median plane up to 12th orbit. The magnetron current was measured using a calibrated wide-band current transformer. The measured shape of the top of the magnetron macro-pulse current is shown in Fig. 2(c).

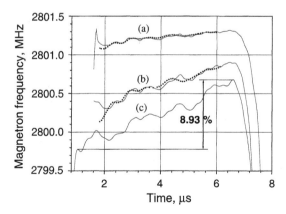

Fig. 2. Magnetron frequency deviations with (a) and without (b) stabilization by the frequency pulling. Curves (a) are respective to the accelerating cavity loading of the magnetron, curves (b) are respective to the passive matched load. Curve (c) demonstrates the shape of the top of the magnetron pulse current (relative units).

The result of the numerical simulation for the measured increase of the magnetron anode current is shown in Fig. 1, curve 3. The result has a satisfactory coincidence with the measured shape of the accelerated current, Fig. 1, curve 4. Increasing the magnetron anode current during the macro-pulse provides a full-scale FIR FEL operation during the beam current.

3. Stabilization of the magnetron frequency

The stabilization was developed for lasing to provide acceptable stability of the bunch repetition rate during the macro-pulse with a simple, reliable and inexpensive magnetron generator, which feeds the microtron cavity. The electron frequency drift caused by the incremental anode current in the magnetron and deteriorating the bunch repetition rate stability was suppressed by means of a simplified scheme. The scheme is based on the frequency pulling in the magnetron through the wave reflected from the accelerating cavity, which also serves as an external stabilizing resonator. The reflected wave passes through a ferrite insulator, having inverse losses of $\approx 18\,\mathrm{dB}$ by the power of 1.7–$2\,\mathrm{MW}$, providing an acceptable level of the passing wave for the frequency pulling. The microtron–microwave system has been optimized in the length.

To determine the coefficient of the magnetron frequency stabilization, the simulations and the measurements have been conducted for both cases of the magnetron load: with the accelerating cavity loaded by the electron current and with a passive matched waveguide load. For the simulations we used the considered abridged equations system describing the transient state in the magnetron and in the microtron cavity and measured time-dependent magnetron and emission currents.

Results of the simulations for the magnetron–microtron cavity system and the magnetron feeding the passive matched load are plotted in Fig. 2 by solid lines (a) and (b), respectively.

Measurements of the magnetron frequency deviations during the macro-pulse in the coupled magnetron–microtron cavity system and the magnetron-passive load system were done using a temporal heterodyne method [6], providing the relative inaccuracy of $\approx 10^{-6}$ in the time interval of ≈ 100 ns. The deviations were measured in the forward wave using the 20-dB directional coupler. The results are shown in Fig. 2 by dotted lines (a) and (b), respectively. The shape of the top of the magnetron macro-pulse current is plotted in Fig. 2(c) in relative units.

In both cases, the simulated and measured results have good coincidence and demonstrate correctness of the simulations and measurements. The value of the stabilization coefficient of ≈ 3.5 was determined from the obtained results. As shown in Fig. 2, the deviations of the magnetron frequency retrace the curve of the shape of the magnetron current, but with the developed stabilization using the accelerating cavity as the stabilizing resonator they are noticeably suppressed.

In Fig. 2 one can see that the "slow" oscillation of the magnetron frequency, having the period of $\approx 0.6 \, \mu s$, takes place in both cases: in the magnetron–microtron cavity system, and by the passive load of the magnetron. This phenomenon is caused by the oscillation in the magnetron current. The magnetron current oscillation has a relative value of $\approx 1\%$ in the amplitude, that corresponds to the oscillation in the magnetron pulse voltage with the level of $\leqslant 0.2\%$, and causes the "slow" oscillation in the magnetron frequency with the relative value of $\approx 10^{-5}$ and $\approx 4 \times 10^{-5}$ by operation with the stabilizing cavity and the passive load, respectively.

4. Pulse stabilization of the emission current

To provide maximal accelerated current by the maximal capture coefficient, we are forced to operate on the left slope of the microtron volt–ampere characteristic in a neighborhood of a main maximum. In this case, a small decrease of the emission current leads to a stripping of the acceleration, a leap of the cavity voltage and break-downs in the cavity.

To avoid this problem, we developed stabilization of the pulse emission current using a computer-controlled system. The system is based on a fast ADC measuring the emission pulse current. The system averages the data for a few pulses, analyzes the data and tunes the cathode filament stabilizer. By the randomized stripping of acceleration, the stabilization system tunes the cathode filament stabilizer to prevent breakdowns. The system provides long-time pulse-to-pulse instability of the emission current $\leqslant 1\%$. The instability in the accelerated current is approximately the same. Measured pulse-to-pulse instability of the extracted current has a level of $< 2.5\%$ by operation of the stabilization system.

5. Effect of the microtron stabilization on a lasing

The effect of stabilization on the FIR lasing has been investigated with parameters of the microtron optimized for long-time operation of the wide-band FIR FEL. The initial magnetron–microtron cavity detuning was chosen to provide minimal range of the bunch repetition rate deviations by the accelerated current at the 12th orbit of ≈ 45 mA in the macro-pulse. The range of the bunch repetition rate deviations measured by the temporal heterodyne method with the monitoring cavity [6] had a relative value of $\leqslant 8 \times 10^{-5}$. The emission current was chosen to provide operation in the neighborhood of the maximum of the microtron volt–ampere characteristic. This provides the value of the beam current in the beamline at the undulator entrance in the range of 40–42 mA by extraction of electrons from the 12th orbit and in the range of 44–46 mA by extraction from the 10th orbit. Measured pulse-to-pulse instability of the current had a value of $\leqslant 1$ mA.

Note that the values of the macro-pulse current provide deep saturation of the FIR FEL generation, small dependence of the FIR macro-pulse energy on the current and, as result, stable enough FIR power during the macro-pulse. Measured dependence of the FIR macro-pulse energy on the beamline current and the FIR power shape by the FIR radiation wavelength of $\approx 110 \, \mu m$ are shown in Figs. 3(a) and (b), curve 2, respectively. The measurements were done using the pyroelectric detector and the wide-band Schottky-barrier detector.

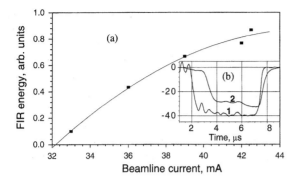

Fig. 3. (a) Dependence of the macro-pulse FIR energy on the beam current at the undulator entrance, (b) 1—The macro-pulse shape of the beamline current, vert. scale is 20 mA/div.; 2—the macro-pulse shape of the FIR power, vert. scale is in arbitrary units.

By the optimized parameters of the microtron, the pulse-to-pulse instability of the FIR energy measured using the pyroelectric detector has a range of $\leqslant 10\%$ during long-time operation and the FIR FEL provides a radiated power of 40–50 W during the macro-pulse having a duration of 3–4 μs in the wavelength range of 100–200 μm.

6. Summary

Simple, inexpensive and reliable classical magnetron-driven microtron has been upgraded to use as an injector of the compact wide-band FIR FEL. The developed stabilization of the microtron operation provides long-time stability in the lasing with a radiated power of 40–50 W in the macro-pulse having a duration of 3–4 μs in the wide-band range of the wavelength.

References

[1] G.M. Kazakevitch, et al., Nucl. Instr. and Meth. A 475 (2001) 599.
[2] G.M. Kazakevitch, et al., Proceedings of the 2001 Particle Accelerator Conference, V4, 2001, pp. 2739–2741.
[3] Joint US–CERN–JAPAN International School on Frontiers in Accelerator Technology, 9–18 September 1996, World Scientific, Singapore ISBN 981-02-3838-X.
[4] L.B. Lugansky, V.N. Melekhin, High Power Electronics, Nauka, Moscow, 1965, pp. 238–256 (in Russian).
[5] V.N. Zavorotilo, O.S. Milovanov, Accelerators, Vol. 16, 1977, Moscow, Atomizdat., pp. 34–37 (in Russian).
[6] G.M. Kazakevitch, et al., Nucl. Instr. and Meth. A 483 (2002) 331.

Available online at www.sciencedirect.com

SCIENCE @ DIRECT°

ELSEVIER Nuclear Instruments and Methods in Physics Research A 528 (2004) 120–124

**NUCLEAR
INSTRUMENTS
& METHODS
IN PHYSICS
RESEARCH**
Section A

www.elsevier.com/locate/nima

Measurements of pulse modulation in an ECM

K. Ronald*, A.W. Cross, A.D.R. Phelps, W. He, C.G. Whyte, J. Thomson, E. Rafferty, I.V. Konoplev

Department of Physics, University of Strathclyde, Glasgow G4 0NG, Scotland, UK

Abstract

We report on experiments which have recently been conducted at the University of Strathclyde investigating rapid amplitude modulations occurring in the microwave output radiation of an electron cyclotron maser (ECM). The experiment used an electron beam injected from a co-axial diode with knife-edged graphite cathode in the fringing field of an adjustable magnet system producing a beam of up to 175 kV and 140 A. The time evolution of the electron beam was measured as the cathode plasma expanded using a Faraday cup in conjunction with upstream beam interceptors as a function of the magnetic compression ratio. The ECM cavity was configured so that its length and the length of the interaction space could be readily adjusted. The microwave output signal was studied using special fast rectifying diode detectors, a high performance deep memory oscilloscope and cut-off filters. Steady-state output was observed at high magnetic compression ratios (16:1) at a frequency of 16 GHz corresponding to cyclotron resonant maser (CRM) coupling between the beam and the radiation in the expected $TE_{1,2}$ mode. At lower compression ratios modulation was observed after an initial steady-state period and shown by antenna pattern measurements to be associated with transverse mode competition in the microwave cavity.
© 2004 Elsevier B.V. All rights reserved.

PACS: 84.40.Ik; 52.59.Rz

Keywords: ECM; Pulse modulation; Gyrotron; Microwave source

1. Introduction

Electron cyclotron masers (ECMs) are microwave sources and amplifiers based on the cyclotron resonant maser (CRM) interaction between electrons gyrating in a static magnetic field and an electromagnetic wave [1–3]. They have advantages over conventional microwave sources and ampli-

fiers at high power levels (> 1 kW) and frequencies (> 10 GHz) as they require no delicate circuit components in high field regions. An example is the gyrotron oscillator producing up to 1 MW CW at 170 GHz for fusion heating.

In normal conditions a single radiation mode dominates the interaction [3] with the electron beam resulting in steady-state saturation. Sometimes however more than one longitudinal radiation mode can be excited leading to rapid amplitude modulations of the microwave output radiation [4–6]. Modulation can also be observed

---
*Corresponding author. Tel.: +44-141-548-3484; fax: +44-141-552-2891.

E-mail address: K.Ronald@strath.ac.uk (K. Ronald).

0168-9002/$ - see front matter © 2004 Elsevier B.V. All rights reserved.
doi:10.1016/j.nima.2004.04.031

when transverse mode competition occurs [7]. Similar modulations have been noticed in preliminary experiments at Strathclyde [8–10] when using explosive electron emission cathodes [11]. We present results of the behaviour of a similar experiment to understand how the pulse modulation arises.

2. Experimental configuration

The experiment used a co-axial diode with a centrally located knife-edged cathode and conical anode to form an electron beam. Graphite was chosen as the emitter material due to its ease of conversion into the explosive emission mode and the resultant slow expansion of the cathode plasma. The configuration of the diode is illustrated in Fig. 1. The cathode voltage of up to 175 kV was applied using a capacitor bank in a Marx configuration. An electron beam of up to 900 A was created in the diode although only a fraction of this beam passed into the ECM cavity. The maser was formed from a long piece of drawn oxygen-free high conductivity copper tube of 37 mm diameter which provided an envelope which could be evacuated to 10^{-6} mbar and also served as the output waveguide. The cavity diameter was chosen to be 32 mm to allow resonance between the electron beam and the $TE_{1,2}$ waveguide mode at a frequency of 16 GHz. The cavity was supported in the output waveguide

by a beam tunnel insert (of variable length) and output taper so that the length of the cavity and its overlap with the plateau region of the main magnet coil could be varied. The beam tunnel diameter was chosen to mitigate against any undesired fundamental resonances and was painted with microwave absorbing material to suppress harmonic coupling. The microwave radiation was coupled out from the experiment using a cylindrical horn antenna with an output diameter of 70 mm. The cavity arrangements are illustrated in Fig. 1. The cavity magnetic field was provided by a superconducting solenoid and the compression ratio was controlled by an additional water-cooled solenoid in the vicinity of the diode as illustrated in Fig. 2.

Diode voltage was monitored using a hybrid divider with copper sulphate solution high-voltage section and metal film resistors on the low-voltage stage. The diode current was measured using a Rogowski belt wound on a nylon former and calibrated against a resistive current shunt. The electron beam in the interaction cavity was measured using a Faraday cup to determine the magnitude of the electron beam passing an upstream 'interceptor' consisting of a metal plate with a hole drilled through its centre. The microwave output amplitude was recorded by sampling a fraction of the output power from the experiment using receiving horn antenna to direct the radiation to a rectifying diode. The detector calibration was checked with a fast pulse

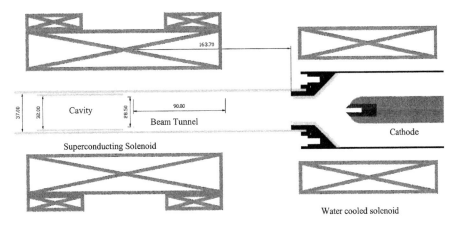

Fig. 1. Partial assembly drawing of the experiment showing the diode, cavity and magnet coils.

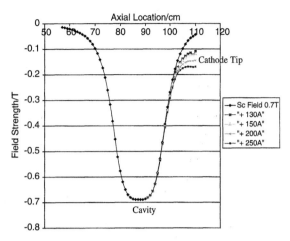

Fig. 2. Magnetic field profile on experiment axis as a function of the current in the diode solenoid.

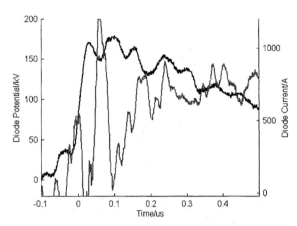

Fig. 3. Electrical evolution of the diode due to the formation of the cathode plasma, the voltage trace is black, current is grey.

microwave generator and the video output of the microwave envelope compared to the AC waveform observed by a 20 GHz digital oscilloscope and found to be in good agreement for pulses rising in 500 ps. Frequency was measured using cylindrical waveguide cut-off filters. The operating mode was checked by measuring the antenna pattern launched from the output of the oscillator and comparing with the operating frequency. By accounting for the effective antenna aperture at the measured frequency and the settings of the attenuators (calibrated by a scalar network analyser) the antenna patterns were integrated to estimate the output power and efficiency.

3. Measurements

Fig. 3 shows the evolution of the diode as accelerating voltage is applied from the Marx generator. A peak voltage of up to 175 kV could be attained with a rise time of ∼50 ns. The diode current shows some initial displacement spikes caused by the capacitance between the cathode and anode assemblies. The peak diode current was up to 900 A. The electron beam diagnostic revealed that the fraction of the diode current coupled into the interaction space varied as a strong function of the magnetic compression factor. At a maximum compression factor of

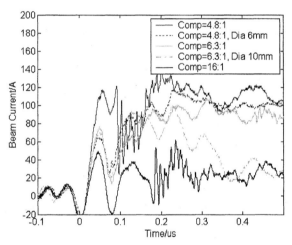

Fig. 4. Electron beam current as a function of time at different compression ratios illustrating the evolution of the beam diameter due to the cathode plasma.

16:1 the beam current was 20 A with a diameter in the cavity region of up to 14 mm. For a compression of 6.3:1 the beam current was 100 A with a diameter <8 mm but expanded to >10 mm after 300 ns. For compression factors <6:1 the beam diameter was initially <6 mm but exceeded 8 mm after 200 ns and 10 mm after 300 ns. At magnetic compression <4.8:1 the current exceeded 100 A and the beam diameter did not exceed 6 mm for the duration of the pulse. Fig. 4

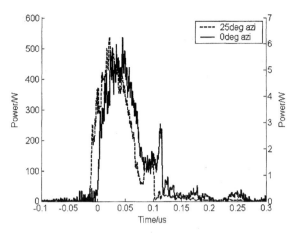

Fig. 5. Single mode output at a compression ratio of 16:1, dashed line on the right hand y-axis, receivers in azimuthal polarisation.

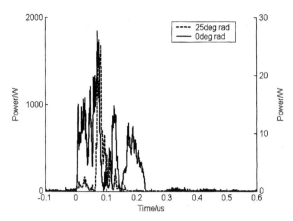

Fig. 6. Spiking output at a compression ratio of 4.8:1, dashed line on the right hand y-axis, receivers in radial polarisation.

illustrates some measurements of the electron beam current. Again a significant distortion occurs at the start of the pulse due to the capacitance between the Faraday cup and the anode plus the inductance of the sensor output wire.

The stability of the microwave output pulse amplitude was studied by varying the length of the interaction space and cavity, the detuning between the microwave field and the electron cyclotron frequency and the compression ratio of the magnetic field profile. It was found that the compression ratio had a very significant effect on the output radiation behaviour. Results are illustrated in Fig. 5 showing the stable output obtained with a 0.25 m long cavity and a 9 cm long interaction space (set by the plateau of the main solenoid) and a 16:1 compression ratio. The frequency was measured to be 16 GHz with the cut-off filters and the antenna patterns were consistent with expectations for the $TE_{1,2}$.

Maintaining all other parameters it was found that reducing the compression ratio to 4.8:1 resulted in a marked change in the output signal. After a period of initially stable output the experiment exhibited strong amplitude variations. Furthermore the time evolution of the amplitude of the microwave signal became a strong function of the position of the receiving antenna, behaviour that is only possible if the output signal contains more than one transverse mode. Analysis of the

data showed that the onset of the pulse modulations observed at $0°$ in the antenna pattern concurred with the appearance of a strong spike at $25°$ in the radial polarisation of the antenna pattern (a location where the antenna pattern of the $TE_{1,2}$ is weak), Fig. 6. The frequency of this spike was measured and found to also be 16 GHz. This strongly indicated that the competing mode is the $TE_{4,1}$ of the interaction cavity ($TE_{1,2}$ and $TE_{4,1}$ cut-off frequencies are, respectively, 15.91 and 15.87 GHz).

In the steady-state regime of operation with 16:1 compression ratio, the experiment had an output power of 300 kW and an electronic efficiency of 10%. The initial steady-state operation at a compression ratio of 4.8:1 had a power of >650 kW corresponding to an efficiency of 4%, the peak power in the spiking regime could be more than double this value.

4. Conclusions

New experiments to investigate the stability of the output radiation from an ECM oscillator with a knife-edged cathode have been conducted. The impact of the length of the cavity, interaction space and degree of magnetic compression were investigated. The greatest effect was found to be due to the magnetic compression ratio. At high levels (16:1) of magnetic compression between the diode and the cavity, the beam current was 20 A and short 300 kW pulses of stable, single mode ($TE_{1,2}$), single frequency (16 GHz), output radiation were obtained. Increasing the diode magnetic field (reducing the compression ratio) to 4.8:1 gave a beam current > 100 A and produced longer pulses of output radiation which were initially in single mode but which broke into a modulated regime where two transverse modes operated. This is consistent with ongoing 2D simulations undertaken with the PIC code KARAT which have hitherto illustrated automodulation only with the excitation of multiple transverse modes.

References

[1] A.V. Gaponov, Izv. VUZ. Radiofizika 2 (1959) 450 and 837.

[2] J.L. Hirshfield, V.L. Granatstein, IEEE Trans. Microwave Theory Tech. MTT-25 (1977) 522.

[3] P. Sprangle, A.T. Drobot, IEEE Trans. Microwave Theory Tech. MTT-25 (1977) 528.

[4] N.S. Ginzburg, et al., Int. J. Electronics 61 (1986) 881.

[5] M.I. Airila, et al., Phys. Plasmas 8 (2001) 4608.

[6] T.H. Chang, et al., Phys. Rev. Lett. 87 (2001) 064802.

[7] O. Dumbrajs, et al., IEEE Trans. Plasma Sci. PS-27 (1999) 327.

[8] K. Ronald, et al., J. Phys. D 34 (2001) L17.

[9] K. Ronald, et al., Proceedings of the ITG Conference on Vacuum Electronics and Displays, 2001, pp. 159–162.

[10] K. Ronald, et al., Proceedings of the 27th International Conference on Infrared and Millimeter Waves, 2002, pp. 41–42.

[11] K. Ronald, et al., IEEE Trans. Plasma Sci. PS-26 (1998) 375.

Available online at www.sciencedirect.com

SCIENCE @ DIRECT®

ELSEVIER   Nuclear Instruments and Methods in Physics Research A 528 (2004) 125–129

NUCLEAR
INSTRUMENTS
& METHODS
IN PHYSICS
RESEARCH
Section A

www.elsevier.com/locate/nima

Generation of coherent far infra-red radiation utilising a planar undulator at the 4GLS prototype

Christopher Gerth[a],*, Brian McNeil[b]

[a] ASTeC, Daresbury Laboratory, Warrington WA4 4AD, UK
[b] Department of Physics, University of Strathclyde, Glasgow G2 0NG, UK

Abstract

An Energy Recovery Linac (ERL) Prototype facility to be built at Daresbury Laboratory serves as a testbed for the study of beam dynamics and accelerator technology important for the design and construction of the 4th Generation Light Source (4GLS). This paper describes the possibility of utilising a planar undulator at the 4GLS prototype facility for the generation of coherent synchrotron radiation in the far infra-red region.
© 2004 Elsevier B.V. All rights reserved.

PACS: 41.60.Cr

Keywords: Coherent spontaneous emission; CSR; THz radiation; T-rays

1. Introduction

Recent advances in accelerator technology have led to steady progress in the generation of high-brightness electron bunches (short bunch length, high peak-current, small transverse emittance) much sought after for driving free-electron lasers (FELs). During circular motion, an electron bunch starts to emit Coherent Synchrotron Radiation (CSR) at wavelengths equal to or longer than the bunch length. The emission of CSR can lead to severe degradation of the beam quality and much effort has been devoted to mitigating the CSR

effects for transport of electron beams in accelerators [1].

CSR is emitted in the far-infrared (FIR) spectral range, i.e. terahertz (THz) region. This region is also referred to as the "THz gap" since it is not directly accessible with both electronic and photonic devices, and thermal sources are very weak. Accelerator-based facilities are promising sources to overcome this limitation. The first evidence of CSR was reported in 1989 [2]. CSR emission has also been observed at several electron storage rings (Refs. [3,4] and references therein). Recently, broadband high-average power CSR from sub-picosecond electron bunches has been measured at the last dipole magnet of the magnetic bunch compressor of the Jefferson Laboratory FEL [5].

Coherent, high-power, narrow-band FIR radiation can be produced by passing sub-picosecond

*Corresponding author. Tel.: +44-1925-603739; fax: +44-1925-603192.

E-mail address: c.gerth@dl.ac.uk (C. Gerth).

URL: http://www.astec.ac.uk.

electron bunches through an undulator [6,7]. In this report, we discuss the predicted performance of a planar wiggler proposed as part of the Energy Recovery Linac Prototype (ERLP) facility at Daresbury Laboratory.

2. Undulator parameters

Daresbury Laboratory has been given funding for the construction of an Energy Recovery Linac Prototype (ERLP) facility [8] that operates at a target electron beam energy of 30–50 MeV. The injector comprises a DC photocathode gun and a super-conducting booster cavity for high repetition rates. The ERLP serves as a testbed for the R&D neeeded for the design study of the 4th Generation Light Source (4GLS) [9].

The main objectives for the ERLP are the demonstration of energy recovery from an electron bunch with an energy spread induced by a radiation source and the development of expertise in such technology. Another aim is the simultaneous operation and synchronisation of multiple radiation sources. To achieve this it is proposed to pass the electron beam through both an IR oscillator FEL [10] (based on a wiggler on loan from Jefferson Laboratory which has previously been used in the IR Demo FEL [11]) and a planar undulator [12] for the generation of CSR which will be de-commissioned from the SRS soon. The latter undulator is a pure-permanent magnet undulator with a variable gap. For a gap of 20–80 mm, the on-axis magnetic peak field $\hat{B}$ and the undulator parameter K can be calculated from Eq. (4) given in Ref. [12]. The wavelength of the fundamental radiated on-axis is then given by the resonance condition:

$$\lambda = \frac{\lambda_w}{2\gamma^2}\left(1 + \frac{K^2}{2}\right) \tag{1}$$

where λ_w is the period length and $\gamma = E/(m_e c^2)$ the electron beam energy in units of the electron rest mass. For the given parameters, the wavelength range 10–300 μm would be covered. The main parameters of the undulator and the proposed ERLP design are listed in Table 1.

Table 1
Main parameters of the ERL prototype design and undulator

Parameter	
Electron beam	
Beam energy E	30–50 MeV
Bunch charge Q	>80 pC
Bunch length (FWHM)	<0.5 ps
Undulator	
Period length λ_w	100 mm
Number of periods N_w	10
Length	1.0 m
Gap	20–80 mm
Peak magnetic field $\hat{B}$	0.7–0.1 T
Undulator parameter K	6.4–1.0
Wavelength range at 30 MeV	300–16 μm
Wavelength range at 40 MeV	175–12 μm
Wavelength range at 50 MeV	113–10 μm

3. Undulator output

The total power P radiated at a frequency ω by an electron bunch moving through an undulator in the absence of any FEL interaction is given by [7,13]

$$P(\omega) = p(\omega)[N_e + N_e(N_e - 1)|\bar{F}(\omega)|^2] \tag{2}$$

where $p(\omega)$ is the radiation power emitted by one electron and N_e the number of electrons in the bunch. The form factor $\bar{F}(\omega)$ is the Fourier transform of the longitudinal charge distribution of the electron bunch and describes the effects of CSR. The emitted spectrum depends on the form factor $\bar{F}(\omega)$ and, hence, on the particular charge distribution in the electron bunch.

For simplicity we assume a Gaussian charge distribution. This approximation is not strictly valid in practice but is useful for estimating the total power and spectral distribution. The Fourier transform then takes the form

$$\bar{F}(\omega) = e^{-(1/2)\omega^2\sigma^2} \tag{3}$$

where σ is the rms electron bunch length in the time domain. The form factor has been calculated for the wavelength range that is spanned by the undulator at a beam energy of 40 MeV. The results are shown in Fig. 1 for four different bunch lengths (0.1, 0.2, 0.3 and 0.4 ps (FWHM)). For the radiation emitted into the cone of half angle $\theta_{cen} = \sqrt{1 + K^2/2}/\sqrt{\gamma^2 N_w}$ the relative spectral bandwidth (FWHM) is given by $\Delta\lambda/\lambda = 0.89/N_w = 0.089$ [13]. The normalized spectral distribution for a fixed gap of 27 mm is also given in Fig. 1. The energy radiated into the central cone θ_{cen} by a single electron passing through a planar undulator is given, in SI units, by [7,13]

$$E_{cen} \simeq \frac{2\pi e^2 f_B^2 \xi}{\varepsilon_0 \lambda_0} \qquad (4)$$

where λ_0 is the resonance wavelength (Eq. (1)) and $f_B = J_0(\xi) - J_1(\xi)$ a parameter that accounts for the additional axial electron motion in a planar undulator with J_n the Bessel function of nth order and $\xi = K^2/(4 + 2K^2)$. The pulse energy (radiation energy per electron bunch) radiated at each resonance wavelength over the entire wavelength range of the undulator can then be calculated from Eq. (2). The result is shown in Fig. 2 for a beam energy of 40 MeV, a bunch charge of 80 pC and four different electron bunch lengths which are in accordance with the form factor calculations in Fig. 1. The strong effect of the form factor on the radiated pulse energy can clearly be seen. The

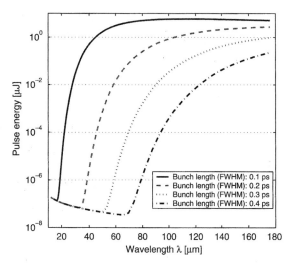

Fig. 2. Variation of the pulse energy over the wavelength range of the undulator calculated for four different electron bunch lengths (bunch charge: 80 pC; beam energy $E = 40$ MeV).

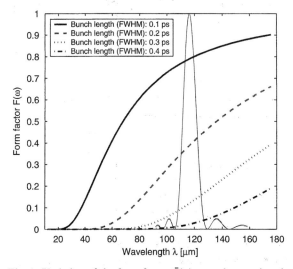

Fig. 1. Variation of the form factor $\bar{F}(\omega)$ over the wavelength range covered by the undulator at a beam energy of 40 MeV. A Gaussian longitudinal bunch profile was assumed. In addition, the normalized spectral distribution for a fixed gap of 27 mm is shown.

pulse energies vary by several order of magnitude over the accessible wavelength range of the undulator. At a given wavelength, i.e. for a fixed gap, the pulse energy is very sensitive to the bunch length and, hence, the undulator may be able to be utilised for the measurement of ultra-short bunch lengths [13,14].

Due to slippage in the undulator—the radiation advances by one wavelength per undulator period—the pulse duration in the CSR regime is $\approx N_w \lambda_0/c$ and the resulting pulse duration is about 1–10 ps.

The radiated pulse energy depends strongly on the bunch charge. This is demonstrated in Fig. 3 for four different bunch charges and a bunch length (FWHM) of 0.2 ps. For bunch charges over 200 pC and wavelengths longer than 120 μm the pulse energy surpasses the 10 μJ level.

4. 1-D time-dependent studies

The previous analysis neglects the effects of any collective FEL interaction that may occur between the emitted radiation and the electrons. Such interaction may effect the charge distribution

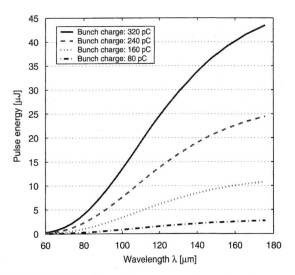

Fig. 3. Variation of the pulse energy over the wavelength range of the undulator calculated for four different electron bunch charges (bunch length (FWHM): 0.2 ps; beam energy $E = 40$ MeV).

within the electron pulse and so alter the characteristics of the radiation emission. In an attempt to model such effects the various parameters used to characterise the FEL interaction were calculated and the 1-D numerical model of Ref. [15] was employed to give estimates of the emitted radiation energy. This model uses unaveraged equations of motion for the electrons and is able, under the 1-D assumption validity, to describe radiation evolution at the sub-wavelength scale so enabling modelling of CSR effects in ultrashort electron pulses.

We assumed an 80 pC Gaussian electron pulse of duration 0.2 ps FWHM with beam energy 40 MeV, and tuned the undulator to give a resonant wavelength of $\sim 175\,\mu$m. Assuming the undulator has natural focussing in both planes and a normalised beam emittance of 10 mm mrad, the matched beam radius is $\sim 223\,\mu$m. The 1-D FEL parameter of Ref. [15] is then $\rho \approx 0.11$ and the gain length is $l_g \approx 0.075$ m. It is immediately apparent from the relatively large value of ρ that the 1-D model of Ref. [15] is close, if not beyond, the Compton limit of validity in which space charge

effects may be neglected [16]. Furthermore, the gain length is significantly longer than the Rayleigh range $Z_r \approx 894\,\mu$m and it would be expected that radiation diffraction effects would dominate any longitudinal FEL interaction. The model of Ref. [15] could not therefore be expected to model in any meaningful way the emission of radiation from the short electron pulse as described. This has been confirmed in simulations which yield a radiation pulse energy of approximately two orders of magnitude above that of the analysis of the previous section.

It would be of interest to extend the model of Ref. [15] to include transverse effects and so be able to model any collective FEL effects in systems such as that described in this paper. Work is currently underway to develop such a model.

5. Conclusions

An existing planar undulator could be employed at the ERL Prototype facility for the generation of coherent FIR radiation in the spectral range 50–300 μm. For bunch charges above 80 pC and electron bunch lengths shorter than 0.4 ps (FWHM), this undulator will emit 1–10 ps long pulses with energies of up to several μJ with a relative spectral bandwidth of $\simeq 9\%$ (FWHM). As the requirements on the beam quality in terms of energy spread and emittance are not stringent for the generation of CSR, it may be possible to operate the undulator following the exit of an IR oscillator FEL and use the radiation of both devices for synchronisation or pump–probe experiments. The undulator could also be used for electron pulse shape/length measurements. Further work is required to describe the CSR emission in time dependent models.

Acknowledgements

The authors are grateful to E. Saldin, E. Schneidmiller, M. Yurkov and N. Thompson for stimulating discussions.

References

[1] J.S. Nodvick, D.S. Saxon, Phys. Rev. 96 (1954) 180.

[2] T. Nakazato, et al., Phys. Rev. Lett. 63 (1989) 1245.

[3] J.M. Byrd, et al., Phys. Rev. Lett. 89 (2002) 224801.

[4] M. Abo-Bakr, et al., Phys. Rev. Lett. 90 (2003) 094801.

[5] G.L. Carr, et al., Nature 420 (2002) 153.

[6] D. Bocek, M. Hernandez, SLAC-PUB-7106 (1995).

[7] B. Faatz, et al., Nucl. Instr. and Meth. A 475 (2001) 363.

[8] M.W. Poole, et al., Proc. PAC 2003, Portland.

[9] M.W. Poole, B.W.J. McNeil, Nucl. Instr. and Meth. 507 (2003) 489.

[10] N.R. Thomson, Nucl. Instr. and Meth. A, (2004) these proceedings, Part II.

[11] G.R. Neil, et al., Phys. Rev. Lett. 84 (2000) 662.

[12] M.W. Poole, et al., Nucl. Instr. and Meth. 208 (1983) 143.

[13] G. Geloni, et al., DESY-Report 03-031 (2003);
G. Geloni, et al., Nucl. Instr. and Meth. A, (2004) these proceedings.

[14] C.P. Neuman, et al., Phys. Rev. STAB 3 (2000) 030701.

[15] B.W.J. McNeil, G.R.M. Robb, Phys. Rev. E 65 (2002) 046503;
B.W.J. McNeil, G.R.M. Robb, Phys. Rev. E 66 (2002) 059902(E).

[16] R. Bonifacio, C. Pelligrini, L.M. Narducci, Opt. Commun. 50 (1984) 373.

Available online at www.sciencedirect.com

Nuclear Instruments and Methods in Physics Research A 528 (2004) 130–133

ELSEVIER

NUCLEAR
INSTRUMENTS
& METHODS
IN PHYSICS
RESEARCH
Section A

www.elsevier.com/locate/nima

High-intensity far-infrared light source using the coherent transition radiation from a short electron bunch

Shuichi Okuda[a,*], Takao Kojima[a], Ryoichi Taniguchi[a], Soon-Kwon Nam[b]

[a] The Research Institute for Advanced Scienece and Technology, Osaka Prefecture University, 1-2 Gakuen-cho, Sakai, Osaka 599-8570, Japan

[b] Department of Physics, Kangwon National University, Chunchon Kangwon-do 200-701, Republic of Korea

Abstract

The characteristics of the far-infrared coherent transition radiation light source used for absorption spectroscopy have been investigated. A single-bunch electron beam of an L-band linear accelerator has been used in the experiments. The energy and the electron charge in a bunch of the beam have been 27 MeV and 13.5 nC/bunch, respectively. The accelerator has been operated to obtain high-intensity radiation. The spectrum of the coherent transition radiation from the beam has been measured at wavelengths of 0.4–2 mm and the electron bunch form factor has been given from the spectrum. For the typical geometrical configurations of the light source the light spectrum has been obtained by using the bunch form factor. The characteristics of the light source have been compared with those of the coherent synchrotron radiation.
© 2004 Elsevier B.V. All rights reserved.

PACS: 41.60.−m; 42.25.Kb; 29.17.+w; 42.72.Ai

Keywords: Coherent radiation; Transition radiation; Submillimeter to millimeter wave light source; Bunch form factor; Electron linear accelerator

1. Introduction

Synchrotron and transition radiation emitted from a bunched electron beam becomes coherent and highly intense at wavelengths about or longer than the bunch length. The radiation has continuous spectrum in a submillimeter to millimeter wavelength range. Recently, far-infrared coherent radiation light sources have been applied to absorption spectroscopy [1–4].

One of the main factors determining the intensity of the coherent radiation is the electron bunch shape. In our previous work a monitoring system has been developed [5] to evaluate the bunch shape of the single-bunch electron beam of the L-band linear accelerator (linac) in Osaka University. The operational conditions of the linac have been investigated for the single-bunch and multibunch beams to generate the high intensity coherent radiation over a relatively wide wavelength range [6]. Thus investigating the beam

*Corresponding author. Tel.: +81-72-254-9846; fax: +81-72-254-9938.

E-mail address: okuda@riast.osakafu-u.ac.jp (S. Okuda).

conditions, the light sources using the coherent synchrotron and transition radiation have been developed. In most of the recent experiments for absorption spectroscopy the coherent transition radiation light source has been used because the system of the source is relatively simple. However, the characteristics of the light source have not been investigated.

The present work has been performed to investigate the characteristics of the coherent transition radiation light source. In the experiment the electron bunch form factor has been obtained from the spectrum of the coherent transition radiation measured with a bunch shape monitoring system. From the results the spectrum of the coherent transition radiation used for light sources has been evaluated.

2. Transition radiation from an electron and a short electron bunch

The intensity of the coherent radiation emitted from an electron bunch is approximately given by $p(\lambda)N^2f(\lambda)$, where $p(\lambda)$ is the intensity of radiation emitted from an electron, N the number of electrons in the bunch, λ the wavelength of radiation. In this expression $f(\lambda)$ is the longitudinal bunch form factor given by $|\int S(z)\exp(i2\pi z/\lambda)\,dz|^2$, where $S(z)$ is the normalized distribution of electrons in the bunch as a function of the length on the beam axis. The bunch form factor has a value below 1 and is proportional to the intensity of the coherent radiation. In the previous work the bunch form factor of the single-bunch beam of the linac in Osaka University has been found to agree with that for a triangular bunch shape [5]. The intensity of radiation from such a triangular bunch is relatively high at wavelengths shorter than the electron bunch length. Hence, in order to expand the wavelength range of the light source, it is important to monitor and control the bunch shape.

The transition radiation emitted from an electron passing through two radiators is schematically shown in Fig. 1. In this figure L is the distance between the radiators and θ is the angle of emission at the radiator 1. In general cases the

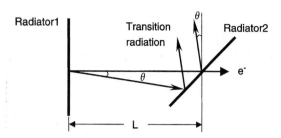

Fig. 1. Schematic diagram showing the transition radiation emitted from an electron passing through two radiators.

direction of emission is symmetrically distributed around the axis of the electron trajectory. When the electron crosses the radiator, transition radiation is emitted backward and forward from the surfaces of the radiator. Transition radiation emitted from the two radiators shown in Fig. 1 is superposed. The spectrum of the transition radiation emitted from the electron is obtained theoretically [7].

In the present work the bunch form factor is obtained from the spectrum of the coherent transition radiation measured by using a bunch shape monitoring system, according to the theory on the emission of transition radiation described above. Thereafter, the spectrum of the coherent transition radiation for the typical geometrical conditions of the light source is obtained from the form factor by calculation.

3. Method for measuring a radiation spectrum

The single-bunch electron beam at an energy of 27 MeV generated with an L-band linac in Osaka University [8] is used in the present experiment. The electron charge of the beam is 13.5 nC/bunch ($N = 2.2 \times 10^{10}$). The linac is operated so as to obtain high intensity radiation. In this operational mode of the linac the energy spread of the electron beam has increased from about 1% to above 10% FWHM, and the bunch length has decreased from 29 to 13 ps FWHM, compared to the normal operational mode.

The spectrum of the coherent transition radiation is measured with the bunch shape monitoring system schematically shown in Fig. 2. In this figure a metal foil and a metal-coated mirror are used for

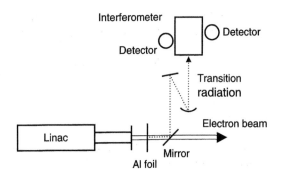

Fig. 2. Schematic diagram of the bunch shape monitoring system used for measuring the spectrum of the coherent transition radiation.

the radiator 1 and 2 shown in Fig. 1, respectively. The surface of the foil is normal to the beam axis and the mirror surface is oriented at an angle of $\pi/4$ from the axis. The coherent transition radiation from the foil in the forward direction and that from the mirror in the backward direction is superposed. The distance between the two radiators is 150 mm. The beam trajectory observed is relatively stable, which is probably due to the fact that the beam is transported straight from the accelerating waveguide and the trajectory is not affected by the change in the beam energy. The radiation spectrum is measured with a Martin-Pupplet interferometer and liquid-helium cooled silicon bolometers in a wavelength range of 0.4–2 mm at a wavenumber resolution of $0.2\,\mathrm{cm}^{-1}$. In the measurement the intensity fluctuation of the incident light due to the change in the electron beam conditions is monitored by measuring the intensity of light from the first splitter in the interferometer. This measurement system is calibrated by using a high-pressure mercury lamp. In the geometrical configurations the radiation at angles of emission smaller than 21 mrad is accepted by the interferometer. These relatively simple configurations have been determined by the requirement of accuracy in the measurement.

4. Experimental results and discussion

The wavelength dependence of the electron bunch form factor obtained from the spectrum

of the coherent transition radiation measured in the present experiment is shown in Fig. 3. The results show that the intensity of radiation is enhanced by 5–7 orders of magnitude by the coherence effect. The relatively large fluctuation observed on the curve is due to the specific bunch shape. The values of the form factor are about two orders of magnitude (by 390 times at $\lambda = 1$ mm) larger than those previously obtained in the normal operational mode [5]. These results and relatively large energy spread of the beam suggest that relatively strong bunch compression has been made in the accelerating waveguide in the present operational mode. At the longer wavelengths the form factor is estimated to increase with the wavelength and close to unity at wavelengths longer than the bunch length (3.9 mm FWHM).

In the present measurement system only a small part of radiation has been measured by the reason mentioned before. Now, typical geometrical configurations for a light source are considered, where the coherent transition radiation emitted from the two radiators shown in Fig. 2 at emission angles θ smaller than 50 mrad is transported and is used for a light source. In this case the spectrum of the coherent radiation obtained by calculation for the same electron bunch is shown in Fig. 4. In the wavelength range of the present measurement (0.4–2 mm) the total energy of the radiation is

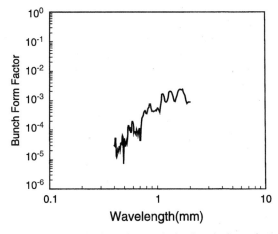

Fig. 3. Wavelength dependence of the bunch form factor of a 27 MeV single-bunch beam at a charge of 13.5 nC/bunch obtained from the spectrum of the coherent transition radiation.

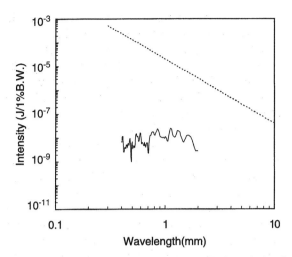

Fig. 4. Spectrum of the coherent transition radiation calculated from the results shown in Fig. 3 for the typical light source configurations where radiation at emission angles θ smaller than 50 mrad is used: the dotted line shows the spectrum obtained by calculation in the case of the bunch form factor of unity.

1.6×10^{-6} J. In this figure the dotted line shows the spectrum when the bunch form factor is unity.

In the previous work the spectrum of the coherent synchrotron radiation used for a light source has been measured [3]. While the beam conditions and the geometrical conditions in the measurement are different from those of the present work, the intensity of the light is roughly 1–2 orders of magnitude larger than that of the coherent transition radiation. The main causes of the difference in the radiation intensity are probably the higher intensity of the incoherent synchrotron radiation and the additional bunch compression in a bending magnet in the case of the synchrotron radiation. In most of our experiments for absorption spectroscopy, the coherent transition radiation has been used because the radiator system is simple and the radiation is relatively stable (electron beam trajectory is not affected by the energy fluctuation in the beam).

The intensity of the incoherent transition radiation in a relatively long wavelength region is less dependent on the electron beam energy compared

to that of the synchrotron radiation. A compact coherent radiation light source using electron beams at energies of about 10 MeV will be established in Osaka Prefecture University.

5. Conclusions

The characteristics of the far-infrared coherent transition radiation light source were investigated. At wavelengths of 0.4–2 mm the spectrum of the light source was obtained for the typical configurations of light source. The energy of radiation was found to be 1–2 orders of magnitude lower than that of the coherent synchrotron radiation light source. The coherent transition radiation light source was characterized by the simple system and the relatively stable radiation intensity.

Acknowledgements

The authors are indebted to Ms. M. Nakamura and Mr. M. Takanaka for their help in the experiments, and S. Nam wishes to acknowledge the financial support of Korea Research Foundation Grant.

References

[1] T. Takahashi, et al., Rev. Sci. Instr. 69 (1998) 3770.
[2] K. Yokoyama, et al., Proceedings of the 20th International Free-Electron Laser Conference, Williamsburg, USA, 1998, pp. II-17.
[3] S. Okuda, et al., Nucl. Instr. and Meth. A 445 (2000) 267.
[4] M. Takanaka, et al., Proceedings of the 23rd International Free-Electron Laser Conference, Darmstadt, Germany, 2001, pp. II-95.
[5] M. Nakamura, et al., Nucl. Instr. and Meth. A 475 (2001) 487.
[6] S. Okuda, et al., Proceedings of the 23rd International Free-Electron Laser Conference, Darmstadt, Germany, 2001, pp. II-49.
[7] Y. Shibata, et al., Phys. Rev. E 49 (1994) 785.
[8] S. Okuda, et al., Nucl. Instr. and Meth. A 358 (1995) 248.

Available online at www.sciencedirect.com

Nuclear Instruments and Methods in Physics Research A 528 (2004) 134–138

NUCLEAR INSTRUMENTS & METHODS IN PHYSICS RESEARCH
Section A

ELSEVIER

www.elsevier.com/locate/nima

Simulations of beam–beam interaction in an energy recovery linac

K. Masuda*, S. Matsumura, T. Kii, K. Nagasaki, H. Ohgaki, K. Yoshikawa, T. Yamazaki

Institute of Advanced Energy, Kyoto University, Gokasho, Uji, Kyoto 611-0011, Japan

Abstract

Interactions between accelerated beams and reinjected spent beams in an energy recovery linac (ERL) were studied numerically aiming at an enhanced efficiency and a reduced shielding load. With low injector energy, 1 MeV, the orthodox scheme for reinjecting spent beams, on one hand, is found to result in undesirable beam-loading deviations in an ERL owing to velocity variations of the beams. On the other hand, a counter reinjection scheme is found to be effective for suppressing the deviations, but it is also found to result in an appreciable degradation in a fresh beam's properties due to enhanced beam–beam interactions. To realize the advantages of both of these two schemes, a new combined scheme for an ERL-based FEL is proposed. Also the effects of spent beams' properties on fresh beams' emittances are discussed.
© 2004 Elsevier B.V. All rights reserved.

PACS: 41.60.Cr; 29.27.Bd

Keywords: Energy recovery; Beam direct energy conversion; Linac; Free-electron laser

1. Introduction

For both reducing the shielding load and further enhancing the electrical utility power service efficiency in an energy recovery linac (ERL) for an FEL [1,2], reducing injector energy is effective and essential, since the dumped beam energy is reduced accordingly.

At low-energy injection to an ERL shown schematically in Fig. 1(a), however, velocity

variations of the electron beams cease being negligible in terms of beam-loading deviations along the accelerator tube. To compensate for these adverse effects, the counter reinjection scheme of the spent beam [3], i.e., reinjecting the spent beam into the tube in the opposite direction to the fresh beam from the injector as shown in Fig. 1(b), was found effective [4]. However, this scheme would also result in a considerable degradation of the fresh beam's emittance exclusively due to enhanced space charge forces through interaction with the spent beam [4].

From these results, a combination of the orthodox and counter reinjection schemes shown

*Corresponding author. Tel.: +81-774-38-3442; fax: +81-774-38-3449.

E-mail address: masuda@iae.kyoto-u.ac.jp (K. Masuda).

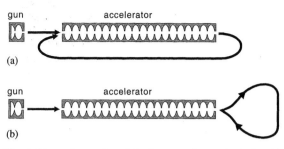

Fig. 1. Two schemes for reinjecting spent beams in ERLs: (a) orthodox and (b) counter reinjection schemes.

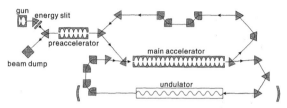

Fig. 2. An ERL-based FEL in a combined scheme of orthodox and counter reinjections.

Table 1
Design parameters of the 1- and 4-m accelerator tubes

Frequency	2.856 GHz
Number of cavities	20/80
Tube inner diameter	81.1 mm
Disk aperture diameter	20.0 mm
Disk spacing	49.5 mm
Disk thickness	3.00 mm
Gradient	7.2 MeV/m
Shunt impedance per cavity	2.80 MΩ/cell

in Fig. 2 arises to minimize the spent beam energy directed to the dump with the least degradation in the fresh beam's properties, i.e., to adopt the counter and orthodox reinjection schemes for the lower and higher energy sections, respectively.

In this study, beam–cavity and beam–beam interactions in an ERL in the combined scheme were studied by use of a two-dimensional time-dependent particle-in-cell (PIC) code [5,6].

2. Numerical model

Fig. 2 shows the combined scheme, consisting of a microwave gun having a 1.0-MeV output, a preaccelerator tube 1 m in length, up to around 8-MeV beam energy, and a 4-m main accelerator tube. Both tubes are S-band standing-wave tubes whose design parameters are listed in Table 1. The resultant beam of around 37 MeV is reinjected into the main tube in the same direction as the fresh beam (orthodox reinjection) to reduce its energy to about 8 MeV. It is then led to the 1-m tube in the counter reinjection scheme and is finally dumped with an energy of around 1 MeV.

In the simulations, the beam optics in the bending or focusing magnets were neglected in order to focus attention on the beam dynamics in the tubes with the least computational effort. In the PIC simulations in the accelerator tubes, the beam-induced fields and cavity fields are handled two-dimensionally in space, and, thus, the transverse misalignment of the two beams with respect to the symmetry axis of the tubes was neglected. Also neglected was the beam-loading effect on the cavity fields, i.e., a constant acceleration gradient in the tubes was assumed, while eigenmode patterns in the cavities were also provided by a two-dimensional eigenmode solver [7].

First, the beam-loading variation and the degradation of the fresh beam's emittance in the tubes were evaluated in comparison with a higher injection energy, 5.5 instead of 1.0 MeV, and with the orthodox reinjection scheme as well.

Second, the time-evolution of the output beam emittance from the 1-m tube was simulated, and the effect of the spent beam's properties on the fresh beam's emittance growth was taken into account. The effect of the emittance and energy spread induced by the undulator were also studied.

For the fresh beams to be injected into the 1-m tube, the output beams from 1.5- and 4.5-cavity S-band microwave guns were calculated by the PIC code and the eigenmode solver. The geometry of the cavities in the gun [6,8] used in the KU-FEL project [9] was adopted for both guns. The highly energetic components then were used as the input fresh beams. The three beams listed in Table 2 were obtained, namely a 5.5-MeV beam by the 4.5-cavity gun, and two 1.0-MeV beams by the 1.5-cavity gun with different emission currents from

Table 2
Properties of the beams from the microwave guns

Number of cavities	4.5	1.5	1.5
Averaged energy (MeV)	5.5	1.0	1.0
Current (A)	0.25	0.56	0.087
Energy spread (%)	2.6	7.6	3.4
Normalized rms emittance (π mm mrad)	3.6	4.3	0.69

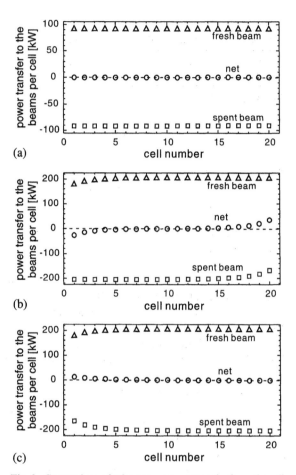

Fig. 3. Comparison of microwave power transfer from the cells to the beams along the 1-m tube, among the injection energies from the gun, and the reinjection schemes: (a) 5.5 MeV, orthodox, (b) 1.0 MeV, orthodox, and (c) 1.0 MeV, counter.

the cathode and different acceptances of the energy slit.

3. Results and discussions

3.1. Comparisons among reinjection schemes, beam energies, and currents

Fig. 3 shows the power transferred from the cavities to the beams along the 1-m preaccelerator tube, comparing the reinjection schemes and injection energies. In these comparisons, the accelerated output beams from the 1-m tube without spent beam reinjection were calculated by separate runs of the PIC code and used as the input conditions of the reinjected spent beams, omitting their time-variation or their degradation induced by the undulator. These omissions will be discussed in the next subsection.

In the orthodox reinjection scheme, the 1.0-MeV (0.56 A) injection is found to result in net beam-loading deviations in the cavities at the both tube ends as seen in Fig. 3(b). These deviations are comparable to a wall loss of around 40 kW per cavity for a case with a quality factor of 10,000, for example. In Fig. 3(a) no net power transfer is seen with the 5.5-MeV injection. This is entirely due to the longitudinal velocity variations of the fresh and spent beams, as clearly seen in Fig. 3(b). The counter reinjection scheme shown in Fig. 3(c), reveals that the beam-loading variations of the two beams compensate each other, and, as a consequence, the net power transfer can be suppressed to a great extent, as expected.

Table 3 shows comparison of 8.2-MeV output beams' properties for the 1.0-MeV injection. The beam–beam interactions in the counter scheme are

found to result in growths of beam size and emittance, especially with the high injection current, 0.56 A, while no growth is seen in the orthodox scheme.

Finally, the resultant beam of 0.087 A was injected into the 4-m main accelerator tube with its spent beam reinjected in the orthodox scheme as shown in Fig. 2. Again, degradation of beam properties induced by the undulator was neglected. As expected, the fresh beam is found to be unaffected by the spent beam reinjected in the orthodox scheme even in the longer 4-m tube, and the properties of the resultant beam to be fed to the undulator are found to be 37.0 MeV in

Table 3
Comparisons of output beams' properties from the 1-m preaccelerator tube with an input beam energy of 1.0 MeV

| Input beam current (A) | 0.56 | 0.56 | 0.56 | 0.087 | 0.087 | 0.087 |
Reinjection scheme	w/o reinjection	orthodox	counter	w/o reinjection	orthodox	counter
Energy spread (%)	2.1	2.1	2.1	5.8	5.8	5.8
Diameter (mm)	3.6	3.6	4.7	5.2	5.2	5.3
Normalized rms emittance (π mm mrad)	2.8	2.8	5.7	1.2	1.2	1.4

averaged energy, 2.0% in energy spread, and 1.7 π mm mrad in normalized emittance.

3.2. Effect of spent beam properties on interaction with fresh beam

Since the fresh beam affected by a spent beam in the 1-m preaccelerator tube ultimately will be reinjected into the tube as an updated spent beam, the output beam properties from the tube will have some time-evolution since the degree of beam–beam interaction depends on the reinjected spent beam's properties. This was omitted in the discussion in the previous subsection.

To study this time-evolution of the output beam emittance from the 1-m tube, the following procedure was adopted. In the PIC simulation, every particle exiting the simulation area from the downstream boundary was reinjected into the boundary with an assigned time delay. The delay time was chosen as 42-rf periods, which corresponds to a round-trip distance of 4.4 m. Though this is much shorter than that in an actual device, it was selected simply to minimize CPU time. Also at the reinjection with the delay, one of the following three processes was applied to simulate degradation in the emittance or energy spread by the undulator, namely (i) to double every particle's radial distance r from the symmetry axis, (ii) to double its angle dr/dz with respect to the axis, or (iii) to double its energy difference ΔE_k from the averaged energy.

The solid line in Fig. 4 shows the emittance time-evolution without artificial induction of emittance or energy spread degradation. At $N_{rf} = 1$, fresh beam injection from the gun starts, and then at $N_{rf} = 11$, the leading particles reach the downstream end of the 1-m tube with an emittance

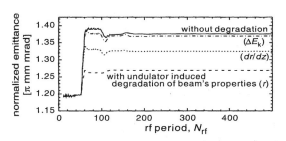

Fig. 4. Time evolutions of output beam emittance from the 1-m preaccelerator tube, with artificially induced degradation in the spent beam's energy spread, divergence, or radius.

around 1.2 π mm mrad. At $N_{rf} = 53$, the leading particles reach the reinjection boundary as a spent beam, and thus the emittance of the output fresh beam starts to increase due to interaction with the spent beam. At around $N_{rf} = 100$, the emittance is clearly seen to decrease. This is because the reinjected beams' emittances, and accordingly their radii as well, are getting large, since the beams affected by the beam–beam interactions in the tube return as updated spent beams, resulting in weaker interactions with the fresh beams. An opposite situation is seen at around $N_{rf} = 140$.

With the undulator-induced degraded beam properties, especially with the doubled beam radius (the broken line in Fig. 4), the emittance growth is found to be small, again resulting from weaker interactions with the less intense beams.

4. Conclusions

The counter reinjection scheme is found to be effective in suppressing the beam loading deviation along the ERL even with the low injection energy, 1.0 MeV. The emittance growth is found not so significant for a beam current of 0.087 A in the present combined scheme, while, with a higher current, 0.56 A, the growth is still considerable. It is also found that the growth is strongly dependent on the reinjected spent beams' properties, especially on the transverse size, indicating that the emittance growth could be suppressed to some extent by defocussing the spent beams prior to their reinjection.

References

[1] G.R. Neil, et al., Nucl. Instr. and Meth. A 445 (2000) 192.

[2] R. Hajima, et al., Proceedings of the 23rd International FEL Conference 2002, II27.

[3] E.J. Minehara, et al., Proceedings of the 22nd International FEL Conference 2001, II47.

[4] K. Masuda, et al., Nucl. Instr. and Meth. A 507 (2003) 133.

[5] K. Masuda, Ph.D. Thesis, Kyoto University, Japan, 1996.

[6] Y. Yamamoto, et al., Nucl. Instr. and Meth. A 393 (1997) 443.

[7] K. Masuda, et al., IEEE MTT 46 (1998) 1180.

[8] K. Masuda, et al., Nucl. Instr. and Meth. A 483 (2002) 315.

[9] T. Yamazaki, et al., Proceedings of the 23rd International FEL Conference 2002, II13.

Available online at www.sciencedirect.com

Nuclear Instruments and Methods in Physics Research A 528 (2004) 139–145

NUCLEAR
INSTRUMENTS
& METHODS
IN PHYSICS
RESEARCH
Section A

www.elsevier.com/locate/nima

Study of partial-waveguide rf-linac FELs for intense THz-pulse generation

M. Tecimer[a,*], D. Oepts[b], R. Wuensch[c], A. Gover[a]

[a] Physical Electronics, Tel Aviv University, 69978 Ramat Aviv, Tel-Aviv, Israel
[b] FELIX-FOM, Rijnhuizen, 3430 BE Nieuwegein , The Netherlands
[c] FZR, 1314 Dresden, Germany

Abstract

In this paper, we present a time-domain analysis of a short pulse partial-waveguide FEL oscillator employing toroidal mirrors and a hole outcoupling. The use of toroidal mirrors with optimized radius of curvatures helps to reduce cavity losses arising from the mismatch of the free space propagating optical field into a waveguided one. We introduce semi-analytical expressions for the calculation of the scattering matrix elements describing the loss and mode coupling mechanism as well as the amount of the extracted power from the cavity. The formulation is implemented in a time-domain FEL code based on modal expansion approach. The described model is applied to a partial-waveguide FEL system and simulation results are compared with measurements.
© 2004 Elsevier B.V. All rights reserved.

PACS: 41.60.Cr

Keywords: Free electron laser; Partial waveguide FEL; THz-pulse generation

1. Introduction

THz technology is being increasingly adopted in a wide variety of applications. The quest for tunable, coherent radiation sources in the THz region of the electromagnetic spectrum (~ 0.1–$10\,THz$) stimulate further developments and research activities in this area. Owing to the increased availability, compactness and conveni-ence in use, many of the applications utilize short pulse, broad-band THz-sources based on conventional laser technology. The available output power in those systems remain however low. In contrast, radiation sources based on accelerated electron devices, such as the (superconducting) rf-linac FEL systems can meet the needs of relatively narrow band, high-power (peak and average) coherent THz radiation offering a wide range of tunability over this spectrum. Due to the large diffraction losses inherent in this spectral region, the operation of rf-linac driven FELs often leads to the use of partial-waveguide resonators [1]. In

---

*Corresponding author.

E-mail addresses: tecimer@post.tau.ac.il, mtecimer@ hawaii.edu (M. Tecimer).

0168-9002/$ - see front matter © 2004 Elsevier B.V. All rights reserved.
doi:10.1016/j.nima.2004.04.034

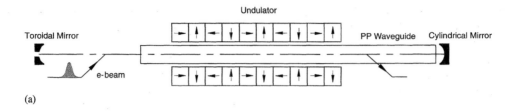

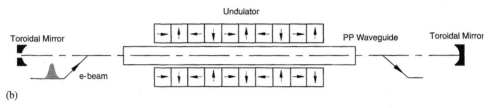

Fig. 1. Illustration of partial-waveguide resonator geometries employed in rf-linac FELs. (a) FELIX-FEL1's cavity utilizing a combination of a cylindrical and a toriodal mirror, (b) CLIO's two-sided partial-waveguide cavity.

Fig. 1a and b, two types of partial waveguide geometries are shown that are common in use. The resonator configuration shown in Fig. 1a has been realized in FELIX's FEL1 that is serving as a powerful, tunable THz-radiation source covering a spectrum range from ∼30 to 250 μm. The geometry shown in Fig. 1b has been implemented in CLIO's rf-linac FEL extending its wavelength range up to 120 μm. Other rf-linac FEL systems such as FELICE ($\lambda \approx 3{-}100\,\mu m$) and (superconducting) FZR-U90 ($\lambda \approx 30{-}150\,\mu m$) will adopt a partial waveguide cavity design in order to reduce the diffraction losses and at the same time to achieve an increase in the filling factor when operating in the long wavelength region above ∼30 μm compared to an open resonator.

In this paper, an overview of the physical model presented that is implemented into the (partial-) waveguide FEL code "wgfel". As in the "Elixer" code on which the "wgfel" bases [2], the longitudinal motion is calculated for each macroparticle separately whereas the transverse motion is accounted for by the beam envelope equations. Finite pulse effects related to the slippage and cavity desynchronization are modeled taking into account the waveguide dispersion for each of the hybrid waveguide modes. To model losses and

mode-coupling effects due to the partial waveguiding a scattering matrix method is implemented into the oscillator simulation code which maps the amplitudes of the outgoing transverse modes from the waveguide aperture to the ones reflected and coupled back into it [3].

To verify the validity of the model employed, simulation results are compared with the experimental data obtained at the FELIX-FEL1.

2. Theoretical description of the model

In the following the field equation that governs the FEL dynamics in a (partially) waveguided resonator is presented. The normalized radiation vector potential $\vec{A_r}$ is expressed by the superposition of the cold cavity eigenmodes:

$$\vec{A_r}(\vec{x}_\perp, z, t) = \frac{e}{m_e c} \sum_{m,n} u_{mn}(z, t)\psi_{mn}(\vec{x}_\perp, z)$$
$$\times \exp[\mathrm{i}(k_{zn}z - \omega t)]\hat{e}_x. \quad (1)$$

In Eq. (1), $u_{mn}(z, t)$ is a time- and space-dependent complex envelope of a modulated wave train with carrier frequency ω and axial wave number k_z. The relation between ω and k_z is established by the waveguide dispersion equation. The transverse profile of the optical field is defined by the transverse modes ψ_{mn} which are excited in a parallel plate waveguide (PPW) resonator [4]. Adopting slowly varying amplitude and phase approximation second-order inhomogeneous wave equation simplifies into

$$\partial_z u_{mn}(z, z') = \frac{\mathrm{i} f_B \mathbf{F} A_{u0}}{2 k_{zn}} \frac{\omega_p^2}{c^2} \zeta_{mn}(z, \bar{z}) \frac{\chi(z, \bar{z})}{N_{\bar{z}}(z, \bar{z})}$$
$$\times \sum_{j}^{N_{\bar{z}}(z, \bar{z})} \mathrm{e}^{-\mathrm{i}\theta_{jn}(z, \bar{z})} \qquad (2)$$

where θ_{jn} is the ponderomotive phase with regard to nth mode. Eq. (2) results from transformations $z' = z - v_g t$ carried out for the radiation pulse longitudinal coordinate and $\bar{z} = z - v_z t$ for the electron beam pulse longitudinal coordinate. The longitudinal distribution of the particles is represented by the dimensionless function $\chi(z, \bar{z}) = N_{\bar{z}}(z, \bar{z})/n_0 \pi r_b^2 \Lambda$ [5,6]. In Eq. (2), F is the filling factor. ζ_{mn} results from averaging the source term over the transverse Gaussian density profile of the electron beam. It can be expressed by

$$\zeta_{mn}(z, \bar{z}) = \frac{\sqrt{1/\mu\varepsilon}}{8\sigma_x(z)\sigma_y(z)} exp(\xi) \, \mathrm{erf}(\sqrt{\mu}X_0)$$
$$\times \{\exp(\theta^2/4\varepsilon)[\mathrm{erf}(\kappa_+) + \mathrm{erf}(\kappa_-)]$$
$$+ \exp(\theta^{*2}/4\varepsilon)[\mathrm{erf}^*(\kappa_+) + \mathrm{erf}^*(\kappa_-)]\} \qquad (3)$$

with the following definition of the parameters:

$$X_0 = 5\sigma_{x0}^2, \quad Y_0 = 5\sigma_{y0}^2, \quad \xi = -\frac{\mathrm{i}}{2}\tan^{-1}(z/z_R)$$
$$\mu = 1/w^2(z) - \mathrm{i}k_{zn}/2R(z) + 1/2\sigma_x^2(z)$$
$$\varepsilon = 1/2\sigma_y^2(z), \quad v = (2n+1)\pi/g, \quad n = 1, 3, 5, \ldots$$
$$\theta = k_u + \mathrm{i}v, \quad \kappa_\pm = \sqrt{\varepsilon}Y_0 \pm \theta/2\sqrt{\varepsilon} \qquad (4)$$

where σ_{x0} and σ_{y0} denote the rms beam waist in the transverse dimensions. $\sigma_x(z) = (\langle x^2 \rangle)^{1/2}$ and $\sigma_y(z) = (\langle y^2 \rangle)^{1/2}$ are the rms transverse dimensions of the beam that are defined by the solution of the beam envelope equations in terms of the longitudinal position "z" [2].

2.1. Slippage and feedback loop in a short pulse (partial-) waveguide oscillator

Due to the waveguide dispersion, slippage effects in short pulse waveguide FELs exhibit differences in comparison with open resonator FELs. The slippage length L_{sp} for the resonant radiation wavelength λ_r is defined as

$$L_{sp} = [(\beta_g/\beta_z) - 1]N\lambda_u$$
$$= N\lambda_r[1 - (k_\perp/\beta_z\gamma_z k_u)^2]^{1/2} \qquad (5)$$

where N and $k_{\perp n}$ are, respectively, the number of the undulator periods and the cutoff wave number of the mnth transverse mode. L_{sp} is mode dependent since each transverse mode overtakes the electron beam with a different group velocity.

The boundary conditions imposed by the resonator mirrors are included into the feedback of the radiation field. For the transverse modes the feedback is described by

$$u_{m''n''}^{(p+1)}(z = 0, t) = R^{(1)}R^{(2)} \times \sum_{m',n'}\sum_{m,n} \mathbf{P}_{m''n''m'n'}^{(1)} \mathbf{P}_{m'n'mn}^{(2)}$$
$$\times \mathrm{e}^{\mathrm{i}\Phi_{m''n''}(\omega)} u_{mn}^{(p)}(L_u, t - t_r) \qquad (6)$$

where $u_{mn}^{(p)}(L_u, t')$ denotes the complex amplitude of the mnth mode at the exit of the interaction region and the suffix "p" the number of passes. $t_R = 2(L_{wg}/v_{gn} + L_f/c)$ is the cavity roundtrip time, L_{wg} and L_f correspond to the waveguided and free space parts of the resonator, respectively. The cavity roundtrip phase shift is given by

$$\Phi_{m''n''}(\omega) = 2[k_{zn}(\omega)L_{wg} + k_s L_f] + \Delta\Phi_G \qquad (7)$$

including the Gouy phase shift $\Delta\Phi_G$ of the hybrid mode. The resonance frequency condition of the cavity is satisfied when $\Phi_{m''n''}(\omega)$ is a multiple of 2π. The intermode scattering between the reflected modes due to the partial waveguiding and the outcoupling hole are expressed by the scattering matrices $\mathbf{P}^{(j)}$ ($j = 1, 2$) where the suffix "j" signifies upstream and downstream mirrors. $R^{(j)}$ denotes the amplitude reflection coefficient accounting for the ohmic losses on the mirror surfaces.

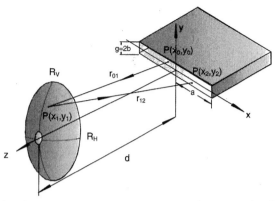

Fig. 2. After interacting with a short electron pulse, the parallel plate waveguide modes Ψ_{mn} propagate in free space to a toroidal mirror.

2.2. Partial PPW resonator

In the hole outcoupled partial-waveguide resonator with toroidal mirrors, cavity loss and mode coupling have been determined by calculating the scattering matrix elements. Here, we evaluate Huygens–Fresnel diffraction integrals describing the free space propagation of the optical field at the exit of a overmoded waveguide to finite size toroidal mirrors ($R_V \neq R_H$), outcoupling through a hole and matching of the back-reflected field into the waveguide aperture (Fig. 2). In the Fresnel approximation, the diffraction integral describing the propagation of the mode $\psi_{mn}(\vec{x}, z)$ to the toroidal mirror is given by

$$
E_{(mn)}(\vec{x}_1, d) = \left(\frac{1}{i\lambda d N_{mn}}\right) \Lambda_{01}
$$
$$
\times \int_{-b}^{b} \int_{-a}^{a} \exp(-ik((x_1 x_0 + y_1 y_0)/d))
$$
$$
\times \exp(ik(x_0^2 + y_0^2)/2d)\psi_m(x_0, z_0)
$$
$$
\times \psi_n(y_0)\, dx_0\, dy_0 \tag{8}
$$

where a and b are the half-width and height of the waveguide aperture, $\Lambda_{01} = \exp(ikd)\exp(ikx_1^2(1 - d/R_H)/2d)\exp(iky_1^2(1 - d/R_v)/2d)$ and R_H and R_v are the horizontal and vertical mirror radius of curvatures, respectively (Fig. 2). The integral in

Eq. (8) can be expressed by

$$
E_{mn}(\vec{x}_1, d) = \left(\frac{1}{i\lambda d N_{mn}}\right) \Lambda_{01} \frac{\sqrt{\pi}\xi(1+i)}{8\sqrt{2\beta\mu}} \exp\left(-\frac{k^2}{4\mu d}\right)
$$
$$
\times \left\{ \operatorname{erf}\left(\sqrt{\mu}a + \frac{ikx_1}{2d\sqrt{\mu}}\right) \right.
$$
$$
\left. + \operatorname{erf}\left(\sqrt{\mu}a - \frac{ikx_1}{2d\sqrt{\mu}}\right) \right\}
$$
$$
\times \left\{ \exp\left(-\frac{i\pi f_{y(-)}^2}{\beta}\right)[\operatorname{erf}(\Theta\Gamma_1) \right.
$$
$$
+ \operatorname{erf}(\Theta\Gamma_2)] + \exp\left(-\frac{i\pi f_{y(+)}^2}{\beta}\right)[\operatorname{erf}(\Theta\Gamma_3)
$$
$$
\left. + \operatorname{erf}(\Theta\Gamma_4)]\right\} \tag{9}
$$

where we define $\Theta = \sqrt{\pi\beta/2}(1+i)$, $\beta = 1/\lambda d$, $\alpha = (n+1/2)/g$, $\gamma = \alpha/\beta$, $f_{y(\pm)} = \beta y_1 \pm \alpha$, $\Gamma_{1,2} = b \mp (y_1 - \gamma)$, $\Gamma_{3,4} = b \mp (y_1 + \gamma)$, and the normalization factor $N_{mn} = (2^m m!\sqrt{\pi/8}w(z)g)^{1/2}$. Eq. (9) allows to calculate the outcoupled radiation power through the hole at any z' within the pulse:

$$
P_{\text{out}}(d, z')
$$
$$
= \frac{kk_{zn}}{2Z_0}(m_e c^2/e)^2
$$
$$
\times \sum_{mn}\sum_{m'n'} \int_{-h_r}^{h_r} dy_1 \int_{-\sqrt{1-y^2}}^{\sqrt{1-y^2}} u_{mn}(L_{\text{wg}}, z')\, u_{m'n'}^*(L_{\text{wg}}, z')
$$
$$
\times E_{mn}(\vec{x}_1, d, z')E_{m'n'}^*(\vec{x}_1, d, z')\, dx_1 \tag{10}
$$

$u_{mn}(L_{\text{wg}}, z')$ being the complex field amplitude of the mode "mn" at the exit of the waveguide aperture. The complex field amplitude of the back reflected field at the waveguide aperture is defined by

$$
E_{mn}(\vec{x}_2, 2d)
$$
$$
= \left(\frac{1}{i\lambda d N_{mn}}\right) \exp(ikd)\exp(ik(x_2^2 + y_2^2)/2d)
$$
$$
\times \left[\int_{-R_y}^{R_y} dy_1 \int_{-R_x\sqrt{1-y_1^2/R_y^2}}^{R_x\sqrt{1-y_1^2/R_y^2}} \exp(-ik((x_2 x_1 + y_2 y_1)/d)) \right.
$$
$$
\times \exp(ik(x_1^2 + y_1^2)/2d)E_{mn}(\vec{x}_1, d)\, dx_1
$$

$$-\int_{-h_r}^{h_r} dy_1' \int_{-\sqrt{h_r^2-y_1'^2}}^{\sqrt{h_r^2-y_1'^2}} \exp(-ik((x_2 x_1' + y_2 y_1')/d)$$

$$\times \exp(ik(x_1'2 + y_1'2)/2d) E_{mn}(\vec{x}_1', d) \, dx_1' \Bigg]. \quad (11)$$

In Eq. (11), limits of the first integration define toroidal mirror's transverse extension and h_r is the outcoupling hole radius. The projection of the back reflected field $E_{m'n'}(x^{\rightharpoonup}_2, 2d)$ into the mnth PPW mode is defined by

$$P_{m'n'mn} = \frac{1}{N_{mn}} \int_{-b}^{b} \int_{-a}^{a} E_{m'n'}(\vec{x}_2, 2d)\psi_{mn}^*(\vec{x}_2, 2d) \, d^2\vec{x}_2. \quad (12)$$

$P_{m'n'mn}$ being the scattering matrix element. The complex field amplitude u_{mn} at the exit of the PPW following the interaction section is related to the backreflected $u_{m'n'}$ containing the cavity loss by

$$u_{m'n'} = \sum_{m,n} P^T_{m'n'mn} u_{mn}. \quad (13)$$

3. Numerical results

Experimental data obtained at the partial waveguide THz-FEL in FELIX by the cavity ring down measurements, output pulse energies and pulse shape measurements is used to verify the validity of the model employed. An overall system

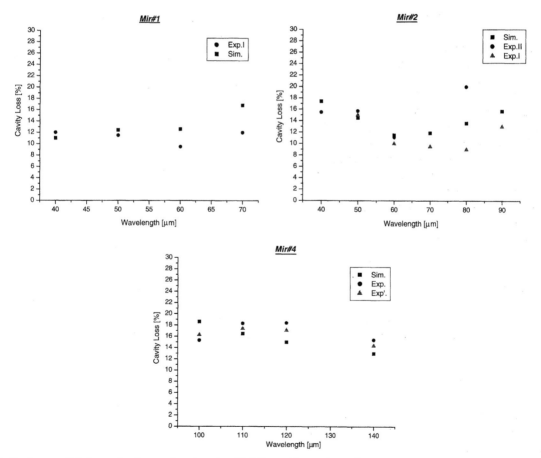

Fig. 3. Cavity roundtrip losses for the wavelength region 40–140 μm using various mirrors at FELIX's partial-waveguide resonator. The loss data is deduced from the cavity ring-down measurements.

description of FEL1 can be found in Ref. [1]. Fig. 3 shows the cavity losses deduced from the decay rate of the extracted radiation power versus radiation wavelength for individual toroidal mirrors. The measurements include two sets of independently taken experimental data. The simulated cavity loss L_{mn} due to the outcoupling hole, finite size toroidal mirror, clipping off at the waveguide aperture and mode-conversion is given by $L_{mn} = 1 - |P_{m'n'mn}|$. In the simulations, combination of TE_{01}, TE_{03}, TE_{05} composes the startup field's transverse mode pattern. The scattering from TE_{01} mode into higher-order modes leads merely to a loss channel of energy since the latter are not sustained by the cavity nor by the gain medium.

The presented numerical results rely on the assumption that the electron injection into the undulator occurs without inducing beam-misalignments with respect to the optical axis of a perfectly aligned resonator. Although they mimic with reasonable agreement the measured wavelength dependent behavior of cavity losses shown in Fig. 3, it is also apparent from the data obtained for mirrors #1–#2 that some of the measured cavity losses happen to be smaller than the computed ones, particularly when the investigated wavelength increases. The latter might be originating from electron beam/resonator-axis misalignment by matching and subsequent steering of the beam through the undulator as the gap is changed to vary the undulator parameter K_{rms}. In the lateral, where the waveguide width is seven times larger than the height, the excitation of the next order free space mode and its coupling to a slightly off-axis steered beam in the same plane is facilitated. In analogy to the cases reported in Refs. [7,8], the maximum intensity of the excited asymmetrical field pattern does not coincide with the hole aperture axis, leading to a reduction in the outcoupling fraction, consequently also to a lower total cavity loss. In Fig. 4, extracted micropulse energies obtained at various toroidal mirror geometries and wavelengths in the region between 40 and 140 μm are compared with the simulated ones. Here, the employed cavity detuning has been optimized to obtain the maximum outcoupled pulse energies. In Fig. 5, detected time structure of

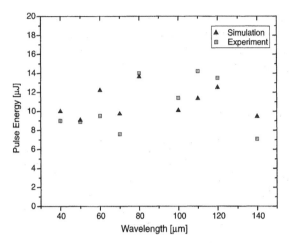

Fig. 4. The outcoupled micropulse energies versus wavelength. The beam energy (14–21 MeV) and the undulator parameter (∼0.8–1.5) are varied to adjust for the wavelength.

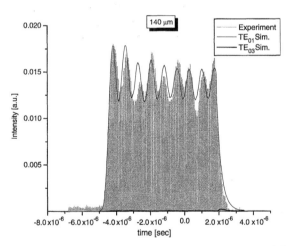

Fig. 5. Detector signal of a macropulse at 140 μm, outcoupled from mirror#4. Simulated macropulse intensity is normalized to the first peak (at saturation) of the measured signal.

a ∼8 μs long macropulse at 140 μm, coupled out from toroidal mirror #4, is compared with the simulated one. Note that, here, the measured output power is scaled to arbitrary units since the detector used in this measurement (P5-01, Molectron) was not calibrated for absolute power measurements. The contribution of TE_{03} mode to the total radiated intracavity power remains according to the simulations around noise level of the detected signal. The measured macropulse shape exhibits $\sim 8\frac{1}{2}$ limit cycle oscillations for a cavity desynchronization set to -0.8 to -0.9λ · The 'wgfel' code uses -1.2λ to fit the simulated pulse shape to the measured one.

4. Conclusion

The physical model presented in this paper provides a means to describe two important aspects of partially waveguided, short pulse FEL oscillators and the interplay between them during the radiation buildup process in multiple passes; short pulse effects in a dispersive medium, and the dynamics of mode-conversion/cavity losses due to abrupt transitions caused by various apertures used in the adopted resonator geometry. The described model is applied to the partial-waveguide THz-FEL in FELIX, making comparisons with the experimental results. Cavity losses and extracted pulse energies at various toroidal mirror

geometries and beam energy/undulator settings are presented. Discrepancies exist between the simulation results and the experimental observations in the cases where the axially symmetric simulation model cannot handle misaligned resonators and/or off-axis electron beams.

References

[1] K.W. Berryman, T.I. Smith, Nucl. Instr. and Meth. A 318 (1992) 885;
FELIX, http://www.rijnh.nl, Characteristics of Felix;
L.Y. Lin, A.F.G. van der Meer, Rev. Sci. Instrum. 68 (1997) 4342.
[2] G. van Werkhoven, et al., Phys. Rev. E 50 (1994) 4063.
[3] M. Tecimer, Nucl. Instr. and Meth. A 483 (2002) 521.
[4] L. Elias, J. Gallardo, Appl. Phys. B 31 (1983) 229.
[5] R. Bonifacio, et al., Phys. Rev. A 44 (1991) 3441.
[6] B.W.J. McNeil, Opt. Commun. 165 (1999) 65.
[7] G.A. Barnett, et al., Nucl. Instr. and Meth. A 358 (1995) 311.
[8] V.I. Zhulin, Nucl. Instr. and Meth. A 375 (1996) ABS 85.

Available online at www.sciencedirect.com

Nuclear Instruments and Methods in Physics Research A 528 (2004) 146–151

NUCLEAR INSTRUMENTS & METHODS IN PHYSICS RESEARCH
Section A

ELSEVIER

www.elsevier.com/locate/nima

Variable height slot-outcoupling for the compact UH THz-FEL

M. Tecimer[a],[*],[1], H. Jiang[b], S. Hallman[a], L. Elias[a]

[a] *Physics and Astronomy, University of Hawaii at Manoa, Honolulu, HI 96822, USA*
[b] *Physics and Astronomy, Clemson University, Clemson, SC 29634, USA*

Abstract

The quest for tunable, compact coherent radiation sources in the THz region of the electro-magnetic spectrum is growing and stimulating further developments and research activities in this area. The THz-FEL Group of the University of Hawaii (UH) pursues the concept of minimizing size and cost of a recirculating-beam electrostatic accelerator THz-FEL. The system employs a low voltage (1.7 MeV) pelletron accelerator in conjunction with a short period ($\lambda_u = 8$ mm) hybrid undulator and miniature beam line components. The continuous tunability offered by the system over a spectral range between 230 and 640 µm necessitates a broad-band outcoupling mechanism. Holes on the axis in resonator mirrors have been previously employed in long wavelength FELs as a means to couple out optical radiation from the cavity. The UH THz-FEL adopted a parallel plate waveguide resonator design introducing a height adjustable slot-aperture on one of the cylindrical metal mirrors. Based on the mode-matching method employed in treating waveguide discontinuities, we examine the characteristics of the chosen outcoupling mechanism and its influence on the intracavity optical fields.
© 2004 Elsevier B.V. All rights reserved.

PACS: 41.60.−m; 41.60.Cr

Keywords: Electrostatic THz-FEL; Waveguide discontinuity; Variable outcoupling

1. Introduction

THz radiation sources based on accelerated electron devices, such as the recirculating-beam electrostatic accelerator FEL can provide narrow band (Fourier-transform limited relative bandwidth ~10^{-7}–10^{-8}), high power coherent THz radiation, operating over the entire THz region [1]. Based on the technology developed at the THz-FEL Group at UH pursues the concept of minimizing size and cost of an electrostatic accelerator THz-FEL. The system employs a low voltage (1.7 MeV) pelletron accelerator in conjunction with a short period ($\lambda_u = 8$ mm) hybrid undulator [2] and miniature beam line components such as an electron gun, beam transport components and a five-stage collector for beam recovery. The matched beam at the entrance to the

*Corresponding author.

E-mail address: mtecimer@hawaii.edu (M. Tecimer).

[1] On leave from Tel Aviv university.

0168-9002/$ - see front matter © 2004 Elsevier B.V. All rights reserved.
doi:10.1016/j.nima.2004.04.035

undulator traverses through a 3.7 mm (parallel plate) waveguide gap along the 185 period long, short period hybrid undulator. An accurate magnetic field tuning scheme enables us to obtain field uniformity within ~0.1% rms [2]. The FEL device, which is being assembled currently, will provide up to 0.6 kW output power in the 230–640 μm region generating several tens of microsecond long narrow band THz-pulses. The final configuration of the improved recirculating beam transport system is expected to enable kW level output power with the prospect of seconds long pulsed- or even a cw-operation.

The large diffraction loss inherent in long wavelength radiation implies the use of waveguide resonators in THz-FELs, which can substantially improve the overlap between the optical and electron beams, and consequently the FEL gain as compared to open resonators. On the other hand, continuous tunability over a wide range of wavelengths makes it necessary to have mirrors that provide broad-band feedback and outcoupling. The adopted height adjustable slot-aperture (Fig. 1) permits a variable, well-defined fraction of the radiation power stored in the cavity to be coupled out. In this paper, we examine the reflection and transmission characteristics of the chosen outcoupling mechanism and its influence on the intracavity mode structure at the presence of the gain medium, as well as the evolution of the extracted power and its mode composition until the FEL saturation sets in.

2. Theoretical description of the model

2.1. FEL equations

This section provides a brief outline of the theoretical model used to describe the FEL interaction of a long electron beam (beam length ≫ slippage length) with radiation fields excited in a parallel plate waveguide (PPW) resonator with slot outcoupling. The formulation is based on the representation of the radiation fields in terms of transverse eigenmodes of an unloaded PPW. Utilizing the orthogonality properties of the transverse cavity eigenmodes, the mode-matching method is adopted to model the outcoupling mechanism. The latter is treated as a waveguide discontinuity problem describing the coupling of a transverse slot on the common cylindrical wall between two parallel plate waveguides with different heights (Fig. 2). For the used resonator configuration, the normalized radiation vector potential is well modelled as a superposition of the hybrid waveguide eigenmodes ψ_{mn}

$$\vec{A}_r(\vec{x}_\perp, z, t)$$

$$= \frac{e}{m_e c} \sum_{m,n} u_{mn}(z, t) \psi_{mn}(\vec{x}_\perp, z) \exp[i(k_{zn}z - \omega t)]\hat{e}_x$$

(1)

where $\psi_{mn}(\vec{x}_\perp, z) = \psi_m(x, z)\psi_n(y)$ is a combination of the Hermite–Gaussian modes in the x

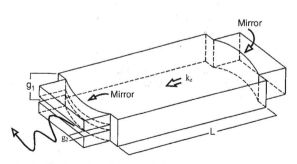

Fig. 1. A schematic of the parallel plate waveguide resonator with a height adjustable slot-aperture on one of the cylindrical metal mirrors.

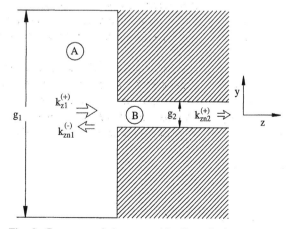

Fig. 2. Geometry of the waveguide discontinuity associated with the slot outcoupling.

direction and TE_{0n} waveguide modes in the y direction [3], which are the only ones to be coupled to the transverse oscillations of the electrons; $\psi_n(y) = \cos(n\pi y/g)$, $n = 1, 3, 5, \ldots$ with $k_{zn}(\omega_s) = [(\omega_s/c)^2 - k_{\perp n}^2]^{1/2}$ and $k_{\perp n} = n\pi/g$ denoting the longitudinal and cutoff wavenumbers of the nth waveguide mode, respectively. In Eq. (1), a steady-state sinusoidal dependence with angular frequency ω_s is assumed. The excitation equation for the slowly varying complex amplitude u_{mn} is given by

$$\partial_z u_{mn}(z) = \frac{\mathrm{i} f_B F_m A_{u0}}{2 k_{zn}(\omega_s)} \frac{\omega_p^2}{c^2}$$

$$\times \frac{1}{N} \sum_J^N \frac{\cosh(k_u y_j)}{\beta_{zj}} \psi_{mn}^*(\vec{x}_{\perp j}, z) \mathrm{e}^{-\mathrm{i}\theta_{jn}(z)} \delta(\vec{x}_\perp - \vec{x}_{\perp j}(z)). \tag{2}$$

The right-hand side of Eq. (2) results from averaging the source term over one cycle of the signal $T = 2\pi/\omega_s$. N is the number of particles crossing the transverse plane at a given z-position during the time T, each carrying the partial beam current I/N. The ponderomotive phase is $\theta_{jn} = (k_{zn} + k_u)z - \omega_s t_j(z)$, where t_j denotes the time taken by the jth particle to arrive at position z. The filling factor is introduced as $F_m = \pi \sigma_x(z) \sigma_y(z) / \Sigma_m(z)$, and the relativistic plasma frequency $\omega_p^2 = e^2 n_0 / \varepsilon_0 m_e \gamma$. $\sigma_x(z)$ and $\sigma_y(z)$ denote the beam envelopes in the transverse dimensions along the interaction region, and $\Sigma_m(z)$ is the hybrid mode area. The equations of electron motion closely follow the KMR (Kroll, Morton, Rosenbluth) formalism [4]. The interaction between the quasi-continuous electron beam and the radiation field is described by solving the wave equation for the amplitudes of the excited waveguide modes in conjunction with a set of six ordinary differential equations. These equations determine the Lorentz factor γ, the electron phase θ, the transverse coordinates x and y, and the dimensionless canonical momenta P_x and P_y.

2.2. Mode-matching at the outcoupler

In an oscillator FEL, the temporal and spatial structure of the radiation field is strongly influenced by the reflection and transmission charac-

teristics of the cavity, since multiple roundtrips are required until saturation is reached. The geometry of the outcoupling aperture determines the extraction ratio, which is the distribution of the transverse modes inside the resonator affecting the coupling of the optical field to the gain medium. A mode-matching procedure is incorporated into the FEL model used in order to determine the influence of the outcoupling mechanism on the amplification process at each pass as well as the fraction of the power transmitted through the aperture.

In the presence of the step discontinuity, boundary conditions at the interface A/B can be formulated as follows:

- The tangential component of the electric and magnetic fields are required to be continuous across the aperture, and
- The tangential component of the electric field must vanish at the perfectly conducting part of the interface.

The above conditions can be cast into the following mathematical formulation. First, the electric fields in both regions A and B are expressed based on the expansion given in Eq. (1):

$$E_x^{(A)}(x, y, z) = \sum_{n=1}^{\infty} (A_n^+ \exp(\mathrm{i} k_{zn_1} z) + A_n^- \exp(-\mathrm{i} k_{zn_1} z))$$

$$\times (2/g_1)^{1/2} \cos(n\pi y/g_1)$$

$$\times (\sqrt{2/\pi}/w(z))^{1/2} \psi_0^{(1)}(x, z) \tag{3}$$

$$E_x^{(B)}(x, y, z) = \sum_{n=1}^{\infty} (B_n^+ \exp(\mathrm{i} k_{zn_2} z)(2/g_2)^{1/2}$$

$$\times \cos(n\pi y/g_2)$$

$$\times (\sqrt{2/\pi}/w(z))^{1/2} \psi_0^{(2)}(x, z) \tag{4}$$

where the eigenmodes modes $\psi_{mn}(\vec{x}_\perp, z)$ are normalized using the orthogonality relation. In Eqs. (3) and (4) A_n^+ is the field amplitude coefficient of the nth transverse mode incident to the outcoupling mirror. A_n^- and B_n^+ are the unknown coefficients of the scattered modes in regions A and B, respectively. The suffixes "+" or "−" account for a field traveling in the positive or negative direction on the z-axis. Apart from propagating modes, the presence of the discontinuity also gives rise to the excitation of evanescent modes with an imaginary propagation constant. The amplitudes of these modes decay exponentially according to the term $\exp(-k_{zn}z)$, and are appreciable only close to the discontinuity. In the next step, the aforementioned boundary conditions are enforced at the cylindrical interface for the electric and magnetic fields. By requiring that the tangential components E_x and H_y be continuous in the aperture, one can determine the scattering coefficients A_n^- and B_n^+ in terms of the incident field amplitudes A_n^+. For the electric field, the condition $E_x^{(A)}(x, y, z) = E_x^{(B)}(x, y, z)$ using Eqs. (3) and (4) can be formulated:

$$
\begin{aligned}
A_m^+ + A_m^- &= \sum_{n=1}^{N} B_n^+ \sqrt{4/g_1 g_2} \int_{-g_1/2}^{g_1/2} \mathrm{Cos}(m\pi y/g_1) \\
&\quad \times \mathrm{Cos}(n\pi y/g_2)\, \mathrm{d}y \\
&= \sum_{n=1}^{N} B_n^+ \sqrt{4/g_1 g_2} \int_{-g_2/2}^{g_2/2} \mathrm{Cos}(m\pi y/g_1) \\
&\quad \times \mathrm{Cos}(n\pi y/g_2)\, \mathrm{d}y
\end{aligned}
\tag{5}
$$

where the infinite series in Eqs. (3) and (4) are truncated to finite summations with only the first P modes in region $\boldsymbol{A}$, and N modes in region $\boldsymbol{B}$. Eq. (5) can be written as a matrix equation:

$$
A^+ + A^- = \boldsymbol{F} \cdot B^+
\tag{6}
$$

where A^+, A^-, and B^+ are vectors with components representing the mode amplitudes of the incident, reflected, and transmitted fields, respectively. The $(P \times N)$ matrix elements $\boldsymbol{F}_{nm}$ are defined by

$$
\begin{aligned}
\boldsymbol{F}_{mn} = \frac{2}{\sqrt{g_1 g_2}} & \left[\frac{\mathrm{Sin}\left(\dfrac{\pi}{2}\left(m\dfrac{g_2}{g_1} - n\right)\right)}{\pi\left(\dfrac{m}{g_1} - \dfrac{n}{g_2}\right)} \right. \\
& \left. + \frac{\mathrm{Sin}\left(\dfrac{\pi}{2}\left(m\dfrac{g_2}{g_1} + n\right)\right)}{\pi\left(\dfrac{m}{g_1} + \dfrac{n}{g_2}\right)} \right]
\end{aligned}
\tag{7}
$$

where $m = 1, 3, 5, \ldots, P$; $n = 1, 3, 5, \ldots, N$ and $\boldsymbol{F}_{mn} = \sqrt{g_2/g_1}$ for the case $m/g_1 = n/g_2$. The magnetic counterpart to Eq. (6) is obtained using $H_y = -(1/i\omega\mu_0)\partial E_x/\partial z$ and the condition $H_y^{(A)}(x, y, z) = H_y^{(B)}(x, y, z)$ on the surface of the cylindrical slot:

$$
\boldsymbol{G} \cdot \boldsymbol{C}_p \cdot (A^+ - A^-) = \boldsymbol{C}_N \cdot B^+
\tag{8}
$$

where $\boldsymbol{C}_P$ and $\boldsymbol{C}_N$ are diagonal matrices with elements k_{zn1} and k_{zn2}, respectively. $\boldsymbol{G}$ is a $N \times P$ matrix with elements

$$
\begin{aligned}
G_{nm} = \sqrt{4/g_1 g_2} \int_{-g_2/2}^{g_2/2} \mathrm{Cos}(m\pi y/g_1) \\
\times \mathrm{Cos}(n\pi y/g_2)\, \mathrm{d}y
\end{aligned}
\tag{9}
$$

where $n = 1, 3, \ldots, N$; $m = 1, 3, \ldots, P$ is transpose of $\boldsymbol{F}_{mn}$. The two mode-matching matrix Eqs. (6) and (8) at the interface provide a system of $P + N$ linear equations for the $P + N$ unknown amplitude coefficients A^- and B^+.

The outcoupling mechanism described by the mode-matching method is incorporated into the FEL formulation to model the reflection, transmission and mode conversion effects on the intracavity radiation build-up in successive round-trips. The complex field amplitude of each transverse mode is determined by the coupled-mode analysis of the FEL amplifier at the output of the interaction region. The calculated mode amplitudes constitute the field A^+, which is incident on the outcoupler mirror. The solution of the mode-matching matrix equations provides the transmitted field B^+ through the outcoupler and the reflected field A^-, which is fed back to the input of the FEL amplifier for the next pass. The matrix equations produce inter-mode scattering, resulting in cross-coupling between the reflected modes.

The presented model is used to study the UH THz-FEL. The results of the simulations are discussed in Section 3.

3. Numerical results

In this section simulation results are presented which describe the characteristics and the performance of the waveguide resonator FEL employing variable slot outcoupling at the UH THz-FEL project. The electrostatic accelerator-based FEL system is driven by a low norm emittance ($\varepsilon_n \sim 2.5 \times 10^{-6}\,\pi\,\mathrm{m\,rad}$), low current (0.2 A) electron beam with a beam energy varying between 1.0 and 1.7 MeV. The accurate magnetic field tuning scheme allows the assumption of perfect undulator fields in the trajectory calculations. The study covers a number of wavelengths; however, here, we present the results of a representative wavelength at 230 μm for the operation of the FEL. We start the multimode analysis of the waveguide outcoupling from spontaneous emission radiated into the TE_{0n} modes excited in the parallel plate waveguide resonator. The associated field amplitudes are inserted as start-up field in A^+. In the simulations, apart from propagating modes, evanescent modes are taken into account that represent local fields existing in the close vicinity of the slot. The field amplitude coefficients A^+, A^- and B^+ can be used to obtain a useful check on the numerical accuracy of the mode-matching algorithm, by evaluating the power conservation where the sum of the reflected and transmitted power equals incident power on perfectly conducting surfaces. Fig. 3 illustrates the verification of the mode-matching alghorithm and the associated boundary conditions at the interface of the step discontinuity. Fig. 4 shows the power transmission, which exhibits stepwise increase at particular values of g_2, namely at those corresponding to the cutoff height for the next order TE_{0n} mode above which the associated k_{zn2} (Fig. 2) becomes real.

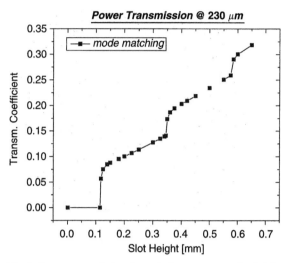

Fig. 4. Power transmission through the aperture vs. slot height. Increasing the slot height $g_2 > 0.35$ mm allows TE_{03} mode to propagate within the small waveguide in addition to TE_{01}.

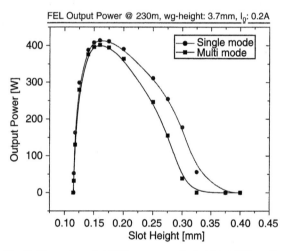

Fig. 5. The outcoupled FEL power vs. slot height at 230 μm. In the single mode case, the TE_{01} mode interacts with the electron beam, and it is the only mode incident on the slot. In the multimode case, the incident field contains the propagating modes.

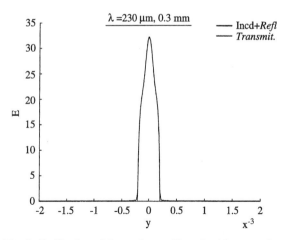

Fig. 3. Verification of the mode-matching algorithm at a slot height of 0.3 mm. Comparison of the incident and reflected field amplitude with the transmitted one across the interface. Fields vanish at the perfectly conducting part of the interface.

The extracted FEL power at 230 μm versus slot height is shown in Fig. 5, for the "single-" and "multi-mode" cases. For the range 0.115–0.35 mm, TE_{01} is the only propagating mode within the small waveguide. Thus, for the aperture range relevant to the FEL operation the transmitted field consists of a single TE_{01} mode.

Fig. 6. The constructed variable slot height mirror is incorporated into the waveguide resonator.

tion and the mode structure evolution of the intracavity field. For the aperture range relevant to the FEL operation it is found that the composition of the outcoupled, propagating field is limited to a single TE_{01} mode.

The presented outcoupler scheme is constructed (Fig. 6) with the aim of providing a broad-band, variable and well-defined output coupling for the UH THz-FEL covering a wide spectral range between 230 and 640 μm. Making use of the total reflexion at cutoff, the scheme may well be suited to the injection of the electron beam into the interaction region allowing the use of straight forward beam optics.

4. Conclusion

We have investigated the reflection and transmission properties of a height adjustable outcoupling slot in a parallel plate waveguide resonator FEL and its influence on the amplifica-

References

[1] G. Ramian, Nucl. Instr. and Meth. A 318 (1992) 225.
[2] M. Tecimer, L.R. Elias, Nucl. Instr. and Meth. A 341 (1994) 126 (ABS).
[3] L.R. Elias, J. Gallardo, Appl. Phys. B 31 (1983) 229.
[4] A. Kroll, et al., J. Quantum Electron. 29 (1981) 2969.

Available online at www.sciencedirect.com

SCIENCE @ DIRECT®

NUCLEAR INSTRUMENTS & METHODS IN PHYSICS RESEARCH
Section A

ELSEVIER   Nuclear Instruments and Methods in Physics Research A 528 (2004) 152–156

www.elsevier.com/locate/nima

Generation of short-pulse far-infrared light from electrons undergoing half-cyclotron rotation

M.R. Asakawa[a],*, T. Marusaki[a], M. Hata[b], Y. Tsunawaki[b], K. Imasaki[c]

[a] Institute of Free Electron Laser, Osaka University, 2-9-5 Tsudayamate, Hirakata, Osaka 573-0128, Japan
[b] Department of Electrical Engineering and Electronics, Osaka Sangyo University, Nakagaito, Daito, Osaka 574-8530, Japan
[c] Institute for Laser Technology, Yamadaoka, Suita, Osaka 565-0871, Japan

Abstract

A novel radiation source in the terahertz wave spectral region is being developed. The radiation is produced by electrons, which undergo a half-cyclotron rotation. Injecting the electron bunch with the energy of 150 MeV into the 3 m-long solenoid field, we observed that the radiation had the intense millimeter spectral component when the cyclotron resonant frequency was set to be 4 THz. The numerical calculation revealed that the radiation could be regarded as a half-cycle radiation with the pulsewidth approximately equaled to that of the electron bunch. Also, the calculation pointed out the possibility to amplify such radiation in the optical resonator.
© 2004 Elsevier B.V. All rights reserved.

PACS: 41.60.−m

Keywords: Cyclotron radiation

1. Introduction

Motivated by recent vital activities in the terahertz wave research, a unique radiator is being developed at iFEL Osaka University. This radiator generates an ultra-short radiation pulse, that is a radiation with a half-cycle; the light pulse is emitted by the electron bunch undergoing a half-cyclotron rotation. Whereas the spectral property of this radiator is similar to that of the widespreading terahertz sources fed by the femtose-

cond lasers [1], the radiator provides higher peak power by the virtue of the high peak power of the electron bunch (150 MeV × 80 A = 1.2 GW). Electrons passing through the bending magnet also radiate a non-periodic short pulse radiation. In the spectral range below terahertz, the collective effect of the electron bunch on the radiation increases the radiation power and such kind of radiation source is developed by using the continuous electron beam accelerator. Compared to the bending magnet radiation source, the cyclotron radiator has an advantage that it has the long interaction length between the electron and radiation field. Our program aims to the amplification of the half-cycle radiation.

---

*Corresponding author.
  E-mail address: asakawa@fel.eng.osaka-u.ac.jp
(M.R. Asakawa).

0168-9002/$ - see front matter © 2004 Elsevier B.V. All rights reserved.
doi:10.1016/j.nima.2004.04.036

In this paper, the preliminary results about the generation of the cyclotron radiation from the electron undergoing a half-cyclotron rotation will be reported. The experiment on the generation of the spontaneous emission will be presented in the next section and then the property of such radiation will be discussed.

2. Experiment

The experiments are underway on the 150 MeV beam line, which was used for the demonstration of ultraviolet free-electron laser [2]. An S-band rf-linac produces the electron bunch of 0.25 nC, 3 ps at a repetition rate of 22.3 MHz (1/128 of S-band frequency) for 25 μs. Thus, the macropulse contains 550 electron bunches. The UV undulator was replaced with 3 m-long solenoid coil wound around a drift tube with an inner radius of 8.5 mm. This solenoid is located between two pairs of the bending magnets to detour the electron beam around the cavity mirrors. The coil is fed by a capacitor bank (0.1 F/500 V) and generates a magnetic field of 1.0 T with a current of 1.5 kA. The discharge time constant is 30 ms so that the change in the solenoid field is negligibly small (less than 0.2%) during the macropulse duration of 25 μs. The incident angle of the bunch is controlled by a set of 'kicker coil' installed at the entrance of the solenoid. The effect of the beam emittance (10 πmm mrad in normalized emittance) on the beam trajectory is also negligibly small, because the incident angle ranges from 1 to 5 mrad in the experiment. The optical resonator consists of a pair of the Au-coated concave mirrors spaced 6.72 m apart. At the 22.3 MHz bunch repetition, the only one radiation pulse is stored in the resonator. Thus, the radiation pulse can interact with the electron bunch for 550 times during the macropulse.

The intensity of the spontaneous emission was measured for various field strength of the solenoid by removing the downstream cavity mirror. The cyclotron resonance condition

$$\omega = \omega_{co}/(1 - \beta_z) \tag{1}$$

where ω_{co}, γ and β_z are the cyclotron frequency, the electron relativistic factor and the normalized

axial velocity, gives a resonant frequency of 4.0 THz for typical experimental condition, an electron energy of 150 MeV, an incident angle of 5 mrad and a solenoid field of 1 T. The Larmor radius and frequency are 2.6 mm and 93 MHz, respectively, in this case. Under this conditions, the electron undergoes a half-cyclotron rotation while it passes through the solenoid. Fig. 1 shows the time traces of the response of the microwave diode with a cutoff frequency of 100 GHz. The detector signal is the pulses synchronized to the electron bunches and thus appears as it is hatched in the figure. The radiation signal when the solenoid field was off, (Fig. 1a), is used by the bending magnets. The radiation power increased when increasing the magnetic field up to 1 T. The radiation power with the solenoid field of 1 T exceeded that without the solenoid field by a factor of 10. In addition, the preliminary calibration of the detector indicated that the power of this cyclotron radiation was greater than that generated by the broad-band terahertz source fed by the femtosecond laser with the peak power of 2 kW at least by a factor of 50.

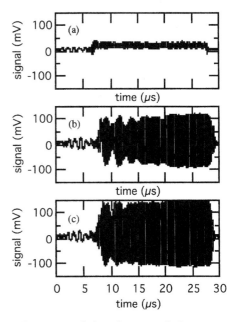

Fig. 1. Time trace of the microwave diode response. The solenoid field is 0, 0.6 and 1.0 T for (a)–(c), respectively.

3. Discussion

In the experiment, we measured only the power in the millimeter part of the spectral region rather than the power of whole spectrum. Because the resonant frequency is in terahertz region, the radiation is expected to have board-band spectrum from terahertz to millimeter wave range, i.e. the radiation is a short pulse.

The electric field produced by the electron is calculated from the Liénard–Wiechert potential given by

$$\mathbf{E}(\mathbf{x}, t) = \frac{q}{4\pi\varepsilon_0}\left[\frac{R(t')}{s(t')^3}(\mathbf{n}(t') - \boldsymbol{\beta}(t'))(1 - \boldsymbol{\beta}(t')^2)\right.$$
$$\left. + \frac{R(t')^2}{cs(t')^3}n(t') \times ((\mathbf{n}(t') - \boldsymbol{\beta}(t')) \times \dot{\boldsymbol{\beta}}(t'))\right] \quad (2)$$

where q is the charge of the particle, c the light velocity, ε_0 the vacuum dielectricity, $\dot{\boldsymbol{\beta}}$ is the normalized acceleration vector, $\mathbf{R}$ and R are the vector and distance from the particle to the observation point, $\mathbf{n}$ is a unit vector in the direction of $\mathbf{R}$, and s is defined as

$$s(t') = R(t') - \frac{\mathbf{R}(t') \cdot v(t')}{c}. \quad (3)$$

Quantities on right-hand side in Eq. (3) are to be evaluated at the retarded time

$$t' = t - \frac{R(t')}{c}. \quad (4)$$

The magnetic field is, then, given by

$$\mathbf{B}(\mathbf{x}, t) = \frac{1}{c}\mathbf{n}(t') \times \mathbf{E}(\mathbf{x}, t) \quad (5)$$

and then the magnitude of the Poynting vector is given by

$$|S(\mathbf{x}, t)| = \varepsilon_0 E^2. \quad (6)$$

The electric field produced by an electron with the initial angle $\theta_{0x} = 3$ mrad and the center of a circular motion $(x_{c0}, y_{c0}) = (0, 0)$ is shown in Fig. 2(a). Initial position of the electron at the entrance of the solenoid is $(x_0, y_0, z_0) = (0, -r_L, 0)$, where r_L is the Lamor radius which is to be 1.5 mm in this case. The coordinates of the observation points, (x_{ob}, y_{ob}, z_{ob}), is $(0, 0, 5$ m$)$, corresponds to

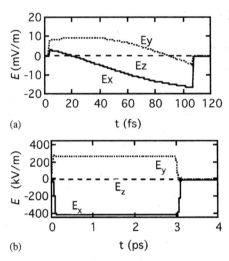

(a)

(b)

Fig. 2. Calculated time trace of the electric field of the radiation at the center of the resonator mirror: (a) shows the field generated by an electron and (b) shows that generated by the electron bunch.

the center of the downstream cavity mirror. The resonant wavelength is to be 65 μm, and corresponding optical period is 217 fs. During passing through the 3 m-long solenoid field, the electron with $\gamma = 300$ executes a half-circulation (166° in phase). As expected, the phases of the fields, E_x and E_y, do not vary by 2π: the fields alter its phase only by π. It depends on the observation point that if the field starts from negative or positive value. The z-component of the field, E_z, is negligibly small comparing to E_x, E_y.

In a practical situation, the electron bunches have a finite size in radial and axial direction. The electric field radiated by an electron bunch is evaluated by adding up the field produced by individual electron included in the bunch. The uniform electron distribution in a block, $4 \times 4 \times 0.9$ mm^3, is assumed. The bunch length of 0.9 mm corresponds to the pulsewidth of 3 ps. The electrons are located at $x_0 = [-2$ mm, 2 mm$]$ and $y_0 = [-2$ mm$-r_L, 2$ mm$-r_L]$, so that the center of the circular electron motion is located at $x_{c0} = y_{c0} = [-2$ mm, 2 mm$]$. Fig. 2(b) shows the time trace of the electric field observed on the downstream mirror. The pulsewidth of the field is almost same as that of electron bunch.

The amplitudes of the fields are much higher than that radiated by electrons undergoing a circular motion. This is because the fields produced by each electron are added up without canceling out each other in the case that the electrons undergo the half-rotation. The enhancement factor of the field is, then, roughly the number of electrons contained in the duration of the radiation field produced by single electron (~ 100 fs).

Similar field can be produced by an electron bunch passing through the bending magnet. However, in that case, the radiation spreads into a wide spatial region along the electron trajectory. Also, the interaction length is too short to store the radiation in the optical cavity in usual cases. In the solenoid field, the interaction length can be longer and the intense radiation may be emitted along the axis of the cyclotron motion. So there is the possibility to gain such half-cycle radiation field via the interaction with successively injected electron bunches.

To confirm if the cyclotron radiation from the electron undergoing a half-rotation propagates along the axis, the contour plot of the energy flux of radiation, $\int dt' |S(x, t')|$, are shown in Fig. 3 for the same condition used in Fig. 2(b). The energy flux has a peak at (0.15, 0.2 cm) and falls to 10% of the peak at a distance of 1.6 cm from the axis of the cyclotron motion and the resonator. The contours tend to be displaced from the axis toward first quadrant in the graph. This is because that electrons undergo the half-rotation mainly in the region $x > 0$ in this calculation. This tendency become more remarkable as larger the initial electron angle. With the initial angle up to 5 mrad, the radiation approximately propagates along the axis.

The radiation field, therefore, can be stored in the optical resonator. Furthermore, the scalar product, $\boldsymbol{\beta} \cdot \mathbf{E} > 0$, is positive in the region where the radiation is intense (first quadrant), while that is small in the forth quadrant where the radiation intensity is relatively weak and the velocity vectors intersect the polarization vectors with the large angles, In such situation, electrons lose their energy by the interaction with the radiation. Thus, there is possibly to amplify the

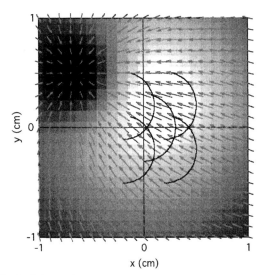

Fig. 3. Calculated energy flux distribution and the polarization vectors on the resonator mirror. The flux is more intense in brighter region. Arrows are the polarization vectors and arcs show the sample of the electron trajectories. Electrons move from $y < 0$ to $y > 0$ region.

radiation by the successive interaction with the electron bunches from the linac in the optical resonator.

4. Conclusion

The cyclotron radiator with the electron undergoing a half-cyclotron motion has potential to generate the intense and short terahertz radiation. By combining it with the electron bunching technology, this radiator can cover the shorter wavelength range. The radiation is regarded as the half-cycle radiation. There is no light source capable to generate such an exotic radiation at this time. We are expecting that this type of the radiator will be a powerful source in many scientific fields, especially in the fields that requires the high-time resolution with far infrared light. The experiment is now underway to investigate how the half-cycle radiation gains its amplitude by the successive interaction with the electron bunches in the optical resonator.

Acknowledgements

This work is supported by a Grant-in-Aid for Scientific Research from Japan Society for the Promotion of Science.

References

[1] B. Ferguson, X.-C. Zhang, Nature Mater. 1 (2002) 26.
[2] M. Asakawa, H. Nishiyama, K. Goto, N. Ohigashi, Y. Tsunawaki, K. Imasaki, Nucl. Instr. and Meth. A 438 (2002) 286.

Available online at www.sciencedirect.com

Nuclear Instruments and Methods in Physics Research A 528 (2004) 157–161

NUCLEAR
INSTRUMENTS
& METHODS
IN PHYSICS
RESEARCH
Section A

www.elsevier.com/locate/nima

Temporal structure of resonator output in a millimeter-wave prebunched FEL

Yukio Shibata[a],*, Kimihiro Ishi[a], Toshiharu Takahashi[b], Tomochika Matsuyama[b]

[a] *Institute of Multidisciplinary Research for Advanced Materials, Tohoku University, Katahira, Sendai 980-8577, Japan*
[b] *Research Reactor Institute, Kyoto University, Kumatori, Osaka 594-0494, Japan*

Abstract

A short-bunched beam of electrons of an L-band linac of 38 MeV is guided to a cylindrical-closed resonator. Wavepackets of coherent transition radiation (CTR) generated from bunches passing through the mirrors are coherently summed up in the resonator. The amplification of the output of the resonator is confirmed in comparison with spontaneous CTR. The temporal structure of the output, observed with a hot-electron bolometer having the response time of 0.35 μs, shows oscillating characteristic of about 1 μs in time scale. The result strongly suggests the inter-bunch distance varies rapidly over the duration of a macropulse, and it consequently results in degradation of the coherent summing up in the resonator.
© 2004 Elsevier B.V. All rights reserved.

PACS: 41.60.Cr; 41.60.Ap; 42.72.Ai; 29.17.+w

Keywords: Prebunched FEL; Coherent transition radiation; Millimeter wave

1. Introduction

In a prebunched FEL, electromagnetic impulses of coherent radiation emitted from a short-bunched beam of electrons are coherently summed up in a resonator [1–5]. The output of the resonator depends on the quality of the resonator and on the quality of the electron beam. However, research on the prebunched FEL is limited and its properties are not well studied.

To elucidate the properties of the prebunched FEL, we lead an electron beam of a linac to a closed resonator. When electron bunches pass through the mirrors of the resonator, coherent transition radiation (CTR) is emitted in the millimeter wavelength region [6]. The electromagnetic impulses of CTR are coherently summed up in the resonator, provided that the length of the resonator is well adjusted to match with the inter-bunch distance of the linac. Here, to elucidate the process of the coherent superposition of the impulses, we observe growth of the resonator output with a hot electron bolometer having the response time of 0.35 μs.

2. Experiment

The experimental setup is schematically shown in Fig. 1. The electron beam generated with the

---

*Corresponding author. Tel./fax: +81-22-217-5350.
E-mail address: shibatay@tagen.tohoku.ac.jp (Y. Shibata).

0168-9002/$ - see front matter © 2004 Elsevier B.V. All rights reserved.
doi:10.1016/j.nima.2004.04.037

L-band linac of Research Reactor Institute, Kyoto University, was guided to a closed resonator. The wavepackets of CTR emitted from bunched electrons passing through the mirrors were coherently summed up in the resonator. The output of the resonator was guided to a grating-type far-infrared spectrometer equipped with several gratings and assembly of many filters and polarizers, which covered wavelength range from 0.1 to 6 mm. In usual case, the output was detected with a liquid-helium-cooled Si bolometer.

The cylindrically symmetric-closed resonator, shown in the inset of Fig. 1, was composed of a polished aluminum pipe and two plane mirrors: The size of the aluminum pipe was (ϕ114 mm × 456 mm) in (inner diameter × length). The plane mirror M1 was made of 15 μm-thick Al-foil and was attached to the aluminum pipe. The round plane mirror M2 was Al-evaporated-fused quartz with a size of ϕ130 mm × 1 mm (diameter × thickness), with diameter larger than that of the resonator pipe.

The central part of ϕ20 mm of M2 was not coated and transparent for the millimeter wave. The output radiation was observed through the coupling window whose power reflectance and transmittance were 21% and 79% at $\lambda = 1$ mm, respectively. The mirror M2 was put on a slide table and the length of the resonator was controlled with a stepping motor around 461 mm, the two times inter-bunch distance.

The conditions of the electron beam were as follows: The radio frequency was 1.3008 GHz and the inter-bunch distance was 230.5 mm. The energy and its spread were 38 MeV and 11%, respectively. The duration of a macropulse was 3 μs and its repetition was 17 Hz. The average beam current was 11.5 μA. Hence the one macro pulse was composed of about 3900 bunches and the average number of electrons was 1.1×10^9 per bunch. The transverse cross-section of the beam was nearly circular and its diameter was about 12 mm at the mirror M1 of the resonator.

3. Results and discussion

3.1. Detuning curve

The detuning curve or the variation of the output of the resonator with the variation of the position of the mirror M2 was observed at the wavelength λ of 2.3 mm and is shown in Fig. 2. The observed curve shows periodic structure composed of the main peak, the secondary peak and a few weak sub-peaks with the period of $\lambda/2$.

The amplitude reflectivity r of the resonator per trip around was derived from the bandwidth of the maximum peak to be 0.87 at $\lambda = 2.3$ mm. Using the reflectivity, the quality factor of the resonator is obtained from the relation, $r = \exp(-2\pi/Q)$, to be 45 at $\lambda = 2.3$ mm. The value of the quality factor is higher than that of the open resonator used in our previous experiments [1].

3.2. Output spectrum and spontaneous CTR

We observed the spectrum of the output, keeping the resonator length at the maximum

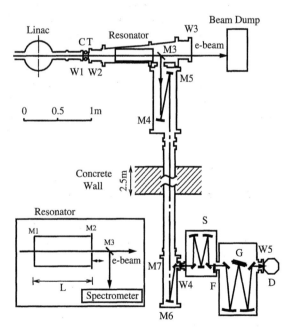

Fig. 1. Experimental layout. CT: a current transformer; W1, W2: titanium windows; W3: an aluminum window; M1–M3, M4, M7: plane mirrors; M5, M6, concave mirrors; F: filters; G: a grating; D: a detector. The inset shows the cylindrical closed resonator with the plane mirrors M1 and M2 which has a coupling window of ϕ20 mm in center.

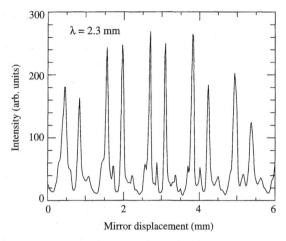

Fig. 2. Detuning curve observed at $\lambda = 2.3$ mm.

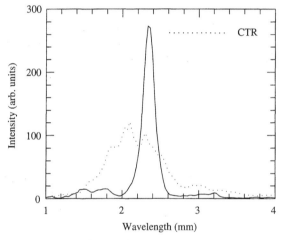

Fig. 3. Output spectrum observed at the maximum peak of detuning curve of $\lambda = 2.3$ mm. The dotted curve shows the spectrum of CTR observed after taking away the mirror M2 in Fig. 1.

peak of the detuning curve of $\lambda = 2.3$ mm. The results are shown by the solid curve in Fig. 3.

According to the theory of the prebunched FEL, the output is composed of higher harmonics of the radio frequency of the accelerator. The resolution of the grating spectrometer was 0.1 mm at $\lambda = 2.3$ mm and was too low to resolve components of the harmonics. The observed spectrum hence appeared to be continuous.

To confirm the amplification of radiation in the resonator, we took away the mirror M2 in Fig. 1,

and then observed spectrum. The result is shown by the dotted curve in Fig. 3. The observed spectrum was superposition of forward CTR emitted from the mirror M1 and backward CTR from the mirror M3. On the other hand, the wavepackets coherently summed up in the resonator were superposition of CTR emitted from the mirrors M1 and M2. The distance between the mirrors M2 and M3 was 8 cm and was shorter than the resonator length. Hence, as the first approximation, we can regard the observed spectrum as the spontaneous CTR generated within the resonator.

The comparison of the solid curve with the dotted curve shows that the output intensity is higher than that of the spontaneous CTR around the wavelength λ of 2.3 mm and that the peak intensity is amplified by a factor of 2.8 compared with the spontaneous CTR. The result confirms the amplification of radiation in the resonator.

Theoretically, the amplification factor is given by $(1 + r)/(1 - r)$ for lossless resonator, where r is the amplitude reflectivity of the resonator per trip around. The reflectivity, derived from the half-width of the main peak, results in the theoretical amplification factor of 14.4, which is much larger than the observed one. The difference will be partly explained from the loss of the coupling window (see for example [7]). The rough analysis of the influence of the window, however, suggests that there remains other reason which depresses the resonator output.

3.3. Temporal structure of resonator output

To observe the temporal structure of the resonator output, we replaced the Si bolometer to an InSb hot-electron bolometer with the response time of 0.35 μs. We then observed the resonator output, keeping the resonator length to the maximum peak of the detuning curve of $\lambda = 2.3$ mm. The result is shown by the solid curve in Fig. 4. The temporal structure of the electron current was also measured with a current transformer having the response time of about 10 ns, and the result is shown in Fig. 5 by the curve of label 4.

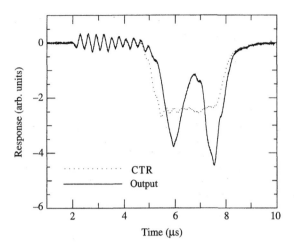

Fig. 4. Temporal structure of the resonator output (solid curve) and CTR (dotted curve) observed at $\lambda = 2.3$ mm with the InSb detector.

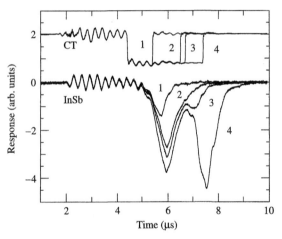

Fig. 5. Temporal structure of the resonator output at $\lambda = 2.3$ mm (lower curves) and of the beam current (upper curves). The figures show duration of the macropulse; 1 = 1.0, 2 = 2.1, 3 = 2.3, 4 = 3.0 μs.

The resonator output varies rapidly in the time scale of about 1 μs. On the other hand, the electron current shows a little variation over the 3 μs duration. To make comparison clear, the temporal structure of CTR was also measured with the InSb bolometer at λ of 2.3 mm, after taking away the mirror M2 from the resonator. The result is shown by the dotted curve in Fig. 4; the temporal structure of CTR does not show such the rapid variation as that of the output.

In addition, we altered the duration of the macro pulse from 3.0 to 2.3, 2.1 and 1.0 μs, keeping the other beam conditions unchanged. The result is shown by the solid curves in Fig. 5, where the figures attached to the curve show the pulse duration; (1) = 1.0, (2) = 2.1, (3) = 2.3 and (4) = 3.0 μs, respectively. The upper curves also show the temporal structure of the beam current measured with the current transformer. These results show that in the resonator it occurred not only coherent summation but coherent subtraction of the radiation field, even if the resonator length was not altered. In a bunch train the phase relation between the bunches, which relation assures coherent summing up of the wavepackets in the resonator, fluctuates over the duration of the macropulse.

The comparison of the temporal structure of the resonator output with that of CTR in Fig. 4 shows that initial increase of the resonator output was delayed by about 0.3 μs from that of CTR. Theoretical simulation of the temporal structure showed that the delay time of 0.3 μs was not explained from the reflectivity of the resonator derived from the band width of the main peak of the detuning curve. The delay suggests that in the bunch train the fluctuation of the phase relation is large at the initial stage of the macropulse.

In Fig. 5, when the duration of the macro pulse increases, the resonator output also increases even the early stage of the macro pulse. The output intensity at $t = 5.5$ μs, for example, increases with the increase of the duration of the pulse. On the other hand, the electric current changes a little with the increase of the duration. When we carefully observe the current, however, the peak current increases with the decrease of the pulse duration.

These responses suggest the effect of the beam loading: a passing beam charge induces wakefields in the RF cavity and the effective RF voltage varies rapidly. The beam loading probably caused the following two influences on the bunch train: (1) The inter-bunch distance fluctuates over the duration of the macropulse. The delay of the resonator output mentioned above will be corresponding to violent fluctuation of the inter-bunch distance in the initial stage of the beam loading. (2) The characteristics of the bunch structure, such as density distribution of the electrons in a bunch, ejected in the initial stage of the bunch train differ probably from those in the middle or final stages of the bunch train. These influences degrade the coherent summing up of the radiation field in the resonator, and will probably explain one of the reasons of the difference in the amplification factor described earlier.

Acknowledgements

The authors thank Dr. E. Bessonov of Lebedev Physical Institute, RAS, for fruitful discussion, and the members of the coherent radiation group of Tohoku University for helpful discussion.

References

[1] Y. Shibata, et al., Phys. Rev. Lett. 78 (1997) 2740.
[2] Y. Shibata, et al., Nucl. Instr. and Meth. B 282 (1998) 49.
[3] V.I. Alexeev, et al., Nucl. Instr. and Meth. A 282 (1989) 436.
[4] H.C. Lihn, et al., Phys. Rev. Lett. 76 (1996) 4163.
[5] S. Sasaki, et al., Nucl. Instr. and Meth. A 483 (2002) 209.
[6] Y. Shibata, et al., Phys. Rev. E 49 (1994) 785.
[7] G.T. McNice, V.E. Derr, IEEE J. Quantum Electron. QE 5 (1969) 569;
    T. Li, H. Zucker, J. Opt. Soc. Am. 57 (1967) 984.

Available online at www.sciencedirect.com

Nuclear Instruments and Methods in Physics Research A 528 (2004) 162–166

**NUCLEAR
INSTRUMENTS
& METHODS
IN PHYSICS
RESEARCH**
Section A

www.elsevier.com/locate/nima

A prebunched FEL using coherent transition radiation in the millimeter wave region

Yukio Shibata[a],*, Kimihiro Ishi[a], Toshiharu Takahashi[b],
Tomochika Matsuyama[b], Fujio Hinode[c], Yasuhiro Kondo[d]

[a] *Institute of Multidisciplinary Research for Advanced Materials, Tohoku University, Katahira, Sendai 980-8577, Japan*
[b] *Research Reactor Institute, Kyoto University, Kumatori, Osaka 594-0494, Japan*
[c] *Laboratory of Nuclear Science, Tohoku University, Mikamine, Sendai 982-0826, Japan*
[d] *Graduate School of Engineering, Tohoku University, Aramaki, Sendai 980-8579, Japan*

Abstract

Using a short-bunched beam of electrons of a linear accelerator, the electromagnetic impulses of coherent transition radiation (CTR) are superposed on the subsequent impulses coherently. The output of a closed resonator is compared with spontaneous CTR to verify amplification of radiation. The main cavity mode is assigned to be TM_{02} from observation of the radial distribution of the intensity. Experiment on the influence of the size of the resonator shows that the output intensity increases with the volume of the resonator.
© 2004 Elsevier B.V. All rights reserved.

PACS: 41.60.Cr

Keywords: Prebunched FEL; Coherent transition radiation; Millimeter wave

1. Introduction

From a short-bunched beam of electrons of a linac, coherent radiation such as synchrotron radiation and transition radiation is emitted in the millimeter wavelength region [1,2]. The coherent radiation emitted from every bunch of the linac interferes coherently with one another [3,4]. The inter-bunch coherence has the potential to amplify the coherent radiation in a resonator: the wavepackets of the coherent radiation circulate in the resonator and superpose on the subsequent bunches to stimulate radiation. In other words, the output of the resonator is amplified due to the coherent summing up of the electromagnetic impulses. In addition, the output has a quasi-continuous broad-band spectrum [5]. We can regard the amplification of radiation as a prebunched FEL, because the electron beam of the linac is already bunched in scale of wavelength.

The prebunched FEL has been studied a little [5–12]. In this work, we have focused on the prebunched FEL based on coherent transition radiation (CTR). To elucidate the properties of the

---

*Corresponding author. Tel./fax: +81-22-217-5350.

E-mail address: shibatay@tagen.tohoku.ac.jp (Y. Shibata).

prebunched FEL, we have made a variety of experiments using several resonators.

2. Experiment

Our experiment was carried out at two facilities; Research Reactor Institute, Kyoto University (RRIKU) and Laboratory of Nuclear Science, Tohoku University (LNSTU). The beam parameters of the linac are listed in Table 1. The bunch length was estimated from the spectrum of CTR to be about 0.3 mm (1 ps) for LNSTU [2] and 6 mm (20 ps) for RRIKU, respectively. Except for the beam conditions, experimental arrangement including a resonator and a spectroscopic system was nearly the same in both facilities.

The electron beam of a linac was guided to a closed resonator (see Fig. 1). The electromagnetic impulses of CTR emitted from bunches passing through the resonator were superposed on the subsequent impulses in the resonator. The output

of the resonator was guided to a grating-type far-infrared spectrometer and was detected with a low-temperature Si bolometer.

The main part of our experiment was carried out at RRIKU, where we fabricated a cylindrically symmetric closed resonator composed of a polished aluminum pipe and two plane mirrors; the size of the aluminum pipe was ($\varnothing 114$ mm × 456 mm) in (inner diameter × length). The M1 was made of a 15 μm-thick Al-foil, and M2 was Al-evaporated fused quartz with the size of ($\varnothing 130$ mm × 1 mm) in (diameter × thickness). The mirror M2 had a coupling window of $\varnothing 20$ mm in center. The mirror M2 was movable and the length of the resonator was controlled with a stepping motor around 461 mm, two times the inter-bunch distance.

When we examined the influence of the size of the resonator on its output, we replaced the resonator with the other one, as described later.

3. Results and discussion

3.1. Detuning curve

We observed the variation of the output intensity as a function of the position of the mirror M2. The results observed at $\lambda = 2.3, 2.8$, and 3.4 mm are shown in Fig. 2. The detuning curves have a periodic structure composed of the main peak, the secondary peak and a few weak peaks with the period of $\lambda/2$. The intensity ratio of the secondary peak to the main peak increased with decrease of the wavelength.

The amplitude reflectivity r of the resonator was derived from the bandwidth of the maximum peak to be 0.87 and 0.88 at $\lambda = 2.3$ and 2.8 mm, respectively. The quality factor of the resonator was obtained from the relation, $r = \exp(-2\pi/Q)$, to be 45 and 50 at $\lambda = 2.3$ and 2.8 mm, respectively. These values are higher than that of the open resonator in our previous experiments [5].

3.2. Spectrum

We observed the spectrum of the resonator output, keeping M2 at the maximum peak of the

Table 1
The beam conditions of the linac

Facility	LNSTU	RRIKU
RF (GHz)	2.856	1.3008
E (MeV)	150	37
$\Delta E/E$ (%)	2	7
Macropulse (μs × Hz)	0.8 × 16.67	1.9 × 13
Current (μA)	0.4	11
$N_{electron}$/bunch	6.6×10^7	2.1×10^9
Beam size (mm × mm)	7 × 8	12 × 12

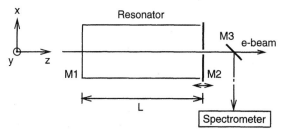

Fig. 1. Schematic layout of a closed resonator. The center of M2 has a coupling window of $\varnothing 20$ mm.

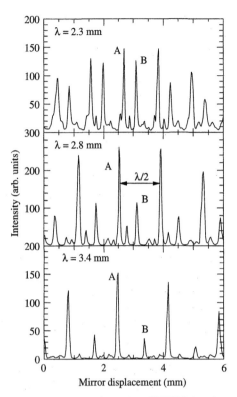

Fig. 2. Detuning curves observed at RRIKU for the $\varnothing 114$ resonator.

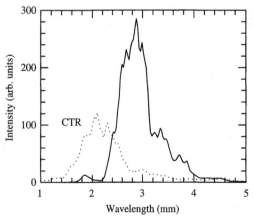

Fig. 3. Spectrum of the output observed at the maximum peak of the detuning curve of $\lambda = 2.8$ mm. The dotted curve shows the spectrum of spontaneous CTR.

detuning curve of $\lambda = 2.8$ mm. The results are shown by the solid curve in Fig. 3. The output has a continuous spectrum with a broad peak.

To confirm the amplification of radiation due to the coherent summing up, we observed the spectrum of the spontaneous CTR, by taking away the mirror M2 from the resonator in Fig. 1. The observed spectrum is shown by the dotted curve in Fig. 3. In comparison with the CTR, the output has high intensity in a wide wavelength range and its peak intensity was amplified by a factor of 14. The theoretical amplification of the coherent summing up is given by $(1 + r)/(1 - r)$ for lossless resonator. Using the value of r derived from the detuning curve, we obtain the amplification factor of 16, which is in agreement with the observed one.

3.3. Transverse mode

The detuning curve shows that there exist a few transverse modes. To identify the mode, we observed the radial distribution of the intensity of radiation by the following way.

At first, in Fig. 1 we replaced the mirror M2 with another plane mirror which had a slit-like window (window size; 90 mm $\times$ 4 mm in width $\times$ height). We also replaced the mirror M3 with a small plane mirror (M3′), which was put on a slide table. The resonator length was controlled to the main peak of the detuning curve, and then the radial distribution of the intensity was observed by moving the mirror M3′ along the x-axis. The results observed at $\lambda = 2.3, 2.8$ and 3.4 mm are shown in Fig. 4.

The observed distributions are nearly symmetric with respect to the center of the mirror. However, the position of the central minimum was shifted by about 8 mm from the mirror center. The shift was probably caused from experimental errors in control of the electron orbit and errors in the optical alignment of the mirrors.

CTR is emitted conically, and its radiation field is consistent with the TM_{0n} modes. The size of the resonator used was not large enough to allow the lowest mode TM_{01}. We hence calculated the theoretical distribution of the TM_{02} mode, and the result is shown by the dotted curve in Fig. 4. The theory is in accordance with the experimental distribution.

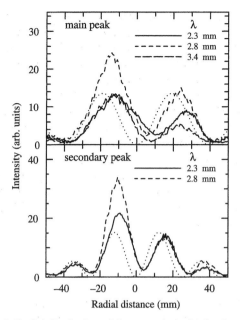

Fig. 4. Radial distribution of the output intensity for the main peak (upper) and the secondary peak (lower). The dotted curves show the theoretical distribution of the TM_{02} mode (upper) and the TM_{03} mode (lower).

The radial distribution of the secondary peak was also observed at $\lambda = 2.3$ and 2.8 mm. The results are shown in Fig. 4. The theoretical distribution of the TM_{03} mode is also shown by the dotted curve and is in accordance with the observed distribution. We therefore assigned the main peak to the TM_{02} mode and the secondary peak to the TM_{03} mode.

3.4. Influence of the volume of resonator

Since CTR is emitted conically with respect to the electron orbit, we expect that the prebunched FEL should have the optimum size in cross-section and length, where the radiation field of CTR is most effectively converted to the main cavity mode. From this view point, we have examined the influence of the size of the resonator on its output. The influence was complicated and difficult to analyze. Here, we show briefly our results and the details will be reported elsewhere.

(A) In RRIKU, we examined the influence of the cross-sectional size of the resonator on its

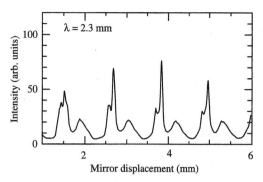

Fig. 5. Detuning curve of $\lambda = 2.3$ mm for the $\varnothing\,54$ resonator. The intensity is shown by the same scale as in Fig. 2.

output. We replaced the resonator of $\varnothing\,114$ mm in inner diameter to another one of $\varnothing\,54$ mm, keeping the resonator length unchanged. We observed the detuning curve at a few wavelengths, and the result of $\lambda = 2.3$ mm is shown in Fig. 5.

In comparison with the detuning curve of the $\varnothing\,114$ resonator, the output intensity of the $\varnothing\,54$ mm resonator was weak. Fig. 5 also shows that the depression of the secondary peak was severe in comparison with that of the main peak.

(B) In LNSTU, we examined the influence of the resonator length on its output. We prepared two rectangular closed resonators; one had the size of (80 mm × 60 mm × $2L_b$) in (width × height × length) and the other of (80 mm × 60 mm × $3L_b$), where L_b is 105 mm, the inter-bunch distance of the linac of LNSTU.

Using the resonators, we observed detuning curves at several wavelengths and the output spectra at its maximum peaks. The output spectrum observed at the maximum peak of the detuning curve of $\lambda = 2.8$ mm is shown in Fig. 6 by the solid curve for the $3L_b$ resonator and by the dashed curve for the $2L_b$, respectively. The output intensity of the $3L_b$ resonator was higher than that of the $2L_b$. To confirm the amplification of radiation due to coherent summing up, we also observed the spectrum of spontaneous CTR after taking away the downstream mirror M2 of the resonator. The result is shown by the dotted curve in Fig. 6. The comparison of the dotted curve with the output confirms the amplification of radiation in the resonator.

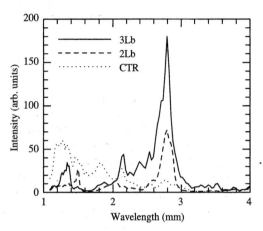

Fig. 6. Output spectrum observed at the maximum peak of the detuning curve of $\lambda = 2.8$ mm. The dotted curve shows the spectrum of spontaneous CTR.

4. Summary

A bunch train of the linac was guided to a closed resonator and CTR was generated in the resonator. The wavepackets of CTR were superposed on the subsequent bunches to stimulate CTR. The output intensity was compared with the spontaneous CTR to verify amplification of radiation in the resonator. Since the electric field of CTR is axially symmetric with respect to the electron beam and the velocity of the electron has no transverse component, no interaction is expected between the electron bunch and CTR. The amplification of radiation hence is due to the coherent summing up of the radiation field of CTR, and it depends on the quality of the resonator.

On the other hand, in usual FEL, the bunching process is essential, in which the interaction of electrons with electromagnetic wave causes modulation of the electron distribution in scale of the wavelength. In our case, the bunching process is not expected, but we consider that the electron bunches can emit much more energy in the resonator due to stimulated emission than the bunches do without resonator.

In the millimeter wavelength region, coherent radiation emitted from a short-bunched beam of a linac is drastically enhanced by several orders of magnitude, in comparison with usual (incoherent) radiation. The present work showed that the CTR was amplified by an order of magnitude in the resonator. This prebunched FEL is simple and will be applicable to various kinds of coherent radiation such as synchrotron radiation, Smith–Purcell radiation and diffraction radiation, to obtain a radiation source with high brightness in the millimeter or longer wavelength region.

Acknowledgements

We appreciate Prof. T. Ohsaka of IMRAM, Tohoku University, for his encouragement and support. We also thank Mr. T. Tsutaya of IMRAM, Tohoku University, for his technical support and assistance.

References

[1] T. Nakazato, et al., Phys. Rev. Lett. 63 (1989) 1245.
[2] Y. Shibata, et al., Phys. Rev. E 49 (1994) 785.
[3] Y. Shibata, et al., Phys. Rev. A 44 (1991) R3445.
[4] T. Takahashi, et al., Rev. Sci. Instrum. 69 (1998) 3770.
[5] Y. Shibata, et al., Phys. Rev. Lett. 78 (1997) 2740.
[6] V.I. Alexeev, et al., Nucl. Instr. and Meth. A 282 (1989) 436.
[7] V.A. Alexeev, E.G. Bessonov, A.V. Serov, Nucl. Instr. and Meth. A 282 (1989) 439.
[8] A. Serov, Nucl. Instr. and Meth. A 308 (1991) 144.
[9] A. Serov, Nucl. Instr. and Meth. A 359 (1995) 70.
[10] Y. Shibata, et al., Nucl. Instr. and Meth. B 282 (1998) 49.
[11] H.C. Lihn, et al., Phys. Rev. Lett. 76 (1996) 4163.
[12] S. Sasaki, et al., Nucl. Instr. and Meth. A 483 (2002) 209.

Available online at www.sciencedirect.com

SCIENCE DIRECT®

ELSEVIER

Nuclear Instruments and Methods in Physics Research A 528 (2004) 167–171

NUCLEAR
INSTRUMENTS
& METHODS
IN PHYSICS
RESEARCH
Section A

www.elsevier.com/locate/nima

Single-mode simulations of a short Rayleigh length FEL

W.B. Colson*, J. Blau, R.L. Armstead, P.P. Crooker

Physics Department, Naval Postgraduate School, Monterey, CA 93943, USA

Abstract

Free electron lasers can make use of a short Rayleigh length optical mode in order to reduce the intensity on resonator mirrors. A simulation method is used that includes the dynamics of this rapidly focusing optical mode and the macroscopic and microscopic electron evolution. The amplitude and phase of the optical fields are represented by a single Gaussian mode. The simulation runs in seconds on small laptop computers and can be used for system analysis. Published by Elsevier B.V.

PACS: 41.60Cr

Keywords: Free electron laser; Short Rayleigh length

1. Introduction

In a high-power free electron laser (FEL), a short Rayleigh length resonator can be used to reduce the optical intensity on the mirrors. Typically, the short optical Rayleigh length requires the use of a short undulator so that the expanding optical beam does not scrape power on the magnets. This paper develops a simple simulation model that describes the energy extraction in a high-power FEL using a short Rayleigh length optical mode with a deterministic form that tracks the rapid diffraction of the laser beam. Thousands of sample electrons interact with the laser field in the middle of the rapidly diffracting wave front. The simulation method can be used over a wide range of parameters to study FEL characteristics more efficiently.

2. The optical mode

The optical mode in an FEL is determined by the electron beam interaction, diffraction, and by the resonator mirrors. In most FEL oscillators, it has been observed that more than 90% of the optical power remains in the fundamental mode at saturation where the gain is reduced to equal the resonator output coupling $\alpha_n = 1/Q_n$. Higher-order modes have little impact on the far-field limit and the FEL interaction. The optical electric field in the fundamental mode [1] is

$$E(r, z) = E_0(\lambda Z_0/A)^{1/2} \, e^{i(kz - \omega t + \phi)} \exp(-\pi r^2/A) \quad (1)$$

*Corresponding author. Tel.: +1-831-656-2765.
E-mail address:* colson@nps.edu (W.B. Colson).

0168-9002/$ - see front matter Published by Elsevier B.V.
doi:10.1016/j.nima.2004.04.039

where $\phi(r,z) = -\tan^{-1}[(z-z_w)/Z_0] + \pi r^2(z-z_w)/AZ_0$ is the optical phase, $A = \lambda Z_0[1+(z-z_w)^2/Z_0^2]$ is the optical mode area, $\lambda = 2\pi/k$ is the FEL optical wavelength, k is the optical wave number, ω is the optical frequency, z is the position along the undulator ($z = 0$ at the beginning, $z = L$ at the end of the undulator), r is the radial position in the mode, z_w is the position of the optical mode waist, Z_0 is the mode's Rayleigh length, and E_0 is the optical electric field amplitude at the mode waist $z = z_w$. The optical electric field is determined by the intra-cavity power, and is given by $E_0^2 = (8\pi Q_n P_{out})/(cZ_0\lambda D_{duty})$ where the FEL average output power is P_{out}, the laser beam duty factor is $D_{duty} = l_b\Omega/c$, l_b is the electron micropulse length, Ω is the pulse repetition frequency, and c is the speed of light [2].

3. The electron beam

The electron beam's initial energy is $E_b = \gamma_0 mc^2$ with a Lorentz factor of γ_0, where m is the electron mass. A realistic beam has a small spread in energies, but in the short undulator this is generally of no consequence. Each micropulse of electrons contains charge q with a repetition frequency of Ω, giving an average electron beam current of $I_{av} = q\Omega$. The average power flowing in the electron beam is $P_b = I_{av}E_b/e$ where e is the electron charge magnitude.

The electron beam's waist radius is r_b at the beam focus, located along the undulator at z_b. A sample electron's initial transverse positions (x_0, y_0) and angles (θ_x, θ_y) are distributed as Gaussians. Consistent with the normalized beam emittance, $\varepsilon_n = \gamma_0 r_b\theta_b$, the resulting electron beam's angular spread is θ_b. For simplicity, the beam is taken to be round here, but the simulation can handle beam asymmetries in x and y. There is no significant betatron focusing in the short undulator, so the electron's injection angles remain constant along the undulator in each direction. An electron's transverse position at distance z along the undulator is given by $x(z) = x_0 + \theta_x(z - z_b)$ and $y(z) = y_0 + \theta_y(z - z_b)$. The

laser interaction does not affect the transverse positions of the electrons, but their transverse positions significantly affect the electron's microscopic longitudinal position, and therefore bunching, energy extraction, and gain.

4. The FEL interaction

An electron's interaction with the transverse fields of the laser light is made possible by passing the beam through the transverse fields of the undulator. The undulator field is linearly polarized with the form $\mathbf{B} = B(0, \sin k_0 z, 0)$ on the undulator axis where the electrons travel. The peak magnetic field in the undulator is B, and the undulator period is $\lambda_0 = 2\pi/k_0$. The electron's transverse motion in the undulator is $\boldsymbol{\beta}_\perp = -(\sqrt{2}K/\gamma)(\cos k_0 z, 0, 0)$ where the undulator parameter is $K = eB\lambda_0/(2\sqrt{2}\pi mc^2)$, and the electron's velocity is $\mathbf{v} = c\boldsymbol{\beta}$. The linearly polarized electric and magnetic optical fields describing the optical mode above are $\mathbf{E}_s = E(\cos\psi, 0, 0)$, $\mathbf{B}_s = E(0, \cos\psi, 0)$ where $\psi = kz - \omega t + \phi$, and the optical mode's amplitude E and phase ϕ are given earlier. Substituting $\boldsymbol{\beta}_\perp$, $\mathbf{E}_s$, and $\mathbf{B}_s$ into the Lorentz force equations, an electron's Lorentz factor, $\gamma = (1 - \boldsymbol{\beta}^2)^{-1/2}$, evolves according to

$$\gamma' = (eKE/\gamma mc^2)[J_0(\xi) - J_1(\xi)]\cos(\zeta + \phi) \quad (2)$$

where J_0 and J_1 are Bessel functions, $\xi = K^2/2(1+K^2)$, $\zeta = (k+k_0)z - \omega t$ is the electron phase, $\gamma' = d\gamma/dz$, and integration along the undulator is in small steps dz instead of time, using $dz = cdt$. "Fast" longitudinal motion in the linearly polarized undulator has been averaged away, giving rise to the Bessel function factors $J_0(\xi) - J_1(\xi)$ [3]. During the integration of the dynamic Lorentz factor $\gamma(z)$ along the undulator, $z = 0 \rightarrow L$, the optical mode's amplitude E and phase ϕ and the electron's transverse positions, x and y, are simply evaluated at each step. The evolution of the electron phase $\zeta = (k+k_0)z - \omega t$ is crucial to FEL performance, and describes electron bunching, gain, and extraction [4].

The electron phase ζ is a microscopic variable measuring the electron position on the optical wavelength scale. The initial electron phases ζ_0 are uniformly spread over a 2π range. When there is no interaction ($\gamma' = \mathrm{d}\gamma/\mathrm{d}z = 0, \gamma = \gamma_0$), the electron phase is determined by $\zeta' = v_0/L$ where the initial electron phase velocity is $v_0 = L[(k + k_0)\beta_z(0) - k]$ and $c\beta_z(0)$ is the electron's initial velocity. Integration gives $\zeta = \zeta_0 + zv_0/L$. When there is an interaction ($\gamma' \neq 0$), the Lorentz factor $\gamma(z)$ evolves by integrating γ' in (2) including self-consistent changes in the electron phase determined by $\zeta' = [v_0 + 4\pi N(\gamma - \gamma_0)/\gamma_0]/L$.

The simulation uses a large number of sample electrons, ranging from 10^3 to 10^4 as more are needed for a larger radial spread r_b. After integration of all the sample electrons through the undulator length ($z = 0 \to L$ in a few hundred small steps $\mathrm{d}z$), the electron beam's extraction is given by $\eta = \langle \gamma_0 - \gamma \rangle/\gamma_0$ where $\langle \ldots \rangle$ is an average over the sampled electrons. If the self-consistent changes in ζ are not included above, $\langle \ldots \rangle$ and η are always zero. In an FEL oscillator, the optical wavelength evolves freely to the value for maximum gain in weak optical fields, or maximum energy extraction in strong optical fields at saturation. The waveform described by E and ϕ above has a single, fixed wavelength that determines the value of the initial phase velocity v_0. The simulation searches through values of v_0 (typically from $v_0 = 0$ to 16) and finds the maximum extraction η (typically around $v_0 = 8-9$). At the best value of v_0 found, the simulation calculates the FEL's final extraction and the resulting output optical power $P_{calc} = P_b\eta$. The value of the optical power that would lead to steady-state saturation can be determined by iteration. At the end of each iteration, the calculated output power is used as the input for the next iteration. If the initial power is above or below the steady-state value, it decreases or increases appropriately over a few iterations until the stable, steady-state power is found.

5. Simulation results

As an example, take the initial electron beam energy to be $E_b \approx 100\,\mathrm{MeV}$ ($\gamma_0 \approx 197$) with micropulse peak current $I_{peak} \approx 1500\,\mathrm{A}$, micropulse length $l_b \approx 0.3\,\mathrm{mm}$ (1 ps duration), micropulse

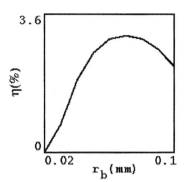

Fig. 1. The FEL extraction $\eta(r_b)$ shows optimum electron beam focal radius r_b.

charge $q \approx 1.5\,\mathrm{nC}$, electron beam radius $r_b \approx 70\,\mu\mathrm{m}$ at the beam focus $z_b = 0.5L$, and normalized emittance $\varepsilon_n \approx 9\,\mathrm{mm\text{-}mradians}$. The undulator has $N = 14$ periods each of length $\lambda_0 \approx 2.7\,\mathrm{cm}$ (total length $L \approx 37\,\mathrm{cm}$) with undulator parameter $K \approx 1.4$ resulting in optical wavelength $\lambda \approx 1\,\mu\mathrm{m}$. Resonator mirrors have an output coupling of $\alpha_n = 1/Q_n \approx 50\%$ ($Q_n \approx 2$), and their curvature determines the Rayleigh length $Z_0 \approx 2.6\,\mathrm{cm}$, dimensionless Rayleigh length $z_0 = Z_0/L \approx 0.07$, creating an optical mode waist $w_0 \approx 90\,\mu\mathrm{m}$ at the center of the undulator $\tau_w = 0.5L$.

Fig. 1 shows the FEL extraction $\eta(r_b)$ as the electron beam focal radius is varied from $r_b = 0.02\,\mathrm{mm}$ up to $r_b = 0.1\,\mathrm{mm}$ with fixed emittance ε_n. As r_b is increased from $0.02\,\mathrm{mm}$, the optical output increases from $\eta \approx 0$ extraction to a peak of $\eta \approx 3.1\%$ at $r_b = 0.07\,\mathrm{mm}$. As the electron beam focal radius is increased further from $r_b = 0.07\,\mathrm{mm}$ to $0.1\,\mathrm{mm}$, the extraction decreases to $\eta \approx 2.3\%$. At large focal radius near $r_b = 0.1\,\mathrm{mm}$, some of the electron beam is outside the optical waist radius $w_0 \approx 0.09\,\mathrm{mm}$, thereby reducing the extraction. For a small focal radius near $r_b = 0.02\,\mathrm{mm}$, the increased angular spread causes some of the beam to diverge outside the optical mode at each end of the undulator. The balance of these two competing effects gives an optimum electron beam focal radius of $r_b \approx 0.07\,\mathrm{mm}$ for these parameters.

Fig. 2 shows the FEL extraction $\eta(z_0)$ as the dimensionless Rayleigh length is varied from $z_0 =$

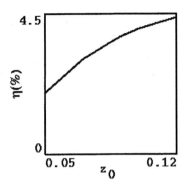

Fig. 2. The FEL extraction $\eta(z_0)$ increases monotonically with increasing Rayleigh length z_0.

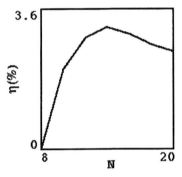

Fig. 3. The FEL extraction $\eta(N)$ shows an optimum number of undulator periods N.

0.05 to 0.12, corresponding to actual Rayleigh lengths $Z_0 = 1.85\,\mathrm{cm}$ to $Z_0 = 4.44\,\mathrm{cm}$ ($Z_0 = Lz_0$). A smaller Rayleigh length increases the mode area on the mirrors, but reduces the electron–optical interaction because of the rapidly changing optical phase, rapidly changing optical amplitude, and short interaction length. For a small Rayleigh length of $z_0 = 0.05$, the extraction is only $\eta \approx 2\%$. As z_0 increases to 0.12, the extraction increases to $\eta \approx 4.5\%$. The extraction steadily increases with z_0, but mirror intensity increases rapidly for two reasons: (i) the increasing extraction, and (ii) the decreasing mode area at the mirrors. The lower values of z_0 significantly reduce the intensity on the mirrors with only a small reduction in the FEL energy extraction η.

Fig. 3 shows the FEL extraction $\eta(N)$ as the number of undulator period is varied from $N = 8$

to 20. With a short Rayleigh length ($Z_0 \ll L$), the optical mode expands significantly along even the shortest undulator and care must be taken that the laser energy "scraped" off the optical beam does not heat or damage the undulator walls. For all the values examined here, the scraped energy remains small (less than a Watt) for an undulator gap of $1\,\mathrm{cm}$. For a shortest undulator of $N = 8$ periods, the interaction length is so small that the FEL is below threshold (output/pass exceeds gain/pass), and extraction is zero. As N increases, the extraction increases to a peak value of $\eta = 3\%$ at $N = 14$ periods. At larger values of N, the extraction decreases slightly down to $\eta = 2.5\%$ at $N = 20$ periods. After reaching the optimum undulator length, there is no further advantage in increasing the undulator length. In fact, there is a penalty since the FEL extraction tends to decrease with increasing N.

6. Conclusions

A simulation method, capturing the important physics of the short Rayleigh length FEL, has been described. The method is then used to explore FEL energy extractions $\eta(r_b)$, $\eta(z_0)$, and $\eta(N)$. The extraction $\eta(r_b)$ shows an optimum electron beam focal radius r_b. A more accurate determination of the optimum value for r_b should be obtained by more sophisticated simulations, or better yet, experiments. Increasing values of the Rayleigh length z_0 steadily improves the FEL interaction and single-pass extraction $\eta(z_0)$, but focuses the power to a smaller mirror spot that can exceed the damage limit. The extraction $\eta(N)$ increases rapidly with N up to an optimum value. Further increases in the undulator length do not improve the interaction because of the diminished optical field strength at both ends of the undulator.

Acknowledgements

The authors are grateful for the support from NAVSEA and the JTO.

References

[1] A. Yariv, Quantum Electronics, Wiley, New York, 1975, p. 112 (Chapter 6).

[2] J.D. Jackson, Classical Electrodynamics, Wiley, New York, 1962, p. 205.

[3] W.B. Colson, G. Dattoli, F. Ciocci, Phys. Rev. A 31 (1985) 828.

[4] W.B. Colson, C. Pellegrini, A. Renieri (Eds.), Free Electron Laser Handbook, North-Holland Physics, Elsevier Science Publishing Co. Inc., The Netherlands, 1990 (Chapter 5).

Available online at www.sciencedirect.com

SCIENCE $\bigcirc$ DIRECT°

ELSEVIER   Nuclear Instruments and Methods in Physics Research A 528 (2004) 172–178

**NUCLEAR
INSTRUMENTS
& METHODS
IN PHYSICS
RESEARCH**
Section A

www.elsevier.com/locate/nima

Consideration on the BPM alignment tolerance in X-ray FELs

T. Tanaka*, H. Kitamura, T. Shintake

The Institute of Physical and Chemical Research, Harima Institute, Koto 1-1-1, Mikazuki, Sayo, Hyogo 679-5148, Japan

Abstract

The effects of the trajectory error on the FEL amplification process are studied analytically and numerically in order to estimate the alignment tolerance of the beam position monitor (BPM) installed in the undulator line of the SASE-based FEL. Analytical investigations show that an angle error on the electron beam disturbs the FEL amplification process by two mechanisms: one is a decrease in radiation efficiency, and the other is smearing of microbunch. Simple formulae to denote these two effects are derived, and a critical angle is introduced as a criterion for the single kick error to be allowed. Then, numerical simulations are performed to estimate the tolerance of the BPM alignment to be required practically. These are in good agreement with the results of the analytical investigations.
© 2004 Elsevier B.V. All rights reserved.

PACS: 41. 60. Cr

Keywords: Gain analysis; FEL simulation; BPM alignment tolerance

1. Introduction

In most X-ray FEL facilities under proposal or construction, a long undulator ranging from 30 to 100 m is to be installed. For technical requirements, the long undulator is divided into several segments between which focusing magnets and beam position monitors (BPMs) are located. The recent progress of the technologies for undulator fabrication enables us to construct an undulator segment with negligible magnetic field errors, in terms of the orbit deviation and optical phase error. Thus, the major source which causes an orbit error is misalignment of the focusing magnets. The beam-based alignment technique [1,2] is proposed which estimates the transverse offset of the BPM origins and thus corrects the orbit error. The disadvantage of this technique is that it requires a change in the electron-beam energy to measure the responses of the BPMs. The latter are affected by the beam halo, dark current from the RF accelerating cavity, and stability of the accelerator. An alternative way is to use the cavity-type BPM, in which the mechanical and electrical origins coincide within the accuracy of several microns, and to perform a fine mechanical alignment of the BPM components. In this case, it is important to know how accurately we should align the BPMs, and this is investigated analytically and numerically in the following sections.

---

*Corresponding author.

E-mail address: ztanaka@srping8.or.jp (T. Tanaka).

0168-9002/$ - see front matter © 2004 Elsevier B.V. All rights reserved.
doi:10.1016/j.nima.2004.04.040

2. Analysis on a single-kick error effect

In order to investigate analytically the effects of the trajectory error on the FEL amplification process, we should solve the three-dimensional FEL equations with the error taken into account, which is usually bulky. Instead, let us consider the simple case when the electron beam is kicked by an error dipole field (single kick error: SKE), and study the FEL amplification process driven by the electron beam with an error angle. This considerably simplifies the problem and enables us to derive analytical expressions to denote the effects of the trajectory error.

2.1. FEL equations

Before moving to the analytical formulation, basic equations to describe the FEL process are derived. To simplify the problem, all equations are written in the time-independent regime. The wave equation of the electric field of FEL radiation is given as [3]

$$\nabla_\perp^2 \tilde{E} + 2i\frac{\omega}{c}\frac{\partial \tilde{E}}{\partial z} \propto b(r, z)$$
$$= \left\langle \frac{\gamma_0}{\gamma_j} e^{-i\psi_j} \delta(r - r_j) \right\rangle, \tag{1}$$

where $\tilde{E}$ is the complex amplitude of the radiation field, ω the frequency of radiation, c the speed of light, γ_j and r_j the Lorentz factor and transverse position of the jth electron, and γ_0 the initial value of the average Lorentz factor. The quantity ψ_j is denoted as the phase of the jth electron and given by the differential equation

$$\frac{d\psi_j}{dz} = \frac{2\pi}{\lambda_u}\left(2\frac{\Delta\gamma_j}{\gamma_0} + \frac{\Delta\lambda}{\lambda_1}\right) - \frac{\omega}{2c}\beta_{j\perp}^2,$$

with

$$\Delta\gamma_j = \gamma_j - \gamma_0, \quad \Delta\lambda = \lambda - \lambda_1,$$

where λ_u is the periodic length of the undulator, λ_1 the fundamental wavelength of the undulator radiation, and $\beta_{j\perp}$ the transverse velocity of the jth electron.

The wave equation (1) can be solved by means of the temporal and spatial Fourier transforms as follows:

$$\tilde{E} \propto \hat{E} = \hat{E}_1 + \hat{E}_2$$

with

$$\hat{E}_1 = -\frac{2\pi ic}{\omega}\int_{z_0}^{z} dz' \mathscr{F}_s^{-1}\left[\mathscr{F}_s[b(r, z')\right.$$
$$\left. \times \exp\left[\frac{ic}{2\omega}k^2(z' - z)\right]\right]$$

$$\hat{E}_2 = \mathscr{F}_s^{-1}\left[\exp\left[\frac{ic}{2\omega}k^2(z_0 - z)\right]\right.$$
$$\left. \times \mathscr{F}_s\left[\hat{E}(r, z_0)\right]\right], \tag{2}$$

where $\hat{E}_{1,2}$ and $\hat{E}$ are normalized (dimensionless) electric fields of radiation, and $\mathscr{F}_s$ denotes the spatial Fourier transform defined as

$$\mathscr{F}_s[f(r)] = \frac{1}{2\pi}\int f(r)\exp(-ik \cdot r)\,dr.$$

The component $\hat{E}_2$ can be easily modified to the Fresnel–Kirchhoff's diffraction integral [5] and is therefore regarded as the electric field of radiation propagating in the free space. The component $\hat{E}_1$ describes the radiation emitted by electrons while they travel from z_0 to z. The quantity b denotes the spatial density of the complex bunching factor having the unit of (length)$^{-2}$.

2.2. Mechanisms of gain degradation

Now, let us consider the SKE effects quantitatively. Fig. 1 shows the schematic illustration of the electron beam with microbunch, which is kicked by an error dipole at a longitudinal position of z_0, resulting in an error angle θ. We can consider a plane nearby in which most electrons exist. It is natural to regard this plane as a wavefront of the microbunch on the electron beam. After the electron beam is kicked, the direction of the beam trajectory changes, while the wavefront orientation does not. As a result, we have a discrepancy between directions of the electron motion and wavefront normal. This disturbs the amplification process by two mechanisms: (a) decrease in radiation efficiency and (b) smearing of microbunch, which will be discussed in the following sections.

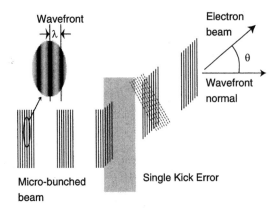

Fig. 1. Electron beam passing through a misaligned quadrupole (dipole). The propagation axis of the electron beam is deflected, while the wavefront orientation is preserved.

2.3. Decrease in radiation efficiency

The intense radiation from the FEL has two optical components. One is that of the undulator radiation and the other is the coherent radiation. The former is emitted towards the electron motion, while the latter towards the wavefront normal of the microbunch. Thus, the discrepancy between the two directions considerably decreases the radiation efficiency.

For analytical formulation of the radiation efficiency loss, let us make an assumption that the amplitude of b does not change significantly over the distance of $\Delta z = z - z_0$. Then the bunching factor as a function of z can be approximated by

$$b(\boldsymbol{r}, z) = \exp\left(i\frac{\omega}{2c}\theta^2\Delta z\right)b(\boldsymbol{r} - \boldsymbol{\theta}\Delta z, z_0). \qquad (3)$$

The first factor denotes the phase variation over the distance of Δz with the error angle error taken into account, while the second denotes the displacement of the spatial profile of the bunch factor brought by the error angle. Substituting Eq. (3) into Eq. (2) and performing integration, we have

$$\hat{E}_1 = 4\pi\mathscr{F}_s^{-1}\left[\exp\left(-i\frac{c}{2\omega}k^2\Delta z\right)\mathscr{F}[b(\boldsymbol{r}, z_0)]\right.$$
$$\left.\frac{1 - \exp\left[i(c/2\omega)\left(\boldsymbol{k} - (\omega/c)\boldsymbol{\theta}\right)^2\Delta z\right]}{\left(\boldsymbol{k} - (\omega/c)\boldsymbol{\theta}\right)^2}\right].$$

Let us assume that the spatial profile of the bunching factor is close to that of the electron beam and has a Gaussian shape with the standard deviation of σ_b. Then the spatial Fourier transform of the bunching factor is given by

$$\mathscr{F}[b(\boldsymbol{r}, z_0)] = b_0\exp\left(-\frac{k^2\sigma_b^2}{2}\right),$$

where b_0 is a constant. Let us calculate the normalized power $\hat{P}$ of radiation emitted while the electron travels the distance of Δz. Using the Parseval's theorem, we have[1]

$$\hat{P} = \int d\boldsymbol{r}|\hat{E}_1|^2 \propto \int d\boldsymbol{k}S_1\left(\boldsymbol{k} - \frac{\omega}{c}\boldsymbol{\theta}\right)S_2(\boldsymbol{k}), \qquad (4)$$

with

$$S_1(\boldsymbol{k}) = \sin c^2\left(\frac{c\Delta z}{4\omega}k^2\right) \qquad (5)$$

$$S_2(\boldsymbol{k}) = \exp(-\sigma_b^2 k^2), \qquad (6)$$

where $\sin c(x) = \sin(x)/x$ is a sinc function which can be seen in the formula of undulator radiation, while the function $S_2(\boldsymbol{k})$ has a Gaussian shape with the standard deviation of $\sigma_2 = (\sqrt{2}\sigma_b)^{-1}$.

Fig. 2 shows typical profiles of $S_1(\boldsymbol{k} - \omega\boldsymbol{\theta}/c)$ and $S_2(\boldsymbol{k})$. From the figure, the radiation power $\hat{P}$, being the area of $S_1 \times S_2$, is found to be a decreasing function of $|\boldsymbol{\theta}|$. Let us approximate the function $S_1(\boldsymbol{k})$ by a Gaussian shape

$$S_1(\boldsymbol{k}) \sim \exp\left(-\frac{k^2}{2\sigma_1^2}\right). \qquad (7)$$

The standard deviation, σ_1, can be determined so that integrations of Eqs. (5) and (7) over the solid angle give the same results. Thus, we have

$$\sigma_1 = \sqrt{\frac{\pi\omega}{c\Delta z}}.$$

Using formulae (6) and (7), we can perform analytically the integral in Eq. (4) to obtain

$$\hat{P} \propto \exp\left[-\frac{(\omega\boldsymbol{\theta}/c)^2}{2(\sigma_1^2 + \sigma_2^2)}\right] = \exp\left(-\frac{\theta^2}{2\sigma_c^2}\right)$$

---

[1] It should be noted that $\hat{P}$ is not equal to the variation of the radiation power between $z = z_0$ and $z = z_0 + \Delta z$, which is in fact expressed as $|\hat{E}_1 + \hat{E}_2| - |\hat{E}_1|^2$. Nevertheless, the dependence of $\hat{P}$ on $\boldsymbol{\theta}$ is expected to be a good measure to denote the SKE effect.

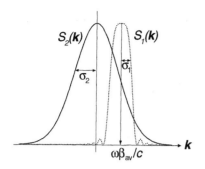

Fig. 2. Typical profiles of the functions of $S_1(k - \omega\theta/c)$ and $S_2(k)$. The error slope causes a peak shift in the function S_1.

$$\sigma_c = \frac{c}{\omega}\sqrt{\sigma_1^2 + \sigma_2^2} = \sqrt{\sigma_{r'}^2 + 2\sigma_{b'}^2}$$

$$\sigma_{r'} = \sqrt{\frac{\lambda}{2\Delta z}}, \quad \sigma_{b'} = \frac{\lambda}{4\pi\sigma_b}, \quad (8)$$

where $\sigma_{r'}$ is the typical angular divergence of the (spontaneous) undulator radiation emitted while the electron travels the distance of Δz, while $\sigma_{b'}$ can be regarded as that of the coherent radiation emitted by the microbunch because it satisfies the diffraction-limit relation $\sigma_{b'}\sigma_b = \lambda/4\pi$. It is worth noting that the condition, $\sigma_{r'} \gg \sigma_{b'}$, is satisfied in most cases. Thus, we have

$$P \propto \exp(-\theta^2 \Delta z/\lambda).$$

This formula shows the effect of the error angle on the amplification of radiation over the distance Δz, which results in lengthening of the saturation length. Let us introduce the new gain length L'_g with

$$\exp(\Delta z/L_g)\exp(-\theta^2\Delta z/\lambda) = \exp(\Delta z/L'_g),$$

where L_g is the gain length without the error angle.[2] Solving the above equation, we have

$$L'_g = \frac{L_g}{1 - \theta^2/\theta_c^2}, \quad \theta_c = \sqrt{\lambda/L_g}. \quad (9)$$

The quantity θ_c is a critical angle to denote the effect of the error angle on the radiation amplification.

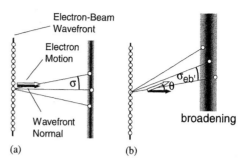

Fig. 3. Microbunch smearing by the finite angular divergence and tilt of the propagation axis with respect to the wavefront normal.

2.4. Smearing of microbunch

Fig. 3 shows the microbunch wavefront and electrons travelling a certain distance with and without the error angle. The finite angular spread of the electron beam results in a difference in time when each electron arrives at the same longitudinal position, and this spoils the phase coherence as shown in Fig. 3(a). This is the so-called debunching effect. The time difference is enhanced by the error angle as shown in Fig. 3(b), which leads to a significant degradation of the bunching factor. Let us call this mechanism smearing of microbunch and distinguish from the normal debunching.

The effect of the microbunch smearing can be estimated as the variation of the total bunching factor $B(z)$ defined as

$$B(z) = \int dr\, b(r, z) \sim \langle e^{-i\psi_j} \rangle.$$

Here, we assumed that the energy deviation of each electron was not so large and omitted the factor γ_j/γ_0 in equation (1).[3]

Using a transverse velocity distribution function $F(\boldsymbol{\beta}_\perp)$, we have

$$B(z) \propto \int d\boldsymbol{\beta}_\perp F(\boldsymbol{\beta}_\perp)\exp\left[i\frac{\omega\Delta z}{2c}(\theta + \boldsymbol{\beta}_\perp)^2\right].$$

Assuming a round Gaussian shape for the distribution function $F(\boldsymbol{\beta}_\perp)$ with the standard

---

[2] This definition is different from that used in Ref. [4], where the radiation power scales as $\exp(\sqrt{3}z/L_g)$.

[3] In fact, $\Delta\gamma_j/\gamma_0$ is of the order of ρ (Pierce parameter), which is usually much less than unity.

deviation σ', we can perform analytically the above integration to obtain

$$|B(z)| \propto \frac{1}{1+(\omega\sigma'^2\Delta z/c)^2} \exp\left(-\frac{\theta^2}{2\sigma_s^2}\right)$$

with

$$\sigma_s = \sqrt{\sigma_{eb'}^2 + \left(\frac{c}{\omega\Delta z\sigma_{eb'}}\right)^2}.$$

The first factor denotes the debunching effect [6,7], while the second denotes the effect of the microbunch smearing. It is easy to show

$$\sigma_s \geqslant \sqrt{\frac{\lambda}{\pi\Delta z}}.$$

Using this relation, we can estimate the maximum effect of the microbunch smearing

$$\begin{aligned}|B(z)| &= |B(z_0)| \exp\left(\frac{\Delta z}{2L_g}\right)\exp\left(-\frac{\theta^2}{2\sigma_s^2}\right)\\ &\geqslant |B(z_0)| \exp\left(\frac{\Delta z}{2L_g}\right)\exp\left(-\frac{\pi\theta^2}{2\lambda}\Delta z\right)\\ &= |B(z_0)|\left(\frac{\Delta z}{2L_g''}\right)\end{aligned}$$

with

$$L_g'' = \frac{L_g}{1 - \pi\theta^2/\theta_c^2},$$

where L_g'' is a newly defined gain length to denote the growth of microbunch. We find that the microbunch smearing effect is also described by the critical angle θ_c.

3. Numerical simulation

The critical angle derived in the preceding section, $\theta_c = \sqrt{\lambda/L_g}$, can be used to roughly estimate the maximum kick error (and the trajectory error) to be allowed in the undulator line. For more precise evaluation of the tolerances of the BPM alignment, we need to perform numerical simulations with the accelerator and undulator parameters of the SCSS project [8], which are summarized in Table 1. All the simulations have been done under the SASE

Table 1
Electron beam and undulator parameters used in the simulation

Phase	I	II
Electron energy (GeV)	1	6
Peak current (kA)	2	4
Normalized emittance (π mm mrad)	2	0.5
Energy spread	2×10^{-4}	
Average betatron function (m)	7	30
Period length (mm)	15	
K value	1.3	
Length per segment (m)	4.5	6
Length of drift section (m)	0.5	
λ_1 (nm)	3.6	0.1
Critical angle θ_c (μrad)	61	6.4

(time-dependent) regime with SIMPLEX [9], an FEL simulation code developed at SPring-8.

3.1. Simulation model

We make several assumptions to simplify the problem. Firstly, the BPMs are located at each drift section and their origins have transverse offsets brought not only by the mechanical alignment but also by the discrepancy between the electrical and mechanical origins of BPMs. Secondly, the electron beam is steered so that it goes through the electrical origins of BPMs, meaning that the electron orbit is specified by the offsets of the BPM origins. Thirdly, each undulator segment is ideal, i.e. it has negligible phase and orbit errors. Under such assumptions, we consider two models. One is that the electrical origins of BPMs distribute randomly within the range of $\pm\Delta d$ in both the horizontal and vertical directions. The other is that all the origins have the same offset in amplitude but the origins of two adjacent BPMs have opposite sign, which result in a zigzag electron orbit. This probably denotes the worst case of alignment of BPMs with the installation tolerance of Δd, which is referred to as a saw-tooth configuration in the following discussions.

3.2. Result of simulation

Fig. 4 shows several typical results of simulations for the SCSS phase I ($\lambda = 3.6$ nm) in

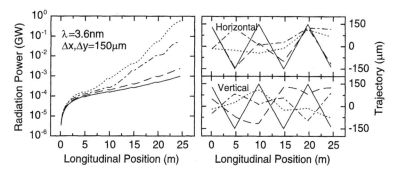

Fig. 4. Results of simulations. Left: radiation power as a function of the distance from the undulator entrance. Right: horizontal and vertical electron orbits. The solid line shows the simulation with the saw-tooth configuration while the dash lines show those with the random BPM distributions. The same line type corresponds to the same simulation.

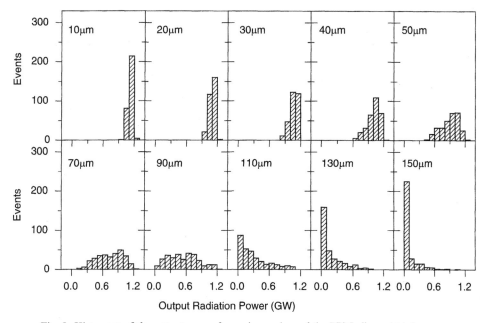

Fig. 5. Histogram of the output power for various values of the BPM alignment tolerances.

terms of the gain curves and orbit trajectories. The gain curves are strongly dependent on the trajectory straightness and the saw-tooth configuration shows the least amplification as expected.

In order to specify the BPM alignment tolerance to be acceptable, we created 300 configurations of BPM origins placed randomly within the square region determined by $|x|, |y| \leqslant \pm \Delta d$ and then performed the FEL simulation with each configuration. Using these results, the histogram of the output power has been calculated and shown in Fig. 5. We have also performed the simulation with the saw-tooth configuration and verified that it has given the minimum output power. This means that the saw-tooth configuration is an appropriate simulation model in order to look for the tolerance of the BPM alignment to ensure a desirable output power. In the case of SCSS phase I, we find the BPM alignment tolerance to be $\pm 40\ \mu$m when we accept half the saturation power as the output. This corresponds to the kick angle of 32 μrad for the saw-tooth configuration, being about half the critical angle θ_c.

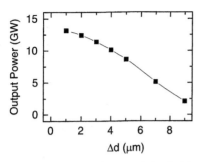

Fig. 6. Output power as a function of the BPM alignment tolerance for the SCSS phase II ($\lambda = 0.1$ nm).

Fig. 6 shows the output power as the function of the BPM tolerance Δd for the saw-tooth configuration obtained by the simulations with the parameters of SCSS phase II ($\lambda = 0.1$ nm). We can find in this case that the allowable tolerance is about $\pm 5\ \mu$m if we accept half the saturation power as the output, which is correspondent to the kick angle of 3.1 μrad. This is again close to half the critical angle. Thus, we find a good agreement between the analytical derivation and numerical simulations as for the criterion of the BPM tolerance.

4. Conclusions

We have studied analytically and numerically the effects due to the trajectory error on the FEL amplification process. In the analytical study, the critical angle has been introduced as a criterion for the BPM alignment. The results of numerical simulations performed to estimate the tolerance of the BPM alignment have shown the validity of the critical angle as a criterion.

It should be noted that since the critical angle is proportional to the square root of the wavelength of radiation, i.e. $\theta_c \propto \sqrt{\lambda}$, the tolerance for trajectory alignment will be more stringent for shorter wavelength FELs.

References

[1] P. Emma, R. Carr, H.D. Nuhn, Nucl. Instr. and Meth. A 429 (1999) 407.
[2] U. Hahn, J. Pflüger, G. Schmidt, Nucl. Instr. and Meth. A 429 (1999) 276.
[3] W.B. Colson, J.L. Richardson, Phys. Rev. Lett. 50 (1983) 1050.
[4] T. Tanaka, H. Kitamura, T. Shintake, Phys. Rev. ST. AB 5 (2002) 040701.
[5] See for example, A. Yariv, Optical Electronics in Modern Communications, Oxford University Press, NY, 1997.
[6] K.J. Kim, Nucl. Instr. and Meth. A 375 (1996) 314.
[7] K.J. Kim, Nucl. Instr. and Meth. A 407 (1998) 126.
[8] T. Shintake, et al., Nucl. Instr. and Meth. A, these Proceedings.
[9] T. Tanaka, to be submitted. See also: http://radiant.harima. riken.go.jp/simplex/

Available online at www.sciencedirect.com

Nuclear Instruments and Methods in Physics Research A 528 (2004) 179–183

NUCLEAR INSTRUMENTS & METHODS IN PHYSICS RESEARCH
Section A

www.elsevier.com/locate/nima

On-line SASE FEL gain optimization using COTRI imaging ☆

A.H. Lumpkin*, J.W. Lewellen, W.J. Berg, Y.-C. Chae, R.J. Dejus, M. Erdmann, Y. Li, S.V. Milton, D.W. Rule[1]

Advanced Photon Source, Bldg. 401, Argonne National Laboratory, 9700 S. Cass Avenue, Argonne, IL 60439, USA

Abstract

Overlap of the particle and the photon beams in a self-amplified spontaneous emission (SASE) free-electron laser (FEL) is one of the keys to optimizing gain. We have now directly demonstrated an on-line method for gain improvement at 540 nm on the Advanced Photon Source FEL. This was achieved by steering the e-beam with correctors before an undulator based on the fringe symmetry in coherent optical transition radiation interference (COTRI) images observed after that undulator. For these conditions we determined that both the SASE and COTR image intensities were improved by about a factor of three in one 2.4-m-long undulator section. The initial tuning had been based on RF beam position monitor readings and the maximization of SASE image intensity in the cameras.
Published by Elsevier B.V.

PACS: 41.60. Cr; 41.60. Ap

Keywords: Free-electron lasers; Coherent optical transition radiation; Microbunching; Beam overlap

1. Introduction

Optimization of the gain in a self-amplified spontaneous emission (SASE) free-electron laser (FEL) is a critical step in the experiment. Besides generating and preserving the ultra-bright electron beam during transport, one must maximize the trajectory overlap of the particle and photon beams within the undulator where the exponential gain process takes place. In the course of performing SASE FEL high-gain experiments at 540 nm at the Advanced Photon Source (APS) low-energy undulator test line (LEUTL) [1], we have used a combination of trajectory adjustments based on RF beam position monitor (BPM) readings, an alignment laser, reference holes in a pick-off mirror, and the intensities observed for the SASE output after each of the eight undulators. Due to the inherent fluctuations in the SASE process and the onset of saturation effects, these techniques have some practical limitations.

We have now directly demonstrated a complementary method for gain improvement by steering the e-beam based on the observed fringe symmetry in the coherent optical transition radiation

☆ Work supported by the US Department of Energy, Office of Basic Energy Sciences, under Contract No. W-31-109-ENG-38.

*Corresponding author. Tel.: +1-630-252-4879; fax: +1-630-252-4732.

E-mail address: lumpkin@aps.anl.gov (A.H. Lumpkin).

[1] NSWC, Carderock Division, West Bethesda, MD, USA.

0168-9002/$ - see front matter Published by Elsevier B.V.
doi:10.1016/j.nima.2004.04.042

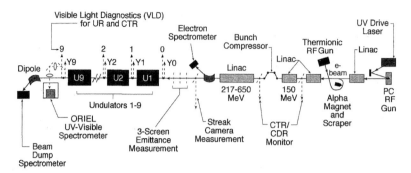

Fig. 1. A schematic of the APS SASE FEL facility showing the PC gun, linac, bunch compressor, diagnostics stations, and the LEUTL undulator hall.

interferometry (COTRI) images obtained after the last of the eight undulators. The intensity of these images is related to the microbunching fraction. We had previously noted the dependence of the theta-x and theta-y planes' symmetries on steering [2,3], but these data are the first quantitative results for gain enhancement. For these experimental conditions we determined that both the SASE and COTRI image intensities improved by a factor of three in one 2.4-m-long undulator section when using this technique. Complementary SASE and COTR z-dependent gain measurements are also reported, and in addition we note further results on transverse profile effects.

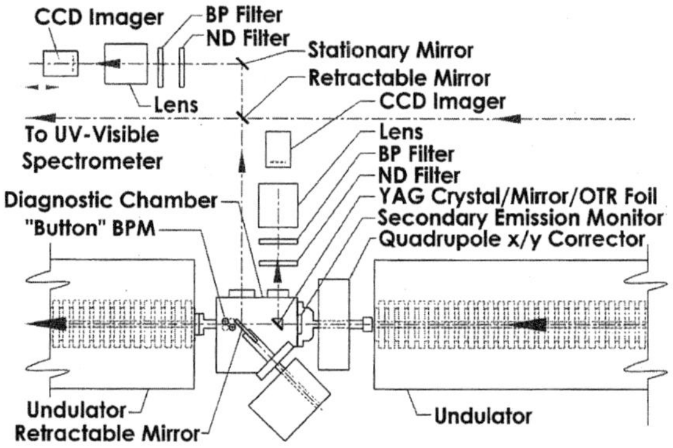

Fig. 2. A schematic of the diagnostics stations located before the first undulator and after each of the eight undulators.

2. Experimental background

The APS SASE FEL experiments were implemented by using a photocathode (PC) RF gun [4], an S-band linac, a chicane bunch compressor [5], a betatron function matching section, and the LEUTL hall where a set of eight undulators are currently installed. These components are shown schematically in Fig. 1. Before and after each of the undulators is located a comprehensive set of diagnostics with a quadrupole magnet and corrector magnets. For completeness, a schematic of the UV–visible diagnostics is shown in Fig. 2 as described previously [6].

The first actuator has positions for a YAG:Ce converter screen with a 45° mirror for SASE light redirection and the 6-μm-thick Al foil used for COTR experiments. A CCD camera with lens

views the selected screen for e-beam or SASE light evaluation. At a second location 63 mm downstream, a 45° mirror on a stepper motor can be used to intercept the SASE light and redirect it to another CCD camera, lens, and filter wheel configuration. The lens was upgraded to fused silica to allow UV viewing down to about 220 nm. The camera is on a translation stage so that the distance to the lens can be adjusted for either near-field or far-field imaging. With the insertion of the upstream thin foil that blocks the SASE light, these systems can be used to view the OTR or COTR from the 45° mirror surface in either the near field or far field. In the latter case, the COTRI images are observed as there is interference between the forward COTR from the foil and the backward COTR from the mirror.

Fundamental to these experiments is the ability to use these imaging stations to make SASE gain measurements and to assess microbunching effects via COTR. For the latter, the spectral angular distribution function can be written as

$$\frac{d^2 N_{ph}}{d\omega\,d\Omega} = |r_{\perp,\parallel}|^2 \frac{d^2 N_1}{d\omega\,d\Omega} I(k)\Im(k), \tag{1}$$

where

$$\frac{d^2 N_1}{d\omega\,d\Omega} = \frac{e^2}{\hbar c} \frac{1}{\pi^2 \omega} \frac{(\theta_x^2 + \theta_y^2)}{(\gamma^{-2} + \theta_x^2 + \theta_y^2)^2}$$

and

$$I(k) = 4\sin^2\left[\frac{kL}{4}(\gamma^{-2} + \theta_x^2 + \theta_y^2)\right].$$

In Eq. (1) $r_{\perp,\parallel}$ are the reflection coefficients, $d^2 N_1/d\omega\,d\Omega$ is the single-electron spectral angular distribution, $I(k)$ is the interference term, and $\Im(k)$ is the coherence function. The latter depends on N_B, the number of microbunched particles in the beam, and the charge/bunch form factors [7]. The shape of this function in θ-space directly multiplies the single-electron spectral angular distribution. Mathematically, we can use an asymmetric, split-Gaussian profile or shift the origin or centroid of this function relative to the single-electron function and produce asymmetric product functions in θ_x and θ_y. Physically, we postulate that if the photon and electron beams are not coaligned, then asymmetric microbunching could occur. Currently, none of the SASE FEL codes address this practical issue to our knowledge.

3. Experimental results and discussion

The fundamental z-dependent intensity gain measurements were performed as a reference for the experiments. Transport through the undulators was first done using minor bunch compression so that saturation was avoided and the trajectory effects would be visible. The 330-pC bunch was then compressed strongly such that a bunch length of 0.25 ps (rms) indicated a peak current of 560 A. The measured normalized emittances after the compressor were 9.8 and 5.9 mm mrad for the horizontal and vertical planes, respectively.

3.1. Angular alignment and gain

As shown in Fig. 3, however, the gain is lower than expected in the first four undulators. Although we do approach the saturation level at a gain of 10^5 compared to the signal after undulator 1, we do not see the long saturated intensity plateau as reported in earlier experiments [1,3]. We also note that the 14.4-m data point is an estimate due to a filter wheel error. The SASE image intensities are about 1000 times stronger than the COTR image intensities. Under these operating conditions the far-field images were reviewed as a function of z. In almost all cases the COTRI images lacked symmetry in the θ_x and θ_y planes. We isolated our attention at undulator 8. Fig. 4a shows the COTRI image from VLD-8 as we found it. We then used the horizontal and vertical correctors before undulator 8. Horizontal corrections immediately moved the observed fringe pattern towards a more symmetric double-lobe in θ_y as seen in Fig. 4b. There also was some effect observed by using the vertical corrector after the undulator (but still located before the 45° mirror in the diagnostics station). We then acquired another 100 images from the VLD-8 camera at this steering for both the SASE and COTRI cases. In post-analysis, the images from before this final steering and after were processed using the same regions of interest, and the

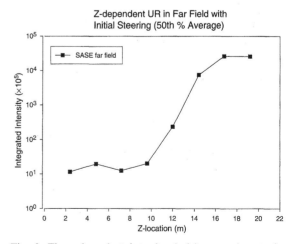

Fig. 3. The z-dependent intensity (gain) measurements for SASE using the far-field focus.

integrated intensities were determined. The 50th percentile averages are provided in Table 1, and the ratio of after/before steering is provided in column 4. The SASE gain improved by 3.7 and the COTRI gain by 2.8, just by steering for appro-

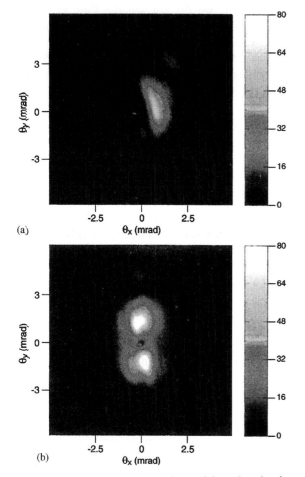

(a)

(b)

Fig. 4. The COTRI image taken after undulator 8 under the gain conditions of Fig. 3, which indicates beam nonalignment (a), and the COTRI image taken after undulator 8 after using the symmetry to guide corrector adjustments (b).

priate fringe symmetry after undulator 8. It is very clear that the charge form factor effectively changed in the COTRI, and mathematically this is simplest to explain as an asymmetric form factor and/or a shift of the centroid location in θ-space. We suggest that the microbunching form factor is dynamically altered when the electron beam and photon beam are not coaligned. Beam time did not permit our repeating the steering technique on the other five undulators where an asymmetric image was found, but we would expect similar improvements up to gain saturation.

3.2. Transverse profile observations

We also investigated the observed e-beam size using OTR and COTR in the first five VLD cameras using the near-field focus as addressed at last year's FEL conference [8]. The COTR beam images were generally two times smaller than the OTR-based beam sizes (which are presumably the real beam distributions). This effect is consistent with our explanation that microbunching is identifiable with a core of beam in the initial gain process, and we have low gain in this particular case. In principle, the SASE beam centroid and the e-beam-generated COTR centroid could be located at each station and steering done to overlay them. In practice, the various sets of ND filters employed for the range of intensities seem to steer the light beams differently. More effort is needed in this area.

4. Summary

In summary, we have reported a novel technique using COTRI for fine-tuning the electron and photon beam trajectories so that gain is improved.

Table 1
Comparison of integrated image intensities obtained before and after the COTRI-guided steering for VLD-8

Radiation type	Before intensity units 10^5	After intensity units 10^5	Ratio after/before
SASE	2.61×10^5	9.56×10^5	3.7
COTR	2.41×10^2	6.78×10^2	2.8

The 50th-percentile integrated intensities are used.

COTRI has the advantage of being more directly related to the SASE process than conventional alignment methods. A systematic application of this technique may lead to optimization of the z-dependent gain process. We have also added to available data in transverse profile effects that may impact the SASE process. We are now planning to extend these studies into the VUV regime, and preliminary results are reported elsewhere [9].

Acknowledgements

The authors acknowledge the support of R. Klaffky and R. Gerig of APS.

References

[1] S.V. Milton, et al., Science 292 (2002) 2037.

[2] A.H. Lumpkin, et al., Proceedings of the 2001 Particle Accelerator Conference, Chicago, IL, June 18–22, 2001, pp. 550.

[3] A.H. Lumpkin, et al., Nucl. Instr. and Meth. A 483 (2002) 394.

[4] S. Biedron, et al., in: A. Luccio, W. MacKay (Eds.), Proceedings of the 1999 Particle Accelerator Conference, New York, NY, 1999, pp. 2024.

[5] M. Borland, et al., in: A.W. Chao (Ed.), Proceedings of the 2000 Linear Accelerator Conference, Monterey, CA, SLAC R561, 2001, pp. 863.

[6] E. Gluskin, et al., Nucl. Instr. and Meth. A 429 (1999) 358.

[7] D.W. Rule, A.H. Lumpkin, Proceedings of the 2001 Particle Accelerator Conference, Chicago, IL, June 18–22, 2001, pp. 1288.

[8] A.H. Lumpkin, et al., Nucl. Instr. and Meth. A 507 (2003) 200.

[9] A.H. Lumpkin, et al., First observations of COTR due to a microbunched beam in the VUV at 157 nm, Nucl. Instr. and Meth. A, these Proceedings.

Available online at www.sciencedirect.com

SCIENCE @ DIRECT°

ELSEVIER   Nuclear Instruments and Methods in Physics Research A 528 (2004) 184–188

NUCLEAR
INSTRUMENTS
& METHODS
IN PHYSICS
RESEARCH
Section A

www.elsevier.com/locate/nima

A method for ultra-short pulse-shape measurements using far infrared coherent radiation from an undulator

G. Geloni[a], E.L. Saldin[b],*, E.A. Schneidmiller[b], M.V. Yurkov[c]

[a] Department of Applied Physics, Technische Universiteit Eindhoven, Eindhoven, The Netherlands
[b] Deutsches Elektronen-Synchrotron (DESY), Notkestrasse 85, Hamburg 22607, Germany
[c] Joint Institute for Nuclear Research, Dubna, 141980 Moskow region, Russia

Abstract

In this paper, we discuss a method for non-destructive measurements of the longitudinal profile of sub-picosecond electron bunches for X-ray free electron lasers. The method is based on the detection of the coherent synchrotron radiation (CSR) produced by a bunch passing through an undulator. Coherent radiation energy within a central cone turns out to be proportional, per pulse, to the square modulus of the bunch form-factor at the resonant frequency of the fundamental harmonic. An attractive feature of the proposed technique is the absence of any apparent limitation which would distort measurements. Indeed, the radiation process takes place in vacuum and is described by analytical formulae. CSR propagates to the detector placed in vacuum. Since CSR energy is in the range up to a fraction of mJ, a simple bolometer is used to measure the energy with a high accuracy. The proposed technique is very sensitive and it is capable of probing the electron bunches with a resolution down to a few microns.
© 2004 Elsevier B.V. All rights reserved.

PACS: 41.60.Cr; 42.55.Vc; 41.75.Ht; 41.85.Ew

Keywords: Free electron lasers; X-ray lasers; Relativistic electron beams

1. Introduction

Effective operation of self-amplified spontaneous emission-free electron laser (SASE FEL) requires high-peak current electron bunches which are produced in the laser-driven RF-gun with subsequent compression in magnetic compressors. The project value of the rms pulse duration for

TTF FEL, Phase 2 is about 150 fs [1]. For a number of FEL application, tailoring of the bunch profile is required to produce shorter radiation pulses. Such a technique has been used at TTF FEL, Phase I to produce radiation pulses with duration down to 30 fs [2,3]. There exists general consensus that the technique for production of even shorter pulses should be developed at TTF FEL, Phase 2 [4]. The femtosecond time scale is beyond the range of standard electronic display instrumentation and the development of non-destructive methods for the measurement of the

*Corresponding author. Tel.: +49-40-8998-2676; fax: +49-40-8998-4475.

E-mail address: saldin@mail.desy.de (E.L. Saldin).

0168-9002/$ - see front matter © 2004 Elsevier B.V. All rights reserved.
doi:10.1016/j.nima.2004.04.043

longitudinal beam current distribution in such short bunches is undoubtedly a challenging problem. Several methods to measure the longitudinal charge distribution of the electron bunch are under development: coherent radiation interferometry [5], longitudinal phase space tomography [6], and technique based on measurements of coherent off-axis undulator radiation [7]. In this paper, we propose a concept of electron bunch diagnostics based on measurement of infrared coherent radiation of electron bunch passing through an undulator [8]. Realization of this method would provide bunch profile measurements with femtosecond accuracy. An important feature of the proposed diagnostics is that it is a non-destructive one. Also, characterization of bunches with strongly non-Gaussian shape is possible [8].

2. Principle of operation

The scheme for electron bunch length diagnostics is sketched in Fig. 1. The electron bunch passes through the undulator and produces a CSR pulse at some specific resonant frequency $\omega_0(K)$. The signal (energy of the CSR pulse within central cone) is recorded by a bolometer. Repetitions of this measurement with different undulator resonant frequencies allow one to reconstruct the modulus of the bunch form-factor.

Since the principle of operation of the proposed scheme is essentially based on the spectral properties of the undulator radiation, we recall some of them. Undulator is described by three parameters: period λ_w, undulator parameter K_w, and number of periods N_w. Single electron passing an undu-

lator radiates electromagnetic wave with N_w cycles. For the radiation within the cone of half angle

$$\theta_{cen} = \frac{\sqrt{1 + K_w^2/2}}{\gamma\sqrt{N_w}} \tag{1}$$

the relative spectral FWHM bandwidth is $\Delta\omega/\omega = 0.89/N_w$ near central frequency

$$\omega_0 = \frac{4\pi c\gamma^2}{\lambda_w[1 + K_w^2/2]} \tag{2}$$

where γ is relativistic factor. Spectral density of the radiation has a sharp central maximum with very weak subsidiary maxima on the sides. The intensity at the next maximum is less than 5% of the first one. If the frequency is increased by any multiple, $\omega = n\omega_0$, we again get other maxima of interference function. The energy at fundamental frequency radiated into the central cone by a single electron is given by [8]

$$\Delta W_{cen} \simeq \frac{\pi e^2 A_{JJ}^2 \omega_0 K_w^2}{c(1 + K_w^2/2)} \tag{3}$$

where $A_{JJ} = [J_0(Q) - J_1(Q)]$, and $Q = K_w^2/(4 + 2K_w^2)$. The energy at fundamental frequency radiated into the central cone by an electron bunch is given by

$$\Delta W_{CSR} = \Delta W_{cen} N[1 + (N - 1)|\bar{F}(\omega_0)|^2] \tag{4}$$

where $F(\omega) = \int dt\, F(t)\exp(-i\omega t)$ is bunch form-factor. This result presents the main essence of our proposal: the CSR radiation energy per pulse is proportional to the square modulus of the bunch form-factor at the resonant frequency. This fact allows one to reconstruct such quantity by repeated measurements of ΔW_{CSR} at different resonant frequencies ω_0 by simple scan of the undulator field. Reconstruction of the bunch profile can be performed with different techniques. For instance, the constrained deconvolution technique fits perfectly for this purpose [8].

In our scheme the undulator has a large value of magnetic field and period. At $\lambda_w = 40$ cm and $H_w = 1.3$ T, undulator parameter is $K_w \simeq 30$, and fundamental wavelength is about 100 μm for $\gamma \simeq 10^3$. A sample of undulator radiation spectrum at zero angle is shown in Fig. 2. The distribution of the radiated energy within different harmonics

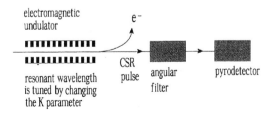

electromagnetic undulator

e⁻

CSR pulse

resonant wavelength is tuned by changing the K parameter

angular filter

pyrodetector

Fig. 1. Scheme for electron bunch length diagnostics based on CSR from an undulator.

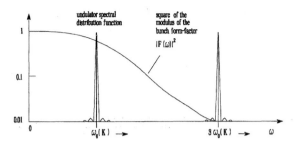

Fig. 2. Some typical spectra of the longitudinal electron distribution and undulator spectral distribution function. The bandwidth of the undulator radiation $(\Delta\omega/\omega_0 = 0.89/N_w)$ must always be as small as the desired spectral resolution.

depends on the value of K_w. For $K_w \gg 1$ we will get strong undulator maxima at $\omega = 3\omega_0$, $5\omega_0$, and so forth. The energy measured by the detector is proportional to the convolution of square modulus of the bunch form-factor and spectral line of the undulator spectrum as it is illustrated with Fig. 2. Reasonable question arises at which conditions measured energy is mainly defined by fundamental harmonic (4). Note that the bunch form-factor is the exponential function at high frequencies, and falls off rapidly for wavelengths shorter than the effective bunch length. For wavelength about three times shorter than the effective bunch length, the radiation power is reduced to about 1% of the maximum pulse energy at the fundamental harmonic. As a consequence, sharp changes of the bunch form-factor result in an attenuation of the high undulator harmonics. In general, $|\bar{F}(\omega)|^2$ varies much more slowly in ω than the sharp resonance term, and we can replace $|\bar{F}(\omega)|^2$ by the constant value $|\bar{F}(\omega_0)|^2$ at the center of the sharp resonance curve.

3. Characterization of bunches with strongly non-Gaussian shape

Let us consider practically important case of an electron bunch with strongly non-Gaussian shape similar to that used to drive TTF FEL, Phase 1 in the femtosecond mode of operation with GW-level radiation pulses [2,3]. Production of such bunches is also planned at Phase 2 of TTF FEL operation

[4]. Bunch compression for relativistic situations can effectively be achieved, at the time being, only by inducing a correlation between longitudinal position and energy offset with an RF system and taking advantage of the energy dependence of the path length in a magnetic bypass section (a so-called magnetic chicane). Downstream of the chicane the charge distribution is strongly non-Gaussian with a narrow leading peak and a long tail, as it is seen from the simulation results in Figs. 3 and 4. Parameters of the leading peak in the bunch (peak current and duration) are of great practical interest, since only this part of the bunch produces powerful FEL radiation with ultra-short pulse duration. In the first order of beam dynamics these parameters are defined by local energy spread in the electron beam before compression. Parameters of the bunch can be also affected by CSR effects in the bunch compressor, and by space charge fields after bunch compressor [9]. However, the shape of the bunch remains similar to that shown in Fig. 4.

Fig. 4 shows plots of the bunch profile versus time for several values of the local energy spread. By comparing these curves, some feeling can be obtained about the effective reduction in peak current. As the initial energy spread is increased from 5 to 15 keV, the maximum peak current falls from 3.5 to 2 kA.

Analysis of plots in Fig. 5 shows that energy within coherent angle at short wavelengths is a strong function of the bunch shape since it changes

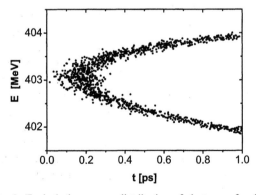

Fig. 3. Typical phase space distribution of electrons after full compression with a single bunch compressor. The head of the bunch is at the left-hand side.

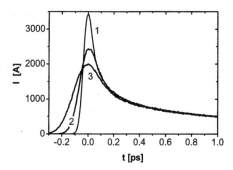

Fig. 4. Current distribution along the bunch after full compression with a single bunch compressor. Curves 1, 2, and 3 correspond to local energy spread of 5, 10, and 15 keV at the entrance of the bunch compressor. The charge of the bunch is 3 nC.

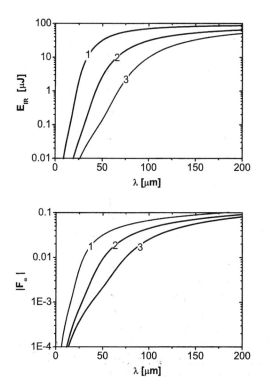

Fig. 5. Energy in the radiation cone and modulus of the bunch form-factor versus radiation wavelength for bunches shown in Fig. 4.

by more than an order of magnitude while peak current only changes by a factor of 1.5. However, the significance of the proposed scheme cannot be fully appreciated until we determine typical values of the energy in the radiation pulse detected within the central cone that can be expected in practice. In the case of TTF FEL, the bunch form-factor modulus falls off rapidly for wavelengths shorter than 40 μm. At the opposite extreme, the dependence of the form-factor on the exact shape of the electron bunch is rather weak and can be ignored in the wavelength range $\lambda > 100$ μm. In this case the energy in the radiation pulse can be estimated simply as in Eq. (4). It is quite clear that in the 40–100 μm wavelength range of the CSR spectrum any bunch with local energy spread smaller than 20 keV produces CSR pulses with energies larger than 0.1 μJ. Radiation energy at such level can be measured with a high accuracy. The values of the modulus of the bunch form-factor are also obtained with a high accuracy, since precise measurements of additional parameters (energy of electron beam, bunch charge and magnetic field entering (4)) is not a problem.

In conclusion, we should notice that measurements of the modulus of the bunch form-factor in the spectral range of 40–100 μm is sufficient for precise reconstruction of the bunch profile. This can be done by means of constrained deconvolution method described in Ref. [8]. An idea consists in finding the best estimate of the bunch profile function, $F(t)$, for a particularly measured form-factor modulus, $|\bar{F}(\omega)|$, including independent measurements of a long tail of the bunch with picosecond resolution (this can be routinely done with streak camera).

Acknowledgements

We thank R. Brinkmann, J. Feldhaus, O. Grimm, M. Koerfer, O.S. Kozlov, J. Krzywinski, E.A. Matyushevskiy, T. Moeller, D. Noelle, J. Pflueger, and J. Rossbach, for many useful discussions. We thank J.R. Schneider and D. Trines for interest in this work.

References

[1] Deutsches Elektronen-Synchrotron, DESY, A VUV Free Electron Laser at the TESLA Test Facility at DESY, Conceptual Design report, DESY Print TESLA-FEL 95-03, Hamburg, 1995.

[2] V. Ayvazayn, et al., Phys. Rev. Lett. 88 (2002) 104802.

[3] V. Ayvazayn, et al., Eur. Phys. J. D 20 (2002) 149.

[4] The TESLA Test Facility FEL Team, SASE FEL at the TESLA Test Facility, Phase 2, DESY print TESLA-FEL 2002-01, Hamburg, 2002.

[5] M. Geitz, et al., Nucl. Instr. and Meth. A 445 (2000) 343.

[6] M. Huening, Proceedings of the Fifth European Workshop on Diagnostics and Beam Instrumentation, May 2001, Grenoble, France, p. 56.

[7] C.P. Neuman, et al., Nucl. Instr. and Meth. A 429 (1999) 287.

[8] G. Geloni, et al., preprint DESY 03-031, Hamburg, 2003.

[9] M. Dohlus, et al., HASYLAB annual report 2003; Nucl. Instr. and Meth., (2004) these proceedings.

Available online at www.sciencedirect.com

SCIENCE  DIRECT°

ELSEVIER   Nuclear Instruments and Methods in Physics Research A 528 (2004) 189–193

NUCLEAR
INSTRUMENTS
& METHODS
IN PHYSICS
RESEARCH
Section A

www.elsevier.com/locate/nima

Longitudinal phase space tomography at the SLAC gun test facility and the BNL DUV-FEL

H. Loos[a],*, P.R. Bolton[b], J.E. Clendenin[b], D.H. Dowell[b], S.M. Gierman[b],
C.G. Limborg[b], J.F. Schmerge[b], T.V. Shaftan[a], B. Sheehy[a]

[a] *National Synchrotron Light Source, Brookhaven National Laboratory, Upton, NY 11973, USA*
[b] *Stanford Linear Accelerator Center, Menlo Park, CA 94025, USA*

Abstract

The Gun Test Facility and the accelerator at the DUV-FEL facility are operated as sources for high brightness electron beams; the first for the Linac Coherent Light Source project and the latter driving an FEL using High Gain Harmonic Generation in the UV. For both accelerators, projections of the longitudinal phase space on the energy coordinate were obtained by varying the phase of an accelerating structure after the gun and measured with a downstream spectrometer dipole. Using an algebraic reconstruction technique, the longitudinal phase space at the entrance to the varied accelerating structure could be reconstructed over a large range of charge from 15 to 600 pC.
© 2004 Elsevier B.V. All rights reserved.

PACS: 41.75.Ht; 29.27.Bd; 42.30.Wb

Keywords: Relativistic electron beams; Beam dynamics; Tomography

1. Introduction

Accelerators for existing and future high brightness electron beam sources require high peak current usually achieved by magnetic bunch compression. A measurement of the longitudinal electron phase space can be useful in understanding and optimizing both injector dynamics and bunch compression. Tomographic methods have been employed to reconstruct the phase space

in recent years [1–3] using measurements of energy spectra with varying longitudinal chirp and dispersion. The technique provides a model independent determination of the phase space as well as a measurement of the slice energy spread of the bunch.

The experiments reported in this paper were done at the Gun Test Facility (GTF) at SLAC and the DUV-FEL at BNL. The GTF accelerator [4] consists of a BNL/SLAC/UCLA 1.6 cell gun with a Nd:Glass drive laser (2 ps FWHM), a 25 MeV s-band accelerating structure, and a dipole energy spectrometer about 3 keV resolution, where the energy spectra were taken while varying the linac

---

*Corresponding author. Tel.: +1-631-344-6110; fax: +1-631-344-3029.

E-mail address: loos@bnl.gov (H. Loos).

phase. The DUV-FEL accelerator [5] shares the gun and accelerating structure with GTF, but the cathode drive laser is Ti:Sapphire-based and there are three additional accelerating structures for up to 200 MeV beam energy and a chicane bunch compressor behind the second linac tank.

2. Reconstruction technique

Various methods have been developed in the past decades in the field of computerized tomographic imaging to retrieve the original two-dimensional distribution from a set of measured projections of this distribution. The simultaneous algebraic reconstruction technique (SART) [6] uses an iterative method to solve the set of linear equations coupling the image and its projections. If the original image is described as discrete pixels g_l, then the ith projection histogram $p_{i,j}$ can be calculated as a sum of the image along the rays of the given projection with the weight coefficients $a_{i,jl}$ according to

$$p_{i,j} = \sum_l a_{i,jl} g_l. \tag{1}$$

Starting from an image $g_q^{(0)}$ as initial guess an iteration step

$$g_q^{(k+1)} = g_q^{(k)} + \sum_j \frac{a_{i,jq}\left(p_{i,j} - \sum_l a_{i,jl} g_l^{(k)}\right)}{\sum_{nl} a_{i,nl}^2} \tag{2}$$

uses one projection p_i after the other until all are utilized once. This procedure is then iterated itself until convergence is reached, which usually occurs within a couple of iterations.

In order to apply this method to the reconstruction of the longitudinal electron phase space distribution $g(\tau, \delta)$, with τ as time coordinate and δ the energy deviation of a particle from the beam centroid energy, the phase space is segmented into discrete bins g_l with size $\Delta\tau$ and $\Delta\delta$. Accelerating the electron beam with initial energy E_0 in a linac section with energy gain V results in a beam energy of $E = E_0 + V\cos(\phi)$ and an energy deviation of $\delta' = \delta + k\tau$ with the chirp $k = -V\omega\sin(\phi)$. A set of energy spectra is obtained at different phases ϕ_i. The spectra with E_i

subtracted are binned into histograms $p_{i,j}$ with the same bin size $\Delta\delta$ as the phase space. The weight coefficients $a_{i,jl}$ for the projections are determined by tracing each phase space pixel according to the applied chirp onto the energy axis. Using the coefficient matrix and measured energy spectra, the phase space can be reconstructed with Eq. (2).

Since this implementation of the tomographic reconstruction only uses an off-crest acceleration and no preceding dispersive element, the set of projections do not include a full 90° rotation of the phase space, i.e., a projection of the time coordinate onto the energy axis. Therefore, the temporal resolution of the reconstructed phase space will be limited due to the maximum chirp k_{max} applied to the beam. In this case, temporal structures converted into an energy distribution are folded with the slice energy spread $\sigma_{\delta,0}$. Considering both a correlated energy modulation and a temporal density modulation the highest frequency that can be resolved is given by

$$f_{res} = k_{max}/(2\pi\sigma_{\delta,0}). \tag{3}$$

3. GTF measurements

The analysis is based on energy spectra measurements presented in Ref. [4]. A fit to the beam energy and energy spread determines the accelerating voltage of the linac structure, the phase difference between maximum energy and minimum spread, and parameters of the longitudinal phase space ellipse at the linac entrance at 5 MeV beam energy. To minimize the rectangular phase space region of the reconstruction, the chirp k is calculated in respect to the phase with minimum spread, thus ignoring the linear chirp of the beam entering the linac. In calculating the weight coefficients, the effects of RF-curvature, wakefields, and low electron energy are not yet taken into account, since they are not expected to significantly influence the slice energy spread.

The phase range of usually 60° corresponds to a maximum chirp of $k_{max} = 220$ keV/ps. A simulation of the experiment with a time and energy modulated phase space was performed to determine the time and energy resolution achievable for the measured energy spectra. The estimate of Eq. (3) was confirmed, giving a resolution of 150–300 fs for a slice energy spread of 5–10 keV.

The reconstructed phase space distributions at the linac entrance, excluding the linear chirp, are shown in Fig. 1. A 7% intensity cut was applied to remove residual artifacts from linac phase and amplitude drifts during the measurement. The bunch current profile obtained from the reconstructions is shown in Fig. 2. Compared with the cathode laser duration of 2 ps (FWHM) the

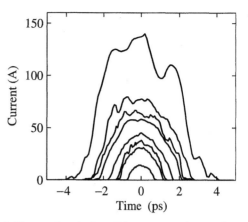

Fig. 2. Temporal projection of the reconstructions as shown in Fig. 1. The bunch charge for the different curves with increasing peak current is 15, 50, 66, 100, 175, 215, 290, and 600 pC.

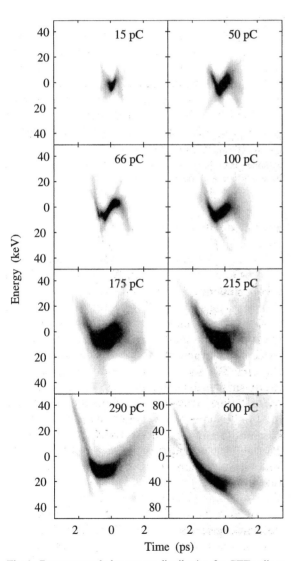

Fig. 1. Reconstructed phase space distribution for GTF at linac entrance.

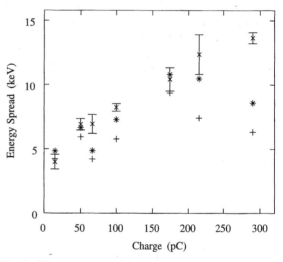

Fig. 3. Slice energy spread from reconstructions as shown in Fig. 1. The "×" symbols are the projected, the "∗" symbols are the averaged slice, and the "+" symbols represent the slice energy spread at maximum peak current.

electron beam is compressed below 100 pC and elongated above. The profiles show some structure which can be attributed to remaining artifacts. The analysis of the energy spread is shown in Fig. 3. The projected energy spread is taken from the longitudinal beam ellipses, whereas the slice data are obtained by a Gaussian fit to the energy profile of each time slice. The average slice energy spread is growing almost as fast with charge as the projected due to increasing spread in the bunch tails. However, the slice energy spread taken at the peak of the current distribution exhibits a significant smaller growth with increasing charge.

4. DUV-FEL measurement

The measurement at the DUV-FEL accelerator was done by varying the phase of the second accelerating structure while the chicane and tanks 3 and 4 were not used. The maximum chirp was 200 keV/ps, resulting in a time resolution of 250 ps for 8 keV energy spread. Fig. 4 shows the reconstructed phase space at the entrance of tank 2 with bunch charge of 200 pC and energy of 38 MeV. Due to a three times longer laser pulse of 2.35 ps compared to GTF, the electron bunch is still ballistically compressed to 1.35 ps. The corresponding drive laser time profile in Fig. 5 exhibits three main substructures as a consequence of the large bandwidth of the Ti:Sapphire oscillator. The structures are not visible in the time distribution of the electron beam, however, they can be recognized in the phase space where they appear as energy–time correlation. This can be

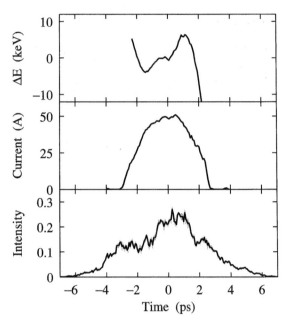

Fig. 5. Temporal distributions corresponding to the reconstruction in Fig. 4. From bottom to top is shown the measured cathode drive laser distribution, the reconstructed electron beam current, and the energy–time correlation.

attributed to longitudinal space charge (LSC) effects [7].

5. Conclusions

An algebraic reconstruction technique was used to reconstruct the longitudinal electron phase space at GTF and DUV-FEL for different bunch charges. The slice energy spread can be significantly smaller than the projected due to nonlinear energy time correlations. Simulations of the injector dynamics to better understand these observations are in progress for both facilities. The implementation of the technique to the compressed bunch at DUV-FEL is planned for the future.

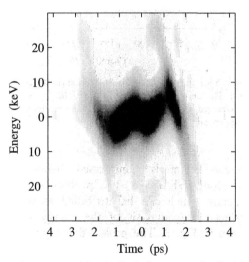

Fig. 4. Reconstructed longitudinal phase space distribution of the DUV-FEL accelerator at the entrance of the second linac tank for a charge of 200 pC.

Acknowledgements

This work was supported by DOE contracts DE-AC03-76SF00515 and DE-AC02-98CH10886.

References

[1] E.R. Crosson, et al., Nucl. Instr. and Meth. A 375 (1996) 87.

[2] M. Hüning, Proceedings of DIPAC 2001, ESRF, Grenoble.

[3] S. Kashiwagi, et al., Proceedings of the XX International Linac Conference, Monterey, CA, 2000, p. 149.

[4] D.H. Dowell, et al., Nucl. Instr. and Meth. A 507 (2003) 331.

[5] W.S. Graves, et al., Proceedings of the 19th Particle Accelerator Conference, Chicago, IL, June 2001, p. 2224.

[6] A.C. Kak, M. Slaney, Principles of Computerized Tomographic Imaging, IEEE Press, New York, 1979.

[7] T. Shaftan, et al., Nucl. Instr. and Meth. A, (2004) these proceedings, TU-P-66.

Available online at www.sciencedirect.com

SCIENCE DIRECT®

ELSEVIER   Nuclear Instruments and Methods in Physics Research A 528 (2004) 194–198

NUCLEAR
INSTRUMENTS
& METHODS
IN PHYSICS
RESEARCH
Section A

www.elsevier.com/locate/nima

First observations of COTR due to a microbunched beam in the VUV at 157 nm ☆

A.H. Lumpkin*, M. Erdmann, J.W. Lewellen, Y.-C. Chae, R.J. Dejus,
P. Den Hartog, Y. Li, S.V. Milton, D.W. Rule[1], G. Wiemerslage

Advanced Photon Source, Bldg. 401, Argonne National Laboratory, 9700 S. Cass Avenue, Argonne, IL 60349, USA

Abstract

The self-amplified spontaneous emission free-electron laser experiments at the Advanced Photon Source are now operating in the VUV at 157 nm for a user experiment. In conjunction with these runs, we have obtained the first coherent optical transition radiation data due to the microbunching of the electron beam in the VUV. We have used both near- and far-field focusing by selecting the spherical mirror with the appropriate focal length for the distance to the CCD chip. The optics are such that much higher resolution than our visible system is attained with calibration factors of 11 μm/pixel and 10 μrad/pixel, respectively. Localized effects in the distributions in both focal conditions are being addressed.
Published by Elsevier B.V.

PACS: 41.60 Cr; 41.60 Ap

Keywords: SASE FEL; VUV; Coherent optical transition radiation; Microbunching

1. Introduction

Following the success of our self-amplified spontaneous emission (SASE) free-electron laser (FEL) high-gain experiments in the UV-visible regime [1,2], the project has pushed on to shorter wavelength work in the VUV near 150 nm. Work

☆ Work supported by the US Department of Energy, Office of Basic Energy Sciences, under Contract No. W-31-109-ENG-38.

*Corresponding author. Tel.: +1-630-252-4879; fax: +1-630-252-4732.

E-mail address: lumpkin@aps.anl.gov (A.H. Lumpkin).

[1] NSWC, Carderock Division, West Bethesda, MD, USA.

0168-9002/$ - see front matter Published by Elsevier B.V.
doi:10.1016/j.nima.2004.04.045

in this regime is strongly driven by the interests of a user program based on the single-photon ionization and resonant ionization to threshold (SPIRIT) techniques [3]. In order to operate successfully in the VUV, a new set of intraundulator diagnostics was proposed, designed, and installed [4]. These diagnostics are basically analogous to the UV-visible stations only with the camera sensor in vacuum and the filter materials adjusted for the VUV regime. Reflective optics are used in these stations, which are currently located after undulators 2, 4, and 6 in the eight-undulator string. A VUV light transport to the endstation where the SPIRIT experiment resides has also been commissioned and is discussed elsewhere [3].

As a complement to the SASE optimization tests, we have preserved the option to perform coherent optical transition radiation (COTR) tests by using a thin foil at a location upstream of the 45° pick-off mirror to block SASE and to serve as one element of the COTR interferometer. We are also taking advantage of the much higher resolution in the system to identify localized effects in the electron beam distributions and in microbunching.

2. Experimental background

These experiments are conducted at the low-energy undulator test line (LEUTL) at the Advanced Photon Source (APS). The facility has been described previously [1], and a schematic is shown in Fig. 1 of our accompanying paper in these proceedings [5]. The RF photocathode (PC) gun is used as the source of bright electron beams. For these experiments the linac beam energy is near 400 MeV with normalized emittances of 6–8π mm-mrad for peak currents of 400–500 A. The FEL is generally operated in the 157 nm regime for the user experiments.

In support of the thrust to the VUV, members of the Experimental Facilities Division (XFD) of APS upgraded the diagnostics. The initial plan was

to use a Kirkpatrick–Baez (K–B) optics design with imaging capability down to 50 nm [4], but a compromise in the minimum operating wavelength specification for the FEL to 120 nm allowed the use of near normal incidence focusing mirrors without the alignment complications of the K–B mirrors. The final implementation is based on two simple reflective flat mirrors and two selectable spherical mirrors of focal length 900 and 2000 mm that provide both near- and far-field imaging, respectively. The distance between the spherical mirrors and the CCD sensor plane is 2.0 m. This results in calibration factors of 11 μm/pixel and 10 μrad/pixel, respectively. The near-field value was checked by stepping the position of a mask with a known hole pattern through the object plane and recording the hole image positions. A schematic of this portion of the diagnostics is shown in Fig. 1. In addition to the thermoelectrically cooled frame transfer Roper Scientific CCD camera with a 512×512 pixel array sensor, a full suite of bandpass filters, neutral density filters, and attenuators is used. The cameras actually image from the VUV (120 nm) through the visible wavelength regime with a 120,000-electron well depth, and they are used with a 16-bit video digitizer. The camera is based on the Marconi VUV EEV57-10 back-thinned chip with no antireflection coating and can transfer the full image at a maximum frame rate of 3 Hz. A faster rate can be obtained by using a software-selectable region of interest or by binning the pixels. One set of VUV filters is characterized at the 120-nm regime based on metallic films on a MgF_2 substrate (Acton Research Corp.), and another set of metallic fused-silica filters is used in the UV-visible regime from 200–800 nm. Seven filter carriers are used to provide the attenuation flexibility needed to cover the OTR, COTR, and SASE intensity ranges. Bandpass filters can also be selected for harmonics measurements.

3. Initial experimental results and discussion

Beam time for commissioning the diagnostics stations has been limited in the last year due to the focus on user experiments. However, in the process

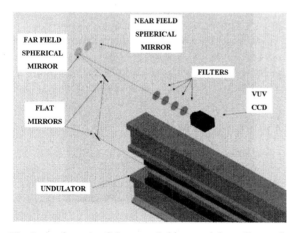

Fig. 1. A schematic of the upgraded intraundulator diagnostic stations for the VUV experiments. Reflective optics, filters, and the VUV camera are shown.

of using the VUV camera to optimize the SASE FEL gain, some ancillary data have been obtained on the microbunched electron beam.

3.1. Near-field focus

From the electron beam diagnostics point of view, we have documented that our limiting resolution in the UV-visible system in the near field of 80 and 90 μm for the x- and y-axes, respectively, was too coarse for the structures we believe are occurring within the photo-injected, bunch-compressed, and microbunched beam. During these experiments we also tried to elucidate the effects of density variations in the beam on final FEL performance. Previously, we reported evidence for microbunching variations in the transverse plane [6], and we suggested that vertical or horizontal charge density variations generated in the PC gun or the bunch compressor could be reflected in the SASE process. Our work in the past using COTR interference images consistently required the use of beam sizes smaller than the predicted matched beam size of 200 μm in the x-plane and 100 μm in the y- plane to explain fringe visibility [2]. We have now used a high-resolution imaging system [7] at a location located after the chicane. As shown in Fig. 2, the 2-D spatial image of the electron beam after the

chicane has several striations in intensity. These are similar to the variations seen by using the `elegant` particle tracking code [8]. Such striations can be generated in the bunch compressor due to coherent synchrotron radiation (CSR) effects depending on the degree of compression or in the photoinjector itself. We have suggested that these localized areas of more intense beam would be more likely to give SASE gain.

As part of this investigation, we show an example near-field image of the VUV SASE beam in Fig. 3a and the COTR image in Fig. 3b from the VUV camera after undulator 4. The x-axis is the horizontal bend plane of the chicane. The thin foil is inserted to allow us to see the VUV COTR

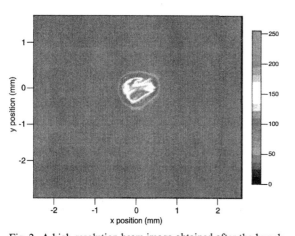

Fig. 2. A high-resolution beam image obtained after the bunch compressor during a SASE FEL run. The striations in intensity are visible and may be subsequently accentuated in microbunching as viewed with COTR.

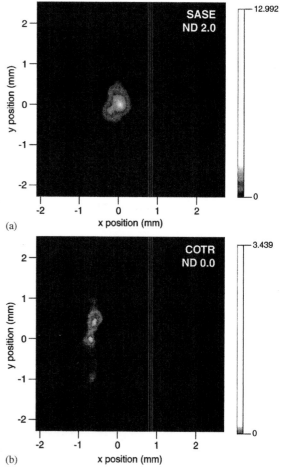

(a)

(b)

Fig. 3. Examples of near-field focus images at 157 nm for (a) SASE VUV light and (b) COTR VUV light.

emitted from it. The localized intensities in the COTR would imply some hot spots in microbunching. Some features do appear to be sub-100 μm with a separation of about 400 μm vertically. On recent runs we have viewed the high-resolution images at the chicane and then the COTR images for the same setup. The localized structures seen at the chicane seem to be carried over into the COTR microbunching images.

3.2. Far-field focus

The far-field focus is achieved in principle by selecting the spherical mirror with appropriate focal length for the CCD camera sensor distance

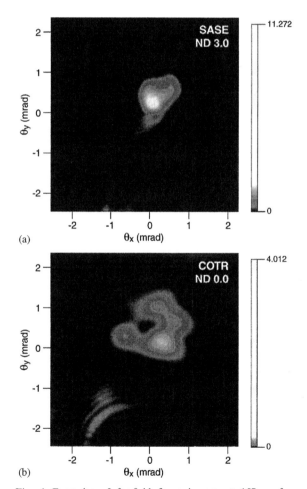

(a)

(b)

Fig. 4. Examples of far-field focus images at 157 nm for (a) SASE VUV light and (b) COTR VUV light.

using a remotely controlled translation stage. For comparison purposes, the SASE far-field image is shown in Fig. 4a. Its shape is not ideal, but it is relatively smooth. In Fig. 4b the COTR interferometric (COTRI) image is shown. At this beam energy, the first OTR lobes are much narrower than the 540 nm case in angle at about ±1 mrad, but the overlap of the bunch form factor and the single-electron interference pattern results in narrower peak positions. The calculated calibration factor seems reasonable since we can reproduce the inner lobe positions by using a split-Gaussian beam distribution of 43 and 25 μm for the left and right side, respectively. The θ_x lobe peaks at −0.5 and +0.7 mrad are then reproduced as well as the relative intensity. The θ_y lobes are at ±0.5 mrad which is reproduced with an rms beam size of 50 μm in the calculation.

4. Summary

In summary, we have begun our experiments on SASE and COTR in the VUV with newly installed reflective optics and VUV-sensitive cameras. The COTR images are the first taken in the VUV, and the high spatial resolution available has facilitated our search for localized, microbunching structure. We are hopeful of cross-comparing beam sizes implied by the COTRI images and these direct measurements. Additional stations after undulators 5 and 7 are being installed, and we anticipate providing more complete z-dependent SASE FEL evaluations and microbunching tests in the VUV in the coming year.

Acknowledgements

The authors acknowledge the support of Kwang-Je Kim and Efim Gluskin of APS as well as the technical support on the VUV cameras by Brian Tieman of XFD.

References

[1] S.V. Milton, et al., Science 292 (2002) 2037.

[2] A.H. Lumpkin, et al., Phys. Rev. Lett. 88 (2002) 234801.

[3] J.F. Moore, Photoionization studies with SPIRIT at the Advanced Photon Source Free Electron Laser, Nucl. Instr. and Meth. A, these Proceedings.

[4] P. Den Hartog, et al., Nucl. Instr. and Meth. A 483 (2002) 407.

[5] A.H. Lumpkin, et al., On-line SASE gain optimization using COTRI imaging, Nucl. Instr. and Meth. A, these Proceedings.

[6] A.H. Lumpkin, et al., Nucl. Instr. and Meth. A 507 (2003) 200.

[7] B.X. Yang, E. Rotela, S. Kim, R. Lill, S. Sharma, Proc. of BIW 2002, AIP Conf. Proc. 648 (2002) 393.

[8] M. Borland, private communication, August 2002.

Available online at www.sciencedirect.com

Nuclear Instruments and Methods in Physics Research A 528 (2004) 199–202

**NUCLEAR
INSTRUMENTS
& METHODS
IN PHYSICS
RESEARCH**
Section A

www.elsevier.com/locate/nima

Optical resonator of powerful free-electron laser

Vitaly V. Kubarev*, Boris Z. Persov, Nilolay A. Vinokurov, Aleksey V. Davidov

Budker Institute of Nuclear Physics, Lavrent'ev av. 11, Novosibirsk 630090, Russia

Abstract

The basic principles of calculation and design of an optical resonator for the high-power free-electron laser of the Siberian Center for Photochemical Research are presented. The resonator has two a time stage of development. The first is a so-called start optical resonator. The main task of the resonator is the determination of fundamental FEL parameters and search of optimal regimes. This resonator has adjustable output coupling and inner vacuum calorimeters for the measurement of high power. Only a small amount of the whole intra-cavity power is output to the atmosphere for diagnostic purposes. The second stage of the resonator has optimal output coupling and maximum output power for user applications.

A method for a simple analytical calculation of losses in the laser resonator is described. Resonator losses measured by a fast Schottky detector are in reasonable agreement with calculated values.
© 2004 Published by Elsevier B.V.

PACS: 42.60.Da; 42.25.Fx

Keywords: Optical resonator; Free-electron laser

1. Design of the optical resonator

The optical resonator of the powerful free-electron laser of the Siberian Center for Photochemical Researches [1] has two time stages of development (Fig. 1). The so-called start optical resonator has recently been created. It is a symmetric open stable resonator. The distance between the resonator mirrors $L_0 = (4/f_0)(c/2) = 26.582$ m is found from the condition of synchronism for the repetition frequency of electron bunches $f_0 = 22.55$ MHz. The radius of curvature of the resonator mirrors $R = 15$ m is defined by the minimization of diffraction losses. At the centers of the mirrors, there are output holes of diameter $d_h = 3.5$ mm for a constant small diagnostic output of radiation and alignment of the resonator. Control blocks are needed for the alignment and monitoring of the main mirrors.

The regulated power output from the resonator into the vacuum calorimeters is carried out by the movable scrapers in the block of calorimeters. When the scrapers are removed, losses in the resonator are minimal, which makes it much easier to obtain lasing. Variable power output is used to find the optimal output coupling of FEL in different modes and measure its parameters. It is

*Corresponding author..

E-mail address: v.v.kubarev@inp.nsk.su (V.V. Kubarev).

0168-9002/$ - see front matter © 2004 Published by Elsevier B.V.
doi:10.1016/j.nima.2004.04.046

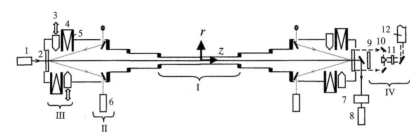

Fig. 1. Scheme of optical resonator: I—undulators, II—control block, III—calorimeter, IV—output system of second stage; 1—HeNe-laser, 2—main mirror, 3—scraper, 4—expander cone, 5—absorber, 6—telescope, 7—Fabry-Perot interferometer, 8—detectors, 9—output mirror of second stage, 10—beam former, 11—diamond window, 12—optical channel.

important for design of the second stage of the optical resonator (Fig. 1). At this stage, maximally possible FEL power in the form of qualitative beam will be output through an optical channel of length 40 m to the radiation-safe part of our building for users.

The calorimeters consist of conic beam expanders and ceramic (Al_2O_3) absorbers deposited on copper heat sinks using a gas dynamic method. The type and thickness of the absorber were chosen in experiments with an universal teraherz gas laser [2]. The same coating was applied to all the transitions of the resonator diameters, which is normal to the beam, for the absorption of scattered radiation.

The FEL mirrors and scrapers are controlled remotely by step motors; the synchronism between the light pulses and the electron bunches is achieved by changing of the f_0 frequency.

2. Calculation of losses in the optical resonator

The calculation of losses and design of the optical resonator were made in 2001 on the basis of a simple analytical method described in Ref. [3]. The method allows us to obtain losses as obvious analytical dependencies on the geometrical resonator sizes. In contrast to numerical methods, the inverse problem of synthesis with an accuracy of 15–20%, sufficient for practice, can be solved easily by this method. It should be noted that the actual accuracy of numerical methods is, as a rule,

of approximately the same order because of the inaccuracy of the simplified calculation model.

According to Ref. [3], the diffraction losses caused by various small perturbations at the center of a Gaussian mode (hole) and at its periphery (mirrors, diaphragms, scrapers) are equal, to double "geometrical" losses, which are a part of the mode power overlapped by a perturbation. The total losses of the resonator can be evaluated as the sum of separate losses. This property of additivity for losses at the holes and apertures of mirrors is shown in a previous paper [3]. For one-type aperture losses at the periphery of a beam (mirrors, diaphragms, scrapers), the additivity condition is satisfied, if the elements with losses are divided by a distance exceeding the length $L_a = d \cdot \delta / \lambda$, where d and δ are present sizes of the mode and perturbation, and λ is the wavelength. The mode fills the cut off part of its periphery at this distance. This condition is satisfied for the main components of the resonator losses. Losses in other sections of the resonator can be ignored because of the exponential sensitivity of the Gaussian beam to narrower diaphragms.

Thus, the resonator losses for round trip are as follows [4]:

$$c \approx 2(c_{mo} + c_{md} + c_{mh} + 2c_{dw} + 2c_{dp}) + c_s \qquad (1)$$

where c_{mo} and c_{md} are the ohmic and aperture losses on the main mirror, respectively, c_{mh} is the hole losses of main mirror, c_{dw} and c_{dp} are the losses on the undulator diaphragm, respectively, and pick-up sensor diaphragm, and c_s is the scraper losses. From the well-known experi-

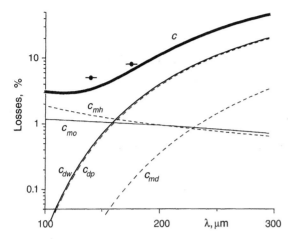

Fig. 2. The power losses in optical resonator per round trip as function of a wavelength: curves—theory, points—experiment.

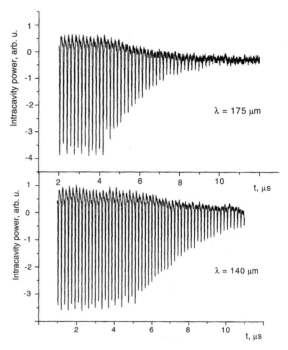

Fig. 3. Signals of output power after switching off the electron beam.

mental data for gold, we can obtain: $c_{mo} = 2 \times 10^{-2}(0.71 - 1.2\lambda[\text{mm}])$.

The curvature of the mirrors was chosen from the minimization $(c_{do} + c_{dp})$: $R = L[1 + (2L_f/L)^2]/2$; where $L_f = 5\,\text{m}$ is Rayleigh length. The diameter of the main mirrors $d_m = 190\,\text{mm}$ was large enough. The diffraction losses on the aperture of the mirror were small compared to other losses in the entire FEL wavelength range.

For losses of the Gaussian TEM_{00}-mode with the radial field distribution $E \sim \exp(-r^2/r_0^2)$; $r_0 = \{\lambda L_f/\pi[1 + (z/L_f)^2]\}^{1/2}$ one can obtain the following formulas:

$$c_{md,dw,dp} = 2\exp[-2\pi d_{m,w,p}^2 L_f/\lambda(4L_f^2 + L_{0,w,p}^2)] \quad (2)$$

$$c_{mh} = \pi d_h^2/\{\lambda L_f[1 + (L/2L_f)^2]\} \quad (3)$$

$$c_s = 2[1 - \text{erf}(2^{1/2}r_s/r_0)] \quad (4)$$

where d_w (78 mm) and d_p (105 mm) are the diameters of the undulator diaphragm and the pick-up sensor diaphragm, respectively, L_w (9 m) and L_p (7.5 m) are distances between these diaphragms, and r_s is the radial position of the scrapers. The minimum losses of the resonator for $c_s = 0$ as a function of the wavelength are shown in Fig. 2.

3. Experiments with the optical resonator

The measurement of losses in the optical resonator also allows us to determine the gain of the small FEL signal and its maximum output power. For instance, the presence of lasing at frequencies of electron bunches constituting 1, $\frac{1}{2}, \frac{1}{3}, \frac{1}{4}$ part of the light pulse round trip and the absence of lasing at $\frac{1}{5}$ part of this frequency implies that the gain exceeds losses by factor of 4–5. The output FEL power can be measured not only by using the direct method but also in other ways. This is carried out by simultaneous measurement of the resonator losses and intra-cavity power, passing through the mirror hole, at different positions of the scrapers.

The measurement of resonator losses was carried out by a fast Schottky diode detector [5]. The time resolution of $\sim 20\,\text{ns}$ was limited by a signal amplifier, but it was sufficient for these experiments. The time history of the FEL radiation before and after the turn off of the electron beam is shown in Fig. 3. The resonator losses $c =$

$1 - \exp(-T/\tau)$, where $T = 2L/c$ is the period of the round trip of the light pulse, τ is the decay time of the light pulse in e time after the turning off of the electron beam. For the wavelengths of 140 and 175 μm, the resonator losses are 5.1% and 8.0%, respectively. The FEL gain constitutes 30–36% in different modes of the first experiment. A comparison of the experimental and theoretical losses is shown in Fig. 2. One can see that the experimental points are close to the calculated curve.

In the first experiments [1], when scrapers were used in only one calorimeter, the wavelength was 150 μm, and the repetition frequency of electron bunches was 5.6 MHz ($\frac{1}{4}$ part of the maximum value) the power in calorimeters was 40 W (the total output power was 50 W). This power can also be calculated by using the following formula:

$$P_c = (1/2)P(c_1 - c_2) \tag{5}$$

where P is the intra-cavity power with scrapers, c_1 the losses with scrapers, and c_2 the losses without scrapers. The number "2" in Eq. (5) reflects the fact of equality between the power, output in calorimeters, and the power scattered by them and absorbed by the diaphragms [3]. For $P = 1\,\text{kW}$, $c_1 = 10.2\%$, and $c_2 = 6.0\%$ we obtain $P_c = 42\,\text{W}$, which is close to the calorimeter measurement. The pulse power of FEL is greater by factor of 2×10^3, that is $\sim 80\,\text{kW}$.

Previous experiments have shown that a simple phenomenological theory of a homogeneously broadened gas laser (see, for example Ref. [6]) can be applied to our FEL. According to this theory, the laser gain $K = \exp(g) - 1$ and logarithmic gain is

$$g = g_0(1 + I/I_0)^{-1} \tag{6}$$

where g_0 is the logarithmic gain of a small signal, I is the intra-cavity intensity, and I_0 is the saturation intensity. Rewriting Eq. (6) as $I = I_0(g_0/g - 1)$ and taking into account that at stationary conditions $g = -\ln(1 - c)$, we obtain

$$I = I_0\left[-g_0/\ln(1 - c) - 1\right] \tag{7}$$

The dependence $I(c)$ calculated by this formula and the experimental points obtained at different repetition frequencies of electron bunches $f = f_0/4n$; $n = 1, 2, 3, 4, 5$, is shown in Fig. 4. A good

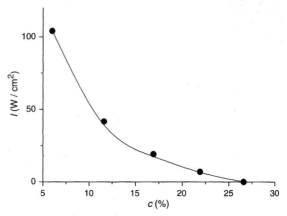

Fig. 4. The intra-cavity intensity vs. resonator losses: solid curve—theory, points—experiment.

agreement is visible. Therefore, we obtain the following for a typical mode: $\lambda = 150\,\mu\text{m}$, $c = 6\%$, $K_0 = 36\%$, and $I_0 = 26\,\text{W/cm}^2$. The greatest power output by the scrapers (S is effective mode cross section)

$$P_{c,\max} \approx (1/2)I_0 S(g_0^{1/2} - c^{1/2})^2 = 49\,\text{W} \tag{8}$$

is achieved at the optimal useful output coupling

$$(1/2)c_s + c_{mh} = (1/2)[(g_0 c)^{1/2} - c] = 4.3\%. \tag{9}$$

This power value within the measurement accuracy, is in agreement with direct measurements (40 W). In a common case, one can easily find an optimum output coupling and a maximal power for any experimental dependence $I(c)$.

The possibility of using Eq. (6) for not strongly saturated FEL was shown theoretically in Ref. [7].

References

[1] N.A. Vinokurov, et al., Proceedings of Free Electron Laser Conference FEL-2003, Tsukuba, Japan, 2003.

[2] V.V. Kubarev, Quantum Electron. 25 (1995) 1141.

[3] V.V. Kubarev, Quantum Electron. 30 (2000) 824.

[4] V.V. Kubarev, Report BINP, 2000.

[5] V.V. Kubarev, et al., Nucl. Instr. and Meth. A 507 (2003) 523.

[6] A. Maitland, M.H. Dunn, Laser Physics, North-Holland Publishing Co., Amsterdam-London, 1969;
V.V. Kubarev, E.A. Kurenskij, Quantum Electron. 26 (1996) 303.

[7] G. Dattoli, et al., Phys. Rev. A 45 (1992) 8842.

Available online at www.sciencedirect.com

SCIENCE DIRECT®

ELSEVIER   Nuclear Instruments and Methods in Physics Research A 528 (2004) 203–207

NUCLEAR INSTRUMENTS & METHODS IN PHYSICS RESEARCH
Section A

www.elsevier.com/locate/nima

Development of the edge-focusing wiggler for SASE

Shigeru Kashiwagi[a,*], Akihito Mihara[a], Ryukou Kato[a], Goro Isoyama[a], Shigeru Yamamoto[b], Kimichika Tsuchiya[b]

[a] *Institute of Scientific and Industrial Research, Osaka University, 8-1 Mihogaoka, Ibaraki, Osaka 567-0047, Japan*
[b] *Institute of Materials Structure Science, High Energy Accelerator Research Organization, 1-1 Oho, Tsukuba, Ibaraki 305-0801, Japan*

Abstract

We propose a novel concept to make a low-error wiggler to be used for self-amplified spontaneous emission. Magnetization errors of permanent magnet blocks can be reduced by using a pair or quartet of magnet pieces quarried out of a single raw block of permanent magnet. The concept has been tested experimentally using magnet blocks for a trial model of the edge-focus wiggler, which can produce a focusing field as high as 1 T/m in the wiggler. The errors in the magnetization of magnet blocks were considerably reduced. This was calculated with the vertical magnetization axis with which it is easier to measure three components of the magnetic field, and the measured values are reliable. It has been experimentally confirmed that the concept works well.

PACS: 41.60.Cr; 41.85.Lc; 07.55.−w

Keywords: FEL(s); SASE; Wiggler; Edge-focusing

1. Introduction

Self-amplified spontaneous emission (SASE) is one of the most promising candidates for the fourth generation light source in the vacuum ultraviolet and X-ray regions. Several SASE experiments in spectral regions from infrared to vacuum ultraviolet have been conducted worldwide in order to study technical feasibility in shorter wavelength regions and to understand the physics involved. Several SASE projects have been proposed in the short-wavelength regions, and technical investigations are being conducted on key components. One of the most critical components is the undulator, also known as the wiggler, which is designed to have a length of 100 m or more in the proposed SASE projects in the hard X-ray region [1]. In such a long wiggler system, a transverse focusing force has to be provided in order to keep the beam size small over the whole wiggler length. Two schemes, the separation-function focusing scheme and the integrated-focusing scheme, have been proposed and are under consideration to focus the electron beam in

---

*Corresponding author. Tel.: +81-6-6879-8486; fax: +81-6-6879-8489.

E-mail address: shigeruk@sanken.osaka-u.ac.jp (S. Kashiwagi).

the wiggler [2–4]. As a new method in the integrated-focusing scheme, we proposed the edge-focus (EF) wiggler, which can produce the high magnetic field gradient for focusing superimposed on the ordinary wiggler field [5].

In order to confirm our proposal, we are currently making an experimental model of the EF wiggler. It is a 5-period wiggler with a period length of 6 cm and an edge angle of 2°. These wiggler parameters were chosen to meet requirements from our SASE experiment, in the far-infrared region, being conducted at the Institute of Scientific and Industrial Research (ISIR), Osaka University [6]. In the fabrication of this model wiggler, we are developing a new method, by trial and error, to make a low-error wiggler without adjustment of the magnetic field. In this paper, we propose a novel concept to reduce errors originating from imperfections of permanent magnet materials and report a preliminary result of the test made for our model of the EF wiggler.

2. Model of the EF wiggler

The EF wiggler is a Halbach-type wiggler made only of permanent magnet blocks, which are cut with an edge angle ϕ, as shown schematically for one period in Fig. 1. The magnet blocks with vertical magnetization are trapezoid shaped, while those with longitudinal magnetization are parallelogram shaped. The focusing force along the wiggler can be varied with the edge angle ϕ, since the field gradient is approximately

proportional to the edge angle as [5]

$$\frac{\mathrm{d}B_y}{\mathrm{d}x} = 4\left(\frac{B_0}{\lambda_W}\right)\phi.$$

The model EF wiggler, which we are fabricating, consists of five periods with the period length $\lambda_W = 60$ mm and the edge angle $\phi = 2°$. The magnetic gap is mechanically fixed at $g = 30$ mm, but it can be easily changed with different columns supporting the upper magnet array. The permanent magnet is Nd–Fe–B with a residual induction of 1.32 T, and the standard dimensions of the magnet blocks are $2a \times 2b \times 2c = 100 \times 20 \times 15$ mm^3. The peak magnetic field is 0.43 T at $g = 30$ mm. The field gradient of the model wiggler is about 1 T/m and the field gradient produces the focusing force $k_x = 26$ m^{-2} at 11.5 MeV energy.

3. Concept for the low-error wiggler

The error magnetic field in a wiggler is produced by imperfections of permanent magnet materials and mechanical inaccuracies of magnet blocks. It is difficult to control errors that can be attributed to characteristics of the materials. In order to maintain uniformity of the magnetization, only one magnet block is usually quarried out of the central part of a raw magnet block. If the uniformity of magnetization is satisfactory, the remaining errors arise due to the magnetization strength and magnetization angle. Magnet blocks provided by manufacturers usually have strength errors up to 1% and angular errors up to 1°. In the standard tuning procedure of the wiggler magnetic field, the strength and the angle of magnetization are measured for each magnet block. Magnet blocks are sorted according to the measured values, and then arranged such that error components are cancelled out with those in the neighbouring blocks. However, in this procedure, it is not guaranteed that all the magnet blocks can be best matched.

Here, we propose a method to spontaneously obtain best matches for cancellation. We assume a magnet block with errors in strength and angle, and the block is cut into two identical pieces along the direction of the magnetization axis, as shown

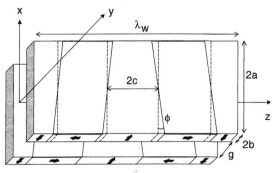

Fig. 1. Schematic drawing of the EF wiggler.

schematically in Fig. 2. If one of the two pieces is rotated by 180° around the main magnetization axis, and then the two pieces are glued to form a single block, then the error components cancel out each other, while the main component remains

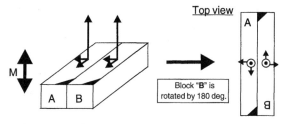

Fig. 2. Method for making magnet blocks with no angular errors. One of the two pieces is rotated by 180° around the main magnetization axis and the error components cancel out each other.

unchanged as shown in Fig. 2. The electron beam detects the magnetic field integrated along its trajectory; therefore, this is practically an ideal magnet block with no angular errors. If the finished block is cut anew into two identical blocks along the plane perpendicular to the magnetization axis, then we have two identical magnet blocks with no angular errors and they can be paired in the wiggler to cancel out the effects of errors in magnetization strength on the electron beam. This is the method for developing a low-error wiggler.

The same method can be applied to the EF wiggler, but the procedure is slightly complicated. We have to consider two cases separately, one for blocks with the vertical magnetization and the other for those with the longitudinal magnetization, as shown in Fig. 3. In the vertical

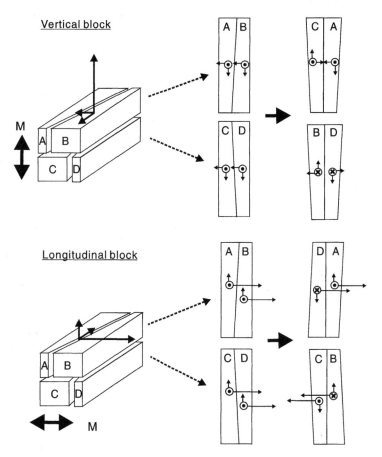

Fig. 3. Method for making magnet blocks with no angular errors for the EF wiggler. Refer to the text for details.

magnetization case, a magnet block is divided into two identical blocks on the plane perpendicular to the main magnetization axis, and then each block is cut with an oblique line into two pieces. These four pieces are labelled as A, B, C, and D. The piece C is rotated by 180° around the magnetization axis and is paired with piece A to make a trapezoid-shaped magnet block without angular errors. Piece B is rotated around the longitudinal axis and the piece D is rotated around the horizontal axis to make the other half of the paired magnet blocks. These pairs should be used with opposite poles placed next to each other in order to cancel out magnetization strength errors. In the longitudinal magnetization case, the original block is divided into two identical blocks on the plane containing the magnetization axis, and each magnet block is cut with an oblique line, and then similarly labelled A, B, C, and D. Piece D is rotated around the main magnetization axis and glued with piece A to make a parallelogram-shaped magnet block without angular errors. Similarly, piece B is glued with piece C, as shown in Fig. 3, so that the other half of the pair is made. This pair of magnet blocks should be used with opposite poles next to each other but in different arrays, that is, the upper and lower arrays.

4. Magnetic field measurement

The magnetic field near a magnet block surface was measured for all the 40 blocks for the model EF wiggler using a Hall probe, which was set 10 mm away from the geometrical centre of the surface on the perpendicular line. The magnetic field for the magnet pair (north- and south-pole magnets) of the surfaces perpendicular to the main magnetization axis were measured for each magnet block. Fractional values of the measured magnetic field with respect to a mean value were separately plotted with the magnet number for the vertical magnetization blocks (hereafter we call them vertical blocks) and for the longitudinal magnetization blocks (longitudinal blocks), as shown in Fig. 4. The solid circles connected with the solid line show measured values for the north-pole magnet surface, and the open circles connected

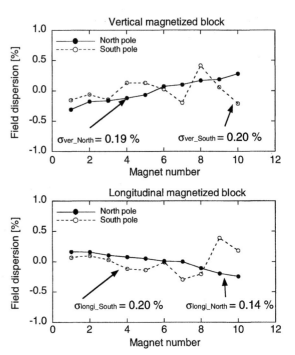

Fig. 4. Surface magnetic field measured for the vertical and longitudinal magnetization blocks for the experimental model of the EF wiggler. Refer to the text for details.

with a dotted line show those for the south-pole magnet surface. The difference between these values is small for each magnet block, indicating that the magnet materials are considerably uniform. Standard deviations of the field strength are also shown in Fig. 4; they range from 0.14% to 0.20% and maximum deviations from the mean values are well below the specification ±1%.

Three components of the magnetic field have been measured for all the blocks. The angular errors of these components have been cancelled with the proposed method, using a special device consisting of a vibrating table and sensor coils. The angle θ, between the mechanical and the magnetization axes, is derived from three components of the magnetic field and is shown in Fig. 5. The angular errors are well reduced for the vertical blocks (shown by the solid circles connected with the solid line) but not as much for the longitudinal blocks (shown with the open circles connected with the broken line). An average value and the

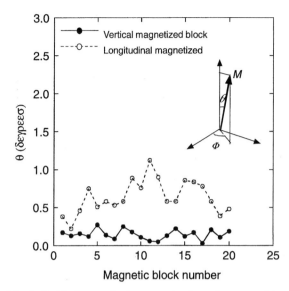

Fig. 5. Residual errors in the magnetization axis measured for low-error magnet blocks made for the experimental model of the EF wiggler. Refer to the text for details.

standard deviation of the angular errors are $0.15 \pm 0.06°$ for the vertical blocks and $0.64 \pm 0.22°$ for the longitudinal blocks. The angular errors for the vertical blocks are reduced to 15% of the specifications, $1°$, while those for the longitudinal blocks are of the same order as the specifications. We suspect that there may have been errors in the measurement for the longitudinal blocks. Since the main magnetization axis is not perpendicular to the mechanical surface due to the edge angle, it was difficult to set the longitudinal blocks precisely to mechanical references of the measurement device. We will continue to further study these points.

5. Conclusions

We proposed a new concept for making a low-error wiggler, and the concept was tested with magnet blocks to be used for the trial model of the EF wiggler. The errors in the magnetization direction were considerably reduced in the vertical magnetization axis where it was easier to measure three components of magnetic field; therefore, the measured results are reliable. Some questions regarding the magnet blocks remain unanswered. We have experimentally confirmed that the concept works well.

Acknowledgements

The authors would like to thank Mrs. T. Koda and K. Okihira of Sumitomo Special Metals Co. Ltd. for their help in magnetic field measurement of the EF wiggler. This research was partially supported by the Joint Development Research at High Energy Accelerator Research Organization (KEK), 2003-17, 2003, and the Ministry of Education, Science, Sports and Culture, Grant- in-Aid for Exploratory Research, 15654036, 2003.

References

[1] LCLS Design Study Group, SLAC-R-0521, 1998, p. 381.
[2] J. Pfluger, Yu.M. Nikitina, Nucl. Instr. and Meth. A 381 (1996) 554.
[3] A.A. Varfolomeev, et al., Nucl. Instr. and Meth. A 381 (1995) 70.
[4] R.D. Schlueter, Nucl. Instr. and Meth. A 381 (1995) 44.
[5] G. Isoyama, et al., Nucl. Instr. and Meth. A 507 (2003) 234.
[6] R. Kato, et al., Nucl. Instr. and Meth. A 445 (2003) 164.

Available online at www.sciencedirect.com

Nuclear Instruments and Methods in Physics Research A 528 (2004) 208–211

ELSEVIER

NUCLEAR
INSTRUMENTS
& METHODS
IN PHYSICS
RESEARCH
Section A

www.elsevier.com/locate/nima

Development of a high-resolution measurement technique of the interval between bunches in a linac with coherent transition radiation

T. Takahashi[a,*], T. Matsuyama[a], Y. Shibata[b], K. Ishi[b]

[a] Research Reactor Institute, Kyoto University, Kumatori, Osaka 590-0494, Japan
[b] Institute of Multidisciplinary Research for Advanced Materials, Tohoku University, Katahira, Aoba-ku, Sendai 980-8577, Japan

Abstract

A new technique of measurement of the inter-bunch distance has been developed. The interval between bunches in the L-band linear accelerator was measured with the cross-correlation interferogram of coherent transition radiation emitted from successive bunches in the millimeter wave region. The radiation was detected by a fast far-infrared bolometer and a fast-gated integrator after passing through an interferometer. The periodic fluctuation of the inter-bunch distance was observed in the time structure of a 4-μs macro-pulse. The maximum amplitude of the fluctuation was 60 fs and corresponds to the resolution of 0.01% compared to the interval between bunches of 769 ps which was calculated from the accelerating radio frequency of 1.3 GHz (L-band).
© 2004 Elsevier B.V. All rights reserved.

PACS: 41.60.Cr; 42.25.Kb; 07.57.Hm; 41.85.Qg

Keywords: Beam diagnostics; Coherent radiation; Transition radiation; Bunch interval; Interferometer; Cross correlation

1. Introduction

The coherent radiation from short bunches of electrons is a new intense light source in the millimeter and sub-millimeter wave regions [1]. It is also well known that coherent radiation is useful for the high-resolution diagnosis of an electron beam. The longitudinal distribution of electrons in a bunch was first analyzed with a spectrum of coherent synchrotron radiation in 1999 [2]. Then, a

method of analysis of the asymmetric bunch shape [3] and a technique of diagnosis with a polychromator [4] were developed. The divergence and the transverse size, i.e., the emittance, of an electron beam can be estimated with an angular distribution of coherent transition radiation [5].

In recent years, we have studied a broadband micro-bunch free electron laser using a pre-bunched electron beam for more intense light sources [6–8]. Since this type of FEL uses the superposition of wave packets from successive bunches, the accuracy of the inter-bunch distance is important. In this paper, we discuss the

---
*Corresponding author.
E-mail address: takahasi@rri.kyoto-u.ac.jp (T. Takahashi).

0168-9002/$ - see front matter © 2004 Elsevier B.V. All rights reserved.
doi:10.1016/j.nima.2004.04.048

development of a new technique of a high-resolution measurement of the interval between bunches in a linac with coherent transition radiation.

2. Theoretical consideration

Since the wave packets of coherent radiation from successive bunches interfere with each other [9], we obtain the cross-correlation interferograms in addition to an autocorrelation one in a long range of the optical path difference (OPD) when the coherent radiation is observed by an interferometer. The electric field of coherent radiation from successive bunches is given by

$$E_A(v) = \sum_{j=1}^{N} \sum_{n=1}^{N_B} E_0 \exp\left[i2\pi v \{x_j - (n-1)L_B\}\right] \tag{1}$$

where N, N_B, L_B, E_0, v, and x are the number of electrons in a bunch, the number of bunches, the inter-bunch distance, the electric field of radiation from a bunch, the wavenumber, and the coordinate along the electron orbit, respectively. In Eq. (1) the emittance of the electron beam was neglected. Then the power spectrum of coherent radiation is calculated as follows:

$$P_A(v) = |E_A(v)|^2 = P(v)K(v) \tag{2}$$

where $P(v)$ represents the spectrum of coherent radiation from a bunch, and $K(v)$ is a function of interference effect due to successive bunches which is represented by

$$K(v) = \left|\sum_{n=1}^{N_B} \exp\left[i2\pi(n-1)L_B v\right]\right|^2$$
$$= \left[\frac{\sin \pi N_B L_B v}{\sin \pi L_B v}\right]^2. \tag{3}$$

The interferogram is given by the Fourier transform of the power spectrum $P_A(v)$ as follows

$$J_A(x) = \sum_{n=-(N_B-1)}^{N_B-1} (N_B - |n|)J(x + nL_B). \tag{4}$$

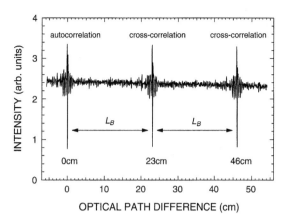

Fig. 1. Interferogram of coherent transition radiation from successive bunches. The key L_B denotes the inter-bunch distance.

This equation shows that the interference pattern around zero OPD, i.e., the autocorrelation interferogram $J(x)$, is repeated at every L_B.

An example of the interferogram is shown in Fig. 1. The first structure around 0 cm in Fig. 1 is the autocorrelation interferogram. The second and third structures around 23 and 46 cm are the cross-correlation interferograms between the wave packet from the first bunch and that from the second one, and between the wave packet from the first bunch and that from the third one, respectively. Since the RF frequency of the linac was 1.3 GHz in this experiment, the theoretical interval between the bunches was 23 cm. Therefore, the interference pattern in Fig. 1 is consistent with Eq. (4) and the inter-bunch distance can be obtained from the interval between the autocorrelation interferogram and the cross-correlation one. Since the resolution of this method depends on the accuracy of the interferometer which can be controlled in a few μm, the high-resolution measurement of the inter-bunch distance is possible.

3. Experimental procedures

The experiment was performed at the facility of the L-band linac at the Research Reactor Institute, Kyoto University. The width of a macro-pulse was 4 μs. The energy, the repetition rate and the average beam current were 33 MeV, 27 Hz and

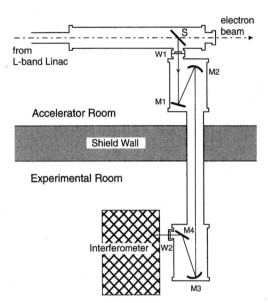

Fig. 2. The schematic diagram of the experimental setup. Keys are: (S) a flat aluminum foil 15 μm thick as a source of transition radiation; (W1) a Kapton window 50 μm thick; (W2) a polyethylene window 0.3 mm thick; (M1, M4) plane mirrors; and (M2, M3) spherical mirrors. The diameter of the parallel light beam between M2 and M3 is 150 mm. The thickness of the shield wall is 2.5 m.

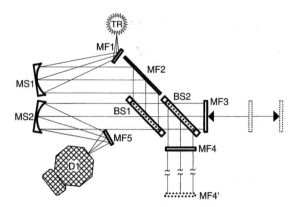

Fig. 3. The details of the interferometer. Keys are (MF1, MF2) plain mirrors; (MS1, MS2) spherical mirrors; (MF3) a movable mirror; (MF4) a fixed mirror; (BS1, BS2) wire-grid beam splitters; (D1) an InSb hot-electron bolometer.

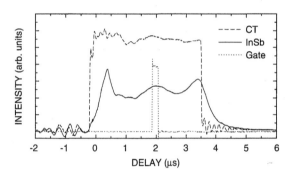

Fig. 4. The waveform of the electron beam (a dashed curve) and the output signal of the detector (a solid curve). The gate of the fast-gated integrator (a dotted curve) was scanned in the 4-μs macro-pulse.

57 μA, respectively. The schematic diagram of the experiment is shown in Fig. 2. The coherent transition radiation in the millimeter wave region was emitted when the electron beam passed through an aluminum foil 15 μm thick. The radiation was detected by an InSb hot-electron bolometer (QMC Instruments Ltd.) through a hand made Martin–Puplett-type polarizing interferometer. The acceptance angle of the mirror M2 was 75 mrad, i.e., the f-number 13.3.

The details of the interferometer are shown in Fig. 3. The two wire-grid beam splitters were made of 10 μm diameter tungsten wire wound with 25 μm wire spacing. The movable mirror was scanned by a stepping motor. The maximum OPD was 55 cm. The fixed mirror MF4 is usually set near the beam splitter BS2. In this experiment, however, the fixed mirror was placed at MF4′ 2.3 m apart from MF4 so that the interference between a wave packet of one bunch and the 21st bunch can be observed in order to clarify the small deviations of the inter-bunch distance.

The output signal from the bolometer was scanned by a fast-gated integrator (SRS250, Stanford Research Systems Inc.) at intervals of 200 ns, as shown in Fig. 4. The wave form of the signal is shown by a solid curve. The frequency response of the InSb hot-electron bolometer was 350 ns. The waveform of the electron beam measured by a current transformer (CT) is represented by a dashed curve. The gate width (dotted curve) of the integrator was 200 ns and 260 bunches were included in the width. Therefore, the average of the inter-bunch distance within 200 ns was measured in this experiment.

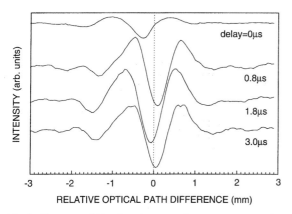

Fig. 5. Cross-correlation interferograms between a wave packet of one bunch and the 21st bunch. The values of the delay represent the position of the gate in the macro-pulse.

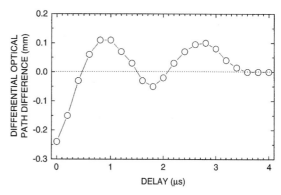

Fig. 6. The fluctuation of the inter-bunch distance in the macro-pulse. The ordinate represents the total amount of the deviation of 20 intervals.

4. Results and discussion

Fig. 5 shows 21st cross-correlation interferograms measured at the gate width of 200 ns at around 4.6-m OPD of the interferometer. The OPD of the abscissa is the relative value because the fixed mirror could not be placed accurately on the order of 10 μm. The relative value, however, has an accuracy of 10 μm since the movable mirror was scanned every 10-μm step. The insetted numerical values of the delay in this figure represent the position of the gate in the macro-pulse. The positions of the minimum bottoms on the interferogram is shown in Fig. 5, where the wave packet from one bunch and that from the 21st bunch are superposed perfectly, vary for every time delay in the 4-μs macro-pulse. This shift is not due to the incorrect action of the stepping motor of the movable mirror, because we did not observe the variation of the bottom position in the autocorrelation interferogram when the fixed mirror was placed at MF4 in Fig. 3.

The deviations of OPD at the minimum bottoms are summarized in Fig. 6. The abscissa represents the position of the gate in the macro-pulse. The ordinate shows the differential OPD. The inter-bunch distance becomes short as the differential OPD decreases. The periodic fluctuation was observed in the 4-μs macro-pulse and the maximum amplitude is 0.36 mm. Since the amplitude of this fluctuation is the total amount of the interval between the first bunch and the 21st one, the inter-bunch distance has the fluctuation of 0.018 mm, i.e., 60 fs. The observed fluctuation corresponds to the time resolution of the order of 10^{-5}, i.e., 0.01%, as compared to the theoretical interval between bunches of 769 ps.

Acknowledgements

This work was partially supported by a Grant-in-Aid for Young Scientists (B) from the Ministry of Education, Culture, Sports, Science and Technology.

References

[1] T. Takahashi, et al., Rev. Sci. Instrum. 69 (1998) 3770.
[2] K. Ishi, et al., Phys. Rev. A 43 (1991) 5597.
[3] R. Lai, U. Happek, A.J. Sievers, Phys. Rev. E 50 (1994) R4294.
[4] T. Watanabe, et al., Nucl. Instr. and Meth. A 480 (2002) 315.
[5] Y. Shibata, et al., Phys. Rev. E 50 (1994) 1479.
[6] Y. Shibata, et al., Phys. Rev. Lett. 78 (1997) 2740.
[7] Y. Shibata, et al., Nucl. Instr. and Meth. B 145 (1998) 49.
[8] S. Sasaki, et al., Nucl. Instr. and Meth. A 483 (2002) 209.
[9] Y. Shibata, et al., Phys. Rev. A 44 (1991) R3445.

Available online at www.sciencedirect.com

SCIENCE ⒟ DIRECT°

ELSEVIER   Nuclear Instruments and Methods in Physics Research A 528 (2004) 212–214

**NUCLEAR
INSTRUMENTS
& METHODS
IN PHYSICS
RESEARCH**
Section A

www.elsevier.com/locate/nima

A composite open resonator for a compact X-ray source

E.G. Bessonov*, R.M. Fechtchenko

P.N. Lebedev Physical Institute of the Russian Academy of Sciences, 119991, Leninsky prospect 53, Moscow 117234, Russia

Abstract

The fineness of a composite open resonator for a compact X-ray source is calculated. The region of the resonator parameters has been found where its fineness is unessentially changed, the excited mode remains Gaussian (with maximum at the axis of the laser beam) and the effective head-on collision of laser and electron beams is realized.
© 2004 Elsevier B.V. All rights reserved.

PACS: 41.60.Cr

Keywords: FEL; Optical resonator; Compton scattering

1. Introduction

Recently, there has been considerable progress in the development of super-reflection mirrors, high-fineness optical resonators (super-resonators) and frequency stabilized high-power cw and pulsed mode-locked lasers. Optical resonators can be used as a photon reservoir to accumulate very high laser power. By this method, a fineness of 10^6 can be achieved for optical resonators, which means the laser power can be enhanced by this order [1]. The high reactive power stored in the super-resonators can be used in gravitational experiments in astrophysics, in Laser-Electron Storage Rings [1–4], in the gravitational analogue of lasers (grasers) [5] and in other applications.

A decrease in mirror performance is a source of concern in open resonators. The main cause of mirror degradation of high fineness dielectric open resonators of free-electron lasers using storage rings is the deposition of chemical components on the mirror surfaces by X-ray synchrotron radiation [6]. The same degradation can occur in Laser-Electron Storage Rings where the backward Compton scattering of laser photons by electron beams is used for production of X-ray and γ-ray radiation [2,3]. A possible solution for the mirror degradation problem is the use of a composite resonator. Below, the calculation of the dependence of fineness on the parameters of such a composite resonator are presented and analyzed.

2. Quality of the composite open resonator

Let a composite open resonator consists of two mirrors (see Fig. 1). The reflectivity of the first

---
*Corresponding author. Tel.: +7-95-3340119; fax: +7-95-9382251.

E-mail address: bessonov@x4u.lpi.ruhep.ru (E.G. Bessonov).

0168-9002/$ - see front matter © 2004 Elsevier B.V. All rights reserved.
doi:10.1016/j.nima.2004.04.049

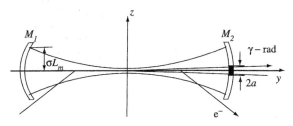

Fig. 1. A composite open resonator.

mirror M_1 and the main part of the second mirror M_2 is R_1. The second mirror has a circular insertion area at the axis of the resonator made from another material having reflectivity R_2. The radius of the insertion area is "a" and the rms radius of the fundamental mode of the resonator at the surface of the mirror is "σ_{Lm}". Both mirrors can be dielectric multilayer mirrors and the insertion area in the second mirror can be Au coated.

The intensity of the laser beam in this case is distributed by the Gaussian law as

$$I_L = \frac{P_0}{2\pi\sigma_L^2} e^{-r^2/2\sigma_L^2} \tag{1}$$

where $P_0 = \int I_L \, dS$ is the initial power of the laser beam stored in the resonator; dS is the area of the element; $\sigma_L = \sigma_{L0}\sqrt{1 + s^2/l_R^2}$ is the dispersion of the laser beam at a point s; $\sigma_{L0} = \sigma_L$ ($s = 0$); the point $s = 0$ corresponds to the waist of the laser beam; $l_R = 4\pi\sigma_{L0}^2/\lambda_L$ is the Rayleigh length; λ_L is the laser wavelength.

In a composite resonator some part of the laser power ($P_1 = P_0 R_1$) is reflected by the first mirror. Part of this power (($P_1 - \Delta P_1)R_1$) is reflected by the external portion of the second mirror and another part is reflected by the internal portion of the mirror ($\Delta P_1 R_2$), where $\Delta P_1 = P_1[1 - \exp(-a^2/2\sigma_{Lm}^2)]$ is the energy incident upon the insertion portion of the second mirror. The round trip reflected power ($P_2 = (P_1 - \Delta P_1)R_1 + \Delta P_1 R_2$) can be expressed in the form of

$$P_2 = P_0 R_1^2\left[1 - \frac{R_1 - R_2}{R_1}(1 - e^{-a^2/2\sigma_{Lm}^2})\right]. \tag{2}$$

After n round trips the laser light power can be expressed as

$$P(t) \simeq P_0\left(\frac{P_2}{P_0}\right)^n = P_0 e^{n \ln(P_2/P_0)}$$

$$= P_0 e^{n \ln(R_1^2[1-(R_1-R_2)/R_1(1-e^{-a^2/2\sigma_{Lm}^2})])} \tag{3}$$

where $n = ct/2L$ and L is the resonator length.

The fineness of the open resonator can be determined by the equation $F = -(2\pi/T)[P/(\partial P/\partial t)]$. In this case, according to Eq. (3), the fineness of the composite resonator can be expressed in the form

$$F = F_0 \frac{\ln R_1^2}{\ln\{R_1^2[1 - (R_1 - R_2)/R_1(1 - e^{-a^2/2\sigma_{Lm}^2})]\}} \tag{4}$$

where $F_0 = F|_{R_1=R_2} = -\pi/\ln R_1 \simeq \pi/(1 - R_1)$.

According to Eq. (4), in the approximations $(1 - R_1)/(1 - R_2) \ll 1$, $1 - R_1 \ll 1$, $1 - R_2 \ll 1$, $a/\sigma_{Lm} \ll 1$, the fineness of the resonator will be decreased F_0/F times when the radius

$$a \simeq \sqrt{2}\sigma_{Lm}\sqrt{\frac{1 - R_1}{1 - R_2}\left(\frac{F_0}{F} - 1\right)}. \tag{5}$$

Example. The relativistic factors of electrons are $\gamma = 10^3$, $\lambda_L = 10\ \mu m$, $\sigma_{L0} = 50\ \mu m$, $L = 2$ m, $F/F_0 = 0.5$, $(1 - R_2)/(1 - R_1) = 10^2$.

In this case, according to Eqs. (1) and (5), the Rayleigh length is $l_R = 3.14$ mm, the mode size is $\sigma_{Lm} = 15.9$ mm, the radius of the X-ray beam at the mirrors is $\sigma_{X\text{-ray}} \simeq L/2\gamma = 1$ mm and the radius of the insertion is $a = 0.1\sqrt{2}\sigma_{Lm} \simeq 2.25$ mm $> \sigma_{X\text{-ray}}$. The magnification M, defined as the ratio of the mode size at the mirrors to the mode size at the waist in the resonator, must be chosen to be high ($\sim 10^2$–10^3), which will be a compromise between mirror degradation, resonator fineness and the source issues.

3. Conclusion

We have shown that if the dimension of the insertion is larger than the effective transverse dimension of the X-ray beam and much less than the diameter of the fundamental Gaussian mode, the mirror degradation time is then greatly increased and the fineness of the resonator is essentially not decreased. Such conditions can be fulfilled when the magnification $M \gg 1$. Further, the effective head-on collision of laser and electron beams is realized.

Acknowledgements

This work was supported by the Russian Foundation for Basic Research, Grant No. 02-02-16209.

References

[1] J. Chen, K. Imasaki, M. Fujita, et al., Nucl. Instr. and Meth. A 341 (1994) 346.

[2] Zh. Huang, R.D. Ruth, Phys. Rev. Lett. 80 (5) (1998) 976.

[3] E.G. Bessonov, Proceedings of the 23rd International ICFA Beam Dynamic WS on Laser–Beam Interactions, Stony Brook, NY, June 11–15, 2001, http//atfweb.kek.jp/icfa/2001/box/index.html; physics/0111084; physics/0202040.

[4] J. Urakawa, M. Uesaka, M. Hasegawa, et al., Proceedings of the 23rd International ICFA Beam Dynamic WS on Laser–Beam Interactions, Stony Brook, NY, June 11–15, 2001, http://atfweb.kek.jp/icfa/2001/box/index.html.

[5] E.G. Bessonov, Proceedings of the International Advanced ICFA Beam Dynamic WS on Quantum Aspects of Beam Physics, Monterey, CA, January 4–9, 1998, World Scientific, USA, p. 330; physics/9802037.

[6] M. Yasumoto, T. Tomimasu, S. Nishihara, N. Umesaki, Proceedings of 21st International Free Electron Lasers Conference, August 23–28, 1999, Hamburg, Germany, pp. II-109.

Available online at www.sciencedirect.com

SCIENCE ⓓ DIRECT°

ELSEVIER   Nuclear Instruments and Methods in Physics Research A 528 (2004) 215–219

**NUCLEAR
INSTRUMENTS
& METHODS
IN PHYSICS
RESEARCH**
Section A

www.elsevier.com/locate/nima

Coherent diffraction radiation diagnostics for charged particle beams

H.K. Avetissian*, G.F. Mkrtchian, M.G. Poghosyan, Kh.V. Sedrakian

Department of Quantum Electronics, Plasma Physics Laboratory, Yerevan State University, 1 A. Manukian, Yerevan 375025, Armenia

Abstract

In this paper, we have presented the properties of coherent diffraction radiation (CDR) for analyzing the general geometry of a charged particle beam passing above a semi-infinite ideally conducting screen, and have discussed the CDR for the measurement of the beam bunch shape.
© 2004 Elsevier B.V. All rights reserved.

PACS: 41.60.Cr; 41.75.Ht; 29.27.Fh

Keywords: Beam; Diagnostic; Superradiant

1. Introduction

A recently developed interest in "diffraction radiation" (DR) is related to both the exploration of the possibility of using it to generate intense radiation from the millimeter to UV and X-ray regions and the implementation of it in non-destructive beam diagnostics [1–7]. For the first problem, one should consider the radiation from many periodically spaced interfaces (known as coherent or resonant diffraction radiation). In both cases, the influence of factors such as transverse beam size, angular beam divergence, monochromaticity, etc., on the radiation characteristics should be taken into account. Consequently, one needs to know about a particle's radiation characteristics from a particle for general geometry and for various screens. Besides, when the observed DR wavelengths are larger or comparable to the transverse or longitudinal size of an electron beam bunch, the collective effects may play a significant role leading to superradiant DR. Such "coherent diffraction radiation" (CDR) can be observed with intensity, which is much higher than of incoherent DR, and can be utilized in beam diagnostics [7].

In this paper, we are focusing on the analysis of CDR for further beam diagnostics, particularly for the measurement of the beam bunch length. In this paper, in order to emphasize the chief principles of electron beam diagnostic method, based on the superradiant DR, we shall present the case of a

*Corresponding author. Department of Quantum Electronics, Plasma Physics Laboratory, Yerevan State University, 1 A. Manukian, Yerevan 375025, Armenia. Tel./fax: +3741-570-597.

E-mail address: avetissian@ysu.am (H.K. Avetissian).

charged-particle beam passing above a semi-infinite ideally conducting screen.

2. Theory of CDR for a particle beam passing above a semi-infinite ideally conducting screen (general geometry)

To calculate the CDR characteristics of a charged-particle beam passing above a semi-infinite ideally conducting screen, it is necessary to know a single particle diffraction radiation field for general geometry as displayed in Fig. 1. The cases for $\Phi = \pi/2$ [8] and $\theta = \pi/2$ [9] have been considered in previous works. However, the fields and spectral–angular distributions of DR for general geometry have not been previously reported. Therefore, in this paper we present solutions for the fields employing Huygens–Fresnel diffraction approach [10] with the Kirchhoff condition on the radiation field

$$\mathbf{E}|_{y=0, x<0} = 0. \tag{1}$$

The x and z components of the radiation field can be expressed as

$$E_{\omega x,z} = \frac{i\omega n_y}{2\pi c R} \int_{S_{\text{screen}}} E^e_{\omega x,z}\, e^{-i(\omega/c)(n_x x + n_z z)}\, dS \tag{2}$$

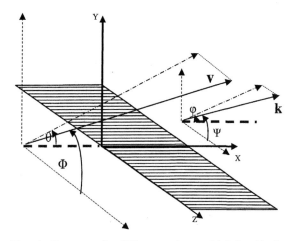

Fig. 1. Geometry for DR near the semi-infinite ideally conducting screen. The screen plane, XZ. Φ represents the angle between the particle velocity $\mathbf{v}$ and OZ axis, θ is the angle between the projection of particle velocity $\mathbf{v}$ onto XZ plane and OX axis. The same holds true for the vector $\mathbf{k}$ and angles Ψ and φ.

where, $E^e_{\omega x,z}$ are the components of the field of an incoming charged particle, $\mathbf{R}$ is the radius vector of an observation point,

$$\mathbf{n} = (n_\perp \cos\varphi, n_\perp \sin\varphi, \cos\Psi)$$

$$n_\perp = \sin\Psi$$

is the unit vector in the radiation direction. The y component of the radiation field can be found from the condition $\mathbf{n}\mathbf{E}_\omega = 0$. On integrating Eq. (2), the radiation field will be obtained as follows:

$$\mathbf{E}_\omega(\mathbf{r}_0, \mathbf{v}, \mathbf{n}) = \frac{e^{ikR}}{R} e^{-i\omega\mathbf{q}\mathbf{r}_0}\mathbf{G}(\mathbf{v}, \mathbf{n}) \tag{3}$$

where, $\mathbf{G}(\mathbf{v}, \mathbf{n})$ is determined by the following formulas as a function of the radiation angle and particle velocity:

$$G_x = \pm \frac{ev_y n_y\left(\frac{v_x}{c^2} - b - iB\right)}{2\pi B c v_\perp^2\left(B + i\left(\frac{n_x}{c} - b\right)\right)}$$

$$G_z = \pm \frac{ev_y n_y\left(\frac{v_z}{c} - n_z\right)}{2\pi c^2 B v_\perp^2\left(B + i\left(\frac{n_x}{c} - b\right)\right)}$$

$$G_y = -\frac{1}{n_y}(n_x G_x + n_z G_z). \tag{4}$$

Here,

$$\mathbf{v} = (v_\perp \cos\theta, v_\perp \sin\theta, v\cos\Phi)$$

$$v_\perp = v\sin\Phi$$

is the particle velocity, $\mathbf{r}_0$ is the initial position of a charged particle, which along with the complex vector $\mathbf{q}$

$$q_x = b + iB \quad q_y = \frac{1}{v_y}(1 - n_z v_z - v_x q_x)$$

$$q_z = \frac{n_z}{c} \tag{5}$$

will take into account phase relations between the different particles of Eq. (3). The signs $+$ and $-$ in Eq. (4) correspond to forward and backward DR, respectively. Then

$$B = \frac{v_y}{v_\perp^2}\left[\left(1 - n_z\frac{v_z}{c}\right)^2 - n_\perp^2\frac{v_\perp^2}{c^2}\right]^{1/2}$$

$$b = \frac{v_x}{v_\perp^2}\left(1 - n_z\frac{v_z}{c}\right). \tag{6}$$

On summing the field of all particles Eq. (3), one can obtain the following expression for the spectral–angular density of DR intensity:

$$I_{n\omega} = \frac{d\varepsilon_{n\omega}}{d\omega \, d\Omega} = cR^2 \left| \sum \mathbf{E}_\omega(\mathbf{r}_0, \mathbf{v}, \mathbf{k}) \right|^2 .$$

By neglecting correlation effects, the latter can be expressed by the one-particle distribution function $f(\mathbf{r}_0, \mathbf{p})$ as follows:

$$I_{n\omega} = cN \int |e^{-i\omega\mathbf{q}\mathbf{r}_0} \mathbf{G}(\mathbf{v}, \mathbf{n})|^2 f(\mathbf{r}_0, \mathbf{p}) \, d\mathbf{r}_0 \, d\mathbf{p}$$
$$+ cN^2 \left| \int e^{-i\omega\mathbf{q}\mathbf{r}_0} \mathbf{G}(\mathbf{v}, \mathbf{n}) f(\mathbf{r}_0, \mathbf{p}) \, d\mathbf{r}_0 \, d\mathbf{p} \right|^2 \quad (7)$$

where, N is the number of particles in the bunch. The first term of Eq. (7) is the incoherent part of the radiation, while the second term of Eq. (7) is the coherent/superradiant part. The radiation density from one particle attains a maximum value for the forward DR (FDR), which coincides with the direction of the particle motion

$$\varphi = \theta, \quad \Psi = \Phi \quad (8)$$

and at

$$\varphi = -\theta, \quad \Psi = \Phi \quad (9)$$

for the backward DR (BDR), which corresponds to the mirror reflection from the screen. The maximal densities of BDR and FDR coincide.

For beam diagnostics, the BDR is more favorable and there is a growing interest in using BDR characteristics to obtain precise information about the beam parameters. Hence, in the following study, we shall be considering only BDR. The parameters $\mathbf{q}$ and $\mathbf{G}(\mathbf{v}, \mathbf{n})$ in Eq. (7) depend on the particle velocity. Since the magnitude of the velocity spread is given by $\delta v/v \simeq \delta\varepsilon/(\varepsilon\gamma^2)$, it can be concluded that for the ultrarelativistic particle beam, the energy spread $\delta\varepsilon$ does not play a significant role in the spectral and angular distributions of CDR. Thus, the main factor would have to be the beam divergence, and we can consider the values corresponding to the average energy of the beam for these parameters. For the factorized distribution function $f(\mathbf{r}_0, \theta, \Phi) = n(\mathbf{r}_0)D(\theta, \Phi)$, we need to integrate the Eq. (3) over $\mathbf{r}_0$, which can be expressed as follows for the Gaussian beam with isotropic transverse

distribution of characteristic width $\sigma_\perp$:

$$I_{n\omega} = cN \int |\mathbf{G}(\mathbf{v}, \mathbf{n})|^2 \, e^{-2B\omega a_0} \, e^{(2\sigma_\perp^2 \omega^2 B^2/\sin^2 \theta)}$$
$$\times D(\theta, \Phi) \, d\theta \, d\Phi + I_{n\omega}^{(c)} \quad (10)$$

where, the coherent part of DR is given by the formula

$$I_{n\omega}^{(c)} = cN^2 e^{(\sigma_\perp^2 \omega^2/\bar{\gamma}^2\bar{v}^2) - (\sigma_1^2 \omega^2/\bar{v}^2)}$$
$$\times \left| \int e^{ia_0\omega(q_x - (n_x/c))} \mathbf{G}(\mathbf{v}, \mathbf{n}) D(\theta, \Phi) \, d\theta \, d\Phi \right|^2 . \quad (11)$$

In this case, σ_1 is the longitudinal characteristic size and the quantity a_0 characterizes the mean position of the beam from the screen edge in the target plane. The bar denotes the average value. The influence of the transverse distribution of the beam on the coherent part of the spectrum is much smaller than the longitudinal one and can be neglected for the relativistic particles as seen in Eq. (11). Hence, the spectrum of superradiation is mainly defined by the longitudinal bunch form factor $\exp(-\sigma_1^2 \omega^2/2\bar{v}^2)$.

As seen from Eq. (11), if we take into account the beam divergence, the CDR cannot be factorized as

$$I_{n\omega}^{(c)} = N^2 e^{-(\sigma_1^2 \omega^2/\bar{v}^2)} I_{1n\omega} \quad (12)$$

as previously shown in Ref. [7], which is crucial for the measurement of the bunch shape. In Eq. (12)

$$I_{1n\omega} = ce^{-2B\omega a_0} |\mathbf{G}(\mathbf{v}, \mathbf{n})|^2$$

is the radiation density from one particle. To clarify the beam divergence effect on the spectral distribution of CDR, the normalized density $F(\omega) = I_{n\omega}^{(c)}/N^2 I_{1n\omega}$ of CDR at the maximum is plotted for various angular spreads as shown in Fig. 2. For the cold beam, it coincides with the modulus squared of the longitudinal form factor of the beam. The beam energy is taken to be 500 MeV, $\Phi_0 = \pi/2$, $\theta_0 = \pi/4$, $a_0 = 3$ mm, the longitudinal characteristic size $\sigma_1 = 0.03$ mm, and transverse size $\sigma_\perp = 0.1$ mm. We considered the isotropic beam divergence, which is described by

$$F(\omega) = I_{n\omega}^{(c)} / I_{1n\omega}(\omega) N^2$$

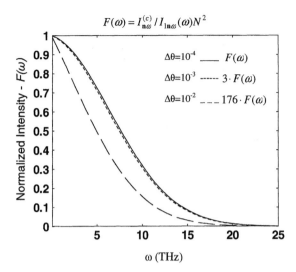

Fig. 2. Effect of the beam divergence on the spectral distribution of coherent BDR at different values of beam divergence. The beam energy is taken to be 500 MeV, $\Phi_0 = \pi/2$, $\theta_0 = \pi/4$, $a_0 = 3$ mm, the longitudinal characteristic size $\sigma_1 = 0.03$ mm, and transverse size $\sigma_\perp = 0.1$ mm. The solid line corresponds to $\Delta\theta = 10^{-4}$ and coincides with the modulus squared for the beam form factor. For visual convenience, we have multiplied $F(\omega)$ in cases $\Delta\theta = 10^{-3}$ (dotted line) and $\Delta\theta = 10^{-2}$ (dashed line) by 3 and 176, respectively.

the Gaussian distribution function

$$D(\theta, \Phi) = \frac{1}{2\pi\Delta\theta^2} e^{-((\theta-\theta_0)^2+(\Phi-\Phi_0)^2)/2\Delta\theta^2} \quad (13)$$

where $\Delta\theta$ is the angular spread of the beam.

Fig. 2 shows that the radiation density gets diminished in magnitude and the spectral width is altered due to the angular divergence. As observed in Eq. (11), the factorized formula shown in Eq. (12) is valid when

$$\Delta\theta \ll \min\left\{\frac{1}{\gamma}, \frac{\sigma_1}{a_0}\right\} \quad (14)$$

which corresponds to the solid curve in Fig. 2. One can recover the beam form factor only at such a beam divergence via Eq. (12) by measuring the quantity $I_{n\omega}^{(c)}$. Furthermore, by inversing the corresponding Fourier transformation, the bunch shape can be restored. Hence, the diagnostic method developed in previous researches [7] is

applicable only if the conditions are satisfied, as in Eq. (14).

For the beam with the divergence $\Delta\theta \sim 1/\gamma$ but $\Delta\theta \ll \sigma_1/a_0$ (dotted curve in Fig. 2), the exponent in the integral of Eq. (11) can be taken at average angles and the density of CDR can then be factorized as

$$I_{n\omega}^{(c)} \simeq cN^2 e^{-2\bar{B}\omega a_0} e^{-(\sigma_1^2\omega^2/\bar{v}^2)}$$

$$\times \left| \int \mathbf{G}(\mathbf{v}, \mathbf{n}) D(\theta, \Phi) \, d\theta \, d\Phi \right|^2. \quad (15)$$

The Eq. (15) suggests that only the maximum radiation density is diminished due to angular divergence. Therefore, in contrast to previous experiments [7] one should also measure in the low-frequency domain to renormalize the longitudinal form factor.

For the beams with $\Delta\theta \sim 1/\gamma$, $\Delta\theta \sim \sigma_1/a_0$ (dashed curve in Fig. 2) the angular distribution of the particles are required to measure the beam longitudinal form factor using the general formula (11).

3. Conclusion

We have presented the theoretical treatment of DR for the general geometry of a charged-particle beam passing above a semi-infinite ideally conducting screen. It is shown that if one takes into account the beam divergence, the frequency dependence of the radiation density of CDR does not get factorized to a single particle radiation density multiplied by the beam form factor, which is important for the bunch shape measurements. We have clarified the beam divergence effect on the spectral distribution of CDR and formulated the criteria of validity for various approaches.

Acknowledgements

This work was supported by the National Foundation of Science and Advanced Technologies (NFSAT) Grant No. NFSAT PH 082-02/ CRDF 12023.

References

[1] M.J. Moran, B. Chang, Nucl. Instr. and Meth. B 40/41 (1989) 970.

[2] M. Castellano, Nucl. Instr. and Meth. A 394 (1997) 275.

[3] D.W. Rule, R.B. Fiorito, W.D. Kimura, in: A.H. Lumpkin, C. Eyberger (Eds.), AIP Conf. Proc. 390 (1997) 510.

[4] R.B. Fiorito, D.W. Rule, Nucl. Instr. and Meth. B 173 (2001) 67.

[5] A.P. Potylitsyn, N.A. Potylitsyna, preprint, lanl.arXiv physics/0002034 (2000).

[6] Y. Shibata, et al., Phys. Rev. E 52 (1995) 6787.

[7] B. Feng, et al., Nucl. Instr. and Meth. A 475 (2001) 492.

[8] A.P. Kazantsev, G.I. Surdutovich, Sov. Phys. Dokl. 7 (1963) 990.

[9] D.M. Sedrakian, Izv. AN ArmSSR 17 (N4) (1964) 103 (in Russian).

[10] M.L. Ter-Mikaelian, High Energy Electromagnetic Processes in Condensed Media, Wiley, New York, 1972.

Available online at www.sciencedirect.com

NUCLEAR
INSTRUMENTS
& METHODS
IN PHYSICS
RESEARCH
Section A

ELSEVIER   Nuclear Instruments and Methods in Physics Research A 528 (2004) 220–224

www.elsevier.com/locate/nima

Waveguide-coupled cavities for energy recovery linacs

S.S. Kurennoy*, D.C. Nguyen, L.M. Young

Los Alamos National Laboratory, Los Alamos, NM 87545, USA

Abstract

A novel scheme for energy recovery linacs used as FEL drivers is proposed. It consists of two parallel beam lines, one for electron beam acceleration and the other for the used beam that is bent after passing through a wiggler. The used beam is decelerated by the structure and feeds the cavity fields. The main feature of the scheme is that RF cavities are coupled with waveguides between these two linacs. The waveguide cut through the two beam pipes provides an efficient mechanism for energy transfer. The superconducting RF cavities in the two accelerators can be shaped differently, with an operating mode at the same frequency. This provides HOM detuning and therefore reduces the beam break-up effects. Another advantage of the proposed two-beam scheme is easy tuning of the cavity coupling by changing the waveguide length.
© 2004 Elsevier B.V. All rights reserved.

PACS: 41.60.Cr; 29.17.+w; 29.27.Bd; 41.75.Fr

Keywords: FEL; Energy recovery; CW; High current; High power

1. Introduction

Energy recovery (ER) will be an important feature of many, if not all, future high-power FELs. This is due to the fact that in FELs an electron beam passing through a wiggler radiates a rather small fraction of its energy, typically 1–2% in oscillator FELs and up to 5–7% in amplifier FELs. For an FEL system to be efficient, the remaining significant beam power must be recovered. Moreover, if this power is not recycled in the system, disposing of the spent beam could be a

difficult problem: the high-power beam dumping becomes a serious challenge.

There are different options for energy recovery. One is reusing the electron beam after it has passed through the wiggler. After conditioning and acceleration, the beam is returned to the wiggler again and again. This scheme is more appropriately called "beam recovery." It is better suited for oscillator-type FELs, where the beam quality degrades less after passing through the wiggler. In a sense, storage-ring FELs are the extreme example of this approach.

Another ER option is to recover only the energy contained in the spent beam, not the beam itself. This can be achieved by two different methods. The beam, after it has passed through the wiggler,

---
*Corresponding author. Tel.: +1-505-665-1459; fax: +1-505-665-2904.

E-mail address: kurennoy@lanl.gov (S.S. Kurennoy).

(1) is sent again through the same accelerating structure as the initial beam, but in a decelerating phase, or (2) is directed through a separate decelerating structure. The last structure should be strongly coupled to the accelerating structure to transfer the recovered energy efficiently. These two methods are classified as the "same-structure (or same-cell) ER" and "different-structure ER."

The idea of energy recovery in accelerators has a long history. Starting from a first proposal by Tigner [1], this idea led to the Reflexotron [2], and then to experiments on the different-structure ER in two copper linacs with bridge couplers at LANL [3], and the same-cell ER in a superconducting (SC) linac at SLAC [4]. The development of ER culminated recently in a record high average power of 1.7 kW CW FEL (2.3 kW by now) achieved in the SC IR Demo FEL at JLab [5] with the same-cell ER.

Obviously, all ER schemes have the common advantage of making an FEL system more effective overall. An important consideration of the FEL system accelerator efficiency, given fixed beam energy at injection and at the wiggler, is related to the ratio P_b/P_w, where P_b is the beam power and P_w is the power dissipated in accelerator structures. The higher this ratio, the more efficient the ER can be. SC systems certainly have a big advantage here, both for "same-cell" and "different-structure" ER. Disadvantages of ER schemes include generally more complex systems than those without ER, except for the beam dump. Additional problems in beam dynamics include possible instabilities at high currents, such as beam break-up (BBU). The transverse BBU appears to be the most dangerous in recirculating ER linacs [6]. From this viewpoint, normal-conducting structures, where damping of high-order modes (HOM) is easier than in SC structures and Q-factors are lower, can be better. For the same reason, SC "different-structure" ER has an advantage over SC "same-cell" ER with respect to the beam stability: one can detune HOM in its decelerating structure.

The "same-cell" ER seems to be a favorite choice today, in large part because of economical reasons. We would like to propose here a simple option for coupling RF cavities in two linacs of the "different-structure" ER that can be useful and important for future really high-power FELs.

2. Waveguide-coupled cavities for ER

2.1. Waveguide coupling

Three-cell bridge couplers were used for energy transfer between two copper linacs at LANL [3]. Their central cell served also for the RF power input. Four tuning posts per bridge allowed changing the field in the decelerator from 15% to 120% of that in the accelerator. Unfortunately, such bridges are too complicated to be implemented in SC linacs. It would be good to have couplers that are simpler.

We have explored an option with a coaxial coupler and found that getting the coupling coefficient on the order of a few percent this way was very difficult. On the other hand, a coupler consisting of a waveguide (WG) section that cut through two beam pipes was found to give a strong coupling; it is easily tunable as well. The model coupler layout is shown in Fig. 1.

2.2. Model of two waveguide-coupled cavities

To demonstrate the idea, we intentionally chose a simple model with two differently shaped

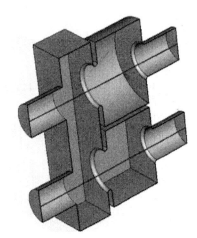

Fig. 1. Model coupler layout. Two beam paths are shown by the lines.

cavities—one in an accelerating linac and the other in a decelerator. The first cavity is a cylindrical pillbox (C), while the second one has a square-shaped cross section (S), but they both have the same fundamental frequency at 700 MHz. The WG coupler here is a section of a standard WR1150, but it can be some other WG. The coupling is regulated by the size of irises connecting the WG with the cavities.

We studied our simple model with the Micro-Wave Studio (MWS) [7] to calculate the mode frequencies and coupling. The working mode

frequency can easily be tuned to 700 MHz by small changes of the transverse dimensions of the two cavities. The electric field of this mode is shown in Fig. 2. If we consider a three-cell system consisting of two linac cells, C and S with the coupling cell WG in between, the mode can be classified as a $\pi/2$-mode. Obviously, the coupling cell is not excited in this mode.

Small changes of the WG length make two neighbor modes (0- and π-type, Fig. 3) separated from the working mode by the same shift, but in opposite directions: at $f_0 = 694$ MHz and $f_\pi = 706$ MHz. Hence, the coupling is $\kappa = (f_\pi - f_0)/f_{\pi/2} = 12/700 = 0.017$ for the model in Fig. 2, where the radii of the coupling apertures are equal to that of the beam pipe, 65 mm. For the model in Fig. 1, where the cavity–WG coupling irises are larger ($r = 80$ mm), the coupling is 3.5%. The coupling also depends on the separation between the cavity and WG, i.e., on the iris thickness.

The HOMs in the model are uncoupled because the cavities in two linacs have different shapes. This is clearly illustrated in Fig. 4, where two dipole modes are plotted. The mode fields are contained mostly in their 'native' cavities.

The WG length L_{wg} in our model was chosen to be slightly larger than one full WG wavelength at 700 MHz, $\lambda_{wg} = 63.02$ cm. The WG ends are at about $L_{wg}/4$ from the cavity axes, and the axis separation is close to $L_{wg}/2$. This makes a compact structure that can fit, for example, in a single

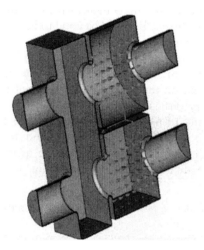

Fig. 2. Electric field of the fundamental $\pi/2$ mode at 700 MHz.

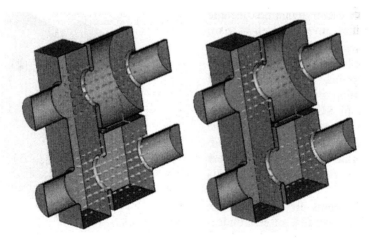

Fig. 3. Electric fields of 0- and π-modes at 694 and 706 MHz.

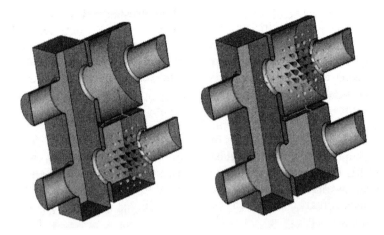

Fig. 4. Electric fields of two HOMs at 804 and 822 MHz.

cryomodule. However, the WG length can be increased by an integer number of $\lambda_{wg}/2$, if needed. Slightly changing the WG section length gives us an effective tuning option. The flat ends of the WG section can be made thin enough to implement mechanical push–pull tuners.

2.3. Coupled multi-cell cavities in linacs

The same approach can be applied for multi-cell cavities in linacs. Fig. 5 illustrates the working mode at 700 MHz for a model with 2-cell cavities. One can see that again the WG coupler is not excited in this mode, while the cavities have field patterns typical for 2-cell π-modes. From the viewpoint of mode classification in a 5-cell cavity consisting of cells C_1, C_2, WG coupler, S_1, and S_2, this working mode has the structure $(-1,1,0,-1,1)$. It can be called $3\pi/4$-mode, as the fourth mode in the 5-mode pass band. This pass band has its lowest, 0-mode at 687.5 MHz and the highest, π-mode at 704 MHz. The cell coupling coefficient here is 2.1%.

3. Discussion

A simple model with simplified cavities was presented to illustrate the idea of WG-coupled linacs for energy recovery in FELs. The WG

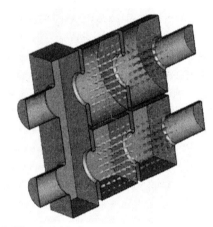

Fig. 5. Electric field of the ER working mode at 700 MHz.

coupling between cavities in the accelerator and decelerator structures appears to be simple, effective, and easily tunable. It is important to note that this scheme can be implemented in SC linacs as well. Of course, for SC linacs the cavities should be, most likely, of the multi-cell elliptical type, but the WG coupling will work for them too, as one can see from Section 2.3.

A rectangular WG was used in our simple model. Possible multipacting effects in such a WG can be mitigated by choosing a different WG shape, e.g., a WG with an elliptical cross-section.

4. Conclusions

A simple scheme for coupling RF cavities in the "different-structure" energy recovery linacs used as high-power FEL drivers is proposed. The RF cavities are coupled between the two linacs—an accelerator and decelerator—with waveguides that cut through two beam pipes, providing an efficient mechanism for energy transfer. The RF cavities in the two linacs can be shaped differently to detune HOMs and increase BBU thresholds.

Using a simple model, we have shown that the WG coupling is simple, tunable, and can be implemented for multi-cell cavities in SC RF linacs. It can be a useful option on a path to high-power ER FELs.

Acknowledgements

The authors would like to acknowledge useful discussions with F.L. Krawczyk, D.L. Schrage, and R.L. Wood (LANL).

References

[1] M. Tigner, Nuovo Cimento 37 (1965) 1228.
[2] S.O. Schriber, E.A. Heighway, IEEE NS-22 (1975) 1060.
[3] D.W. Feldman, et al., Nucl. Instr. and Meth. A 259 (1987) 26.
[4] T.I. Smith, et al., Nucl. Instr. and Meth. A 259 (1987) 1.
[5] G.R. Neil, et al., Phys. Rev. Lett. 84 (2000) 662.
[6] L. Merminga, Nucl. Instr. and Meth. A 483 (2002) 107.
[7] MicroWave Studio, CST, Darmstadt, Germany.

Available online at www.sciencedirect.com

NUCLEAR
INSTRUMENTS
& METHODS
IN PHYSICS
RESEARCH
Section A

ELSEVIER   Nuclear Instruments and Methods in Physics Research A 528 (2004) 225–230

www.elsevier.com/locate/nima

Test facility for investigation of heating of 30 GHz accelerating structure imitator for the CLIC project

A.V. Elzhov[a], N.S. Ginzburg[b], A.K. Kaminsky[a,*], S.V. Kuzikov[b], E.A. Perelstein[a], N.Yu. Peskov[b], M.I. Petelin[b], S.N. Sedykh[a], A.P. Sergeev[a], A.S. Sergeev[b], I. Syratchev[c], N.I. Zaitsev[b]

[a] *Laboratory of Particle Physics, Joint Institute for Nuclear Research, 6 Joliot-Curie Street, Dubna, Moscow Region 141980, Russia*
[b] *RAS Institute of Applied Physics, 46 Vlyanova St., Nizhny Novgorod, 603600, Russia*
[c] *CERN, Geneva, Switzerland*

Abstract

Since 2001 an experimental test facility for investigation of lifetime of a copper material, with respect to multiple RF pulse actions, was set up on the basis of the JINR (Dubna) FEM oscillator, in collaboration with IAP RAS (Nizhny Novgorod). A high-Q copper cavity, which simulates the parameters of the accelerating structure of the collider CLIC at an operating frequency of 30 GHz, is used in the investigation. The experimental setup consists of a wavebeam injector—FEM oscillator (power of ~ 25 MW, pulse duration up to 200 ns, spectral bandwidth not higher than 0.1%), a quasi-optic two-mirror transmission line, a wave-type converter, and a testing cavity. The frequency and transmission features of the components of the quasi-optic line were analyzed.
© 2004 Elsevier B.V. All rights reserved.

PACS: 41.60.Cr; 84.40.Ik

Keywords: Free-electron maser; Oscillator; Bragg resonator; High-power microwaves

1. Introduction

A high-energy (0.5–5 TeV center-of-mass), high-luminosity $(10^{34}$–10^{35} cm^{-2} s$^{-1})$ e$^+$e$^-$ Compact Linear Collider (CLIC) [1] is being studied at CERN as a possible new high-energy physics facility for the post-LHC era.

Accelerating gradient at room temperature [2] is limited by RF breakdown, dark current trapping, and pulsed heating. In 1996, Wilson developed parameterizations for these limits with wavelength using the geometry of an NLC traveling wave structure as a reference. The formulas and plots of the limits are shown in Fig. 1.

At frequencies over 10 GHz in collider accelerating structures, the pulsed heating limitation becomes the most significant.

---

*Corresponding author. Tel.: +7-09621-65443; fax: +7-09621-65767.
E-mail address: alikk@sunse.jinr.ru (A.K. Kaminsky).

0168-9002/$ - see front matter © 2004 Elsevier B.V. All rights reserved.
doi:10.1016/j.nima.2004.04.052

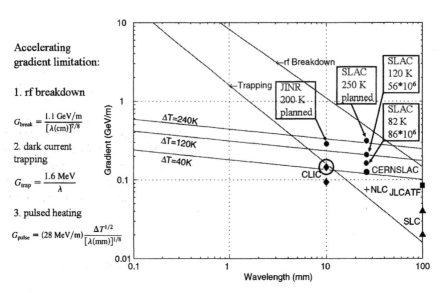

Accelerating
gradient limitation:

1. rf breakdown

$$G_{break} = \frac{1.1\ \text{GeV/m}}{[\lambda(\text{cm})]^{7/8}}$$

2. dark current
trapping

$$G_{trap} = \frac{1.6\ \text{MeV}}{\lambda}$$

3. pulsed heating

$$G_{pulse} = (28\ \text{MeV/m})\frac{\Delta T^{1/2}}{[\lambda(\text{mm})]^{1/8}}$$

Fig. 1. Gradient limits due to RF breakdown, dark current trapping, and pulsed heating.

Under RF pulsed heating over a short time, the inertia of the material prevents expansion and thermal stresses are induced. If these stresses are larger than the elastic limit, known as yield strength, then damage in the form of microcracks on the surface may occur after many pulses. This type of damage is known as cyclic fatigue. Experimental investigation of this limit at the frequency of 11.4 GHz is carrying out now in SLAC [2].

The experiment at JINR has been proposed and prepared to investigate the lifetime of a copper material undergoing power pulsed heating up to 200 K at 30 GHz frequency.

2. Test cavity design

The available RF power source (FEM oscillator, see Section 3) possesses considerably lower power than the value required for powering the CLIC full-scale accelerating structure. Therefore, in order to investigate the copper degradation undergoing 30 GHz pulsed heating, a special high-Q cavity with a large local surface magnetic field was selected. Various profiles of the cavity surface were studied, and a form composed of two

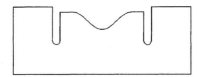

Fig. 2. Contour of the longitudinal cross-section of the test cavity.

diaphragms with a sinusoidal section between them (Fig. 2) was chosen.

In order to determine the geometry of the experimental cavity, a series of estimations and simulation have been carried out.

After preliminary cavity matching and temperature rise evaluation, a step-by-step corrected simulation has been made, taking into account the transient heating process during the filling of the cavity, increase of resistance with temperature, spectral losses, and the effect of loaded Q-factor reduction. The results of the last two factors for injected power of 20 MW are plotted in Fig. 3.

The expected temperature rise that might be achieved with the existing RF source is about 200 K, provided the power is properly delivered and injected into the cavity.

3. RF source: FEM oscillator

The 30 GHz free-electron maser (FEM) in oscillator configuration with the Bragg resonator

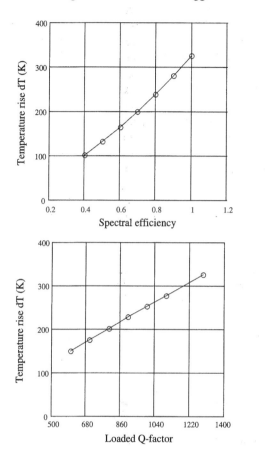

Fig. 3. The influence of spectral losses (top) and loaded Q-factor (bottom) on temperature rise.

is under development by the LPP JINR in collaboration with the IAP RAS, Nizhny Novgorod [3]. We used an induction accelerator LIU-3000 (0.8 MeV, 200 A, 200 ns, 0.5 Hz), a helical electromagnetic wiggler, and a guide solenoid magnetic field. A modern variant of the FEM oscillator can produce radiation with an output power of about 20–25 MW, pulse duration up to 200 ns, and a spectral bandwidth of about 30 MHz.

The experimental results on the stability of radiation parameters were reported in Ref. [4]. The obtained RF pulse duration is close to the duration of the flat top of the linac voltage pulse. In short series (500 pulses), the FWHM of the measured pulses keeps within the required 10% interval in amplitude and duration values. The spectral bandwidth of the measured signals was about 30–40 MHz, i.e., about 0.1%. The long-term stability should be improved.

To optimize power extraction from the FEM with respect to electric breakdown (see Fig. 4) and provide the proper wavebeam profile, different geometries of the FEM output waveguide and the position of the vacuum window in it were investigated both in cold and hot experiments.

We use wavebeam self-reproduction in long oversized waveguides (Talbot effect) to optimize output wavebeam and reduce the field strength at the window. Reinstalling the vacuum window into the region of minimal electric field did not facilitate the complete solution of the breakdown problem. Therefore, we had to seek an optimum

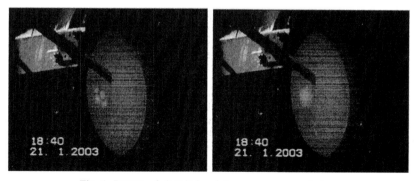

Fig. 4. Images of electric breakdown on vacuum window.

Axial dist. in tube (diameter 75 mm)

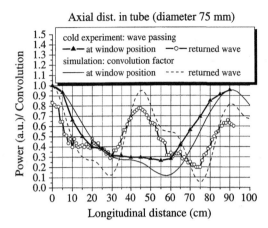

Fig. 5. Forward and reflected wave in FEM output waveguide: simulation and cold measurement.

Fig. 6. Test cavity overview and cross-section.

over two parameters, minimal values of the electric field and the reflection coefficient. For this purpose, cold measurements with a radiating horn and an oversized waveguide, the same as that used in hot experiments, were carried out. The RF power radiated through the horn was reflected from the metallic plate at the tube exit and returned partially to the horn. The dependence of the incident power at the tube exit and power part accepted by the horn, on the distance between the plate and the horn is presented in Fig. 5 (dotted lines).

The power amplitude evolution of the wavebeam with an initially Gaussian profile (convolution parameter) was studied with numerical simulation as well. In Fig. 5, the results for the forward wave at the window position and the reflected wave accepted by the horn are plotted as solid and dashed lines, respectively.

4. Experimental facility

In order to apply the JINR-IAP FEM radiation to the experimental investigation of the lifetime of an accelerating structure material, a special single cell high Q-factor copper resonator has been developed and manufactured (see Fig. 6). This test cavity simulates conditions of

the accelerating structure of the CLIC, in terms of electromagnetic field intensity at the metal surface.

The experimental scheme for investigation of the pulsed heating consists of an electron linac, an FEM oscillator, a test cavity, a transportation channel for the RF beam from the FEM to the cavity, and electron beam/RF power measuring systems (see Fig. 7).

To measure spatial distributions of RF power at a megawatt level in the near-field zone, a unique detector based on a dielectric waveguide with adjustable attenuation was developed and manufactured.

In order to simplify the alignment of the RF transmission line, a monitor of wavebeam position and size has been made and tested. The results of the measurement are illustrated in Fig. 8. The images of the wavebeam cross-section obtained at the position of the first mirror (a) and behind the test cavity (c) indicate that the intensity distribution transforms from H_{11}- to H_{01}-like one.

5. Conclusions

An experimental facility for investigating the effect of copper degradation due to pulsed heating by 30 GHz radiation had been set up.

The 30 GHz FEM was upgraded. The registration and analysis of more than 10^4 pulses of the microwave radiation of the FEM oscillator have been carried out. The required parameters, namely, frequency band—0.1%, pulse duration ~ 150 ns, and good short-term stability, were attained. However, the long-term stability is under consideration. A special high-Q cavity

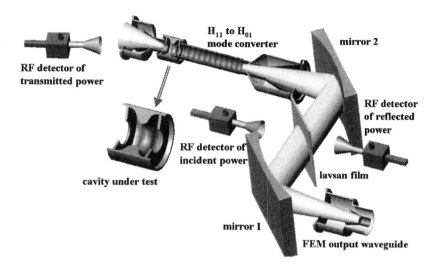

Fig. 7. Layout of experimental facility for test cavity heating.

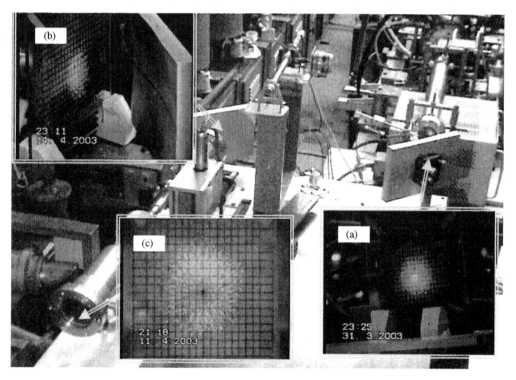

Fig. 8. Cross-section images of the wavebeam at different positions.

was designed and manufactured. The results of the cold measurements demonstrate the suitable parameters of the whole system. Furthermore, the diagnostic complex has been extended.

Acknowledgements

This work is supported by Grants ## 03-02-16530, 02-02-17438 of the Russian Foundation for Basic Research.

References

[1] J.P. Delahaye, et al., Proceedings of the European Particle Accelerator Conference (EPAC '98), Stockholm, 1998, p. 58.

[2] D.P. Pritzkau, Phys. Rev. Spectrosc. Topics—Accel. Beams 5 (2002) 112002.

[3] N.S. Ginzburg, A.A. Kaminsky, A.K. Kaminsky, et al., Phys. Rev. Lett. 84 (2000) 3574.

[4] A.V. Elzhov, I.N. Ivanov, A.K. Kaminsky, et al., Proceedings of the European Particle Accelerator Conference (EPAC 2002), Paris, 2002, p. 2311.

Available online at www.sciencedirect.com

SCIENCE  DIRECT®

ELSEVIER   Nuclear Instruments and Methods in Physics Research A 528 (2004) 231–234

NUCLEAR INSTRUMENTS & METHODS IN PHYSICS RESEARCH
Section A

www.elsevier.com/locate/nima

Optical resonator optimization of JAERI ERL-FEL

R. Nagai*, R. Hajima, M. Sawamura, N. Nishimori, N. Kikuzawa, E. Minehara

Free-Electron Laser Laboratory, Advanced Photon Research Canter, JAERI 2-4, Shirakata-Shirane, Tokai, Ibaraki 319-1195, Japan

Abstract

The optical resonator geometry of the JAERI ERL-FEL is optimized to ensure high-power operation. The FEL power strongly depends on the performance of the optical resonator including output efficiency, round-trip loss, and interaction mode volume. The performance of the optical resonator is evaluated by a simulation code using the Fox–Li procedure. The optimum resonator parameters for the JAERI ERL-FEL are determined through a numerical evaluation of the optical resonator.

© 2004 Elsevier B.V. All rights reserved.

PACS: 41.60.Cr; 42.60.Da

Keywords: Free-electron laser; Optical resonator; Far-infrared; Efficiency

1. Introduction

A free-electron laser based on a superconducting energy recovery linac at the Japan Atomic Energy Research Institute (JAERI ERL-FEL) operating in the far-infrared region has been demonstrated [1]. An R&D program aimed at a 10-kW ERL-FEL is in progress. This includes injector upgrades [2], HOM analysis [3], return-path optimization [4], and optical resonator optimization. In this paper, the optical resonator optimization for the JAERI ERL-FEL is presented.

The optical resonator consists of metal-coated end mirrors and a partially cut-away output coupler. This allows wide broadband operation [5] and ultrashort optical pulse generation [6]. To ensure high-power FEL operation, the optical resonator geometry is optimized about the FEL efficiency, the usability of the FEL optical beam, and the stability of the FEL. The FEL extraction efficiency is inversely proportional to the square root of the interaction mode volume and the square root of the round-trip loss of the optical resonator [7]. The interaction mode volume and the round-trip loss should be minimized by optimization of the curvature radii of the optical resonator end mirrors. The curvature radii of the end mirrors are optimized through a numerical evaluation of the optical resonator by an eigenmode calculation code using the Fox–Li procedure. The code has been improved, using

*Corresponding author. Tel.: +81-29-282-6752; fax: +81-29-282-6057.

E-mail address: r_nagai@popsvr.tokai.jaeri.go.jp (R. Nagai).

fast-Fourier-transform (FFT), to rapidly find the solution [8]. The output coupler introduces additional diffractive loss and expansion of the interaction mode volume due to scattering at the edge of the coupler. In the numerical evaluation of the optical resonator, the output coupler should be taken into account. The output FEL optical beam is transported to an experimental room for various applications. The output coupler should be matched to a long-path optical transport system. There is the misalignment of the optical resonator end mirrors due to changes in environment. The stability of the FEL depends on the tolerance of the mirror alignment. The tolerance should be taken into account in the optical resonator optimization.

2. Optimization model

An inserted scraper coupler system is a very efficient coupler in the far-infrared region [9]. But the scraper coupler is not well matched to a long-path optical transport system. To use the FEL optical beam for various applications, the optical beam is transported to experimental rooms by an optical transport system with a He–Ne laser alignment system. The coupling efficiency of the scraper coupler for the He–Ne laser is very low because the scraper is located at the offset position of the optical resonator axis. Since the He–Ne laser alignment system is not suitable for the scraper, a center-hole output coupler has been selected for the JAERI ERL-FEL.

The FEL efficiency is defined as $\eta_{fel} = \eta_{out}\eta_{ext}$, where η_{out} is the output coupling efficiency of the optical resonator, and η_{ext} is the FEL extraction efficiency. The output coupling efficiency is defined as $\eta_{out} = \alpha_{out}/\alpha_{loss}$, where α_{out} is the output power of the resonator, and α_{loss} is the total round-trip loss of the resonator. The total round-trip loss consists of the reflective loss of 1.2%, the diffractive loss, and output power. If the FEL is lasing in the spiking-mode, the FEL extraction efficiency is $\eta_{ext} \propto 1/\sqrt{\alpha_{loss} \cdot V}$, where V is the interaction mode volume [7]. The FEL efficiency is then represented by the following

equation:

$$\eta_{fel} \propto \frac{\alpha_{out}}{\alpha_{loss}^{3/2} \cdot V^{1/2}}. \tag{1}$$

With assumed operation in the Gauss mode and with the waist position at the center of undulator, the optimum curvature radii of the end mirrors are well understood as follows. The optical beam radius at the ends of undulator duct is $\omega^2 = (\lambda z_R/\pi)(1 + (L_d/2z_R)^2)$, where λ is the wavelength, z_R is the Rayleigh range, and L_d is the undulator duct length. Thus the beam radius is a minimum when $z_R = L_d/2$. The optimum curvature radius for the round-trip loss is therefore represented by the following equation:

$$R_{loss\_opt} = \frac{1}{2}\left(d + \frac{L_d^2}{d}\right) \tag{2}$$

where d is length of the optical resonator. The interaction mode volume is given by $V = \lambda(z_R L_u + L_u^3/(12z_R))$, where L_u is the undulator length. Then the interaction mode volume is a minimum when $z_R = L_u/\sqrt{12}$. The optimum curvature radius for the interaction mode volume is therefore represented by the following equation:

$$R_{vol\_opt} = \frac{1}{2}\left(d + \frac{L_u^2}{3d}\right) \tag{3}$$

which is not equal to R_{loss_opt}. The optimum mirror curvature radius for FEL efficiency is roughly between R_{loss_opt} and R_{vol_opt}. The diffractive loss and the interaction mode volume are enlarged due to the scattering at the center-hole coupler. Hence, the curvature radii of the end mirrors can be optimized through numerical evaluation of the optical resonator by an eigenmode calculation code using the FFT Fox–Li procedure. The eigenmode of the optical resonator at a wavelength of 22 μm is calculated to evaluate the performance of the optical resonator. The output coupling efficiency, round-trip loss, and interaction mode volume with the eigenmode of the resonator are used to estimate the FEL efficiency.

As shown in Fig. 1, the optimization model contains two end mirrors, four apertures, and a center-hole output coupler. The effects of the undulator duct and bending magnet ducts are

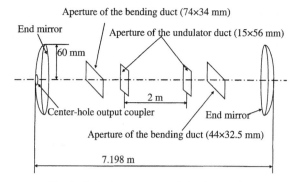

Fig. 1. Optical resonator optimization model.

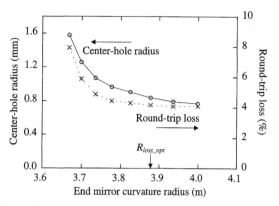

Fig. 2. Center-hole radius and round-trip loss for an output coupling efficiency of 0.35. The circles and crosses represent the center-hole radius and round-trip loss, respectively.

Table 1
Optimization model parameters

Length of the resonator	7.198 m
Radius of the end mirror	60 mm
Curvature radius of the end mirror	3.669–3.999 m
Reflectivity of the end mirror	99.4%
Radius of the center-hole	0.0–2.0 mm
Length of the undulator duct	2 m
Aperture of the undulator duct	15×56 mm
Aperture of the bending duct (upstream)	44×32.5 mm
Aperture of the bending duct (downstream)	74×34 mm
Distance of undulator to upstream bending aperture	1.824 m
Distance of undulator to downstream bending aperture	1.724 m
Distance of bending aperture to upstream end mirror	0.75 m
Distance of bending aperture to downstream end mirror	0.9 m

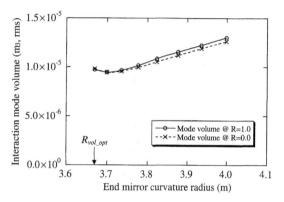

Fig. 3. Interaction mode volume with a 1.0 mm radius center-hole and without a center-hole. The circles and crosses represent the interaction mode volume with and without a center-hole coupler.

taken into account by the four apertures, one at each end of the undulator and one at the outer side of each bending magnet duct. In this model, the optical guide effect of the electron beam is not taken into account. Parameters of the optimization model are listed in Table 1.

3. Optimization results

The center-hole radius and the round-trip loss for an output coupling efficiency of 0.35 are shown in Fig. 2 as a function of the mirror curvature radius. For the same output coupling efficiency,

the center-hole radius decreases with an increase in the end mirror curvature radius, since the optical beam radius on each end mirror decreases with the end mirror curvature radius. The round-trip loss increases rapidly as the mirror curvature radius grows smaller, since the optical beam radius is enlarged at the ends of undulator duct. The mirror curvature radius for the minimum round-trip loss is slightly larger than $R_{\text{loss\_opt}}$ due to the diffraction of the apertures and the center-hole coupler.

The interaction mode volumes with a 1.0 mm radius center-hole and without a center-hole are shown in Fig. 3 as a function of the mirror

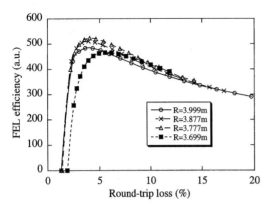

Fig. 4. FEL efficiencies for various end mirror curvature radii.

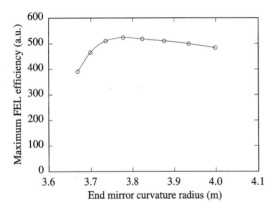

Fig. 5. Maximum FEL efficiency as a function of the end mirror curvature radius.

curvature radius. The mirror curvature radius for the minimum mode volume is slightly larger than R_{volOpt} due to the diffractions of the apertures. Since the mode volume with the center-hole differs little from the mode volume without the center-hole, the mode degeneracy resulting from the center-hole coupler is satisfactorily small for the FEL interaction.

The FEL efficiency for various end mirror curvature radii is presented as a function of the round-trip loss in Fig. 4. The efficiencies are maximized at a round-trip loss of about 4%. The maximum FEL efficiency as a function of mirror curvature radius is shown in Fig. 5. The optimum mirror curvature radius for the JAERI ERL-FEL is 3.777 m and the FEL efficiency is almost flat in

the range of 3.75 to 3.90 m. On the other hand, the tolerance of the mirror alignment is restricted by

$$\Delta\Theta_M \ll \left(\frac{2\lambda}{\pi d}\right)^{1/2} (1 - g)^{1/4}(1 + g)^{3/4} \qquad (4)$$

where $g = 1 - d/R$ is the stability parameter, and R is the curvature radius of the end mirror. Eq. (4) infers that the optical resonator is more stable for a large curvature radius end mirror. Thus, the end mirror curvature radius of 3.877 m has been chosen for the JAERI ERL-FEL. This curvature radius corresponds with a Rayleigh range that is half the undulator duct length. For this end mirror curvature radius, the optimum center-hole radius is 0.8 mm.

4. Conclusions

The optical resonator geometry for the JAERI ERL-FEL with a wavelength of 22 μm is optimized about the FEL efficiency, the usability of the FEL optical beam at the experimental room, and the stability of the FEL. The optimum resonator geometry is composed of a center-hole output coupler, an end mirror curvature radius of 3.877 m, and a center-hole radius of 0.8 mm.

References

[1] R. Hajima, et al., Nucl. Instr. and Meth. A 507 (2003) 115.
[2] N. Nishimori, et al., Proceedings of the 28th Linear Accelerator Meeting in Japan, July 30–August 1, Tokai, 2003, pp. 159–161 (in Japanese).
[3] M. Sawamura, et al., Proceedings of the 2003 Particle Accelerator Conference (PAC2003), May 12–16, 2003, Portland.
[4] R. Hajima, E. Minehara, Nucl. Instr. and Meth. A 507 (2003) 141.
[5] R. Nagai, et al., Proceedings of the 13th Symposium on Accelerator Science and Technology, October 29–31, Suita, 2001, pp. 455–457.
[6] R. Nagai, et al., Nucl. Instr. and Meth. A 483 (2002) 129; R. Hajima, R. Nagai, Phys. Rev. Lett. 91 (2003) 024801.
[7] N. Piovella, et al., Phys. Rev. E 52 (1995) 5470.
[8] R. Nagai, et al., Proceedings of the 28th Linear Accelerator Meeting in Japan, July 30–August 1, Tokai, 2003, pp. 381–383 (in Japanese).
[9] R. Nagai, et al., Nucl. Instr. and Meth. A 475 (2001) 519.

Available online at www.sciencedirect.com

SCIENCE DIRECT®

Nuclear Instruments and Methods in Physics Research A 528 (2004) 235–238

NUCLEAR
INSTRUMENTS
& METHODS
IN PHYSICS
RESEARCH
Section A

www.elsevier.com/locate/nima

ELSEVIER

Simulation of the resonator with inserted scraper output coupler for the CAEP far-infrared FEL

Yuhuan Dou[a],*, Xiaojian Shu[b], Yuanzhang Wang[b]

[a] Graduate School, China Academe of Engineering Physics, P.O .Box 2101, Beijing 100088, China
[b] Institute of Applied Physics and Computational Mathematics, P.O. Box 8009, Beijing 100088, China

Abstract

To improve the quality of the resonator, the simulation and design of the CAEP FIR FEL with inserted scraper output coupler has been made using our 3D FEL oscillator code. Using inserted scraper output coupler, the optical cavity Q can be tuned conveniently. The transverse modes of the optical cavity are calculated. The saturated power, the net gain and the coupling efficiency are evaluated as a function of the scraper radius and the distance between scraper and the center of mirror.
© 2004 Elsevier B.V. All rights reserved.

PACS: 41.60.Cr

Keywords: Free-electron laser; Far-infrared; Optical resonator; Coupler

1. Introduction

Usually, a resonator with an on-axis hole in one of the mirrors is an effective method to reduce space and weight, and eliminate the cost of bending magnets for directing the electron beam into the resonator of a free-electron laser (FEL). So-called hole coupling allows a wide broadband operation and higher damage threshold. The performance of the hole-coupled resonator has been studied extensively [1–3]. For the far-infrared FEL, this output coupler has some disadvantages, however. So the inserted scraper output coupler has been used in FEL systems [4,5]. It was proved

that the inserted scraper coupler was the most suitable for the far-infrared FEL using Fox–Li procedure with the cold-cavity case [6]. In this paper, the interaction of optical radiation with electrons is taken into account and compared with Ref. [6].

In the CAEP far-infrared FEL, it has been found that the diffractive loss is so large that it is half of the total loss due to the longer wavelength while using hole-coupled resonator [7]. By changing the size of the hole, the total loss of the hole-coupled resonator cannot be reduced effectively, since this would result in a low hole-coupling efficiency and output power because the presence of the hole would cause a large gradient in the optical wave front and thus a large diffraction loss [8]. To improve the quality of the resonator, the

*Corresponding author.
E-mail address: yuhuan_dou@yahoo.com.cn (Y. Dou).

0168-9002/$ - see front matter © 2004 Elsevier B.V. All rights reserved.
doi:10.1016/j.nima.2004.04.054

simulation and design of the CAEP FIR FEL with inserted scraper output coupler was performed with the help of our 3D FEL oscillator code [8,9]. The transverse modes of the optical cavity are calculated. The result shows that the fundamental mode is dominant. The output power, the net gain, the total optical loss and the coupling efficiency are evaluated as a function of the scraper radius and the distance between scraper and the center of mirror. It is proved that the inserted scraper is effective and applicable for the CAEP FIR FEL. Using inserted scraper output coupler, the optical cavity Q can be tuned conveniently so that it can satisfy the different demands in the cases of different experiments. The results will be helpful in designing the far-infrared FEL device.

2. Computational code

The simulation is based on our 3D FEL oscillator code [7–9], which has been modified to accommodate the inserted scraper output coupled resonator. The electron-beam and wiggler parameters used in the simulation are listed in Table 1. In the simulations, the distribution functions of transverse position and velocity and energy of the

Table 1
Parameters of the CAEP FIR FEL used in the simulations

Electron beam	
Energy (MeV)	6.5
Peak current (A)	5
Micro bunch (ps)	20
Emittance (π mm mrad)	1.5
Energy spread (%)	0.5
Wiggler	
Period (cm)	3.0
Peak field strength (kG)	3.0
Number of periods	50
Optical	
Wavelength (μm)	110
Cavity length (m)	2.536
Mirror curvature (m)	1.768
Beam duct	
Length (m)	1.5
Cross section	3.0 cm × 1.5 cm

electrons are assumed to be Gaussian. The corresponding initial values of the sample electrons are determined by a Monte Carlo method, and the initial phases are loaded according to the 'quiet start' scheme to eliminate numerical noise. The energy spread and emittance are specified as FWHM and RMS, respectively. The emittance in the y-direction is the same as that in the x-direction and the initial size of the electron beam is chosen to obtain a circular cross section at the centre of the wiggler [7]. The optical field intercepted by the beam-duct wall is assumed to be entirely absorbed by the wall, which is reasonable [10]. The insertion direction of the scraper is parallel to the wiggling plane of the electron beam. The crossing angle between the scraper and the cavity mirror is 45°.

3. The analysis of transverse optical modes

The transverse modes are composed of Hermite–Gaussian modes. If the electron beam is injected without misalignment, the overlapping factor between odd-order modes and electron beams will be zero because of the asymmetry of the Hermite polynomial, and the odd-order modes cannot be stimulated. When the output mirror is shifted from the center to the edge, the symmetry will change, and the proportion of the odd-order modes will become large. Using the code OSIFEL, the transverse construction of optical modes in the middle of the wiggler is calculated, as shown in Table 2, where f_{mn} is the ratio of the TEM$_{mn}$ modes to the intracavity power, and rs0 expresses

Table 2
Transverse construction of optical modes in the middle of the wiggler, where f_{mn} is the ratio of the TEM$_{mn}$ modes to intracavity power

rs0 (cm)	f_{00} (%)	f_{01} (%)	f_{10} (%)	f_{02} (%)	f_{20} (%)	f_{22} (%)
0	87.39	3.11×10^{-6}	2.48×10^{-5}	4.26	4.17	0.209
0.2	86.98	2.01×10^{-3}	2.01×10^{-5}	4.53	4.22	0.225
0.3	86.94	2.56×10^{-3}	1.82×10^{-5}	4.66	4.27	0.240
0.5	87.27	3.77×10^{-3}	9.94×10^{-6}	4.85	4.42	0.268
0.7	87.59	1.91×10^{-3}	4.30×10^{-6}	4.95	4.53	0.292

the distance between the inserted scraper output coupler and the center of the end mirror in the y-direction. It can be seen that the fundamental mode is still dominant when the output mirror is shifted from the center to the edge due to the gain and optical guiding effects from the electron beam that travels in a small region near the axis of the wiggler and the optical cavity. Therefore, it is reasonable to use the fundamental mode to estimate the output coupling and the diffractive loss of the optical cavity.

4. The characteristics of resonator

First, we simulated the case of the hole-coupled resonator. Table 3 shows the characteristics of the hole-coupled resonator as a function of the hole radius (rhb), where P is the saturated intracavity power, P_{out} the output power, G_{net} the net gain, Tloss the total loss of the resonator, Touth the coupling efficiency, which is defined as the ratio of the useful loss through the hole to the total loss. It can be seen that the total loss of the resonator decreases with a reduction in the size of the hole; however, the output power and coupling efficiency is reduced even more. Therefore, the total loss cannot be effectively reduced [8]. If the size of the hole is increased, the interaction of the optical radiation with the electrons will become weaker so the net gain of the resonator is decreased.

Next, the characteristics of the resonator with inserted scraper output coupler are simulated as shown in Tables 4 and 5. The descriptions of the parameters are the same as in Table 3. Table 4 shows the edge-coupled resonator as a function of

Table 3
Characteristics of the hole-coupled resonator as a function of the hole size

rhb (cm)	P (MW)	P_{out} (kW)	G_{net} (%)	Tloss (%)	Touth (%)
0.03	1.20	3.97	28.96	18.95	1.84
0.06	1.07	14.14	26.62	21.34	6.55
0.09	0.89	24.74	23.15	24.94	11.83
0.12	0.74	31.34	19.67	28.39	16.12
0.15	0.47	32.93	12.51	35.42	22.78

Table 4
Simulation results of the edge-coupled resonator as a function of the scraper size when rs0 is 0.25 cm, where rs0 expresses the distance between the inserted scraper output coupler and the center of the end mirror

rsa (cm)	P (MW)	P_{out} (kW)	G_{net} (%)	Tloss (%)	Touth (%)
0.05	1.22	2.09	29.24	18.61	1.08
0.1	1.15	8.65	27.67	20.09	4.40
0.2	0.90	27.08	21.74	25.55	14.02
0.25	0.74	34.55	17.34	29.57	19.21
0.3	0.55	36.69	12.04	34.37	24.40

Table 5
Characteristics of the edge-coupled resonator as a function of the parameter rs0 when the radius of the scraper is 0.2 cm

rs0 (cm)	P (MW)	P_{out} (kW)	G_{net} (%)	Tloss (%)	Touth (%)
0	0.55	33.74	14.60	33.47	20.92
0.2	0.82	29.91	20.25	27.45	15.66
0.4	1.09	14.78	25.35	21.66	8.70
0.6	1.20	4.34	28.15	19.26	3.26
0.9	1.24	0.53	29.74	18.37	0.34

the scraper size when rs0 is 0.25 cm. Table 5 lists the characteristics of the edge-coupled resonator as a function of the parameter rs0 when the radius of the scraper is 0.2 cm. From Tables 4 and 5, it can been seen that the characteristics of the resonator, such as the output power, the wiggler gain, the net gain, the total loss of the resonator, and the coupling efficiency, are determined by the radius and the position of the scraper. As the radius of the scraper (rsa) increases, the output power and the coupling efficiency increase; however, the net gain is small and the total loss is large due to the output optical proportion being too large. When rs0 becomes large, the net gain will increase and the total loss will decrease; however, the output power and the coupling efficiency will decrease because the optical density at the edge is lower than that at the center. So the optimum radius (rsa) and distance (rs0) must be selected. According to Tables 4 and 5, we may select the size of rsa as 0.15–0.20 cm and the size of rs0 as 0.1–0.3 cm when operating the FEL system. Although there are some differences between the

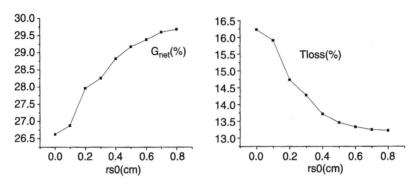

Fig. 1. Net gain (left) and total (right) loss as the 0.09-cm radius output mirror is shifted from the center to the edge.

results of the simulation and the results obtained experimentally, these values of rsa and rs0 will be references for the experimental system.

When the radius of the output mirror is selected as 0.09 cm, making the output mirror shift from the center to the edge, we can obtain the net gain and total loss shown in Fig. 1. It can be seen that the net gain increases and the total loss decreases with the mirror shifted to the edge. The main difficulty of the CAEP FEL is that the total loss is too large to allow start up, as mentioned in Ref. [7]. We can reduce the total loss by making use of the output mirror shift to the edge, which is an easy adjustment and can be achieved conveniently.

5. Conclusions

In general, the simulation and design of the CAEP FIR FEL with inserted scraper output coupler has been made with the help of our 3D FEL oscillator code. The transverse construction of optical modes is calculated. The optimum radius (rsa) and distance (rs0) have been chosen. We have shown that the edge-coupled resonator can provide satisfactory performance for the CAEP FIR FEL. The coupler is flexible to use and can be fabricated conveniently. Using the inserted scraper output coupler, the Q value of the optical cavity can be tuned conveniently in many ways, such as by shifting or by replacing the

scraper, and changing the radius and the position of the scraper, all of which will be helpful in the experiments. The Q value of the optical cavity can be improved to obtain higher net gain to reach first lasing in the initial stage of experiments, such as that of the CAEP FEL device. Moreover, the output power and efficiency of the FEL oscillator can be optimized to a maximum by increasing the output-coupling factor and the coupling efficiency in the upgrade stage of experiments.

References

[1] M. Xie, K.J. Kim, Nucl. Instr. and Meth. A 318 (1992) 877.
[2] B. Faatz, R.W.B. Best, D. Oepts, et al., IEEE J. Quantum Electron. QE-29 (1993) 2229.
[3] M. Keselbrener, S. Ruschin, B. Lissak, A. Gover, Nucl. Instr. and Meth. A 304 (1991) 782.
[4] E. Minehara, et al., Nucl. Instr. and Meth. A 445 (2000) 183.
[5] E.A. Antokhin, et al., First lasing at the high power free electron laser at Siberian Center for Photochemistry Research, The 25th FEL Conference, Japan, 2003.
[6] R. Nagai, et al., Nucl. Instr. and Meth. A 475 (2001) 519.
[7] Xiaojian Shu, Yuanzhang Wang, Nucl. Instr. and Meth. A 483 (2002) 205.
[8] Xiaojian Shu, Yuanzhang. Wang, Yunqing Jiang, et al., Opt. Eng. 39 (2000) 1543.
[9] Xiaojian Shu, Yuanzhang Wang, et al., Nucl. Instr. and Meth. A 407 (1998) 76.
[10] M. Sobajima, et al., Nucl. Instr. and Meth. A 407 (1998) 121.

Available online at www.sciencedirect.com

Nuclear Instruments and Methods in Physics Research A 528 (2004) 239–243

NUCLEAR INSTRUMENTS & METHODS IN PHYSICS RESEARCH
Section A

www.elsevier.com/locate/nima

Improvement of the PFN control system for the klystron pulse modulator at LEBRA

K. Yokoyama[a],*, I. Sato[a], K. Hayakawa[a], T. Tanaka[a], Y. Hayakawa[a], K. Kanno[b], T. Sakai[b], K. Ishiwata[b], K. Nakao[b]

[a] Laboratory for Electron Beam Research and Application, Nihon University, 7-24-1 Narashinodai, Funabashi 274-8501, Japan
[b] Graduate School of Science and Technology, Nihon University, 7-24-1 Narashinodai, Funabashi 274-8501, Japan

Abstract

The short-wavelength free electron laser (FEL) requires a stable electron beam with long pulse duration, narrow energy spread, and high current density. The pulse forming network (PFN) for the klystron pulse modulator at LEBRA was designed to realize pulse flatness error within 0.05%. The PFN consists of 30 capacitors and adjustable inductors. A PFN adjustment was performed to suppress changes of the RF phase difference between the input and the output of the klystron. The flatness of the pulse voltage applied to the klystron has been adjusted to an error within 0.06% as a result of the phase flatness error within 0.3° achieved by the PFN adjustment. Furthermore, the time dependence of the energy spectrum in the pulse duration has also been suppressed.
© 2004 Elsevier B.V. All rights reserved.

PACS: 41.60.Cr.; 29.27.Fh.; 29.17.+w.; 29.27.Bd.

Keywords: FEL; Infrared; Electron linac; RF; Modulator

1. Introduction

The quality of FEL is seriously affected by the performance of the accelerator in terms of the stability of the RF power and phase, the bunch length, the beam emittance, the current density and the energy spectrum. Essentially, RF linac-based FELs require a stable and long RF pulse from klystrons [1,2]. Generally, the fractional change in the electron energy gained in an accelerator section due to the fractional change in voltage of the klystron is expressed as

$$\frac{\delta V_m}{V_m} = \frac{5}{4}\frac{\delta V_k}{V_k} \tag{1}$$

where V_m is the maximum possible electron energy that can be obtained from a given length of accelerator supplied with a given level of RF power, and V_k is the voltage applied to the klystron [3]. The phase shift that arises in the klystron, which can occur both during a pulse and

---

*Corresponding author. Tel.: +81-47-469-5489; fax: +81-47-469-5490.

E-mail address: k_yokoyama@lebra.nihon-u.ac.jp (K. Yokoyama).

from pulse-to-pulse, is caused by modulation of the klystron beam-voltage pulse. This follows that

$$\Delta\theta_k = -2\pi f \frac{l_k}{c} \frac{1}{\left(\sqrt{1 - 1/(1 + M_k)^2}\right)^3}$$

$$\times \frac{1}{(1 + M_k)^3} M_k \frac{\Delta V_k}{V_k} \qquad (2)$$

where M_k is substituted for eV_k/m_0c^2, f is the RF frequency, l_k is the klystron drift tube length, m_0 is the rest mass of the electron, c is the velocity of light, and e is the electron charge [3].

A change in the applied voltage is a serious problem in the LEBRA linac since the accelerating RF is provided by two klystrons driven with separate pulse modulators. The different phase shift between the klystrons causes fluctuations in the beam energy at the exit of the linac, resulting in the orbit length variation of the beam in the momentum analysing magnet system. Therefore, the bunched electron beam can be easily mistuned to the light pulse accumulated in the optical cavity [4]. The LEBRA FEL requires voltage fluctuation within 0.05% at 240 kV, which corresponds to the phase fluctuation of 0.26° for PV3030 klystron, as deduced from Eq. (2).

The inductances in the PFN have been optimized to reduce the time dependence of the RF phase difference between the input and the output of the klystron in the pulse duration. Then, the voltage fluctuation was deduced from the phase fluctuation.

2. The RF system for the LEBRA linac

2.1. Performance of new RF amplifiers

The klystron output power is nearly independent of small fluctuations in the input RF power. But phase fluctuations are introduced through the drive amplifier. A new solid-state RF amplifier was developed to obtain a pulsed high-power microwave with constant phase to drive a klystron. Since the phase of the amplifier output RF changes rapidly at the head of the pulse, the RF power within the last 20 μs in a total 50 μs pulse duration

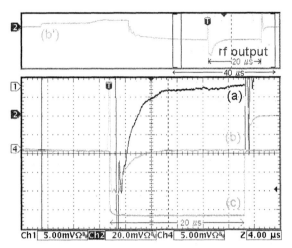

Fig. 1. The phase and the amplitude of the RF amplifiers. (a) RF phase of amplifier #1 (0.3°/div), (b) and (b′) are the RF phase of amplifier #2 (0.7°/div) that differ in horizontal scale and that are equal in vertical full scale, (c) RF amplitude of amplifier #2 (120 W/div).

is coupled to the output port of the amplifier as shown in Fig. 1(b′). The resultant phase shift of the RF amplifier output is less than ±0.5° over 18 μs at the maximum output power of 400 W, which was measured for each klystron driving RF amplifier using the double-balanced mixer (DBM) as shown in Fig. 1. The RF amplifier input is a cw RF from a master oscillator (Agilent Technologies E4425B-ATO-11188). Both amplifiers have similar phase shift characteristics at the start of the RF output. The phase shift during the pulse duration is compensated with a function generator and the RF amplifier output is supplied to a klystron [5]. Therefore, the effect of these phase shifts on the beam energy will be reduced; however, the effect on FEL lasing is not negligible.

2.2. The PFN control system

The pulse modulator consists of a PFN circuit with 30 LC sections. Each variable inductor system consists of an air-core coil, a motor-driven aluminum slug, and a potentiometer for readout of the slug position. The tuning slug is controlled by a personal computer (PC) through an I/O board [6].

2.3. The method of PFN adjustment

The optimal value of each inductance, which produces the flat-pulsed voltage, is deduced from the following calculation [6]. If the target waveform, which could have a high flatness voltage or phase signal, is denoted as $g(t_i)$ in a simplified manner (i.e., a function of the inductances L_0-L_{29} and time t_i); the initial waveform is denoted as $f(t_i)$; the reference voltage is V; the jth inductance is L_j and the variation of L_j is ΔL_j, where t_i is ith time in the pulse duration partitioned into n ($i = 0-n$); and j is the channel number of the inductor ($j = 0-29$), then $g(t_i)$ is approximated as the first order expansion by

$$g(t_i) = f(t_i) + \frac{\partial f(t_i)}{\partial L_0}\Delta L_0 + \cdots + \frac{\partial f(t_i)}{\partial L_j}\Delta L_j. \quad (3)$$

Thus, when the sum of squares of deviations $\Sigma(V - g(t_i))^2$ is minimum, the function $g(t_i)$ represents the high flatness pulse [3]. It means

$$\frac{\partial}{\partial L_j}\sum_{i=0}^{n}\{V - g(t_i)\}^2 = 0. \quad (4)$$

Indeed, the differential function can be replaced with

$$\frac{\partial f}{\partial L_j} \cong \frac{f(L_j + \delta L_j) - f(L_j)}{\delta L_j} \quad (5)$$

where δL_j is the change in L_j. ΔL_j should be within the range of 2.0–3.6 µH and t_i should lie within the pulse duration. The value of δL_j is set so that the change in the waveform signal is sufficiently larger than the noise level. From the measurement of the waveform dependence on each change δL_j, Eq. (4) is solved for ΔL_j. Then, the target function $g(t_i)$ is constructed. The value of ΔL_j is set on the PFN inductor using a PC. The fluctuation of the modulator output voltage is not large enough to analyze for deriving the differential coefficient in Eq. (5) due to a poor signal-to-noise ratio. Instead, the RF phase difference between the input and the output of the klystron is useful for this method as can be expected from Eq. (2).

3. Improvement of energy spectrum

Waveforms of the RF phase difference between the input and the output of the klystron before (A)

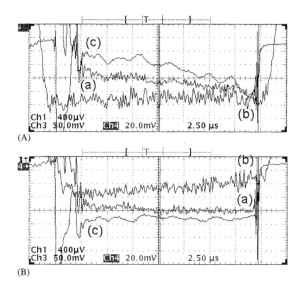

(A)

(B)

Fig. 2. The RF flatness of klystron #1. (A) Before adjusting the PFN inductance, (B) after adjusting the PFN inductance. (a) RF amplitude (0.5%/div), (b) waveform of the PFN (0.2%/div), (c) RF phase (0.6°/div).

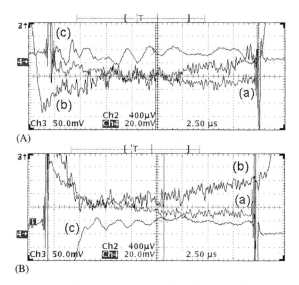

(A)

(B)

Fig. 3. The RF flatness of klystron #2. (A) Before adjusting the PFN inductance, (B) after adjusting the PFN inductance. (a) RF amplitude (0.5%/div), (b) waveform of the PFN (0.2%/div), (c) RF phase (0.5°/div).

and after (B) the optimization of the PFN inductances carried out by this method are shown in Figs. 2 and 3 for klystron Nos. 1 and 2, respectively. Before adjustment, the phase shift in the RF pulse duration of 17 μs was greater than 2° for klystron No.1 and greater than 1° for klystron No.2. Each phase shift has been reduced to around 0.3° over 17 μs by the optimization. This corre-

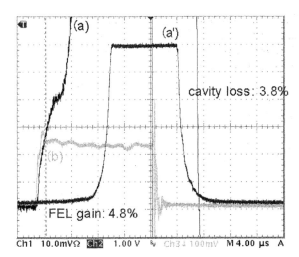

Fig. 5. FEL oscillation. (a) FEL pulse shape, (a') is (a) displayed on a different vertical scale, (b) electron beam current (50 mA/div). The IR detector is saturated.

sponds to the modulator output voltage fluctuation of approximately 0.06%.

The difference of the electron beam energy spectra between the two situations is compared in Fig. 4. Waveforms of the electron beam current and the FEL power are shown in Fig. 5. The uniformity of the central energy over the pulse duration has been considerably improved. The average energy spread (FWHM) over the pulse duration is 0.5%. The drift of central energy during the pulse duration is also much improved. As a result of this improvement, the growth of the FEL optical power is found to be initiated at the head of the beam pulse. However, energy fluctuation is still observed in the pulse duration due to residual phase fluctuation of the relative phase between the two klystrons. The fine fluctuation of the central energy possibly has a considerable effect on the intensity of the FEL or SASE [4].

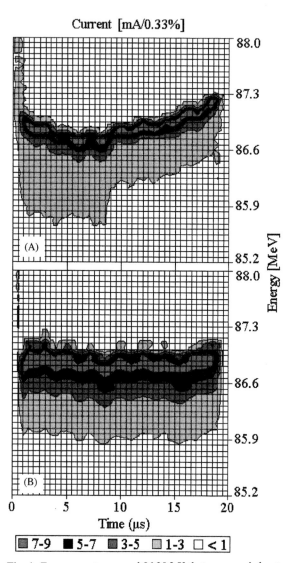

Fig. 4. Energy spectra around 86.8 MeV that were carried out, utilizing the first 45° bending magnet as a spectrometer. (A) Before adjusting the PFN inductance, (B) after adjusting the PFN inductance.

4. Conclusion

The method of PFN adjustment described in this report is shown to be simple and effective to obtain uniform energy spectra over the RF pulse duration. The set of proper PFN inductances depends on the voltage applied to the klystron.

Therefore, this method will be useful to develop a database of various sets of proper inductances for different PFN output voltages in order to realize a reliable wavelength variability of FEL.

References

[1] R. Chaput, et al., Proceedings of the Fourth European Particle Accelerator Conference (EPAC'94), London, England, 2001, p. 728.

[2] E. Oshita, et al., Proceedings of the 16th IEEE Particle Accelerator Conference (PAC'95), Dallas, Texas, 1995, p. 1608.

[3] P.M. Lapostolle, A.L. Septier, Linear Accelerators, North-Holland Publishing Company, Amsterdam, 1970, p. 324.

[4] T. Tanaka, et al., Nucl. Instr. and Meth. A, (2004) these Proceedings.

[5] K. Yokoyama, et al., Jpn. J. Appl. Phys. 41 Pt.1 (7)A (2002) 4758.

[6] K. Yokoyama, et al., Proceedings of the 28th Linear Accelerator Meeting, Tokai, Japan, 2003, p. 464 (in Japanese).

Available online at www.sciencedirect.com

SCIENCE @ DIRECT°

ELSEVIER   Nuclear Instruments and Methods in Physics Research A 528 (2004) 244–248

**NUCLEAR
INSTRUMENTS
& METHODS
IN PHYSICS
RESEARCH**
Section A

www.elsevier.com/locate/nima

Stability analysis of the RF linac based on an AR model

R. Kato[a,*], S. Isaka[a], H. Sakaki[b], S. Kashiwagi[a], G. Isoyama[a]

[a] *Institute of Scientific and Industrial Research, Osaka University, 8-1 Mihogaoka, Ibaraki, Osaka 567-0047, Japan*
[b] *Japan Atomic Energy Research Institute, 2-4 Shirakata-Shirane, Tokai, Naka, Ibaraki 319-1195, Japan*

Abstract

We are pursing operational stability of the RF linac with a method of system analysis. The RF phase and the power for the pre-buncher have been measured, together with environmental parameters including temperatures of the main components, the room temperatures and the AC line voltage. The measured data have been analyzed with the autoregressive model, and the feedback structure of the system has been derived from noise contribution ratios and impulse responses analysis. It was found that the AC line voltage and the temperatures of the klystron room significantly affected operational stability of the linac.
© 2004 Elsevier B.V. All rights reserved.

PACS: 29.17.+w; 43.50.Rq; 05.45.Tp

Keywords: Linear accelerators; Noise; Time-series analysis

1. Introduction

We are conducting experimental studies on free electron laser (FEL) and Self-Amplified Spontaneous Emission (SASE) in the infrared region using the L-band linac at the Institute of Scientific and Industrial Research (ISIR), Osaka University. Although stability of the electron beam is essential for these studies, we observe fluctuations of the electron trajectory and the electron energy, so that the experiments are sometimes interrupted. To solve the problem, we have begun study on instability of the linac. The electron beam accelerated with the linac is affected by fluctuations of

the RF field for acceleration after all and the RF field is influenced by varying surroundings. In order to identify sources of the instability, we have measured environmental conditions for a long time, which may affect the accelerating RF field, including the AC line voltage and temperatures at various points. We have analyzed the measured data using the autoregressive (AR) model, which has been successfully applied to analysis of instability of the RF linacs of SPring-8 and JAERI [1]. Preliminary analysis results were already reported, which were obtained only using noise contribution ratios of the AR model [2]. In this paper, we have analyzed the measured data using an impulse response analysis in addition to the noise contribution ratios, and showed the feedback structure of the linac system.

*Corresponding author. Tel.: +81-6-6879-8486; fax: +81-6-6879-8489.

E-mail address: kato@sanken.osaka-u.ac.jp (R. Kato).

0168-9002/$ - see front matter © 2004 Elsevier B.V. All rights reserved.
doi:10.1016/j.nima.2004.04.056

2. Linac and the measurement systems

The measurement system for the accelerator and facilities are shown schematically in Fig. 1. The RF power from the 5 MW klystron is provided to the pre-buncher (PB) and the buncher, while the power from the 20 MW klystron is provided to the accelerating tube. In this study, we measured the RF power and the phase supplied to PB. The cooling system for the linac has a cooling tower (CT3) for discharging heat to the outside world. Temperatures at the various locations shown in Fig. 1 were measured with thermistors.

We measured the RF phase and the power for PB, which affect the electron beam most significantly, together with the AC line voltage and temperatures of the main components and the room temperatures. Results of the measurement are shown in Fig. 2. The RF phase and the power are shown in the upper panel, and the AC line voltage and temperatures are shown in the lower panel of Fig. 2. The RF phase shown in Fig. 2 contains high-frequency noise components, but the periodical variations are similar to those of the temperatures of the klystron room and CT3.

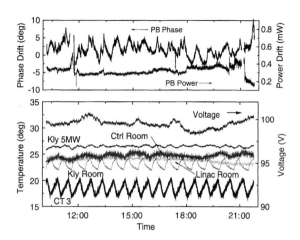

Fig. 2. RF phase and the power for PB measured together with temperatures of the main components, the room temperatures and the AC line voltage. The RF power is the power measured with the directional coupler.

3. Autoregressive model

Stability or instability of the linac system is determined by its constituent components. If all the components work independently, stability analysis would be very simple. The components are, however, connected to each other with facilities, such as water-cooling systems and electrical power lines. As a result, the interrelated components form a complicated system. In such a system, fluctuations produced in a component will be propagated to others, and circulated among the components. The AR model is a method for time-series analysis [3]. It may be applicable in analysis of the feedback structure in a complicated system consisting of mutually interacting elements.

An example of the feedback system with two parameters is schematically shown in Fig. 3. For the analysis with the AR model, the present data $X(n)$ can be expressed by the linear combination of the past data $X(n-m)$, $Y(n-m)$ and the white noise $e_x(n)$:

$$
\begin{bmatrix} X(n) \\ Y(n) \end{bmatrix} = \sum_{m=1}^{M} \begin{bmatrix} a_{xx}(m) & a_{xy}(m) \\ a_{yx}(m) & a_{yy}(m) \end{bmatrix} \cdot \begin{bmatrix} X(n-m) \\ Y(n-m) \end{bmatrix} + \begin{bmatrix} e_x(n) \\ e_y(n) \end{bmatrix}
$$

(1)

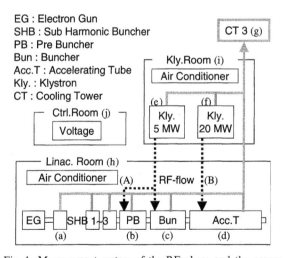

Fig. 1. Measurement system of the RF phase and the power together with environmental parameters. Letters (A) and (B) show locations of the directional couplers, and (a)–(j) are measurement points for temperature. In this measurement, the RF phase and the power of the pre-buncher were measured at point (A). The AC line voltage is also measured in the control room. The air conditioners in the linac room and the klystron room exchange heat directly with the outside air.

where M is a regressive number. Fitting Eq. (1) to time-series fluctuations, we obtain the factor a_{ij} and the noise e_i. These factors and noises show the feedback system modelled on the computer.

Assuming that the noise $e_x(n)$ and $e_y(n)$ are equal to zero, the system will either converge or diverge. When an impulse noise is applied to the feedback system in the stationary state without noise, it will converge soon for stable system, because there is no noise except for the impulse. By

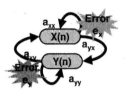

Fig. 3. Schematic diagram of the feedback system for two parameters.

analyzing the impulse response of the system, it is possible to know how the fluctuation propagates temporally among the components, and whether the system is stable or not. This method is called "impulse response analysis" and gives physical images of the feedback structure in the temporal region. Although the given impulse is a delta function, since the component undergoes influence by its own past value and the other components, it has the temporal tail unlike the delta function.

The power spectrum is defined as the Fourier transform of an auto-covariance of the fluctuation data. Fractional factors of the influences due to the each intrinsic noise power contributing to the power spectrum are called the noise (power) contribution ratios. By analyzing the noise contribution ratios, we can know which is the most influential component in fluctuating the system.

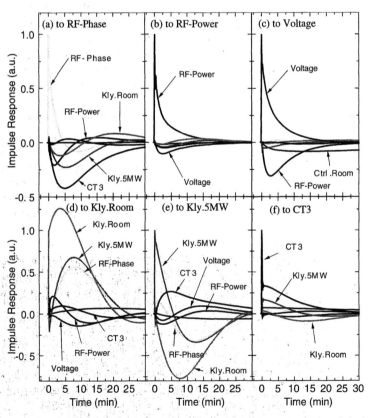

Fig. 4. Impulse responses for the RF phase, the RF power, the AC line voltage, the temperatures of the klystron room, the 5 MW klystron, and the cooling tower 3.

4. Analysis with the AR model

4.1. Impulse response

Results of impulse response analysis, in which an impulse of magnitude 1 is given at time 0, are shown in Fig. 4. The vertical units in the figure are the physical units which were used when measuring. For example, all the temperatures are shown in degrees, the AC line voltage is in V, the RF phase in degrees, and the RF output is mW, which is output value from the directional couplers. The analysis shows:

(a) Fluctuation of the RF phase influences temperatures of CT3 and the 5 MW klystron, and the RF power.
(b) Fluctuation of the RF power does not affect the others and converges in a short time.
(c) Fluctuation of the AC line voltage influences the RF power.
(d) Temperature fluctuation of the klystron room affects the RF phase and the temperature of the 5 MW klystron.
(e) Temperature fluctuation of the 5 MW klystron affects the RF phase and the temperature of the klystron room.
(f) Temperature fluctuation of CT3 influences the temperature of the 5 MW klystron.

All these items except (a) are reasonable and consistent with our experiences. However, if interpreted as follows, we can understand that item (a) is not to be contrary to the experience. First, the flucuation of the RF phase changes the beam loading. This changes the heat load of the dummy load and the temperature of CT3. Since the temperature shift of CT3 changes the temperature of the klystron cavity, the RF power changes.

4.2. Noise contribution ratio

Noise contribution ratios for the RF phase, the RF power, and the AC line voltage are shown in Fig. 5. In the noise contribution ratios for the RF phase, contributions from temperatures of the klystron room and the 5 MW klystron can be seen. In those for the RF power, a contribution from the

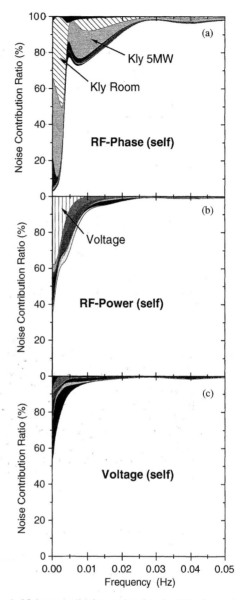

Fig. 5. Noise contribution ratios for the RF phase, the RF power, and the AC line voltage.

AC line voltage can be seen. The AC line voltage is not affected from other things.

4.3. Feedback structure of the system

The feedback structure of the system is shown in Fig. 6, which is derived from the impulse response analysis and major noise contribution ratios of the

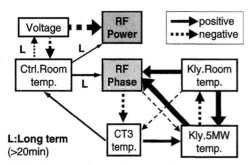

Fig. 6. Feedback structure of the L-band linac system derived with the AR model.

system. An arrow shows the direction of the noise contribution and its width expresses the noise contribution ratio. A dashed arrow indicates that the impulse response is negative. We see in Figs. 4–6 that the main contribution to fluctuation of the RF phase for PB is the temperature of the 5 MW klystron and the next one is the temperature of the klystron room. Especialy the RF phase follows the change of the temperature of the 5 MW klystron immediately. The fluctuation of the RF power for PB is produced by the AC line voltage. The voltage fluctuation gives the RF power, the opposite change, after a delay of a few minutes.

5. Summary

In order to find sources of instability of the L-band linac at ISIR, Osaka University, we measured various parameters affecting the RF phase and the RF power for PB and analyzed the measured data with the AR model. The results of the analysis show that instability of the linac is produced by fluctuations of the temperatures of the 5 MW klystron and the klystron room and also by fluctuation of the AC line voltage.

Based on the results of this study, we are now replacing and modifying the accelerator components as well as facilities in order to make the linac stabler.

References

[1] H. Sakaki, et al., T. SICE 35 (10) (1999) 1283.
[2] S. Isaka, et al., Proceedings of the Second Asian Particle Accelerator Conference, Beijing, China, 2001, pp. 660–662.
[3] H. Akaike, Ann. Inst. Statist. Math. 20 (1968) 425.

Available online at www.sciencedirect.com

SCIENCE ⓓ DIRECT®

Nuclear Instruments and Methods in Physics Research A 528 (2004) 249–253

**NUCLEAR
INSTRUMENTS
& METHODS
IN PHYSICS
RESEARCH**
Section A

www.elsevier.com/locate/nima

Design of a dimple mirror system for uniform illumination from an MIR-FEL

Manabu Heya[a],*, Taku Saiki[b], Koji Tsubakimoto[c], Kunio Awazu[a], Masahiro Nakatsuka[c]

[a] Institute of Free Electron Laser, Graduate School of Engineering, Osaka University, 2-9-5 Tsuda-Yamate, Hirakata, Osaka 573-0128, Japan
[b] Institute of Laser Technology, Osaka University, 2-6 Yamada-Oka, Suita, Osaka 565-0871, Japan
[c] Institute of Laser Engineering, Osaka University, 2-6 Yamada-Oka, Suita, Osaka 565-0871, Japan

Abstract

In order to generate a flat-topped beam profile from a tunable, broadband, mid-infrared free-electron laser (MIR-FEL), we have proposed and designed a dimple mirror system (DMS), using a laser optics calculation code LOCCO. The principles of the DMS are similar to those of a micro-lens array system involving a principal focusing lens and a two-dimensional (2D) array of micro-lenses located ahead of the principal lens. The DMS is composed of a parabolic mirror, which acts as the principle lens and a 2D array of micro-dimples, which splits the incident FEL beam into many partial beams. All beamlets overlap near the focal point of the parabolic mirror, resulting in the generation of an approximately flat-topped intensity profile due to the interference of light. We proved that the DMS can generate a flat-topped profile even over a wide MIR waveband. Thus, we showed the applicability of the DMS for uniform illumination from a tunable, broadband MIR-FEL.
© 2004 Elsevier B.V. All rights reserved.

PACS: 42.79.Bh

Keywords: Laser processing; Uniform illumination; Broadband lasers; MIR-FEL; Dimple mirror system; Flat-topped beam

1. Introduction

In laser processing of tissue and material, uniform illumination of the target surface is essential. There are a variety of uniform illumination techniques, such as those using a micro-lens array (MA) [1], a diffractive optical element [2], or a

deformable mirror [3]. These techniques are severely limited by the laser wavelength used. Therefore, these are not applicable to a mid-infrared free-electron laser (MIR-FEL), which is both tunable and broadband. In this paper, we propose a dimple mirror system (DMS), which is a reflective type of the MA, for MIR-FEL uniform illumination. Reflective optics must be used to avoid the effects of focusing performance on color aberration.

We designed the specifications of the DMS using a one-dimensional (1D) laser optics calcula-

*Corresponding author. Tel.: +81-72-897-6415; fax: +81-72-897-6419.
E-mail address: heya@fel.eng.osaka-u.ac.jp (M. Heya).

0168-9002/$ - see front matter © 2004 Elsevier B.V. All rights reserved.
doi:10.1016/j.nima.2004.04.057

tion code (LOCCO) around a laser wavelength (λ) of 6.0 μm. This waveband is one of the candidates for non-invasive bio-tissue treatment [4]. We proved the applicability of the DMS for the generation of a flat-topped profile over a wide MIR waveband.

2. Principles of the DMS

The DMS is a reflective version of the MA [1]. Figs. 1(a) and (b) show the principles of the MA and DMS, respectively. The MA is composed of a principal focusing lens with a focal length F and a two-dimensional (2D) lens array of many micro-lenses, each with a focal length f. The size of each micro-lens, d, is determined by $d = D/N$, where D is the diameter of the incident beam and N is the number of micro-lenses per array. The micro-lenses split the incident beam into N partial beams. Each of these beamlets focuses on the compound focal surface (E), diverges, and illuminates the target. This compound focal length, F_c, is given by

$F_c = (1/F + 1/f)^{-1}$ [5]. Each elementary focal spot on E can be regarded as a coherent point light source. A focused pattern is formed near the focal point of the principal focusing lens (G) as a multiple beam interference pattern, and it can be carefully controlled by moving the target surface away from the original focal point G by a backward moving distance, d_b (Fig. 1(a)). Let us consider 1D interference patterns with three point sources A, B, and C at $d_b = 0$, 1, and 2, respectively, where $d_b = 0$ is at the position of the original focal plane (G). Gaussian-shaped, approximately flat-topped, and non-uniform profiles are generated at $d_b = 0$, 1, and 2, respectively, as shown by the dotted lines. Thus, the MA can generate a flat-topped profile at the optimum backward moving distance.

The DMS has a 2D micro-dimple array on a parabolic mirror, which acts as the principal focusing lens of the MA (Fig. 1(b)). θ represents the off-axis angle of the parabolic mirror. This complex configuration with one mirror surface can be treated as a combination of the principal lens and the micro-lens array. It should be noted that each dimple is parabolic but not concave, and the focal length of each dimple is F_c but not f.

The features of the DMS are as follows: (1) Since the DMS is composed of reflective optics and can eliminate the effects of focusing performance on color aberration, it is applicable for broadband and tunable lasers. (2) The use of ultra-fast pulse lasers is possible because there is no group velocity dispersion in the DMS. (3) Speckles due to diffraction effects at the edges between the dimples can be reduced with the use of a broadband laser. Thus, the DMS is a fairly suitable focusing system for broadband, tunable, and ultra-fast MIR-FEL uniform illumination.

3. Design of the DMS and discussion

3.1. Laser Optics Calculation Code (LOCCO)

We designed the specifications of the DMS using the 1D LOCCO for broadband laser systems [6]. Using a simulation based on a configuration of the MA, we calculated and estimated the focusing

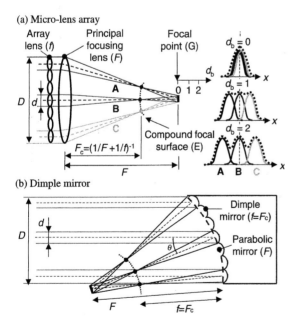

Fig. 1. Principles of (a) the MA and (b) the DMS. For simplicity, both of these systems are drawn in one dimension. Conceptual 1D patterns near the focal point are shown in (a).

performance of the DMS. The profile of the incident beam was treated as Gaussian-shaped [7]. In Section 3.2, the laser wavelength and the number of dimples per array were $\lambda = 6.0$ μm and $N = 12$, respectively, and in Section 3.3 these were varied to investigate the dependencies of the focusing performance on λ and N. The bandwidth was fixed at 1% ($\Delta\lambda/\lambda$ for the FWHM), which corresponds to that of our MIR-FEL system [7]. The focal length of the parabolic mirror was fixed at $F = 30$ cm. The simulation parameters are as follows: (1) Incident beam: the FWHM beam diameter ($D = 2.0$–8.0 cm). (2) Dimple array: the focal length ($f = 63.5$–1000 cm, resulting in a compound focal length (F_c) of 20.37–29.13 cm). We calculated 1D interference profiles near the focal plane (G) for the respective parameters by changing the backward moving distance (d_b).

3.2. D and f dependency

Fig. 2 shows 1D spatial profiles as a function of d_b at $\lambda = 6.0$ μm, $D = 4$ cm, $F = 30$ cm, $f = 200$ cm ($F_c = 26.09$ cm), and $N = 12$ ($d = D/N = \frac{4}{12} = 0.3333$ cm). All the intensity profiles were normalized by the corresponding maximum values. The closer it was to $d_b = 135$–143 μm, the more uniform was the intensity profile. The intensities gradually decrease as d_b is increased (that is, as the target surface is moved away from the parabolic mirror).

Figs. 3(a)–(c) show the dependencies of d_b, the flat-spot size, and the flatness on D and f, respectively. The flat-spot size, S_{flat}, is defined by the spatial width wherein the intensity is above 97.5% of the maximum value. The flatness, R_{flat}, is obtained from the ratio of S_{flat} to the spatial width wherein the intensity is above 10% of the maximum value ($R_{flat} = 11\%$ for a Gaussian-shaped beam). A higher R_{flat} indicates a more uniform intensity distribution in the focused pattern.

In Fig. 3(a), we plotted the optimum backward moving distance, d_b, for flat-topped pattern generation ($d_b = 135$–143 μm in the case of Fig. 2). d_b strongly depends on f in the smaller D range (<3 cm), while d_b is approximately constant in the larger D range (> 6 cm). S_{flat} does not depend

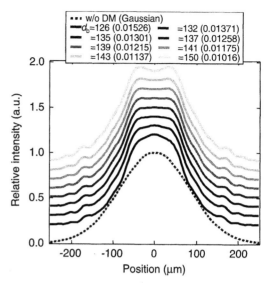

Fig. 2. Normalized 1D interference patterns as a function of d_b. For reference, a Gaussian-shaped profile without the DMS at $d_b = 150$ μm is also shown by the dotted line. The values in parentheses represent the maximum values in energy density (J/cm^2).

significantly on f at a given D (Fig. 3(b)). However, S_{flat} decreases appreciably with increasing D, since an increase in D leads to a reduction of d_b (signifying a smaller spot size of the focused beam), as shown in Fig. 3(a). R_{flat} becomes higher as D increases (Fig. 3(c)).

Using the DMS condition in the simulations shown in Fig. 2, we can obtain a uniform intensity profile within the optimum backward moving distance range of 135–143 μm. This d_b range, at which a flat-topped beam can be obtained, depends on the DMS conditions. Error bars indicated in Fig. 3(a) represent the corresponding d_b ranges for flat-topped beam generation but not the statistical errors in the calculation. Moreover, since d_b affects S_{flat} and R_{flat}, the latter show the variation according to the corresponding d_b ranges, as shown by error bars in Figs. 3(b) and (c).

3.3. N and λ dependency

In Section 3.2, it was found that a higher flatness above 20% was obtained at $\lambda = 6.0$ μm, $N = 12$,

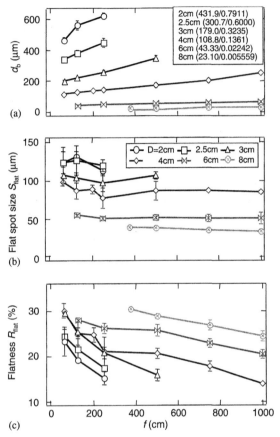

Fig. 3. Calculated results as functions of D and f. (a) d_b dependency. The two values (b/m) in parentheses show intercept/slope values for the corresponding fitted lines. (b) Flat-spot size dependency. (c) Flatness dependency.

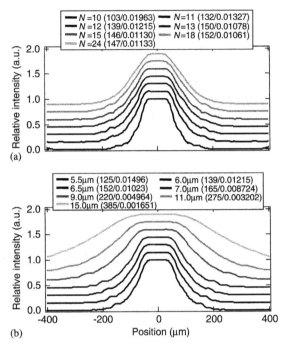

Fig. 4. Normalized 1D spatial profiles as functions of (a) N and (b) λ at $D = 4$ cm, $F = 30$ cm, $f = 200$ cm, and $F_c = 26.09$ cm. The two values in parentheses show (d_b (μm)/maximum energy density (J/cm^2)) for each calculation.

3.4. Discussion

As described in Section 1, the DMS was designed for an MIR waveband of 5.75–6.45 μm. In addition, the MIR-FEL at our facility has a FWHM beam diameter of $D = 4$–6 cm in the user room [7]. Therefore, we should choose the following specifications for the DMS (to realize higher flatness): $F = 30$ cm, $f = 200$ cm, $F_c = 26.09$ cm, $N = 12$, and $d = 0.3333$–0.5 cm. θ is selectable within the range of 30–90°. We can obtain an approximately flat-topped profile with a flat-spot size of ~ 50–90 μm and with a flatness of $\sim 25\%$ at $d_b = \sim 50$–150 μm by the use of this DMS. This DMS is applicable in the longer waveband region up to $\lambda = 9.0$ μm; however, it causes a decrease of energy density. It should be noted that the focusing performance of this DMS is similar to that of smaller or larger DMSs with an identical design.

$f = \sim 150$–250 cm, $D = 4$–6 cm, and $F = 30$ cm. In this section we examine the N and λ dependencies at $D = 4$ cm, $F = 30$ cm, $f = 200$ cm, and $F_c = 26.09$ cm. Figs. 4(a) and (b) show the normalized 1D profiles as functions of $N(= 10$–24) and $\lambda(= 5.5$–15.0 μm), respectively. As N increases, the profiles gradually change from an approximately flat-topped to a Gaussian-like profile. $N = 10$–13 (including $N = 12$) allows uniform illumination. Flat-topped patterns are also affected by an increase in λ as was the case with increasing N. $\lambda = 5.5$–7.0 μm still allows a uniform intensity distribution but $\lambda > 9.0$ μm becomes a Gaussian-like profile.

Each focal spot on the compound focal plane (E) must be a coherent point light source to generate a fairly uniform distribution. This requires a relatively larger D, smaller N, and shorter λ, since the size of each focal spot on E is proportional to $(F_c N/D)\lambda$. On the contrary, a smaller D, larger N, and longer λ result in a smoother profile with lower flatness (Figs. 3(c) and 4) due to diffraction effects.

4. Conclusions

We have proposed and designed a DMS using the 1D LOCCO for obtaining uniform illumination from a tunable and broadband MIR-FEL. We have proved that the DMS can generate an approximately uniform profile over a wide MIR waveband by moving the target surface slightly away from the original focal point. We will further develop the DMS and utilize it in a TOF mass spectrometer system.

References

[1] X. Deng, et al., Appl. Optics 25 (1986) 377.
[2] Y. Kato, et al., Phys. Rev. Lett. 53 (1984) 1057.
[3] J. Feinleib, et al., Appl. Phys. Lett. 25 (1974) 311.
[4] M. Heya, et al., Nucl. Instr. and Meth. A 507 (2003) 564.
[5] M. Born, E. Wolf, Principles of Optics I, Orion Press, Tokyo, 1974, pp. 184–276 (in Japanese).
[6] Opt Electronics Laboratory Ltd. LOCCO-1.1 (in Japanese).
[7] K. Awazu, et al., Nucl. Instr. and Meth. A 507 (2003) 547.

Available online at www.sciencedirect.com

SCIENCE DIRECT®

ELSEVIER   Nuclear Instruments and Methods in Physics Research A 528 (2004) 254–257

NUCLEAR
INSTRUMENTS
& METHODS
IN PHYSICS
RESEARCH
Section A

www.elsevier.com/locate/nima

Development of MCP-based photon diagnostics at the TESLA Test Facility at DESY

A. Bytchkov[a], A.A. Fateev[a], J. Feldhaus[b], U. Hahn[b], M. Hesse[b],
U. Jastrow[b], V. Kocharyan[b], N.I. Lebedev[a], E.A. Matyushevskiy[a],
E.L. Saldin[b], E.A. Schneidmiller[b], A.V. Shabunov[a], K.P. Sytchev[a],
K. Tiedtke[b], R. Treusch[b], M.V. Yurkov[a],*

[a] *Joint Institute for Nuclear Research, Dubna 141980, Moscow Region, Russia*
[b] *Deutsches Elektronen-Synchrotron (DESY), Hamburg, Germany*

Abstract

A nondestructive radiation detector unit is under construction at the TESLA Test Facility at DESY providing monitoring of the radiation (pulse energy, transverse intensity distribution, and statistical properties). The concept of the detector design is based on experience obtained during operation of TTF FEL, Phase 1. Key element of the detector is a wide dynamic range micro-channel plate (MCP) which detects scattered radiation from a thin gold wire crossing photon beam. Operating wavelength range of the detector is from 6 to 100 nm. An important feature of the MCP-based detector is that it is capable to cover all dynamic range of the radiation intensity, from the level of spontaneous emission up to the saturation level of SASE FEL.
© 2004 Elsevier B.V. All rights reserved.

PACS: 41.60.Cr; 42.55.Vc; 41.75.Ht; 41.85.Ew

Keywords: Free electron lasers; X-ray lasers; Relativistic electron beams

1. Introduction

The free electron laser at the TESLA Test Facility, Phase 2 will cover a wavelength range 6–100 nm. Radiation of this band is strongly absorbed by any material, and all devices for beam diagnostics must operate under UHV con-

ditions. A set of different detectors is foreseen for characterization of properties of the photon beam: PtSi-photodiodes, thermopiles based on YBCO high-T_c-superconductors (HTSCs), a gas ionization detector, and a detector based on a microchannel plate (MCP) [1–5]. A VUV/soft X-ray monochromator equipped with an intensified CCD camera will be used to record single-shot spectra. All these detectors were tested during Phase 1 of TTF FEL operation with unique VUV radiation pulses of GW-level, ultra-short

---

*Corresponding author. Tel.: +7-9621-62154; fax: +7-9621-65767.

E-mail address: yurkov@mail.desy.de (M.V. Yurkov).

(30–100 fs) duration. There was a special concern about radiation characterization, since calibration of the detectors was performed only with conventional pulsed sources (lasers and UV flash-lamps) having different wavelength band and pulse duration. Thorough cross-check of all detectors gave us a more reliable base for precise characterization of the energy in the radiation pulse.

TTF FEL, Phase 2 will mainly serve user's experiments which requires to develop such a monitoring of the photon beam that allows to deliver the most of the photon beam to the user's stations without significant interference. A nondestructive, MCP-based radiation detector was successfully used during Phase 1 of TTF FEL operation, and it was decided that similar detector will be used as one of the primary diagnostics at TTF FEL, Phase 2.

2. Principle of operation and calibration

The principle of nondestructive photon beam monitoring with a MCP detector is illustrated in Fig. 1. A thin gold wire scatters a tiny fraction of the incident radiation onto a micro-channel plate (MCP). The MCP consists of a large number of thin conductive glass capillaries which work as an independent secondary-electron multipliers and form a two-dimensional secondary-electron multiplier as a whole. The MCP amplification coefficient can be easily tuned in a wide range by changing the voltage which is applied to the MCP assembly. These features make an MCP a perfect detector for monitoring VUV radiation.

A specific feature of MCPs is a nonlinear, nearly exponential dependence of the gain on the applied voltage, so special efforts have been directed to a calibration procedure (the manufacturer of the MCP does not provide the calibration characteristics with required accuracy). Here we describe a calibration procedure proven to be very effective during operation of TTF FEL, Phase 1. The calibration procedure is performed with detector assembled in a photon beamline and proceeds as follows. We use a window in a vacuum chamber to measure relative gain of the MCP with Xe flash lamp in a voltage range from 1500 to 2600 V. The gain curve outside calibrated region is extrapolated from experimental points by

$$\log_{10}(G) = a_0 + a_1 V + a_2 V^2 + a_3 \times V^3. \qquad (1)$$

General gain characteristic of the MCP-based detector is shown in Fig. 2. The next step is absolute calibration of the detector. The signal of the detector is proportional to the number of photons scattered onto MCP. The procedure for absolute calibration is based on measurements of incoherent radiation from the electron bunch passing the undulator. The energy of electrons and bunch charge are known with a high accuracy, and the angular acceptance of the detector is known, too. Since the properties of the undulator radiation are well known, we can calculate the absolute value of the radiation energy (number of photons) passing through the detector acceptance.

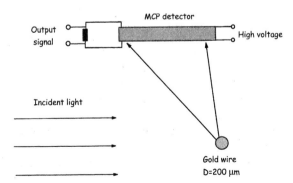

Fig. 1. Scheme of nondestructive radiation detector.

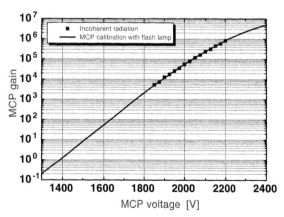

Fig. 2. MCP gain versus applied voltage. The points are calibration with spontaneous radiation from TTF FEL undulator.

As a result, we can put a corresponding point on the relative gain curve in Fig. 2. Variation of the beam energy changes the radiation wavelength, thus allowing to measure frequency response of the detector under actual experimental conditions. The TTF accelerator allows to change the bunch charge in wide limits, from a fraction of nC to 5 nC. Insertion of diaphragms with different aperture between the TTF FEL undulator and the detector extends the range of intensities by an order of magnitude more. As a result, we have absolute calibration of the detector in a wide dynamic range. Note that such an absolute calibration is performed with an assembled detector in the beamline, and can be easily checked at any time. A question arises if extrapolation (1) works well outside the calibrated range. Our answer is a positive one. For large intensities of the SASE radiation we can use other detectors. For instance, a gas ionization detector has absolute calibration, and during cross-measurements at saturation difference in the results was less than 20% only [5]. That is, described procedure gives rather good calibration in a wide dynamic range of the intensities, about six orders of magnitude.

3. Design of MCP detector unit for TTF FEL, Phase 2

MCP detector unit (see Fig. 3) will be used as a primary photon diagnostic system for nondestructive measurements of the radiation energy, photon beam profile, and for statistical measurements. We use the MCP of F4655-type manufactured by the Hamamatsu Corporation. This device is directly sensitive to UV/VUV radiation with high detection efficiency. Installation of two-stage MCP allows to cover a wide dynamic range of intensities (about six orders of magnitude). Detector will allow to determine individual parameters of the radiation pulses with 9 MHz repetition rate.

The detector unit consists of three parts. A thin gold wire and a mesh are mounted in the first (second) parts providing the possibility to perform scans of the photon beam in the horizontal (vertical) plane. The MCPs and photodiodes are

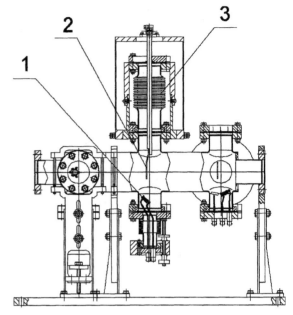

Fig. 3. Layout of MCP diagnostic unit for TTF FEL, Phase 2. Here 1 is MC-detector, 2 is target, and 3 is target mover.

mounted in the first and second parts allowing to detect the radiation scattered at large angles (this configuration is optimal for long wavelengths). A similar set is mounted in the second and third parts for detection of the radiation scattered at small angles (this is optimal for short wavelengths).

A set of apertures (0.5–10 mm diameter) is installed in front of the MCP detector unit. An aperture of a large size (10–15 mm) must be installed when using the MCP as a calorimeter. This is necessary for reducing the background, while catching all SASE radiation. A small aperture (0.5–1 mm) is used for scanning photon beam profiles.

4. Experimental experience from TTF FEL, Phase I

The MCP-based photon detectors were in operation at the TTF FEL, Phase 1 for 2 years without any signature of degradation of performance. With this detector we were able to perform precise measurements of the radiation starting from spontaneous emission up to the saturation

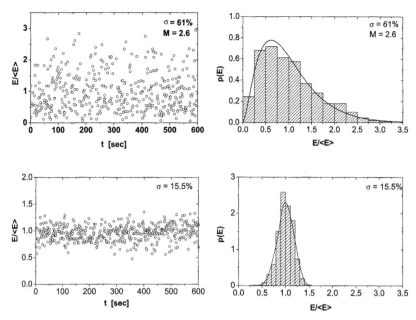

Fig. 4. Online monitoring of the radiation energy with MCP detector at TTF FEL, Phase I. Upper plots: measurements for linear regime at active length of undulator $z = 9$ m. Lower plots: measurements for saturation regime at active length of undulator $z = 14.2$ m. Left column: measured energy versus time. Right column: probability distributions of the radiation energy.

regime of SASE FEL. MCP detectors were used for online, nondestructive monitoring of radiation intensity in parallel with a pilot user experiment [6]. Our experience shows that at normal operating regime of the accelerator, the background (electron beam and dark current losses) almost does not disturb the signal which allows us to monitor very low levels of VUV radiation intensities. For example, we can detect simultaneously the direct SASE radiation pulse from the undulator and the same pulse passed complete round-trip in the optical feedback system of the RAFEL [4]. Our experience also shows that use of the MCP-based detector significantly simplifies tuning the SASE regime, since it detects reliably very small increase of the radiation intensity at a level of a few percent when normalized to the bunch charge. Fig. 4 shows typical online monitoring of TTF FEL radiation. Plots at the right side show probability distributions of the radiation energy derived from data shown at the left side. The relative accuracy of these measurements is better than 5%. One

should not wonder that large fluctuations of the shot-to-shot radiation energy occur. These is intrinsic property of the single-pass FEL amplifier starting from shot noise. The use of such a precise tool as MCP-based radiation detector gave us the possibility to perform precise characterization of the radiation pulse energy and study statistical properties of the SASE FEL radiation [7,8].

References

[1] R. Treusch, et al., Nucl. Instr. and Meth. A 445 (2000) 456.
[2] R. Treusch, et al., Nucl. Instr. and Meth. A 467–468 (2001) 30.
[3] Ch. Gerth, et al., Nucl. Instr. and Meth. A 475 (2001) 481.
[4] B. Faatz, et al., Nucl. Instr. and Meth. A 483 (2002) 412.
[5] M. Richter, et al., Appl. Phys. Lett. 83 (2003) 2970.
[6] H. Wabnitz, et al., Nature 420 (2002) 482.
[7] V. Ayvazyan, et al., Phys. Rev. Lett. 88 (2002) 104802;
    V. Ayvazyan, et al., Euro. Phys. J. D 20 (2002) 149.
[8] V. Ayvazyan, et al., Nucl. Instr. and Meth. A 507 (2003) 368.

Available online at www.sciencedirect.com

SCIENCE ⬧ DIRECT°

Nuclear Instruments and Methods in Physics Research A 528 (2004) 258–262

ELSEVIER

NUCLEAR INSTRUMENTS & METHODS IN PHYSICS RESEARCH
Section A

www.elsevier.com/locate/nima

Tolerance studies for the BESSY FEL undulators ✩

B. Kuske[a],*, M. Abo-Bakr[a], J. Bahrdt[a], A. Meseck[a], G. Mishra[b], M. Scheer[a]

[a] *BESSY, Albert-Einstein-Strasse 15, 12489 Berlin, Germany*
[b] *DA University, Khandra Road, Indore 452017, India*

Abstract

A multiuser, soft X-ray FEL facility is currently planned at BESSY. In preparation for the technical design report, tolerance studies have been performed for the undulator section adjacent to the 2.25 GeV linac. Deficiencies of the incoming beam as well as undulator imperfections have been investigated. Special aspects of the planned APPLE II type undulators have been considered.
© 2004 Elsevier B.V. All rights reserved.

PACS: 41.60.Cr; 07.85.Fv; 41.60−m

Keywords: Free electron laser; Tolerance; SASE; Undulators

1. Introduction

BESSY is presently preparing a technical design report for the planned multiuser, soft X-ray FEL facility [1]. A photoinjector followed by a 2.25 GeV superconducting linac, including two bunch compressors, will provide macrobunches at a repetition rate of 1 kHz and above 4 kA peak current. A fast kicker will distribute the pulses between three different undulator lines operating at 1.88 and 2.25 GeV in SASE operation, see Table 1. The undulators will provide light in the energy range from 20 to 1000 eV of variable polarization. All undulators will be variable gap APPLE II type

devices about 60 m in length [2]. The performance of the FELs in terms of peak power and saturation length depends strongly on the parameters of the incoming beam, such as emittance, bunch length, peak current, beam size, beam offset, and the like, as well as on the quality of the undulators themselves, which is expressed in terms of field errors, gap errors, and offsets in the focussing structure. The influence of these effects have been investigated using the FEL-code GENESIS [3] for all three undulator lines. All tolerance studies presented were performed for SASE-undulators, although alternative FEL schemes like high gain harmonic generation (HGHG) and high harmonic gain (HHG) were considered as well. Modeling these schemes with the existing codes is extremely complicated and automated runs are not yet feasible. The results presented though, can atleast be partly extrapolated to other schemes.

✩ Funded by the 'Bundesministerium für Bildung, Wissenschaft, Forschung und Technologie', the 'Land Berlin', and the 'Zukunftsfond des Landes Berlin'.

*Corresponding author.

E-mail address: bettina.kuske@bessy.de (B. Kuske).

0168-9002/$ - see front matter © 2004 Elsevier B.V. All rights reserved.
doi:10.1016/j.nima.2004.04.059

Table 1
Parameters of three different undulators lines

	*UE*66	*UE*36.5	*UE*27.5
Energy range [eV]	20–300	270–550	500–1000
Period length λ_0 [mm]	66	36.6	27.5
Periods/module	52	96	130
Number of modules	15	15	15
γ	3669	3669	2930–4403
K-value	4.92–0.82	1.54–0.81	0.86–0.74
Gap$_{min}$ [mm]	10	10	10

2. Procedures and influence of errors

All calculations presented here were performed with the GENESIS code in the time-independent mode, where the bunch is assumed to be infinitely long without any temporal structure. This reduces the computation time to acceptable limits. External UNIX-shell script were used to automate parameter scans and error runs. C-programs were written that allow the automated evaluation of GENESIS output files. They determine the saturation length, saturation power, gain length, rms-beam wander, and the like. The determination of the saturation point turned out to be nontrivial in the presence of errors. The point where, shortly after saturation, energy of the radiation field is transferred back to the beam, i.e., the first minimum of the total beam energy function, turned out to be the most reliable. Still, there is an uncertainty in this evaluation of a few meters. For errors that cause offsets in the beam trajectory, it is essential to correct the trajectory, otherwise the impact of the errors will be totally over estimated. A C-code has been developed, that reads the GENESIS error distribution from the appropriate file, calculates the distorted trajectory and the necessary corrector settings using a SVD-algorithm (singular-value decomposition). All rms-values of trajectories given are residual offsets after the correction has been applied.

Two basic mechanisms deteriorate the amplification process in the undulators: Any offset in the trajectory of the beam will lead to a reduction of the overlap between the electron beam and the radiation field. 'Beam wander', defined as the rms-value of the trajectory offset, leads to a smaller bunching rate and thus mainly to a reduction in the output power, rather than to an increased gain length. Trajectory offsets are caused by field errors in the undulators, quadrupole offsets, and fluctuations in the offset and angle of the incoming beam. Although an orbit correction system is foreseen, no BPMs can be placed within the undulators due to the variable of undulator gaps. Correctors and BPMs will be placed between the undulator modules leaving ≈ 3.5 m of uncontrolled space for trajectory drifts.

The second mechanism is called 'phase shake' and refers to a deviation between the electron phase and the ponderomotive potential. Phase shake is caused, e.g., by gap or tilt errors of the undulator gaps, by phase mismatch between the undulators or by correlated field errors. Large trajectory offsets also lead to phase errors, as longitudinal velocity of the particles gets shifted into the transverse plane. Phase errors primarily lead to a longer gain length. The original power level will finally be reached, but the necessary undulator length increases.

Trajectory offsets at the undulator entrance and reduced beam quality from the linac usually affect both output power and saturation length.

3. Incoming beam

The effects of errors in the BESSY linac still have to be evaluated, but studies performed at SLAC [4] for the LCLS project might serve as a guideline. While most static or slowly varying linac errors might be counteracted by feedback or other correction systems, there is no cure for fast jitters that change the beam properties on a shot-to-shot basis.

Gun timing, initial bunch charge, and RF phase jitters will lead to a spread in the initial beam conditions for the FEL process at the entrance of the undulators. The energy, peak current, emittances, arrival time, energy spread, and position and angle of the bunches will jitter from shot-to-shot. Table 2 lists the design parameters at the end of the BESSY linac and the rms value of the jitter expected for the LCLS linac.

Table 2
Beam design parameters at the entrance of the BESSY undulators and LCLS jitter estimates

	Design value	LCLS jitter estimate
Energy [GeV]	1.87, 2.25	0.04%
Energy spread	0.04–0.07	10%
Emittance [πmm mrad]	1.5	5%
σ_x [μm]	75	—
σ_y [μm]	30	—
Bunch length [μm]	35	—
Peak current [kA]	3.5	10%
Offset$_{x,y}$ [μm]	0, 0	9, 1.5
Angle$_{x,y}$ [μrad]	0, 0	55, 21

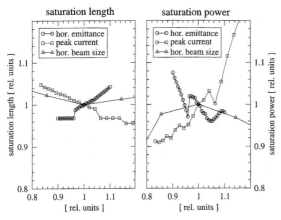

Fig. 1. While the saturation length (left) is mainly affected by emittance changes, the power reduction (right) is greatest for lower peak currents. Beta function mismatch plays a minor role.

Fig. 1 shows how deviations from the design horizontal emittance, peak current, and horizontal beam size translate to changes in saturation lengths and output power. The lacking smoothness of the curves results from ambiguities in the interpretation of the GENESIS output.

While an enlarged emittance (+10%) mainly increases the saturation length (+4%), reduced peak current (−10%) strongly reduces the output power (−8%). A beam mismatch though, seems to be of minor importance as long as the 10% margin is not exceeded.

4. Effects of undulator errors

BESSY chose APPLE II type undulators for the FEL as they provide the highest magnetic field among all polarizing devices. A modified design with even higher fields at the expense of reduced horizontal access is in discussion [2]. In these devices, the magnet arrays have a slit (1 mm) in beam direction to allow a horizontal shifting of the array halves to create polarized light. This leads to strong, shift dependent, horizontal focussing, see Fig. 2, and differs greatly from the dominantly vertical focussing properties of conventional planar undulators. While these focussing characteristics can be modeled by GENESIS, the independence of horizontal and vertical field errors cannot. Also quadrupole and undulator positioning errors remain for further studies.

The performance was evaluated for each of the three undulator lines for low and high K-values and for uncorrelated Gaussian distributed errors with four different σ_K values (0.0002–0.0008). Two short wavelength, high K-value results are presented in Fig. 3. A perfect correction of the

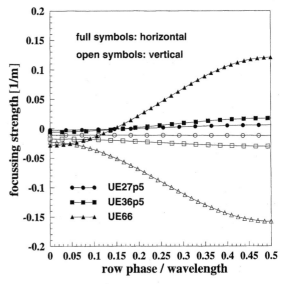

Fig. 2. Focussing strength of APPLE II undulators as a function of the shift parameter that determines the degree of polarization.

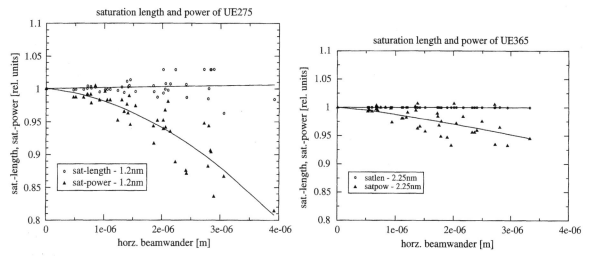

Fig. 3. Smaller radiation wavelengths lead to a higher sensitivity to the reduced overlap between the photon and the electron beam. A horizontal beam wander of 4 μm rms reduces the output power by 20%.

trajectory between the modules is assumed without BPM reading or positioning errors.

The degradation of the FEL performance due to field errors is mainly due to a reduction in the overlap between the photon and electron beam, and thus results primarily in power reduction. The effect increases with decreasing optical wavelength. It is displayed as a function of the residual horizontal beam wander after correction, which roughly correlates with σ_K.

Experience at BESSY shows that sorting and shimming of the undulator magnets allows to confine the electron trajectory offset in the undulator to fractions of a wiggle amplitude (6–10 μm). The resulting reduction in output power is only $\approx 2\%$. On the other hand, with today's BPM resolution, trajectory offsets due to jitters in the linac and in the extraction kicker easily reach the 10 μm level. In this case, the output power is typically reduced by 50%, and the saturation length increases by 20% at the shortest wavelength.

Preliminary studies of the high energy undulator with random gap settings, but perfect fields, show no power reduction; but, as expected, an increase in saturation length of 10% for a $\sigma_{K_{\mathrm{gap}}} = 0.0018$.

5. Conclusion

The influences of different error sources that will lead to a degradation of the FEL performance have been studied. Depending on which of the two possible mechanisms, phase shake or beam wander, is triggered by the error, either the saturation length or the output power deteriorates. As the effects act independently, it is expected that they will add up, the dominant problem being the stability and alignment of the trajectory inside the undulators. Possible error sources in the modeling procedure are the correlation between the horizontal and vertical field errors (provided by GENESIS), the lack of shimming procedures in the undulator model or overestimation of the effect of linac jitters. Further studies will include quadrupole and undulator offsets. Time-dependent calculations will be performed to estimate the effects of errors on the pulse structure and the spectrum.

Acknowledgements

The authors would like to thank Sven Reiche and Bart Faatz for the very productive help and collaboration in all questions concerning GENESIS.

References

[1] D. Krämer et al., Closing in on the Design of the BESSY-FEL, Proceedings of the PAC 2003, Portland, p. 1083

[2] J. Bahrdt et al., Undulators for the BESSY SASE-FEL Project, Proceedings of the SRI 2003, San Francisco.

[3] S. Reiche, Nucl. Instr. Meth. A 429 (1999) 243.

[4] P. Emma, S2E in the LCLS Linac, Talk during the Start-to-End X-ray SASE FEL Workshop, Zeuthen 2003.

Available online at www.sciencedirect.com

SCIENCE  DIRECT®

Nuclear Instruments and Methods in Physics Research A 528 (2004) 263–267

NUCLEAR INSTRUMENTS & METHODS IN PHYSICS RESEARCH
Section A

www.elsevier.com/locate/nima

Stabilization of the naturally pulsed zones of the Super-ACO Storage Ring Free Electron Laser

M.E. Couprie[a,*], C. Bruni[a], G.L. Orlandi[a], D. Garzella[a], S. Bielawski[b]

[a] *Service de Photons, Atomes et Molécules, CEA/DSM/DRECAM, bât. 522, 91 191 Gif-sur-Yvette,
and LURE, bat. 209 D, Université de Paris-Sud, BP 34, 91 898 Orsay Cedex, France*
[b] *Laboratoire de Physique des Lasers, Atomes et Molecules (UMR 8523)
and CERLA, Bat P5, Université des Sciences et Technologies de Lille, F-59655 Villeneuve d'Ascq Cedex, France*

Abstract

A Storage Ring Free Electron Laser presents different regimes versus the synchronization between the optical pulses bouncing in the optical cavity and the electron bunches stored in the ring. Both experiments and simulations show that, on Super-ACO, the so-called detuning curve presents five zones: a "CW" regime at the ms time scale in the central and lateral ones, and a pulsed behavior at intermediate detunings. Apart from a longitudinal feedback allowing the FEL to be maintained in the central "cw" region for ensuring a high level of stability for the users, a recent feedback allowed the pulsed regions to be stabilized. The new detuning curves are presented.
© 2004 Elsevier B.V. All rights reserved.

PACS: 41.60.Cr; 42.65.Sf; 42.60Rn

Keywords: Storage Ring FEL; Feedback control; Stability; Relaxation oscillations

1. Introduction

Free Electron Laser (FEL) oscillators can be viewed as a particular non-linear dynamical system. In this perspective, chaos has been studied on different FEL sources, such as Raman devices [1], and LINAC-based infra-red FEL oscillators [2]. On Storage Ring Free Electron Lasers (SRFEL), a macrotemporal structure at the millisecond time-

scale can appear, for particular detunings (i.e. synchronization between the electron bunches stored in the ring and the optical pulses bouncing in the optical resonator) in addition to the microtemporal structure, corresponding to the temporal pattern of the electron bunches (at a high-repetition rate) [3]. The theoretical and experimental SRFEL responses to a detuning modulation have been studied and they can lead to deterministic chaos [4]. We consider here the control of the naturally pulsed regimes that have been performed on the Super-ACO FEL [5], following the approach developed in the dynamical systems fields applied to conventional lasers, such

---

*Corresponding author. Tel.: +33-1-64-46-80-44; fax: +33-1-64-46-41-48.

E-mail address: marie-emmanuelle.couprie@lure.u-psud.fr (M.E. Couprie).

0168-9002/$ - see front matter © 2004 Elsevier B.V. All rights reserved.
doi:10.1016/j.nima.2004.04.060

as Nd-YAG lasers [6], Nd-doped optical fiber lasers [7], passively mode locked laser [8], and in class B (pump modulated) lasers [9]. The theoretical and experimental control of the pulsed zones of the Super-ACO Free Electron Laser is here presented.

2. The detuning curve of the Super-ACO FEL

2.1. Experimental behavior of the Super-ACO FEL versus detuning

In Fig. 1a, the Super-ACO FEL [10] intensity response for a slow sweep of the detuning shows different regimes. Around perfect tuning and for a small detuning interval (zone 3), the FEL is "cw" at the ms time scale, with the shortest widths of the temporal and spectral distributions, the FEL being closed to the Fourier limit [11]. For intermediate values of the detuning (zones 2 and 4), the FEL presents systematically a pulse structure at the ms time scale, with slightly larger temporal and spectral widths. The detuning curve shows the minima and

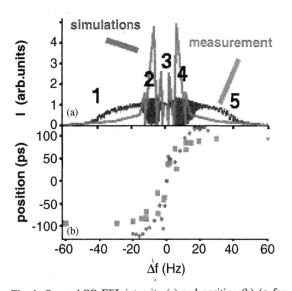

Fig. 1. Super-ACO FEL intensity (a) and position (b) (● for measurements and squares for LAS simulations) versus detuning. Energy: 800 MeV; mirror losses: 1%; gain: 2%; wavelength: 350 nm; TEM00 transverse mode. Main RF cavity operating at 100 MHz. A modification of 100 Hz corresponds to a round trip mismatch of 120 ns, or to a cavity length detuning of 18 μm.

the maxima of this macrotemporal structure. For even larger detunings (zones 1 and 5), the laser presents again a "cw" temporal structure at the ms time scale, with larger distributions and reduced power. Fig. 1b shows the evolution of the FEL pulse position with respect to the position of the synchronous electron. This "arctang" like function shows rapid changes around perfect detuning. Clearly, zone 3 is the most suitable one for user applications [12]. Such a type of detuning curve represents a rather general situation for the dynamics of SRFELs, even though the noise can mask the "cw" central region [13].

2.2. Modelling of the detuning curve

First theoretical representation of the behavior of a storage ring FEL was performed according to a phenomenological model in the temporal domain [3]. It describes the evolution of the intensity distribution and temporal position of the FEL pulse (with respect to the electron bunch center of mass), the increase of the electron beam energy spread ("heating") induced by the FEL interaction. Using the "LAS" code based on this approach, one can reproduce the FEL intensity and position versus detuning, as illustrated in Fig. 1 [14]. Analogous results can be obtained, for example, with the numerical codes "SRFELn" [15] or "STOK_2D" based on FEL equations [14], or with "SRFEL" in the frequency domain [16], or with Ref. [17]. The pass-to-pass phenomenological model can be simplified for small variations of the detuning by rewriting it in a form of partial differential equations of the normalized laser longitudinal profile Y and the electron beam energy spread σ, since the pulse shape varies slowly from pass-to-pass, in the case of a low gain and losses system. It leads to the following equations, at the first order of the cavity losses:

$$\partial_T Y - \Delta\Omega \partial_\theta Y = -Y + G[Y + \eta] \tag{1}$$

$$d\sigma^2/dT = \alpha[\sigma_0^2 - \sigma^2 + (\sigma_e^2 - \sigma_0^2)I] \tag{2}$$

where T is the continuous time, expressed in units of cavity photon lifetime (usual time multiplied by the cavity losses and divided by the cavity round trip time τ_R). θ, the new dimensionless fast time, synchronous with the electron bunch passage, is the sum of the temporal coordinate τ and the delay between the electron bunch and the laser pulse at each round trip $\delta\tau_n$, normalized with respect to τ_R. $\Delta\Omega$ represents the normalized detuning, G is the laser gain, and ηG is the spontaneous emission. $\alpha = 2\tau_R/\tau_s$ is typically much smaller than 1, τ_s being the synchrotron damping time. σ_0, σ, σ_e, respectively, refer to the initial energy spread, the energy spread at time τ, and at equilibrium. I indicates the laser intensity integrated over the longitudinal profile Y. Eq. (1) expresses the laser intensity evolution. Eq. (2) describes the laser heating damped via the synchrotron oscillations. Fig. 2 shows a typical detuning curve obtained using this simplified model. It is in good agreement with the experimental results and with the previous pass-to-pass simulations.

3. The control of the Super-ACO FEL

3.1. A longitudinal feedback for maintaining the Super-ACO FEL in zone 3

Since a Fourier limit powerful stable laser source is required for the user applications, a feedback aiming to keep the FEL in zone 3 was first developed on Super-ACO [18]. With the position of the FEL pulse changing very rapidly near perfect tuning, it provides a good parameter to evaluate the drift in synchronism. It is deduced from the measurement of the FEL pulse distribution, the drift in synchronism is compensated via the frequency of the RF cavity. Such a feedback system allows the intensity fluctuations to be damped to 1%, the spectral drift to be limited to less than 0.001% and the temporal jitter of the FEL pulse to be reduced, and compensated up to 1 ms, as illustrated in Fig. 3.

3.2. The stabilization of the pulsed regimes of the Super-ACO FEL in zones 2 and 4

The FEL can be viewed as a spatio-temporal system, with relevant coordinates T and θ, the effect of the detuning appearing in the form of an advection term $\Delta\Omega\partial_\theta Y$. The detuning curve (see

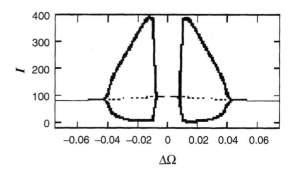

Fig. 2. Detuning curve derived from the simplified model in the case of the Super-ACO FEL. Synchrotron damping time: 8 ms, cavity losses 1%, gain: 2%, cavity round trip time: 120 ns.

Fig. 3. Streak camera image showing the compensation of the FEL jitter with the longitudinal feedback system.

Fig. 2), plotted using Eqs. (1) and (2), can also be considered as a bifurcation diagram. Stable stationary states exist around perfect tuning and for large detunings. Between these zones, a limit cycle regime leads to the pulsed behavior of the FEL, together with an unstable stationary state. Then, the control of the pulsed macrotemporal structure of the Super-ACO storage ring FEL becomes possible, by forcing the laser to operate onto the unstable stationary state. As in Refs. [7,8], a signal proportional to the derivative of the laser intensity (with a gain β) is used, and applied to the RF frequency pilot, as for the longitudinal feedback. However, the reference position is set in the centre of zone 2 or 4, and no more in the centre of zone 3. A photomultiplier, whose bandpass is smaller than the revolution frequency, and much larger than the macropulse frequency, is used for the detection. Simulations using the simplified model showed that the pulsed regime of the FEL can be stabilized.

Experimentally, the control has been achieved by choosing empirically the feedback gain ($\beta = 0.1$–$0.5\ \mu W^{-1}$). Fig. 4 shows the controlled laser intensity and the feedback signal with the FEL operation in zone 4. The transient regime

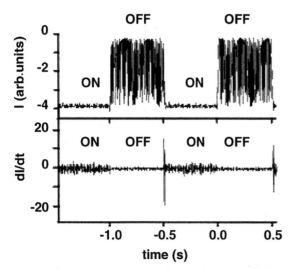

Fig. 4. Super-ACO FEL intensity I and control signal when the feedback is established and then switched off. The black zones correspond to the pulsed regime of the SUPER-ACO FEL (the macropulses are not resolved at this time scale).

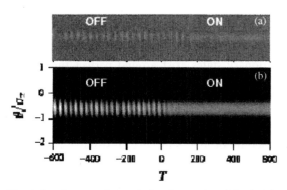

Fig. 5. Streak camera image during the transient following the application of the feedback (a) experimental results (b) simulation with the typical values of the Super-ACO FEL, corresponding to the present experimental conditions.

lasts a few tens of ms leading to the stabilization of the pulsed zones. The resulting laser intensity fluctuations depend on the value of the applied gain β. Fig. 5 shows the behavior of the FEL micropulse observed with a double sweep streak camera, while the control is established. The experimental results in (a) are in good agreement with the simulated ones in (b). The sign of the reaction on the RF frequency pilot is opposite for zones 2 and 4, leading either to a stabilization of the pulsed zones or to an induced pulsed regime

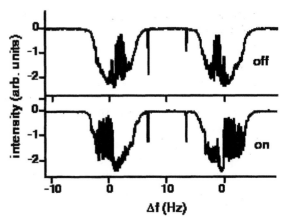

Fig. 6. Detuning curve plotted with a slow triangle function of frequency change applied on the RF pilot, without (off) and with control (on).

starting from original "cw" behavior. In Fig. 6, one observes a stabilization of the pulse region of one half of the detuning curve, and an increase of the pulses amplitude in the other half.

4. Conclusion

Following the work performed on conventional lasers, a control of the Super-ACO FEL has been applied to stabilize the naturally pulsed regimes of operation, allowing the width of "cw" operation of the FEL to be widened. It is of great importance for the user applications, and for SRFELs for which the central "cw" region is extremely tiny, because of the combination of various parameters such as the synchrotron frequency, synchrotron damping time, gain and cavity losses. It could also probably be very useful for storage ring FEL where line instabilities prevent a good stability in zone 3.

References

[1] E.H. Park, et al., Nucl. Instr. and Meth. A 358 (1995) 448;
    S. Kawasaki, et al., Nucl. Instr. and Meth. A 358 (1995) 114.
[2] P. Chaix, et al., Phys. Rev. E 48 (1993) 3259;
    D.A. Jaroszynski, et al., Nucl. Instr. and Meth. A 407 (1998) 407;

R. Hajima, et al., Nucl. Instr. and Meth. A 475 (2001) 270;
W.P. Leemans, et al., Nucl. Instr. and Meth. A 358 (1995) 208.

[3] M. Billardon, et al., Phys. Rev. Lett. 69 (1992) 2368.

[4] M. Billardon, Phys. Rev. Lett. 65 (1990) 713;
G. De Ninno, et al., Eur. Phys. J. D 22 (2003) 269;
M.E. Couprie, Nucl. Instr. and Meth. A 507 (2003) 1;
C. Bruni, et al., Nucl. Instr. and Meth. A, (2004) these proceedings.

[5] S. Bielawski, C. Bruni, G.L. Orlandi, D. Garzella, M.E. Couprie, Phys. Rev. E 69 (2004) 045502 (R).

[6] R. Roy, et al., Phys. Rev. Lett. 68 (1992) 1259.

[7] S. Bielawski, et al., Phys. Rev. A 47 (1993) 3276.

[8] N. Joly, S. Bielawski, Opt. Lett. 26 (2001) 692.

[9] M. Ciofini, et al., Phys. Rev. E 60 (1999) 398;
R. Meucci, R. Mc Allister, R. Roy, Phys. Rev. E 66 (1999) 026216.

[10] M.E. Couprie, et al., Nucl. Instr. and Meth. A 358 (1995) 374.

[11] M.E. Couprie, et al., Nucl. Instr. and Meth. A 475 (2001) 229.

[12] M.E. Couprie, et al., Nucl. Instr. and Meth. A 375 (1996) 639.

[13] M.E. Couprie, P. Elleaume, Nucl. Instr. and Meth. A 259 (1987) 77;
H. Hama, et al., Nucl. Instr. and Meth. A 375 (1996) 39;
N. Sei, et al., Nucl. Instr. and Meth. A 375 (1998) 187;
V. Litvinenko, et al., Nucl. Instr. and Meth. A 475 (2001) 240;
R.P. Walker, et al., Nucl. Instr. and Meth. A 475 (2001) 20.

[14] C. Bruni, G. De Ninno, "LAS", R. Bartolini, "STOK_2D", Proceedings of the Brainstorming Workshop, Europ. TMR network "Storage ring FEL towards 200 nm", Maratea, Italy, 14–19/04/2002.

[15] C.A. Thomas, et al., Nucl. Instr. and Meth. A 507 (2003) 281.

[16] T. Hara, et al., Nucl. Instr. and Meth. A 375 (1996) 39.

[17] V. Litvinenko, et al., Nucl. Instr. and Meth. A 358 (1995) 53.

[18] M.E. Couprie, et al., Nucl. Instr. and Meth. A 358 (1995) 374.

Available online at www.sciencedirect.com

SCIENCE @ DIRECT°

Nuclear Instruments and Methods in Physics Research A 528 (2004) 268–272

NUCLEAR INSTRUMENTS & METHODS IN PHYSICS RESEARCH
Section A

www.elsevier.com/locate/nima

Lasing below 200 nm in the NIJI-IV compact storage-ring-based free electron laser

K. Yamada*, N. Sei, H. Ogawa, M. Yasumoto, T. Mikado

Photonics Research Institute, AIST, Tsukuba Central 2, 1-1-1 Umezono, Tsukuba, Ibaraki 305-8568, Japan

Abstract

Laser gain of the NIJI-IV compact storage-ring-based Free Electron Laser (FEL) system at AIST reached ∼9% at 200 nm for an average beam current of 16.3 mA. In addition to such a drastic gain enhancement, improvement of the laser-cavity performance has led to successful FEL lasing in the vacuum ultraviolet below 200 nm even in a compact system. Temporal and spectral characteristics of the NIJI-IV FEL were examined around 200 nm. FEL performance will be briefly discussed from the application point of view.
© 2004 Elsevier B.V. All rights reserved.

PACS: 41.60.Cr; 52.59.Rz; 52.59.Ye

Keywords: VUV free-electron laser; Storage ring; Ring impedance; Beam instability; FEL gain; Cavity loss

1. Introduction

Storage ring Free Electron Lasers (SRFELs) are excellent light sources with a very good optical quality in principle. At AIST (National Institute of Advanced Industrial Science and Technology), efforts have been made to shorten the FEL wavelength down to the Vacuum Ultraviolet (VUV) range on the compact storage ring NIJI-IV. In addition to renewal of the RF cavity [1] and installation of thin sextupole magnets [2], we recently replaced the ring vacuum chambers with low-impedance-type ones to suppress the longitudinal microwave instability, which enabled us to

drastically enhance the FEL gain. Such a gain enhancement and reduction of the optical-cavity loss brought successful FEL lasing in the VUV range, shorter than 200 nm, in a compact FEL system. From the results of analytic calculation [3], we can expect to obtain much higher gain and to achieve lasing even below 190 nm, which is the world's shortest lasing wavelength [4] in SRFELs. Here, we report distinctive FEL characteristics in the deep UV (DUV) and VUV range, recently obtained with the NIJI-IV FEL system.

2. Impedance reduction and gain enhancement

Since there were many vacuum components, which have quite different inner cross-section

---

*Corresponding author. Tel.: +81-29-861-5679; fax: +81-29-861-5683.

E-mail address: k.yamada@aist.go.jp (K. Yamada).

0168-9002/$ - see front matter © 2004 Elsevier B.V. All rights reserved.
doi:10.1016/j.nima.2004.04.061

shapes, in a very limited space, broadband impedance of the NIJI-IV was considerably large compared with medium-sized modern storage rings. This caused a strong longitudinal microwave instability [5] that limited the peak beam current and FEL gain through bunch lengthening for higher beam current regions. So, we replaced old vacuum chambers with new low-impedance-type ones over 75% of the circumference. The new vacuum chambers include some key components, for example, low-impedance bellows chambers with RF contact fingers inside, smooth transition chambers to change the cross-section shapes and bend chambers whose inner surfaces are as smooth as possible. Inside the kicker, one of the most crucial components, a DELTA-type low-impedance coil [6] was adopted. For chamber connection, VAT-type flat flanges were mainly used to minimize the inner surface discontinuity.

From the bunch-length measurement as a function of average beam current, ring impedance was estimated for old and new vacuum chambers [3,7]. Since the NIJI-IV is operated in one-bunch mode, the bunch length is mainly affected by potential well distortion and longitudinal microwave instability. For old chambers, a clear onset of steep bunch lengthening due to microwave instability was observed at a beam current of ~ 2 mA. From this threshold current, the broadband impedance of old chambers was calculated to be $\sim 20\ \Omega$. For new chambers, on the other hand, any onset of microwave instability was not found and the bunch length slowly increased only due to potential well distortion at least up to 16.3 mA. Assuming this value to be a new threshold for microwave instability, broadband impedance was calculated to be 1.8 Ω, which is smaller by an order of magnitude than before.

Taking the measured bunch lengths into account, FEL gain was estimated at 200 nm using the well-known analytic formula [8]. In Fig. 1, solid circles and open squares indicate the gains estimated for old high-impedance and new low-impedance chambers. Dashed and solid curves show predicted gains for chamber impedance of 2 and 1 Ω, respectively. In this figure, the gain is found to be limited below $\sim 2.5\%$ for high-impedance (20 W) chambers due to microwave

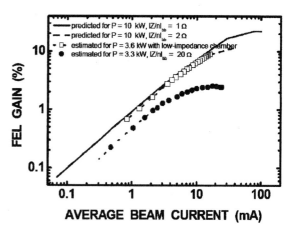

Fig. 1. FEL gain estimated at 200 nm for high-impedance (solid circles) and low-impedance (open squares) chambers, taking the reduced bunch lengths into account. Dashed and solid lines show the results of calculation for ring impedances of 2 and 1 Ω.

instability, while it does not have any limitation and can reach $\sim 9\%$ at a beam current of 16.3 mA for low-impedance chambers. According to the prediction curves, a higher gain of at least 10% or hopefully more than 20% is achievable for higher current regions, which will bring FEL lasing at much shorter wavelength.

3. Lasing characteristics in the DUV/VUV range

Due to the drastic gain enhancement and improved cavity-mirror performance, lasing wavelength was successfully reached in the VUV range. Fig. 2 shows FEL gains, cavity losses and FEL line spectra in the DUV/VUV range. Since our previous gain was 2.5% at maximum, the shortest laser wavelength obtained was 211 nm, considering the cavity-loss curve around 215 nm. For the wavelength shorter than this, cavity loss was comparable to the gain, which made it difficult to obtain lasing. However, recently we have got two major improvements, that is, gain enhancement mentioned above and a reduction of minimum loss of the laser cavity, composed of two Al_2O_3/SiO_2 mirrors, down to $\sim 0.6\%$ in a shorter wavelength region around 200 nm. These improvements enabled us to obtain FELs below 200 nm.

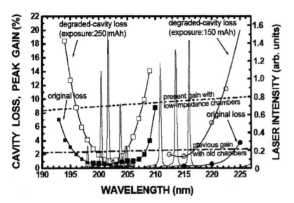

Fig. 2. FEL gains, cavity losses and FEL line spectra in the DUV/VUV range.

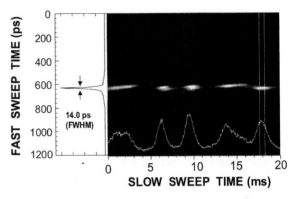

Fig. 3. Time structure of the NIJI-IV FEL observed with a dual-sweep streak camera near the best cavity-tuning condition.

Though the lasing range lies from 198 to 205 nm at present, it should be wider from the relation between the gain and cavity-loss curves. This will be achieved by careful alignment among electron beam, laser axis and vacuum chambers connected with cavity-mirror vessels. Since Al_2O_3/SiO_2 mirrors whose optical loss is 1.4% at 193 nm have already been prepared, we expect to obtain lasing at a shorter wavelength even below 190 nm.

Time structure of the NIJI-IV FEL was observed with a dual-sweep streak camera around 200 nm. Fig. 3 shows a typical result near the best cavity-tuning condition. Due to relatively short bunch length and mechanically unstable structure of mirror holders, the lasing mode was not fixed at a stable CW mode but in a little fluctuating CW-like mode. In Fig. 3 the laser micro-pulse width is observed to be 14 ps in FWHM. Since this value is almost comparable to the temporal resolution for the measuring sweep speed, the micro-pulse width is expected to be much shorter than this. Actually, a pulse width of 6.7 ps in FWHM was observed with the fastest sweep speed of the streak camera, though the lasing was not in a CW but in a pulsed mode.

Fig. 4 (a) and (b) show the temporal change of macro-temporal laser intensity and corresponding laser-line width at 202 nm for different cavity-detuning frequencies of ~ 0 Hz and $+20$ Hz, respectively. These data were obtained by relaying the output images of a monochromator with a

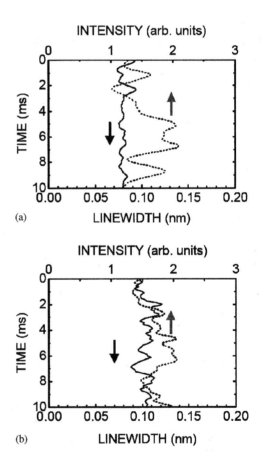

Fig. 4. Temporal change of macro-temporal laser intensity (dashed lines) and corresponding laser-line width (solid lines) for different cavity-detuning frequencies of ~ 0 Hz (a) and $+20$ Hz (b). A 20-Hz detuning is equivalent to a cavity-detuning length of 3.6 μm in the NIJI-IV system.

resolution of 0.04 nm onto the entrance slit of the streak camera [9]. In case (a), where the laser cavity length is near the best tuning condition and the lasing mode should be near CW, the line width is observed to be relatively constant around 0.08 nm within 10-ms time scale, in spite of a large intensity fluctuation. This is probably because the laser starts up from nearly single seed light and its peak intensity already reaches a saturation level where the line width should approach a certain value determined by the transform-limited condition. On the other hand, in case (b), where the lasing mode should be quasi-CW [9], the line width is found to fluctuate irregularly around a little larger value of 0.1 nm. This suggests that the laser micropulses include some components originating from different seed lights and the peak intensities of these components have never reached a saturation level yet. From these results, we find that a very precise cavity-length tuning of the order of 0.1 μm or a few Hz in RF frequency is necessary to obtain lasing with a stable and sharp line width.

The average FEL output power was measured to be 100 μW per two ports at 202 nm, since the transmission of the cavity mirrors was only 0.05%. Considering an enough margin of the gain against the cavity loss, one-order increase of the mirror transmission is acceptable, which will bring an average power of the order of 1 mW. Assuming the diffraction-limited divergence angle, 1 mW at 200 nm is equivalent to a net average spectral brightness of $\sim 1.7 \times 10^{16}$ photons/s/mrad^2/0.05%BW which can be actually utilized without a monochromator. Further increase of the power up to 10 mW level can be achieved by increasing the beam energy. We plan to increase the beam energy from 0.3 to 0.5 GeV in the next phase upgrade of the NIJI-IV system. According to the lasing performance mentioned above, we started to use the NIJI-IV FEL as a light source for photoelectron emission microscopy (PEEM). By using VUV FELs with an average power of 1–10 mW level, real-time images of collective motions in surface chemical reactions are expected to be observed with video-rate time resolution.

4. Conclusions

The ring impedance of the NIJI-IV compact storage ring was successfully reduced from 20 down to less than 2 Ω by replacing vacuum chambers. This effectively suppressed the microwave instability and led to a drastic gain enhancement up to $\sim 9\%$. Through such a successful gain enhancement and an improved performance of the cavity mirrors, the NIJI-VI FEL has reached the entrance of the VUV range. Since a gain of at least 10% or hopefully more than 20% is obtainable in higher current regions, we can expect to obtain lasing below 190 nm. Though the average FEL power is 100 μW per two ports at present, one-order increase up to 1 mW level is achievable by increasing the cavity-mirror transmission. Further increase up to 10 mW level will also be possible by raising the beam energy in the next phase upgrade of the NIJI-IV. To obtain FELs at much shorter wavelengths in the extreme ultraviolet (EUV) range, shorter than 100 nm, we are also studying a feasibility of the intra-cavity coherent harmonic generation scheme in the NIJI-IV system [10,11]. We are sure that the NIJI-IV FEL will become a high-quality VUV coherent light source for various research fields, especially in fine surface measurement.

Acknowledgements

We would like to thank Dr. K. Watanabe for useful discussion about the PEEM measurement. This work was supported by the Budget for Nuclear Research of the Ministry of Education, Culture, Sports, Science and Technology, based on the screening and counscling by the Atomic Energy Commission.

References

[1] N. Sei, et al., Nucl. Instr. and Meth. A 445 (2000) 437.
[2] N. Sei, et al., Nucl. Instr. and Meth. A 429 (1999) 185.
[3] K. Yamada, et al., Nucl. Instr. and Meth. A 475 (2001) 205.
[4] M. Trovo, et al., Nucl. Instr. and Meth. A 483 (2002) 157.

[5] M. Furman, J. Byrd, S. Chattopadhyay, Beam instabilities, in: H. Winick (Ed.), Synchrotron Radiation Sources—A Primer, World Scientific, Singapore, 1994, pp. 306–343.

[6] B. Baasner, et al., Nucl. Instr. and Meth. A 331 (1993) 163.

[7] K. Yamada, et al., Proceedings of the 24th International FEL Conference, Argonne, IL, 2002, II-41.

[8] D.A.G. Deacon, et al., IEEE Trans. Nucl. Sci. NS-28 (1981) 3142.

[9] K. Yamada, et al., Nucl. Instr. Meth. A 483 (2002) 162.

[10] V.N. Litvinenko, Nucl. Instr. and Meth. A 507 (2003) 265.

[11] H. Ogawa, et al., Abstract book of 25th International FEL Conference, Tsukuba, Japan, 2003, We-P-40.

Available online at www.sciencedirect.com

SCIENCE  DIRECT°

Nuclear Instruments and Methods in Physics Research A 528 (2004) 273–277

**NUCLEAR
INSTRUMENTS
& METHODS
IN PHYSICS
RESEARCH**
Section A

www.elsevier.com/locate/nima

Chaotic nature of the super-ACO FEL

C. Bruni[a,b,*], D. Garzella[a,b], G.L. Orlandi[a,b], M.E. Couprie[a,b]

[a] CEA/DRECAM/SPAM, bât 522, 91 191 Gif-sur-Yvette, France
[b] LURE, CEA Université Paris Sud, bât 209D BP 34, 91 898 Orsay, France

Abstract

The response of the Free Electron Laser (FEL) to an external perturbation can be studied by applying a laser gain modulation. When its frequency is close to the laser resonance one, a non-linear gain variation of the Super-ACO storage ring FEL can lead to a chaotic dynamic behaviour of the laser macro-temporal structure. At a fixed frequency of the modulation, and increasing its amplitude, the systematic analysis of the macro-temporal structure of the laser, shows that chaos appears between two periodic regimes. The attractors, i.e. the phase-space representation of the macro-temporal structure, characterize the regime as periodic, quasi-periodic, or chaotic. The bifurcation scheme evolves with the frequency of the modulation. The experimental macro-temporal sequences can be reproduced by LAS the numerical empirical code developed in the Super-ACO FEL group. Here the comparison between the experimental and numerical results obtained on the non-linear dynamics of the Super-ACO FEL is presented.
© 2004 Elsevier B.V. All rights reserved.

PACS: 41.60.Cr; 29.20.Dh; 42.65.Sf; 05.45.Tp; 05.45.Pq

Keywords: FEL Chaos; Gain modulation; Attractors; Lyapunov exponents

1. Introduction

The storage ring FEL dynamics acts on different time scales. A perturbation with a characteristic time of the order of the laser rise time ($10–60\,\mu s$) would prevent the laser from starting. The electron beam coherent synchrotron oscillations [1] ($35–70\,\mu s$) modify the FEL gain at a time scale of the order of its rising time and affect its stability. The

FEL is also sensitive to perturbations close to its natural resonant frequency f_r ($2–5$ ms). A modulation of the laser gain around f_r allows the chaotic dynamics of the laser to be underlined. First studies on chaos on the ACO and Super-ACO storage ring FEL showed that the laser intensity macrotemporal structure follows the laws of deterministic chaos [2]. The theoretical and experimental responses of the storage ring FEL to a detuning modulation has been performed [3]. Further theoretical investigation carried out with a simplified model allowed the Lyapunov characteristic exponents to be determined [4]. First the experimental gain modulation and the measurable

*Corresponding author. Université de Paris Sud, BP 34, LURE Bt 209D, 91 898 Orsay Cedex, France. Tel.: +33-1-64-46-80-96; fax: +33-1-64-46-41-48.
E-mail address: christelle.bruni@lure.u-psud.fr (C. Bruni).

0168-9002/$ - see front matter © 2004 Elsevier B.V. All rights reserved.
doi:10.1016/j.nima.2004.04.069

variables of the laser are introduced. Then, the deterministic character of the chaotic dynamics observed on the Super-ACO FEL is pointed out. Finally, the chaotic regime is presented in terms of phase-space diagram with its calculated embedding dimension (phase-space dimension).

2. Experimental set-up

2.1. Gain modulation

The Super-ACO FEL can present a chaotic answer to a non-linear modulation of its gain. The gain depends on the detuning between the pass frequency of the electrons in the optical klystron and the frequency of back and return of the light pulse in the optical cavity. For a Gaussian longitudinal distribution of the electron bunches in the storage ring, the detuning and the gain are linked by the following relation:

$$g(s) = g_0 \exp\left(-\frac{(s+\delta)^2}{2\sigma_\tau^2}\right) \qquad (1)$$

g being the gain, s the longitudinal coordinate, g_0 the small initial signal gain, δ the detuning and σ_τ the rms width of the longitudinal distribution of the electron bunches. A sinusoidal modulation of the detuning leads to a non-linear variation of the laser gain versus time t : $\delta(t) = A\sin(2\pi f t) + B$, A and f being, respectively, the amplitude and the frequency of the modulation, and B a constant. Usually, the modulation is applied for the FEL near the perfect tuning ($B = 0$). The detuning is experimentally applied on Super-ACO by changing the frequency of the 100 MHz RF (radio frequency) cavity.

2.2. Measurable variables

The laser intensity $I(t)$ is measured with a photomultiplier. $X(t)$, the centre of mass of the laser intensity distribution with respect to the longitudinal distribution of the bunch one is measured using the longitudinal feedback [5] in open loop. It measures $X(t)$ with respect to the one at zero detuning and compensates the difference by changing the RF frequency. $I(t)$ and $X(t)$,

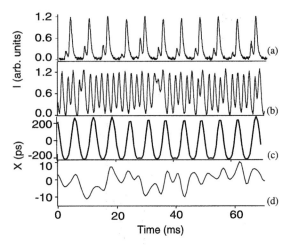

Fig. 1. (a) $2T$ periodic regime, $f = 320$ Hz, $A = 100$ Hz (b) chaotic regime, $f = 520$ Hz, $A = 10$ Hz: $I(t)$ oscilloscope record (for figure clarity there's only 25 points), (c) $f = 320$ Hz, $A = 100$ Hz, (d) $f = 520$ Hz, $A = 10$ Hz: $X(t)$ double sweep streak camera image analysis.

which composes the time series, are recorded by an oscilloscope and sampled at 50,000 points on 2 s (see Figs. 1a,b). The analysis of the double sweep streak camera images [6] gives these variables too on a maximum time scale of 1 s (see Figs. 1c,d).

3. Deterministic chaos

3.1. Time series and frequency spectrum

Besides, the intensity is no more periodic, the laser intensity changes on every macropulse in Fig. 1b contrarily to a periodic regime of Fig. 1a. In Fig. 1a, the maximum of the macropulse intensity I_{max} fluctuates about 10%, and its minimum I_{min} is always zero. On the other hand, in Fig. 1b, I_{max} changes to about 35%, and I_{min} to about 60%. These features (aperiodicity and intensity fluctuations) are characteristics of a chaotic regime.

The aperiodicity of a time series can be further analysed in the Fourier space. Fig. 2b shows the frequency spectrum rise up for a chaotic regime [7], in comparison to a limit cycle regime (see Fig. 2a). As the chaoticity is forced by a periodic modulation, the frequency of the modulation dominates and some subharmonics can appear.

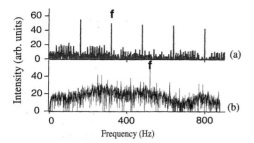

Fig. 2. Fast Fourier transform of (a) Fig. 1a, (b) Fig. 1b.

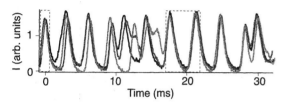

Fig. 3. $I(t)$ for three trajectories with a close initial condition, $f = 320$ Hz, $A = 198$ Hz.

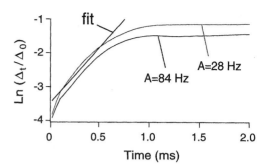

Fig. 4. Calculation of $\ln(\Delta_t/\Delta_0)$ versus time for experimental time series for $f = 320$ Hz. The numerical code used is described in Ref. [9]. The fit is done on the linear part of calculations.

3.2. Lyapunov exponents

Fig. 3 presents three intensity series, where the macropulses, shown in the squares, are identical, and then differs both in period and intensity. Two close initial conditions (e.g. $I(t_1) \approx I(t_2)$), whose difference in intensity is $\Delta_0(I(t_1) \approx I(t_2))$, can be considered as a small perturbation. If it $(\Delta_t \approx I(t_1 + t)) - I(t_2 + t))$ grows exponentially on time, one can find the maximum Lyapunov exponent λ as: $|\Delta_t| \approx \Delta_0 \exp(\lambda t)$, which characterizes the divergence of the trajectories in the phase-space [8]. It illustrates the exponential growth versus time of the difference of intensity between nearby trajectories.

Different numerical computations for obtaining the maximum Lyapunov exponent of a time series exist. Fig. 4, deduced from the method described in Ref. [10], represents the exponential growth in time of the difference of nearby trajectories. It is characterized by a linear increase, whose slope is the maximum Lyapunov exponent of $3.38\,\mathrm{ms}^{-1}$. This positive Lyapunov exponent, estimated from the Super-ACO FEL time series, is a signature of its chaotic deterministic nature.

4. Phase-space representation

4.1. Las model

The numerical model, LAS [3], reproduces the dynamics pass per pass in the optical klystron of the laser intensity versus the detuning. It is based on the evolution of the laser intensity distribution and of the energy spread. A third coupled equation of the detuning modulation is added. So the theoretical phase-space has three dimensions. One can access experimentally on the second time scale to the 2-D phase-space $I(t)$, $\delta(t)$, but not directly to the third dimension given by the energy spread.

4.2. Delay coordinates

The experimental phase-space is built with the delay coordinates [11]. The phase-space is given by the independent coordinates $[I(t), I(t+\delta\tau), I(t+2\delta\tau), \ldots, I(t+m\delta\tau)]$, where $(m-1)$ is called the embedding dimension, and $\delta\tau$ the delay. For a better phase-space representation, the autocorrelation function[1] of the time series is calculated [12]. For a high autocorrelation function, $I(t+\delta\tau)$ can

---

[1] The autocorrelation function is defined as follows:

$$C(\delta\tau) = \frac{1}{N - \delta\tau/\Delta t} \frac{\sum_{i=0}^{N-1}(I(t_i) - \langle I \rangle)(I(t_i + \delta\tau) - \langle I \rangle)}{s^2(I)}$$

with N the number of data points, $\delta\tau$ the delay, $\langle I \rangle$ the intensity average, s its standard deviation, Δt the sampling time.

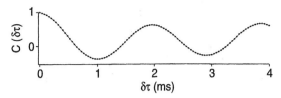

Fig. 5. Autocorrelation function versus delay. The optimized delay for the phase–space diagram of $I(t)$ in Fig. 1b is 0.64 ms.

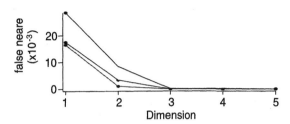

Fig. 6. Number of false nearest versus the dimension for three experimental time series $I(t), f = 320$ Hz.

be deduced from $I(t)$, and they are independent for a small one (see. Fig. 5).

4.3. Embedding dimension

The phase-space dimension of a time series can be determined by the false nearest method [13]. A data point $I(t_1)$ is the nearest of $I(t_2)$ if the distance R_i between these two points does not exceed a heuristic threshold R_t. If the fraction of points for which $R_i > R_t$ is zero, or at least sufficiently small, then the embedding dimension is considered to be high enough. The method consists in calculating the number of nearest n_1 in the 1D-space $I(t)$, and n_2 in the 2D-space $I(t), I(t + \delta\tau)$. The number of false nearest in the 2D-space is n_1-n_2. The number of false nearest falls to zero for the embedding dimension of the time series. Fig. 6 shows that the embedding dimension for the Super-ACO FEL is three: $I(t), I(t + \delta\tau), I(t + 2\delta\tau)$. After the calculation of the delay, one can reconstruct the phase-space diagram with the appropriate embedding dimension.

Fig. 7 compares the experimental and theoretical intensity $I(t)$ and position $X(t)$ phase-space diagram. Fig. 7a1 exhibits an oyster-shaped surface. It has been realized using an experimental

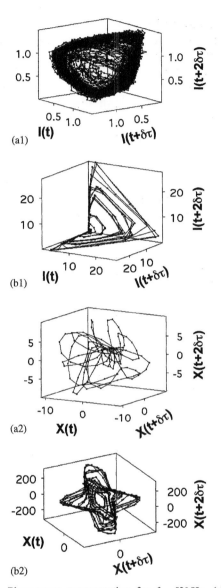

Fig. 7. Phase-space representation for $f = 520$ Hz, (a) $A = 10$ Hz, experimental (a1) $I(t)$, $\delta\tau = 0.64$ ms, (a2) $X(t)$, $\delta\tau = 6.9$ ms (b) $A = 70$ Hz, theoretical (b1) $I(t)$, $\delta\tau = 0.12$ ms, (b2) $X(t)$, $\delta\tau = 0.48$ ms.

time series recorded on a digital oscilloscope (50,000 samples on 2 s). The topological character of a chaotic regime is verified, in a 3-D space: the surface [7] does not fill all the space. The theoretical attractor of Fig. 7b1, reconstructed with 16,500 points on 60 ms, is in agreement with the experimental one (see. Fig. 7a1), presenting also an oyster-like surface. The theoretical attractor of the position $X(t)$ looks like a symmetrical star with four branches. The surface is not exactly filled because of the number of iterations used (16,500 points on 60 ms), which is a compromise with the time of calculation. The corresponding experimental attractor in Fig. 7a2 is limited by the number of points of the time series obtained with a double sweep streak camera (128 points on 70 ms). It seems to correspond qualitatively to Fig. 7b2, even though the very limited number of points makes this analysis very imprecise.

The phase-space diagrams of the theoretical time series are in qualitative agreement with the experimental one. The LAS code reproduces qualitatively the longitudinal dynamics versus detuning observed at Super-ACO.

5. Conclusion

In this paper, it has been demonstrated that the Super-ACO FEL follows the deterministic chaotic dynamics. This means that some control [14] could be applied to such a FEL, which would present some undesired chaotic dynamics resulting from a perturbation on the resonant laser frequency time scale. This deterministic character has allowed to control the pulsed zones on the Super-ACO detuning curve [15]. There is good agreement between the experimental and theoretical phase-space representation. Further investigations would be done with a topological approach [16].

References

[1] M.E. Couprie, et al., Nucl. Instr. and Meth. A 429 (1999) 165.
[2] M. Billardon, Phys. Rev. Lett. 65 (1990) 713.
[3] G. De Ninno, et al., Eur. Phys. J. D 22 (2003) 269.
[4] W. Wang, et al., Phys. Rev. E 51-1 (1995) 653.
[5] M.E. Couprie, et al., Nucl. Instr. and Meth. A 258 (1995) 374.
[6] R. Roux, et al., Nucl. Instr. and Meth. A 393 (1997) 33.
[7] D. Dangoisse, et al., Phys. Rev. A 36 (1987) 4775.
[8] G. Benettin, et al., Phys. Rev. A 14 (1976) 2338.
[9] H. Kantz, T. Schreiber, Cambridge University Press, Cambridge, 1997.
[10] H. Kantz, Phys. Rev. A 185 (1994) 77.
[11] J.D. Farmer, Physica D4 (1982) 336.
[12] P. Grassberger, et al., Int. J. Bifurcation Chaos 1 (1991) 521.
[13] M.B. Kennel, et al., Phys. Rev. A 45 (1992) 3403.
[14] E. Ott, et al., Phys. Rev. Lett. 64 (1990) 1196.
[15] S. Bielawski, et al., Phys. Rev. E 69 (2004) R045502;
     M.E. Couprie, et al., Nucl. Instr. and Meth. A, (2004) these Proceedings.
[16] C. Letellier, et al., Phys. Rev. E 52-5 (1995) 4754.

Available online at www.sciencedirect.com

ELSEVIER   Nuclear Instruments and Methods in Physics Research A 528 (2004) 278–282

NUCLEAR
INSTRUMENTS
& METHODS
IN PHYSICS
RESEARCH
Section A

www.elsevier.com/locate/nima

Q-switching regime of the ELETTRA storage-ring free-electron laser

G. De Ninno*, M. Trovò, M. Danailov, M. Marsi, B. Diviacco

Sincrotrone Trieste, Area Science Park, 34012 Basovizza, Trieste, Italy

Abstract

Different methods can be used to concentrate the power generated by a storage-ring free-electron laser into a series of giant pulses. One of these techniques, Q-switching, is based on a periodically induced detuning between the electron bunches circulating into the ring and the light pulse stored in the optical cavity. Such a method has been recently implemented at ELETTRA. This new regime significantly improves the peak power and the reproducibility of laser signal making the source more reliable. A full characterization of such an operation mode is presented in this paper.
© 2004 Published by Elsevier B.V.

PACS: 29.20.Dh; 41.60.Cr

Keywords: Storage ring; Free electron laser

1. Introduction

The maximum average lasing power that can be achieved using a Storage-Ring Free-Electron Laser (SRFEL) is limited by the heating of the electron beam induced by the laser onset [1,2]. The increase of the electron-beam energy spread is indeed responsible for the diminution of the optical gain while, at saturation, the latter reaches the level of the optical cavity losses. However, for applications requiring a high peak power, the FEL power can be "concentrated" into a series of giant pulses. In this case, the peak power is considerably

enhanced (one to few orders of magnitude) while the average power is only slightly reduced [3].

The giant-pulse operation on a SRFEL can be obtained by means of a radio-frequency (RF) modulation, the so-called Q-switching, or using a gain-switching technique [3]. The first method, currently used at Super ACO [4], UVSOR [5] and NIJI-IV [6], and recently implemented at ELETTRA, is based on a periodically induced longitudinal detuning between the electrons circulating into the ring and the light pulse stored in the optical cavity. When the system is detuned, the lasing process is stopped. Maintaining the detuned condition for a time long enough allows the electron beam to cool down and the gain of the amplification process to recover its initial (i.e. laser-off) maximum value. Once this situation is reached, the system is led back to the perfect

*Corresponding author. Tel.: +39-040-375-8008; fax: +39-040-375-8565.

E-mail address: giovanni.deninno@elettra.trieste.it (G. De Ninno).

0168-9002/$ - see front matter © 2004 Published by Elsevier B.V.
doi:10.1016/j.nima.2004.04.070

tuning condition, which is maintained for a period of few milliseconds (typically) so to induce the giant pulse onset. Then, the system is detuned again and the process repeated.

As it is shown in the following, the implementation of this technique at ELETTRA has led to a significant improvement of the stability and reproducibility of the laser signal over the standard operation mode. To achieve this result a systematic study has been carried out of both the electron-beam sensitivity to the choice of the modulation parameters and the correlated source performance.

2. A significant improvement of the ELETTRA FEL performance

Fig. 1 shows an example of the different macro temporal structures displayed by the ELETTRA FEL when operated in standard (Fig. 1a) or Q-switching (Fig. 1b) mode. The aperiodic character of the "natural" regime is a direct consequence of the instabilities perturbing the electron beam dynamics. The origin of these instabilities can be traced back both to the electron's interaction with the ring environment (microwave instability) and to external perturbations, such as mechanical vibrations and line-induced modulations. In particular, the laser dynamics shows to be particularly reactive to the systematic presence of a 50 Hz perturbation (and its harmonics), whose origin is presently under investigation. This slow instability may manifest in different non-predictable ways, e.g. as a kick or as a continuous modulation, and it

can be at the origin of fast synchrotron-like instabilities [8]. This lack of reproducibility reflects in the noisy temporal structure of the laser signal (see Fig. 1a). It is also worth mentioning that the increased sensitivity to electron-beam instabilities, which is proper to "high-gain" SRFELs with respect to lower-gain devices, is one of the reasons why SRFELs like ELETTRA and DUKE do not show a "cw" regime of the laser intensity around the perfect light-electron beam tuning condition [7,9]. Such a behaviour is instead characteristic of lower-gain devices like Super-ACO [10] and UVSOR [11].

When operated in Q-switched mode, the FEL appears to be much less affected by the imperfect electron-beam stability. Fig. 1b shows the significant improvement of the ELETTRA FEL performance in terms of peak power (enhancement of about a factor 70 with respect to the standard operation mode) and stability (fluctuation of the peak power of the order of only few percents). This achievement represents an important step towards the possibility of fully exploiting the source for pilot user experiments in photoelectron microscopy and time-resolved spectroscopy [12].

3. FEL efficiency versus the modulation parameters

The FEL performance in terms of (peak and average) power and stability when the system is operated in Q-switched regime is strongly correlated with the choice of the modulation parameters, i.e. the repetition rate and slope of the RF jump leading the system back to the perfect

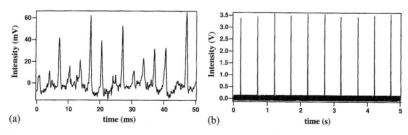

Fig. 1. ELETTRA FEL signal in standard (a) and Q-switched operation mode (b). The experimental setting is the following: beam energy: 900 MeV, beam current (four bunches): $\simeq 10$ mA, laser wavelength: 250 nm, optical gain: $\simeq 7\%$, cavity losses: $\simeq 5\%$. For a complete list of the ELETTRA parameters which are relevant for the FEL operation see, e.g., Ref. [7].

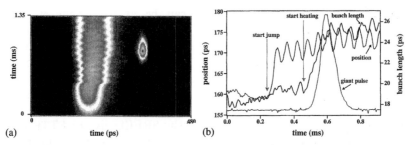

Fig. 2. (a) Streak camera image of the electron beam (left trace) and of the laser macropulse (right trace) in Q-switched operation mode. A horizontal cut provides the longitudinal distributions while on the vertical axis one can follow the evolution in time of the distributions profile. (b) Analysis of (a). Beam current: $\simeq 8$ mA, slope of the RF jump: 40 Hz/μs.

synchronism. These parameters have an influence on the status of the electron beam at the start of successive giant pulses. The electron-beam reaction to the RF jump has been monitored using a double-sweep streak camera [13]. For this purpose, the signal produced by a pulse generator has been used both for driving the RF modulation and as a trigger for the slow sweep of the streak-camera. Fig. 2 shows an image in which are displayed, at the same time, the electron beam (perturbed by the modulation) and the generated giant pulse. One can clearly notice (see Fig. 2b) the main phenomena involved in the process: the excitation of synchrotron oscillations due to the RF jump, the electron-beam heating and the slightly delayed start of the giant pulse.

3.1. Efficiency versus repetition rate

Fig. 3 shows the behaviour of the average power, rise time and temporal stability (with respect to a given reference, see caption for details) of the giant pulse as a function of the repetition rate of the modulation. The comportment of the average power is as qualitatively expected: it grows almost linearly up to a given repetition frequency, above which the time left to the electron beam to cool down between the generation of two successive pulses becomes too short. As a consequence, for higher frequencies the process starts to loose efficiency. A surprising result comes from the threshold value, around 15 Hz, which is much higher than expected. In fact, one could presume the time necessary for the electron beam to cool down to be of the order of few synchrotron

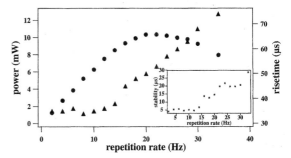

Fig. 3. Average power (circles), rise time (triangles) and stability (see inset) of the giant pulse as a function of the repetition rate of the modulation. The stability has been characterized by measuring the rms jitter of the pulse centroid with respect to the start of the modulation. Beam current: $\simeq 7$ mA, slope of the RF jump: 10 Hz/μs.

damping times, i.e. few hundred milliseconds for the actual experimental setting, and not, as shown in Fig. 3, around 65 ms. The observed phenomenon, the origin of which is under investigation, seems to suggest that the electron-beam cooling between successive giant pulses is driven by a mechanism which is different and faster with respect to the (standard) synchrotron damping. As for the laser rise time, which is inversely proportional to the net optical gain, it stays almost constant while the growth of the average power is linear with the frequency. Starting slightly before threshold, an evident increase is observed, corresponding to a diminution of the net gain. The behaviour of the temporal stability (see inset of Fig. 3) shows that the time jitter between two successive giant pulses is smaller when the system is operated well inside the regime of maximum efficiency. Note the correlation between such a behaviour and that of the rise time.

3.2. Efficiency versus the slope of the RF jump

The most delicate point of the Q-switching technique resides in the process leading the system back to the perfect synchronism. In fact, such a process has to be quick enough to prevent any significant heating of the electron beam before the perfect tuning is re-established. On the other hand, the restoring process has to be not so quick to induce strong synchrotron oscillations, which may both reduce the net gain at the laser start and spoil the stability of the giant pulse to be generated.

A systematic study has been carried out aimed to characterize the amplitude of the induced synchrotron oscillations as a function of the slope of the RF jump. This has been done by making use of a double-sweep streak camera, similarly to what is shown in Fig. 2. As a general result, it has been found that the induced oscillations starts to be significant for slopes steeper than 20 Hz/μs (see Fig. 4). This in practice means that performing a RF jump of 400 Hz (which is the typical RF offset necessary to completely switch off the laser) in about 40 μs (or even slightly less) would not induce significant synchrotron oscillations. Although comparable with the laser rise time (see Fig. 3), such a duration is short enough to prevent a significant heating of the electron beam before the start of the giant pulse. Indeed, as it is shown

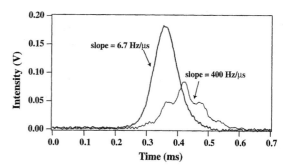

Fig. 5. Giant pulse profile for different slopes of the RF jump. Beam current: ≃ 8 mA; modulation rate: 3 Hz.

in Fig. 2, there is a delay between the RF jump and the start of the beam heating which is of the order of 200 μs. The fact that the best performance in terms of power and stability of the giant pulse is obtained for slopes smaller than 10 Hz/μs is confirmed by the results shown in Figs. 4 and 5. In particular, the curve in Fig. 5 obtained for the larger slope shows how sensitive is the stability of the laser profile (which appears to be modulated at the synchrotron frequency) to the oscillations induced on the electron beam by the RF jump.

4. Conclusions

The implementation of the Q-switching operation mode at ELETTRA allowed a significant improvement of the FEL peak power and reproducibility. A careful study has been carried out of the source characteristics depending on the choice of the RF modulation parameters. As a result, a reliable criterion has been established to control and minimize the electron beam reaction to the modulation and, thus, to achieve the best FEL performance.

Acknowledgements

The support given by EUFELE, a Project funded by the European Commission under FP5 Contract No. HPRI-CT-2001-50025, is acknowledged. It is also a pleasure to thank the accelerator staff at Sincrotrone Trieste (in particular

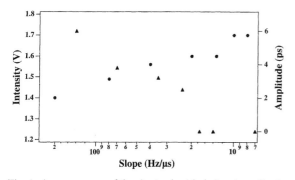

Fig. 4. Average power of the giant pulse (circles) and amplitude of the induced synchrotron oscillations (triangles) as a function of the slope of the RF jump. Beam current: ≃ 10 mA; modulation rate: 3 Hz.

F. Iazzourene and L. Tosi) for continuous support, and M.E. Couprie, L. Giannessi and G. Orlandi for their participation in FEL shifts.

References

[1] N.A. Vinokurov, et al., preprint INP77.59 Novosibirsk, (unpublished) 1977.

[2] A. Renieri, Il Nuovo Cimento 35 (1979) 161.

[3] I.V. Pinayev, et al., Nucl. Instr. and Meth. A 475 (2001) 222.

[4] T. Hara, et al., Nucl. Instr. and Meth. A 341 (1994) 21.

[5] M. Hosaka, et al., Nucl. Instr. and Meth. A 507 (2003) 289.

[6] K. Yamada, et al., Nucl. Instr. and Meth. A 483 (2002) 162.

[7] G. De Ninno, et al., Nucl. Instr. and Meth. A 507 (2003) 274.

[8] G. De Ninno, et al., PAC Conference Proceedings, 2003.

[9] G. De Ninno, et al., ELETTRA Internal Report, ST/SL-03-01, 2003.

[10] M.E. Couprie, et al., Nucl. Instr. and Meth. A 331 (1993) 37.

[11] H. Hama, et al., Nucl. Instr. and Meth. A 375 (1996) 32.

[12] B. Diviacco, et al., Nucl. Instr. and Meth. A, (2004) these proceedings.

[13] M. Ferianis, M. Danailov, in: G.A. Smith, T. Russo (Eds.), AIP Conference Proceedings, Vol. 648, 2002, BNL, Upton, NY, USA.

Available online at www.sciencedirect.com

SCIENCE DIRECT®

**NUCLEAR
INSTRUMENTS
& METHODS
IN PHYSICS
RESEARCH**
Section A

ELSEVIER   Nuclear Instruments and Methods in Physics Research A 528 (2004) 283–286

www.elsevier.com/locate/nima

Observation of a giant pulse time structure produced by a storage ring FEL ☆

I.V. Pinayev*, V.N. Litvinenko, K. Chalut, E.C. Longhi, S. Roychowdhury

FEL Laboratory, Duke University, P.O. Box 90319, Durham, NC 27708, USA

Abstract

A fast steering magnet (gain modulator) allows redistribution of free electron laser (FEL) power into a series of high peak power pulses. In this paper, we report experimental study of properties of giant pulses depending on the FEL settings and pulse repetition rate. Localized growth of the electron energy spread, caused by the interaction of the electron beam and the optical pulse, is experimentally demonstrated.
© 2004 Elsevier B.V. All rights reserved.

PACS: 41.60.Cr; 41.85.Ew

Keywords: Free electron laser; Storage ring; Giant pulse; Ok-4/Duke FEL

1. Introduction

Giant pulse generation proved to be a useful tool for storage ring free electron lasers (SRFEL) [1]. It allows redistribution of the optical power into a series of reproducible high-power pulses thus considerably expanding SRFEL capabilities. Initially designed to satisfy user applications demanding high peak power, the technique was utilized for generation of harmonics [2,3] and also for reducing background in the nuclear physics experiments using free electron lasers (FEL) generated γ-ray beam [4].

In addition to the advantages for user operation, giant pulses provide valuable means for measuring FEL performance and studying SRFEL dynamics.

2. FEL power time structure

The experiments described in this paper were performed on the Duke storage ring capable of operating in 250–1200 MeV energy range. It has dedicated straight section for the OK-4/Duke FEL. Measurements were performed with 5–6 mA stored electron beam current at 450 MeV. The lasing wavelength was 710 nm. The value of the stored current was chosen in such a way that the electron beam energy spread induced by the FEL remains within present energy acceptance of the Duke storage ring and this current is much lower than 15 mA which is achievable for the

☆ This work is supported by AFOSR Grant under Contract F49620-10370.

*Corresponding author. Tel.: +1-919-660-2657; fax: +1-919-660-2671..

E-mail address: pinayev@fel.duke.edu (I.V. Pinayev).

single bunch operation. The gain modulator provides displacement of the electron beam in the FEL region away from the optical axis and stops lasing. With the arrival of the trigger pulse the disturbed orbit adiabatically returns to the lasing one and the giant is generated [1].

Part of the extracted optical power is directed to the silicon photodiode with an amplifier. The second-order low-pass filter eliminates the ripples associated with the electron beam revolution period. The output signal is observed on the LeCroy LC574A digital oscilloscope. Because the time constant of the electronics used for diagnostics is much larger than the micropulse duration, the instantaneous value of intensity corresponds to the micropulse energy.

Typical envelope of the giant pulse is shown in Fig. 1. The area under the curve is proportional to a sum of all micropulses and therefore corresponds to the macropulse energy. The giant pulse on the positive slope has exponential growth and from the time constant one can obtain the value of net gain. Net optical cavity losses were estimated from the falling edge time constant and are around 0.5% per roundtrip. Real losses of the optical cavity are obtained directly from the ring-down time of the optical cavity and are about 0.8% per roundtrip. Lower value of the net optical loss may be attributed to the residual non-zero gain of the optical klystron after the giant pulse. Adding net gain and losses of the optical cavity one can obtain the values of the SRFEL gain.

Initially, we measured dependence of the peak micropulse energy on the buncher setting. Increase in the buncher current leads to the initial growth of net gain and enhances micropulse power. However, higher values of N_D makes SRFEL gain more sensitive to the energy spread of the electron beam and therefore, lowers the level of the induced energy spread (and hence the optical power) at which net gain reaches zero. The results shown in Fig. 2 are in agreement with theory [5], with maximum reached at $N_D \sim 20$ which is close to the value from the previous experiments [1].

The dependence of the micropulse energy versus giant pulse repetition rate is shown in Fig. 3. The synchrotron damping time at this energy is 94 ms. The pulse shape and amplitude of giant pulses remain unchanged at repetition rates with period less than 3 synchrotron damping times. So, the average power is directly proportional to the frequency of the giant pulses. Further increase of the repetition rate leads to the "residual" energy spread demonstrating itself in decrease of the initial gain and the maximal micropulse energy.

The dependence of the macropulse energy versus giant pulse repetition rate is shown in Fig. 4. When giant pulse repetition period exceeds two synchrotron damping times, the macropulse energy is not changing. With increase of giant pulse frequency the energy starts declining. Also, it takes more time for the FEL to reach maximal

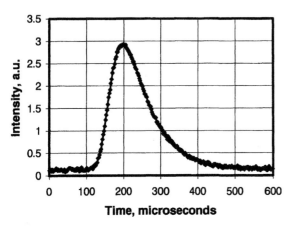

Fig 1. Giant pulse profile.

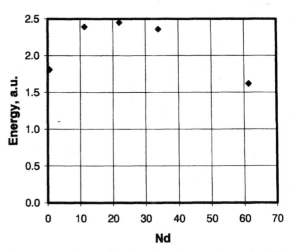

Fig. 2. Dependence of peak micropulse energy on buncher setting.

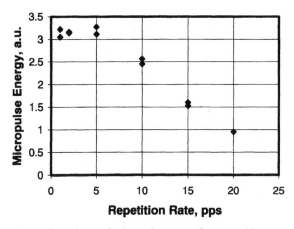

Fig. 3. Dependence of micropulse energy from repetition rate of giant pulses.

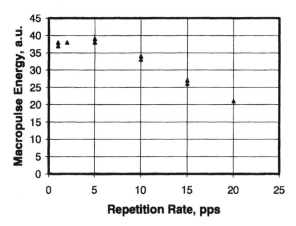

Fig. 4. Dependence of macropulse energy from repetition rate of giant pulses.

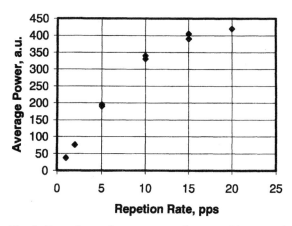

Fig. 5. Dependence of average power from repetition rate of giant pulses.

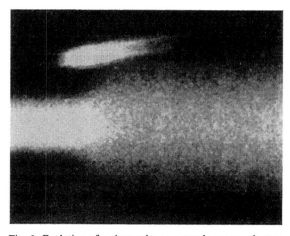

Fig. 6. Evolution of a giant pulse on a streak camera, electron bunch on the bottom, FEL light on the top. The slow horizontal sweep is 500 μs, the fast vertical sweep in 1.3 ns.

peak power and the giant pulses become longer. This explains the steeper decline of the micropulse energy in comparison with macropulse energy.

For many experiments the average power is more significant than peak power. For example, utilizing the giant pulse operation mode with time gate improved signal-to-noise ratio in experiments with γ-rays [4]. The average power continues to grow because the macropulse energy drop is compensated by the increase of the repetition rate (see Fig. 5). It reaches saturation at repetition rate that corresponds to the 0.5–1 damping times. Such unexpected result may be attributed to the fact that FEL induces energy spread, which suppresses microwave instability. Therefore, one of the

sources of the energy spread is eliminated and electron beam is cooled faster.

3. Electron beam phase space dynamics

For the numerical analysis we used code #VUVFEL [6]. The results of the modeling include Poincaré phase–space plot of the electron beam and optical energy in the pulse. Set of phase–space plots at different stages of giant pulse indicates that "heating" of the electron beam takes place in the center of the electron bunch where lasing power is maximal and electrons with energy below

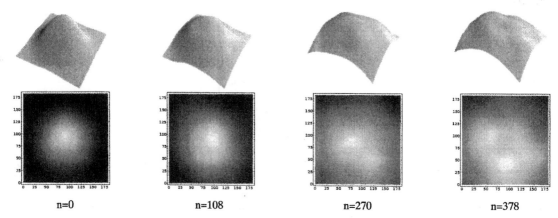

Fig. 7. Evolution of the electron beam profile: 3D plot of distribution functions (top) and the density pictures (bottom). The numbers indicate the quantity of electron bunch revolutions after the start of the giant pulse. Time scale (horizontal) is 0.71 ns, relative energy spread (vertical) $\sigma_E/E = 0.105\%$.

equilibrium are affected. The experiment described below is aimed to observe such a behavior. There is no way to obtain directly the distribution of the electron beam in the energy-time phase space. Fortunately, the synchrotron motion rotates the distribution thus enabling phase–space tomography by observing the evolution of the temporal structure of the electron beam.

The dual sweep streak camera (Hamamatsu, model C5860) provides the sequence of equidistant one-dimensional (time) projections of the longitudinal distribution of the electrons in the two-dimensional phase space. The synchrotron frequency was 25.9 kHz (one synchrotron period is 38.6 μs) and therefore one may neglect the evolution of the phase–space distribution and consider obtained projections as equivalent to the ones observed at different angles. Using Radon transforms [7] we are able to reconstruct the phase space distribution of the electrons.

The giant pulse data from the streak camera operating in the double-sweep mode are shown in Fig. 6. The top trace shows the profile of the FEL light and the lower trace corresponds to the electron beam. There are 640 one-dimensional time projections (dual sweep) of the electron bunch, taken over about 13 synchrotron periods. There are about 50 projections in each synchrotron oscillation. Since the one-dimensional projection functions are symmetric, we need only one-half period's worth of data. The results of the

reconstruction are shown in Fig. 7. As one can see the obtained two-dimensional profiles indicate good agreement with results of modeling; however, further studies and technique development are required.

4. Conclusions

Utilization of the gain modulator provides useful tools for user experiments and SRFEL diagnostic. Giant pulses are well described by the existing theoretical models. For the first time it is demonstrated experimentally that heating of the electron beam with optical radiation occurs locally in the phase space. We plan more detailed study of the electron beam dynamics.

References

[1] I.V. Pinayev, et al., Nucl. Instr. and Meth. A 475 (2001) 222.
[2] V.L. Litvinenko, Nucl. Instr. and Meth. A 507 (2003) 265.
[3] E.C. Longhi, et al., Nucl. Instr. and Meth., A (2003), these Proceedings.
[4] M.W. Ahmed, et al., Nucl. Instr. and Meth. A 516 (2004) 440.
[5] V.L. Litvinenko, et al., Nucl. Instr. and Meth. A 475 (2001) 65.
[6] V.L. Litvinenko, et al., Nucl. Instr. and Meth. A 358 (1995) 334.
[7] S.R. Deans, The Radon Transform and its Applications, Wiley, New York, 1983.

Available online at www.sciencedirect.com

SCIENCE $\bigcirc$ DIRECT°

ELSEVIER Nuclear Instruments and Methods in Physics Research A 528 (2004) 287–290

**NUCLEAR
INSTRUMENTS
& METHODS
IN PHYSICS
RESEARCH**
Section A

www.elsevier.com/locate/nima

Control of the electron beam vertical instability using the super-ACO free electron laser

M.E. Couprie[a,b,*], C. Bruni[a], G.L. Orlandi[a], D. Garzella[a], G. Dattoli[b]

[a] *Service de Photons, Atomes et Molécules, CEA/DSM/DRECAM, bât. 522, 91 191 Gif-sur-Yvette, France Université de Paris-Sud, bat. 209 D, BP 34, 91 898 Orsay cedex, France*
[b] *ENEA, Divisione Fisica Applicata, Centro Ricerche Frascati, Roma, Italy*

Abstract

The operation of the Super-ACO storage ring FEL using a harmonic RF cavity exhibits vertical instabilities, which are generally compensated with an increase of the chromaticity of the operating point. It has been previously reported that the FEL itself can compete, and even damps various types of electron beam instabilities. Here, we show that, when the FEL is operating, the threshold of appearance of the vertical instability is modified and that the FEL enlarges the zone of stable regime of the beam operation. Experiments showing the evolution of the chromaticity allowing the beam to be stable versus current are compared, in the presence of the FEL or for the natural beam operation.
© 2004 Elsevier B.V. All rights reserved.

PACS: 41.60.Cr; 41.75.Fr

Keywords: Storage ring FEL; Head–tail instability; Stability

1. Introduction

Free Electron Laser (FEL) sources result from the interaction of an optical wave and a relativistic electron bunch, leading to an energy exchange between the light pulses and the relativistic electrons. The light amplification is accompanied by an increase in the energy spread of the beam,

---
*Corresponding author. Laboratoire pour l'Utilisation du, Rayonnement Electromagnetique, Universite Paris Sud, Bat. 209D, BP 34, Orsay Cedex 91898, France, Tel.: 33-1-64-46-80-44; fax: +33-1-64-46-41-48.

E-mail address: marie-emmanuelle.couprie@lure.u-psud.fr (M.E. Couprie).

the so-called laser heating. On recirculating based accelerators such as the storage ring, this FEL induced modification of the electron bunch kept on many turns, and then, can compete with the various collective effects affecting the electron beam stability. Generally, the FEL has a stabilising effect on the electron bunch dynamics [1]. Longitudinally, the FEL can compete with different types of instabilities such as microwave instability [2], saw-tooth instability [3], coherent synchrotron oscillations [4]. Transversally, the FEL can also damp instabilities, such as the saw-tooth one [5]. The FEL can even be considered as a perturbation of the electron beam dynamics and

0168-9002/$ - see front matter © 2004 Elsevier B.V. All rights reserved.
doi:10.1016/j.nima.2004.04.072

lifetime such as the ones provided by the various types of instability [6]. The FEL being established, it can even modify the injection current threshold [7]. Here, we examine how the Super-ACO FEL in Orsay can affect the mode coupling instability.

The storage ring Super-ACO in Orsay is operated for the FEL at 800 MeV, with two bunches, with a main RF cavity at 100 MHz or with an additional 500 MHz cavity for bunch shortening [8,9].

2. Super-ACO beam vertical stability

2.1. Super-ACO operation with the main 100 MHz cavity

A current limiting single bunch vertical instability (SBVI) is observed with a threshold of about 30 mA at nearly zero absolute vertical chromaticity ($\xi_z = 0$) increasing to over 120 mA for $\xi_z = +2$. This instability has been attributed to the transverse mode-coupling (TMC) phenomenon [10].

2.2. Super-ACO operation with two rf cavities, at 100 and 500 mhz

A 500 MHz cavity with an excitation voltage ranging up to 300 kV is used on Super-ACO in the bunch shortening mode [9] in order to enhance the FEL gain. Two regimes of a new single bunch vertical instability : NSBVI-I ($v_z = 0.698$) and NSBVI-II ($v_z = 0.794$) can be distinguished, as shown in Fig. 1.

The first regime, the so-called NSBVI-I, presents a threshold close to 10 mA/bunch which is independent of the charge configuration (1 bunch, 2 equidistant bunches, 2 adjacent bunches) or the presence of non-zero dispersion in the harmonic cavity. The beam presents a vertical excitation and a self-excited line appears in the transverse beam spectrum at a frequency slightly above the one of the vertical tune (v_z). This frequency does not correspond to that of a head-tail mode even taking into account the current variation of the frequencies of these modes. The amplitude is not constant but oscillates at a low frequency. The threshold

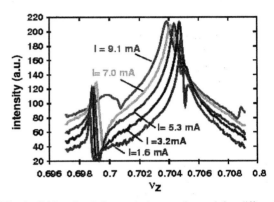

Fig. 1. Sidebands of the vertical tune observed for different beam currents.

current always lies in the 8–10 mA range. It is rather insensitive to various parameters, such as the absolute value of tune, on both sides of the integer, the chromaticity for values between 0 and +2.5, the transverse non-linearity, the harmonic voltage. However, the distance to the coupling resonance (v_z-v_x) appears to influence the intensity of the instability. Very strong values of the chromaticity (of the order of 5) suppresses it but it may result from a change in tune spread due to the very strong sextupole excitation rather than from the chromaticity. Different mechanisms, such as an interaction between the beam and charged particles, a synchro-betatron instability or an instability due to the narrow band resistive wall impedance cannot explain such a phenomenon. This instability is nevertheless strongly affected by the presence of four Super-ACO insertion devices. It disappears when all the undulators are not in place. It might be caused by the effective octupolar component introduced by the insertion devices, that could tend to reduce the overall octupolar component and thus the non-linear tune spread, leading to a lower Landau damping and a greater tendency to instability. Practically, this instability is damped with a very strong increase in vertical chromaticity to the detriment of the dynamic aperture and the beam lifetime.

The second regime, NSBVI-II , appears at a threshold current close to 30 mA/bunch. It is characterised by the self-excitation of the mode head–tail −1 (v_z-v_x), corresponding to the mixing of the 0 and −1 head–tail modes, with vs the synchrotron tune. At a current of 40 mA/bunch, the two lines become indistinguishable and the injection saturates. In this situation, increasing the vertical chromaticity from 2.5 to 4 leads to a sudden separation of the two lines and a damping of the instability giving place to NSBVI-I. The level of the needed chromaticity increase depends nevertheless on the current. These observations seem compatible with the Transverse Mode Coupling mechanism already observed on the Super-ACO in normal mode. The only difference is the need for higher values of the chromaticity in order to calm the instability at a given current.

3. Influence of the FEL on the Super-ACO beam vertical stability

3.1. the Super-ACO FEL

Generally, for the FEL operation with the harmonic RF cavity, the chromaticity is set to a value of 4, leading to a suppression of the vertical instability. For the experimental sessions reported here, the Super-ACO FEL was operated with only the FEL insertion device (SU7) in place, at 350 nm. The minimum current for which the FEL can oscillate is of the order of 10 mA/bunch, for mirror cavity losses of 1%. The FEL can be operated in presence of the NSBVI-II mode of the vertical instability, which corresponds to the Transverse Mode Coupling mechanism. Here, we try to analyse how the FEL can modify the instability threshold itself.

3.2. FEL effect on the instability threshold for Super-ACO operated with the harmonic cavity at 90 kV

Experimentally, the chromaticity is set to a value of 4, which is sufficient to completely annihilate the vertical instability. Then, the chromaticity is reduced step by step, until the instability sidebands appears in the vertical tune. This procedure is applied starting with or without the FEL. The chromaticity difference is then plotted versus current in Fig. 2, for a voltage on

the harmonic cavity of 90 kV. Without laser, when the current decreasing, the value of the chromaticity where the instability starts decreases, but it increases at 12 mA/bunch, which corresponds to coherent synchrotron oscillations whose amplitude depends on the electron phase with respect to the synchronous one [11]. When the FEL is in operation, the threshold appears systematically for smaller values of the chromaticity. It means that the laser modifies the instability threshold and provides a more stable electron beam. The Super-ACO FEL induced heating induces a competition with the microwave instability, so the bunch length remains practically unchanged (as shown in Fig. 3). The suppression of the instability can not be attributed to a change in ring peak current, and results in the interplay between the FEL and the vertical instability.

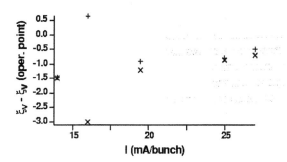

Fig. 2. Chromaticity variation from the usual operating point corresponding to the threshold of the instability for the laser off (+) and on (x) with the 500 MHz cavity at 90 kV. cavity losses : 1.12%.

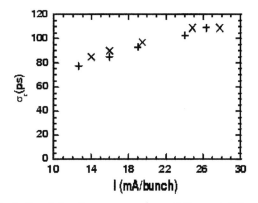

Fig. 3. Bunch length measurements : x FEL on, + FEL off, with the 500 MHz cavity at 90 kV. cavity losses: 1.12%.

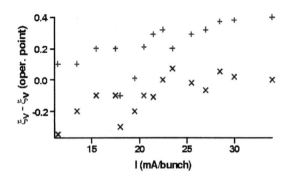

Fig. 4. Chromaticity variation from the usual operating point corresponding to the threshold of the instability for the laser off (+) and on (x) with the 500 MHz cavity at 120 kV.cavity losses : 1.12%

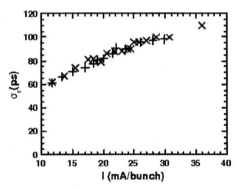

Fig. 5. Bunch length measurements : x FEL on, + FEL off, with the 500 MHz cavity at120 kV.cavity losses: 1.12%.

3.3. FEL effect on the instability threshold for Super-ACO operated with the harmonic cavity at 120 kV

When the voltage of the harmonic cavity is higher, the required chromaticity allowing a stabilisation of NSBVI-II is larger. It is confirmed by the measurements performed without laser, as shown in Fig. 4. Again, we observe an intermediate current zone where the instability is stronger, now in the 24–15 mA / bunch range. Here it also leads to a stabilisation of the instability due to the presence of the FEL. As noted (see Fig. 5), there is no significant bunch lengthening, which could affect the peak current and the vertical instability.

4. Conclusion

In consequence, all the above data, mainly taken in the mode coupling regime show that the FEL improves the beam stability and modifies the instability threshold. The FEL can either stabilise the electron beam when a particular instability is established, or it can even modify the instability current threshold. It seems that injecting the beam in presence of the FEL would allow to further push the limit of injected current.

Acknowledgements

The support given by EUFELE, a Project funded by the European Commission under FP5 Contract No. HPRI-CT-2001-50025, is acknowledged.

References

[1] M.E. Couprie, Nuovo Cimento A 112 (5) (1999) 475.
[2] A. Renieri, Nucl. Instr. and Meth. A 375 (1996) 1.
[3] R. Bartolini, G. Dattoli, L. Mezi, A. Renieri, M. Migliorati, M.E. Couprie, R. Roux, G.De. Ninno, Phys. Rev. Lett. 87 (2001) 134801.
[4] M.E. Couprie, et al., Nucl. Instr. and Meth. A 304 (1991) 58.
[5] G. Dattoli, L. Mezi, A. Renieri, M. Migliorati, M.E. Couprie, R. Roux, D. Naturelli, M. Billardon, Phys. Rev. E 58 (1998) 6570.
[6] R. Bartolini, G. Dattoli, L. Gianessi, L. Mezi, A. Renieri, M. Migliorati, C. Bruni, M.E. Couprie, D. Garzella, G.L. Orlandi, Phys. Rev. E, 69 (2004) 036501.
[7] K. Casarin et al., in J. Chew, P. Lucas, S. Web (Eds), PAC 03 p. 2309.
[8] M. Billardon, M.E. Couprie, D. Nutarelli, G. Flynn, P. Marin, R. Roux, M. Sommer, Proceedings of the EPAC '98, Institute of Physics Publishing, Philadelphia, PA, p. 954.
[9] M. Billardon, M.E. Couprie, L. Nahon, D. Nutarelli, B. Visentin, A. Delboulbé, G. Flynn, R. Roux, Proceedings of the EPAC '98, Institute of Physics Publishing Bristol and Philadelphia, p. 670.
[10] M.-P. Level, et. al., EPAC'88, Rome, June 1988.
[11] M.E. Couprie, D. Nutarelli, R. Roux, L. Nahon, B. Visentin, A. Delboulbé, G. Flynn, M. Billardon Nucl. Instr. and Meth. A 407 (1998) 215.

Available online at www.sciencedirect.com

SCIENCE DIRECT®

ELSEVIER Nuclear Instruments and Methods in Physics Research A 528 (2004) 291–295

NUCLEAR INSTRUMENTS & METHODS IN PHYSICS RESEARCH
Section A

www.elsevier.com/locate/nima

Upgrade of the UVSOR storage ring FEL

M. Hosaka[a,*], M. Katoh[a], A. Mochihashi[a], J. Yamazaki[a], K. Hayashi[a], Y. Takashima[b]

[a] *UVSOR Facility, Institute for Molecular Science, Myodaiji, Okazaki 444-8585, Japan*
[b] *Materials Processing Engineering, Nagoya University Graduate School of Engineering, Furo, Chikusa, Nagoya 464-8603, Japan*

Abstract

The UVSOR electron storage ring has been shutdown since April till August 2003 for upgradation of its performance. This upgrade includes high brightness lattice, longer straight sections for in-vacuum undulators and high-quality vacuum system, and it can increase the brightness of synchrotron radiation from undulators with reduced beam emittance. The improved quality of the electron beam is also of great advantage to the storage ring FEL. To further improve the performance of the FEL and deliver shorter wavelength laser to the user, we are considering a project, which can upgrade the FEL system. One of the solutions is to construct a new in-vacuum optical klystron type undulator in the new long straight section, which is optimized for lasing in the deep UV and VUV region.
© 2004 Elsevier B.V. All rights reserved.

PACS: 29.20.Dh; 41.60.Cr

Keywords: Storage ring; Free electron laser

1. Introduction

Small-beam emittance, short bunch, and long straight section for a long undulator are key issues involved in lasing of a storage ring free electron laser in deep UV and VUV regions. The third generation light sources, such as Elettra and Duke meet these requirements and have recently succeeded in lasing at these wavelength regions [1,2]. At the UVSOR storage ring, a second generation VUV light source has been operational since 1983; an upgrade project was started in 2000. The chief aim of this upgrade is to provide users with brighter SR from undulator, to modify the magnetic lattice configuration in order to reduce the beam emittance, and to extend the straight sections for undulators. The transverse emittance is reduced to $\frac{1}{6}$ of the present value and is close to that of third generation light sources. As these changes are very favourable to the SRFEL, we expect successful lasing in shorter wavelength. From April to August, 2003, the storage ring was shut down for the upgrade; most of the elements were replaced with new ones and it was successfully commissioned in July.

*Corresponding author. Tel.: +81-564-55-7401; fax: +81-564-54-7079.
E-mail address: hosaka@ims.ac.jp (M. Hosaka).

0168-9002/$ - see front matter © 2004 Elsevier B.V. All rights reserved.
doi:10.1016/j.nima.2004.04.073

In this paper, we describe the upgraded storage ring (UVSOR-II) and present the expected performance of the SRFEL. We have also described the future upgrade of the storage ring and the FEL system.

2. Upgrade of the storage ring

The magnetic lattice of the UVSOR storage ring was reconstructed to optimize the storage ring for use of undulator radiation. Positions of bending magnets were not changed in order to minimize the influences on the existing beam-lines and the ring circumference was also not changed. The quadrupole triplets between the bending magnets were replaced with two quadrupole doublets and a short straight section was introduced between them. The doublets at the longer straight sections were also replaced to maximize the length of the straight sections. The old sextupole magnets were removed, and the sextupole field needed for chromaticity correction was integrated with the quadrupole magnets [3]. As a result, the ring now has eight straight sections; four of them are 4 m long and the remaining four are 1.5 m long.

The optical functions before and after the upgrade are compared in Fig. 1. In the new lattice, the horizontal focusing has increased and the dispersion function has a non-zero value (0.8 m) across all the straight sections. Both contribute to reduce the emittance by a factor of 6 from the previous values: 27.4 nm-rad for SR users (750 MeV) and 17.5 nm-rad for FEL experiment (600 MeV). The vertical betatron function at each straight section is optimized by making it small for installing narrow-gap insertion devices, such as in-vacuum short period undulators. The parameters of the storage ring are summarized in Table 1.

At the end of March 2003, we terminated the operation of UVSOR-I and started the reconstruction work. All the magnets and vacuum components, except for the bending sections, were removed. The bending magnets, RF cavities, injection magnets, and undulators were retained. Next, new components were installed. Most of the vacuum components were replaced with new ones; since their structures are much smoother than the

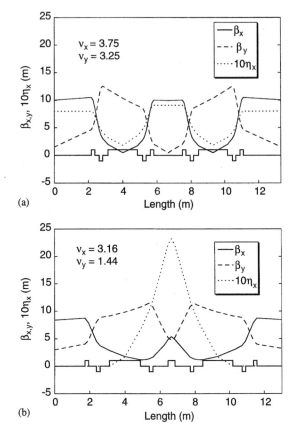

Fig. 1. Lattice functions of (a) UVSOR-II and (b) UVSOR-I.

previous ones, impedance is expected to be reduced considerably. The entire reconstruction work was completed by the end of June. All the vacuum components were baked in the first week of July. We started commissioning in the second week of July under the operating conditions in which the betatron tunes and the optical functions were very similar to those before the reconstruction had started. On 14th July, we succeeded in storing the beam in the storage ring. One week later, the maximum beam current reached 500 mA.

On 30th July, we succeeded in storing the electron beam in the low emittance mode. We faced no difficulties in injection and storage. We measured the horizontal and vertical betatron tunes and the dispersion function. The measured values were consistent with the designed ones and this result confirmed that the design goal was achieved.

Table 1
Parameters of UVSOR- I and II for the FEL experiment

	UVSOR-I	UVSOR-II
Electron energy (MeV)	600	600
Circumference (m)	53.2	53.2
Number of super-periods	4	4
Straight sections (m^2)	3×4	$4 \times 4, 1.5 \times 4$
Emittance (nm-rad)	106	17.5
Energy spread	3.4×10^{-4}	3.4×10^{-4}
Betatron tunes (v_x, v_y)	(3.16, 1.44)	(3.75, 3.20)
Natural chromaticity (ξ_x, ξ_y)	(−3.4, −2.5)	(−8.1, −7.3)
Momentum compaction factor	0.026	0.028
XY coupling (presumed) (%)	10	10

3. Expected performance of the SRFEL

The reduced beam emittance of UVSOR-II is of great advantage to the FEL because the FEL gain strongly depends on it. For the FEL experiment at the electron energy of 600 MeV, the expected horizontal emittance is 17.5 nm-rad, and the expected FEL gain is more than 2 times higher compared with the previous one. As mentioned previously, in order to reduce the beam emittance we have chosen non-zero dispersion at the straight section where the optical klystron is placed. It is essential to know the effect of this non-zero dispersion. The horizontal beam size at a position, where the betatron function is β_x and the horizontal dispersion is η_x, is given by

$$\sigma_x = \sqrt{\varepsilon_x \beta_x + \eta_x^2 \left(\frac{\sigma_\gamma}{\gamma}\right)^2} \tag{1}$$

where ε_x and σ_γ/γ are the horizontal emittance and

the relative energy spread, respectively. In a storage ring FEL, the energy spread is not constant but increases with power evolution. Therefore, the effect of the non-zero dispersion increases the electron beam size considerably and leads to a reduction in the FEL gain. However, another effect from the non-zero dispersion

increases the FEL gain. Due to finite dispersion, the electrons whose energy is higher or lower than the central energy are removed from the horizontal centre of the electron beam, where the optical wave overlaps with the electron beam. Assuming the Gaussian distribution of beam energy and horizontal position, the distribution function of an electron whose horizontal position is x and relative energy is ε/γ is given by

$$P(x, \varepsilon/\gamma) = \frac{1}{\sqrt{2\pi \varepsilon_x \beta_x}} \frac{1}{\sqrt{2\pi \sigma_\gamma/\gamma}}$$
$$\times \exp\left[-\frac{(\eta_x \varepsilon/\gamma - x)^2}{2\varepsilon_x \beta_x} - \frac{(\varepsilon/\gamma)^2}{2(\sigma_\gamma/\gamma)^2}\right]. \tag{2}$$

Multiplying Eq. (2) by a laser power distribution $\left(\sqrt{(2/\pi)}/w\right)\exp(-2x^2/w^2)$ and integrating it with x, one can deduce the electron energy spread seen by the laser which is given by

$$\sigma_{\gamma\text{eff}}/\gamma = \sqrt{\left(\varepsilon_x^2 \beta_x^2 + (w/2)^2\right)/\left(\eta_x^2 (\sigma_\gamma/\gamma)^2 + \varepsilon_x^2 \beta_x^2 + (w/2)^2\right)} \, \sigma_\gamma/\gamma \tag{3}$$

where w is the waist size of the laser. In an SRFEL using an optical klystron, the dependence of the gain on the energy spread can be expressed by the so-called modulation factor

$$f = \exp\left(-8\pi\left((N + N_d)\sigma_\gamma/\gamma\right)^2\right) \tag{4}$$

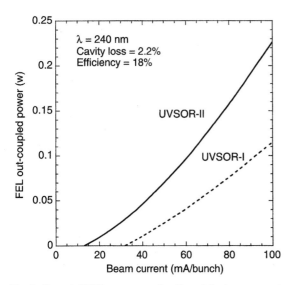

Fig. 2. Expected FEL power as a function of the beam current.

where N is the number of periods in the undulator section and N_d is the number of optical waves passing over an electron in the dispersive section. If there is finite dispersion, the σ_γ/γ in Eq. (3) can be replaced by $\sigma_{\gamma eff}/\gamma$ in Eq. (2) and this leads to enhanced FEL gain. In the case of UVSOR, the combined effects of the gain degradation and enhancement almost seem to cancel out; however, it leads to a slight increase in the threshold current for the lasing (or decrease the initial FEL gain) and in the FEL out-coupled power at the saturation. Taking into account these effects, the expected out-coupled power at a wavelength of 240 nm is calculated and shown in Fig. 2. The value of cavity loss used in the calculation was obtained from a previous experiment. As seen in the figure, high-power lasing of more than 100 mW can be obtained.

4. Future improvement of the storage ring and the FEL

The beam lifetime due to intra-beam scattering (Touscheck lifetime) is short when beam emittance is small because the lifetime is proportional to the bunch volume. At the UVSOR-II, where the emittance is reduced to $\frac{1}{6}$ of its previous value,

the lifetime will be shorter and disadvantageous for SR uses. We therefore, decided to employ the third harmonic cavity to increase the bunch length [4] and increase the XY coupling to enlarge the bunch transverse volume. But the most efficient way to improve the lifetime is to increase the voltage of acceleration of RF cavity. In the present system the voltage is only 46 kV, and it is apparently too low. It is already known that the performance of the presently employed RF cavity is not good enough to be used with high voltage. Hence we are considering upgrading the RF cavity. In our estimate, cavity voltage of 200 kV can be obtained with the present power supply system if the old cavity is replaced by a new one. The high voltage of the RF cavity has a five times longer Touscheck lifetime. This high cavity voltage is also advantageous to the SRFEL. Since the FEL gain is proportional to its electron density in a bunch, the reduced bunch length due to the high cavity voltage will enhance the FEL gain by a factor of 2.

We are considering constructing a new optical klystron for FEL lasing in the VUV region in our future projects for UVSOR-FEL. Since the FEL gain using an optical klystron is proportional to the square of the number of undulator periods, a long optical klystron is favourable for an SRFEL. As mentioned earlier, the straight section has been elongated from 3 to 4 m in UVSOR-II, which allows the installation of a longer optical klystron. However, its length is much smaller than those of the third generation light sources such as Duke and Elettra. Therefore, the new optical klystron should have as many number of periods as possible within the limited length, and also the optical klystron should have sufficient K-value. In addition the polarity should be helical because of high gain and absence of higher harmonics on the axis. These requirements can be fulfilled if one uses an in-vacuum short period helical optical klystron. The narrow gap of the in-vacuum optical klystron allows a sufficiently high K-value. We have successfully installed two in-vacuum undulators at UVSOR; therefore, we have adequate experience in installation.

We began calculations using the code Radia [5] and one of the results is shown in Fig. 3. The

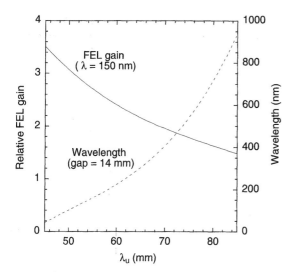

Fig. 3. FEL gain and wavelength at the minimum gap as a function of the undulator period length.

calculation was based on the assumption that the helical undulator was of Apple-2 type. Although the calculation is in progress, the immediate solution is to employ a period of 65 mm to compensate the FEL gain and the longest wavelength at the minimum gap length (14 mm). In this case the expected FEL gain is about 5 times higher as compared to the case using helical optical klystron. With all these improvements, including high cavity voltage and in-vaccum helical optical klystron, the expected FEL gain at a beam current of 20 mA/bunch is 30% at a wavelength of around 200 nm.

References

[1] M. Trovò, et al., Nucl. Instr. and Meth. A 483 (2002) 157.
[2] V.N. Litvinenko, et al., Nucl. Instr. and Meth. A 475 (2001) 195.
[3] J. Yamazaki, et al., UVSOR-29, UVSOR Activity Report 2001, 2002, p. 39.
[4] M. Hosaka, et al., Proceedings of 25th ICFA Advanced Beam Dynamics Workshop, 2002, p. 17.
[5] P. Elleaum, et al., Proceedings of the IEEE Particle Accelerator Conference, Vancouver, BC, Canada, May 1997, p. 3509.

Available online at www.sciencedirect.com

SCIENCE ⓓ DIRECT°

ELSEVIER   Nuclear Instruments and Methods in Physics Research A 528 (2004) 296–300

NUCLEAR
INSTRUMENTS
& METHODS
IN PHYSICS
RESEARCH
Section A

www.elsevier.com/locate/nima

Simulation for spectral evolution of a storage ring free electron laser in a macropulse zone

N. Sei*, K. Yamada, M. Yasumoto, H. Ogawa, T. Mikado

Photonics Research Institute, National Institute of Advanced Industrial Science and Technology, 1-1-1 Umezono, Tsukuba, Ibaraki 305-8568, Japan

Abstract

A computer simulation for evolution of a storage ring free electron laser (SRFEL) micropulse in a macropulse zone has been performed. The one-dimension simulation based on the bunch heating theory could describe dependence of the macropulse period on the longitudinal detuning length and evolution of the pulse width. We modified the simulation code so as to describe the evolution of the SRFEL spectrum. It can explain the characteristics of the NIJI-IV SRFEL spectrum in the macropulse zone. The evolution of the spectral width and wavelength in the macropulse zone is discussed.

© 2004 Elsevier B.V. All rights reserved.

PACS: 41.60.Cr

Keywords: Free electron laser; Storage ring; Spectral width; Macropulse zone; Micropulse

1. Introduction

It is recognized that longitudinal detuning between an electron bunch and an optical pulse forms macrotemporal structure of a storage ring free electron laser (SRFEL) [1]. When the detuning is comparatively small (∼1 μm), the SRFEL has stable macropulses. Evolution of the SRFEL intensity can be described by one-dimension finite difference equations based on a theory of the bunch heating [2]. The simulation with using the finite difference equations also explains the characteristics of the pulse width in the macropulse zone [3]. However, the evolution of the SRFEL spectrum has not been investigated much. Then we improve the one-dimension finite difference equations to consider the temporal characteristics of the SRFEL spectrum in the macropulse zone. The improved simulation can qualitatively describe the characteristics of the evolution of the SRFEL spectrum. In this article, we explain the improved finite difference equations. Then, the numerical simulations are compared with measurements performed on the NIJI-IV FEL.

2. Basic equations for the SRFEL evolution

The evolution of the SRFEL micropulse is explained by a theory of the bunch heating [4].

---
*Corresponding author. Tel.: +81-29-861-5680; fax: +81-29-861-5683.

E-mail address: sei.n@aist.go.jp (N. Sei).

0168-9002/$ - see front matter © 2004 Elsevier B.V. All rights reserved.
doi:10.1016/j.nima.2004.04.074

According to this theory, energy spread of an electron bunch σ_γ is given by the following equation as a function of the SRFEL intensity I_F:

$$\frac{\mathrm{d}\sigma_{\gamma,n}^2}{\mathrm{d}n} = -2\frac{T_0}{\tau_s}(\sigma_{\gamma,n}^2 - \sigma_{\gamma,i}^2) + \alpha_\sigma I_{F,n} \qquad (1)$$

where T_0 and τ_s are round-trip time of the micropulse and synchrotron dumping time, respectively. The suffix i means the initial state where the electron beam and light pulse do not interact, and the suffix n means the round-trip number. Although the symbol α_σ is a function of the energy spread and the wavelength, we adopt the simplest model and consider it to be a constant which is derived from the saturation condition of SRFEL power [2]. When the macropulse grows up enough, this assumption would be appropriate due to spectral narrowing. We express longitudinal and spectral distribution of I_F as $I_F(\tau, \lambda)$, namely,

$$I_F = \int_{\lambda_R(1-\Delta_\lambda)}^{\lambda_R(1+\Delta_\lambda)} \mathrm{d}\lambda \int_{-T_0/2}^{T_0/2} I_F(\tau, \lambda)\,\mathrm{d}\tau \qquad (2)$$

where λ_R is a fundamental resonant wavelength of an undulator on the axis of the electron-beam trajectory and τ is local time of which period is equal to T_0. We assume that Δ_λ is equal to reciprocal of the number of periods in the undulator. When there is longitudinal detuning between the electron bunch and the SRFEL micropulse, the intensity of the micropulse shifts from the synchronous position. The evolution of the intensity can be described with using evolution of the SRFEL gain distribution $G(\tau, \lambda)$ by the following finite difference equation:

$$\frac{\mathrm{d}I_{F,n}(\tau, \lambda)}{\mathrm{d}n} = [G_n(\tau, \lambda) - l_c(\lambda)]I_{F,n}(\tau + \delta\tau, \lambda) \\ + I_{S,n}(\tau, \lambda) \qquad (3)$$

where $I_S(\tau, \lambda)$ is the intensity distribution of the spontaneous emission on the axis of the electron-beam trajectory, $\delta\tau$ is the detuning length per a round trip divided by the speed of light. Since cavity mirrors causes narrowing of the bandwidth due to exposure to the undulator radiation, the cavity loss $l_c(\lambda)$ strongly depends on the wavelength [5].

The electron bunch and the optical cavity might have periodic vibrations. In this article, we consider the simplest model that the electron bunch vibrates longitudinally in a sine wave as the NIJI-IV FEL system [6]. Then $I_S(\tau, \lambda)$ and $G(\tau, \lambda)$ are given by

$$I_{S,n}(\tau, \lambda) = \frac{I_S(\lambda)}{\sqrt{2\pi}\sigma_{\tau,n}} \exp\left(-\frac{(\tau - \Delta\tau_n)^2}{2\sigma_{\tau,n}^2}\right) \qquad (4)$$

$$G_n(\tau, \lambda) = G_n(\lambda) \exp\left(-\frac{(\tau - \Delta\tau_n)^2}{2\sigma_{\tau,n}^2}\right) \qquad (5)$$

with

$$\Delta\tau_n = A_b \sin(2\pi f_b n T_0) \qquad (6)$$

where f_b and A_b are frequency and amplitude of the vibration of the electron bunch, respectively. To simplify the calculation, the electron-bunch shape is assumed to be Gaussian with the bunch length divided by the speed of light σ_τ. The SRFEL gain spectrum $G(\lambda)$ can be calculated from the intensity spectrum of the spontaneous emission $I_S(\lambda)$.

Calculating the finite difference equations in order of the round trip, the evolution of the micropulse can be described. A mesh size for the time and spectrum should be set to be larger than the Fourier limit in the simulation. The SRFEL spectrum is obtained from the integration of $I_F(\tau, \lambda)$ over τ.

3. Comparison with the spectral evolution in the experiments

The spectral evolution of the micropulse was observed with the NIJI-IV FEL at wavelengths around 300 nm before the improvement of the vacuum chambers [7]. The optimum wavelength at which a loss of the cavity mirrors became the minimum was 305 nm. It was observed that the cavity loss increased except the optimum wavelength as the exposure to the optical klystron radiation increased [5]. The cavity losses at the wavelengths of 305 and 300 nm were estimated to be 0.9 and 1–2% in measurements of the spectral evolution. The electron-beam current was 14–21 mA in the experiments, so that the SRFEL gain was 4.3–4.6%. The synchrotron damping time was about 40 ms. The NIJI-IV electron bunch longitudinally vibrated by the period of 100 Hz and the amplitude was about 4.5 mm. This vibration caused uneven intensity of the macropulse. The characteristics of the NIJI-IV electron beam were fully reported in Ref. [6].

The spectral evolution of the micropulse was measured by a combination of a dual-sweep streak camera with a monochromator [8]. The resolution of the monochromator was 0.04 nm in FWHM and the time resolution of the streak camera in the slow-sweep direction was about 0.1 ms in FWHM. The spectral resolution of the system measured with mercury spectral lines at 313.155 and 313.183 nm was ~0.06 nm in FWHM. Examples of the spectral evolution of the SRFEL micropulse with longitudinal detuning of ~0.8 μm (~5 Hz) are shown in Fig. 1. Because the evaluation of the detuning length was not perfect due to the instable optical cavity, there was a gap in the macropulse periods of the experiment and the simulation. Typical shapes of the SRFEL spectral lines are illustrated in Fig. 2. The spectrum was asymmetry and had a long foot of the pulse to the short wavelength. Such the spectrum was clearly observed in a Q-switch operation [9]. In the cw zone, the spectrum was closed to Gaussian. This fact supports that the SRFEL raises not from a white noise but from resonate light of the spontaneous emission spectrum in the macropulse mode.

Evolution of the spectral width can be evaluated from Fig. 1. Although the shape of the SRFEL spectral line is not Gaussian, we define one standard deviation of a Gaussian fitting to the actual SRFEL spectrum as the spectral width. As shown in Fig. 3, the spectral width started to shorten from the rise of the SRFEL micropulse and kept shortening even if the intensity passed over the peak. The phase to which the spectral width is minimized would depend on the effective gain and the detuning length. The value of the spectral width in the experiment was about 50% larger than that in the simulation. The reason is that two or more peaks grow up in the beginning of the macropulse around the wavelength where the gain is the maximum. However, the simulation can qualitatively describe the evolution of the spectral width in the macropulse zone well.

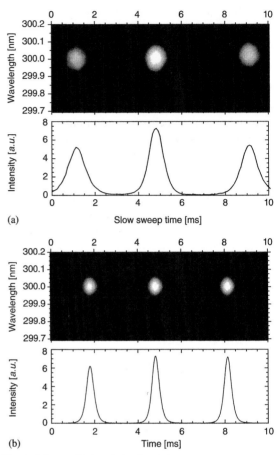

Fig. 1. Measured (a) and simulated (b) spectral evolution with longitudinal detuning of ~0.8 μm at the electron-beam current of ~17 mA. The cavity loss at the wavelength of 300 nm was assumed to be 1.4% in the simulation.

The SRFEL wavelength shifts slightly in the macropulse. It varies periodically but the evolution is complex due to the influence of $I_S(\tau, \lambda)$ and $l_c(\lambda)$. Fig. 4 shows examples of the evolution of the SRFEL wavelength in the macropulse zone. Because signal-to-noise ratio of the streak camera data was low in regions of the low SRFEL intensity, the SRFEL wavelength in the experiment changed irregularly. However, it is noted that the evolution around the largest peak in the experiment was qualitatively similar to that in the simulation. The FEL wavelength has the extrema just before the extrema of the FEL intensity. Generally, it is difficult to measure the shift

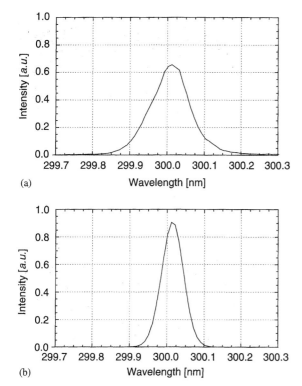

(a)

(b)

Fig. 2. Measured (a) and simulated (b) spectral lines. They are the data on the largest peaks of the SRFEL intensity in Fig. 1.

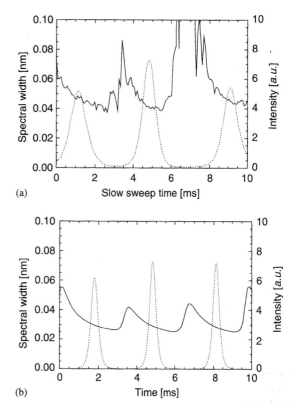

(a)

(b)

Fig. 3. Typical evolution of the spectral widths evaluated from Fig. 1 for the experiment (a) and the simulation (b). The experimental data are evaluated in consideration of the resolution of the measurement system. The FEL intensity is also shown in the dotted lines for the comparison.

accurately because the typical gap of the wavelength is extremely small (~ 0.01 nm). However, we observed a relatively large gap of the wavelength (~ 0.05 nm) in a two-color SRFEL oscillation. In this case, we also observed that the FEL wavelengths mutually shifted in the opposite direction. This fact suggests that the shift is not caused by a change in the electron-beam energy. We have not analyzed the evolution of the two-color SRFEL micropulse in detail yet. We will clarify conditions of the shift in the wavelength and the two-color oscillation with using the simulation.

4. Summary and discussion

The evolution of the SRFEL spectrum was simulated by using the simple one-dimension finite difference equations based on the theory of the bunch heating. This simulation showed the asymmetric shape of the spectral line and the shift of the wavelength in the macropulse zone. It also showed that the spectral width started to shorten from the rise of the SRFEL micropulse and kept shortening even if the intensity passed over the peak. The evolution of the SRFEL spectrum in the experiment was qualitatively in accord with that in the simulations.

Not only the macropulse structure of the intensity but also the evolution of the pulse width and spectral width should be considered in classification of the SRFEL macrotemporal mode [3]. In the case of the NIJI-IV FEL, the intensity was not stable even if the micropulse completely

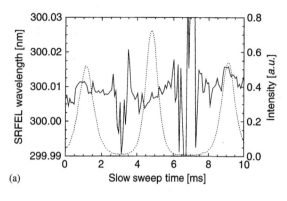

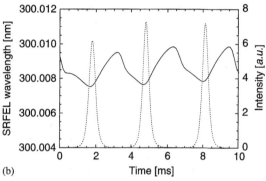

Fig. 4. Evolution of the SRFEL wavelength evaluated from Fig. 1 for the experiment (a) and the simulation (b). The FEL intensity is also shown in the dotted lines for the comparison.

synchronized with the electron bunch. However, we observed that the pulse width and spectral width were almost constant in the perfect tuning [8]. We can consider that such SRFEL oscillations belong to the cw mode. To explain the instable intensity in the perfect tuning, it would be necessary to introduce effects of local behavior of the electrons in the bunch to the simulation [10].

Acknowledgements

This work was supported by the Budget for Nuclear Research of the Ministry of Education, Culture, Sports, Science and Technology of Japan.

References

[1] M.E. Couprie, et al., Nucl. Instr. and Meth. A 331 (1993) 37.
[2] T. Hara, et al., Nucl. Instr. and Meth. A 375 (1996) 67.
[3] N. Sei, et al., Jpn. J. Appl. Phys. 43 (2004) 577.
[4] P. Elleaume, J. Phys. 45 (1984) 997.
[5] K. Yamada, et al., Nucl. Instr. and Meth. A 393 (1997) 44.
[6] N. Sei, et al., Jpn. J. Appl. Phys. 42 (2003) 5848.
[7] H. Ogawa, in preparation for publication.
[8] K. Yamada, et al., Nucl. Instr. and Meth. A 483 (2002) 162.
[9] M. Hosaka, et al., Nucl. Instr. and Meth. A 507 (2003) 289.
[10] G. De Ninno, et al., Phys. Rev. E 65 (2002) 056504.

Available online at www.sciencedirect.com

SCIENCE @ DIRECT®

Nuclear Instruments and Methods in Physics Research A 528 (2004) 301–304

ELSEVIER

NUCLEAR
INSTRUMENTS
& METHODS
IN PHYSICS
RESEARCH
Section A

www.elsevier.com/locate/nima

Simulation study on the coherent harmonic generation at the NIJI-IV FEL with a hole-coupled resonator

H. Ogawa*, K. Yamada, N. Sei, M. Yasumoto, T. Mikado

Photonics Research Institute, National Institute of Advanced Industrial Science and Technology (AIST), 1-1-1 Umezono, Tsukuba, Ibaraki 305-8568, Japan

Abstract

Generation of the VUV radiation has been numerically studied based on coherent harmonic generation (CHG) scheme at NIJI-IV with a hole-coupled resonator. The optical mode of the light pulse inside the optical cavity in a hole-coupled resonator was simulated by using the code GENESIS 1.3 and its modified version including resonator configurations. In this paper, the expected harmonic radiation power for the different configurations of the hole-coupled resonator combined with a 6.3-m optical klystron ETLOK-II is discussed.
© 2004 Elsevier B.V. All rights reserved.

PACS: 41.60.Cr; 52.59.Ye

Keywords: Free-electron laser; Storage ring; Optical klystron; Harmonic generation; Q-switched mode

1. Introduction

Coherent harmonic generation (CHG) is an attractive method as a source of tunable coherent radiation in the vacuum ultraviolet (VUV) range [1–5]. Taking advantage of the CHG scheme which can eliminate the difficulty in obtaining low loss laser cavity in the VUV, we are preparing to produce FELs at the wavelength shorter than 150 nm in the storage ring NIJI-IV FEL [6] by using either an external laser focused into an optical klystron [7] or an FEL oscillator itself [3,8]. In this paper, we focus on the investigation of the

harmonic generation for a Q-switched FEL with a hole-coupled resonator combined with a 6.3-m optical klystron ETLOK-II [6] using a Monte Carlo simulation code.

2. Simulation procedures and results

The time evolution of the fundamental wave by FEL process was numerically calculated by using the code GENESIS 1.3 [9] and its extended version including resonator configurations. The code also takes into account storage ring FEL (SRFEL) saturation mechanism through a change of the beam energy spread. The detail of this simulation is described in Section 2.1. The calculated optical

*Corresponding author. Tel.: +81-29-861-3424; fax: +81-29-861-5683.

E-mail address: ogawa.h@aist.go.jp (H. Ogawa).

0168-9002/$ - see front matter © 2004 Elsevier B.V. All rights reserved.
doi:10.1016/j.nima.2004.04.075

wave and the electron beam are interacted in the first undulator as shown in Fig. 1 and, consequently, the harmonic radiation is produced. The expected intensity of the third harmonic is presented in Section 2.2.

2.1. Procedures and results of simulation for the fundamental wave

The electron beam dynamics and the fundamental at 300 nm in the first undulator were simulated along the axis by using the code GENESIS 1.3 [9], giving the 6D phase space distribution $(x, y, p_x, p_y, \gamma, \theta)$ of the electron beam and radiation field at the exit of the first undulator. The parameters used for the simulation are listed in Table 1 and more detailed ones are described in Ref. [10].

In the dispersive section, the phase shift is induced by the different transit time depending on the difference of the electron energy. We took the phase energy relation [11] between the phase of the electron θ_j and Ψ_j at the entrance and the end of the dispersive section for jth electron, respectively,

$$\Psi_j = \theta_j + 2\pi N_d \left[-1 + 2\left(\frac{\gamma_j - \gamma_0}{\gamma_0}\right) \right]. \tag{1}$$

Here $\gamma_j - \gamma_0$ is the energy deviation from the mean energy of the electron beam in terms of the Lorentz factor, and the parameter N_d represents the average number of optical wavelengths passing over an electron in the dispersive section. In this simulation $N_d = 60$ was taken so as to obtain the maximum third harmonic intensity. Applying the phase distribution at the end of the first undulator to Eq. (1), the 6D phase space distribution $(x, y, p_x, p_y, \gamma, \Psi)$ at the exit of dispersive section

was determined. Thus, the obtained 6D parameters were written into a file as an input to GENESIS 1.3 for calculation in the second undulator and, consequently, we got the 6D parameters and the radiation distribution at the exit of the optical klystron.

In a storage ring the laser induces energy spread during passing through the optical klystron, calculated by the above simulation, accumulates from turn to turn and the energy spread σ_γ decay also occurs due to synchrotron radiation [12]

$$\frac{d\sigma_\gamma^2}{dt} = -\frac{2}{\tau_s}(\sigma_\gamma^2 - \sigma_{\gamma 0}^2). \tag{2}$$

Table 1
Simulation parameters for the NIJI-IV FEL

Electron beam	
Energy	310 MeV
Relative energy spread	3.3×10^{-4}
Revolution frequency	10.1 MHz
Beam current	15 mA
Bunch length	18.9 mm
Beam size	
σ_x	0.85 mm
σ_y	0.23 mm
Optical klystron	
Magnetic period	
Undulator section	72 mm
Dispersive section	216 mm
Number of period	42×2
Total length	6.288 m
K-value	2.03
Optical cavity	
Wavelength λ_L	300 nm
Cavity length	14.8 m
Mirror radius	15 mm
Cavity loss	1.0%

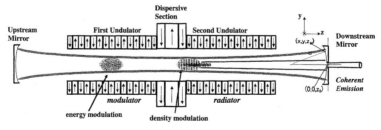

Fig. 1. Schematic view of the setup for CHG scheme with a hole-coupled resonator in the optical klystron.

where τ_s is the synchrotron damping time (40 ms for the NIJI-IV) and $\sigma_{\gamma 0}$ is the rms energy spread without an FEL. At this time the peak current of the electron beam decreases since the electron bunch length σ_l is proportional to energy spread. The resulting σ_γ and peak current in a single turn calculation were used for an initial γ distribution and peak current in the succeeding turn simulation.

As regards a round-trip simulation for the electromagnetic wave, in order to calculate resonator configurations, implementation of free space and mirror, which has been studied by Faatz et al. [13], was performed as follows: the propagation of an electromagnetic wave in the free spaces on both sides of the undulators was simulated by solving the paraxial wave equation

$$\left(\Delta_\perp + 2ik_s\frac{\partial}{\partial z}\right)a_s e^{i\phi} = 0 \tag{3}$$

with making use of existing subroutines. The reflected field $a_{\text{ref}}(x, y, z_0)$ from the mirror was calculated as

$$a_{\text{ref}}(x, y, z_0) \approx a_{\text{inc}}(x, y, z_0)e^{2ik_s z_M} \tag{4}$$

where the $a_{\text{inc}}(x, y, z_0)$ represents the incident field and the surface of a spherical mirror with radius of curvature R_M was parameterized as $z_M(r = \sqrt{x^2 + y^2}) = z_0 - r^2/2R_M$. In this simulation, the reflection of a hole on the mirror axis was assumed a smooth-edge function as $R(r) = 2(r/R_a)^4 - (r/R_a)^8$ because a hole with a sharp-edge would cause numerical problems in calculating the field derivatives [12]. The effective radius of the hole, $R_{\text{eff}} = 0.74R_a$, is defined as the radial position where the reflection R becomes 0.5.

Fig. 2(a) shows the simulation results of the intra-cavity peak power with a 0.16 mm effective diameter hole at 15 mA using the NIJI-IV FEL parameters. The power of $R_M = 8$ m is larger than that of 10 m because the loss of a hole-coupled mirror for $R_M = 8$ m is smaller than that for 10 m due to the difference of waist size on the mirror (the gain of a single pass is almost the same in both R_M in this simulation).

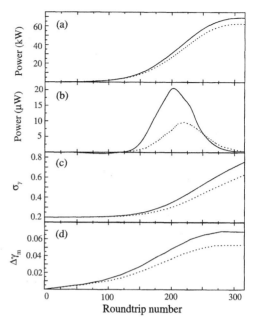

Fig. 2. Fundamental (a) and third harmonic (b) intra-cavity peak power with radius of curvature $R_M = 8$ m (solid line) and 10 m (dotted line), and the corresponding electron energy spread (c) and modulation (d), as a function of the number of round-trips through the optical cavity. The downstream mirror has a hole with an effective diameter of 0.16 mm.

2.2. Harmonic generation

In this study, we made an analytical approach to calculate the harmonic generation. The coherent emission from the number of electron N_e in the straightforward direction at the harmonic n is related to the emission $d^2I_0/d\omega\,d\Omega$ of a single electron [14–16] by

$$\frac{d^2I_n}{d\omega\,d\Omega} = \frac{1}{2}N_e^2 a_n^2 \frac{d^2I_0}{d\omega\,d\Omega} \tag{5}$$

where

$$a_n = 2J_n\left(4\pi n N_d\frac{\Delta\gamma_m}{\gamma}\right)f_n. \tag{6}$$

J_n is Bessel function of the order n and the modulation factor is defined as

$$f_n = \exp\left[-8\pi^2 n^2(N_u + N_d)^2\left(\frac{\sigma_\gamma}{\gamma}\right)^2\right] \tag{7}$$

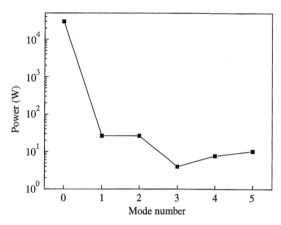

Fig. 3. Optical mode distribution obtained by decomposing the fundamental in the cavity at 200 round-trips. The downstream mirror has a hole with an effective diameter of 0.16 mm.

and the coherent light is emitted into a solid angle $d\Omega = \lambda_L^2/2\pi n^2 \sigma_x \sigma_y$ with the line width $d\omega/\omega = \lambda_L/2n\sqrt{\pi}\sigma_1$ [16].

The energy modulation $\Delta\gamma_m$ induced by optical pulse, which was simulated in Section 2.1, was numerically calculated along the first undulator axis by GENESIS 1.3. As shown in Fig. 2(d), the energy modulation increases with the growth of the fundamental peak power in a Q-switched regime. The third harmonic intra-cavity power calculated from Eq. (5) was plotted in Fig. 2(b). As the laser pulse intensity grows in the cavity, the harmonic production increases, while the harmonic production reaches maximum intensity before the end of the growth of laser pulse due to reduction of the modulation factor by induced energy spread. As a result, maximum peak power, 20 μW, was obtained with a 0.16 mm effective diameter hole.

In order to study transverse optical mode, the fundamental at 200 round-trips where third harmonic became maximum intensity was decomposed by Gauss–Lagueree function. In this situation, the optical mode was found to be mainly fundamental Gaussian and the presence of higher-order modes was much smaller as shown in Fig. 3.

The intra-cavity peak power of third harmonic at 100 nm was also simulated for the different size

of hole. The calculated power of $R_M = 8$ m was 12 and 29 μW at an effective diameter of 0.20 and 0.12 mm, respectively.

3. Conclusion

The harmonic generation at NIJI-IV was numerically investigated by taking into account optical mode of light pulse. The calculated intra-cavity peak power of third harmonic at 100 nm was 20 μW with a 0.16 mm effective diameter hole in the Q-switched FEL. The optical mode inside the optical cavity was mainly fundamental Gaussian with a peak intensity of the third harmonic.

Acknowledgements

This study was financially supported by the Budget for Nuclear Research of the Ministry of Education, Culture, Sports, Science and Technology, based on the screening and counseling by the Atomic Energy Commission, Japan.

References

[1] J.M. Ortega, et al., IEEE J. Quantum Electron. QE-21 (1985) 909.
[2] R. Prazeres, et al., Europhys. Lett. 4 (1987) 817.
[3] V.N. Litvinenko, Nucl. Instr. and Meth. A 507 (2003) 265.
[4] R. Bonifacio, et al., Nucl. Instr. and Meth. A 293 (1990) 627.
[5] H.P. Freund, et al., IEEE J. Quantum Electron. 36 (2000) 275.
[6] T. Yamazaki, et al., Nucl. Instr. and Meth. A 331 (1993) 27.
[7] K. Yamada, et al., Nucl. Instr. and Meth. A 407 (1998) 193.
[8] C. Rippon, et al., Nucl. Instr. and Meth. A 507 (2003) 299.
[9] S. Reiche, Nucl. Instr. and Meth. A 429 (1999) 243.
[10] N. Sei, et al., Jpn. J. Appl. Phys. 42 (2003) 5848.
[11] C.M. Tang, W.P. Marable, Nucl. Instr. and Meth. A 318 (1992) 675.
[12] P. Elleaume, J. Phys. (France) 45 (1984) 997.
[13] B. Fattz, et al., J. Phys. D 26 (1993) 1023.
[14] P. Elleaume, Nucl. Instr. and Meth. A 250 (1986) 220.
[15] J.M. Ortega, Nucl. Instr. and Meth. A 250 (1986) 203.
[16] B. Kincaid, Nucl. Instr. and Meth. A 250 (1986) 212.

Available online at www.sciencedirect.com

NUCLEAR
INSTRUMENTS
& METHODS
IN PHYSICS
RESEARCH
Section A

ELSEVIER   Nuclear Instruments and Methods in Physics Research A 528 (2004) 305–311

www.elsevier.com/locate/nima

Superconducting RF guns for FELs

D. Janssen[a],*, H. Büttig[a], P. Evtushenko[a], U. Lehnert[a], P. Michel[a],
K. Möller[a], C. Schneider[a], J. Stephan[a], J. Teichert[a], S. Kruchkov[b], O. Myskin[b],
A. Tribendis[b], V. Volkov[b], W. Sandner[c], I. Will[c], T. Quast[c], K. Goldammer[d],
F. Marhauser[d], P. Ylä-Oijala[e]

[a] FZ-Rossendorf, Zentralabteilung Strahlungsquelle Elbe, Postfach 510119, Dresden 01314, Germany
[b] Budker Institute of Nuclear Physics, Novosibirsk, Russia
[c] Max-Born-Institut, Berlin, Germany
[d] BESSY, Berlin Germany
[e] Helsinki University of Technology, Helsinki, Finland

Abstract

This paper provides an overview of the advantages and problems of superconducting RF guns. The results of the Rossendorf experiments are presented here. These results are integrated in the design of a new 3.4 cell superconducting RF gun. The beam parameters of this gun correspond to the demands for the new generation of high current, high brightness injectors.
© 2004 Elsevier B.V. All rights reserved.

PACS: 41.60.Cr; 41.75.Fr; 42.55.Xi; 82.25.-j

Keywords: Cavity; Superconductivity; Photocathode; Laser; Electron gun

1. Introduction

In recent years, a number of projects for linacs with high average current and high brightness have been started at different laboratories. These linacs are drivers for free electron lasers (FELs). One has to distinguish two types of linacs in the dependence on the wavelengths of the FELs. The first type is for FELs in the infrared region. In this case, the linac is designed for a small bunch charge but high average current, and a transversal emittance $\varepsilon \leq 10\,\mathrm{mm\,mrad}$ is sufficient. In the second case, the users of FEL demand wavelengths in the X-ray region. Here, a large bunch charge and a small emittance at high energy are necessary, and the average current is smaller as in the first case. Examples for these two types of linacs are given in Table 1.

In contrast to the storage ring accelerators, the beam parameters of the linac are mainly determined by the injector. Table 1 shows the injector parameters that correspond to the two types of linacs. Therefore, the injector determines the

---

*Corresponding author.
E-mail address: d.janssen@fz-rossendorf.de (D. Janssen).

0168-9002/$ - see front matter © 2004 Elsevier B.V. All rights reserved.
doi:10.1016/j.nima.2004.04.076

Table 1
Parameters of linac projects

Linac parameters	
BESSY-FEL	$E = 0.3$–$2.25\,\text{GeV}$, $Q = 1\,\text{nC}$, $I_{av} = 1.25\,\text{mA}$, $\varepsilon = 1.5\,\text{mm mrad}$, $\sigma_z = 265$–$85\,\text{fs}$
TJNAF-FEL	$E = 200\,\text{MeV}$, $Q = 135$–$200\,\text{pC}$, $I_{av} = 5$–$100\,\text{mA}$, $\varepsilon < 10\,\text{mm mrad}$, $\sigma_z = 5\,\text{ps}$
Rossendorf-FEL	$E = 40\,\text{MeV}$, $Q = 77\,\text{pC}$, $I_{av} = 1\,\text{mA}$, $\varepsilon = 1\,\text{mm mrad}$, $\sigma_z = 2\,\text{ps}$
Injector parameters	
1. Mode	$Q = 1\,\text{nC}$, $I_{av} \sim 1\,\text{mA}$, $\varepsilon = 1.5\,\text{mm mrad}$, $\sigma_z = 10\,\text{ps}$, $E = 10\,\text{MeV}$, $\sigma_E = 10\,\text{keV}$
2. Mode	$Q = 80\,\text{pC}$, $I_{av} > 1\,\text{mA}$, $\varepsilon < 10\,\text{mm mrad}$, $\sigma_z = 2\,\text{ps}$, $E = 10\,\text{MeV}$, $\sigma_E = 5\,\text{keV}$

average current, the bunch charge, and the emittance of the final beam. The bunch lengths is not defined by the injector parameters. Using cavities with higher harmonics for linearization of the longitudinal phase space and chicanes for bunch compression, it is possible to shorten the bunch length inside the linac.

2. Status of the development of superconducting RF guns

An overview of high brightness, high average current photoinjectors has been given in Ref. [1]. The results for normal conducting RF guns and DC-SC photo-injectors were last reported in Ref. [2]. Therefore, we will restrict our discussion to superconducting RF guns.

The main components of a superconducting RF gun are the cryostat, the cavity with a special gun cell, the drive laser, and the RF power supply. The laser is synchronized with the RF frequency and determines the time structure of the electron beam.

Some special features of the superconducting RF gun are discussed in greater detail, following the discussion on the high average power and the superconductivity. Drive lasers are widely used for normal conducting RF guns [3]. The additional feature of lasers for superconducting guns is the cw-mode and the high average power of the laser beam. The principle scheme of the laser is shown in Fig. 1.

The time and pulse structure is created at an infrared wavelength ($\lambda = 1047\,\text{nm}$). This wavelength is then converted by two nonlinear crystals to 262 nm. This conversion is the critical

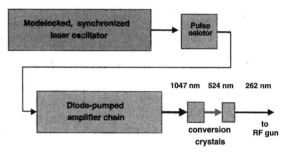

Fig. 1. Scheme of the drive laser.

point of a high average power laser for the following reasons:

- High power of IR pulses enhances the conversion efficiency. If the peak power of the IR pulses is limited to several 100 kW, then the conversion efficiency is only 5–10%.
- Significant absorption of UV radiation in the second conversion crystal heats the crystal and limits the beam quality (thermal lensing).
- The heat can damage the crystal and leads to insufficient long-term stability.

The presently available parameter range (Max Born Institute Berlin, FZR) is as follows: the pulse shape is Gaussian with a pulse duration $\tau \geq 3\,\text{ps}$ in the UV region, the repetition rate varies between 10 and 30 MHz, and the average power is $P_{IR} = 20\,\text{W}$ and $P_{UV} = 0.8$–$1.2\,\text{W}$.

We now discuss the properties of photo-cathodes, which lie inside the superconducting cavity. The illumination by a laser beam could lead to evaporation of particles. These particles could reduce the high-quality factor of the superconducting cavity. The simplest way to avoid this

Fig. 2. Cavity of the Brookhaven gun.

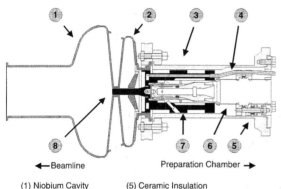

← Beamline   Preparation Chamber →

(1) Niobium Cavity   (5) Ceramic Insulation
(2) Choke Flange Filter   (6) Thermal Insulation
(3) Cooling Insert   (7) 3 Stage Coaxial Filter
(4) Liquid Nitrogen Tube   (8) Cathode Stem

Fig. 3. Cavity with the cathode and cooling insert of the Rossendorf gun.

problem is to consider the back wall of the cavity as the cathode ([4] and Fig. 2).

In this case, the cathode material is niobium, and no contamination occurs. However, the quantum efficiency (QE) of niobium is only about 5×10^{-5}; therefore, for a reasonable laser power, the average electron current is limited to approximately $10 \, \mu A$.

For this reason, we decided to use Cs_2Te photocathodes in the Rossendorf project [5], where the QE value is three orders of magnitude larger. The scheme of the gun cavity is shown in Fig. 3.

The cathode is separated from the cavity by a vacuum gap of 1 mm. A special support structure thermally and electrically isolates the cathode from the surrounding cavity, and a four-step coaxial filter prevents the leakage of RF power. The

cathode itself is maintained at liquid nitrogen temperature by a special cooling insert. For the first time, we have obtained a stable working superconducting RF gun using this cavity. Fig. 4 shows the set-up of the experiment.

The results of the measurements are published in Ref. [5]. At 4.2 K, the quality factor of the cavity, $Q = 2.5 \times 10^8$, remained constant, during a 7-week period of operation. The field strength at the cathode was $22 \, MV/m$, and the electron beam was accelerated to an energy of 900 keV. The bunch charge of 20 pC was limited by the laser power and the low quantum efficiency of the used photocathode. Fig. 5 presents the cathode emission and

Fig. 4. Set up of the Rossendorf experiment.

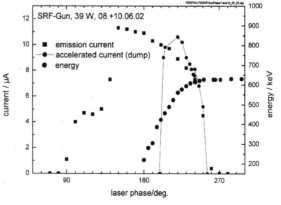

Fig. 5. Energy and current as a function of the laser phase.

accelerated current together with the corresponding electron energy as a function of the laser phase. Complete transmission was obtained within a phase window of 60°. Energy width and transverse emittance were measured for small bunch charges.

These results, especially the constancy of the Q value, has stimulated the start of the development of a new 3.4 cell superconducting RF gun together with a new cryostat specially designed for this gun.

3. Design of a new 3.4 cell superconducting RF gun

The new RF gun should work with a frequency of 1300 MHz as in the first case. In order to reduce space charge effects in the following drift space, an energy of 10 MeV is required. Using the available RF power and main coupler, the beam power is limited to 10 kW, which corresponds to an average current of 1 mA. These parameters can be obtained if the gun cavity consists of three TESLA cells [6] and a specially designed gun cell. The RF field is limited using a condition, wherein the accelerating field strength in the TESLA cells is 25 MV/m, which seems to be a conservative value. With this limit, the maximum surface fields in the TESLA cells are given by $B_S = 0.11$ T and $E_S = 52$ MV/m. Fig. 6 shows the geometry of the new gun cavity.

The parameters of the elliptic arc between the gun cell and the first TESLA cell were optimized to obtain the minimum surface field of the gun cell. As a result, the maximum surface field value B_S of the gun cell agrees with the corresponding value of the TESLA cell; therefore, the quench probability should be the same. In contrast to the normal conducting RF guns, the length of the gun cell is smaller than $\lambda/4$. Due to this, the field strength at the cathode becomes smaller, but the optimum start phase of the electron bunch increases; as a result, the emission field strength $E_{cath} \times \sin\varphi$ also increases. We obtain the length of the gun cell shown in Fig. 7 by calculating the beam properties of the gun for different lengths of the cell. The cone angles of 1° and 6° in the geometry of the gun cell are required for the cleaning procedure of superconducting cavities. Some results of field calculation are presented in Figs. 7 and 8.

In correspondence with Ref. [7], we put the cathode 1.5 mm behind the back wall of the gun cell. This geometry enhances the focusing effect of

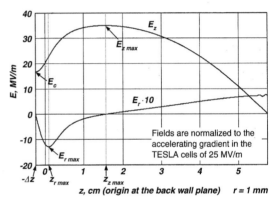

Fig. 7. The RF field inside the gun cell.

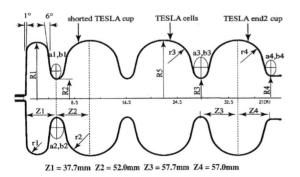

Z1 = 37.7mm Z2 = 52.0mm Z3 = 57.7mm Z4 = 57.0mm

Fig. 6. Geometry of the new 3.4 cell cavity.

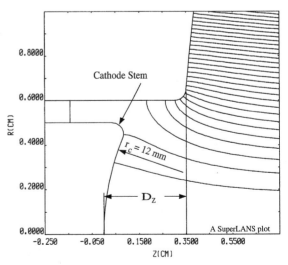

Fig. 8. The RF field near the cathode.

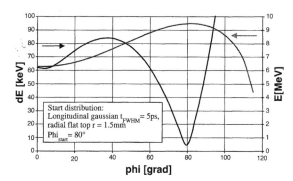

Fig. 9. Dependence of energy and energy spread from the laser phase.

the RF field. In the superconducting case, the RF focusing partially replaces the effect of the static magnetic field applied in normal conducting RF guns.

Calculations with the code MULTIPAC [8] show a probability for multipacting effects in the gap between the cathode and the cavity at low field strengths. This can be suppressed by a DC voltage between the cathode and cavity.

4. Tracking calculation

For bunch charges of 80 pC and 1 nC, we have performed tracking calculations using the ASTRA code, in order to find the beam parameters. The results are presented in Figs. 9–11. Fig. 9 shows the dependence of the energy on the start phase and gives the optimal start value around 80° . This value results from the short length of the gun cell.

Fig. 10 shows the calculated transverse emittance and the beam diameter for a bunch charge of 80 pC, using the start distribution given in Fig. 9, which conforms with the laser parameters of Section 1.

In the tracking calculation, for a bunch charge of 1 nC, we added a solenoid 1 m behind the cathode and additional 8 TESLA cavities for accelerating the bunch to the final energy of

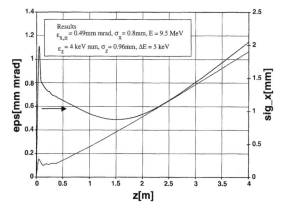

Fig. 10. Results of tracking calculation for $Q = 80$ pC.

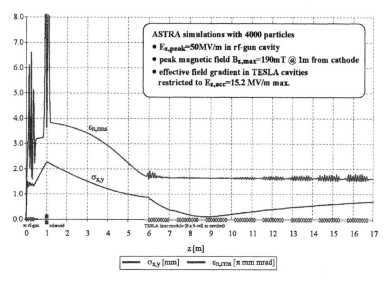

Fig. 11. Emittance and beam radius for $Q = 1$ nC.

Table 2
Calculated beam parameters for a bunch charge of 1 nC

Initial distribution	Comments	Results	
Pulse length 20 ps(flat top) Radial uniform distribution $r = 1.5$ mm	Magnetic field: $B_{max} = 0.19$ T 1 m after the cathode	$\varepsilon_n = 1.7$ mm mrad rms spot size 0.7 mm rms bunch length 4.1 mm	BESSY
Pulse length 20 ps FWHS(Gaussian distribution) Radial uniform distribution $r = 2.8$ mm	No magnetic field $z_{min} = 1.8$ m	$\varepsilon_n = 2.5$ mm mrad rms spot size 1.8 mm rms bunch length 3.5 mm	Rossendorf

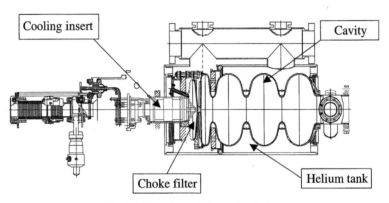

Fig. 12. First draft of the cavity design.

138 MeV. Fig. 11 shows the development of the transverse emittance and the beam diameter for 1 nC in dependence on the distance from the cathode.

The influence of the magnetic field on those values is demonstrated. For an initial distribution, using a flat top profile, a 20 ps length, and a magnetic field of 0.19 T, an emittance of 1.7 mm mrad was obtained. In a second calculation, we have used a Gaussian start distribution of the same length, but without a magnetic field and additional accelerating cavities (see Table 2). The emittance increased to 2.5 mm mrad. Therefore, the flat top profile of the laser pulse together with the external magnetic field reduced the emittance for the bunch charge of 1 nC.

5. Design work

Fig. 12 shows a first draft of the superconducting RF gun.

The cathode holder, cooling insert, and RF-filter are taken from the tested half-cell variant of the RF gun. The helium tank surrounding the cavity is designed for a temperature of 2 K. Due to the special shape and stiffness of the gun cell and the TESLA cells, we need two tuners and two pick up probes for measuring the field amplitudes.

6. Conclusion

After successful experiments with a half cell superconducting RF gun in Rossendorf, we are convinced that the superconducting RF gun is the first candidate for a high current and high brightness injector. Numerical calculations and design work show that a 3.4 cell superconducting RF gun fulfils the demands for an injector in the current FEL projects. After the start of the technical development, we hope to obtain the first beam of the new gun in 2–3 years.

References

[1] S.J. Russel, Nucl Instr., Meth. A 507 (2003) 304.

[2] Zaho Kui, et al., Nucl. Instr. and Meth. A 483 (2002) 125; D.C. Nguyen, et al., Proceedings of the International FEL conference 2003, Nucl. Instr. and Meth. A, (2004) these Proceedings; D.C. Nguyen, et al., Nucl. Instr. and Meth. A (2004), in print.

[3] K.J. Snell, et al., Topical Meeting on Advanced Solid-State Laser, Davos, Switzerland, February 2000.

[4] T. Srinivasan-Rao, et al., MOPB010, PAC 2003.

[5] D. Janssen, et al., Nucl. Instr. and Meth. A 507 (2003) 314.

[6] B. Aune, et al., DESY 00-031, (2000).

[7] D. Janssen, et al., Nucl. Instr. and Meth. A 452 (2000) 34.

[8] P. Yl ä-Oijala, Part. Accel. 63 (1999) 105.

Available online at www.sciencedirect.com

SCIENCE @ DIRECT°

ELSEVIER   Nuclear Instruments and Methods in Physics Research A 528 (2004) 312–315

NUCLEAR
INSTRUMENTS
& METHODS
IN PHYSICS
RESEARCH
Section A

www.elsevier.com/locate/nima

CeB$_6$ electron gun for the soft X-ray FEL project at SPring-8

K. Togawa[a,*], H. Baba[a], K. Onoe[a], T. Inagaki[a], T. Shintake[a], H. Matsumoto[b]

[a] SPring-8/RIKEN Harima Institute, Hyogo 679-5148, Japan
[b] High Energy Accelerator Research Organization (KEK), Ibaraki 305-0801, Japan

Abstract

A pulsed high-voltage electron gun with a thermionic cathode is under development for the injector system of the soft X-ray FEL project at SPring-8 (SCSS project). A CeB$_6$ single crystal of 3 mm diameter was chosen as a thermionic cathode because of its excellent emission properties, i.e., high resistance against contamination, uniform emission density and smooth surface. The CeB$_6$ cathode can produce a 3 A beam with 2 μs FWHM. A gun voltage of -500 kV was chosen as a compromise between the need for controlling emittance growth and minimizing the risks of high-voltage arcing. We have constructed a 500 kV electron gun test stand and have begun performance tests. This paper describes the basic design and the current status of the hardware R&D on the CeB$_6$ gun.
© 2004 Elsevier B.V. All rights reserved.

PACS: 41.60.Cr

Keywords: FEL; X-ray laser; Low emittance; Electron gun; CeB$_6$ cathode

1. Introduction

The SPring-8 Compact SASE Source (SCSS) is a high peak-brilliance soft X-ray free-electron laser project [1]. The machine consists of a low-emittance injector, a high gradient C-band accelerator, and an in-vacuum short-period undulator. For the injector system, we chose a pulsed high-voltage electron gun with a thermionic cathode instead of a photocathode RF-gun. It is well known in the SASE-FEL theory that the quality of the internal structure of the bunched beam dominates the FEL gain, that is, the sliced emittance of the beam should be very low to saturate SASE-FEL. Moreover, from the FEL application point of view, the FEL light should be stable for longer periods of operation. Therefore, stability is essential for the injector system and the electron gun. We believe that a cathode made from a single crystal of CeB$_6$ is suitable to produce an extremely stable beam with low emittance, because its surface remains fairly flat due to material evaporation, and it possesses better resistance against carbon contamination.

This paper describes the basic design and the current status of the hardware R&D on the CeB$_6$ gun.

2. Electron injector

The CeB$_6$ cathode of 3 mm diameter can produce a 3 A beam with 2 μs FWHM. A gun

---

*Corresponding author. Tel.: +81-791-58-2929; fax: +81-791-58-2840.

E-mail address: togawa@spring8.or.jp (K. Togawa).

0168-9002/$ - see front matter © 2004 Elsevier B.V. All rights reserved.
doi:10.1016/j.nima.2004.04.077

Table 1
Design beam parameters at gun exit.

Beam energy	500 keV
Peak current	3 A
Pulse width	2 μs FWHM
Repetition rate	60 Hz
Normalized RMS emittance	0.4 πmm mrad

Fig. 1. CeB$_6$ cathode assembly.

voltage of -500 kV was chosen as a compromise between HV breakdown technical problems and emittance growth due to space charge. Design beam parameters at the gun exit are summarized in Table 1 [2].

A beam chopper that can select a 2 ns part of the 2 μs long pulse would be installed. A 476 MHz sub-harmonic buncher modulates the beam energy to form a short bunch. A subsequent energy filter removes the energy tails (top and bottom). Finally, an L-band accelerator captures the bunch and accelerates it to 20 MeV. PARMELA simulation predicts that a single bunch of 0.9 nC/bunch, 13 ps FWHM, and normalized RMS emittance of 1.9 πmm mrad is feasible. This emittance does not exceed a requirement of 2 πmm mrad for FEL operation at the undulator section.

3. The cathode assembly

The normalized RMS emittance of electrons emitted from a hot cathode is described by

$$\varepsilon_{n,\text{RMS}} = \frac{r_c}{2}\sqrt{\frac{k_B T}{mc^2}}$$

where r_c is the cathode radius, k_B is Boltzman's constant, and T is the cathode temperature. It is obvious from the above relation that a small cathode is necessary to obtain a low-emittance beam. The single-crystal CeB$_6$ cathode with a 3 mm diameter can produce a 3 A peak current when heated to 1450°C. In this case, the beam emittance becomes as low as 0.4 πmm mrad.

Fig. 1 shows the cathode assembly. The CeB$_6$ crystal is mounted on a graphite sleeve. This produces a uniform electric field on the whole cathode surface [3]. This is quite important for the elimination of a halo in beam emission from

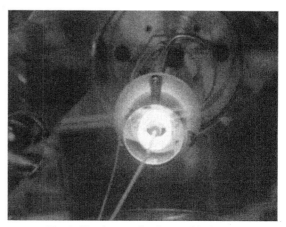

Fig. 2. Heating test for the graphite heater.

the cathode edge, which can cause damage to the undulator magnets.

We used a graphite heater rather than the conventional metallic filament made of tungsten or the like. Graphite is mechanically and chemically stable even at very high temperatures and does not evaporate like other metals. Since its electric resistance does not change much as a function of temperature, it is easy to control the heater power. Fig. 2 shows the graphite heater in operation in the vacuum chamber. The cathode was heated up to more than 1450°C with 400 W of heater power. To avoid over-heating problems, we are developing a cathode assembly which operates at lower heater power.

4. Design of the gun geometry

We chose a flat Wehnelt rather than the common Pierce-type electrode. The reasons for this are as follows: (1) The Pierce electrode was originally designed to produce a parallel beam whose space charge field is balanced by a focusing electric field. However, if the cathode is not exactly centered due to misalignment of cathode mount or shifts in cathode position due to heating, the asymmetric focusing field acts on the beam. This may cause emittance growth. The flat Wehnelt does not have such an effect. (2) We planned to vary the beam current over a wide range in order to tune the accelerator system. The gun would be operated in a temperature limited region. The Pierce electrode is not suitable for such an operation, because at a low current, the beam is over-focused; however, the flat Wehnelt does not over-focus the beam.

In order to check emittance growth of the beam from the flat Wehnelt, we performed a computer simulation using the EGUN code [4]. As shown in Fig. 3, the beam trajectory does not diverge much at the electrode gap. We have also simulated phase space plots for various mesh sizes and checked emittance values [5]. The thermal emittance was not considered in the simulation. The estimated emittance for a rough mesh sizes (Fig. 4) is deemed

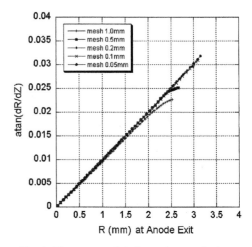

Fig. 4. Phase space plots for various mesh sizes.

to be non-physical and to be caused by a simulation error. When the mesh size becomes small, the phase space plot becomes a straight line, and the emittance converges to $0.1\,\pi\mathrm{mm}\,\mathrm{mrad}$ below a mesh size of 0.05 mm. We conclude that the emittance growth due to space charge in the gun region is negligible.

5. The 500 kV gun test stand

We have constructed a 500 kV electron gun test stand and have begun performance tests. A side view of the test stand is shown in Fig. 5. It consists of the 500 kV electron gun and an emittance measurement system.

Since we need to apply a −500 kV pulse voltage to the cathode, all the high-voltage components, namely, the ceramic insulator, pulse transformer, dummy load, etc., are immersed in insulating oil to eliminate discharge problems.

We have used the same model of the C-band klystron modulator to feed −24 kV pulsed voltage to a pulse transformer, which steps-up the input voltage to −500 kV. In order to match the impedance of the gun to the modulator PFN circuit, a 1.9 kΩ dummy load is connected in parallel with the cathode. Fig. 6 shows a gun voltage waveform measured by a capacitive voltage divider. The high voltage ceramic was

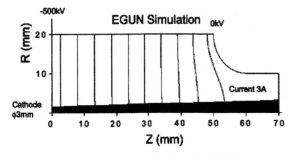

Fig. 3. Beam trajectory simulated by the EGUN code.

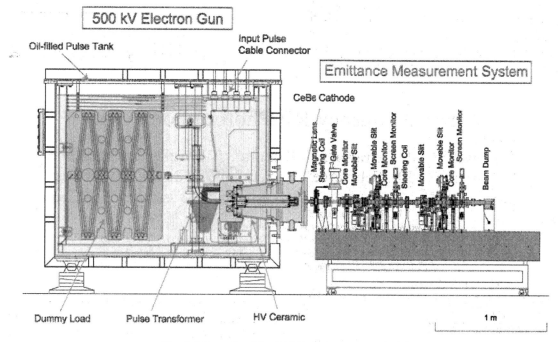

Fig. 5. Schematic of the 500 kV electron gun test stand.

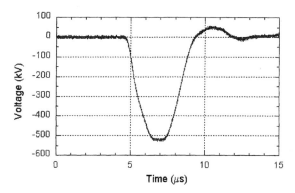

Fig. 6. Gun voltage waveform.

vertical emittance and the other for horizontal. The width and position of the slit gate could be controlled with an accuracy of 10 μm. The temporal profiles of beam currents were measured by means of core monitors. We can also measure a time-resolved beam profile using a YAP:Ce fluorescent screen. This year, we plan to measure emittance of a 500 keV beam from the CeB_6 cathode.

not installed during this test. The measured flat-top portion of the pulse is about 500 ns, which is enough to generate a 2 ns bunch.

We can measure the beam emittance by the so-called double slit method. Two sets of movable slits were prepared in the beam line. One was for

References

[1] http://www-xfel.spring8.or.jp.
[2] T. Shintake, et al., SPring-8 Compact SASE Source (SCSS), SPIE2001, San Diego, USA, June 2001.
[3] H. Kobayashi, et al., Emittance measurement for high-brightness electron guns, Linear Accelerator Conference, Ottawa, Canada, August 1992.
[4] W.B. Herrmannsfeldt, SLAC-Report 331, 1988.
[5] A.D. Yeremian, private communication.

Available online at www.sciencedirect.com

SCIENCE DIRECT®

ELSEVIER   Nuclear Instruments and Methods in Physics Research A 528 (2004) 316–320

NUCLEAR
INSTRUMENTS
& METHODS
IN PHYSICS
RESEARCH
Section A

www.elsevier.com/locate/nima

A two-Frequency RF Photocathode Gun

D.H. Dowell[a],*, M. Ferrario[b], T. Kimura[c], J. Lewellen[d], C. Limborg[a],
P. Raimondi[b], J.F. Schmerge[a], L. Serafini[e], T. Smith[c], L. Young[f]

[a] Stanford Linear Accelerator Center, Mail Stop 18, 2575 San Hill Rd., Menlo Park, CA 94025-7015, USA
[b] INFN-Frascati, via E. Fermi 40, 00040 Frascati, Italy
[c] Hansen Labs, Stanford University, Stanford, CA 94305-4085, USA
[d] Advanced photron Source, Argonne National Laboratory, Argonne, IL 60439, USA
[e] INFN-Milan-LASA, via Fratelliceni 201, 20090 Segrate (MI), Italy
[f] Los Alamos National Laboratory, P.O. Box 1663, Los Alamos, NM 87545, USA

Abstract

In this paper we resurrect an idea originally proposed by Serafini (Nucl. Instr. and Meth. A 318 (1992) 301) in 1992 for an RF photocathode gun capable of operating simultaneously at the fundamental frequency and a higher frequency harmonic. Driving the gun at two frequencies with the proper field ratio and relative phase produces a beam with essentially no RF emittance and a linear longitudinal phase space distribution. Such a gun allows a completely new range of operating parameters for controlling space charge emittance growth. In addition, the linear longitudinal phase space distribution aids in bunch compression. This paper will compare results of simulations for the two-frequency gun with the standard RF gun and the unique properties of the two-frequency gun will be discussed.
© 2004 Elsevier B.V. All rights reserved.

PACS: 29.25.Bx; 29.27.Ac; 41.60.Cr; 41.85.Ar

Keywords: Electron beam; Photocathode RF gun; RF harmonics; Emittance

1. Introduction

The RF gun transverse emittance is predominately due to space charge forces, RF fields and thermal emittance. Emittance compensation does well to remove the linear space charge contribution, but the gun parameters are still constrained to operate between the space charge and RF limits.

This is illustrated in Fig. 1, where the uncompensated emittance at the gun exit is plotted in the plane defined by beam size and bunch length. The operating region is shown for the LCLS gun and is typical of most guns. They perform best in the saddle between the space charge and RF dominated regimes. Emittance compensation extends the region into the space charge regime, and does a remarkably good job of recovering the emittance where the correct combination of solenoid, drift and linac gradient and phase, reduce the projected emittance from three microns to one micron after

---
*Corresponding author. Tel.: +1-650-9262494; fax: +1-650-926-8533.

E-mail address: dowell@slac.stanford.edu (D.H. Dowell).

0168-9002/$ - see front matter © 2004 Elsevier B.V. All rights reserved.
doi:10.1016/j.nima.2004.04.078

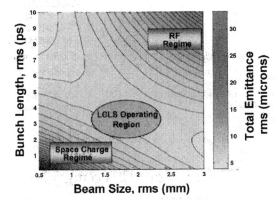

Fig. 1. Contour plot of the total emittance at the gun exit in the plane of bunch size and length. The bunch charge is 1 nC. The formulas of Travier [1] were used to compute the RF and space charge effects.

acceleration to high energy. However, further improvement requires one to advance from compensating to eliminating the sources of the emittance.

It has been shown [2] that the addition of a harmonic to the RF fundamental field would greatly reduce the RF emittance, allowing complete freedom of beam size and length to control the space charge forces. This work showed a harmonic RF field in the same $N + \frac{1}{2}$ cells as the fundamental will linearize the RF force to fourth order, both transversely and longitudinally. This condition is achieved when the bunch exit phase is $\pi/2$, $n = 3,7,11\ldots$ and the nth harmonic field, E_n, in terms of the fundamental field, E_0, is given by

$$E_n = (-1)^{(n-3)/2} E_0/n^2.$$

This work resurrects these ideas and applies them to the BNL/SLAC/UCLA gun. The space charge and RF-dominated regimes are simulated with a new version of Parmela [3] capable of superimposing the RF fields at two frequencies in the same cell. Short bunches are used to explore the space charge dominated regime while long bunches investigate the RF-dominated regime. In addition, the two-frequency gun is combined with emittance compensation to achieve very low transverse emittance. Comparisons of the transverse and longitudinal emittances and phase space distributions are presented and discussed.

2. Superfish fields

The standard 1.6 cell BNL/SLAC/UCLA shape was modified to have the desired fundamental and 3rd harmonic frequencies while keeping the 1.6 $\lambda/2$ length. The field shapes for the Superfish model are shown in Fig. 2. The gun's fundamental mode is unbalanced due to the shape change used to obtain the harmonic. The cathode cell 3rd harmonic is a TM021-like mode and the full cell mode is TM012-like, which should not significantly affect our results since the beam samples only the on-axis fields.

3. The parmela simulations

The simulations were designed to explore the two-frequency gun's properties in both the space charge and RF regimes. The study compares emittances for short, space charge dominated bunches with emittances for long, RF-dominated bunches. The longitudinal electric field is assumed to be the sum of the fundamental and the 3rd harmonic fields given by

$$E_z = E_0 \cos(kz)\sin(\omega t + \phi_0)$$
$$+ E_3 \cos(3kz)\sin(3(\omega t + \phi_3)).$$

In each case, the parameters ϕ_0, E_3 and ϕ_3 are varied to obtain the lowest projected emittance.

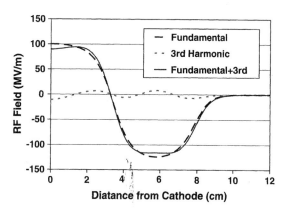

Fig. 2. The Superfish fields used in the Parmela simulations. The full cell field is approximately 20% higher than the cathode cell field.

For both the short and long bunch simulations, the beam size on the cathode is 2 mm radius, flat top distribution and the results are given at the gun exit, with no solenoid field or any emittance compensation. The space charge regime (short bunch) uses a 10 ps full-width, square bunch shape. The RF regime (long bunch) is studied with a 40 ps long bunch. Simulations are performed with 0 and 1 nC to separate the RF and space charge contributions.

For the emittance compensated case the beam radius is 1 mm and the full-width bunch length is 30 ps. In all cases, the thermal emittance is zero.

3.1. Longitudinal phase space

Fig. 3 shows that even for short bunches 3rd harmonic linearization improves the longitudinal phase space. The full-width, correlated energy spread is reduced from 100 to 40 keV in the presence of space charge.

The short bunch case for 0 nC, RF only, is shown in Fig. 4. The addition of the 3rd harmonic not only makes the distribution more linear, but also flips the sign of the correlation. This is not observed in the analytic theory [2], and is due to the 1.6 cell length of the gun used in these simulations. The original theory is for a 1.5 cell

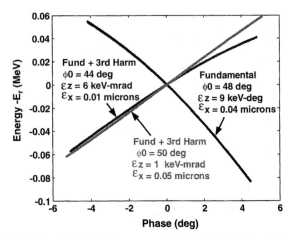

Fig. 4. Longitudinal phase space with 0 nC of charge, 10 ps full-width for fundamental only and fundamental + 3rd harmonic: $E_0 = 82\,\text{MV/m}$, $E_3 = -28\,\text{MV/m}$, $\phi_3 = 15°$, ϕ_0 as shown.

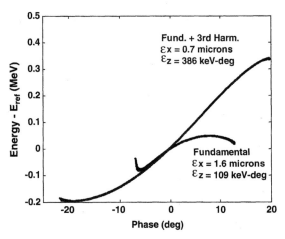

Fig. 5. The 40 ps long bunch longitudinal phase space at 1 nC with and without the 3rd harmonic. For fundamental only: $E_0 = 82\,\text{MV/m}$, $\phi_0 = 20°$. For fundamental + harmonic: $E_0 = 82\,\text{MV/m}$, $\phi_0 = 35°$, $E_3 = -31.5\,\text{MV/m}$, $\phi_3 = 17°$.

gun. In addition, the figure shows the fundamental phase which produces the lowest transverse emittance is slightly different than that which best linearizes the longitudinal phase space.

The 40 ps long bunch for a 1 nC bunch is shown in Fig. 5. With only the fundamental, the bunch has been compressed approximately by a factor of two. However, with the 3rd harmonic, the bunch length is unchanged.

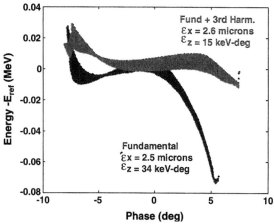

Fig. 3. The short bunch longitudinal phase space at 1 nC with and without the 3rd harmonic: $E_0 = 82\,\text{MV/m}$, $\phi_0 = 25°$, $E_3 = -21\,\text{MV/m}$, $\phi_3 = 17°$.

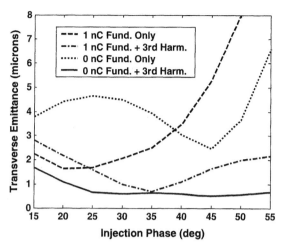

Fig. 6. Transverse projected emittance for a 40 ps long square bunch. For 0 nC: $E_0 = 82$ MV/m, $\phi_0 = 35°$, $E_3 = -28$ MV/m, $\phi_3 = 16°$. For 1 nC: $E_0 = 82$ MV/m, $\phi_0 = 35°$, $E_3 = -31.5$ MV/m, $\phi_3 = 18°$. Although the minimum projected emittance is larger for 0 nC at 45° injection phase than for 1 nC at 25°, the 0°nC slice emittance is smaller over more of the bunch. Slice mismatch makes the 0°nC emittance larger than 1 nC emittance.

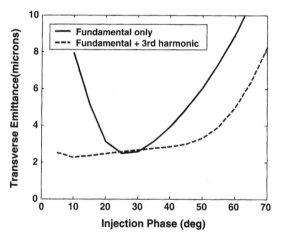

Fig. 7. Short bunch transverse emittance for 1 nC at the exit of RF gun without and with the 3rd harmonic field: $E_0 = 82$ MV/m, $\phi_0 = 25°$, $E_3 = -21$ MV/m, $\phi_3 = 17°$.

3.2. Transverse emittance

It is expected that the 3rd harmonic should mostly benefit long bunches in the RF dominant regime. In this case, the RF emittance is expected to be small over a wide range of injection phases as given in the

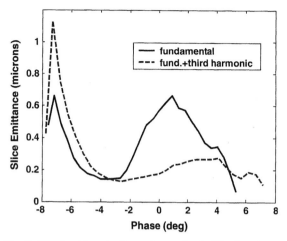

Fig. 8. The short-bunch, slice emittance for 1 nC is plotted as a function of position along the bunch. The head of the bunch is to the left and 1° at s-band = 1.05 ps. $E_0 = 82$ MV/m, $\phi_0 = 35°$, $E_3 = -21$ MV/m, $\phi_3 = 17°$.

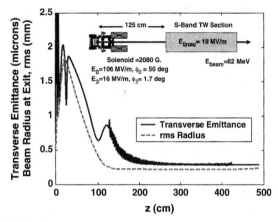

Fig. 9. The projected transverse emittance and beam size of the two-frequency gun with emittance compensation. The final bunch length is 29 ps full-width.

original work [2]. Fig. 6 shows this is verified by the Parmela simulations, reducing the RF emittance a factor of four or more for injection phases from 25° to 55°. With the addition of space charge, the fundamental + 3rd harmonic emittance is approximately half that of the fundamental only value.

The results of both frequencies upon the emittance for a short, space charge dominated bunch is given in Fig. 7. While the fundamental + 3rd minimum projected emittance is not any lower than that with the fundamental alone, the beam quality is good over a wider range of injection phase, even indicating some improvement down to $10°$.

The reduction in the short bunch slice emittance is larger. Fig. 8 gives the slice emittance plotted along the length of the bunch in degrees of RF and indicates that except for the head slices, the emittances are reduced from 0.6 to $0.3\,\mu m$ or less over the main body of the bunch.

The configuration and simulation results for the emittance compensation case are shown in Fig. 9 for a 30 ps long, 1 nC bunch. The projected emittance equilibrates to $0.28\,\mu m$ and the slices (not shown) are very well aligned with emittances of $0.2\,\mu m$ over more than 95% of the bunch.

Acknowledgements

SLAC is operated by Stanford University for the Department of Energy under contract number DE-AC03-76SF00515.

References

[1] C. Travier, Nucl. Instr and Meth. A 340 (1994) 26.
[2] L. Serafini et al, Nucl. Instr and Meth. A 318 (1992) 301; T.I. Smith, Proceedigs of the Linear Account Conference, SLAC PUB-303.
[3] L.M. Young, "PARMELA," Los Alamos National Laboratory Report LA-UR-96-1835(revised July17, 2003).

Available online at www.sciencedirect.com

SCIENCE @ DIRECT°

ELSEVIER   Nuclear Instruments and Methods in Physics Research A 528 (2004) 321–325

NUCLEAR INSTRUMENTS & METHODS IN PHYSICS RESEARCH
Section A

www.elsevier.com/locate/nima

Experimental investigations of DC-SC photoinjector at Peking University ☆

Rong Xiang, Yuantao Ding, Kui Zhao*, Xiangyang Lu, Shengwen Quan, Baocheng Zhang, Lifang Wang, Senlin Huang, Lin Lin, Jia'er Chen

RF SC Group, SRF Laboratory, Institute of Heavy Ion Physics, Peking University, Beijing 100871, China

Abstract

We report the progress of the DC-SC photoinjector at Peking University. This is a compact electron gun integrating a DC Pierce gun with a 1.3 GHz superconducting cavity. The photoinjector is designed to provide an electron beam having an average current of 1 mA with the energy of 2.61 MeV and normalized rms transverse emittance of 3 mm mrad. The test facility has been completely installed in our laboratory. The photocathode preparation chamber can produce Cs_2Te and Cs_3Sb cathodes, and the laser system can provide laser pulses with 532 or 266 nm wavelength at an 81.25 MHz repetition rate. The timing jitter of less than 1 ps between the laser and RF power has been achieved by using a timing stabilizer. A new method using "duo image pattern" of Cherenkov radiation will be commissioned to measure beam emittance. This paper summarizes some of the ongoing experimental activities.
© 2004 Elsevier B.V. All rights reserved.

PACS: 41.60.Cr; 41.75.Fr; 42.55.Xi; 07.77.Ka

Keywords: Photoinjector; Superconducting cavity; FEL

1. Introduction

It was proposed that a free-electron laser facility (FEL) [1] based on superconducting accelerators be built at Peking University. For this program, a new type of superconducting electron gun, named DC-SC photoinjector, is being developed

☆ Supported in part by Chinese Department of Science and Technology under the National Basic Research Projects (No. 2002CB713602) and by National Natural Science Foundation of China (10075006)(19985001).

*Corresponding author. Tel.: +86-10-627-57195; fax: +86-10-627-51875.

E-mail address: kzhao@pku.edu.cn (K. Zhao).

[2]. The chief goal is to produce minimum transverse emittance beams at a high average current (1–5 mA).

The DC-SC photoinjector test facility was constructed in January 2003. This is a compact system, integrating a DC Pierce gun with a superconducting $1 + \frac{1}{2}$ niobium cavity. Fig. 1 shows a schematic overview of the facility, including the cathode preparation chamber, the cryostat housing the DC gun and the SC cavity, the 100-kV high-voltage source for the cathode of DC gun, the RF main coupler and the 4.5 kW solid-state power amplifier, the drive laser system, and the beam diagnostic.

The first test of the injector is in process. The cathode preparation has been built and operated. The UV laser of 1.2 W at 266 nm has been achieved. According to the early performance analysis for the $1 + \frac{1}{2}$ cell SC cavity, we have improved some sub-systems. We have also developed a new method using "duo image pattern" of Cherenkov radiation to measure beam emittance. Details of the hardware components are described below.

2. DC Pierce gun and $1 + \frac{1}{2}$ cell SC cavity

2.1. Design considerations

As shown in Fig. 2, the core elements of this electron gun are the DC Pierce gun and the $1 + \frac{1}{2}$ cell superconducting cavity. The photocathode is placed at the cathode of the Pierce structure, and the anode makes up the bottom of the $1 + \frac{1}{2}$ cell cavity.

The structure of DC-SC injector has several advantages [2];

(1) The effect of the photocathode on the SC cavity can be avoided because the photocathode is placed outside the superconducting cavity. The cathode plug with the photolayer can also be operated at low temperatures, and the good vacuum conditions should increase the life span of the highly sensitive photocathode.

(2) The back wall of the half-cell has a conical geometry, which leads to an RF focusing of the electron bunch.

Code PARMELA [3] is used to simulate the performance of the whole injector. The optimized results at the exit of the injector are listed in Table 1.

The choke point of this design is its narrow neck between the DC Pierce gun and the SC cavity. From the PARMELA simulation, we found that when the electron bunches with low energy (70 keV) pass through this neck, the space-charge effect leads to a rise in the emittance. However, before the charge per bunch increases to 100 pC, the simulation results display good potential to meet our requirements. Moreover, the RMS

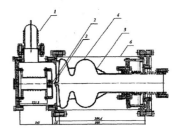

(1) Ceramic insulation   (2) Photocathode

(3) Pierce DC gun   (4) Niobium cavity

(5) Stiffening ring   (6) LHe tank

Fig. 2. Draft of the $1 + \frac{1}{2}$ cell cavity.

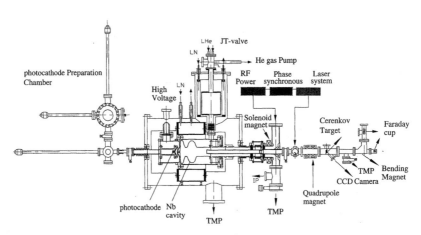

Fig. 1. Schematic overview of the current set-up.

Table 1
Simulation results of the DC-SC injector

Electron bunch	
Length (ps)	7.8
Energy spread (rms) (%)	1.16
Energy (MeV)	2.61
I_{ave} (mA)	1
Emittance (rms) (mm mrad)	3
Radius (mm)	2.8
SC cavity	
E_{acc} (MV/m)	15

Fig. 3. Image of the $1 + \frac{1}{2}$ cell niobium cavity.

emittance in Table 1 is as good as 3 mm mrad at the end of the whole injector.

2.2. Cavity preparation and test results

The whole procedure of preparing $1 + \frac{1}{2}$ cell cavity was carried out at Peking University. Cups are made of 2.5-mm thick sheet niobium (RRR = 250) by spinning, followed by trimming and electron beam welding. After heat treatment, the cavity undergoes mechanical polishing, electric etching, buffered chemical polishing (BCP), and ultraclean water rinsing. The assembled cavity (Fig. 3) is mechanically tuned to adjust the resonance frequency to the design value.

The first cold RF-test of the SC cavity has been recently carried out. The cavity was tested without the cathode in it to evaluate the cavity and to prove the compatibility of the superconducting cavity and the RF input coupler. A resonant at 1300 MHz at 4.2 K was achieved. The first test showed that the unloaded Q value of $1 + \frac{1}{2}$ cell was

$\sim 10^8$ and the average gradient was about 4–5 MV/ m, limited by field emission. A test on the DC gun has been performed, in which the current reached 100 μA.

3. Drive laser and photocathode preparation

A photocathode preparation chamber has been designed and manufactured at Peking University. This chamber is bakeable, and the vacuum can reach values of 10^{-6} Pa in the course of the evaporation. Cs_2Te and Cs_3Sb are optionally fabricated. One layer of Cs_2Te on a stainless plug is excited by 266 nm UV laser, and its quantum efficiency is above 5% for several days. The "Yo–Yo" technique is used in the growth of Cs_3Sb, and the QE at 532 nm is measured to 1%.

A schematic view of the laser system currently in operation is shown in Fig. 4. GE-100-XHP is a high-power mode-locked and diode-pumped $Nd:YVO_4$ picosecond laser, and its average output power is 10.4 W at 1064 nm. Using a timing stabilizer (CLX-1100) it can provide a full synchronization to the RF cavity with a time jitter of less than 1 ps. The pulse duration is 10 ps. KTP is used as a frequency doubler and a CLBO crystal, developed in China, is used to obtain 266 nm UV laser. After frequency doubling and quadrupling, it is able to generate 5 W at 532 nm or 1.2 W at 266 nm with the repetition of 81.25 MHz. The size of the RMS laser spot was measured to be about 6 mm.

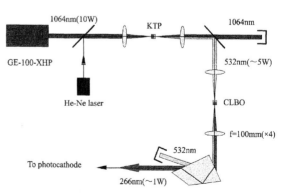

Fig. 4. Optical scheme of the current drive laser.

4. Main power coupler and RF system

The cavity is powered by a 4.5 kW solid-state power amplifier via a coaxial input coupler. A 50 Ω coaxial line is adopted. The transition between the waveguide from the generator and the coaxial line of the coupler is realized by a doorknob configuration. To decrease the thermal loading and the RF loss, the inner conductor is made of thin stainless steel, coated with copper on the outside, and the metal of outer conductor is stainless steel, coated with copper on the inside as well.

5. Beam diagnostic

A beam diagnostic system has been researched and commissioned in relation to the beam line. In Fig. 1, a dipole magnet and two Faraday cups are installed to detect the energy and the energy spread.

A new measure of electron beam emittance with Cherenkov radiation "duo image pattern" has been researched at Peking University. The Cherenkov radiation, excited by the electron beams traversing the Cherenkov crystal target, passes through two thin convex lenses, and two images at the focal plane and the image plane are captured by a CCD camera. Using an image-processing technique, the emittance can be directly obtained. The detailed process is described elsewhere [4].

6. Summary and outlook

We have shown that the DC-SC photoinjector test facility at Peking University has been developed. The performance in our current test shows good potential as an injector for the electron source with moderate average currents at low charge per bunch but at a very high repetition rate. First results from the cold test of superconducting cavity have been presented. The beam test on the DC gun has been conducted, and the electron current of 100 µA has been accelerated. This is a first stage, and the next step, to prepare all the equipment for the beam tests, is under way.

Table 2
Beam parameters of the $2 + \frac{1}{2}$ gun

$2 + \frac{1}{2}$ cavity	
E_{acc} (MV/m)	15
Drive laser	
Pulse length (ps)	10
Spot radius (mm)	3
Repetition rate (Hz)	50
Electron bunch	
Charge/bunch (pC)	500
Energy (MeV)	3.72
Energy spread (rms) (%)	1.68
Emittance (rms) (mm mrad)	6.1

In parallel to the experimental effort on DC-SC injector, we are also studying the other type of superconducting RF gun, such as the gun for a higher peak current [5]. We are working on the design of a 1.3 GHz $2 + \frac{1}{2}$ cell SRF gun for possible application at infrared SASE FEL [6]. To share the existing SC cavity technology, the geometry of the full cells is based on the TESLA design [7]. The half cell is similar to that of the DC-SC cavity, in which the conical geometry leads to RF focusing. An RF filter replaces the DC Pierce gun because of the beam loading, which is as high as 500 pC/bunch, and the cathode is installed on a stem in the center of the filter. The results of PARMELA calculations are given in Table 2.

Acknowledgements

We thank our many colleagues who are supporting our work with their advice and helping us in the experiments. Special thanks are due to Dr. Proch from DESY and Dr. Xie Ming from LBL for their support to our work. We also wish to thank Dr. Peter Michel for several helpful discussions.

References

[1] Zhao Kui, et al., Nucl. Instr. and Meth. A 483 (2002) 125.
[2] Zhao Kui, et al., Nucl. Instr. and Meth. A 475 (2001) 564.

[3] L. Young, PARMELA, LA-UR-96-1835, LANL, 1996.

[4] Jia'er Chen, et al., Proceedings of PAC 2003, Portland, OR, USA.

[5] E. Barhels, et al., Nucl. Instr. and Meth. A 445 (2000) 408.

[6] Ding Yuantao, et al., Proceedings of FEL2003, Tsukuba, Ibaraki, Japan.

[7] B. Aune, et al., Phys. Rev. Spec. Topics-Accel. Beams 3 (2000) 092001.

Available online at www.sciencedirect.com

Nuclear Instruments and Methods in Physics Research A 528 (2004) 326–329

NUCLEAR
INSTRUMENTS
& METHODS
IN PHYSICS
RESEARCH
Section A

www.elsevier.com/locate/nima

Temporal properties of coherent synchrotron radiation produced by an electron bunch moving along an arc of a circle

G. Geloni[a], E.L. Saldin[b], E.A. Schneidmiller[b], M.V. Yurkov[c],*

[a] Department of Applied Physics, Technische Universiteit Eindhoven, The Netherlands
[b] Deutsches Elektronen-Synchrotron (DESY), Hamburg, Germany
[c] Particle Physics Laboratory, Joint Institute for Nuclear Research, Dubna 141980, Moskow Region, Russia

Abstract

In the limit for a large distance between bunch and detector and under the assumption that the entire process, i.e. radiation and detection, happens in vacuum, one can use the well-known Schwinger formulas in order to describe the single-particle radiation in the case of circular motion. Nevertheless, these formulas cannot be applied for particles moving in an arc of a circle. In this paper, we present a characterization of coherent synchrotron radiation (CSR) pulses in the time-domain as they are emitted by an electron bunch moving in an arc of a circle. This can be used in order to give a quantitative estimation of the effects of a finite bending magnet extension on the characteristics of the CSR pulse.
© 2004 Elsevier B.V. All rights reserved.

PACS: 41.20.Jb; 41.60.Ap; 41.75.Ht

Keywords: Synchrotron radiation; Coherent radiation; Coherent synchrotron radiation; Relativistic electron and positron beams

1. Introduction

Short bunches of rms length of order of 100 fs are needed for XFEL applications [1,2], which is achieved by compression chicanes. The development of nondestructive methods for the measurement of the longitudinal beam current distribution in such short bunches is a challenging problem: one possible solution is based on measurements of coherent synchrotron radiation (CSR) produced by a bunch passing a magnetic system. CSR is also

*Corresponding author. Tel.: +7-09621-62154; fax: +7-09621-65767.

E-mail address: yurkov@mail.desy.de (M.V. Yurkov).

interesting as concerns its application for the design of radiation sources in the mid-far infrared [3].

Standard theory of synchrotron radiation and CSR is based on Schwinger formulas [4], which cannot be used to describe the CSR pulse from a dipole magnet. In fact, while these are valid in order to describe radiation from a dipole in the X-ray range they are useless when it comes to characterize the long-wavelength asymptote from a particle in a bend, which is the only interesting spectral region concerning CSR effects. In Section 2, we derive a new analytical expression for the CSR pulse from a bunch moving in an arc of a circle. Our conclusions are summed up in

0168-9002/$ - see front matter © 2004 Elsevier B.V. All rights reserved.
doi:10.1016/j.nima.2004.04.080

Section 3, while Schwinger formulas cannot be used directly in order to characterize the CSR pulse from a bending magnet, their long-wavelength asymptote can still be used in an elegant way to derive an analytical description in the far field limit.

2. CSR pulse characterization

Let us consider an electron bunch moving in a hard edge bending magnet characterized by an angular extension ϕ_m. We will focus only on the radiation seen by an observer located at large distance from the sources, on the tangent to the electrons orbit at the middle point of the magnet.

In this case we can use the fixed coordinate system (x, y, z) shown in Fig. 1, suited for the circular motion. The observation point and the vector $\vec{n}$ are within the (y, z)-plane and radiation is emitted at an angle θ with respect to the z-axis. Let us start expressing the total CSR pulse as a superposition of single particle fields at the given position in the far zone. In the case of an arc of a circle the CSR field can be expressed as

$$\vec{E}_{CSR}(t) = \int_{t-T}^{t+T} \vec{E}_r(t - \tau)NF(\tau)\,d\tau. \tag{1}$$

Here $\vec{E}_r$ is the single-particle radiation field, N is the number of particles and F is the bunch density function. We stated in the Introduction, that Schwinger formulas fail to describe the long-wavelength radiation limit and that cannot be

directly used when an expression for the CSR pulse from a bend is looked for: in this regard, note how the time T in the integration limits is in loco of a window function in the integrand, which cuts the contributions of the single particle radiation pulse when the electron is not in the arc.

Eq. (1) contains the observation time T, which should be replaced by the retarded time t_e'. The two times are related by

$$2T = 2t_e' + \frac{1}{c}|\vec{r}(t_e')| - \frac{1}{c}|\vec{r}(-t_e')| \tag{2}$$

where $t_e' = \phi_{2m}/(\omega_0)$ and ϕ_m is the bending magnet angular extension. Our analysis focuses on the case of a long bending magnet, $\gamma\phi_m \gg 1$.

The single-particle radiation field obeys the following relation:

$$\int_{-\infty}^{\infty} \vec{E}_r\,dt = 0. \tag{3}$$

Using Eq. (3) and assuming

$$\frac{R}{c\gamma^3}\frac{dF(t)}{dt} \ll F(t) \tag{4}$$

the field of the CSR pulse is readily shown to be

$$\begin{aligned}
\vec{E}_{CSR}(t) = &\int_{t-T}^{t-\delta} \vec{E}_r(t - \tau)NF(\tau)\,d\tau \\
&- NF(t)\int_{-\infty}^{t-\delta} \vec{E}_r(t - \tau)\,d\tau \\
&+ \int_{t+\delta}^{t+T} \vec{E}_r(t - \tau)NF(\tau)\,d\tau \\
&- NF(t)\int_{t+\delta}^{\infty} \vec{E}_r(t - \tau)\,d\tau
\end{aligned} \tag{5}$$

where δ satisfies

$$\delta \gg \frac{R}{c\gamma^3} \quad \delta\frac{dF(t)}{dt} \ll F(t). \tag{6}$$

These limitations define a "slowly" varying function of the time and we simply take $F(\tau)$ outside the integral sign and call it $F(t)$ when calculating the 2nd and the 4th integral of Eq. (5). Adding and subtracting suitable edge terms one can perform

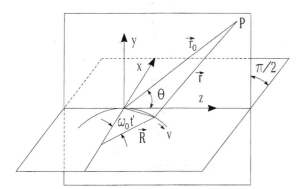

Fig. 1. Geometry for synchrotron radiation from circular motion.

integration by parts, thus obtaining

$$
\begin{aligned}
\vec{E}_{CSR}(t) = {} & NF(t+T) \int_{-\infty}^{t+T} \vec{E}_r(t-\tau)\,d\tau \\
& - NF(t-T) \int_{-\infty}^{t-T} \vec{E}_r(t-\tau)\,d\tau \\
& - \int_{t-T}^{t-\delta} \Phi[\vec{E}_r](t-\tau) N \frac{dF(\tau)}{d\tau}\,d\tau \\
& - \int_{t+\delta}^{t+T} \Phi[\vec{E}_r](t-\tau) N \frac{dF(\tau)}{d\tau}\,d\tau.
\end{aligned}
\tag{7}
$$

Here $\Phi[\vec{E}_r]$ is a primitive of the single-particle radiation field. Since conditions (6) hold for δ we may substitute the 3rd and the 4th integral in Eq. (7) with a single integral in which the primitive, $\Phi[\vec{E}_r]$, is replaced by its asymptotic for large values of the argument, $\Phi[\vec{E}_r^A]$. Under the assumption of a long bunch ($\omega_c T \gg 1$) the 1st and the 2nd integral can be expressed by means of the primitive asymptotic too. Moreover, we can perform a change of variables in all the integrals $t - \tau \to \tau$. As a result Eq. (7) can be written in the form:

$$
\begin{aligned}
\vec{E}_{CSR}(t) = {} & - NF(t+T) \int_{-T}^{\infty} \vec{E}_r^A(\tau)\,d\tau \\
& - NF(t-T) \int_{\infty}^{T} \vec{E}_r^A(\tau)\,d\tau \\
& - \int_{-T}^{T} \Phi[\vec{E}_r^A](\tau) N \frac{dF(t-\tau)}{d\tau}\,d\tau.
\end{aligned}
\tag{8}
$$

To calculate the primitive of $\vec{E}_r$ we use the well-known expressions for the Lienard–Wiechert fields, where the electric field is expressed in terms of quantities at the retarded time t'. The calculation is simplified if we use the following, widely used consideration: since, in general $dt/dt' = (1 - \vec{n} \cdot \vec{\beta})$ and

$$
\frac{d}{dt'}\left[\frac{\vec{n} \times [(\vec{n} \times \vec{\beta})]}{(1 - \vec{n} \cdot \vec{\beta})} \right] = \left[\frac{\vec{n} \times [(\vec{n} - \vec{\beta}) \times \dot{\vec{\beta}}]}{(1 - \vec{n} \cdot \vec{\beta})^2} \right]
\tag{9}
$$

we also have

$$
\frac{d}{dt}\left[\frac{\vec{n} \times [(\vec{n} \times \vec{\beta})]}{(1 - \vec{n} \cdot \vec{\beta})} \right] = \left[\frac{\vec{n} \times [(\vec{n} - \vec{\beta}) \times \dot{\vec{\beta}}]}{(1 - \vec{n} \cdot \vec{\beta})^3} \right].
\tag{10}
$$

Because the angles are very small and the relativistic γ factor is very large, it is useful to express Eq. (8), using the Lienard–Wiechert fields

and Eq. (10) in a small angle approximation. The triple-vector product is calculated from Fig. 1

$$
\vec{n} \times [\vec{n} \times \vec{\beta}] = \omega_0 t' \vec{e}_x + \theta \vec{e}_y.
\tag{11}
$$

Here θ is the vertical angle, $\omega_0 = \beta c / R$ is the revolution frequency and $\vec{e}_{x,y}$ are unit vectors directed along the x- and y-axis of the fixed Cartesian coordinate system (x, y, z) shown in Fig. 1. The definition of $\vec{n}$ and $\vec{R}$ can be used to compute the scalar product in the denominator in Eq. (10) so that

$$
\begin{aligned}
\vec{n} \cdot \vec{\beta} &= \beta \cos \theta \cos \omega_0 t' \\
&\simeq \beta(1 - \theta^2/2)(1 - (\omega_0 t')^2/2).
\end{aligned}
\tag{12}
$$

We assume, here, that the vertical angle is small enough to leave out the cosine factor.[1] In this situation, we have

$$
\frac{\vec{n} \times [(\vec{n} \times \vec{\beta})]}{(1 - \vec{n} \cdot \vec{\beta})} \simeq \frac{\omega_0 t' \vec{e}_x + \theta \vec{e}_y}{(\omega_0 t')^2/2}.
\tag{13}
$$

This quantity must be evaluated at the retarded time $t' \simeq [6\tau/\omega_0^2]^{1/3}$. Substitution of these expressions in Eq. (8) gives

$$
\begin{aligned}
\vec{E}_{CSR}(t) = {} & \frac{-2eN}{4\pi\varepsilon_0 c |\vec{r}_0|} \left\{ \int_{-T}^{T} \left[\frac{\epsilon(\tau)\vec{e}_x}{[6\omega_0 |\tau|]^{1/3}} \right. \right. \\
& \left. + \frac{\theta \vec{e}_y}{[6\omega_0 |\tau|]^{2/3}} \right] \frac{dF(t-\tau)}{d\tau}\,d\tau \\
& + [F(t+T) + F(t-T)] \frac{\vec{e}_x}{(6\omega_0 T)^{1/3}} \\
& \left. - [F(t+T) - F(t-T)] \frac{\theta \vec{e}_y}{(6\omega_0 T)^{2/3}} \right\}
\end{aligned}
\tag{14}
$$

where $T = \phi_m^3/(48\omega_0)$. Eq. (14) is a new, elegant result which allows one to characterize the far field CSR pulse from a bunch moving in an arc of a circle.

As an example of the application of this expression, consider the situation when $\theta = 0$ and the bunch profile is a Gaussian. According to Eq. (14) the CSR field in this case is given by

$$
E_x(t) = G_1(t) + G_2(t)
\tag{15}
$$

---

[1] This assumption can be readily relaxed.

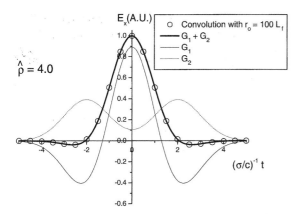

Fig. 2. Time structure of the CSR pulse from a Gaussian electron bunch moving along an arc of a circle. Here $\theta = 0$, $\hat{\rho} = \phi_m^3/(24\omega_0\sigma_T) = 4$. Circles present the results obtained from direct superposition of single particle pulses. The solid line corresponds to the shape calculated by means of Eq. (14).

where

$$G_1 = \frac{2eN}{4\pi\varepsilon_0(2\pi)^{1/2}6^{1/3}\sigma_T^3\omega_0^{1/3}c|\vec{r}_0|}$$

$$\times \int_{-T}^{T} \frac{\epsilon(\tau)(t-\tau)}{|\tau|^{1/3}} e^{(-(t-\tau)^2/2\sigma_T^2)} \, d\tau \qquad (16)$$

$$G_2 = \frac{-2eN}{4\pi\varepsilon_0(2\pi)^{1/2}6^{1/3}\sigma_T(\omega_0 T)^{1/3}c|\vec{r}_0|}$$

$$\times [e^{(-(t+T)^2/2\sigma_T^2)} + e^{(-(t-T)^2/2\sigma_T^2)}]. \qquad (17)$$

Fig. 2 presents the results of calculations obtained using Eq. (14). As expected, a simple cross-check with numerical results shows a very good agreement. The finite magnet length chosen in Fig. 2,

$\hat{\rho} = \phi_m^3/(24\omega_0\sigma_T) = 4$, substantially modifies the shape of the CSR pulse with respect to the case of a circular motion. As one can see from Ref. [5], the tails shrink and the integral of the total field deviates from zero (which is the expected result for the circular case) as $\hat{\rho}$ becomes smaller and smaller. This, of course, can be directly ascribed to the fact that Schwinger formulas are not valid to describe the long-wavelength asymptote of the single particle radiation.

3. Conclusion

While Schwinger formulas cannot be used directly in order to characterize the CSR pulse from a bending magnet, their long-wavelength asymptote can still be used in an elegant way to derive an analytical description in the far field limit. In this paper, we derived such a new expression. This can be used in order to give a quantitative estimation of the effects of a finite bending magnet extension on the characteristics of the CSR pulse, as has been done in Ref. [5].

References

[1] F. Richard, et al. (Eds.), TESLA Technical Design Report, DESY 2001-011.
[2] The LCLS Design Study Group, LCLS Design Study Report, SLAC reports SLAC-R521, Stanford (1998).
[3] G.L. Carr, et al., Nature 420 (2002) 153.
[4] J. Schwinger, Phys. Rev. 75 (1949) 1912.
[5] G. Geloni, et al., preprint DESY 03-031, DESY, Hamburg, 2003.

Available online at www.sciencedirect.com

SCIENCE  DIRECT®

**NUCLEAR
INSTRUMENTS
& METHODS
IN PHYSICS
RESEARCH**
Section A

ELSEVIER   Nuclear Instruments and Methods in Physics Research A 528 (2004) 330–334

www.elsevier.com/locate/nima

Application of constrained deconvolution technique for reconstruction of electron bunch profile with strongly non-Gaussian shape

G. Geloni[a], E.L. Saldin[b], E.A. Schneidmiller[b], M.V. Yurkov[c],*

[a] *Department of Applied Physics, Technische Universiteit Eindhoven, The Netherlands*
[b] *Deutsches Elektronen-Synchrotron (DESY), Notkestrasse 85, 22607 Hamburg, Germany*
[c] *Joint Institute for Nuclear Research, Dubna 141980 Moskow Region, Russia*

Abstract

An effective and practical technique based on the detection of the coherent synchrotron radiation (CSR) spectrum can be used to characterize the profile function of ultra-short bunches. The CSR spectrum measurement has an important limitation: no spectral phase information is available, and the complete profile function cannot be obtained in general. In this paper we propose to use constrained deconvolution method for bunch profile reconstruction based on a priori-known information about formation of the electron bunch. Application of the method is illustrated with practically important example of a bunch formed in a single bunch-compressor. Downstream of the bunch compressor the bunch charge distribution is strongly non-Gaussian with a narrow leading peak and a long tail. The longitudinal bunch distribution is derived by measuring the bunch tail constant with a streak camera and by using a priory available information about profile function.
© 2004 Elsevier B.V. All rights reserved.

PACS: 41.60.cr; 42.55.Vc; 41.75.Ht; 41.85.Ew

Keywords: Free electron lasers; X-ray lasers; Relativistic electron beams

1. Introduction

The bunch length for XFEL applications is of order of 100 femtoseconds [1–3]. Time scale of fine structure of the electron bunch is beyond the range

*Corresponding author. Deutsches Elektronen Synchrotorn (DESY), Notkestrasse 85, 22607 Hamburg, Germany. Tel.: +49-40-8998-2676; fax: +49-40-8998-4475.
E-mail address: yurkov@mail.desy.de (M.V. Yurkov).

of standard electronic display instrumentation, and different techniques based on the detection of the coherent synchrotron radiation (CSR) spectrum are developed. If both modulus and phase of the bunch form-factor were somehow measured, we would also know the Fourier spectrum of the bunch; an inverse Fourier transform of the measured data $\bar{F}(\omega)$ would then yield the desired profile function $F(t)$ with a resolution limited by the maximum achievable frequency

0168-9002/$ - see front matter © 2004 Elsevier B.V. All rights reserved.
doi:10.1016/j.nima.2004.04.081

range. Unfortunately, in practice, it is impossible to extract the phase information from the CSR measurement. A more realistic task is to measure the modulus of the bunch form-factor only. In Ref. [4] we proposed the method for measurements of the modulus of the bunch form-factor based on the detection of the Coherent Synchrotron Radiation (CSR) spectrum produced by a bunch passing an undulator. Relevant scientific background of this method is described in Ref. [5]. However, independently on the technique selected, the following problem arises, to study what information about the bunch profile can be derived from the collected experimental data: the problem is generally referred to as phase retrieval.

In the particular case of interest, namely an electron bunch originating by the XFEL driver accelerator, we can combine a priori knowledge about the bunch density distribution with measured information about the bunch form-factor. We know that, in the case under examination, the bunching process in zero order approximation can be treated by single particle dynamical theory. We cannot neglect the CSR induced energy spread and other wake fields when we calculate the transverse emittance, but we can neglect them when we calculate the longitudinal current distribution, simply because their contribution to the longitudinal dynamics is small. As a result, we will show that we can define a general temporal structure of the electron bunch after compression without any measurements. The purpose of the measurement is then to determine the numerical value of the parameters on which such temporal structure depends. In this case, sufficient information is provided by the modulus of the bunch form-factor, allowing us to ignore the phase information. The process of finding the best estimate of the profile function $F(t)$ for a particular measured modulus of the bunch form-factor, including utilization of any a priori information available about $F(t)$ refers to a technique called "constrained deconvolution", which will be exemplified below in detail. Although the context in which the suggestion of this method was made first was spectroscopy [6], the ideas apply equally well in the present context of bunch profile reconstruction.

2. Constrained deconvolution

Let us consider an example which shows the deconvolution process in easy-to-understand circumstances. For emphasis we choose, here, an experiment aimed at measuring the strongly non-Gaussian electron bunch profile at TTF, Phase 2 in femtosecond-mode operation. In order to study the CSR diagnostic for a compressed bunch, first let us find the longitudinal charge distribution for our bunch model when it is fully compressed by a chicane (here we assume that the perturbation of the bunching process by CSR is negligible). Downstream of the chicane the charge distribution is strongly non-Gaussian with a narrow leading peak and a long tail. Our attention is focused on determining the longitudinal density distribution after compression. As first was shown in Ref. [7] there is an analytical function which provides quite a good approximation for the simulation data. The current distribution along the beam after a single compression is given by

$$I(t) \simeq \begin{cases} I_0 \exp\left(-\dfrac{t^2}{2\tau_0^2}\right) & \text{for } t_1 > t > -\infty, \\[2mm] \dfrac{A\exp(-t/\tau_1)}{\sqrt{(t+t_0)/\tau_1}} & \text{for } t > t_1 > 0. \end{cases} \quad (1)$$

To illustrate how good an approximation to the actual current distribution is provided by Eq. (1) we show, in Fig. 1, the current as a function of time, first from numerical simulation data, then as

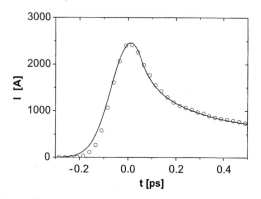

Fig. 1. Current distribution along the bunch reconstructed from an ideal set of experimental data for the form-factor and correct value of time constant for the bunch tail (solid curve). Circles present data from a tracking simulation code.

computed from Eq. (1). Eq. (1) involves four independent parameters in the calculation of longitudinal density distribution: τ_0, which characterizes the local charge concentration (non-Gaussian feature) of the compressed density, τ_1, which characterizes the bunch tail and fitting parameters, t_0 and t_1. In this case detailed information on the longitudinal bunch distribution (actually constants: τ_0, τ_1 and t_0, t_1) can be derived from the particular measured square modulus of the bunch form-factor by using an independently obtained tail constant τ_1 (for example by means of a streak camera). A quantity of considerable physical interest is the parameter τ_0 which characterizes the leading peak. The value of τ_0 can be related to the value of the initial local energy spread $\Delta\gamma/\gamma$ in the electron beam if the compaction factor R_{56} of the magnetic chicane is known. Fig. 2 shows plots of the peak current

versus time for several values of the local energy spread. By comparing these curves, some feeling can be obtained about the effective reduction in peak current.

The highest time resolution obtainable by means of a streak camera is in the sub-picosecond range. By means of coherent radiation, the resolution can be made comparatively high because it is determined by the spectral range in the measurement. Hence, this method will be useful for monitoring an ultra-short electron bunch at TTF FEL. In the TTF FEL Phase 2 the measurement of CSR spectrum can be made by using long period electromagnetic undulator [5]. As already said, for the complete determination of the bunch profile function, downstream of the first TTF FEL bunch compressor, a measurement of the bunch length can be made independently by using a sub-picosecond resolution streak camera. The limiting resolution of the streak camera will prevent the measurement of the spike width. Nevertheless, the streak camera can detect the long time constant characterizing the tail, which is outside the CSR detector response.

The significance of the proposed scheme cannot be fully appreciated until we determine typical values of the energy in the radiation pulse detected within the central cone that can be expected in practice. In the case of TTF FEL, the bunch form-factor modulus falls off rapidly for wavelengths shorter than 40 μm. At the opposite extreme, the dependence of the form-factor on the exact shape of the electron bunch is rather weak and can be ignored outside the wavelength range 40 μm < λ < 100 μm. This point is made clear upon examination of the curves presented in Fig. 2. The energy in the radiation pulse can be estimated simply as in Ref. [5]: in the 40–100 μm wavelength region of the CSR spectrum any local energy spread in the electron beam smaller than 10 keV produces energies larger than 1 μJ in the radiation pulse.

To conclude this section we present examples of deconvolution of computer generated data. By simulating a spectrum rather than using an actual CSR spectrometer to record it, we have a complete control over such factors as resolution, shape and noise. Also, we know what the perfectly

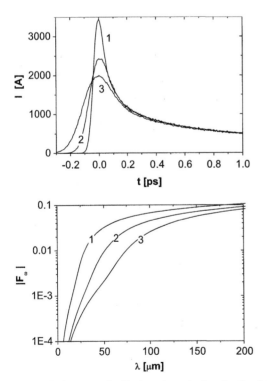

Fig. 2. Typical current distribution along the bunch after full compression with a single bunch compressor (upper plot), and form-factor of the electron bunch (lower plot). Curves 1, 2, and 3 correspond to local energy spread of 5, 10, and 15 keV, respectively, at the entrance of the bunch compressor. The head of the bunch is at the left side. The charge of the bunch is 3 nC.

deconvolved profile function would look like. In the deconvolution of actual CSR spectral data, the presence of noise is usually the main limiting factor: the effects of noise on deconvolution are demonstrated in the following. For the purpose of examining the deconvolution process, we begin with noisy data, which of course, can be realized in a simulation process. When other aspects of deconvolution, such as errors in the prior information are examined, noiseless data are used. One important question to answer is about how well the modulus of bunch form-factor must be known in order to perform a successful deconvolution. To obtain the sensitivity to error effects, we use a set of input error data for a simulated bunch profile spectrum (see Fig. 3). A 40% rms error (variable over the undulator scan) is included. The sensitivity of the final bunch profile function to the simulation input errors are summarized in Fig. 4. Only one random seed has been presented here as an example, but several seeds have been run with similar success. Fig. 4 shows that rms of CSR measurement should be within $\pm 50\%$. It is seen that the constrained deconvolution method is insensitive to the errors in the bunch form-factor.

As shown above, the bunch profile function is given by Eq. (1). An important issue to deal with consists in determining the dependence of the deconvolved function on the tail constant. One way to investigate this matter is to use a set of tail constants which differs from the real one in the

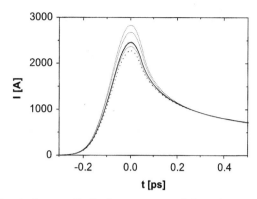

Fig. 4. Current distribution reconstructed from five sets of measurements with relative error $\pm 40\%$ (see Fig. 3). The solid line corresponds to an ideal measurement. The tail time constant in all cases is 9 ps.

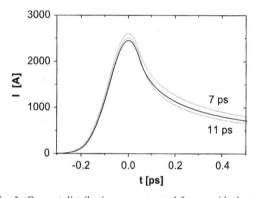

Fig. 5. Current distribution reconstructed from an ideal set of experimental data for the form-factor (see Fig. 3), but with different values of time constants, 7 and 11 ps, for the bunch tail (dotted lines). The solid line corresponds to the correct value of the time constant, 9 ps.

deconvolution of a simulated spectrum. Fig. 5 shows the results of such a test. At first glance all three traces of Fig. 5 are practically identical, indicating that the tail parameter is less important than it would have probably been expected. We can therefore conclude that the width of the tail parameter should be within $\pm 50\%$ of the proper value.

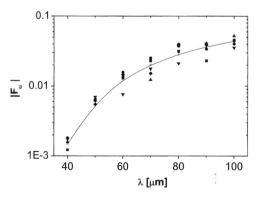

Fig. 3. Electron bunch form-factor corresponding to the case of 10 keV energy spread at the entrance of the bunch compressor. The solid curve presents an ideal measurement. Symbols present five sets of measurements with relative error $\pm 40\%$.

References

[1] F. Richard, et al., (Eds.), TESLA Technical Design Report, DESY 2001-011.

[2] The LCLS Design Study Group, LCLS Design Study Report, SLAC reports SLAC- R521, Stanford, 1998.

[3] V. Ayvazyan, et al., Phys. Rev. Lett. 88 (2002) 104802.

[4] G. Geloni, et al., Nucl. Instr. and Meth. A, (2004) these proceedings.

[5] G. Geloni, et al., preprint DESY 03-031, DESY, Hamburg, 2003.

[6] P.A. Jansson (Ed.), Deconvolution with Application in Spectroscopy, Academic Press, Inc., New York, 1984.

[7] R. Li, Nucl. Instr. and Meth. A 475 (2001) 498.

Available online at www.sciencedirect.com

SCIENCE ⒹDIRECT°

NUCLEAR INSTRUMENTS & METHODS IN PHYSICS RESEARCH
Section A

Nuclear Instruments and Methods in Physics Research A 528 (2004) 335–339

www.elsevier.com/locate/nima

Emittance compensation in a return arc of an energy-recovery linac

R. Hajima*

Japan Atomic Energy Research Institute, 2-4 Shirakata-Shirane, Tokai-mura, Ibaraki 319–1195, Japan

Abstract

Dilution of the electron beam emittance through a return arc is one of the critical problems in the design of an energy-recovery linac for a high-power FEL and synchrotron light sources. This study reports that the emittance dilution can be compensated by matching a beam envelope to the CSR-induced dispersion. In linear regime, this optimized design of an achromatic cell is obtained from a simple matrix calculation.
© 2004 Elsevier B.V. All rights reserved.

PACS: 29.27.Bd; 41.60.Ap; 41.60.Cr

Keywords: FEL; Energy recovery linac; CSR; Emittance

1. Introduction

An energy-recovery linac (ERL) is considered to be a promising device for a high-power free-electron laser (FEL) [1] and a next-generation synchrotron light source [2]. Since the generation of electron bunches with durations shorter than picoseconds is an essential requirement for such devices, emittance dilution arising from coherent synchrotron radiation (CSR) is one of the critical issues in the design of ERLs.

For an electron bunch of Gaussian temporal distribution, rms energy spread caused by CSR is

$$\Delta E_{\mathrm{rms}} = 0.2459 \frac{eQL_{\mathrm{b}}}{4\pi\varepsilon_0\rho^{2/3}\sigma_{\mathrm{s}}^{4/3}}, \qquad (1)$$

*Tel.: +81-29-282-6315; fax: +81-29-282-6057.
E-mail address: hajima@popsvr.tokai.jaeri.go.jp (R. Hajima).

where Q is the bunch charge, L_{b} is the bending path length, ρ is the bending radius, and σ_{s} is the rms bunch length. This energy spread induced in an achromatic cell results in the growth of projection emittance at the cell exit. It is known that this emittance growth can be compensated by setting the cell-to-cell betatron phase advance at an appropriate value [3].

The present paper reports that the emittance compensation can be achieved in a single achromatic cell by matching a beam envelope to the CSR-induced dispersion.

2. Weak CSR, linear analysis

When the electron energy E_0 is much larger than the CSR-induced energy spread, $\Delta E/E_0 \ll 1$, a linearized approximation can be adopted to solve the electron dynamics. The CSR-induced energy spread results in the displacement of bunch slices in (x, x') phase space at the end of an achromatic cell. In the linear regime, it can be assumed that all the bunch slices align on a single line as shown in Fig. 1, which also indicates that the projection emittance depends on the orientation of the CSR kick and the phase ellipse, and has a minimum value, when the achromatic cell is designed so that the CSR kick coincides with the orientation of the phase ellipse at the cell exit. This study shows that an achromatic cell for such minimum emittance can be designed by a simple matrix calculation.

A first-order equation of electron motion in a uniform field of a dipole magnet is

$$x'' = -\frac{x}{\rho^2} + \frac{1}{\rho}(\delta_0 + \delta_{csr} + \kappa[s - s_0]), \qquad (2)$$

where the variable x denotes the deviation of an electron from the ideal path in the bending plane, x'' is the second-order derivative of x with respect to a coordinate along the ideal path s, ρ is the bending radius, and δ_0 is the initial momentum deviation at $s = 0$ normalized by the reference momentum. The last two terms on the right-hand side have been added to calculate the CSR effect: δ_{csr} is the normalized momentum deviation caused by CSR in the upstream path $(0 < s < s_0)$, $s = s_0$ is the entrance of the bending magnet, $\kappa = W/E_0$ is

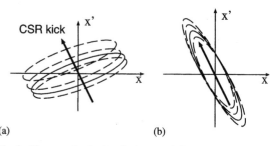

Fig. 1. Electrons in the (x, x') plane and displacement of beam slices due to the CSR kick. The emittance growth depends on the CSR kick direction and the phase ellipse orientation.

following conditions: (1) all the dipole magnets have the same bending radius, (2) the electron bunch does not change its longitudinal profile, and (3) the transient CSR effect at the entrance and exit of the magnet is not large. The first two conditions are reasonable for a return arc of high-energy ERLs, but the last condition is delicate to apply. The effect of transient CSR will be examined later in a numerical example.

In the constant CSR wake regime, Eq. (2) can be solved analytically, and electron dynamics through an achromatic cell is calculated using the extended R-matrix [4]. A vector was defined to express electron motion: $\vec{x} = (x \ x' \ \delta_0 \ \delta_{CSR} \ \kappa)^T$ and an R-matrix for a sector magnet

$$R_{bend} = \begin{pmatrix} \cos\theta & \rho\sin\theta & \rho(1-\cos\theta) & \rho(1-\cos\theta) & \rho^2(\theta-\sin\theta) \\ -\frac{1}{\rho}\sin\theta & \cos\theta & \sin\theta & \sin\theta & \rho(1-\cos\theta) \\ 0 & 0 & 1 & 0 & 0 \\ 0 & 0 & 0 & 1 & \rho\theta \\ 0 & 0 & 0 & 0 & 1 \end{pmatrix}, \qquad (3)$$

the normalized CSR wake potential in the bending path determined by CSR wake potential W and the reference energy E_0.

Although the wake potential is a function of the bending radius, the bunch profile, and the position of an electron in the bunch, in the following discussion, it was assumed that each electron experienced constant CSR wake through the entire bending path. This assumption is valid under the

and similar R-matrices for a drift and a quadrupole, by which the electron motion through a beam transport can be tracked,

$$\vec{x}(s_1) = R_{0 \to 1} \ \vec{x}(s_0). \qquad (4)$$

Following the momentum dispersion function to calculate the effect of initial momentum error, (η, η'), the CSR wake dispersion function, (ζ, ζ'), defined below was introduced:

$$(\zeta_x(s_1) \ \zeta'_x(s_1) \ 0 \ L_b(s_1) \ 1)^T$$
$$= R_{0 \to 1} \ (\zeta_x(s_0) \ \zeta'_x(s_0) \ 0 \ L_b(s_0) \ 1)^T \quad (5)$$

where $L_b(s_1)$ is the total bending path length for $0 < s < s_1$.

Using the CSR wake dispersion function, it is possible to calculate the displacement of electrons in the (x, x') phase space, and the emittance growth due to the CSR effect. Unnormalized emittance at an arbitrary position along the beam transport is obtained by

$$\varepsilon^2 = (\varepsilon_0 \beta_x + D^2)(\varepsilon_0 \gamma_x + D'^2) - (\varepsilon_0 \alpha_x - DD')^2 \quad (6)$$

where $(\alpha_x, \beta_x, \gamma_x)$ are Courant–Snyder parameters, ε_0 is the initial emittance and (D, D') is the rms spread of the CSR wake dispersion obtained by

$$(D, D') = \Delta\kappa_{rms}(\zeta_x, \zeta'_x). \quad (7)$$

$$\Delta\kappa_{rms} = \Delta E_{rms}/E_0/L_b. \quad (8)$$

Terms related to the initial momentum error have been omitted from Eq. (6), because the emittance is usually evaluated where the momentum dispersion is zero, $\eta = \eta' = 0$.

As an example of the linear analysis, the emittance growth in a triple-bend achromatic cell (TBA) was calculated: bending radius $\rho = 25$ m, bending angle $\theta = 3 + 6 + 3 = 12°$, and the quadrupoles inside the cell were chosen as isochronous parameters. Fig. 2 shows the betatron envelope, the momentum dispersion and the CSR dispersion functions for the bending plane, in which the strength of the quadrupoles outside the cell has been determined so that β_x and ζ_x have same envelope after the cell. This matched envelope corresponds to the coincidence of the CSR kick with the beam ellipse in the (x, x') phase space, which results in the minimum emittance.

The emittance compensation in the optimized TBA cell has been confirmed by particle tracking code ELEGANT including CSR calculation [5]. The initial electron bunch parameters were assumed to be as follows: central energy $E_0 = 3.07$ GeV, bunch charge $Q = 770$ pC, normalized rms emittance $\varepsilon_{n,0} = 0.1$ mm-mrad, longitudinal size $\sigma_s = 30$ μm (100 fs), uncorrelated energy spread

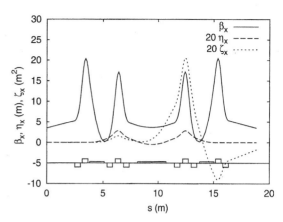

Fig. 2. Betatron function β_x, momentum dispersion η_x and CSR wake dispersion ζ_x of an optimized triple-bend achromatic cell.

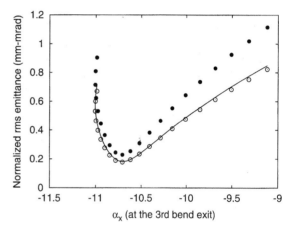

Fig. 3. Emittance growth as a function of α_x at the 3rd bend exit. First-order matrix (solid line), particle tracking without transient CSR (○), particle tracking with transient CSR (●).

$\sigma_E/E_0 = 0.02\%$, and Gaussian distribution of electrons in the 6D phase space. Projection emittance at the cell exit is calculated with scanning Courant–Snyder parameters at the cell entrance: $-5.0 < \alpha_x < 3.0$, $\gamma_x = 0.29$ m^{-1}.

Fig. 3 shows the calculated projection emittance as a function of α_x at the third bend exit. It has been confirmed that the projection emittance is a function of the beam envelope inside the TBA cell and can be estimated by matrix calculation. The matrix approach gives an optimized design for emittance compensation even when the transient CSR is included.

3. Strong CSR, nonlinear regime

When the CSR-induced energy spread is relatively large, higher-order terms should be taken into account. It was observed that the emittance growth due to the CSR in the nonlinear regime was related to the phase ellipse parameters at the arc exit as well as the linear analysis.

Emittance growth in the JAERI-ERL arc is calculated, where an electron bunch, $E_0 = 17$ MeV, $Q = 0.5$ nC, $\sigma_s = 660$ μm (2.2 ps), $\sigma_E/E_0 = 0.2\%$, and $\varepsilon_n = 20$ mm-mrad, travels through a triple-bend arc, $\rho = 0.23$ m, $\theta = 60 \times 3 = 180°$. The CSR-induced energy spread after the arc was found to be 38 keV (0.22%) from Eq. (1).

From the simulations by ELEGANT, it can be seen that the emittance at the arc exit is a function of the beam envelope inside the arc. Figs. 4 and 5 show the optimized beam envelope for the minimum emittance growth and the emittance values for various beam envelope, respectively. In the emittance calculation, the ellipse parameters at the arc entrance were scanned: $-2 < \alpha_x < 2$, $\gamma_x = 3.6$ m^{-1}. The emittance growth by the second-order aberration for the initial rms energy spread, $\sigma_E/E_0 = 1\%$, was also plotted, where the CSR effect was absent.

Fig. 5 shows that the emittance growth due to the transient CSR in the JAERI-ERL arc cannot

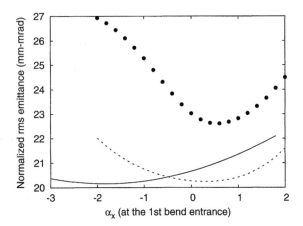

Fig. 5. Emittance growth as a function of α_x at the arc entrance. First-order matrix (solid line), particle tracking with transient CSR (●), Second-order aberration for large initial energy-spread, $\sigma_E/E_0 = 1\%$, without CSR (dashed line).

be estimated by linear matrix analyses. This is because the horizontal beam size is not small enough compared with the bending radius. Moreover, the CSR-induced energy spread is also relatively large, thus, the first-order approximation is not adequate in this case. However, there exists an optimum beam envelope for the minimum emittance growth. Further investigations on the correlation between the CSR-induced emittance growth and the second-order aberration shown in Fig. 5 should be conducted in future.

4. Conclusions

This study shows that CSR-induced emittance growth in an achromatic cell of an ERL return arc can be minimized by choosing an appropriate beam envelope inside the cell. The optimum cell design is derived from matrix calculations in the linear regime, and from particle tracking in the nonlinear regime.

Acknowledgements

This work was supported in part by Japan Society for the Promotion of Science, Grant-in-Aid for Scientific Research (B) 15360507.

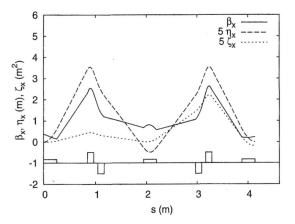

Fig. 4. Betatron function β_x, momentum dispersion η_x, and CSR wake dispersion ζ_x of the optimized JAERI-ERL arc.

References

[1] G.R. Neil, et al., Phys. Rev. Lett. 84 (2000) 662;
R. Hajima, et al., Nucl. Instr. and Meth. A507 (2003) 115.

[2] S.M. Gruner, et al., Rev. Sci. Instrum. 73 (2002) 1402.

[3] D. Douglas, Thomas Jefferson, National Accelerator Laboratory Technical Note, JLAB-TN-98-012 (1998);
J.H. Wu, Proceedings of the PAC-2001, Chicago, USA, IEEE, New Jersey, 2001, p. 2866.

[4] R. Hajima, Jpn. J. Appl. Phys. 42 (2003) L974.

[5] ELEGANT version 14.8, Nov. 26, 2002;
M. Borland, Argonne National Laboratory Advanced Photon Source Report No. LS-287, 2000.

Available online at www.sciencedirect.com

Nuclear Instruments and Methods in Physics Research A 528 (2004) 340–344

ELSEVIER

NUCLEAR INSTRUMENTS & METHODS IN PHYSICS RESEARCH
Section A

www.elsevier.com/locate/nima

A full-DC injector for an energy-recovery linac

R. Hajima*, E.J. Minehara, R. Nagai

Japan Atomic Energy Research Institute, 2-4 Shirakata-Shirane Tokai-mura, Ibaraki 319-1195, Japan

Abstract

We propose a full-DC injector for an energy-recovery linac. The injector is based on a 2 MeV DC accelerator that is commercially available (Dynamitron type), combined with a photo cathode. A bunched beam from the DC accelerator is merged with a recirculating beam and injected into a superconducting buncher-booster, which accelerates the beam up to ~20 MeV. The buncher-booster is operated in partial energy-recovery mode, in which we can reduce the capacity of the RF generators to 10 kW even for a high-average current operation, 100 mA. We present results of beam dynamics simulation and RF system optimization.

PACS: 29.27.Ac; 41.60.Ap; 41.60.Cr

Keywords: FEL; Energy recovery linac; Injector; DC accelerator

1. Introduction

An energy-recovery linac (ERL) is expected as a driver of high-power free-electron lasers [1] and next generation light sources [2]. To realize such ERL machines in the future, one of the most critical issues is the development of an injector that fulfills the requirements of high-brightness and a high-average current, simultaneously. Two types of injectors have been proposed for future ERLs: a photo-cathode RF gun and a photo-cathode DC gun. The latter one, a DC gun combined with a GaAs cathode has been used in JLAB/IR-demo

with a 5 mA current, and attempts to increase the average current towards 100 mA are underway [3].

An electron bunch generated from a GaAs cathode has a temporal duration of more than 10 ps, which is determined by the laser absorption depth of the cathode surface. In order to obtain a subpicosecond electron bunch at the undulator, the electron bunch must be compressed to a few picoseconds before the main linac. In previous studies, the bunch compression before the main linac was conducted by a superconducting booster [4], which was located before the merger chicane. The adoption of the booster cavity introduces a practical problem—a large amount of RF power consumption, and a technological challenge—the development of a high-power RF coupler for superconducting cavities. Using the typical configuration of a booster, we need to feed an RF of

---

*Corresponding author. Tel.: +81-29-282-6315; fax: +81-29-282-6057.

E-mail address: hajima@popsvr.tokai.jaeri.go.jp (R. Hajima).

0168-9002/$ - see front matter © 2004 Elsevier B.V. All rights reserved.
doi:10.1016/j.nima.2004.04.064

100 kW (CW) through the main coupler, which is much larger than the TESLA design parameter with an average power of 1.4 kW [5].

In the present paper, we propose a full-DC injector for future high-average current ERLs. The injector is based on a Dynamitron type DC accelerator, which is a commercial product for electron beam irradiation facilities [6]. In this study, we use velocity bunching inside an ERL loop, where partial energy-recovery is established, and the RF input power of the booster cavities is largely reduced to 10 kW.

2. Design of a full-DC injector

In this section, we show a sample design of a full-DC injector for a high-average current ERL, based on TESLA superconducting cavities. The basic parameters of the injector are listed in Table 1.

We use a DC accelerator of a commercial standard size, and a 200 kW beam power. A 250 kV DC gun with a GaAs cathode and a solenoid focusing magnet are installed in the DC accelerator as shown in Fig. 1. The solenoid magnet can be replaced by an electrostatic lens. An electron bunch from the DC accelerator is

Table 1
Parameters for the sample design

Gun	
Cathode	GaAs
Anode voltage (kV)	250
Charge (pC)	77
RMS bunch length (ps)	17
Repetition rate (GHz)	1.3
Average current (mA)	100
DC accelerator	
Acc. voltage (MV)	2
Gradient (MV/m)	2
Merger dipole	
Bending angle (deg.)	15
Bending radius (m)	1
Drift between dipoles (m)	0.82

merged with the ERL loop by a 3-dipole chicane and compressed via velocity bunching in super-conducting buncher-booster cavities consisting of 3-cell and 9-cell cavities as shown in Figs. 2 and 3. The accelerating voltage is chosen as 1.2 MV for each 3-cell cavity and 20 MV for the 9-cell cavity. All the cavities are assumed to have a structure that is similar to the TESLA cavity, and $Q_0 = 5 \times 10^9$.

Quadrupole magnets before and after the merger were installed to provide envelope matching at the beam injection into the ERL loop. A quadrupole doublet inside the buncher-booster produces a small size beam at the high-gradient 9-cell cavity to maintain small beam emittance.

3. Beam dynamics simulations

Electron beam motion in the designed full-DC injector was studied by particle tracking code, PARMELA [7]. Field distribution for the DC accelerator and the superconducting cavities were prepared by POISSON and SUPERFISH, respectively [8]. The focusing strength of the solenoid and the quadrupoles were determined to minimize the transverse emittance after the bunch compression. The initial bunch parameters at the exit of the 250 kV gun were assumed as a bunch length of 17 ps (rms) and normalized rms transverse emittance of $\varepsilon_{n,rms} = 0.5$ mm-mrad including the thermal emittance for a laser spot size of 1 mm radius, $\varepsilon_{n,rms,th} = 0.13$ mm-mrad.

The simulation results for transverse and longitudinal beam dynamics are plotted in Fig. 4. The electron bunch parameters after the bunch compression are $E = 23$ MeV, $\varepsilon_x = 1.5$ mm-mrad, $\varepsilon_y = 1.5$ mm-mrad, and $\sigma_z = 3.3$ ps (rms).

The transverse emittance is diluted by various sources along the beam transport. In the DC accelerator, radial expansion of the beam caused by space charge force results in the emittance growth. It is known that this type of emittance growth can be compensated by applying an external solenoid field which makes the laminar flow of the beam [9]. In the present design, however, only a single solenoid is installed just

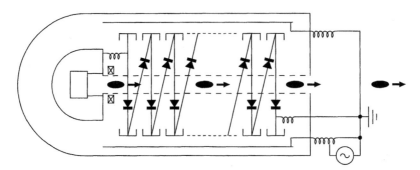

Fig. 1. Dynamitron type DC accelerator equipped with a photo-cathode gun.

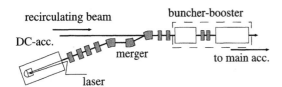

Fig. 2. The layout of a full-DC injector.

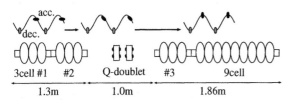

Fig. 3. The configuration of the buncher-booster for velocity bunching.

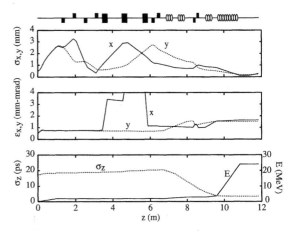

Fig. 4. Simulation results: transverse beam envelope, emittance, bunch length, and energy of electrons.

after the anode and the emittance compensation is imperfect.

The emittance growth noted in the merger chicane is due to the energy redistribution during a dispersive path and appears in x-plane only. There are two sources of energy redistribution, longitudinal space charge force and coherent synchrotron radiation. However, in our calculation, only the longitudinal space charge force is taken into account. Simulations including coherent radiation will be considered in future work.

The emittance growth during the bunch compression appears in both x and y planes. The radial electric force due to the off-crest acceleration combined with the longitudinal mixing of the

bunch slices results in the transverse emittance growth.

Figs. 5 and 6 shows a longitudinal phase plot and temporal profile of an electron bunch at the exit of the 9-cell cavity. The bunch duration is 3 ps as an rms value, but has a narrow peak and a rather long tail because of the nonlinearity in the compression scheme. Such a steep profile is not preferred for high-current operation of ERLs, because it contains a higher frequency component of the beam current in comparison with a bunch having a smooth profile such as Gaussian, and induces larger HOM power in the superconducting accelerators. Possible linear compression with an additional third-harmonic cavity remains to be studied in a future work.

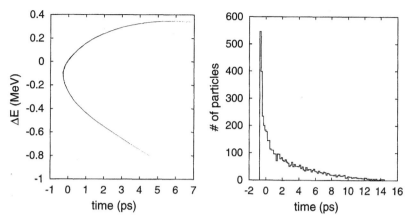

Fig. 5. Simulation results: longitudinal phase plot and temporal profile of an electron bunch at the exit of the 9-cell cavity.

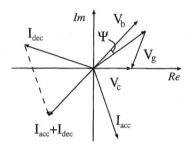

Fig. 6. RF vector diagram for the 3-cell cavity. Accelerating beam current, I_{acc}; decelerating beam current, I_{dec}; cavity voltage, V_c; beam induced voltage, V_b; generator voltage, V_g; and cavity detuning angle, Ψ.

4. Optimization of the RF parameters

The RF power balance in the superconducting cavities is determined from the current and phase of the accelerated and the decelerated beams, the Q value and the detuning angle of the cavities under load [10]. The phase difference between the two beams is not 180° in 3-cell cavities that are optimized for velocity bunching. The vector sum of the two beams is, however, in the decelerating direction. Therefore, ideally, it is possible to achieve perfect energy-recovery and make the RF power supply zero, by choosing an appropriate loaded Q value and detuning. It should be noted that the perfect energy-recovery requires a high-speed cavity tuner to change the detuning angle from zero at beam-off to a certain value at beam-on.

Table 2
Optimized RF parameters for the 3-cell cavities: bunch phase for acceleration Ψ_{acc} and deceleration Ψ_{dec} (degree), loaded Q, cavity detuning δf (Hz), and generator power P_g (kW)

	Ψ_{acc} / Ψ_{dec}	Q_L	δf	P_g
#1	−70/162	4.8×10^4	2700	9.9
#2	−67/150	6.5×10^4	1700	6.2
#3	77/241	1.3×10^5	400	1.8

To avoid a feasibility discussion on such a high-speed tuner, we fix the detuning angle and the loaded Q value of each cavity in the present design, so that equal amounts of RF power is consumed in both cases, beam-off and beam-on. Table 2 shows the optimized RF parameters for the 3-cell cavities under these conditions. The generator power fed through the RF coupler is less than 10 kW, i.e. one-order smaller than the accelerating beam power. This reduction of the coupler power in the 3-cell cavities is due to the partial energy-recovery. The generator power for the 9-cell cavity, in which perfect energy-recovery is expected, is determined by the control margin of the RF stability and estimated to be about 8 kW [4].

5. Conclusions

This paper presents a full-DC injector for an energy-recovery linac. A Dynamitron type

DC accelerator equipped with a GaAs photo cathode gun enables the generation of a 100 mA CW electron beam with small emittance, $\varepsilon_{n,rms} < 2$ mm-mrad. The electron bunch from the DC accelerator is compressed to 3 ps (rms) via velocity bunching in a superconducting buncher-booster after it is merged with a recirculating beam. The amount of RF power supplied to the buncher-booster is greatly reduced by partial energy-recovery operation. Estimation of the emittance growth by coherent synchrotron radiation, calculation of the effect of voltage ripple in the DC accelerator, and further optimization of bunch compression by adding a third-harmonic cavity, remains to be studied in a future work.

References

[1] G.R. Neil, et al., Phys. Rev. Lett. 84 (2000) 662;
    R. Hajima, et al., Nucl. Instr. and Meth. A 507 (2003) 115.
[2] S.M. Gruner, et al., Rev. Sci. Istrum. 73 (2002) 1402.
[3] T. Siggins, et al., Nucl. Instr. and Meth. A 475 (2001) 549;
    C. Sinclair, in Proceedings of PAC-2003, Portland, 2003.
[4] S.M. Gruner, M. Tigner (Eds.), Cornell University, CHESS Technical Memo 02-003, JLAB-ACT-01-04, 2001.
[5] B. Aune, et al., Phys. Rev. ST-AB 3 (2000) 092001.
[6] C.C. Thompson, M.R. Cleland, Nucl. Instr. and Meth. B 40/41 (1981) 1137.
[7] L.M. Young, PARMELA ver. 3.26, Los Alamos National Laboratory, LA-UR-96-1835.
[8] J.H. Billen, L.M. Young, POISSON SUPERFISH ver. 6.12, Los Alamos National Laboratory, LA-UR-96-1834.
[9] L. Serafini, J.B. Rosenzweig, Phys. Rev. E 55 (1997) 7565.
[10] H. Padamsee, et al., RF Superconductivity for Accelerators, Wiley, New York, 1998.

Available online at www.sciencedirect.com

SCIENCE  DIRECT®

ELSEVIER   Nuclear Instruments and Methods in Physics Research A 528 (2004) 345–349

NUCLEAR
INSTRUMENTS
& METHODS
IN PHYSICS
RESEARCH
Section A

www.elsevier.com/locate/nima

Impact of beam energy modulation on rf zero-phasing microbunch measurements

Z. Huang[a],*, T. Shaftan[b]

[a] Stanford Linear Accelerator Center, Stanford, CA 94309, USA
[b] Brookhaven National Laboratory, Upton, NY 11973, USA

Abstract

Temporal profile of a simple bunch distribution may be obtained by measuring the horizontal density profile of an energy-chirped electron beam at a dispersive region using the rf zero-phasing technique. For an energy-modulated beam, the horizontal profile obtained by this technique is also modulated with an enhanced amplitude. We study the microbunching experiment at the NSLS source development laboratory and show that the horizontal modulation observed by the rf zero-phasing technique can be explained by the space–charge-induced energy modulation in the accelerator.
© 2004 Elsevier B.V. All rights reserved.

PACS: 29.27.Bd; 29.30.Aj

Keywords: RF zero-phasing; Space–charge oscillation; High-brightness electron beam

1. Introduction

Time-resolved measurements of very short electron bunches are essential for free-electron lasers (FEL), linear colliders and other advanced accelerators. Diagnostic techniques that have femtosecond resolutions are of particular interests (see, e.g., Ref. [1]). Among them, the rf zero-phasing technique [2] is relatively straightforward to implement since accelerating cavities and regions of nonzero dispersion are usually available at linear accelerators. Knowledge of the horizontal density profile at the dispersive region determines

the energy profile of the bunch, which in turn can be related to the current profile for a linearly chirped beam. Recent applications of such a technique at the NSLS source development laboratory (SDL) yield unexpected information about the longitudinal bunch distribution [3]: the measured horizontal density profiles show large high-frequency modulations after the beam is compressed by a bunch compressor chicane. Analysis based on the recently developed coherent synchrotron radiation (CSR) instability in the bunch compressor [4,5] does not yield micro-bunching that supports the observed structures [6]. Note that similar structures in energy spectra of chirped electron beams due to off-crest rf acceleration have been reported in Refs. [7–9].

---

*Corresponding author.
E-mail address: zrh@slac.stanford.edu (Z. Huang).

0168-9002/$ - see front matter © 2004 Elsevier B.V. All rights reserved.
doi:10.1016/j.nima.2004.04.065

Motivated by these experimental observations, we perform a thorough analysis of the rf zero-phasing technique for a general beam distribution. We show that the horizontal density modulation obtained by this method is significantly enhanced due to the beam energy modulation. For the SDL experiment, the energy modulation can be induced by the longitudinal space charge (LSC) force in the linac and can dominate the horizontal spectrum.

2. Analysis of rf zero-phasing technique

The rf zero-phasing technique uses one or several rf cavities operated at the zero accelerating phase to impart a large energy–time correlation to the beam. The energy-chirped beam is then dispersed horizontally by a spectrometer dipole and intercepted by a measurement screen. The horizontal profile of the beam at the screen is used to determine its energy profile, which can be related to its temporal profile for a beam with a smooth, linear energy–time phase space distribution (see Fig. 1(a)).

In general, we consider a distribution function $f(x, x', z, \delta; s)$ in a dispersive region at a distance s from the beginning of the spectrometer dipole. Here x and $x' \equiv dx/ds$ are the horizontal phase space coordinates, z is the longitudinal (temporal) coordinate centered in the bunch ($z > 0$ for the head of the bunch), and $\delta = \Delta E/E_0$ is the relative energy deviation for a beam with the average energy $E_0 = \gamma mc^2$. The Fourier transformation of

the horizontal beam profile is

$$a(k_m; s) = \frac{1}{N} \int dX e^{-ik_m x} f(X; s)$$
$$= \frac{1}{N} \int dX_d e^{-ik_m x(X_d)} f_d(X_d), \qquad (1)$$

where $X = (x, x', z, \delta)$, $N = \int dX f(X; s)$ is the total number of electrons, k_m is the measured modulation wave-number, $f_d(X_d)$ is the beam distribution at the entrance of the dipole (referred to by the subscript d), and we have applied the Liouville theorem in transforming the phase space from s to the beginning of the dipole. (Collective effects such as CSR in the spectrometer dipole is small due to Landau damping from beam emittance [4,5,10].)

Let us assume that the beam at the entrance of the dipole has both a small longitudinal density variation $\Delta n(z_d)$ as well as a small energy variation $\Delta\delta(z_d)$, i.e.,

$$f_d(X_d) = \frac{n_0 + \Delta n}{2\pi\epsilon_x \sqrt{2\pi}\sigma_\delta} \exp\left[-\frac{(\delta_d - hz_d - \Delta\delta)^2}{2\sigma_\delta^2}\right]$$
$$\times \exp\left[-\frac{x_d^2 + (\beta_d x_d' + \alpha_d x_d)^2}{2(\sigma_x)_d^2}\right]. \qquad (2)$$

where n_0 is the average line density, ϵ_x is the horizontal emittance, σ_δ is the rms incoherent energy spread, α_d and β_d are the twiss parameters, $(\sigma_x)_d = \sqrt{\epsilon_x \beta_d}$ is the rms horizontal beam size, and

$$h = \frac{eV_{rf} k_{rf} \cos\phi}{E_0} \qquad (3)$$

is the energy chirp generated by an accelerating voltage V_{rf} with the rf wavelength $\lambda_{rf} = 2\pi/k_{rf}$ at a phase ϕ ($\phi = 0$ or π for zero-crossing). Note that the horizontal position x at s is

$$x(X_d) = C(s)x_d + S(s)x_d' + \eta(s)\delta_d, \qquad (4)$$

where $C(s), S(s), \eta(s)$ are the cosine-, sine-like, and dispersion functions, respectively. Inserting Eq. (4) into Eq. (1) and keeping only first-order terms in Δn and $\Delta\delta$, we obtain

$$a(k_m; s) = \left[b_d(k) - i\frac{k}{h}p_d(k)\right] e^{-\frac{k^2\sigma_\delta^2}{2h^2} - \frac{k^2\sigma_x^2(s)}{2\eta^2 h^2}}, \qquad (5)$$

where $k = k_m(s)\eta(s)h$ is the initial modulation wave number, $b_d(k)$ and $p_d(k)$ are the current

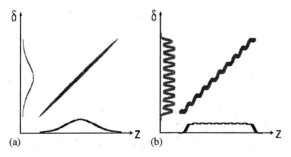

(a)   (b)

Fig. 1. Current and energy profiles of a chirped beam (a) without energy modulation, (b) with energy modulation.

and energy modulations at the beginning of the dipole, i.e.,

$$b_d(k) = \frac{1}{N} \int dz_d \, \Delta n(z_d) e^{-ikz_d},$$

$$p_d(k) = \frac{1}{N} \int dz_d \, \Delta \delta(z_d) e^{-ikz_d} \qquad (6)$$

and $\sigma_x(s)$ is the rms horizontal beam size at s. Note that Eq. (5) is valid when $|b_d(k)| \ll 1$ and $|p_d(k)| \ll h/k$. In the absence of any initial energy modulation (i.e., when the longitudinal phase space correlation is linear), and if

$$\lambda = \frac{2\pi}{k} \gg \max\left(\frac{\sigma_\delta}{|h|}, \frac{\sigma_x(s)}{\eta|h|}\right), \qquad (7)$$

then

$$a[k_m(s); s] \approx b_d(k),$$

$$\int dx' dz \, d\delta f(\mathbf{X}; s) \approx \Delta n\left(\frac{x}{\eta(s)h}\right) \qquad (8)$$

reproduces the longitudinal density variation (i.e., the current profile) at the entrance of the dipole. The right-hand side of Eq. (7) determines the temporal resolution of the rf zero-phasing technique due to finite beam size and energy spread. For a high-brightness electron beam generated from a photocathode rf gun, the incoherent energy spread σ_δ is typically very small, and the horizontal beam size σ_x can be focused down to 100 μm or less at the measurement screen. If we take $|h| \approx 20$ m^{-1} and $\eta \approx 1$ m, Eq. (7) indicates a temporal resolution of about 5 μm or 17 fs.

However, if the longitudinal phase space distribution has a higher-order correlation, e.g., if the beam energy is modulated at a modulation wavelength λ, the energy modulation can be converted into horizontal density modulation through the dispersion and hence distort the simple relation given by Eq. (8). From Eq. (5), we see that the amplitude of the horizontal modulation is magnified by a factor

$$\frac{k}{|h|} = \frac{E_0}{eV_{rf}|\cos\phi|} \frac{\lambda_{rf}}{\lambda} \gg 1 \qquad (9)$$

for $\lambda_{rf} \gg \lambda$ above the resolution limit, even if $eV_{rf} \approx E_0$ (i.e., 100% correlated energy spread). As illustrated in Fig. 1(b), the very rapid but small energy modulation on top of a linear chirp

redistributes the electrons in the energy space, causing a very large density modulation in the energy spectrum.

Therefore, the horizontal (or energy) profile obtained by the rf zero-phasing technique is sensitive to even very small high-frequency energy modulation of the beam. For a current-modulated bunch, the energy modulation can be induced by wakefields in the accelerator. An important source of energy modulation for a high-brightness electron beam is the LSC force in the linac. We study its effect on rf zero-phasing measurements in the next two sections.

3. Energy modulation due to space charge

There is no LSC force if the bunch current profile is uniform. However, if there is a density clustering, the LSC force tends to push electrons away from each other, accelerating the front electrons and decelerating the back electrons to give rise to the energy modulation. For a sinusoidal current modulation characterized by the bunching parameter $b(k)$, the on-axis longitudinal electric field in free space is [11]

$$E_z(k) = -\frac{4ien_0 b(k)}{kr_b^2}\left[1 - \frac{kr_b}{\gamma} K_1\left(\frac{kr_b}{\gamma}\right)\right], \qquad (10)$$

where r_b is the beam radius for a circular cross-section, K_1 is the modified Bessel function, and the velocity of the electrons is taken to be the speed of light c. We have neglected a small transverse variation of E_z. Thus, the LSC impedance per unit length is

$$Z_{LSC}(k) = \frac{4i}{kr_b^2}\left[1 - \frac{kr_b}{\gamma} K_1\left(\frac{kr_b}{\gamma}\right)\right]$$

$$\approx \begin{cases} \dfrac{4i}{kr_b^2}, & \dfrac{kr_b}{\gamma} \gg 1, \\[2mm] \dfrac{ik}{\gamma^2}\left(1 + 2\ln\dfrac{\gamma}{r_b k}\right), & \dfrac{kr_b}{\gamma} \ll 1, \end{cases} \qquad (11)$$

The free space approximation is satisfied when the beam pipe radius is much larger than the reduced modulation wavelength in the beam's rest frame $\gamma\lambda/(2\pi)$. For the SDL microbunching experiment,

$\gamma \approx 130$, and the free space approximation is valid up to $\lambda \sim 100$ μm.

In the zero-phasing accelerating cavities, the slip factor is simply $1/\gamma^2$ for a small energy deviation. A standard instability analysis [13] shows that the beam is stable for the LSC impedance with an oscillation frequency

$$\Omega = c \left[\frac{I_0}{\gamma^3 I_A} k |Z_{LSC}(k)| \right]^{1/2}, \qquad (12)$$

where $I_0 = n_0 ec$ is the peak electron current, and $I_A = 17,045$ A is the Alfven current. In the limit that the transverse beam size is much larger than the reduced modulation wavelength in the beam's rest frame (i.e., $kr_b/\gamma \gg 1$), Ω becomes the plasma frequency.

In this paper, we focus on the longitudinal beam dynamics in the SDL rf zero-phasing section. Since the upstream energy modulation can be effectively converted into current modulation through the chicane [12], we assume the beam has only the current modulation $b_0(k)$ and no energy modulation prior to the rf zero-phasing section (right after the bunch compressor). Thus, the current and energy modulations after a zero-phasing section of length ΔL are

$$b_d(k) = b_0(k) \cos(\Omega \Delta L / c),$$

$$p_d(k) = -i \left[\frac{\gamma I_0}{I_A} \frac{|Z_{LSC}(k)|}{k} \right]^{1/2} \sin \left(\frac{\Omega \Delta L}{c} \right) b_0(k), \quad (13)$$

where $-i$ means 90° phase shift between the current and energy modulations in the longitudinal z-coordinate. Landau damping in the straight section is ignored because of the negligible path length difference for the very small energy spread and the emittance considered here. Since $(\Omega \Delta L / c) \sim 1$ over a wide range of the modulation wavelength for the compressed bunch (with a large peak current), the space–charge-induced energy modulation is close to its maximum at the entrance of the spectrometer dipole.

4. Gain in horizontal modulation

As discussed in Section 2, the LSC induced energy modulation can be converted to an enhanced horizontal modulation in the spectro-

meter dipole. As a result, the horizontal modulation measured by the rf zero-phasing technique is much larger than the current modulation and may be quantified by a "gain" factor as

$$G_m \equiv \left| \frac{a_m(k_m)}{b_0(k)} \right|, \qquad (14)$$

where the subscript m refers to the measurement screen. Inserting Eq. (13) into Eq. (5), we obtain

$$G_m = \left| \cos \left(\frac{\Omega \Delta L}{c} \right) - \frac{k}{h} \left[\frac{\gamma I_0}{I_A} \frac{|Z_{LSC}(k)|}{k} \right]^{1/2} \sin \left(\frac{\Omega \Delta L}{c} \right) \right|$$
$$\times \exp \left[-\frac{k^2 \sigma_\delta^2}{2h^2} - \frac{k^2 (\sigma_x)_m^2}{2h^2 \eta_m^2} \right], \qquad (15)$$

where the rms beam size at the screen is

$$(\sigma_x)_m = (\sigma_x)_d \left[\left(C_m - \frac{\alpha_d S_m}{\beta_d} \right)^2 + \frac{S_m^2}{\beta_d^2} \right]. \qquad (16)$$

For a measurement screen located at a distance L_s behind a sector dipole with a bending radius ρ and a magnetic length L_b, the final lattice functions at the screen are

$$C_m = \cos \frac{L_b}{\rho} - \frac{L_s}{\rho} \sin \frac{L_b}{\rho},$$

$$S_m = \rho \sin \frac{L_b}{\rho} + L_s \cos \frac{L_b}{\rho},$$

$$\eta_m = \rho \left(1 - \cos \frac{L_b}{\rho} \right) + L_s \sin \frac{L_b}{\rho}. \qquad (17)$$

Eq. (15) gives the apparent "gain" of the horizontal density modulation to the longitudinal density modulation as a function of the modulation wavelength λ. We apply this formula to study the SDL microbunching experiment with the typical parameters listed in Table 1. As shown in Fig. 2, the gain is much larger than 1 for a wide range of modulation wavelengths above the resolution limit, indicating that the horizontal spectrum is dominated by the space–charge-induced energy modulation. In addition, Fig. 2 shows that the horizontal modulation is sensitive to the average beam size in the zero-phasing linac, in agreement with the experimental observations [14].

Table 1
Parameters for the SDL experiments

Parameters	Symbol	Value
Beam energy	E_0	65 MeV
Peak current	I_0	200 A
Normalized emittance	$\gamma\epsilon_x$	2.5 μm
Horizontal beta at dipole	β_d	8.5 m
Incoherent energy spread	σ_δ	1×10^{-4}
Chirp	h	20 m^{-1}
Zero-phasing linac length	ΔL	15 m
Dipole bending radius	ρ	0.79 m
Dipole magnetic length	L_b	1 m
Dipole to screen length	L_s	0.31 m

[15]. Therefore, the rf zero-phasing technique can be used to extract the beam energy modulation instead of the current modulation.

Acknowledgements

We thank A. Chao, D. Dowell, P. Emma, and C. Limborg for useful discussions. This work was supported by the Department of Energy contracts DE-AC03-76SF00515 and DE-AC02-76CH00016.

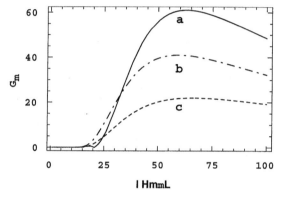

Fig. 2. Gain of the horizontal density modulation relative to the current modulation as a function of the modulation wavelength at the entrance of the dipole for (a) $r_b = 0.5$ mm (solid line), (b) $r_b = 1.2$ mm (dashed line), and (c) $r_b = 2.4$ mm (dotted line).

References

[1] P. Krejcik, SLAC-PUB-9527, 2002.
[2] D.X. Wang, G.A. Krafft, C.K. Sinclair, Phys. Rev. E 57 (1998) 2283.
[3] W.S. Graves, et al., in: Proceedings of the 2001 Particle Accelerator Conference, 2001, p. 2224.
[4] S. Heifets, G. Stupakov, S. Krinsky, Phys. Rev. ST Accel. Beams 5 (2002) 064401.
[5] Z. Huang, K.-J. Kim, Phys. Rev. ST Accel. Beams 5 (2002) 074401.
[6] H. Loos, et al., in: Proceedings of the 2002 European Particle Accelerator Conference, 2002, p. 814.
[7] H.H. Braun, et al., Phys. Rev. ST Accel. Beams 3 (2000) 124402.
[8] Ph. Piot, et al., in: Proceedings of the 2000 European Particle Accelerator Conference, 2000, p. 1546.
[9] M. Huning, Ph. Piot, H. Schlarb, Nucl. Instr. and Meth. A 475 (2001) 348.
[10] E.L. Saldin, E.A. Schneidmiller, M.V. Yurkov, Nucl. Instr. Meth. A 490 (2002) 1.
[11] J. Rosenzweig, et al., DESY-TESLA-FEL-96-15, 1996.
[12] E.L. Saldin, E.A. Schneidmiller, M.V. Yurkov, TESLA-FEL-2003-02, 2003.
[13] A. Chao, Physics of Collective Beam Instabilities in High Energy Accelerators, Wiley, New York, 1993.
[14] T. Shaftan, et al., Nucl. Instr. and Meth. A (2003), these Proceedings.
[15] T. Shaftan, L.-H. Yu, BNL-71490, 2003.

Finally, when $G_m \gg 1$ and $|p_d(k)| \ll |h|\lambda/(2\pi)$, we have from Eq. (5)

$$p_d(k) \approx i\frac{h}{k}a_m(k_m) \qquad (18)$$

above the resolution limit. When $|p_d(k)| \sim |h|\lambda/(2\pi)$, other methods can be used to obtain p_d

Available online at www.sciencedirect.com

Nuclear Instruments and Methods in Physics Research A 528 (2004) 350–354

NUCLEAR
INSTRUMENTS
& METHODS
IN PHYSICS
RESEARCH
Section A

www.elsevier.com/locate/nima

Sensitivity studies for the LCLS photoinjector beamline

C. Limborg, P. Bolton, D.H. Dowell, S. Gierman, J.F. Schmerge*

SLAC, MS 69, 2275 Sand Hill Road, Menlo Park, CA 94025, USA

Abstract

The LCLS photoinjector beamline is now in the Design and Engineering stage. The fabrication and installation of this beamline is scheduled for the summer 2006. The photoinjector will deliver 10 ps long electron bunches of 1 nC with a normalized transverse emittance of less than 1 mm mrad for 80% of the slices constituting the core of the bunch at 150 MeV. Based on PARMELA simulations, stability requirements for all the physical parameters were derived. In this paper, we discuss the regulation requirements of the solenoid fields, accelerating fields and the phase stability of RF components. We also describe how specifications for the laser pulse were defined, in particular for the longitudinal and transverse uniformities. Finally, requirements for alignment of components and steering of the electron beam along the beamline are presented.
© 2004 Elsevier B.V. All rights reserved.

PACS: 29.25.Bd; 29.27.Ac; 41.60.Cr; 41.85.Ar

Keywords: Photoinjector; Stability; Wakefield

1. Introduction

The LCLS photoinjector has been designed to deliver bunches with 100 A current from a 1 nC 10 ps pulse with slice emittances smaller than 1 mm mrad and a projected emittance smaller than 1.2 mm mrad at the entrance of the main linac at 150 MeV. Computations had been performed for an ideal laser beam and perfectly tuned and stable accelerator. The beam quality for this ideal tuning gives slice emittances around 0.85 mm mrad and a projected emittance of around 1 mm mrad. Errors on physical parameters of components, time-dependent drifts and misalignment will deteriorate the beam quality. Specifications on machine and laser components have been derived to assure continued SASE FEL operation by maintaining a high beam quality, with the emittance, energy and peak current kept stable. In this paper, we review the possible sources of beam quality deterioration and discuss tolerances.

In the following, we refer to the 80% emittance. It corresponds to the projected emittance for a series of 80 out of 100 adjacent slices of the bunch centered on the middle slice. It does not include the head and tails which significantly increase the

*Corresponding author. Tel.: +1-650-926-2320; fax: +1-650-926-4100.

E-mail address: schmerge@slac.stanford.edu (J.F. Schmerge).

0168-9002/$ - see front matter © 2004 Elsevier B.V. All rights reserved.
doi:10.1016/j.nima.2004.04.066

projected emittance but do not contribute to the SASE FEL. This quantity is more conservative than the slice emittance.

2. Sensitivity of LCLS photoinjector beamline

2.1. LCLS photoinjector beamline

The LCLS photoinjector beamline has been described in Ref. [1]. It consists of a 1.6 cell S-Band gun with a copper cathode, an emittance compensation solenoid, two SLAC S-Band linac structures and a second solenoid wrapped around the first third of the first linac section. The gun will be operated with 120 MV/m peak field. Operating parameters are described in Table 1. The thermal emittance was assumed to be of 0.6 mm mrad per mm radius and so 0.72 mm mrad for our 1.2 mm radius, following Ref. [2]. With this large thermal emittance, the best optimisation for a perfect beamline gives an 80% emittance of 0.9 mm mrad. With all possible errors included, this quantity should not exceed 1 mm mrad.

2.2. Tolerance budget

The tolerances on parameters are given in Table 1. The last column gives the maximum variation on the parameter which increases the 80% emittance from 0.9 to 0.95 mm mrad when that single parameter is varied. Obviously, those tolerances are too large to assure good beam

quality. Indeed when those errors are combined, the 80% emittance gets much larger than 1 mm mrad. The regulation requirements were determined to be physically realistic values and such that the combination of errors would not lead to an 80% emittance larger than 1 mm mrad and such that the overall FEL performance would not be deteriorated. The tolerance budget for the entire accelerator is such that the peak current does not vary by more than 12% and the energy by more than 0.1% at the undulator entrance [3].

The power supply of the solenoid will be maintained stable to $\pm 2 \times 10^{-4}$ around its operating point. A gun field stability of 0.5 MV/m around 120 MV/m is required. An amplitude stability of 0.1% rms is routinely achieved at the Gun Test Facility at 110 MV/m for periods of 20 min without feedback. The gun field ratio between the two cells ("gun balance") can be maintained to $\pm 3\%$ for weeks. It was also checked that if the gun is unbalanced by 20%, a tuning of the beamline can be found which meets the beam quality requirements. Around that working point, similar regulation tolerances are needed.

The phase stability for the gun and the two linac sections is of 0.1° S-Band. These tolerances are imposed by the timing constraints of the entire accelerator. They are the most challenging ones, but appears to be realistic [4]. The charge stability of 2% rms is necessary to maintain the peak current stable to 12% after the two compressors.

The two linac gradients will be easily maintained to ± 0.5 MV/m of their nominal value. Running

Table 1
Operating parameters, (2) nominal value, (3) regulation requirements, (4) increase of 80% emittance from 0.9 to 0.95 mm mrad

Parameter	(2)	Unit	(3)	(4)
Solenoid 1*	2.7235	kG	$\pm 0.02\%$	$\pm 0.3\%$
Solenoid 2*	0.748	kG	$\pm 1°$	$\pm 20\%$
Gun phase*	27.45	deg/0cross	$\pm 0.1°$	$\pm 2.5°$
Gun field*	120	MV/m	± 0.5	$\pm 0.5\%$
$Gun_{Balance}$	1	E_{cell1}/E_{cell2}	$\pm 3\%$	$\pm 3\%$
$R_{Laser\ spot}$	1.2	mm	$\pm 2\%$	5%
Charge*	1	nC	2% rms	5%
$Field^*_{Linac1}$	19.8	MV/m	± 0.5	12%
$Field_{Linac2}$	28.7	MV/m	± 0.5	
$Phase_{Linac1}$	-2	deg/crest	$\pm 0.1°$	
$Phase_{Linac2}$	-10	deg/crest	$\pm 0.1°$	

the second linac gradient as high as 28.7 MV/m, it is feared that too much dark current will be generated. We recently verified that operating this section at 24 MV/m would be viable. There will be no emittance growth in the matching section even if the final energy will now be 135 MeV. The operation will be more stable and regulation tolerances similar.

In Fig. 1, we show the results from PARMELA simulations in which multiple errors were assumed simultaneously. In those 64 simulations, the 6 main parameters took either one of the two extreme values given by the regulation requirements. The 80% and projected emittances re-

mained, respectively, below 1 and 1.2 mm mrad. (Fig. 2)

3. Alignment and steering requirements

Additional PARMELA simulations were done to determine requirements for alignment of components. Our criteria was set to have no more than 1–2% growth of the 80% emittance at the end of the beamline, as very little margin is left when multiple errors on components are combined.

3.1. Transverse wakefields in S-band structures

If a beam enters a linac structure with some transverse offset or angle, it will then be subject to transverse wakefields. At the end of the accelerating structure, the beam will be skewed from head to tail. The slice emittance will not be deteriorated but the 80% emittance will be damaged. The transverse wakefield kick is particularly strong at low energy, so our study was only done for the first linac section. The wakefield kick was applied at the end of every cell for the first meter of section and every other cell further downstream. Those computations showed that the 80% emittance would grow from 0.9 to 0.91 mm mrad if the beam was offset by 100 μm or if it had an entrance angle of 120 μrad. The growth from 0.9 to 1 mm mrad would appear for offsets of 300 μm and entrance angle of 360 μrad. Correctors and Beam Position Monitors in the drift from Gun to linac have been

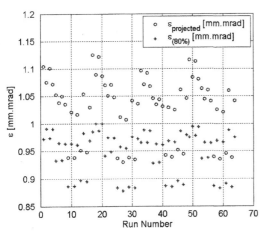

Fig. 1. PARMELA simulations results for combined errors on the 6 parameters* from Table 1.

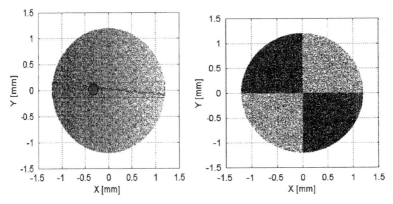

Fig. 2. (1) Slope; (2) CheckerBoard.

designed to give steering well within a few tens of micrometers in position and a few tens of microradians in angle. Steering will be available at the end of the first linac section and at the entrance of the second one. Anyway, alignement tolerances are large for the second section.

3.2. Solenoid alignment

An angular error on the solenoid position can generate head to tail offsets, which then deteriorates the 80% emittance. The transverse kicks are second-order effects. They vary slowly along the bunch which carries an energy chirp from longitudinal space charge effects. The tolerance is as large as 1.5 mrad for the angular position and 500 μm for the transverse position. The criteria was to keep the growth of the 80% emittance smaller than 1% at the entrance of the linac.

3.3. Laser steering on cathode

If the laser is not perfectly steered on the electrical center of the gun, radial fields at the exit of the gun will produce different transverse kicks from the head to tail of the energy-chirped bunch. Computations show that the 80% emittance at the entrance of the linac will not increase by more than 1% if the beam is steered within 100 μm of the electrical center.

4. Laser specifications

4.1. Transverse uniformity

It can be checked that the most damaging density irregularities across the transverse profiles are low-frequency structures. The high-frequency structures are more easily diluted. Tolerances for the high-frequency structure has been described in Ref. [1]. Two types of low-frequency structures are studied here.

4.1.1. Slope across transverse profile
For a linear increase of charge density across the transverse laser spot, the center of gravity will be displaced. For a displaced center of gravity, the

three possible misalignment studied above could damage the emittance. The two last types of misalignment are of second order, only the transverse wakefield in the linac is of concern here. A ±15% linear variation of charge density across the transverse pushes the center of gravity to 100 μm at the entrance of the linac section.

4.1.2. "Lowest order checkerboard"
In this case, the center of gravity is on axis, but the structure generates ellipticity to the beam. The local increase in charge density produces slice emittance growth. A ±20% variation from low- to high-charge density region will increase the 80% emittance by 6%, a ±10% will leave the emittance unchanged.

The transverse uniformity of the emitted electron beam should then have a maximum of ±10% peak-to-peak. This uniformity results from the product of the laser intensity by the cathode quantum efficiency.

4.2. Longitudinal uniformity

The temporal distribution of the laser corresponds to a flat top with rise and fall times of 0.7 ps. The irregularities on the flat top are characterized by their amplitudes and wavelengths.

4.2.1. Emittance criteria
We first computed what would be the increase in slice emittance due to some undesired irregularities. The increase in emittance gets larger for the longer wavelengths. A 20% modulation at a wavelength of 480 μm increases the 80% emittance by 5%. For amplitudes below 10%, and all wavelengths up to 480 μm, the increase in 80% emittance is smaller than 1%. A ±10% had been specified for the laser temporal profile. But, irregularities on the current line density can in fact be more damaging.

4.2.2. Longitudinal space charge instability
A modulation of the current density generates local longitudinal space charge forces which push particles away, diluting the local high density regions. Once the current density structure is washed out, particles have acquired or lost locally some

small amounts of energy. The energy distribution is now modulated. This corresponds to a quarter of a cycle of a plasma oscillation. If this instability grows, either as an increase of the current density modulation or of the energy modulation, some emittance growth will appear in the bunch compressors. Preliminary computations show that a 0.1% current modulation at the end of the photoinjector beamline with wavelength shorter than 40 µm would grow in the main accelerator and result in an uncorrelated energy spread larger than 5×10^{-4} [5] after the second bunch compressor, preventing the saturation of the FEL. For the LCLS photoinjector beamline, PARMELA computations show that for a 10% initial modulation of the laser temporal distribution, the residual current density will be at least as large as 2% rms at the end of the beamline for all wavelengths ranging from 50 to 500 µm. The associated rms energy modulation will be of 2–4 keV depending on the wavelength. Even if the modulation is strongly attenuated inside the gun, the residual amplitude at the end of the photoinjector beamline is still too large. The use of a "beam heater" as suggested in Ref. [6] is under discussion. The inverse FEL mechanism would warm up the beam energy spread to 75 keV rms, preventing the instability to gain amplitude in the main accelerator.

5. Conclusion

We have computed regulation requirements to maintain the emittances, the timing and the energy at the exit of the photoinjector beamline within the needs of the overall LCLS accelerator. These requirements seem realistic to meet. Alignment tolerances are not challenging as they are well above what is commonly achieved. Great care is presently taken to characterize definitely the needs of the temporal laser pulse and of a "beam heater".

References

[1] C. Limborg, New Optimization for the LCLS Photoinjector, EPAC 2002, SLAC-Pub 9555.
[2] W. Graves, et al., Measurement of thermal emittance for a Copper PhotoCathode, Proceedings PAC 2001, 2001.
[3] P. Emma, LCLS CDR, Chapter 7, http://www-ssrl.slac.stanford.edu/lcls/CDR/.
[4] R. Akre, Measurements on SLAC linac RF system for LCLS Operation, Proceedings PAC 2001, 2001.
[5] M. Borland, Private Communications.
[6] E.L. Saldin, E.A. Schneidmiller, M.V. Yurkov, Longitudinal space charge driven microbunching instability in TTF2 linac, TESLA-FEL-2003-02, Nucl. Instr. and Meth. A, (2004) these Proceedings.

Available online at www.sciencedirect.com

SCIENCE  DIRECT°

ELSEVIER   Nuclear Instruments and Methods in Physics Research A 528 (2004) 355–359

NUCLEAR INSTRUMENTS & METHODS IN PHYSICS RESEARCH
Section A

www.elsevier.com/locate/nima

Longitudinal space charge-driven microbunching instability in the TESLA Test Facility linac

E.L. Saldin[a], E.A. Schneidmiller[a],*, M.V. Yurkov[b]

[a] *Deutsches Elektronen-Synchrotron (DESY), Notkestrasse 85, Hamburg, Germany*
[b] *Joint Institute for Nuclear Research, Dubna 141980, Moscow region, Russia*

Abstract

In this paper, we study a possible microbunching instability in the TESLA Test Facility linac. A longitudinal space charge is found to be the main effect driving the instability. Analytical estimates show that initial perturbations of beam current in the range 0.5–1 mm are amplified by a factor of a few hundred when the beam passes two bunch compressors. A method to suppress the instability is discussed.
© 2004 Elsevier B.V. All rights reserved.

PACS: 41.60.Cr; 42.55.Vc; 41.75.Ht; 41.85.Ew

Keywords: Free electron lasers; X-ray lasers; Relativistic electron beams

1. Introduction

Magnetic bunch compressors are designed to produce short electron bunches with a high peak current for linac-based short-wavelength free electron lasers (FELs) [1–4]. Microbunching instabilities of bright electron beams in linacs with bunch compressors are of serious concern [5–10]. A simple physical picture of the instability [6,11,12] suggests that high-frequency components of the beam current spectrum cause energy modulations at the same frequencies due to collective fields (longitudinal space charge, CSR, geometrical wakefields etc.). The energy modula-

tion is converted into an induced density modulation while the beam is passing the bunch compressor. If the high-frequency self-fields are strong enough, the induced density modulation can be much larger than the initial one. In other words, the system can be treated as a high-gain klystron-like amplifier. An amplification of parasitic modulations due to the longitudinal space charge (LSC) field was considered [6,12] in the context of phase 1 of the TESLA Test Facility (TTF1) [1]. The aim of this paper is to study a possible LSC-driven instability in TTF2 linac [2].

2. Gain in a single bunch compressor

Let us assume that at some point of the beam-line upstream of the bunch compressor there is a

---

*Corresponding author. Tel.: +49-40-8998-2676; fax: +49-40-8998-4475.
E-mail address: schneidm@mail.desy.de
(E.A. Schneidmiller).

0168-9002/$ - see front matter © 2004 Elsevier B.V. All rights reserved.
doi:10.1016/j.nima.2004.04.067

small current perturbation ρ_i at some wavelength $\lambda = 2\pi/k$:

$$I(z) = I_0[1 + \rho_i \cos(kz)]. \tag{1}$$

The amplitude of energy modulation $\Delta\gamma$ (in units of electron rest energy) due to collective self-fields can be described by longitudinal impedance $Z(k)$:

$$\Delta\gamma = \frac{|Z(k)|}{Z_0} \frac{I_0}{I_A} \rho_i \tag{2}$$

where $Z_0 = 377\ \Omega$ is the free-space impedance and $I_A = 17\ \text{kA}$ is the Alfven current. If self-fields inside bunch compressor can be neglected, one can calculate the amplitude of the induced density modulation ρ_{ind} at the end of compressor. In general case, to find final density modulation ρ_f one should sum up induced modulation and (transformed to the end of compressor) initial one, taking care of phase relations. But in this paper we use approximation $\rho_i \ll \rho_{\text{ind}} \ll 1$. In other words, $\rho_f \simeq \rho_{\text{ind}}$ and the gain in density modulation $G = \rho_f/\rho_i \simeq \rho_{\text{ind}}/\rho_i$ is assumed to be high, $G \gg 1$. Under this approximation the gain depends neither on phase of $Z(k)$ nor on sign of momentum compaction factor R_{56} of the bunch compressor and is found to be [6,12]:

$$G = Ck|R_{56}|\frac{I_0|Z(k)|}{\gamma_0 I_A Z_0} \exp\left(-\frac{1}{2}C^2 k^2 R_{56}^2 \frac{\sigma_\gamma^2}{\gamma_0^2}\right). \tag{3}$$

Here C is the compression factor, γ_0 is relativistic factor, and σ_γ is rms uncorrelated energy spread (in units of rest energy) before compression.

The influence of beam emittance ε on longitudinal dynamics is negligible when $Ck\varepsilon S_c/\beta \ll 1$, where S_c is the length of a path through compressor and β is the beta-function. Formula (3) can be used when the effect of wakefields in front of compressor is much stronger than any collective effects inside the bunch compressor. In particular, CSR-related effects are strongly suppressed [7–9] when $k\sqrt{\varepsilon\beta}\theta_d \gg 1$, where θ_d is the bending angle in a dipole of the bunch compressor. Both conditions for emittance can be met simultaneously.

3. Longitudinal space charge impedance

Let us consider longitudinal space charge field of a modulated beam (1) propagating in free space. This field can easily be calculated in the rest frame of the beam. We consider here the limit of "pencil" beam, when

$$p = \gamma/(k\sigma_\perp) \gg 1. \tag{4}$$

Here, $\sigma_\perp$ is the rms transverse size of the beam, and γ is relativistic factor. The impedance of a drift with the length L_d is then given by (with logarithmical accuracy):

$$\frac{|Z(k)|}{Z_0} \simeq \frac{2kL_d \ln p}{\gamma^2}. \tag{5}$$

The influence of vacuum chamber can be neglected when $\gamma/(kb) \ll 1$, b being the transverse size of the vacuum chamber. In the opposite case the parameter p in (5) should be substituted by $b/\sigma_\perp$.

Let us now estimate an effective LSC impedance for an accelerating structure with the length L_A and the average accelerating gradient $d\gamma/ds$. Considering γ in (5) as a function of s, neglecting change of $\ln p$ in the structure, and making integration, we can generalize (5) as follows: $|Z(k)|/Z_0 \simeq 2kL_A \ln p/(\gamma_i \gamma_f)$, where γ_i and γ_f are relativistic factors before and after acceleration, respectively. When $\gamma_f \gg \gamma_i$, this formula can be rewritten in the form:

$$\frac{|Z(k)|}{Z_0} \simeq \frac{2k \ln p}{\gamma_i(d\gamma/ds)}. \tag{6}$$

4. Gain estimation for TTF2

The TESLA Test Facility Free Electron Laser (phase 2) is supposed to produce powerful coherent soft X-ray radiation [2]. A nominal design of TTF2 [13] assumes linearized compression of the electron beam in two bunch compressors. The laser driven RF gun produces low-emittance, flat-top electron bunch (total length is 7 mm) with the current of about 40 A, which is accelerated off-crest in the TESLA module (ACC1). A third harmonic cavity is then used to linearize longitudinal phase space. At the energy of

120 MeV the beam is compressed by a factor of $C_1 = 8$ (the current increases up to 320 A) in the 4-bend chicane BC1 with the $R_{56} = 17$ cm. After that the beam passes a 14 m long drift, and is then accelerated off-crest in two TESLA modules (ACC2 and ACC3) up to 438 MeV. In an S-type chicane BC2 ($R_{56} = 4.9$ cm) the peak current increases by another factor of $C_2 = 8$, reaching the required level of about 2.5 kA. According to start-to-end simulations [13], the local energy spread after compression is below 200 keV, and the normalized slice emittance is 1–1.5 mm mrad in the lasing area of the bunch.

Our estimates of the gain of LSC-driven microbunching instability will be based on these parameters of the machine and of the electron beam. A very important parameter is the local energy spread before compression which is assumed to be 3 keV. We perform the estimates using the following scheme. We do not analyze the processes in the gun and assume that the beam with some small amplitude of density modulation and a period λ_1 enters accelerating module ACC1. There the beam is quickly accelerated, the longitudinal motion gets frozen, and we can calculate the energy modulation (and, therefore, impedance in accordance with (2)). Then we compute the gain in BC1 using formula (3). A compressed beam (with the enhanced density modulation with a period $\lambda_2 = \lambda_1/C_1$) moves in the long drift and in the modules ACC2 and ACC3. We calculate impedance for this part of the machine and use again (3) to calculate the gain in BC2. The total gain is a product of partial gains in two compressors. We study pure effect of the space charge and neglect CSR effects inside bunch compressors.

Analyzing the above-mentioned parameters, we come to the conclusion that the cut-off wavelength is defined by the smearing effect due to the energy spread in BC2. Corresponding initial modulation wavelength λ_1 is in the range 0.5–1 mm. Analytical calculations of the impedance in ACC1 would be difficult because the condition (4) is not well satisfied at the entrance to ACC1 (rms transverse size at this point is about 0.8 mm, and $\gamma_i \simeq 9$). Besides, it is not clear a priori, whether or not the density modulation stays constant while the energy modulation is growing (taking into account low initial energy and reduced accelerating gradient in

the first half of ACC1, $d\gamma/ds = 24$ m^{-1}). For these reasons we performed numerical simulations of this part of the machine with the help of the code Astra [14], introducing small density modulations at different wavelengths and getting corresponding energy modulations at the exit of ACC1. The impedance is then calculated using formula (2). The results of these calculations are presented in Fig. 1. Note that at the shortest wavelength on this plot, the density modulation slightly decreases while beam is moving in ACC1.

As for the part of the machine between BC1 and BC2, the analytical approach is justified there. With the help of formulas (5) and (6) we calculate the total impedance, defining transverse size of the beam from the design optics and taking the nominal accelerating gradient in ACC2 and ACC3, $d\gamma/ds = 37$ m^{-1}. The resulting impedance is plotted in Fig. 2, where the wavelength is scaled to that before compression.

Finally, using formula (3) and Figs. 1 and 2, we calculate the total gain which is plotted in Fig. 3 versus initial modulation wavelength. In the given wavelength range the high-gain condition is well satisfied for each stage of compression. The maximum of the gain curve corresponds to the wavelength 10 μm after two-stage compression.

After compression the beam is accelerated up to 1 GeV (for the shortest FEL wavelength 6.4 nm) and then moves about 80 m to the undulator. The LSC impedance for the part of the machine between BC2 and the undulator at the wavelength

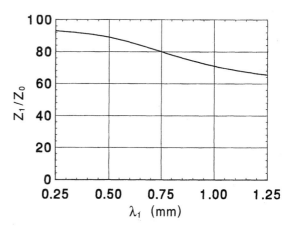

Fig. 1. The LSC impedance in ACC1 (numerical simulations).

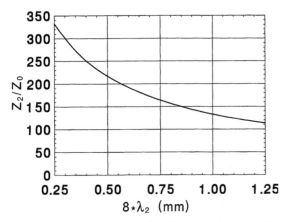

Fig. 2. The LSC impedance for the part of the machine between BC1 and BC2 (analytical calculations).

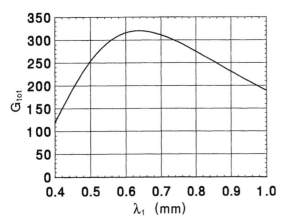

Fig. 3. Total gain versus initial modulation wavelength.

10 μm can be estimated at $|Z|/Z_0 \simeq 200$. According to formula (2), a 30% density modulation at the exit of BC2 would result in more than 4 MeV energy modulation at the undulator entrance. A reliable operation of the facility would require to keep the density modulation at the entrance to ACC1 well below the level of 10^{-3} (or to suppress the amplification mechanism).

5. Discussion

An effective way to suppress the instability is to increase local energy spread. For instance, increase of the energy spread at TTF2 up to 15–20 keV

would eliminate the instability. At a high energy, the use of superconducting wiggler, where the energy spread increases due to quantum fluctuations of synchrotron radiation, was suggested [3]. A simple method to control the energy spread at low energy would be to use FEL-type modulation of the beam in optical wavelength range by a laser pulse in an undulator. Then the beam goes through the bunch compressor where these coherent energy modulations are quickly dissipated, leading to the effective "heating" of the beam (similar mechanism takes place in storage ring FELs). For illustration we present here a numerical example for TTF2. The undulator with ten periods, a period length 3 cm, and a peak field 0.49 T is located in front of BC1. A fraction of power in the second harmonic ($\lambda = 0.52$ μm) of the Nd:YLF laser is outcoupled from the photoinjector laser system and is transported to the undulator. For a transverse size of the laser beam 0.5 mm (Rayleigh length is 1.5 m) and a power 300 kW, the amplitude of energy modulation will be about 20 keV (rms energy spread is smaller by $\sqrt{2}$). Of course, the required value of the modulation should be defined experimentally.

Generally speaking, our studies have led us to the conclusion that LSC effect on microbunching instability is stronger than the effects of CSR and of geometrical wakefields for typical parameters of linacs with bunch compressors for FELs. Of course, CSR field is much stronger at the energy of compression, but LSC effect is accumulated at much lower energies and over much longer distances. For example, as one can see from Fig. 2, the LSC impedance at the optimal wavelength ($\lambda_2 = 0.6/8 \simeq 0.08$ mm) is about 200, while CSR impedance in the first dipole of BC2 is 10. In addition, a suppression of CSR instability due to finite emittance effects does not play any role for LSC-driven instability. Our conclusion is that for the studies of microbunching instabilities the space charge effect should be included into start-to-end simulations and analytical calculations.

References

[1] V. Ayvazyan, et al., Phys. Rev. Lett. 88 (2002) 104802.

[2] J. Rossbach, Nucl. Instr. and Meth. A 375 (1996) 269.

[3] Linac Coherent Light Source (LCLS) Design Report, SLAC-R-593, 2002.

[4] TESLA Technical Design Report, DESY 2001-011, 2001.

[5] M. Borland, et al., Nucl. Instr. and Meth. A 483 (2002) 268.

[6] E.L. Saldin, E.A. Schneidmiller, M.V. Yurkov, Nucl. Instr. and Meth. A 483 (2002) 516.

[7] E.L. Saldin, E.A. Schneidmiller, M.V. Yurkov, TESLA-FEL-2002-02, 2002;
E.L. Saldin, E.A. Schneidmiller, M.V. Yurkov, Nucl. Instr. and Meth. A 490 (2002) 1.

[8] S. Heifets, G. Stupakov, S. Krinsky, Phys. Rev. ST Accel. Beams 5 (2002) 064401.

[9] Z. Huang, K.-J. Kim, Phys. Rev. ST Accel. Beams 5 (2002) 074401.

[10] Z. Huang, et al., SLAC-PUB-9818, 2003.

[11] T. Limberg, Ph. Piot, E.A. Schneidmiller, Nucl. Instr. and Meth. A 475 (2001) 353.

[12] E.L. Saldin, E.A. Schneidmiller, M.V. Yurkov, DESY-01-129, 2001.

[13] T. Limberg, Ph. Piot, F. Stulle, in: Proceedings of EPAC2002, 2002, p. 811.

[14] K. Flöttmann, Astra User Manual, http://www.desy.de/mpyflo/Astra_dokumentation/

Available online at www.sciencedirect.com

SCIENCE DIRECT®

ELSEVIER   Nuclear Instruments and Methods in Physics Research A 528 (2004) 360–365

**NUCLEAR
INSTRUMENTS
& METHODS
IN PHYSICS
RESEARCH**
Section A

www.elsevier.com/locate/nima

Characterization of the electron source at the photo injector test facility at DESY Zeuthen

K. Abrahamyan[b,1], W. Ackermann[a], J. Bähr[b], I. Bohnet[b], J.P. Carneiro[c], R. Cee[a], K. Flöttmann[c], U. Gensch[b], H.-J. Grabosch[b], J.H. Han[b], M.V. Hartrott[d], E. Jaeschke[d], D. Krämer[d], M. Krasilnikov[b,*], D. Lipka[b], P. Michelato[e], V. Miltchev[b], W.F.O. Müller[a], A. Oppelt[b], C. Pagani[e], B. Petrossyan[b,1], J. Roßbach[c], W. Sandner[f], S. Schreiber[c], D. Sertore[e], S. Setzer[a], L. Staykov[g], F. Stephan[b], I. Tsakov[g], T. Weiland[a], I. Will[f]

[a] *TU Darmstadt, Darmstadt 64289, Germany*
[b] *DESY, Zeuthen 15738, Germany*
[c] *DESY, Hamburg 22603, Germany*
[d] *BESSY, Berlin 12489, Germany*
[e] *INFN Milano, Segrate 20090, Italy*
[f] *Max-Born-Institute, Berlin 12489, Germany*
[g] *INRNE Sofia, Sofia 1784, Bulgaria*

Abstract

The Photo Injector Test Facility at DESY Zeuthen (PITZ) was built to test and optimize electron sources for Free Electron Lasers and future linear colliders. The focus is on the production of intense electron beams with minimum transverse emittance and short bunch length as required for FEL operation. The experimental setup includes a 1.5 cell L-band gun cavity with coaxial RF coupler, a solenoid for space charge compensation, a laser capable to generate long pulse trains with variable temporal and spatial pulse shape, an UHV photo cathode exchange system, and an extensive diagnostics section. This contribution will give an overview on the facility and will mainly discuss the measurements of the electron beam transverse phase space. This will include measurements of the transverse and longitudinal laser profile, beam charge as a function of RF phase, and transverse emittance as a function of different parameters. The corresponding measurements of momentum and momentum spread as well as the RF commissioning results will be summarized. As a first application of the PITZ electron source it will be installed at the TESLA Test Facility Free Electron Laser at DESY Hamburg in autumn 2003.
© 2004 Elsevier B.V. All rights reserved.

PACS: 41.75.Ht; 41.85.Ar

Keywords: Photo injector; Emittance

---

*Corresponding author. Tel.: +49-33762-77213; fax: +49-33762-77330.
E-mail address:* mikhail.krasilnikov@desy.de (M. Krasilnikov).
[1] on leave from YERPHI, Yerevan 375036, Armenia.

0168-9002/$ - see front matter © 2004 Elsevier B.V. All rights reserved.
doi:10.1016/j.nima.2004.04.068

1. Introduction

First beam measurements at the photo injector test facility at DESY Zeuthen (PITZ) have been presented at earlier conferences [1,2]. Since then, the photo cathode laser system has been upgraded significantly and several operation periods have followed. The current near term goal of PITZ is to do a full characterization of the existing electron source and then to install it at the VUV-FEL at TTF2 in Hamburg in autumn 2003. The experience gained with the PITZ gun at Zeuthen and Hamburg will be the basis for the development of a more demanding injector for the recently approved X-FEL project. In this paper, an overview of the achieved experimental results is given and detailed measurements of the transverse beam emittance as a function of injector parameters are presented.

A schematic overview of the current PITZ installation is given in Fig. 1.

2. Achievements on RF commissioning

A smooth commissioning procedure yielded an operation with up to 900 μs long RF pulses at 10 Hz repetition rate and a gradient at the cathode of about 40 MV/m. That corresponds to a maximum average power of 27 kW in the gun cavity with 0.9% duty cycle. This long pulse operation fulfils the TTF2 requirements and no fundamental limit on the gradient has been detected yet.

The dark current in the gun cavity has been measured as a function of the accelerating gradient, main solenoid and bucking magnet currents [2,3]. Measurements were performed using Mo and Cs_2Te cathodes. A maximum dark current of 180 μA has been measured with a gradient on the cathode of about 40 MV/m and a main solenoid current of 200 A. In standard operation at higher solenoid fields the dark current is over focussed and the amount transported downstream is at least a factor of 2 smaller.

3. Upgrade of the laser system

The main achievement of the cathode laser upgrade at PITZ is a stable production of long laser pulse trains where each micro pulse has a flat top longitudinal profile. The laser is based on a diode-pumped pulsed oscillator synchronized with the RF. A diode-pumped amplifier chain and two flash-lamp-pumped booster amplifiers follow. A pulse shaper inserted between the oscillator and the diode-pumped amplifier chain allows for generation of temporal flat top pulses. The laser material is Nd:YLF operated at a wavelength of 1047 nm. Since the photo cathode requires ultraviolet radiation, the infrared laser pulses are converted to the fourth harmonic (262 nm) by means of two nonlinear crystals. The rising edge of the flat top pulses after conversion to the UV is presently in the order of 4–6 ps. A typical laser micro pulse longitudinal profile is shown in Fig. 2a. The laser is able to generate trains of micro pulses of up to 800 μs length.

The transverse profile of the laser beam is controlled by imaging a diaphragm onto the photo cathode. The RMS spot size can be varied from 0.3 to 1 mm. For the emittance measurements shown later, the measured RMS laser spot

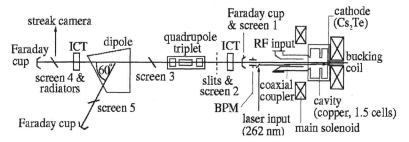

Fig. 1. Schematics of the current set-up. The beam goes from right to left, the total length is about 6 m.

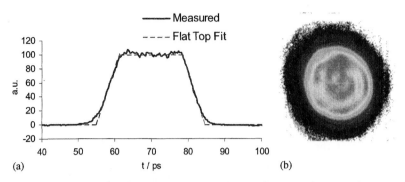

Fig. 2. (a) Temporal profile of the laser micro pulse measured at a wavelength of 524 nm using a streak camera. Ten laser pulses have been integrated. The dashed line shows a flat top fit with ~24 ps FWHM and 6 ps rise/fall time. (b) Transverse laser intensity distribution on the virtual cathode. Modulation depth ~20%.

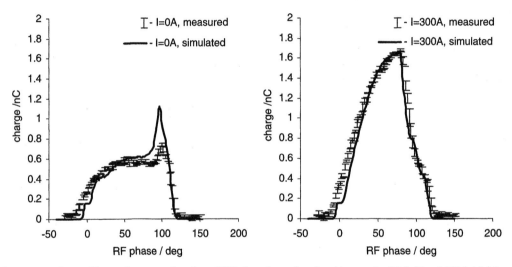

Fig. 3. Phase scans—detected beam charge as a function of RF phase for main solenoid currents of 0 (left) and 300 A (right), compared with simulations. Gradient at the cathode ~40 MV/m.

size on the cathode was $\sigma_x = 0.45 \pm 0.03$ mm and $\sigma_y = 0.52 \pm 0.03$ mm (see e.g. Fig. 2b).

4. Beam charge measurements

The charge of the electron bunch is measured with Faraday Cups and integrating current transformers (ICT). A basic measurement is the so called phase scan: the accelerated charge downstream of the gun measured as a function of launch phase, the relative phase of the laser pulses with respect to the RF.

The space charge effects on and near the cathode which depend on the laser pulse shape, the solenoid position and strength, and the acceleration gradient at the cathode determine the shape of the phase scan. Fig. 3 shows two-phase scans for main solenoid currents of 0 and 300 A. The data are compared to simulations [4]. The general agreement is fairly good, but there are differences in detail.

In order to have a defined comparison between measurements and simulations a corresponding reference RF phase Φ_0 has to be settled. To not rely on details of the comparison between the measured and simulated phase scans, the reference RF phase in our paper is chosen as the phase with maximum mean energy gain. This is easy and reliably defined using a beam transverse size vs. RF-phase measurement, as described in Ref. [2]. In the rest of this paper the RF phases will always be quoted with respect to this reference phase Φ_0.

5. Longitudinal phase-space measurements

The mean momentum and the momentum distribution of the electron beam were measured using the dipole spectrometer. For an accelerating gradient of $\sim 42\,\text{MV/m}$ and a beam charge of 1 nC a maximum mean momentum of $\sim 4.7\,\text{MeV}/c$ and a minimum momentum spread of $\sim 30\,\text{keV}/c$ were obtained. An electron bunch length of $\sim 20\,\text{ps}$ FWHM was measured using a streak camera and a Cherenkov radiator. For detailed results see Ref. [5].

6. Beam emittance measurements

Before starting emittance measurements numerous beam dynamics simulations have been performed. The emittance has been simulated for a 1 nC beam with a laser longitudinal profile of 25 ps FWHM and 5 ps rise/fall time. A homogeneous transverse laser profile with $\sigma_{x,y} = 0.45\,\text{mm}$ and an accelerating gradient at the cathode of $40\,\text{MV/m}$ were assumed. The results for the above mentioned flat top laser profiles are shown in Fig. 4, the simulated minimum emittance is $\sim 1.6\pi\,\text{mm\,mrad}$. Only changing the transverse laser shape to a Gaussian with the same $\sigma_{x,y}$ yields an emittance minimum of $\sim 5\pi\,\text{mm\,mrad}$.

Measurements of the transverse emittance were performed using a single-slit scan technique. Beamlet profiles were observed 1010 mm downstream of a single-slit mask (1 mm thick tungsten plate, 50 μm slit opening) at screen 3 (see Fig. 1). Beamlets from three slit positions were taken into account for the emittance calculation (see Fig. 5).

Horizontal and vertical emittances were measured as a function of main solenoid current for a gradient at the cathode of $\sim 40\,\text{MV/m}$. The RF phase has also been varied by $\pm 10°$ from the

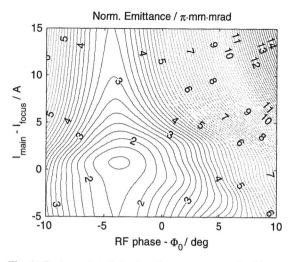

Fig. 4. Contour plot of simulated transverse normalized beam emittance as a function of RF phase and main solenoid current. The reference solenoid current corresponds to the conditions of electron beam focusing at screen 2.

phase of maximum mean energy gain Φ_0. The laser power was tuned for 1 nC beam charge, readjusted for each RF phase chosen. The bucking coil was off during these measurements, so the magnetic field at the cathode is supposed to be small but not zero. Results of the measurements are plotted in Fig. 6.

The normalized beam emittance has been simulated using ASTRA for injector parameters close to the ones observed during emittance measurements. This includes the modelling of the measured transverse and longitudinal laser profiles. Results are shown in Fig. 7.

The coarse agreement between measurement and simulation (minimum emittances between 2 and $5\,\pi\text{mm\,mrad}$ for different transverse laser profiles) is good. In detail, the usage of the measured transverse laser distribution as an input for the simulations results in a rotational asymmetry. At least a part of the disagreement between simulated and measured emittances can be explained by the fact, that the space charge routine used in ASTRA is based on a cylinder symmetric beam model. Another probable explanation comes from possible imperfections in slit orientation, which causes X–Y coupling resulting in increased measured emittances.

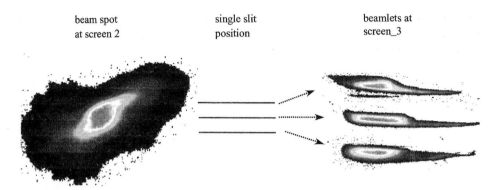

Fig. 5. Schematics of the single slit scan technique for the beam emittance measurement. After measurements of the beam position and beam size at screen 2, the size of the beamlets is measured at screen 3 for three slit positions: $y_n = \langle Y \rangle^{\text{screen2}} + n \cdot 0.7\sigma_y^{\text{screen2}}; n \in \{-1, 0, 1\}$.

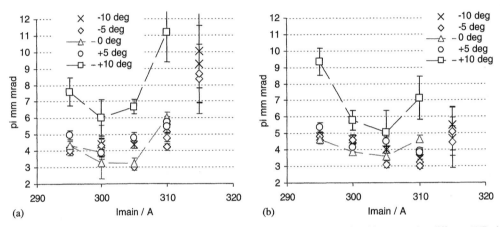

Fig. 6. Measured horizontal (a) and vertical (b) beam emittance as function of main solenoid current for different RF phases with respect to Φ_0.

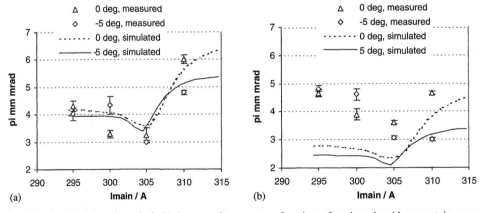

Fig. 7. Simulated horizontal (a) and vertical (b) beam emittance as a function of main solenoid current in comparison with measurements. RF phases are given with respect to Φ_0.

The next steps for optimizing the electron source include further improvement of the transverse laser profile and investigating the influence of the residual magnetic field at the cathode.

7. Conclusions

The experimental optimization and full characterization of the electron source at PITZ is ongoing. The current optimum machine parameters have been found. A normalized transverse beam emittance of 3π mm mrad for 1nC electron beam is measured with high stability and reproducibility. The simulations for the phase scan show reasonable agreement, while emittance simulations need further studies.

References

[1] J. Bakker, et al., Nucl. Instr. and Meth. A 507 (2003) 210.
[2] M. Krasilnikov, et al., Experimental characterization of the electron source at the photo injector test facility at DESY Zeuthen, Proceedings of the PAC 2003, Chicago, May 2003.
[3] J.H. Han, et al., Dark current measurements at the PITZ gun, Proceedings of the DIPAC2003, Mainz, May 2003.
[4] ASTRA, see http://www.desy.de/~mpyflo/.
[5] D. Lipka, et al., Measurements of the longitudinal phase space at the photo injector test facility at DESY Zeuthen, FEL2003.

Available online at www.sciencedirect.com

SCIENCE ⓓ DIRECT°

Nuclear Instruments and Methods in Physics Research A 528 (2004) 366–370

NUCLEAR INSTRUMENTS & METHODS IN PHYSICS RESEARCH
Section A

www.elsevier.com/locate/nima

Measurements of the beam quality on KU-FEL linac

H. Ohgaki*, S. Hayashi, A. Miyasako, T. Takamatsu, K. Masuda,
T. Kii, K. Yoshikawa, T. Yamazaki

Institute of Advanced Energy, Kyoto University, Gokasho, Uji, Kyoto, Japan

Abstract

An infrared FEL facility for advanced energy research is under construction at the Institute of Advanced Energy, Kyoto University. An S-band linac with a thermionic RF gun have been installed and measurement of beam quality has been performed until the entrance of the accelerator tube. The energy spread of below 5% and above 100 mA beam current was measured. The emittance of 8 πmm mrad in horizontal and 15 πmm mrad in vertical at the gun section was obtained by the tomographic method. PARMELA calculation well reproduced both in the energy spectrum and in the emittance.
© 2004 Elsevier B.V. All rights reserved.

PACS: 41.60.cr; 29.17.+w

Keywords: Infrared free electron laser; Linac; Rf-gun; Emittance; Simulation

1. Introduction

We have started construction of an IR-FEL facility (KU-FEL [1]) based on a conventional S-band linac for research on an advanced energy system, especially for the sustainable energy system (bio energy, solar energy, and so on). Since a tunable coherent light can selectively induce a specific molecular reaction, efficiencies of biological or chemical reactions are enhanced by a radiation of IR-FEL. A thermionic RF gun, a 3-m accelerator tube, and a beam transport system have been installed and are waiting for the permission of the radiation-regulation. Although

*Corresponding author. Tel.: +81-774-38-3421; fax: +81-774-38-3426.

E-mail address: ohgaki@iae.kyoto-u.ac.jp (H. Ohgaki).

we have evaluated the electron beam parameters and expected FEL performance with a start-to-end simulation [2], it is crucial to measure the real beam parameters, such as beam emittance, energy spread, and peak current. A system diagnostic can be done with a comparison between simulation and measurement. In this paper, we will describe the electron beam measurement of the KU-FEL system from the gun to the entrance of the accelerator tube. Comparison between measurements and simulation will be described.

2. Measurement of the electron beam parameters

The KU-FEL system consists of an injector, an accelerator tube, a beam transport system, and an undulator. Fig. 1 shows a schematic view of the system. A 4.5-cell thermionic RF gun has been installed for the injector, because it has features of compactness and high brightness of the output beam. An energy dispersive bend with an energy-filtering slit is located between the gun and the accelerator tube. This section consists of two 45° bending magnets and 3 quadrupole magnets and works as an achromatic transport system (called 'Dog-Leg'). A 3-m S-band accelerator tube has been installed and we are waiting for the permission of radiation-regulation. Thus, the measurement of the electron beam parameters has been performed from the gun to the entrance of the accelerator tube.

A 3D-emittance, transverse and longitudinal emittance, is a crucial parameter for lasing FEL. We have measured the transverse emittance, energy spread, and current of the electron beam. Unfortunately, the micro-bunch duration, whose measurement system is under construction, was not measured.

The macro-pulse duration is also important for the operation of the FEL. However, the back-streaming electron in the thermionic RF gun reduces the macro-pulse duration and degraded the beam quality. Thus, we use a short macro-pulse duration (2 μs) during the measurement. The cathode temperature, which was measured by an IR-pyrometer, was fixed to 1258 K and the corresponding surface current was 10 A/cm².

2.1. Energy spread

The energy spread of the electron beam from the gun has been measured with the bending magnet (B-1 in Fig. 1) and a beam current monitor (Faraday cup 2, FC2). Fig. 2(a) shows the results of the measurement as a solid line. The energy spread of 10% was obtained with an RF input power of 4.5 MW. To compare the experimental data with the simulation, PARMELA [3] calculation has been performed. It should be noted that the energy measured at the FC2 was filtered by the vacuum duct of the Dog-Leg section. Fig. 2(a) also shows the energy spectra of the non-filtered (GUN) and filtered (FC2 calc.) beams calculated by PARMELA. We obtained a good agreement between the measurement and the simulation.

In order to reduce the energy spread, we used the slit system. The PARMELA simulation showed that 3.4% energy spread was reduced to 1.4% with a 7 MeV beam by using the 5 mm slit [2]. Thus, the slit width was fixed to 5 mm (half-width) throughout the measurement. The bending magnet downstream the slit system (B-2) and the calibrated current transformer (CT-3) were used to measure the energy spread of the electron beam which was filtered by the slit system. The result of the measurement is shown in Fig. 2(b). It is clear that the slit system reduced the energy spectrum and we obtained 4.0% energy spread of the beam. The energy spectrum given by the simulation is also shown in Fig. 2(b) as a dotted line. One can

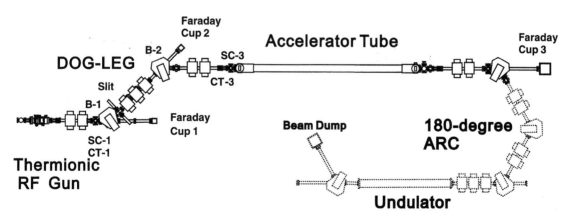

Fig. 1. Schematic view of the KU-FEL system. Dotted parts are under construction.

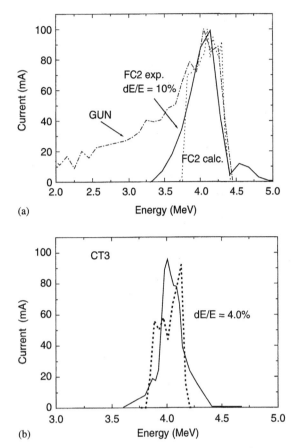

(a)

(b)

Fig. 2. Energy spectra of the electron beam (a) before and (b) after the slit. The solid lines represent the measured spectra and dotted lines of simulation. The half-dotted line in (a) shows the bare spectrum of the electron beam from the gun (simulation).

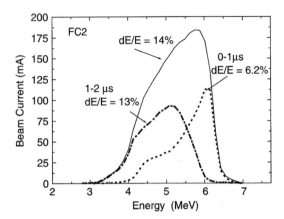

Fig. 3. Energy spectra of the electron beam measured with 5.4 MW RF power. The dotted line shows the energy distribution of the electron beam in the front half of the macro-pulse and half-dotted line shows that of the rest half.

also see a reasonable agreement between the measurement and the simulation.

PARMELA calculation shows that the energy spread of the electron beam from the RF gun is reduced when the RF input power is increased. When the thermal electrons emitted from the cathode are accelerated with the higher RF field, the space charge effect which degrades the beam quality works less effectively. The energy spread of the electron beam at the gun with the RF input power of 5.4 MW was measured and the result is displayed in Fig. 3. Contrary to the simulation ($dE/E = 6.2\%$), the measured energy spread was a broad one, 14%. The reason for this broadness is

due to the back-streaming electron. We also measured the time-dependent energy distribution and found that the front half of the macro-pulse (0–1 μs), where the cathode was not heated by the back-streaming electrons, consists of high-energy electrons. On the other hand, the rear part of the macro-pulse (1–2 μs), where the cathode was overheated by back-streaming electrons, consists of low-energy electrons. Although we could pick up the high-energy component of the electron beam by using the slit system, the macro-pulse duration would be shorter than 1 μs. Actually, the energy spread of 4.8%, which was still larger than that of simulation (3.5%), and macro-pulse duration of 650 ns was observed at the entrance of the accelerator tube (CT-3). It should be noted that the measured beam current was 180 mA with 5.4 MW RF power at the entrance of the accelerator tube. On the other hand, 110 mA beam current was measured with 4.5 MW. According to the PARMELA calculation, we can expect the micro-pulse duration of 2.1 ps, 18 A peak current, at the entrance of the accelerator tube.

2.2. Transverse emittance

The tomography technique has been applied for the precise measurement of the beam emittance [4]. This method brings us a detailed phase space information and is suitable for the thermionic RF gun whose phase space distribution is non-Gaussian. Therefore, we tried to measure the transverse beam emittance with the phase space tomographic method at the gun section. The beam profile was measured with a phosphor screen and a CCD camera (SC-1). The CCD signal was captured by a frame grabber board with an accurate timing trigger. To reconstruct the phase space ellipse, 40 shots of the screen frames with different quadrupole settings were taken for the horizontal and vertical planes. The rotation angles in the phase space were 0.4π in horizontal and 0.2π in vertical which were limited by the distance from the scanning quadrupoles to the phosphor screen. Although, an extrapolation process has been introduced to reconstruct the phase space image, a large error was predicted in the vertical plane. Fig. 4 shows a typical tomographic image

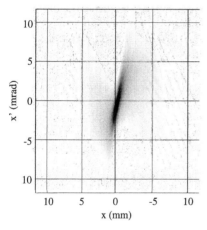

Fig. 4. A reconstructed image of the phase space ellipse measured in the SC-1 screen in the horizontal plane with the RF power of 4.5 MW.

reconstruction. As is shown in Fig. 4, we got a clear image of the phase space distribution. The measured emittances are listed in Table 1. For a comparison, PARMELA calculation was performed and the results are listed in Table 1. The calculation well reproduces the reconstructed image both in the shape and in the value. However, we found that the rotational angles of the reconstructed image were slightly different from the calculation. This suggests the quadrupole settings were slightly different from the calculation. A conventional method of the emittance measurement, the quadrupole scanning and quadratic fitting method, was also performed and the results are showed in Table 1. A reasonable agreement between two methods was found with 4.5 MW RF power.

We tried to measure the emittance at the entrance of the accelerator tube, where PARMELA calculation predicted small values, $3.8\,\pi$mm mrad in horizontal and $0.88\,\pi$mm mrad in vertical. The phosphor screen (SC-3) and a CCD camera were used for the measurement. The measured emittance with the tomographic method were $13.6\,\pi$mm mrad in horizontal and $9.3\,\pi$mm mrad in vertical, This disagreement is mainly due to the reconstruction error with the thick grid drawing on the screen. We noticed that the reconstructed images were separated several islands by the grid drawing and this degraded the read-out emittance. However, we roughly conclude that the emittance of about $10\,\pi$mm mrad or less beam was obtained at the entrance of the accelerator tube.

The RF power dependency on the emittance was also measured. A better emittance can be obtained by simulation with 5.4 MW RF input power

Table 1
Unnormalized emittance (πmm mad)

| Gun RF input (MW) | Direction | Measurement | | Simulation |
		Tomography	Quadratic fit	
4.5	Horizontal	7.5	5.3	7.51
	Vertical	15.3	27.2	21.7
5.4	Horizontal	7.2	4.1	3.3
	Vertical	10.3	34.6	11.8

(Table 1). 5.4 MW power was applied for the gun and the measurement of the emittance has been performed at the gun section. The results are also listed in Table 1. Almost the same emittances as the 4.5 MW case were observed with 5.4 MW RF power. This could be due to the heavy back-streaming effect at the 5.4 MW RF power.

3. Conclusion

We have measured the electron beam parameters, energy spread and transverse emittance from RF gun to the entrance of the accelerator tube. The energy spread of 5% and above 100 mA beam current was obtained at the entrance of the accelerator tube. The emittance was measured to be 8 πmm mrad in horizontal and 15 πmm mrad in vertical at the gun section by using the tomographic method. The PARMELA simulation well reproduced the beam parameters measured with 4.5 MW RF power, but failed with 5.4 MW RF power because of heavy back-streaming effect.

References

[1] T. Yamazaki, et al., Proceedings of the International Free Electron Laser Conference 2001, 2002, pp. II-13.
[2] H. Ohgaki, et al., Nucl. Instr. and Meth. A 507 (2003) 150.
[3] L.M. Young, J.H. Billen, PARMELA, LA-UR-96-1835, 2001.
[4] C.B. McKee, et al., Nucl. Instr. and Meth. A 358 (1995) 264;
X. Qiu, et al., Phys. Rev. Lett. 76 (1996) 3723.

Available online at www.sciencedirect.com

SCIENCE  DIRECT®

ELSEVIER   Nuclear Instruments and Methods in Physics Research A 528 (2004) 371–377

**NUCLEAR
INSTRUMENTS
& METHODS
IN PHYSICS
RESEARCH**
Section A

www.elsevier.com/locate/nima

3-D simulation study for a thermionic RF gun using an FDTD method

H. Hama*, F. Hinode, K. Shinto, A. Miyamoto, T. Tanaka

Laboratory of Nuclear Science, Graduate School of Science, Tohoku University, 1-2-1 Mikamine, Taihaku-ku, Sendai 982-0826, Japan

Abstract

Beam dynamics in a thermionic RF gun for a new pre-injector in a future synchrotron radiation facility at Tohoku university has been studied by developing a 3-D Maxwell's equation solver. Backbombardment (BB) effect on a cathode, which is a crucial problem for performance of the thermionic RF gun, has been investigated. It is found that an external dipole magnetic field applying around the cathode is effective to reduce high-energy backstreaming electrons from the accelerating cell. However, the low-energy electrons coming back from the first cell inevitably hit the cathode, so that characteristics of the cathode material seems to be crucial for reduction of the BB effect.
© 2004 Elsevier B.V. All rights reserved.

PACS: 41.60.Cr

Keywords: Thermionic RF gun; FDTD method; Backbombardment

1. Introduction

Accelerator complex for a future project of synchrotron radiation facility has been under design at Laboratory of Nuclear Science, Tohoku university [1]. The project will contain a storage ring FEL on a 1.5 GeV storage ring and an infrared FEL driven by a pre-injector linac. A newly designed pre-injector linac will provide a 150 MeV beam into a 1.2 GeV booster synchrotorn that is already operational. We are going to choose a thermionic RF gun for the electron injector to extend potential ability of the pre-injector linac.

The thermionic RF gun is quite attractive because of simple configuration with no high voltage stage. In addition, the beam emittance is possibly very low and the bunch length can be compressed to have sufficient peak current to drive FEL [2]. Currently there are not so many linacs equipped with thermionic RF guns because the BB effect strongly restrains further development as the electron source. Most of facilities has been empirically reducing the BB power by applying dipole magnetic field to deflect the backstreaming aside [3]. In order to minimize the BB power in a designing work, we have developed a 3-D simulator code employing finite difference time domain (FDTD) method. At the moment, we investigate spatial distribution of the backstreaming electrons and effectiveness of the external dipole field.

*Corresponding author. Tel.: +81-22-743-3432; fax: +81-22-743-3401.

E-mail address: hama@lns.tohoku.ac.jp (H. Hama).

0168-9002/$ - see front matter © 2004 Elsevier B.V. All rights reserved.
doi:10.1016/j.nima.2004.04.083

2. Remarks in the FDTD code

A standard method of the finite difference in the time domain as well as in the spatial domain is employed. Basic electromagnetic field equations for the time domain are

$$E^n = E(t = t_0 + n\Delta t) \tag{1}$$

$$\varepsilon_0 \dot{E}^{n-1/2} = \nabla \times H^{n-1/2} - j^{n-1/2} \tag{2}$$

$$\mu_0 \dot{H}^n = -\nabla \times E^n \tag{3}$$

where Δt is a time interval between steps and n is the number of steps. To avoid unstable solution, the time interval has to be chosen as

$$(c\Delta t)^{-1} \geqslant \sqrt{\Delta x^{-2} + \Delta y^{-2} + \Delta z^{-2}} \tag{4}$$

where Δx, Δy and Δz are spatial steps of 3-D grid. Perfect electric conductor is applied for the boundary condition at the surface wall of the gun cavity. For the gun exit, the first order Mur's absorbing boundary condition is employed [4].

Number of simulation codes for the beam dynamics in the RF gun have been developed so far. The most conventional way of the silumation is that the equation of motion of electrons is solved in an intrinsic mode of the RF field calculated by a separated code. Since the FDTD code does not separate the electron motion and the microwave propagation, the space charge effect and the beam loading are able to be fully taken into account. Because the thermionic RF gun generates a multi-bunch beam which is extracted from the cathode immediately feeding the RF, the beam loading is supposed to be particularly important. Although it may take a very long cpu time to calculate "start-to-end" simulation of one macro-pulse, the FDTD code has a potential ability for such a simulation. In addition, the code can be easily extended for the beam transport line to investigate effects of wakefields upon the beam property. At the moment, the main purpose of the simulation study is, however, to obtain overall properties of the beam dynamics in a few RF cycle, and relatively large step sizes (Δx, y, $z = 2$ mm and $\Delta t = 3$ ps) are used to save the cpu time sacrificing accuracy.

3. RF gun

An RF gun structure of on-axis coupled 3-cell ($\frac{1}{2} + \frac{1}{2} + 1$) cavity is chosen [5]. The cavity geometry has been determined to satisfy following conditions: (i) resonant frequency of accelerating mode ($\pi/2$-mode) is close to 2.856 GHz, (ii) frequencies of other modes should be sufficiently distant, (iii) electric field on the cathode surface is rather low ($\sim$ half of the maximum one on the beam axis) to reduce the BB power, (iv) kinetic energy gain is around 1 MeV, which is also aiming low back-bombardment (BB) power.

Resonant frequencies depend on the steps of the grid and the time, so that the intrinsic TM modes of the standing wave in the RF gun are excited by feeding a wide-band (1 GHz) Gaussian pulse on a certain proper point, which is a common way in the FDTD method to obtain the electromagnetic response of the system. Well-separated three TM modes are found as shown in Fig. 1(a). The $\pi/2$-mode is then excited by using the derived resonant frequency and the spatial distribution of E_z when that on the cathode surface is at the maximum is shown in Fig. 1(b).

4. Beam simulation and results

4.1. Properties of extracted beam

Equation of motion of electrons

$$\frac{d\vec{\beta}}{dt} = \frac{\sqrt{1 - \beta^2}\,e}{m_0 c}[\vec{E} + c\vec{\beta} \times \vec{B} - (\vec{E} \cdot \vec{\beta})\vec{\beta}] \tag{5}$$

is numerically solved using conventional Runge–Kutta method, where e and m_0 are the electron charge and the rest mass, β is the relative velocity. The electromagnetic field between each time step in the FDTD calculation is interpolated for the electron motion, and then new set of field is again calculated including the new current density.

An impregnated dispenser cathode is supposed to be used and the constant emission current density over a 6 mm diameter cathode is assumed. One of calculated results is shown in Fig. 2, in which the emission current density is 30 A/cm² and

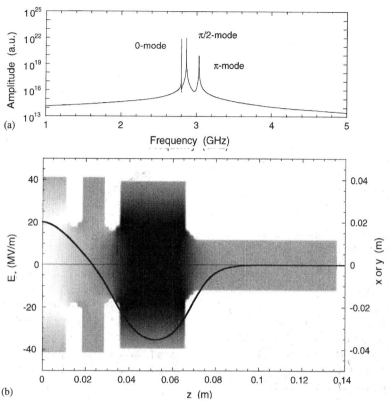

Fig. 1. (a) Frequency spectrum of standing wave in the gun excited a wide-band Gaussian pulse. (b) Calculated longitudinal electric field E_z on the beam axis and its distribution in a transverse plane excited by a narrow-band pulse whose frequency is corresponding to the $\pi/2$-mode.

the maximum accelerating field on the cathode surface is chosen to be 20 MV/m. As one can see in Fig. 2(a), the kinetic energy linearly depends on the exit time which is quite preferable for the energy selection and the bunch compression. The peak kinteic energy is consequently ~1 MeV, which is much lower than that of usual photo-cathode injectors. However, high energy gain in the gun is considered to be dangerous for the BB. It is noticed that the electrons are well concentrated into the nucleus of the beam, so that widths of both the energy and the time are very narrow.

4.2. Transverse emittance and the beam current

A beam current dependence on the transverse emittance is investigated by varying the electron current density from the cathode. The normalized

rms emittance is defined as

$$\varepsilon_x = \beta\gamma\sqrt{\langle x^2\rangle\langle x'^2\rangle - \langle xx'\rangle^2}. \tag{6}$$

Since the extracted beam energy is a continuum spectrum, the beam current is of course depending on the width of the energy selection, the emittance as well.

The beam current averaged over one RF period and the normalized emittance are shown in Fig. 3 for the energy widths $\Delta\beta\gamma = 1\%$, 5% and 10% from the maximum $\beta\gamma$. For the larger energy aperture, $\Delta\beta\gamma = 10\%$, the emittance is much increased, which implies the time-dependent trans-verse electric field considerably affects the emit-tance. The space charge force can be seen as the gradients in Fig. 3(b). For the actual operation, the medium aperture around $\Delta\beta\gamma = 5\%$ is pretty

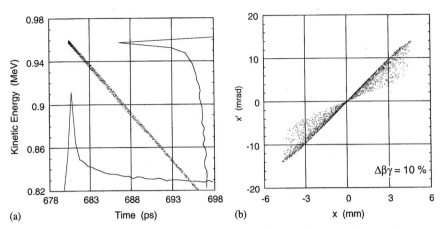

Fig. 2. (a) Longitudinal beam phase space around the peak kinetic energy at the exit of the gun with projections onto the time axis and the energy axis. (b) The transverse phase space is plotted for electrons whose total energies are more than 90% of the maximum. The accelerating phase of the RF field is started at $t = 0$.

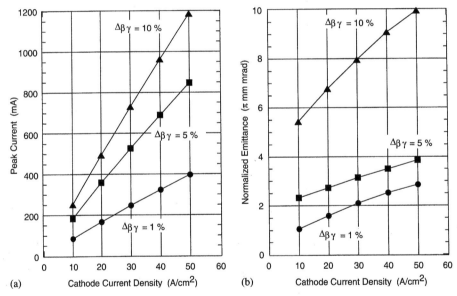

Fig. 3. (a) The beam current and (b) the beam emittance for three different energy apertures plotted as a function of the cathode current density.

desirable because the higher beam current is obtained though the emittance is still lower.

4.3. Backbombardment

The simulation result shows there are two groups of the backstreaming electrons. One group bombards the cathode with relatively low kinetic energy,

which immediately comes back from the first cell. Another one hits the cathode with higher energy. The electrons of this group come back from the accelerating (third) cell, and are focused onto the center of the cathode. Though small number of electrons, the bombarding power of the high-energy group is much higher than that of the low-energy group as shown in Fig. 4 [6]. In the case

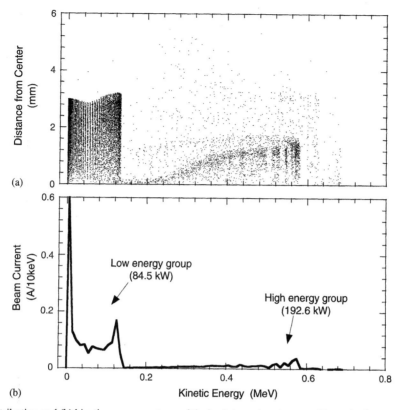

Fig. 4. (a) Radial distribution and (b) kinetic energy spectrum of the backstreaming electrons. The cathode current density is 30 A/cm^2.

of the current density of 30 A/cm^2, a total BB power averaged over one RF period is 277 kW, meanwhile the extracted total beam power is 1.21 MW.

Reduction of the BB power by means of applying external dipole magnetic field is also examined. In the simulation, an ideal dipole field of 50 G covers the first cells and fringing field is not taken into account. The BB power integrated over the cathode area is reduced to 81 from 264 kW by applying the dipole field as shown in Fig. 5. This is evidently caused by which the high-energy group is turned aside. In other words, the low-energy group mostly remains the same because trajectory path of the low-energy group is too short to be sufficiently deflected.

4.4. Small size cathode

For further reduction of the BB effect, a small size cathode such as LaB$_6$ is often used in the operational machines [7]. For the simulation a 2 mm diameter cathode is employed. The total emission current is assumed to be same as that of the 6 mm diameter cathode, so that the current density is 9 times higher. The spatial distribution of the high-energy group is spreading, which is probably due to the space-charge effect because of the higher charge density. On the other hand, most of the low-energy group is still hitting the cathode as shown in Fig. 6(a), and the total BB power on the cathode is 119 kW. By applying the dipole field, the total power is reduced to 60 kW, which is not so much different from that of the 6 mm diameter cathode. The BB power density on the 2 mm diameter cathode is therefore much more higher than that on the 6 mm one. Accordingly, the advantage of the LaB$_6$ cathode against the BB effect does not come from the area size. Since operating temperature of low work function cathode material is higher, we suppose that the

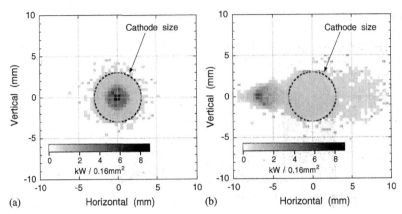

Fig. 5. Spatial distribution of the BB power (a) without and (b) with the dipole field. The cathode current density is $30\,A/cm^2$.

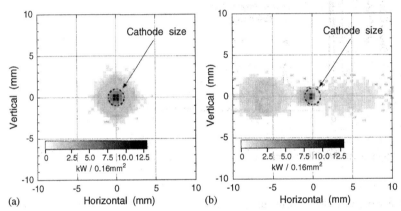

Fig. 6. Spatial distribution of the BB power (a) without and (b) with the dipole field for the 2 mm diameter cathode. The cathode current density is $270\,A/cm^2$.

BB heat capacity may be relatively small for the LaB_6 cathode.

5. Summary and future prospect

Using a 3-D FDTD code, numerical simulations for an on-axis coupled 3-cell RF gun show the excellent beam performances as an FEL driver, i.e., high beam current, low emittance and short pulse length. At the moment the external dipole magnetic field may be effective to reduce the high-energy backstreaming electrons. However, the BB effect of the low-energy one seems to be not completely eliminated. It is desired to find a way out of the difficulty for further development of the thermionic RF gun keeping its simple structure.

Coming assignments are to investigate the beam dynamics in the transient time of microwave feeding and the beam loading effect. Since the FDTD method can involve all of the electromagnetic wave propagation and the charge motion, the simulator code will be further improved. On the other hand, we are going to manufacture a prototype and install it on a test bench soon. Progress on the gun performance will be facilitated by direct comparison with simulation results.

Acknowledgements

One of authors (H.H.) wishes to thank Dr. Kai Masuda of Kyoto University for valuable discussions and suggestions on the simulator code.

References

[1] H. Hama, et al., Design Report of LNS Storage Ring, Tohoku university, http//www.lns.tohoku.ac.jp/.

[2] M. Yokoyama, et al., Nucl. Instr. and Meth. A 475 (2001) 38;
S.V. Benson, et al., Nucl. Instr. and Meth. A 250 (1986) 39.

[3] C.B. MacKee, J.M.J. Maday, Nucl. Instr. and Meth. A 296 (1990) 716.

[4] G. Mur, IEEE Trans. Electromagn. Compat. EMC-23 (1981) 377.

[5] E.A. Knapp, et al., Rev. Sci. Instr. 39 (1968) 979.

[6] Y. Yamamoto, et al., Nucl. Instr. and Meth. A 393 (1997) 443.

[7] M. Kawai, private communication.

Available online at www.sciencedirect.com

NUCLEAR
INSTRUMENTS
& METHODS
IN PHYSICS
RESEARCH
Section A

ELSEVIER   Nuclear Instruments and Methods in Physics Research A 528 (2004) 378–381

www.elsevier.com/locate/nima

Cherenkov interaction and post-acceleration experiments of high brightness electron beams from a pseudospark discharge

H. Yin*, A.W. Cross, A.D.R. Phelps, W. He, K. Ronald

Department of Physics, University of Strathclyde, Glasgow G4 0NG, UK

Abstract

A pseudospark-sourced electron beam has two phases, an initial hollow cathode phase (HCP) beam followed by a conductive phase (CP) beam. The beam brightness was measured by a field-free collimator to be 10^9 and $10^{11}\,A\,m^{-2}\,rad^{-2}$ for HCP beam and CP beam, respectively. The initial HCP beam from an eight-gap pseudospark discharge was applied in a Cherenkov interaction between the electron beam and the TM_{01} mode of a 60-cm long alumina-lined waveguide. While the CP beam from a three-gap pseudospark discharge chamber was propagated and post-accelerated from about 200 V to more than 40 kV.
© 2004 Elsevier B.V. All rights reserved.

PACS: 41.75.Ht

Keywords: Pseudospark discharge; High brightness; Electron beam; Cherenkov interaction; Post-acceleration

1. Introduction

Pseudospark discharge [1–3] experiments to generate a low-temperature plasma for use as a copious source of electrons have been carried out at the University of Strathclyde. The plasma can be regarded as a low work function surface that facilitates electron extraction. Initial study of electron beam production was carried out on a single-gap pseudospark system for a wide range of parameters, including cathode cavity length, cathode hole size, applied voltage, external capacitance

and the inductance in the discharge circuit [4]. Higher-energy electron beam production, more suitable for high power microwave generation, was studied using multi-gap pseudospark systems. Electron beam pulses of duration of tens of ns, current density ($>10^8\,A\,m^{-2}$), brightness of up to $10^{12}\,A\,m^{-2}\,rad^{-2}$ and emittance of tens of mm mrad were measured from a pseudospark discharge [5]. This beam has a higher combined current density and brightness compared to electron beams formed from any other known type of electron source. It was applied in a Cherenkov maser interaction and also transported in a plasma-induced ion background and simultaneously accelerated by an accelerating potential. This article will present some results from these experiments.

*Corresponding author. Tel.: +44-141-548-3355; fax: +44-141-552-2891.

E-mail address: cabs58@strath.ac.uk (H. Yin).

0168-9002/$ - see front matter © 2004 Elsevier B.V. All rights reserved.
doi:10.1016/j.nima.2004.04.084

2. Cherenkov maser interaction experiment

A schematic outline of the pseudospark-based Cherenkov maser amplifier is shown in Fig. 1. The main components of the experiment are the pseudospark-based electron beam source, the magnetic field for beam transport, the Cherenkov interaction region, electrical/beam diagnostics and the microwave launching/diagnostic system. In the maser system, the presence of the dielectric in the waveguide reduces the phase velocity of the electromagnetic waves, allowing a resonant inter- action to occur between a TM or HE waveguide mode and the rectilinear electron beam. Coherence of the generated radiation arises due to bunching of electrons in phase with respect to the electro- magnetic wave.

Microwave radiation was detected successfully from the pseudospark-based dielectric Cherenkov maser amplifier. The temporal profile of the microwave output radiation from the maser is shown in the lower part of Fig. 2, time correlated with the electron-beam current and voltage pro- files.

The frequency range of the microwave radiation from the Cherenkov maser amplifier was measured to be between 25.5 and 28.6 GHz by applying different cylindrical cut-off filters in the wave- guide. The observed frequency was found to be 20% higher than that predicted by the resonance condition for 70–80 keV beam energy. This dis- crepancy is probably due to charging of the dielectric liner [6].

The output antenna pattern associated with the azimuthal E-field component was measured to be independent of the presence of the dielectric and close to zero, confirming the operation of a TM mode. The measured pattern was in good agree-

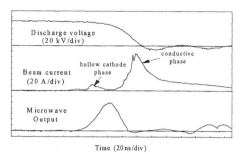

Fig. 2. Typical waveforms of pseudospark discharge voltage, the beam current and the microwave pulse.

ment with the results from bench experiments in which a 27 GHz TM_{01} microwave signal was launched using the same horn. This confirms the operation mode to be TM_{01} mode. The peak power was measured to be around 2 ± 0.2 kW and the gain to be 29 ± 3 dB. A relative spectral energy distribution was obtained and approximately 65% of the radiation was found to lie in 25–28.6 GHz frequency band.

To complement the experimental investigations of the Cherenkov maser, a three-dimensional numerical simulation code was developed [7,8]. The simulations show that the TM_{01} mode at ~ 21 GHz is amplified strongly, attaining a power of ~ 3.4 kW at $z = 60$ cm, whereas the power in the TM_{02} mode at ~ 55 GHz remains around its initial level. These simulations support the inter- pretation of the experimental results as microwave amplification via a Cherenkov interaction between the high-quality electron beam and the TM_{01} mode of the dielectric-lined waveguide.

3. Post-acceleration experiment

A schematic outline of the experimental setup for the study of the propagation and the post- acceleration of the pseudospark-sourced beam is shown in Fig. 3. The electron beam was extracted from a three-gap pseudospark discharge chamber. The single Perspex disc of inner diameter 5 mm, outer diameter 300 mm and 26-mm thickness was used for acceleration gap insulation. It was made with a recess to achieve a 5-mm acceleration gap separation.

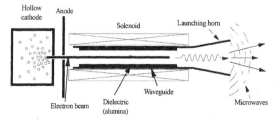

Fig. 1. Cherenkov maser experimental configuration.

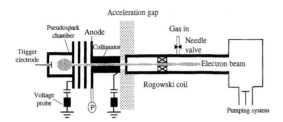

Fig. 3. Schematic diagram for the post-acceleration experiment of the pseudospark electron beam.

The pseudospark discharge was powered by a DC power supply. The hollow cathode was connected through a 30 MΩ charging resistor to a negative voltage source (−30 kV, 1 mA DC power supply) and the charging voltage was measured by a capacitive voltage probe. The external energy storage capacitance was 600 pF. The acceleration unit was driven by a 40 kV, 125 ns voltage pulse produced by a cable Blumlein and the acceleration voltage was measured by another capacitive voltage probe. The pseudospark discharge and the cable Blumlein were triggered by two sets of trigger signals controlled by two delay units and one trigger source. Both trigger signals for the pseudospark and the cable Blumlein were 15 kV pulses. Careful adjustment of the delay units ensured the beam acceleration voltage was applied at the right time.

The propagation of the electron beam from a three-gap pseudospark discharge chamber was studied as a function of the length of a collimator of 3.5-mm internal diameter. Rogowski coil beam current measurements 150 mm downstream from the anode showed that with no magnetic guiding field, 70% and 50% of the beam propagated through 30 and 60 mm long collimators, respectively. A magnetic field free collimator technique also enabled a beam brightness of up to $10^{11-12}\,A\,m^{-2}\,rad^{-2}$ to be measured from a 3-gap pseudospark discharge.

In the post-acceleration experiment a 30 mm long collimator of 3.5 mm internal diameter was used after the anode of the pseudospark discharge chamber and the acceleration unit was located immediately after the collimator to achieve a gas pressure gradient and to optimize beam current. The beam acceleration experiments showed that

careful adjustment of the trigger system could ensure synchronization between the beam propagation and the application of the acceleration voltage. A 100 A, 40 kV electron beam pulse was measured at a distance of 120 mm from the acceleration gap without a magnetic guiding field, as shown in Fig. 4. However, in Fig. 4 some beam loading effect is evident where it can be seen that as the beam current increases the post acceleration voltage decreases and the flat top of the voltage signal becomes shorter. During the experiments, the beam loading effect was mitigated by reducing the internal impedance of the cable pulser from 50 to 14 Ω. It is possible to further reduce the beam loading effect by continuing to decrease the internal impedance of the cable Blumlein or by using an alternative lower impedance pulse forming line.

Beam propagation across the acceleration gap and further along the beam channel was simulated using the electromagnetic PIC code MAGIC. The simulations show that a 200 V, 200 A beam in a pseudospark CP phase will propagate across a 40 kV post-acceleration gap of 5 mm separation in an ion background of certain densities. About 10% and 70% of the beam would propagate across the gap when the ion densities are 1×10^{12} and $6 \times 10^{12}\,cm^{-3}$, respectively. Fig. 5 shows a simulated beam profile and current variation during its propagation across the acceleration gap in an ion background of density $6 \times 10^{12}\,cm^{-3}$.

The simulation also shows that with ion background, the shapes of both the cathode and anode of the acceleration gap have little effect on beam propagation.

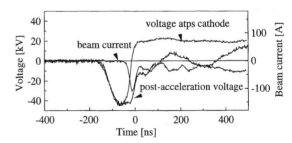

Fig. 4. Typical record of the time-correlated pseudospark discharge voltage, beam current and the acceleration voltage pulse.

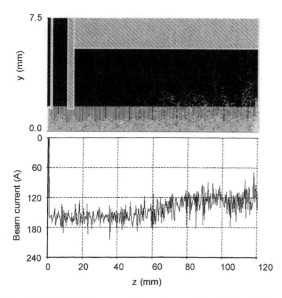

Fig. 5. Simulated beam profile and current variation during propagation across the acceleration gap and further along the beam channel with an ion background of density $6 \times 10^{12}\,cm^{-3}$ (K, cathode; A, anode; z, beam channel axis; vertical dense lines, plasma zone; and cloud, electrons).

4. Discussion and conclusions

In conclusion, we have presented measurements of coherent electromagnetic radiation generation in a free electron maser using an electron beam from a pseudospark discharge. The microwave radiation was generated by Cherenkov amplification of the broadband emission from the pseudospark discharge. The frequency of the microwave output after the Cherenkov maser interaction was measured to be mainly around 25.5 GHz and the dominating mode was identified as being TM_{01}. The peak power is of around 2 ± 0.2 kW and the gain was measured as 29 ± 3 dB.

A 100 A, 40 kV electron beam pulse was measured at a distance of 120 mm from the acceleration gap without a magnetic guiding field. Comparing this with the simulation implies that a favorable ion background exists along the beam channel and the acceleration gap. The ion background can be formed from the background gas ionization by the initial high energy HCP beam during the pseudospark discharge, which also expels electrons in the ionized gas media. In summary, the beam in the pseudospark conductive phase was successfully accelerated from about 200 V to more than 40 kV.

Acknowledgements

The authors would also like to thank the Engineering and Physical Sciences Research Council (EPSRC) for supporting this work.

References

[1] J. Christiansen, C. Schultheiss, Z. . Phys. A290 (1979) 35.
[2] M.A. Gunderson, G. Schaefer, NATO ASI Ser. B, Plenum, New York, 1990.
[3] K. Frank, J. Christiansen, IEEE Trans. Plasma Sci. 17 (1989) 748.
[4] H. Yin, W. He, A.W. Cross, A.D.R. Phelps, K. Ronald, J. Appl. Phys. 90 (2001) 3212.
[5] H. Yin, G.R.M. Robb, W. He, A.D.R. Phelps, A.W. Cross, K. Ronald, Phys. Plasmas 7 (2000) 5195.
[6] E. Garate, R. Cook, P. Heim, R. Layman, J.E. Walsh, J. Appl. Phys. 58 (1985) 627.
[7] H. Yin, A.D.R. Phelps, W. He, G.R.M. Robb, K. Ronald, P. Aitken, B.W.J. McNeil, A.W. Cross, C.G. Whyte, Nucl. Instr. And Meth. A 407 (1998) 175.
[8] H.P. Freund, Phys. Rev. Lett. 65 (1990) 2989.

Available online at www.sciencedirect.com

SCIENCE  DIRECT®

ELSEVIER   Nuclear Instruments and Methods in Physics Research A 528 (2004) 382–386

**NUCLEAR
INSTRUMENTS
& METHODS
IN PHYSICS
RESEARCH**
Section A

www.elsevier.com/locate/nima

Photocathode RF gun using cartridge-type electric tubes

Jun Sasabe[a],*, Hirofumi Hanaki[b], Takao Asaka[b], Hideki Dewa[b],
Toshiaki Kobayashi[b], Akihiko Mizuno[b], Shinsuke Suzuki[b], Tsutomu Taniuchi[b],
Hiromitsu Tomizawa[b], Kenichi Yanagida[b], Mitsuru Uesaka[c]

[a] *Hamamatsu Photonics K.K., Japan*
[b] *JASRI/SPring-8, Japan*
[c] *University of Tokyo, Japan*

Abstract

This report describes an S-band photoinjector with replaceable Cs_2Te photocathodes. The photoinjector is equipped with a cathode-module, which is composed of four cartridge-type electric tubes with Cs_2Te photocathodes. The photocathodes can be inserted into a single cell pillbox-type RF cavity. The maximum electric field strength on the photocathode surface, 90 MV/m, has been achieved for the pillbox-type cavity in only 2 h of RF processing following the exchange of the photocathode. The temporal behavior of the photocathode's quantum efficiency was measured following application of the electric field strength of 90 MV/m. The initial quantum efficiency of 3% was halved to 1.5% in 5 h and finally settled to 1% after 15 h. Analysis of the surface suggested the possibility of oxidization on the cathode surface rather than damage by vacuum discharges.
© 2004 Elsevier B.V. All rights reserved.

PACS: 29.25.Bx; 29.27.Ac; 41.60.Cr

Keywords: Photoinjectors; Photocathode; Cs_2Te; RF gun; Free-electron lasers

1. Introduction

High-brightness X-ray sources, such as the SASE FEL (Self-Amplified Spontaneous Emission Free Electron Laser), for future fourth-generation light sources and inverse Compton scattering sources are being developed throughout the world. An electron beam injector with high-brightness and high-quality electron beams is necessary for these light sources. Accordingly, there is a strong demand for practical applications of a photocathode RF gun that produces an electron beam with a short pulse length and low emittance. There are several other demands, such as high Quantum Efficiency (QE), long lifetime, stability of QE, and low field emission ("dark current"). At present, copper (Cu) is mainly used as the photocathode for a RF gun, in spite of its low QE (typically less than 10^{-4}). However, it is more desirable to use photocathodes with a high QE. Cesium telluride (Cs_2Te) with a high QE (typically several percent) and a long lifetime (several hundred hours) has

*Corresponding author. Tel: +81-53-586-7111; fax: +81-53-586-6180.

E-mail address: sasabe@crl.hpk.co.jp (J. Sasabe).

0168-9002/$ - see front matter © 2004 Elsevier B.V. All rights reserved.
doi:10.1016/j.nima.2004.04.085

been studied as a promising cathode material [1,2]. However, unlike Cu, Cs_2Te requires a deposition chamber with a complex load-lock system. This imposes technical difficulties related to coating, long deposition time, and variations in QE values. To solve such problems, we started developing of a new photocathode RF gun using cartridge-type electric tubes instead of a cathode deposition chamber. The Cs_2Te photocathode is deposited on a Mo plug in the tube. After pushing the cathode out of the tube in a vacuum, the cathode is inserted into a center hole in the back plate of the RF gun. A cathode-module (a revolver-type cartridge holder) can accommodate up to four cartridge-type electric tubes.

In this paper, we report on the cartridge-type electric tubes and a RF cavity with the cathode-module. Then we describe a high-power test using the RF gun as well as a QE lifetime test.

2. Photocathode RF gun

2.1. Cartridge-type electric tubes

A photograph of the cartridge-type electric tube is shown in Fig. 1. The electric tube consists of a UV-transparent glass tube, flanges made of Kovar welded on both sides of the glass tube, a Mo cathode plug with a surface layer of Cs_2Te, and a vacuum bellows to move the cathode plug. A thin Kovar foil is welded to the flange on the cathode side. A RF contact made of Cu–Be is fitted around

Fig. 1. Photograph of cartridge-type electric tube.

the shaft of the cathode plug. To reduce the dark current, the surface of the cathode plug is polished with a diamond polishing compound. The surface has a mirror-like appearance, and the surface roughness is about 10 nm (Ra value). The process of fabricating the Cs_2Te photocathode is based on a first evaporation of 300 Å of Cr followed by a second evaporation of 500 Å of Te onto the Mo cathode plug. We then weld it to the electric tube in air. After the electric tube is exhausted and baked at 170°C, Cs is deposited on the cathode plug while monitoring the photo-current generated by UV light from a mercury lamp. Thus, the Cs reacts with the Te at 170°C to produce the Cs_2Te film. A bias voltage of 90 V is applied between the cathode and anode when the photocurrent is measured. The initial QE is routinely in the range of 5–10% at 260 nm (DC measurement).

2.2. RF cavity with a cathode-module

We have designed a single cell pillbox-type RF cavity. The back plate of this cavity, which is made of OFC (oxygen-free copper), is used as a Cu photocathode to produce a low-emittance electron beam [3,4]. To make it possible to use the cartridge-type cathode with a pillbox-type cavity, we have recently developed a new pillbox-type cavity with a back plate that has a hole 7.8 mm in diameter at its center. Fig. 2 is a schematic drawing of the RF cavity with a cathode-module. After the cartridge-type electric tube is moved from the cathode-module and positioned just before the back plate, a pair of cutters cut the Kovar foil. The cathode (diameter: 7.2 mm) then is inserted into the hole. At this time, the top of the RF finger is in contact with the back plate. While exchanging the photo cathode, the RF cavity is not exposed to ambient air, since the gate valve between the RF cavity and the cathode-module is closed. The main specifications of the RF cavity are almost the same as those of the previous one [3].

3. High-power test

An 80-MW klystron was used to feed a 2856-MHz microwave to the RF cavity. Its RF pulse

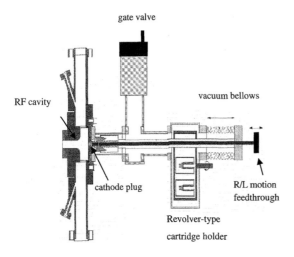

Fig. 2. Schematic drawing of RF cavity with a cathode module.

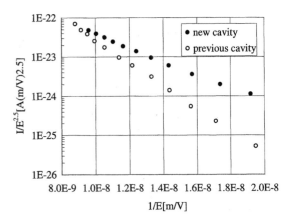

Fig. 3. Fowler and Nordheim (F–N) plots for the dark current of the new cavity and the previous one, where E is the maximum field gradient on the cathode and I is the peak current

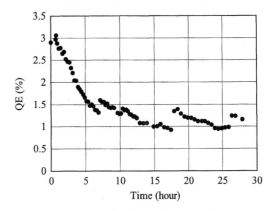

Fig. 4. Temporal behavior of the QE during the beam test.

duration was 500 ns (FWHM), and its repetition rate was typically 10 Hz. The initial RF processing of the RF cavity was carried out with an OFHC copper cathode plug in place of the Cs_2Te photocathode. The charge per bunch, including the dark current, was measured with a Faraday cup at the exit of the gun. The electric field strength on the surface of the copper cathode at 100 MV/m was achieved after 70 h of RF processing. Fowler and Nordheim (F–N) plots, shown in Fig. 3, compare the dark current of the new cavity (Fig. 2) with that of the previous one. The results indicate that the dark current of the new cavity is about 1.5 times higher than that of the previous one at 100 MV/m. This increase in dark current can probably be attributed to the gap between the cathode plug and the hole in the back plate. At present, no damage to the RF contact has been observed, so it is likely that it performs correctly.

4. Beam test

After 70 h of RF processing of the cavity, we replaced the copper cathode with the Cs_2Te photocathode (initial QE was 8.3% by DC measurement), and after another 2 h of RF processing of the photocathode, we started an RF gun beam test. The electric field strength on the cathode surface was 90 MV/m and the pressure

in the gun was 1.2×10^{-7} Pa. The cathode was illuminated by an UV laser with a wavelength of 263 nm, a pulse length of 8 ps (FWHM), a repetition rate of 10 Hz, and an average energy per pulse of 120 nJ. The QE for each pulse was deduced from the ratio of the charge per bunch to the UV energy per pulse. Fig. 4 shows the temporal behavior of the QE during the beam test. The initial QE of 3% was halved to 1.5% in 5 h and finally, settled to around 1% in 15 h. The initial QE of 3% for each pulse is rather low compared with initial DC QE of 8.3%. In addition, four cycles of vacuum discharges followed by rapid QE recovery were observed during the beam test. To investigate the reason for rapid QE degradation and recovery, after the beam test the cathode surface was examined with an optical microscope, a scanning electron microscope (SEM), and an electron probe X-ray micro-analyzer (EPMA). Damage was observed on the cathode surface. There was a crater about 100 μm in diameter at the center of the damage (Fig. 5), and photocathode materials were dislodged by the vacuum discharges (Fig. 6). Two large sites of damage were detected, but they were not more than 5% of the area of the entire region. It is unlikely that the rapid QE degradation resulted

from this damage. The degradation may be attributed to oxidization of the cathode surface by outgassing during the RF processing before the beam test. This outgassing is probably due to discharges between the cathode plug and the hole in the back plate. The QE recovery may be due to the reduction of the oxidized surface by a vacuum discharge. In the future, we will experimentally examine the possibility of this hypothesis. Moreover, the dark current from the cathode was simultaneously measured. The dark current from the Cs_2Te cathode plug was about twice that from the Cu cathode plug. We measured a dark current value of 1.3 nC/pulse at 90 MV/m.

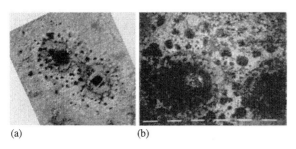

(a)   (b)

Fig. 5. Microscopic views of the damage (a) and SEM images (b).

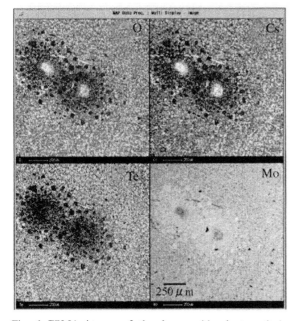

Fig. 6. EPMA images of the damage (signal strength is expressed as a gray-scale intensity; white > black).

5. Concluding remarks

We developed a new photocathode RF gun using cartridge-type electric tubes instead of a cathode deposition chamber and conducted a high-power test for this RF gun as well as a QE lifetime test. We obtained an electric field strength at 90 MV/m on the surface of the Cs_2Te photo-cathode and a QE as much as 1% after a one-day lifetime test. A rapid decrease in QE was observed after applying a high RF field. The results of the surface analysis showed the possibility of oxidiza-tion on the cathode surface instead of damage by vacuum discharge. In the future, we will examine the possibility of this hypothesis, and study the photocathode material and the structure of a cathode plug to achieve stable QE and lower dark current. Our intention is to eventually employ the Cs_2Te photocathode RF guns in practical applications.

Acknowledgements

The authors are particularly thankful to Dr. M. Hagino, M. Tachino, M. Sugiyama, K. Okano, and N. Suzuki for helpful discussions; to Y. Iigami, H. Fujimatsu, S. Shigeji, T. Iida, T. Mitsunari, and M. Kawahara (VALQUA) for technical assistance; and to H. Watanabe for EPMA analysis.

References

[1] S.H. Kong, et al., J. Appl. Phys. 77 (11) (1995) 6031.
[2] P. Michelato, Nucl. Instr. and Meth. A 393 (1997) 455.

[3] T. Taniuchi, et al., Proceedings of the LINAC Conference, Korea, August 2002, p. 685.
[4] H. Tomizawa, et al., Proceedings of the European Particle Accelerator Conference, Paris, July 2002, p. 1819.

Available online at www.sciencedirect.com

SCIENCE DIRECT·

ELSEVIER   Nuclear Instruments and Methods in Physics Research A 528 (2004) 387–391

NUCLEAR
INSTRUMENTS
& METHODS
IN PHYSICS
RESEARCH
Section A

www.elsevier.com/locate/nima

Three-dimensional simulation code for SPring-8 RF gun system

A. Mizuno[a,*], T. Asaka[a], H. Dewa[a], T. Kobayashi[a], S. Suzuki[a], T. Taniuchi[a],
H. Tomizawa[a], K. Yanagida[a], H. Hanaki[a], M. Uesaka[b]

[a] Japan Synchrotron Radiation Research Institute (SPring-8), Mikazuki, Sayo, Hyogo 679-5198, Japan
[b] Nuclear Engineering Research Laboratory, University of Tokyo, Tokai, Naka, Ibaraki 319-1188, Japan

Abstract

To analyze the beam characteristics of an RF gun system, we developed a true three-dimensional beam tracking code. All forces imposed on each macrocharged particle are included in this code. Asymmetrical figures of beams and RF fields can be calculated with this code. In this paper, we describe the details of the developed code, and show comparisons between our code and PARMELA. In addition, we compare the results of our code and the experimental results of the SPring-8 S-band single-cell RF gun system. Using this code, significant differences in measured emittance values in horizontal and vertical directions resulted from an oblique incidence of a laser, that is, the difference of arrival time of the laser wavefront on the cathode.
© 2004 Elsevier B.V. All rights reserved.

PACS: 02.60.Pn; 29.25.Bx; 29.27.Bd

Keywords: RF gun; Beam tracking; Emittance; PARMELA

1. Introduction

Future light sources based on electron linear accelerators require electron guns that can generate a high-brightness, low-emittance electron beam. To date, thermionic electron guns have been used as linac injectors. Conventional thermionic electron guns cannot be used for this purpose because the space charge effect is too strong. To reduce the space charge effect, the electron beam needs to be accelerated more than several MeV in a short distance. As a candidate device, development of a photocathode RF gun has been widely pursued.

A photocathode RF gun with a 2856 MHz single-cell pillbox type cavity has been under development in SPring-8 since 1996. To study the fundamental physics of RF guns and high-gradient acceleration, we chose a single cell cavity.

To design an RF gun system that has a low-emittance beam, several analytical theories [1,2] involving beam dynamics need to be investigated. However, some assumptions are necessary to

---

*Corresponding author.

E-mail address: mizuno@spring8.or.jp (A. Mizuno).

shape and reconcile these theories. Indeed, an electron beam has nonuniform density distributions and an asymmetrical cross-section. These beam characteristics bestow a nonlinear space charge effect and complicated dynamics. Therefore, a beam tracking simulation code is required.

A simulation code for this purpose, PARMELA [3], is used worldwide. However, it cannot calculate an asymmetrical system, such as the spatial and temporal asymmetrical beam shapes and asymmetrical RF fields that are often found in actual experiments.

To investigate the asymmetrical effects, we have been developing a true three-dimensional tracking code. In this code, the forces imposed on each macroparticle are calculated to determine a space charge force. Using this code, we can explain measured emittance behavior with asymmetrical conditions such as the oblique incidence of a laser.

2. Tracking code

2.1. Outline of code

This tracking code treats each charged particle, which is actually a cluster of electrons, as a macroparticle. The charge of the macroparticle is equal to the sum of electron charges in the cluster, and is constant throughout one calculation process.

Instead of creating a mesh to calculate the space charge force as is done in PIC codes, we calculate the forces between all macroparticles.

If we assume macroparticle B is undergoing uniform motion, both the electric and magnetic fields at point A, which is produced by macroparticle B, are calculated as follows:

$$\mathbf{E}_A = \frac{1}{4\pi\varepsilon_0\gamma_B^2} \frac{-Q\mathbf{r}}{[|\mathbf{r}|^2 - \frac{|\mathbf{v}_B\times\mathbf{r}|^2}{c^2}]^{3/2}},$$

$$\mathbf{B}_A = \frac{1}{c^2}\mathbf{v}_B \times \mathbf{E}_A \tag{1}$$

where $\mathbf{r}$ is the vector from B to A, $\mathbf{v}_B$ is the velocity of the macroparticle B, γ_B is the Lorentz factor of B, and Q is the charge of the macroparticle B.

These fields act on the macroparticle A as follows:

$$M_0\frac{d(\gamma_A\mathbf{v}_A)}{dt} = -Q(\mathbf{E}_A + \mathbf{v}_A \times \mathbf{B}_A) \tag{2}$$

where M_0 is the rest mass of the macroparticle A. From the equation of motion above, following equation can be derived:

$$\frac{d\mathbf{v}_A}{dt} = -\frac{1}{\gamma_A}\frac{Q}{M_0}\left(\mathbf{v}_A \times \mathbf{B}_A + \mathbf{E}_A - \frac{(\mathbf{v}_A \cdot \mathbf{E}_A)}{c^2}\mathbf{v}_A\right). \tag{3}$$

Note that, Q/M_0 is equivalent to a specific charge of electron.

Based on the Eq. (3), the tracking of macroparticles is numerically performed using the fourth-order Runge–Kutta method with rectangular coordinates. The z-axis is defined as being along the beamline, the x-axis is the horizontal direction, and the y-axis is the vertical direction. Note that x and y components of $\mathbf{v}_B$ in Eq. (1) are ignored to reduce the calculation time, which is permissible because they are negligible in comparison with the z component.

Initially, a macrocharge bunch having a length equivalent to a laser pulse length is positioned behind the cathode. Random numbers determine each particle position. When the tracking starts, each macroparticle runs parallel to the z-axis with the velocity of light until it reaches the cathode surface. Once it leaves the cathode, its velocity is established by the electromagnetic field in the cavity, space charge, image charge of the cathode, and magnetic fields of the solenoid coils. Near the surface of the cathode, we assume an image charge as if there were a mirror particle of each macroparticle. Each mirror particle has the opposite polarity as its macroparticle. The value of velocity is opposite to that of a z component; this also holds for the x and y components. There is a shield section for image force calculation on the cathode surface with a thickness of several μm. When a macroparticle is located inside the shield section, the calculation process of the image charge is not applied. This is because the motion equation shown in Eq. (3) does not depend on the rest mass of the macroparticle, but rather on the specific charge of the electrons. Therefore, if a macroparticle is close to its image particle, a significantly strong force is produced between them, which drives the macroparticle far away by a single tracking step.

Time step tracking durations of 0.1 ps were chosen near the cathode and in the RF cavity because velocity is changed significantly in this section. On the other hand, it is changed to 1.0 ps after the cavity section because of small changes of the electromagnetic field.

Our cavity has two RF ports in the horizontal direction as shown in Fig. 2, thus the horizontal and vertical field distributions are slightly different. Therefore, the code MAFIA [4] was used to calculate the three-dimensional fields. The field data obtained are imported into our simulation code. The defined field in the cavity is expressed by the following equations:

$$\mathbf{E}_{cavity} = -\mathbf{E}_0 \cos(\omega t - \phi_i)$$
$$\mathbf{B}_{cavity} = \mathbf{B}_0 \sin(\omega t - \phi_i) \qquad (4)$$

where ϕ_i is the initial phase of RF, and E_0 and B_0 are calculated matrix field data using MAFIA. We define $t = 0$ when an electron at the head of the bunch leaves the cathode surface.

The fields of solenoid coils that are located after the cavity are calculated in this simulation code.

The rms normalized transverse emittance of the beam is calculated using the following formula:

$$\varepsilon_x = \langle \gamma \rangle \langle \beta \rangle \sqrt{\langle x^2 \rangle \langle x'^2 \rangle - \langle x \cdot x' \rangle^2} \qquad (5)$$

where the symbol $\langle \rangle$ is an average of all macroparticles. Emittance ε_x in Eq. (5) is equivalent to the Courant–Snyder invariant, if x and x' are Gaussian distributions. Therefore, we express the emittance value as ε_x πmrad. For ε_y, a formula can be obtained from Eq. (5), by replacing x with y.

2.2. Comparison of beam envelope with PARMELA

The behaviors of beam orbit extensions were calculated in free space using our code and PARMELA to examine space charge effects. We used PARMELA with a two-dimensional space charge calculation mode. Fig. 1 shows plots of the beam envelope as a function of the longitudinal beam position with various initial beam energies.

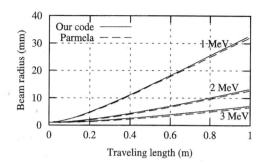

Fig. 1. Calculated beam envelopes in free space using our code and PARMELA with charge per bunch of 1 nC.

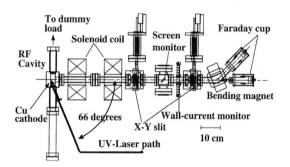

Fig. 2. Schematic of our RF gun system.

We set an electron bunch at $z = 0$ with initial conditions of charge per bunch of 1 nC, a beam radius of 1.0 mm, beam bunch length of 6.0 mm, etc. Both spatial and temporal initial beam distributions are almost uniform. The number of particles was 5000 for both codes. We obtained good agreement between our code and PARMELA.

3. Emittance comparison with measurement value

We have compared our simulation code to experimental results. A schematic of our system is shown in Fig. 2.

The copper cathode is part of the cavity wall and generates electrons by an illuminating laser pulse with a wavelength of 263 nm, which is the third harmonic of a Ti:Sa laser. The electrons produced are accelerated to 3.1 MeV in the cavity, which corresponds to a maximum field gradient on the cathode of 135 MV/m. Two solenoid coils located after the cavity focus the electrons.

For emittance measurements, two slits and a Faraday-cup are located downstream from the solenoid coils. The first slit is located 670 mm downstream from the cathode surface. The second slit is located 460 mm downstream from the first slit. Each slit has two sets of copper blocks to scan horizontally and vertically, respectively. There is a gap of 0.3 mm at the center of the blocks.

Measurements are obtained by scanning the first and second slit independently. The transverse emittance, derived from the x–x' (or y–y') matrix data set, is acquired as output signals from the Faraday cup. Therefore, we measure the emittance value at the first slit position.

Fig. 3 shows horizontal and vertical normalized emittance as a function of charge per bunch. It consists of two data series measured on different experimental runs and has reproducibility. The minimum normalized x-emittance value we measured was 2.3 πmm mrad with a net charge of 0.1 nC/bunch.

Simulations for these measurements were performed with the same parameters as for the measurements [5]. Both the measured and calculated y-emittances are larger than the x-emittances. These phenomena are due to the large incident angle of the laser. Our cavity has a laser port with an incident angle of 66°, as shown in Fig. 2. Therefore, the entire wavefront of a laser pulse does not reach the cathode surface simultaneously. The initial beam shape then becomes asymmetrical in shape, both spatially and temporally. Since our simulation code is a true three-dimensional code, these effects can be calculated. The simulation results can explain the measurement results with a charge of less than 0.8 nC/bunch.

4. Comparison of emittance with PARMELA

Our code predicts that a lower emittance value can be achieved by means of a right angle incidence of the laser. We surveyed optimum parameters for the lowest emittance using our code. In these optimizations, the maximum electric field on the cathode was held at 135 MV/m, and the charge remained at 0.1 nC/bunch. The optimum parameters are listed in Table 1.

These are symmetric conditions so simulations of PARMELA are applicable. In Fig. 4, values of calculated emittance as a function of charge are shown using our code and PARMELA. For calculations of high charge, over 0.1 nC/bunch, the parameters listed in Table 1 are used.

Table 1
Optimum parameters for the lowest emittance value

Initial RF phase	85°
Temporal and spatial laser profile	uniform
Laser spot size and pulse length	ϕ1.6 mm, 20 ps
Fields of two solenoid coils	1680, 1190 G

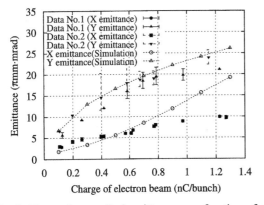

Fig. 3. Measured normalized emittance as a function of a charge per bunch. Calculated data are also plotted.

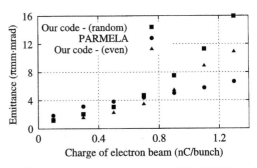

Fig. 4. Comparison between our code and PARMELA using optimum parameters which are surveyed by our code.

Calculated y-emittances are almost the same as the x-emittances since there were symmetrical conditions; as a result only x-emittances were plotted.

The results of our code and PARMELA nearly agreed at 0.7 nC/bunch and below, but not over this value.

During the tracking process, if two particles are close to each other, a strong field will interact with these particles and disturb their orbits. Therefore, the emittance is exaggerated in this case. To avoid this effect, the calculations were repeated. Instead of using random numbers to define the initial particle positions, the particles were placed at even intervals to avoid particles in proximity. The results of these new calculations are also plotted in Fig. 4. The emittance differences over 0.7 nC/bunch are decreased, although they still remain.

There is a possibility that the emittance difference over 0.8 nC/bunch in Fig. 3 is also due to these effects. It is necessary to investigate the cause of the difference.

5. Summary

We developed a true three-dimensional beam-tracking code to investigate asymmetrical conditions of the RF gun. The envelope calculations in free space show good agreement with our code and PARMELA. Our code can explain the difference between the x- and y-measured emittance caused by the difference of arrival times of the laser wavefront on the cathode. In the case of an asymmetrical system, the measured- and calculated-emittance values also agree when the charge is less than 0.8 nC/bunch.

References

[1] B.E. Carlsten, Nucl. Instr. and Meth. A 285 (1989) 313.

[2] K.J. Kim, Nucl. Instr. and Meth. A 275 (1989) 201.

[3] L.M. Young, J.H. Billen, Los Alamos National Laboratory, LA-UR-96-1835, 2000.

[4] M. Bartsch, et al., Comput. Phys. Commun. 72 (1992) 22.

[5] H. Tomizawa, et al., Proceedings of the European Particle Accelerator Conference, Paris, July 2002, p. 1819.

Available online at www.sciencedirect.com

Nuclear Instruments and Methods in Physics Research A 528 (2004) 392–396

ELSEVIER

NUCLEAR
INSTRUMENTS
& METHODS
IN PHYSICS
RESEARCH
Section A

www.elsevier.com/locate/nima

Development of photoinjector RF cavity for high-power CW FEL

S.S. Kurennoy[a,*], D.L. Schrage[a], R.L. Wood[a], L.M. Young[a], T. Schultheiss[b], V. Christina[b], J. Rathke[b]

[a] Los Alamos National Laboratory, Los Alamos, NM 87545, USA
[b] Advanced Energy Systems, Medford, NY 11763, USA

Abstract

An RF photoinjector capable of producing high continuous average current with low emittance and energy spread is a key enabling technology for high-power CW FEL. A preliminary design of the first, and the most challenging, section of a 700-MHz CW RF normal-conducting photoinjector—a 2.5-cell, pi-mode cavity with solenoidal magnetic field for emittance compensation—is completed. Beam dynamics simulations demonstrate that this cavity with an electric field gradient of 7 MV/m will produce an electron beam at 2.7 MeV with the transverse rms emittance 7 mm mrad at 3 nC of charge per bunch. Electromagnetic field computations combined with a thermal and stress analysis show that the challenging problem of cavity cooling can be successfully resolved.

We are in the process of building a 100-mA (3 nC of bunch charge at 33.3 MHz bunch repetition rate) photoinjector for demonstration purposes. Its performance parameters will enable a robust 100-kW-class FEL operation with electron beam energy below 100 MeV. The design is scalable to higher power levels by increasing the electron bunch repetition rate and provides a path to a MW-class amplifier FEL.
© 2004 Elsevier B.V. All rights reserved.

PACS: 41.60.Cr; 29.17.+w; 29.27.Bd; 41.75.Fr

Keywords: FEL; Photoinjector; High current; High brightness; CW

1. Introduction

For a MW-class FEL, a high-current emittance-compensated photoinjector is a key element. We present results on the RF cavity design and cooling schemes for a demo 700-MHz CW RF photo-injector (PI) accelerating 100 mA of the electron beam (3 nC per bunch at 33.3-MHz repetition rate) to 5.5 MeV [1]. The transverse rms emittance is below 10 mm mrad at the wiggler, with an energy spread below 1% to allow bunch compression before the wiggler. The PI is scalable to higher beam currents, up to 1 A, by increasing the bunch repetition rate. To keep the beam emittances low, high electric-field gradients are needed, especially near the photocathode, along with an external

*Corresponding author. Tel.: +1-505-665-1459; fax: +1-505-665-2904..
E-mail address: kurennoy@lanl.gov (S.S. Kurennoy).

0168-9002/$ - see front matter © 2004 Elsevier B.V. All rights reserved.
doi:10.1016/j.nima.2004.04.087

magnetic field to compensate space charge. Wall power losses increase as cavity fields squared, for a given cavity shape, leading to challenges in cavity cooling and increasing the required RF power. Therefore, an optimal design must be a trade-off between the requirements of beam dynamics, RF power, and feasibility of the cavity cooling.

It is important to emphasize that many of the required beam parameters have already been achieved in normal-conducting photoinjectors. A beam charge up to 5–7 nC per bunch as well as a bunch repetition rate of 108 MHz in macropulses have been demonstrated in the AFEL [2]. However, the AFEL duty factor was rather low. The highest average beam current of 32 mA was achieved in 1992, by the Boeing photoinjector at 25% duty factor [3]. For a CW operation of high-current normal-conducting photoinjectors, one must solve the challenging problem of photoinjector thermal management.

2. RF cavity design

2.1. Split photoinjector

After studying various options, we arrived at the photoinjector design that can be called a "split" design. The photoinjector has a two-stage configuration consisting of a 2.5-cell cavity with the accelerating gradient $E_0 = 7$ MV/m (the PI proper) that brings the electron beam energy to 2.7 MeV, followed immediately by a conventional 4-cell booster with $E_0 = 4.5$ MV/m, where the beam is accelerated to 5.5 MeV. Starting with a $\left(n + \frac{1}{2}\right)$-cell, π-mode RF structure, with cell-to-cell coupling of 0.03–0.05, we studied a few options for cell coupling [4], including magnetic coupling via slots and on-axis electric coupling through apertures. Using slots provides effective structures with high shunt impedance but leads to very high power-loss densities near the slots. As a result, the on-axis coupled cavity structure was chosen [4]. It has an additional advantage: the septa (cell-separating walls) are easier to cool than those with coupling slots.

A long structure, $n = 6$ or 7, which would be needed to achieve the beam energy of 5 MeV, while

keeping the heat flux manageable, gives a rather small mode separation. Our solution, therefore, was to split it. The 2.5-cell PI cavity produces a good quality beam due to its higher accelerating gradient and a matching emittance-compensating external solenoid, with a bucking coil that cancels the magnetic field near the cathode. The magnet designs are similar to those for the 1.3-GHz AFEL [2]. The booster then raises the electron beam energy to 5.5 MeV. The wall power density in the 4-cell booster, due to its lower gradient, is below 35 W/cm^2; its cooling is less of a problem.

The challenging part of the PI is the 2.5-cell RF cavity. Its schematic, including cooling and tapered ridge-loaded RF waveguides, is shown in Fig. 1. With an aperture radius of 65 mm and septum thickness of 20 mm, the cell coupling in the 2.5-cell structure is near 0.03, so that the π-mode at 700 MHz is well separated from its nearest neighbors. The cavity design is a result of a few iterations made with Superfish (SF) [5] and MicroWave Studio (MWS) [6], followed by a thermal analysis.

Beam dynamics was simulated with PARMELA [5] to select field configurations minimizing the emittance for a given cavity shape. We found that a flat electric field profile in the 2.5-cell PI cavity—equal field amplitudes in all 3 cells—combined with a proper focusing gives the best transverse emittance.

Results of PARMELA simulations confirm that our split design satisfies all beam requirements. The photoinjector can produce the beam transverse rms emittance of 7 mm mrad at 3 nC per bunch, see Ref. [1].

2.2. Electromagnetic modeling

The on-axis electric field of the π-mode in the PI 2.5-cell RF cavity has a flat distribution. The field alternates its direction from one cell to the other and reaches about 10 MV/m near the cathode for the required gradient $E_0 = 7$ MV/m. Fields in the vacuum plenum are small, because it would resonate at a lower frequency of about 650 MHz. Wall power densities are calculated using both SF and MWS. The surface current distribution inside the cavity is illustrated in Fig. 2. The highest power

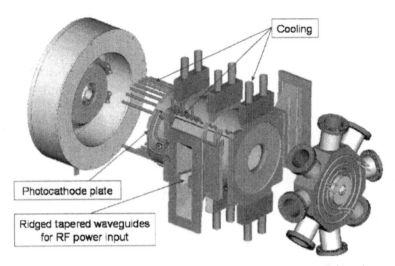

Fig. 1. Schematic of the 2.5-cell RF cavity (center) with emittance-compensating magnets (left) and vacuum plenum (right).

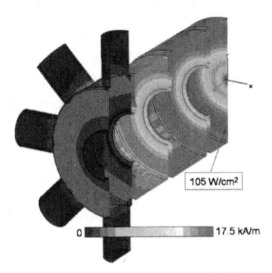

Fig. 2. Surface current distribution in the π-mode of the 2.5-cell cavity.

density is on the cell septa and particularly on the first half-cell end wall, where it reaches $105\,\mathrm{W/cm^2}$ for $E_0 = 7\,\mathrm{MV/m}$.

The RF feeds are in the third (the second full) cell of the 2.5-cell PI cavity; cf. Fig. 1. Two symmetrically placed ridge-loaded tapered waveguides are connected to the cavity via "dog-bone" coupling irises—long narrow slots with two circular holes at their ends—in a 0.5 in.-thick copper wall. This design is based on the LEDA RFQ and SNS high-power RF couplers [7]. The required cavity–waveguide coupling is given by the coefficient $\beta_c = (P_w + P_b)/P_w \approx \frac{4}{3}$, where $P_w = 780\,\mathrm{kW}$ is the CW wall power loss, and $P_b = 270\,\mathrm{kW}$ is the beam power of a 100-mA beam at the 2.5-cell cavity exit. The coupling is adjusted by changing the radius of the holes at the ends of the coupler slot.

The waveguide–cavity system was modeled [8] using a pillbox cavity (Fig. 3—a slice of the third cell in the 2.5-cell PI cavity), with two short ridge-loaded waveguides. The required waveguide–cavity coupling for the pillbox is $\beta_{pb} = \beta_c(W_c/W_{pb})(Q_{pb}/Q_c)$, where W_i, Q_i are the field energies and unloaded quality factors of the pillbox TM_{010} mode and of the π-mode in the 2.5-cell cavity. We assume that W_{pb} is scaled such that the surface magnetic field far from the coupler iris in the transverse cross-section at the coupler location is equal to one at the same location in the third PI cell (14.8 kA/m) for the nominal PI gradient. This gives the waveguide–pillbox coupling $\beta_{pb} \approx 6.1$.

Time-domain simulations with MWS to adjust coupling and calculate the power density distribution around the coupler are described in Ref. [8]. We simulate the RF feeding of the model via the waveguides with a TE-mode at the pillbox–waveguide resonance frequency, $f_r = 699.575\,\mathrm{MHz}$.

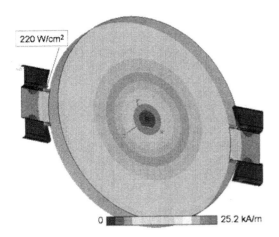

Fig. 3. Surface currents near the RF coupler irises in the model.

While the waveguide input remains constant, the output decreases, reaching a point where it vanishes, and subsequently increases. The point of zero reflected power corresponds to an exact match. Field pictures at that moment give us field distributions for the matched case, as shown in Fig. 3. The highest surface power densities of 220 W/cm^2 are localized near the ends of the slot aperture. This is a rather high value, but the hot spots are very small and can be cooled efficiently using dedicated cooling channels placed near the coupler iris in the thick cavity wall. The average RF power transferred through each of the two waveguides is about 525 kW, as also follows from the energy balance for the matched case.

One can calculate that our RF cavity without beam (unmatched) will reflect only 2.7% of the input power, with the output amplitude equal to 16.5% of the input one. We can observe such a situation in initial thermal tests of the 2.5-cell cavity with the full RF load but without beam. Field snapshots at this point give us field distributions for the unmatched case. The maximal power density near the iris is about the same as that for the matched case, but the average input RF power is now 400 kW through each waveguide.

3. Thermal management

The required high-field operation of this cavity results in high RF losses on the surface. While cooling with liquid nitrogen (LN$_2$) would lower the wall loads significantly [1], the cost of a high-pressure LN$_2$ system was prohibitive for the demo PI. The operating cost of CW LN$_2$ system also makes it very unattractive for PI applications. Therefore, a water-cooled design was pursued. This design, with some tuning to adjust the frequency (e.g., with slug tuners in the second cell of the 2.5-cell cavity), would also work with high-pressure LN$_2$ if desired.

About 800 kW of RF power are deposited on the walls of the water-cooled PI 2.5-cell cavity. This requires intricate cooling channels and a relatively large flow of cooling water. The cooling design is substantially complete. Flow requirements based on RF heat loads at surface temperature are reasonable: 36 l/s at 5 m/s with inlet water temperature of 20°C and mean outlet temperature 26°C; see details in Refs. [1,9].

An axisymmetric model of the cavity was developed using the finite element code ANSYS. RF loads were mapped on the surface from SF output and modified based on the surface temperature. Significant stresses generally occur between the heated RF surface and cooling channels. Glidcop®, a dispersion-strengthened copper, is used throughout this cavity. It has a yield strength of 270 MPa and provides significant margin for this design. The steady-state temperature distribution was mapped onto a structural finite element model, using the same nodes and element connectivity as the thermal model. The resulting steady-state von Mises local stresses do not exceed 70 MPa [1,9].

4. Conclusions

The physical design of the PI RF cavity is completed. This design allows us to surpass the required beam parameters while addressing the key issue for a high-current normal conducting CW RF photoinjector, which is the structure cooling. It provides a path forward to a very high-power amplifier FEL.

The cooling of the regions around the coupling irises is a challenging thermal management issue due to surface power density exceeding 200 W/cm^2

at some points. However, the hot spots are localized in small regions around the slot ends and can be cooled effectively with cooling channels placed properly around the slots in the thick cavity wall.

The cold model manufacturing is nearly completed at AES. Our plan is to construct the full-power prototype and install it in the existing facilities at LANL to perform RF and thermal tests without beam, by late 2004. Once the RF/thermal management with high heat flux has been demonstrated, the system will be upgraded to operate with electron beam, by adding a real photocathode and drive laser, a second RF power source, and a short beam line with diagnostics.

Acknowledgements

The authors would like to acknowledge useful discussions with D.C. Nguyen and S.J. Russell (LANL).

References

[1] S.S. Kurennoy, et al., Photoinjector RF cavity design for high power CW FEL, Proceedings of PAC, 2003, p. 920.

[2] R.L. Sheffield, et al., Nucl. Instr. and Meth. A 318 (1992) 282.

[3] D.H. Dowell, et al., Appl. Phys. Lett. 63 (1992) 2035.

[4] S.S. Kurennoy, et al., CW RF cavity design for high-average-current photoinjector for high power FEL, FEL 2002, Elsevier, Amsterdam, 2003, p. II-53.

[5] J.H. Billen, L.M. Young, Reports LA-UR-96-1834, 1835 (revised 2003).

[6] MWS v.4.3, CST GmbH, http://web.cst-world.com.

[7] L.M. Young, et al., High power RF conditioning of the LEDA RFQ, Proceedings of PAC 1999, NY, p. 881.

[8] S.S. Kurennoy, L.M. Young, RF coupler for high-power CW FEL photoinjector, Proceedings of PAC 2003, p. 3515.

[9] A.M.M. Todd, et al., High-power electron beam injectors for 100 kW free-electron lasers, Proceedings of PAC 2003, p. 977.

Available online at www.sciencedirect.com

SCIENCE ⓓ DIRECT°

ELSEVIER   Nuclear Instruments and Methods in Physics Research A 528 (2004) 397–401

**NUCLEAR
INSTRUMENTS
& METHODS
IN PHYSICS
RESEARCH**
Section A

www.elsevier.com/locate/nima

Experiments with electron beam modulation at the DUVFEL accelerator

T. Shaftan[a],*, Z. Huang[b], L. Carr[a], A. Doyuran[a], W.S. Graves[c], C. Limborg[b], H. Loos[a], J. Rose[a], B. Sheehy[a], Z. Wu[a], L.H. Yu[a]

[a] *BNL-NSLS, Upton, NY 11973, USA*
[b] *SLAC, 2575 Sand Hill Road, Menlo Park, Ca 94025, USA*
[c] *MIT-Bates, Middleton MA 01949, USA*

Abstract

Recently strong modulation of the electron bunch longitudinal phase space was obtained in the DUVFEL accelerator at NSLS. The chirped beam energy spectra exhibit a spiky structure with a subpicosecond spike separation. A model based on the longitudinal space-charge effect has been suggested as an explanation of the observed effect. For characterization of the structure and comparison with the model we performed several experiments. In this paper we present experimental data and discuss them in relation to the model.
© 2004 Elsevier B.V. All rights reserved.

PACS: 41.75.Fr; 52.59.Sa; 41.60.Cr

Keywords: Electron beam; Collective effects; Energy spectrum

1. Introduction

In our previous papers we described a spiky structure in the chirped beam energy spectra observed while measuring bunch length [1,2]. The horizontal beam profile in the energy spectrometer (6 and 7 on Fig. 1) exhibited modulation with as much as 100% depth and period of 100 μm (compare with DUVFEL slippage of 70 μm for the lasing at 266 nm). The interpretation of the

spiky structure as the longitudinal density modulation led to the conclusion about inevitable degradation of the FEL performance. However, no real degradation of the FEL output was observed which motivated investigation of an effect responsible for this phenomenon.

A model based on longitudinal space-charge effect was suggested in [3]. According to this model small density clusters in the electron bunch create fluctuations in the static space charge field. Field fluctuations accelerate particles at the head of the cluster and decelerate particles at the tail, creating energy modulation along the bunch. This initiates plasma oscillations in the bunch during its travel

---

*Corresponding author. Tel.: +1-631-344-5144; fax: +1-631-344-3029.

E-mail address: shaftan@bnl.gov (T. Shaftan).

0168-9002/$ - see front matter © 2004 Elsevier B.V. All rights reserved.
doi:10.1016/j.nima.2004.04.119

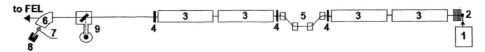

Fig. 1. The DUVFEL layout. 1—RF gun drive laser, 2—RF gun with focusing solenoid, 3—3 GHz linac tank, 4—focusing triplets, 5—magnetic chicane, 6—spectrometer dipole, 7—beam position monitor, 8—beam dump, 9—CTR station.

along the accelerator. As a result, the phase space distribution of the electron bunch at the end of the accelerator is strongly distorted.

Another possible mechanism driving the structure is coherent synchrotron radiation (CSR) instability in the chicane-compressor [5,6]. In this case, coherent radiation emitted by electrons while passing through the chicane creates density bunching along the beam. For DUVFEL experimental conditions, both numerical simulation [2] and analytical estimates [3] give a small value of the gain for CSR microbunching instability.

The layout of the DUVFEL accelerator is depicted in Fig. 1. For energy measurements, we use a spectrometer dipole located at the end of the accelerator. A beam position monitor intercepts the beam 30 cm downstream of the dipole, the image on the monitor is recorded, and the transverse distribution in the spectrometer dispersion direction measures the electron energy distribution.

The spectrometer set-up also allows us to measure the bunch length, using the "zero-phasing" technique [7]. In this mode the last linac tank is running off-crest, imparting an energy-time correlation in the electron beam. Measuring a chirped beam's energy spectrum, one can estimate the bunch length [1].

A coherent transition radiation (CTR) station [8] is located upstream of the spectrometer (Fig. 1). The station consists of an in-vacuum copper mirror mounted on a retractable support. The infrared light is outcoupled into a nitrogen-cooled bolometer by a 3-mirror collection system.

The described experimental set-up gives us an opportunity to characterize the longitudinal phase space only at the end of the accelerator. In our studies, we varied the beam parameters in the beam line upstream and studied the impact of these variations on the measured projections of the longitudinal phase space [4]. In the experiments,

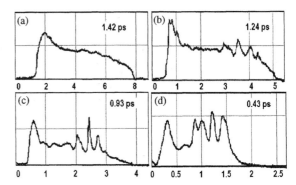

Fig. 2. Development of the structure during compression process. Horizontal axis is scaled in picoseconds.

we used beam parameters close to regular DUVFEL operations (charge of ∼300 pC, bunch length of 4 ps, normalized emittances of ∼4 mm mrad).

2. Gradual bunch compression

As discussed in [1], structure appears in an initially smooth chirped beam energy profile after the beam gets compressed. In order to study the dynamics of modulation during the compression process we performed the following experiment. The second linac section was gradually dephased from the on-crest position, providing the energy-time correlation required for compression. The chirped beam energy spectra for every compression ratio are shown in Fig. 2. The first profile corresponds to an uncompressed bunch with a charge of 300 pC. The spike at the head of the bunch (on the left side) is caused by nonlinearity of the phase space due to RF curvature. The beam profile exhibits small amount of modulation with average period of ∼900 fs.

In frames (b)–(d) we provided compression ratios of 1.15, 1.53 and 3.30 respectively, by changing the phase offset in the second tank.

The small modulation in the chirped beam profile evolved into a highly pronounced structure with a depth of more than 50%. Another important observation, derived from the profiles, is in the evolution of the modulation wavelength. Though structure shapes differ for consecutive profiles, the average spike separation tends to scale with compression ratio. For instance, in (d), the spike separation is about 250 fs.

This experiment demonstrates the enhancement of the structure during the compression process. The compression decreases the average period of the structure while increasing the peak current in the bunch.

3. Experiment with the chicane strength

The result of the previous experiment leads us to the question of the sensitivity of the structure to the chicane strength. The impedance [9] of CSR in the chicane depends on the bending radius in the chicane magnets and, for the DUVFEL configuration, it is comparable to the length of the bends. Therefore, for a CSR-mediated effect, the observed structure should be sensitive to the chicane settings.

Changing the chicane strength has two major consequences. First, it changes the compression ratio and, therefore, the bunch peak current. Second, due to the edge focusing of the chicane dipoles, it affects the transverse beam envelope along the accelerator. Thus, changing the chicane strength significantly impacts the beam dynamics, requiring the consideration of many other effects.

In order to create a "clean" experiment we exploit the fact that the compression ratio $1-h \cdot R_{56}$ depends on both the chicane strength R_{56} and the energy chirp h. This allows us to maintain the compression ratio (and the peak current of the postcompressed bunch) while simultaneously changing the bend angle with the chirping tank phase.

In order to maintain the postcompressed beam envelope independent on the chicane settings we designed lattices for any particular chicane settings.

For the experiment we picked three isolines (lines of the constant bunch length on (h, R_{56})

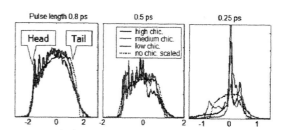

Fig. 3. Chirped beam energy spectra for different isolines.

plane) at low (1.8), medium (2.5) and strong (4.5) compression ratio (Fig. 3). The results of the experiment are shown on Fig. 3. For a low compression ratio structure seems enhanced only at the highest chicane strength. At the same time there is no difference observed for the medium compression. For high compression, profiles are dominated by the structure regardless to the chicane strength. We may conclude that, in general, the structure is not sensitive to the chicane strength.

Another important result of this experiment was the close correspondence of the measured values of the bunch length to the calculated ones in any compression scenario. It has also been observed that the measured uncorrelated energy spread is constant along the isoline.

4. Dependence on transverse beam size

As shown in [3], the longitudinal space charge effect has a strong dependence on the beam radius. In the next experiment we tested the sensitivity of the structure to the variation of transverse beam size in the accelerator.

Varying quadrupole triplets, we created three different focusing scenarios. The RMS beam sizes, averaged over the length of the accelerator, were measured as 0.25, 0.5 and 1 mm RMS. Special care was taken to insure that spectrometer resolution stayed constant during the beam size change.

The result of the experiment is shown in Fig. 4. The deep modulation on the chirped beam profile for the smallest beam (0.25 mm) almost vanishes for the largest beam (1 mm) profile. The average

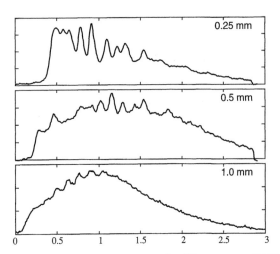

Fig. 4. Chirped beam energy spectra for different average RMS beam sizes.

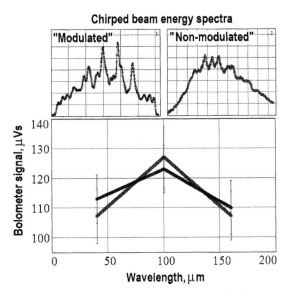

Fig. 5. IR measurements: bolometer signal (μVs) versus wavelength (μm).

modulation period stays nearly constant. This result supports the interpretation of the structure as dominated by energy modulation, caused by the longitudinal space-charge effect.

5. Coherent transition radiation measurements

The spectrometer set-up does not allow us to study the bunching content in the longitudinal beam density. Spiky structure in the chirped energy spectrum can be interpreted as a combination of energy modulation and bunching (local longitudinal density fluctuations). We used the CTR station to measure the IR power generated by the compressed beam.

Varying the transverse beam size we created two cases of "strongly modulated" and "non-modulated" beam profiles (Fig. 5). All other beam parameters (bunch length, peak current) were maintained constant. The characteristic modulation wavelength was estimated as 90 μm, using chirped bunch profiles. Using an IR bolometer together with low-pass IR filters (cut-off at 40, 100, 160 μm) we measured CTR.

The result of the experiment is shown in Fig. 5. If the observed structure were dominated by the longitudinal bunching we would have to expect enhancement of the measured IR power at the wavelength region of about 90 μm. However, the measurements show no significant difference between these two bunch conditions, indicating that the spiky structure is associated with a modulation in energy, not in density.

6. Conclusion

In this paper we discussed several experiments performed for the characterization of the spiky structure observed in the energy spectra of the compressed bunch. The longitudinal space charge model [3,10] is found to be in a qualitative agreement with the results of experiments. The interpretation of the structure follows as dominated by the energy modulation.

Acknowledgements

Authors wish to thank D. Dowell, P. Emma, S. Krinsky, J. B. Murphy, and X.J. Wang for many useful discussions. This work is performed under DOE contract DE-AC02-76CH00016.

References

[1] W. Graves, et al., Proceedings of PAC-2001, Chicago, June 2001, p. 2224

[2] H. Loos, et al., Proceedings of EPAC-2002, Paris, June 2002, p. 814.

[3] Z. Huang, T. Shaftan, SLAC-PUB-9788, p. 329

[4] T. Shaftan, et al., Proceedings of PAC-2003, Portland, June 2003.

[5] S. Heifets, S. Krinsky, G. Stupakov, Phys. Rev. ST AB 5 (2002) 064401.

[6] Z. Huang, K.-J. Kim, Phys. Rev. ST AB 5 (2002) 074401.

[7] D.X. Wang, et al., Phys. Rev. E 57 (1998) 2283.

[8] G.L. Carr, et al., Proceedings of PAC-2001, Chicago, June 2001, p. 2608.

[9] M. Borland, Proceedings of LINAC-2000, Gyeongju, Korea, p. 833.

[10] Z. Huang, et al., Nucl. Math. and Mech. A (2003), these Proceedings.

Available online at www.sciencedirect.com

Nuclear Instruments and Methods in Physics Research A 528 (2004) 402–407

ELSEVIER

NUCLEAR INSTRUMENTS & METHODS IN PHYSICS RESEARCH
Section A

www.elsevier.com/locate/nima

Development of far-infrared FEL with needle photo-RF-gun

T. Inoue*, S. Miyamoto, S. Amano, T. Mochizuki

Laboratory of Advanced Science and Technology for Industry (LASTI), 3-1-2 Kouto, Kamigori, Akou, Hyogo 678-1205, Japan

Abstract

Far-infrared free electron lasers (FELs) with central wavelengths of 11.6 and ~70 μm were obtained by using an S-band RF linac (LEENA) at electron energies of 14.4 and ~5.5 MeV, respectively. Small-signal gain and cavity loss were measured to be 8.2% and 3.3%, respectively, at a peak current of 4.4 A, an electron energy of 15 MeV, and a central wavelength of 10 μm. The pulsewidth of the FEL signal could be lengthened by tailoring the waveform of klystron drive voltage, since uniformity of electron energy in a macropulse was improved. A tungsten needle photo-RF gun was investigated using the field-dependence of QE for a third-harmonics Nd:YAG laser (wavelength of 355 nm). Numerical calculations of electron beam performances in a needle photo-RF-gun were made. A tungsten needle photocathode, with a tip radius of 10 μm, was introduced into the RF-gun cavity of LEENA.
© 2004 Elsevier B.V. All rights reserved.

PACS: 29.25.Bx; 41.60.Cr

Keywords: Needle photocathode; Laser-excited RF-gun; Quantum efficiency; High electric field; Schottky effect; Free electron laser

1. Introduction

Free electron laser (FEL) is a coherent light source with broad-band wavelength tunability. FEL is the only tunable laser in the far-infrared (FIR) region. FIR FEL can selectively photo-excite a particular molecule in materials (condensed matter, medical tissue, protein, etc.). An S-band RF electron linac (LEENA) can generate FIR FEL at a wavelength region of 10–100 μm by changing electron energy from 4 to 15 MeV. LEENA has a Halbach-type undulator and an optical resonator.

The RF-gun is driven by using a mode-locked Nd:YLF laser. The fundamental parameters of LEENA are shown in Table 1.

2. Far-infrared FEL of LEENA

Far-infrared signals were detected with a mercury–cadmium–telluride (MCT) detector, or a germanium–gallium (Ge:Ga) detector in wavelength regions of 2–12 and 52–120 μm, respectively. Measured spectra of FEL signals at central wavelengths of 11.6 and 70 μm, where electron beam energies were 14.4 and 5.58 MeV, respectively, are shown in Fig. 1. These central wavelengths λ_c are considered to be 11.8 and 72.3 μm,

---

*Corresponding author. Tel: +81-791-58-0249, fax: +81-791-58-0242.

E-mail address: inoue@lasti.himeji-tech.ac.jp (T. Inoue).

0168-9002/$ - see front matter © 2004 Elsevier B.V. All rights reserved.
doi:10.1016/j.nima.2004.04.120

Table 1
Fundamental parameters of LEENA

RF frequency	(MHz)	2856
Electron beam		
Energy	(MeV)	4–15 (controllably)
Micropulse	(ps)	10–20
Macropulse	(μs)	5
Repetition rate	(Hz)	10
Halbach-type undulator		
Period/period	(mm/-)	16/50
Undulator parameter		0.1–0.98 (controllably)
Resonant λ	(μm)	10–100
Optical resonator		
Mirror		Au-coated Cu
Reflectivity	(%)	98.7 ± 0.7
Mode-locked Nd:YLF laser		
Wavelength	(nm)	351 (THG), 266 (FHG)
Pulse width	(ps)	10
Pulse energy	(μJ/pulse)	1–3
Repetition rate	(MHz)	89.25 (2856/32)

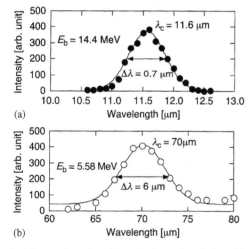

Fig. 1. Measured spectra: (a) central wavelength $\lambda_c = 11.6\,\mu m$ for electron energy $E_b = 14.4\,MeV$, undulator parameter $K_u = 0.720$, and spectral bandwidth $\Delta\lambda = 0.7\,\mu m$, (b) $\lambda_c = 70\,\mu m$ for $E_b = 5.58\,MeV$, $K_u = 0.754$, and $\Delta\lambda = 6\,\mu m$.

respectively, by using the following equation:

$$\lambda_c = \left(\frac{\lambda_u}{2\gamma^2}\right)\left(1 + \frac{K_u^2}{2}\right), \tag{1}$$

where γ is the Lorentz factor, λ_u is the undulator period, and K_u is the undulator parameter. Ratios

of spectral bandwidth to central wavelength are less than 10%.

A photocathode RF-gun driven by a mode-locked laser can synchronously control an electron emission phase against an RF field phase; thus, it is effective for improving an FEL gain. Waveforms of a photocurrent signal (electron beam energy of 15 MeV) and an FEL signal (central wavelength of 10 μm) are illustrated in Fig. 2. The FEL signal increased from the middle point of the photocurrent waveform and decreased thereafter. The FEL peak power was estimated to be 70 W.

Small-signal gain and cavity loss were evaluated from the FEL waveform. The rising part and sinking part could be fit to two exponential functions, respectively. Small-signal gain and cavity loss were evaluated to be 8.2% and 3.3%, respectively. The net gain is thus 4.9%. The calculated cavity loss, including diffraction, outcoupling, and reflection losses, is 5.2%. The peak photocurrent was evaluated to be 4.4 A, where the ratio of a peak photocurrent (measured using a streak camera) to a time-averaged photocurrent (shown in Fig. 2) is 20.6. The obtained small-signal gain could be represented by a numerical calculation with a peak current of 4.4 A, a normalized emittance of 24.3 π mm mrad, an energy spread of 0.2%, and a pulse width of 16.5 ps [1,2]. The emittance was measured using a quadrupole scanning method, the energy spread was obtained using a diagram of electron energy distribution,

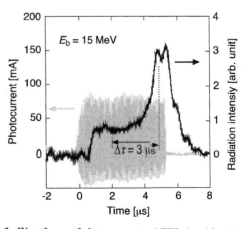

Fig. 2. Waveforms of photocurrent and FEL signal for electron energy of 15 MeV. Rise-up time of FEL signal $\Delta\tau = 3\,\mu s$.

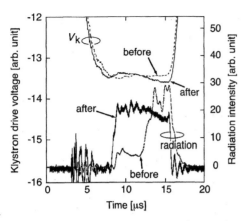

Fig. 3. Tailoring klystron derive voltage waveform can improve uniformity of electron energy in a macropulse, and pulsewidth of an FEL signal is increased. (a) Klystron drive voltage before/after tailoring, (b) FEL signal before/after tailoring.

and the pulse width was investigated by using a streak camera.

Uniformity of electron energy in a macropulse is required to improve FEL gain and to obtain a high-power FEL. Beam loading is a cause of reduction of electron energy. Waveforms of klystron drive voltage V_k and FEL signal are shown in Fig. 3. Before tailoring V_k, electron energy reduced due to beam loading, and pulsewidth of FEL signal was limited to 4 µs. After tailoring V_k, the energy reduction was compensated, and FEL pulsewidth was increased to 7 µs. The energy spread in a macropulse should be made more uniform in order to efficiently improve FEL gain. However, tailoring V_k is one of the effective methods for increasing FEL power.

3. Design of needle photo-RF-gun

A high-brightness electron source is a critical element in improving FEL performances. In order to obtain a high-power FEL in an RF-linac, an RF-gun is required to provide a high peak current, a low emittance, a low energy spread, and synchronization of an electron bunch using an RF wave. A photocathode with high quantum efficiency (QE) and high stability is required to increase FEL power. It is well known that QEs of several metal photocathodes increased with the electric field due

to the Schottky effect [3]. Photoelectron emissivity is improved since the effect reduces a potential barrier close to a metal photocathode surface. Therefore, a high electric field is required to obtain a high-QE metal photocathode.

A tungsten needle was fabricated with an electrolytic etching method. Tungsten, which has a high melting point, was selected as the material of the needle photocathode to prevent the needle tip from being damaged due to intense laser focusing or high current emission. Needle tip radii could be controlled between 0.1 and 1 µm by changing the applied voltage and the density of the etching solution. Dependence of QE on the high electric field was investigated with the tungsten needle photocathode irradiated by a third-harmonics of Nd:YAG laser at 355 nm [4]. The peak intensity on the needle tip was evaluated to be 2.3×10^3 W/cm². The measured dependence is shown in Fig. 4. The QE was calculated by dividing an emitted electron number from the needle tip by an incident photon number upon one. The QE exponentially increased with the electric field from 600 to 800 MV/m, and it reached 3% at ~800 MV/m. The QE value is comparable to the high-QEs of semiconductor photocathodes [5]. This result indicates that a tungsten needle photocathode is a candidate of high-QE photocathodes.

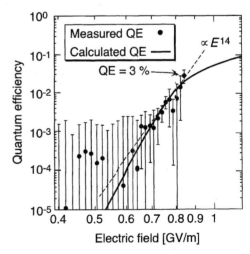

Fig. 4. Dependence of QE on electric field. The solid line represents the calculated dependence of QE using a photo-assisted field-emission model [4].

Electric field on a needle tip is one of the most important parameters for a needle photo-RF-gun, since it directly influences the QE of the needle tip. The field-enhancement factor on the needle tip can be calculated by using SUPERFISH, which is a simulation code to calculate RF electromagnetic field distribution in a resonant cavity or waveguide. The calculated electric field distribution in an RF-gun cavity of LEENA including a needle photocathode is illustrated in Fig. 5, where the RF cavity is a standing-wave type with the maximum electric field strength of 40 MV/m [6]. Electric field lines concentrate on the needle tip. The electric field on the z-axis (a dashed line in Fig. 5) increases around the needle tip due to field concentration. Field-enhancement factor can be approximately represented as $\beta \approx 3.3 + 1400/(r_0\,[\mu m])$, where r_0 is the needle tip radius. For a tip radius $< 66\,\mu m$, the electric field strength on the needle tip is calculated to be $> 800\,MV/m$.

Electric field strength in an RF-gun cavity changes temporally and spatially, as a result, the QE of a needle tip also changes. Furthermore, bunch length at a cavity exit varies due to bunch-compression (or expansion) in acceleration under RF field. The bunching property depends on an initial phase where electrons are extracted from a cathode surface against the RF field phase. Thus, the peak photocurrent at a cavity exit depends on an initial phase (a laser injection phase φ upon a photocathode). Electric field strength starts in-creasing at $\varphi = 0°$, and it reaches up to the maximum value at $\varphi = 90°$. Calculated dependences of the peak photocurrent on a laser injection phase and a tip radius are shown in Fig. 6, where an incident laser has a peak intensity of $500\,MW/cm^2$ and a pulsewidth of 10 ps. The peak laser intensity was determined using numerical calculations on laser heating so that a peak temperature of a needle tip is lower than one-third of the melting-point of tungsten (1228 K) [6]. This calculation includes both space-charge limitation and field-dependence of QE, where QE-saturation was assumed as a solid line shown in Fig. 4 [4]. The calculated peak photocurrent is larger than 90 A (pulsewidth $\leqslant 10\,ps$) for tip radii of 5–10 μm, laser intensity of $500\,MW/cm^2$, and electric field of 40 MV/m. Peak photocurrent density is evaluated to be $\geqslant 100\,MA/cm^2$; however, increment in the temperature of a needle tip due to joule heating is $\ll 1\,K$.

A tungsten needle photocathode was introduced into the RF-gun of LEENA, as shown in Fig. 7, for the purpose of investigating performances of a needle photo-RF-gun and demonstrating the generation of a far-infrared FEL. Fig. 7(b) illustrates a scanning microscopy image of the needle tip (the tip radius is 10 μm), and Fig. 7(c) is a monitored image of a heated needle cathode. A preliminary experiment using a needle-RF-gun has

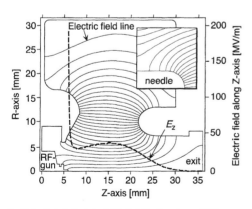

Fig. 5. Calculated electric field distribution in an RF-gun cavity of LEENA including a needle photocathode using a simulation code SUPERFISH. The dashed line represents the electric field strength along the z-axis.

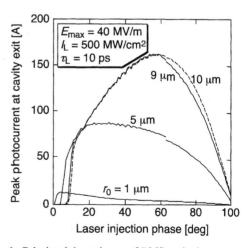

Fig. 6. Calculated dependences of RMS peak photo-current of a photo-electron beam on laser injection phase at a cavity exit for tip radii of 1, 5, 9, and 10 μm.

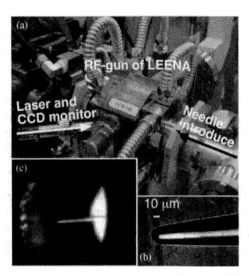

Fig. 7. Tungsten needle photocathode was introduced into an RF-gun of LEENA. (a) Image of an RF-gun, (b) a scanning electron microscope image of the needle cathode, (c) image of the needle cathode heated in the RF-gun cavity.

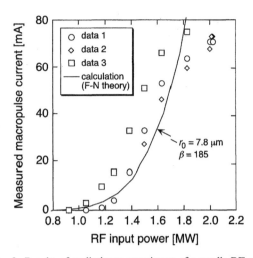

Fig. 8. Results of preliminary experiment of a needle-RF-gun. Dependence of macropulse current (field-emission) on the RF power of an RF-gun cavity was measured. The measured dependence is in good agreement with a calculation using Fowler–Nordheim theory for a tip radius of 7.8 μm and a field enhancement factor of 185.

been demonstrated in LEENA. The field-emission current of the needle-RF-gun was measured by changing the RF power of the RF-gun cavity. The measured dependence is shown in Fig. 8, where the solid line indicates a theoretical line calculated using the Fowler–Nordheim theory [7,8]. The effective tip radius was obtained as 7.8 μm by fitting the theoretical line to the measured data, and a field-enhancement factor was evaluated to be 185. The peak field-emission current was evaluated to be 1.3 A, and the electric field on the needle tip was estimated to be ∼6 GV/m. Therefore, this result indicates that the electric field is enough to obtain a high-QE ⩾3% in the RF-gun of LEENA.

4. Summary

Far-infrared FELs at a wavelength region of 11–70 μm were obtained by using an S-band RF-linac (LEENA) at an electron energy region of 5.6–15 MeV. Measured small-signal gain of 8.2% and cavity loss of 3.3% agreed with the numerical calculations using experimental parameters. Electron energy was reduced due to the beam loading

in a macropulse. The energy reduction was compensated by tailoring the waveform of klystron drive voltage, and a pulsewidth of an FEL signal could be inceased from 4 to 7 μs.

A needle photo-RF-gun was proposed for a high-brightness electron source from the experimental investigation and numerical calculations of its performances. The QE of a tungsten-needle photocathode reached 3% at 800 MV/m for third-harmonics of the Nd:YAG laser. The peak photocurrent was calculated to be >90 A for tip radii of 5–10 μm, a laser intensity of 500 MW/cm², and an electric field of 40 MV/m. A tungsten needle, with a tip radius of 10 μm, has been introduced into the RF-gun of LEENA, and a dependence of a field-emission current on RF power of the RF-gun cavity was investigated. The maximum electric field of a needle tip was estimated to be ∼6 GV/m using the Fowler–Nordheim theory, and the field strength is enough to obtain a high-QE of ⩾3%.

References

[1] G. Dattoli, T. Letardi, J.M.J. Madey, A. Renieri, IEEE J. Quantum Electron. 20 (1984) 637.

[2] AESJ, In: Introductory Book on Free Electron Laser, Tomimasu T, (Ed) AESJ, Tokyo, 1995, p.112 (in Japanese).

[3] T. Srinivasan-Rao, J. Fischer, T. Tsang, J. Appl. Phys. 69 (1998) 3291.

[4] T. Inoue, S. Miyamoto, S. Amano, M. Yatsuzuka, T. Mochizuki, Jpn. J. Appl. Phys. 41 (2002) 7402.

[5] S.H. Kong, J. Kinross-Wright, D.C. Nguyen, R.L. Sheffield, Nucl. Instr. and Meth. A 358 (1995) 272.

[6] T. Inoue, S. Miyamoto, S. Amano, M. Yatsuzuka, T. Mochizuki, Jpn. J. Appl. Phys. 42 (2003) 311.

[7] E.L. Murphy, R.H. Good Jr., Phys. Rev. 102 (1956) 1464.

[8] R.D. Young, Phys. Rev. 113 (1959) 110.

Available online at www.sciencedirect.com

SCIENCE @ DIRECT°

ELSEVIER   Nuclear Instruments and Methods in Physics Research A 528 (2004) 408–411

NUCLEAR
INSTRUMENTS
& METHODS
IN PHYSICS
RESEARCH
Section A

www.elsevier.com/locate/nima

Improvement of beam macropulse properties using slim thermionic cathode in IAE RF gun

Toshiteru Kii*, Atsushi Miyasako, Syusuke Hayashi, Kai Masuda,
Hideaki Ohgaki, Kiyoshi Yoshikawa, Tetsuo Yamazaki

Institute of Advanced Energy, Kyoto University, Gokasyo, Uji, Kyoto 611-0011, Japan

Abstract

A long beam macropulse is strongly required for free-electron lasers. RF guns can potentially produce high brightness electron beam using a simple and compact system. However, due to a back-bombardment, a cathode surface is overheated. Thus, it is difficult to maintain a constant beam current and beam energy during a macropulse. The use of a photo cathode with a short-pulsed laser is one of the solutions, but it affects the simplicity and the compactness of the RF guns. We studied a mechanism of back-bombardment and experimentally and numerically found that a low energy component of the back-streaming electrons plays an important role in cathode surface heating.
© 2004 Elsevier B.V. All rights reserved.

PACS: 41.60.Cr

Keywords: Free-electron laser; RF gun; Electron injector

1. Introduction

RF guns are widely used as electron injectors of accelerators. However, electrons that drop from the accelerating phase are decelerated and hit the cathode. Consequently, the cathode surface is overheated and the surface temperature increases. Then, the extraction of constant electron beam from the RF gun becomes difficult.

To avoid the back-bombardment problem, a group at Stanford University proposed the appli-

cation of a transverse magnetic field and success in the production of a longer macropulse [1]. A group at the Science University of Tokyo was also successful in producing a long macropulse by using the slim LaB6 cathode [2].

We developed a two-dimensional (2-D) particle simulation code KUBLAI [3] and a 1-dimensional (1-D) heat conduction model [4] to study the back-bombardment problem in our 4.5 cell thermionic RF gun. We found that the low energy back-streaming electrons, which distributed widely on the cathode, had a serious influence on the beam quality [5]. In this study, an effect of the magnetic field on back-streaming electrons has been analysed by measuring macropulse duration of

*Corresponding author. Tel.: +81-774-38-3422; fax: +81-774-38-3426.

E-mail address: kii@iae.kyoto-u.ac.jp (T. Kii).

0168-9002/$ - see front matter © 2004 Elsevier B.V. All rights reserved.
doi:10.1016/j.nima.2004.04.121

extracted electron beam and beam current. We also studied an effect of the cathode diameter by a particle simulation and time evolutions of the cathode surface temperature.

2. Experiment

Extracted beam current and effective beam macropulse duration were measured. Fig. 1 shows an experimental setup. A dipole magnet is located just behind a cathode to reduce back-streaming electrons. A disk-shaped dispenser cathode with a 6 mm diameter was used. The surface temperature of the cathode was kept at about 1000°C. Fig. 2 shows the arrangement of the dipole magnet attached to the RF gun. Field distribution along z-axis is shown in Fig. 3. In this paper, field strength is defined as the maximum field strength along the z-axis ($z = 0$).

A typical signal from a current transformer (CT) and an input and a reflected RF power are shown in Fig. 4. Due to the back-bombardment effect, beam loading increases during macropulse. Thus, the resonant frequency changes and reflected RF increases and extracted electron beam drops. Here, effective macropulse duration is defined as the time from the leading edge of the macropulse to the time when signals from the CT start swinging.

Fig. 5 shows measured pulse duration with and without the magnetic field. As shown in Fig. 5, the pulse duration is improved with an increase in the magnetic field.

On the other hand, by applying the magnetic field, extracted beam current decreases because the beam halo is stopped by the aperture of the RF gun (Fig. 6). In our case, about 20 G of a magnetic

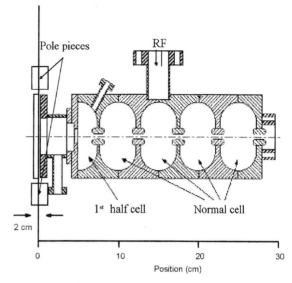

Fig. 2. Arrangement of the dipole magnet.

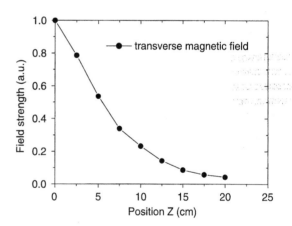

Fig. 3. Field distribution along a beam axis.

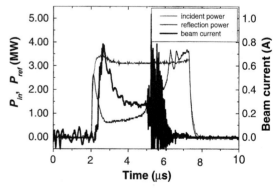

Fig. 4. Input and reflected RF power and extracted beam current when no magnetic field is applied.

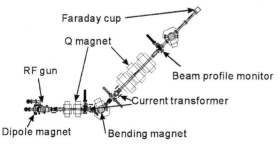

Fig. 1. Experimental setup.

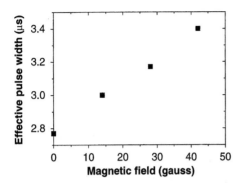

Fig. 5. Measured effective beam macropulse duration as a function of the field strength of the magnetic field.

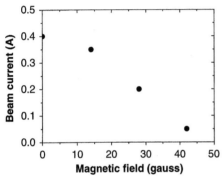

Fig. 6. Extracted beam current as a function of applied field strength.

field is acceptable because the main component of extracted beam tends to be obstructed by the aperture. In this case, the improvement in macropulse duration was about 10%.

3. Evaluation

To evaluate the effect of the dipole magnetic field to the beam properties, we used a particle simulation code PARMELA (version 3.30) [6] and the 1-D thermal conduction model [4]. The input parameters are shown in Table 1. Energy spectrum of the back-streaming electrons is shown in Fig. 7. A main component of back-streaming electrons lower than 250 keV is reduced by the dipole magnetic field. However, most of the low energy component remains when magnetic field is applied,

Table 1
Input parameters for calculations

Frequency (MHz)	2856
Input RF power (MW)	3.0
Current density on the cathode surface (A/cm^2)	10
Cavity voltage (1st half cell) (MV/m)	15
Cavity voltage (2nd–5th cell) (MV/m)	∼28

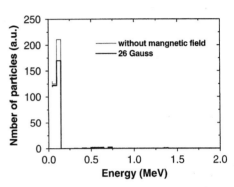

Fig. 7. Energy spectrum of back-streaming electrons.

because very low energy component, which is very sensitive to the surface temperature, tends to change its direction close to the cathode, and returns to the almost same position on the cathode.

Time evolution of the cathode surface temperature is also estimated by using the 1-D heat conduction model. Fig. 8 shows time evolutions when the magnetic field is applied and magnetic field of 20 G goes on and off.

As shown in Fig. 8, when the magnetic field is applied, the gradient is 10% smaller than that without the magnetic field. This is consistent with the experimental results.

To reduce low energy back-streaming electrons, using a thermionic cathode with smaller diameter appears to be effective, because the ratio of the electrons wiped out to those outside the cathode will increase. Thus, we also estimated an effect of a slim cathode with a 2 mm diameter by using the PARMELA and the 1-D heat conduction model.

In our design, injection energy to an accelerator tube was designed to be 7 MeV [7], and input parameters were chosen as shown in Table 2.

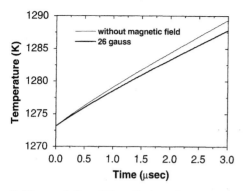

Fig. 8. Time evolution of the cathode surface temperature.

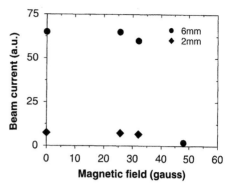

Fig. 9. Extracted beam current as a function of applied field strength.

Table 2
Input parameters for calculations

Frequency (MHz)	2856
Input RF power (MW)	7.4
Current density on the cathode surface (A/cm^2)	10
Cavity voltage (1st half cell) (MV/m)	22
Cavity voltage (2nd–5th cell) (MV/m)	∼40

The estimated extracted beam current and time evolutions of cathode surface temperature are shown in Figs. 9 and 10.

In case of the cathode with a 2 mm diameter, the reduction ratio when magnetic field is applied is larger compared to when the cathode with a 6 mm diameter is used. Thus, the combination of a slim cathode and dipole magnetic field will improve the macropulse duration of electron beam.

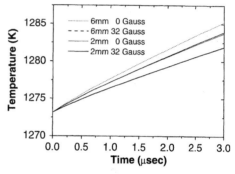

Fig. 10. Time evolution of the cathode surface temperature.

slim cathode with a 2 mm diameter is used. It was found that the temperature growth during macropulse was effectively suppressed by the magnetic field.

4. Summary and conclusion

To extract a long pulse from the IAE 4.5 cell thermionic RF gun with thermionic cathode with a 6 mm diameter, we tested the effect of a magnetic field applied close to the cathode surface. It was experimentally found that the magnetic field can improve pulse duration, which is shown by the PARMELA and 1-D thermal conduction model. However, the improvement was not sufficient for FEL experiments.

We estimated the time evolution of the cathode surface temperature during macropulse when a

References

[1] C.B. McKee, J.M.J. Madey, Nucl. Instr. and Meth. A 296 (1990) 716.
[2] F. Oda, et al., Nucl. Instr. and Meth. A 483 (2002) 45.
[3] K. Masuda, Development of Numerical Simulation Codes and Application to Klystron Efficiency Enhancement, Ph.D.Thesis, Kyoto University, 1997.
[4] T. Kii, et al., Nucl. Instr. and Meth. A 483 (2002) 310.
[5] T. Kii, et al., Nucl. Instr. and Meth. A 507 (2003) 340.
[6] L.M. Young, J.H. Billen, PARMELA, LA-UR-96-1835, 2001.
[7] H. Ohgaki, et al., Nucl. Instr. and Meth. A 507 (2003) 150.

Available online at www.sciencedirect.com

SCIENCE @ DIRECT°

ELSEVIER   Nuclear Instruments and Methods in Physics Research A 528 (2004) 412–415

NUCLEAR
INSTRUMENTS
& METHODS
IN PHYSICS
RESEARCH
Section A

www.elsevier.com/locate/nima

Photo-injector study for the ELETTRA linac FEL

V.A. Verzilov*, R.J. Bakker, C.J. Bocchetta, P. Craievich, M. Danailov,
G. D'Auria, G. De Ninno, S. Di Mitri, B. Diviacco, M. Ferianis, A. Gambitta,
F. Mazzolini, G. Pangon, L. Rumiz, L. Tosi, D. Zangrando

Sincrotrone Trieste, S.S. 14 Km. 163.5 Area Science Park, 34012 Basovizza, Trieste, Italy

Abstract

A new injector is one of the necessary upgrades to be made in the existing ELETTRA linac to improve its performance to the level required for FEL operation. To this end, the design of an RF photo-injector is in progress at ELETTRA. Several types of RF guns are currently under consideration. The drive laser system, beam optics, diagnostics, and other relevant issues are also being addressed.
© 2004 Elsevier B.V. All rights reserved.

PACS: 41.60.Cr; 41.75.Fr; 52.59.Rz

Keywords: Photo-injector; RF gun; FEL

1. Introduction

The ELETTRA single pass free electron laser (FEL) project [1] aims to build a high brilliant, tunable source of coherent radiation in the wavelength range from 100 to 1.2 nm. In the first stage, in which lasing down to 10 nm is anticipated, it will make use of the existing 1 GeV S-band linac and infrastructure. Certain modifications, in particular a new low emittance electron source, are required to transform the present ELETTRA storage ring injector into an

FEL driver. Our choice for the new electron source, an RF photo-injector, has been governed by the recent progress in the technology and operation of these devices, which has made it possible to conduct a number of successful FEL experiments in the visible and UV spectral range. Our design of the RF photo-injector is based on the LCLS studies and is developed to meet the requirements and constraints of the existing infrastructure. In particular, the new injector design is fully compatible with the continued use (roughly 1.5 h of operation per day) of the present thermionic gun, since the linac still has to serve as a reliable injector for the ELETTRA storage ring. The latter task will remain necessary until completion of a new full energy injector, i.e. a booster synchrotron.

*Corresponding author. Tel.: +39-040-3758079.
E-mail address: victor.verzilov@elettra.trieste.it
(V.A. Verzilov).

0168-9002/$ - see front matter © 2004 Elsevier B.V. All rights reserved.
doi:10.1016/j.nima.2004.04.122

2. Overview

2.1. Layout

The new injector section, the layout of which is shown in Fig. 1, has been developed on the basis of the following criteria:

- Optimum performance for the new RF photo-injector
- Reasonably good transport and acceleration of the beam generated by the old thermionic gun
- Fitting the available space in the existing tunnel
- Incorporation of all necessary optics, RF, vacuum, and diagnostic elements.

For these reasons, the RF gun is installed in place of the old gun, which is relocated 3 m upstream. To bypass the RF cavity and the compensation solenoid, a magnetic chicane is used. For the integrated energy spread of 0.5%, the small generated momentum compaction, $R_{56} < 1$ mm, has a negligible effect on the longitudinal beam dynamics. The last dipole of the chicane is also used as the RF gun beam energy and energy spread analyzer . The vacuum system is designed to provide safe operation of either of the two injectors.

The drive laser room will be built in the existing shielded area located next to the linac tunnel. From there, the laser pulses will be transported over several meters to the gun and will illuminate the cathode at either normal or grazing incidence.

2.2. RF gun

Three options for the S-band RF gun are under consideration: a 1.6-cell BNL/SLAC/UCLA type [2], a 2.6-cell gun of TUE [3], and a 1-cell higher-order mode gun [4]. Of these, the first type of gun is currently the most developed and is considered to be ahead of the others.

Since the outer dimensions are very similar, in all probability no serious modifications to the present design would be needed to install any of these options.

2.3. Laser system

The laser system with the required parameters [1] will be built on the basis of a commercially available Ti:Sapphire laser and amplifier modules updated with additional units and control loops. Proper transverse and longitudinal laser pulse-shaping are required for optimum performance. This will be done by shape-controlled elements capable of forming nearly uniform temporal (<0.5 ps rise time) and transverse profiles. Proper stability of the system will be obtained by implementation of phase and amplitude control loops, using a fast laser energy trimming and exploiting the virtual cathode.

2.4. Booster

According to novel ideas in the operation of photo-injectors [5], the best performance is obtained when the beam is properly accelerated up to energies of more than 100 MeV, after which space charge forces have an almost negligible effect on the beam dynamics. In our design, the energy booster is formed by the first two sections (S0A and S0B) of the present linac (only S0A is shown in Fig. 1). These sections are 3.2 m long, SLAC type S-band TW structures, each embedded in its own solenoid with a maximum magnetic field of 0.25 T. Both structures are powered by a single 45 MW klystron.

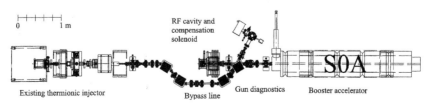

Fig. 1. The new injector layout.

2.5. Diagnostics

The laser diagnostics will measure and stabilize the transverse shape and position on a virtual cathode. In addition, the pulse energy and phase will be controlled on a pulse-to-pulse basis. The pulse duration will be measured by cross-correlation with the oscillator output.

The diagnostic section in the drift between the gun and booster will provide an essential control over the main beam parameters and will also fit in the space required for compensation of the emittance. Beam position monitors, current transformers, Faraday Cups, and view screens will be used to measure the beam position, profile, and charge. Emittance will be measured using single slit or multi-slit methods. Given the short distance to the view screen a spatial resolution of 10–15 μm is required. The beam energy and energy spread will be measured in the 60° dispersive bend. The longitudinal beam profile will be measured by a streak-camera in case a suitable light emitter is available.

Detailed beam characterization will be performed just behind the booster, in the bunch compressor area.

3. Simulated performance

The performance of the photo-injector is determined by the whole system along the beam path from the cathode to the booster exit. Multi-parameter scans using the ASTRA [7], a beam dynamics simulation tool, were performed for several injector configurations to optimize parameters and evaluate the overall performance.

3.1. Simulation model

Three configurations are considered: the BNL/SLAC/UCLA gun operated with an on-axis field gradient of 140 MV/m, the same gun with a reduced field gradient of 120 MV/m, and the TUE gun operated at 100 MV/m. The idea behind lowering the operating gradient is to improve gun reliability.

Intensive parameter optimizations have been previously performed at LCLS for the 1.6-cell BNL/SLAC/UCLA gun in both the cases, i.e. 140 MV/m [6] and 120 MV/m [8]. Although our design is similar to that of LCLS, independent parameter optimizations were necessary for at least two reasons:

- Different RF frequency: 2998 MHz vs. 2856 MHz
- Different simulation tool: ASTRA vs. PARMELA

Beam dynamic simulations were preceded by detailed electromagnetic studies of both these guns using SUPERFISH. In the course of these studies, the 1.6-cell gun was scaled to the ELETTRA linac operating frequency of 2998 MHz. Both guns are shown in Fig. 2 and some basic parameters are reported in Table 1. In contrast to the BNL/SLAC/UCLA gun, the TUE gun has an ideal rotational symmetry due to axial coupling of RF power.

Up to six parameters were varied in the beam dynamics simulations. The figure of merit for the optimization was the smallest transverse emittance at the booster exit. The charge of electron bunches was 1 nC. The flat-top-shaped laser pulse duration was fixed at 10 ps FWHM with a 1 ps rise time as accepted in the LCLS design. Given that the cathode-to-booster distance is fixed (1.5 m), this might not be the optimum pulse length for every configuration [9]. Therefore, the laser pulse length optimization will be also necessary in the following steps. The transverse profile of the laser beam was a uniform radial distribution with a variable radius. No thermal emittance was included in

1.6-cell BNL/SLAC/UCLA Gun   2.6-cell TUE Gun

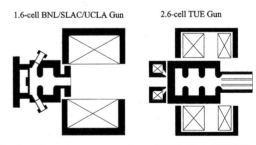

Fig. 2. Schematic representation of BNL/SLAC/UCLA and TUE guns combined with compensation solenoids.

Table 1
Injector parameters for three gun configurations: 1.6-cell BNL/SLAC/UCLA gun operated at the field of 140 MV/m (BNL[1]), 120 MV/m (BNL[2]), and 2.6-cell TUE gun

Parameter (unit)	BNL[1]	BNL[2]	TUE
Gun field (MV/m)	140	120	100
Eff. shunt impedance R_s (MΩ)	1.85	1.85	2.87
Char. shunt impedance R_s/Q (Ω)	138	138	208
RF peak power (MW)	11.1	8.2	8.2
Average heat flux at 4μs × 10 Hz (W/cm^2)	1.56	1.14	0.77
Laser spot radius (mm)	1.0	1.1	1.0
Gun phase (deg.)	34	28	29
Solenoid field(kG)	3.02	2.64	4.01
S0A average gradient (MV/m)	16.5	14.3	18
S0A focusing field (kG)	0.6	0.6	0.7
S0B average gradient (MV/m)	22	22	22

Table 2
Output beam characteristics for the same gun configurations as in Table 1

Beam parameter (unit)	BNL[1]	BNL[2]	TUE
Gun field (MV/m)	140	120	100
Beam size (rms)(mm)	0.34	0.33	0.28
Norm. emittance (rms) (μm)	0.57	0.63	0.57
Electron energy (MeV)	119	111	123
Bunch length (rms)(mm)	0.90	0.93	1.0
Energy spread (rms)(%)	0.21	0.27	0.36
Peak current (A)	104	102	93

simulations for the sake of convenience and ease of comparison among different options.

3.2. Simulation results

The main injector parameters obtained as a result of the optimization procedure are listed in Table 1, and the output beam parameters are presented in Table 2. The phases of the two booster TW sections were tuned to get the maximum beam energy at the injector exit. All numbers refer to the projected values.

Simulations show that the estimated performance of the new electron source is adequate for the ELETTRA linac FEL throughout the entire wavelength range.

Comparing the parameter list and output beam characteristics to LCLS simulations [6,8], we note that there is generally an agreement between both simulations. This is what we expected given a few small differences between the two designs. However, ASTRA systematically predicts somewhat (10–20%) smaller beam emittances as compared to PARMELA.

The output beam energy may not be sufficient for the complete suppression of the space charge effects (150 MeV is required [5]). Additional beam acceleration before bunch compression may be necessary.

From the simulations, the 2.6-cell TUE gun looks very attractive since it develops nearly the same performance as the 1.6-cell gun at 140 MV/m, while it is operated at a lower gradient; therefore it has a substantially lower heat load per unit area. A slight increase in the bunch length and consequently in energy spread is accounted for by a stronger effect of space charge forces at lower accelerating gradients. This should not affect the slice energy spread.

4. Conclusion

The design of an RF photo-injector is in progress for the ELETTRA linac FEL. The technical solutions to main subsystems have been developed. Several types of RF guns are presently under investigation. Parameter optimizations have been performed using ASTRA code.

References

[1] G. D'Auria, Proceedings of LINAC 2002, pp. 643.
[2] D. Palmer, The next generation photo-injector, Ph.D. Thesis, Stanford University, 1988.
[3] M.J. de Loos, Proceedings of EPAC 2002, pp. 1831.
[4] J.W. Lewellen, Proceedings of EPAC 2002, pp. 673.
[5] M. Ferrario, SLAC-PUB-8400, 2000.
[6] P.R. Bolton, et al., SLAC-PUB-8962, 2001.
[7] K. Flöttmann, ASTRA User Manual, 2000.
[8] C. Limborg, et al., SLAC-PUB-9555, DESY, Hamburg, Germany, unpublished, 2002.
[9] M. Ferrario, et al., Report LNF-03/6(p), 2003.

Available online at www.sciencedirect.com

SCIENCE DIRECT•

Nuclear Instruments and Methods in Physics Research A 528 (2004) 416–420

**NUCLEAR
INSTRUMENTS
& METHODS
IN PHYSICS
RESEARCH**
Section A

www.elsevier.com/locate/nima

Design and optimization of IR SASE FEL at Peking University ✩

Yuantao Ding, Senlin Huang, Jiejia Zhuang, Yugang Wang, Kui Zhao*, Jiaer Chen

Institute of Heavy Ion Physics, RF SC Group, Peking University, Beijing 100871, China

Abstract

We present the layout of the Peking University FEL facility (PKU-FEL), to be driven by a superconducting accelerator, which is currently under construction. PKU-FEL is a test facility designed to study the physical and technical issues of single pass SASE FEL and high-average power FEL. During the intial stage of the IR SASE FEL, the first goal is to demonstrate a high-brightness electron beam and set up a SASE FEL platform to study some novel aspects in physics. In this paper, the study of theoretical and numerical simulations required to design the PKU SASE FEL has been described. A superconducting RF injector and a main linac composed of two 9-cell superconducting cavities are used to produce a high-quality electron beam at about 40 MeV. The SASE FEL is designed to reach saturation at 7 μm with a 5-m long hybrid undulator.
© 2004 Elsevier B.V. All rights reserved.

PACS: 41.60.Cr; 29.25.Bx; 29.17.+w

Keywords: SASE FEL; Superconducting accelerator; Planar undulator; Saturation length

1. Introduction

Free-electron lasers (FELs) have developed rapidly over the past 30 years. At present, there are two principle areas for future FEL development: shorter wavelength and higher-average powers [1]. Short-wavelength FELs are primarily directed toward an X-ray regime with wavelengths down to 1 Å. Such a source of coherent laser-like X-rays would have many applications and therefore, the SASE mode has been chosen. SASE FEL requires a high-quality electron beam and long undulators. To generate high-average power FELs, a superconducting radio-frequency (SRF) technology is a good driver for the high-duty factor in continuous-wave operation [2].

The PKU-FEL facility is an ideal platform to study the physical and technical issues of single-pass SASE FEL and high-average power FEL. The Peking University Superconducting

---

✩ Work supported by Chinese department of Science and Technology under the National Basic Research Projects No. 2002CB713600.

*Corresponding author. Tel.: +86-10-627-571-95; fax: +86-10-627-518-75.

E-mail address: kzhao@pku.edu.cn (K. Zhao).

0168-9002/$ - see front matter © 2004 Elsevier B.V. All rights reserved.
doi:10.1016/j.nima.2004.04.123

Accelerator Facility (PKU-SCAF) [3], which comprises a superconducting RF photoinjector and a superconducting main linac, will be used to drive FELs. The main coupler and cryostats are designed to work at CW mode. In the first stage, a SASE FEL experiment with a wavelength of 7 μm will be studied. The accelerator will be operated to produce an electron beam with a peak current of about 200 A at about 35–40 MeV with micropulse repetition frequency of 10–30 Hz. The first goal at this stage is to demonstrate a high brightness electron beam and set up the SASE FEL platform, after which we will focus on some novel physical aspects in SASE FEL experiments, including harmonic generation after saturation and short bunch effect in the SASE process. In this paper, the optimization of SASE FEL design based on a superconducting accelerator has been described.

2. PKU SASE FEL platform

The PKU SASE FEL platform mainly consists of PKU-SCAF, a bunch compressor, and an undulator. Fig. 1 is a sketch of the PKU SASE FEL platform. The main parameters of the SASE FEL platform are given in Table 1.

2.1. PKU-SCAF

PKU-SCAF mainly comprises a superconducting RF photoinjector, a superconducting main linac, an RF power system, and a drive laser system. The injector is made up of newly developed superconducting $2 + \frac{1}{2}$ cavities operated at 1.3 GHz which is under design [4]. It will be operated in CW mode, with a micropulse repetition frequency of 10–30 Hz for the SASE experiment. The drive laser system includes a diode

pumped Nd:YVO$_4$ laser and a regenerative amplifier. A Cs$_2$Te photocathode has also been adopted. Currently, great progress has been achieved in areas such as laser testing and photocathode preparation.

The main accelerator includes two 9-cell TESLA type 1.3 GHz superconducting cavities. The cavities are operated below 2 K in liquid helium. Each cavity is driven by a 10 kW klystron amplifier. The two cavities are housed in one cryostat, similar to that of the superconducting linear accelerator of the ELBE project at Rossendorf [5]. For the SASE FEL experiment, electron energy from the main accelerator is about 40 MeV.

2.2. Compressor

High electron density is one of the most stringent requirements for high gain and short saturation length of SASE FEL. A bunch compressor, located behind the main accelerator, is used to get short bunch length. By means of an off-crest RF acceleration in the main accelerator, a

Table 1
Parameters of PKU SASE FEL

Electron beam energy (MeV)	40
Peak current (A)	200
RMS emittance (mm mrad)	5
RMS energy spread (keV)	80
Bunch duration (ps)	~ 1
Undualtor	
Period length (mm)	27
K parameter	1.5
Average β function (cm)	26
Undulator length (m)	5
FEL	
Wavelength (μm)	7
Saturation length (m)	~4
Saturation power (MW)	~50

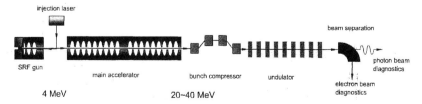

Fig. 1. Sketch of PKU SASE FEL platform.

chirping of the electron energy over the bunch is produced, correlating the individual electron's longitudinal position within the bunch with its energy. Next, the electrons move through a non-isochronous magnetic section, where the path length depends on the electron energy, and the bunch is compressed. Optimization of a magnetic chicane for the compressor is under way.

2.3. Optimized hybrid planar undulator

Undulator is the most prominent FEL-specific component. Given the electron beam and undulator parameters, the radiation wavelength is determined by a resonance condition

$$\lambda = \frac{\lambda_w(1 + a_w^2)}{2\gamma_0^2}$$

where $a_w = K/\sqrt{2}$ for the planar undulator, γ_0 is related to the average beam energy. The undulator has to provide a sinusoidal magnetic field so that FEL process can take place, and it also has to keep the beam size small across the whole undulator length. A planar hybrid permanent magnet undulator combined with a superimposed periodic quadrupole lattice structure [6] is adopted at PKU SASE FEL. We optimized the undulator parameters according to the fitting numerical solutions of the coupled Maxwell–Vlasov equations [7]. Fig. 2 is a contour plot of saturation length L_{sat} versus λ_w and K for FEL wavelength at $\lambda = 7\,\mu m$. The electron beam parameters I, ε_n, σ_e are fixed at nominal values given in Table 1, and the average beta function $\beta = 26\,cm$. According to our design, the FODO period length is $8\lambda_w$ with focusing field gradient of $17\,T/m$ and dispersing field gradient of $14\,T/m$ in the x direction. With this FODO lattice and natural focusing of the planar undulator, an average beta function of $26\,cm$ in, both, x and y directions are obtained, along with the ratio $\beta_{max}/\beta_{min} = 2$. Fig. 3 presents the beta functions along the undulator. The accessible area in the λ_w-K space is limited by practical constraints. According to the limits of electron beam energy and the undulator gap, parameters of $\lambda_w = 2.7\,cm$, $K = 1.5$ for the PKU SASE experiment are chosen. This requires beam energy of $32.7\,MeV$

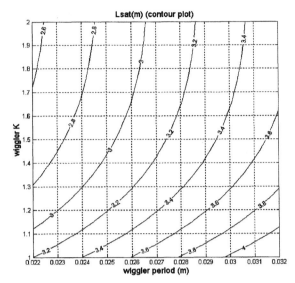

Fig. 2. L_{sat} vs. λ_w and K for the undulator at a wavelength of $7\,\mu m$.

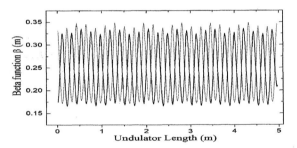

Fig. 3. Horizontal (dotted line) and vertical beta function (solid line) along the undulator length.

at $7\,\mu m$ FEL. This undulator is now being designed.

3. Calculations of SASE FEL performance

With the chosen electron beam and undulator parameters given in Table 1, the SASE FEL process can be studied analytically and with numerical simulation codes.

Fig. 4 shows the influence of the beam parameters on the FEL saturation length, calculated using Xie's fitting formula method [6] and the GENESIS1.3 code in steady-state mode. When one parameter is changed, the others are used from

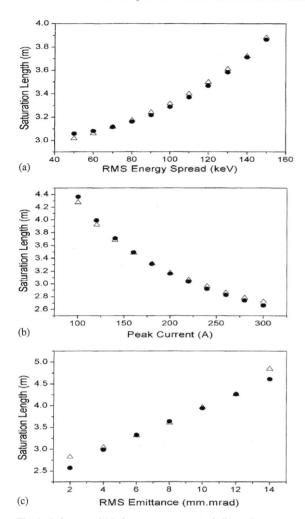

Fig. 4. Influence of (a) electron energy spread, (b) peak current, and (c) emittance on saturation length. Circles are analytical results, triangles are simulation results using GENESIS1.3 code in steady-state mode.

the list of Table 1. From these results, we can control the beam qualities within a reasonable range during the experiment so as to reach saturation before the end of the undulator.

For PKU SASE FEL, short bunch effect could not be neglected. With the electron bunch length (FWHM) at about 1 ps, it is of the order of radiation cooperation length $2\pi l_c$. In this short bunch limit, only a single temporal spike is typically present, and the spectral profile is smooth [8]. The fluctuation of saturation length can be as

large as a few gain lengths for PKU SASE FEL. Using the GENESIS1.3 code in a time dependent mode, we get a saturation length of about 4 m. We will change the bunch compressor parameters to get an electron beam with different bunch lengths. The short bunch effect in the SASE process can be studied in detail using this platform which will provide a benchmark for theoretical study. The undulator is designed to be 5 m long. About 1 m longer than saturation length, the harmonic generation after saturation will be studied experimentally.

4. Scientific merit and development plan of the project

The IR range PKU SASE FEL platform is a small-scale high-gain FEL system with a full range of diagnostics to effectively study the critical physical issues involved in SASE FELs. This facility is flexible and convenient to test new ideas and instrumentation techniques. A long undulator is used to study the harmonic generation after saturation. The short bunch effect in the SASE process is expected to be studied under experimental conditions. Based on the superconducting injector and the accelerator, it is potentially superior for driving future sources of high-average power and attaining high-efficiency. High-average power FEL will be studied after the SASE FEL stage.

In this paper, design and optimization of IR SASE FEL at Peking University has been described. As a platform, the physical process of SASE FEL will be studied in detail under experimental conditions. This facility has been developed to provide an experimental benchmark for designing large-scale SASE FEL at X-ray wavelengths in the future in China.

Acknowledgements

We thank Dr. Ming Xie for his useful suggestions and help in design optimization.

References

[1] P.G. O'Shea, H.P. Freund, Science 292 (2001) 1853.
[2] G.R. Neil, L. Merminga, Rev. Mod. Phys. 74 (2002) 685.
[3] K. Zhao, S.W. Quan, J.K. Hao, et al., Nucl. Instr. and Meth. A 483 (2002) 125.
[4] R. Xiang, et al., Proceedings of FEL 2003, Tsukuba, Ibaraki, Japan.

[5] A. Buchner, F. Gabriel, P. Michel, et al., The ELBE-project at Dresden-Rossendorf, Proceedings of EPAC 2000, p.732.
[6] J. Plufger, Nucl. Instr. and Meth. A 445 (2002) 366.
[7] X. Ming, Nucl. Instr. and Meth. A 445 (2002) 59.
[8] R. Bonifacio, L.D. Salvo, P. Pierini, et al., Phys. Rev. Lett. 73 (1994) 70.

Available online at www.sciencedirect.com

Nuclear Instruments and Methods in Physics Research A 528 (2004) 421–426

**NUCLEAR
INSTRUMENTS
& METHODS
IN PHYSICS
RESEARCH**
Section A

www.elsevier.com/locate/nima

Bunch compressor for the SPring-8 compact SASE source (SCSS) project

Yujong Kim[a,*], T. Shintake[b], H. Kitamura[b], H. Matsumoto[c], D. Son[d], Y. Kim[d]

[a] *Deutsches Elektronen-Synchrotron DESY, D-22603 Hamburg, Germany*
[b] *RIKEN Harima Institute, SPring-8, Hyogo 679-5148, Japan*
[c] *KEK, High Energy Accelerator Research Organization, Ibaraki 305-0801, Japan*
[d] *The Center for High Energy Physics, Daegu 702-701, Republic of Korea*

Abstract

At the SPring-8 compact SASE source (SCSS), a high-quality beam with high peak current and low emittance should be supplied to saturate the SASE mode within a 22.5 m long undulator. In this paper, we describe the design concepts of an SCSS bunch compressor and the manner in which the CSR-induced emittance growth can be reduced in the bunch compressor.
© 2004 Elsevier B.V. All rights reserved.

PACS: 29.17.+w; 41.85.Lc; 41.60.Ap; 41.60.Cr

Keywords: Bunch compressor; SASE-FEL; CSR; CSR microbunching instability

1. Introduction

The SPring-8 compact SASE source (SCSS) project is divided into two phase stages according to its installed components and the radiation wavelength of the SASE source [1,2]. During the Phase-I stage (2001–2005), an injector, a C-band main linear accelerator, and a bunch compressor (BC) will be installed to generate a 40 nm wavelength radiation. During the Phase-II stage (2005–2007), three additional C-band main linear accelerators will be added to generate a wave-

length radiation of about 3.6 nm. Detail design beam parameters for the two stages are summarized in Table 1 of Ref. [2]. At the Phase-I stage, a peak current of about 500 A is required to saturate the SASE mode within a 13.5 m long undulator. However, no present injector technology can directly supply the required high-quality electron beam. Since the peak current is inversely proportionate to the bunch length, the required peak current can be obtained by compressing the bunch length. After upgrading the injector system, a 4 ps (FW) long bunch will be directly supplied to the bunch compressor at the Phase-II stage. And by retuning the operation conditions of the bunch compressor, our required high peak current of 2 kA can be obtained using the same bunch

---

*Corresponding author.
E-mail address: yujong.kim@desy.de (Yujong Kim).

0168-9002/$ - see front matter © 2004 Elsevier B.V. All rights reserved.
doi:10.1016/j.nima.2004.04.124

compressor. In this paper, we describe the design concepts of the SCSS bunch compressor.

2. The principle of a bunch compressor

When electrons in a bunch pass through a magnetic chicane, their traveling path lengths or times are changed according to their energies or energy spreads, which implies a bunch length change [2,3]. In a bunch compressor, the longitudinal coordinate relation to the second order is given by

$$dz_f = dz_i + R_{56}(dE/E)_i + T_{566}(dE/E)_i^2 \qquad (1)$$

$$R_{56} \approx 2\theta_B^2(\Delta L + \tfrac{2}{3}L_B) \approx -\tfrac{2}{3}T_{566} \qquad (2)$$

where dz_f (dz_i) is the longitudinal position deviation from the bunch center after (before) the chicane, $(dE/E)_i$ is the relative energy deviation before the chicane, which is supplied by a precompressing linac, R_{56} is the momentum compaction factor, which is supplied by the chicane, T_{566} is the second-order momentum compaction factor due to the second-order dispersion in the chicane, θ_B is the bending angle in the radian, ΔL is the drift space between the first dipole and the second dipole in the chicane, and L_B is the effective length of the dipole in the chicane. Here, we assume that the chicane consists of four rectangular dipoles, and the head electrons have positive dz and negative (dE/E).

3. Design concepts of bunch compressor

When the bunch length is compressed in a bunch compressor, the bunch length may become smaller than the radiation wavelength. In this case, CSR can be generated. Since CSR from the tail electrons can overtake the head electrons after traveling the overtaking length, the head electrons will be accelerated by CSR, and the tail electrons will be decelerated due to their own CSR loss. The electrons will be transversely kicked at the nonzero dispersion region due to the CSR-induced correlated energy spread along the bunch. It must be noted that the projected emittance can be diluted due to CSR in the bunch compressor while the slice emittance dilution is sufficiently small [3]. We have reduced the emittance dilution due to CSR in the BC using the following methods:

First, after considering the slice emittance growth in a double-chicane due to the CSR microbunching instability, the slice emittance growth in a wiggler-combined single chicane due to the spontaneous radiation in the wiggler, and the projected and slice emittance growths in an S-chicane due to the residual dispersion and CSR microbunching instability, we chose a normal single chicane with four rectangular magnets for our bunch compressor, as shown in Fig. 1 [3].

Second, we have reduced CSR by a weak strength chicane or small R_{56} [3]. According to Eqs. (1) and (2), this is possible by choosing a somewhat large $(dE/E)_i$, a small bending angle, and a somewhat large drift space between the first and second dipoles. Optimized parameters of the

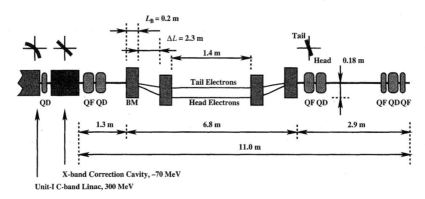

Fig. 1. Layout of the SCSS bunch compressor.

Table 1
Parameters of the SCSS BC in Phase-I and Phase-II stages

Parameter	Phase-I	Phase-II
Beam energy E (MeV)	230	218
Initial bunch length Δz_i (mm)	2.4	1.2
Final bunch length Δz_f (mm)	0.6	0.15
Initial projected relative energy spread σ_δ (%)	2.14	1.98
Initial slice rms relative energy spread $\sigma_{\delta s}$ (10^{-5})	~ 5	~ 5
Initial maximum relative energy deviation $(\mathrm{d}E/E)_i$ (10^{-2})	3.6	3.5
Beam phase at the C-band linac ϕ_c (deg)	12.5	24
Momentum compaction factor R_{56} (mm)	24.8	15.0
Second-order momentum compaction factor $|T_{566}|$ (mm)	37.2	22.5
Effective dipole length L_B (m)	0.2	0.2
Drift length between first and second dipoles ΔL (m)	2.3	2.3
Dipole bending angle θ_B (mrad)	71	56
Dipole magnetic field $|B|$ (T)	0.27	0.20
Maximum horizontal dispersion $\eta_{x,\max}$ (m)	0.18	0.14
Initial slice emittance before the BC ε_{ns} (μm)	1.50	1.50
Initial projected emittance before the BC ε_n (μm)	1.58	1.54
Change in slice emittance $\Delta\varepsilon_{ns}/\varepsilon_{ns}$ (%)	0.04	0.18
Change in projected emittance $\Delta\varepsilon_n/\varepsilon_n$ (%)	2.3	63.2
Change in energy spread due to CSR $\Delta\sigma_{\delta,\mathrm{CSR}}$ (10^{-4})	2.9	8.7
Change in energy spread due to ISR $\Delta\sigma_{\delta,\mathrm{ISR}}$ (10^{-8})	6.2	3.8

SCSS BC for the Phase-I and Phase-II stages are summarized in Table 1, where ε_{ns} (ε_n) is the slice (projected) transverse normalized rms emittance and $\sigma_{\delta s}$ (σ_δ) is the slice or local (projected) rms relative energy spread [2]. Since electrons separated by more than one cooperation length ($\sim \mu$m) will not interact with each other, we should focus on the slice parameters to predict FEL performance [3]. Although the normal operational R_{56} can be reduced further by choosing higher σ_δ, we have chosen $\sigma_\delta \sim 2\%$ to keep the chromaticity-induced projected emittance dilution in linacs within 5%.

Third, we have reduced CSR further by installing a higher harmonic X-band correction cavity before the BC to compensate the nonlinearities in the longitudinal phase space. The X-band correction cavity compensates the second-order nonlinearities due to the RF curvature of the C-band linac, the second-order path dependence on particle energy in the chicane T_{566}, and the short-range wakefields in linacs, as shown in Figs. 1 and 2 [3,4]. Here, the residual weak nonlinearity after the X-band correction cavity is to compensate T_{566}

of the chicane. Since the beam energy at the SCSS BC is sufficiently high, the nonlinearity due to the space charge force is negligible. If those nonlinearities are not properly compensated, then a local charge concentration or spike is generated during the compression process. This local charge concentration or spike amplifies CSR effects in the bunch compressor [2,4]. By compensating the nonlinearities with the X-band correction cavity, the final obtainable minimum bunch length can be also reduced further [2]. However, beams are decelerated by about 70 MeV in the X-band correction cavity to compensate the nonlinearities [2,4]. The projected emittance dilution due to the transverse short-range wakefield in the X-band correction cavity is within 7% if its misalignment is smaller than 50 μm [4].

Fourth, the projected emittance dilution due to CSR can be reduced further by forcing the beam waist close to the fourth dipole, where α-function ($= -0.5\beta'$) is zero, and the β-function is minimum [2,3]. The optimized β-functions are shown in Fig. 3.

Fifth, although the final bunch length at the Phase-II stage is four times smaller than that at the

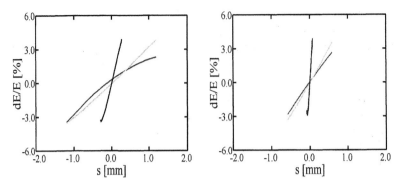

Fig. 2. ELEGANT simulation results of the longitudinal phase space distribution at Phase-I (left) and Phase-II (right) stages. The red, green, and blue lines indicate the longitudinal phase space distribution after the C-band linac, after the X-band correction cavity, and after the BC, respectively. The negative s $(= |dz|)$ indicates the bunch head.

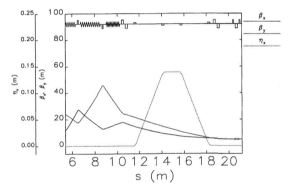

Fig. 3. β-functions at the SCSS BC Phase-II.

Phase-I stage, we can control CSR effects by reducing the chicane strength or R_{56} to 15 mm. This R_{56} can be obtained by maintaining σ_δ at the BC at about 2% and by supplying 4 ps (FW) long bunch from the injector. We can reduce the energy spread of 2% to the design value by the longitudinal short-range wakefields in three additional C-band main linacs and by operating those linacs close to the RF crest.

4. Investigation of CSR wakefield effects

We have investigated CSR wakefield effects in the bunch compressor for the SCSS Phase-I and Phase-II stages using ELEGANT and TraFiC[4] codes. As shown in Figs. 2 and 4, the final bunch length and peak current are 0.6 mm and about 500 A at the Phase-I stage, respectively, and those

of the Phase-II stage are 0.15 mm and about 2 kA, respectively. Since the SCSS injector has a beam deflector, a beam chopper, and an energy filter, we have assumed that the initial linear density before the BC is a somewhat flat-top shape, as shown in Fig. 4 [1]. With the help of the X-band correction cavity, the linear densities are fairly symmetric around the bunch center, and there is no strong spike or local charge concentration after the BC, as shown in Fig. 4. The CSR induced correlated energy spread along the bunch is increased as the bunch length is compressed, as shown in Fig. 5. Here, the head electrons with negative s are accelerated by CSR, and the tail electrons are decelerated by their CSR loss. At the Phase-II stage, there is one small special spike around the tail region in the energy change, as shown in Fig. 5 (right). This is due to the uncompensated higher order nonlinearities such as the third order term in the longitudinal short-range wakefields and the third order term in the momentum compaction.

Since the longitudinal space charge force and the intrabeam scattering increase the slice energy spread in the linac, and $\sigma_{\delta s}$ in the BC is increased further by the incoherent synchrotron radiation (ISR), CSR, and the longitudinal space charge force, the estimated gain of the CSR microbunching instability is less than 4, as shown in Fig. 6 [5,6]. Therefore, the CSR microbunching instability in the SCSS BC can be prevented to a certain extent.

After the SCSS BC, the final slice (projected) normalized rms emittances at the Phase-I and

Phase-II stages are about 1.501 μm (1.6 μm) and 1.503 μm (2.5 μm), respectively, all of which are within our design values as summarized in Table 1 of Ref. [2].

5. Summary

We have designed a single chicane bunch compressor for the SCSS project. By compensating

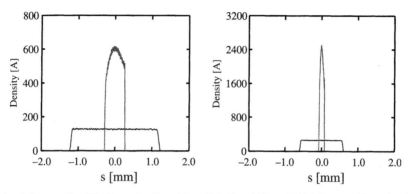

Fig. 4. ELEGANT simulation results of the linear density at Phase-I (left) and Phase-II (right) stages. The red and green lines indicate the linear density at the ends of the first and last dipoles, respectively.

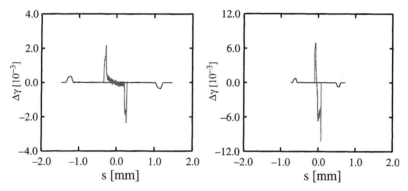

Fig. 5. ELEGANT simulation results of the energy change $\Delta\gamma$ due to CSR at Phase-I (left) and Phase-II (right) stages. The red and green lines indicate $\Delta\gamma$ at the ends of the first and last dipoles, respectively.

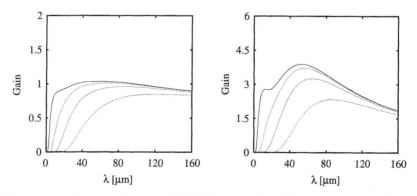

Fig. 6. Gain of the CSR microbunching instability at the SCSS BC Phase-I (left) and Phase-II (right) stages. Here, $\sigma_{\delta s}$ indicated by red, green, blue, and magenta lines are 9.5×10^{-6}, 1.25×10^{-5}, 2.5×10^{-5}, and 5.0×10^{-5}, respectively.

nonlinearities in the longitudinal phase space with the X-band correction cavity, using the weak strength chicane under a somewhat large energy spread condition, and by optimizing Twiss parameters around the bunch compressor, we can reduce the CSR-induced emittance growth and the microbunching instability in the bunch compressor.

Acknowledgements

The authors sincerely thank M. Borland, Ph. Piot, Z. Huang, and P. Emma for their useful discussions and recommendations.

References

[1] T. Shintake, et al., Proceedings of LINAC 2002, Gyeongju, Korea, 2002, pp. 526–530.
[2] Yujong Kim, et al., Proceedings of LINAC 2002, Gyeongju, Korea, 2002, pp. 100–102;
    Yujong Kim, et al., Proceedings of EPAC 2002, Paris, France, 2002, pp. 808–810.
[3] LCLS Conceptual Design Report, SLAC-R-593, 2002.
[4] P. Emma, LCLS-TN-01-1, 2001.
[5] Yujong Kim, et al., Proceedings of PAC 2003, Portland, USA, 2003.
[6] Z. Huang, et al., Phys. Rev. ST Accel. Beams 5 (2002) 074401.

Available online at www.sciencedirect.com

SCIENCE  DIRECT®

ELSEVIER   Nuclear Instruments and Methods in Physics Research A 528 (2004) 427–431

**NUCLEAR
INSTRUMENTS
& METHODS
IN PHYSICS
RESEARCH**
Section A

www.elsevier.com/locate/nima

Start-To-End (S2E) simulations on microbunching instability in TESLA Test Facility Phase 2 (TTF2) project

Yujong Kim[a],*, D. Son[b], Y. Kim[b]

[a] *Deutsches Elektronen-Synchrotron DESY, D-22603 Hamburg, Germany*
[b] *The Center for High Energy Physics, Daegu 702-701, South Korea*

Abstract

Microbunching instability in the FEL driving linac is induced by collective self-fields such as longitudinal space charge, coherent synchrotron radiation, and geometric wakefields. Our simulations are the first start-to-end simulations including all important collective self-fields from the cathode to the end of linac. In this paper, we estimate the gain of the microbunching instability in TESLA Test Facility Phase 2 linac by applying current density modulations at the cathode, and describe several new smearing effects of the microbunching instability which are not considered in the analytical gain estimation formula.
© 2004 Elsevier B.V. All rights reserved.

PACS: 29.17.+w; 52.59.Sa; 41.60.Ap; 41.60.Cr

Keywords: Bunch compressor; SASE-FEL; CSR; Microbunching instability; Longitudinal space charge; TESLA Test Facility Phase 2

1. Introduction

Microbunching instability is induced by collective self-fields when an electron bunch has a current density modulation before a bunch compressor (BC) [1,2]. The initial density modulation is converted into the energy modulation by longitudinal space charge (LSC) and geometric wakefields in the linac and by coherent synchrotron radiation (CSR) in the bunch compressor. In the BC with nonzero dispersion, the initial density modulation is amplified by this induced energy modulation, and the microbunching instability

occurs if the collective self-fields are strong enough. Generally, the microbunching is only due to LSC does not give a large emittance growth in the BC. But if the CSR is added to the LSC driven microbunching in the BC, the slice emittance, the projected emittance, and the local energy spread are largely increased. Therefore FEL performance is severely influenced by the microbunching instability. Recently, it was reported that the microbunching instability can be induced in the TESLA Test Facility Phase 2 (TTF2) linac by density modulation in the gun driving laser pulse, and LSC is the main source of the instability [1]. To estimate the strength of the instability, we define gain parameter as the ratio of the normalized amplitude of a density modulation

---

*Corresponding author.
E-mail address: yujong.kim@desy.de (Yujong Kim).

0168-9002/$ - see front matter © 2004 Elsevier B.V. All rights reserved.
doi:10.1016/j.nima.2004.04.125

at the final position to that of the initial modulation. In the case of TTF2, the analytically estimated maximum gain of the microbunching instability after the second BC is about 320, when an initial density modulation with about 2.0 ps period is applied before the first TESLA accelerating module (ACC1) [1,3]. Since this analytically estimated gain is large, we perform start-to-end (S2E) simulations with consideration of all important collective self-fields to estimate a more realistic maximum gain in TTF2 linac. We use a code, ELEGANT to consider the CSR at two bunch compressors and the geometric short-range wakefields at the ends of all TESLA accelerating modules [4,5]. And we use ASTRA to consider the space charge force through entire regions from the cathode to the end of TTF2 linac [6,7]. A detail simulation layout and machine parameters of TTF2 are described in Refs. [3,6].

2. S2E simulation results

2.1. Amplification in the first BC (BC2)

To investigate the amplification of an initial modulation at the end of the first bunch compressor (BC2), three different initial density modulations with 1.0, 2.0, and 4.0 ps periods and $\pm 10\%$ amplitude are assumed at the cathode as shown in Fig. 1(left) [6]. Although $\pm 10\%$ amplitude is not small, the initial density modulation with 2.0 ps period is well damped by the fast plasma oscillation and the strong Landau damping at the low-energy region as shown in Fig. 1(right) [2,6,8]. Here fifty thousand macroparticles are tracked in

the ASTRA simulations, and ten radial grid rings and fifty longitudinal grid cells are used for the space charge calculation. We also find that this strong damping in the initial density modulation appears for 1.0 ps (4.0 ps) period, where we use one hundred thousand (fifty thousand) macroparticles and one hundred (fifty) longitudinal grid cells [6].

According to the theory, the LSC induces the energy modulation in a linac or in a drift space if the current density is modulated in the upstream path [1,2]. In our calculations, however, the induced energy modulation before BC2 is small enough because the amplitude of the initial density modulation is not constant and reduced due to the fast plasma oscillation and the strong Landau damping at the low energy region as shown in Figs. 1(right) and 2(left). With 1.5 million macroparticles, we also observe the same damping at the low-energy region for initial modulations with 2.0 ps period and $\pm 1.25\%$, $\pm 2.5\%$, $\pm 5.0\%$, $\pm 7.5\%$, and $\pm 10\%$ amplitudes [9]. In the BC, density modulation is converted into energy modulation by the CSR, and the energy modulation is also converted to the density modulation via the nonzero dispersion, even if those modulations before the BC are small. The energy modulation after BC2, consequently, grows with the same frequency as that of the initial current modulation as shown in Figs. 2(right) and 3(left), and the current density modulation after BC2 is also amplified as shown in Fig. 3(right).

However, amplification of the current density modulation after BC2 is weak as shown in Fig. 3(right), where the gain of the current density modulation after BC2 is about 1.6 with respect to

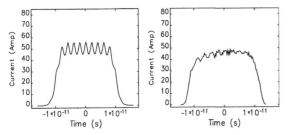

Fig. 1. Current profile at the cathode (left) and at 0.5 m downstream from the cathode (right) for an initial modulation with 2.0 ps period and $\pm 10\%$ amplitude.

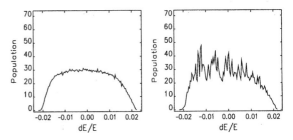

Fig. 2. Energy profiles before BC2 (left) and after BC2 (right) for an initial modulation with 4.0 ps period and $\pm 10\%$ amplitude at the cathode.

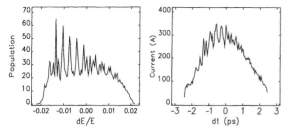

Fig. 3. After BC2, energy profile (left) and current profile (right) for an initial modulation with 2.0 ps period and ±10% amplitude at the cathode.

an initial modulation with ±10% amplitude at the cathode. For different initial modulations with 1.0 and 4.0 ps periods, similar weak density modulations are generated after BC2 [6]. Therefore, a single stage bunch compressor is safe from the strong current density modulation, even if the gun driving laser pulse has a current density modulation with ±10% amplitude. Recently, we observed this strong energy modulation after BCs at SDL of BNL and TTF1 of DESY [2,6,10]. Since the density modulation in the current profile is amplified in BC2, the longitudinal electric field due to the space charge force is also amplified and modulated after BC2 with the same frequency as that of the initial modulation frequency [6]. After BC2, this modulated nonlinear longitudinal electric field due to the space charge increases the energy spread in the small local area [6]. Therefore, the microbunching instability in the second bunch compressor (BC3) will be somewhat smeared out by this increased local energy spread before BC3 [6]. If we neglect the space charge force from the first bunch compressor, the local energy spread is not increased at the downstream of BC2 and the microbunching instability will be somewhat overestimated at BC3. Note that in the case of S2E simulations for the LCLS project, the space charge force is not included from the first bunch compressor.

2.2. Gain after the second BC (BC3)

According to the analytical gain estimation, an initial modulation with about 2.0 ps period gives the maximum gain of the microbunching instabil-

ity in TTF2 linac [1]. To estimate its magnitude with S2E simulations, four different initial density modulations with 2.0 ps period and ±2.5%, ±5.0%, ±7.5%, and ±10% amplitudes are assumed at the cathode and tracked down to the end of TTF2 linac. After the first bunch compressor, strong numerical noise appears for the initial modulations with ±2.5% and ±5.0% amplitudes, while the noise becomes small for the initial modulation with ±10% amplitude [6,9]. This numerical noise artificially increases the gain of the microbunching instability at BC3 [6]. We can reduce this noise by increasing the simulated macroparticles or by increasing the initial modulation amplitude [9].

Since the plasma oscillation and the Landau damping are weak after BC2, the density modulation after BC2 is frozen up to BC3. Hence the energy modulation is continuously accumulated by LSC up to the BC3 entrance, and the energy modulation is converted into the density modulation in BC3 via the nonzero dispersion. Therefore, the current density modulation is strongly amplified after BC3, even if the CSR is not considered in BC3, as shown in Fig. 4. Note that for all the cases, LSC is included in the upstream of BC3. Since the modulation strength of the ±10% amplitude case is much stronger than that of the ±5.0% amplitude case as shown in Fig. 4, the modulation with 2.0 ps period and ±10% amplitude at the cathode is not saturated at BC3 by LSC.

Since the CSR becomes stronger as the modulation amplitude or nonlinearity in the current

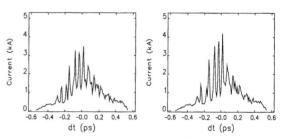

Fig. 4. After BC3, current profile without consideration of CSR in BC3 for an initial modulation with 2.0 ps period and ±5.0% amplitude at the cathode (left) and for an initial modulation with 2.0 ps period and ±10% amplitude at the cathode (right).

profile is increased, the local energy spread, the slice normalized emittance, and the projected normalized emittance are largely increased due to the CSR as shown in Figs. 5 and 6(left) [11]. In Fig. 5, the local relative rms energy spread in ± 0.1 mm core region is 3.5×10^{-4} without consideration of CSR in BC3, while the spread becomes 9.8×10^{-4} with consideration of CSR in BC3. In the left plot of Fig. 6, red and green (blue and magenta) lines indicate the projected emittance and the slice emittance with (without) consideration of CSR in BC3, respectively. If CSR is considered in BC3, the LSC driven microbunching instability is smeared by the increased local energy spread and the increased emittance by the Landau damping as shown in Figs. 4(right), 5(right), 6(left), and 7. This CSR induced smearing effect becomes larger for the modulations with $\pm 7.5\%$ and $\pm 10\%$ amplitudes as shown in Fig. 6(right), where red (green) line indicates the gain of the microbunching instability

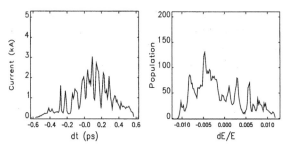

Fig. 7. After BC3, current profile (left) and energy profile (right) with consideration of CSR for an initial modulation with 2.0 ps period and $\pm 10\%$ amplitude at the cathode.

after BC3 with (without) consideration of CSR in BC3. Since the numerical noise is strong for the initial modulations with $\pm 2.5\%$ and $\pm 5.0\%$ amplitudes, their gains after BC3 are somewhat overestimated as shown in Fig. 6(right), while the noise is reduced for the initial modulations with $\pm 7.5\%$ and $\pm 10\%$ amplitudes [6]. Therefore, the maximum gain of TTF2 after BC3 with respect to the initial density modulation at the cathode is smaller than 10. Since LSC impedance becomes smaller after BC3 due to the higher energy and the higher gradient, the gain at the end of TTF2 linac is not increased largely [1,6]. Although the maximum gain of the microbunching instability is not so high in TTF2 linac, the slice emittance is largely increased once the instability is induced. Therefore, to keep safety margin, we should try to reduce the modulation in the gun driving laser pulse or to damp the possible microbunching instability by increasing the local energy spread at the injector system or at the bunch compressors [1].

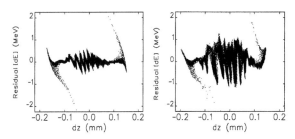

Fig. 5. After BC3, local energy spread without consideration of CSR in BC3 (left) and with the consideration of CSR in BC3 (right) for an initial modulation with 2.0 ps period and $\pm 10\%$ amplitude at the cathode.

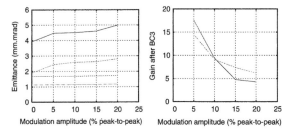

Fig. 6. After BC3, normalized horizontal emittances (left) and gains of the microbunching instability in TTF2 linac with respect to initial modulations at the cathode (right). Here modulation period is 2.0 ps for all amplitudes.

3. Summary

If the current density at the cathode has the modulation initially, the energy modulation due to LSC is amplified after the first bunch compressor, while the amplification of the density modulation is weak enough. Since the energy modulation after the first bunch compressor is continuously accumulated in the linac by LSC, strong density modulation appears after the second bunch compressor. The maximum gain of the micro-bunching instability after the TTF2 second bunch compressor is smaller than 10, because several smearing effects such as the fast plasma oscillation and strong Landau damping at the low energy region, the local energy spread growth in the linac due to LSC, and the local energy spread and transverse emittance growth in the bunch compressor due to CSR are not included in the analytical gain estimation.

Acknowledgements

Yujong Kim sincerely thanks K. Flöttmann, S. Schreiber, Z. Huang, E.A. Schneidmiller, P. Emma, M. Borland, M. Dohlus, T. Limberg, Prof. J. Rossbach, Prof. K.-J. Kim, and Prof. W. Namkung for their encouragements, many useful discussions, and comments on this work.

References

[1] E.L. Saldin, et al., TESLA-FEL-2003-02, 2003.
[2] Z. Huang, et al., SLAC-PUB-9788, 2003.
[3] The TESLA Test Facility FEL team, TESLA-FEL-2002-01, 2002.
[4] M. Borland, Advance Photon Source Report No. LS-287, 2000.
[5] T. Weiland, et al., TESLA-2003-19, 2003.
[6] http://www.desy.de/~yjkim/FEL03P02.pdf
[7] K. Flöttmann, ASTRA manual, http://www.desy.de/~mpyflo/Astra_dokumentation/
[8] J. Rosenzweig, et al., TESLA-FEL-96-15, 1996.
[9] http://www.desy.de/~yjkim/FEL03P03.pdf
[10] M. Hüning, et al., Nucl. Instr. and Meth. A 475 (2001) 348.
[11] R. Li, Nucl. Instr. and Meth. A 475 (2001) 498.

Available online at www.sciencedirect.com

SCIENCE  DIRECT°

ELSEVIER   Nuclear Instruments and Methods in Physics Research A 528 (2004) 432–435

**NUCLEAR
INSTRUMENTS
& METHODS
IN PHYSICS
RESEARCH**
Section A

www.elsevier.com/locate/nima

Performance simulation and design of the magnetic bunch compressor for the infrared free-electron laser at the Pohang light source test linac

Eun-San Kim

Pohang Accelerator Laboratory, Pohang University of Science and Technology, Pohang, KyungBuk 790-784, Republic of Korea

Abstract

A chicane bunch compressor was designed for the infrared free-electron laser project at the Pohang Light Source test linac. The bunch compressor has triplet quadrupoles before and after the chicane to obtain a transversely uncorrelated beam at the center of the bunch compressor. In this paper, we describe the design considerations of the chicane that provides a compression from 3 ps rms to 250 fs rms a beam energy of 100 MeV. Simulations provide predictions of emittance variation with the chicane parameters and the linac RF phase in the precompressor. The simulations also indicate that the normalized emittance in the designed bunch compressor increases from 5 mm mrad to about 5.5 mm mrad.

PACS: 29.27.Bd; 41.60.Ap; 29.17.+w

Keywords: Free electron laser; Coherent synchrotron radiation; Bunch compressor

1. Introduction

An important requirement for the self-amplified spontaneous emission (SASE) free-electron laser (FEL) is to provide a beam with low-emittance and high peak beam current. The design specification for the SASE-FEL requires a photocathode RF gun that can produce the required emittance, and a magnetic bunch compressor to reduce the bunch length, which will increase the peak beam current. Coherent synchrotron radiation (CSR) induced by a short bunch in the magnetic bunch

compressor becomes a major issue because it may cause emittance growth. When an electron bunch is compressed in a bunch compressor, the CSR induces an energy redistribution along the bunch [1,2]. In the dispersive region, the chromatic transfer function will couple this longitudinal energy redistribution into a betatron oscillation motion, thus it leads to the growth of transverse emittance [3]. The infrared FEL system at the Pohang Light Source is being designed by utilizing the test linac facility. The system is based on the existing electron linac that consists of a thermionic RF gun, an alpha magnet, and two S-band linac sections to provide electron energies from 60 to

E-mail address: eskim1@postech.ac.kr (E.-S. Kim).

0168-9002/$ - see front matter © 2004 Elsevier B.V. All rights reserved.
doi:10.1016/j.nima.2004.04.126

100 MeV. A chicane bunch compressor is required to shorten the bunch length in the system. Since the CSR may cause degradation in beam quality, the CSR effects have to be carefully analyzed for the design of the bunch compressor. Accordingly, the chicane parameters in our system are chosen to minimize the CSR effects. In this paper, we describe the design consideration of the bunch compressor for the infrared FEL and the performance of the bunch compressor. A simulation code [4] is utilized to investigate the effects of the coherent synchrotron radiation on beam quality. The CSR algorithm in the code is based on an analytical model for the longitudinal and transverse CSR wake [5,6].

2. Design of the chicane bunch compressor for the infrared FEL

A four-dipole chicane is being designed for the infrared FEL system. Fig. 1 shows a detailed scheme of the bunch compressor with its Twiss parameters. The dispersion function (η_x) and its derivative, and α_x and α_y have a zero value as their initial values at the entrance of the bunch compressor. The triplet quadrupoles before and after the chicane are included to obtain a transversely uncorrelated beam, that is $\alpha_{x,y} = 0$,

Table 1
Beam and four-dipole chicane bunch compressor parameters

Parameter	Symbol	Value	Unit
Electron beam energy	E	100	MeV
Bunch charge	Q	0.3	nC
Total length	L_t	6.1	m
rms energy spread	σ_δ	0.1	%
Initial rms bunch length	σ_z	3	ps
Initial normalized emittance	$\varepsilon_{n(x,y)}$	5	μm
1st momentum compaction	R_{56}	−69	mm
2nd momentum compaction	T_{566}	105.6	mm
Linac RF phase in precompressor	ϕ	14	deg
Bend angle per dipole	$|\theta|$	10.3	deg
Bend length per dipole	L	0.22	m
Bend radius per dipole	$|\rho|$	1.2	m
Field of bending magnets	B	0.27	T
Maximum horizontal dispersion	η	0.205	m

in the center of the bunch compressor. Table 1 shows the main parameters of the beam and the designed bunch compressor. The effects of coherent synchrotron radiation in the bunch compressor are estimated.

For a bunch with a Gaussian distribution, the energy gain due to the CSR along the bunch is given by [5]

$$\frac{dE}{dz} = \frac{-2Ne^2}{\sqrt{2\pi}(3\rho^2\sigma_z^4)^{1/3}} \int_{-\infty}^{\xi} \frac{d\xi'}{(\xi - \xi')^{1/3}} \frac{d}{d\xi'} e^{-\xi'^2/2} \quad (1)$$

where N is the bunch population, ρ is the bending radius, σ_z is the rms bunch length, and $\xi = z/\sigma_z$. When Eq. (1) is plotted by using the parameters from Table 1, the head particles in the bunch gain energy while most particles in the tail and center lose energy, as shown in Fig. 2 (top). A beam through the RF linac ahead of the crest induces an energy chirp into the beam, thus the particles in the tail of the beam have higher energy than the head particles. As a result, the higher energy electrons in the tail catch up with the lower energy particles of the head in the bunch compressor, giving a shorter bunch. To compress the bunch, before the chicane, we introduced a chirp in the RF linac cavity at a chirping angle $\phi = 14°$ (in our notation $\phi = 0$ means on-crest). The chicane has an $R_{56} = 0.069$ m. When we consider energy dependence of path length with the relation $R_{56} = -E/(dE/dz)$, the energy slope in the

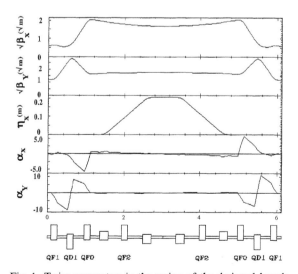

Fig. 1. Twiss parameters in the region of the designed bunch compressor.

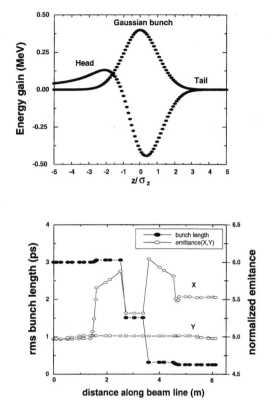

Fig. 2. (Top) Energy gain and loss due to the CSR along the Gaussian beam distribution. (Bottom) Evolution of the rms bunch length, and normalized x and y emittances along the beam line of the bunch compressor.

precompressor dE/dz has a value of -1.45×10^9 eV/m.

3. Performance of the bunch compressor for the infrared FEL

Our simulation shows that a bunch in the designed bunch compressor is compressed from 3 ps to 250 fs with a beam energy of 100 MeV. A nominal set of parameters to fulfill the above bunch compression is: the beam energy has $E_0 = 100$ MeV, $Q = 0.3$ nC, $\sigma_\delta = 0.1\%$ (before the chirping) and $\varepsilon_{n(x,y)} = 5$ mm mrad. The rms bunch length and the beam emittance growth induced by the CSR, shown in Fig. 2 (bottom), are plotted along the beam line of the chicane. A drift and matching 2 m section is added after the chicane to monitor the rms bunch length and beam emittance after the final bend. The horizontal emittance increases about 10% due to the coherent synchrotron radiation effect. Fig. 3 shows the difference of the path length versus the momentum difference. The second-order momentum compaction factor of the bunch compressor has a value of 105.6 mm.

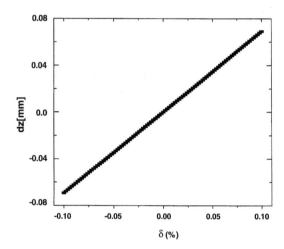

Fig. 3. Variation of the path length versus momentum difference.

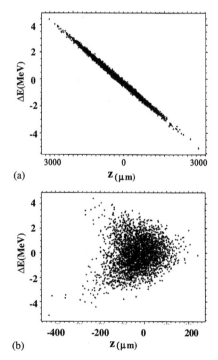

Fig. 4. Longitudinal phase spaces (a) at entrance and (b) exit of the bunch compressor.

Fig. 4 shows the longitudinal beam phase spaces at the entrance (top) and at the exit (bottom) of the bunch compressor.

The longitudinal resistive wall wakefield in the beam chamber may induce the energy spread. The rms energy spread generated in a beam chamber of length L_B is given by

$$\sigma_{\delta RW} \sim 0.22 \frac{e^2 c N L_B}{\pi^2 b E \sigma_z^{3/2}} \sqrt{Z_o/\sigma} \qquad (2)$$

where $Z_o = 377\,\Omega$, b is the radius and σ is the conductivity. For an aluminum chamber ($\sigma \sim 3.6 \times 10^7\,\Omega^{-1}\mathrm{m}^{-1}$), $b = 5$ cm, and the bunch length $\sigma_z = 75\,\mu\mathrm{m}$, the rms energy spread due to the resistive wake gives $\sim 4 \times 10^{-7}$, which is a smaller value than the energy spread due to the CSR in the bend.

4. Conclusions

In this paper, we presented the design and the performance results of the bunch compressor for the infrared SASE-FEL at the PLS. The CSR effects in the bunch compressor are also discussed. It is shown that an electron bunch in the designed bunch compressor can be compressed from 3 ps to 250 fs and horizontal emittance is increased by 10%. The simulations also give predictions of emittance variation with the parameters of the bunch compressor and chirping angle in the RF linac.

References

[1] Linac Coherent Light Source Conceptual Design Report, 2002, http://www-ssrl.slac.stanford.edu/lclc/CDR.
[2] TESLA Technical Design Report, DESY TESLA-FEL 2001-05, 2001.
[3] M. Borland, Phys. Rev. Spec. Top. Accel. Beams 4 (2001) 074201.
[4] K. Yokoya, unpublished.
[5] J. Murphy, S. Krinsky, R. Gluckstern, in: Proceedings of PAC, 1995, p. 2980.
[6] Y. Derbenev, et al., DESY TESLA-FEL 95-05, 1995.

Available online at www.sciencedirect.com

SCIENCE  DIRECT®

**NUCLEAR
INSTRUMENTS
& METHODS
IN PHYSICS
RESEARCH**
Section A

ELSEVIER   Nuclear Instruments and Methods in Physics Research A 528 (2004) 436–442

www.elsevier.com/locate/nima

Ultraviolet high-gain harmonic-generation free-electron laser at BNL

L.H. Yu*, A. Doyuran, L. DiMauro, W.S. Graves[1], E.D. Johnson, R. Heese,
S. Krinsky, H. Loos, J.B. Murphy, G. Rakowsky, J. Rose, T. Shaftan, B. Sheehy,
J. Skaritka, X.J. Wang, Z. Wu

National Synchrotron Light Source, Brookhaven National Laboratory, AFT, Bldg 725C, Box 5000, Upton, NY 11973, USA

Abstract

We report the first experimental results on a high-gain harmonic-generation (HGHG) free-electron laser (FEL) operating in the ultraviolet. An 800 nm seed from a Ti-Sapphire laser has been used to produce saturated amplified output at the 266 nm third-harmonic. The results confirm the advantages of the HGHG FEL: stable central wavelength, narrow bandwidth and small pulse energy fluctuation. The harmonic output at 88 nm, which accompanies the 266 nm radiation, has been used in an ion pair imaging experiment in chemistry.
© 2004 Elsevier B.V. All rights reserved.

PACS: 41.60.Cr

Keywords: Free-electron laser; High-gain harmonic-generation

1. Introduction

There is great interest in utilizing a high-gain single-pass free-electron laser (FEL) to generate intense, short pulse radiation in the spectral region from the deep ultraviolet down to hard X-ray wavelengths [1]. The most widely studied approach has been self-amplified spontaneous-emission (SASE). In a SASE FEL [2–5], the amplifier is seeded by the shot noise in the electron beam.

SASE can produce short wavelength radiation with high peak power and an excellent spatial mode. However, the output has limited temporal coherence (coherence time much shorter than the pulse duration) and chaotic shot-to-shot intensity variations.

An alternate single-pass FEL approach is high-gain harmonic-generation (HGHG) [6–7], which is capable of producing temporally coherent pulses. In HGHG: (1) a small energy modulation is imposed on the electron beam by interaction with a seed laser in a short undulator (the modulator) tuned to the seed frequency ω; (2) the energy modulation is converted to a coherent longitudinal density modulation as the electron beam traverses

---

*Corresponding author. Tel.: +1-631-344-5012.

E-mail address: lhyu@bnl.gov (L.H. Yu).

[1]Present address: MIT-Bates Linear Accelerator Center Middleton, MA 01949, USA.

0168-9002/$ - see front matter © 2004 Elsevier B.V. All rights reserved.
doi:10.1016/j.nima.2004.04.127

a three-dipole chicane (the dispersion magnet); (3) in the second undulator (the radiator), tuned to the nth harmonic of the seed frequency, the micro-bunched electron beam emits coherent radiation at the harmonic frequency $n\omega$, which is then amplified until saturation. In HGHG, the light output is derived from a coherent sub-harmonic seed pulse; consequently, the optical properties of the HGHG FEL are a map of the characteristics of the high-quality seed laser. This has the benefit of providing better stability and control of the central wavelength, narrower bandwidth, and smaller energy fluctuations than SASE. Furthermore, HGHG has the potential to produce light pulses with duration much shorter than the electron bunch length.

The basic principle of the HGHG has been demonstrated at IR with the second harmonic [7]. Recently we reported the HGHG experiment in the UV region [8]. Here we give a more detailed report on the results obtained with the deep ultraviolet FEL (DUV-FEL) at the National Synchrotron Light Source of Brookhaven National Laboratory. The DUV-FEL design and commissioning are discussed in Refs [9–11]. In October 2002, we achieved the third harmonic HGHG starting from a seed at 800 nm with saturated output at 266 nm. We have measured HGHG output as a function of the undulator length for the first time. Since January 2003, the harmonic output at 88 nm accompanying the 266 nm fundamental has been used in an ion pair imaging experiment [12] in chemical physics as its first user application [13]. The experiment benefited from the high stability of the HGHG output. Here we report observations of the HGHG process: initial coherent generation, exponential amplification and saturation. We describe the key properties of HGHG, emphasizing its high stability and narrow bandwidth.

2. Basic description of the NSLS DUV-FEL

The layout of the facility is as illustrated in Fig. 1. The injector is comprised of a photo-cathode RF gun, illuminated by a frequency-tripled Ti:Sa laser at 266 nm, producing a 300 pC, 4.5 MeV, 4 ps (FWHM) electron bunch with normalized emittance of 3–5 μm. Two SLAC-type 2.856 GHz linac sections accelerate the electron beam up to 77 MeV. The second linac tank provides an energy chirp for the bunch compressor [14] (a four-magnet chicane). The third linac tank, located after the chicane, removes the residual energy chirp, with additional acceleration. The last tank is used to complete the acceleration to the desired energy. It is also used in combination with the downstream spectrometer magnet for bunch length measurement, employing the "zero-phasing" technique [15,16].

To allow injection of the seed, a combination of four dipole trims produces a "local bump" of the trajectory to bend the electron beam around the laser seeding mirror. Then follows the energy modulating undulator which has an 8 cm period with $K = 1.67$, so it is resonant to 800 nm at 177 MeV. Following the modulator is a 30 cm long dispersion magnet which converts the energy modulation to microbunching of the electron beam. Next is the 10 m long "NISUS" undulator [17] with 3.89 cm period, 0.31 T peak field, and equal focusing in the horizontal and vertical planes by means of canted poles. Since NISUS was not designed for the DUV-FEL, its parameters are not ideal for this application. Its period is longer and the electron transverse focusing is weaker than optimum—25 m betatron wavelength at 177 MeV. The resulting gain length of $L_G \sim 0.8$ m is too long for the system to reach SASE saturation. However, there is sufficient gain to saturate as an HGHG FEL. Every section of the long undulator is equipped with horizontal and vertical dipole correctors as well as quadrupole trims based on a 4-wire system. The electron trajectory and the transverse beam sizes are measured using Cerium-doped YAG-crystal profile monitors [18]. More details about the DUV-FEL system can be found in [9–11].

3. The experimental results

The photo-cathode laser with a pulse energy of 60 μJ at 266 nm is set at 60° before the RF crest, and the tank 2 phase is set at 23° before the RF crest. The charge in a typical electron bunch is

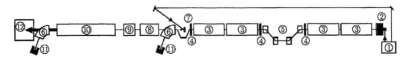

Fig. 1. The NSLS DUV-FEL layout: 1—gun and seed laser system, 2—RF gun, 3—linac tanks, 4—focusing triplets, 5—magnetic chicane, 6—spectrometers dipoles, 7—seed laser mirror, 8—modulator, 9—dispersive section, 10—radiator (NISUS), 11—beam dumps, 12—FEL radiation measurements area.

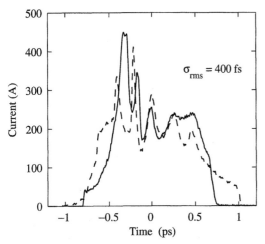

Fig. 2. Current profile measured by zero-phasing. The solid is for the electron bunch at zero phase with positive slope, i.e., the electron energy is chirped so that the head energy is reduced while the tail increased. The head is at the left side. For the dashed curve, the phase is shifted 180°, so the head gain energy, but we flipped the energy coordinate and shifted so the head is still at the left, and matched to the first curve to show the precision of the measurement. The ripples at the peak are due to small energy modulation, they do not represent spikes in time domain [10].

300 pC. As shown in Fig. 2, the current profile as measured using the zero-phasing method has a FWHM pulse length after compression of 1 ps with a current averaged within the pulse to be 300 pC/1 ps = 300/A. The measured projected normalized emittance after bunch compression is about 4.7 μm. The slice emittance is smaller, measured to be between 2.5 and 3.5 μm.

The trajectory in the NISUS undulator was corrected [19] using the trim dipoles to within 200 μm peak to peak about a straight line determined by referencing the pop-in monitors to a HeNe alignment laser. The beam size as measured from the pop-in monitors provides reliable data for matching the electron beam into

the NISUS (see Fig. 3), and for measurement of the projected emittance, with results exhibiting excellent agreement with the emittance as measured by a quadrupole scan. The projected energy spread is estimated to be 0.05% rms for the compressed bunch.

The 800 nm seed input is derived from the same chirped-pulse-amplified Ti:Sapphire laser system that drives the photocathode RF gun [20]. A separate compressor leaves a residual chirp in the 9 ps (FWHM) seed pulse, so that the bandwidth seen by the 1 ps electron bunch is only 0.8 nm. A few nanometer tuning range can thus be obtained by varying the delay of the seed pulse relative to the electron bunch. This also varies the seed power since the stretched pulse is not a flat-top.

The synchronization between the 1 ps electron bunch and the 9 ps seed laser was achieved first using a streak camera at the end station measuring both the seed at 800 nm and SASE at 266 nm. Later, it was improved by using the HGHG signal. During the HGHG operation, the dispersion magnet current and the electron beam energy are varied to optimize the output. When the dispersion magnet current is 110 A, the maximum excursion in the dispersion magnet is measured to be $x_m = 2.1$ mm by a monitor at its center, the dispersion [7] is found to be $d\psi/d\gamma = 32\pi x_m^2/(3s\lambda_s\gamma) \cong 5.4$, where ψ is the ponderomotive phase in the NISUS, $s = 30$ cm is the dispersion section length, $\lambda_s = 266$ nm, and $\gamma = 346$ the normalized beam energy. When the dispersion is optimized for maximum bunching, i.e., maximum initial coherent generation in the first part of NISUS, its value can be used to calculate the energy modulation. When combined with the measured seed laser Rayleigh range of 2.4 m and the modulator parameters, this in turn provides information about the laser intensity at its overlap with the electron bunch.

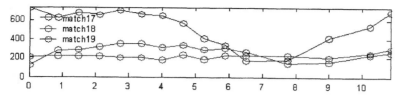

Fig. 3. Use pop-in monitors to find beam matrix and match into NISUS wiggler. The three sets of data of beam size measured along the wiggler sequentially improved the beam profile matching until the last one "match19" is nearly uniform with 200 μm beam size. The resulting Twiss parameters agree with quad scan results.

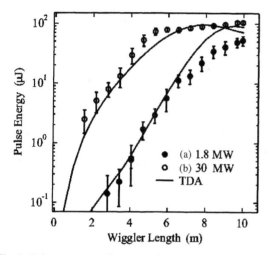

Fig. 4. Pulse energy vs. distance in the radiator for two values of the seed laser input power: (a) 1.8 MW and (b) 30 MW. The solid curves are simulation results by the TDA code.

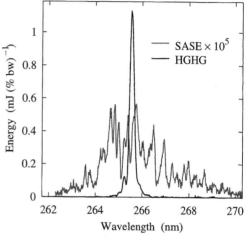

Fig. 5. Single shot HGHG spectrum for 30 MW seed power, exhibiting a 0.1% FWHM bandwidth. The grey line is the single shot SASE spectrum far from saturation when the 30 MW seed was removed. This spectrum serves as the background of the HGHG output. The average spacing between spikes is used to estimate the pulse length.

The output pulse energy versus distance for two different seed powers: (a) $P_{in} = 1.8$ MW and (b) 30 MW is presented in Fig. 4. The data was taken using a single downstream detector and sequentially kicking the beam away from the undulator axis using the 16 correctors uniformly distributed along NISUS. As a check, this data was compared with a set of measurements by 5 photodiode detectors installed on the side of NISUS. Gain lengths measured by these two methods agree. At $P_{in} = 1.8$ MW, with $d\psi/d\gamma = 8.7$, the gain length is found to be 0.8 m.

For $P_{in} = 30$ MW, $d\psi/d\gamma = 3$, the output single shot spectrum of HGHG is shown in Fig. 5, together with the single shot SASE spectrum when the seed was turned off are presented. The average spacing between the SASE spectral spikes is used

to estimate the pulse length [21] as $T_b = \dfrac{\lambda^2}{0.64c\Delta\lambda} = 0.9$ ps, about equal to the result of 1 ps electron bunch length obtained from the zero-phasing technique. Note that the HGHG spectral width of Fig. 5 is very nearly equal to the width of a single spike in the SASE spectrum. This is evidence of the very high temporal coherence of the HGHG output. The HGHG spectral brightness is 2×10^5 times larger than the SASE as shown in Fig. 5, because NISUS is too short to achieve SASE saturation. So this is not an appropriate comparison. If the NISUS length was doubled to 20 m, the SASE would reach saturation, but because of its broader bandwidth it would still have an order of magnitude lower brightness than the HGHG, as calculated by the code GENESIS [24] (see Fig. 6). A histogram of the shot-to-shot HGHG output pulse energy for a 30 MW seed is shown in Fig. 7 (lower right plot). The rms intensity fluctuation is

seen to be small, only 7%, mostly due to variation of the electron beam parameters. In Fig 7 it is visible that between about 14–20 s after counting started there was a minor glitch in the system parameters that contributes to the increased base size of the histogram, thus increases the fluctuation from 3.8% to 7%. The HGHG output energy can also exhibit a slow drift if the electron beam parameters change slowly. But we have observed the HGHG output to be stable, with typical output energy of 100 μJ Since the slippage of laser pulse relative to the electron bunch over the whole NISUS (256 periods long) is $256 \times 0.266\,\mu m/c \cong 200\,fs$ which is 5 times smaller than the 1 ps electron pulse length, the SASE fluctuation would be $1/\sqrt{5} \approx 44\%$ for an idealized electron beam.

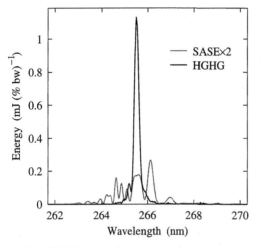

Fig. 6. The HGHG single shot spectrum versus a simulated saturated SASE spectrum for a NISUS wiggler with doubled length of 20 m to reach saturation.

4. Analysis

The time-independent approximation as used by the code "TDA" is valid, because the slippage is much smaller than the electron bunch length. Furthermore, as a rough approximation, we

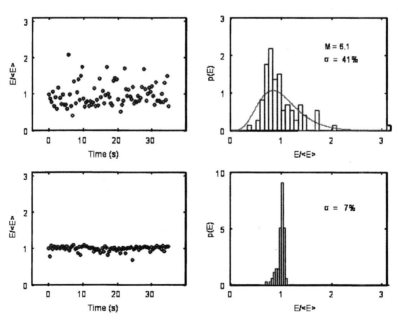

Fig. 7. Histogram of HGHG output pulse energy with 30 MW seed power (lower right) versus a SASE histogram (upper right) obtained under the same condition. The two plots at the left show the shot to shot fluctuation within about 40 s.

neglect the detailed time-structure of the electron bunch. When the seed power is low, as in the $P_{in} = 1.8$ MW data of Fig. 4a, there is a significant exponential growth. From the measured gain length of 0.8 m, we can estimate the electron beam parameters. Since the current is ~ 300 A, an analytical gain length calculation [22] indicates that the slice emittance is below 3 μm. Since the measured slice emittance is between 2.5 and 3.5 μm, the analytical solution also indicates the local rms energy spread should be smaller than the measured projected value of 510^{-4}. In fact, if we assume the local rms energy spread to be 110^{-4} and the emittance to be 2.7 μm, the simulation by a modified TDA code [7] reproduces the measured gain length of 0.8 m and predicts saturation at the end of the NISUS—as observed. The Pierce parameter [23] for this case is $= 310^{-3}$. Since the simulation predicts an output peak power of 130 MW and we measure the average pulse energy to be 60 μJ, we estimate the radiation pulse length in this case ($P_{in} = 1.8$ MW) to be 0.5 ps, t which is consistent with the pulse length we measured by an auto-correlation method (see Fig. 8). This suggests that only the part with highest peak current of the 1 ps electron bunch significantly contributes to the output. The simulated pulse energy vs. distance curve in Fig. 4a shows good agreement of the simulation with the measured data.

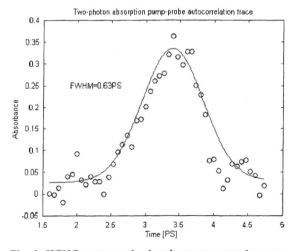

Fig. 8. HGHG output pulse length measurement by auto-correlation method in the case corresponding to Fig. 4a.

At the higher seed power of 30 MW, the auto-correlation of the 266 nm output indicates the pulse length is ~ 1 ps (in agreement with the SASE output pulse length of 0.9 ps estimated from the spacing of the spikes in Fig. 5), showing that in this case the whole electron bunch contributes. For the 30 MW seed, the coherent radiated energy in the initial part of the NISUS undulator is more than a factor 50 greater than that for the low power seed of 1.8 MW, and saturation is reached at 5 m after amplifying the initial coherent radiation by only a factor of 10. The fact that we need a gain of only 10 to bring the initial coherent emission to saturation shows that coherent radiation is the dominant feature in the HGHG process. This greatly reduces the required undulator length and makes the system much less sensitive to electron beam parameter variation. Hence the HGHG is very stable, as illustrated in Fig. 7. Analysis suggests that the apparent slow growth after the saturation at 5 m instead of the drop of power as indicated by the simulation is due to the fact that the whole bunch contributes to the output, and individual slices having different currents reach saturation at different rates. Since the whole bunch is contributing to the output, part of the beam is mismatched. To take this into account we do not use the slice emittance in the simulation, rather the emittance is approximated by the measured projected emittance of 4.7 μm. The rms energy spread is assumed to be 1×10^{-4}. The simulation results show reasonable agreement with the pulse energy vs. distance data shown in Fig. 4b.

The bandwidth within a 1 ps slice of the chirped seed is 0.8 nm (0.1% bandwidth) and the chirp in the HGHG output is expected to be the same, i.e., $0.1\% \times 266$ nm $= 0.26$ nm. This is in agreement with the measured FWHM bandwidth of 0.23 nm observed in Fig. 5. A Fourier-transform limited flat-top 1 ps pulse would have a bandwidth of 0.23 nm, while a FWHM 1ps Gaussian pulse would have bandwidth of 0.1 nm.

5. Summary

The coherent generation and the ensuing exponential growth of the HGHG light at 266 nm has been observed to be in agreement with theory. The output exhibits the predicted and designed high intensity stability and a nearly Fourier-transform limited bandwidth.

Acknowledgements

This work was supported by US Department of Energy, Office of Basic Energy Sciences, under Contract Nos. DE-AC02-98CH10886 and by Office of Naval Research Grant no. N00014-97-1-0845.

References

[1] S. Leone, Report of the BESAC Panel on Novel Coherent Light Sources, US Department of Energy, Washington, DC, 1999.

[2] J. Galayda, (Ed.), Linac Coherent Light Source Conceptual Design Report, SLAC-R-593, 2002.

[3] V. Ayvazyan, et al., Phys. Rev. Lett. 88 (2002) 104802.

[4] S.V. Milton, et al., Science 292 (2001) 2037.

[5] A. Tremaine, et al., Phys. Rev. Lett. 88 (2002) 204801.

[6] I. Ben-Zvi, L.F. Di Mauro, S. Krinsky, M. White, L.H. Yu, Nucl. Instr. and Meth. A 304 (1991) 181; L.H. Yu, Phys. Rev A 44 (1991) 5178.

[7] L.H. Yu, et al., Science 289 (2000) 932.

[8] L.H. Yu, A. Doyuran, L. DiMauro, W.S. Graves, E.D. Johnson, R. Heese, S. Krinsky, H. Loos, J.B. Murphy, G. Rakowsky, J. Rose, T. Shaftan, B. Sheehy, J. Skaritka, X.J. Wang, Z. Wu, Phys. Rev. Lett. 91 (2003) 074801.

[9] L.H. Yu, et al., The DUV-FEL development program, Proceedings of the PAC 2001, Chicago 2001, p. 2830.

[10] L. DiMauro, et al., Nucl. Instr. and Meth. A 507 (2003) 15.

[11] A. Doyuran, et al., Nucl. Instr. and Meth. A 507 (2003) 392.

[12] X. Liu, R.L. Gross, A.G. Suits, Science 294 (2001).

[13] Wen Li, A.G. Suits, manuscript in preparation (2003).

[14] T. Shaftan, et al., Bunch compression in the SDL linac. Proceedings of the EPAC2002, Paris, June 3–7, 2002, p. 834.

[15] D.X. Wang, G.A. Krafft, C.K. Sinclair, Phys. Rev. E 57 (1998) 2283.

[16] W.S. Graves, et al., Proceedings of 2001 Particle Accelerator Conference, Chicago, pp. 2224–2226.

[17] K.E. Robinson, et al., Nucl. Instr. and Meth. A 291 (1990) 394.

[18] W.S. Graves, E.D. Johnson, S. Ulc, AIP Conf. Proc. 451 (1998) 206.

[19] H. Loos, et al., Beam-based trajectory alignment in the NISUS Wiggler, Proceedings of EPAC 2002, Paris, June 3–7, 2002, p. 837.

[20] W. Graves, et al., Proceedings of 2001 Particle Accelerator Conference, Chicago, pp. 2860–2862.

[21] S. Krinsky, R.L. Gluckstern, Nucl. Instr. and Meth. A 483 (2002) 57.

[22] L.H. Yu, S. Krinsky, R.L. Gluckstern, Phys. Rev. Lett. 64 (1990) 3011.

[23] R. Bonifacio, C. Pellegrini, L.M. Narducci, Opt. Commun. 50 (1984) 373.

[24] S. Reiche, Nucl. Instr. and Meth. A 429 (1999) 243.

Available online at www.sciencedirect.com

SCIENCE  DIRECT®

ELSEVIER   Nuclear Instruments and Methods in Physics Research A 528 (2004) 443–447

**NUCLEAR
INSTRUMENTS
& METHODS
IN PHYSICS
RESEARCH**
Section A

www.elsevier.com/locate/nima

Evolution of transverse modes in a high-gain free-electron laser

S.G. Biedron[a,*], H.P. Freund[b], S.V. Milton[a,c], G. Dattoli[d],
A. Renieri[d], P.L. Ottaviani[e]

[a] *MAX-Laboratory, University of Lund, Lund, Sweden SE-22100*
[b] *Science Applications International Corporation, McLean, VA 22102, USA*
[c] *Advanced Photon Source, Argonne National Laboratory, Argonne, IL 60439, USA*
[d] *ENEA, Unite Tecnico Scientifica Tecnologie Fisiche Applicate, Centro Ricerche Frascati, C.P. 65, 00044 Frascati, Rome, Italy*
[e] *ENEA, Divisione Fisica Applicata,Centro Ricerche E. Clementel, Via Don Fiammelli 2, Bologna, Italy*

Abstract

At the point of saturation in a high-gain free-electron laser (FEL) the light is fully transversely coherent. The number and evolution of the transverse modes is important for the effective tune-up and subsequent operation of FELs based on the photon beam characterization and in designing multi-module devices that rely on relatively stable saturation distances in each module. In the latter, this is particularly critical since each section will seed another module. Overall, in a single- or multi-module device, experimental users will desire stability in power and in photon beam quality. Using a numerical simulation code, the evolution of the transverse modes in the high-gain free-electron laser (FEL) is examined and is discussed. In addition, the transverse modes in the first few higher nonlinear harmonics are investigated.
© 2004 Elsevier B.V. All rights reserved.

PACS: 41.60; 42.65.Ky; 52.59.-f

Keywords: Free-Electron Lasers; Harmonic Generation; Frequency Conversion; Intense Particle Beams and Radiation Sources; Coherence

1. Introduction

A full understanding of the transverse coherence in a high-gain (HG), single-pass (SP), free-electron laser (FEL) as a function of distance is essential. If the gain process progresses differently than that designed, e.g. the FEL saturates before or after the end of the undulator, the transverse coherence properties will be affected. Knowledge of any such deviation from that expected is important for the user who requires knowledge of the coherence correlated to their data. A further challenge is the determination and control of the transverse coherence of the nonlinear harmonics, which do not necessarily saturate at the same longitudinal position as the fundamental and could also be more susceptible than the fundamental to the electron beam emittance and transverse electron beam distribution Nonlinear harmonics, see for example Ref. [1]. To predict the evolution of the transverse coherence of the fundamental and

*Corresponding author. Tel.: 46-46-222-3069; fax: 46-46-222-4710.

E-mail address: sgbiedron@elementaero.com (S.G. Biedron).

nonlinear harmonics as a function of distance through a HG SP FEL, the simulation code MEDUSA [2] was modified to calculate and read out the intensity map for as many transverse modes as one requires to attain convergence of the transverse intensity profile. In this study, the development of the transverse intensity profile of two similar systems were studied: the SPARC project, managed by different Italian institutions [3] and the operational Low-Energy Undulator Test Line (LEUTL) at the Advanced Photon Source (APS) at Argonne National Laboratory (ANL) [4].

2. The simulation code—MEDUSA

MEDUSA is a 3D, multifrequency, macroparticle simulation code that represents the electromagnetic field as a superposition of Gauss–Hermite modes and uses a source-dependent expansion to determine the evolution of the optical mode radius. The field equations are integrated simultaneously with the 3D Lorentz force equations. MEDUSA differs from the other nonlinear simulation codes in that no undulator-period average is imposed on the electron dynamics. It is capable of treating quadrupole and corrector fields, magnet errors, and multiple segment undulators of various quantities and types. MEDUSA is able to treat the fundamental and all harmonics simultaneously. Because of these features, MEDUSA is capable of following the FEL evolution of the fundamental and all harmonics. Most recently, diagnostic output of the mode pattern as a function of axial position has been added to MEDUSA.

3. Cases under study

Using MEDUSA, the fundamental and nonlinear harmonics were simulated to investigate the evolution of the transverse modes and intensity profile in the theoretical cases of the SPARC project in Italy and the LEUTL at the APS at ANL. The parameters for the SPARC project are from an earlier iteration of the present design parameters. These SPARC and LEUTL parameters are listed in Table 1.

Table 1
Parameters of the SPARC Project in Italy and those of LEUTL at the APS ANL

Parameters	SPARC	LEUTL
Electron beam energy (MeV)	150	219.5
Normalized emittance (π mm-mrad)	2	5
Peak current (A)	150	150
Undulator period (m)	0.033	0.033
Undulator strength (K)	1.886	3.1
Energy spread (%)	0.1	0.1
Fundamental wavelength (nm)	529.6	518.8

4. Results

In both cases, the fundamental and the third and fifth harmonics were simulated and the mode structure was mapped out after each undulator segment in the conceptual design of the SPARC system and in the actual LEUTL system. A flat-pole-face wiggler model was used with quadrupole focusing between each undulator section. The fundamental was seeded with 10 W of optical power and the harmonics were started at zero power. Note that since the fundamental was seeded with 10 W of optical power, we ran MEDUSA in the amplifier mode of operation. The difference between SASE and amplifier operation in FELs is largely (1) the start-up from noise region, and (2) the spectral bandwidth. Once the start-up regime is passed, however, the fields grow exponentially as in an amplifier. Further, once saturation is reached in a SASE FEL, the spectral narrows about the wavelength of the fastest-growing mode. Hence, an amplifier model can give a reasonable simulation of a SASE FEL as long as the fastest-growing mode is used along with a good estimate of the start-up noise power. The self-amplified spontaneous emission (SASE) case, or start up from noise case, now requires investigation. Twenty five modes were used with a total of 34,992 particles.

4.1. SPARC

In the SPARC case, the powers of the fundamental, third harmonic, and fifth harmonic are

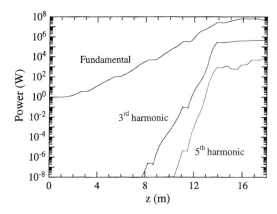

Fig. 1. Fundamental, 3rd harmonic, and 5th harmonic FEL power (W) growth as a function of distance, z (m), through the SPARC undulator system.

plotted in Fig. 1 as a function of distance through the HG SP FEL. Note the points of saturation vary for the wavelength (or harmonic), as evidenced in Fig. 1 as well as in Figs. 2 and 3 in greater detail. In these latter figures, the horizontal (x) and vertical (y) intensity cross the sections for $y = 0$ and $x = 0$, respectively, as a function of distance for the fundamental, third harmonic, and fifth harmonic are shown. The cross-sections are those following each undulator section (every 70 periods), the points at which MEDUSA was programmed to write out the extensive modal information. All cross-sections have been normalized to a peak intensity of 1 and offset in both x and y to allow easier viewing of the propagation along the length of the undulator. The fundamental saturation occurs between the 4th and 5th undulators, the third harmonic saturation occurs between the 5th and 6th undulators, and the fifth harmonic saturation occurs between the 6th and 7th undulators. To determine the exact positions of saturation of each wavelength, an extensive numerical undertaking is required with the code to read out many more modal maps in z. At the points of near saturation, the narrowing of the modes is clear and the mode narrowing increases as a function of harmonic number. Note that the maximum output power does not necessarily provide the "cleanest" mode structure—closest to TEM_{00}—for the FEL user, when comparing Fig. 1 with Figs. 2 and 3. In other words—it is insufficient to only examine the output

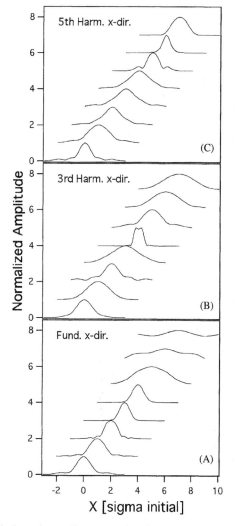

Fig. 2. Intensity profile cross-sections in the x-direction (wiggle plane) for $y = 0$. The lower left is following the first undulator and the profiles toward the 8th undulator is offset up and to the right for clarity. (A) Fundamental wavelength (532.1 nm), (B) 3rd harmonic (177.4 nm), and (3) 5th harmonic (106.4 nm).

power as a function of distance in planning for future user facilities that intend to employ nonlinear harmonic FEL radiation—the beam physicist must *also* examine the mode structure of each wavelength. Please also note that when designing and optimizing an FEL, a much higher resolution of the transverse mode development for the fundamental and harmonics must be simulated. In this paper, however, we intend only to outline the importance of investigating the evolution of

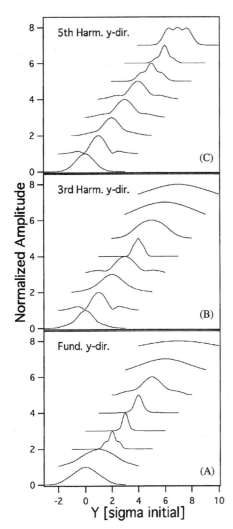

Fig. 3. Intensity profile cross-sections in the *y*-direction (opposite to the wiggle plane) for *x* = 0. The lower left is following the first undulator and the profiles toward the 8th undulator is offset up and to the right for clarity. (A) Fundamental wavelength (532.1 nm), (B) 3rd harmonic (177.4 nm), and (C) 5th harmonic (106.4 nm).

the fundamental and harmonics for the design of future user facilities.

Saturation of the fundamental occurs near the end of the 5th undulator. In both the *x* and *y* directions the cross-sections tend toward a true TEM$_{00}$ mode as the intensity grows toward saturation and beyond. At saturation, this mode size begins to grow since the so-called gain guiding is no longer effective. Also following saturation, additional modes tend to gain

on the TEM$_{00}$ mode and the intensity profile loses its strictly Gaussian shape.

The intensity profile of the harmonics behave differently due to the shorter gain length. There seems to be no true mode that the profile settles down to. There are differences in *x* and *y*. These nonlinear harmonics are very much a product of the fundamental interaction and are entirely dependent upon it. Furthermore, although the beam emittance is below the diffraction limit in the case of the fundamental this is not necessarily the case for the harmonics. The 5th harmonic, for example, does not meet the diffraction limit criteria of $\varepsilon < \lambda/4\pi$. The harmonics can be there-fore, much more susceptible to the actual electron beam distribution.

4.2. LEUTL

Similar results were obtained for the LEUTL case. The fundamental transverse intensity dis-tribution tends toward a true TEM$_{00}$ mode at saturation, Fig. 4, and begins to deviate from this beyond saturation. The harmonics, as in the case of the SPARC, have a very rich mode content as shown in Fig. 5; the third harmonic beyond the saturation point in the LEUTL case.

5. Conclusions

We have added diagnostics to the MEDUSA simulation code in order to explicitly study the evolution of the transverse amplitude profile in a future light source, and have applied the code to study this evolution in the SPARC and LEUTL projects. This approach can be used to determine the profile evolution and transverse coherence in future light sources. The transverse mode evolu-tion of the fundamental and harmonics vary with longitudinal position and differ based upon specific electron beam and undulator properties. In the case of an operational facility, the transverse coherence and intensity profile at the exit of the undulator line will be important for some of the user experiments. Small variations in electron beam quality may lead to unwanted variations shot-to-shot. This is particularly true for the

Fundamental after 6th Wiggler

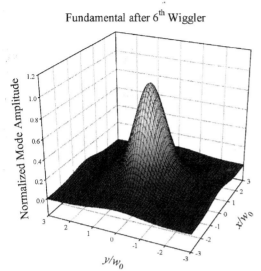

Fig. 4. LEUTL fundamental signal after saturation.

3rd Harmonic after 6th Wiggler

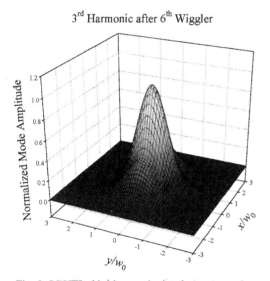

Fig. 5. LEUTL third harmonic signal near saturation.

ous more longitudinal positions, with start-up from noise to simulate SASE operation, and to simulate the far-field behaviour of the radiation. These three extensions will greatly benefit the future FEL users. It is hoped that a comparison can be made with experiment in the near future after these additional goals are achieved.

Acknowledgements

The activity and computational work for H. P. Freund is supported by Science Applications International Corporation's Advanced Technology Group under IR&D subproject 01-0060-73-0890-000. The work of S.V. Milton is supported at Argonne National Laboratory by the US Department of Energy, Office of Basic Energy Sciences under Contract No. W-31-109-ENG-38. The work of G. Dattoli, P.L. Ottaviani, and A. Renieri is supported by ENEA the Italian Agency for New Technologies, Energy, and the Environment.

We wish to thank our colleagues and collaborators for encouraging us to pursue these transverse mode efforts, albeit in our spare time: B. Faatz, R. Bartolini, S. Benson, F. Ciocci, R. Dejus, W.M. Fawley, G. Felici, L. Gianessi, Z. Huang, K-.J. Kim, J. Lewellen, Y. Li, A. Lumpkin, G. Neil, and J. Rocca.

harmonics and even partly true for the fundamental beyond saturation. These effects may be more pronounced as the wavelengths get shorter and the electron beam emittance approaches or exceeds the optical mode emittance such as in an X-ray SASE source.

The next step in the investigation of the transverse modes is to examine a system more thoroughly by analyzing the evolution at numer-

References

[1] S.G. Biedron, Z. Huang, K.-J. Kim, S.V. Milton, G. Dattoli, A. Renieri, W.M. Fawley, H.P. Freund, H.-D. Nuhn, P.L. Ottaviani, Phys. Rev. St. Accel. Beams 5 (2002) 030701 and references therein.

[2] H.P. Freund, T.M. Antonsen Jr.,, Principles of Free-electron Lasers, 2nd Edition, Chapman & Hall, London, 1986;
H.P. Freund, Phys. Rev. E 52 (1995) 5401;
S.G. Biedron, H.P. Freund, S.V. Milton, Development of a 3D FEL code for the simulation of a high-gain harmonic generation experiment, in: H.E. Bennett, D.H. Dowell (Eds.), Free Electron Laser Challenges II, Proceedings of SPIE, Vol. 3614, 1999, p. 96;
H.P. Fruend, et al., IEEE J. Quant. Electron. 36 (2000) 275.

[3] A. Renieri, Nucl. Instr. and Meth. A 507 (2003) 507.

[4] S.V. Milton, et al., Science. 292 (5524) (2001) 2037. (Originally published in Science Express as 10.1126/science.1059955 on May 17, 2001).

Available online at www.sciencedirect.com

SCIENCE @ DIRECT°

ELSEVIER Nuclear Instruments and Methods in Physics Research A 528 (2004) 448–452

NUCLEAR INSTRUMENTS & METHODS IN PHYSICS RESEARCH
Section A

www.elsevier.com/locate/nima

Start-to-end simulations of SASE FEL at the TESLA Test Facility, Phase I: comparison with experimental results

M. Dohlus[a], K. Flöttmann[a], O.S. Kozlov[b], T. Limberg[a], Ph. Piot[c], E.L. Saldin[a], E.A. Schneidmiller[a],*, M.V. Yurkov[b]

[a] Deutsches Elektronen-Synchrotron (DESY), Notkestrasse 85, Hamburg 22607, Germany
[b] Joint Institute for Nuclear Research, Dubna, 141980 Moscow Region, Russia
[c] Fermilab, Batavia, IL 60510, USA

Abstract

VUV SASE FEL at the TESLA Test Facility (Phase 1) was successfully running and reached saturation in the wavelength range 80–120 nm. We present a posteriori start-to-end simulations of this machine. The codes Astra and elegant are used to track particle distribution from the cathode to the undulator entrance. An independent simulation of the beam dynamics in the bunch compressor is performed with the code CSRtrack. SASE FEL process is simulated with the code FAST. Simulation results are in a good agreement with the measured properties of SASE FEL radiation.
© 2004 Elsevier B.V. All rights reserved.

PACS: 41.60.Cr; 42.55.Vc; 41.75.Ht; 41.85.Ew

Keywords: Free electron lasers; X-ray lasers; Relativistic electron beams

1. Introduction

SASE FEL at the TESLA Test Facility (TTF), Phase I reached saturation in the wavelength range 80–120 nm and produced GW level radiation pulses with a duration of 30–100 fs [1]. Previous analysis of the experiment used a simplified model of the electron bunch (the lasing part of the bunch was approximated by Gaussian) [1]. Although that

model allowed us to describe main properties of the radiation, some important features were not understood well. In this paper we perform comprehensive studies of the beam dynamics (including, for instance, space charge effects after compression), as well as SASE FEL simulations. We compare simulation results with a set of experimental data taken at similar tuning of TTF FEL in a femtosecond mode of operation and published earlier in Refs. [1,2]. Not all of these characteristics were measured at the same time, so that the electron beam parameters may be slightly different for different measurements. For start-to-end simulations we took settings for one of typical

---

*Corresponding author. Tel.: +49-40-8998-2676; fax: +49-40-8998-4475.

E-mail address: schneidm@mail.desy.de (E.A. Schneidmiller).

0168-9002/$ - see front matter © 2004 Elsevier B.V. All rights reserved.
doi:10.1016/j.nima.2004.04.129

accelerator tuning (settings for magnetic elements and RF system from logbook), and found very good agreement with experimental results.

2. Beam dynamics simulations

The description of TTF accelerator, operating under standard lasing conditions, can be found in Ref. [1]. The details of beam dynamics simulations are presented in Ref. [3]. Start-to-end simulation of the beam dynamics were performed in the following way. The initial part of the machine, from the cathode to the bunch compressor entrance, was simulated with Astra [4]. Then the beam was tracked through the bunch compressor with the code elegant taking into account a simplified model of coherent synchrotron radiation (CSR) wake [5]. The bunch compression at TTF was strongly nonlinear, resulting in "banana-like" shape of the longitudinal phase space. In time domain we obtained a short, high-current leading peak, and a long, low-current tail. In order to check that we did not miss any important CSR-related effect, we performed independent simulations of the bunch compressor with a newly

developed code CSRtrack [6]. The main results were in agreement with the results of elegant. From the bunch compressor exit to the undulator entrance we used Astra again, because the space charge effects are very important for a short, high-current leading peak. A part of the longitudinal phase space at the undulator entrance as well as the slice parameters in the head of the bunch, that were used in SASE FEL simulations are presented in Fig. 1. One can notice a strong energy chirp $\Delta\gamma$ induced by the longitudinal space charge field on the way from the bunch compressor to the undulator

$$\frac{\mathrm{d}(\Delta\gamma)}{\mathrm{d}z} \simeq \frac{I}{I_A} \frac{\ln(\gamma\sigma_z/\sigma_r)}{\gamma^2\sigma_z}.$$

Here I is peak current, I_A is Alfven's current, σ_z and σ_r are the longitudinal and transverse size of the spike.

3. SASE FEL simulations

SASE FEL simulations were performed with three-dimensional time-dependent FEL code FAST [7]. Since SASE FEL radiation is a

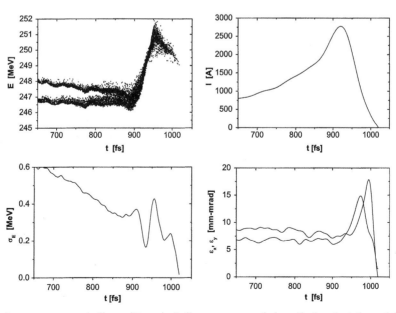

Fig. 1. Longitudinal phase space, current, slice emittance and slice energy spread along the bunch at the undulator entrance. Bunch head is at the right side.

statistical object (due to the start up from shot noise), we performed 200 statistically independent runs with code FAST.

In Fig. 2 we present average energy in the radiation pulse and the rms energy fluctuations versus position in the undulator. One can notice a reasonable agreement between measurement and simulations.

In Fig. 3 we present measured [8] and simulated averaged spectrum of SASE FEL operating in linear regime. One can see a little bump on the left—it is due to the energy chirp in the electron bunch.

Single-shot spectra as well as the averaged one are shown in Fig. 4 for the saturation regime [1]. The agreement between simulations and measurements is quite good.

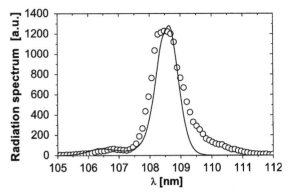

Fig. 3. Average spectrum of TTF FEL operating in the high-gain linear regime. Circles are experimental data [8], and curve represents results of simulations with code FAST.

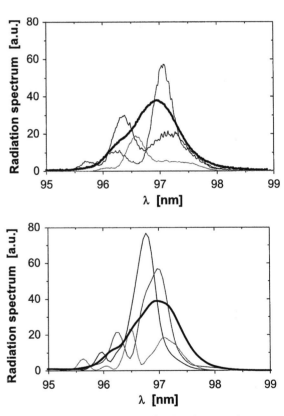

Fig. 4. Single shot spectra (thin lines) and averaged spectrum (bold lines) of TTF FEL operating in the nonlinear regime. Upper plot shows experimental data [1] and lower plot shows results of simulations with code FAST.

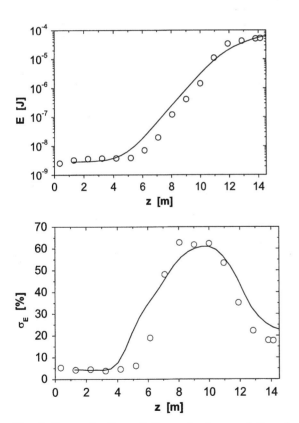

Fig. 2. Energy in the radiation pulse (upper plot) and fluctuations of the energy in the radiation pulse (lower plot) versus undulator length. Circles are experimental data [1], and curves represent results of simulations with code FAST.

A comparison between measured [1] and simulated angular distribution of the radiation intensity in far zone is presented in Fig. 5, and in Fig. 6 we

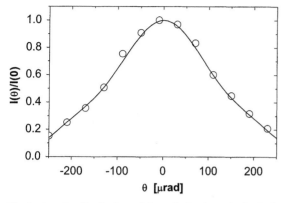

Fig. 5. Angular distribution of the radiation intensity in the far zone. TTF FEL operates in the nonlinear regime. Circles are experimental data [1], and curves represent results of simulations with code FAST.

compare probability distributions of the total energy in the radiation pulse and after narrow band monochromator. TTF FEL operates in the nonlinear regime. Note that these are relative characteristics independent of absolute calibration of the radiation detector. Perfect agreement between experimental and simulated results is a strong argument in favor of correct simulations.

4. Summary

Detailed analysis of beam dynamics leads us to the conclusion that TTF FEL, Phase I was driven by strongly non-Gaussian bunch with short leading peak having current of about 3 kA. Space charge is the main physical effect for beam dynamics after the bunch compressor: a large value of energy chirp in the leading spike is gained in a long drift spaces. Good agreement between experimental and simulation results is an encouraging message that physical models realized in codes ASTRA-elegant-FAST do not miss important physical effects, at least for parameter range of TTF FEL, Phase I. This allows us to determine the parameters of the SASE FEL which are not directly accessible to measurement. First of all this refers to the temporal structure of the radiation pulse (see Fig. 7): the computed FWHM pulse duration is about 40 fs, and averaged peak power is about 1.5 GW.

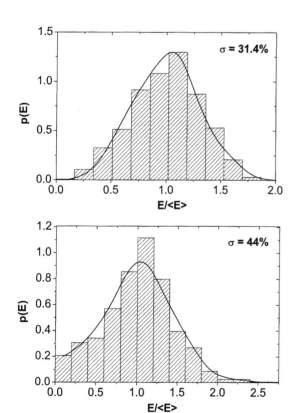

Fig. 6. Probability distributions of the total energy in the radiation pulse (upper plot) and after narrow band monochromator (lower plot) [2]. TTF FEL operates in the nonlinear regime. Solid lines represent simulations with code FAST.

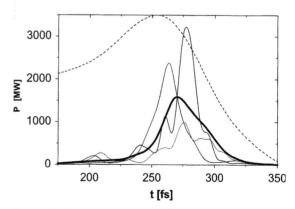

Fig. 7. Radiation power along the bunch for TTF FEL operating in the nonlinear regime. Thin curves: single shots; bold curve: averaged profile; dashed curve: profile of electron bunch. Simulations are performed with code FAST.

References

[1] V. Ayvazayn, et al., Phys. Rev. Lett. 88 (2002) 104802;
V. Ayvazayn, et al., Eur. Phys. J. D 20 (2002) 149.

[2] V. Ayvazayn, et al., Nucl. Instr. and Meth. A 507 (2003) 368.

[3] M. Dohlus, et al., Start-to-end simulations of SASE FEL at the TESLA test facility, Phase 1, Preprint DESY 03-197, November 2003

[4] K. Flöttmann, Astra User Manual, http://www.desy.de/mpyflo/Astra_dokumentation/.

[5] M. Borland, Elegant: a flexible SDDS-compilant code for Accelerator Simulations, Preprint APS LS-287, September 2000.

[6] M. Dohlus, private communication.

[7] E.L. Saldin, E.A. Schneidmiller, M.V. Yurkov, Nucl. Instr. and Meth. A 429 (1999) 233.

[8] J. Andruszkow, et al., Phys. Rev. Lett. 85 (2000) 3825.

Available online at www.sciencedirect.com

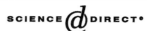

NUCLEAR
INSTRUMENTS
& METHODS
IN PHYSICS
RESEARCH
Section A

ELSEVIER   Nuclear Instruments and Methods in Physics Research A 528 (2004) 453–457

www.elsevier.com/locate/nima

Two-color FEL amplifier for femtosecond-resolution pump-probe experiments with GW-scale X-ray and optical pulses

J. Feldhaus[a], M. Körfer[a], T. Möller[a], J. Pflüger[a], E.L. Saldin[a,*],
E.A. Schneidmiller[a], S. Schreiber[a], M.V. Yurkov[b]

[a] *Deutsches Elektronen-Synchrotron (DESY), Notkestrasse 85, Hamburg 22607, Germany*
[b] *Joint Institute for Nuclear Research, Dubna 141980, Moskow Region, Russia*

Abstract

The paper describes a scheme for pump-probe experiments that could be performed at the soft X-ray SASE FEL at the TESLA Test Facility (TTF) at DESY and determines what additional hardware developments will be required to bring these experiments to fruition. Pump-probe experiments combining pulses from a XFEL and optical femtosecond laser are very attractive for sub-picosecond time-resolved studies. Since the synchronization between the two light sources to an accuracy of 100 fs is not yet solved, it is proposed to derive both femtosecond radiation pulses from the same electron bunch but from two insertion devices. This eliminates the need for synchronization and developing tunable, high power femtosecond quantum laser. In the proposed scheme for pump-probe experiments, GW-level soft X-ray pulse is naturally synchronized with his GW-level optical pulse and cancel jitter. The concept is based on generation of the optical radiation in the master oscillator-power FEL amplifier configuration. An attractive feature of the FEL amplifier scheme is the absence of limitation which would prevent operation in the femtosecond regime in a wide (200–900 nm) wavelength range. The problem of tunable quantum seed laser can be solved with commercially available long pulse dye laser. An important feature of the proposed scheme is that optical radiator uses the spent electron beam. As a result, saturation mode of operation of the optical FEL does not interfere with the main mode of the soft X-ray SASE FEL operation.
© 2004 Elsevier B.V. All rights reserved.

PACS: 41.60.Cr; 42.55.Vc; 41.75.Ht; 41.85.Ew

Keywords: Free electron lasers; X-ray lasers; Relativistic electron beams

1. Introduction

Time-resolved experiments are used to monitor time-dependent phenomena. In a typical pump-probe experiment a short probe pulse follows a short pump pulse at some specified delay.

*Corresponding author. Tel.: +49-40-8998-2676; fax: +49-40-8998-4475.

E-mail address: saldin@mail.desy.de (E.L. Saldin).

0168-9002/$ - see front matter © 2004 Elsevier B.V. All rights reserved.
doi:10.1016/j.nima.2004.04.130

Femtosecond capabilities have been available for some years at visible wavelengths. However, there is a strong interest in extending these techniques to X-ray wavelengths because they allow to probe directly structural changes with atomic resolution.

Recent progress in free electron laser (FEL) techniques have paved the way for the production of GW-level, sub-100 fs, coherent X-ray pulses. This has recently been demonstrated experimentally at the TESLA Test Facility (TTF) at DESY [1], although only at vacuum ultraviolet (VUV) wavelengths down to 80 nm. First user experiments have led to exciting results [2]. A SASE X-ray FEL will be commissioned as a user facility in 2004 at DESY, covering the VUV and soft X-ray range down to 6 nm wavelength. The short-wavelength, GW-level radiation pulses with sub-100 fs duration are particularly interesting for time-resolved studies of transient structures of matter on the time scale of chemical reactions.

One of the main technical problems for the pump-probe facility, combining X-ray SASE FEL and conventional optical lasers, is the synchronization between two radiation pulses. A new method proposed in this paper is an attempt to get around this obstacle by using a two step FEL process, in which two different frequencies (colors) are generated by the same femtosecond electron bunch. It has the further advantage to make a wide frequency range accessible at high peak power and high repetition rates not so easily available from conventional lasers.

2. A two-color FEL amplifier concept

The concept of the proposal is schematically illustrated in Fig. 1. Two different frequencies (colors) are generated by the same electron bunch, but in different insertion devices. The optical radiation is generated in a master oscillator-power FEL amplifier (MOPA) configuration. The X-ray radiation is generated in a X-ray undulator inserted between modulator and radiator sections of the optical MOPA scheme. The scheme operates as follows: the electron beam and the optical pulse from the seed laser enter the modulator radiator. Due to the FEL process the electron bunch gains energy modulation at the optical frequency which is then transformed to a density modulation in the dispersion section. The density modulation exiting the modulator (i.e. the energy-modulation undulator and the dispersion section) is about 10–20%. Thus, the optical seeding signal is imprinted in the electron bunch. Then the electron bunch is directed to the X-ray undulator. The process of amplification of the radiation in the main (soft X-ray) undulator develops in the same way as in the conventional SASE FEL: fluctuations of the electron beam current density serve as input signal.

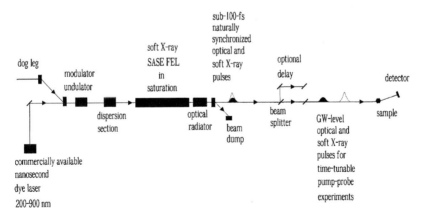

Fig. 1. Scheme for pump-probe experiments employing an optical pulse as a pump and a soft X-ray pulse as a probe or vice versa. A very long laser pulse is used for modulation of the energy and density of the electrons at the optical frequency. Optical photons for pump-probe experiments are generated by an additional insertion device (optical radiator) using the same electron bunch.

At the chosen level of density modulation the SASE process develops nearly in the same way as with an unmodulated electron beam because of the large ratio of the cooperation length to the optical wavelength [3]. As a result, at the exit of the X-ray undulator the electron bunch produces a GW-level X-ray pulse. A GW-level optical pulse is then produced when the electron bunch passes the optical radiator. The optical radiator is a conventional FEL amplifier seeded by the density modulation in the electron bunch. Although the electron beam leaving the soft X-ray FEL has acquired some additional energy spread, it is still a good "active medium" for an optical radiator at the end. Approximately 20% of density modulation is sufficient to drive the optical FEL amplifier in the nonlinear regime and to produce GW-level optical pulses in a short undulator. An important feature of the proposed scheme is that the optical radiator uses the spent electron beam. As a result, the optical FEL can operate in saturation mode without interfering with the soft X-ray SASE FEL operation.

3. Parameters of the two-color FEL amplifier at the TESLA Test Facility

We illustrate the two-color FEL amplifier scheme for the parameters of the TESLA Test Facility at DESY [4]. The technical details of the two-color amplifier can be found in [5]. A commercially available dye laser, for example, can be used as a seed laser. The typical pulse energy of a dye laser system is in the range of 2 to 10 mJ with a pulse duration of 5–10 ns, and the peak power is in the range of 1 MW which gives us sufficient safety margin for operation of the modulator. The viewport for a seed laser has been already foreseen [4]. Also it is very important, that there is free space upstream and downstream of the main undulator available for the optical undulators. Both modulator and radiator undulators are identical tunable-gap devices similar to those used at DORIS. The dispersion section is composed of four standard bending magnets similar to those used at the HERA storage ring. Therefore, this optical radiation source could be

realized at the TTF rather quickly and with minimum cost expenses.

A detailed description of the physical processes in both optical and X-ray undulators can be found in [5]. The parameters of the facility have been optimized using the time-dependent FEL simulation code FAST [6]. Two modes of TTF operation were analyzed in [5]: the femtosecond mode and the short wavelength mode. We describe here the femtosecond mode of which feature is that both optical and X-ray radiation are produced by a sharp current spike (with the duration of the order of 100 fs) at the head of the electron bunch. The bunch profile at the entrance of the undulator is shown as a solid line in Fig. 2. The peak current is about 2 kA, and the normalized emittance in the leading spike is about 7π mm-mrad in our simulations. An important feature is a pronounced decrease of the local energy spread in the head of the bunch [5].

The performance of the two-color facility is illustrated for an optical wavelength of 400 nm and an X-ray wavelength of 30 nm. The seed laser pulse interacts with the electron beam in the 4.5 m long modulator undulator and produces an energy modulation of about 100 keV in the electron bunch. The amplitude of the induced beam modulation is less than a percent. Then the electron beam passes through the dispersion

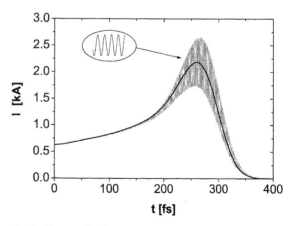

Fig. 2. Current distribution along the bunch after the dispersion section. The bunch is modulated with a period equal to the optical wavelength. The solid line shows the bunch profile at the entrance to the optical modulator.

section where the energy modulation is converted to a density modulation (20%) at the optical wavelength. The grey line in Fig. 2 shows the bunch profile after the dispersion section. The beam modulation is non-uniform due to the strongly varying energy spread along the bunch.

Upon leaving the dispersion section, the electron beam passes the X-ray undulator. Although the electron bunch density is strongly modulated (see Fig. 2), the SASE FEL process in the X-ray undulator remains almost the same as for an unmodulated electron beam. Of course, the output radiation has contents of the sideband harmonic [3], but its contribution to the total radiation energy is tiny, of the order of 10^{-4}. Fig. 3 shows the time structure of the radiation pulse.

The electron bunch leaving the X-ray undulator has a large induced energy spread of about 1 MeV but is still "cold" enough for the generation of optical radiation. Since the bunch is strongly modulated at the optical wavelength λ_{opt}, it readily starts to produce powerful optical radiation when it enters the optical radiator resonant at λ_{opt}. The temporal structure of the optical pulse at the exit of the optical radiator (at $z = 4.5$ m) is shown in Fig. 4. It should be noted that the optical pulse is completely coherent.

The main parameters of the radiation pulses for the femtosecond mode are summarized in Table 1. Note that in a short wavelength mode of operation

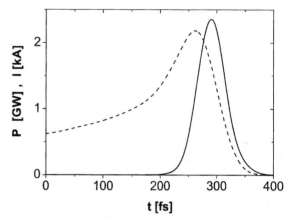

Fig. 4. Time structure of radiation pulse at the exit of optical radiator for a radiation wavelength of 400 nm. The dashed line indicates the profile of electron bunch.

Table 1
Properties of the radiation pulses of the two-color pump-probe facility

Optical pulse	
Wavelength	200–900 nm
Pulse energy	0.1–0.5 mJ
Pulse duration (FWHM)	30–100 fs
Peak power	2 GW
Spectrum width	Fourier-limited
Spot size (FWHM)	150–200 μm
Angular divergence (FWHM)	100–500 μrad
Repetition rate	10 (10^4) Hz
X-ray SASE pulse	
Wavelength	30–120 nm
Pulse energy	0.1 mJ
Pulse duration (FWHM)	30–100 fs
Peak power	2 GW
Spectrum width	0.4–0.6%
Spot size (FWHM)	350–1400 μm
Angular divergence (FWHM)	40–150 μrad
Repetition rate	10 (10^4) Hz

the wavelength of X-ray pulses will be extended down to 6 nm [4,5].

4. Conclusion

A novel two-color FEL amplifier for pump-probe experiments has been described combining sub-100 fs optical and X-ray pulses. The proposed

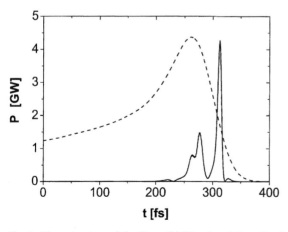

Fig. 3. Time structure of the X-ray SASE pulse at the exit of the undulator at a radiation wavelength of 30 nm. The dashed line shows the bunch profile.

facility has unique features: both pulses have very high peak power in the GW range. The wavelengths of both radiation sources are continuously tunable in a wide range: 200–900 nm for the optical pulses, and 6–120 nm for the X-ray pulses. Both pulses have diffraction limited angular divergence. The spectral width of the optical pulse is transform limited. Finally and most important, optical and X-ray pulses are precisely synchronized at a femtosecond level, since they both are produced by the same electron bunch, and there are no reasons for any time jitter between the pulses. Based on these unique features a pump-probe facility could be built with unique possibi-lities for studying time-dependent processes on the time scale of chemical reactions.

References

[1] V. Ayvazyan, et al., Phys. Rev. Lett. 88 (2002) 10482.
[2] H. Wabnitz, et al., Nature 420 (2002) 482.
[3] E.L. Saldin, E.A. Schneidmiller, M.V. Yurkov, Opt. Commun. 205 (2002) 385.
[4] SASE FEL at the TESLA Test Facility, Phase2, preprint TESLA-FEL 2002-01.
[5] J. Feldhaus, et al., preprint DESY 03-091.
[6] E.L. Saldin, E.A. Schneidmiller, M.V. Yurkov, Nucl. Instr. and Meth. A 429 (1999) 233.

Available online at www.sciencedirect.com

SCIENCE @ DIRECT°

Nuclear Instruments and Methods in Physics Research A 528 (2004) 458–462

NUCLEAR INSTRUMENTS & METHODS IN PHYSICS RESEARCH
Section A

www.elsevier.com/locate/nima

Femtosecond X-ray pulses from a spatially chirped electron bunch in a SASE FEL

P. Emma, Z. Huang*

Stanford Linear Accelerator Center, Stanford University, Stanford, CA 94309, USA

Abstract

We propose a simple method to produce short X-ray pulses using a spatially chirped electron bunch in a SASE FEL. The spatial chirp is generated using an RF deflector which produces a transverse offset (in y and/or y') correlated with the longitudinal bunch position. Since the FEL gain is very sensitive to an initial offset in the transverse phase space at the entrance of the undulator, only a small portion of the electron bunch with relatively small transverse offset will interact significantly with the radiation, resulting in an X-ray pulse length much shorter than the electron bunch length. The X-ray pulse is also naturally phase locked to the RF deflector and so allows high precision timing synchronization. We discuss the generation and transport of such a spatially chirped electron beam and show that tens of femotsecond long pulse can be generated for the linac coherent light source.
© 2004 Elsevier B.V. All rights reserved.

PACS: 41.60.cr

Keywords: Femtosecond X-ray; Spatially chirped electron bunch; Self-amplified spontaneous emission

1. Introduction

There is presently a great deal of interest in producing very short duration photon pulses at the femtosecond time scale for future X-ray free-electron lasers (FEL) based on self-amplified spontaneous emission (SASE). In typical SASE FEL designs the photon pulse is similar in duration to the electron bunch length, which is usually limited to 100–200 fs due to short-bunch collective effects in the accelerator. One possible method to shorten the photon

pulse with respect to the electron bunch is to suppress the FEL gain for most of the bunch by somehow degrading or distorting the electron bunch quality over all but a very short duration.

To this end, we note that the FEL gain can be suppressed when an electron beam has an initial offset in the transverse coordinates x, y, x' or y', at the entrance of the undulator. Thus, by giving the electron beam a transverse offset (spatial or angular) correlated with the longitudinal bunch position, a shorter X-ray pulse can be generated because only the portion of the electron bunch with little transverse offset contributes significantly to the FEL process.

---
*Corresponding author.
E-mail address: zrh@slac.stanford.edu (Z. Huang).

0168-9002/$ - see front matter © 2004 Elsevier B.V. All rights reserved.
doi:10.1016/j.nima.2004.04.131

Such a spatially (or angularly) chirped electron bunch may be produced by a transverse RF deflecting cavity or by residual dispersion and time-correlated energy spread at the exit of a bunch compressor. For the linac coherent light source (LCLS) [1], transverse RF deflecting cavities are incorporated in the baseline design, hence no additional hardware is required for this short-pulse scheme. In the following section, we discuss the generation and transport of such a beam and show that a 30-femtosecond long pulse (fwhm) can be generated for the LCLS.

2. Method

2.1. General description

A transverse RF deflecting structure is included in the LCLS design to provide absolute bunch length and slice emittance measurements [2]. The deflector is oriented such that it kicks (pitches) the bunch in the vertical direction. Near the zero-crossing RF phase, the head of the bunch is kicked (for example) in the positive vertical direction ($+y$) and the tail of the bunch in the negative vertical direction ($-y$). This produces a spatially chirped electron bunch which is immediately dumped onto an off-axis screen using a horizontal kicker magnet. The screen image is used to analyze the vertical beam size (i.e., the bunch length) and the horizontal beam size at various vertical positions (i.e., the horizontal slice emittance). If the kicker magnet is switched off, the spatially chirped bunch will propagate on axis through the accelerator (for reasonable chirp levels) and pass through the undulator. Since the head and tail of the bunch have large vertical offsets, only the central core of the beam will continually interact with the radiation field and produce exponential FEL gain. This short central core will generate a short X-ray pulse with duration set by the level of spatial chirp.

2.2. RF deflector

As a numerical example, Fig. 1 shows that a 1-m long, 5-MV (10 MW) vertical S-band RF deflector (2856 MHz) prior to the second bunch compressor

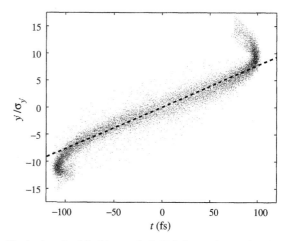

Fig. 1. Angular (y') chirp on the LCLS electron bunch, in units of rms angular beam size ($\sigma_{y'}$), at the undulator entrance (14.3 GeV), induced by a 5-MV S-band RF deflector prior BC2.

(BC2), at 4.5 GeV, generates about $\pm 10\,\sigma_{y'}$ chirp over the 210-fs bunch length at the LCLS undulator entrance. In this case the RF deflector was placed at a vertical betatron phase advance of $\Delta\psi_y \approx n\pi$ ($n = 1, 2, 3,...$) with respect to the undulator entrance, resulting in an initial angular (y') chirp and no significant spatial (y) chirp. It is also possible to set $\Delta\psi_y \approx (2n + 1)\pi/2$, by relocating the deflector or adjusting the linac phase advance, to get a pure spatial chirp. FEL simulations indicate that the output power is similarly sensitive to spatial chirp at the undulator entrance as it is to angular chirp (when each is normalized to its un-chirped rms beam size). Therefore, a similar pulse length will be produced, for the same normalized chirp, with $\Delta\psi_y \approx (2n + 1)\pi/2$, as compared to $\Delta\psi_y \approx n\pi$.

The RF deflection (in units of rms angular beam size) is given by

$$\frac{y'}{\sigma_{y'}} = \sqrt{\frac{\beta_y}{\gamma_0 \varepsilon_N} \frac{eV_0}{mc^2}} \cos\varphi \cdot \omega t \qquad (1)$$

where β_y (40 m) is the vertical beta function in the RF deflector, γ_0 (8880) is the electron energy in the deflector (in rest mass units), ε_N (1 μm) is the normalized vertical emittance, e is the electron charge, V_0 (5 MV) is the RF deflector voltage at crest phase, mc^2 is the electron rest mass, φ is the

RF phase (zero-crossing at $\varphi = 0$), and ω is the RF frequency ($2\pi \cdot 2856$ MHz).

The deflector is placed prior to the BC2 bunch compressor where the bunch length is longer than in the undulator (1.8 ps fwhm). Setting $t = (1.8\,\text{ps})/2$, as the half width, the normalized maximum chirp amplitude from Eq. (1) is ~ 10, which is the case of Fig. 1. Note that the BC2 bunch compressor then compresses the bunch length to 210 fs fwhm, but does not alter the vertical chirp amplitude. (The curved head and tail sections in Fig. 1 are a normal feature of the LCLS bunch and are produced by the linac longitudinal wakefields).

This tracking example also includes the transverse linac wakefield, which (for this well aligned linac example) does not appreciably alter the spatial chirp.

2.3. FEL interaction

Three-dimensional SASE code *GENESIS 1.3* [3] is used to study the FEL interaction of the spatially chirped electron bunch. Since the slippage length is about 1 fs for a 100-m long, 3-cm period undulator, the FEL interaction depends only on the local properties of the electron bunch. Thus, we model the spatially chirped electron bunch as consisting of many thin time slices that have the same peak current and the same rms beam parameters but an initial angular offset correlated with the longitudinal position of each slice. By studying the dependence of the FEL output power as a function of the initial angular offset, we can trace out the time-dependence of the radiation power as well as the expected X-ray pulse length. Although the curved head and tail sections shown in Fig. 1 produce larger current spikes, they do not contribute to the FEL gain because of their relatively large transverse offsets generated by the RF deflector.

Fig. 2 shows *GENESIS* simulation results and indicate that a $2\sigma_{y'}$ angular offset degrades the final radiation power by more than one order of magnitude at the radiation wavelength 1.5 Å. Thus, we expect the final X-ray pulse will have a fwhm value of 29 fs for the given spatial chirp of the electron bunch described in Section 2.2 (see Fig. 3). Since the maximum saturation power is

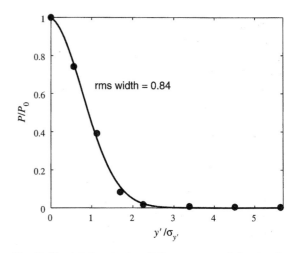

Fig. 2. Simulated output radiation power, relative to the maximum saturated power P_0, as a unction of the normalized angular offset at the undulator entrance. The solid line is a Gaussian fit with rms width of 0.84.

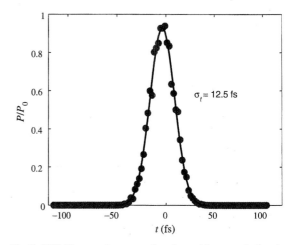

Fig. 3. FEL X-ray pulse versus time formed by convoluting the spatial chirp in Fig. 1 with the power sensitivity curve of Fig. 2. The rms pulse length is 12.5 fs and the fwhm value is 29 fs.

$P_0 \approx 12$ GW, about 2.7×10^{11} X-ray photons are contained in this short pulse.

2.4. Timing stability

The FEL gain with a spatially chirped electron bunch only produces X-rays within about ± 15 fs of its zero-crossing point. Even if the RF deflector phase or bunch arrival time varies from shot-to-shot, as long as the electron bunch overlaps the zero-crossing phase, the short X-ray pulse is naturally phase locked to the RF deflector. Thus, the X-ray timing can be determined by the phase of the RF deflector alone, which may be synchronized to an external laser for pump-probe experiments [4].

The RF deflector phase jitter tolerance is determined by the electron bunch length. As long as the RF phase (or the bunch arrival time) varies from shot-to-shot by less than the rms bunch length, there will always be a portion of the electron bunch in the undulator which crosses at $y' = 0$, or $y = 0$ (see Fig. 1 where this crossing occurs at $t \approx 0$). For the RF deflector located prior to BC2 where the bunch length is 1.8 ps fwhm, the RF phase jitter tolerance is about 0.6 ps ($\approx 0.6°$ S-band), which is quite reasonable and has already been demonstrated [2]. Whereas placing the deflector after BC2 requires a much tighter RF phase tolerance of 0.07° S-band since the bunch length is 210 fs fwhm.

Placing the RF deflector before BC2 creates another source of X-ray timing jitter because the electron energy jitter becomes arrival time jitter after the chicane. BC2 has a net momentum compaction $|R_{56}| \approx 25$ mm, therefore rms shot-to-shot electron energy stability of $\sigma_\delta \approx 0.1\%$ will introduce additional timing jitter of $\sigma_t = |R_{56}|\sigma_\delta/c \approx 80$ fs. However, a beam position monitor (BPM) located at the center of the chicane can be used to monitor the shot-to-shot energy fluctuations and provide a delayed signal to accurately reconstruct the arrival time per pulse. With a BC2 maximum momentum dispersion of $\eta_x \approx 340$ mm, and a BPM resolution of $\sigma_{BPM} \approx 20$ μm, this delayed signal has a timing resolution of $\sigma_\tau = \sigma_{BPM} |R_{56}/(c\eta_x)| \approx 5$ fs. The timing signal can then be provided to pump-probe experiments and used to accurately bin the experimental coincidence data with timing precision possibly approaching 10–20 fs, which is similar to the minimum X-ray pulse length achieved with this method.

2.5. Aperture restrictions

The minimum pulse length obtainable with this method is ultimately limited by the chirped beam

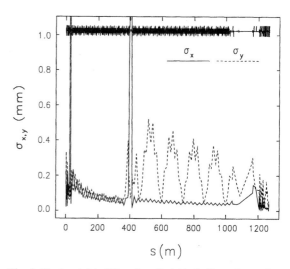

Fig. 4. Horizontal (solid) and vertical (dashed) rms beam sizes along LCLS accelerator with RF deflector switched on and located at $s = 360$ m, prior to BC2. The large horizontal beam sizes at $s = 25$ and 400 m are in the two compressor chicanes (BC1 and BC2).

size generated in the accelerator and the undulator. In the above numerical example the RF deflector has been limited to 5 MV based on the rms beam size in the accelerator, which has increased to 10-times its nominal size due to the large spatial chirp. This produces a maximum rms beam size in the linac of 0.5 mm (see Fig. 4), which is small compared with the 12-mm iris radius of the accelerator structures. It may be possible to increase the RF deflector power to further reduce the X-ray pulse length, but this is better determined experimentally given realistic apertures and shot-to-shot phase jitter.

3. Summary

We have described a very straightforward way to produce SASE X-ray pulses with fwhm duration of ~30 fs, which are 7-times shorter than the LCLS electron bunch length. The method uses an S-band RF deflecting structure already included in the LCLS design (although not optimally located for this purpose). It may be possible to push this method to ~10 fs fwhm depending on aperture restrictions. For example, an octupole magnet located after the deflector might be used to roll over the large transverse amplitude at the head and tail sections and provide an increased spatial chirp in the beam core without exceeding aperture limits. To take full advantage of these naturally phase-locked X-ray pulses, a dedicated X-band RF deflector (four times the S-band frequency) located just before the undulator entrance may provide a similar level of spatial chirp and a closer distance to the experimental stations for better synchronization with an external laser [4].

The spatial chirp method might also be implemented, without the need for a transverse RF deflector, by taking advantage of the transverse wakefields of the linac accelerating structures or by generating a large dispersion function in the linac in the presence of a time-correlated energy spread, which already exists in the BC2. The dispersion can easily be added with adjustable gradient quadrupole magnets inside the bunch compressor chicanes, which are standard components in short-wavelength FELs. Thus, this method will be useful for SASE FELs to produce bright X-rays with pulse length much shorter than the electron bunch length.

Acknowledgements

This work is supported by Department of Energy Contract No. DE-AC03-76SF00515.

References

[1] LCLS Conceptual Design Report, SLAC-R-593, April 2002, http://www-ssrl.slac.stanford.edu/lcls/CDR/.
[2] R. Akre, et al., EPAC'02, Paris, France, June 3–7, 2002, pp. 1882, http://accelconf.web.cern.ch/accelconf/e02/PAPERS/THPRI097.pdf.
[3] S. Reiche, Nucl. Instr.and Meth. A 429 (1999) 243.
[4] J. Galayda, J. Hastings, private communications.

Available online at www.sciencedirect.com

Nuclear Instruments and Methods in Physics Research A 528 (2004) 463–466

NUCLEAR INSTRUMENTS & METHODS IN PHYSICS RESEARCH
Section A

www.elsevier.com/locate/nima

Chirped pulse amplification at VISA-FEL

R. Agustsson[a], G. Andonian[a], M. Babzien[b], I. Ben-Zvi[b], P. Frigola[a], J. Huang[c], A. Murokh[a,*], L. Palumbo[d], C. Pellegrini[a], S. Reiche[a], J. Rosenzweig[a], G. Travish[a], C. Vicario[d], V. Yakimenko[b]

[a] *Department of Physics Astronomy, University of California, 405 Hilgard Ave., Los Angeles, CA 90095-1547, USA*
[b] *Brookhaven National Laboratory, Upton, NY 11973, USA*
[c] *Phoang Accelerator Laboratory, South Korea*
[d] *University of Rome, "La Sapienza", Italy*

Abstract

Chirped beam manipulations are of the great interest to the free electron laser (FEL) community as potential means of obtaining ultra short X-ray pulses. The experiment is under way at the accelerator test facility (ATF) at Brookhaven National Laboratory (BNL) to study the FEL process limits with the under-compressed chirped electron beam. High gain near-saturation SASE operation was achieved with the strongly chirped beam (~2.8% head-to-tail). The measured beam dynamics and SASE properties are presented, as well as the design parameters for the next round of experiment utilizing the newly installed UCLA/ATF chicane compressor.
© 2004 Elsevier B.V. All rights reserved.

PACS: 41.60.Cr

Keywords: Short pulse; SASE FEL

1. Introduction

With the prospects of the X-ray free electron laser (FEL) becoming a reality within the next decade [1], there is a great interest in a broader scientific community towards shortening the pulse duration of the anticipated X-ray FEL pulses down to 10-femtosecond range. For example, it would allow singular molecule diffraction experiments in the time interval shorter than it takes for the Coulomb explosion to destroy the sampled molecule [2]. Recently it was proposed [3] to use a chirped electron bunch to drive the self-amplified spontaneous emission (SASE) FEL so that the resulted SASE frequency modulation could allow separating single lasing spike from the rest of the bunch, or even using the optical gratings to longitudinally compress the X-ray beam. The limits associated with the chirped electron beam based SASE process have been studied theoretically and through simulations [4,5], and one can

*Corresponding author.
E-mail address: murokh@physics.ucla.edu (A. Murokh).

0168-9002/$ - see front matter © 2004 Elsevier B.V. All rights reserved.
doi:10.1016/j.nima.2004.04.132

approximately state, that the characteristic chirp value of interest is given by

$$\left(\frac{\Delta p}{p}\right)\frac{\lambda_c}{L_b} \sim \rho \qquad (1)$$

where λ_c and L_b are the FEL cooperation length [6] and the electron beam bunch length respectively, and ρ is the 3-D FEL parameter. The amount of electron beam chirp indicated in Eq. (1) is sufficient to separate individual spikes in time and frequency domain, while still preserving high gain SASE operation (which gain length generally increases with the higher uncorrelated energy spread).

2. Description of the VISA experimental system

The experiment is under way at VISA-FEL (visible to infrared SASE amplifier) to demonstrate a proof of principle, that the efficient chirped SASE amplification is possible with the electron beam chirp value of interest as is indicated in Eq. (1). The experiment is driven by the high brightness ATF photoinjector beam of about 71 MeV, matched into the strong focusing VISA undulator ($\beta_x, \beta_y \sim 30$ cm). The original VISA-I experiment in 2001 demonstrated SASE high gain and saturation within 4-m undulator at 840 nm [7,8].

To achieve high gain SASE lasing, an unusual bunch compression mechanism is used, utilizing nonlinear properties of the long dispersive section in the ATF beam line (Fig. 1); which is taking advantage of the large negative second order compression coefficient T_{566}. The initial electron bunch (Table 1) with the current of 55 A is

Table 1
Electron beam characteristics and FEL gain length measured without compression (A) and with compression (B)

	Case A	Case B
Electron beam energy	71.2 MeV	70.4 MeV
Beam charge, Q	300 pC	300 pC
Horizontal emittance, ε_n	$2.1 \pm 0.2\,\mu$m	$\sim 4\,\mu$m
Peak current, I_p	55 A	~ 300 A
Sliced energy spread, $\Delta\sigma_\gamma$	0.05%	0.45%
FEL gain length, L_g	~ 30 cm	~ 20 cm

compressed longitudinally up to the peak current of 300 A, through manipulating the beam line tune settings, electron beam RMS linear chirp, σ_γ, and central momentum offset of the electron bunch, $\Delta p/p$:

$$\Delta\zeta = \left(T_{566}\frac{\Delta p}{p}\right)\sigma_\gamma. \qquad (2)$$

The typical chirp value to obtain maximum compression was of the order of $\sigma_\gamma \sim 0.17\%$, during VISA-I experiment. Even though such a compression process allowed obtaining short FEL gain length and corresponding studies of the SASE properties at saturation, it had significantly restricted controllability of the electron beam parameters.

On the other hand, in accordance with the Eq. (1), an electron beam chirp of about 2% is required to demonstrate chirped beam amplification of interest, which is an order of magnitude larger value than used during VISA-I. The new ongoing round of experiments (VISA-II) [9] has two stages: (1) initial optimization of the beam line tune and beam running conditions to achieve SASE operation with the large chirp (Table 1B); and (2) the main part involving utilization of the magnetic chicane compressor at ATF and electron beam transport linearization to achieve independent control over the electron beam chirp and longitudinal bunch profile.

3. Initial experimental results on the chirped beam amplification

One of the challenges of running a strongly chirped electron beam through the ATF beam line

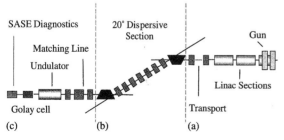

Fig. 1. Experimental layout of the ATF Beam line III: gun and linac area (a); 20° double-bend dispersive section (b); and VISA experimental area (c), including the undulator and diagnostics.

is to control the size of the beam, through the dispersion section of the beam line. The chirped beam transmission through the beam line is controlled by the high-energy slit (HES), which is an adjustable collimator located at the beginning of the dispersive section. In the recent experimental round, the 500 pC electron beam (Fig. 1a) has a 2.8% energy chirp after the linac exit, of which only 1.7% (330 pC) can propagate through the fully open HES (Fig. 2b).

The compression process in the dispersive section is monitored by the Golay cell installed in front of the undulator to measure the coherent transition radiation (CTR) intensity [8]. When the beam central momentum is chosen to optimize the compression in the dogleg, the CTR energy is peaked. Then, by closing the HES one can determine the part of the beam that contributes to the compression process (Fig. 2c). The start-to-end PARMELA-ELEGANT simulations reproduced the compression process as observed by the

combined slit and Golay cell measurements showing that the bunch peak current enhancement can reach up to 300 A.

With such a current very high SASE gain is achieved, despite the large sliced energy spread and degraded sliced emittance in the compressed part of the beam. The average SASE radiated energy around 2 μJ was recorded, which is within the order of magnitude from the saturation level. The spectrum of the FEL radiation (Fig. 3) have unusual width of up to 10%, and a characteristic dual spike structure, indicating two lasing modes. It should be noted, that by closing the HES, one could eliminate one of the spectral spikes, without degrading the second spike intensity (Fig. 2d), indicating longitudinal mapping of the SASE wavelength to the electron beam chirp. Yet, the detailed GENESIS simulations, and further measurements are necessary to confidently reproduce the SASE process under these non-trivial experimental conditions.

Another interesting observation is the unusual stability of the SASE process. Because the beam chirp is very linear and only 60% propagates through the HES, the FEL performance becomes insensitive to the RF phase jitter. Unlike in VISA-1, where small drifts in the RF phase would degrade lasing significantly, in this new, large chirp regime, the same beam fraction propagates through HES, regardless of the beam centroid jitter, thus making the lasing much more stable.

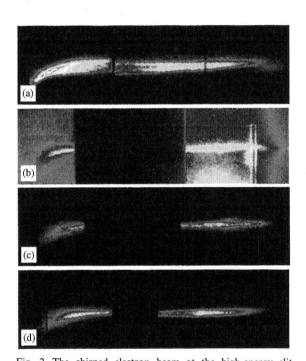

Fig. 2. The chirped electron beam at the high-energy slit monitor: [a] closed slit (500 pC, 2.8% chirp); [b] fully open slit (60% transmission); [c] compressed fraction of the beam (1.5% chirp); and [d] the fraction of a beam generating a single SASE spike (0.8% chirp).

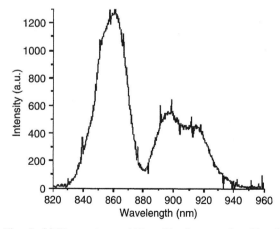

Fig. 3. SASE spectrum width—with the strongly chirped beam—reaches up to 10%.

4. Beam line improvements and outlook for the future

While the studies shown above are still under way to investigate the limits of the chirped beam amplification, there steps have been undertaken to improve the VISA experimental system in the near future. First and foremost, the magnetic chicane has been built and installed into the H-line [10] at the ATF (Fig. 4). Numerical simulations show that using the chicane one can increase the peak current in the electron beam by an order of magnitude without the need to use nonlinear properties of the dispersive section. Furthermore, the second order effects in the dispersive section utilized during the VISA-I and initial period of VISA-II experiments may become parasitic with the introduction of the chicane and would degrade the performance of the system. To avoid these problems, a set of sextupoles is being installed into the dispersive section [11], which according to simulations can linearize the transport even for the strongly chirped beams.

The chicane and sextupoles will be commissioned before the end of 2003. In addition to beam line modifications, the diagnostics at VISA are being upgraded. The longitudinal electron beam diagnostics, including the CTR interferometer, is under development before and after the dispersive section. Also, additional beam profile monitors are being installed in the beam line 3, as well as upgraded SASE optical diagnostics. In addition there are plans for autocorrelation of the SASE light at the undulator exit.

In conclusion, here we report operation of the high gain SASE FEL driven by the strongly chirped electron beam. SASE spectra indicate very large energy spread in the lasing portion of the beam, as well as short pulse duration. In parallel, the ongoing upgrades to VISA beam line will improve in the near future the control over the electron beam longitudinal phase space, as well as

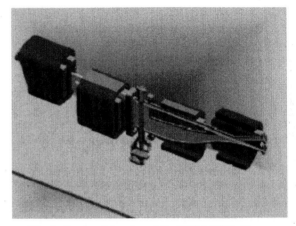

Fig. 4. Rendered drawing of the UCLA/ATF chicane.

transport efficiency, which will enable unambiguous chirped pulse amplification studies throughout the next year.

Acknowledgements

The authors thank M. Woodle and the ATF staff for technical support, and also BNL Cottage #30 for inspiration. The work is performed under ONR Grant # N00014-02-1-0911.

References

[1] M. Cornacchia, Nucl. Instr. and Meth. A, (2004) these Proceedings.
[2] R. Neutze, et al., Nature 406 (2000) 752.
[3] C. Pellegrini, unpublished technote, 1999.
[4] C. Schroeder, et al., J. Opt. Soc. Am. B 19 (2002) 1782.
[5] S. Krinsky, Z. Huang, Phys. Rev. STAB 6 (2003) 050702.
[6] R. Bonifacio, et al., Phys. Rev. Lett. 73 (1994) 70.
[7] A. Tremaine, et al., Phys. Rev. Lett. 88 (2002) 204801.
[8] A. Murokh, et al., Phys. Rev. E 67 (2003) 066501.
[9] G. Andonian, et al., Proceedings of PAC 2003, Portland, April 2003.
[10] R. Agustsson, UCLA Master Thesis, 2004.
[11] J. England, private communication.

Available online at www.sciencedirect.com

SCIENCE  DIRECT°

ELSEVIER   Nuclear Instruments and Methods in Physics Research A 528 (2004) 467–470

NUCLEAR INSTRUMENTS & METHODS IN PHYSICS RESEARCH
Section A

www.elsevier.com/locate/nima

Chirped pulse amplification of HGHG-FEL at DUV-FEL facility at BNL

Adnan Doyuran[a,*], Louis DiMauro[a], W. Graves[b], Richard Heese[a],
Erik D. Johnson[a], Sam Krinsky[a], Henrik Loos[a], James B. Murphy[a],
George Rakowsky[a], James Rose[a], Timur Shaftan[a], Brian Sheehy[a],
Yuzhen Shen[a], John Skaritka[a], Xijie Wang[a], Zilu Wu[a], Li Hua Yu[a]

[a] *Brookhaven National Laboratory, NSLS, Upton, New York, NY 11973, USA*
[b] *MIT, Bates Linear Accelerator Center Middleton, MA 01949, USA*

Abstract

The DUV-FEL facility has been in operation in the High Gain Harmonic Generation (HGHG) mode producing a 266-nm output from 177-MeV electrons for 1 year. In this paper, we present preliminary results of the Chirped Pulse Amplification (CPA) of the HGHG radiation. In the normal HGHG process, a 1-ps electron beam is seeded by a chirped 9 ps long, 800-nm Ti:Sapphire laser. The electron beam sees only a narrow fraction of the seed laser bandwidth. However, in the CPA case, the seed laser pulse length is reduced to 1 ps, and the electron beam sees the full bandwidth. We introduce an energy chirp on the electron beam to match the chirp of the seed pulse, enabling the resonant condition for the whole beam. We present measurements of the spectrum bandwidth for various chirp conditions.
© 2004 Elsevier B.V. All rights reserved.

PACS: 41.60.Cr

Keywords: CPA; FEL; HGHG; Seed laser; Accelerator; High gain

1. Introduction

There is great interest in producing lasers with a pulse length in the femtosecond region. In general, to serve FELs, the photo-cathode electron guns produce 4–5-ps beams. These beams can be compressed to 1 ps or less; however, it is not trivial to compress the beams to femtosecond level.

By producing a frequency chirp along the pulse, we can use conventional laser pulse compression to shorten the pulse duration to the femtosecond level [1].

The accelerator consists of a Ti:Sapphire laser system, a 1.6 cell RF photo-cathode gun, four SLAC-type linac tanks, and a four-dipole chicane (Fig. 1). A photo-cathode RF gun is illuminated by the tripled Ti:Sapphire laser at 266 nm producing a 300-pC charge, a 4–5-ps FWHM bunch length, and a 4.5-MeV energy electron beam with

---

*Corresponding author.

E-mail address: doyuran@physics.ucla.edu (A. Doyuran).

0168-9002/$ - see front matter © 2004 Elsevier B.V. All rights reserved.
doi:10.1016/j.nima.2004.04.133

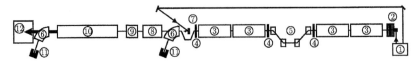

Fig. 1. The NSLS DUV-FEL layout: (1) gun and seed laser system, (2) RF gun, (3) linac tanks, (4) focusing triplets, (5) magnetic chicane, (6) spectrometer dipoles, (7) seed laser mirror, (8) modulator, (9) dispersive section, (10) radiator (NISUS), (11) beam dumps, and (12) FEL measurement area.

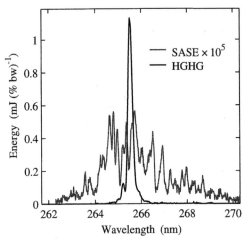

Fig. 2. The spectrum of HGHG (black) at normal operating conditions (no electron energy chirp and long seed pulse) and SASE spectrum (gray).

normalized emittance of 4–5 mm mrad. The first two linac tanks accelerate the electron bunch with the second tank off-crest by 23° to introduce an energy chirp. A four-dipole chicane compresses the bunch to about 1 ps FWHM. During the past year, the HGHG-FEL reached saturation at 266 nm with the seed at 800 nm. The typical output energy is about 100 μJ with a 1-ps pulse length. Energy fluctuation is measured to be about 7%, which is mostly due to the machine performance. The spectrum is measured to be very narrow (0.23 nm) (Fig. 2) and the output is transversely and temporally Fourier-transform limited [2]. The third harmonic at 88 nm accompanied by the 266-nm fundamental mode has been used in a novel chemistry experiment since January 2003 [3].

2. Experimental procedure for CPA

In this experiment the seed laser pulse length is adjusted to 1 ps FWHM, which is chosen to be the same as the electron bunch length. A proper delay is introduced to the seed laser pulse to establish synchronization with the electron beam. This way the electron bunch sees the full bandwidth of the seed laser. However, the usual 1% bandwidth of the seed laser is not supported by the FEL process. The resonant condition for an FEL is

$$\lambda = \frac{\lambda_w}{2\gamma^2}(1 + K^2/2) \tag{1}$$

To satisfy the resonant condition along the bunch we need to introduce an energy chirp so that every slice within the electron bunch will be resonant. Eq. (1) yields the relationship between the wavelength chirp of the seed laser and the energy chirp of the electron as

$$\frac{\Delta\gamma}{\gamma} = -\frac{\Delta\lambda}{2\lambda} \tag{2}$$

Thus, we need to introduce an energy chirp that is equal to half the laser wavelength chirp.

The energy of the electron beam after the linac tanks is

$$E = E_{T1} + 34\cos(\phi_2) + 52\cos(\phi_3) + A_{Tank4}\cos(\phi_4) \tag{3}$$

where each term represents the energy gain at each linac tank, and the angles are the tank wave phases measured with respect to the crest of the RF. The wave phase of tank 2 is set to −22° to −26° by the compression requirements. Tank 3 is operated on a crest and tank 4 is varied to produce different energy chirps along the beam. The wave amplitude of tank 4 is adjusted so that the total energy is the resonant energy for the HGHG-FEL. The energy chirp can be expressed from Eq. (3) as

$$\frac{\Delta E}{E} = -\frac{34\omega\sin(-22)\Delta t_{uncomp}}{E} - \frac{A_{Tank4}\omega\sin(\phi_4)\Delta t_{comp}}{E} \tag{4}$$

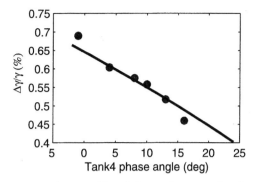

Fig. 3. FWHM energy chirp of electron beam as a function of tank 4 phase. Blue curve is calculation and red dots are measurement.

where ω is the RF frequency, and $\Delta t_{uncomp} \cong 5 \, ps$ and $\Delta t_{comp} \cong 1 \, ps$ are uncompressed and compressed electron bunch lengths, respectively. Fig. 3 shows the FWHM percentage of energy chirp as a function of the phase angle of tank. We observe good agreement between the measurement and the calculation. We measure the spectrum of the HGHG output at each electron energy chirp condition.

3. Spectrum measurements

We scan the tank 4 phase from $24°$ to $-2°$ and measure the spectrum of HGHG output at each case. This way the chirp per picosecond slope is varied, and we expect to have the widest bandwidth when the chirp slope is half the seed laser wavelength chirp slope. Fig. 4 shows the spectra for different chirps. Note that the smoothness of the chirped spectra indicates that the electron density profile is smooth. We plot bandwidth as a function of chirp in Fig. 5.

We see a clear peak in Fig. 5, which shows that when the chirp of the electron beam is matched to the seed laser chirp, the HGHG bandwidth is the widest. The widest spectrum was measured at a chirp of 0.58%. The typical bandwidth of the seed laser is about 1%. The measured seed laser FWHM bandwidth during the experiment was 5.5 nm, which is about 0.7%. Thus, we expect to have the largest bandwidth at a 0.35% chirp, and the bandwidth should be one-third of the seed

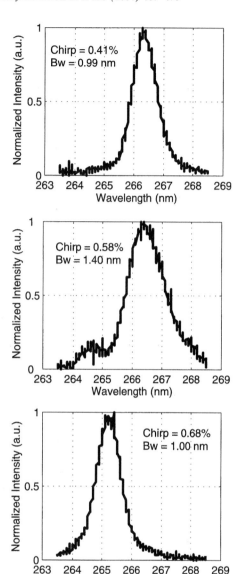

Fig. 4. Spectra of CPA-HGHG for different chirps.

bandwidth, which is 1.8 nm. The fact that we introduced more chirp to the electron beam than 0.35% might suggest that the electron bunch length is longer than we measured by the zero-phasing technique [4]. The widest measured bandwidth of 1.4 nm is not far from the expected value of 1.8 nm. One of the critical issues during the experiment is the longitudinal jitter of the system. We estimate this time jitter is about 0.3 ps

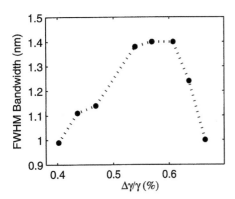

Fig. 5. HGHG bandwidth vs. percent energy chirp of the electron beam.

between the electron beam and the seed laser pulse. Considering 1 ps long electron and photon beams, 0.3-ps jitter is significant for the performance of the system. The fluctuations were larger than the usual HGHG condition for this experiment. This jitter could also be the reason for not having the full 1.8-nm bandwidth, because the electron beam would be seeing a different bandwidth at every shot. This is also consistent with the wavelength fluctuations that were observed. We accumulated a number of spectrum data for the same chirp conditions and chose those with high output. Currently we are planning a method that would reduce this jitter by illuminating the cathode with the 266-nm HGHG output. This would reduce the jitter by the compression ratio, which is about 4–5 times.

We are in the process of building a compressor to shorten the CPA-HGHG output and a SPIDER to analyze the output in more detail.

4. Conclusion

The first steps toward the Chirped Pulse Amplification of HGHG-FEL have been taken. Spectrum widening has been observed when the electron beam is properly chirped. The results are encouraging for the future compression of the CPA-HGHG output, which should produce pulses with pulse lengths in the femtosecond region.

Acknowledgements

This work was supported by US Department of Energy, Office of Basic Energy Sciences, under Contract No. DE-AC02-98CH10886 and by Office of Naval Research Grant No. N00014-97-1-0845.

References

[1] L.H. Yu, et al., Phys. Rev. E 49 (1994) 4480.
[2] L.H. Yu, et al., Phys. Rev. Lett. 91 (2003) 074801.
[3] X. Liu, R.L. Gross, A.G. Suits, Science 294 (2001) 2527.
[4] D.X. Wang, G.A. Kraft, C.K. Sinclair, Phys. Rev. E 57 (1998) 2283.

Available online at www.sciencedirect.com

NUCLEAR INSTRUMENTS & METHODS IN PHYSICS RESEARCH
Section A

ELSEVIER   Nuclear Instruments and Methods in Physics Research A 528 (2004) 471–475

www.elsevier.com/locate/nima

Efficient frequency doubler for the soft X-ray SASE FEL at the TESLA Test Facility

J. Feldhaus[a], M. Körfer[a], T. Möller[a], J. Pflüger[a], E.L. Saldin[a], E.A. Schneidmiller[a], M.V. Yurkov[a,b],*

[a] *Deutsches Elektronen-Synchrotron (DESY), Notkestrasse 85, Hamburg 22607, Germany*
[b] *Joint Institute for Nuclear Research, Dubna 141980, Moscow region, Russia*

Abstract

This paper describes an effective frequency doubler scheme for SASE free electron lasers (FEL). It consists of an undulator tuned to the first harmonic, a dispersion section, and a tapered undulator tuned to the second harmonic. The first stage is a conventional soft X-ray SASE FEL. Its gain is controlled in such a way that the maximum energy modulation of the electron beam at the exit is about equal to the local energy spread, but still far away from saturation. When the electron bunch passes through the dispersion section this energy modulation leads to effective compression of the particles. Then the bunched electron beam enters the tapered undulator and produces strong radiation in the process of coherent deceleration.

We demonstrate a frequency doubler scheme that can be integrated into the SASE FEL at the TESLA Test Facility at DESY, and will allow to reach 3 nm wavelength with GW-level of output peak power. This would extend the operating range of the FEL into the so-called water window and significantly expand the capabilities of the TTF FEL user facility.
© 2004 Elsevier B.V. All rights reserved.

PACS: 41.60.cr; 42.55.Vc; 41.75.Ht; 41.85.Ew

Keywords: Free electron lasers; X-ray lasers; Relativistic electron beams

1. Introduction

The soft X-ray FEL at the TESLA Test Facility at DESY (TTF FEL) will cover a spectral range between approximately 60 and 6 nm wavelength. The minimum wavelength of 6 nm is determined by the maximum electron beam energy of 1 GeV [1]. It would be extremely interesting to extend this range into the so-called water window, i.e. the range between the K-absorption edges of carbon and oxygen at 4.38 and 2.34 nm, respectively. There are basically two methods to up-convert the fundamental radiation frequency via nonlinear harmonics: the generation of third harmonic radiation in a planar SASE undulator through a nonlinear mechanism driven by bunching at the fundamental frequency [2,3], and a two-undulator (second) harmonic generation scheme, also referred to as the "after-burner" method [4–6].

*Corresponding author. Deutsches Elektronen-Synchrotron (DESY), Notkestrasse 85, Hamburg 22607, Germany. Tel.: +49-40-8998-2676; fax: +49-40-8998-4475.

E-mail address: yurkov@mail.desy.de (M.V. Yurkov).

0168-9002/$ - see front matter © 2004 Elsevier B.V. All rights reserved.
doi:10.1016/j.nima.2004.04.134

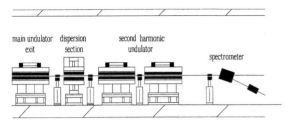

Fig. 1. Side view of the electron beam transport system, showing the location of the frequency doubler.

In this paper, we propose an effective frequency doubler scheme for a SASE FEL [7]. It consists of an undulator tuned to the first harmonic, a dispersion section, and a tapered undulator tuned to the second harmonic (see Fig. 1). The first stage is a conventional soft X-ray SASE FEL. The gain of the first stage is controlled in such a way that the maximum energy modulation of the electron beam at the FEL exit is almost equal to the local energy spread, but still far away from saturation. When the electron bunch passes through the dispersion section this energy modulation leads to the effective compression of the particles. Then the bunched electron beam enters a tapered undulator, and from the very beginning produces strong radiation because of the large spatial bunching. The strong radiation field produces a ponderomotive well which is deep enough to trap the particles, since the original beam is relatively cold. The radiation produced by these captured particles increases the depth of the ponderomotive well, and they are effectively decelerated.

We illustrate the operation of the proposed frequency doubler for the parameters of the TTF FEL. The result of our study is that the TTF FEL would be able to produce radiation down to 3 nm wavelength with an output peak power in the GW range, and with excellent spectral properties. The power of the third harmonic from this device (i.e. at 1 nm wavelength) is still in the 10 MW-level power range, exceeding any other pulsed radiation source presently available at 1 nm, and sufficiently high for novel applications.

2. Operation of a frequency doubler

In this section, we illustrate the operation of the frequency doubler for the parameters of the TTF

FEL, Phase 2 [1,8]. The beam parameters, and the parameters of the doubler undulator and the dispersion section are listed in Table 1. The resonance wavelength in the main X-ray undulator is equal to 6 nm, and the doubler undulator is tuned to the resonance wavelength of 3 nm.

The frequency doubler scheme operates as follows. The electron bunch enters the main X-ray undulator and produces SASE radiation at the wavelength of 6 nm. During the amplification process the radiation power grows exponentially with the undulator length. Simultaneously, the energy and density modulation of the beam are growing. At the end of the X-ray undulator the

Table 1
Parameters of the frequency doubler for the TESLA Test Facility FEL

Electron beam	
Energy	1000 MeV
Peak current	2.5 kA
Normalized rms emittance	2π mm-mrad
Rms energy spread	0.2 MeV
Rms bunch length	50 μm
External β-function	4.5 m
Rms beam size	68 μm
Undulator (1st harmonic)	
Type	Planar
Period	2.73 cm
Gap	12 mm
Peak magnetic field	0.47 T
Segment length	4.5 m
Undulator length	13.5 m
Dispersion section	
Net compaction factor	1.5 μm
Undulator (2nd harmonic)	
Type	Planar, tapered
Period	1.95 cm
Gap	10 mm
Peak magnetic field	0.39 T
Segment length	4.5 m
Undulator length	13.5 m
Undulator tapering	−0.14%/m
Coherent radiation	
Wavelength	3 nm
Energy per pulse	180 μJ
Peak power	1.5 GW
Bandwidth (FWHM)	0.2%
Pulse duration (FWHM)	130 fs

beam energy modulation is comparable with the local energy spread. In the present example this takes place at an undulator length of 12.3 m. The upper left plot in Fig. 2 shows the phase space distribution of particles for the slice corresponding to 330 fs time coordinate along the bunch. Such a picture is typical for every spike. We see that the modulation amplitude at the second harmonic is small (see lower left plot in this figure). On the other hand, there is visible energy modulation with an amplitude of about the same value as that of the local energy spread. When the electron bunch passes through the dispersion section this energy modulation leads to effective compression of the particles as it is illustrated with plots in the right column in Fig. 2. When the bunched beam enters the undulator tuned to the second harmonic, it

immediately starts to produce powerful radiation at the second harmonic. That is important, but not the main feature of our proposal. Let us study more closely the phase space distribution of the particles at the exit of the dispersion section. We see that this distribution is double-periodic with respect to the second harmonic, i.e., only each second bucket is populated with particles. If we trace the evolution of the FEL process in the uniform undulator, we find that "thermalization" takes place. The population of particles in the originally filled buckets is reduced, and some of these particles travel into the ponderomotive well of originally empty buckets. Simultaneously, the energy spread in the bunch grows due to the FEL process. Finally, saturation occurs.

Thus, another key idea of our proposal is to preserve the original bunching of the beam and to organize an effective extraction of the energy from the electron beam. This idea is simply realized by using a tapered undulator. The process of amplification proceeds as follows. The bunched beam (double-periodic) produces strong radiation from the very beginning because of the large spatial bunching. The strong radiation field produces a ponderomotive well which is deep enough to trap particles, as the original beam is relatively cold. The radiation produced by these captured particles increases the depth of the ponderomotive well, and they are effectively decelerated. The undulator tapering preserves the synchronism of trapped particles and radiation, and a significant fraction of the energy can be extracted. Simulations using the time-dependent FEL code FAST [9] confirm this qualitative picture. Fig. 3 shows the evolution of the energy in the radiation pulse in the tapered undulator. Despite the original spiking seeding the process of the second harmonic, we effectively trap a significant fraction of the particles, and can achieve much higher power than for the case of an untapered undulator. Fig. 4 shows the temporal and spectral structure of the radiation pulse after 13.5 m of second-harmonic undulator. Note that the radiation pulse length is about a factor of two shorter than that of a traditional SASE FEL, due to nonlinear transformation provided by the dispersion section. Another important feature of the radiation from a tapered undulator is the significant

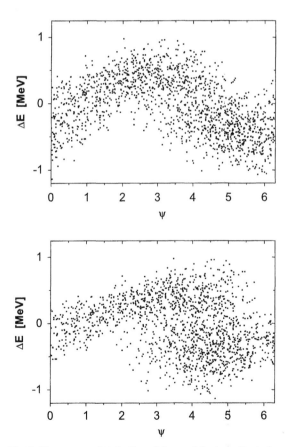

Fig. 2. Phase space distribution of the particles in a slice before (upper plot) and after (lower plot) the dispersion section. The radiation wavelength is equal to 6 nm.

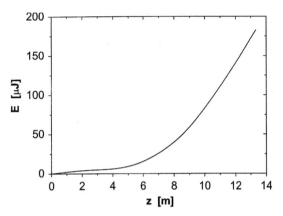

Fig. 3. Energy in the radiation pulse versus undulator length in the frequency doubler.

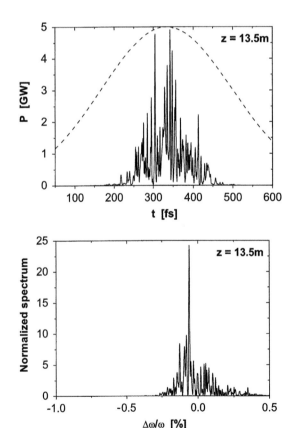

Fig. 4. Time structure (upper plot) and spectral structure (lower plot) of the radiation pulse at a length of the frequency doubler of 13.5 m. The radiation wavelength is equal to 3 nm. The dashed line shows the bunch profile.

suppression of the sideband growth in the nonlinear regime. This means that in the proposed scheme the spectral brightness of the radiation is increased proportionally to the radiation power, while in the case of a uniform undulator the peak brightness is reached at the saturation point and is then reduced due to the sideband growth [10].

The present design of the TESLA Test Facility assumes 12 m free space downstream of the main undulator which is sufficient for a 9 m long (2 m × 4.5 m sections) second harmonic radiator. It is planned now that the main X-ray undulator will consist of six sections in order to have sufficient safety margin during the first commissioning of the SASE FEL. With project parameters of the electron beam, five undulator modules will be sufficient. This means that in the future there will be space for three modules of the frequency doubler. This will allow to realize the full frequency doubler and to extend the wavelength range of the TTF FEL down to 3 nm with GW-level power without changing the present TTF layout.

3. Summary

Proposed frequency doubler scheme has several significant advantages over traditional SASE FELs with uniform undulators: shorter total magnetic length; a possibility to attain higher output power and brightness of the radiation; shorter radiation pulse duration. The realization of a frequency doubler at the TTF is considered as extremely important not only for reaching the water window, but also for realizing the X-ray FEL user facility in the 0.1 nm range. The use of frequency doublers at a large-scale facility could result in a significant reduction of the project costs. In addition, the safety margin for facility operation becomes more relaxed.

References

[1] SASE FEL at the TESLA Test Facility, Phase 2, DESY print TESLA-FEL 2002-01, Hamburg, 2002.

[2] H.P. Freund, S.G. Biedron, S.V. Milton, IEEE J. Quant. Electron. 36 (2000) 275;
H.P. Freund, S.G. Biedron, S.V. Milton, Nucl. Instr. and Meth. A 445 (2000) 53;
H.P. Freund, S.G. Biedron, S.V. Milton, Nucl. Instr. and Meth. A 445 (2000) 95.

[3] Z. Huang, K.-J. Kim, Phys. Rev. E 62 (2000) 7295.

[4] R. Bonifacio, L. De Salvo, P. Pierini, Nucl. Instr. and Meth. A 293 (1990) 627.

[5] F. Ciocci, et al., IEEE J. Quantum Electron. 31 (1995) 1242.

[6] W.M. Fawley, et al., Proceedings of the IEEE 1995 Particle Accelerator Conference, 1996, p. 219.

[7] J. Feldhaus, et al., preprint DESY 03-092, Hamburg, 2003.

[8] T. Limberg, Ph. Piot, F. Stulle, Proceedings of the EPAC 2002, Paris, France, p. 811.

[9] E.L. Saldin, E.A. Schneidmiller, M.V. Yurkov, Nucl. Instr. and Meth. A 429 (1999) 233.

[10] E.L. Saldin, E.A. Schneidmiller, M.V. Yurkov, The Physics of Free Electron Lasers, Springer, Berlin, 1999.

Available online at www.sciencedirect.com

SCIENCE d DIRECT°

Nuclear Instruments and Methods in Physics Research A 528 (2004) 476–480

ELSEVIER

NUCLEAR
INSTRUMENTS
& METHODS
IN PHYSICS
RESEARCH
Section A

www.elsevier.com/locate/nima

Start to end simulations for the BESSY FEL project ☆

M. Abo-Bakr*, M.v. Hartrott, J. Knobloch, D. Krämer, B. Kuske,
F. Marhauser, A. Meseck

BESSY GmbH, Albert Einstein Strasse 15, Berlin, Germany

Abstract

BESSY plans to construct a FEL multi user facility for the Soft-X-Ray spectral range from 20 eV to 1 keV. Central part of the project is a superconducting linac, based on the TESLA technology but operating in CW mode. A photo injector and a bunch compression scheme is introduced, allowing to reach the required electron beam parameters at the undulators beginning: up to 4 kA peak current (1 nC bunch charge), 1.5π mm mrad normalized transverse emittance and an energy spread below 1%. Extensive Start to End (S2E) simulations have been performed, using the programs ASTRA for the injector part, ELEGANT for the linac and bunch compression studies and GENESIS for the FEL process. Results of these simulations are presented and discussed.
© 2004 Elsevier B.V. All rights reserved.

PACS: 41.60.Cr; 41.85.Ja

Keywords: Free electron lasers; Start to end simulations S2E

1. Introduction

The Berliner Elektronenspeicherring-Gesellschaft für Synchrotronstrahlung (BESSY) operates storage ring based synchrotron light source facilities since 1981. The most recent one, BESSY II [1], started its user operation in January 1999, delivering high brilliance photon beams in the VUV to XUV spectral range. As an addition to the existing light-source BESSY proposes the con-

struction of a linac-based single-pass free-electron laser (FEL) user-facility for photon energies from 20 eV to 1 keV (62 nm $\geqslant \lambda \geqslant$ 1.2 nm). A peak-brilliance of 10^{31} photons/(s mm^2 mrad^2 0.1%BW) is aspired, i.e., a peak power up to 10 GW at pulse-durations less than 200 fs (FWHM) initially and ultimately down to 20 fs. Presently a three year design-phase, funded as a collaboration between BESSY, DESY and the Max-Born-Institute Berlin, is in its final stage.

In this paper the present basic machine layout as well as procedure and results of Start to End (S2E) simulations for the BESSY FEL are described. From the various layout options under investigation, the high energy SASE-FEL scheme with $\lambda = 1.24$ nm, which is currently the most

---

☆ Funded by the Bundesministerium für Bildung, Wissenschaft, Forschung und Technologie, the Land Berlin and the Zukunftsfonds des Landes Berlin.

*Corresponding author.

E-mail address: michael.abo-bakr@bessy.de
(M. Abo-Bakr).

demanding case, has been chosen to be presented here.

2. General machine layout

Central part of the project will be a 2.25 GeV superconducting linac, based on DESY's 1.3 GHz TESLA technology [2], operating in CW mode, allowing a free choice of the timing pattern. For the first project phase the use of the existing PITZ photo injector [3] design from DESY is foreseen, operating at 1 kHz macro pulse repetition rate with a pulse duration of about 10 µs. Later on an upgrade to a superconducting injector cavity is planed, allowing to use all benefits of a superconducting linac. From two extraction sections in the linac electrons are guided to the 3 undulator beamlines, suited to cover the desired wavelength range. A schematic drawing of the BESSY FEL scheme is shown in Fig. 1.

2.1. Injector

Most of the injector parameters have been adopted from the PITZ design. With respect to the higher average power the gradient at the cathode has been reduced to 40 MV/m. The main parameters are listed in Table 1. Adjusting carefully the solenoid and booster module position and the field strength, the emittance compensation is very efficient and a normalized projected value of 0.77π mm mrad with 3 keV rms energy spread can be reached for a 1 nC bunch of about 2.0 mm rms length. Simulations for the injector part have been performed using ASTRA [4], a 3D particle tracking code, taking into account space charge forces.

Results of these simulations—energy, rms-beam size and emittance evolution over the initial injector part—are shown in Fig. 2.

2.2. Linac and bunch compressors

The BESSY FEL linear accelerator will consist of eighteen TESLA modules, each equipped with eight 9-cell cavities with a length of about 12 m. Operating at an average accelerating gradient of 15–16 MeV/m a maximum energy of 2.25 GeV can be reached. Two magnetic, S-shape chicanes, compressing the bunches to the required peak currents in the kA range, are integrated in the linac after the second and sixth module, respectively. Main parameters of the bunch compressors are summarized in Table 2. Simulations were executed with ELEGANT [5], a 3D nonlinear tracking program which includes the wake field and 1D longitudinal CSR effects. Fig. 3 shows the optical functions over the entire linac, beginning with the 3rd harmonic module, including the bunch com-

Table 1
Injector parameters

Laser:	Type	TiSa
	Wavelength	266 nm
	Flat top duration	20 ps
	Rise/fall time	2 ps
Cathode:	Material	Cs_2Te
	Quantum eff.	(8–10)%
Cavity:	Type	1 1/2 cell
	Max. gradient	40 MeV/m
	Heat load	$\leqslant$100 kW
Solenoid:	Magnetic field	180 mT
Booster:	Type	TESLA 8 cav.
	Grad. cav. 1–4	11.25 MeV/m
	Grad. cav. 5–8	15.25 MeV/m

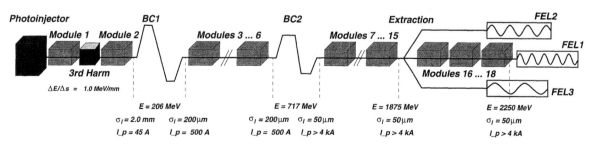

Fig. 1. General Layout of the BESSY FEL, equipped with three SASE-FEL beamlines.

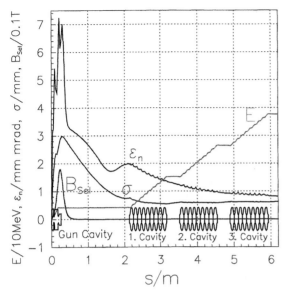

Fig. 2. Injector: energy E, transverse, normalized emittance ε_n and beam size σ evolution up to No. 3 booster cavity (gun and cavity shapes as well as solenoid field B_{Sel} are also plotted).

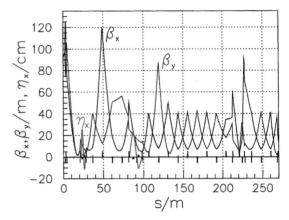

Fig. 3. Optical function over the entire linac: β_x: red line; β_y: blue line; η_x: green line.

Table 2
Bunch compressor parameters

	BC1	BC2
Type	6 dipole S-shape chicane	
Energy	206 MeV	717 MeV
$\sigma_{E,corr}$	1.45%	0.46%
R_{56}	−12.8 cm	−3.5 cm
T_{566}	20.0 cm	5.2 cm
Total length	5.7 m	15.1 m
ρ_{dipole}	1.1/1.4/2.8 m	16.5/32.1 m
L_{dipole}	0.4 m	0.4 m

Table 3
Undulator parameters

Type	Variable gap Apple-II device
λ_u	2.75 cm
K	0.865
g	10 mm
Lattice	FODO, external focusing
L_{total}	60.0 m

pressor chicanes and the (disabled) low energy extraction sections and ending before the high energy FEL's undulator section (collimators are not included here).

2.3. Undulator

For the high energy SASE beamline a variable gap "Apple II" type device of about 60 m total length with a period length $\lambda_U = 2.75$ cm is planed. Tuning of the FEL wavelength can be achieved either by gap changes or by changing the

gradient in the last accelerator modules (downstream of the low energy extraction section) and thus adjusting the electron energy. A modular construction with a block length of 3.5 m is foreseen. A Drift space of 0.70–0.75 m between two modules will contain quadrupoles, steering magnets, phase shifters and diagnostics equipment. The main undulator parameters are summarized in Table 3. The FEL process has been studied with GENESIS [6], a fully 3D time-dependent FEL simulation code.

3. Start to end simulations

Usually several programs are used to simulate a full FEL device from the injector's cathode to the undulator end. Since each code requests different type of inputs, often a parametrization of the electron bunch is necessary. For these purposes projected mean and rms values are calculated. Due

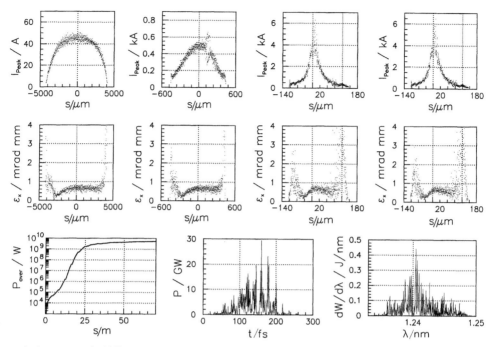

Fig. 4. Slice analysis: upper and middle row: peak current and normalized emittance at end of the injector (ASTRA output), BC1, BC2 and linac (GENESIS input); bottom row: power growth in the undulator and radiation power, both in spectral and time domain.

to the non-standard particle distributions and varying sliced values over the bunch length, this usually results in losing information. To avoid this, in our S2E simulations we abandon any parametrization and pass the 6D phase space directly between the programs.

For the injector part, 100,000 particles have been tracked with ASTRA up to the first bunch compressor. The ASTRA output has been converted and passed to ELEGANT. The calculated phase space distribution at the linac's end has been analysed and transformed to the sliced input format of GENESIS. Fig. 4 shows the slice analysis of the S2E simulation results: peak current and normalized emittance, taken after the injector (module 1), BC1, BC2 and at the linac's end. The radiation power growth along the undulator and the temporal and spectral properties of the SASE output radiation power are also displayed in the figure.

At the linac's end a bunch peak current clearly above the desired 4.0 kA is reached. The projected horizontal emittance increases to 0.83π mm mrad

after the first and 1.45π mm mrad after the second bunch compressor, caused by varying horizontal shifts of the sliced bunch centroids. The horizontal sliced emittance in the high peak current area is close to the undistorted value around 0.8π mm mrad. The vertical emittance, which is not affected by CSR effects, is fully conserved. The sliced energy spread is below 1×10^{-4} and thus sufficiently small for the FEL interaction. The SASE process saturates at about 35 m with an output power of 6 GW. From the power density distribution in time domain the rms pulse duration can be estimated as 45 fs whereas the inspection of the spectral domain yields a bandwidth of 2%.

4. Conclusion

We presented Start to End Simulations for the BESSY FEL High Energy SASE case. The results of these studies are well within our FEL performance aims. The tools, used for the S2E simulations, are suited for automated runs. In a next step tolerance studies are needed to perform.

References

[1] D. Krämer, et al., BESSY II: exceeding design parameters—the first year of user service, Proceedings of the European Particle Acceleration Conference, Wien, 2000, p. 640.

[2] TESLA TDR, DESY 2001-011, ECFA 2001-209, CD-ROM, March 2001.

[3] F. Stephan, et al., Photo injector test facility under construction at DESY Zeuthen, FEL 2000, Durham.

[4] K. Flöttmann, ASTRA user manual, http://www.desy.de/~mpyflo.

[5] M. Borland, Elegant: a flexible SDDS-compliant code for accelerator simulation, Advanced Photon Source LS-287, September 2000.

[6] S. Reiche, Nucl. Instr. and Meth. A 429 (1999) 243.

Available online at www.sciencedirect.com

Nuclear Instruments and Methods in Physics Research A 528 (2004) 481–485

ELSEVIER

NUCLEAR INSTRUMENTS & METHODS IN PHYSICS RESEARCH
Section A

www.elsevier.com/locate/nima

Accurate synchronization between short laser pulse and electron bunch using HGHG output at 266 nm as photo-cathode laser for RF-gun

L.H. Yu[a,*], Hanwei Yi[b]

[a] National Synchrotron Light Source, Brookhaven National Laboratory, ATF, Bldg 725c, Box 5000, Upton, NY 11973, USA
[b] Deptartment of Physics, Science College , Changchun University of Science and Technology, Changchun 130022, P.R. China

Abstract

We report a new scheme to realize accurate synchronization between the electron bunch and seed laser by using HGHG output at 266 nm as photo-cathode laser for the RF gun. One by-product of this method is the reduction of the energy fluctuation in addition to the reduced time jitter.
© 2004 Published by Elsevier B.V.

PACS: 41.60.Cr

Keywords: Free electron laser

1. Introduction

Synchronization between seed laser pulse and electron bunch to bellow 50 fs is an important issue in carrying out cascading HGHG stages to achieve intense temporally coherent X-ray FEL. The single-stage HGHG experiment in the IR [1] and recently in UV region [2] has achieved significant success.

Hence, to demonstrate the cascading of two stages of HGHG becomes important in confirming the viability of this scheme. Here we describe a simple way to achieve the required highly accurate synchronization.

During the HGHG experiment, we obtained consistently more than 100 μJ output at 266 nm [2] with 7% rms fluctuation. The seed pulse at 800 nm is about 9 ps long, while the electron pulse is 1 ps long. Thus time jitter which is less than 1 ps does not affect the HGHG output, and the HGHG output is perfectly synchronized with the electron bunch.

Now the idea of accurate synchronization is to split this HGHG output into two beams as shown in Fig. 1, send 90% of the intensity back to the photo-cathode of the RF gun to generate a second electron bunch, and use the rest of the 10% as a seed for the second electron bunch with appropriate delay. The crucially important point is now that both the new seed and the second electron bunch are highly accurately synchronized with the RF phase, as we shall explain now.

---

*Corresponding author. Tel.: + 1-631-344-5012.
E-mail address: lhyu@bnl.gov (L.H. Yu).

0168-9002/$ - see front matter © 2004 Published by Elsevier B.V.
doi:10.1016/j.nima.2004.04.136

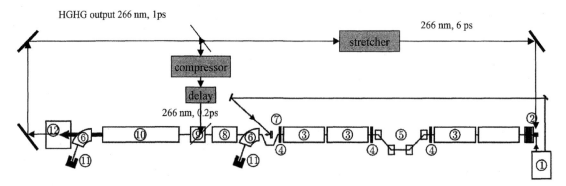

Fig. 1. The NSLS DUVFEL layout: 1—gun and seed laser system, 2—RF gun, 3—linac tanks, 4—focusing triplets, 5—magnetic chicane, 6—spectrometers dipoles, 7—seed laser mirror, 8—modulator, 9—dispersive section, 10—radiator (NISUS), 11—beam dumps, 12—FEL radiation measurements area.

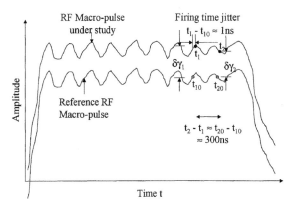

Fig. 2. Effect of the laser pulse firing time jitter relative to the klystron firing time ($t=0$).

The time interval between the Ti–saphire photo-cathode laser pulse reaching the cathode and the HGHG pulse reaching the cathode again is about 300 ns, corresponding to the round trip distance from the RF gun to the end of the NISUS undulator exit, about 100 m. During the 300 ns, the phase and amplitude of the RF system do change. From shot-to-shot the amplitude of the klystron output fluctuates and the firing time relative to laser pulse jitters about 1 ns (see Fig. 2). However, as we shall quantitatively analyse in the following, in spite of these effects, in our scheme the jitter between the electron bunch and laser pulse is reduced by the compression ratio.

2. Reduction of the shot-to-shot energy fluctuation

The macro-pulse of the RF is more than $2\,\mu s$ long, Usually we operate the cathode laser pulse at the peak of this pulse, which is near the end of a nearly flat top. For our case, we shall send the first Ti–saphire laser pulse about 300 ns earlier, so that the second pulse will arrive at the cathode near the end of the macro-pulse to optimize the second electron bunch for use in cascading HGHG, as shown in Fig. 2.

Now let us analyse the reduction of the time jitter quantitatively. We first assume a reference shot (or, the reference RF macro-pulse). For example, we denote the arrival time of the Ti–saphire pulse at the cathode as t_{10} while the arrival time of the split HGHG output pulse at the cathode as t_{20} (see Fig. 2). During the next macro-pulse, separated by several hundreds of milliseconds, the laser arrival times are t_1 and t_2, respectively. For convenience we always use the starting time of the macro-pulse as zero time. According to this convention, t_1-t_{10} is not of the order of hundreds of milliseconds, but is of the order of 1 ns, which is the klystron firing time jitter relative to the laser. Then we study the jitter and fluctuation of the next RF macro-pulse relative to this reference shot. Assume the acceleration is on a slope h, and during the next shot the acceleration before the compressor increases by $\delta\gamma_1/\gamma_0$, while the cathode laser is delayed by τ_1 (notice that τ_1 is the laser jitter relative to the RF phase, which is of the order of picosecond, rather than the klystron firing time jitter t_1-t_{10} of order of 1 ns). Then, the energy of the electron bunch increases by

$$\frac{\Delta\gamma_1}{\gamma_0} = h\tau_1 + \frac{\delta\gamma_1}{\gamma_0}. \qquad (1)$$

Let the complete accelerator system has $R_{56} > 0$ (main contribution is from compressor though there is a smaller contribution from the RF gun), when the energy increases by $\Delta\gamma_1/\gamma_0$, the time delay after the compressor relative to the reference shot is

$$
\begin{aligned}
\tau_2 &= -R_{56}\frac{\Delta\gamma_1}{\gamma_0} + \tau_1 \\
&= -R_{56}\left(h\tau_1 + \frac{\delta\gamma_1}{\gamma_0}\right) + \tau_1 \\
&= (1 - R_{56}h)\tau_1 - R_{56}\frac{\delta\gamma_1}{\gamma_0} \qquad (2)
\end{aligned}
$$

Since the HGHG output is synchronized with the electron bunch, when feedback to the cathode, during the same RF macro-pulse it will generate the second-electron bunch also delayed by τ_2. The time is later than the first electron bunch by $t_2 - t_1 \cong 300ns$. We assume that due to this time laps, relative to the reference shot, the acceleration before the compressor changes by $\delta\gamma_2/\gamma_0$, then the second electron bunch energy, relative to the reference shot, increases by

$$
\begin{aligned}
\frac{\Delta\gamma_2}{\gamma_0} &= h\tau_2 + \frac{\delta\gamma_2}{\gamma_0} = h(1 - R_{56}h)\tau_1 - hR_{56}\frac{\delta\gamma_1}{\gamma_0} + \frac{\delta\gamma_2}{\gamma_0} \\
&= (1 - R_{56}h)\left(h\tau_1 + \frac{\delta\gamma_1}{\gamma_0}\right) + \frac{\delta\gamma_2 - \delta\gamma_1}{\gamma_0} \\
&= (1 - R_{56}h)\frac{\Delta\gamma_1}{\gamma_0} + \frac{\delta\gamma_2 - \delta\gamma_1}{\gamma_0}. \qquad (3)
\end{aligned}
$$

The first term means that the electron beam energy fluctuation is suppressed by a factor $(1-R_{56}h)$, i.e. the compression ratio. We shall show later that the fluctuation of the second term can be made negligibly small.

3. Analysis of the time jitter reduction

After the compressor, the second electron bunch is delayed by

$$\tau_3 = -R_{56}\frac{\Delta\gamma_2}{\gamma_0} + \tau_2 = (1 - R_{56}h)\tau_2 - R_{56}\frac{\delta\gamma_2}{\gamma_0} \qquad (4)$$

The second-seed pulse, obtained by splitting from the HGHG output, is delayed by τ_2. Hence the jitter between the second-seed laser and the second electron bunch is

$$
\begin{aligned}
\tau_3 - \tau_2 &= -R_{56}h\tau_2 - R_{56}\frac{\delta\gamma_2}{\gamma_0} \\
&= -R_{56}\left(h\tau_2 + \frac{\delta\gamma_1}{\gamma_0}\right) - R_{56}\left(\frac{\delta\gamma_2}{\gamma_0} - \frac{\delta\gamma_1}{\gamma_0}\right) \\
&= -R_{56}\left(h(1 - R_{56}h)\tau_1 - hR_{56}\frac{\delta\gamma_1}{\gamma_0} + \frac{\delta\gamma_1}{\gamma_0}\right) \\
&\quad - R_{56}\frac{\delta\gamma_2 - \delta\gamma_1}{\gamma_0} \\
&= -(1 - R_{56}h)\left(R_{56}h\tau_1 + R_{56}\frac{\delta\gamma_1}{\gamma_0}\right) \\
&\quad - R_{56}\frac{\delta\gamma_2 - \delta\gamma_1}{\gamma_0} \\
&= (1 - R_{56}h)(\tau_2 - \tau_1) - R_{56}\frac{\delta\gamma_2 - \delta\gamma_1}{\gamma_0} \qquad (5)
\end{aligned}
$$

where we used Eq. (2).

The first term means that the time jitter without feedback, $\tau_2-\tau_1$, is reduced by a factor $(1-R_{56}h)$, which is just the well-known compression ratio. The second term is due to the RF system amplitude change during the time laps $t_2 - t_1 \cong 300$ ns.

4. Analysis of the klystron variation

We shall show now that the second term of Eq. (5) can be made negligibly small. In Fig. 2 we show the acceleration before the compressor as a function of the injection time of the cathode laser. This macro-pulse lasts about 2–3 μs. By carefully tuning the RC circuit in the modulator, the top of the pulse can be made as flat as 0.1%, with residual ringing of about ten "periods". The shape of the ringing is not very regular, but is reproducible, and can be adjusted by tuning the PFN circuit. Even though it is not sinusoidal, we approximately represent the last part of the marco-pulse by a sinusoidal function as a rough estimate: $\delta\gamma(t) = A \sin(\omega t + \varphi)$. Consulting Fig. 2, we have

$$
\begin{aligned}
\delta\gamma_2 - \delta\gamma_1 &= \delta\gamma(t_2) - \delta\gamma(t_1) \\
&= (\delta\gamma(t_2) - \delta\gamma(t_{20})) - (\delta\gamma(t_1) - \delta\gamma(t_{10})) \\
&\quad + (\delta\gamma(t_{20}) - \delta\gamma(t_{10})) \\
&\cong A\omega(t_2 - t_{20})\cos(\omega t_{20} + \varphi) \\
&\quad - A\omega(t_1 - t_{10})\cos(\omega t_{10} + \varphi) + \Delta\gamma_0 \\
&\cong A\omega(t_1 - t_{10})(\cos(\omega t_{20} + \varphi) \\
&\quad - \cos(\omega t_{10} + \varphi)) + \Delta\gamma_0
\end{aligned}
\tag{6}
$$

where $\Delta\gamma_0 \equiv (\delta\gamma(t_{20}) - \delta\gamma(t_{10}))$ is a constant determined by the reference macro RF pulse, which does not contribute to any fluctuation. We have used $(t_2 - t_{20}) - (t_1 - t_{10}) = (t_2 - t_1) - (t_{20} - t_{10}) \cong 0$.

The klystron firing time jitter $t_1 - t_{10}$ is of the order of 1 ns, the ringing amplitude is taken as $A/\gamma_0 = 0.5\%$ (a conservative estimate)[3], the ringing period [3] is roughly 200 ns so $\omega \cong 2\pi/200$ ns. Under the worst scenario we can take the timing of the first- and second-cathode laser pulses to be such that $(\cos(\omega t_{20} + \varphi) - \cos(\omega t_{10} + \varphi)) \cong 2$. With these parameters we find the fluctuation of $(\delta\gamma_2 - \delta\gamma_1)/\gamma_0 \cong 3 \times 10^{-4}$. In the DUVFEL system $R_{56} = 5$ cm, hence we find the contribution of the second term of Eq. (5) as

$$
\begin{aligned}
R_{56}(\delta\gamma_2 - \delta\gamma_1)/\gamma_0/c &= 5 \text{ cm} \times 3 \times 10^{-4}/c \\
&= 15 \text{ μm}/c = 50 \text{ fs}
\end{aligned}
\tag{7}
$$

Actually, the laser firing time and the optical path length can be adjusted, while the ringing curve can be tuned such that $\cos(\omega t_{20} + \varphi) \cong \cos(\omega t_{10} + \varphi) \cong 0$. This corresponds to selecting the laser arriving time at times when the time derivative of the ringing curve is zero. Then the effect of the second term can be neglected. Thus the time jitter is reduced by the compression ratio. As an important by product, we also reduce the energy fluctuation before the compressor by the compression ratio, as discussed following Eq. (3).

5. Numerical example in DUV FEL

Combined with the ballistic compression in the RF gun, we have the effective slope $h \approx 0.17/\text{cm}$. Thus the compression ratio is $1 - R_{56}h \cong 0.15$. The time jitter between the Ti–Saphire laser and the electron bunch is 0.3 ps FWHM, as recently measured by the electro-optic detection method [4]. Thus the jitter between the second electron bunch and the second seed from the HGHG output is reduced to 0.3 ps × 0.15 = 45 fs ≈ 50 fs.

6. Requirement on the laser system

To have the RF gun operating correctly, we need to stretch the 1 ps HGHG pulse to about 6 ps, hence in Fig. 1 we show a stretcher after the beam splitter. On the other hand, to carry out a cascading experiment using the "fresh bunch technique", we need to use a seed pulse much shorter than the electron pulse. Hence we show a compressor. In this example of a proof-of-principle experiment, we compress the seed pulse to 0.2 ps. Also shown in the figure is a delay line for the seed laser because we need to arrange the new seed and the second electron bunch to arrive at the NISUS at the same time. To accomplish this compression, we need to operate the HGHG in a chirped pulse amplification (CPA) mode. The CPA mode operation is being planned at the DUVFEL to be carried out in near future. Analysis showed that with the 0.25% bandwidth it is possible to compress to about 200 fs. Thus we need to reduce the seed pulse of 1% bandwidth from 9 to 4 ps so that the 1 ps electron bunch has 0.25% chirp.

The optical path length of the second cathode laser can be arranged so that it will reach the cathode at about the same RF phase as the first electron bunch. It is important to stretch the pulse length in such a way that wavelength jitter would not cause time jitter. The important point is that this stretcher should be independent of wavelength chirp. One example of the method is to expand the laser beam and send to a grating, the output pulse length from the first diffraction line is then determined by the ratio of the transverse width of the expanded laser beam versus the transverse beam size of the output. There may be other methods to achieve the same goal though.

The loss in the stretcher of a grating at 266 nm is of order of 30%, hence the second cathode laser pulse energy is about 60 μJ, if we use part of the present HGHG output number of 100 μJ as the input to the stretcher. This is about what we need for the copper cathode now: about 60 μJ. To shape the pulses there may be more losses. There are two solutions to the problem. One is to use the magnesium cathode to increase the cathode photo efficiency. The other is to improve the HGHG output. Actually, so far the maximum output of the HGHG experiment is 200 μJ when the 800 nm seed is optimally maintained. We expect the HGHG output will reach steady output of more than 200 μJ in near future.

7. Possible applications

The new seed and the second electron bunch are accurately synchronized. Hence the 266 nm FEL amplifier using them can be used to study the FEL performance as has never been done before: we can move the seed to different parts of the second electron bunch and measure the gain for different slices of the bunch. Since the second electron bunch has a very small jitter relative to the RF gun, its performance will be very stable, and can be used to optimize the electron gun operating conditions.

As a first step to test this idea, we do not need the stretcher and the compressor at the first stage of the experiment. We only need an optical transport line for the HGHG output to be sent back to the cathode. The second electron bunch will reach the detector 11 at the end of the NISUS

after passing through the energy spectrometer 6, and observed together with the first electron bunch. Since the first electron bunch has been used by the HGHG process, it has large energy spread, while the second one will have small energy spread. So it is easy to distinguish them. Since the second electron bunch has much smaller time jitter than the first one we should see much smaller energy fluctuation compare with the first one. This will be a proof of the success of this scheme.

Since the second electron pulse is separated from the first by 300 ns, there are many ways to separate them in the future plan. For example, the magnetic kicker used in storage rings with speed faster than a few tens of ns can be used for this purpose.

A possible residual jitter may come from the vibration of the mirrors in the long distance transport of the HGHG output to the cathode. However, since most vibration on the floor is order of sub microns, it will contribute to the jitter by only about a few fs at the most.

Acknowledgements

The author thanks Dr. Jim Rose, Dr. Timur Shaftan and Dr. A. Doyuran for discussions on various subjects about the RF system and the compressor crucially important for this work.

References

[1] L.H. Yu, et al., Science 289 (2000) 932.
[2] L.H. Yu, et al., Phys. Rev. Lett. 91 (2003) 074801.
[3] J. Rose, private communications.
[4] H. Loos, et al.,Proceedings of PAC2003 2003.

Available online at www.sciencedirect.com

SCIENCE ⓓ DIRECT°

ELSEVIER   Nuclear Instruments and Methods in Physics Research A 528 (2004) 486–490

**NUCLEAR
INSTRUMENTS
& METHODS
IN PHYSICS
RESEARCH**
Section A

www.elsevier.com/locate/nima

Observation of SASE in LEBRA FEL system

T. Tanaka*, K. Hayakawa, I. Sato, Y. Hayakawa, K. Yokoyama

Laboratory for Electron Beam Research and Application (LEBRA), Nihon University, 7-24-1 Narashinodai, Funabashi 274-8501, Japan

Abstract

A large enhancement of spontaneous undulator radiation has been observed during FEL lasing experiments at LEBRA. The enhancement has been observed only with the detector for the infrared fundamental radiation. The detector output signal showed spikes during the electron beam pulse, yet no apparent enhancement was observed with a CCD camera monitoring the visible harmonic radiations. An enhancement factor greater than 10 has been obtained with a 2.4 m long undulator with a completely detuned FEL optical cavity length and depends strongly on the parameters of the linac RF system. This implies that the SASE operation is possible even with a conventional electron beam by achieving suitable bunch compression.
© 2004 Published by Elsevier B.V.

PACS: 41.60.Cr; 41.85.Ja

Keywords: Infrared; FEL; SASE; Bunch compression

1. Introduction

A self-amplified spontaneous emission (SASE) device has been considered as the next generation high-power radiation source, especially in the vacuum ultraviolet (VUV) to X-ray ranges where high-reflectance optical mirrors are unavailable. Typical X-ray SASE projects proposed so far [1] assume a combination of a high-energy electron linac and an extremely long undulator magnet. Meanwhile, recent experimental results on SASE in the near infrared (IR) to visible spectral ranges have shown that the SASE devices in these spectral ranges can reach optical power saturation within

several meters along the undulator path by implementing a high charge density electron beam of medium energy [2,3].

The linac-based FEL system of the Laboratory for Electron Beam Research and Application (LEBRA) in Nihon University [4] was designed to generate FEL in the near IR to UV spectral ranges by using an optical cavity having high-reflectance mirrors. The present undulator consists of 50 periods of a Halbach-type permanent magnet array having a period of 4.8 cm, which was designed for lasing in the near IR range with the maximum RMS K-value of 1.2 [5]. Although the LEBRA FEL system was not designed to be used for SASE experiments, a large enhancement of the fundamental spontaneous radiation was observed at a detuned optical cavity length by adjusting the

*Corresponding author.

E-mail address: tanaka@lebra.nihon-u.ac.jp (T. Tanaka).

0168-9002/$ - see front matter © 2004 Published by Elsevier B.V.
doi:10.1016/j.nima.2004.04.137

linac operating condition. This suggests that the SASE mode operation is possible even with a relatively low-current electron beam from a conventional linac.

This paper reports the behavior of the optical power enhancement observed at the LEBRA FEL system in terms of the linac operating condition and the bunch compression effect in the achromatic bending system.

2. LEBRA FEL system

The schematic layout of the LEBRA FEL system, upgraded since its first lasing, is shown in Fig. 1.

Since the electron beam is ejected continuously from the 100-kV thermionic gun over a linac RF pulse duration, the electron bunch is accelerated during every period of the 2856 MHz RF. The electron beam accelerated in the linac is transported to the FEL undulator through the 90° achromatic bending and analyzing magnet system (BM1 to BM2). The energy spread of the electron beam is defined with a horizontal slit placed in the dispersive section, where the energy spread is restricted to 0.3–2% of the central energy depending on the purpose of experiment. The transverse beam positions at the entrance and the exit of the undulator are adjusted by monitoring both the

horizontal and vertical positions of the beam with strip-line type beam position monitors [6,7].

The optical cavity consists of two silver-coated copper mirrors (M1, M2) having a reflectance of approximately 99.3%. The radiation is extracted from the cavity through a coupling hole of 0.15-mm radius in the mirror M2. The extracted divergent optical beam then is converted into a parallel beam by the beam expander optics consisting of an ellipsoidal mirror (M5) and a parabolic mirror (M6). The parallel optical beam is transported to the next room and then is sent to the experimental facility. In the next room, the optical beam is detected with either an InSb IR detector or a CCD camera after being separated by a cold mirror (M11) into IR and other higher harmonic components.

The latest operation of the LEBRA FEL system demonstrated the FEL wavelength tunability from 1.4 to 5 μm with electron beams ranging from 53 to 86 MeV in combination with various K-values. The FEL output power of approximately 10 mJ/macropulse has been obtained at 1.6 μm, which has been measured by means of an IR power meter placed at the same position as the IR detector in Fig. 1. Considerable enhancement of the harmonic components in the visible range also has been observed with the CCD camera, following the intense lasing of the fundamental FEL [5].

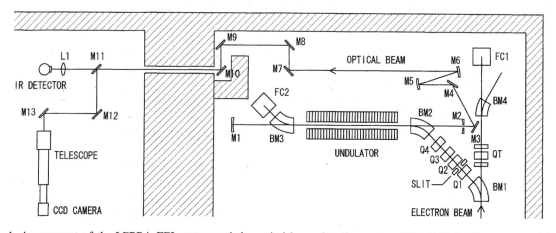

Fig. 1. Arrangement of the LEBRA FEL system and the optical beam detecting system. BM1-BM4: bending magnets; Q1–Q4: quadrupole magnets; FC1, FC2: Farady cups; M1, M2: cavity mirrors; M3, M4, M7-M10, M12, M13: total reflection plane mirrors; M11: cold mirror.

3. Observation of SASE

An enhancement of the spontaneous radiation in the LEBRA FEL system was initially observed during an adjustment of the electron beam performed to search for a beam condition that allowed maximum FEL power. A large enhancement of the optical power in the cavity has been reproduced by the experiments performed at various combinations of the electron energy and the undulator radiation wavelength. Here, we discuss the results of the experiment performed at an electron energy of 86 MeV.

Typical behavior patterns of the optical power buildup observed at the fundamental radiation of 1.6 μm are shown in Fig. 2, together with the macropulse shape of the electron beam current. In the experiment, the cavity mirrors were aligned to build up the optical power, which allowed us to observe the large output signal from the InSb detector. The cavity length was completely detuned by shifting the distance between the two mirrors by 1 mm to eliminate the FEL gain, which is sufficient to avoid an overlap of a beam bunch and a previously generated photon pulse considering the RF phase stability [8]. In this case, only the

fundamental component of the spontaneous emission was injected into the InSb detector due to the property of the cold mirror M11. Thus, the buildup waveform of the spontaneous emission in the cavity usually takes the shape shown in Fig. 2B, corresponding to a high reflectance of the mirrors.

The enhancement of the spontaneous emission power shown in Fig. 2A was achieved by adjusting the RF phase for the buncher and the regular accelerating tubes. The enhanced power is not uniform over the entire macropulse; a large enhancement seems to occur intermittently. In the experiment the waveform has varied drastically from pulse to pulse, showing that the quality of the electron beam is critical in satisfying a condition to enhance the power. The enhancement has also been strongly dependent on the RF phase and the magnetic focusing. In some cases, the enhanced power was in excess of 100 times the common level of buildup.

In order to investigate the property of the enhanced optical power, dependence of the optical power on the deflection of the electron beam in the central plane of the undulator field was measured by exciting a steering coil placed at a distance of 0.95 m downstream the undulator entrance. In this measurement, the power buildup was suppressed by deflecting the upstream mirror M2 by 1.5 mrad, in addition to the cavity length detuning. Then, instead of the buildup and decay curves shown in Fig. 2A, we observed the optical power shape as a sum of a square shape due to the contribution from the spontaneous undulator radiation and many unstable, strong peaks with amplitude five times that of the square shape.

If the beam is sufficiently deflected with the steering coil, the amplitude of the spontaneous emission power has to be reduced to less than $\frac{1}{6}$, considering the shortening of the effective undulator length [9]. However, the amplitude of the square shape was only reduced to nearly $\frac{1}{2}$ even with a maximum beam deflection of about 2 mrad, which is due to a wide angular distribution of the spontaneous emission compared with the deflection angle. On the other hand, the amplitude of the peaks in the optical power shape was strongly reduced by the increase in the deflection angle of

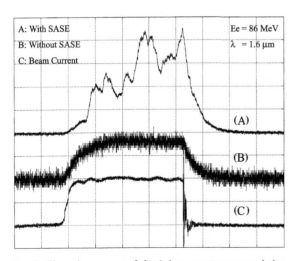

Fig. 2. The enhancement of the 1.6 μm spontaneous emission power observed using the InSb detector (A, vertical scale is 200 mV/div), compared with the buildup shape without amplification (B, vertical scale is 20 mV/div). The electron beam pulse shape is shown at the same timing (C, vertical scale is 20 mA/div). The horizontal scale is 4 μs/div.

the electron beam; the peaks disappeared with the maximum beam deflection. Therefore, evidently the peaks were due to a strong emission of radiation in the last half of the undulator. Furthermore, the large decrease in the amplitude of the peaks that resulted from the beam deflection cannot be explained by the property of the spontaneous emission or a coherent radiation that may be emitted by a fine structure of the longitudinal beam bunch profile. The result suggests the existence of an amplified radiation by the interaction between the electron beam and the undulator radiation. Thus, we conclude that the strong peaks in the optical power shape were due to the amplification of radiation in the undulator by the SASE process.

These facts remind us of the possibility of strong bunch compression in the bending magnet system, which must be sufficient to achieve the bunch length of around 0.3 ps or shorter based on the peak beam current [2,5,10].

On the other hand, however, no indication of enhancement in the higher harmonics has been found on the CCD image monitoring the visible components.

4. Consideration of bunch compression

The linac consists of the prebuncher, buncher, and regular 4-m long accelerating tubes ACC1, ACC2 and ACC3. The phase of the RF fed in each component is adjusted independently, so that the electron bunch can be accelerated at any RF phase in the regular accelerating tubes, which allows a wide variability of the output energy and a fine adjustment of the longitudinal beam property.

As described with the open circles and solid lines in Fig. 3, the energy spectrum of the electron beam, which was measured by means of the 90° bending system that worked as a spectrometer, had a broad peak of FWHM 2%. Despite the reduction of beam current in the undulator beam line, SASE has been observed only with this beam condition. Another spectrum described with solid circles and solid lines corresponds to the operation when the maximum beam current was obtained in the undulator beam line, which has a narrow peak

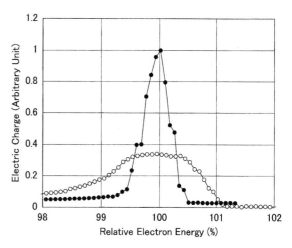

Fig. 3. Differences in the energy spectra, which depend on the linac operation, measured at the exit of the linac. SASE was observed at the spectrum described with open circles. The solid circles correspond to the operation providing the maximum current in the undulator beam line. The spectra were obtained by averaging the data from 2 to 18 μs in a total 20 μs beam pulse duration.

of FWHM 0.5% that represents the approximate minimum energy gradient at the exit of the linac. However, SASE has not been observed with this condition.

The phase of the electron bunch when self-amplification occurred has been deduced from the measurement of the RF phase shift in terms of the peak phase (90°) that results in the maximum beam loading. The deduced values are 81.5° in ACC1, 81.4° in ACC2, and 35.1° in ACC3, respectively. For the beam with 86 MeV at the exit of the linac, the longitudinal energy gradient in the bunch, approximately 0.60 MeV/deg or 0.70%/deg, has been deduced from the bunch phase and the accelerating field in all the regular tubes with linear approximation. The result is in agreement with the value, 0.73%/deg, deduced from another SASE experiment performed with the same procedure at 69 MeV.

According to the design calculation based on the first order beam matrix theory [11], the 90° achromatic bending system causes a path difference factor of −0.86 mm/%, or −2.9°/% for the displacement of the electron energy. The total path difference, the sum of the path difference factor and the inverse number of the energy gradient, is $-1.5°/\%$. The result suggests that considerable bunch compression is possible if the energy gradient at the exit of the buncher is a negative value, e.g. around $-0.35\%/\text{deg}$. Thus, it is necessary to precisely simulate the bunch formation process in the prebuncher and the buncher for further investigation of the SASE phenomenon observed with the low-current beam from the conventional linac.

5. Summary

The LEBRA FEL system demonstrated the FEL tunability from 1.4 to 5 μm. SASE in the near IR range has been observed using the electron beam with a low macropulse beam current, 40 mA, from the conventional linac of LEBRA, which suggests the formation of a very short bunch with considerable bunch compression in the achromatic bending system. The behavior of the energy distribution in the bunch was estimated in terms of the bunch compression, however, the preliminary data obtained from the experiment is insufficient to prove the achievement of bunch compression.

References

[1] G. Dattoli, A. Renieri, Nucl. Instr. and Meth. A 507 (2003) 464.
[2] A. Murokh, et al., Nucl. Instr. and Meth. A 507 (2003) 417.
[3] A. Doyuran, et al., Nucl. Instr. and Meth. A 507 (2003) 392.
[4] T. Tanaka, et al., Proceedings of APAC'98 (Tsukuba, Japan March 1998) 722.
[5] Y. Hayakawa, et al., Nucl. Instr. and Meth. A507 (2003) 404.
[6] T. Suwada, et al., AIP Conference Proc. 319 (1993) 334.
[7] K. Ishiwata, et al., Proceeding of LINAC2002, Gyeongju, Korea, 2002, p179.
[8] K. Yokoyama, et al., Proceeding of FEL2003, Tsukuba, Japan, 2003, Nucl. Instr. and Meth. A (2003), these Proceedings.
[9] C.A. Brau, Free-Electron Lasers, Academic Press, Inc., San Diego, 1990.
[10] J.N. Galayda, Proceeding of LINAC2002, Gyeongju, Korea, August 2002, p. 6.
[11] K.L. Brown, SLAC-75 (1972).

Available online at www.sciencedirect.com

SCIENCE DIRECT•

Nuclear Instruments and Methods in Physics Research A 528 (2004) 491–496

NUCLEAR
INSTRUMENTS
& METHODS
IN PHYSICS
RESEARCH
Section A

www.elsevier.com/locate/nima

High-gain ring FEL as a master oscillator for X-ray generation

Nikolay A. Vinokurov*, Oleg A. Shevchenko

Budker Institute of Nuclear Physics, 11 Acad. Lavrentyev Prosp., 630090 Novosibirsk, Russian Federation

Abstract

High-gain free electron laser (FEL) with bends between undulator sections is discussed. Such FEL configuration may be used for the mirror-free master oscillator in X-ray band. The oscillator linewidth is estimated. Results of computation for the state-of-art electron beam parameters show the feasibility of the X-ray oscillator FEL.
© 2004 Elsevier B.V. All rights reserved.

PACS: 41.60.Cr

Keywords: Free electron laser

1. Introduction

One of the main problems of modern X-ray free electron laser (FEL) projects is poor radiation quality. The reason for it is in the SASE principle of operation—noise amplification at high-gain amplifier. The solution of the problem—the use of master oscillator—is wellknown in laser physics. The lack of proper mirrors in the X-ray band makes such oscillator not so simple, as it is in longer-wavelength bands. The possible mirror-free FEL configuration is discussed in this paper. It consists of several undulator sections separated with almost isochronous bends. As the electron beam bunching is partly conserved in bends, the bunching growth in such system is possible. After

360° rotation, radiation from the last undulator section illuminates the beam in the first section and induces bunching in the incoming fresh beam (see Fig. 1). This way the feed-back loop is closed, and for high-enough gain the signal growth until saturation takes place. The electron beam parameters, which are acceptable to satisfy the oscillation condition are calculated with modified code RON [1].

2. Analytical estimation

The scheme of ring FEL was proposed first in paper [2]. In that paper the break between two undulator sections with magnetic buncher inside was considered using the wide electron beam approximation. The length of undulator sections was significantly more than the gain length. It was shown that if one absorbs (makes zero) the radiation field in the break (and therefore at the

*Corresponding author. Tel.: +7-3832-394003; fax: +7-3832-342163.

E-mail address: n.a.vinokurov@inp.nsk.su (N.A. Vinokurov).

0168-9002/$ - see front matter © 2004 Elsevier B.V. All rights reserved.
doi:10.1016/j.nima.2004.04.138

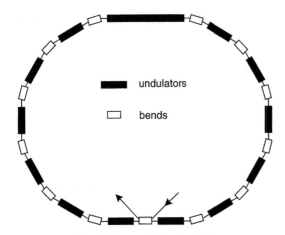

Fig. 1. Scheme of the ring FEL.

entrance to the second undulator section), the signal at the exit of in the second undulator section will be reduced, although not so much. The cause of the signal existence at the end of the second undulator section is the electron beam bunching at the entrance of the second undulator section. For the simplest case without buncher for the cold electron beam the signal amplitude reduction factor is $\frac{2}{3}$. The non-zero beam energy spread comparable with $\lambda_u/(4\pi L_g)$ (L_g is the power gain length and λ_u is the undulator period), will decrease this factor. Therefore to keep the signal amplification along the ring FEL, it is necessary to provide the undulator length more than two power gain length. Similar results were obtained in the paper on undulator alignment tolerances [3] using both analytical and numerical methods.

Finite energy spread and transverse emittances decrease the beam bunching at the bend exit. The particle slippage (delay) with respect to the reference particle can be expanded over energy, coordinate and angle deviations. To obtain an isochronous bend, one has to make zero the expansion coefficient up to some order. The first-order compensation requires to make zero three elements of 6×6 matrix of the bend—R_{51}, R_{52} and R_{56} (if the bend is planar). This compensation can be easily obtained in symmetric three-bend achromat. It is worth noting that bunching conservation at visible wavelength in achromatic bend was demonstrated in Ref. [4]. The second-order coeffi-

cients T_{516}, T_{526} and T_{566} may be made zero by sextupoles (so-called second-order achromat). But the coefficients T_{511}, T_{512}, T_{521} and T_{522} typically stay non-zero (positive). For the straight-line section of length L, the particle average delay may be expressed by simple equation

$$c\langle\Delta t\rangle = \frac{\varepsilon_x}{2}\int_0^L \gamma_x\,\mathrm{d}s + \frac{\varepsilon_y}{2}\int_0^L \gamma_y\,\mathrm{d}s$$

where ε_x and ε_y are beam emittances, γ_x and γ_y are Twiss parameters. This expression may be used for rough estimation of debunching in the second-order achromatic bend.

To estimate the linewidth of the ring FEL consider the equivalent electric circuit shown in Fig. 2.

Variable component of beam current I_2 in the last undulator emit light, which induces variable component of the beam current I_1 in the first undulator. This dependence is linear

$$I_1 = \beta I_2. \tag{1}$$

The ring FEL amplify this current together with shot noise I_s

$$u(t) = \int_0^\infty K(\tau)[I_1(t-\tau) + I_s(t-\tau)]\,\mathrm{d}\tau. \tag{2}$$

To take into account a saturation, nonlinear element $F(u)$

$$I_2 = F(u) \tag{3}$$

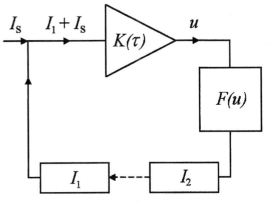

Fig. 2. The ring FEL flowchart.

is included to the circuit. Eqs. (1)–(3) give non-linear transformation

$$u(t+T) = \int_0^\infty K(\tau)\{\beta F[u(t-\tau)] + I_s(t-\tau)\}\,d\tau \tag{4}$$

where T is the time of particle circulation over ring FEL. T is relatively large, therefore no phase correlations exist over this time. In other words, we just study the properties of transformation

$$u_{out}(t) = \int_0^\infty K(\tau)\{\beta F[u_{in}(t-\tau)] + I_s(t-\tau)\}\,d\tau. \tag{5}$$

Assuming that $F(u) = u - \alpha u^3$ (the quadratic term can be skipped for narrow-band system) and defining slow complex amplitudes as $u_{in}(t) = a_{in}(t)e^{-i\omega t} + \text{c.c.}$ and $u_{out}(t) = a_{out}(t)e^{-i\omega t} + \text{c.c.}$, one can neglect oscillating terms in Eq. (5):

$$a_{out}(t) = \int_0^\infty K(\tau)e^{i\omega\tau}\{\beta[a_{in}(t-\tau) \\ - 3\alpha a_{in}(t-\tau)|a_{in}(t-\tau)|^2] \\ + I_s(t-\tau)e^{i\omega(t-\tau)}\}\,d\tau. \tag{6}$$

For small signal one can easily get from Eq. (6), the well-known oscillation condition

$$\left|\int_0^\infty \beta K(\tau)e^{i\omega\tau}\,d\tau\right| > 1. \tag{7}$$

To simplify notations, it is convenient to change variables as $b = a\sqrt{3\alpha}$, then

$$b_{out}(t) = \int_0^\infty k(\tau)\Big\{ b_{in}(t-\tau)\big[1 - |b_{in}(t-\tau)|^2\big] \\ + I_s(t-\tau)e^{i\omega(t-\tau)}\frac{\sqrt{3\alpha}}{\beta} \Big\}\,d\tau \tag{8}$$

where $k(\tau) = \beta K(\tau)e^{i\omega\tau}$. The stationary-amplitude solutions of homogeneous transformation are

$$|b_s|^2 = 1 - \frac{1}{\left|\int_0^\infty k(\tau)\,d\tau\right|}. \tag{9}$$

Linearization of transformation Equation (8) near b_s $b = b_s + c$ gives

$$c_{out}(t) = \int_0^\infty k(\tau)\Big\{ c_{in}(t-\tau)\big(1 - |b_s|^2\big) \\ - 2b_s Re\big[b_s^* c_{in}(t-\tau)\big] \\ + I_s(t-\tau)e^{i\omega(t-\tau)}\frac{\sqrt{3\alpha}}{\beta} \Big\}\,d\tau. \tag{10}$$

To simplify further consideration, one can choose $\arg\left[\int_0^\infty k(\tau)\,d\tau\right] = 0$ and then $(b_{in})_s = (b_{out})_s = |b_s| = b_s$. It can be shown that the main contribution to the linewidth is given by phase diffusion $\langle[\varphi(t) - \varphi(t-\theta)]^2\rangle$, where $\varphi = \arg(b) \approx Im(c)/b_s$. Neglecting term with small $Re(c_{in})$, the equation for phase transformation can be simplified to

$$\varphi_{out}(t) = \int_0^\infty \Big\{ Re[k(\tau)]\varphi_{in}(t-\tau)(1-b_s^2) \\ + Im\Big[k(\tau)I_s(t-\tau)e^{i\omega(t-\tau)}\frac{\sqrt{3\alpha}}{\beta b_s}\Big] \Big\}\,d\tau. \tag{11}$$

Corresponding equation for spectral density is

$$\big|\varphi_{out}(\Omega)\big|^2 = \big|\varphi_{in}(\Omega)\big|^2(1-b_s^2)^2|(Re\,k)(\Omega)|^2 \\ + \frac{3\alpha}{2b_s^2}|I_s(\omega)|^2|K(\omega)|^2 \tag{12}$$

where $(Re\,k)(\Omega)$ is the Fourier transform of $Re\,k(\tau)$. The steady-state solution of Eq. (12) is

$$|\varphi(\Omega)|^2 = \frac{3\alpha}{2b_s^2}|I_s(\omega)|^2|K(\omega)|^2 \\ \times \frac{|(Re\,k)(0)|^2}{|(Re\,k)(0)|^2 - |(Re\,k)(\Omega)|^2}. \tag{13}$$

One can choose the carrier frequency ω in such a way that $|(Re\,k)(0)|^2 = \max_\Omega|(Re\,k)(\Omega)|^2$. Then the natural definition of the round-trip bandwidth $\Delta\omega$ is

$$|(Re\,k)(\Omega)|^2 \approx |(Re\,k)(0)|^2\left[1 - \left(\frac{\Omega}{\Delta\omega}\right)^2\right]. \tag{14}$$

The phase spectral density frequency dependence is $2D/\Omega^2$. Then

$$\langle[\varphi(t) - \varphi(t-\theta)]^2\rangle = 2D\theta \tag{15}$$

where the phase diffusion coefficient (and therefore linewidth $\delta\omega$) is

$$\delta\omega = D = \frac{3\alpha}{4b_s^2}|I_s(\omega)|^2|K(\omega)|^2(\Delta\omega)^2. \qquad (16)$$

The linewidth is equal to the phase diffusion coefficient as $1/D$ is the signal dephasing time. The spectral density of shot noise is $|I_s(\omega)|^2 = 2Ie$, I is the beam current, and e is the charge of electron. The stationary amplitude corresponds to almost full bunching, therefore $b_s^2 = 3\alpha a_s^2 \sim \alpha I^2$. Then

$$\frac{\delta\omega}{\omega} \sim \frac{e\omega}{I}\left(\frac{\Delta\omega}{\omega}\right)^2|K(\omega)|^2. \qquad (17)$$

The typical value for $(\Delta\omega/\omega)^2$ is 10^{-7}, for $e\omega/I \sim 10^{-3}$ assuming $K = 10$ we have $\Delta\omega/\omega \sim 10^{-8}$. For wavelength $\lambda = 2\,\text{Å}$ it corresponds to the coherence length 1 cm. Typical bunch length in from RF linacs is much less, therefore the linewidth value will be limited by the Fourier-transform limit (10^{-6} or more).

3. X-ray FEL

To make calculations for the electron beam with finite energy spread and transverse emittances, we used the linear time-independent code RON [1]. The list of input parameters is shown in Table 1.

Table 1
X-ray ring FEL parameters

Energy GeV	14.35
Peak current (kA)	2
Relative energy spread (%)	0.008
Normalized rms emittance (μm)	1.2
Undulator period (m)	0.03
Undulator deflection parameter K	3.71
Radiation wavelength (Å)	1.5
Undulator section length (m)	18
Undulator first and last section length (m)	18
Bend angle (deg)	30
Bend length (m)	6
Bend $\int \gamma_x \, ds$	0.864
Bend $\int \gamma_y \, ds$	1.245
Distance between first and last undulator ends (m)	2

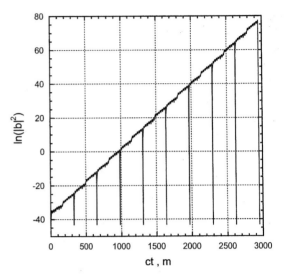

Fig. 3. Time dependence of logarithm of the bunching amplitude square. Deep drops correspond to the bunching transfer to the fresh (new-coming) electron beam.

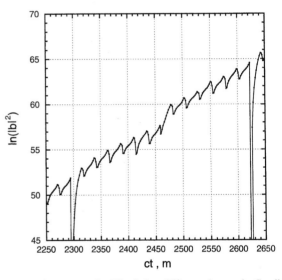

Fig. 4. The same as for Fig. 3, but different time scale. Small drops correspond to the bunch passing through the bends. Higher growth rate at the center is due to lack of bend between two central undulator sections.

To use RON the bends were substituted by long straight sections. The radiation field was cancelled at each straight section. The ring FEL was substituted by the sequence of several ring equivalents. The bunching at the entrance of the

Table 2
Soft X-ray ring FEL parameters

Energy (GeV)	0.485
Peak current (kA)	0.3
Relative energy spread (%)	0.05
Normalized rms emittance (μm)	5
Undulator period (m)	0.03
Undulator deflection parameter (K)	2
Radiation wavelength (Å)	500
Undulator section length (m)	12
Undulator first and last section length (m)	5
Bend angle (deg)	60
Bend length (m)	10
Bend $\int \gamma_x \, ds$	3.1
Bend $\int \gamma_y \, ds$	6.22
Distance between first and last undulator ends (m)	2

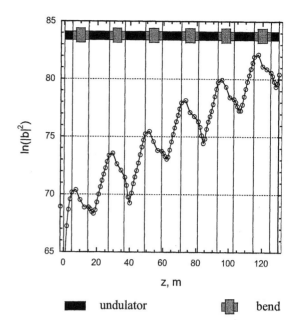

Fig. 6. The same as for Fig. 4, but for the prototype. The decrease of bunching not only at the bends, but at the beginnings of undulator sections is seen.

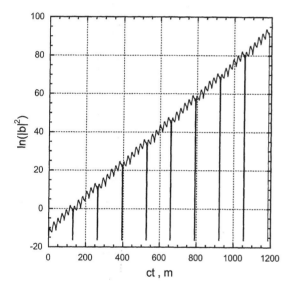

Fig. 5. The same as for Fig. 3, but for the low-energy prototype.

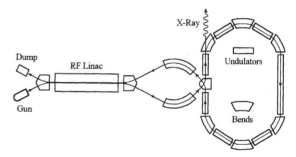

Fig. 7. Possible layout of ring FEL with energy recovery linac.

first undulator was made zero. After few passes through the ring the main eigenmode has set in. The typical bunching dependence on the longitudinal coordinate (and, correspondingly, time) is shown in Figs. 3 and 4.

4. Soft X-ray prototype

To test the ring FEL concept the small-size prototype with moderate electron beam para-

meters may be useful. Moreover, there are lots of interesting applications for the soft X-ray ring FEL. Its parameters are listed in Table 2. The results of calculations are presented in Figs. 5 and 6.

5. Accelerator

The ring FEL requires CW beam with high-enough peak current. Average current may be rather low, as the minimum bunch repetition rate is few MHz (corresponding to the ring circumference). On the other hand, the beam average power is significant. Therefore an energy recovery linac (probably, multiturn) seems to be a good candidate as the source of electron beam with required parameters. One of the possible schemes, which looks prospective for lower energies [5], is shown in Fig. 7.

6. Conclusion

In this paper we tried to "prove an existence theorem" for the high-gain ring FEL. Real design and after that a low-energy prototype are necessary to explore this concept further. Some combinations with seeded higher power FEL and harmonics generation are possible.

References

[1] R.J. Dejus, O.A. Shevchenko, N.A. Vinokurov, Nucl. Instr. and Meth. A 429 (1999) 225.
[2] N.A. Vinokurov, Nucl. Instr. and Meth. A 375 (1996) 264.
[3] T. Tanaka, H. Kitamura, T. Shintake, Phys. Rev. STAB 5 (2002) 040701.
[4] N.G. Gavrilov, G.N. Kulipanov, V.N. Litvinenko, I.V. Pinayev, V.M. Popik, I.G. Silvestrov, A.S. Sokolov, P.D. Vobly, N.A. Vinokurov, IEEE J. Quantum Electron. 27 (1991) 2569.
[5] E. Minehara, Nucl. Instr. and Meth. A 483 (2002) 8.

Available online at www.sciencedirect.com

SCIENCE  DIRECT°

ELSEVIER

Nuclear Instruments and Methods in Physics Research A 528 (2004) 497–501

NUCLEAR
INSTRUMENTS
& METHODS
IN PHYSICS
RESEARCH
Section A

www.elsevier.com/locate/nima

Future concept for compact FEL using a field emission micro-cathode

H. Mimura[a,b,*], H. Ishizuka[c], K. Yokoo[b]

[a] Research Institute of Electronics, Shizuoka University, 3-5-1 Johoku, Hamamatsu 432-8011, Japan
[b] Research Institute of Electrical Communication, Tohoku University, 2-1-1 Katahira, Aoba-ku, Sendai 980-8577, Japan
[c] Faculty of Engineering, Fukuoka Institute of Technology, Higashi-ku Fukuoka 811-0295, Japan

Abstract

The paper proposes the concept of the compact free electron laser using Smith–Purcell radiation and field emitters, and describes preliminary experiments associated with the generation of a modulation beam directly from the cathodes. The emission characteristics of the GaAs emitter designed for the Gunn effect strongly suggest the generation of a modulation beam. The illumination of a pulse laser on a p-type emitter also generates a modulation beam. However, for a high-frequency modulation beam, the suppression of the diffusion of the photogenerated carriers in the emitter is necessary.
© 2004 Elsevier B.V. All rights reserved.

PACS: 85.45.Db; 41.60.Cr; 85.30.Fg

Keywords: Smith–Purcell radiation; Field emitter array; Fabry–Perot resonator; Gunn effect; Modulation beam; Photo-response

1. Introduction

Field emitter arrays (FEAs) have been developed as a key element of vacuum nanoelectronics. Their applications to RF tubes such as a gyrotoron and a travelling wave tube have already been demonstrated [1]. Potential applicability of FEAs to compact free electron lasers (FELs) using Smith–Purcell or Cherenkov devices has been pointed out by several authors [2–4] for FEA's capability of yielding low-emittance high brightness electron beams.

Recently, we have successfully demonstrated Smith–Purcell radiation (SPR) at visible wavelength by using Si FEAs [5]. SPR covers a broad band ranging from visible light to millimeter waves in accordance with the grating constant, the electron beam velocity, and the angle of observation of the light.

In this paper, we propose the concept of the compact FEL using SPR and FEAs, and describe preliminary experiments for the concept.

---

*Corresponding author. Research Institute of Electrical Communication, Tohoku University, 2-1-1 Katahira, Aoba-ku, Sendai 980-8577, Japan. Tel.: +81-22-217-5511; fax: +81-22-217-5513.

E-mail address: mimura@riec.tohoku.ac.jp (H. Mimura).

2. Concept of the FEL

Fig. 1 shows the concept of the compact FEL. The modulation electron beam generated directly from FEAs is passed over the surface of a metal grating to emit SPR. It is expected that the adjusting of the frequency and velocity of the modulation beam enhances the SPR by the coherent effect [6] and that the changing of the mirror spacing of the Fabry–Perot resonator produces a stimulated radiation.

3. Modulation electron beam

3.1. Electron emission from Gunn domain

When a voltage higher than the threshold voltage is applied on GaAs or InP diodes, it is well known that travelling dipole domains are generated in the diode and the Gunn oscillation takes place [7]. If FEAs are fabricated of light-doped GaAs and the electric field inside the emitter exceeds 3.2 kV/cm, the travelling domains could be generated in the emitter and the accumulated charges in the domains could be periodically emitted into vacuum when they reach the emitter surface. The frequency of the modulation beam is determined by the emitter size and covers microwave and millimeter wave frequencies, similar to that in a conventional Gunn diode.

We fabricated a GaAs field emitter with a high aspect ratio using both anisotropic and isotropic etching [8]. The doping concentration of the GaAs emitter is 1×10^{15} cm^{-3}, which satisfies formation of the travelling dipole domains. A scanning electron microscope (SEM) image of the GaAs emitter is shown in Fig. 2. The emitter with an area of $1\,\mu m \times 1\,\mu m$ and a height of $14\,\mu m$ was realized. The radius of the tip curvature was less than 20 nm. The expected frequency of the modulation beam for the emitter is 10 GHz. The emission characteristic of the emitter is shown in Fig. 3. The emission current was measured at a pressure of 2×10^{-8} Torr using a diode configuration with an anode–cathode distance of $10\,\mu m$. The emission current is observed at the anode voltage of 200 V and the emission current increases with increasing anode voltage up to 600 V. The emission characteristic below 600 V is the same as that of a conventional field emitter. With

Fig. 2. SEM image of the GaAs emitter with a high aspect ratio.

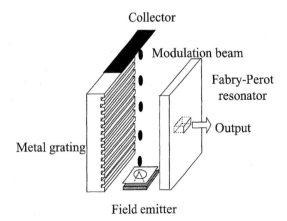

Fig. 1. Concept of the compact FEL.

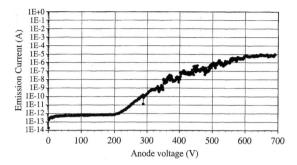

Fig. 3. Emission characteristic of the GaAs emitter.

increasing anode voltage beyond 600 V the emission current saturates at the current level of 10 μA. For a conventional Gunn diode with an area of 1 μm × 1 μm, the diode current saturates at the order of 10 μA, indicating that the electric field inside the GaAs active-layer reaches 3.2 kV/cm at this current level and the travelling dipole domain is formed, that is, the Gunn effect takes place. Therefore, the current saturation observed at the current level of 10 μA in the GaAs emitter strongly suggests that the travelling dipole domain is formed and the modulation beam is emitted into vacuum.

3.2. Photoexcited electron emission from a p-type Si field emitter

The emission current from a p-type semiconductor field emitter saturates in dark with increase of the gate voltage, because the supply of minority carriers (electrons) for field emission is limited in the depletion region. When a p-type emitter is illuminated by light with photon energy larger than the band gap energy, the emission current remarkably increases because of the additional supply of the photoexcited carriers. It suggests that the modulation beam could be obtained, when the p-type emitter is irradiated with a pulse laser. In addition, the applying of the photo-mixing technique [9] to a p-type semiconductor cathode could generate an ultra-high frequency modulation beam in excess of terahertz (THz). Two continuous frequency-offset pump lasers are used in the photo-mixing technique and the frequency of the modulation beam is widely tunable by changing

the frequency of one of the frequency-offset pump lasers. In these schemes, the frequency of the modulation beam is limited by the photo-response of the emitter. As preliminary experiments, we investigated the photo-response of a p-type Si field emitter under the illumination of a GaAlAs pulse laser.

The p-type single tip Si emitter was mounted in a high vacuum chamber (1×10^{-8} Torr), and the anode voltage was biased at 1 kV. The emitter was irradiated with a GaAlAs pulse laser having a wavelength of 830 nm and an optical power of 30 mW. The repetition frequency was 2 kHz, and the original time of rise and fall of the laser pulse was less than 1 μs. A convex lens focused the laser beam to a spot of about 1 mm diameter on the cathode surface, and a part of the laser beam was introduced to a conventional Si photodiode using a beam splitter. The emission currents flowing into an anode and a photodiode were converted to voltage using a current–voltage (I–V) conversion amplifier. The rise and fall times of the I–V conversion amplifier was about 4 μs. Therefore, the minimum rise and fall times of the measurement system was about 4 μs.

Fig. 4 shows the dependence of the emission current waveform on the gate voltage under the illumination of a GaAlAs pulse laser. The waveform denoted as photodiode in the figure is obtained from the Si photodiode. Pulsed emission currents, whose profiles are similar to that of the photodiode, are observed, indicating that a modulated electron beam is generated directly from the emitter under the illumination of the GaAlAs pulse laser.

Figs. 5(a) and (b) show the rising and falling characteristics of the photo-response in the emitter, respectively, at the gate voltage of 93 V. Since the waveforms were taken in the AC mode of an oscilloscope, the current values of the photo-response were the residuals after the subtraction of the DC current. The large noise components in the emission waveforms are not probably due to the instability of field emission, but due to the measurement system. The figure shows that the rise time of the photo-response in the emitter is about 4 μs, which is similar to that of the photodiode. This indicates that the real rise time

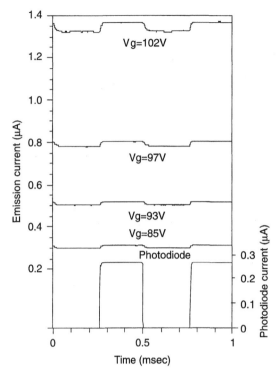

Fig. 4. Dependence of the emission current waveform on the gate voltage.

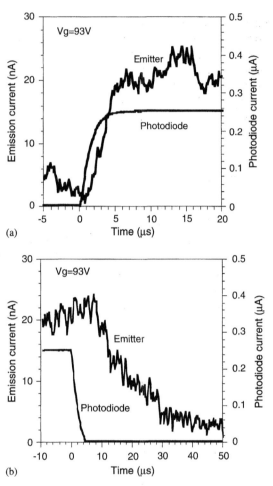

Fig. 5. Rising (a) and falling (b) characteristics of the photo-response at the gate voltage of 93 V.

of the photo-response in the emitter is at least less than 4 µs and the measured rise time is determined by the rise time of the amplifier used in this experiment. On the other hand, the fall time of the photo-response in the emitter is about 40 µs, which is much larger than that of the photodiode. Photogenerated electrons in the depletion layer are readily collected by the high field and emitted into vacuum with a quick response to an optical pulse. However, electrons generated in a field-free layer diffuse to the surface and then are emitted into vacuum. The diffusion process is much slower than the drift process. In this experiment, the depletion layer thickness is much less than the photon absorption depth because of the long wavelength (830 nm) of the GaAlAs laser. The bulk of photoelectrons were generated outside the depletion layer. Therefore, this slow fall time is determined by the diffusion process of the electrons photogenerated outside the depletion layer in the p-type emitter.

4. Conclusions

The paper proposes the concept of the compact FEL using SPR and FEAs, and describes preliminary experiments associated with the generation of a modulation beam directly from the cathodes. The GaAs emitter designed for the Gunn effect shows a clear saturation of electron emission, suggesting the generation of the modulation beam are caused by the travelling dipole domains in the emitter. We also measured the emission characteristics of a p-type Si emitter under the illumination of a GaAlAs pulse laser. A modulation electron beam is generated directly from the emitter under the illumination of the pulse laser. However, to generate a high-frequency modulation beam, we should design the emitter structure where the depletion layer is deep enough for almost all electrons to be generated inside the depletion layer, or the emitter where the electrons photogenerated outside the depletion layer was blocked not to diffuse to the surface.

Acknowledgements

A part of this work was supported by Grant-in-Aids for Top Priority R&D to be Focused Frequency Resources Development from the Ministry of Public Management, Home Affairs, Posts and Telecommunications, and for Scientific Research from the Ministry of Education, Culture, Sports, Science and Technology, Japan.

References

[1] H. Makishima, H. Imura, M. Takahashi, H. Fukui, A. Okamoto, Technical Digest of the 1997 IVMC, Kyongju, 1997, p. 194.

[2] C.M. Tang, M. Goldstein, T.A. Swyden, J.E. Walsh, Nucl. Instr. and Meth. A 358 (1995) 7.

[3] T. Taguchi, K. Mima, Nucl. Instr. and Meth. A 341 (1994) 322.

[4] H. Ishizuoka, Y. Nakahara, S. Kawasaki, N. Ogiwara, K. Sakamoto, A. Watanabe, M. Shiho, Nucl. Instr. and Meth. A 331 (1993) 577.

[5] H. Ishizuka, Y. Kawamura, K. Yokoo, H. Shimwaki, A. Hosono, Nucl. Instr. and Meth. A 445 (2000) 276.

[6] K. Ishi, Y. Shibata, T. Takahashi, S. Hasebe, M. Ikezawa, K. Takami, T. Matsuyama, K. Kobayashi, Y. Fujita, Phys. Rev. E 51 (1995) R5212.

[7] J.B. Gunn, Solid State Commun. 1 (1963) 88.

[8] H. Hasegawa, H. Mimura, K. Yokoo, Jpn. J. Appl. Phys. 42 (2003) 4051.

[9] E.R. Brown, F.W. Smith, K.A. Mcintosch, J. Appl. Phys. 73 (1993) 1480.

Available online at www.sciencedirect.com

Nuclear Instruments and Methods in Physics Research A 528 (2004) 502–505

**NUCLEAR
INSTRUMENTS
& METHODS
IN PHYSICS
RESEARCH**
Section A

www.elsevier.com/locate/nima

Using VUV high-order harmonics generated in gas as a seed for single pass FEL

D. Garzella[a,b,*], T. Hara[c], B. Carré[a], P. Salières[a], T. Shintake[c], H. Kitamura[c], M.E. Couprie[a,b]

[a] CEA, DSM/SPAM, 91 191 Gif-sur-Yvette, France
[b] LURE, 91 898 Orsay Cedex, France
[c] SPring-8/RIKEN Harima Institute, Hyogo 679-5148, Japan

Abstract

In this paper we illustrate the main features of a new scheme which aims at obtaining coherent radiation at very short wavelengths (down to 1.5 nm) by using the Harmonics of an intense laser, generated in a gas, as a seed to inject in a long undulator, either in the pure SASE or in the HGHG configuration. The unique spatial and temporal coherence properties of the High-Order Harmonics are then conserved, and the High Gain FEL amplifier system saturates up to the GW level. An experiment is here proposed, to be performed on the new compact LINAC-based source SCSS, currently under development in SPring-8.
© 2004 Elsevier B.V. All rights reserved.

PACS: 41.60.Cr; 42.65.Ky

Keywords: Harmonics Generation in gas; VUV radiation; Free Electron Laser; Seeding

1. Introduction

During the last years, new schemes other than the Self-Amplified Spontaneous Emission (SASE) have been proposed for achieving very short wavelengths in Free Electrons Laser (FEL)-based systems [1–3], where there is a need for more compact and fully temporally coherent sources. Solutions based on Harmonics Generation and the seeding of coherent light pulses (e.g. from lasers)

have appeared to be the most competitive and affordable. Thus, a High Gain Harmonics Generation (HGHG) experiment has been performed and obtained first results in the visible and near UV [4,5]. Moreover, the theory of Non-Linear Harmonics Generation (NHG) in SASE sources has been completed and the first validity tests are currently being performed [6]. In parallel to these studies, progress in strong laser–matter interaction have resulted in the generation of the high harmonics of intense laser pulses in rare gases. Very recently, during a Franco-Japanese Workshop [7] aiming at reinforcing close collaboration between the research agencies and teams of the

*Corresponding author. CEA/SPAM, bât 522, Centre d'Etudes de Saclay, 91191 Gif-sur-Yvette, France.
E-mail address: garzella@drecam.cea.fr (D. Garzella).

0168-9002/$ - see front matter © 2004 Elsevier B.V. All rights reserved.
doi:10.1016/j.nima.2004.04.089

two countries in FEL science, a new idea has been proposed to use the High-Order Harmonics of a laser, generated in a gas (HHG), as the seed to inject a High Gain FEL Amplifier. The experiment proposed here on the SCSS project aims at exploring the opportunity of using HHG in gas either as a seeding in SASE regime, or in the HGHG or NHG schemes for extracting harmonics of higher orders. In the present work, we describe the properties of the coherent radiation produced with such a scheme.

2. High-order harmonics generation

HHG in gas [8] is a linearly polarized source in the VUV–XUV region between 100 and 3 nm (12–400 eV), of high temporal [9] and spatial [10] coherence, emitting very short pulses (less than 100 fs), with a relatively high repetition rate (*up to few* kHz). The emission process is directly related to the interaction between the electromagnetic field of a focused intense laser, the "pump", and rare gas atoms (Ar, Xe, Ne and He the most used). A simple but physically meaningful picture of HHG is given by the three-step semi-classical model [11,12]: in the interaction region close to laser focus, the external electromagnetic field strength is comparable to the internal static field of the atom; thus atoms ionize by tunneling of the electron. The ejected free electrons are then accelerated in the external field. Those which are driven back close to the core can either be scattered or recombine to the ground state emitting a burst of XUV photons every half-optical cycle. The quasi-periodical train of XUV bursts is a superposition of the High-Order odd Harmonics of the fundamental frequency; a typical spectrum measured in Ne is shown in Fig. 1. The spectrum illustrates the characteristic distribution of intensities into the "plateau" (intensities almost constant with order) and "cutoff" region, where intensities decrease rapidly. The cut-off law gives the upper spectral limit of the harmonic spectrum as $E_{cutoff} = I_p + 3.2U_p$, where E_{cutoff} is the "cutoff" photon energy, I_p is the gas medium Ionization Potential and $U_p \propto I_{pump}\lambda^2_{pump}$, respectively, the focused intensity and the wavelength of the pump. As a conse-

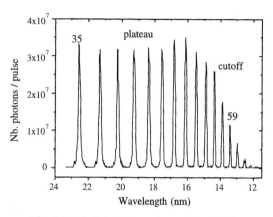

Fig. 1. High-Order Harmonics spectrum in Ne.

Table 1
Beam characteristics

Beam energy E (GeV)	0.5–1
FWHM bunch length (fs)	250
Norm. emitt. (π mm mrad)	2
Peak current $\mathfrak{I}_p$ (kA)	2
β function (m)	10
Trans. dim. $\sigma_{x,y}$ (μm)	100
Pulse rep. rate (Hz)	60

quence, the lighter the gas, the higher the ionization potential and the laser intensity which can be used without ionizing the atoms, therefore the higher the cutoff energy. Furthermore, in the VUV–XUV region, the radiation spectrum is completely tunable by means of frequency-mixing techniques applied on the pump laser [13].

3. Experimental set-up

The main electron beam characteristics are summarized in Table 1, while the experiment layout is shown in Fig. 2 [14].

Before the first undulator section, a Ti:Sa laser system at a wavelength $\lambda_{ph} = 800$ nm ($E_{ph} = 1.55$ eV) is intended to deliver 60 fs long pulses, with a repetition rate up to 60 Hz, as for the SCSS electron pulse. The laser pulse energy must be about 15 mJ in order to reach an intensity between 1×10^{14} and 3×10^{14} W/cm^2 in the interaction region, for a laser focal spot diameter $\cong 500\,\mu$m [14].

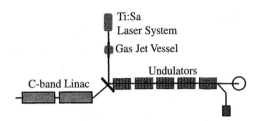

Fig. 2. Layout of the proposed experiment. The C-band linac modules are the two last units of the SCSS whole accelerator.

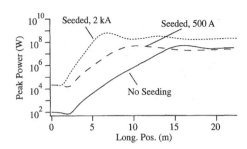

Fig. 3. SASE emitted peak power at 14 nm vs. z calculated with SRW. Solid line: starting from noise, 500 A beam peak current; dashed line: with $\lambda_{H57} = 14$ nm seeding, 500 A; dotted line: with seeding and 2 kA beam peak current.

Under these conditions, the 19th ($\lambda_{H19} = 42$ nm) and 57th ($\lambda_{H57} = 14$ nm) harmonics are generated in Ar and Ne, with efficiencies ranging between 10^{-5} and 10^{-7}, respectively, thus giving $\cong 10^{10}$ ph/pulse at 42 nm and $\cong 10^{8}$ ph/pulse at 14 nm [15]. For sake of simplicity, we assume the same pulse duration for all harmonics, say 30 fs [16], a focal spot diameter half of the laser one and a harmonic beam divergence of $\cong 1$ mrad [17]. The beam is then injected in the undulator by means of a multidielectric spherical mirror. In the 50–40 nm spectral region, Sc/Si mirrors are available with a reflectivity from 30% up to 50%, while Mo/Si can be manufactured at 14 nm with even higher reflectivity [18,19]. The maximum overlap between the harmonic pulse and the electron beam is obtained by imposing that the radiation waist be $w_0 = 2\sigma_{x,y}$, where $\sigma_{x,y}$ are the electron beam transverse dimensions. For SCSS, $w_0 = 240$ μm and the beams can be matched using a spherical mirror with a radius of curvature of 10 m at 42 nm.

4. Simulation results

Several configurations have been tested by calculating the SASE with the code SRW, a 3D code based on Genesis [20], while HGHG has been calculated via 0D macros based on the analytical theories developed by Dattoli [1,21] and Yu [22,23]. All the codes are implemented on the IGOR® software.

4.1. SASE

Fig. 3 shows the SASE emission at 14 nm, calculated with SRW starting from noise, then by

Table 2
Undulators characteristics

	Modulator	Radiator
Period λ_0 (mm)	22.5	15
K-value	2.33	1.3
Per./Section N	200	300
No. of sections	1	4–5

a HHG seeding pulse with 22 kW peak power, and finally by injecting the same seeding pulse, but with an electron beam peak current $\Im_p$ of 2 kA, the ultimate value for SCSS [14]. In the first two cases, the same peak power P_p at saturation, about 50 MW, is obtained, but use of the seeding allows to shorten the saturation length L_s from about 16 m to less than 10 m. Then, with a higher $\Im_p$, both an effective increase of P_p ($\cong 600$ MW) and a further shortening of L_s ($\cong 6$ m) at saturation are observed.

4.2. HGHG

The use of a modified first undulator section as modulator (cf. Table 2), together with the use of a 42 nm pulse as a seeding, leads to even better results, as shown in Fig. 4.

At 42 nm, 750 kW can be injected as seeding power in the modulator. Calculations with the Dattoli and Yu methods agree on the obtention of a high value for the bunching parameter b_n on the third harmonics ($b_3 = 0.34$ and 0.22, respectively). The behaviour of the emitted P_p vs. the

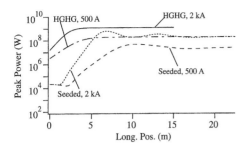

Fig. 4. Comparison between SASE and HGHG emission at 14 nm. The horizontal axis indicates the longitudinal position from the entrance of the radiator section. Dashed line: seeded SASE, 500 A peak current; dotted line: seeded SASE, 2 kA; solid-dotted line: HGHG, 500 A; solid line: HGHG, 2 kA.

longitudinal position z in the radiator, compared to that in the "seeded" SASE configuration, shows a significant increase of P_p by more than a factor 3, with again L_s reduced. Finally, taking $\Im_p = 2$ kA, the 1 GW saturation level at 14 nm seems to be obtained within only one radiator section.

4.3. Non-linear harmonic generation

From the results given by the MEDUSA code, based on the theories on NHG [24,25] and the recent experimental results obtained at LEUTL [6] and VISA [26], it is possible to get rough estimates of the higher harmonics content at the end of the SCSS radiator, in the HGHG configuration. Peak power $P_p \cong 1$ MW in the "water window" at $\lambda = 2.8$ nm (5th Harmonic) and $P_p \cong 60$ kW at $\lambda = 1.55$ nm (9th Harmonic) are estimated for a seeding at 14 nm. The latter value corresponds to $\cong 10^7$ ph/pulse.

5. Perspectives and conclusions

Using state-of-the-art High Order Harmonics in gas for seeding High Gain FEL amplifiers appears very interesting, because the seed radiation is fully coherent and tunable in the VUV–XUV range. Such a seeding can reduce the saturation lengths, thus giving a more compact source. Moreover, higher peak power can be obtained with this scheme than with a pure SASE source or multiple stages. Finally, the temporal and spatial coherence properties of HHG in gas should be essentially conserved. This calls for further investigations to be carried out with time-dependent 3D codes.

References

[1] F. Ciocci, et al., IEEE J. Quant. Electron. 31 (1995) 1242.
[2] L.H. Yu, I. Ben-Zvi, Nucl. Instr. and Meth. A 393 (1997) 96.
[3] S.G. Biedron, et al., Nucl. Instr. and Meth. A 475 (2001) 401.
[4] L.H. Yu, et al., Science 289 (2000) 932.
[5] L.H. Yu, et al., Phys. Rev. Lett. 91 (2003) 074801.
[6] S.G. Biedron, et al., Nucl. Instr. and Meth. A 483 (2002) 100.
[7] http://www.jp.cnrs.fr/local/workshop/fel02/index.html.
[8] P. Salières, et al., Adv. At. Mol. Opt. Phys. 41 (1999) 83.
[9] P. Salières, et al., Science 292 (2001) 902.
[10] L. Le Déroff, et al., Phys. Rev. A 61 (2000) 043802.
[11] P.B. Corkum, Phys. Rev. Lett. 71 (1993) 1994.
[12] M. Lewenstein, et al., Phys. Rev. A 49 (1994) 2117.
[13] H. Eichmann, et al., Phys. Rev. A 51 (1995) R3414.
[14] T. Shintake, et al., Proceedings of SPIE Conferences, 2001.
[15] J.F. Hergott, et al., Phys. Rev. A 66 (2002) 021801.
[16] C. Lyngå, et al., Phys. Rev. A 60 (1999) 4623.
[17] L. Le Déroff, et al., Opt. Lett. 23 (1998) 1544.
[18] Yu.A. Uspenskii, et al., Nucl. Instr. and Meth. A 448 (2000) 147.
[19] T. Feigl, et al., Nucl. Instr. and Meth. A 483 (2002) 351.
[20] S. Reiche, Nucl. Instr. and Meth. A 429 (1999) 243.
[21] G. Dattoli, P.L. Ottaviani, J. Appl. Phys. 86 (1999) 5331.
[22] L.H. Yu, Phys. Rev. A 44 (1991) 5178.
[23] L.H. Yu, J. Wu, Nucl. Instr. and Meth. A 293 (2002) 493.
[24] H.P. Freund, Phys. Rev. E 52 (1995) 5401.
[25] H.P. Freund, et al., IEEE J. Quant. Electron. 36 (2000) 275.
[26] A. Tremaine, et al., Phys. Rev. Lett. 88 (2002) 204801.

Available online at www.sciencedirect.com

SCIENCE @DIRECT•

ELSEVIER   Nuclear Instruments and Methods in Physics Research A 528 (2004) 506–510

**NUCLEAR
INSTRUMENTS
& METHODS
IN PHYSICS
RESEARCH**
Section A

www.elsevier.com/locate/nima

New concept of the gamma-ray source: coherent stimulated annihilation of relativistic electrons and positrons in a strong laser field

I.V. Smetanin*, K. Nakajima

Advanced Photon Research Center, JAERI Umemidai 8-1, Kizu-cho, Soraku-gun, Kyoto 619-0215, Japan

Abstract

The process of stimulated coherent annihilation of relativistic electron–positron pairs in the strong laser field is examined. It is shown that if the energy of laser photon in the beam (center-of-mass) reference frame exceeds mc^2, coherent stimulated generation of γ-ray photons becomes possible. Electron–positron pair in the electromagnetic field can be considered as an inverted two-level quantum system, in which the two-photon transition is allowed with the stimulated emission of one γ-quantum and one photon of the laser mode. In analogy with the effect of self-induced transparency in the conventional nonlinear optics, the nonlinear oscillation regime arises at high laser intensities, in which inversion can be removed in one pass and the π-pulse of coherent γ-ray radiation is generated. Estimates show the scheme proposed is a promising high-brightness source of coherent γ-radiation.
© 2004 Elsevier B.V. All rights reserved.

PACS: 41.75.Fr; 42.55.Vc; 52.38.Ph.

Keywords: Stimulated annihilation; Coherent gamma-ray; Coherent π-pulse; Two-photon resonance

1. Introduction

The problem of electron–positron annihilation and pair creation in a strong electromagnetic field is under intense investigation since the early Dirac [1] and Shwinger [2] works. Detailed study of annihilation in the super-strong external field has been done in Ref. [3] where the non-perturbative theory of electron–positron interactions has been developed for various field configurations.

Recent progress of high-power femtosecond laser technologies [4] has renewed interest in the annihilation assisted by a strong field. Self-action of intense electromagnetic wave in electron–positron vacuum has been studied in Ref. [5]. Photon–photon scattering in the waveguides has been proposed recently for detection of QED vacuum nonlinearities [6].

With an increase in the field strength, coherent interaction effects has come into play, leading to a dramatic enhancement of pair creation and annihilation rates. The mechanism of coherent

*Corresponding author. Tel.: +81-774713369; fax: +81-774713316.

E-mail addresses: smetanin@apr.jaeri.go.jp, smetanin@sci.lebedev.ru (I.V. Smetanin).

pair creation in electron–positron collisions in linear colliders has been examined in Ref. [7]. The idea of stimulated amplification and generation of coherent resonant γ-ray photons (i.e., γ-laser) in the electron–positron plasma has been proposed for the first time in Ref. [8].

In this paper, using a deep physical analogy between the Compton and electron–positron annihilation processes [9], we propose a new concept of coherent generation of γ-ray radiation, the mechanism of which is the stimulated non-linear coherent annihilation of electron–positron pairs in a strong laser field. In our concept, the relativistic beam of electrons and positrons inter-acts with the counter-propagating high-power laser radiation. We show that if the energy of laser photon in the beam (center-of-mass) refer-ence frame exceeds mc^2, m is the rest mass of an electron, coherent stimulated generation of γ-ray photons becomes possible. The energy of photon emitted is then; $\hbar\omega_\gamma \approx 4\gamma_0^2\hbar\omega_i \approx (mc^2)^2/\hbar\omega_i$; γ_0 is the beam energy and $\hbar\omega_i$ is the laser quantum. Electron–positron pair in the electromagnetic field can be considered as an inverted two-level quantum system, in which the upper state consists of one electron, one positron and N laser photons, and the lower state $(N+1)$ laser photons and one gamma-ray photon, and the lower state $(N+1)$ photon pairs. In analogy with the effect of self-induced transparency in conventional nonlinear optics, the nonlinear oscillation regime arises at high laser intensities, in which inversion can be removed in one pass and the π-pulse of coherent γ-ray radiation is generated. Estimates show the scheme proposed is a promising high-brightness source of coherent γ-radiation.

In our analysis, the Klein–Gordon field theory is used to describe interaction of electrons and positrons with the radiation fields, the general approach can be found in Ref. [3], for example. The deep analogy between annihilation and Compton processes [9] allows to suppose a similarity in the dynamics of the proposed regime of coherent stimulated annihilation and the Compton free-electron lasers. It is worth mention-ing that theoretical studies of free-electron lasers originally employed the quantum representation of physical processes in an FEL [10–14]. The

Klein–Gordon theory of the Compton FEL has been developed in Ref. [15]. It has been shown that with an expansion of oscillation frequency into the X-ray region, the quantum recoil effect becomes crucial for the lasing in the Compton FEL with an optical undulator [16], and the FEL becomes the two-level quantum generator with a completely inverted active medium [17]. In this paper, we construct the theory of the proposed coherent stimulated electron–positron annihilation by analogy with that of two-level Compton X-ray FEL [17].

2. Electron–positron pair as a two-level quantum system

Let us consider the relativistic electron–positron beam interacting with a strong counter-propagat-ing laser pulse. If the energy of laser photon in the beam reference frame exceeds mc^2, m is the rest mass of electron, coherent stimulated generation of γ-ray photons becomes possible. Let the vector potential of the γ-ray photons be $A_1 \exp(i[k_1 z - \omega_1 t])$, and that of the laser pulse be $A_2 \exp(i[k_2 z + \omega_2 t])$. In the Klein–Gordon theory, the annihila-tion Hamiltonian can be written in the form [9] ($\hbar = c = 1$)

$$H = \sum_p \{g_p^* a_p^+ b_{p'}^+ + g_p b_p a_{p'}\}$$
$$\times \delta(p + p' - (k_1 - k_2)). \qquad (1)$$

Here a_p^+, b_p^+ are the creation operators for elec-trons and positrons of the momentum p and energy ε_p, respectively, and a_p, b_p are the corre-sponding annihilation operators, $g_p = e^2 A_2 A_1^*/4\sqrt{\varepsilon_p \varepsilon_{p'}}$ is the coupling constant. Delta function expresses the energy and momentum conservation laws for the pair creation and annihilation, $p + p' = k_1 - k_2 \equiv q$ and $\varepsilon_p + \varepsilon_{p'} = \omega_1 + \omega_2$, respectively.

We intend to show in this section that the system of electrons and positrons in the field of two strong electromagnetic field satisfying the above energy and momentum conservation laws can be con-sidered as a conventional two-level system with a two-photon resonance transition.

Let us consider Bogolubov's transformation

$$c^+ = \alpha_{11}a^+ + \alpha_{12}b,$$
$$d^+ = \alpha_{21}a - \alpha_{22}b^+. \tag{2}$$

Here operators a and b correspond to the momentum state p and $q-p$, respectively. The new field operators should obey the commutation rules $[c^+, c] = [d^+, d] = 1, [c, d] = [c, d^+] = [c^+, d] = [c^+, d^+] = 0$. For the following values of the ratio $x = \alpha_{11}/\alpha_{12}$

$$x = \frac{(\varepsilon_p + \varepsilon_{q-p}) \pm \Omega_p}{2g_p} \tag{3}$$

$\Omega_p = \sqrt{(\varepsilon_p + \varepsilon_{q-p})^2 + 4|g_p|^2}$ is the Rabi frequency for the system under consideration, Bogolubov's transform diagonalizes Hamiltonian (1)

$$H = \sum_p E_{1p}c_p^+ c_p + E_{2p}d_p^+ d_p \tag{4}$$

where the spectrum of new quasi-particles describing operators c and d is

$$E_{1p} = \frac{\varepsilon_p|x|^2 + \varepsilon_{q-p} - 2g_p x}{|x|^2 - 1},$$
$$E_{2p} = \frac{\varepsilon_{q-p}|x|^2 + \varepsilon_p - 2g_p x}{|x|^2 - 1}. \tag{5}$$

Note, the difference in energy of levels is $E_{1p} - E_{2p} = \varepsilon_p - \varepsilon_{q-p}$ and vanishes in the case of symmetric photon momentums $q = 0$. Spectrum (5) indicates that the Rabi oscillations occur in this two-level electron–positron system driven by the given external field, accompanied by the coherent generation of γ-ray photons in the resonant two-photon transition.

3. Nonlinear dynamics of coherent stimulated pair annihilation and γ-ray generation

To describe the nonlinear evolution of the electron–positron pair in the interaction region and coherent generation of coherent γ-rays, we will use the Klein–Gordon equation and seek its solution for electrons and positrons in the form

$$\Psi_e(z, t)$$
$$= \sum_p B_p u_p^{(-)}(z) \exp\left(\frac{i}{\hbar}[pz - \varepsilon_p t]\right)$$
$$+ B_{q-p} v_{q-p}^{(-)}(t) \exp\left(-\frac{i}{\hbar}[(q - p)z - \varepsilon_{q-p}t]\right),$$

$$\Psi_p(z, t)$$
$$= \sum_p B_p v_p^{(+)}(z) \exp\left(-\frac{i}{\hbar}[pz - \varepsilon_p t]\right)$$
$$+ B_{q-p} u_{q-p}^{(+)}(z) \exp\left(\frac{i}{\hbar}[(q - p)z - \varepsilon_{q-p}t]\right) \tag{6}$$

where the momentum and the energy of particles satisfy the above conservation law. The normalization coefficients are $B_p = (mc^2/2V\varepsilon_p)^{1/2}$, V is the quantization volume. The amplitudes of the upper $u(z)$ and lower $v(t)$ states of the two-level electron–positron system are assumed to be slow varying functions, and after substitution in the Klein–Gordon equation we have

$$\frac{\partial u_p^{(\pm)}}{\partial z} = -i\frac{e^2(A_1 A_2^*)}{4\hbar\sqrt{\varepsilon_p \varepsilon_{-p-q}}} v_p^{(\pm)},$$
$$\frac{\partial v_p^{(\pm)}}{\partial z} = -i\frac{e^2(A_1^* A_2)}{4\hbar\sqrt{\varepsilon_p \varepsilon_{-p-q}}} u_p^{(\pm)}. \tag{7}$$

We will assume that the electromagnetic modes have equal polarizations, and the laser (counter-propagating) wave is sufficiently strong to neglect its depletion. Evolution of the second high-frequency electromagnetic wave is described by the Maxwell equations with the current density of the Klein–Gordon field

$$\left(\frac{\partial}{\partial z} + \frac{1}{c}\frac{\partial}{\partial t}\right)A_1 = i\frac{2\pi e^2 A_2 n}{\omega_1\sqrt{\varepsilon_p \varepsilon_{q-p}}}$$
$$\times \int d\varepsilon_p(u_p^{(-)}v_p^{(-)} + u_p^{(+)}v_p^{(+)}) \tag{8}$$

where n is the density of electron–positron pairs. Eqs. (7) and (8) are analogous to that for the X-ray Compton FEL in the quantum regime [17], and form closed self-consistent system of Maxwell–Bloch equations, thus allowing solutions of the self-induced transparency type with generation of the π-pulses of coherent γ-rays.

By analogy with the theory of two-level quantum FEL [17], we introduce the pulse area

$$\chi(z) = \frac{\mu_1}{c\hbar} \int_0^z E_1(z) \, \mathrm{d}z \qquad (9)$$

with the characteristic dipole momentum of the two-level system under consideration

$$\mu = \frac{e^2 c^2 E_2}{2\omega_1 \omega_2 \sqrt{\varepsilon_p \varepsilon_{-p-q}}}. \qquad (10)$$

Assuming boundary conditions $u_p^{(\pm)}|_{z=0} = 1, v_p^{(\pm)}|_{z=0} = 0$, which correspond to the initially inverted two-level electron–positron system, we have solutions to Eqs. (7)

$$u_p^{(\pm)} = \cos(\chi/2), \quad v_p^{(\pm)} = -i \sin(\chi/2) \qquad (11)$$

i.e. the amplitudes oscillate with the pulse area.

Evolution of the pulse area for the amplified γ-ray signal is guided by the pendulum equation

$$\frac{\partial^2 \chi}{\partial z^2} = \alpha^2 \sin(\chi) \qquad (12)$$

with the linear gain coefficient

$$\alpha = \sqrt{\frac{\pi e^4 E_2^2 nc^2}{\hbar \omega_1 \omega_2^2 \varepsilon_p \varepsilon_{q-p}}}. \qquad (13)$$

Intensity of the coherently generated γ-ray field in the process of stimulated annihilation oscillates with the interaction distance z

$$I_1(z) = \mathrm{dn}^{-2}\left(\frac{\alpha z}{\kappa}, \kappa\right) I_1(0) \qquad (14)$$

where parameter $\kappa = \sqrt{1 + I_1(0)/I_m}$ and $I_1(0)$ is the initial intensity of the γ-ray photon flux, determined by the spontaneous pair annihilation. For the process of stimulated pair creation, the argument in this solution has to be shifted by the half of the period of the Jacobi dn function, $K(\kappa)$. The parameter

$$I_m = \hbar \omega_1 nc \qquad (15)$$

is the maximum intensity of the amplified wave which can be attained as a result of annihilation of positrons and electrons having the equal beam densities n. The maximum intensity can be achieved for the interaction distance

$$L_{\max} = \alpha^{-1} \kappa K(\kappa). \qquad (16)$$

$K(x)$ is the complete elliptic integral of the first kind,

$$I_1(L_{\max}) = I_1(0) + I_m \qquad (17)$$

which corresponds to the formation of the π-pulse of the amplified wave, i.e. the pulse area becomes $\chi(L_{\max}) = \pi$.

Let us conclude with some estimates. The optimum interaction conditions correspond to the case when, in the center-of-mass reference frame, both photons have equal energies $\hbar \omega_1' \approx \hbar \omega_2' \approx mc^2$ and electron and positron are non-relativistic and have small kinetic energies. Let the second electromagnetic field be formed by the super-strong laser pulse. The gamma factor for this moving reference frame is $\gamma_0 \approx mc^2/2\hbar\omega_2 \sim 2.5 \times 10^5$ for the frequency $\hbar \omega_2 \sim 1$ eV in the laboratory frame, which corresponds to the energy of co-propagating electron and positron beams of ~ 125 GeV. The linear gain coefficient is $\alpha \approx (\pi r_0 na^2/2\gamma_0^3)^{1/2} \approx 0.53 \times 10^{-2} a(n/10^{24} \text{ cm}^{-3})^{1/2} \text{ cm}^{-1}$ where $a = eA_2/mc^2$ is the normalized vector potential amplitude of the laser field. The coherent γ-ray field generated has the frequency $\hbar \omega_1 \approx 4\gamma_0^2 \hbar \omega_2 \approx 2mc^2 \gamma_0 \approx 250$ GeV and π-pulse intensity $I_m \sim 1.2 \times 10^{27}(n/10^{24} \text{ cm}^{-3})\text{W/cm}^2$. The length of formation of coherent π-pulse of γ radiation $L_{\max} \sim \alpha^{-1}$ scales as $\sim \gamma_0^{3/2}$ and can be significantly reduced by using a shortwavelength laser radiation produced by an X-ray FEL. Using estimates $\hbar \omega_2 \sim 125$ eV photons to stimulate pairs annihilation, we have the beam threshold energy reduced to GeV level $\gamma_0 \sim 2 \times 10^3$, and the linear gain coefficient becomes $\alpha \approx 7.4a(\text{cm}^{-1})$ for the above pair densities. However, the gamma-photon yield is also reduced to the level $I_m \sim 10^{25}$ W/cm² due to the decrease in photon's energy $\hbar \omega_1 \approx 2mc^2 \gamma_0 \sim 2$ GeV. This estimates show that the scheme proposed is a prospective way to generate intense high-energy coherent gamma rays.

References

[1] P.A.M. Dirac, Proc. Cambridge Phil. Soc. 26 (1930) 361.
[2] J. Shwinger, Phys. Rev. 82 (1951) 664.

[3] A.I. Nikishov, Problems of an intense external field in QED, in: N.G. Basov (Ed.), Proceedings of P. N. Lebedev Physics Institute, Vol. 111, Nauka, Moscow, 1979, pp. 152–278 (in Russian).

[4] M.D. Perry, et al., Opt. Lett. 24 (1999) 160.

[5] N. Rozanov, JETP 86 (1998) 284.

[6] G. Brodin, M. Marklund, L. Stenflo, Phys. Rev. Lett. 87 (2001) 171801.

[7] P. Chen, V. Telnov, Phys. Rev. Lett. 63 (1989) 1796.

[8] L. Rivlin, Sov. J. Quantum Electron. 3 (1976) 2413 (in Russian).

[9] V.B. Berestetskii, E.M. Lifshitz, L.P. Pitaevskii, Quantum Electrodynamics, Pergamon, Oxford, 1978.

[10] R.H. Pantell, G. Soncini, H.E. Puthoff, IEEE J. Quant.-Electron. 4 (1968) 905.

[11] J.M.J. Madey, J. Appl. Phys. 42 (1971) 1906.

[12] J.M.J. Madey, H.A. Schwettman, W.M. Fairbank, IEEE Trans. Nucl. Sci. NS-20 (1973) 980.

[13] V.P. Sukhatme, P.W. Wolf, J. Appl. Phys. 44 (1973) 2331.

[14] W.B. Colson, Phys. Lett. A 59A (1976) 187.

[15] M.V. Fedorov, Progr. Quant. Electron. 7 (1981) 73; M.V. Fedorov, J. McIver, Sov. Phys. JETP 49 (1979) 1012.

[16] P. Dobiash, P. Meystre, M.O. Scully, IEEE J. Quantum Electron. 19 (1983) 1812; J. Gea-Banacloche, et al., IEEE J. Quantum Electron. 23 (1987) 1558.

[17] I.V. Smetanin, Laser Physics 7 (1997) 318; E.M. Belenov, et al., JETP 105 (1994) 808.

Available online at www.sciencedirect.com

SCIENCE @ DIRECT°

ELSEVIER   Nuclear Instruments and Methods in Physics Research A 528 (2004) 511–515

**NUCLEAR
INSTRUMENTS
& METHODS
IN PHYSICS
RESEARCH**
Section A

www.elsevier.com/locate/nima

Methods of electron beam bunching

E.G. Bessonov*

P.N. Lebedev Physical Institute of the Russian Academy of Sciences, Leninsky prospect 53, Moscow 119991, Russia

Abstract

A review of electron beam-bunching methods is presented. A method to create trains of short electron microbunches by creating an electron multilayer mirror in the longitudinal phase space and then rotating it in fields of radiofrequency accelerators or Free-Electron Laser amplifiers is proposed.
© 2004 Elsevier B.V. All rights reserved.

PACS: 07.85.Fv; 29.25.Bx; 41.60.Cr; 41.85.Ct

Keywords: Free electron laser; Buncher; Chopper; Accelerator

1. Introduction

The production of short period trains of electron bunches (electron multilayer mirrors) is very important for the solution of the problem of compact, stable, monochromatic, diffraction-limited prebunched FELs in the mm to X-ray regions.

Prebunched FELs were suggested by V.L. Ginsburg in 1947 and named "Sources of Coherent Undulator Radiation (UR)" [1]. The terms parametric and prebunched FELs appeared later [2]. Motz performed the first experiments on generation of coherent UR in 1953 [3]. Phillips generated stimulated UR in a conventional FEL in the radiofrequency (RF) region in 1960 [4]. In the early 1970s interest in prebunched FELs was renewed [5,6].

Bunched electron beams of length ~ 0.1–1 mm are produced in klystron bunchers (KB) and RF accelerators [7]. The basic components of a KB is an RF cavity followed by a drift space. The cavity is excited to the TM_{010} mode. The electron beam at the exit of the buncher receives sinusoidal energy modulation. After the drift space the electron beam will be bunched due to the acquired velocity modulation.

A simple way to produce a bunched beam is to pass a continuous beam through a chopper, where the beam is deflected across a narrow slit or a system of slits, resulting in a pulsed beam behind the slit. The chopper includes an RF cavity excited like the buncher cavity but with the beam port offset by a distance from the cavity axis.

The chopper mode of bunching is rather wasteful. An RF buncher, which concentrates electrons from a large range of phases towards a particular phase is more efficient for electron bunching. However, a chopper that produces a

*Tel.: +7-95-3340119; fax: +7-95-9382251.
E-mail address: bessonov@x4u.lpi.ruhep.ru
(E.G. Bessonov).

0168-9002/$ - see front matter © 2004 Elsevier B.V. All rights reserved.
doi:10.1016/j.nima.2004.04.091

beam with higher depth of modulation (clear bunches) does not introduce additional energy spread and can be used for specific reasons. Modern choppers work effectively in the RF region [8].

Short-period trains of bunches are produced in undulator klystron bunchers (UKB). The energy modulation occurs in the undulator and electromagnetic wave fields of the UKB (FEL amplifier configuration) and bunching takes place in a free space [4]. The bunching length in UKB can be shortened if a dispersion section (chicane magnet) follows the undulator. Such a system was called an optical klystron buncher (OKB) [9].

Undulators with high deflection parameters (wigglers) possess high dispersion. If the UKB works in the regime of high deflection parameter and high intensity of the wave (HHKB), the energy modulation and bunching can occur in the same undulator [10]. In HHKB a high degree of bunching occurs if electrons produce a quarter of a period of phase oscillations. The bunching length in this case can be short (~ 1 m in the optical region, see Section 3).

A scheme of bunching based on convergent waves and/or undulators with variable parameters (VPKB) was developed in Ref. [11]. In this scheme an electron beam is trapped as a whole in regions less than those limited by separatrices of the FEL amplifier for about $\frac{3}{4}$ of a period of phase oscillations (Fig. 1). Clear bunched beam can be used for high-efficiency generation on the fundamental or higher harmonics in the same undulator or in an undulator radiator. A similar scheme of bunching can occur in cases being developed in Refs. [12–15].

A bunching scheme based on the extraction of a series of microbunches from an electron beam located in storage ring buckets, when an undulator with variable parameters and electromagnetic wave are used, was considered in Ref. [16]. Extracted microbunches stay in the storage ring buckets.

Shock acceleration of electron bunches or trains of bunches by wavepackets with a longitudinal component of electric field can lead to their compression. Such schemes can be realized in linear accelerators [17] and laser wake-field accelerators.

2. A method of electron beam bunching

The proposed method is based on the production of electron beams in the form of trains of electron bunches, composed in the longitudinal plane in N energy layers and followed by rotation of the bunches at a right angle [18]. Rotation converts the bunches into short-period ($< \lambda_L/N$) trains of micro bunches with layers transverse to the beam propagation (Fig. 2).

FEL amplifier configuration can be used for rotation of bunches. In the case of a helical undulator, the wavelength of the circular polarized laser wave

$$\lambda_L = \frac{\lambda_u}{2\gamma_s^2}(1 + p_\perp^2),\tag{1}$$

where $p_\perp = \sqrt{p_{\perp u}^2 + p_{\perp L}^2}\big|_{p_{\perp L} \ll p_{\perp u}} \simeq p_{\perp u}$; $p_{\perp u (L)} = eB_{u (L)}\lambda_{u (L)}/2\pi mc^2$ are the transverse electron momenta determined by the electromagnetic fields of the undulator and the electromagnetic wave; λ_u, B_u are the period and the magnetic

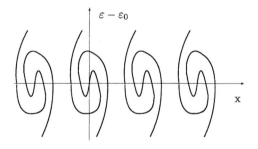

Fig. 1. Electron beam at the exit of VPKB in the phase space ($\varepsilon - \varepsilon_0, x$). The amplitude of the laser electric field is $E_{Lm} = E_0(1 + \alpha z)$. $E_0 = 180$ kV/cm, $\alpha = 0.8$ m^{-1}, $\lambda_u = 1.5$ cm, the undulator length $L = 5$ m, $B_u = 3 \times 10^3$ Gs, $\lambda_L = 1.06$ mkm.

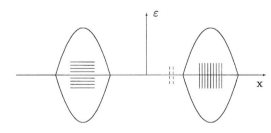

Fig. 2. An electron bunch arranged in energy layers in the left bucket is converted after its rotation in the right bucket to a train of micro bunches.

field strength of the undulator; B_L, the magnetic field strength of the wave; $\gamma_s = 1/\sqrt{1 - \beta_s^2} = \sqrt{\lambda_u(1 + p_\perp^2)/4\lambda_L}$, the relative electron equilibrium energy; $\beta_s = \sqrt{\beta_{\perp s}^2 + \beta_{z s}^2}$; $\beta_{\perp s} = \beta_{\perp us} + \beta_{\perp Ls} = (p_{\perp u} + p_{\perp L})/\gamma_s$; $\beta_{zs} = \lambda_u/(\lambda_u + \lambda_L)$; $2\pi mc^2/e \simeq 10,700$ Gs cm; e and m, the electron charge and mass [10].

The frequency of phase oscillations of electrons in a helical undulator and a circular polarized laser wave $\omega_\varphi = \omega_u\sqrt{B_u B_L(1 + \beta_{zs})}/\beta_{zs}\gamma_s B_c$, where $\omega_u = 2\pi\beta_{zs}c/\lambda_u$; $B_c = 2\pi mc^2/e\lambda_u$ [10].[1]

Electrons produce a quarter of the period of phase oscillations inside the length

$$l_{1/4} = \frac{\pi c \beta_{zs}}{2\omega_\varphi} = \frac{\beta_{zs}\gamma_s}{4}\sqrt{\frac{\lambda_u \lambda_L}{(1 + \beta_{zs})p_{\perp u}p_{\perp L}}}$$

$$= \frac{\lambda_u}{8}\sqrt{\frac{1 + p_\perp^2}{p_{\perp u}p_{\perp L}}}. \qquad (2)$$

The maximum deviation of the electron energy in the bucket determined by a separatrix is

$$\frac{\Delta\gamma_{sep}}{\gamma_s} = \frac{2\beta_{zs}\omega_\varphi}{(1 + \beta_{zs})\omega_u}$$

$$= 4\beta zs\sqrt{\frac{p_{\perp u}p_{\perp L}}{(1 + \beta_{zs})(1 + p_\perp^2)}}. \qquad (3)$$

The magnetic field strength of the circular polarized Gaussian laser beam

$$B_L(Gs) = \sqrt{\frac{2P_L}{c\sigma_L^2}} \simeq 2.58 \times 10^{-2}\sqrt{\frac{P_L\ (W)}{\sigma_L^2\ (cm^2)}} \qquad (4)$$

where P_L is the power of the laser wave; σ_L.

Example. The parameters of the undulator and electromagnetic laser wave are the following:

laser power $\qquad\qquad P_L = 1.6 \times 10^9$ W,
laser wavelength $\qquad\quad \lambda_L = 10^{-4}$ cm,
laser beam dispersion $\quad \sigma_L = 2 \times 10^{-2}$ cm,
undulator magnetic field $\ B_u = 10,700$ Gs,

undulator period $\qquad\qquad \lambda_u = 3$ cm,
deflecting parameter $\qquad\ p_\perp = 3$,
duration of laser wavepacket $\tau_l = 2.5 \times 10^{-11}$ s,
energy of laser wavepacket $\ \varepsilon_l = P_l\tau_l = 0.04$ J.

In this case $\gamma_s = 274$, $p_{\perp u} = 3$, $p_{\perp w} = 4.82 \times 10^{-4}$, $l_{1/4} = 31.2$ cm, $B_L = 5.16 \times 10^4$ Gs, $l_R = 50.3$ cm $(2l_R > l_{1/4})$, $\Delta\gamma_{sep}/\gamma_s = 1.07 \times 10^{-2}$.

2.1. Arrangement of electron beams with energy layers

Scheme 1: A system of low emittance electron beams produced by N electron guns of energy $\varepsilon_n = \varepsilon_0 + n\Delta\varepsilon$ can be arranged in one beam with energy layers by means of a bending magnet (Fig. 3) [18]. Here $n = 0, 1, 2, 3, \ldots, N - 1$, $\Delta\varepsilon$ is the interval between energy layers. The energy spread of the electron guns $\delta\varepsilon \ll \Delta\varepsilon$.

If the guns produce short bunches, the beam at the exit of the system will be arranged in layers and bunched. The control systems of guns must have delay lines for phase adjustment to match produced bunches at the exit of the system in the longitudinal direction. The beam can be bunched at the exit of the bending magnet as well.

Scheme 2: A thin electron beam with moderate energy spread $\Delta\varepsilon$ and low transverse emittance passes through a system of a dispersive magnet[2] and a metal lattice with N slits (Fig. 4). After the lattice, the electron beam is arranged in N energy layers and has increased transverse dimension.

If $N\Delta\varepsilon/\varepsilon = 10^{-2}$, $N = 5$, the average bending radius $R = 30$ cm, the increase of the transverse dimension of the beam due to the energy spread is $\Delta r = 2RN\Delta\varepsilon/\varepsilon = 6$ mm and the lattice period $d = 1.2$ mm. The beam dimension at the entrance of the bending magnet must be smaller than the lattice period and the transverse emittance of the beam must be small ($<10^{-7}$ m).

The increase of the transverse dimension of the beam at the exit of the lattice can be removed by using, after the lattice, another bending magnet with the same parameters.

The beam can be bunched at the exit of the bending magnet by a chopper.

---

[1] We suppose that the laser beam intensity is constant along the interaction region of the laser and electron beams ($l_{1/4} < 2l_R$, where $l_R = 4\pi\sigma_L^2/\lambda_L$ is the Rayleigh length; σ_L, the rms transverse laser beam dimension).

---

[2] More perfect systems of focusing bending magnets, quadrupole lenses and α-magnets can be used.

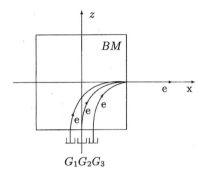

Fig. 3. The scheme of arranging an electron beam in energy layers. G_1, G_2, G_3 are the electron guns, BM is the bending magnet.

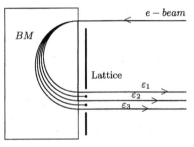

Fig. 4. The scheme of production of the electron beam arranged in layers. BM is the bending magnet.

After rotation at the bunching frequency, the beam in schemes 1 and 2 will be a train of bunches separated by a distance λ_L, where λ_L is the wavelength of the laser beam in the buncher. Bunches will consist of N microbunches (see Fig. 2).

The maximum rate of the electron energy loss in parametric (prebunched) FELs is $d\varepsilon/dy = (\pi^2 mc^2/\lambda_u)(p_\perp^2/(1+p_\perp^2))(i/i_A)$, where $i_A = mc^3/e$, $\pi^2 mc^2/i_A \simeq 0.3$ MeV/kA. This expression is valid when short $(<\lambda_r)$, narrow $(\sigma_{e\perp} \ll \lambda_r \gamma/\sqrt{1+p_\perp^2})$ microbunches are situated in sequence at distances λ_L, where λ_r, the wavelength of the radiation emitted in the radiator (λ_r multiple to λ_L). In the case of a wide beam the rate tends to the previous one at the distance from the origin of the radiator $l_e = 2\pi\sigma_{e\perp}^2/\lambda_r$, where $\sigma_{e\perp}$ is the rms transverse electron beam dimension. A circularly polarized laser beam and a helical undulators were selected.

The power of radiation emitted in the prebunched FEL $P = K\lambda_u(d\varepsilon/dy)(i/e)$ or [19]

$$P = \frac{\pi^2 K i^2}{c} \frac{p_\perp^2}{1+p_\perp^2} \simeq 300 K i^2 \frac{p_\perp^2}{1+p_\perp^2} \,(\text{W/A}) \qquad (5)$$

where K is the number of the undulator periods.

The laser beam power (5) is high. If $i = 1$ A, $p_\perp = 1$, $K = 30$, the power $P = 9$ kW. It means, that the currents ~ 100 mA–1 A in the trains of microbunches are enough for experiments. There is no space charge problems in this case if relativistic electron beams are used. FELs work with such currents. The total current from electron guns in Example 1 can be accelerated after the exit from the bending magnet if low energy guns are used.

Power (5) does not depend on the emitted wavelength. The emitted radiation is diffraction-limited and the linewidth is determined by the length of the train of the electron microbunches. The main problem in prebunched FELs is the problem of beam bunching.

Scheme 3: If a thin electron beam has a small energy spread, low transverse emittance and energy modulation, Scheme 2 will work in a chopper mode at RF, optical and harder regions. The electron beam at the exit of the lattice will be bunched at Nth harmonic.

If the ionization losses of electrons in the medium of a thin lattice are equal to $\Delta\varepsilon_l/2$, then electrons of the incoming beam will be arranged in layers without losses.

Scheme 4: Two counter-propagating short laser beams stored in an open resonator perform an interference lattice in a storage ring at a position of low-beta and high dispersion function. An electron beam having low transverse emittance intersects the lattice at a 90° angle [20].

In this case, the energy losses of electrons depend on their position in the laser beam. Electrons moving in the bright zones of fringes lose more energy than those in the dark zones. Their energy (closed orbits) will tend to the energy (closed orbits) of electrons in the nearest dark zones if the energy of scattered quanta $\hbar\omega_{max} \ll \Delta\varepsilon_l$. As a result, the electron beam will be arranged in energy layers if the interaction time will be much less than the period of phase oscillations of electrons in the storage ring.

More effective interaction of counterpropagating laser and electron beams can be produced in straight sections of storage rings if $TEM_{n,0}$ laser mode is used ($n > 1$).

This scheme of bunching can be realized in ion-storage rings using present-day technology (Rayleigh cross-section is 7–10 orders higher than Compton one) [18]. Realization of the electron version of this scheme depends on the possibility of storing high-power laser beams in high-finesse optical resonators.[3]

3. Conclusion

Linear accelerator, storage ring and buncher technologies offer high-beam quality and short bunches necessary for operation of high efficiency, high-power, high-degree monochromatic pre-bunched FELs in optical and X-ray regions.

Acknowledgements

This work was supported partly by the Russian Foundation for Basic Research, Grant No. 02-02-16209.

References

[1] V.L. Ginzburg, Izv. Acad. Sci. USSR Ser. Phys. 11 (2) (1947) 165 (in Russian).

[2] E.G. Bessonov, Parametric free-electron lasers, Nucl. Instr. and Meth. A 282 (1989) 442.

[3] H. Motz, W. Torn, R.N. Whitehurst, J. Appl. Phys. 24 (7) (1953) 826.

[4] R.M. Phillips, Trans. IRE Electron Devices 7 (4) (1960) 231.

[5] D.F. Alferov, Yu.A. Bashmakov, E.G. Bessonov, Part 1 Sov. Phys. Tech. Phys. 23 (8) (1978) 902; D.F. Alferov, Yu.A. Bashmakov, E.G. Bessonov, Part 2 Accel. 9 (4) (1979) 223.

[6] E.G. Bessonov, Proceedings of the Fourth General Conference of the European Physics Society, York, UK, 1979, p. 471 (Chapter 7).

[7] B.E. Carlsten, D.W. Feldman, J.M. Kinross-Wright, Proceedings of the MICRO BUNCHES WS, Upton, New York, September 1995, p. 21.

[8] H. Wiedemann, Particle Accelerator Physics, Vol. 1, Springer, Berlin, 1998.

[9] N.A. Vinokurov, A.N. Skrinsky, preprint, Inst. Nucl. Phys. No. 77-59, Novosibirsk, 1977.

[10] D.F. Alferov, E.G. Bessonov, preprint FIAN No. 162, Moscow, 1977; D.F. Alferov, E.G. Bessonov, Sov. Phys. Tech. Phys. 23 (4) (1979) 450.

[11] E.G. Bessonov, A.V. Serov, preprint No. 87, Lebedev Physical Institute, USSR, 1980; E.G. Bessonov, A.V. Serov, Sov. Phys. Tech. Phys. 27 (12) (1982) 245.

[12] J. Blau, T. Campbell, W.B. Colson, et al., Nucl. Instr. and Meth. A 483 (2002) 142.

[13] W. Colson, A. Todd, G.R. Neil, in: M. Brunken, H. Genz, A. Richter (Eds.), Proceedings of the International Conference on Free Electron Lasers, 2001, II-9.

[14] A.V. Savilov, Nucl. Instr. and Meth. A 483 (2002) 200.

[15] G.N. Neil, L. Merminga, Rev. Mod. Phys. 74 (2002) 685.

[16] E.G. Bessonov, Proceedings of the 21st International Free Electron Lasers Conference, Hamburg, Germany, August 23–28, 1999, II-51.

[17] E.G. Bessonov, V.G. Kurakin, A.V. Serov, Sov. Phys. Tech. Phys. 21 (9) (1976) 1158; E.G. Bessonov, V.G. Kurakin, A.V. Serov, Doklady Acad. Nauk USSR 280 (4) (1985) 843 (in Russian).

[18] E.G. Bessonov, Proceedings of the MICRO BUNCHES WS, Upton, New York, September 1995, p. 367.

[19] E.G. Bessonov, Nucl. Instr. and Meth. A 358 (1995) 204.

[20] T. Shintake, Nucl. Instr. and Meth. A 311 (1992) 453.

[21] E.G. Bessonov, K. Je Kim, Phys. Rev. Lett. 76 (3) (1995) 431; E.G. Bessonov, K. Je Kim, Proceedings of the Fifth European Particle Acceleration Conference, Sitges, Barcelona, 10–14 June 1995, Vol. 2, p. 1196.

---

[3] Note that prebunched Free-Ion Lasers present the ultimate in the capabilities of lasers [19]. Relativistic ion beams can be cooled to a high degree in 6D space [21].

Available online at www.sciencedirect.com

NUCLEAR
INSTRUMENTS
& METHODS
IN PHYSICS
RESEARCH
Section A

ELSEVIER   Nuclear Instruments and Methods in Physics Research A 528 (2004) 516–519

www.elsevier.com/locate/nima

Experiment on gamma-ray generation and application

D. Li[a,*], K. Imasaki[a], M. Aoki[a], S. Miyamoto[b], S. Amano[b], K. Aoki[b], K. Hosono[b], T. Mochizuki[b]

[a] *Institute for Laser Technology, 2-6 Yamada-oka, Suita, Osaka 565-0871, Japan*
[b] *Lasti, Himeiji Institute of Technology, 3-1-2 Koto, Kamigori, Hyogo 678-1201, Japan*

Abstract

An experimental setup of gamma-ray generation through laser Compton scattering has been built on the NewSUBARU storage ring. The aim is to study nuclear transmutation, which is regarded as the first stage to explore the feasibility of developing a nuclear waste disposal system based on the concept of irradiating long-lived fission products by laser Compton scattering gamma ray. In this paper, the gamma-ray generation facility is presented, and some experimental results such as gamma-ray energy spectrum, intensity distribution, and the coupling efficiency of nuclear transmutation, are given. The experimental data is in good agreement with the analytic calculation or simulation analysis.
© 2004 Elsevier B.V. All rights reserved.

PACS: 07.85.FV

Keywords: Gamma ray; Laser Compton scattering; Storage ring; Nuclear transmutation; Long-lived fission product

1. Introduction

In order to dispose nuclear waste composed of long-lived fission products, a proposal discussing the concept of shortening their long radioactive life by transmuting their nuclei to an unstable isotope through irradiation by laser Compton scattering gamma ray, which is based on a storage ring, was presented in recent years [1]. The concept points at the conversion efficiency [2] of nuclear transmutation, induced by the gamma-ray photons coupling to nuclear giant resonance of a certain material. In accordance with the proposed scheme, experiments are being conducted to investigate the fundamental issues concerning nuclear transmutation. We have successfully developed a laser Compton scattering setup on the NewSUBARU storage ring providing electron beam of energy 1 GeV, to produce gamma-ray photons with maximum energy of 17.6 MeV by a low power Nd:YAG laser. Results obtained in experiments including gamma-ray energy spectrum and intensity distribution have been given in this paper, and the application of the produced gamma ray to nuclear transmutation have also been reported.

*Corresponding author.
E-mail address:* dazhi_li@hotmail.com (D. Li).

0168-9002/$ - see front matter © 2004 Elsevier B.V. All rights reserved.
doi:10.1016/j.nima.2004.04.092

2. Gamma-ray generation

2.1. Experimental setup

One of the straight sections of the NewS-UBARU storage ring was chosen to realize the laser Compton scattering chamber, as shown in Fig. 1, where the electron beam collides with the incoming laser beam in a head-to-head manner. Thus the collisions between electrons and laser photons would give rise to higher energy photons, gamma-ray photons in our case, going along the incident electron moving direction in a forward cone of angle $1/\gamma$ where γ is the relativistic factor of electron, namely, 0.5 mrad in our experiment for the 1 GeV electron beam. A reflected mirror is located at the downstream end to guide the laser light travelling along the beam line through the interaction point designed at the center of the straight section, and the light is reflected out of the chamber by another upstream mirror. The produced gamma-ray photons would go through the downstream mirror and reach the detector or irradiate the nuclear sample.

Fig. 2 shows the transport system of the laser light coming from a Nd:YAG laser operating at CW mode with a wavelength of 1.064 μm and a

power of 0.67 W. The laser light is guided into the vacuum chamber by five mirrors arrayed deliberately and a convex lens with focal length of 5 m in a well-designed position, 7.5 m away from the YAG laser and 15 m away from the center point of the straight section. This results in a focused spot of light with radius of 0.82 mm. Taking into account the loss of reflection and diffraction, the laser power at the interaction point is expected to be 0.35 W.

A collimator, which is also a sample holder, is set just before a High-Purity Germanium Coaxial Photon Detector, produced by ORTEC, with a crystal measuring 64.3 mm in diameter and 60.0 mm in length, exhibiting an efficiency of 45%.

The electron beam size is determined by the β function and emittance. For the NewSUBARU storage ring, at the center point of the straight section, these parameters are characterized as $\beta_x = 2.3$ m, $\beta_y = 9.3$ m, $\varepsilon_x = 40$ nm, and $\varepsilon_y = 4$ nm, resulting in the electron beam size of 0.30 mm for the horizontal direction and 0.19 mm for the vertical direction. Consequently, the size of electron beam is smaller than that of the laser beam at the interaction point.

2.2. Gamma-ray energy spectrum

The average current of the electron beam supplied by the NewSUBARU storage ring ranges from several milliamperes up to 200 mA. The measurement of gamma rays is carried out at a lower current of several milliamperes, lest it saturates the Germanium detector. The experimental results for Laser On and Laser Off, detected by a collimator of 6 mm diameter are shown in Fig. 3. The apparent separation between

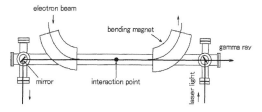

Fig. 1. Collision system.

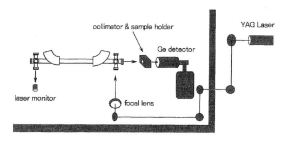

Fig. 2. Laser and detector system.

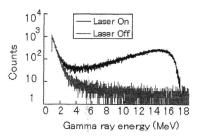

Fig. 3. Gamma-ray energy spectrum detected by Ge detector.

the two signals of laser Compton scattering gamma ray and the background presents a good signal-to-noise ratio. The maximum energy appears around 17 MeV, which is in agreement with the theoretical prediction.

We simulated the whole process of generated gamma-ray photons passing through the reflected mirror, output window, collimator, and being detected by the germanium detector, by employing the EGS4 code [3]. The EGS4 code is well known and widely used in the field of interaction of particles and material, taking into account many physics processes such as Bremsstrahlung production, pair production, Compton scattering, and photoelectric effect. The simulation curve is well consistent with the experimental data as shown in Fig. 4. After processing the experimental data, we achieved the actual gamma-ray photons' luminosity of 2.5×10^5 counts/A/W/s. In conclusion, the gamma-ray photons' yield of 1.75×10^4 counts/s can be accomplished by our facility under the running condition of $I_e = 0.2$ A and $P = 0.35$ W.

2.3. Spatial distribution of intensity

The theoretical analysis predicts that the spatial distribution of intensity of laser Compton scattering gamma-ray is in connection with the polarization of initial laser photons. Circularly polarized or unpolarized initial photon gives rise to an azimuthally symmetric pattern in transverse distribution of intensity, whereas linearly polarized initial photon results in azimuthal modulation. In our experiment, the incoming laser photons were of linear polarization, and an image plate was placed 15 m away from the interaction point to detect the spatial distribution of produced gamma

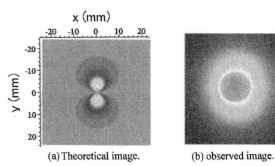

(a) Theoretical image.　　(b) observed image.

Fig. 5. Image of gamma-ray indicates its spatial distribution of intensity.

ray. The results of the analytic calculation as well as the experiment are shown in Fig. 5. Though the experimental image is not so clear we can identify the modulated pattern around the central area.

3. Application on nuclear transmutation

The final goal of this research is to understand the feasibility of transmuting the long-lived fission products in nuclear wastes such as [129]iodine and [135]cesium, which have a long radioactive life of more than million years. There is a possibility of water transporting the nuclear waste from deep underground to the surface [1]. As our first step, we aim at exploring the coupling efficiency of gamma rays to nuclear giant resonance, which is defined as the transmutation rate by per gamma-ray photon. Coupling efficiency was derived by improving the one described in reference [2] by considering geometrical structure of a cylindrical target as

$$\eta = \frac{N_0 \int_0^b \int_0^a \sigma_L(E)\sigma_g(E)e^{-\mu z} \cdot 2\pi r \, dr}{\int \sigma_L(E) \cdot 2\pi r \, dr} \quad (1)$$

where N_0 is the number of atoms per volume, $\sigma_L(E)$ is the cross section of laser Compton scattering defined by Klein–Nishina formula, $\sigma_g(E)$ is the cross section for nuclear giant resonance, μ is the total linear attenuation coefficient including the effects of photoelectron, Compton, and pair production as expressed in Ref. [2], a and b represent the radius and length of

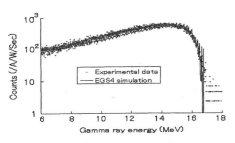

Fig. 4. Simulation and detected curves of energy spectrum.

Fig. 6. Transmutation of Au target.

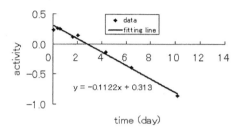

Fig. 7. The exponential decay of activity (semi-logarithm plot).

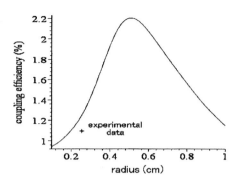

Fig. 8. Coupling efficiency vs. Au rod radius with a fixed length of 5 cm (theoretical curve and experimental data).

the cylindrical target, respectively, and E indicates the gamma-ray energy.

A 5 cm long gold rod with a radius of 0.25 cm was adopted as the nuclear target in the present experiment and irradiated for a duration of 8 h on-axis, 15 m away from the interaction point. The transmutation process of this target is as shown in Fig. 6, and the main decay occurs from Au-196 to Pt-196, giving rise to radioactivity in the form of gamma-ray photons with a peak energy of 355.73 keV in the energy spectrum. This radioactivity was measured by a NaI(Tl) detector, and the activity line is given in Fig. 7 with semilogarithm coordinates, presenting a good agreement with the acknowledged half-life of 6.183 D. Through data processing, we concluded that the number of transmuted nuclei was 3.165×10^6 at the moment the irradiation was complete. On the other hand, by the germanium detector, the absorbed laser Compton gamma-ray photons by the target during the irradiation was determined as 2.95×10^8. Hence, the coupling efficiency of gamma ray to nuclear giant resonance was derived as 1.1%. Actually, this value should be lower than the real value because the attenuation of gamma ray from the radioactivity inside the target was not involved, and future experiment would provide a more accurate estimation. However, the experimental result is close to the theoretical analysis as shown in Fig. 8. It is predicted that a maximum efficiency $\sim 2.2\%$ can be reached when an Au rod of radius 0.5 cm is used.

4. Conclusion

We built a storage ring-based laser Compton scattering setup and successfully produced gamma-ray photons. The gamma rays were applied to nuclear transmutation research for exploring new means to dispose nuclear waste. The theoretical evaluation and the observed results of the experiment are consistent with each other.

References

[1] K. Imasaki, A. Moon, High brightness γ ray and its application for nuclear transmutation, SPIE 3886 (2000) 721.

[2] D. Li, K. Imasaki, M. Aoki, J. Nucl. Sci. Tech. 39 (2002) 1247.

[3] W.R. Nelson, H. Hirayama, W.O. Roger, The EGS4 code system, SLAC-Report, 1985, p. 265.

Available online at www.sciencedirect.com

Nuclear Instruments and Methods in Physics Research A 528 (2004) 520–524

**NUCLEAR
INSTRUMENTS
& METHODS
IN PHYSICS
RESEARCH**
Section A

ELSEVIER

www.elsevier.com/locate/nima

On limitations of Schwinger formulae for coherent synchrotron radiation produced by an electron bunch moving along an arc of a circle

G. Geloni[a], E.L. Saldin[b],*, E.A. Schneidmiller[b], M.V. Yurkov[c]

[a] *Department of Applied Physics, Technische Universiteit Eindhoven, The Netherlands*
[b] *Deutsches Elektronen-Synchrotron (DESY), (MPY) Notkestrasse 85, 22607 Hamburg, Germany*
[c] *Joint Institute for Nuclear Research, Dubna, 141980 Moskow region, Russia*

Abstract

Re-examination of dogmatic "truths" can sometimes yield surprises. For years we were led to believe that famous Schwinger's formulas are directly applicable to describe synchrotron radiation from dipole magnet and even now no attention is usually paid to the region of applicability of these expressions. While such formulas are valid in order to describe radiation from a dipole in the X-ray range, their long-wavelength asymptote are not valid, in general. In the long-wavelength region, Schwinger's formulas must be analyzed from a critical viewpoint, and corrections must be discussed when one is looking for an application to CSR-based diagnostics. In this paper, we perform such a task by means of a consistent use of similarity techniques, discussing the limits of validity of Schwinger's formulas which arise from a finite magnet length, from a finite distance of the detector to the sources and from diffraction effects (due to the presence of vacuum pipe and aperture limitations).
© 2004 Elsevier B.V. All rights reserved.

PACS: 41.20.Jb; 41.60.Ap; 41.75.Ht

Keywords: Synchrotron radiation; Coherent radiation; Coherent synchrotron radiation; Relativistic electron and positron beams

1. Introduction

Short bunches of RMS length of the order of 100 fs are needed for XFEL applications [1,2], which is achieved by compression chicanes. The development of nondestructive methods for the measurement of the longitudinal beam current

distribution in such short bunches is a challenging problem: one possible solution is based on measurements of Coherent Synchrotron Radiation (CSR) produced by a bunch passing a magnetic system. CSR is also interesting as a radiation source in the far infrared wavelength range [3].

Standard theory of synchrotron radiation (and CSR) is based on Schwinger's formulas [4], which rely upon several approximations. In general, due to the presence of vacuum pipes and aperture limitations, diffraction effects must be taken into

*Corresponding author. Tel.: +49-40-8998-2676; fax: +49-40-8998-4475.

E-mail address: saldin@mail.desy.de (E.L. Saldin).

0168-9002/$ - see front matter © 2004 Elsevier B.V. All rights reserved.
doi:10.1016/j.nima.2004.04.093

account when studying radiation emission. The solution of the diffraction problem involves, in its turn, the characterization of the fields at the aperture position which can not fulfill, for practical reasons, the usually assumed condition of an observer at infinite distance from the sources. However, both finite distance and diffraction effects are usually ignored when it comes to the problem of characterizing radiation from moving electrons. Moreover, as pointed out in Ref. [5], the CSR pulse form a dipole magnet is not properly described by means of the usual synchrotron radiation theory, since the spectral region of interest is at long (far-infrared) wavelengths. All the effects introduced above must be carefully taken into account when designing a diagnostic method or a radiation source based on CSR.

2. Limitations of standard results

In this section, we address the temporal structure of the CSR pulse from a Gaussian bunch studying how it is altered by the failure of the assumptions in Schwinger formulas. The CSR pulse is described by a characteristic time of the order of the inverse of the frequency spread in the CSR pulse. We assume $\sigma_T \gg R/(c\gamma^3)$, σ_T being the RMS electron bunch duration and R being the magnet radius. We will focus on the radiation pulse seen about the tangent to the orbit at the middle point of magnet. When the vertical angle is zero, the normalized coherent field amplitude $E_x(t)/E_{max}$ is a function of six dimensional parameters:

$$t, \quad v, \quad R, \quad \phi_m, \quad \sigma_T, \quad |\vec{r}_0| \qquad (1)$$

where v is the electron velocity, ϕ_m is the magnet angular dimension and $|\vec{r}_0|$ is the distance between source and observer. After appropriate normalization, it is a function of four dimensionless parameters only

$$E_x(t)/E_{max} = \hat{E}_x = D(\hat{t}, \hat{\sigma}, \hat{\rho}, \hat{r}_0) \qquad (2)$$

where $\hat{t} = t/\sigma_T$ is the dimensionless time, $\hat{\sigma} = \omega_0\sigma_T\gamma^3$ is the dimensionless electron pulse duration, $\hat{r}_0 = |\vec{r}_0|(c\sigma_T)^{-1/3}R^{-2/3}$ is the dimensionless distance between source and observer, $\hat{\rho} =$

$\phi_m^3/(24\omega_0\sigma_T)$ is the magnet length parameter (ω_0 being the particle revolution frequency). In the general case the universal function D should be calculated numerically by means of strict Lienard–Wiechert formulas. It is easy to show that the CSR pulse duration is given by $\sigma_T \simeq R\phi^3/(24c)$, while the radiation source extends over some finite length $R\phi \sim L_f = (c\sigma_T)^{1/3}R^{2/3}$ along the particle path. We see that the reduced distance can be expressed as $\hat{r}_0 = |\vec{r}_0|/L_f$. The ratio $(R\phi_m)^3/(6L_f^3)$ is equal to $\hat{\rho}$, which we use now as a measure of finite magnet length effects.

2.1. Diffraction effects

In realistic situations, the long-wavelength synchrotron radiation from bending magnets passes through many different vacuum chamber pieces with widely varying aperture, and this may perturb the CSR spectrum. Consider an electron moving in a circle, as in Fig. 1. Between the observer and the source there is an aperture with a characteristic dimension d. Qualitatively, an observer looking at a single electron is presented with a cone of radiation characterized by an aperture angle of order $\theta \simeq \sqrt{2d/R}$. Fig. 1 shows part of the trajectory of an electron travelling along an arc of a circle of radius R. The presence of a finite aperture introduces diffraction effects specific to the geometry and clearly dependent on the wavelength. For structures such as pinholes it is found that these diffraction patterns propagate away from the structure at angles of order $\theta_d \simeq \lambda/d$, where d is the characteristic aperture dimension. The region of applicability for the far diffraction zone is given by the relation $L_p \gg L_d \simeq d^2/\lambda$, where L_p is the distance between observer and aperture and L_d the typical diffraction

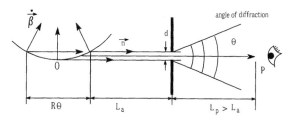

Fig. 1. Depiction of the effects of an aperture limitation.

distance. When the wavelength is about $R\theta^3/24 \simeq \sqrt{2}(d^3/R)^{1/2}/12$, the latter condition transforms to $L_p \gg L_d \simeq 6R\theta$. For example, if $R = 3$ m and $\lambda = 100$ μm, $\theta \simeq 0.08$, a 1 cm diameter hole will significantly perturb the SR spectrum at wavelengths in this region for distances greater than about 1 m. In order to solve the diffraction problem we must characterize the field at the aperture position: since the aperture is expected to be only a couple of centimeters away from the sources, finite distance effect play an important role in the overall determination of the CSR pulse at the detector position.

2.2. Finite distance effects: acceleration field

Consider an electron bunch moving in a circle, without any aperture limitation. In this situation we can address the contribution to the field detected by an observer given by the acceleration field alone and then focus on the contributions by the velocity field. In this case the spectrum is independent of the $\hat{\rho}$ parameter and, in the long-wavelength asymptote $(\hat{\sigma} \gg 1)$, the acceleration field is described by $\hat{r}_0$ only. The region applicability of Schwinger formulas requires $\hat{r}_0 \gg 1$, since the radiation source extends over some finite length $R\phi \simeq L_f$, corresponding to a transverse size of the radiation source $h \simeq R\phi^2$. The vector $\vec{n}$ changes its orientation between point A and point B of an angle of order h/r_0, where r_0 is the distance between source and observer. Therefore, when $h/r_0 \ll \phi$, the vector $\vec{n}$ in Lienard–Wiechert formula is almost constant when the electron moves along the formation length $R\phi$, and the unit vector $\vec{n}$ can be considered constant throughout the electron evolution only if $r_0 \gg L_f$. The results of numerical calculations for several values of $\hat{r}_0$ are presented in Fig. 2. Calculations have been performed using the general Lienard–Wiechert formula. At large distance the CSR pulse profile is simply the far zone profile predicted by Schwinger formulas. One can see that they work well at $\hat{r}_0 = 100$, but already at $\hat{r}_0 = 10$ the CSR pulse envelope is visibly modified. Note that, for example, for a RMS bunch length of about 100 μm and a radius $R = 3$ m the value $\hat{r}_0 = 10$ corresponds to $r_0 \simeq 1$ m. If we detect CSR in

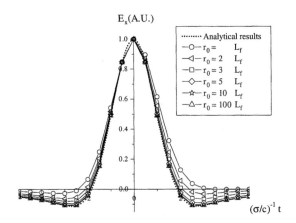

Fig. 2. Time structure of a CSR pulse from a Gaussian electron bunch moving in a circle at different reduced distance between source and observer.

vacuum at a distance smaller than 1 m, or if we have an aperture at a distance smaller than 1 m, this effect is important.

2.3. Finite distance effects: velocity field

In the long-wavelength asymptote, the velocity part of the coherent electric field from a particle in a circle is a function of two dimensionless parameters

$$E_{(v)}(t)/E_{\max} = \hat{E}_{(v)} = D(\hat{t}, \hat{\sigma}, \hat{r}_0) \qquad (3)$$

where the normalization of the velocity field is performed with respect to the maximal acceleration field amplitude. Figs. 3 and 4 show the dependence of the normalized velocity field amplitude on the value of the reduced electron pulse duration for different values of the reduced distance. It is seen from the plots in Fig. 4 that in the near zone we cannot neglect the influence of the velocity field on the detector. In the far zone the velocity field from one electron is close to an antisymmetric function: when the electron bunch is long the velocity field from one electron can be substituted by its asymptotic behavior hence, the normalized amplitude of the coherent velocity field is found by dividing the asymptotic behavior of the velocity field kernel by the asymptotic behavior of the acceleration field kernel; in the far zone, the normalized velocity field is of order

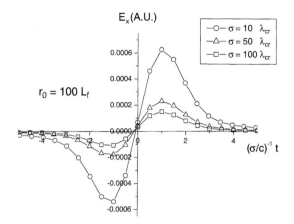

Fig. 3. Electric field pulse due to velocity term from a Gaussian electron bunch moving in a circle at different reduced bunch length. The reduced distance is held constant, $\hat{r}_0 = 100$. Here the normalization is performed with respect to the maximal acceleration field amplitude.

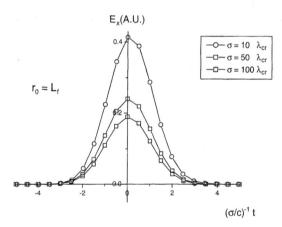

Fig. 4. Electric field pulse due to velocity term from a Gaussian electron bunch moving in a circle at different reduced bunch length. The reduced distance is held constant, $\hat{r}_0 = 1$. Here the normalization is performed with respect to the maximal acceleration field amplitude.

$E_v/E_{acc} \simeq R/(\gamma^2 \phi^2 r_0)$, where $\phi \simeq (c\sigma_T/R)^{1/3}$ is the natural synchrotron radiation opening angle with frequency $\omega \simeq \sigma_T^{-1} \ll c\gamma^3/R$. Using normalized variables we get

$$\hat{E}_{(v)} \simeq \hat{r}_0^{-1}\hat{\sigma}^{-2/3} \quad \text{for } \hat{\sigma}, \hat{r}_0 \gg 1. \tag{4}$$

As we can see from Fig. 3, numerical calculations in the far zone confirm this simple physical consideration. The value of $\hat{E}_v$ that we can expect is found remembering that, in the example given in

Fig. 3, $\hat{r}_0 = 100$, $\hat{\sigma} = 100$; therefore, the normalized velocity field would be about 0.0004, which is the same order of magnitude as results of numerical calculations (0.0002). Also, $\hat{E}_v$ varies roughly as $\hat{\sigma}^{-2/3}$. The normalized velocity field amplitude decreases, as we see from our estimations, linearly with distance, which means that if we are in the near zone at $\hat{r}_0 = 1$, our model predicts $\hat{E}_v \simeq 0.04$, in contrast with numerical calculation results, which give $\hat{E}_v = 0.2$ (again, note that in the case for $R = 3$ m and bunch length of 100 μm this case corresponds to $r_0 \simeq 10$ cm): our approximate treatment of coherent velocity field breaks down once source and observer get as close as they are at $\hat{r}_0 \simeq 1$. We conclude that, in addition to the antisymmetric part of the single-particle field, there is also a symmetric one which becomes dominant at close distances and which also varies with the separation.

2.4. Finite magnet length effects

Until now we have considered the case in which electrons move in a circle. Here we generalize this situation to the case where the electrons move along an arc of a circle. We obtain that, in the far-field approximation, the CSR pulse profile is a function of only one dimensionless parameter $\hat{\rho} = \phi_m^3/(24\omega_0\sigma_T)$, where ϕ_m is the magnet angular extension. The region of applicability of the

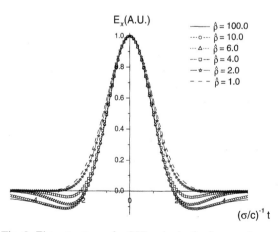

Fig. 5. Time structure of a CSR pulse in the far zone from a Gaussian electron bunch moving along an arc of a circle at different reduced magnet lengths.

Schwinger formulas is for $\hat{\rho} \gg 1$. Using the plots presented in Fig. 5, one can characterize quantitatively the region of applicability of the circular motion model. Note that the field tail is completely modified for $\hat{\rho} = 6$, which, for our example of $R = 3$ m, RMS bunch length of 100 μm means $\phi_m \simeq 0.17$. This means that a 50 cm magnet is short enough to give nonnegligible spectral distortions. This corresponds to a view from a hole of about $d \simeq 1$ cm.

References

[1] F. Richard, et al. (Eds.), TESLA Technical Design Report, DESY 2001-011.
[2] LCLS Design Study Report, SLAC Reports SLAC-R521, Stanford, 1998.
[3] G.L. Carr, et al. Nature November (2002) 153.
[4] J. Schwinger, Phys. Rev. 75 (1949) 1912.
[5] G. Geloni, et al., preprint DESY 03-031, DESY, Hamburg, 2003.

Available online at www.sciencedirect.com

SCIENCE DIRECT°

ELSEVIER   Nuclear Instruments and Methods in Physics Research A 528 (2004) 525–529

NUCLEAR
INSTRUMENTS
& METHODS
IN PHYSICS
RESEARCH
Section A

www.elsevier.com/locate/nima

Free-electron lasers as pumps for high-energy solid-state lasers

G. Travish[a],*, J.K. Crane[b], A. Tremaine[b]

[a] UCLA Department of Physics and Astronomy, Los Angeles, CA 90095, USA
[b] Lawrence Livermore National Laboratory, Livermore, CA 94551, USA

Abstract

High average-power free-electron lasers may be useful for pumping high peak-power solid-state laser-amplifiers. At very high peak-powers, the pump source for solid-state lasers is non-trivial: flash lamps produce thermal problems and are unsuitable for materials with short florescence-times, while diodes can be expensive and are only available at select wavelengths. FELs can provide pulse trains of light tuned to a laser material's absorption peak, and florescence lifetime. An FEL pump can thus minimize thermal effects and potentially allow for new laser materials to be used. This paper examines the design of a high average-power, efficient high-gain FEL for use as pump source. Specifically, the cases of a 100 J class pump, and a 100 TW-class laser at a planned fourth-generation light-source are considered.
© 2004 Elsevier B.V. All rights reserved.

PACS: 52.59.−f; 41.60.Cr; 42.55.Xi

Keywords: Lasers; FEL; Pump; High energy; High power

1. Introduction

High energy and high power lasers have found wide usage in research and application including material processing, solid-state research, particle physics, and as drivers for other radiation sources such as X-ray (K-alpha) production. From early proposals of laser driven accelerators [1] to recent applications such as high-field physics [2], nuclear physics, [3] fusion science [4], proton-beam generation [5], and radiography [6], the demand for high power and high-energy lasers has increased. Still, lasers in the greater than 100 TW or 10 J class

are challenging to build, involve large investments, and can only be developed for a limited set of wavelengths. While thermal limits and optical-damage thresholds constraint the selection of gain media, the lack of viable pump-sources further restricts usable materials. Flashlamps produce large thermal-loads and are not suited to materials with short fluorescence-times (such as Ti:S). Diodes remain expensive, difficult to use on very large crystals, and are more practical at longer wavelengths.

Free-electron lasers (FELs) have long promised to provide high average optical-powers, and recent work indicates that this promise can be delivered upon by increasing the efficiency, duty factor and bunch repetition-rate of existing designs. FELs can

---

*Corresponding author. Tel.: +1-310-206-1677.
E-mail address: gil.tavish@physics.ucla.edu (G. Travish).

0168-9002/$ - see front matter © 2004 Elsevier B.V. All rights reserved.
doi:10.1016/j.nima.2004.04.094

provide pulse trains of light tuned to a laser material's absorption peak, and duration. An FEL pump would thus minimize thermal-effects and potentially allow for new laser-materials to be used. For instance, a high-power FEL could pump a high-energy Ti:S amplifier and produce PW-class pulses. Moreover, FEL light-source facilities are being designed and built already; a high-energy laser at such a fourth-generation light-source could be used for novel pump-probe experiments while taking advantage of existing infrastructure.

This paper proposes the use of a high average-power, efficient, high-gain FEL for use as a pump source. Specifically, two cases are considered: a 1 kJ-class pump for a 100 J high-energy laser; and, a 25 J pump for a 100 TW-class high peak-power laser.

2. The FEL as a pump

Existing solid-state lasers are pumped by flash lamps, diodes or other lasers. A comparison of existing pump-sources with the type of FEL considered here is shown in Table 1 and reveals the regime of applicability for novel pumps: at high pump-energies and shorter wavelengths conventional pump-sources are inadequate or do not exist. While flashlamps produce unwanted heat due to their broadband spectrum, they are the most mature and most widely used type of pump—the world's most powerful laser, NIF, is flashlamp pumped [7]. Diodes are becoming widely used because they are reliable, stable, tuneable to match common laser-material absorption-bands and

Table 1
A comparison of existing laser pump sources with the FEL-based pump. The FEL is suited to high energy and short wavelength applications

	Pump source			
	Flashlamp	Diode	Laser	FEL
Avg. energy	Very high	High	Low	High
Peak energy	Medium	Low	High	Very high
Heat load	High	Low	Low	Very low
Wavelength	VIS	IR–NIR	IR–UV	IR–UV

therefore are thermally favourable, and are becoming economically viable (est. at $5/W). However, diodes are still difficult to couple to large crystals, and are not practical (efficient) for wavelengths below about 800 nm.

2.1. The ideal pump

An ideal laser-pump is matched to the gain medium in wavelength, bandwidth, time structure and size [8]. The wavelengths of available pumps have, in many ways, determined which materials are investigated for use in lasers. New pump sources especially in the UV would make possible the use of new materials offering new lasing-wavelengths and new operating-regimes (e.g. Ce^{3+}:Li-CAF). The ideal pump source is also stable shot-to-shot; is electrically efficient and has a low cost per Watt. The optical properties of pump sources is also an issue for large amplifier crystals as they require many sources to cover the material.

The FEL is able to provide most of the above characteristics with a few possible exceptions: the stability, efficiency, and cost of the FEL warrant discussion. A seeded high-gain FEL driven to saturation is stable to the extent that the electron-beam parameters do not fluctuate significantly. Operating as a pump source, the integrated energy delivered to the laser crystal is more significant than shot–shot fluctuations; thus, a feedback system could be used to assure that the integrated energy is indeed stable. The efficiency of conventional high-gain FELs is low (varying as the FEL parameter ρ; approximately 0.1–1% in typical systems). Tapering of the undulator, along with bunch compression, has been shown to produce efficiencies approaching 10% in already-achieved undulator lengths. Finally, the cost of free-electron lasers is inherently high due to the attendant infrastructure. At lower pump energies (<10 J), the costs of building an FEL seems unlikely to compete with conventional lasers or diodes. However, at pump energies around 100 J, the cost analysis is vague. Moreover, large diode-arrays costing millions of dollars can only operate at one wavelength (or at best, over a narrow wavelength-range), and have a limited lifetime: an FEL can be tuned and thus pump multiple types of laser amplifiers. (of course, time structure is constrained to what the RF system can deliver).

2.2. Description of a pump FEL

The components of an (high gain) FEL acting as a pump are a high average-power accelerator with a high-brightness injector; a long tapered-undulator; and, a high repetition-rate seed-laser. (Alternative configurations such as FEL oscillators or regenerative amplifiers—RAFEL—are not considered here.) The "gain medium" in an FEL is the electron beam. Because the electron beam is continuously refreshed, there are no thermal-load problems in an FEL, and hence, a high average-power device has long been promised [9]. Still, production of high average-power, high-brightness electron-beams has been a challenge and a number of approaches have been considered including energy-recovery linacs, and superconducting systems. In addition, conventional accelerating systems can be and have been used at high duty-cycles. The salient point here is that the electron beam needs to provide a high integrated-energy only over the fluorescence time of the crystal to be pumped.

3. Challenges and opportunities

The production of a high-brightness high average-power beam presents significant technical challenges including thermal management, efficient production of high-power RF, and development of a suitable drive-laser (assuming a photoinjector is used). In the accelerator, correction for beam loading and mitigation of wakefields are issues. The design of a high-efficiency FEL is straightforward if a tapered undulator is employed; however, such an undulator would introduce a large energy-spread on the beam making recirculation or energy-recovery schemes difficult. Finally, tailoring the repetition rate, microbunch current (charge), and macropulse duration to deliver the desired pump-energy during the gain crystal's fluorescence time may be difficult. The

above significant challenges face most high average-power devices in some form.

4. Example systems

We consider two examples of an FEL pumped solid-state laser. In the first example, we attempt to match the performance of the state-of-the-art MERCURY diode-pumped laser system while the second example is based on a system operating parasitically at the Linac Coherent Light Source (LCLS).

4.1. A MERCURY-like pump

The MERCURY laser utilizes arrays of diodes (over 6000 in total) to drive a 4-pass gas-cooled amplifier system using ytterbium-doped strontium fluorapatite crystals [10]. The project goals include 10% electrical efficiency at 10 Hz and 100 J with a 5-ns pulse length, and represents the state-of-the-art in diode pumped solid-state lasers (DPSSL). The peak power of the diode arrays is 640 kW. Yb:S-FAP has a fluorescence time of about 1.1 ms with an absorption peak of 905 nm. The light from the complex diode-arrays is funnelled into the disk amplifier through multiple reflections in a polished chamber (non-imaging optic).

We consider a MERCURY-class pump that can deliver 1 kJ of 905 nm light in 1.1 ms. A superconducting linac is selected to take advantage of the long fluorescence-time. For simplicity, we assume a TTF-based linac capable of generating about 3×10^5 bunches of 1 nC each—filling 1 in 5 RF buckets—at about 60 MeV [11]. Production of the beam may be challenging due to the long RF-pulse and the long drive-laser train: an RF thermionic-gun with a compression alpha-magnet could be considered [12] due to the long FEL wavelength. The resulting beam is sent though an undulator (≈ 10 m) with a period of 2 cm and a $K = 1$ to produce a 5%-efficient FEL. Each electron bunch would produce some 3 mJ of optical energy, yielding 1 kJ of optical power for the train. Of course, this is an un-optimized design; one could, for instance, consider filling every RF bucket, and shortening the train to 200 μs.

4.2. A 100 TW-class laser for LCLS

The LCLS is one of a few fourth-generation light-sources being constructed [13]. Such a facility will provide X-ray pulses with unprecedented brightness and brilliance. The intensity of the X-ray pulses is sufficient that single-shot measurements are possible and necessary—the target is vapourized. Pump-probe experiments have long been a standard at existing light-sources, and will likely continue at fourth-generation sources. The destruction of the target after a single "probe" pulse implies that new methods must be used to obtain time-delay probe information (i.e. chirped pulses, multiple samples, etc.). The need for multiple laser-sources synchronized to the X-ray pulse seems clear from the user demands [14]. The availability of a 100 TW-class laser may allow for novel experiments (i.e. pumping of high density materials and non-linear techniques) using the intense X-ray pulse as a probe of sufficient intensity to image the target.

We envision an FEL driven by the front end of the LCLS (Fig. 1) including the photoinjector, low-energy linac and bunch compressor. The photoinjector is assumed to have a high quantum-efficiency cathode and be driven by a multipulse laser (e.g. LANL AFEL or the DESY TTF). The resultant beam parameters are assumed to be 2000 bunches (filling every 5th RF bucket) each of 1 nC. At the end of the first bunch compressor (BC-1), the beam is 250 MeV with a peak current of 500 A. Such a beam, sent into an FEL with 5% efficiency can deliver 25 J of pump light (at 490 nm) over the fluorescence time of Ti:S (about

3.5 μs). While these parameters appear ambitious, and the beam-loading compensation may require additional linac-sections downstream (running all sections at a lower gradient); the necessary beam-brightness is considerably lower than that required for the LCLS, and the undulator is a long but straightforward design (e.g. 5 cm period, $K \approx 2.5$, 10–20 m un-optimized) — such undulators have already been built and used in FELs [15]. A final-amplifier driven by this pump could produce over 10 J of 800 nm light, and could easily be compressed below 100 fs, yielding over 100 TW peak power at the experimental hall.

5. Future work

The use of an high energy, high-efficiency FEL as a pump for a solid-state lasers may find application in existing facilities as well as purpose-built machines. Ultimately, the practicality of such a system may be an economic decision as diodes become more affordable. However, the flexibility of the FEL to pump at multiple wavelengths and to act as a useful source in its own right may prevail over a simple cost-analysis. Work remains to find laser-materials better suited to the FEL-based pump-source, optimize the FEL design, and consider a realistic accelerator. Finally, accelerator-based alternatives to FEL pumping need to be considered such as direct electron-beam excitation of a gain material, optical pumping of laser diodes, and FEL-assisted mixing using an optical parametric amplifier (OPA).

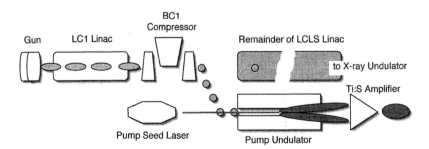

Fig. 1. A schematic of the front end of the LCLS with a multibunch beam diverted into a seeded FEL acting as a pump for a Ti:S amplifier.

References

[1] T. Tajima, J.M. Dawson, Phy. Rev. Lett. 43 (1979) 267.

[2] M.D. Perry, G. Mourou, Science 264 (1994) 917.

[3] T.E. Cowan, et al., Laser Particle Beams 17 (1999) 773.

[4] M.H. Key, Nature 412 (2001) 775.

[5] Y. Sentoku, et al., Phys. Plasmas 10 (2003) 2009.

[6] M.D. Perry, et al., Rev. Sci. Instrum. 70 (2) (1999) 265.

[7] J.A. Paisner, et al., SPIE Proceedings, Series 2633, Bellingham, WA, 1995, p. 2.

[8] W. Koechner, Solid-State Laser Engineering, Springer, Berlin, 1999, 312pp.

[9] J.T. Weir, et al., Proc. SPIE 1133, 1989, 97.

[10] A.J. Bayramian, et al., Proc Adv. Solid State Photon. 83 (2003) 268.

[11] V. Ayvazyan, et al., Phys. Ref. Lett. 88 (2002) 104802.

[12] J. Lewellen, et al., Proceedings of the 1998 Linac Conference, ANL-98/28, 1999, pp. 863–865.

[13] Linac Coherent Light Source (LCLS), SLAC-R-521, UC-414, 1998.

[14] J. Als-Nielsen, Proceedings of Workshop on 4th Generation LS, ESRF Report, Grenoble, 1996.

[15] I.B. Vasserman, et al., Proceedings Particle Accelerator Conference 1999.

Available online at www.sciencedirect.com

Nuclear Instruments and Methods in Physics Research A 528 (2004) 530–533

NUCLEAR
INSTRUMENTS
& METHODS
IN PHYSICS
RESEARCH
Section A

www.elsevier.com/locate/nima

X-ray laser on the relativistic ion and pump laser beams

H.K. Avetissian*, G.F. Mkrtchian

Plasma Physics Laboratory, Department of Quantum Electronics, Yerevan State University, 1 A. Manukian, Yerevan 375025, Armenia

Abstract

In this paper, the possibility of lasing in the X-ray domain by means of a relativistic high-density ion beam and a counter-propagating pump laser field in a high-gain regime is investigated. This deliberation is based on the self-consistent set of Maxwell and relativistic quantum kinetic equations describing the X-ray generation process at which a pump wave resonantly couples two internal ionic levels and the necessity of initial inverse population for lasing vanishes.
© 2004 Elsevier B.V. All rights reserved.

PACS: 41.60.Cr; 42.55.Vc

Keywords: Ion; X-ray; FEL

1. Introduction

The recent advancement of high brightness particle and laser beam technology makes achievable nonlinear schemes for the generation of short-wave coherent radiation in induced free–free transitions [1]. In the current stage of study the problem of creation of short-wave coherent radiation sources in general aspects reduces to implementation of Free Electron Lasers (FEL) [2]. For the realization of an X-ray FEL, one should overcome two problems: (1) compensate for the small electron–photon interaction cross-section

due to various resonances or to provide a larger length of coherency and (2) provide electron beams with probable high brightness. From this point of view, the relativistic ion beams are considered as a new generation of light sources of short-wave radiation [3], especially in the X- and γ-ray regions, where, due to the existence of bound states, the ion–photon interaction cross-section is resonantly enhanced by several orders with respect to that of electron–photon scattering. Hence, stimulated radiation of relativistic ion beams is of interest for the potential synthesis of conventional Quantum Generators and FEL in the X-ray domain.

In this paper, the possibility of lasing in the X-ray domain by means of a relativistic high-density ion beam and counter-propagating pump laser field is investigated. This consideration is based on the self-consistent set of Maxwell and relativistic

---

*Corresponding author: Department of Quantum Electronics, Plasma Physics Laboratory, Yerevan State University, 1 A. Manukian, Yerevan 375025, Armenia. Tel./fax: +3741-570-597.

E-mail address: avetissian@ysu.am (H.K. Avetissian).

0168-9002/$ - see front matter © 2004 Elsevier B.V. All rights reserved.
doi:10.1016/j.nima.2004.04.095

quantum kinetic equations describing the X-ray generation process at which a pump wave resonantly couples two internal ionic levels. As a result, in the considering scheme, the necessity of initial inverse population of levels for lasing vanishes. Besides, the cross-section of the laser–ion interaction process and, consequently, the X-ray generation gain is resonantly enhanced by several orders with respect to the Compton-laser scheme on the free electrons.

2. Self-consistent set of equations

We consider as our model a relativistic beam of two ion levels, a co-propagating (Z-axis) probe electromagnetic (EM) wave with a frequency ω and wave vector $\mathbf{k}$ and counter-propagating strong pump EM wave of frequency ω_0 and wave vector $\mathbf{k}_0$. The EM waves are treated as classical fields and the total electrical field is given by

$$\mathbf{E}(\mathbf{r}, t) = \tfrac{1}{2}\mathbf{e}_0 E_0 e^{i\omega_0 t - i\mathbf{k}_0 \mathbf{r}}$$
$$+ \tfrac{1}{2}\mathbf{e} E_e(t, \mathbf{r}) e^{i\omega t - i\mathbf{k}\mathbf{r}} + c.c.. \tag{1}$$

The probe wave is characterized by a slowly varying amplitude $E_e(t, \mathbf{r})$ and unit polarization vector $\mathbf{e}$, while a pump wave of a given amplitude E_0 and polarization vector $\mathbf{e}_0$ (both waves are linearly polarized). We assume that an internal ionic electron is nonrelativistic and the transition takes place from an S state to a P state. The Hamiltonian governing the evolution of the ion beam in the field (1) takes the following second quantized form in the resonant approximation:

$$\hat{H} \simeq \sum_{\mathbf{p}, s=1,2} \varepsilon_s(\mathbf{p})\hat{a}_{s,\mathbf{p}}^+ \hat{a}_{s,\mathbf{p}}$$
$$+ \sum_{\mathbf{p}} [\hbar\Omega_{0\mathbf{p}} e^{i\omega_0 t}\hat{a}_{1,\mathbf{p}-\hbar\mathbf{k}_0}^+ \hat{a}_{2,\mathbf{p}}$$
$$+ \hbar\Omega_{\mathbf{p}}(\mathbf{r}, t)e^{i\omega t}\hat{a}_{1,\mathbf{p}-\hbar\mathbf{k}}^+ \hat{a}_{2,\mathbf{p}} + c.c.]. \tag{2}$$

Here $\varepsilon_s(\mathbf{p}) = \sqrt{c^2\mathbf{p}^2 + (mc^2 + w_s)^2}$ $(s = 1, 2)$ is the total energy of the ion with momentum $\mathbf{p}$ of the center-of-mass motion (CMM) of the ion and w_1, w_2 are the binding energies of the internal ionic electron in the ground and excited states, respectively ($\hbar$ is the Plank constant, c is the light speed in vacuum, m is the ion mass). Then $\hat{a}_{s,\mathbf{p}}^+$, $\hat{a}_{s,\mathbf{p}}$

denote ionic creation and annihilation operators for the internal states $s = 1, 2$ with CMM momentum $\mathbf{p}$. They satisfy either the usual bosonic- or fermionic-type equal time commutation rules. The couplings

$$\Omega_{0\mathbf{p}} = \frac{E_0 \mathbf{e}_0 \mathbf{d}_{12}}{2\hbar}\left(1 - \frac{\mathbf{v}\mathbf{k}_0}{\omega_0}\right) \tag{3}$$

$$\Omega_{\mathbf{p}}(\mathbf{r}, t) = \frac{E_e(t, \mathbf{r})\mathbf{e}\mathbf{d}_{12}}{2\hbar}\left(1 - \frac{\mathbf{v}\mathbf{k}}{\omega}\right) \tag{4}$$

take into account the dipole interaction as well as the interaction of magnetic moment $[\mathbf{d}_{12} \times \mathbf{v}]/c$ (because of moving electric dipole) with the magnetic field of the waves. In Eq. (4) $\mathbf{v} = \mathbf{p}/m\gamma$ is the ion velocity, γ is the Lorenz factor, $\mathbf{d}_{12}$ is the ionic transition dipole moment.

We will use the Heisenberg representation, where the evolution of operators is given by the following equation:

$$i\hbar\frac{\partial\hat{L}}{\partial t} = [\hat{L}, \hat{H}] \tag{5}$$

and expectation values are determined by the initial density matrix $\hat{D}: \langle\hat{L}\rangle = Sp(\hat{D}\hat{L})$.

Eq. (5) should be supplemented by the Maxwell equation for slowly varying amplitude $E_e(t, \mathbf{r})$. The resonant current is defined by the non-diagonal element of the single particle density matrix $\rho_{12\mathbf{p},\mathbf{p}+\hbar\mathbf{k}}(t) = \langle\hat{a}_{2,\mathbf{p}+\hbar\mathbf{k}}^+ \hat{a}_{1,\mathbf{p}}\rangle$. Hence, for the determination of the self-consistent field we need the evolution equation for the single particle density matrix $\rho_{ij\mathbf{p},\mathbf{p}'}(t) = \langle\hat{a}_{j,\mathbf{p}'}^+ \hat{a}_{i,\mathbf{p}}\rangle$. We will assume that initially ions are in the ground state and the pump laser field is not so strong or it is far off resonance and consequently, the population of the exited state remains small. Then introducing the functions $\rho_{11\mathbf{p},\mathbf{p}+\hbar\mathbf{k}-\hbar\mathbf{k}_0}(t) = \rho_{11\mathbf{p},\mathbf{p}+\hbar\mathbf{k}-\hbar\mathbf{k}_0}(0) + (2\pi\hbar)^3 e^{i(\omega-\omega_0)t}\delta n(\mathbf{p}, t)$, $\rho_{12\mathbf{p},\mathbf{p}+\hbar\mathbf{k}}(t) = \rho_{12\mathbf{p},\mathbf{p}+\hbar\mathbf{k}}(0) + (2\pi\hbar)^3 e^{i\omega t}J(\mathbf{p}, t)$, from the Heisenberg and Maxwell equations one can obtain the self-consistent set of equations which determines the evolution and dynamics of the considered system

$$\frac{\partial E_e}{\partial t} + \frac{c^2\mathbf{k}}{\omega}\frac{\partial E_e}{\partial \mathbf{r}} - i\Delta E_e$$
$$= 4\pi i\omega \mathbf{e}\mathbf{d}_{12}^* \int\left(1 - \frac{\mathbf{v}\mathbf{k}}{\omega}\right)J(\mathbf{p}, t)\,d\mathbf{p} \tag{6}$$

$$\frac{\partial J}{\partial t} + \mathbf{v}_0 \frac{\partial J}{\partial \mathbf{r}} + i\Gamma_1(\mathbf{p})J = i\Omega_{0p}\delta n(\mathbf{p}, t) \tag{7}$$

$$\frac{\partial \delta n}{\partial t} + \mathbf{v}_0 \frac{\partial \delta n}{\partial \mathbf{r}} + i\Gamma_0(\mathbf{p})\delta n$$
$$= i\Omega_{0p}^* \Omega_p \left[\frac{N_{\mathbf{p}+\hbar\mathbf{k}-\hbar\mathbf{k}_0}}{\Gamma_1(\mathbf{p}) - \Gamma_0(\mathbf{p})} - \frac{N_{\mathbf{p}}}{\Gamma_1(\mathbf{p})} \right]. \tag{8}$$

To take into account the pulse propagation effects, we have replaced the time derivatives $\partial/\partial t \rightarrow \partial/\partial t + \mathbf{v}_0 \partial/\partial \mathbf{r}$, where $\mathbf{v}_0$ is the mean velocity of the electron beam. Here, it is assumed that the initial beam is uniform and consequently

$$\rho_{11\mathbf{p},\mathbf{p}'}(0) = (2\pi\hbar)^3 N_{(\mathbf{p}+\mathbf{p}')/2}\delta_{\mathbf{p},\mathbf{p}'} \tag{9}$$

where $N_{\mathbf{p}}$ is the classical momentum distribution function, $\delta_{\mathbf{p},\mathbf{p}'}$ is the Kronecker symbol (summation is replaced by integration). Then

$$\Delta = \frac{2\pi\omega|\mathbf{ed}_{12}|^2}{\hbar} \int d\mathbf{p} \frac{\left(1 - \frac{\mathbf{vk}}{\omega}\right)^2 N_{\mathbf{p}}}{\Gamma_1(\mathbf{p})} \tag{10}$$

is the frequency shift because of the ion beam polarization (refractive index caused by ion beam) and

$$\Gamma_0(\mathbf{p}) = \hbar^{-1}(\varepsilon_1(\mathbf{p}) - \varepsilon_1(\mathbf{p} + \hbar\mathbf{k} - \hbar\mathbf{k}_0)$$
$$+ \hbar\omega - \hbar\omega_0) \tag{11}$$

is the resonance detuning for the Compton scattering, while

$$\Gamma_1(\mathbf{p}) = \hbar^{-1}(\varepsilon_1(\mathbf{p}) + \hbar\omega - \varepsilon_2(\mathbf{p} + \hbar\mathbf{k})) \tag{12}$$

is the resonance detuning for absorption/emission of the probe wave's quanta.

3. High gain regime

Our goal is to determine the conditions under which we will have collective instability, which causes exponential growth of the probe wave. We assume steady-state operation, i.e., dropping of all partial time derivatives in Eqs. (6)–(8). Making a Laplace transformation

$$F(q) = \int_0^\infty F(z)e^{-qz} \, dz$$

in Eqs. (6)–(8) we arrive at the following characteristic equation for variable q:

$$q - i\Delta = \int \frac{K(\mathbf{p}) \, d\mathbf{p}}{(q + i\Gamma_0(\mathbf{p}))(q + i\Gamma_1(\mathbf{p}))} \tag{13}$$

with the

$$K(\mathbf{p}) = \frac{2\pi i\omega|\mathbf{ed}_{12}|^2|\Omega_{0p}|^2}{\hbar v_{0z}^2 c}\left(1 - \frac{\mathbf{vk}}{\omega}\right)^2$$
$$\times \left[\frac{N_{\mathbf{p}}}{\Gamma_1(\mathbf{p})} - \frac{N_{\mathbf{p}+\hbar\mathbf{k}-\hbar\mathbf{k}_0}}{\Gamma_1(\mathbf{p}) - \Gamma_0(\mathbf{p})}\right].$$

This is a transcendental equation, which allows one to determine a small signal gain in various regimes. In this paper, we will investigate only the cold beam regime, when $N(\mathbf{p}) = N_0\delta(\mathbf{p} - \mathbf{p}_0)$, where N_0 is the ion beam density and $\delta(\mathbf{p} - \mathbf{p}_0)$ is the Dirac delta function. Taking into account Eqs. (11), (12) and neglecting the quantum recoil, as well as the conditions $|q| \gg |\Gamma_0(\mathbf{p}_0)|, |\Delta|$ (high gain regime) and $|q| \ll |\Gamma_1(\mathbf{p}_0)|$ from Eq. (13) one can obtain the exponential growth rate, which can be expressed in the following form:

$$G = \frac{\sqrt{3}}{2}\left[\frac{\Omega_r^2}{\delta^2}\frac{2\pi\omega^3|\mathbf{ed}_{12}|^2}{v_{0z}^2\gamma_0^5 mc^3}N_0\right]^{1/3}. \tag{14}$$

Here $\Omega_r = E_0\mathbf{e}_0\mathbf{d}_{12}/(2\hbar)$ is the Rabi frequency associated with the pump wave, $\delta = \omega_{12} - \omega_0\gamma_0(1 + v_{0z}/c)$ is the resonance detuning, and $\omega_{12} = \hbar^{-1}(w_2 - w_1)$ is the transition frequency.

Formula (14) defines the exponential growth rate of X-rays at the induced "Compton" scattering of a strong pump laser radiation on the ion beam at the resonance.

Let us make some estimates for hydrogen-like ions with a charge number of nuclear $Z_i = 3$ and specify the requirements on the ion and pump laser beams. In this case $\hbar\omega_{12} \simeq 90$ eV. We assume a KrF pump laser $\hbar\omega_0 \simeq 5$ eV with a peak power of 10^4 W focused to a Rayleigh range of the order of saturation length, which is assumed to be of the order of $L_{sat} \sim 10$ cm. For the resonance $\omega_{12} \simeq 2\gamma_0\omega_0$, one should have an ion beam with $\gamma_0 \simeq 9$. In this case, the frequency of amplifying radiation will be $\hbar\omega \simeq 2\gamma_0\omega_{12} \simeq 1.6$ keV. For an ion beam with a peak density $N_0 \simeq 10^{16}$ cm^{-3} at $\hbar\delta \simeq 10^{-3}$ eV the gain will be $G \sim 1$ cm^{-1}. For

H.K. Avetissian, G.F. Mkrtchian / Nuclear Instruments and Methods in Physics Research A 528 (2004) 530–533 533

comparison, note that at the same parameters the small signal gain for the Compton laser on free electrons will be $G_c \sim 10^{-2}$ cm^{-1}.

Acknowledgements

This work is supported by International Science and Technology Center (ISTC) Project No. A-353.

References

[1] H.K. Avetissian, G.F. Mkrtchian, Phys. Rev. E 65 (2002) 046505.

[2] L.R. Elias, et al., Phys. Rev. Lett. 36 (1976) 771; D.A. Deacon, et al., Phys. Rev. Lett. 38 (1977) 892.

[3] E.G. Bessonov, K.-J. Kim, Sources of coherent X-rays based on relativistic ion beams, W2-108, in: Proceedings of the 10th ICFA Beam Dynamics Panel Workshop on 4th Generation Light Source, Grenoble, France, January 22–25, 1996.

Available online at www.sciencedirect.com

NUCLEAR
INSTRUMENTS
& METHODS
IN PHYSICS
RESEARCH
Section A

ELSEVIER   Nuclear Instruments and Methods in Physics Research A 528 (2004) 534–538

www.elsevier.com/locate/nima

Superradiant X-ray laser on the channeled in a crystal ion beam

H.K. Avetissian*, G.F. Mkrtchian

Plasma Physics Laboratory, Department of Quantum Electronics, Yerevan State University, 1 A. Manukian, Yerevan 375025, Armenia

Abstract

In this paper, we present a new scheme for superradiant X-ray radiation generation on fast, multiply charged, channeled ion beams in a crystal. It is shown that X-ray superradiation occurs due to macroscopic coherent excitation of an ionic beam by a crystal's electrostatic field.
© 2004 Elsevier B.V. All rights reserved.

PACS: 42.50.Fx; 42.55.Vc; 41.60.Cr

Keywords: Superradiation; X-ray; Ion

1. Introduction

As a potential source of shortwave coherent radiation, the non-linear regimes of stimulated channeling radiation of electron or positron beams have been considered. These are the analogue of free electron laser (FEL) on channeled particles beams [1]. In this paper, we consider a new scheme of a coherent X-ray radiation generated by a fast, multiply charged, channeled ion beam in a crystal. Here X-ray transitions involving K or L shell electrons in ions can be resonantly excited by the periodic crystal potential seen by fast channeled

ions [2,3]. The emission frequencies in this case are determined by the discrete spectrum of the electron states in ions and by the Doppler shift due to the ion center of mass motion. With respect to the moving ions, the crystal potential plays the role of an effective pumping field with the Rabi frequency corresponding to a high-power "X-Ray laser". By varying the crystal thickness, one can obtain diverse equivalent "X-Ray pulses" leading to various coherent superposition states, from which one can obtain coherent X-ray radiation from the ion beam spontaneous superradiation (SR).

Below, we will investigate superradiant coherent X-ray radiation generation when an ion beam moves close to the crystal-lattice axis. This radiation is predicted by the second quantized Maxwell and Quantum equations governing the motion of an ion beam in a crystal.

---

*Corresponding author. Department of Quantum Electronics, Plasma Physics Laboratory, Yerevan State University, 1 A. Manukian, Yerevan 375025, Armenia. Tel./fax: +3741-570-597.

E-mail address: avetissian@ysu.am (H.K. Avetissian).

0168-9002/$ - see front matter © 2004 Elsevier B.V. All rights reserved.
doi:10.1016/j.nima.2004.04.096

2. Resonant excitation of fast ions in a crystal

For channeling an ion beam in a crystal, we assume that the incident angle of ions (with a charge number of the nucleus Z_i) with respect to a crystalline axis (OZ) is smaller than the Lindhard angle. Then the potential of the atomic chain, which governs the ion motion, can be represented in the following form [3]:

$$V(z, r_\perp) = \sum_n V_n(r_\perp) \exp\left[i\frac{2\pi n}{d}z\right] \qquad (1)$$

where d is the crystal lattice period along the channel axis, $V_n(r_\perp)$ is defined by the single atomic potential of the crystal, which is given by the screening Coulomb potential with the radius of screening r_c and charge number of the nucleus Z_c that has the following form:

$$V_n(r_\perp) = \frac{2eZ_c}{d} K_0(r_\perp q_n)$$

$$q_n = \sqrt{\frac{1}{r_c^2} + \left(\frac{2\pi n}{d}\right)^2} \qquad (2)$$

where K_0 is a modified Bessel function.

The potential (1) acts on the internal electron as well as on the ion center-of-mass motion (CMM), providing channeling. The CMM of the ion represents slow oscillations in the transversal direction ($r_\perp$) and free motion (on average) along the crystalline axis. For the ionic electron the atomic chain potential acts as an exciting field [3]. The latter is obvious in the rest frame of reference (FR) of the ion (neglecting transversal oscillations) where there is an oscillating time/space electromagnetic field with a fundamental frequency $2\pi \gamma v_z/d$ (γ is the Lorenz factor, v_z is the ion longitudinal velocity). If one of the harmonics (n) of this frequency is close to the frequency ω_{12} associated with the energy difference of the ionic electron levels

$$2\pi n \gamma v_z/d \simeq \omega_{12} \qquad (3)$$

we can expect resonant excitation of ions. The latter represents the conservation law for the total energy (neglecting quantum recoil) in the laboratory FR.

As the physical picture of the considered process is more evident in the FR connected with the ion beam and the problem becomes nonrelativistic in this FR, then it is more convenient to pass to the FR of the ion beam (moving with the mean velocity v_0 of the beam). If the resonance condition holds (3), we can keep only the resonant harmonic in the potential (1) and the Hamiltonian describing the quantum kinetics of the channeled ion beam takes the following second quantized form in the resonant approximation:

$$\hat{H}_{ic} \simeq \sum_{\mathbf{p}, s=1,2} \varepsilon_s(\mathbf{p}) \hat{a}^+_{s,\mathbf{p}} \hat{a}_{s,\mathbf{p}}$$
$$+ \sum_{\mathbf{p}} [\hbar \Omega_c e^{i\omega_c t} \hat{a}^+_{1,\mathbf{p}-\hbar g_n} \hat{a}_{2,\mathbf{p}} + \text{c.c.}]. \qquad (4)$$

Here, we have introduced the lattice vector $\mathbf{g}_n = (0, 0, -2\pi n \gamma_0/d)$, where $\gamma_0 = (1 - v_0^2/c^2)^{-1/2}$, and $\hat{a}^+_{s,\mathbf{p}}$, $\hat{a}_{s,\mathbf{p}}$ denote ionic creation and annihilation operators for the states $s = 1, 2$ with CMM momentum $\mathbf{p}$ and energy $\varepsilon_s(\mathbf{p}) = \mathbf{p}^2/2m + w_s$ (w_s is the binding energy of the ionic electron). They satisfy either the usual bosonic or fermionic type equal time commutation rules. The coupling is

$$\Omega_c = \frac{2eZ_c\gamma_0}{\hbar d}\left\{-ig_n f_z K_0(\bar{r}_\perp q_n) + \frac{\mathbf{f r}_\perp}{r_\perp} q_n K_1(\bar{r}_\perp q_n)\right\} \qquad (5)$$

where $\mathbf{f}$ is the ionic transition dipole moment, which represents the Rabi frequency, with the assumption that the crystal potential acts as a quasi-monochromatic wave with the frequency

$$\omega_c = v_0 g_n; \quad g_n = 2\pi n \gamma_0/d \qquad (6)$$

In Eq. (5) we have neglected the ion transverse oscillations, since they are much slower than the frequency of collisions of ions with the atoms of the crystalline axis. Here $\bar{r}_\perp$ is the ion mean transverse displacement.

The full Hamiltonian describing also the radiation processes will be

$$\hat{H} = \hat{H}_{ic} + \sum_{\mathbf{k}, \lambda=1,2} \hbar \omega \hat{c}^+_{\mathbf{k}, \lambda} \hat{c}_{\mathbf{k}, \lambda}$$
$$+ \sum_{\mathbf{p}, \mathbf{k}, \lambda} [\hbar \Omega_{\mathbf{k}, \lambda} \hat{a}^+_{1, \mathbf{p}-\hbar \mathbf{k}} \hat{a}_{2, \mathbf{p}} \hat{c}_{\mathbf{k}, \lambda} + \text{c.c.}] \qquad (7)$$

where the second term is the Hamiltonian of the photon field with the creation and annihilation operators $\hat{c}^+_{\mathbf{k},\lambda}$, $\hat{c}_{\mathbf{k},\lambda}$ of photons with the momentum $\hbar\mathbf{k}$ and linear polarization $\mathbf{e}_\lambda$ ($\lambda = 1,2$). The last term is the Hamiltonian of interaction of ions with the photon field and

$$\Omega_{\mathbf{k},\lambda} = [2\pi\hbar\omega]^{1/2}(\mathbf{e}_\lambda\mathbf{f}) \tag{8}$$

is the Rabi frequency for the quantized photon field (the quantization volume is taken to be $V = 1$). We will use the Heisenberg representation where evolution of operators is given by the following equation:

$$i\hbar\frac{\partial\hat{L}}{\partial t} = [\hat{L}, \hat{H}] \tag{9}$$

and expectation values are determined by the initial density matrix $\hat{D}$: $\langle\hat{L}\rangle = Sp(\hat{D}\hat{L})$.

If the effective Rabi frequency is large enough and the crystal length is short enough, the spontaneous emission and the relaxation processes may be neglected during the time of interaction of ions with the crystal. In this case, the Heisenberg equation (9) for the operators $\hat{a}_{s,\mathbf{p}}$ may be solved analytically. This gives the following solution:

$$\hat{a}_{1,\mathbf{p}} = e^{-(i/\hbar)\varepsilon_1(\mathbf{p})t}e^{-i\frac{1}{2}\delta_{v_z}\tau}\bigg\{\cos\Omega\tau$$

$$+ i\frac{\delta_{v_z}}{2\Omega}\sin\Omega\tau\bigg\} \cdot \hat{a}^{(0)}_{1,\mathbf{p}}$$

$$\hat{a}_{2,\mathbf{p}} = -ie^{-(i/\hbar)\varepsilon_2(\mathbf{p})t}e^{i\frac{1}{2}\delta_{v_z}\tau}\frac{\Omega_c}{\Omega}$$

$$\times \sin\Omega\tau \cdot \hat{a}^{(0)}_{1,\mathbf{p}-\hbar g_n}. \tag{10}$$

Here $\hat{a}^{(0)}_{1,\mathbf{p}}$ is the initial operator, τ is the ion interaction time with the crystal.

$$\delta_{v_z} = \omega_{12} - \omega_c - g_n v_z \tag{11}$$

is the resonance detuning, and

$$\Omega = \sqrt{|\Omega_c|^2 + \delta^2_{v_z}/4} \tag{12}$$

is the effective Rabi frequency. We assume that initially ions are in the ground state, so that in Eq. (10) we have not written the terms with the operator $\hat{a}^{(0)}_{2,\mathbf{p}}$. As we see, the population of electrons oscillates coherently between the states depending on the crystal length, $L_c \simeq v_z\tau$. If $|\Omega_c| \gg |\delta_{v_z}|$ and the crystal length corresponds to "pulse area" $|\Omega_c|\tau = j\pi/4$ ($j = 1, 2, \ldots$), the ion

beam will then have the maximal polarization (macroscopic dipole moment).

3. Superradiation of ion beam

To investigate the properties of the ion beam radiation (in free space) we come back to the full Hamiltonian and perturbatively calculate the photonic operators $\hat{c}_{\mathbf{k},\lambda}(t)$

$$\hat{c}_{\mathbf{k},\lambda}(t) = -i\pi\hbar\Omega_{\mathbf{k},\lambda}\sum_{\mathbf{p}}\hat{a}^+_{1,\mathbf{p}-\hbar\mathbf{k}}\hat{a}_{2,\mathbf{p}}$$

$$\times \delta(\hbar\omega + \varepsilon_1(\mathbf{p} - \hbar\mathbf{k}) - \varepsilon_2(\mathbf{p})). \tag{13}$$

The output spectrum consists of coherent and incoherent radiation. The coherent/superradiant radiation is defined by the mean value of the photonic operators $\langle\hat{c}_{\mathbf{k}\alpha}(t)\rangle$; i.e. it is proportional to the Fourier transform of the mean ionic polarization $\langle\hat{a}^+_{1,\mathbf{p}-\hbar\mathbf{k}}\hat{a}_{2,\mathbf{p}}\rangle$. In this work, we will represent only the coherent radiation by assuming that the total number of photons is much smaller than the number of ions, $N_{ph} \ll N$. In accordance with this assumption, one can neglect the retro radiation effects. Otherwise, ions would respond collectively, and as is well known [4] the N-particle spontaneous emission rate might be much larger than a particle spontaneous emission rate, consequently considering equations for the photons and ions operators should be solved self consistently.

Considering Eq. (13), we obtain the following equation for the total number of emitted photons, with momentum $\hbar\mathbf{k}$ and polarization λ per unit time

$$\frac{\partial N^{(coh)}_{\mathbf{k}\lambda}}{\partial t} = 2\pi\hbar|\Omega_{\mathbf{k},\lambda}|^2\sum_{\mathbf{p}_1,\mathbf{p}}\text{Re}\{\rho_{12\mathbf{p}_1-\hbar\mathbf{k},\mathbf{p}_1}$$

$$\times \rho_{21\mathbf{p},\mathbf{p}-\hbar\mathbf{k}}\delta(\hbar\omega + \varepsilon_1(\mathbf{p} - \hbar\mathbf{k}) - \varepsilon_2(\mathbf{p}))\} \tag{14}$$

where $\rho_{12\mathbf{p},\mathbf{p}'}(t) = \langle\hat{a}^+_{2,\mathbf{p}'}\hat{a}_{1,\mathbf{p}}\rangle$ is the non-diagonal element of the single particle density matrix defined by the operators (10). By summing over photon polarization and integrating over frequency one can obtain the following expression for the angular distribution of superradiant power:

$$\frac{dI_{coh}}{do} = N^2 I_1(\hat{\mathbf{k}}) \left| G\left(\hat{\mathbf{k}}\frac{\omega_{12}}{c} - \mathbf{g}_n\right)\right|^2$$

$$\times \left| \int e^{i(\hat{\mathbf{k}}\mathbf{v}/c)\omega_{12}t} P(v_z) f(\mathbf{v})\, d\mathbf{v}\right|^2 \qquad (15)$$

where

$$G(\mathbf{q}) = \frac{1}{N}\int n(\mathbf{r})e^{i\mathbf{q}\mathbf{r}}\, d\mathbf{r} \qquad (16)$$

is the beam form-factor with $n(\mathbf{r})$ being the ion beam density function, $f(\mathbf{v})$ is the velocity distribution function of ions, $I_1(\hat{\mathbf{k}})$ is the single particle radiation power with the unit vector $\hat{\mathbf{k}}$ in the radiation direction and

$$P(v_z) = e^{-ig_n v_z\tau}\frac{\Omega_c}{\Omega}\sin\Omega\tau\left\{\cos\Omega\tau + i\frac{\delta_{v_z}}{2\Omega}\sin\Omega\tau\right\}. \qquad (17)$$

For the beam spatial and velocity distributions, we will assume Gaussian functions with isotropic transverse distributions. Then, from Eq. (15) we obtain

$$\frac{dI_{coh}}{do} = N^2 I_1(\hat{\mathbf{k}})e^{-(\delta v_\perp^2/c^2)\omega_{12}^2 t^2\sin^2\vartheta}|P(t,\vartheta)|^2$$

$$\times e^{-(l_\perp^2\omega_{12}^2/c^2)\sin^2\vartheta - l_z^2 g_n^2(1+(\omega_{12}/cg_n)\cos\vartheta)^2} \qquad (18)$$

where

$$P(t,\vartheta) = \int P(v_z)e^{i(v_z\cos\vartheta/c)\omega_{12}t - (v_z^2/2\delta v_z^2)}\frac{dv_z}{\sqrt{2\pi}\delta v_z}. \qquad (19)$$

Here $l_\perp$, l_z are the transverse and longitudinal bunch sizes of the beam with the transverse and longitudinal velocity spreads $\delta v_\perp$, δv_z. As is seen from Eq. (18), if the observed wavelengths are much smaller than the transverse size of an ion beam, the SR from the ion beam will occur primarily along the Z-axis and will cover only a tiny solid angle, which will be defined by the transverse size of the ion beam

$$\Delta o \simeq \pi\frac{c^2}{l_\perp^2\omega_{12}^2}. \qquad (20)$$

The superradiant pulse duration depends on velocity spreads of the beam and will be defined by the function $P(t,\vartheta)$. The analyses of Eq. (18) shows the existence of two superradiant regimes of X-ray generation. For the first regime when the phase matching condition holds

$$\omega_{12} = cg_n \qquad (21)$$

the SR from the ion beam may occur primarily in the backward direction and the longitudinal bunch size l_z of the ion beam should not be smaller than the wavelength of SR. On the other hand, for the resonant excitation it should be fulfilled the condition $|\delta_0| \ll \omega_{12}$. Then taking into account the phase matching condition (21), for the detuning (11) we have

$$\delta_0 \simeq \omega_{12}(1 - v_0/c). \qquad (22)$$

The latter means that for the backward SR it is necessary for a relativistic ion beam to satisfy the resonance condition $\delta_0 \simeq \omega_{12}/2\gamma_0^2 \ll \omega_{12}$.

In the opposite case when resonance condition holds $\omega_{12} = \omega_c = v_0 g_n$, one can easily fulfill the condition for maximal dipole moment ($|\Omega_c| \gg \delta_0$) for the light ion beams ($Z_i < 10$, $\gamma_0 \simeq 1$), but since the phase matching condition (21) is violated: $\omega_{12} < cg_n$, the SR will take place if the longitudinal bunch size of the ion beam is smaller than the crystal lattice period d.

Let us make some estimations for the mean power of backward SR. From the Eq. (18) one can obtain the following approximate formula for the mean power

$$I_{mean} \simeq N^2 I_1|P(0,\pi)|^2\Delta o. \qquad (23)$$

We will consider hydrogen-like ions with a charge number of nucleus $Z_i \sim 20$. Regarding the crystal we will assume $Z_c \sim 60$. In this case the characteristic transition frequencies $\hbar\omega_{12} \sim 5$ keV. Then we suppose an ion beam with $\gamma_0 \simeq 2$, $l_\perp \simeq 10^{-3}$, $N \simeq 10^9$. In this case $\Delta o \simeq 10^{-10}$ rad., $I_1 \simeq 10^{-2}$ W, $|P(0,\pi)|^2 \sim |\Omega_c|^2/\delta_0^2 \simeq 10^{-3}$ and, consequently, $I_{mean} \simeq 1$ kW. Note that these formulas have been obtained in ion beam FR. In the laboratory FR the wavelength of backward SR $\lambda = 2d \sim$ Å and $I_{\text{mean}}/2\gamma_0^2 \sim 100$ W.

Acknowledgements

This work was supported by National Foundation of Science and Advanced Technologies (NFSAT) Grant No. NFSAT PH 082-02/CRDF 12023 and by the International Science and Technology Center (ISTC) Project No. A-353.

References

[1] H.K. Avetissian, K.Z. Hatsagortsian, G.F. Mkrtchian, Kh.V. Sedrakian, Phys. Rev. A 56 (1997) 4121;
    H.K. Avetissian, G.F. Mkrtchian, Nucl. Instr. and Meth. A 507 (2003) 479.
[2] S. Shindo, Y.H. Ohtsuki, Phys. Rev. B 14 (1976) 3929.
[3] V.A. Bazylev, N.K. Zhevago, Zh. Eksp. Teor. Fiz. 77 (1979) 312 (Sov. Phys. JETP 50 (1979) 161).
[4] L. Allen, J.H. Eberly, Optical Resonance and Two Level Atoms, Wiley-Interscience, New York, 1975.

Available online at www.sciencedirect.com

SCIENCE @ DIRECT®

ELSEVIER

Nuclear Instruments and Methods in Physics Research A 528 (2004) 539–543

NUCLEAR
INSTRUMENTS
& METHODS
IN PHYSICS
RESEARCH
Section A

www.elsevier.com/locate/nima

Evolution of the energy and momentum distribution of electron beam in the gamma-ray laser synchrotron sources

I.V. Smetanin*, K. Nakajima

Japan Atomic Energy Research Institute, Umemidai 8-1, Kizu-cho, Soraku-gun, Kyoto 619-0215, Japan

Abstract

Evolution of the relativistic electron beam in the laser-beam Compton scattering was investigated. The quantum kinetic equation for the electron beam distribution function is derived in the frame of the Klein–Gordon field theory. When the beam energy spread exceeds the energy of scattered photons, this equation is reduced to the diffusion-like kinetic equation which describes the laser cooling of electron beam and includes quantum fluctuations limiting the cooling process. In the opposite limit of high beam energies, we describe the evolution of transverse beam emittance in the laser-beam Compton scattering.
© 2004 Elsevier B.V. All rights reserved.

PACS: 41.75.Fr; 41.60.Cr; 42.55.Vc

Keywords: Laser-beam Compton scattering; Laser cooling; Quantum fluctuations; Beam emittance

1. Introduction

JAERI with Osaka University and JASRI has carried out laser-beam Compton scattering experiments under the leadership of Prof. M. Fujiwara. Recently, the discovery of a new elementary particle consisting of five quarks has been reported by this group [1]. This $S = +1$ baryon resonance has been found in photoproduction reaction from the neutron. This discovery was enabled by the usage of laser Compton gamma-rays in the above-mentioned team effort. There are other numerous applications of laser Compton gamma-rays including nuclear transmutation [2].

In the concept of the Laser Synchrotron Source [3], high-brightness X-ray and γ-radiation is produced in the scattering of intense laser pulse by the counter-propagating relativistic electron beam. The energy of scattered photons rapidly increases with the energy of electrons $\hbar\omega \approx 4\gamma^2\hbar\omega_i$, γ is the Lorentz factor of beam electrons, ω_i is the frequency of laser radiation. The character of the laser-beam interaction depends strongly on the ratio between the beam energy spread $\Delta\gamma mc^2$ and the quantum emitted $\hbar\omega$. At low energies of electrons,

$$\Delta\gamma/\gamma > 4\gamma\hbar\omega_i/mc^2 \qquad (1)$$

*Corresponding author.
E-mail address:* smetanin@sci.lebedev.ru (I.V. Smetanin).

laser-beam scattering is accompanied by the cooling of electron beam [4,5]. In the opposite situation, when the quantum emitted exceeds the beam energy spread, the beam emittance increases as a result of laser-beam scattering. For typical energy spread $\Delta\gamma/\gamma \sim 10^{-3}$ and CO_2 laser radiation $\hbar\omega_i \sim 0.1$ eV, we have the boundary value $\gamma^* \sim 10^3$ which corresponds to the sub-GeV level of electron energy. The energy of scattered photons are in the γ-ray spectral region but small enough $\hbar\omega_i\gamma/mc^2 \ll 1$ to satisfy the low-frequency limit of the Klein–Nishina theory [6]. In the laser-beam Compton X- and γ-ray generation experiments, both the above limiting cases can be realized [7].

Here, we intend to describe the evolution of electron beam in the process of Compton laser-beam interaction. Using the Klein–Gordon field theory, we derive the quantum kinetic equation for the momentum distribution function of beam electrons. In the limit of low beam energies (1), this equation describes both the laser-beam cooling [4,5] and quantum fluctuations which results in a diffusion-like growth of the beam energy spread and thus restrict the cooling process. In the limit of high electron energies, when the quantum emitted exceeds the initial beam energy spread, we use the derived kinetic equation to study evolution of the beam emittance.

2. Quantum kinetics of relativistic electron beam under spontaneous Compton scattering

To describe the Compton laser-beam interaction, we start with the Klein–Gordon scalar field Hamiltonian [6]. Assuming the linear (i.e., single-photon) Compton scattering, we have derived the following equation for the electron momentum distribution function $f(\mathbf{p}) = \langle a_{\mathbf{p}}^+ a_{\mathbf{p}} \rangle (\hbar = c = 1)$

$$\frac{df(\mathbf{p})}{dt} = 2\sum_{\mathbf{k},\lambda}\{G_{\mathbf{k},\lambda}(\mathbf{p}+\mathbf{q})f(\mathbf{p}+\mathbf{q})$$
$$- G_{\mathbf{k},\lambda}(\mathbf{p})f(\mathbf{p})\}, \qquad (2)$$

with the coefficients

$$G_{\mathbf{k},\lambda}(\mathbf{p}) = \left| g_{\mathbf{k}}(\mathbf{p}-\mathbf{q})(\mathbf{e}_{\mathbf{k},\lambda}^*, \mathbf{e}_i) \right.$$
$$- \frac{\mu_i(\mathbf{p}-\mathbf{k})\mu_{\mathbf{k}}(\mathbf{p}-\mathbf{k})(\mathbf{e}_i, \mathbf{p}-\mathbf{k})(\mathbf{e}_{\mathbf{k},\lambda}^*, \mathbf{p})}{\varepsilon_{\mathbf{p}-\mathbf{k}} - \varepsilon_{\mathbf{p}-\mathbf{q}} + \omega_i}$$
$$- \left. \frac{\mu_{\mathbf{k}}(\mathbf{p}-\mathbf{q})\mu_i(\mathbf{p})(\mathbf{e}_i, \mathbf{p})(\mathbf{e}_{\mathbf{k},\lambda}^*, \mathbf{p}+\mathbf{k}_i)}{\varepsilon_{\mathbf{p}+\mathbf{k}_i} - \varepsilon_{\mathbf{p}-\mathbf{q}} - \omega_k} \right|^2 n_i$$
$$\times \frac{\sin(\varepsilon_{\mathbf{p}} - \varepsilon_{\mathbf{p}-\mathbf{q}} - \omega_k + \omega_i)t}{\varepsilon_{\mathbf{p}} - \varepsilon_{\mathbf{p}-\mathbf{q}} - \omega_k + \omega_i}. \qquad (3)$$

Here $a_{\mathbf{p}}^+, a_{\mathbf{p}}$ are the field operators for the electron of the momentum $\mathbf{p}$ and the energy $\varepsilon_{\mathbf{p}}$, $\{\mathbf{k},\lambda\}$ represents the scattered mode of electromagnetic field of a wave vector $\mathbf{k}$ and polarization number λ, and $\mathbf{e}_{\mathbf{k},\lambda}$ is the polarization vector. The incident laser pulse is modelled by the single mode $\{\mathbf{k}_i, \lambda_i\}$, with the polarization vector $\mathbf{e}_i$, $n_i = I_i/c\hbar\omega_i$ is the density number of photons in the laser wave of intensity I_i. The coefficients in Eq. (3) are $\mu_{\mathbf{k}}(\mathbf{p}) = e(2V\omega_k\varepsilon_{\mathbf{p}+\mathbf{k}}\varepsilon_{\mathbf{p}})^{-1/2}$, $g_{\mathbf{k}}(\mathbf{p}) = e^2(4V^2\omega_k\omega_i\varepsilon_{\mathbf{p}+\mathbf{q}}\varepsilon_{\mathbf{p}})^{-1/2}$, $\mathbf{q} = \mathbf{k} - \mathbf{k}_i$ is the total change in momentum of an electron, V is the quantization volume, and the sum is over scattered modes $\{\mathbf{k},\lambda\} \neq \{\mathbf{k}_i, \lambda_i\}$. The factor 2 on the right-hand side of Eq. (2) is conditioned by the spin of electron. Eq. (2) is a typical "decay" kinetic equation, the right-hand side of which (i.e., the collision integral) represents the balance of electrons in the state $|\mathbf{p}\rangle$. The term $\sin(\Delta t)/\Delta$ corresponds to the energy conservation law.

The character of evolution of the beam distribution function depends on the relation between the recoil momentum q and the characteristic beam momentum spread. Under condition (1), the right-hand side of Eq. (2) becomes diffusion-like. In the opposite limiting case, when the recoil momentum q well exceeds the distribution's width, the dynamics of the beam electrons is reduced to the exponential decay in a multilevel quantum system.

3. Quantum diffusion in the laser cooling of electron beam

Under condition (1), expanding right-hand side of Eq. (2) to the second-order of small parameter

$|k|/|p|$, we get a diffusion-like kinetic equation

$$\frac{d}{dt}f(\mathbf{p}) = A(p)f(p) + \sum_v B_v \frac{\partial f}{\partial p_v}$$
$$+ \frac{1}{2}\sum_{v,\xi} C_{v,\xi}\frac{\partial^2 f}{\partial p_v \partial p_\xi} + \cdots . \qquad (4)$$

Here the coefficients are

$$A = 2\sum_{k,\lambda}\left[G_{k,\lambda}(p+q) - G_{k,\lambda}(p)\right]$$

$$B_v = 2\sum_{k,\lambda} G_{k,\lambda}(p+q)q_v$$

$$C_{v,\xi} = 2\sum_{k,\lambda} G_{k,\lambda}(p+q)q_v q_\xi . \qquad (5)$$

The first two terms on the right-hand side of Eq. (4) are classical and describe the beam cooling and decrease in its energy. The term with the second derivative contains the Plank constant and describes the diffusive growth of the beam energy spread. To illustrate competition between these two processes, let us consider the 1D approximation, keeping only the longitudinal momentum p_z. Eq. (4) becomes (in dimension units)

$$\frac{d}{dt}f(p_z) = smc^2\gamma^2\left[\frac{c}{\varepsilon_p}f + \frac{\partial f}{\partial p_z} + \frac{88}{15}\hbar k_i\gamma^2\frac{\partial^2 f}{\partial p_z^2}\right]. \qquad (6)$$

Here $s = 8/3\pi r_0^2 I_i/mc^2$ is the characteristic cooling rate. In the classical cooling regime, when quantum fluctuations are negligible, the second term on the right-hand side of Eq. (6) dominates, and the solution can be written in the form

$$f(p_z, t) = \int dp' F_0(p')\delta$$
$$\times \left(\frac{p'}{mc} - \frac{(p_z/mc) + \tan(st)}{1 - (p_z/mc)\tan(st)}\right) \qquad (7)$$

where $F_0(p)$ is the initial distribution function of beam electrons, and the delta function contains the particle's trajectory. According to this solution, the average energy of electrons (corresponding to the maximum of the distribution function) decreases with time as $\langle p_z \rangle \approx p_0/(1 + (p_0/mc)\tan(st))$, p_0 is the initial average longitudinal momentum of electrons, whereas the width of the distribution function decreases as $\sigma(t) \sim \sigma_0/(1 + p_0 \tan(st))^2$. The beam energy spread $\Delta\gamma/\gamma$ is thus vanishing in time, and the beam

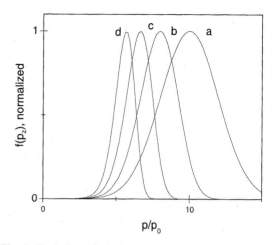

Fig. 1. Evolution of the beam distribution function in the process of laser cooling: (a) initial distribution $\gamma_0 = 10$, $st = 0$; (b) $st = 0.025$; (c) $st = 0.005$; (d) $st = 0.075$.

quality increases as a result of laser-beam Compton scattering [4,5]. With a decrease in the beam energy spread, the third term on the right-hand side of Eq. (6) comes into play. The quantum diffusion in the momentum space leads to an effective heating of the electron beam and thus terminate the process of laser Compton cooling. It is easy to estimate from Eq. (6) that the limiting beam energy spread is $\sigma_q \sim \left(\frac{88}{15}\right)\gamma^2\hbar\omega_i$ which is of the order of the energy of scattered photon. The laser cooling of relativistic electron beam is illustrated by Fig. 1, in which evolution of the beam distribution function in time is shown, the initial distribution $F_0(p_z)$ is assumed to be Gaussian.

4. Evolution of beam emittance in the γ-ray laser synchrotron source

In the opposite case of high-energy electron beam, when the energy of the scattered photon well exceeds the beam energy spread, we seek solution to the quantum kinetic equation (2) as a sequence of distribution functions

$$f(\mathbf{p}) = f_0(\mathbf{p}) + f_1(\mathbf{p}) + f_2(\mathbf{p}) + \cdots \qquad (8)$$

where $f_n(\mathbf{p})$ corresponds to the momentum distribution of electrons, which exhibited n Compton scattering events. Functions $f_n(\mathbf{p})$ are well separated from each other and located in momentum space in the vicinity of the following discrete set of values of electron's momentum and energy:

$$\varepsilon_n = \varepsilon_{n-1} - \hbar\omega_n$$
$$\mathbf{p}_n = \mathbf{p}_{n-1} - \hbar\mathbf{k}_n \qquad (9)$$

which are separated by a sequence of quanta $\hbar\omega_n \approx 4\gamma_{n-1}^2 \hbar\omega_i$, $\gamma_n = \varepsilon_n/mc^2$. Substituting Eq. (8) into Eq. (2), we get the chain of coupled equations, which describe the distributions on each energy level

$$\frac{\partial f_0(\mathbf{p}, t)}{\partial t} = -\Gamma(\mathbf{p})f_0(\mathbf{p}, t)$$

$$\frac{\partial f_1(\mathbf{p}, t)}{\partial t} = 2\sum_{\mathbf{k},\lambda} G_{\mathbf{k},\lambda}(\mathbf{p}+\mathbf{q})f_0(\mathbf{p}+\mathbf{q}) - \Gamma(\mathbf{p})f_1(\mathbf{p})$$

$$\cdots \quad \cdots \quad \cdots \quad \cdots$$

$$\frac{\partial f_n(\mathbf{p}, t)}{\partial t} = 2\sum_{\mathbf{k},\lambda} G_{\mathbf{k},\lambda}(\mathbf{p}+\mathbf{q})f_{n-1}(\mathbf{p}+\mathbf{q})$$
$$- \Gamma(\mathbf{p})f_n(\mathbf{p}). \qquad (10)$$

Here, the coefficient $\Gamma(\mathbf{p}) = 2\sum_{\mathbf{k},\lambda} G_{\mathbf{k},\lambda}(\mathbf{p})$ is the total rate for the drain of electrons in the state characterized by momentum $\mathbf{p}$,

$$\Gamma(\mathbf{p}) \approx \frac{4\pi}{3} r_0^2 \frac{I_i}{\hbar\omega_i} (1 + \beta\cos\Theta_i)\left[1 - \cos^2\psi_i \sin^2\Theta_i\right] \qquad (11)$$

where ψ_i is the angle between the polarization vector of the laser mode $\mathbf{e}_i$ and the plain which is formed by vectors $\mathbf{p}$ and $\mathbf{k}_i$.

At sufficiently low laser intensities, population of the states with $n \geq 2$ can be neglected, and the beam dynamics is reduced to the evolution of the distribution functions f_0 and f_1. The first equation of system (10) describes exponential decay of the initial beam distribution,

$$f_0(\mathbf{p}, t) = F_0(\mathbf{p})\exp(-\Gamma(\mathbf{p})t). \qquad (12)$$

One can easily see that when the recoil momentum q well exceeds the initial momentum spread of electron beam σ, the characteristic angle width of the function $f_0(\mathbf{p}+\mathbf{q})$, $\delta\Theta \sim (2k_i\sigma/k^2)^{1/2}$ is much less than the width of the rate coefficient $G_{\mathbf{k},\lambda}(\mathbf{p})$, $\delta\Theta \leq 1/\gamma$. As a result, the momentum spread of scattered electrons is determined by the function $G_{\mathbf{k},\lambda}(\mathbf{p})$, and the initial distribution f_0 on the right-hand side of the second equation in system (10) can be considered as the δ-function.

The transverse beam emittance is characterized by the mean transverse momentum spread $\varepsilon_{tr} = (1/p^2)\int d\mathbf{p} f_1(\mathbf{p})\mathbf{p}_\perp^2$, where $\mathbf{p}_\perp$ is the component of electron's momentum which is perpendicular to the average initial momentum $\mathbf{p}_0$. One can split the transverse momentum $\mathbf{p}_\perp$ in two components, one of each ("horizontal") is in the plane formed by the laser wave vector k_i and $\mathbf{p}_0$, the second ("vertical") is normal to that plane. The transverse momentum spread becomes thus the sum of correspondent horizontal and vertical components, $\varepsilon_{tr} = \varepsilon_h + \varepsilon_v$, for which we have found the following equations:

$$\frac{\partial \varepsilon_h}{\partial t} \approx \Sigma_h \exp(-\Gamma_0 t) - \Gamma_0 \varepsilon_h,$$

$$\frac{\partial \varepsilon_v}{\partial t} \approx \Sigma_v \exp(-\Gamma_0 t) - \Gamma_0 \varepsilon_v. \qquad (13)$$

Here, the decay rate Γ_0 is determined by Eq. (11) at $\mathbf{p} = \mathbf{p}_0$, and the emittance growth rate functions $\Sigma_{h,v}$ are

$$\Sigma_{h,v} = \frac{4\pi}{15} r_0^2 \frac{I_i\hbar\omega_i}{(mc^2)^2} (1 + \beta\cos\Theta_i)^3 \Phi_{h,v}(\psi_i, \Theta_i) \qquad (14)$$

The geometrical factors $\Phi_{h,v}$ are different for horizontal and vertical components

$$\Phi_h = 1 + \sin^2\psi_i - (1 - \zeta)\cos^2\psi_i \sin^2\Theta_i$$
$$\Phi_v = 2 - \sin^2\psi_i - (1 - \zeta)\cos^2\psi_i \sin^2\Theta_i \qquad (15)$$

where the small correction factor is $\zeta = 5(5 + 15\beta^{-3}(\ln[(1 + \beta)\gamma] - \beta))/(\gamma\beta)^2 \ll 1$. The factor Φ_v reaches its maximum value when the polarization vector $\mathbf{e}_i$ of the laser mode is in the plane $(\mathbf{p}_0, \mathbf{k}_i)$, while Φ_h becomes maximal when $\mathbf{e}_i$ is normal to the plane $(\mathbf{p}_0, \mathbf{k}_i)$. This dependence is in agreement with the general rule for the spatial distribution of photon number density, $dN/d\Omega d\omega \sim |\mathbf{R} \times (\mathbf{R} \times \mathbf{p}_\omega)|^2$, according to which the maximum number of scattered photons is in the direction perpendicular to the direction of oscillations.

As one can see from Eqs. (13) the normalized transverse momentum spread varies in time as $\varepsilon_{v,h} \approx \Sigma_{v,h} t \exp(-\Gamma_0 t)$ and its maximum value can be estimated as $\varepsilon_{v,h} \sim \Sigma_{v,h}/2.7\Gamma_0 \sim (4/15)[\hbar\omega_i/mc^2]^2$. Taking into account the transverse vertical and horizontal beam sizes, Eqs. (11)

and (13)–(15) determine the evolution of beam emittance in the process of Compton laser-beam scattering.

References

[1] T. Nakano, et al., Phys. Rev. Lett. 91 (2003) 012002.

[2] T. Tajima, Photonuclear reactions and nuclear transmutations, in: LEPS Workshop, SPring-8, March 3–4, 2003.

[3] P. Sprangle, A. Ting, E. Esarey, A. Fisher, J. Appl. Phys. 72 (1992) 5032.

[4] P. Sprangle, E. Esarey, Phys. Fluids B 4 (1992) 2241.

[5] V. Telnov, Phys. Rev. Lett. 80 (1997) 4757.

[6] V.B. Berestetskii, E.M. Lifshitz, L.P. Pitaevskii, Quantum Electrodynamics, Pergamon, Oxford, 1978.

[7] I.V. Pogorelsky, I. Ben-Zvi, X.J. Wang, T. Hirose, Nucl. Instr. and Meth. A 455 (2000) 176;
I. Sakai, T. Hirose, K. Dobashi, et al., in: Proceedings of the 21st ICFA Beam Dynamics workshop on laser-beam interactions, Stony Brook, New York, 2001.

Available online at www.sciencedirect.com

Nuclear Instruments and Methods in Physics Research A 528 (2004) 544–548

ELSEVIER

NUCLEAR INSTRUMENTS & METHODS IN PHYSICS RESEARCH
Section A

www.elsevier.com/locate/nima

FEL-like interaction regime in the up-shifted scattering of radiation by the laser-plasma wake

I.V. Smetanin[a],*, T.Zh. Esirkepov[a], S.V. Bulanov[b]

[a] *Advanced Photon Research Center, Japan Atomic Energy Research Institute, 8-1 Umemidai, Kizu-cho, Soraku-gun, Kyoto 619-0215, Japan*
[b] *General Physics Institute, Russian Academy of Science, Moscow 117924, Russia*

Abstract

We study the process of resonant scattering of radiation by the laser excited plasma wake in the wavebreaking regime. The resonant interaction arises between the cusp electrons and the slow wave of ponderomotive potential which is formed by the incident and scattered electromagnetic waves. Scattered wave is amplified when the density cusp is moving slightly faster than ponderomotive wave, and is absorbed in the opposite case. This FEL-like interaction leads to nonlinearity in reflection, the intensity-dependent reflection coefficient is calculated.
© 2004 Elsevier B.V. All rights reserved.

PACS: 52.38.Kd; 52.35.Mw; 41.60.Cr

Keywords: Laser-plasma wake; Wavebreaking regime; Frequency up-shift; Resonant scattering; Amplification

1. Introduction

In a free-electron laser [1], coherent radiation is generated as a result of scattering an electromagnetic wave, which represents the undulator field according to the Weizsäcker–Williams approximation, by the counter-propagating beam of relativistic electrons. The frequency of scattered radiation enlarges $\sim 4\gamma^2$ times as a result of the Doppler up-shifting, γ is the beam Lorentz factor. Analogous up-shifting effects were found in a broad variety of configurations, where they were caused by different mechanisms, including the backward Thompson scattering at relativistic electrons [2,3], the reflection at the moving ionization fronts [4], various schemes of the counter propagating laser pulses in underdense plasma and a use of parametric instabilities [5].

Recently, the plasma wakefield in the wavebreaking regime has been proposed as a tool for generating a coherent radiation of ultra-high frequency [6]. In this scheme, a short intense driver laser pulse induces nonlinear wakefield in an underdense plasma, which in the wave-breaking regime leads to the formation of cusps of electron density, moving at the group velocity $v_{ph} = c\beta_{ph}$ of the driver pulse. The cusp acts as a

*Corresponding author.

E-mail address: smetanin@sci.lebedev.ru (I.V. Smetanin).

0168-9002/$ - see front matter © 2004 Elsevier B.V. All rights reserved.
doi:10.1016/j.nima.2004.04.098

semi-transparent mirror for the counter-propagating electromagnetic wave; the frequency of the reflected radiation is then up-shifted by the factor $(1 + \beta_{ph})^2 \gamma_{ph}^2$, where $\gamma_{ph} = (1 - \beta_{ph}^2)^{-1/2}$ is the cusp Lorentz factor. Reflection of radiation by this moving relativistic mirror proceeds under conditions close to that of the free-electron laser interaction, which allows to suggest the possibility of *stimulated* scattering of radiation by the density cusp.

In this paper, we intend to show that the resonant amplification of scattered radiation in the interaction of the moving density cusp in a plasma with counter-propagating electromagnetic wave is possible in analogy with the amplification mechanism in the free-electron laser. The resonant interaction arises between the cusp electrons and the slow wave of the ponderomotive potential, which is formed by the source and scattered electromagnetic waves. The scattered wave is amplified when the cusp's velocity slightly exceeds the velocity of the ponderomotive wave and damped in the opposite case. The gain coefficient, in analogy with the FEL, is proportional to the intensity of the source wave and scales as L^2 with the interaction distance L. Interaction with the slow ponderomotive wave leads to nonlinearity in the reflection of the cusp mirror and the dependence of the reflection coefficient on the wave intensity is investigated.

2. Laser-plasma interaction in the wavebreaking regime

Excitation of the nonlinear plasma wave in the presence of the short driver wave $A_{\perp d}$ and two counter-propagating incident A_i and scattered A_s waves is guided by the system of equations [7]

$$\partial_{xt}\varphi - 4\pi n_e p_\| / m_e c \gamma = 0 \tag{1}$$

$$n_e = n_i + \partial_{xx}\varphi / 4\pi e \tag{2}$$

$$\partial_t p_\| + \partial_x(e\varphi - m_e c^2 \gamma) = -\frac{e^2}{2m_e \gamma c^2} \partial_x A_\perp^2 \tag{3}$$

$$\gamma = (1 + (eA_\perp / m_e c^2)^2 + p_\|^2 / m_e^2 c^2)^{1/2} \tag{4}$$

where $A_\perp$ is the total vector potential, and in the right-hand side of Eq. (3) the ponderomotive force is taken into account. Assuming that incident and scattered waves are sufficiently weak, one can seek solution in the form

$$p_\|(x, t) = p_{\|1}(x, t) + \delta p_\| \tag{5}$$

$$n_e(x, t) = n_1(x, t) + \delta n \tag{6}$$

$$\varphi(x, t) = \varphi_1(x, t) + \delta\varphi \tag{7}$$

where perturbations are generated by the waves A_s and A_i.

For the unperturbed variables we have

$$\partial_{xt}\varphi_1 - 4\pi n_1 p_{\|1} / m_e c \gamma = 0 \tag{8}$$

$$n_{e1} = n_i + \partial_{xx}\varphi_1 / 4\pi e \tag{9}$$

$$\partial_t p_{\|1} + \partial_x(e\varphi_1 - m_e c^2 \gamma) = 0 \tag{10}$$

$$\gamma = (1 + (eA_{\perp d} / m_e c^2)^2 + p_{\|1}^2 / m_e^2 c^2)^{1/2}. \tag{11}$$

According to Eqs. (8)–(11), short intense driver laser pulse induces wakefield in an underdense plasma. Its group velocity equals to zero, and its phase velocity is equal to the group velocity of the laser pulse in the plasma. The corresponding Lorentz factor is $\gamma_{ph} \approx \omega_d / \omega_{pe}$, where ω_d is the *driver* pulse frequency and ω_{pe} is the Langmuir frequency. The nonlinearity of strong wakefield causes a nonlinear wave profile, including the steepening of the wave and formation of the cusps in the electron density [8]. This amounts to the wavebreaking regime [7]. Theoretically the electron density in the cusp depends on the coordinate as $\propto (x - v_{ph}t)^{-2/3}$ and tends to infinity, but remains integrable [7].

3. Resonant scattering of radiation by the plasma wake

Let us consider the interaction of two counter-propagating plane electromagnetic waves characterized by the vector potentials

$$A_s = (mc^2/e)a_s(x, t) \exp(i[k_s x - \omega_s t])$$

$$A_i = (mc^2/e)a_i \exp(i[k_i x + \omega_i t]) \tag{12}$$

with the cusp of electron density moving along the x-axis $n_1(x - v_{ph}t)$. The frequency and wavenumber of the scattered (s) and incident (i) waves are $\omega_{s,i}$ and $k_{s,i}$, respectively. We will neglect the depletion of the incident wave $a_i = const$ and the evolution of the scattered mode is assumed to be sufficiently slow. The electromagnetic waves (12) form the slow wave of the ponderomotive potential which propagates at the phase velocity

$$u = \frac{\omega_s - \omega_i}{k_s + k_i}. \tag{13}$$

The resonant interaction arises when the cusp and ponderomotive wave velocities are close to each other $u \approx v_{ph}$ which corresponds to the frequency up-shifting for the scattered wave [6]

$$\omega_s = (1 + \beta_{ph})^2 \gamma_{ph}^2 \omega_i. \tag{14}$$

It is convenient to consider the laser-cusp interaction in the moving with the plasma wake reference frame. The transformation of coordinates to the moving frame is given by $t' = (t - v_{ph}x/c^2)\gamma_{ph}$, $x' = (x - v_{ph}t)\gamma_{ph}$. In this reference frame, the electron density reads $n_1(x - v_{ph}t) \approx n_0(2/9)^{1/3}\gamma_{ph}(v_{ph}\gamma_{ph}/\omega_{pe}x')^{2/3} \equiv n(x')$ in the vicinity of the cusp [6]. The frequencies and wavenumbers of the electromagnetic waves in this new reference frame are $\omega_s' = (\omega_s - v_{ph}k_s)\gamma_{ph}$, $k_s' = (k_s - v_{ph}\omega_1/c^2)\gamma_{ph}$, $\omega_i' = (\omega_i + v_{ph}k_i)\gamma_{ph}$, $k_i' = (k_2 + v_{ph}\omega/c^2)\gamma_{ph}$.

Assuming the potential perturbation to be small with respect to the ponderomotive potential, the perturbation to the longitudinal motion of cusp electrons is guided, according to Eqs. (3) and (5), by the ponderomotive force

$$\frac{d\delta p_{\parallel}}{dt} = -\frac{e^2}{2mc^3}\frac{\partial}{\partial x}(A_1 + A_2)^2. \tag{15}$$

Here and below throughout this paragraph we will omit the prime assuming that all the variables belong to the moving frame. We assume the amplitude of the scattered wave $a_s(x, t)$ to be the slow varying function. After integration, we have for the longitudinal velocity of cusp electrons

$$\delta v_{\parallel} \approx \frac{kc^2}{\Omega} a_i a_s(x, t) \exp(i[kx - \Omega t]) \tag{16}$$

where $k = k_s + k_i$ is the wavenumber of the slow ponderomotive wave and $\Omega = \omega_s - \omega_i \ll \omega_{s,i}$ is its

frequency in the moving frame. Oscillations in longitudinal current $\delta j_{\parallel} = en(x)\delta v_{\parallel}$ results in the density oscillations according to the continuity equation

$$\frac{\partial}{\partial t}\delta n = -\frac{\partial}{\partial x}\delta j_{\parallel} \tag{17}$$

and, assuming zero perturbation $\delta n|_{t=0} = 0$ as the initial condition, we get for the density oscillation the following relation:

$$\delta n = -i\frac{kc^2}{\Omega^2} a_i a_s \left[ikn + \frac{\partial n}{\partial x}\right]$$
$$\times \exp[ikx][\exp[-i\Omega t] - 1]. \tag{18}$$

The correspondent resonant term in the transverse current perturbation is

$$\delta j_{\perp} = e\delta n v_{\perp} \approx \chi(x, t)|a_i|^2 A_s \tag{19}$$

where the complex coefficient $\chi(x, t)$ is

$$\frac{2m}{ke^2 c}\operatorname{Re}\chi = kn\left(\frac{\sin\eta}{\eta}\right)^2 - \frac{1}{2}\frac{\partial n}{\partial x}\frac{\partial}{\partial \eta}\left(\frac{\sin\eta}{\eta}\right)^2$$

$$\frac{2m}{ke^2 c}\operatorname{Im}\chi = \frac{\partial n}{\partial x}\left(\frac{\sin\eta}{\eta}\right)^2 - \frac{1}{2}kn\frac{\partial}{\partial \eta}\left(\frac{\sin\eta}{\eta}\right)^2. \tag{20}$$

One can find from Eq. (20) that the dependence of χ on the detuning from the exact resonance $\eta = \Omega t/2$ is given by the diffraction-like term $f(\eta) = (\sin\eta/\eta)^2$ and its derivative $df/d\eta$. The function $f(\eta)$ has its maximum value $f = 1$ at the exact resonance $\eta = 0$ and is an even-positive function. The derivative $df/d\eta$ is an even function that vanishes at the exact resonance and reaches its maximum and minimum values at $\eta \approx \pm\pi/2$, respectively.

Imaginary part of the coefficient χ describes the effect of amplification of the scattered signal due to the conventional FEL mechanism. Let us assume that the cusp mirror is transparent and consider amplification of a given external signal at frequency ω_s. Substituting the resonant current term (19) into the right-hand side of the wave equation, we get for the slow-varying amplitude $a_s(x, t)$ the following equation

$$2k_s\frac{\partial a_s}{a_s\partial x} + 2\omega_s\frac{\partial a_s}{a_s\partial t} = -\frac{4\pi}{c}\operatorname{Im}\chi|a_i|^2. \tag{21}$$

Seeking solution in the form $a_s = a_{s0} \exp(G(x, t))$, we get for the signal gain function

$$G \simeq \frac{1}{8} \frac{k^2 \omega_p^2}{k_s} l_c t^2 |a_i|^2 \frac{d}{d\eta} \left(\frac{\sin \eta}{\eta} \right)^2. \tag{22}$$

Here $\omega_p = (4\pi n_0 e^2/m)^{1/2}$ is the plasma frequency, l_c is the characteristic size of the cusp, $l_c = (1/n_0) \int n(x) \, dx$, t is the duration of interaction in the moving frame, and $\eta = \Omega t/2$ is the detuning parameter. One can see that gain (22) has a typical for the FEL gain structure. The gain is proportional to the intensity of incident wave $\sim |a_i|^2$, and increases (as t^2) with the travelled distance, which is just the time t in the moving reference frame, and has a typical diffraction-like term $(d/d\eta)(\sin^2\eta/\eta^2)$ which determines the optimum interaction length at optimum detuning $\eta \sim \pi/2$.

One can see from Eq. (22) that, in accordance with the conventional FEL mechanism, the gain is positive when the detuning is positive $\eta > 0$, i.e. when the cusp is slightly faster than the wave of ponderomotive potential, and negative in the opposite case. In the case of amplification, the scattered radiation mode receives energy from the driver laser pulse which induces the nonlinear wake wave.

4. Nonlinear reflection coefficient

The nonlinear resonant current perturbation (19), (20) leads to the nonlinear corrections to the plasma mirror reflection coefficient [6]. Let us represent the reflected electromagnetic wave as $A_s = b(x, t) \exp(-i\omega_s t)$. In fact, the function $b(x, t)$ is related to the above slow amplitude $a_s(x, t)$ as $b = a_s \exp[ik_s x]$. Let the reflection coefficient be R. Taking into account that in the moving reference frame frequencies and wavenumbers of the incident and scattered waves are close to each other $\omega_i \approx \omega_s$, $k_i = k_s$, one can rewrite Eq. (19) in the form $\delta j_\perp \approx \chi(x, t) R |a_i|^2 A_i$. The wave equation for the scattered wave has the following form:

$$\frac{d^2 b}{dx^2} + \left(k_s^2 - \frac{g}{x^{2/3}} \right) b \approx g a_i \exp(-ik_i x)$$

$$\times \left[\frac{1}{x^{2/3}} \left(1 - \frac{\alpha R |a_i|^2}{2} \right) + i \frac{\alpha^* R |a_i|^2}{3k} \frac{1}{x^{5/3}} \right] \tag{23}$$

where $\alpha = k^2 c^2 t^2 (f(\eta) - (i/2) \, df/d\eta)$ is the complex dimensionless coefficient of nonlinearity, the wavenumber $k_s^2 = \omega_s^2/c^2$ and $g = (2/9)^{1/3} k_p^{4/3} \gamma_{ph}^{2/3}$.

Assuming that the reflection coefficient is sufficiently small, we find an asymptotical solution at $x \to \infty$

$$b(x, t) \approx \frac{i^{4/3} g \Gamma(1/3)}{k_s k^{1/3}} (1 - R |a_i|^2 \mathrm{Re}\,\alpha) a_i \exp(ik_s x) \tag{24}$$

where Γ is the Euler gamma function. Thus the modified reflection coefficient is

$$R = \frac{i^{4/3}(g/k_s k^{1/3}) \Gamma(1/3)}{1 + i^{4/3}(\mathrm{Re}\,\alpha)(g/k_s k^{1/3}) \Gamma(1/3) |a_i|^2}. \tag{25}$$

At low intensities this reflection coefficient coincides with the result of the linear reflection theory [6]. With an increase in the intensity of incident laser pulse, the nonlinear effects come into play. Using detuning from the resonance, one can handle the reflection process and thus optimize the interaction regime.

5. Conclusion

In conclusion, we have shown in this paper that, in analogy to the free-electron laser interaction, the effect of resonant amplification is possible in the scattering of electromagnetic radiation by the plasma wake. The resonant interaction arises between the plasma wake and the slow wave of the ponderomotive potential, which is formed by the incident and scattered radiation. This effect causes nonlinearity in the reflection by the moving plasma mirror, and the reflection coefficient becomes dependent on the intensity of radiation.

References

[1] R.H. Pantell, G. Soncini, H.E. Puthoff, IEEE J. Quant. Electron. 4 (1968) 905;
J.M.J. Madey, J. Appl. Phys. 42 (1971) 1906;
J.M.J. Madey, H.A. Schwettman, W.M. Fairbank, IEEE Trans. Nucl. Sci. NS-20 (1973) 980.
[2] F.R. Arutyunian, V.A. Tumanian, Phys. Lett. 4 (1963) 176.
[3] P. Sprangle, A. Ting, E. Esarey, A. Fisher, J. Appl. Phys. 72 (1992) 5032.

[4] V.I. Semenova, Sov. Radiophys. Quantum Electron. 10 (1967) 599;
W.B. Mori, Phys. Rev. A 44 (1991) 5118;
R.L. Savage Jr., C. Joshi, W.B. Mori, Phys. Rev. Lett. 68 (1992) 946;
W.B. Mori, et al., Phys. Rev. Lett. 74 (1995) 542;
M.I. Bakunov, et al., Phys. Plasmas 8 (2001) 2987;
J.M. Dias, et al., Phys. Rev. E 65 (2002) 036404.

[5] G. Shvets, et al., Phys. Rev. Lett. 81 (1998) 4879;
G. Shvets, A. Pukhov, Phys. Rev. E 59 (1999) 1033;

Z.-M. Sheng, et al., Phys. Rev. E 62 (2000) 7258;
Y. Ping, et al., Phys. Rev. E 62 (2000) R4532.

[6] S.V. Bulanov, et al., Sov. Phys.- Lebedev Inst. Rep. 6 (1991) 9 ISSN 0364-2321.

[7] S.V. Bulanov, et al., in: V.D. Shafranov (Ed.), Reviews of Plasma Physics, Vol. 22, Kluwer Academic Plenum Publishers, NY, 2001, p. 227.

[8] A.I. Akhiezer, R.V. Polovin, Sov. Phys. JETP 30 (1956) 915.

Available online at www.sciencedirect.com

SCIENCE DIRECT°

NUCLEAR
INSTRUMENTS
& METHODS
IN PHYSICS
RESEARCH
Section A

Nuclear Instruments and Methods in Physics Research A 528 (2004) 549–552

www.elsevier.com/locate/nima

Novel scheme of the long wavelength FEL amplifier with a transverse electron current

N.S. Ginzburg[a], R.M. Rozental[a,*], A.V. Arzhannikov[b], S.L. Sinitsky[b]

[a] *Institute of Applied Physics, Russian Academy of Sciences, 46 Ulyanov Str., Nizhny Novgorod, GSP-120, 603950, Russia*
[b] *Budker Institute of Nuclear Physics, Siberian Branch of Russian Academy of Sciences, 11 pr. Akademika Lavrent'eva, Novosibirsk, 630090, Russia*

Abstract

A novel scheme of the long wavelength FEL amplifier with a transverse electron current has been discussed. In this scheme an amplified signal propagates across the electron beam and the operating wavelength is equal to the undulator period. A specific feature of the transverse current amplifier is the absence of saturation, i.e. under special optimization of the undulator field, the output power grows proportionally to interaction distance, which is defined by the beam transverse size. The X-band amplifier should be based on the U-2 accelerator at the Budker Institute of Nuclear Physics and produce 500 kV, 1 kA/cm sheet electron beam with transverse size amounts 140 cm (J. Appl. Phys. 72 (1992) 1657). It has been observed that a gain of ~38 dB corresponding to a 6.5 GW output power and 10% efficiency could be achieved for a 2 kOe undulator field and a 1 MW seed signal. Additional optimization of guiding and undulator fields provides the possibility of increasing the output power up to 18 GW with an efficiency of 27%.
© 2004 Elsevier B.V. All rights reserved.

PACS: 41.60.−m; 41.60.Cr

Keywords: Free-electron lasers; Amplifier; Sheet beam; High-power microwaves; Transverse current

1. Introduction

The main advantage of FEL driven by high current sheet relativistic electron beams is a high total output power under moderate density of electromagnetic flux [2–4]. For sheet electron beams with transverse size of 10^2–10^3 wavelengths, 2-D distributed feedback based on the 2-D planar Bragg resonators have been suggested [5,6] to provide spatial coherence of radiation from different parts of the electron beam. In this paper one more possibility for utilization of hundreds of gigawatts of large-size sheet beams has been discussed. The new scheme is based on the amplification on an electromagnetic signal, which propagates across the sheet electron beam (also referred to as transverse electron current amplifier). However, this scheme can be used only for the centimeter waveband because the operating wavelength is equal to the undulator period.

*Corresponding author. Fax: +7-8312-160616.
E-mail address: rrz@appl.sci-nnov.ru (R.M. Rozental).

0168-9002/$ - see front matter © 2004 Elsevier B.V. All rights reserved.
doi:10.1016/j.nima.2004.04.099

2. Basic equations

Let us assume, that a sheet electron beam with initial velocity $\vec{v}_0 = v_0\vec{z}_0$ passes through a planar undulator with period d in a guiding magnetic field $\vec{H}_0 = H_0\vec{z}_0$. The amplified signal wave propagates in a groove waveguide (Fig. 1) and the electron beam excites only the lowest mode of this waveguide. The vector potentials of the undulator field (subscript u) and signal wave (subscript s) are given by

$$\vec{A}_u = Re\ \{A_u f_u(z)ch(\bar{h}x)e^{i\bar{h}z}\vec{y}_0\},$$
$$\vec{A}_s = Re\ \{A_s(y)f_s(z)e^{i(\omega t - ky)}\vec{x}_0\}$$

where $A_s(y)$ is the slowly varying amplitude, $f_{s,u}(z)$ is the distribution of the fields along z-axis, $\bar{h} = 2\pi/d, k = \omega/c$. The electron beam interacts with the signal wave under the conditions of undulator synchronism $\omega \approx \Omega$ far from the cyclotron resonance $|\omega - hv_z - \omega_H| \cdot T \gg 2\pi, |\Omega - \omega_H| \cdot T \gg 2\pi$, where $\Omega = \bar{h}v_z$, $\omega_H = eH_0/m_0c\gamma_0$ are the bounce and cyclotron frequencies, respectively, v_z is the longitudinal electron velocity, $\gamma_0 = (1 - v_0^2/c^2)^{-1/2}$ is the relativistic mass factor, and T is the interaction time.

According to Ref. [7], the self-consistent system of averaged equations in ultra-relativistic case $\gamma_0 1$ can be written as follows:

$$\frac{da_s}{dY} = -\frac{1}{N_s}\int_0^L\int_0^{2\pi}\frac{g^2}{\varepsilon^2 - g^2}f_s^*(\zeta)a_u(\zeta)e^{-i\theta}d\theta_0 d\zeta$$
$$\frac{d\varepsilon}{d\zeta} = \frac{4g}{\varepsilon^2 - g^2}Re\ \{a_u^*(\zeta)a_s(\zeta)e^{i\theta}\}$$

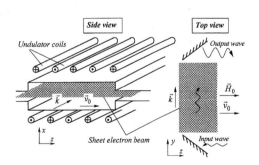

Fig. 1. Scheme of the transverse current FEL-amplifier.

$$\frac{d\theta}{d\zeta} = \frac{1}{\varepsilon^2} - \varDelta + \frac{|ia_s(\zeta)e^{i\theta} + a_u(\zeta)|^2}{(\varepsilon + g)^2}$$
$$+ \frac{|ia_s(\zeta)e^{i\theta} - a_u(\zeta)|^2}{(\varepsilon - g)^2} \tag{1}$$

with the boundary conditions: $a_s|_{Y=0} = a_0$, $\varepsilon|_{\zeta=0} = 1$, $\theta|_{\zeta=0} = \theta_0 \in [0, 2\pi)$.

The following dimensionless variables have been used: $a_{s,u}(\zeta) = f_{s,u}(\zeta) \cdot eA_s/2m_0c^2$ are the amplitudes of the signal wave and the undulator field, $N_s = \int_0^L|f_s(\zeta)|^2 d\zeta$, $\varepsilon = \gamma/\gamma_0$ is the electron energy, $\theta = \omega t - \bar{h}z$ is the electron phase, $\zeta = kz/2\gamma_0^2$ and $Y = ej_0 y/m_0c^2\gamma_0 v_z ha$ are the normalized coordinates, L is the normalized waveguide width, j_0 is the current density, a is the waveguide gap, $g = \omega_H/\Omega$, and $\varDelta = 2\gamma_0^2(1 - \bar{h}/k)$ is the normalized mismatch.

The amplifier efficiency could be defined as $\eta_{total} = B^{-1}\int_0^B \eta_{part}dY$, where B is the normalized beam width and

$$\eta_{part}(Y) = \frac{1}{2\pi} \cdot \frac{\gamma_0}{\gamma_0 - 1}\int_0^{2\pi}(1 - \varepsilon|_{\zeta=L})\,d\theta_0$$

is the partial efficiency of energy extraction from the beam fraction injected at cross-section with coordinate Y.

3. Small signal regime

In the case of uniform distribution of undulator field $a_u(\zeta) \equiv a_u \equiv$ const and the sinusoidal longitudinal distribution of the operating wave $f_s(\zeta) = \sin(\pi\zeta/L)$ from Eqs. (1) we get for the small signal gain

$$\varGamma = \frac{32\pi^3 a_u^2 g^2 L^2}{(1 - g^2)^2}$$
$$\left(\frac{F(\Psi)}{L} - \left(1 - g^2 + 2a_u\frac{1 + 3g^2}{(1 - g^2)^3}\right)\frac{dF(\Psi)}{d\Psi}\right)$$

where $F(\Psi) = (1 + \cos\ \Psi)/2(\pi^2 - \Psi^2)^2$, $\Psi = L(\varDelta - 1 - 2a_u^2(1 + g^2)/(1 - g^2)^2)$.

To find the conditions for maximum gain, the parameters of the U-2 accelerator Budker Institute of Nuclear Physics (BINP) that produces a 500 kV, 1 kA/cm sheet electron beam with transverse size amounts 140 cm have been used [1]. The sheet

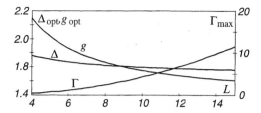

Fig. 2. Optimal parameters for providing the maximum gain Γ. Normalized mismatch Δ and factor g as function of normalized waveguide width L.

electron beam is transported through the guiding magnetic field with the strength of 10–15 kOe (the lowest value is limited by the conditions of the stable beam transportation) and undulator field with on-axis field strength up to 1.6 kOe and period of about 3 cm.

These parameters corresponds to the following values of dimensionless variables: $a_u = 0.2$, $g = 1.6 - 2.4$, $B = 2.2$. Fig. 2 presents the optimal ratio of Δ, g vs. normalized waveguide width L for providing the maximum gain Γ_{max}. One can see, that the gain increases with an increase in the waveguide width. Simultaneously the optimal value of g (i.e., value of guiding field) decreases. As a result, the minimal possible value of g corresponds to the normalized waveguide width of $L = 10$ and real width of about 40 cm. Further, we use these values for the nonlinear simulations.

4. Nonlinear simulations and fields optimization

The power of about 1 MW was chosen for a master oscillator for nonlinear simulations, that corresponded to the normalized value of signal wave of about $a_0 \approx 0.003$. In Fig. 3, the dependencies of the gain and efficiency through the y-coordinate in the case of uniform magnetic fields distributions along y are plotted. Dash lines correspond to the optimal values taken from the linear theory: $\Delta = 1.55$, $g = 1.59$ and solid lines correspond to the optimal values obtained from the nonlinear simulations $\Delta = 1.58$, $g = 2.0$. The maximum gain amounts to 38 dB with a 6.3 GW output power and 9.5% efficiency. Note that in both cases saturation and a subsequent decrease in partial efficiency takes place. Obviously, two

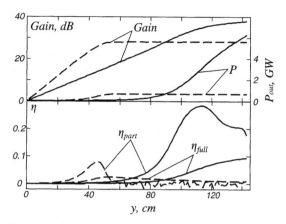

Fig. 3. Nonlinear simulation without optimization: gain, output power and efficiency vs. beam width.

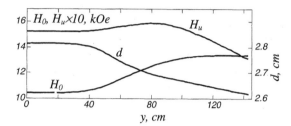

Fig. 4. Nonlinear simulation with optimization: optimal profiles for magnetic fields H_0, H_u and undulator period d.

factors are responsible for the decrease in partial efficiency. The first factor is that the undulator field amplitude becomes too high for the large signal wave amplitude, i.e., overbunching takes place. The second factor is the shift of the amplification band with increasing signal amplitude. It is possible to minimize the influence of these factors by tapering the strength of the undulator and guiding fields over the y-coordinate.

The results of optimization are presented in Fig. 4, where the distributions of magnetic field strength H_0, H_u and the undulator period d vs. the y-coordinate have been presented. The dependencies of the gain, output power, and efficiencies vs. beam width are presented in Fig. 5. One can see that the gain increases up to 42.5 dB with an efficiency of 27% and an output power of about 18 GW. These values exceed those obtained for the uniform field distributions by threefold.

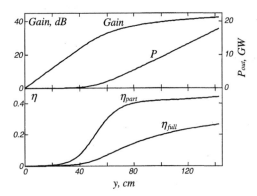

Fig. 5. Nonlinear simulation with optimization: gain, output power and efficiency vs. beam width.

5. Conclusion

The novel scheme of the transverse current FEL amplifier was studied theoretically. The optimal parameters were found both at the linear and nonlinear amplification regimes for the uniform distributions of the undulator and guiding magnetic fields over the transverse coordinate. Additional optimization of the field distributions across the electron beam were carried out to provide nearly threefold increase in the output power and efficiency in comparison with a uniform case.

Acknowledgements

This work was supported by the Russian Fund of Basic Researches (Grant No 01-02-16749).

References

[1] A.V. Arzhannikov, et al., J. Appl. Phys. 72 (1992) 1657.
[2] A.V. Arzhannikov, S.L. Sinitsky, M.V. Yushkov, preprint 91-85 INP SB RAS, Novosibirsk, 1991 (in Russian).
[3] A.V. Arzhannikov, et al., 14th International FEL Conference Kobe, Japan, 1992, p. 214.
[4] S. Cheng, et al., IEEE Trans. Plasma Sci. 24 (1996) 750.
[5] N.S. Ginzburg, N. Yu Peskov, A.S. Sergeev, Sov. Tech. Phys. Lett. 18 (1992) 285.
[6] N.S. Ginzburg, et al., Phys. Rev. E 60 (1999) 935.
[7] N.S. Ginzburg, et al., Nucl. Instr. and Meth. A 483 (2002) 255.

Available online at www.sciencedirect.com

ELSEVIER Nuclear Instruments and Methods in Physics Research A 528 (2004) 553–556

NUCLEAR INSTRUMENTS & METHODS IN PHYSICS RESEARCH
Section A

www.elsevier.com/locate/nima

The MIT bates X-ray laser project

M. Farkhondeh, W.S. Graves, F.X. Kaertner, R. Milner, D.E. Moncton*,
C. Tschalaer, J.B. van der Laan, F. Wang, A. Zolfaghari, T. Zwart

MIT-Bates Laboratory, Middleton, MA 01949, USA

Abstract

MIT and the Bates Linear Accelerator Center are exploring the construction of an X-ray free electron laser user facility. It will be based on a superconducting linac of approximately 4 GeV energy, and produce XUV light in the 0.3–100 nm wavelength range at kilohertz repetition rates. The facility will be a full user facility incorporating multiple beamlines. Conventional lasers that produce the electron beam, seed the FEL, and execute pump–probe experiments are carefully integrated. The current design of the facility is discussed.
© 2004 Elsevier B.V. All rights reserved.

PACS: 41.60.Cr; 41.60.Ap; 29.17.+w; 42.55.Vc

Keywords: X-ray laser; Linac; Free electron laser

1. Introduction

Recent developments in the fields of laser and accelerator technology now allow consideration of a user facility based on highly coherent, powerful and ultra-short pulses of X-ray radiation ranging in wavelength from 100 to 0.1 nm.

Three key elements of the facility we envision would make it unique. First, a 4-GeV linear accelerator with superconducting radio frequency cavities would produce such high electron pulse rates that 20 or more beamlines could be extracted to serve a large user community. Second, integrated high-harmonic generation laser technology would seed the electron beam and generate photon beams with high longitudinal coherence and pulse lengths significantly below 100 fs, perhaps below 1 fs. Third, taking advantage of the ability of linear accelerators to extract beams at different energies, we envision a facility spanning both the traditional extreme ultraviolet and X-ray wavelength range. This approach provides for integration and synergy between the UV and X-ray communities and the laser community, where scientists are anxious to move to wavelengths shorter than conventional table-top technology can provide with high pulse power.

In recent years, a number of short wavelength FEL experiments have demonstrated key technologies and obtained good agreement between experiment and theory. There is a growing consensus in the FEL community that the technology demonstrations and our understanding

*Corresponding author. Tel.: +617-253-8333.
E-mail address: dem@aps.anl.gov (D.E. Moncton).

0168-9002/$ - see front matter © 2004 Elsevier B.V. All rights reserved.
doi:10.1016/j.nima.2004.04.100

of FEL physics have reached sufficient maturity to permit the design of an X-ray laser user facility. The proposed MIT X-ray laser incorporates design features that take advantage of many recent developments. It blends proven technologies into a powerful new instrument that combines the high power, coherence, and ultrashort timescale probe of a laser with the energy reach and spatial resolution of synchrotron X-rays. It is a primary goal to integrate the instruments and experimental methods from both the laser and synchrotron radiation communities at the earliest stages of design.

2. Facility layout and components

A sketch of the layout of the accelerator and experimental halls is shown in Fig. 1. The overall length of the proposed facility is less than 1 km, which fits comfortably on the site of MIT's Bates laboratory. The major components are the superconducting electron linac of length ∼300 m, undulator tunnels for 4–8 undulators each that are ∼50 m long, and three experimental halls (UV, nanometer, and X-ray) following the undulators that are also ∼50 m long. Each hall contains a

number of conventional lasers used for multiple color experiments and seeding of the FEL.

2.1. Room temperature RF photoinjector

The current design calls for a 10 kHz, 2.5 cell copper structure operating at an RF frequency of 1.3 GHz [1]. The major challenge is handling the heat load associated with high duty factor operation while producing a high-brightness beam. The duty factor is 5% for 5 μs long RF pulses. The maximum cathode field is 55 MV/m to keep peak power density to 100 W/cm^2. A Cs$_2$Te photocathode will be used, offering >1% quantum efficiency. The drive laser will be a mode-locked, direct diode pumped Yb:KYW/YAG operating at 10 kHz with flat-top 20 ps long pulses. The full-cell peak gradients are 44 MV/m, and each of the three cells is individually powered and phased. Two modes of operation will be explored: low charge (100–200 pC) per bunch for seeding with ultrashort pulses, and high charge (∼1 nC) for narrow bandwidth seeding.

2.2. Superconducting RF linac

We propose to use a 4-GeV linear accelerator based on the DESY design that will produce such

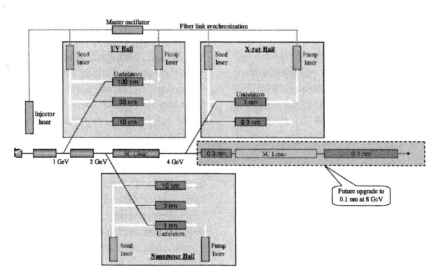

Fig. 1. Layout of facility showing multiple extraction points feeding 3 undulator halls each with multiple undulators. Lasers for seeding and pump–probe experiments are synchronized with the XFEL. The overall facility length is ∼500 m.

high electron pulse rates that 20 or more beamlines can be extracted to serve a large user community. CW operation of the RF is preferred to achieve the required energy and timing stability, and to enable flexibility in pulse timing in response to different user needs. While the cavities will run CW RF, the beam will be pulsed at 10 kHz. The TESLA design is currently pulsed, however Cornell and BESSY are actively modifying the cryomodules for CW, high gradient operation. Optimization of linac length and cryogenic cost will determine the operating gradient, which will be in the range 20–25 MV/m. The electron beam is extracted by fast switches at 1, 2, and 4 GeV to the separate experimental halls.

In addition to the 1.3 GHz accelerating structures, a harmonic cavity at 3.9 GHz will be used to linearize the electron beam's time-energy distribution prior to compression in a magnetic chicane. Other RF structures, possibly pulsed rather than CW, are deflecting cavities at a subharmonic of 1.3 GHz for beam switching, and a RF deflector to allow studies of "slice" electron beam properties.

2.3. Bunch compressors and switchyards

Two stages of bunch compression are performed with magnetic chicanes. The first chicane follows the first two 8-cavity cryomodules and the 3.9 GHz linearizer at an energy of 200 MeV. Following the first stage of compression the beam is accelerated through two more cryomodules to 540 MeV where it is compressed to its final value of approximately 100 fs for the 100 pC case and 1 ps for the 1 nC case. Peak currents are 1–1.5 kA.

At 1 GeV, the first switchyard selects individual pulses for delivery to the UV hall containing approximately four undulators. Both fast ferrite switches and RF deflectors are under study as solutions for the switching needs. An RF switch allows routing of closely spaced bunches to separate halls and so has advantages for seeding at multiple energies, which is under study for production of pulses with relative bandwidth less than 10^{-5}. Similar switchyards are located at 2 and 4 GeV to feed the Nanometer and X-ray halls, respectively.

2.4. Undulators

The initial design calls for 10 total undulator lines, with room for upgrade to 30 undulator lines. The total undulator lengths will reach about 10 m for long wavelength (100 nm) and up to 60 m for short wavelength (0.3 nm). Long undulators will consist of separate segments and short break sections where beam focusing quadrupoles, orbit correction magnets, phase adjusters and diagnostic devices will be located. This approach has been successfully demonstrated at ANL [2], and significantly eases construction of long undulators. For most of the undulators, each segment will most likely be a permanent-magnet planar hybrid device. Electromagnetic undulators are a viable alternative at longer periods, and high-field superconducting undulator technology may also be a useful new technology. It is important to tune the photon energy by tuning the gap, the field strength, or the period of the undulator, rather than the electron beam energy, so that other experiments are not affected. Furthermore, it would be advantageous to adjust the effective undulator length by tuning it in sections to optimize the FEL properties in response to changing electron beam parameters or different seeding configurations.

The layout of the accelerator and experimental halls will allow for a future upgrade to a higher energy, longer linac and the placement of additional long undulators to produce 0.1 nm radiation in the fundamental. Even in the first phase, where lasing is limited to 0.3 nm in the fundamental, significant energy ($\sim 1\,\mu J$) will be produced at the 0.1 nm third harmonic.

3. Laser seeding

Integrated high-harmonic generation laser technology [3,4] will seed the electron beam and generate photon beams with high longitudinal coherence and pulse lengths significantly below 100 fs, perhaps below 1 fs [5]. The FEL itself will use the high gain harmonic generation (HGHG) method [6] to produce multiple harmonics of the tunable input seed. BNL has demonstrated

Table 1
Photon parameters for proposed facility

	MIT Bates X-ray Laser		
	SASE FEL	Min bandwidth seeded FEL	Short pulse seeded FEL
X-rays per pulse (0.1% max BW)	3.E+11	3.E+11	6.E+09
Peak brilliance (p/s/0.1%/mm2)	1.E+33	3.E+35	7.E+33
Peak flux (p/s/0.1%)	6.E+24	6.E+24	1.E+23
Avg. flux (p/s/0.1%)	3.E+14	3.E+14	6.E+12
Average brilliance (p/s/0.1%/mm2)	5.E+22	1.E+25	3.E+23
Degeneracy parameter	4.E+09	3.E+11	6.E+09
Pulse length (fs)	50	50	1
Photon beamlines	10–30		
Wavelength (nm)	0.3–100		
Pulse frequency (Hz)	1000		

saturated output in the HGHG regime at 266 nm [7]. The output radiation has the full longitudinal and transverse coherence and stability of the seed laser, providing substantial improvement over performance based solely on self-amplified spontaneous emission (Table 1).

Different lasers distributed throughout the facility will be used to generate the electron beam, to seed the undulators, and for use in pump–probe experiments. Through seeding, the FEL can generate amplified pulses with desirable properties such as full temporal coherence, extremely short pulses (<1 fs), improved bandwidth, and better energy stability. The FEL output will also be synchronized at femtosecond timescales with conventional lasers for pump–probe studies.

Seed pulses of EUV or soft X-rays will be generated by high harmonic generation (HHG) in a gas, and by nonlinear harmonic crystals for seed wavelengths longer than 180 nm. The HHG method uses millijoule pulses of 800 nm light from a Ti:Sapp fire laser that are compressed in a hollow fiber compressor to 5 fs and focused into a gas jet to produce the high harmonics. For successful seeding, the VUV pulses must be superimposed on the electron bunches before the FEL with a temporal uncertainty of about 10 fs. Rapid advances in frequency metrology based on ultrafast lasers and, therefore, also in laser stabilization and synchronization, show that such low timing jitters between different laser systems can be achieved and maintained over arbitrarily long times [8,9]. The seed wavelength can be made tunable when generated by an optical parametric amplifier and by selecting among closely spaced high harmonics.

A central goal of the seeding program is the development of a multi-kHz source of intense 20–30 fs pulses, as well as phase-controlled few-cycle (sub-5 fs) light pulses and their full characterization and control with respect to intensity, shape, pulse width, and carrier-envelope phase.

References

[1] M. Farkhondeh, et al, Design Study for the RF Photoinjector for the MIT-Bates X-ray Laser, Proceedings of the Particle Accelerator Conference, Portland, 2003.
[2] E. Gluskin, et al., Nucl. Instr. and Meth. A 429 (1999) 358.
[3] M. Schnürer, et al., Phys. Rev. Lett. 80 (1998) 3236.
[4] H. Hentschel, et al., Nature 414 (2001) 509.
[5] W.S. Graves, et al, Seeded X-ray FEL simulations using a high-gain harmonic-generation cascade, Nucl. Instr. and Meth. A, (2004) these proceedings.
[6] L.-H. Yu, et al., Science 289 (2000) 932.
[7] L.-H. Yu, et al., Phys. Rev. Lett. 91 (2003) 074801.
[8] R.K. Shelton, et al., Opt. Lett. 27 (2002) 312.
[9] T.R. Schibli, et al., 300-attosecond active synchronization of passively mode-locked lasers using balanced cross-correlation, Proceedings of Conference on Lasers and Electro-Optics, Baltimore, 2003.

Available online at www.sciencedirect.com

SCIENCE ⬀ DIRECT®

ELSEVIER   Nuclear Instruments and Methods in Physics Research A 528 (2004) 557–561

NUCLEAR
INSTRUMENTS
& METHODS
IN PHYSICS
RESEARCH
Section A

www.elsevier.com/locate/nima

A French proposal for an innovative accelerators based coherent UV–X-ray source

M.E. Couprie[a,b,*], M. Belakhovsky[c], B. Gilquin[d], D. Garzella[a,b], M. Jablonka[e], F. Méot[e], P. Monot[a], A. Mosnier[e], L. Nahon[a,b], A. Rousse[f]

[a] Service de Photons, Atomes et Molécules, CEA/DSM/DRECAM, bât. 522, 91 191 Gif-sur-Yvette, France
[b] Université de Paris-Sud, bat. 209 D, BP 34, 91 898 Orsay cedex, France
[c] CEA/DSM/ DRFMC, DRFMC, CEA-Grenoble 17 avenue des Martyrs, 38054, Grenoble Cedex 9, France
[d] CEA-Département d'Ingénierie et d'Etude des Protéines, Centre d'Etude de Saclay, Bat 152, 91191 Gif sur Yvette, France
[e] CEA/DSM/DAPNIA/SACM, bat. 701, Orme des Merisiers, 91 191 Gif-sur-Yvette, France
[f] Laboratoire d'Optique Appliquée, LOA - ENSTA, Laboratoire d'Optique Appliquée, Chemin de la Hunière, 91761 Palaiseau, France

Abstract

At the initiative of the CEA (Commissariat à l'Énergie Atomique), discussions were conducted in France on fourth generation light sources. A new independent accelerator based radiation facility ARC-EN-CIEL (Accelerator Radiation Complex for ENhanced Coherent Intense Extended Light) is proposed, aiming at providing coherent femtosecond light pulses in the UV- to X-ray range for scientific applications. The project is based on a 700 MeV superconducting LINAC, providing low emittance 200 fs RMS electron bunches. They can be injected in undulators used in the Self Amplified Spontaneous Emission mode (SASE), or in the High Gain Harmonic Generation (HGHG), seeded with high harmonics in gases at 20 nm. The SASE source covers the 100–5 nm spectral range, the HGHG goes down to 0.8 nm. Two optional loops, for Energy Recovery or energy enhancement (1.4 GeV), will accommodate fs synchrotron radiation sources in the IR-, VUV- and X-ray ranges, together with a FEL oscillator providing radiation down to 10 nm, taking advantage of the optical development for lithography. The facility also proposes to test plasma acceleration and to provide a Thomson radiation source. Characteristics of the light source will be described.
© 2004 Elsevier B.V. All rights reserved.

PACS: 41.60.Cr; 41.75.Ht; 41.60.Ap

Keywords: FEL; ERL; Synchrotron radiation; Harmonic generation

*Corresponding author. Laboratoire pour l'utilisation du Rayonnement Electromagnetique, Universite Paris Sud, Bat. 209D, BP 34, Orsay Cedex 91898, France. Tel.: +33-1-64-46-80-44; fax: +33-1-64-46-41-48.

E-mail address: marie-emmanuelle.couprie@lure.u-psud.fr (M.E. Couprie).

1. Introduction

Short pulses Free Electron Laser (FEL) in the VUV-soft X-ray spectral range seem very attractive sources for time-resolved studies in various scientific domains [1]. France developed different

0168-9002/$ - see front matter © 2004 Elsevier B.V. All rights reserved.
doi:10.1016/j.nima.2004.04.101

FEL devices, starting with the first storage ring FEL on ACO [2] in 1983, the UV FEL on Super-ACO [3] and the Linac based Infra-red FEL user facility CLIO in 1992 [4]. First time-resolved pump–probe two-colour experiments were performed, using the Super-ACO FEL and synchrotron radiation [5]. Starting from a scientific case (condensed matter, chemistry, biology, plasma and molecular physics) that cannot be satisfied by the present synchrotron radiation capabilities, discussions were then conducted for the design of a new accelerator based light source, offering tuneability, adjustable polarisation, high brilliance and fs light pulse for scientific applications in the UV–X-ray range. The scheme of the machine is shown in Fig. 1. Intense electron bunches, delivered by a superconducting 700 MeV accelerator, present a very low transverse emittance and a 200 fs RMS duration. Optional Recirculation loops allow the energy to be enhanced (re-acceleration) or the current to be increased (Energy Recovery Mode), as successfully demonstrated on the Jefferson Laboratory FEL [6]. They will comport undulators for the production of fs synchrotron radiation in the X-ray (keV range) and the VUV ranges, and an infra red bending magnet source [7]. A 120–10 nm FEL oscillator, with a 15 m long optical cavity, is installed on the first loop. In the LINAC axis, several undulator sections are planned, for Self Amplified Spontaneous Emission (SASE) in the 200–7 nm range and for HGHG [8] in the 100–0.8 nm range, in particular starting from coherent harmonic generated in gases at 20 nm [9]. A kHz powerful fs laser will be used for the production of harmonics in gases and of fs X-rays by Thomson scattering, for test of plasma electron acceleration, and for time-resolved pump–probe two-colour experiments.

2. The light source

2.1. The accelerator ensemble

Energy recovery justifies the choice of the superconducting technology, which offers a more flexible temporal structure for the users. Injector 1, a 40 MV/m RF gun, equipped with a CsTe photocathode [10], operates with macropulses of 200 μs at 50 Hz or 1 ms at 10 Hz, with either 1 nC and 1 MHz for the SASE/GHC mode or 0.1 nC and 10 MHz for the FEL oscillator. Injector 2, in a CW mode, could be based on an electrostatic gun [11], such as the one developed in TJNAF with 5 mA and 320 kV, now aiming at 10 mA and 500 kV, with 1 μm rad and 0.1 nC or 7 μm rad and 1 nC, or with a RF gun powered by a single short RF pulse, with 1 μm rad and 0.1 nC as in the LBNL proposal [12]. A future injector, for high bunch frequency, charge could be based on a RF superconducting gun [13] (see Fig. 2). A preaccelerating structure raises the energy up to 10–20 MeV. The LINAC (see Table 1) is composed of six TTF like modules, with 8–9 cavities of 9 cells,

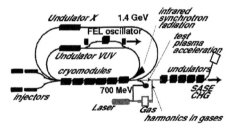

Fig. 1. Layout of the ARC-EN-CIEL proposal. The 72 m LINAC is constituted by 6 modules of 12 m long, 4 m undulator sections are dedicated to SASE and HGHG. Two optional loops will accommodate a FEL oscillator and fs synchrotron radiation sources on undulators and on a dipole.

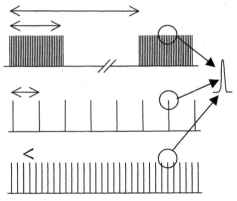

Fig. 2. Temporal structure: (a) injector 1: 1 ms, 10 Hz macropulses, with 200 fs RMS, 10 MHz micropulses (b) injector 2: CW 200 fs, 10 kHz micropulses after compression, (c) future injector : CW 200 fs micropulses at 1 MHz

at 1.3 GHz, with 10^{10} quality factors, powered each by two 100 kW CW klystrons, with 10^7 external coupling. Efficient extraction of high-order modes excited by the beam in the cavities is one of the main concerns. Two optional loops, one for energy recovery and high average current operation, and one for beam energy enhancement at 1.4 GeV, provide a beam up to 2.1 GeV. Beam stability and collective effects remain a difficult issue. Coherent synchrotron radiation and wake-field effects make crucial the bunch compression down to 200 fs, with emittance conservation. A first stage in the injector uses a magnetic compression in a chicane or the velocity bunching scheme, followed by a chicane in the middle of the LINAC. A more complex scheme, using one complete injector module in a separate loop as in the LUX scheme [14] is also considered.

2.2. Sources in the loops

To provide more flexibility for users in terms of light polarisation and spectral range, two 12 m long planar to helical permanent magnets could be placed. Table 2 gives the main undulator characteristics for different beam energies E: period λ_o, deflection parameter K, wavelength λ_1 and energy E_1 of the fundamental. They provide typically a few 10^{12} ph/s/0.1% BW in the VUV and the X-ray for 0.1 mA average current. One order of magnitude would be gained with in emerging vacuum superconducting undulators. An infra-red source, using the central or the edge field of a bending magnet of the loop, could provide 10^{11} ph/s mrad^2 0.1% BW in the 10 mm–0.01 μm range, for two colour experiments coupling the infra-red with the VUV–X-ray radiation.

2.3. Laser sources

A fs Titanium:Sapphire laser source will serve for the seeding of HGHG, Thomson scattering X production, plasma radiation or electron beam-plasma interaction studies. A first part will deliver a few mJ at kHz, sufficient for the production of higher order harmonics in gases. An amplification chain will raise the energy up to a few J at 10 Hz, offering 30 TW power and 10^{20} W/cm^2. A FEL oscillator will be installed in the first loop, and undulators, located in the Linac axis, will be used for SASE and HGHG from higher harmonics in gases. The undulators characteristics are given in Table 3. The SASE radiation covers the 200–7 nm spectral range for beam energies ranging between 135 and 700 MeV, with a Pierce parameter of the order of a few 10^{-3}. The oscillator covers the 120–10 nm, thanks to recent development multilayer mirrors for lithography, and to SiC in normal

Table 1
Main Linac beam characteristics

Energy	MeV	700
RF frequency	GHz	1.3
Gradient	MV/m	15
Intensity per 1 passage	mA	1
Intensity with ER	mA	5–10
Bunch charge	nC	1
Bunch frequency	MHz	$\leqslant 10$
Transverse emittance	m rad	2×10^{-6}
Energy spread		1×10^{-3}

Table 2
Undulators in the loops

E (GeV)	λ_o (mm)	Gap (mm)	K	λ_1 (nm)	E_1 (eV)
XUV undulator in the second loop					
0.7	30	40–10	0.01–2	8–25	155–50
1.4	20	40–10	0.01–0.9	1–2	1200–600
VUV undulator in the first loop					
0.25	30	10–40	2.1–0.1	200–63	6–20
Possible future in vaccum superconducting undulator					
0.7	4		0.1–2	1.1–3.2	1100–400
1.4	4		0.1–2	0.3–0.8	4000–1550

Table 3
Undulators for FELs

	SASE		Oscillator
Period (mm)	30	20	25
Gap (mm)	10–30	10–20	10–24
Magnetic field (T)	0.1–0.75	0.1–0.5	0.1–1.05
Length (m)	4	4	5
Section number	1–2	4	1
Wavelength (nm)	200–8	60–6	120–13
Photon energy (eV)	6–150	20–200	10–90

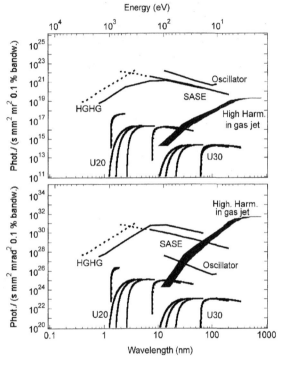

Fig. 3. Peak and average brilliance of ARC-EN-CIEL. High Harm in gas jet stands for the higher harmonics produced in a Xenon gas jet. U20 and U30 refer to the undulators in the loops, respectively, with 20 and 30 mm period.

incidence, offers 0.1-1 kW average power, a 0.1–1% bandwidth. One could also produce mutually coherent FEL sources. For coherent harmonic generation, the seeding will be provided either by the harmonics 2 and 4 of the Ti:Sa laser (1 kHz repetition rate, 5 W, 5 mJ/pulse at 400 nm and 1 W, 1 mJ/pulse at 200 nm) or the harmonics in the gases (n°11 to 31 in the plateau) in the 70–21 nm

spectral range, delivering 2 μJ–0.2 nJ/ pulse (5×10^{11}–2×10^{7} ph/pulse). For the SASE undulators, it leads to 25 MW at 7 nm (FEL harmonics n°3), 1 MW at ~4 nm. The cascade scheme, with one or two sections optimised for 21 nm at 700 MeV and 4 sections designed for 7 nm, will allow the radiation to be extended down to 0.8 nm (1.5 keV, on the 9th harmonics) with a few MW peak power and a flux of 10^{14} ph/s/0.1% BW, with 0.1–1% bandwidth. Harmonics can also be produced from the FEL oscillator, with 500 MW at 4.5 nm and 10 MW at 2.7 nm. Besides, using the beam at 1.4 GeV from the loop will allow to shorten further the radiation wavelength down to 0.4 nm with an additional undulator section. A second cascade module could also be considered. In addition, an 8–10 keV X-ray source, produced by Thomson scattering, will be implemented. Fig. 3 shows the brilliance performances of this light source project.

3. Scientific prospects

This facility proposes a variety of different fs coherent light sources from the infra-red to the X-rays, with an adjustable polarisation, allowing different scientific topics to be investigated. In gas phase, femtochemistry on free species or adsorbates can be studied by photoemission (electronic states) or EXAFS (structure). Photoionisation and alignment dynamics (valence or internal band), spectroscopy on extremely dilute species (radicals, ions), and non-linear physics in the XUV on isolated systems such as clusters can be investigated. In condensed matter, it offers higher resolutions due to high brilliance, new coherence possibilities in imagery, microscopy, holography, especially for nanostructures (spintronics, integrated devices), and the extension of absorption X diffraction, diffusion techniques towards ultimate temporal resolutions for the study of transitory systems out of equilibrium and in situ analysis of ultrafast processes under an external constraint (such as a pulse of magnetic field) due to the fs pulse duration. Optical phonons role in phase transitions, spin polarisation dynamics, identification of complex intermediates in biochemical reactions, study of plasma generation and dynamical behaviour can be analysed. The flux hungry "photon in-photon out" techniques, such as time resolved magnetic holography and resonant inelastic scattering will be developed. In the UV–X domain, protein folding and cell studies using microscopy will benefit from the high brilliance of the source. Exploitation of non-linear techniques opens new prospects for structure and magnetism at interfaces.

4. Conclusion

The short wavelength FEL oscillator and the seeding with harmonics in gas provides a spectral range extension towards short wavelengths, in a much more compact device as compared to a simple SASE scheme, and will offer better source quality (Fourier limit) for users. Moreover, SASE, oscillator and harmonics properties can be compared on the same facility, allowing a step forward in the FEL physics understanding. This project relies also on a strong synergy between the accelerator, the FEL and the laser community. It can help to gather a user community exploiting fs X-ray sources.

References

[1] H. Wabnitz, et al., Nature 420 (2002) 482.
[2] M. Billardon, et al., Phys. Rev. Lett. 51 (1983) 1652.
[3] M.E. Couprie, et al., Phys. Rev. A. 44 (1991) 1301.
[4] R. Prazeres, et al., Phys. Rev. Lett. 78 (1997) 2124.
[5] M. Marsi, et al., Appl. Phys. Lett. 70 (1997) 895.
[6] G. Neil, Phys. Ev. Lett. 84 (2000) 662.
[7] Workshop IRSR, Proceedings ed. LURE, 1998.
[8] L.-H. Yu, et al., Science 289 (2000) 932.
[9] P. Salières, et al., Adv. At. Mol. Opt. Phys. 41 (1999) 83.
[10] D. Dwersteg, et al., Nucl. Instr. and Meth. A 393 (1997) 93.
[11] B.C. Yunn, PAC 2001, p. 2254.
[12] R.A. Rimmer, et al., PAC 2001, p. 1801.
[13] V.N. Volkov, et al., PAC 2001, p. 2218;
T. Schultheiss, et al., PAC 2001, p. 2287.
[14] J.N. Corlett, et. al., PAC 2003.

Available online at www.sciencedirect.com

Nuclear Instruments and Methods in Physics Research A 528 (2004) 562–565

ELSEVIER

NUCLEAR
INSTRUMENTS
& METHODS
IN PHYSICS
RESEARCH
Section A

www.elsevier.com/locate/nima

Spontaneous coherent cyclotron emission from a short laser-kicked electron buncn

A.V. Savilov[a,*], D.A. Jaroszynski[b]

[a] Russian Academy of Sciences, Institute of Applied Physics, 46 Ulyanov Street, Nizhny Novgorod, Russia
[b] TOPS, Strathclyde University, Glasgow, UK

Abstract

Ponderomotive kicking of an electron bunch by the field of a laser pulse is proposed as a method for generating coherent cyclotron emission. It is shown that the imparted gyro-rotation can provide selective RF generation in an oversized microwave system, which can be rapidly tuned over a broad frequency range. A possible realization of a moderately relativistic source of short sub-millimeter wavelength pulses is studied.
© 2004 Elsevier B.V. All rights reserved.

PACS: 41.60.-m; 41.60.Ap

Keywords: Cyclotron radiation; Short electron bunches; Sub-millimeter waves

1. Introduction

Photoinjectors are commonly used as a front end in modern accelerators to produce low-emittance high-charge sub-picosecond bunches of relativistic electrons. The availability of ultra-short e-bunches opens up the attractive possibility for spontaneous coherent radiation in the sub-millimeter and the far-infrared wavelength range [1,2]. Coherent radiation is emitted when all electrons of the bunch have approximately the same phase with respect to the radiated wave. For devices that depend on longitudinal electron bunching, such as

the ubitron free-electron maser, bunching on a wavelength scale or an e-bunch length shorter than half a wavelength is necessary to achieve coherent emission. The latter requirement is difficult to realize for radiation frequencies higher than a few 100 GHz because of the challenge of producing and transporting ultra-short high-charge e-bunches while conserving their duration in the presence of space-charge forces. A completely different scenario exists for cyclotron masers, where electron bunching has a mixed transverse-longitudinal character. This relaxes the need for ultra-short e-bunches because electron gyro-phases are not correlated longitudinally [3]. The ideal e-bunch represents a rotating helix with the rotation frequency and wavenumber matched to the corresponding parameters of the radiation wave.

*Corresponding author. Tel.: +7-8312-384-318; fax: +7-8312-362-061.

E-mail address: savilov@appl.sci-nnov.ru (A.V. Savilov).

0168-9002/$ - see front matter © 2004 Elsevier B.V. All rights reserved.
doi:10.1016/j.nima.2004.04.102

In this work, we propose to use a laser pulse as a kicker to impart transverse (rotatory) motion to the electrons of a short bunch. This provides the proper correlation of phases of electron gyro-rotation and, therefore, the coherent character of the emitted cyclotron radiation.

2. Laser-kicked electron bunch

We consider an e-bunch moving rectilinearly in a uniform magnetic field with a velocity v_0. A short laser pulse propagates parallel to the e-bunch at the speed of light, c (Fig. 1). As an example, we consider a laser pulse with linear polarization, $E_x = B_y = A \sin \theta_L$ where $\theta_L = \omega_L t - k_L z$ and $k_L = \omega_L/c$. A displacement between the e-bunch and the axis of the transverse distribution of laser field, $A(x, y)$, results in a ponderomotive force, F_L, which drives the electrons in the direction of diminishing laser field (Fig. 1a). Thus, the laser pulse initiates electron cyclotron rotation and transforms the rectilinear e-bunch into a helically rotating e-bunch (Fig. 1).

Electrons of the e-bunch following laser kicking have a longitudinal velocity, $v_z < v_0$, and a rotatory velocity, $v_+ = v_x + i v_y = v_\perp \exp i\theta_e$, where $\theta_e = \Omega(t - t_{kick}), \Omega = eB_0/\gamma mc$ is the cyclotron frequency and t_{kick} is the time representing the start of kicking. If at the time $t = 0$ the electrons are distributed over an interval $0 < z_0 < L_{e0}$ and the laser pulse is situated at the point $z = 0$, then $t_{kick} = z_0/(c - v_0)$ and the current electron coordinate is expressed as $z = z_0 + v_0 t_{kick} + v_z(t - t_{kick})$. Thus, the electron gyro-phase can be expressed in the following form:

$$\theta_e = \frac{\Omega}{1 - \beta_z}t - \frac{\Omega/c}{1 - \beta_z}z \qquad (1)$$

with $\beta_z = v_z/c$. Thus, laser kicking provides a spatio-temporal modulation of the e-bunch with the frequency and the wavenumber, which are close to parameters of a cyclotron-resonant wave

$$\omega = \frac{\Omega}{1 - \beta_z/\beta_\phi}, \quad h = \frac{\Omega/(\beta_\phi c)}{1 - \beta_z/\beta_\phi} \qquad (2)$$

where $\beta_\phi = \omega/hc$. If the wave phase velocity is equal to the speed of light, $\beta_\phi = 1$, then exact synchronization appears: all electrons have the same gyro-phases with respect to the wave. This enables spontaneous coherent cyclotron radiation over the whole e-bunch. In a waveguide $\beta_\phi > 1$, and the synchronization is only approximate. The condition of phase synchronization $\Delta\theta < \pi/2$ (here $\theta = \omega t - hz - \theta_e$ is the electron phase with respect to the wave and $\Delta\theta$ is the phase difference between the e-bunch edges) results in limitation of the e-bunch length following kicking:

$$\frac{L_e}{\lambda} < \frac{1 - \beta_z}{4(1 - \beta_\phi^{-1})}. \qquad (3)$$

This condition can be used for controlling the mode of coherent cyclotron emission. If L_e is sufficiently long, then the e-bunch radiates mainly in the lowest possible transverse mode, which has the smallest β_ϕ and, therefore, the highest frequency corresponding to the cyclotron resonance.

If laser frequency is far from cyclotron resonance, then the electron motion consists of fast oscillations in the laser field and the averaged slow motion. The latter is described by the equations

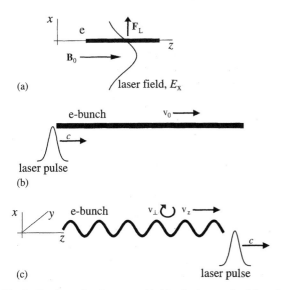

(a)

laser pulse
(b)

(c)   laser pulse

Fig. 1. Pondermotive force provided by the laser pulse (a), and e-bunch before (b) and following (c) laser kicking.

$$\frac{d\bar{v}_x}{dt} = F_L(t) - \bar{v}_y\Omega, \quad \frac{d\bar{v}_y}{dt} = \bar{v}_x\Omega, \qquad (4)$$

where

$$F_L = \frac{-\alpha^2}{4\omega_L^2(1 - \Omega^2/\omega_L'^2)} \frac{\partial A^2}{\partial x}$$

and $\alpha = e/m\gamma$. For a Gaussian shape $F_L(t) = F_0 \exp(-4t^2/t_L^2)$, the solution of Eqs. (4) has the form

$$\bar{v}_+ = -\exp[i\Omega(t - t_{kick})]\frac{F_0}{\Omega}S(\omega) \qquad (5)$$

where $\psi = \Omega t_L'$, $S(\psi) = \psi \exp(-\psi^2/16)\sqrt{\pi}/2$, and $t_L' = t_L/(1 - \beta_z)$. The optimal phase, $\psi \sim \pi$, corresponds to the laser pulse duration

$$t_L \approx \pi(1 - \beta_z)/\Omega \approx \pi/\omega \qquad (6)$$

In the case of a laser pulse with a Gaussian transverse distribution, $E^2(r) \propto \exp(-r^2/r_L^2)$, Eq. (5) leads to the following estimate

$$K = \gamma v_\perp/c \approx 2.5\sqrt[4]{P_L(TW)}\sqrt{\lambda_L/\lambda_c} \qquad (7)$$

where P_L is the laser power, $\lambda_c = 2\pi c^2/\Omega\gamma$ is the wavelength corresponding to the non-relativistic cyclotron frequency, and λ_L is the laser wavelength. Here we have assumed that the transverse electron oscillation is confined within the region of effective force F_L: $2r_c \approx r_L$, where r_c is the electron Larmor radius. It is important that the imparted gyro-velocity weakly depends on the resonant frequency ω, which allows broadband fast frequency tuning.

3. Simulation for the TOPS e-bunch

We study cyclotron emission from a laser-kicked e-bunch in a circular waveguide, and assume that after kicking the particles rotate around the waveguide axis. In this case, only transverse waveguide modes that have the azimuthal index equal to unity can be excited at the fundamental cyclotron resonance [4]. Assuming a small electron gyro-radius compared with the waveguide radius we obtain the motion equations averaged over fast gyro-rotation [5], which are valid after

kicking ($t > t_{kick}$):

$$\frac{d\gamma}{d\tau} = -\beta_\perp e \sum_m Re \, a_m \exp(i\theta_m) - \boldsymbol{\beta}\hat{\mathbf{E}}_{ch} \, ,$$

$$\frac{dp_z}{d\tau} = -\sum_m Re \, \chi_m a_m \exp(i\theta_m) - \mathbf{z}_0\hat{\mathbf{E}}_{ch}$$

Here $\tau = ct$, $\boldsymbol{\beta} = \mathbf{v}/c$, $p_z = \gamma\beta_z$, $\theta_m = \omega_m t - h_m z - \theta_e$, ω_m and h_m correspond to the resonance condition (2), $\hat{\mathbf{E}}_{ch}$ is the normalized space-charge field, $\chi_m = \beta_{\phi_m}^{\mp 1}$ for TE (−) and TM (+) modes, a_m is the m-th transverse mode amplitude described by the following equations

$$(1 - \beta_{z0}\beta_{\phi_m})\frac{\partial a_m}{\partial \zeta} + \frac{\partial a_m}{\partial \tau} = G_m \left\langle \frac{\beta_\perp}{\beta_z}\exp(-i\theta_m)\right\rangle_\varphi$$

Here $\zeta = z - \beta_{z0}\tau$, $G_m = Ie/mc^3 \cdot k_\perp^2 m/S_m\chi_m$, I is the electron current, $k_\perp$ is the transverse wavenumber, and S is the wave norm. We assume that at the moment $t = t_{kick}$ initial electron gyro-phases, φ, are distributed homogeneously over the interval $0 < \varphi < \varphi_0$. In simulation a quite big phase size, $\varphi_0 = 0.2\pi$, is taken.

We study possibility for realizing a moderately relativistic sub-millimeter wavelength frequency-tunable source utilising the TOPS e-bunch source [6]. We consider a 2 nC 1 MeV e-bunch moving in a waveguide with quite a large diameter (5 mm), and study the excitation of the two lowest modes (TE$_{1,1}$ and TM$_{1,1}$). The magnetic field is chosen to correspond to the non-relativistic cyclotron frequency of 100 GHz ($\lambda_c = 3$ mm) so as to study a quite high ($\sim \gamma^2$) Doppler frequency up-conversion. For these parameters and for the IR laser pulse with $\lambda_L = 1$ μm and $P_L = 1 - 3$ TW, estimate (7) gives $K = 0.14$–0.22. Fig. 2 illustrate the shapes of the output RF pulses of the modes in the case of $K = 0.17$. In this case, the length of the operating region is about 20 cm. If the e-bunch length is short, then the both waves are excited. An increase of L_e results in a fast decrease of the TM-wave power and, simultaneously, in a significantly slower decrease of the TE-wave peak, so that at $L_e > 3$ mm practically single-mode emission occurs, which is consistent with the condition (3). Fig. 3 illustrates the possibility for frequency tuning by changing the value of the magnetic field for $L_e = 3$ mm. The TE-mode pulse power depends

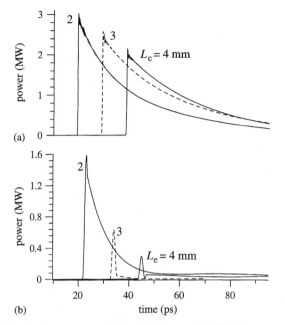

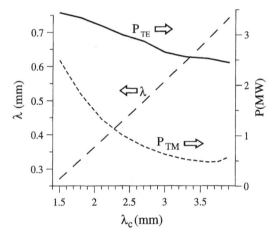

Fig. 2. Output power of the TE (a) and the TM (b) waves versus the time at various e-bunch lengths.

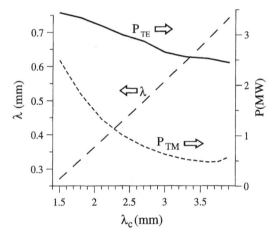

Fig. 3. Wavelength and output peak power of the TE wave, as well as output peak power of the TM wave versus the magnetic field (measured in wavelengths).

on the magnetic field very weakly. However, at short wavelengths, the TM-wave admixture becomes significant.

4. Conclusion

As a conclusion, we discuss the advantages of laser-kicked e-bunch cyclotron emission. First of all, in a relatively long and, therefore, low-density e-bunch, it provides a spatio-temporal modulation, which is almost ideal for coherent cyclotron emission. Second, good mode selectivity can be provided in a simple oversized microwave system. Actually, in addition to azimuthal selectivity imposed by an axis-encircling e-bunch, there is also a mechanism which provides the best condition for radiation in the lowest transverse mode. Third, since the laser kicking has a non-resonant character and, therefore, can be provided in a wide range of parameters, it is possible to provide fast frequency tuning by simply changing the resonant cyclotron frequency.

Acknowledgements

This work is supported by the Russian Science Support Foundation, the Russian Foundation for Basic Research, Grant No. 02-02-17205 and the EPSRC UK.

References

[1] A. Doria, R. Bartolini, J. Feinstein, G.P. Gallernao, R.H. Pantell, IEEE J. Quantum Electron. 29 (1993) 1428.
[2] D.A. Jaroszynski, R.J. Bakker, A.F.G. van der Meer, D. Oepts, P.W. van Amersfoort, Phys. Rev. Lett. 71 (1993) 3798.
[3] D.B. McDermott, N.C. Luhmann Jr., A. Kupiszewski, H.R. Jory, Phys. Fluids 26 (1983) 1936.
[4] J.L. Hirshfield, Phys. Rev. A 44 (1991) 6847.
[5] V.L. Bratman, N.S. Ginzburg, G.S. Nusinovich, M.I. Petelin, P.S. Strelkov, Int. J. Electronics 51 (1981) 541.
[6] D.A. Jaroszynski, B. Ersfeld, G. Giraud, et al., Nucl. Instr. Meth. Phys. Res. A 445 (2000) 317.

Available online at www.sciencedirect.com

ELSEVIER   Nuclear Instruments and Methods in Physics Research A 528 (2004) 566–570

NUCLEAR
INSTRUMENTS
& METHODS
IN PHYSICS
RESEARCH
Section A

www.elsevier.com/locate/nima

Two-color infrared FEL facility at the Saga synchrotron light source in 2003

T. Tomimasu[a],*, S. Koda[a], Y. Iwasaki[a], M. Yasumoto[b], T. Kitsuka[c], Y. Yamatsu[c],
T. Mitsutake[c], M. Mori[c], Y. Ochiai[c]

[a] SAGA-LS, 181-2 Tosu Hokubu Kyuuryou Shintoshi Tosu, Saga 841-0002, Japan
[b] National Institute of Advanced Industrial Science and Technology (AIST), Tsukuba 305-8568, Japan
[c] Saga Prefectural Government, Jonai Saga-shi, Saga 840-8570, Japan

Abstract

The proposed Saga two-color IR-FEL facility employs a part of the electron linac injector of the Saga synchrotron light source that is operated in the fall of 2004 in Tosu by a local Government in Kyushu, Japan. It consists of two sets of undulators and optical cavities in series. Its two IR-FEL facilities independently cover most useful wavelength ranges of 4–10 μm and 7–22 μm for biomedical applications and semiconductor applications, respectively. New radio-frequency (RF) system for the ring cavity and the linac accelerating tube has been designed to synchronize the IR-FEL macropulse with the synchrotron radiation (SR) macropulse as much as possible for new applications such as combination of the molecular bond-selective excitation by the IR-FEL and the SR-based X-ray spectro-microscopy of 500 nm spatial resolution. Beam position monitors for the Saga linac and the beam transport lines are also reported.
© 2004 Published by Elsevier B.V.

PACS: 41.60.Cr; 52.59.Ye

Keywords: Infrared; Free electron laser; Two-color lasing; Synchrotron radiation; Linac injector; Beam position monitor

1. Introduction

Since the Saga synchrotron light source (SAGA-LS) project was started in 1998, we have proposed a two-color IR-FEL facility employing a new linac injector of a Saga storage ring and have discussed beam qualities of the electron injectors suitable for a two-color IR-FEL facility and positron genera-tion [1–3]. Recently, the Saga storage ring and a 250-MeV electron linac injector are being con-structed and will be operated in the fall of 2004 in Tosu by Saga prefectural Government, a local Government in Kyushu, Japan. The ring has eight double-bend cell and 2.93-m long straight sections. The DB cell structure with a distributed dispersion system was chosen to produce a compact design. The circum-ference is 75.6 m and the emittance is 15 nm rad at 1.4 GeV. Six insertion devices includ-ing a 7.5-T wiggler can be installed. The critical energy of synchrotron radiation (SR) from the

*Corresponding author. Tel.: +81-942-83-5017; fax: +81-942-83-5196.
E-mail address: tomimasu@saga-ls.jp (T. Tomimasu).

0168-9002/$ - see front matter © 2004 Published by Elsevier B.V.
doi:10.1016/j.nima.2004.04.103

bending magnet and the wiggler are 1.9 and 9.8 keV, respectively. In the previous papers, we reported design study of SAGA-LS [4], lattice selection [5], control system [6]. We expect that the stored beam current and its lifetime will be 300-mA at 1.4 GeV and 5 h and that the linac beam can be used for two-color IR-FEL generation during the interval of 5 h between electron injections in 2006.

The Saga two-color IR-FEL facility consists of two sets of undulators and optical cavity in series along a 40-MeV beam line. The configuration is similar to the two-color IR-FEL facility at the FELI [7]. However, as available electron macropulse length is reduced from 12 to 9 μs because of the limited budget for SAGA-LS project, the rise time of IR-FEL macropulse should be shortened. Therefore, the first optical cavity length is designed to be a half of the FELI FEL-1 (6.72 m long) and the period number of the second undulator is designed to be more than 45 comparing to 30 periods of the FELI FEL-4 [8].

The Saga two IR-FEL facilities cover most useful wave length ranges of 4–10 and 7–22 μm for biomedical applications and semiconductor applications, respectively [3]. Recently, new applications are proposed such as combination of the SR based X-ray spectro-microscopy of 500 nm spatial resolution and the molecular bond-selective ex-

citation by the IR-FEL [3], and the RF frequencies of the ring cavity and the linac accelerating tube are selected to be 499.8 and 2856 MHz, respectively, to synchronize the IR-FEL macropulse with the synchrotron radiation (SR) macropulse as much as possible.

In this paper, we report the Saga linac beam for the storage ring and FEL oscillation, the RF frequency system and RF power level for new applications, beam position monitors for the Saga linac and the beam transport lines, and recent design study on the two-color IR-FEL facility.

2. Saga linac beam for storage ring injection and FEL oscillation

The Saga linac consists of an FELI type 6-MeV injector [9] and six Electrotechnical Laboratory (ETL) type accelerating tubes [10]. The accelerating tubes with a length of 2.93 m are of linearly narrowed iris type to prevent beam blow up effect.

The schematic layout of the Saga linac and the two-color IR-FEL facility is shown in Fig. 1. The new RF frequency system and power level for the linac accelerating tube and the ring cavity is also shown in Fig. 1. The frequencies of 499.8 and 2856 MHz (89.25 MHz × 32) are selected for the ring cavity and the linac accelerating tube,

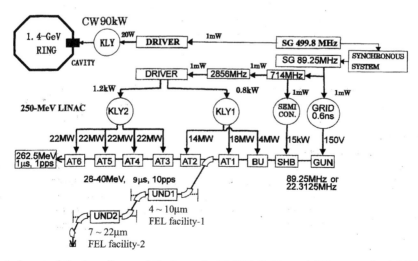

Fig. 1. The schematic layout of the Saga linac and the two-color IR-FEL facility and RF systems for 1.4-GeV storage ring and 262-MeV linac.

respectively, to synchronize the IR-FEL macro-pulse with the synchrotron radiation (SR) macro-pulse as much as possible for new applications. The 89.25-MHz signals for the linac accelerating tube are generated from a 499.8-MHz signal generator for the ring cavity by a synchronous system including a frequency divider.

The injector consists of a 120-keV thermionic triode gun, a 714-MHz prebuncher and a 2856-MHz standing wave-type buncher. The gun with a dispenser cathode (EIMAC Y646B) usually emits 600-ps (FWHM) pulses of 2.3 A at 22.3125 MHz or 89.25 MHz. The grid pulser is supplied by Kentech Instruments, Ltd., UK These pulses are compressed to 60 A × 10 ps by the prebuncher and the buncher. The RF source for the prebuncher is a 714-MHz semiconductor type 25-μs macro-pulse RF source, of which the latter phase stable part is available for beam bunching. A 2856-MHz klystron (Toshiba E3729, 36 MW) is used for the buncher and the first two accelerating tubes (AT1–AT2).

The linac is operated in two modes; 1-μs macropulse 262-MeV electron beam operation for the storage ring injection and 9-μs macropulse 28–36-MeV electron beam operation for two-color FEL oscillation. At the injection mode, a 2856-MHz klystron (Toshiba E3712, 88 MW) is used for the following four accelerating tubes (AT3–AT6). At the FEL mode, the 9-μs macropulse electron beam is accelerated up to 40-MeV at the end of the AT-1.

The electron beam consists of a train of several ps, 0.6-nC microbunch repeating at 22.3125 or 89.25-MHz like the former FELI linac [11]. At the 1-μs macropulse 262-MeV beam operation and an electron charge of 12-nC (0.6-nC × 20 pulses) is injected to the storage ring per second. The beam energy is ramped from 262-MeV to 1.4 GeV in a minute after beam storage, since all storage ring magnets are made of laminations of 1 mm or 0.5 mm thick steel. The RF frequencies of the linac accelerator tube and the ring cavity are selected to be 2856 and 499.8-MHz to synchronize the IR-FEL macropulse with the SR macropulse as much as possible for new applications such as combination of the SR based X-ray spectro-microscopy of 500 nm spatial resolution and the molecular bond-

Table 1
Beam parameters of the Saga linac

Electron energy at injection	262.5 (MeV)
Energy spread (FWHM)	0.5 (%)
Peak current	130 (A)
Beam radius	0.5 (mm)
Normalized emittance	$25\pi \times 10^{-6}$ (m rad)
Micropulse charge	0.6 (nC)
Micropulse duration	4 (ps)
Micropulse separation	44.8 (ns)
Macropulse duration	1 (μs)
Macropulse repetition rate	1 (Hz)
Electron energy at FEL application	~40 (MeV)
Energy spread (FWHM)	~1 (%)
Peak current	60 (A)
Beam radius	0.5 (mm)
Macropulse charge	0.6 (nC)
Micropulse duration	6 (ps)
Micropulse separation	11.2 (ns)
Macropulse duration	9 (μs)
Macropulse repetition‘	10 (Hz)

selective excitation by the IR-FEL. Table 1 shows beam parameters of the Saga linac at injection and FEL application.

3. Beam position monitors for the Saga linac and the beam transport lines

The linac beam size and position is always monitored and controlled to pass through the center of accelerating tubes. A screen monitor is installed at the inlet and outlet of every accelerating tube and quadrupole doublet. The interval of screen monitors is 3.81 m from AT-2 to AT-6. Each screen monitor is an alumina plate and has a 2-mm ϕ aperture where the linac beam can pass through. The center of each 2-mm ϕ aperture is aligned with the fiducial line. The iris center of the accelerating tube is also aligned with the line. The almina plate is set at 45° but the 2-mm ϕ aperture is bored in parallel to the fiducial line. A CCD camera with a zoom lens is set at horizontal side of the almina plate and the fluorescent light from the almina is observed with it. Since the normalized emittance of the Saga linac beam is estimated to be $25\,\pi$ mm mrad, the linac beam is focused to be less than 2-mm ϕ so as to pass through succeeding one or two screen monitors. Further, four screen monitors are installed in S-type BT line [12] for electron energy spectrum measurement. The beam size and position are adjusted to pass through a narrow slit of a septum magnet. The first bending magnet of each S-type BT line and the water absorber are used as an energy spectrometer.

Two Al-foil optical transition radiation (OTR) beam profile monitors [13] will be installed in each narrow vacuum chamber of the two-color IR-FEL facility. Each Al foil has a 1-mm ϕ aperture. The S-type BT line can focus about 80–90% of the electron beam to pass through the aperture. The two-profile monitor emittance measurement method with the OTR beam profile monitor is useful for its simplicity and short time for data acquisition along the linac beam line.

4. Undulators and optical cavities for the Saga two-color IR-FEL facility

At the two IR-FELs operation, first we try to oscillate the second FEL at an IR range at small signal gain of 30%, widening the first undulator gap length to 50 mm (K-value ~ 0). Second, we gradually narrow the first undulator gap length to 24 mm (K-value $= 0.58$) to start up the first oscillation. The macropulse of the first FEL rises near the end of 9-µs electron macropulse length at a small signal gain of several %. The first FEL oscillation gives a small influence near the end of the second FEL macropulse. However, the first FEL macropulse length increases with an increase of the small signal gain. The tail of the second FEL macropulse is influenced with an increase of the first FEL macropulse length.

The two IR-FELs are delivered to two different user groups for the application experiment synchronizing with SR X-rays. Although the design parameters of the two-color IR-FEL were discussed for the 12-µs electron macropulse [3], revised parameters for the 9-µs macropulse are given in Table 2.

The cavity lengths of the first and the second FELs are 3.36 and 6.72 m, respectively. The period number of the second undulator is increased to be more than 45 comparing to 30 periods of the FELI

Table 2
Main parameters of the two-color IR-FEL oscillators

	(4–10 µm)	(7–22 µm)
Electron energy	36–28 (MeV)	
Undulator type	Halbach	Halbach
Undulator period (mm)	32	48
Period number	40	45
Gap length (mm)	15–24	24–36
K value	1.4–0.58	1.93–0.87
Permanent magnet	Nd–Fe–B	Nd–Fe–B
Cavity type	Fabry–Perot	Fabry–Perot
Cavity length (m)	3.36	6.72
Mirror curvature (m)	2.0	3.555
Material of mirror	Au/Cu	Au/Cu
Extraction	Hole	Hole
Aperture dia. (mm)	1.5	2.5
Window for FEL guide	ZnSe	KRS-5

FEL-4 [8] to get higher net gain. The first FEL covers 4–10-µm and the second FEL can cover a broader range of 7–22-µm. Each optical cavity is a Fabry–Perot cavity which consists of two mirror vacuum chambers [13]. The spontaneous radiation and the IR-FEL beam are delivered with Au-coated mirrors through an evacuated optical pipe from the optical cavity to a diagnostic station on the same floor in the experimental hall of the SAGA-LS. Their macropulse shape, power and spectrum are simultaneously observed at the diagnostic station and the linac control room.

5. Summary

The proposed Saga two-color IR-FEL facility employs a part of the 250-MeV linac injector of SAGA-LS operated in the fall of 2004. The Saga two IR-FELs oscillate at useful wavelength range of 4–22-µm in 2006 and are delivered to two different user groups for the application experiment synchronizing with SR X-rays.

References

[1] T. Tomimasu, et al., Proceedings of AFEL'99, KAERI, June 8–10, 1999, p. 340.
[2] T. Tomimasu, et al., Nucl. Instr. and Meth. A 475 (2001) 54.

[3] M. Yasumoto, et al., Jpn. J. Appl. Phys. 41 (Suppl. 41–1) (2002) 44.

[4] T. Tomimasu, et al., Presented at 2003 PAC, Porland, May, 2003.

[5] Y. Iwasaki, et al., Presented at 2003 PAC, Porland, May, 2003.

[6] H. Ohgaki, et al., Presented at 2003 PAC, Porland, May, 2003.

[7] A. Zako, et al., Nucl. Instr. and Meth. A 429 (1999) 136.

[8] T. Takii, et al., Nucl. Instr. and Meth. A 407 (1998) 21.

[9] T. Tomimasu, et al., Nucl. Instr. and Meth. A 407 (1998) 370.

[10] T. Tomimasu, IEEE Trans. Nucl. Sci. NS-28 (3) (1981) 3523.

[11] T. Tomimasu, et al., Nucl. Instr. and Meth. A 429 (1999) 141.

[12] Y. Miyauchi, et al., Proceedings of 9th Symposium on Accelerator Science and Technology, KEK, Aug. 1993, pp.416–418.

[13] K. Saeki, et al., Nucl. Instr. and Meth. A 375 (1998) 10.

Available online at www.sciencedirect.com

Nuclear Instruments and Methods in Physics Research A 528 (2004) 571–576

ELSEVIER

NUCLEAR INSTRUMENTS & METHODS IN PHYSICS RESEARCH
Section A

www.elsevier.com/locate/nima

Design consideration for Tohoku light source storage ring equipped with UV free electron laser

H. Hama*, F. Hinode, K. Shinto, A. Miyamoto, T. Tanaka

Laboratory of Nuclear Science, Graduate School of Science, Tohoku University, 1-2-1 Mikamine, Taihaku-ku, Sendai 982-0826, Japan

Abstract

An integrated photon source facility has been planed at Laboratory of Nuclear Science, Tohoku University. A 1.5 GeV main ring designed as a synchrotron light source of VUV and soft X-ray region contains straight sections with very low beta function to accept high field superconducting wigglers for X-ray. One of two 8-m long straight sections is allocated for storage ring free electron laser (SRFEL) in the UV region. The beam property in the ring is evaluated and then the beam quality for the SRFEL oscillation is also discussed including possibility of coherent higher harmonic generation by showing results of numerical simulation.
© 2004 Elsevier B.V. All rights reserved.

PACS: 41.60.Cr; 29.20.Dh

Keywords: Storage ring; Potential-well distortion; Free electron laser; Coherent harmonic generation

1. Introduction

A new accelerator complex project to provide photons of which the energy range covers far-infrared, VUV, X-ray and hundreds MeV has been developed at Tohoku University. A 150 MeV injector linac may contain a thermionic RF gun [1], and then an infrared FEL is also considered. A 1.5 GeV storage ring is designed to realize relatively low emittance and capability of inserting a couple of superconducting high-field wigglers for X-ray. Furthermore an SRFEL is also taken into account to provide coherent higher harmonic photons and high energy γ-rays via intracavity Compton back-

scattering. At the present, the lattice of the storage ring is designed based on eight-hold Chasman–Green cells and two 8 m long straight sections. Because of limited budget and site area, the storage ring should be small, so that we will employ multipole magnets which produce both quadrupole and sextupole magnetic fields [2]. A planned layout of accelerator complex is shown in Fig. 1.

2. Storage ring

2.1. Lattice design and basic parameters

Lattice design has been done by taking availability of superconducting 7 T wiggler (wavelength shifter) into account, so that a vertical betatron

---

*Corresponding author. Tel.: +81-22-743-3432; fax: +81-22-743-3401.

E-mail address: hama@lns.tohoku.ac.jp (H. Hama).

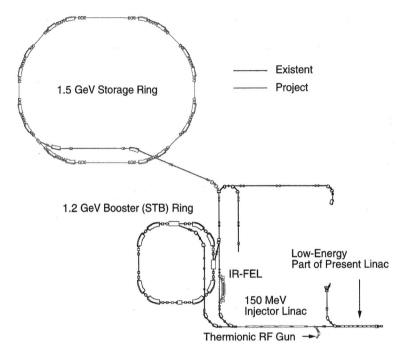

Fig. 1. Accelerator complex at LNS, Tohoku university. Greyish lines and letters indicate future project.

function in the straight section of the normal cell is reduced to 0.8 m. Distortion of the lattice function due to the wigglers is sufficiently small and can be corrected by trifling change of excitation currents for multipole magnets nearby. One of two 8.3 m long straight sections will be dedicated to the SRFEL device. The lattice functions along a half of the ring is shown in Fig. 2, and basic ring parameters are listed in Table 1.

The multipole magnets have been designed to contain sextupole moments so as to increase chromaticity up to $+2$, and additional backleg coils can vary it by ± 2. Detail of designing work of the ring will be reported elsewhere [3]. Since the ring is a moderate focus machine, i.e., the natural emittance is not much small, a large dynamic aperture is secured.

2.2. Bunch lengthening and ring impedance

Designing the machine is recently getting tight to pack electrons in a very small 6-dimensional phase space. Such a high peak current density is preferable for the SRFEL, which is, however, also

taking a risk of strong electromagnetic interaction with the impedance of the beam pipe. On the other hand, very low impedance of the vacuum chambers is realized due to excellent design and fabrication at facilities of 3rd generation light sources.

If the ring impedance does not strongly depend on the frequency, shape of the electron bunch at a current is well predicted by Haissinski equation [4,5],

$$\frac{\mathrm{d}f}{\mathrm{d}t} = -\frac{tf + (-)\xi f^2}{1 + (-)f}, \tag{1}$$

where f is a normalized distribution function of electrons and ξ is a ratio of resistive impedance $R\sigma_{\tau 0}$ to inductive one L. The minus sign in the equation is for negative momentum compaction α. The differential equation can be numerically solved by applying a normalized non-dimensional beam current

$$\Gamma \equiv \int_{-\infty}^{+\infty} f \, \mathrm{d}t = \left(\mathrm{Im}\left| \frac{Z}{n} \right| + \mathrm{Re}\left| \frac{Z}{n} \right| \right) \frac{Ih^2}{2\pi f_{\mathrm{RF}}^3 \hat{V}_{\mathrm{RF}} \sigma_{\tau 0}^3}, \tag{2}$$

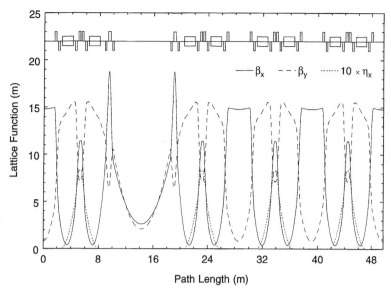

Fig. 2. Lattice functions in a half of the ring.

where $Z = R\sigma_{\tau 0} + iL$ is a modified ring impedance, n is the harmonic number of the impedance at a certain frequency, and I, h, f_{RF}, $\hat{V}_{RF}$ and $\sigma_{\tau 0}$ are the average beam current, the harmonic number, the RF frequency, the peak accelerating voltage and the natural bunch length, respectively. In general the inductive impedance is dominant for the bunch lengthening that is so-called potential-well distortion, and the resistive one makes some asymmetry of the bunch shape [5].

When the beam current increases, the phase stability in the accelerating RF potential is lost due to large wake-potential, which is well known as Boussard criterion [6]. However this criterion does not take potential-well distortion into account. Assuming a parabolic bunch shape as a solution of Haissinski equation, the threshold current of $\Gamma = 17.5$ is obtained [5], which is the upper bound of potential-well distortion. Further increase of the beam current the energy spread increases as well as increase of the bunch length, so that the non-dimensional current stays at the threshold current. This is not turbulent but a kind of microwave instability in general.

Fig. 3 presents the bunch length and the energy spread calculated by Haissinski equation with the phase stability criterion. One notices that if

Table 1
Basic ring parameters

Beam energy	1.2 ~ 1.8 GeV
Circumference	99.503 m
Horizontal emittance	17.2 nmrad @ 1.2 GeV
Straight section	3.1 m × 6, 8.3 m × 2
Betatron tune	(7.80, 3.72)
Momentum compaction factor	0.0064
Chromaticity	$(-20.9, -13.2) \Rightarrow (+2, +2)$
Relative energy spread	0.00052 @ 1.2 GeV

the impedance is able to be reduced to $0.2\,\Omega$, the threshold current will be never met within actual operating beam current. Consequently the ring impedance is crucial to earn a large FEL gain. In the case of the impedance of $0.2\,\Omega$, the peak current reaches 100 A at the average one of 20 mA.

3. SRFEL and coherent harmonic generation

Although an available length for the FEL undulator is more than 8 m, here we suppose a length around 5 m because another undulator as a radiator of coherent harmonic generation (CHG) is considered to put in remaining straight section.

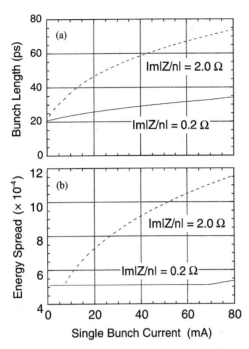

Fig. 3. Calculated (a) bunch length and (b) energy spread for the designed ring operated at 1.2 GeV and a peak RF voltage of 500 kV.

Since SRFEL is a resonator type FEL, performance of optical devices such as cavity mirrors is crucial particularly in UV region. The CHG from microbunches induced by the FEL interaction is possibly a promising method to generate coherent synchrotron radiation in shorter wavelength.

Fig. 4 shows laser intensity dependences of the FEL gain for an optical klystron (OK) configuration and an undulator. The maximum microbunching is occurred at the saturation intensity ($10^{8.5}$ MW/m^2 and 10^9 MW/m^2 for the OK and the undulator, respectively). However such very high intracavity power may not be obtained in ordinal SRFEL oscillation because of increase of the energy spread (*the bunch-heating*) and equilibrium between the effective gain and the cavity loss.

Ignoring the transverse motion, a total radiation intensity from an electron bunch can be expressed as

$$I_{tot}(\omega) \approx I_{in\text{-}coh}(\omega)\{N_e + N_e^2|S(\omega)|^2\}, \tag{3}$$

where N_e is a number of electrons in the bunch and $|S(\omega)|^2$ is the bunch form factor, which is deduced

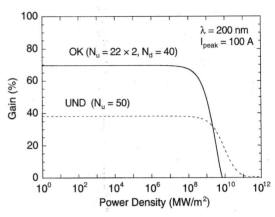

Fig. 4. The calculated FEL gain for a helical optical klystron (OK) and a helical undulator (UND) plotted as a function of the laser power density. Period length of both configurations is 0.1 m. No energy spread taken into account.

by using a distribution function $\rho(z)$,

$$S(\omega) = \int \rho(Z)e^{i\omega z/c}\,dz\left(\int \rho(z)\,dz = 1\right) \tag{4}$$

To evaluate with respect to the FEL frequency ω_0, Eq. (4) is rewritten with a normalized length $u = \omega_0 z/c$,

$$S(\omega/\omega_0) = \int (\omega_0/c)\rho(u)e^{i(\omega/\omega_0)u}\,du. \tag{5}$$

The ratio f of coherent part of the radiation to in-coherent one from one electron bunch can be roughly estimated as

$$f = n_{MB}N_{eMB}^2|S|^2/N_e. \tag{6}$$

Using a peak current of the electron bunch I_p, the number of the microbunches, n_{MB}, the number of electrons in a microbunch, N_{eMB}, and the number of electrons in the bunch, N_e, are expressed as $\sqrt{2\pi}\sigma_{FEL}c/\lambda_{FEL}$, $I_p\lambda_{FEL}/(ce)$ and $I_P\sqrt{2\pi}\sigma_b/e$, respectively, where σ_{FEL} the FEL time duration, c the velocity of light, λ_{FEL} the FEL wavelength, e the electron charge, σ_b the bunch length. Using these rough evaluation, Eq. (6) is reduced to be

$$f = \frac{\sigma_{FEL}}{\sigma_b}\frac{I_P\lambda_{FEL}}{ce}|S|^2. \tag{7}$$

The FEL does not involve the entire electron bunch because of a bell shape of the gain profile, so that the ratio σ_{FEL}/σ_b is assumed to be $10^{-2\sim-1}$. Employing the peak current of 100 A and the FEL wavelength of 200 nm, one obtains $|S|^2 > 10^{-5 \sim -4}$ for which the coherent radiation intensity is larger than the incoherent one ($f > 1$). Assuming the Gaussian shape for the FEL induced microbunch, the form factors are calculated for different microbunch length shown in Fig. 5. If one wants to obtain the 5th harmonics, the microbunch length has to be at least reduced to be the 10% of the wavelength. Fig. 6 shows the longitudinal electron distributions for the undulator and the OK configurations obtained by a one-dimensional simulation. Calculations were done for different initial energy spreads. Ideal case of no energy spread gives a nice micro-

bunching. The microbunch length for the natural energy spread ($\sigma_\gamma/\gamma = 5 \times 10^{-4}$) is still favorable. However even the gain (or Q) switching lasing, the energy spread is no longer the natural one when once the FEL power is evolved.

For the OK case, CHG would be unlikely if the energy spread becomes 3 times larger than the natural one, on the other hand it seems to be still possible for the undulator case.

4. Concluding remarks

A wonderful SRFEL performance using the high brilliant electron beam of a 3rd generation light source was successfully demonstrated on ELETTRA [7]. This achievement will open the next door for further development of the SRFELs. Overall characteristics of the electron beam in the storage ring have to be highly efficient for completion of SRFEL oscillation, which means careful consideration for every possible beam instabilities is indispensable in the designing work.

The longitudinal phase space of the electrons is in transient state during the gain switching lasing. Particularly synchrotron oscillation which is an intrinsic property of the storage ring may much affect both the FEL power and the

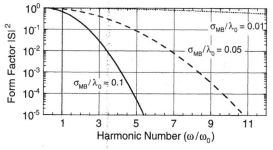

Fig. 5. The form factors for the Gaussian microbunches, where σ_{MB} and λ_0 denote the microbunch length and the FEL wavelength, respectively. The frequency is normalized to the FEL frequency as explained by Eq. (5).

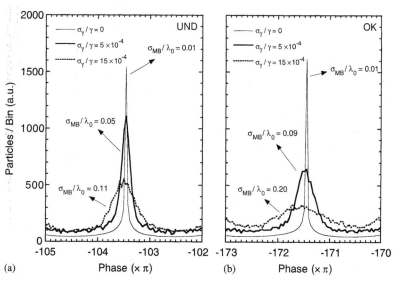

Fig. 6. Microbunch shapes at the exit of the undulator (a) and the OK (b). The laser power is at the respective saturation intensity. The microbunch lengths are deduced from a simple Gaussian fit.

micro-bunching [8]. To investigate the possibility of the FEL induced CHG, precise simulation involving entire beam dynamics of the storage ring and the FEL interaction is definitely required.

References

[1] H. Hama, et al.,3-D simulation study for a thermionic RF gun using an FDTD method, Nucl. Instr. and Meth.

[2] A. Andersson, et al., Design report for the MAX II ring, University of Lund, Lund, 1992.

[3] H. Hama, et al.,Design Report of LNS storage ring, Tohoku university, http://www.lns.tohoku.ac.jp/.

[4] J. Haissinski, Nouvo Cimento 18B (1) (1973) 72.

[5] H. Hama, Proceedings of the ICFA 17th Advance Beam Dynamics Workshop on Future Light Sources, Argonne National Laboratory, Argonne (2000) wg3-05;
H. Hama, et al., Nucl. Instr. and Meth. A 429 (1999) 172.

[6] D. Boussard, Int. CERN Rept. Lab II/RF/Int 75-2, CERN, Geneva, 1975.

[7] M. Trovo, et al., Nucl. Instr. and Meth. A 483 (2002) 157.

[8] V.N. Litvinenko, et al., Nucl. Instr. and Meth. A 358 (1995) 334.

Available online at www.sciencedirect.com

SCIENCE @ DIRECT°

Nuclear Instruments and Methods in Physics Research A 528 (2004) 577–581

**NUCLEAR
INSTRUMENTS
& METHODS
IN PHYSICS
RESEARCH**
Section A

www.elsevier.com/locate/nima

BESSY soft X-ray FEL ☆

A. Meseck[a,*], M. Abo-Bakr[a], D. Krämer[a], B. Kuske[a], S. Reiche[b]

[a] *BESSY, Albert-Einstein-Strasse 15, 12489 Berlin, Germany*
[b] *UCLA, Department of Physics and Astronomy, CA 90095, USA*

Abstract

The BESSY soft X-ray FEL is planned as a linac-based single pass FEL multi user facility covering the VUV to soft X-ray spectral range (0.02 keV $\leqslant \hbar\omega \leqslant$ 1 keV) [1,2]. A Photoinjector and a superconducting CW linac [3] will feed three independent SASE-FEL-lines consisting of APPLE II type undulators providing variable polarized radiation [2,4]. The performance of the SASE-FEL can be enhanced using a coherent seed which dominates the spontaneous emission at the beginning of the process. We present simulation studies for three BESSY-SASE-FELs as well as for possible multistage high-gain harmonic generation (HGHG) FELs and discuss their performance in terms of pulse power, pulse duration, pulse shape and signal-to-noise ratio.
© 2004 Elsevier B.V. All rights reserved.

PACS: 41.60.Cr; 07.85.Fv

Keywords: Free electron laser; HGHG; SASE

1. Introduction

Based on the experience with the third-generation light-source BESSY II delivering high-brilliance photon beams in the VUV to XUV spectral range, BESSY proposes a linac-based free electron laser multi user facility. The scientific case for the BESSY soft X-ray FEL was published in 2001 [1]. The target photon energy range is 20 eV to 1 keV (62 nm $\geqslant \lambda \geqslant$ 1.2 nm) with a peak-brilliance of

about 10^{31} photons/s/mm²/mrad²/0.1% *BW*, i.e. a peak power of up to a few *GW* for pulse lengths less than 100 fs (rms). For a SASE-FEL the duration and power of the radiation pulse are basically determined by the properties of the electron beam. Among other parameters such as beam energy and emittance, the radiation pulse duration depends in particular on the electron pulse length and peak current. The output of a SASE-FEL has a spiky structure in time and frequency, since the electron beam amplifies its own initial spontaneous emission which is inherently stochastic. Using a coherent seed with sufficient initial power to dominate the spontaneous emission in the beginning of the process, the spiky structure can be overcome. In this case the

☆ Funded by the Bundesministerium für Bildung, Wissenschaft, Forschung und Technologie (BMBF), the Land Berlin and the Zukunftsfonds des Landes Berlin.

*Corresponding author.

E-mail address: meseck@bessy.de (A. Meseck).

properties of the seed determines the properties of the amplified radiation. Thus the final pulse duration and pulse shape are dependent basically on the pulse duration and shape of the seed. Conventional lasers however cannot be used as a seed source in the X-ray regime, since they still do not deliver the necessary short wavelengths at the needed power level of a few hundred MW. Therefore, techniques which allow to produce coherent short wavelengths are of major interest. High-gain harmonic generation is a promising method to produce coherent short wavelength from a long wavelength laser seed. In this paper, we present simulation studies for SASE-FELs and for possible multistage HGHG FELs and discuss their performance in terms of pulse power, pulse duration, pulse shape and signal-to-noise ratio. We use the time-dependent 3D simulation code GENESIS [5] for our studies. For simulation of the HGHG-FELs the output routine of GENESIS has been modified to enable harmonic generation calculations.

2. The BESSY-SASE-FELs

To cover the photon energy range from 20 eV to 1 keV three undulator-lines are forseen. The so called "low-energy" and "medium-energy" FELs will operate at an electron beam energy of 1.875 GeV and will deliver photons with energies ranging from 20 to 300 eV and 250 to 550 eV, respectively. The "high-energy" FEL operates at variable electron beam energies of 1.5 to 2.25 GeV and delivers photon energies of 550–1000 eV. The undulator-lines are of a typical length of up to 60 m including the intersection for the superposed FODO-lattice, phase shifter and diagnostic devices. The length of the undulator modules will be about 3.5 m. The average betafunctions amounts to about 10 m. The nominal electron beam parameters at the entrance of the undulators are summarized in Table 1. Even though start to end simulations [3] predict a peak current of about 6 kA at the entrance of the undulators the present simulations assumed a conservative value for the peak current of 3.5 kA.

Table 1
The nominal electron beam parameters for BESSY SASE-FELs at the entrance of the undulators used in the simulation studies

Parameter	High-energy	Medium-/low-energy
ε_n (mm mrad)	1.5	1.5
I_{peak} (kA)	3.5	3.5
E (GeV)	1.5–2.25	1.875
$\Delta E/E$ (%)	0.04–0.07	0.05
σ_t (fs) (rms)	120	120

A summary of undulator parameters and FEL performance for the BESSY-SASE-FELs can be found in Refs. [2,4].

The output power of the high-energy FEL as a function of the time for the shortest wavelength is shown in Fig. 1a. The spiking in time as well as in spectral domain, shown in Fig. 1b, is typical for SASE-FELs. The spiky radiation pulse can be fitted with a Gaussian function. The pulse length of about 50 fs (rms) is short compared to the electron bunch length ($\sigma_t = 120$ fs). The spectral bandwidth amounts to 0.1% corresponding to the ρ-parameter for BESSY soft X-ray FELs.

The pulse shape is dominated by the Gaussian shape of the electron current distribution used in the simulations. The signal to spontaneous radiation ratio is better than four orders of magnitude which is sufficient for most of the user experiments. Note that for the simulation studies the beam parameter and undulator field errors were not taken in to account. They may deteriorate the signal-to-noise ratio [6].

3. HGHG scenario

In the HGHG scheme a first undulator modulates the electron beam energy on the seed frequency while a following dispersion section converts this modulation in to spatial bunching optimized for a particular harmonic. A second undulator which is set in resonance to the harmonic acts as a radiator. Cascading such modulator–dispersion–radiator-structures allows to achieve very short wavelengths from a long-wavelength seed. A theoretical treatment of HGHG can be found for example by YU [7,8].

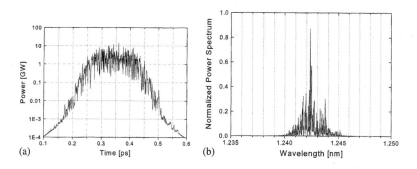

Fig. 1. Time-dependent GENESIS simulation for the shortest wavelength. (a) The time-resolved power distribution and (b) the spectral power distribution are calculated.

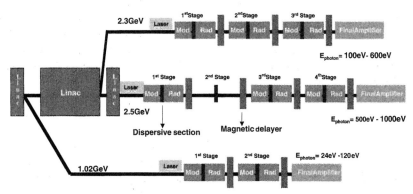

Fig. 2. Schematic view of the HGHG-scenario for the multi user FEL-facility. To cover the target wavelength range of BESSY we need three HGHG-lines. The seed laser is tunable covering the spectral range of 230–460 nm with Gaussian profile, a peak power of 500 MW and a pulse length of about 14 fs (rms).

An efficient coherent harmonic generation however requires an energy spread of the electron beam smaller than the Pierce parameter otherwise the growth rate of the radiation is significantly reduced. In a multistage HGHG-FEL the energy spread of the electron beam increases from stage to stage due to the energy modulation caused by the seed. A promising method to overcome this problem is the "fresh bunch technique" described by Ref. [7].

For the HGHG-scenario of a multi user FEL-facility we assume a tunable seed laser covering the spectral range of 230 to 460 nm with Gaussian profile, a peak power of 500 MW and a pulse length of about 14 fs (rms). To cover the target wavelength range of BESSY we need three HGHG-lines, as shown in Fig. 2. The "low-energy" HGHG-FEL operates in two stages at a beam energy of 1.02 GeV delivering photons in a spectral range of 24–120 eV. An energy of 2.30 GeV is chosen for the "medium-energy" HGHG-FEL. A cascade of three stages covers the energy range of 100–600 eV while the "high-energy" four-stage HGHG-FEL operates at an energy of 2.5 GeV. The delivered photon energy ranges from to 500–1000 eV.

For the simulation studies a box-shaped electron current distribution is assumed with an average current of 1.75 kA and a pulse length of about 600 fs. The normalized transverse emittance and relative energy spread assumed to be 1.5 mm mrad and about 0.02%, respectively. In this case the typical length of the modulators is about 2 m, while the radiator lengths vary between

3.5 and 7 m. The final amplifier following the last HGHG stage reaches saturation within an undulator length of about 10 m. We use the "fresh bunch technique", i.e. magnetic delayers position the seed radiation between HGHG-stages at the undisturbed part of the bunch. GENESIS offers the possibility to generate output files for the radiation field and the electron distribution at the end of the undulator. These files can be used as input files for the next GENESIS run. Using these features and shifting the particle distribution to higher harmonics, time-dependent GENESIS runs for HGHG FELs including all magnetic elements were performed.

The laser seed interacts with electrons at the rear of the bunch. Due to the slippage effect only part of the interacting electrons experience the full power of the Gaussian-shaped seed. Optimizing the modulator length and dispersive section to a somewhat reduced power level the bunching on the harmonics can be enhanced. Thus the electrons in the middle, which experience the full power long enough are somewhat overbunched. This causes a power dip in the radiation pulse provided by the first radiator, as shown in Fig. 3a. Fig. 3b shows the radiation spectrum of the first radiator. For a Gaussian seed the optimal bunching on the high harmonics is box shaped [8]. The convolution of the box-shaped bunching and the frequency causes the side spikes in the spectrum. The side spikes can be avoided by optimizing the first stage for the seed peak power. In this case the bunching factor of the harmonic is reduced. The bunching is of more Gaussian shape. The resulting radiation power and pulse length are reduced compared to the box shaped bunching.

Fig. 4 shows the final output power and spectrum for the low energy HGHG FEL. The

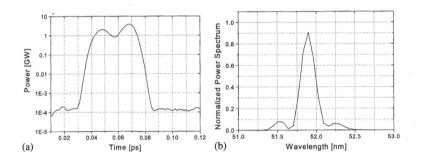

Fig. 3. Time-dependent GENESIS simulation for first radiator. (a) The time-resolved power distribution and (b) the spectral power distribution are calculated.

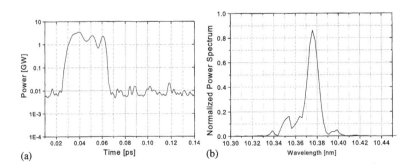

Fig. 4. Time-dependent GENESIS simulation for low energy HGHG FEL. (a) The time-resolved power distribution and (b) the spectral power distribution were calculated for a final wavelength of 10.37 nm. The seed wavelength was 259 nm.

initial wavelength of 259 nm is transferred to a final wavelength of 10.37 nm after two stages. The pulse length of 33 fs is larger than the seed length of 30 fs due to slippage effect, while the output power reaches a few *GW*. The slippage shifts the side spikes to one side of the spectrum.

The Penman algorithm [9] used in GENESIS code for the generation of the initial shot noise statistics is only valid for the fundamental frequency. Therefore, the simulated noise level is overestimated by 1 to 2 orders of magnitude compared to analytical and numerical estimates [10,11].

4. Conclusion

While the SASE radiation has a spiky structure in time and frequency and suffers from its own stochastic character, the HGHG radiation benefits from the seed properties and delivers photon beams superior to the SASE case. Especially the output pulse length is mainly controlled by the seed. However, the HGHG scheme requires a more complex and demanding technical solution than a SASE FEL. The seed power has to be large enough to overcome the spontaneous emission. The initial energy spread has to be smaller than the Pierce parameter for the final harmonic. The time jitter between laser pulse and electron bunch has to be minimized. Further studies are in progress to decide on an optimal FEL scheme.

Acknowledgements

We would like to thank L.H. Yu for very fruitfull discussions concerning the HGHG scheme and optimization.

References

[1] Vision of Science, The BESSY SASE-FEL in Berlin Adlershof, BESSY 2001.
[2] D. Krämer, Closing in on the design of the BESSY-FEL, Proceedings of PAC 2003, Portland.
[3] M. Abo-Bakr, et al., Start to end simulations the for BESSY FEL, Nucl. Instr. and Meth. A, (2004) these proceedings.
[4] J. Bahrdt, et al., Undulators for the BESSY-SASE-FEL project, Proceedings of SRI 2003, San Francisco, Nucl. Instr. and Meth. A, (2004) to be published.
[5] S. Reiche, GENESIS 1.3, Nucl. Instr. and Meth. A 429 (1999) 243.
[6] B. Kuske, et al., Tolerance studies for the BESSY FEL undulators, Nucl. Instr. and Meth. A, (2004) these proceedings.
[7] L. Yu, I. Ben-Zvi, Nucl. Instr. and Meth. A 393 (1997) 96.
[8] L. Yu, J. Wu, Nucl. Instr. and Meth. A 483 (2002) 493.
[9] C. Penman, B.W.J. McNeil, Opt. Commun. 90 (1992) 82.
[10] S. Reiche, Private communication.
[11] E.L. Saldin, et al., Opt. Commun. 202 (2002) 169.

Available online at www.sciencedirect.com

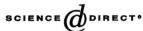

ELSEVIER   Nuclear Instruments and Methods in Physics Research A 528 (2004) 582–585

NUCLEAR
INSTRUMENTS
& METHODS
IN PHYSICS
RESEARCH
Section A

www.elsevier.com/locate/nima

0.3-nm SASE-FEL at PAL

J.S. Oh*, D.E. Kim, E.S. Kim, S.J. Park, H.S. Kang, T.Y. Lee, T.Y. Koo,
S.S. Chang, C.W. Chung, S.H. Nam, W. Namkung

*Pohang Accelerator Laboratory, Pohang University of Science and Technology, San 31, Hyoja-dong, Nam-gu, Pohang 790-784,
Republic of Korea*

Abstract

Recent success in SASE-FEL experiments provides high confidence in achieving FEL radiation wavelengths as short as 0.1 nm. The SASE-XFEL requires multi-GeV electron beams with extremely low-emittance, short bunches, and long undulator systems. PAL is operating a 2.5-GeV electron linac as a full-energy injector to the PLS storage ring. With a proposed energy upgrade to 3.0 GeV and an in-vacuum undulator, PAL will be able to produce coherent X-ray radiation at wavelengths as short as 0.3 nm. This paper presents the preliminary design details for the proposed PAL-XFEL. The required undulator period is 12.5 mm with a 3-mm gap. The third harmonic enhancement technique can be used to obtain radiation wavelengths of 0.1 nm. The technical parameters related to these goals are reviewed.
© 2004 Elsevier B.V. All rights reserved.

PACS: 41.60.Cr

Keywords: PAL; X-ray SASE; In-vacuum undulator; 0.3 nm; Third-harmonic enhancement

1. Introduction

The last two decades have ushered in rapid progress in basic and applied science research using synchrotron radiation and technological advancements in extremely intense electron beams for high-energy physics research. These result in both strong demands for, as well as the possibilities of, fully coherent radiation with high brilliance, called fourth-generation light sources. LCLS [1] at SLAC and TESLA-XFEL [2] at DESY are in the detailed design stages with

government funding approval. These SASE-XFEL facilities are expected to be operational by 2008 and 2012, respectively. Those projects require large-scale accelerators with 14–20-GeV electron beams and very long undulator systems with 113–175-m lengths.

We propose a compact X-ray Free Electron Laser Program at the Pohang Accelerator Laboratory, PAL-XFEL. PAL operates an electron linac, the third largest in the world, 2.5 GeV at present, as a full-energy injector to the PLS storage ring [3]. The linac is regularly injecting these beams to a storage ring twice a day for up to 5 min per injection. In its design stage, PAL was designed to use multi-application beams concurrently. With

*Corresponding author.
E-mail address: jsoh@postech.ac.kr (J.S. Oh).

the proposed upgrade modification for the beam energy of 3.0 GeV and an in-vacuum undulator system, it will be possible to produce coherent X-ray radiation as short as 0.3 nm. To obtain radiation wavelengths of 0.1 nm, we propose to employ the third-harmonic enhancement technique that will utilize an additional undulator system with a shorter periodic length. The design goal is for the undulator to be less than 60 m in total length.

2. PAL X-FEL

The fundamental radiation wavelength λ_x of an undulator is given by

$$\lambda_x = \frac{\lambda_u}{2\gamma^2}(1 + K^2/2)$$

where $\gamma = E_0/0.511$, $K = 0.934 B_u \lambda_u$, E_0 is the beam energy in MeV, B_u is the peak magnetic field of the undulator in Tesla, and λ_u is the undulator period in cm. Either a short-period undulator or a high-energy beam can provide short-wave radiation. However, the value of K should be reasonably large to obtain a short saturation length. An in-vacuum mini-gap undulator can meet this requirement. This concept was introduced by the SCSS project at SPring-8 [4].

We will use a 45° magnetized undulator with $H = \lambda_u/2$. The peak magnetic field on axis is calculated by

$$B_y^{peak} = \frac{4\sqrt{2}B_r}{\pi} \sum_{n=1,5,9}^{\infty} \frac{1}{n}(1 - e^{-2n\pi H/\lambda_u})e^{-n\pi g/\lambda_u}$$

where B_r is assumed to be 1.19 T provided by $Nd_2Fe_{14}B$ magnets, H is the block height, g is full-gap length as shown in Fig. 1 [5].

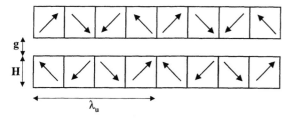

g
H
λ_u

Fig. 1. Undulator geometry with 45° magnetization.

Nominal beam parameters for the PAL X-FEL linac are summarized in Table 1. The emittance can be further optimized by reducing the bunch charge and peak current, which provides the same saturation length with a moderate power reduction.

Fig. 2 shows the saturation lengths and beam energies for 0.3-nm SASE as a function of undulator period and gap size, which are calculated by the use of equations given by Xie [6]. This figure clearly shows that the undulator must have a period of less than 1.3 cm and a gap size of less than 3 mm to limit the saturation length within 50 m for a 3-GeV beam energy.

Table 2 summarizes key parameters of the undulator for a 0.3-nm PAL X-FEL. The undulator beta value is optimized to obtain as short a saturation length as possible. The result shows that we need 13 undulators, each 4.5-m long. Table 3 lists the X-ray radiation parameters. Due to the

Table 1
Beam parameters for PAL X-FEL

Beam energy (GeV)	3.0
Normalized emittance (μm-rad)	1.5
Peak current (kA)	4.0
Bunch charge (nC)	1.0
FWHM bunch length (fs)	235
Energy spread (%)	0.02

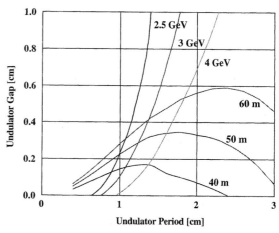

Fig. 2. Undulator period vs. gap length for 0.3-nm SASE with saturation lengths of 40, 50, and 60 m and beam energies of 2.5, 3, and 4 GeV.

Table 2
Undulator parameters for 0.3-nm XFEL

Period (mm)	12.5
Gap (mm)	3.0
Peak magnetic field (T)	0.97
Undulator parameter, K	1.14
Beta (m)	15
Saturation length (m)	52

Table 3
X-ray radiation parameters

FEL parameter	0.00043
1D gain length (m), L_{1d}	1.35
3D gain length correction, η^*	0.97
Gain length (m), L_g	2.67
Peak power (GW)	2.1
Peak brightness ($\times 10^{32}$)**	1.4

*$L_g = (1 + \eta)L_{1d}$.
**Photons/s-mm^2-mrad^2 -0.1%BW.

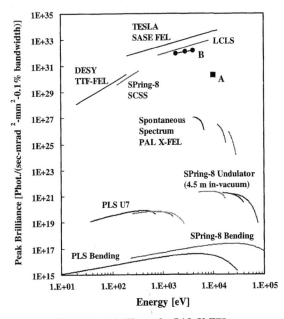

Fig. 3. Peak brilliance for PAL X-FEL.

rather higher emittance, the 3D gain length is twice the 1D gain length.

The solid line 'B' in Fig. 3 shows the expected peak brilliance for the 0.3-nm PAL X-FEL. The three circles on the line correspond to beam energies of 2.0, 2.5, and 3.0-GeV. 'A' denotes the 0.1-nm PAL X-FEL with the third-harmonic enhancement technique employing an additional undulator with a shorter periodic length, as shown in Table 4.

Fig. 4 shows the sensitivity of saturation length on the system performance. The system parameters such as beam emittance, energy spread, bunch length, and undulator beta are normalized by the nominal values given in Table 1. The most sensitive parameter is the emittance of the electron beam for PAL X-FEL. Therefore, the prime effort has to be concentrated on the design of a low emittance gun. Fig. 5 shows a possible upgrade layout of the PAL linac including a new 0.5-GeV injector. The injector consists of a low-emittance laser-driven photocathode-gun, three S-band accelerating modules (X1, X2, X3), and a bunch compressor BC1.

Table 4
Undulator parameters for 0.1-nm XFEL

Period (mm)	6.5
Gap (mm)	3.0
Peak magnetic field (T)	0.5
Undulator parameter (K)	0.3
Beta (m)	30
Undulator length (m)	23

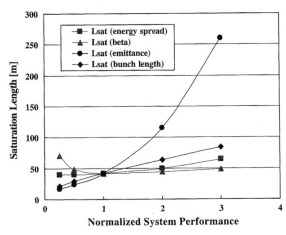

Fig. 4. Saturation length vs. system performance for beam emittance, energy spread, bunch length, and undulator beta.

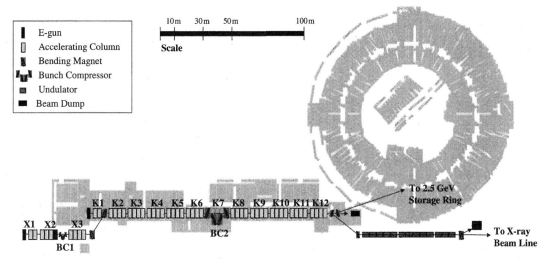

Fig. 5. PAL X-FEL compression and acceleration scheme.

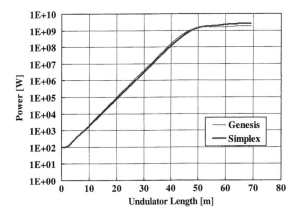

Fig. 6. Saturation curves of X-ray SASE at PAL linac.

12.5 mm with a 3-mm gap. To obtain the 0.1-nm radiation wavelength, we propose to employ the third harmonic enhancement technique utilizing an additional undulator system with a shorter periodic length of 6.5 mm.

Acknowledgements

This work is supported by POSCO and the Ministry of Science and Technology (MOST) of Korea.

The analytic design is confirmed by simulation codes. The FODO layout with a focusing strength of 18 T/m (F) and 12.5 T/m (D) is assumed in the simulation. Fig. 6 shows the saturation curves for the 0.3-nm PAL X-FEL by SIMPLEX [7] and GENESIS [8].

3. Summary

With an upgrade modification for the beam energy of 3.0 GeV and an in-vacuum undulator system, the PAL linac will be able to produce coherent X-ray radiation down to 0.3 nm. The undulator period required to obtain this is

References

[1] LCLS Conceptual Design Report, SLAC-R-593, 2002.
[2] TESLA-XFEL Technical Design Report, Supplement, DESY 2002-167, TESLA-FEL 2002-09, 2002.
[3] http://pal.postech.ac.kr/.
[4] T. Shintake, et al., SPring-8 compact SASE source (SCSS), in: Proceedings of the SPIE2001, San Diego, USA, 2001: http://www-xfel.spring8.or.jp.
[5] X. Maréchal, J. Chavanne, P. Elleaume, On 2D periodic magnetic field, ESRF-SR/ID 90-38, 1990.
[6] M. Xie, Design Optimization for an X-ray Free Electron Laser Driven by SLAC Linac, LBL Preprint No-36038, 1995.
[7] http://radiant.harima.riken.go.jp/simplex/; Takashi Tanaka, The Institute of Physical and Chemical Research.
[8] S. Reiche, Nucl. Instr. and Meth. A 429 (1999) 243.

Available online at www.sciencedirect.com

SCIENCE @DIRECT®

Nuclear Instruments and Methods in Physics Research A 528 (2004) 586–590

NUCLEAR
INSTRUMENTS
& METHODS
IN PHYSICS
RESEARCH
Section A

www.elsevier.com/locate/nima

ELSEVIER

Status of the SPARC project

D. Alesini[a], S. Bertolucci[a], M.E. Biagini[a], C. Biscari[a], R. Boni[a], M. Boscolo[a],
M. Castellano[a], A. Clozza[a], G. Di Pirro[a], A. Drago[a], A. Esposito[a], M. Ferrario[a],
V. Fusco[a], A. Gallo[a], A. Ghigo[a], S. Guiducci[a], M. Incurvati[a], C. Ligi[a],
F. Marcellini[a], M. Migliorati[a], C. Milardi[a], A. Mostacci[a], L. Palumbo[a],
L. Pellegrino[a], M. Preger[a], P. Raimondi[a], R. Ricci[a], C. Sanelli[a], M. Serio[a],
F. Sgamma[a], B. Spataro[a], A. Stecchi[a], A. Stella[a], F. Tazzioli[a], C. Vaccarezza[a],
M. Vescovi[a], C. Vicario[a], M. Zobov[a], F. Alessandria[b], A. Bacci[b], I. Boscolo[b],
F. Broggi[b], S. Cialdi[b], C. DeMartinis[b], D. Giove[b], C. Maroli[b], M. Mauri[b],
V. Petrillo[b], M. Romè[b], L. Serafini[b,*], D. Levi[c], M. Mattioli[c], G. Medici[c],
L. Catani[d], E. Chiadroni[d], S. Tazzari[d], R. Bartolini[e], F. Ciocci[e], G. Dattoli[e],
A. Doria[e], F. Flora[e], G.P. Gallerano[e], L. Giannessi[e], E. Giovenale[e], G. Messina[e],
L. Mezi[e], P.L. Ottaviani[e], S. Pagnutti[e], L. Picardi[e], M. Quattromini[e], A. Renieri[e],
C. Ronsivalle[e], A. Cianchi[f], A.D. Angelo[f], R. Di Salvo[f], A. Fantini[f],
D. Moricciani[f], C. Schaerf[f], J.B. Rosenzweig[g]

[a] INFN-LNF, Via E.Fermi 40, 00044 Frascati (RM), Italy
[b] Università di Milano e INFN-Milano, Via Celoria 16, 20133 Milano, Italy
[c] INFN-Roma I, University of Roma "La Sapienza", p.le A. Moro 5, 00185 Roma, Italy
[d] INFN-Roma II, University of Roma "Tor Vergata", Via della Ricerca Scientifica 1, 00133 Roma, Italy
[e] ENEA-FIS, Via E. Fermi 45, 00040 Frascati (RM), Italy
[f] University of Roma "Tor Vergata", Via della Ricerca Scientifica 1, 00133 Roma, Italy
[g] UCLA—Department of Physics and Astronomy, 405 Hilgard Avenue, Los Angeles, CA 90024, USA

Abstract

In this paper we report on the final design of the SPARC FEL experiment which is under construction at the Frascati INFN Laboratories by a collaboration between INFN, ENEA, ELETTRA, Un. of Rome (Tor Vergata), CNR and INFM. This project comprises an advanced 150 MeV photo-injector aimed at producing a high brightness electron beams to drive a SASE-FEL experiment in the visible using a segmented 12 m long undulator. The project, finally approved and funded early this year, has a 3 year time span, with the final goal of reaching saturation on the fundamental of the SASE-FEL and studying the resonant non-linear generation of harmonics. Peculiar features of this project are the optimized design of the photo-injector to reach minimum emittances by using flat-top laser pulses on the

*Corresponding author.
E-mail address: luca.serafini@mi.infn.it (L. Serafini).

photocathode, and the use of an uncompressed electron beam of 100 A peak current at very low emittance to drive the FEL. Results of start-to-end simulations carried out to optimize the performances of the whole system are presented, as well as the status of the construction and assembly of the system components. Activities planned for a second phase of the project are also mentioned: these are mainly focused on velocity bunching experiments that will be conducted with the aim to reach higher peak currents with preservation of low transverse emittances.

PACS: 41.60.Cr

Keywords: Photo-injector; High brightness electron beam; SASE-FEL; Non-linear resonant higher harmonics

1. Introduction

The overall SPARC Project consists of 4 main lines of activity. A *150 MeV Advanced Photo-Injector* aimed at investigating the generation of high brightness electron beams and their compression via magnetic and/or velocity bunching. This beam will be used to drive a *SASE-FEL Visible-VUV Experiment*: this is aimed to investigate the problems related to the beam matching into a segmented undulator and the alignment with the radiation beam at 500 nm, as well as the generation of non-linear coherent higher harmonics. In parallel, R&D activities are pursued at different sites on *X-ray Optics and Monochromators*, to analyze radiation-matter interactions in the spectral range of SASE X-ray FELs. Studies of *Soft X-ray table-top Sources* are also part of the SPARC program, with an anticipated upgrade of the present compact source at INFM-Politecnico Milano, delivering 10^7 soft X-ray photons in 10–20 fs pulses by means of high harmonic generation in a gas. In the following we present an overview of the system under construction at the Frascati National Laboratories of INFN, aiming at reaching the scientific and technological goals indicated in the first two topics listed above.

A 3D model of the whole system is illustrated in Fig. 1: the photo-injector, the FEL undulator, the beam dump and the undulator by-pass beam line

are hosted inside a dedicated underground bunker which is 36 m long and 14 m wide.

The 150 MeV photo-injector consists of a 1.6 cell RF gun operated at S-band (2.856 GHz, of the BNL/UCLA/SLAC type) and high-peak field on the cathode (120 MV/m) with incorporated metallic photo-cathode (Cu or Mg), generating a 6 MeV beam [1]. The beam is then focused and matched into 3 SLAC-type accelerating sections, which boost its energy up to 150–200 MeV. The first section is embedded in solenoids in order to provide additional magnetic focusing to better control the beam envelope and the emittance oscillations [2]. The photo-cathode drive laser is a Ti:Sa system with the oscillator pulse train locked to the RF. To perform temporal flat top laser pulse shaping we will manipulate frequency lines in the large bandwidth of Ti:Sa, either using a liquid crystal mask in the Fourier plane for nondispersive optic arrangement or a collinear acousto-optic modulator [3] (DAZZLER). We aim achieving a pulse rise time shorter than 1 ps.

The photo-injector design is by now completed and is reported in a dedicated TDR [4] with full specification of each system component: all bids for acquisition of main components have been so far launched, so we expect to be on schedule with delivery of RF gun, laser system, RF sources and linac accelerating sections. The expected start of installation for the photo-injector components is

Fig. 1. A map view of the SPARC photo-injector and FEL undulator systems: from left to right, the RF gun, the first linac section embedded in solenoids, the two additional linac sections, the 6 undulator sections, the beam dump and the undulator by-pass beam line are visible.

confirmed for spring of 2005. The first beam at full energy is expected by the beginning of 2006.

2. Start-to-end simulations for the FEL experiment

A basic choice of SPARC phase 1 is to drive the FEL with an uncompressed beam: it was a decision by the project group to postpone the study of magnetic compression and velocity bunching to phase 2 of the project, as explained in Section 3. As a consequence, in order to make the FEL saturate with the natural beam current produced by the photo-injector one has to deliver at the undulator entrance a very high-quality (brightness) beam, i.e. with very low rms slice emittance and energy spread. To this aim we designed a photo-injector based on the Ferrario working point [2] lay-out to achieve full emittance compensation at the linac exit, where the emittance is no longer sensitive to envelope oscillations [5]. The selected beam parameters are listed in Table 1: they slightly differ from those of a previous analysis [6] because of the desire to keep a reasonable FEL saturation length with a bunch charge close to 1 nC. A peak current in excess of 100 A along a substantial fraction of the bunch is anticipated, despite the slight debunching caused in the gun to linac drift by the longitudinal space charge field. By properly matching the beam into

Table 1
SPARC FEL experiment parameter list

Electron beam energy (MeV)	155
Bunch charge (nC)	1.1
Cathode peak field (MV/m)	120
Laser radius spot size on the cathode (mm, hard edge)	1.13
Laser pulse duration, flat top (ps)	10
Laser pulse rise time (ps) 10% → 90%	1
Bunch peak current at linac exit (A)	100
Rms norm. transv. Emittance at linac exit (µm); includes thermal comp. (0.3)	<2
Rms slice norm. emitt. (300 µm slice)	<1
Rms uncorrelated energy spread (%)	0.06
Undulator period (cm)	2.8
Undulator parameter, K	2.14
FEL radiation wavelength (nm)	499
Average beta function (m)	1.52
Expected saturation length (m)	<12

the linac, set for on-crest acceleration at a gradient of 25 MV/m, the final rms normalized transverse emittance is minimized at the linac exit (155 MeV), with a nominal value of 0.75 µm as predicted by simulations: the corresponding slice emittance is about 0.6 µm over a substantial fraction of the bunch. A detailed analysis of errors and misalignments [4] in the system leads us to evaluate an upper limit for these two quantities as reported in Table 1, i.e. 2 µm for the rms normalized emittance and 1 µm for the slice emittance. The behavior of relevant beam parameters over the bunch slices is shown in Fig. 2: the energy spread is well below 0.06% for all slices and the current is above 100 A for 54% of the bunch slices. Matching this beam into the segmented undulator (parameters listed in Table 1) requires the use of two triplets in the transfer line in order to reduce the beta function of the beam down to about 1.5 m (average): to assure focusing in the horizontal plane along the undulator, single quadrupoles are located in undulator drift sections. Start-to-end simulations were performed using PARMELA [7] and GENESIS [8]: the result on FEL performances is shown in Fig. 3, where the FEL power growth along the undulator is plotted. Saturation power is reached at nearly 10^8 W in about 12 m of total length including drift sections, leaving some margin of extra undulator length. This comes out to be

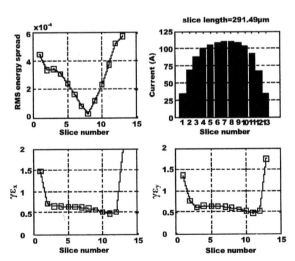

Fig. 2. Computed beam slice parameters at the photo-injector exit (energy spread, current, rms normalized emittance in x and y planes): the slice thickness is about 300 µm.

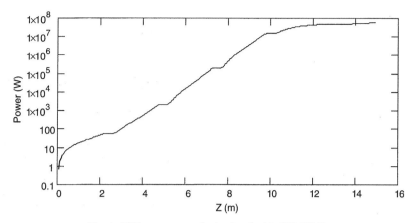

Fig. 3. FEL power growth simulated with GENESIS.

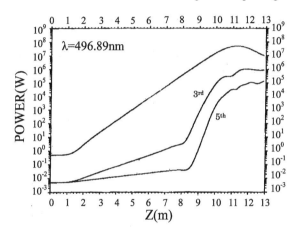

Fig. 4. FEL power growth on fundamental, third and fifth harmonic.

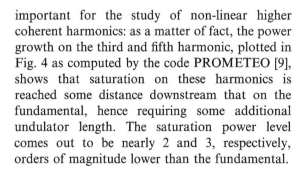

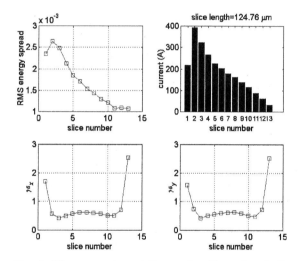

Fig. 5. Slice parameters at linac exit with weak velocity bunching.

important for the study of non-linear higher coherent harmonics: as a matter of fact, the power growth on the third and fifth harmonic, plotted in Fig. 4 as computed by the code PROMETEO [9], shows that saturation on these harmonics is reached some distance downstream that on the fundamental, hence requiring some additional undulator length. The saturation power level comes out to be nearly 2 and 3, respectively, orders of magnitude lower than the fundamental.

3. Phase 2 of the project

As previously mentioned, phase 1 of the SPARC Project was restricted to the use of uncompressed beams to drive the FEL experiment, mainly because of budget limitation issues. Full implementation of beam compression is foreseen in SPARC phase 2. As illustrated in Fig. 1, the second and third accelerating sections in the linac are not embedded in solenoids: in phase 2, expected to start in 2006, we plan to add solenoids to these sections in order to provide additional focusing along all the linac with flexible magnetic field profile. This is one of the most relevant demands of the velocity bunching technique [10]: it is needed to fully control envelope oscillations during compression, which in turns determine the minimum emittance achieved. A magnetic compressor will also be installed on the second beam line to study CSR effects and emittance degradation issues. An example of possible performances that can be obtained by applying velocity bunching to the beam and by properly using the additional solenoid focusing provided in the last two sections, is shown in Fig. 5. By injecting the beam into the first section at $-83°$RF ($0°$RF corresponds to on-crest acceleration) at 25 MV/m, the beam current is raised up to 200 A for 40% of the bunch slices, with a slice at 400 A, while the slice emittance is kept below 0.6 μm for most slices (total rms normalized emittance is about 1.4 μm). We plan to perform a dedicated experimental investigation for characterizing the full potentiality of the velocity bunching technique.

References

[1] D.T. Palmer, The next generation photoinjector, Ph.D.Thesis, Stanford University.

[2] M. Ferrario, et al., Homdyn Study for the LCLS RF Photoinjector, SLAC-PUB 8400.

[3] F. Verluise, et al., Opt. Lett. 25 (2000) 572. See also S. Stagira, Proceedings of the Workshop on Laser Issues for Electron RF Photoinjectors, SLAC-WP-025, 2002.

[4] L. Palumbo, J.B. Rosenzweig (Eds.), Technical Design Report for the SPARC Advanced Photo-Injector, in publication.

[5] L. Serafini, J.B. Rosenzweig, Phys. Rev. E 55 (1997) 7565.

[6] D. Alesini, et al., Nucl. Instr. and Meth. A 507 (2003) 345.

[7] L.M. Young, J.H. Billen, Parmela, LA-UR-96–1835, 2000.

[8] S. Reiche, Nucl. Instr. and Meth. A 429 (1999) 243.

[9] G. Dattoli, et al., ENEA Report RT/INN/93/09 1993.

[10] L. Serafini, M. Ferrario, AIP Conf. Proc. 581 (2001) 87.

Available online at www.sciencedirect.com

SCIENCE @ DIRECT®

ELSEVIER   Nuclear Instruments and Methods in Physics Research A 528 (2004) 591–594

NUCLEAR
INSTRUMENTS
& METHODS
IN PHYSICS
RESEARCH
Section A

www.elsevier.com/locate/nima

The Shanghai high-gain harmonic generation DUV free-electron laser

Z.T. Zhao[a],*, Z.M. Dai[a], X.F. Zhao[a], D.K. Liu[a], Q.G. Zhou[a], D.H. He[b], Q.K. Jia[b], S.Y. Chen[c], J.P. Dai[c]

[a] SSRF, Shanghai Institute of Applied Physics, Shanghai 200800, China
[b] NSRL, University of Science and Technology of China, Hefei 230029, China
[c] Institute of High Energy Physics, Beijing 100084, China

Abstract

The Shanghai deep ultraviolet free-electron laser source (SDUV-FEL) is an HGHG FEL facility designed for generating coherent output with wavelength down to 88 nm. The design and the relevant R&D of this HGHG FEL source have been under way since 2000. Currently, a 150 MeV S-band electron injector is under construction as the first linac section to produce a high brightness beam. The design study and the present R&D status of the SDUV-FEL have been presented in this paper.
© 2004 Elsevier B.V. All rights reserved.

PACS: 41.60.Cr; 52.75.Ms

Keywords: High-gain harmonic generation; Free-electron laser; Deep ultraviolet

1. Introduction

The Shanghai deep ultraviolet free-electron laser source (SDUV-FEL) is a single-pass FEL facility designed for generating coherent radiation in the spectral region from 500 to 88 nm. It consists of a 300 MeV linac for producing a high brightness electron source and an undulator section with a length of about 10 m for FEL interaction. The linac is composed of a low emittance electron injector with photo-cathode RF gun, S-band main linac sections, and bunch compressors. For stabilization of the FEL process, the HGHG seeding scheme [1,2] will be implemented in the SDUV-FEL. Further, the undulator section includes a modulator undulator, a dispersive section, and a radiator undulator.

The design and the relevant R&D of the SDUV-FEL source have been under way since 2000. Fig. 1 shows the optimized layout of SDUV-FEL facility. The construction of the SDUV-FEL building for housing the accelerator, FEL, and the experimental equipment was started in 2002, and it will be completed in November, 2003. For the time being, a 150 MeV S-band electron injector

*Corresponding author.
E-mail address: zhaozt@ssrc.ac.cn (Z.T. Zhao).

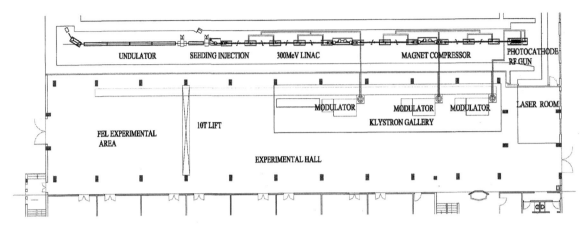

Fig. 1. Layout of SDUV-FEL facility.

is under construction as the first linac section, which is planned to be commissioned by the end of next year. Based on the high brightness electron beam delivered from the 150 MeV injector, it is expected that the SASE FEL and/or seeded FEL experiments will be carried out in the spectral region from 500 to 264 nm from the beginning of 2005.

2. Design study of the SDUV HGHG FEL

In the SDUV HGHG FEL, a seed laser at a wavelength of 264 nm is injected into the first undulator for energy modulation, in which the electron beam is forced to be resonant with the seed laser. Following this modulation, a dispersive section is traversed, in which spatial bunching is induced with a strong higher harmonic content. The modulated electron beam then enters a second undulator, the radiative section, which is tuned to be resonant to a wavelength of 88 nm, the third harmonic of the seed laser. The main parameters of the SDUV HGHG FEL are summarized in Table 1.

To optimize the system, we first optimized the radiator undulator to obtain minimum power e-folding length, L_G at 88 nm. Fig. 2 shows the dependence of the gain length of the FEL at 88 nm on the undulator period and matched beta

function for the planar permanent magnet undulator. This dependence was obtained using the three-dimensional analytical formulae of Ref. [3] for high-gain FEL. Fig. 2 indicates that the design values of 2.5 cm for the undulator period and 2–4 m for the beta function are close to the optimal.

A FODO focusing scheme with quadrupoles installed in the drift space between the undulator sections is introduced in the radiator undulator of the SDUV-FEL. The influence of the section length, drift length between undulator sections, the focusing strength on the FEL gain length, and the saturation power have been studied extensively using the GENESIS 1.3 [4]. The optimal gain length simulated by GENESIS 1.3 is about 0.9 m, which is approximately equal to the analytical result.

The HGHG process is studied using the TDA3D code [5] to determine the parameters of the modulator undulator, dispersive section, radiator undulator, and the seed laser. Fig. 3 illustrates the evolution of HGHG FEL at a wavelength of 88 nm for different slice energy spreads of the electron beam. It indicates that a small slice energy spread is crucial to the HGHG process.

The sensitivity of the FEL performance on electron beam parameters, such as the current, emittance, energy spread, and undulator errors, has also been studied.

Table 1
Main parameters of the SDUV-FEL

FEL parameters	
Wavelength (nm)	88
Gain length L_G (m)	~0.8
Output power (MW)	~100
Electron beam parameters	
Energy (MeV)	276
Bunch charge (nC)	1
Peak current (A)	400
Normalized emittance (mm-mrad)	4
Local energy spread (%)	0.1
Seed laser parameters	
Wavelength (nm)	264
Input power (MW)	20–50
Rayleigh range (m)	0.8
Modulator undulator	
Length (m)	0.781
Period (cm)	3.55
Peak magnetic field (T)	0.778
Dispersive section	
Dispersion $(d\psi/d\gamma)$	1.0–6.0
Radiator undulator (FODO focusing)	
Period (cm)	2.5
Peak magnetic field (T)	0.62
Section number	6
Section length (m)	1.5
Drift length between sections (m)	0.1
Average beta function (m)	~3.5

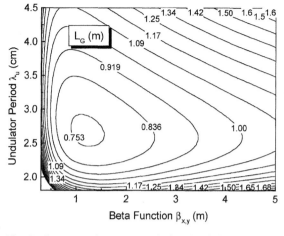

Fig. 2. Contours of constant gain length of the FEL at a wavelength of 88 nm. (The energy of the electron beam is changed as needed to preserve the resonant wavelength).

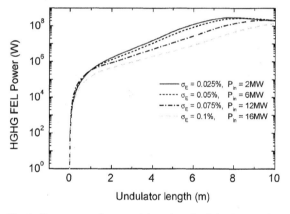

Fig. 3. Power vs. radiator undulator length of the SDUV-FEL for different beam energy spreads.

3. 150 MeV injector and UV FEL experiment

The linac dedicated to the SDUV-FEL possesses the task of producing high brightness electron bunches up to the energy of 300 MeV. The construction of a 150 MeV S-band electron injector as the first linac section and R&D of the undulator are being undertaken by the collaboration of Shanghai Institute of Applied Physics (SINAP), National Synchrotron Radiation Laboratory (NSRL), and Institute of High Energy Physics (IHEP). Its aim is the generation of electron beams with high brightness and the SASE FEL and/or seeded FEL in the spectral region from 500 nm to 264 nm.

The 150 MeV linac (see Fig. 4) consists of a photo-cathode RF gun, five SLAC-type accelerating tubes (2.856 GHz), and a magnetic bunch compressor. The first linac section will deliver high brightness electron bunches with peak current being in excess of 300 A and normalized emittance being lower than 6 mm·mrad. SASE FEL and/or seeded FEL experiments in the spectral region from 500 to 264 nm will be performed with undulator sections which are defined in Table 1.

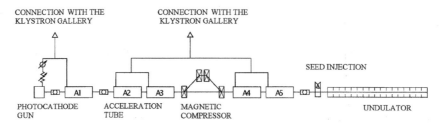

Fig. 4. Schematic layout of the 150 MeV linac and the UV FEL experiment.

The 150 MeV linac will be commissioned by the end of next year, SASE FEL and/or seeded FEL experiments are expected to be carried out from the beginning of 2005.

The BNL type photo-cathode RF gun is operated at S-band (2.856 GHz) and high peak field on the cathode (120–140 MeV/m) with incorporated metallic photo-cathodes (copper or magnesium). The RF gun cavity and the emittance compensation magnet have been manufactured and are being tested. The design of the 150 MeV linac was completed in the summer of 2002, followed by development and procurement of its equipment and components. Meanwhile, an FEL experimental building including a concrete bunker for housing the linac with energy of up to 300 MeV and the HGHG undulator has been designed and is under construction since the beginning of 2003. The construction of the building will be completed in November 2003, and the installation of the first linac section is scheduled to begin from December. While developing the photo-cathode RF gun, this linac will be initially commissioned with thermionic cathode DC electron gun and S-band buncher from April 2004.

4. Conclusion

Single-pass high-gain FEL in short wavelength region, from DUV to X-ray, is of great interest in many fields of applications. As a first step towards the high-gain SDUV-FEL facility, the design and R&D are being carried out in three institutes of the

Chinese Academy of Sciences (SINAP, NSRL and IHEP), and a 150 MeV linac for conducting high-gain FEL experiment is under construction and expected to be commissioned at the end of next year.

Acknowledgements

This work is partially supported by the Major State Basic Research Development Programme of China under Grant No. 2002CB713600, the Chinese Academy of Sciences, and the National Natural Science Foundation of China, and it has been carried on by the SDUV-FEL project team. The SDUV-FEL profits from the collaboration of many scientists worldwide. We are thankful to Drs. L.H. Yu, X.J. Wang, Z.R. Huang, J.H. Wu, B. Faatz, S. Reiche, and others for their valuable consultation, useful comments, helpful suggestions, and discussions on the design and R&D of SDUV-FEL.

References

[1] L.H. Yu, Phys. Rev. A 44 (1991) 5178.
[2] L.H. Yu, et al., Science 289 (2000) 932.
[3] M. Xie, Proceedings of PAC, 1995, p.183.
[4] S. Reiche, Nucl. Instr. and Meth. A 429 (1999) 243;
    S. Reiche, User Manual of Genesis 1.3.
[5] T.M. Tran, J.S. Wurtele, Comput. Phys. Commun. 54 (1989) 263;
    B. Faatz, S. Reiche, User Manual of TDA3D 1.0.

Available online at www.sciencedirect.com

Nuclear Instruments and Methods in Physics Research A 528 (2004) 595–599

NUCLEAR INSTRUMENTS & METHODS IN PHYSICS RESEARCH
Section A

ELSEVIER

www.elsevier.com/locate/nima

Performance simulation of infrared free electron laser at the Pohang Light Source test linac

Eun-San Kim

Pohang Accelerator Laboratory, Pohang University of Science and Technology, Pohang, KyungBuk 790-784, South Korea

Abstract

Numerical simulation studies are performed for the proposed infrared self-amplified spontaneous-emission free-electron laser (SASE-FEL) at the existing PLS test linac. It is shown that a high-gain SASE-FEL with a 1.5 µm radiation wavelength could be driven by a 61-MeV electron beam from the S-band rf linac and a 5-m long undulator. Third-harmonic generation to 0.5 µm radiation wavelength was also investigated to enhance the usefulness of the infrared SASE-FEL facility. Bunching fractions of the nonlinear third harmonic to the fundamental mode in the designed SASE-FEL is estimated and showed that the third-harmonic emission resulted in the same trend as the fundamental. This paper attempts to investigate the effect of the beam parameters such as the emittance, the energy spread, the beam energy, and the peak beam current to the designed infrared SASE-FEL. The wake effects due to the surface roughness of beam chamber in the undulator are also estimated.
© 2004 Elsevier B.V. All rights reserved.

PACS: 41.60.Cr; 42.65.Ky; 52.59.−f

Keywords: Free-electron laser; Nonlinear third-harmonic generation; Undulator

1. Introduction

At the Pohang Light Source (PLS), SASE-FEL system is being designed, which uses the existing test linac facility. The system is based on the existing electron linac which consists of a thermionic RF gun, an alpha magnet for bunch compression, and two S-band linac sections to provide electron energies from 60 to 100 MeV. The design study is focussed on reaching saturation at the radiation wavelength of 1.5 µm. For this design, we consider a 5 m long undulator with a

period length of 15 mm and a peak magnetic field of 1.37 T. In this paper, we investigate the effects of the emittance, the energy spread, the beam energy and the peak beam current on the performance of the designed IR SASE-FEL. In particular, we investigated characteristics of third harmonic because it may be a promising way to produce radiation power in the region of shorter wavelengths [1]. Our simulation results show that the third harmonic experiences a trend in gain and saturation similar to that experienced in the fundamental mode. The GINGER [2] simulation code was used to investigate the influence of electron beam parameters on the field energy of

E-mail address: eskim1@postech.ac.kr (E.-S. Kim).

the fundamental mode, the bunching of the fundamental mode and third harmonic. Longitudinal wakefield effects that are caused by the surface roughness of beam pipe in the undulator are also investigated.

2. Description of simulation

In the simulation, we chose an electron beam with a parabolic distribution in the longitudinal direction and a Gaussian distribution in the transverse direction. We used 2048 macroparticles per slice and 60 slices to represent an electron beam in the simulation. The sensitivity scans in the fundamental and third-harmonic generation are performed by using the nominal parameters given by in Table 1. For simplicity, we adopted a single-segment planar undulator with curved pole-face focusing. We also injected an input power of 1 μW at the fundamental mode. The characteristics of saturation of the system is investigated as a function of the beam current, energy spread, beam emittance and peak beam current. The peak position in the bunching is shown to be consistent with the position that the output field energy saturates.

3. Parameter investigations for the infrared SASE-FEL

3.1. Emittance scans

For the emittance sensitivity scans, the normalized emittance was varied between 1 and

Table 1
Basic parameters of the infrared SASE-FEL at PLS

Electron beam energy	E	61.3 MeV
Rms beam emittance	ε_n	5 πmm mrad
Peak beam current	I_p	300 A
Beam energy spread	$\Delta E/E$	0.1%
Radiation wavelength	λ	1.5 μm
Undulator period	λ_u	15 mm
FEL parameter	ρ	0.0046
Peak undulator magnetic field	B_0	1.37 T
Power gain length	L_G	17.8 cm
Undulator parameter	K	1.918

10 πmm mrad, which resulted in the saturation length approximately doubling. Fig. 1(a) shows the field energy as a function of undulator length for various values of the beam emittance at the fundamental wavelength. The field energies for emittances smaller than 6 πmm mrad show saturation at distances smaller than an approximately 5 m long undulator. The field energies for emittances larger than 6 πmm mrad show slow increases of the field energy up to a 15 m long undulator. Larger emittances require much longer saturation lengths and show decreased outputs, even though the FEL operation is maintained. Over this limited range, the field energy at the fundamental shows a significant sensitivity to the emittance.

3.2. Energy spread scans

For the energy spread scans, the initial beam energy spread was varied between 1% and 0.001%. Fig. 1(b) shows the field energy for various values of the energy spread at the fundamental mode. For energy spreads smaller than 0.1%, the field energy saturates at undulator length smaller than 5 m. For energy spreads larger than 0.5%, the field energies do not show a peak value up to an undulator length of 15 m.

3.3. Peak beam-current scans

For the beam current scans, the peak beam current was varied between 200 and 1000 A. Fig. 1(c) shows the field energy at the fundamental for various values of the peak beam current. For peak beam currents larger than 300 A, the field energies show saturation for undulator distances smaller than 5 m. For a peak beam current of 200 A, the field energy shows a trend of increasing up to 15 m long undulator. Over the limited range, the field energy at the fundamental is found to be significantly sensitive to the peak beam current.

3.4. Beam energy scans

For the electron beam energy scans, beam energy was varied between $\gamma = 120$ and 200. Fig. 1(d) shows the field energy at the fundamental

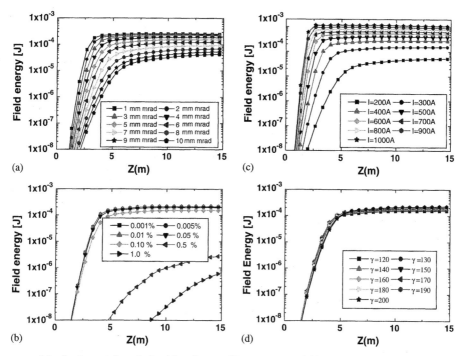

Fig. 1. Field energy of the fundamental mode for (a) emittance (b) energy spread (c) peak beam current and (d) beam energy as a function of undulator length.

mode for various values of the beam energy. Over this limited range, the field energy does not show any significant sensitivity to the beam energy.

3.5. Performance of the IR SASE-FEL at the PLS test linac facility

For the nominal parameters in Table 1, the simulation results show that saturation can be reached for the undulator distance of about 4.5 m, as shown in Fig. 2(a). The peak bunchings at the fundamental mode and the third harmonic also occur at about 4.5 m. Fig. 2(b) shows bunching factor for the fundamental mode and the third harmonic as a function of the distance. It is shown from Fig. 2(b) that the ratio of the bunching fraction of the third harmonic to that of the fundamental mode is about 15% for an undulator distance of 4.5 m. From these simulation results, we note that an undulator of about 5 m long is required to achieve saturation in the IR SASE-FEL.

In the one-dimensional model, the FEL parameter ρ is given by [3]

$$\rho = \left[\frac{I\gamma \Lambda_u^2}{I_A 16\pi^2 \sigma^2} \frac{K^2}{(1 + K^2/2)} (J_0(\xi) - J_1(\xi))^2 \right]^{1/3} \quad (1)$$

where I is the peak beam current, $I_A = 17,045$ A the Alfvên current, J_0 and J_1 Bessel functions, and σ the rms transverse beam size. $\xi = K^2/2(1 + K^2)$, where K is the undulator parameter. Under the nominal design parameters, the FEL parameter is given by 0.0046.

The gain in the SASE-FEL is defined as $G = E/E_0$, where E is the total energy when saturation occurs and E_0 is the initial energy off seeded power. We found the gain G to be $\simeq 1.0 \times 10^7$. The gain length in the one-dimensional model is given by $L_G = \lambda_u/4\sqrt{3}\pi\rho$, where λ_u is fundamental radiation wavelength, the calculated gain length is given by 0.15 m. The three-dimensional gain length obtained from the numerical simulation is

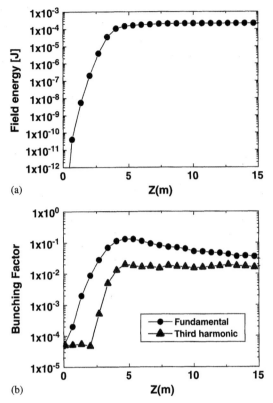

(a)

(b)

Fig. 2. (a) Field energy of the fundamental mode, and (b) bunching factor for the fundamental mode and the third harmonic as functions of the distance.

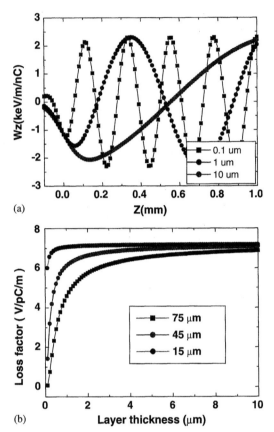

(a)

(b)

Fig. 3. (a) Wake potentials due to a 75 μm bunch length in the beam pipe of the radius $R = 5$ cm for different thickness of surface layer $\delta = 0.1$, 1 and 10 μm. (b) Loss factor versus thickness of surface layer for bunch length of 15, 45 and 75 μm.

0.17 m, which is good agreement with the one-dimensional calculation.

4. Effect of wakefield due to the surface roughness in beam chamber of the undulator

In addition to the resistive wall wakefield, other wake can be generated due to the surface roughness of the vacuum chamber in the undulator. To obtain the wake field, the roughness can be modeled by a thin dielectric layer $\varepsilon = 2$ on the smooth beam pipe [4]. In this model, the wake function is given by

$$W(z) = -\frac{ce^2 Z_0}{\pi R^2} \cos(k_0 z) \qquad (2)$$

defining the wave number $k_0 = \sqrt{4/R\delta}$, δ is the longitudinal rms variation of the roughness, R is the radius of the beam pipe and Z_0 is 377Ω. The convolution of the wake function with the longitudinal Gaussian beam profile gives the wake potential, as shown in Fig. 3(a). The wake potential shows an absolute maximum amplitude of roughly 2 keV/m for a roughness of $\delta = 1$ μm, a bunch charge of 1 nC and a bunch length of 75 μm.

The loss factor per unit length, integral of the bunch wake potential with the charge distribution is given by

$$K_{\text{loss}} = \frac{Z_0 c}{2\pi b^2} \exp - (K_0 \sigma_z)^2. \qquad (3)$$

The dependence of the loss factor on the surface layer thickness is shown in Fig. 3(b) for different bunch lengths σ_z.

5. Conclusions

We have examined the influence of the electron-beam parameters on the performance of the high-gain infrared SASE-FEL by a numerical simulation. The FEL will be a 1.5-μm SASE system driven by a 61 MeV electron beam and an 5 m long undulator. We have also investigated the sensitivities of the radiation output energy to variations in the beam emittance, the energy spread, the beam energy and the peak current centered around the nominal design parameters for the infrared SASE-FEL. It is shown that nonlinear third-harmonic generation in the de-signed infrared SASE-FEL can be used to achieve

shorter wavelengths, which shows the same trends in the growth and saturation characteristics with the fundamental mode. It is also shown that the surface roughness in the beam pipe of the undulator can give a harmful effect for the bunch length of 75 μm and surface layer thickness of 1 μm.

References

[1] H.P. Freund, S. Biedron, S. Milton, Nucl. Instr. and Meth. A 115 (2000) 53;
S.G. Biedron, Z. Huang, K. Kim, S. Milton, G. Dattoli, P. Ottaviani, A. Renieri, W. Fawley, H. Freund, H. Nuhn, Nucl. Instr. and Meth. A 483 (2002) 101;
R. Kato, et al., Nucl. Instr. and Meth. A 475 (2001) 334.
[2] W.M. Fawley, LBNL Report No. LBNL-49625, 2002.
[3] K. Kim, M. Xie, Nucl. Instr. and Meth. A 331 (1993) 1.
[4] A. Novokhatski, A. Mosier, Proceedings of the PAC97 Conference, Vancouver, 1997.

Available online at www.sciencedirect.com

SCIENCE  DIRECT®

ELSEVIER   Nuclear Instruments and Methods in Physics Research A 528 (2004) 600–604

NUCLEAR
INSTRUMENTS
& METHODS
IN PHYSICS
RESEARCH
Section A

www.elsevier.com/locate/nima

Design study of the Compton backscattering photon beam facility at the Pohang light source

J.K. Ahn[a,1], E.-S. Kim[b,*]

[a] *Department of Physics, Pusan National University, Busan 609-735, Republic of Korea*
[b] *Pohang Accelerator Laboratory, Pohang University of Science and Technology, POSTECH, Pohang, KyungBuk 790-784, Republic of Korea*

Abstract

We design a high-intensity Compton-backscattered photon beam facility at the PLS. A laser beam is shot to head-on collide with 2.5 GeV electrons, so that it produces a highly polarized photon beam with the wide energy range spanning between a few MeV and 400 MeV. We study the maximum photon beam intensity and the polarization of Compton backscattered photon beams with three different wavelength lasers (CO_2, Nd-YAG, and Ar-ion). We present design parameters for the proposed Compton beam facility as well as experimental programs. Simulation results on the increase of energy spread due to Compton backscattering in the PLS storage ring are also presented.
© 2004 Elsevier B.V. All rights reserved.

PACS: 06.20.F; 29.20.D; 41.75.H

Keywords: Compton backscattering; Energy spread; Photon beam line

1. Introduction

We design a Compton-backscattered photon beam line at Pohang Light Source (PLS). Compton backscattering with highly relativistic electrons provides a high-energy linearly or circularly polarized photon beam for extensive studies in nuclear and hadron physics. Compton-backscattered photons are produced almost equally over the energy range up to the maximum energy,

thereby providing higher photon flux than bremsstrahlung photons with the flux decreasing as $1/E_\gamma$. Linear polarization is one of the most important feature of the Compton scattered photons, which can be only partially achieved for bremsstrahlung photons by a coherent radiation technique. The PLS circulates 2.5 GeV electrons in the 280 m storage ring, and 15 synchrotron beam lines are currently in operation [1]. A laser beam is shot to head-on collide with a 2.5 GeV electron beam in the storage ring, so that a high-intensity polarized photon beam is generated with a wide energy range from a few MeV to 400 MeV [2–4]. Monte-Carlo simulations on the Compton backscattering have been also performed by using the

---

*Corresponding author.

E-mail address: eskim1@postech.ac.kr (E.-S. Kim).

[1] Supported by Korea Research Foundation Grant (KRF-2003-015-C00130).

PLS storage ring parameters and laser beam parameters.

2. Proposed Compton backscattering beam line

The Compton backscattering beam line is proposed to produce high-energy photon beam at a 6.8 m long straight section. The PLS storage ring consists of 12 cells, two of them are injection and RF parts and the others are composed of the 6.8 m long straight sections between bending magnets. Several sections are already used for undulator-driven X-ray beam lines. Our proposed location of the beam line is shown in Fig. 1.

We propose to build a wide energy-range photon beam line using three different wavelength lasers. A head-on collision between a long-wavelength CO_2 laser (9400 and 10,600 nm) beam

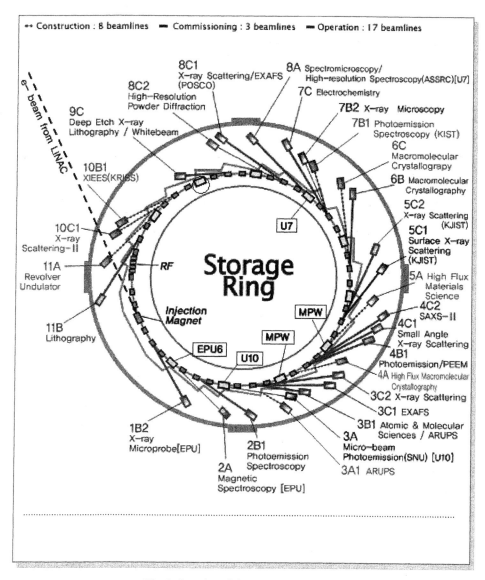

Fig. 1. Location of the proposed beam line.

and 2.5 GeV electrons will produce high-intensity MeV photons with the maximum energy of 15 MeV. This low-energy scattered photons share initial kinetic energy little with recoil electrons, so the recoil electron still lies within the momentum acceptance of the storage ring. It prevents us from tagging photon energies with measuring recoil electrons. Instead, we can use a collimator to define photon energy range, since Compton-scattered photon energies vary with scattering angle.

Mid-energy range photons up to 110 MeV will be available with a Nd-YAG (1064 nm) laser, as shown in Fig. 2. A Ti-Sapphire laser (690–1100 nm) can be also used for production of the mid-range energy photons. High-energy photon production up to 400 MeV needs as short laser wavelength as 250 nm. Among commercially available CW lasers a 351 nm Ar-ion laser or Kr/Ar laser using a frequency doubler (257 nm) is one of the strong candidates. Higher harmonic laser beam can be also an alternative, but we should sacrifice photon flux.

When Compton-backscattered photon energy becomes as high as 100 MeV, we are in a position to determine photon energy by measuring the direction of scattered electrons. Since the recoil electron energy is not much different from the original electron energy, 2.5 GeV, the recoil electron path deviates from the 2.5 GeV one within a few cm at the exit of the bending magnet. In this respect, we propose a minor change of the beam line chamber at the exit of the bending magnet, in order to accommodate a photon-energy tagger. The tagger will be composed of two layers of the plastic scintillating fiber array alternatively stacked by two layers of solid-state tracking device for fast timing and fine tracking. We aim at achieving the energy resolution of less than 1%.

In the high-energy region, the electron beam lifetime is shortened with increasing Compton scattered photon flux. The Compton-scattering associated lifetime τ_C is given by $I/e \cdot T_{rev}/\dot{N}_\gamma$, where I is the electron beam current [A], T_{rev} the revolution time 0.93 μs for the PLS ring. If we keep the lifetime τ_C to be three times as long as the original Touschek-scattering dominated lifetime (25 h for PLS), the maximum allowable photon flux will be 2.2×10^7 [$A^{-1} s^{-1}$]. Since the power of all the above commercial lasers are high enough to generate the photon flux limit, the key parameter is a transmission efficiency of the photon transport line.

The laser-optic system will be equipped in a shielded hutch and the photon path will be kept in vacuum to prevent light scattering with the air. For high-energy photon beam, a permanent magnet will be used to sweep out e^+e^- pairs. Experimental apparatus with low-energy photon beam will be implemented in the laser hutch. Fig. 3 shows a schematic view of the low-energy photon beam line using a CO_2 laser.

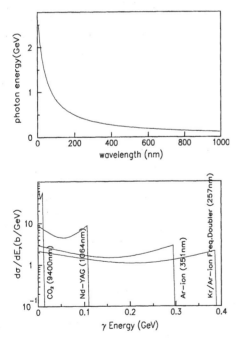

Fig. 2. Backscattered photon energy as a function of laser wavelength (top), cross-sections of Compton backscattering in terms of the choice of laser.

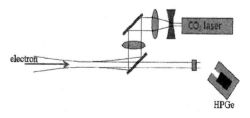

Fig. 3. Schematic view of low-energy photon beam line.

A Compton-supressed high-purity Ge detector system is best suited for high-resolution gamma ray detection.

The experimental apparatus for mid- and high-energy photon beams should be different from that for low-energy beams. For gamma-ray detection, it should be more radiation-resistive and of larger acceptance. NaI(Tl), BGO, and CsI(Tl) detectors can be good candidates. While low-energy charged particle energies are usually measured by time-of-flight and $\Delta E - E$ measurements, high-energy charged particles are magnetic-analyzed by wire chambers and a large electromagnet.

A Monte-Carlo simulation [5] of Compton backscattered photons from 2.5 GeV electrons was performed to see the effects of Compton backscattering on the electron beam parameters. The tracking simulation were performed in six-dimensional phase space. Initial electron beam distribution is chosen randomly from a Gaussian distribution with emittance of 18.9 nm, beam energy of 2.5 GeV, energy spread of 8.6×10^{-4}, bunch length of 8.5 mm and coupling constant of 1%. The simulation results show that the rms natural energy spread of 2.16 MeV due to the Compton backscattering is increased by a factor of 1.9. Laser with a wavelength of 1.053 μm (1.177 eV) and 10 Hz pulse is considered for the simulation. Then the number of scattered photons per bunch is estimated to be 2×10^{8}.

3. Proposed physics programs

The proposed Compton backscattering beam line enables us to study many aspects of nuclear physics in a very wide energy range spanning between a few MeV and 400 MeV. Fig. 4 shows an excitation spectrum measured from inclusive (e, e') scattering, in which one can see how much smaller objects can be detected at a given energy. Low-energy photons see a target nucleus as a whole. In the energy region below 10 MeV one can assign quantum numbers for discrete nuclear bound states, in particular for multi-photon levels and intruder states. One can also study nucleosynthesis s- and p-processes by measuring inverse reactions of $A(n, \gamma)B$ and (p, γ). When the nucleus A is very

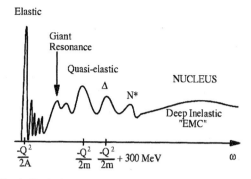

Fig. 4. Excitation spectrum from inclusive (e, e') scattering.

unstable, one can hardly measure the neutron-capture cross-section off A.

In the mid-energy region of 10–100 MeV, nuclear resonance becomes main interest. In particular, there is a huge bump of many nuclear resonances populated in the energy region from 20 to 30 MeV. While the isovector Giant resonance levels are quite well known from the extensive studies in last 50 years, we should note the importance of the study on iso-spin mixing states. One well-known example is the $^{10}B^{*}$ system, where the E1 transition from two 2^{+} levels of 19.3 and 18.43 MeV to a 3^{+} ground state and the M1 transition from 2^{-} level at 18.8 MeV to the ground state coexist. One should clarify this mixing is just due to the existence of iso-spin mixing state or interpret it as a parity-violating process.

While Giant dipole resonance accounts for the relative motion of protons and neutrons in a nucleus, there is a so-called pygmy E1 resonance whose name comes from its small fraction of the total nuclear excitation. The pygmy resonance is known to be associated with a relative motion between neutron–proton core and neutron skin. It is interesting to see how much bigger or smaller the pygmy resonance contribution is as the neutron number increases.

Pion photoproduction at threshold comes to attract much attention to nuclear physicists since it is the first case that Chiral perturbation theory is applied to nuclear system. For an elementary processes $\gamma p \rightarrow n\pi^{+}$, our interest energy region is just 1–2 MeV above threshold. Due to this highly limited phase space, it is very hard to detect

π^+. Therefore, one should detect a neutron, instead. Measurement of nucleon polarizability provides considerable information in understanding the structure of nucleons. High-energy photons are Compton-scattered off a liquid deuterium target to measure a dipole moment of neutron.

4. Summary

We design a high-intensity Compton-backscattered photon beam facility at PLS. A laser beam is shot to head-on collide with 2.5 GeV electrons, so that it produces a highly polarized photon beam with the wide energy range spanning between a few MeV and 400 MeV. We study the maximum photon beam intensity and the polarization of Compton backscattered photon beams with three different wavelength lasers (CO_2, Nd-YAG, and Ar-ion). We present design parameters for the proposed Compton beam facility as well as experimental programs. Simulated results on the increase of energy spread due to the Compton backscattering in the storage ring are also presented.

References

[1] Annual Report, Pohang Accelerator Laboratory, 2002.
[2] T. Ohgaki, E.-S. Kim, J.K. Ahn, Joint 28th ICFA Advanced Beam Dynamics & Advanced & Novel Accelerators Workshop on Quantum Aspects of Beam Physics, January 7–11, Hiroshima, Japan, 2003.
[3] W. Kim, J. Korean Phys. Soc. 26 (316) (1993).
[4] G.N. Kim, D.C. Son, in: H. Bhang, S.W. Hong, H. Shimizu (Eds.), Proceedings of the International Workshop on Nuclear Physics in Different Degree of Freedom, February 19, 2001, Institute of Basic Science, SungKyunKwan University, Suwon, Rep. of Korea.
[5] K. Yokoya, http://www-acc-theory.kek.jp/members/cain/default.html.

Available online at www.sciencedirect.com

SCIENCE  DIRECT®

ELSEVIER Nuclear Instruments and Methods in Physics Research A 528 (2004) 605–608

**NUCLEAR
INSTRUMENTS
& METHODS
IN PHYSICS
RESEARCH**
Section A

www.elsevier.com/locate/nima

Picosecond visible/IR pump–probe dynamics of photoactive yellow protein

Aihua Xie[a], Lorand Kelemen[a], Britta Redlich[b], Lex van der Meer[b], Robert Austin[c],*

[a] *Department of Physics, Oklahoma State University, Stillwater, OK 74078, USA*
[b] *FOM Institute for Plasma Physics, 3430 BE Nieuwegein, The Netherlands*
[c] *Department of Physics, Princeton University, Princeton, NJ 08544, USA*

Abstract

We present 2-color pump/probe experiments on the sensory protein Photoactive Yellow Protein (PYP). We present a snapshot of the transient IR spectrum of PYP at 100 ps after the pump pulse, in the important amide I and amide II region around 6 μm.
© 2004 Elsevier B.V. All rights reserved.

PACS: 87.15.Aa

Keywords: Picosecond; Infrared; Visible; Photoactive yellow protein

1. Introduction

Although X-ray crystallography remains the premier technology to ascertain the structure of proteins, there remain two fundamental problems with this technique. First, the proteins are confined in the crystal lattice so that large-scale conformational transitions are not allowed. These conformational transitions can be quite important, particularly in the case of a signaling protein such as PYP, which is believed to make a very large structural transition to a molten globule-like conformation after the *trans–cis* isomerization of the p-coumaric acid chromophore [1]. The effect of this confinement was strikingly brought out in a time-resolved FTIR experiment by Xie et al. [2], where it was shown that the large changes in the amide I region of the IR spectra of PYP are NOT seen in crystals, implying that these large structural changes cannot occur in the crystalline state of the protein. Second, even granted that X-ray crystallography does give one the time-averaged structure of the protein, many proteins are dynamic entities which undergo a carefully organized series of structural changes which are difficult to do using X-ray techniques.

PYP is an ideal protein for time-resolved structural studies. PYP contains a unique 4-hydroxycinnamic acid (pCA) chromophore. The

---

*Corresponding author.
E-mail address: austin@princeton.edu (R. Austin).

0168-9002/$ - see front matter © 2004 Elsevier B.V. All rights reserved.
doi:10.1016/j.nima.2004.04.111

trans to cis photoisomerization of this chromophore activates a photocycle involving first a short-lived red-shifted intermediate (pR), then a long-lived blue-shifted intermediate (pB), and finally recovery of the original receptor state (pG). The protein in the receptor state is characterized by a strong absorption band centered at 446 nm, resulting in its yellow color. Absorption of a blue photon drives PYP into a cyclic chain of thermal processes. The first observed photointermediate is a red-shifted species pR with a max at 465 nm. On a sub-millisecond time scale pR is converted into a blue-shifted species pB with max at 335 nm. The change in protonation state of pCA has been implied to trigger large protein conformational changes, a "protein quake". Key residues at the active site of PYP include Glu46 and Tyr42, whose OH groups hydrogen bond to the phenolic oxygen of pCA, and Arg52, whose positively charged guanidinium group has been proposed to stabilize and delocalized the buried negative change on pCA [3].

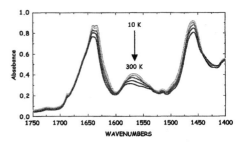

Fig. 1. The temperature dependence of the IR absorbance of PYP.

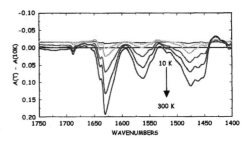

Fig. 2. Absorbance at temperature T (290–50 K)—Absorbance at 10 K.

2. Experimental protocol

PYP samples for both the FTIR and FEL studies were prepared at pH 7.0. A homogeneous PYP film was made by spreading 2 μl of PYP dissolved in D_2O over an area of 20 mm diameter on a calcium fluoride (CaF_2) plate (25.4 mm diameter, 2 mm thickness), with a 12 μm spacer between the two CaF_2 plates. The final photoactive yellow protein concentration was 150 mg/ml or approximately 4 mM, and had a 1 OD absorbance at the peak of the amide I band. The temperature-dependent samples were dissolved in a glass-forming 75% deuterated glycerol-25% D_2O (V/V) solvent. Single-beam FTIR spectra of pG as a function of temperature were collected at 2 cm^{-1} resolution from 800 to 3800 cm^{-1}, using an Oxford Instruments Optistat static gas exchange cryostat. Fig. 1 shows the temperature dependence of the IR absorbance spectrum of PYP from 10 to 300 K. The bands associated with the protein between 6 and 8 μm which are components of the amide I and II show not a shifting of the bands with decreasing temperature but instead either an

increase in the oscillator strength for the amide II bands or the appearance of a new red-shifted band in the amide I. When a protein is placed in a deuterated solvent not all the hydrogens exchange, to deuterons: the inner core of the protein does not exchange its deuterium atoms, resulting in two amide II bands. Since the amide II band is due to a "scissoring" N–X motion in our sample, the scissoring mode frequency is dependent on the presence of a hydrogen or deuterium atom. The band at 6.50 μm is due to the N–H mode in the inner core of the protein, and the band at 6.90 μm is due to the deuterated N–D modes at the surface of the protein where hydrogen exchange has occurred. This is more clearly seen in Fig. 2.

The time resolved two color visible/IR experiments were done using the FELIX FEL as a tunable IR probe with approximately 50 nJ energy in each micro pulse and a 10 μJ visible pulse at 400 nm as a photolyzing pulse. IR pulse widths were set at the longer range of FELIX parameters in order to achieve 7 cm^{-1} spectral resolution (0.025 μm at 6 μm). The 400 nm excitation pulse was provided by doubling the 800 nm light from a

Ti:S seeded regenerative amplifier running at 10 Hz, synced with 10 ps accuracy to the FELIX FEL micropulses. The delay between the 400 nm pump and the IR probe was set at +100 ps for these measurements. The sample, held between the 25.4 mm diameter plates, was rotated at approximately 1 Hz. We ensured that the PYP molecules had sufficient time to recover between excitation by moving the 100 μm diameter focused beams 15 mm from the axis of rotation. This resulted in approximately 1000 distinct irradiation spots, and a time between reexcitation of a given PYP molecule of approximately 100 s. Since the PYP recovery time is approximately 0.2 s this provides enough time for protein recovery between 400 nm pulses. The probe pulse sequence consisted of 2 adjacent micropulses from FELIX separated by 40 ns, each micropulse was split into a pair of pulses separated by 4 ns. The first pair of pulses (reference pair) preceeded the 400 nm pump pulse, in the second pair (measurement pair) the 400 nm pulse was delayed electronically around the second pulse of the pair.

3. Experimental results

Fig. 3 gives the change observed in the amide I–II region of PYP 100 ps after the *trans–cis* isomerization event. We coplot the steady-state difference spectra seen at 80 K for a PYP sample in glycerol–water (Xie, unpublished work). There are two observations to be made. First, the Glu 46 shift in hydrogen bonding observed as a derivative in the 80 K data at 1740 cm^{-1} is not present, implying that within our S/N level of 10^{-4} that the shift in hydrogen bonding in Glu46 has not yet occurred at 100 ps. Second, there is a rather large change in the amide I region, with a maximum at 1640 cm^{-1}. This band is due to C–O stretch, and so it is difficult at present to unambiguously assign the origin of the change to some combination of the *trans–cis* isomerization and changes in the backbone conformation of the protein. While the 80 K FTIR difference spectrum of Xie is somewhat similar to the 100 ps difference spectrum seen here, they are by no means identical.

We have plotted in Fig. 4 the difference spectrum of the IR absorbance of PYP taken from Fig. 2 at 250–290 K and the transient IR difference spectrum at +100 ps. While in principle there is no obvious connection between the temperature dependent absorption of ground state PYP (pG) and the transient spectrum seen 100 ps after the *trans–cis* isomerization one can argue that these temperature accessible conformations could also represent the configurations which a strained configuration of the protein can drop into after the *cis–trans* event. The similarity of the amide I difference band seen +100 ps after *cis–trans* isomerization and the temperature dependent

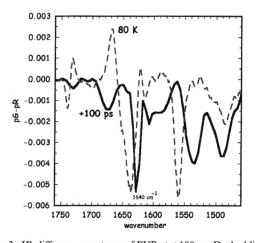

Fig. 3. IR difference spectrum of PYP at +100 ps. Dashed line is pG–pR at 80 K.

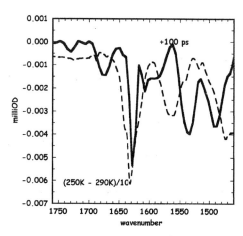

Fig. 4. +100 ps (solid line) and the 250–290 K in PYP (dashed line).

band seen in the ground state protein would perhaps argue for a protein conformation change origin to the prompt changes seen in this experiment. But, more work needs to be done before the important question of the extent of protein conformational changes in signaling proteins can be answered.

Acknowledgements

We thank the staff of FELIX for laser operations, and W.D. Hoff for discussions. Supported by the Office of Naval Research and the Stichting voor Fundamental Onderzoek der Materie (FOM).

References

[1] W.D. Hoff, A. Xie, I.H.M. Van Stokkum, X.-J. Tang, J. Gural, A.R. Kroon, K.J. Hellingwerf, Biochemistry 38 (1999) 1009.

[2] A. Xie, L. Kelemen, J. Hendriks, B.J. White, K.J. Hellingwerf, W.D. Hoff, Biochemistry 40 (2001) 1510.

[3] A. Xie, W.D. Hoff, A.R. Kroon, K.J. Hellingwerf, Biochemistry 35 (1996) 14671.

Available online at www.sciencedirect.com

Nuclear Instruments and Methods in Physics Research A 528 (2004) 609–613

**NUCLEAR
INSTRUMENTS
& METHODS
IN PHYSICS
RESEARCH**
Section A

www.elsevier.com/locate/nima

An improvement of matrix-assisted laser desorption/ionization mass spectrometry using an infrared tunable free electron laser

Yasuhide Naito*, Sachiko Yoshihashi-Suzuki, Katsunori Ishii, Kunio Awazu

Institute of Free Electron Laser, Graduate School of Engineering, Osaka University, 2-9-5 Tsuda-Yamate, Hirakata, Osaka 573-0128, Japan

Abstract

Matrix-assisted laser desorption/ionization (MALDI) combined with time-of-flight mass spectrometry (TOFMS) is a powerful yet robust tool for protein identification, due to its high sensitivity and theoretically unlimited detectable mass range. A large part of functional proteins, such as membrane proteins, are insoluble as native forms in a matrix solution without a strong denaturing condition, hence are not amenable to the conventional MALDI–TOFMS analysis. Aiming at overcoming this difficulty, we have developed a novel MALDI technique (UV/FEL-MALDI). An infrared free electron laser (IR-FEL) has a wide tunability in a mid-IR range and is quite attractive as a source of selective vibrational excitation. The FEL wavelength can be tuned to activate a denaturant, which impedes the conventional MALDI process, without an excess heating of analyte molecules. This scheme lets a dense denaturant to be used for the MALDI sample preparation of insoluble proteins. A simultaneous use of the FEL with a nitrogen pulse laser for MALDI achieves spatially and temporally defined desorption, which is essential to TOFMS detection, while specificity and selectivity owing to an FEL wavelength can be conserved. Some attractive features of the protein clustering have been found in the application of UV/FEL-MALDI to hair keratins, which was chosen as a model of insoluble proteins.
© 2004 Elsevier B.V. All rights reserved.

PACS: 42.62.Be; 82.80.Ms; 82.80.Rt; 87.14.Ee; 36.20.Cw

Keywords: FEL; MALDI; TOFMS; Insoluble protein; Hair keratin

1. Introduction

Matrix-assisted laser desorption/ionization (MALDI) is one of the ionization techniques used in mass spectrometry (MS) [1,2], and particularly suited for detecting labile and nonvolatile macromolecules. Analytes are thoroughly mixed with a matrix compound, whose molar ratio to the analytes is usually more than 10^3. By exposing the matrix-dominated samples to a pulse laser, which is generally nitrogen laser (337 nm), analyte molecules are released from the condensed phase into the gas phase in the forms of protonated or cation-adducted (positive mode) or deprotonated (negative mode) ions. MALDI has been successfully used for the analyses of peptides, proteins, oligonucleotides, oligosaccharides, and synthetic polymers. The roles of matrix are thought to be as

*Corresponding author. Tel.: +81-72-897-6413; fax: +81-72-897-6411.

E-mail address: naito@fel.eng.osaka-u.ac.jp (Y. Naito).

follows: firstly, matrix is a solvating medium, which enfeebles cohesive forces between analyte molecules in the condensed phase. Secondly, matrix is a photon absorber, which prevents analyte molecules being broken by laser and converts photon energy to heat in a good efficiency. Lastly, matrix is a reaction reagent, which gets analyte molecules to be charged via proton-transfer reactions.

Recent years, identifying a large set of proteins has been increasingly emphasized as a global challenge of life science in the post-genome era. The systematic view of the whole proteins expressed from a particular gene is proteome, and a study on a proteome is proteomics. The key technology of proteomics is a high throughput identification of proteins based on MS and searching a protein database. MALDI combined with time-of-flight mass spectrometry (TOFMS) is a powerful yet robust tool for protein identification, due to its high sensitivity and theoretically unlimited detectable mass range.

The ionization efficiency of each analyte in the MALDI process depends on a microscopic crystal formation of the matrix-analyte system, whose mixing condition is thus important to gain the signal yield. A large part of functional proteins, such as membrane proteins, are not amenable to the conventional MALDI–TOFMS analysis, because they are insoluble as native forms in a matrix solution unless using a strong denaturing condition. Aiming at overcoming this difficulty, we have developed a novel MALDI technique (UV/FEL-MALDI). An infrared free electron laser (IR-FEL) has a wide tunability in a mid-IR range and is quite attractive as a source of selective vibrational excitation. The FEL wavelength can be tuned to activate a denaturant, which impedes the conventional MALDI process, while analyte molecules are not heated excessively by the laser. This scheme lets a dense denaturant to be used for the MALDI sample preparation of insoluble proteins. A simultaneous use of the FEL with a nitrogen pulse laser for MALDI achieves spatially and temporally defined desorption, which is essential to TOFMS detection, while specificity and selectivity owing to an FEL wavelength can be conserved.

In this paper, we present that UV/FEL-MALDI TOFMS was successfully applied to analyze an insoluble protein with a dense denaturant. Hard keratin is a primary component of human hair shaft and known as an extremely insoluble protein having a high content of cysteine residues, which form intermolecular disulfide bonds. The molecular weights of hair keratin and its clusters were measured for the first time.

2. Experimental

2.1. Mid-infrared free electron laser

The FEL used in this study covers a mid-infrared range (5–22 µm), where many organic compounds have rich vibrational absorption bands. The diameter of the FEL beam delivered to the user's facility was approximately 10 cm. This was reduced to approximately 5 mm by a confocal minification optical system using a parabola mirror and a ZnSe lens. The minified FEL beam was guided by gold-coated mirrors and focused by a ZnSe lens of 12.7 mm in diameter and 127 mm in focal distance to generate a small exposed spot on a sample. A portion of the FEL beam was separated by a gold-coated mirror inserted in front of the minification optics and received by a MCT infrared detector (R005, Vigo System, Warsaw, Poland). The MCT output was converted to a TTL signal by a delay generator (DG535, Stanford Research Systems, Sunnyvale, CA, USA), which also put an appropriate delay (ca. 4 µs) on the TTL signal to discard an early part of the FEL macropulse suffering from a distortion and a shot-to-shot fluctuation.

2.2. Mass spectrometer

A commercially manufactured MALDI TOF mass spectrometer (Voyager DE Pro, Applied Biosystems, Foster City, CA, USA) was used with minor in-house modifications. The mass spectrometer is equipped with a nitrogen laser (VSL-337ND, Spectra-Physics, Mountain View, CA, USA) for the conventional UV-MALDI technique. The nitrogen laser was disconnected from the original control circuit and externally triggered by the TTL signal generated from the FEL pulse as above described. A CaF_2 window mounted on a 1.33 inch mini conflat flange was attached to the vacuum chamber for making the FEL beam pass into the ion source region. The superposition of the nitrogen laser and FEL spots on the sample plate was ensured by a liquid crystal film sensitive to a small change of the surface temperature.

An operation of the TOF mass spectrometer can be chosen as either linear or reflector modes. The former mode is suitable for detecting heavy ions at the cost of the mass resolving power lower than that can be obtained by the latter one, whose use is limited for less than m/z 10,000. The TOF mass spectrometer was operated in the linear mode throughout this study. Contemporary MALDI TOF mass spectrometers employ the delayed extraction technique, which corrects the spread of ionic velocities and thereby improves the mass resolving power. In the delayed extraction mode, an ionization laser pulse is followed by a short delay; an extraction voltage pulse was applied afterward between a sample plate and an acceleration grid. The delayed extraction parameters were adjusted to maximize the performance in a high mass region ($> m/z$ 50,000).

2.3. Sample preparation

A human hair keratin powder was obtained from Nacalai Tesque (Kyoto, Japan) and used without further purification. The MALDI matrix grade of 3,4-dimethoxy-4-hydroxycinnamic acid (sinapinic acid) was obtained from Fluka (Buchs SG, Switzerland). The matrix solution was prepared by dissolving 10 mg of sinapinic acid into 1 ml of acetonitrile/water (1:1) with 0.05% (v/v) trifluoroacetic acid. The keratin powder was added to a mixture of 50 mM 2-amino-2-(hydroxymethyl)-1,3-propanediol, 8 M urea and 0.1 M 2-mercaptoethanol in the final concentration to be 7 mg/ml. The analyte solution was rigorously vortexed, then the supernatant layer was separated for use. Deposits of the sample on the plate were prepared by two different techniques, the dried-droplet method and the crushed-crystal method. In the former one, a drop of the aqueous analyte

solution (ca. 1 μl) was mixed with the same quantity of the matrix solution and stirred with a pipette tip. A drop of the mixed solution was applied to the sample plate and dried in ambient air. In the latter method, firstly a drop of the matrix solution was applied to the plate and dried in air. Then a slide glass was placed on the matrix deposit and turned laterally with a small force down to the surface. Crushed particles were wiped out with a tissue paper and a smeared spot of the matrix was left. A drop of the analyte/matrix solution prepared in the same way as the former method was applied to the smeared spot. Before the deposit was completely dried, water was applied to wash the spot. Finally, excess water was removed by blotting with a tissue paper and the deposit was dried in ambient air.

3. Results and discussion

The dried-droplet method for the keratin sample showed the morphology forming an opaque amorphous solid, rather than a polycrystalline thin-film desirable to a successful MALDI measurement. Attempts to obtain a MALDI TOF mass spectrum from this solid by exposing to the UV laser resulted in failure (Fig. 1a). In contrast to this, the simultaneous exposure to the UV laser and the FEL (UV/FEL-MALDI) gave a sign of extraordinary heavy ions on the TOF mass spectrum (Fig. 1b). With respect to the FEL wavelength, which was varied in the range of 5.0–7.0 μm, these heavy ions were observed exclusively in tuning at 5.8 or 5.9 μm. These effective FEL wavelengths are coincident with the maximum absorption band of urea in the mid infrared region. The peaks appeared on Fig. 1b are attributable to either keratins themselves or materials originating from the powdery keratin sample, because the blank control mass spectrum which was obtained from a deposit of the same solution but not containing the keratin powder indicates no sign of heavy ions (Fig. 1c).

Although urea dominating the solid deposit is transparent to the UV wavelength, the analyte molecules were deeply embedded in the bulky solid and the UV matrix was diluted by urea. Therefore,

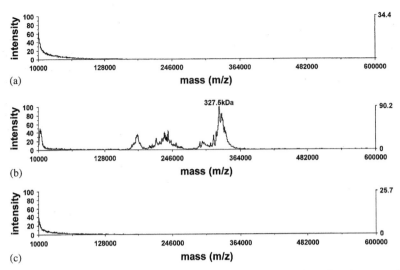

Fig. 1. UV-MALDI (a) and UV/FEL-MALDI (b) TOFMS spectra of a keratin/matrix/urea deposit prepared by the dried-droplet method. (c) Same as (b) but obtained from a matrix/urea deposit.

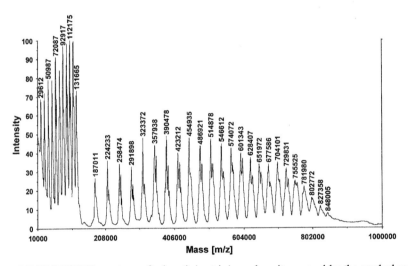

Fig. 2. UV/FEL-MALDI TOFMS spectrum of a keratin/matrix/urea deposit prepared by the crushed-crystal method.

the heat which was converted from the UV photons by the matrix was insufficient to initiate the desorption process. In the case of UV/FEL-MALDI, the bulky solid of urea was presumably activated by the IR photons whose wavelength was tuned at the absorption band of the denaturant. Urea has been examined as a matrix of IR-MALDI using 3 μm region and shown a relatively poor performance [3]. It was possible that urea might act as a matrix and an IR-MALDI process might generate the ions; however, the present sample exposed to the FEL alone did not give a successful mass spectrum (result not shown). It can be suggested that the simultaneous exposure to the UV and IR lasers was engaged in an essential step of the MALDI process, such as a local rapid heating of the sample deposit.

Comparing with the above dried-droplet case, the crushed-crystal method reduced the thickness of the solid deposit and gave the UV/FEL-MALDI TOF mass spectrum allowing an analytical inspection (Fig. 2). Two distinct series of peaks can be found in the mass spectrum. A series in the lower mass region (m/z 20,000–135,000) exhibits almost the same mass difference (ca. 10,200 Da[1]) between every adjacent peaks. The other series (m/z 180,000–850,000) also appears with equal spacing (ca. 28,800 Da). These mass differences are attributable to two types of keratin molecules, presumably γ-keratins, which are amorphous components in hair rod. Native keratin molecules have many intermolecular disulfide bonds to form supramolecular networks. The contents of keratin clusters which can be conserved via the sample preparation and the ionization process of UV/FEL-MALDI most probably elucidate the observation of the peak series.

4. Conclusion

We have developed a novel MALDI technique, UV/FEL-MALDI, in which a sample is exposed to FEL and nitrogen laser pulses simultaneously. The principle of UV/FEL-MALDI was examined by using a hair keratin sample prepared under a strong denaturing condition. The molecule-related ions of keratins, i.e. their total masses, have been analyzed by mass spectrometry for the first time, along with an evidence of their clustering. The remarks are as follows: firstly, the FEL macropulse with 15 µs width was utilized for MALDI–TOFMS without an extra instrumentation, such as the Pockels cell, to obtain a sliced micropulse train; secondly, extremely heavy ions (> m/z 800,000) can be generated in the gas-phase by UV/FEL-MALDI; lastly, the present approach allows a solubilizing agent to be used in a high concentration for sample preparations.

References

[1] K. Tanaka, et al., Rapid Commun. Mass Spectrom. 2 (1988) 151.
[2] M. Karas, et al., Anal. Chem. 60 (1988) 2299.
[3] M. Sadeghi, et al., Rapid Commun. Mass Spectrom. 11 (1997) 393.

---

[1] The abbreviation of "dalton": the unified atomic mass unit defined as the mass of one atom of ^{12}C divided by 12.

Available online at www.sciencedirect.com

SCIENCE @ DIRECT°

ELSEVIER   Nuclear Instruments and Methods in Physics Research A 528 (2004) 614–618

NUCLEAR
INSTRUMENTS
& METHODS
IN PHYSICS
RESEARCH
Section A

www.elsevier.com/locate/nima

The effects of FEL irradiation against a phosphorylated peptide and the infrared spectrographic identification method for a phosphate group

Katsunori Ishii[a,b,*], Sachiko Suzuki-Yoshihashi[b],
Kunihiro Chihara[a], Kunio Awazu[b]

[a] Graduate School of Information Science, Nara Institute of Science and Technology,
8916-5 Takayama-cho, Ikoma, Nara 630-0101, Japan
[b] Institute of Free Electron Laser, Graduate School of Engineering, Osaka University,
2-9-5 Tsuda-Yamate, Hirakata, Osaka 573-0128, Japan

Abstract

Phosphorylation and dephosphorylation, which are the most remarkable post-translational modifications, are considered to be important chemical reactions that control the activation of proteins. First, we examine the phosphorylation analysis method by measuring the infrared absorption peak of the phosphate group that is observed at about $1070\,cm^{-1}$ ($9.4\,\mu m$) with Fourier Transform–Infrared Spectrometer (FT–IR). Next, we attempt to control the quantity of phosphorylation, that is to say an action like dephosphorylation without enzyme reactions, by irradiating $9.4\,\mu m$-Free Electron Laser ($9.4\,\mu m$-FEL). FEL irradiation has an effect of some kind on the organization of infrared absorption of a phosphate group. We would now like to go on to develop this photochemical reaction like dephosphorylation by examining under several conditions and in detail.
© 2004 Elsevier B.V. All rights reserved.

PACS: 41.60.Cr; 42.62.Be

Keywords: Free electron laser (FEL); Phosphorylation and dephosphorylation; Fourier transform–infrared spectrometer (FT–IR); Phosphorylated peptide; Infrared absorption spectrum

1. Introduction

After translation, based on the information in DNA, the active factors and functions of proteins

─────────
*Corresponding author. Institute of Free Electron Laser, Graduate School of Engineering, Osaka University, 2-9-5 Tsuda-Yamate, Hirakata, Osaka 573-0128, Japan. Tel.: +72-897-6410; fax: +72-897-6419.

E-mail address: ishii@fel.eng.osaka-u.ac.jp (K. Ishii).

are regulated by post-translational modifications under several physical conditions. The most well known posttranslational modifications of proteins are phosphorylation and dephosphorylation, and these are considered to be important reactions that control the active and inactive factors of proteins. It is also known that various types of stresses affect the phosphorylation changes, and it is considered that a phosphorylation-induced

0168-9002/$ - see front matter © 2004 Elsevier B.V. All rights reserved.
doi:10.1016/j.nima.2004.04.113

abnormality in the active factor of a protein leads to many chronic ailments and carcinogenesis. Therefore, undoubtedly, an insight into the modification reactions, including phosphorylation and dephosphorylation, is important for a breakthrough in physiologic alignment systems, involving the manufacture of new drugs and analysis of disease determinants.

Presently, many analysis methods of phosphorylated proteins exist, but many established methods require several preparations like antibody or enzyme reactions [1–8].

First in this study, we examine the phosphorylation analysis method by measuring the infrared absorption peak of phosphate group that observed at about $1070 \, cm^{-1}$ (9.4 μm) with Fourier Transform–Infrared Spectrometer (FT–IR). Next, we attempt to control the quantity of phosphorylation, an action such as dephosphorylation without enzyme reactions, by irradiating 9.4 μm-Free Electron Laser (9.4 μm-FEL).

2. Materials and methods

2.1. Sample preparation

We use peptides and phosphorylated peptides (TORAY Research Center, Tokyo, Japan) in this study. Table 1 shows the molecular arrangement and mass of samples.

A 1 mg sample is dissolved into a 500 μl solution mixed with 5 ml ultra pure water, 5 ml methanol (CH_3OH) and 10 μl acetic acid (CH_3COOH). The prepared samples solutions are pipetted onto a BaF_2 crystal plate (13 mm in diameter, 1 mm in thickness) (GL Sciences Inc., Tokyo, Japan) and dried at room temperature. Two microlitre of the sample is pipetted four times at every spot. The subsequent samples are pipetted after drying the previous one.

2.2. Microscopic FT–IR method

The samples are examined by a microscopic FT–IR (FT-520, Horiba, Tokyo, Japan) to measure the infrared absorption spectra. The measurements are conducted by the transmittance method and the size of transmittance area is a 100 μm × 100 μm square. The measured area in wavenumbers is $700–4000 \, cm^{-1}$, the gain is 1 and the resolution is $2 \, cm^{-1}$.

2.3. Experimentation of 9.4 μm-FEL irradiation

A peptide phosphorylated threonine (C02059B) is irradiated by MIR–FEL delivered from iFEL, Osaka University. The wavelength is 9.4 μm corresponding to the infrared absorption peak of phosphate group in phosphorylated peptides. The MIR–FEL at iFEL has a double pulse structure: a macropulse and a micropulse. The pulse widths/time intervals of the macropulse and the micropulse are ∼15 μs/0.1 s and ∼5 ps/44.8 ns, respectively. Each macropulse consists of ∼330 micropulses [9]. The FEL spot size is estimated to be 260 μm × 400 μm in FWHM. The average power is 5–10 mW and the average power density is $6.1–12.2 \, W/cm^2$. The exposure time is 1–10 s. The effects of 9.4 μm-FEL irradiation are estimated by comparing the before and after infrared absorption spectra measured by FT–IR.

Table 1
The details of samples

Sample name	Molecular arrangement	Molecular mass [D]	Spectrum
C02059A	MHRQETVDC	1102	Fig. 2(a)
C02059B	MHRQET(PO_3H_2)VDC	1183	Fig. 2(b)
C02014A	MHRQESVDC	1088	Fig. 2(c)
C02014B	MHRQES(PO_3H_2)VDC	1169	Fig. 2(d)
C02014C	MHRQEYVDC	1164	Fig. 2(e)
C02014D	MHRQEY(PO_3H_2)VDC	1245	Fig. 2(f)

3. Results

3.1. Measured results by FT–IR

Fig. 1 shows the infrared absorption spectra of peptides and phosphorylated peptides (q.v. Table 1). In the infrared absorption spectra of each phosphorylated peptide, the peak of phosphate group are observed at (C02059B): $1061 \, cm^{-1}$ ($9.43 \, \mu m$), (C02014B): $1068 \, cm^{-1}$ ($9.36 \, \mu m$), (C02014D): $1076 \, cm^{-1}$ ($9.29 \, \mu m$) in wavenumbers (wavelength).

Fig. 2 shows the wavenumber distribution in the measured peaks of the infrared absorption peak of phosphate group. The peaks are measured at 1064–$1070 \, cm^{-1}$ (average: $1064 \, cm^{-1}$) when the phosphorylated amino acid is threonine, 1070–$1076 \, cm^{-1}$ (average: $1073 \, cm^{-1}$) for serine, and 1082–$1093 \, cm^{-1}$ (average: $1086 \, cm^{-1}$) for tyrosine.

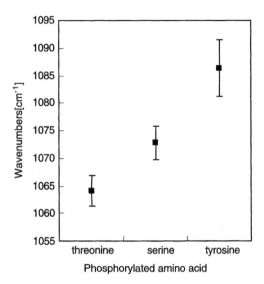

Fig. 2. The wavenumber distribution in the measured peaks of the infrared absorption spectra of phosphate group.

3.2. Results of 9.4 µm-FEL irradiation

Fig. 3 shows the infrared absorption spectra before and after the $9.4 \, \mu m$-FEL irradiation for 10 s. Fig. 3(a) is for non-irradiation; Figs. 3(b) and (c) are for 6.1 and $8.6 \, W/cm^2$ (average power density), respectively. In case of $6.1 \, W/cm^2$, irradiation effects are not observed. In case of $8.6 \, W/cm^2$, the infrared absorption peak of the phosphate group ($1064 \, cm^{-1}$) is reduced.

4. Considerations

4.1. Infrared absorption peak of phosphate group

The peak derived from phosphorylation is measured in 1056–$1093 \, cm^{-1}$. This is the infrared absorption peak of the phosphate group, attributed to mainly the O–P–O–stretching vibration in about 1000–$1100 \, cm^{-1}$ [10].

Phosphorylation causes changes in any remaining groups of threonine, serine, and tyrosine. In case of this study, because the sample's peptides have a similar amino-acid sequence except for the phosphorylated site, it is possible to identify the phosphorylated sites by comparing each peak. For example, the infrared absorption peak tended

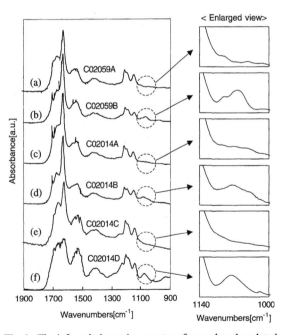

Fig. 1. The infrared absorption spectra of non-phosphorylated and phosphorylated peptides measured by microscopic FT–IR. C02059B, C02014B and C02014D are phosphorylated peptides, and their phosphorylated sites are for threonine, serine and tyrosine, respectively.

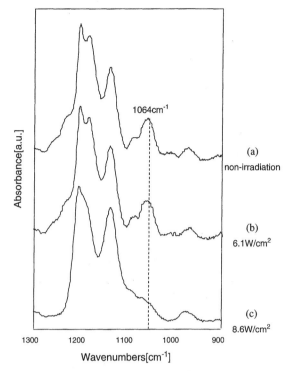

Fig. 3. The infrared absorption spectra before and after FEL irradiation. After FEL irradiation, when average power density is 8.6 W/cm^2, the peak of the phosphate group decreased.

towards high wavenumbers in the case of threonine, is about 9 cm^{-1} if the phosphorylated amino acid is serine, and about 22 cm^{-1} if it is tyrosine. Thus, we can identify phosphorylated amino acids for the same amino acid sequences that have a different phosphorylated site.

4.2. Comparison between FT–IR method and the established one

There are many phosphorylation analysis methods in the life science field [1–8]. The following are representative examples of established phosphorylation analysis. enzyme-linked immunosorbent assay (ELISA) is the method that makes use of antigen–antibody reaction and fluorescence spectrometry. It enables the identification of phosphorylation sites easily by using a specific antibody on the phosphorylated amino acid. Differential display method by mass spectrometry with a stable isotope or radioisotope enables not only identification of phosphorylation but also a quantitative analysis of phosphorylation [2].

This study shows that it is possible to identify phosphorylation and phosphorylated amino acids by comparing similar amino-acid sequences that have a different phosphorylated site, using the FT–IR method. The advantage of this method is that it does not require preparation, such as enzyme and antibody reactions. In addition, the measuring time is comparatively shorter than that of other established methods. However, it is difficult to identify phosphorylated sites under all circumstances and measure them quantitatively.

4.3. Effects of 9.4 μm-FEL irradiation

The infrared absorption peak of the phosphate group, at about 1070 cm^{-1}, is reduced after 9.4 μm-FEL irradiation. The lowering of the infrared absorption peaks of phosphate groups leads to a quantitative decrease in the phosphate group in a phosphorylated peptide, because the level of the infrared absorption peak reflects the quantity of measuring targets such as a molecular bond and a functional group. This effect can be regarded as dephosphorylation without an enzyme reaction. It may be presumed that this reducing is peak shifting to the side of high wavenumbers. We would now like to go on to develop this photochemical or thermal reaction like dephosphorylation by examining under several conditions, for example, by controlling the macropulse width of MIR–FEL in detail.

5. Conclusion

This study indicates that it is possible to identify a phosphorylation by measuring the infrared absorption peak of phosphate group observed at about 1070 cm^{-1} with FT–IR method. In this method, identifying only phosphorylation, is comparatively easier and faster than established methods. As long as target peptides have the same amino-acid sequence, it is possible to identify the phosphorylated sites (threonine, serine and tyrosine). We would now like to go on to examine peptides or proteins that have different amino acid sequences and/or high molecular mass and/or multiple phosphate groups.

The infrared absorption peak of the phosphate group is reduced after the 9.4 μm-FEL irradiation (average power density 8.6 W/cm^2; irradiation time 10 s). The detail of this phenomenon is unclear. We might go on to an even more detailed examination of the effects of 9.4 μm-FEL irradiation.

References

[1] T. Isobe, et al., The analysis methods of proteome, Youdosya (2000).

[2] Y. Oda, et al., Nat. Biotechnol. 19 (2001) 379.

[3] E. Yamauchi, et al., J. Biol. Chem. 273 (1998) 4367.

[4] M. Matsubara, et al., J. Biol. Chem. 271 (1996) 21108.

[5] H. Taniguchi, et al., J. Biol. Chem. 269 (1994) 22481.

[6] T. Sasaki, et al., J. Biol. Chem. 277 (2002) 36032.

[7] Y. Yamakata, Proteins, Nucleic Acids Enzymes 47 (2002) 51 (in Japanese).

[8] H. Yamamoto, Proteins, Nucleic Acids Enzymes 47 (2002) 241 (in Japanese).

[9] H. Horiike, et al., Jpn. J. Appl. Phys. 41 (2002) 10.

[10] H. Horiguchi, The book of infrared absorption spectra, Sankyo (1973) 322.

Available online at www.sciencedirect.com

SCIENCE ⓓ DIRECT•

Nuclear Instruments and Methods in Physics Research A 528 (2004) 619–622

**NUCLEAR
INSTRUMENTS
& METHODS
IN PHYSICS
RESEARCH**
Section A

www.elsevier.com/locate/nima

Isomer separation of bis(trifluoromethyl)benzene by use of mid-infrared free electron laser

Keiji Nomaru[a,b,*], Sergey R. Gorelik[a], Haruo Kuroda[a], Koichi Nakai[a]

[a] *IR FEL Research Centre, Institute for Science and Technology, The Tokyo University of Science, 2641 Yamazaki, Noda, Chiba 278-8510, Japan*
[b] *Technical Institute, Kawasaki Heavy Indstries Ltd., 118 Futatsuzuka, Noda, Chiba 278-8585, Japan*

Abstract

We have demonstrated the possibility of the isomer separation by means of the multi-photon dissociation induced by the irradiation with FEL light, by using a gaseous mixture of 1,3- and 1,4-bis(trifluoromethyl)benzene as the sample. The both isomers have CF_3 antisymmetric stretching vibration mode at $1284\,cm^{-1}$ for the 1,3-isomer and $1324\,cm^{-1}$ for the 1,4-isomer. By the irradiation of FEL light to the mixture at the power of $14\,mJ$, the multiphoton dissociation (MPD) of the mixture has occurred effectively. We have obtained the MPD spectrum of the mixture in the frequency region from 1333 to $1253\,cm^{-1}$. At the frequency region lower than $1290\,cm^{-1}$, the 1,3-isomer was mainly dissociated and the selectivity of the dissociation exceeded 100 when the FEL frequency was tuned to $1253\,cm^{-1}$.
© 2004 Published by Elsevier B.V.

PACS: 33.80.−b

Keywords: FEL; Infrared; Multi-photon; Isomer separation; Bis(trifluoromethyl)benzene

1. Introduction

Mid-infrared free electron laser (MIR-FEL) is an ideal light source for selective photochemical reaction because of the strong monochromatic light pulse and the wide wavelength tuneability. Several researches have been done relating to the isotope separation [1–4] utilizing these advantages.

The product of a chemical synthesis is often composed of a mixture of isomers, and it is one of the important problem in the chemical industry to isolate the specified isomer among the product of a synthetic process. The different isomers (excepting chiral molecules) usually show partly different mid-infrared spectra depending on their molecular structures. Thus, it is possible to remove a specified isomer(s) from the product of chemical synthesis by utilizing the muti-photon dissociation with a strong light beam from an infrared free electron laser.

In this paper, we will report the experimental results of an isomer separation of bis(trifluoromethyl)

*Corresponding author. Technical Institute, Kawasaki Heavy Indstries Ltd., 118 Futatsuzuka, Noda, Chiba 278-8585, Japan. Tel.: +81-4-7124-1501x6200; fax: +81-4-7123-9839.

E-mail address: keiji_aus@com.home.ne.jp (K. Nomaru).

0168-9002/$ - see front matter © 2004 Published by Elsevier B.V.
doi:10.1016/j.nima.2004.04.114

benzene, which were carried out by use of the MIR-FEL at the IR FEL Research Centre of the Tokyo University of Science.

2. Experiments

In case of the MIR-FEL in the IR FEL Research Centre, the wavelength can be tuned from 4 to 16 μm. The maximum macro pulse energy in this wavelength range was 65 mJ. The pulse width and the repetition rate of the macro pulse were 0.5–2 μs and 5 Hz, respectively. The micro pulse composing the macro pulse had the time width of 2 ps and the repetition rate was 2856 MHz [5].

Schematic diagram of an experimental setup used in the present experiment, is shown in Fig. 1. The experiments of the isomer separation were carried out in the gaseous phase. The sample gas, the mixture of 1,3-bis(trifluoromethyl)benzene and 1,4-bis(trifluoromethyl)benzene, was filled in the L130 mm × ∅ 40 mm cylindrical stainless-steel cell with NaCl windows at the both ends. The FEL beam was irradiated through the NaCl window. It was focused by the ZnSe lens with the focal length of 250 mm. The infrared spectra of the mixture before and after the irradiation of the FEL light were measured by a FT-IR spectrometer (FT-IR/615; JASCO).

Both of the isomer were liquid under the normal condition and were used in the gaseous phase by reducing the pressure less than a few hundreds Pa. The infrared spectra of the both isomers at the pressure of 133 Pa were shown in Fig. 2. Although most of the intense peak appears at the similar

position, we can identify two strong absorption peaks with large wavenumber gap near 1300 cm^{-1}. These peaks corresponds to the CF_3 antisymmetric stretching vibration modes of the 1,3-isomer (1284 cm^{-1}) and 1,4-isomer (1324 cm^{-1}), respectively. We can expect a good selectivity because the separation the two peaks is 40 cm^{-1} is significantly larger than the spectral bandwidth of FEL light.

Through the experiments, the 1,3- and 1,4-isomers were filled as a mixture in the cell by the ratio of 3:7 at the total pressure of 13 Pa and the FEL light with the macro pulse energy of 14 mJ were irradiated by the macro-repetition rate of 5 Hz.

3. Results and discussions

The MPD rates of the isomers in the mixture, which is defined by the dissociated ratio of the each isomer per single shot of the FEL macro

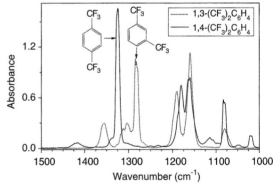

Fig. 2. The infrared spectra of the 1,3- and the 1,4-bis (trifluoromethyl)benzene at the pressure of 133 Pa.

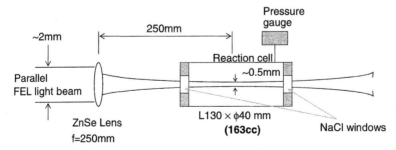

Fig. 1. The experimental setup for the isomer separation.

pulse, were measured for the different irradiation wavenumber of the FEL from 1333 to 1253 cm^{-1}. As shown in Fig. 3(a), the irradiation wavenumber dependence can be seen clearly for both isomers. The MPD rates were highest at $v_{FEL} = 1253$ cm^{-1} for the 1,3-isomer and at $v_{FEL} = 1303$ cm^{-1} for the 1,4-isomer, respectively. Here, we noted that the maximum MPD of the 1,3-isomer was higher than that of the 1,4-isomer notwithstanding the absorbance of the 1,3-isomers smaller than that of the 1,4-isomer. At this moment we do not have a clear explanation for this phenomenon.

The ratio of the MPD rates of the two isomers was expressed as the selectivity ($S = k_{\text{1,3-isomer}}/k_{\text{1,4-isomer}}$) in Fig. 3(b). Around $v_{FEL} = 1313$ cm^{-1}, the 1,4-isomer is dissociated faster than the 1,3-isomer and the selectivity was 0.1. On the other hand, the selectivity exceeded 100 at

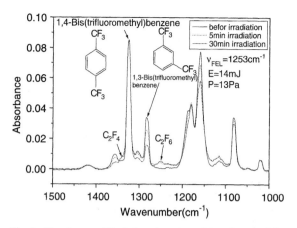

Fig. 4. The change of the infrared spectra of the mixture of the 1,3- and the 1,4-isomers of the bis(trifluoromethyl)benzene during the irradiation of MIR-FEL light at $v_{FEL} = 1253$ cm^{-1}.

$v_{FEL} = 1253$ cm^{-1} and only the 1,3-isomer was selectively dissociated. This can be seen from the spectral data shown in Fig. 4.

Here, we briefly discuss about the reaction mechanism. After the irradiation, the some peaks appeared in the spectra in Fig. 4. The peaks at 1120, 1250 and 1340 cm^{-1} are identified to be C_2F_6, C_2F_4 and C_2F_6, respectively. Although the spectrum of the dissociation of the 1,4-isomer is not shown here, the same products appeared. From these products, the main MPD pathways is considered to be expressed as follows.

$$(CF_3)_2 - C_6H_4 + nh\nu \rightarrow CF_3 - \bullet C_6H_4 + \bullet CF_3, \tag{1}$$

$$2 \bullet CF_3 \rightarrow C_2F_6. \tag{2}$$

Details of the reaction mechanism and the products have not been analyzed yet because of the complexity of the molecule. This will be studied in the next stage of our work.

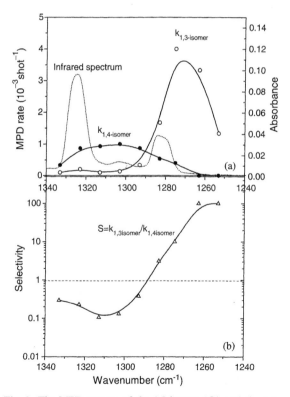

Fig. 3. The MPD spectra of the 1,3-isomer (○) and the 1,4-isomer (●) of the bis(trifluoromethyl)benzene by the irradiation of MIR-FEL (a) and the corresponding selectivity (△) (b). The dotted line represents the infrared spectrum of the mixture of the bis(trifluoromethyl) benzene.

4. Conclusion

The isomer separation of the bis(trifluoromethyl) benzene has been successfully carried out by use of the MIR-FEL. The dissociation rates of the two isomers have shown clear wavelength dependence. The selectivity of the 1,3-isomer exceeded 100 when the irradiated wavenumber of the FEL was $1253\,cm^{-1}$. Further experiments must be done for the understanding of the reaction mechanism and the difference of the dissociation rates between two isomers.

Acknowledgements

This study was carried out by the Grant-in-Aid for the Creative Scientific Research of JSPS, No.11NP0101.

References

[1] A.K. Petrov, E.N. Chesnokov, S.R. Gorelik, Yu.N. Nolin, K.D. Straub, E.B. Szarmes, J.M.J. Maydey, Nucl. Instr. and Meth. B 144 (1998) 203.

[2] S. Kuribayashi, T. Tomimasu, S. Kawanishi, S. Arai, Appl. Phys. B 65 (1997) 393.

[3] M. Hashida, M. Matsuoka, Y. Izawa, Y. Nagaya, M. Miyabe, Nucl. Instr. and Meth. A 429 (1999) 485.

[4] J.L. Lyman, B.E. Newnam, T. Noda, H. Suzuki, J. Phys. Chem. A 103 (1999) 4227.

[5] M. Yokoyama, F. Oda, K. Nomaru, H. Koike, M. Sobajima, M. Kawai, H. Kuroda, K. Nakai, Nucl. Instr. and Meth. A 507 (2003) 261.

Available online at www.sciencedirect.com

Nuclear Instruments and Methods in Physics Research A 528 (2004) 623–626

NUCLEAR INSTRUMENTS & METHODS IN PHYSICS RESEARCH
Section A

www.elsevier.com/locate/nima

Study of electron dynamics in *n*-type GaN using the Osaka free electron laser

T. Takahashi[a], T. Kambayashi[a], H. Kubo[a], N. Mori[a,*], N. Tsubouchi[b], L. Eaves[c], C.T. Foxon[c]

[a] *Department of Electronic Engineering, Osaka University, 2-1 Yamada-oka, Suita City, Osaka 565-0871, Japan*
[b] *Institute of Free Electron Laser, Osaka University, 2-9-5 Tsuda-Yamate, Hirakata, Osaka 573-0128, Japan*
[c] *School of Physics and Astronomy, University of Nottingham, Nottingham NG7 2RD, UK*

Abstract

We conducted an experimental and computational study on electron dynamics in an *n*-type GaN crystal using the Osaka free electron laser. We focus on effects of anisotropy of polar-optical-phonon scatterings on electron distribution under intense mid-infrared irradiation.
© 2004 Elsevier B.V. All rights reserved.

PACS: 78.20.−e; 72.20.Ht; 63.20.Kr

Keywords: Free electron laser; Optical properties; GaN

1. Introduction

Band-gap luminescence from various compound semiconductors under intense mid-infrared (mid-IR) irradiation by a free electron laser (FEL) has been observed [1]. An example of the FEL-induced luminescence spectra is shown in Fig. 1 for a wurtzite GaN crystal. The sample is *n*-type and the electron concentration is 2.2×10^{18} cm^{-3} at room temperature. In Fig. 1, we also plot a photo-luminescence spectrum obtained with a HeCd laser with 325 nm-wavelength for comparison. As can be seen in Fig. 1, the band-gap luminescence induced by the mid-IR FEL pulses is clearly observed. This FEL-induced luminescence is originated in the electron–hole pairs, which are generated through impact-ionization (II) induced by free-carrier heating [2,3]. In the previous full-band Monte Carlo (FBMC) study of the FEL-induced II in a wurtzite GaN crystal [4], the spherical parabolic-band approximation and an isotropic matrix element for the electron–polar-optical-phonon (POP) interaction have been assumed. The electron–POP coupling coefficient, however, is known to be different from a simple cubic form [5,6]. In the present work, we include the full-band and anisotropic nature of a wurtzite GaN crystal for the POP scatterings, and study how the electron-heating process is affected by it.

*Corresponding author. Tel.: +81-6-6879-7766; fax: +81-6-6879-7753.

E-mail address: mori@ele.eng.osaka-u.ac.jp (N. Mori).

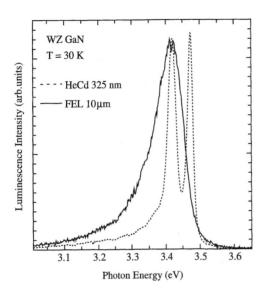

Fig. 1. Luminescence spectra of an *n*-type GaN crystal induced by mid-IR FEL pulses with 10 μm-wavelength (solid line) and that obtained with a 325 nm-wavelength HeCd laser (dotted line).

2. Scattering rates

We assumed the spherical parabolic-band and isotropic interaction approximations for POP scatterings for the Γ valley in the previous FBMC simulation [4]. For inter-valley scatterings we also assumed isotropic interaction. However, the electron–POP coupling coefficient is known to be different from a simple cubic form [5,6], and it is necessary to include the anisotropic nature of POP scatterings in order to obtain a quantitative understanding of electron dynamics in *n*-type GaN under intense FEL pulses. For this purpose, we adopt the Fröhlich-like electron–optical phonon interaction Hamiltonian derived by Lee and co-workers [5] in the present study.

Within Fermi's golden rule, the optical-phonon emission rate from an initial state k to a final state k' is written as

$$W(k, k') = \frac{2\pi}{\hbar} |M_q|^2 \delta(E_{k'} - E_k + \hbar\omega_q), \qquad (1)$$

where $q = k' - k$ is the phonon wavevector, ω_q is the phonon frequency, and M_q is a matrix element for the longitudinal-optical (LO)-like mode. The matrix element, M_q, is given by

$|M_q|^2 = \Delta(k, k')|C_q|^2(N_q + 1)$, where $N_q = \{\exp(\hbar\omega_q/k_B T) - 1\}^{-1}$ is the phonon occupation number, $\Delta(k, k')$ is an overlap parameter evaluated with pseudopotential results, and C_q is a coupling coefficient given by

$$|C_q|^2 = \frac{2\pi e^2 \hbar}{V q^2 \omega_q} \left[\frac{\sin^2\theta}{(1/\varepsilon_\perp^\infty - 1/\varepsilon_\perp^0)\omega_\perp^2} + \frac{\cos^2\theta}{(1/\varepsilon_z^\infty - 1/\varepsilon_z^0)\omega_z^2} \right]^{-1}. \qquad (2)$$

In this equation, $\omega_\perp$ (ω_z) is the phonon frequency perpendicular to (along) the *c*-axis, $\varepsilon_\perp^\infty$ (ε_z^∞) is the high-frequency dielectric constant perpendicular to (along) the *c*-axis, $\varepsilon_\perp^0$ (ε_z^0) is the static dielectric constant perpendicular to (along) the *c*-axis, θ is an angle of the phonon wavevector q with the *c*-axis, and $\omega_q = \{\omega_z^2 \cos^2\theta + \omega_\perp^2 \sin^2\theta\}^{1/2}$. The θ-dependence of the phonon frequency is so weak that the main anisotropic effect comes from the square-bracket term in Eq. (2).

Scattering rate, $W(k) = \sum_{k'} W(k, k')$, is transformed from the summation of delta-functions into integration in the reciprocal space over an isoenergy surface S,

$$W(k) = \int_S dS \frac{W(k, k')}{|\nabla_k E(k')|}. \qquad (3)$$

We applied the tetrahedron integration technique of Lehmann and Taut [7] for the overall computation of the scattering rate [6]. All the scattering paths in the irreducible Brillouin zone are stored in a table, ensuring minimization of errors associated with discretized angles and the singularity of the q^{-2}-term in Eq. (2). In the present study, we focus on effects of the full-band and anisotropic nature of a wurtzite GaN crystal for POP scatterings, and neglect the absorption process of the LO-like phonons and scatterings due to the transverse-optical (TO)-like mode.

Fig. 2 shows the calculated POP emission rate as a function of electron energy along the Γ–K direction. Open circles are results obtained by using Eqs. (1)–(3), and solid circles are those of neglecting the overlap parameter, $\Delta(k, k')$. Results within a parabolic band approximation of $m^* = 0.20m_0$ are also plotted by dotted line. We find that the parabolic band approximation significantly

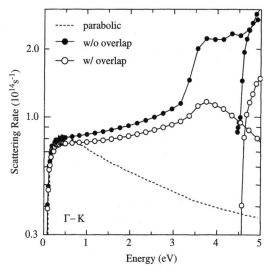

Fig. 2. Polar-optical-phonon emission rate as a function of electron energy along the Γ–K direction at $T = 300$ K.

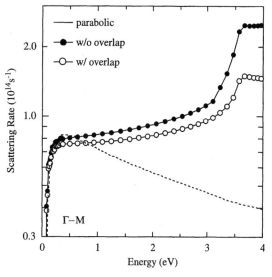

Fig. 3. The same as Fig. 2 but for the Γ–M direction.

deviates from the anisotropic model for electron energy, E, higher than ≈ 0.5 eV and underestimates the scattering rate for high-energy regions. We also find that the overlap parameter plays an important role only for very high electron energies ($E > 4$ eV). Similar features can be seen for the Γ–M direction as shown in Fig. 3.

3. Results

Electron dynamics in wurtzite GaN crystals under mid-IR FEL pulses can be calculated by performing an FBMC simulation [8–10]. Using the scattering rates calculated above, we performed an FBMC simulation with the numerical integration of the equation of motion

$$\hbar \frac{d\boldsymbol{k}}{dt} = -e\boldsymbol{E} \cos \omega t, \qquad (4)$$

where $\boldsymbol{E}$ is the FEL electric field strength, ω is the FEL frequency, and $\boldsymbol{k}$ is the electron wavevector. We assume that electrons continue to move within a band during a free-flight. For the II scattering rate R_{II}, we use an analytical form

$$R_{II} \; [\mathrm{s}^{-1}] = 1.98 \times 10^9 \; (E \; [\mathrm{eV}] - 3.4)^{5.84} \qquad (5)$$

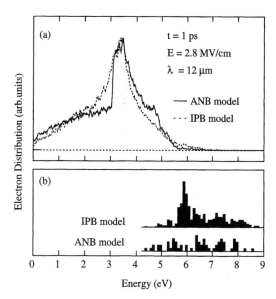

Fig. 4. Electron distribution (a) and histograms showing the impact-ionization occurrences (b) for FEL wavelength, $\lambda = 11$ μm and field strength, $E = 2.8$ MV/cm at $t = 1$ ps.

which is obtained by fitting the numerical results reported in Ref. [11].

Fig. 4 shows the electron distribution (a) and histograms showing the II occurrences (b) for FEL wavelength, $\lambda = 11$ μm and field strength, $E = 2.8$ MV/cm at $t = 1$ ps. The electric field is applied along the Γ–M direction. In Fig. 4(a), we plot the results obtained by the present model, namely the anisotropic non-parabolic-band (ANB) model, by solid lines and those obtained in the previous study [4], namely the isotropic parabolic-band (IPB) model, by dotted lines. The figure clearly shows difference between the ANB and the IPB models. Electrons are mostly confined in a rather narrow energy region between 3 and 5 eV for the ANB model, while the electron distribution spreads more widely in the energy space for the IPB model. This may be attributed to the fact that electrons are more easily scattered into satellite valleys for the IPB model because of large density of states in the satellite valleys and the q-independent final-state selection approximation. This difference in the electron distribution significantly affects the number of II occurrences. Because the II scattering rate exponentially depends on electron energy as given by Eq. (5), a small difference in the electron distribution at higher energy tail results in a large difference in the number of II occurrences as shown in Fig. 4(b).

4. Summary

We have performed FBMC simulation of electron-heating processes in a wurtzite GaN crystal under intense mid-IR irradiation, and find that the anisotropic non-parabolic nature for the POP scatterings significantly affect the steady-state electron distribution. We also find that the ANB model predicts smaller number of II occurrences compared to the IPB model.

References

[1] N. Mori, T. Takahashi, T. Kambayashi, H. Kubo, C. Hamaguchi, L. Eaves, C.T. Foxon, A. Patanè, M. Henini, Physica B 314 (2002) 431.
[2] N. Mori, H. Nakano, H. Kubo, C. Hamaguchi, L. Eaves, Physica B 272 (1999) 431.
[3] H. Nakano, H. Kubo, N. Mori, C. Hamaguchi, L. Eaves, Physica E 7 (2000) 555.
[4] N. Mori, et al., presented at 12th International Conference on Nonequilibrium Carrier Dynamics in Semiconductors, Santa Fe, NM, USA, August 27–31, 2001.
[5] B.C. Lee, K.W. Kim, M. Dutta, M.A. Stroscio, Phys. Rev. B 56 (1997) 997.
[6] C. Bulutay, B.K. Ridley, N.A. Zakhleniuk, Phys. Rev. B 62 (2000) 15754.
[7] G. Lehmann, M. Taut, Phys. Status Solidi B 54 (1972) 469.
[8] H. Shichijo, K. Hess, Phys. Rev. B 23 (1981) 4197.
[9] J.Y. Tang, K. Hess, J. Appl. Phys. 54 (1983) 5139.
[10] M.V. Fischetti, S.E. Laux, Phys. Rev. B 38 (1988) 9721.
[11] J. Kolnik, I.H. Oguzman, K.F. Brennan, J. Appl. Phys. 79 (1996) 8838.

Available online at www.sciencedirect.com

SCIENCE  DIRECT°

Nuclear Instruments and Methods in Physics Research A 528 (2004) 627–631

NUCLEAR INSTRUMENTS & METHODS IN PHYSICS RESEARCH
Section A

www.elsevier.com/locate/nima

The investigation of excited states of Xe atoms and dimers by synchronization of FEL and SR pulses at UVSOR

Tatsuo Gejo[a,*], Eiji Shigemasa[a], Eiken Nakamura[a], Masahito Hosaka[a], Shigeru Koda[a], Akira Mochihashi[a], Masahiro Katoh[a], Jun-ichiro Yamazaki[a], Kenji Hayashi[a], Yoshifumi Takashima[a], Hiroyuki Hama[b]

[a] *Institute for Molecular Science, Myodaiji, Okazaki 444-8585, Japan*
[b] *Laboratory of Nuclear Science, Tohoku University, Sendai 980, Japan*

Abstract

Pump and probe experiments using storage ring free electron laser (SRFEL) and synchrotron radiation (SR) pulses have been tried to perform for the last decade, because the natural synchronization of the SRFEL pulse with the SR pulse makes it easier to do such experiments. Recently, we have successfully carried out the two-photon double-resonant excitation on Xe atoms, utilizing a SR pulse as a pump and an FEL pulse as a probe light. Here, we report the recent experimental results of the investigation for excited states of Xe dimers as well as Xe atoms. The results on the pressure-dependent measurements that were applied to the same system, strongly suggest the formation of Xe dimers in the intermediate states prepared by the primary pump processes.
© 2004 Elsevier B.V. All rights reserved.

PACS: 41.60.Ap; 41.60.Cr

Keywords: Synchronization; Pump and probe technique; Synchrotron radiation; FEL; Xe dimer

1. Introduction

Storage ring free electron laser (SRFEL or FEL) has been developed at many synchrotron radiation (SR) facilities all over the world as a powerful light source, owing to its unique potential capacity such as high power, high coherence and unique

temporal feature [1–5]. Each FEL pulse naturally synchronizes with the SR one, and therefore provides an opportunity for performing two-color pump/probe experiments in a one-to-one shot ratio. In this context, pump and probe experiments using FEL and SR pulses have been tried to carry out for the last decade. The first two-color experiment using SRFEL and SR was performed at Super-ACO in 1995 [2,3]. The surface photo voltage effect induced by FEL UV photons (at 350 nm) on a semiconductor's surface was measured by SR pulses as a function of the FEL/SR

*Corresponding author. Himeji Institute of Technology, Kouto 3-2-1, Kimigori, Hyogo 678-1297, Japan. Fax: +81-791-58-0132.

E-mail address: gejo@sci.himeji-tech.ac.jp (T. Gejo).

delay [4]. Recently, pump/probe experiments combining the SRFEL and the infrared SR pulses are in progress at the same facility [5].

Very recently, as the first gas-phase experiment combining FEL with SR, we have carried out the two-photon double-resonant excitation on Xe atoms with success, utilizing an SR pulse as a pump and an FEL pulse as a probe light. Here, we report the present status and recent results of the combined experiments with FEL and SR at UVSOR. The undulator radiation as well as the conventional SR from a bending magnet were used as a pump source.

2. Experiments

2.1. The performance of SRFEL UVSOR

The performance of FEL at UVSOR is summarized in Table 1. Since the power of one FEL pulse exceeds 100 nJ (CW mode [6]) or 10 μJ (Q-switching mode [6]) around 300 nm wavelength region, one can consider the FEL in UV range to be a useful power source. Note that such a high performance can hardly be realized by any mode-locked laser.

During the experiment, there were serious background signals due to the scattered stray light of SR pulses (typically about 10^5 counts/s), which made it difficult to detect the real ion signals. Therefore, for searching the real signals, we temporarily employed a Q-switching technique [6]. In this technique, much larger peak power of FEL is provided, although the duration of lasing becomes relatively short (~ 0.2 ms). However, if

events are selected only during this duration, the improvement of signal-to-noise ratio (S/N) by a factor of 100 can be achieved. At UVSOR, the Q-switching technique has been performed by a repetitive jump of the RF frequency.

2.2. Pump/probe experiments

The experiments were conducted both on the undulator beamline BL3A1 and on the bending-magnet-based beamline BL7B at UVSOR. At BL3A1, no monochromator is installed. Therefore, an LiF filter was employed to suppress higher-order harmonics of the undulator radiation. The FEL pulses were extracted through the backward mirror and transported to experimental stations through a series of multi-layer mirrors. The flight path of FEL, which was adjusted to synchronize timing between the FEL and the SR pulses, was about 30 m. A focusing mirror ($f = 10$ m) was placed in the center of the flight path to keep the beam size of FEL small throughout the transport. About 69% of the extracted power were transferred to the experimental station. Fine-tuning of the delay timing was made by using a movable optical delay system (50 cm) at the experimental station. The FEL and SR pulses introduced coaxially crossed an effusive jet of Xe atoms from a gas nozzle. The singly charged Xe ions produced in the interaction region were detected by means of a conventional channeltron.

3. Results

3.1. Excited states for Xe atoms

The first experimental results on the two-photon double-resonant excitation of the Xe* $5p^5nf'$ autoionization states using the combination of a mode-locked laser and SR have already been demonstrated by Meyer's group at LURE [7]. In the present work, the combination of FEL and the undulator radiation was chosen in place of the former. The peak position of the fundamental harmonic of the undulator was adjusted to be 10.4 eV, in order to prepare the Xe* $5p^55d$ $[5/2]_1$

Table 1
Performance of UVSOR-FEL

Wavelength	240 nm–590 nm
Average power	~ 0.5 W (max: 1.2)
Repetition rate	11.25 MHz
Temporal width	10 ps
Linewidth ($d\lambda/\lambda$)	10^{-4}
Tunability	10–25 nm
Maximum pulse energy	100 nJ (CW)
	10 μJ (Q-switch)

intermediate states (10.401 eV) as a first step. The undulator radiation has a certain energy spread including 10.593 eV photons, which can excite Xe atoms in their ground state to the Xe* $5p^5 7s$ intermediate states. However, taking account of the oscillator strength for each excitation and the photon intensity of the undulator radiation at each photon energy, it is found that the contribution of the Xe* $5p^5 7s$ states is less than 10% of the Xe* $5p^5 5d$ states. The Xe* $5p^5 4f'$ $[5/2]_2$ autoionization resonance, which lies within the wavelength region covered by FEL, was excited as a second step. Because the lifetime of the intermediate states is quite short (600 ps) [7], the synchronization between the SR and laser pulses is essential in this experiment.

The ion yield spectra for the autoionization Xe* $5p^5 4f'$ $[5/2]_2$ resonance measured with two different operational modes for FEL are shown in Fig. 1. In these measurements, the wavelength of FEL was swept by changing the gap of the helical optical klystron. Using the Q-switching technique, we optimized the spatial and temporal overlap between the FEL and SR pulses to reduce the background (Fig. 1(a)). After the precise alignment of the beams and the gas nozzle position, we have measured the ion yield spectrum using a CW FEL (Fig. 1(b)). During the measurements, a newly developed feedback system [8] was operated to stabilize the lasing. Since the bandwidth of CW FEL is much narrower than that of the Q-switching FEL, the quality of the spectrum was considerably improved. The asymmetric line shape described by the Fano formula [7] has been clearly observed in Fig. 1(b).

3.2. Excited states for Xe dimers

In order to investigate the excited states for Xe dimers, we have performed pressure-dependent measurements for the same system on the beamline BL7B, where a normal incidence monochromators is installed. The base pressure of the experimental chamber was increased by a factor of 10, relative to the measurements in the previous chapter. Fig. 2(a) shows the excitation spectrum near the Xe* $5p^5 5d$ $[5/2]_1$ resonance region obtained by setting the FEL wavelength to the maximum of the 5d →

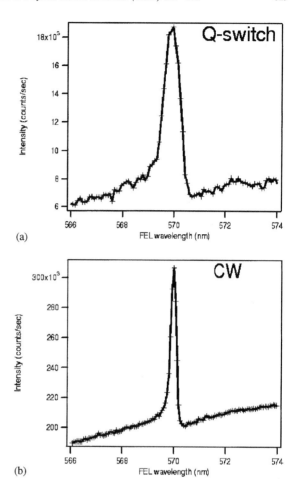

Fig. 1. Two-photon ionization signal from the autoionization Xe* $5p^5 4f'$ $[5/2]_2$ resonance of Xe atoms as a function of the wavelength of FEL, in two different operational modes.

4f' transition. In this measurement, the entrance- and exit-slit openings of the monochromator at BL7B were set at 500 μm. The corresponding resolution ($\lambda/\Delta\lambda$) of the SR pulses was about 500. Fig. 2(b) represents the ion yield spectrum in the vicinity of the autoionization Xe* $5p^5 4f$ resonance state, which was obtained by setting the SR wavelength to the maximum of the 5p → 5d transition. In this measurement, the slit width were kept wide (2000 μm), resulting in very wide energy spread for the pump SR pulses, which enables to excite both Xe atoms and Xe dimers simultaneously.

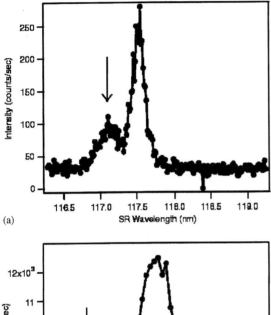

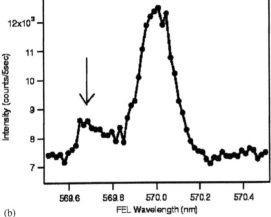

Fig. 2. Two-photon ionization signals as a function of the wavelength of SR (a), and that of FEL (b). The stronger peak in each figure is due to the two-photon double resonance excitation of Xe atoms, while the peak indicated by the arrow may correspond to that of Xe dimers.

The stronger peak in each figure is due to the two-photon double resonance excitation of Xe atoms. The clear enhancement just below the Xe* $5p^5 5d$ resonance indicated by the arrows are observed both in Fig. 2(a) and (b), which have not been detected in the previous measurements at lower pressure. This result strongly suggests that the newly found structures are relevant to the formation of Xe clusters, mainly dimers. The energy difference between two peaks in Fig 2(a) is about $320 \, cm^{-1}$, while that in Fig 2(b) is about $15 \, cm^{-1}$.

The binding energy of the Xe dimers in the ground state is about $196 \, cm^{-1}$ [9]. The value of $320 \, cm^{-1}$ for the energy difference indicates that the energy of the Xe dimers may exceed the dissociation energy for $Xe(^1S_0) + Xe(5d \, [3/2]_1)$. Castex et al. have already observed the same band that locate at $430 \, cm^{-1}$ from the atomic line $5p^5 5d$ by the absorption spectrum of Xe dimer in a pressure of 15 Torr [10,11]. They suggested that there may be a potential hump arising from the 1_u molecular state character [11]. On the basis of the IR wavelengths detected by exciting the Rydberg state in this region with laser, Tsukiyama and Kasuya reported that the predissociating level is a repulsive state dissociating to $Xe + Xe^*(6p)$ [12]. Later on, Mao et al. have measured the same band by using a VUV laser, and have suggested two possible electronic symmetries for the state observed: one is 1_u state which dissociates to Xe^* 7s $[3/2]_1$ and another is 0_u^+ state which dissociates to Xe^* 7s $[3/2]_2$ [13]. The assignment of this band is not well established. In Fig. 2(a), the FEL wavelength was set to the maximum of the $5d \rightarrow 4f'$ transition. This implies that the band near $Xe(5d \, [3/2]_1)$ atomic line is associated with the $Xe(5d \, [3/2]_1)$ dissociation limit, which supports the interpretation by Castex et al. [11]; namely, the band observed is due to the 1_u state dissociating to the limit of $Xe(^1S_0)$ and $Xe(5d \, [3/2]_1)$.

Although the assignment of the small peak is unknown at 569.7 nm in Fig. 2(b), this is probably due to the molecule Rydberg state converging to $Xe(^1S_0) + Xe(5p^5 4f' \, [5/2]_2)$. No potential energy curves for excited states of Xe_2 dissociating to the autoionization Rydberg states have been studied either experimentally or theoretically [14]. To our knowledge, this is the first observation of the Xe dimer states whose dissociation limit is converging to the autoionization Rydberg states of Xe atoms.

We have also investigated in the energy region corresponding to the Xe^* $5p^5 nf'$ ($n > 10$) autoionization states, by scanning the wavelength of FEL from 410 to 430 nm. The 10.4 eV undulator radiation was used as a pump source in order to excite Xe atoms and molecules to the intermediate state as many as possible. Meyer's group observed the Rydberg series from $5p^5 7f'$ to $5p^5 13f'$ by exciting via Xe^* $5p^5 6d$ intermediate state [7].

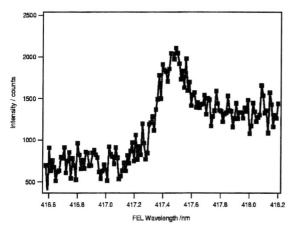

Fig. 3. Two-photon ionization signal of Xe dimers as a function of the wavelength of FEL.

However, we could not observe any atomic line arising from these levels, except for one peak at around 417.5 nm as shown in Fig. 3. This peak is presumably due to Xe dimers because the peak appeared in the same condition as the experiments above. The potential energy curves for several excited states of Xe_2 have been intensively investigated both experimentally and theoretically [14], but no information about the excited states in this energy region is obtainable. Since the pump SR pulses includes the excited states which dissociate into $Xe(^1S_0) + Xe(5d\ [3/2]_1)$, the band observed here may be the autoionization state converging to C $^2\Pi^+_{1/2u}$ (13.28 eV) or $D^2\Sigma^+_{1/2g}$ (13.36 eV) ionic state [15], based on the fact that this peak has a total excitation energy of 13.4 eV.

4. Conclusion

As the first gas-phase experiment combining FEL with SR, we have carried out the two-photon double-resonant excitation on Xe atoms and Xe dimers, utilizing a SR pulse as a pump and an FEL pulse as a probe light. As a result, we have observed for the first time the excited state of Xe dimers whose dissociation limit is converging to the autoionization Rydeberg states. This is one typical example for demonstrating the usefulness of the SR–FEL pump/probe technique, because the further extension of the photon energy region of interest towards the EUV region, which allows us to attain higher energy Rydberg levels, is easy for the SR pulses.

References

[1] M.E. Couprie, R. Bakker, D. Garzella, L. Nahon, M. Marsi, F. Merola, T. Hara, M. Billardon, Nucl. Instr. and Meth. A 375 (1996) 630.

[2] M.E. Couprie, D. Garzella, T. Hara, J.H. Codabox, M. Billardon, L. Nahon, M. Marsi, Nucl. Instr. and Meth. A 358 (1995) 374.

[3] M. Marsi, M.E. Couprie, L. Nahon, D. Garzella, R. Bakker, A. Delboulbé, D. Nutarelli, R. Roux, B. Visentin, C. Grupp, G. Indlekofer, G. Panaccione, A. Taleb-Ibrahimi, M. Billardon, Nucl. Instr. and Meth. A 393 (1997) 548.

[4] M. Marsi, R. Belkhou, C. Grupp, G. Panaccione, A. Taleb-Ibrahimi, L. Nahon, D. Garzella, D. Nutarelli, E. Renault, R. Roux, M.E. Couprie, M. Billardon, Phys. Rev. B 61 (2000) R5070.

[5] L. Nahon, E. Renault, M.E. Couprie, F. Mérola, P. Dumas, M. Marsi, A. Taleb-Ibrahimi, D. Nutarelli, R. Roux, M. Billardon, Nucl. Instr. and Meth. A 429 (1999) 489.

[6] M. Hosaka, S. Koda, M. Katoh, J. Yamazaki, K. Hayashi, K. Takashima, T. Gejo, E. Shigemasa, H. Hama, Nucl. Instr.and Meth. A 483 (2002) 146.

[7] M. Gisselbrecht, A. Marquette, M. Meyer, J. Phys. B 31 (1998) L977.

[8] S. Koda, M. Hosaka, J. Yamazaki, M. Katoh, H. Hama, Nucl. Instr. and Meth. A 475 (2001) 211.

[9] D. Green, S.C. Wallace, J. Chem. Phys. 110 (1994) 6129.

[10] M.-C. Castex, Chem. Phys. 5 (1974) 448.

[11] M.-C. Castex, N. Damany, Chem. Phys. Lett. 13 (1972) 158.

[12] K. Tsukiyama, T. Kasuya, J. Mol. Spectrosc. 151 (1992) 312.

[13] D.M. Mao, X.K. Hu, S.S. Dimov, R.H. Lipson, J. Mol. Spectrosc. 181 (1997) 435.

[14] C. Join, F. Spiegelmann, J. Chem. Phys. 117 (2002) 3059.

[15] Y. Lu, Y. Morioka, T. Matsui, T. Tanaka, H. Yoshii, R.I. Hall, T. Hayaishi, K. Ito, J. Chem. Phys. 102 (1995) 1553.

Available online at www.sciencedirect.com

Nuclear Instruments and Methods in Physics Research A 528 (2004) 632–635

ELSEVIER

NUCLEAR INSTRUMENTS & METHODS IN PHYSICS RESEARCH
Section A

www.elsevier.com/locate/nima

Application of a portable pulsed magnet to magneto-spectroscopy using FEL

Yasuhiro H. Matsuda*, Yuji Ueda, Hiroyuki Nojiri

Department of Physics, Faculty of Science, Okayama University, 3-1-1 Tsushimanaka, Okayama 700-8530, Japan

Abstract

A very small pulsed high magnetic field generator that can be applicable to magneto-spectroscopy using a free electron laser (FEL) has been developed. Although, it is generally difficult to perform a high magnetic field experiment over 20 T since high electric power and large space are required, we can make it rather easy by this small pulsed magnet system. We have made magneto-transmission experiment using a millimeter wave coherent radiation from a linear accelerator and observed the electron spin resonance. Possible study using a combination of the pulsed magnet and a FEL in the terahertz range is also discussed.
ⓒ 2004 Elsevier B.V. All rights reserved.

PACS: 07.55.Db; 07.57.Ty; 78.20.Ls

Keywords: Pulsed high magnetic field; Terahertz radiation; Magneto-spectroscopy

1. Introduction

Magneto-spectroscopy is one of the most powerful techniques to investigate small energy phenomena in solids such as the inter- and intra-band transitions, the cyclotron resonance and the electron spin resonance (ESR), and so on. In high magnetic fields exceeding 20 T the cyclotron energy and the Zeeman energy can be in the terahertz (THz) range and become comparable to other energy parameters such as phonon energies or quantum confinement potential. Therefore, it is envisioned that the new phenomena due to some multiple excitations are observed by the THz spectroscopy in high magnetic fields. Moreover, typical time scale of carrier dynamics, e.g. scattering, tunneling, etc., is in this frequency range. Since ultra-fast electronics, opt-electronics, and possibly spintronics are developing very rapidly, it is also technologically important to understand the THz-dynamics of carriers in solids. Because a free electron laser (FEL) is an only intense, and wavelength-tunable radiation source in the THz range, high field magneto-spectroscopy using THz FEL is one of the most exciting and urgently required experiments.

Although lots of scientists may come up with the idea that a high magnetic field generator can be installed into a THz FEL facility or vice versa, there are only a few reports on this issue [1]. This is

---

*Corresponding author.

E-mail address: ymatsuda@cc.okayama-u.ac.jp (Y.H. Matsuda).

0168-9002/$ - see front matter ⓒ 2004 Elsevier B.V. All rights reserved.
doi:10.1016/j.nima.2004.04.117

partly because it is technically not easy to install a high electric power source that requires large space into an FEL facility for the field generation, and also generally it takes a long period of time and costs a large expenditure of money. A huge power source as large as 8–40 MW is needed to generate a steady state field of 30–45 T [2,3]. For a pulsed magnet the required power is smaller than the steady magnet due to its short pulse duration; the typical energy of a capacitor bank for a non-destructive pulsed magnet is 50 kJ–1 MJ. The maximum field and the pulse duration for a non-destructive pulsed magnet are 30–80 T and 0.1–100 ms, respectively. Although the combination of a pulsed magnet and FEL looks feasible, the dimensions of the power source (the capacitor bank) is still not very small; normally it is several tens of m^3 or larger.

In this paper, we propose a new concept that a very small pulsed magnet can make the magneto-spectroscopy using FEL easy to be performed. The small pulsed magnet system has been made so that it can be carried easily and applied to various kinds of FEL or other advanced light source facilities without any large additional expense. A prototype of the small pulsed magnet system has been completed. We have also done a magneto-transmission experiment using the small pulsed magnet and a millimeter wave coherent radiation from a linear accelerator as a first real experiment.

2. Portable pulsed magnet

The very small pulsed magnet system consists of a capacitor bank as a power source and a miniature coil as a magnet. The parameters and dimensions of the capacitor bank are shown in Table 1. The bore size and the length of a coil is typically 3–6 mm diameter and 10–20 mm long,

respectively. Since the required energy for the field generation is proportional to the volume of the field space, only a few kJ energy is sufficient for a 40-T field, while a 100 kJ class capacitor bank is used for this high field when we use a conventional pulsed magnet [3]. A photograph of the capacitor bank is shown in Fig. 1. We can easily carry the capacitor bank to the place for an experiment, thus call this small pulsed magnet system *a portable pulsed magnet*.

Field wave-forms of a miniature coil (3 mm bore, the inductance is about 25 μH) at different voltages are shown in Fig. 2. We see that only half

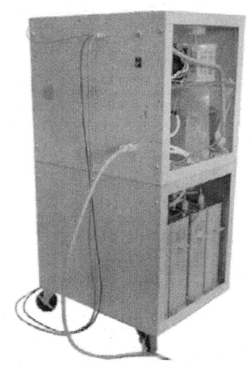

Fig. 1. The capacitor bank of a portable pulsed-magnet system. The bank is composed of electric circuits for a charge and a discharge of the energy (upper part) and six 160 μF-capacitors (lower part).

Table 1
Parameters and dimensions of the small capacitor bank

Energy (kJ)	Capacitance (μF)	Voltage (V)	Height (mm)	Width (mm)	Depth (mm)	Weight (kg)
1.92	960	2000	1234	658	533	∼350

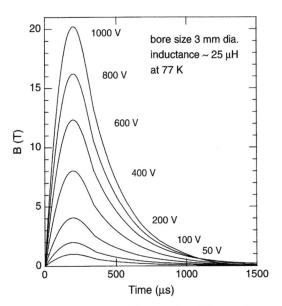

Fig. 2. Pulse shapes of a miniature coil at different voltages.

of the energy of the capacitor bank is used for a generation of 20 T and it was confirmed that we can safely repeat 20-T field generation with an interval of 60 s. The upper limit of the maximum field is determined by the tensile strength of the wire used for winding the coil. (A coil is destroyed when the Maxwell force $B^2/2\mu_0$ exceeds the tensile strength of the wire, where B is the magnetic field and μ_0 is the permeability of vacuum.)

Since we used normal soft Cu wire, the maximum field expected is about 30 T if no additional reinforcement is done to the coil. A record of the non-destructive pulsed magnet in the world is about 80 T at the present [4]. Development of a miniature magnet for higher fields will be carried out by using a wire with higher tensile strength such as the Cu–Ag wire in the near future.

3. Millimeter wave experiment

In order to see if the small pulsed magnet can be used for a real experiment we have made the magneto-transmission experiment. We used a coherent radiation from a linear accelerator in the Research Reactor Institute, Kyoto University

as a light source [5]. The wavelength (frequency) range of the radiation is 1–10 mm (300–30 GHz). The radiation consists of macro pulses with repetition rate of 10–20 Hz and the macro-pulse is composed of many micro-pulses. The pulse duration of the macro-pulse and micro-pulse are 10 ns–4 μs and ∼ps, respectively. This time structure of the radiation is similar to that of a FEL. Actually an application of the coherent radiation to FEL has been reported [6].

A test sample is $RbMg_{0.7}Mn_{0.3}F_3$ and the ESR of Mn ion is expected to be observed. At temperatures higher than the antiferromagnetic exchange constant between Mn ions the standard ESR signal with $g = 2.0$ can be observed, where g is the g-value of the Mn ions.

Fig. 3 shows the ESR spectra observed at 2.5 mm and 77 K. We synchronize the field pulse to the macro-pulse to measure the transmission at high fields; we can tune the field value by changing the time delay for the synchronization or by changing the charge voltage. Although, we have not made use of the advantage of the radiation (i.e. tunability of the wavelength, time structure, and coherence) in this experiment, it was confirmed that the magneto-optical study using FEL or other advanced light sources can be made by use of the very small pulsed magnet.

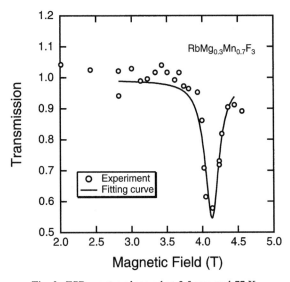

Fig. 3. ESR spectra observed at 2.5 mm and 77 K.

4. Possible THz-magneto spectroscopy

According to previous reports on the magneto-spectroscopy, using FEL by a conventional super-conducting magnet [7,8] and that using pulsed magnet [1], the cyclotron resonance of the semiconductors is the main subject. It is partly because THz carrier dynamics in semiconductors are very important for the future technology. On the other hand, there are few reports on ESR study in spite of the importance of manipulation of spins in solids for developing new spintronics devices.

When we use a 1-THz radiation the resonance line of ESR is observed at 35.7 T if $g = 2.0$. This means that using high magnetic fields are essentially important to study THz ESR, while the cyclotron resonance of the conduction band electrons can be observed at much lower fields because of the small effective mass of electrons in semiconductors.

A possible intriguing ESR study is ongoing on the molecular nanomagnet such as Mn_{12}-acetate [9], because several phenomena related to the quantum tunneling of the magnetization are observed in these materials [10]. The energy levels of each spin state that are separated from each other due to the anisotropy energy cross at a high magnetic field and the spin state of the ground level is changed successively with increasing the magnetic field. Since there is an energy barrier for the reorientation of the magnetization, a spin flip at the level crossing can occur due to the quantum tunneling at low temperatures. Since we can directly observe the energy levels by means of the magneto-spectroscopy, ESR can be a powerful tool to study the spin states at the level crossing. For this experiment, we need to control the magnetic field and the photon energy simultaneously. Using wavelength tunability of a THz FEL and a pulsed high magnetic field is a solution to realize this kind of experiment.

Acknowledgements

We thank T. Takahashi, K. Ishi and Y. Shibata for valuable discussions and technical assistance for the millimeter wave ESR experiment at the Research Reactor Institute, Kyoto University.

This work was supported in part by the REIMEI Research Resources of Japan Atomic Energy Research Institute and by the Ministry of Education, Science, Sports and Culture, Grant-in-Aid for Young Scientists (B), 15740183.

References

[1] L. Van Bockstal, L. Li, C.J.G.M. Langerak, M.J. van de Pol, A. De Keyser, F.G. van der Meer, A. Ardavan, J. Singleton, R.J. Nicholas, F. Herlach, R. Bogaerts, H.U. Mueller, M. von Ortenberg, Physica B 246–247 (1998) 208.

[2] F. Herlach, Physica B 294–295 (2001) 500.

[3] N. Miura, F. Herlach, in: F. Herlach (Ed.), Pulsed and Ultrastrong Magnetic Fields, Springer Topics in Applied Physics, Vol. 57, Springer, Berlin, 1985, p. 247.

[4] K. Kindo, Physica B 294–295 (2001) 585.

[5] T. Takahashi, T. Matsuyama, K. Kobayashi, Y. Fujita, Y. Shibata, K. Ishi, M. Ikezawa, Rev. Sci. Instrum. 69 (1998) 3770.

[6] S. Sasaki, Y. Shibata, K. Ishi, T. Ohsaka, Y. Kondo, F. Hinode, T. Matsuyama, M. Oyamada, Nucl. Instr. and Meth. A 483 (2002) 209.

[7] B.N. Murdin, A.R. Hollingworth, M. Kamal-Saadi, R.T. Kotitschke, C.M. Ciesla, C.R. Pidgeon, P.C. Findlay, H.P.M. Pellemans, C.J.G.M. Langerak, A.C. Rowe, R.A. Stradling, E. Gornik, Phys. Rev. B 59 (1999) R7817.

[8] J. Kono, A.H. Chin, A.P. Mitchell, T. Takahashi, H. Akiyama, Appl. Phys. Lett. 75 (1999) 1119.

[9] T. Lis, Acta Cryst. B 36 (1980) 2042.

[10] L. Thomas, F. Lionti, R. Ballou, D. Gatteschi, R. Sessoli, B. Barbara, Nature 383 (1996) 145.

Available online at www.sciencedirect.com

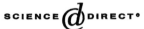

ELSEVIER

Nuclear Instruments and Methods in Physics Research A 528 (2004) 636–640

NUCLEAR
INSTRUMENTS
& METHODS
IN PHYSICS
RESEARCH
Section A

www.elsevier.com/locate/nima

Exploring vibrational coherence of molecular systems with simultaneous excitation of close frequencies using the CLIO-FEL

C. Crépin[a],*, M. Broquier[a], A. Cuisset[a], H. Dubost[a], J.P. Galaup[b], J.M. Ortéga[c]

[a] Lab. de Photophysique Moléculaire, CNRS, Bât. 210, Université Paris-Sud, 91405 Orsay, Cedex, France
[b] L.A.C.-CNRS, Bât. 505, Université Paris-Sud, 91405 Orsay, Cedex, France
[c] L.U.R.E., Bât. 209, Université Paris-Sud, 91405 Orsay, Cedex, France

Abstract

We studied the vibrational coherence of DCl molecules and clusters embedded in solid nitrogen by means of photon echo experiments using the CLIO-FEL. Infrared vibrational spectra of the samples exhibit numerous absorption bands assigned to different species in the 2080–2020 wavenumber range. The tunability of the laser source in terms of spectral range, spectral width of the IR pulses allows to excite coherently these species in different ways. The time-resolved non-linear response of the sample depends on the various experimental conditions. We observe various time features, involving oscillations related to the number of simultaneously excited vibrational modes, and time decays related to the average coherence time of the probed transitions. All these results bring a complete information on vibrational dynamics of the different trapped species.
© 2004 Elsevier B.V. All rights reserved.

PACS: 42.50.Md; 33.20.Tp; 41.60.Cr

Keywords: Vibrational coherence; Photon echo; Free electron laser; Matrices; DCl; Spectral diffusion

1. Introduction

By means of the Free Electron Laser of Orsay (CLIO: Centre Laser Infrarouge d'Orsay) [1,2], we performed time-resolved experiments devoted to the characterization of vibrational dynamics of molecular systems in their ground electronic state [3,4]. In order to probe the influence of the environment on the molecular behaviour, we focused our attention especially on the vibrational coherence. The coherence time or dephasing time (T_2) reflects the effect of all the dephasing processes perturbing the vibration: the relaxation of vibrational populations and the interaction with the environment characterized by time T_2^*, called "pure" dephasing time. The vibrational coherence is studied in "photon echo" experiments

---

*Corresponding author. Tel.: +33-1-69-15-75-39; fax: +33-1-69-15-67-77.

E-mail address: claudine.crepin-gilbert@ppm.u-psud.fr (C. Crépin).

0168-9002/$ - see front matter © 2004 Elsevier B.V. All rights reserved.
doi:10.1016/j.nima.2004.04.118

or one-colour time-resolved Degenerated Four Wave Mixing experiments (denoted hereafter DFWM).

The molecular system, a very simple one, consists of diatomic molecules (DCl) embedded in a well-known lattice structure (solid nitrogen). The sizes of DCl and N_2 being very close, the guest molecules occupy substitutional sites of the N_2 fcc lattice; no rotation of the DCl molecule is observed [5]. In the same sample, we have access to the behaviour of both isolated molecules interacting with the matrix (van der Waals interactions) and DCl dimers or clusters (involving weak hydrogen bonds). The D–Cl stretching mode frequencies differ by tens of wavenumbers in the different species.

From previous studies of such systems [6], vibrational population relaxation time (T_1) of the isolated molecule is known to be very long, in the microsecond range (the main dephasing processes will proceed from the interaction between the molecule and the solid). In this case, we will observe accumulated infra-red photon echoes with our experimental set-up [4], from which another dynamical process can be analyzed: the spectral diffusion. The spectral diffusion consists in erratic jumps of the molecular vibrational frequencies under study; it comes from the fluctuations of the vibrational energies in the condensed phase. In our samples, it is due to intermolecular vibrational energy transfers (V–V transfers) between species trapped in the solid.

The recorded time-resolved signal $S(\tau)$ is related to the third-order polarization of the sample created by the interaction between the molecules and three laser pulses delayed in time, with a tunable delay τ between the first two pulses and a long fixed delay T between the last two pulses ($T \simeq 16$ ns): $S(\tau) = \int_0^\infty |P^{(3)}(t,\tau)|^2 \, dt$. Pulse frequencies are resonant with the vibrational transitions. The first two pulses create a transient vibrational population grating in the sample, the lifetime of this grating is the lifetime of the vibrational populations; the third pulse is diffracted by this grating in a well-defined direction, generating the fourth wave of the DFWM experiment [7]. In a specific direction of the fourth wave detection, the DFWM signal depends on the sign

of τ. In our experimental conditions, $\tau > 0$ corresponds to the case of a photon echo signal: the third pulse creates coherence on the studied vibrational transition with a phase evolution in time exactly opposite to that of the coherence created by the first pulse, a rephasing process occurs at time τ leading to a coherent emission– the photon echo signal [8]. This rephasing process is due to the broad inhomogeneous width of the vibrational transition in the solid: molecules have slightly different vibrational frequencies and the time evolution of the vibrational coherence involve these frequencies, but at time τ after the third pulse, all the molecules are in phase again [9]. The evolution of the photon echo intensity is characterized by time T_2: $S(\tau > 0) \propto \exp(-4\tau/T_2)$.

The rephasing process can be strongly disturbed by spectral diffusion [7]. If spectral diffusion occurs during time T, the population grating is strongly perturbed; when the third pulse arrives on the sample, the produced vibrational coherence has not any clear relation with the coherence produced by the first pulse. In the case $\tau < 0$, the rephasing process is also not observed. The decay of the signal versus $|\tau|$ is then faster than in a photon echo response, because of the dephasing of the coherence due to the wide range of vibrational frequencies in the inhomogeneous distribution. When taking into account a probability of V–V transfers, mainly taking place during the long delay T, one obtains expressions of $S(\tau)$ reflecting the different dynamical processes in good agreement with experimental data [4,10].

On one hand, the tunability of the laser frequency allows the excitation of the different trapped species, giving access to their specific coherence times. On the other hand, the spectral resolution of the laser source (~ 20 cm^{-1}) can be modified in a narrow range, leading to the coherent excitation of a various number of molecular transitions: a collective signal is observed. In particular, these experiments give access to V–V transfer dynamics between all the excited species [11]. Using different laser frequencies (ν_{las}) and spectral widths ($\Delta\nu_{las}$), DFWM experiments on DCl monomer and dimer species in nitrogen matrices widely explore both dephasing and spectral diffusion processes. The time-resolved signals can exhibit complex features. In this paper, simultaneous excitation of monomers and dimers D–Cl stretching modes is described, the results are rationalized in a semi-qualitative way in order to extract the main physical meaning.

2. Experimental set-up

Samples were prepared as described in Ref. [10]. The very efficient D/H exchanges on the walls of the set-up leads to a DCl/HCl ratio about 2. $c =$ [DCl + HCl]/[N_2] is nearly equal to 1%. The final composition of the solid samples was checked and analysed measuring absorption spectra by means of a Mattson FTIR spectrometer with a resolution of 0.25 cm^{-1}.

The CLIO Free Electron Laser frequency was tuned between 2000 and 2100 cm^{-1}. The spectral width was tuned from 12 to 40 cm^{-1}. The average laser power was typically 200 mW. The FEL time sequence was as follows: the rate of sub-picosecond pulses was 62.5 MHz (giving time $T = 16$ ns) in bunches of 10 μs, and the rate of bunches was 25 Hz.

The DFWM experimental set-up was described in details elsewhere [3]. Briefly, the laser beam was separated in two beams by reflection/transmission on a ZnSe window. One of these beams was delayed with respect to the other using a retro-reflector fixed to a computer-driven time delay translation plate (Newport Instrument); the time delay range was −300/ + 300 ps. A parabolic mirror focused the two parallel beams on the sample–the size of the focus point was around 100 μm. The DFWM signal was emitted in a well-defined direction $\mathbf{k}_s$ given by: $\mathbf{k}_s = 2\mathbf{k}_2 - \mathbf{k}_1$ where $\mathbf{k}_{1,2}$ were the directions of the two laser beams in the sample, $\mathbf{k}_2$ corresponding to the beam delayed by time τ from the other. The three beams with $\mathbf{k}_{1,2,s}$ directions were then collected by another parabolic mirror in order to get three parallel beams from which the signal was extracted geometrically with a well-positioned diaphragm. This signal beam was detected on a liquid nitrogen-cooled MCT detector. The DFWM signal $S(\tau)$ was then recorded by the computer.

3. Results and discussion

The same sample contains different guests: HCl and DCl monomers, dimers, mixed dimers for both ^{35}Cl and ^{37}Cl isotopic species in their natural abundance. There are three main features in the infrared absorption spectra (see Fig. 1), one assigned to monomers composed of four bands, two assigned to dimers, corresponding to modes v_1 and v_2 localized, respectively, on the DCl molecules acting as a proton acceptor and as a proton donor in the weak hydrogen bond of the dimer. Table 1 summarizes the observed transitions in the D–Cl stretching mode region and their assignments.

The DFWM signal results from the excitation of all the vibrational transitions lying in the CLIO spectral range. The narrowest spectral width of the source was 12 cm^{-1} so that, in any case, any of the listed transition cannot be excited alone. The results obtained on the simultaneous excitation of the four monomers bands in dilute samples are described in Ref. [10]. Sample concentrations of hydracids of ~1% allow the study of dimers. The DFWM signals were recorded with laser frequencies tuned in the 2060–2030 cm^{-1} region. Depending on the excitation characteristics in terms of laser frequency and spectral width, the time-resolved signals exhibit very different behaviours, as shown in Fig. 2.

In most cases, the DFWM signal included two components: (i) an oscillating structure around the zero temporal delay, which is the signature of the

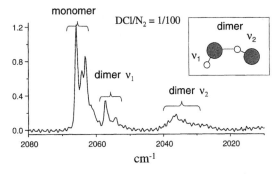

Fig. 1. Absorption spectrum of the 2100–2000 cm^{-1} region recorded for a [DCl + HCl]/N_2 = 1/100 sample at 7.5 K.

Table 1
Vibrational transition frequencies of DCl monomers and dimers in N_2 matrices at $T = 7.5$ K

Wavenumber (cm^{-1})	Assignment [5]
2067.7	$D^{35}Cl$ monomer
2066.0	$D^{35}Cl$ nnn dimer[a]
2064.7	$D^{37}Cl$ monomer
2063.4	$D^{37}Cl$ nnn dimer[a]
2058.9	$D^{35}Cl$-HCl dimer
2058.7	$D^{35}Cl$-DCl dimer, mode ν_1
2055.7	$D^{37}Cl$ dimer
2039.5	HCl-$D^{35}Cl$ dimer
2037.7	DCl-$D^{35}Cl$ dimer, mode ν_2
2034.8	$D^{37}Cl$ dimer

[a] DCl molecules in position of next nearest neighbours (nnn) in the fcc lattice [5].

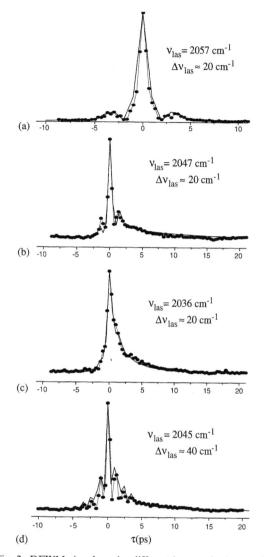

(a) $\nu_{las} = 2057$ cm^{-1} $\Delta\nu_{las} \approx 20$ cm^{-1}

(b) $\nu_{las} = 2047$ cm^{-1} $\Delta\nu_{las} \approx 20$ cm^{-1}

(c) $\nu_{las} = 2036$ cm^{-1} $\Delta\nu_{las} \approx 20$ cm^{-1}

(d) $\nu_{las} = 2045$ cm^{-1} $\Delta\nu_{las} \approx 40$ cm^{-1}

τ(ps)

Fig. 2. DFWM signals under different laser excitation conditions, $c \sim 1\%$, $T = 10$ K; dots: experimental data, solid line: simulation.

spectral diffusion process; (ii) a long and weak component, corresponding to the rephasing process, involving the coherence time of the probed transitions. Such a behaviour has been observed under the excitation of monomers [10]. The periodic time structure of the oscillations is related to the frequency shifts between all the excited transitions affected by the spectral diffusion. It depends clearly on the laser frequency and the laser spectral width as shown in Fig. 2: a period of 3 ps is measured in (a)–corresponding to the coherent excitation of the monomer and the mode ν_1 of the dimer ($|\Delta\omega| \sim 11$ cm^{-1}), a period of 1.33 ps in (b)–corresponding to the coherent excitation of modes ν_1 and ν_2 of the dimer ($|\Delta\omega| \sim 25$ cm^{-1}), and oscillations in (d) involve both periods (3 and 1.33 ps) because in this case, all the three kinds of transitions are excited by the laser. These oscillations can be reproduced using a crude model where the part of the third-order polarization of the sample affected by the spectral diffusion is expressed by

$$P_d^{(3)}(t, \tau) \propto P^{(1)}(-t) \times P^{(1)}(\tau) \qquad (1)$$

with

$$P^{(1)}(t) \propto \exp[\gamma|t|] \sum_{i=1}^{n} N_i \exp[-2i\pi\omega_i t].$$

$P^{(1)}(\pm t)$ reproduces the evolution of the molecular coherence following the interaction with one laser pulse [12]; in the sequence of the three pulses, this evolution is stopped with the interaction with the second pulse delayed by τ, and because of spectral diffusion, there is no memory of this evolution at the arrival of the third pulse which thus creates a "new" evolution of the coherence. In Eq. (1), the laser is assumed to excite n transitions, with frequencies ω_i, weighted by parameters N_i

including the populations of the excited species, the transition dipole moments and the probability of excitation of the transition due to the spectral profile of the laser; $\exp[\gamma|t|]$ is a function reproducing in a rough approximation, by Fourier Transform, the envelope of the part of the absorption spectrum probed by the laser, coherence times (T_2) are hidden within parameter γ. The periods of the oscillations give thus access to the number of excited transitions and involved frequencies, and the contrast of the oscillations gives access to the ratio of the weights N_i/N_j. We write [10] $P^{(3)}(t, \tau) = pP_r^{(3)}(t, \tau) + (1 - p)P_d^{(3)}(t, \tau)$, where the rephasing process is described by $P_r^{(3)}(t, \tau)$ and p is the probability that no $V–V$ transfer affects the non-linear response. $P_r^{(3)}(t, \tau)$ is responsible for the long component (ii) simulated in a simple way by a single exponential decay $\exp[-4\tau/T_2']$, T_2' is a global coherence time reflecting the effects of the coherence times of all the excited transitions.

The case $p = 0$ (no rephasing process) is illustrated in Fig. 2a, involving in a large part the excitation of monomers: from Ref. [10], we know that $V–V$ transfers are very fast between the isolated DCl molecules when the concentration reaches 1%. The ratio $N_{\text{monomer}}/N_{\nu_1 \text{ dimer}} = 0.8$ deduced from the simulation is in very good agreement with the ratio ($= 0.7$) calculated from the absorption spectrum overlapped by the laser pulse, assuming a Gaussian shape of the pulse. In Fig. 2b and c, both modes of the dimers are excited with different weights depending on the laser frequency, as reflected by the contrast of the oscillations. The simulations lead to the values $R_d = N_{\nu_1 \text{ dimer}}/N_{\nu_2 \text{ dimer}} = 1$ (well-defined oscillations) and 0.2 (weak modulation) in Fig. 2b and c, respectively. Following the decrease of ν_{las} (and the decrease of R_d), there is a decrease of T_2' which means that the coherence time of the proton donor mode (ν_2) is shorter than that of the proton acceptor mode (ν_1) – $T_2' = 25$ and 13 ps in Fig. 2b and c, respectively. The coherence times for each

transition will be obtained when taking into account the fitted parameters N_i in the analysis of the long component. At last, p is directly related to the characteristic time of the $V–V$ transfers between all the probed transitions. Its dependence on $\Delta\nu_{\text{las}}$ brings information on the kinetics of $V–V$ transfers between the different kinds of transitions.

The results described in this paper emphasize the possibilities of vibrational studies using CLIO. This laser source provides an excellent compromise between time and spectral resolution.

Acknowledgements

The authors want to acknowledge the members of the CLIO group of the LURE laboratory, and especially F. Glotin and R. Andouart, for their helpful assistance with the experiments.

References

[1] R. Prazeres, F. Glotin, C. Insa, D.A. Jaroszynski, J.M. Ortéga, Eur. Phys. J. D 3 (1998) 87.
[2] F. Glotin, J.M. Ortéga, R. Prazeres, C. Rippon, Nucl. Instr. and Meth. B 144 (1998) 8.
[3] J.P. Galaup, M. Broquier, C. Crépin, H. Dubost, J.M. Ortéga, F. Chaput, J.P. Boilot, J. Lumin. 86 (2000) 363.
[4] C. Crépin, M. Broquier, H. Dubost, J.P. Galaup, J.L. Le Gouët, J.M. Ortéga, Phys. Rev. Lett. 85 (2000) 964.
[5] C. Girardet, D. Maillard, A. Schriver, J.P. Perchard, J. Chem. Phys. 70 (1979) 1511 and references therein.
[6] J.M. Wiesenfeld, C.B. Moore, J. Chem. Phys. 70 (1979) 930.
[7] A.M. Weiner, S. De Silvestri, E.P. Ippen, J. Opt. Soc. Am. B 2 (1985) 654.
[8] T. Joo, A.C. Albrecht, Chem. Phys. 176 (1993) 233.
[9] Y.R. Shen, The Principles of Nonlinear Optics, Wiley–Interscience, New York, 1984.
[10] M. Broquier, C. Crépin, A. Cuisset, H. Dubost, J.P. Galaup, P. Roubin, J. Chem. Phys. 118 (2003) 9582.
[11] C. Crépin, Phys. Rev. A 67 (2003) 013401.
[12] S. Mukamel, Principles of Nonlinear Optical Spectroscopy, Oxford University Press, New York, 1995.

FREE ELECTRON LASERS 2003
E.J. Minehara, R. Hajima and M. Sawamura (Eds.)
Published by Elsevier B.V.

The Dipole Magnet System for the Jlab THz/IR/UV Light Source Facility

G. H. Biallas*, D. R. Douglas, T. Hiatt, R. Wines

Thomas Jefferson National Accelerator Facility, 12000 Jefferson Blvd. Newport News VA 23606

Abstract

Six families of dipoles were designed, manufactured and tested for the THz/IR/UV Light Source Facility at the Thomas Jefferson National Accelerator Facility (Jlab). These magnets obtained core field and field integral uniformities at the parts in 10,000 level. They form the basis of an ERL transport system with considerable operational flexibility (e.g. independently, order-by-order tuneable momentum compaction) and very large momentum acceptance – in the order of 10%. They were modelled in 3D using ANSYS, Radia or Opera 3D. The highlights of dipole magnet system design are presented along with summaries of family characteristics. Operational experience is presented. These dipoles run at two core field ranges, 0.05 T and 0.22 to 0.71 T.

PACS codes: 41.85.Lc, 41.85.Ja

Keywords: dipole, magnet, ERL, transport, modelling, measurement

1. Introduction

Twenty-eight dipoles are required for the injection, extraction, chicanes and arcs of the 10 kW IR/1 kW UV THz/IR/UV Free-Electron Laser Upgrade [1,7] at the Thomas Jefferson National Accelerator Facility (Jefferson Lab). The dipoles run at relatively low fields ranging from 0.05 to 0.7 Tesla. Two sets of 180° "Bates Lab" arcs using three families of dipoles (designated GQ, GX, GY) form the ends of the racetrack beam path. An additional family (GW) of four dipoles form two chicanes as well as the UV Bends. Two low field (0.05 T) families (GU, GV) form injection/extraction lines.

2. Dipole Systems

The three families of the first arc are run in series with either of the two chicanes using one power supply (230A maximum). Shunts on the 180° bend (GY) and across the either chicane (four GW magnets) match those elements to the already matched Bend-Reverse Bend (GX-GQ) dipoles. The second arc operates on a second power supply with a shunt on its GY. The second arc is separate because it

has to be adjusted to a lower field integral to accommodate the momentum decline during lasing. For UV operation, both arcs switch their outboard GXs to half field and switch in their respective UV Bends (GW) adjusted with individual shunts.

We obtained (with one exception) the required specifications that these magnets match to 1 part in 1000 within a family and have core field and integral for any one magnet uniform within a few parts in 10,000 over the good field region. Our tools were 3D magnetic modelling, careful manufacturing and adjustment during magnet measurement.

At the design stage we used the successful window frame with Purcell gap system dipole style of the IR Demonstration FEL [2] as the model for four dipole families. The Arc Bend family (GX) was designed as an "H" style because it has an optical pipe through the return legs of its yokes. The Reverse Bend family (GQ) is an "H" because its fields must track the GX.

The dipoles were modelled in 2001using computer codes to resolve fields in 3D to parts in 10^5. [4,5,7]. We used a simple (non-hysteretic) BH curve for annealed 1006 steel in our models. Residual field added to the gap's field by powering was not

---

* Email: biallas@jlab.org

accounted for.

3. Dipole Family Summaries

Specifics of measurement of the individual dipole families are reported elsewhere [3, 6, 7]

3.1 Injection and Extraction Dipoles (GU/GV)

The Injection line uses three 20 degree wedge dipoles (GV) in a three magnet, 10 MeV/c transport to the first cryomodule. Recirculating beam of up to 210 MeV/c may be recombined with the injected beam using a chicane formed by the single GU dipole (double the field integral of a GV) and a single GV. The extraction beam line inverts the above recombination to a spread using an identical chicane. The small and large injection dipoles (GV & GU) were designed by DULY Research [4] using the 3D code Radia. For the small magnet, they invented a trim coil system (patent applied for) that compensates for core field droop on the short part of the wedge to flatten core field and field integral. The large dipole uses the Purcell gap system to flatten field. Operation shows that the family members match to within 20 G-cm.

3.2 Chicane and UV Bend Dipole (GW)

The GW dipole is a window frame style, rectangular pole magnet with Purcell gap system, modelled using Opera 3D. Magnet measurement revealed no problems.

3.3 Bend/Reverse Bend Dipoles (GX/GQ) and Pi (180°) Bend (GY) for the Arcs

The GX has a large wedge angle (39.4°) and a small good field region (10.2 cm). An enlarged pole in one region allows for an alternate, larger radius beam path. When only half the coil turns are powered, the field reduces to half the IR nominal value, making the magnet a switch wherein the larger radius diverts the electron beam to the UV branch of this Light Source. The GX's twin, the Reverse Bend Magnet (GQ) has a small wedge angle (10.4°) with larger good field region (25.4 cm).

Advanced Energy Systems Corp. (AES) modelled the GX & GQ magnets using ANSYS. [5] Micrometric steps in the pole tip edges flatten the fields of both families. AES also designed the Pi Bend in window frame style with Purcell gap system

using ANSYS. Additional coils built into the ends of its poles, create path length corrector dipoles. Core field flatness is out of specification transverse to the beam. Fortunately, the field error is compensated for by using planned trim quadrupoles and sextupoles.

Operation shows that when operating as a unit, the three families of the arc match to within 20 G-cm, are repeatable to within order 100 G-cm and provide edge focusing equivalent gradient integrals within order 20 G of design.

4. Summary and Conclusions

Six different families of dipoles were designed as part of an energy recovery beam transport system. 3D computer modelling produced designs with excellent performance. Manufacturing with attention to critical parameters yielded magnets that match across families, have flat fields and behave in agreement with design projections. All out-of-tolerance conditions were recoverable by final adjustment at the measurement stage or by use of available trim elements. Injection and recirculator operation uses low trim magnet strengths, verifying the quality of the magnet systems.

5. References

[1] D. Douglas, et.al. "A 10 kW IRFEL Design For Jefferson Lab", Proc. PAC2001, Chicago, June 2001.

[2] G, Biallas, et.al. " Making Dipoles to Spectrometer Quality Using Adjustments During Measurement ", Proc. PAC1999, New York, (1999).

[3] T. Hiatt, et.al. "Magnetic Measurement of the 10 kW, IR FEL Dipole Magnets". Proc. PAC2003, Portland, (2003).

[4] D. Newsham, et. al. "Wedge-shaped, Large-aperture, Dipole Magnet Designs for the Jefferson lab Facility Upgrade", Proc. PAC2003, Portland, (2003).

[5] T. Schultheiss, et al. "Magnetic Analysis of the Arc Dipoles for the JLAB 10 kW Free Electron Laser Upgrade", FEL'02, Argonne, IL, Sep. 2002.

[6] K Baggett, et.al., "Magnetic Measurement of the 10 kW, IR FEL 180 Degree Dipole". Proc. PAC2003, Portland, (2003).

[7] G, Biallas, et.al., "Magnetic Modelling Vs Measurements of the Dipoles for the Jlab 10 KW Free Electron Laser Upgrade". Proc. PAC2003, Portland, (2003).

6. Acknowledgements

Work supported by the US DOE Contract #DE-AC05-84ER40150, the Office of Naval Research, The Air Force Research Laboratory, the Commonwealth of Virginia, and the Laser Processing Consortium.

FREE ELECTRON LASERS 2003
E.J. Minehara, R. Hajima and M. Sawamura (Eds.)
Published by Elsevier B.V.

Tunable Hybrid Helical Wiggler with Multiple-Poles per Period

Yoshiaki Tsunawaki[a*], Nobuhisa Ohigashi[b], Makoto Asakawa[c],
Mitsuhiro Kusaba[a], Kazuo Imasaki[d] and Kunioki Mima[e]

[a] *Department of Electrical Engineering & Electronics, Osaka Sangyo University*
3-1-1 Naka-gaito, Daito, Osaka 574-8530, Japan
[b] *Department of Physics, Kansai University, Yamate-cho, Suita, Osaka 564-8680, Japan*
[c] *Institute of Free Electron Laser, Osaka University, Tsuda-yamate, Hirakata, Osaka 573-0128, Japan*
[d] *Institute for Laser Technology, 2-6 Yamadaoka, Suita, Osaka 565-0871, Japan*
[e] *Institute of Laser Engineering, Osaka University, 2-6 Yamadaoka, Suita, Osaka 565-0871, Japan*

A hybrid helical wiggler with multiple poles per period has been designed for the free electron laser (FEL) at Institute of FEL (iFEL) Osaka University in order to obtain much higher power of FEL in the far-infrared region using an electron beam with the highest energy of 165 MeV [1]. The period of the wiggler (λ_w) must be long due to the high electron beam energy. It is considered to be 14 cm in our study. The necessary gap diameter is assumed to be 2 cm. One period of the wiggler is constructed with some segments to obtain a sinusoidal magnetic field. Each segment consists of two fan-shaped permanent magnets (Nd-Fe-B : PMs) and two fan-shaped ferromagnetic bodies (permendurs : FMs) which are alternately configured to each other in a plane perpendicular to the wiggler axis (z-axis). The PMs are magnetized normal to the surface of the segment. The magnetized direction of a PM is opposite to that of the other PM. A wiggler is constructed in such a way that each segment is stacked along the z-axis and rotated by an angle $\theta_P = 360/N_P$ (degrees) where N_P is the number of

FM's poles per period. It is found that an optimal number of Np is 8 from the viewpoint of easy production although larger Np is a little better [1]. In this study, this type of wiggler with Np=8 was re-designed by using the MAGTZ computational code for the analysis of magnetic field to add the function of a tunability of the wiggler field.

The fundamental structure of a tunable hybrid helical wiggler is shown in Fig.1. It consists of two coaxial helical wigglers. In a normal hybrid helical wiggler the radius of the PM is larger than that of the FM to suppress the leakage field. The projecting portion of the PM along the circumference of a circle is separated from the rest of the wiggler to the outer wiggler. The wiggler field is varied by rotating the outside wiggler. Two kinds of wigglers are considered in this work as shown in Fig. 1; the outside wiggler doesn't have any PM except for PMs magnetized to z direction (type A wiggler), and includes PMs magnetized to the radial direction (PMrad) inserted between the PMs oppositely magnetized to z direction (type B wiggler)

The size of the magnetic materials was decided by similar method to the previous work [1] but FM was divided into smaller elements in the field calculation to analyze it more precisely. In the normal hybrid helical wiggler, the same radii between PM and FM gave the saturated field of about 1.25 T around the radius of 15 cm. The reduction of the radius of only FM to 10 cm brought the maximum field of about 1.50 T due to the suppression of the leakage field. Under these results, the dimensions of the inside and outside wigglers were selected as follows; for the inside wiggler, the inner and outer radii are 2 and 10 cm, respectively, and for the outside wiggler, those radii are 10.6 and 15 cm, respectively. The space between the inside and the outside wigglers is necessary to the holders of magnetic materials in each wiggler.

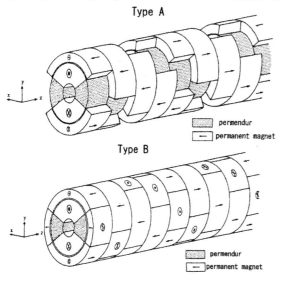

Fig.1 Tunable hybrid helical wiggler with 8 poles per period.

Fig. 2 shows the helical wiggler field depending on the rotational angle (θ r) of the outer wiggler.

It is seen at θ r = 0° that providing a space of 6 mm between the inner and outer wigglers reduces the field from 1.50 to 1.40 T and the addition of PM$_{rad}$s improves it to 1.58 T. Both fields for type A and type B wiggler decrease with the rotational angle and they are approximated to each curve of fourth degree with an inflection point around θ r = 90° where they are almost the same strength. If these wigglers are applied to an electron beam with an energy of 165 MeV, the tunable wavelength range of FEL is estimated to be 72 to 225 μ m and 40 to 288 μ m for type A and type B wigglers, respectively.

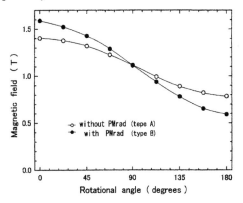

Fig.2 Magnetic field depending on the rotational angle of the outer wiggler.

The magnetic field distribution on the z-axis was expanded in a Fourier series. As is well known, an ideal helical wiggler has only 0th Fourier component because it has a constant helical magnetic field. Both wigglers of type A and type B have 0th Fourier component fluctuating between 99.1 and 99.7 % depending on the θ r except 97.2 % at θ r = 90° in type A wiggler. Fundamental Fourier component of the Bx and By in both wigglers also varies between 96.6 and 98.7 % except 94.8 % at θ r = 90° in type A wiggler and then it was seen that the wiggler of type B has, on the whole, a tendency including the fundamental component higher than that of type A. It will be, therefore, said that PM$_{rad}$s not only contributes strongly to the field strength but also gives rise to a smooth field distribution along the z-axis.

Fig.3 shows the distribution of By along the y-axis where the FM poles are arranged. In type A wiggler, it is seen that the field increases with the distance from the center. With increase of θ r, the curvature approaches to the curve of cosh(k_wy) where k_w is $2\pi / \lambda_w$ and y is the distance from the wiggler axis. It

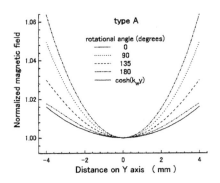

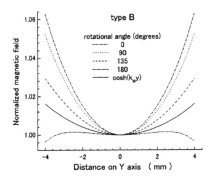

Fig.3 Magnetic field distribution on Y axis where the FM poles are arranged.

will be expected that the focusing power of the wiggler is stronger than that of a usual wiggler. On the other hand, type B wiggler shows a different aspect at θ r = 180° where the field reduces at some distance from the center but it is slightly concave around the center.

In conclusion, two kinds of coaxial hybrid helical wiggler have been analyzed by a computational simulation code. The tunable ranges were estimated to be 1.40 to 0.79 T and 1.58 to 0.59 T for the wiggler without and with PM$_{rad}$s, respectively. PM$_{rad}$s will also give rise to a uniform field distribution along the z-axis, on the whole, with respect to the θ r. However the wiggler without PM$_{rad}$s will maintain a proper focusing power for an electron beam in any θ r. On the other hand, , the wiggler with PM$_{rad}$s has similar focusing power at any θ r except near 180°. We think that the simulation work must be confirmed by the fabrication of a test wiggler before the construction of a real wiggler.

Reference

[1] Y.Tsunawaki, N.Ohigashi, M.Asakawa, K.Imasaki, K.Mima, Nucl. Instr. and Meth. **A507** (2003) 166.

FREE ELECTRON LASERS 2003
E.J. Minehara, R. Hajima and M. Sawamura (Eds.)
Published by Elsevier B.V.

Control System Upgrade of JAERI ERL-FEL

N. Kikuzawa*

Japan Atomic Energy Research Institute, Tokai-mura, Ibaraki 319-1195 Japan

Abstract

The control system used for the JAERI ERL-FEL is a PC-based distributed control system that has been in operation since 1992. Since an interface bus of the PCs is peculiar, interface boards for the PCs are difficult to obtain in recent years. Thus we have been developing the CAMAC controller with μITRON operating system to replace the old PCs connected with CAMAC. We will introduce a Java and CORBA environment in the new control system. The control system upgrade, including hardware upgrading and applications rewriting, is described in this paper.

PACS codes: 41.60.Cr, 07.05.Dz

Keywords: ERL-FEL, μITRON, JAVA, CORBA

1. Introduction

An accelerator control system used for the JAERI ERL-FEL, FELOWS (FEL Operators Window System), is a PC-based distributed control system that has been in operation since 1992 [1]. The system currently consists of 15 PCs grouped into functions of operator consoles, instrumentation nodes, internet servers and database servers, communicating over an Ethernet LAN. The PCs run Windows 95/98 operating system and use NetBIOS and TCP/IP to communicate over the network. A homemade RPC (remote procedure call), based on TCP/UDP protocols, is used for communication between PCs.

These old PCs (NEC PC-98 series 32 bit personal computers) connected with CAMAC were originally localized and de facto standard in Japan in the 1980's. Since an interface bus of the PCs was peculiar, interface boards for the PCs have been difficult to obtain in recent years. The most serious problem is that we need much cost and manpower to maintain these old computers. In order to reduce the cost and labor for constructing the control system for computer upgrade, it is preferable to take over as many CAMAC modules as possible. Therefore, we

have been developing the CAMAC controller with μITRON operating system to replace these obsolete PCs.

In this article, we discuss a newly introduced control system to improve the control functionalities, and/or to enable better maintenance capabilities. The new controllers for the CAMAC crates are described in detail in Section 2. Control software design for the controllers is discussed in Section 3.

2. New Controller

Old PCs and interface boards for the control system, which were made about 10 years ago, could not be reproduced any more, mainly because of production discontinuance of their components. Thus a new controller was designed not to depend on the development of the PC technologies, depending on the needs to take over as many CAMAC modules as possible continuously. The controllers employed an embedded μITRON operating system machine (Nichizo Densi-seigyo Co. ND-MCU) with 10/100Base-T network. The μITRON operating system was designed for an embedded system in all industrial fields as a real time operating system.

---

* Tel: +81-29-282-6752; FAX: +81-29-282-6057; e-mail:
kikuzawa@popsvr.tokai.jaeri.go.jp

Since the controller is equipped with no HDD, it seldom breaks down. It is expected that the controller works in the more stably than PC with MS-Windows. The CAMAC controller has three PCI slots and the controller has flexibilities for enlargement by adding interface cards in the PCI slots.

In order to hold setting parameters during power off, the setting values are kept in battery backup RAM by a control program in the controller. Thus the parameters are restored automatically at the start up of the controller.

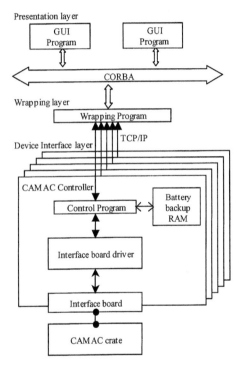

Figure 1. New control system architecture.

As shown in Fig. 1, the control system is divided into three layers, a presentation layer, a wrapping layer and a device interface layer. The device interface layer is connected with CAMAC crates. The wrapping layer is mediation of CORBA (Common Object Request Broker Architecture) protocol and TCP/IP protocol. The presentation layer is a GUI (Graphical User Interface) application to control the devices or to display current status.

3. New control software

Our software developed as a 16-bit application, worked on the Windows3.1/95 operating system. It was difficult to change this software to 32-bit application working on Windows2000/XP. Therefore, we determined to develop software newly.

Recently, the object-oriented technologies for distributed computers, such as Java and CORBA, have been used in the field of accelerator controls as well as in all industrial fields [2]. Java is a platform-independent language with the capability to create an interactive GUI at a web-browser. CORBA has become the standard communication protocol between distributed computers over network. The use of CORBA enables smooth communication between different languages and different operating systems. If an application is written in Java and communicates with the CORBA protocol, it is expected to work on any platform with the same source. Thus, we will introduce a Java and CORBA environment in the new JAERI ERL-FEL control system.

4. Conclusions

We have been developing the CAMAC controller with the μITRON operating system to replace the old PCs. It is expected that the controller works in the more stably than PC with MS-Windows.

We will introduce a CORBA environment in the new JAERI ERL-FEL control system. In this case, we will use Java as a client-side language. We need, however, to confirm whether the existing programs written in Borland Delphi can communicate by using the CORBA protocol.

5. REFERENCES

1. M. Sugimoto, Nucl. Instrum. Meth. A331 (1993) 340-345.
2. S. Kusano et al., Proc. of the 1999 International Conference on Accelerator and Large Experimental Physics Control Systems (ICALEPCS'99), Trieste, Italy, October, 1999.

FREE ELECTRON LASERS 2003
E.J. Minehara, R. Hajima and M. Sawamura (Eds.)
Published by Elsevier B.V.

ENHANCEMENT OF LASER COMPTON GAMMA-RAY BEAM WITH F-P CAVITY IN STORAGE RING TERAS

H. Ohgaki*[1], H.Toyokawa[2], K. Yamada[2], S. Hayashi[1], T. Kii[1], T. Yamazaki[1]

[1]Institute of Advanced Energy, Kyoto University, Gokasho, Uji, Kyoto, JAPAN
[2]National Institute of Advanced Industrial Science & Technology, Umezono, Tsukuba, Ibaraki, JAPAN

Abstract

We have designed and installed a 6-m long, in-vacuum Fabry-Perot cavity in a straight sections of the storage ring TERAS for head-on laser-Compton experiment. The Rayleigh length of the cavity is determined from the Form Factor. A 1 W mode-matched laser light is guided into the cavity from outside the vacuum chamber. We operate the cavity as an amplifier of the injected laser power inside the cavity up to 400, which is limited by the electron life-time, and the gamma-ray enhancement factor of 20 is estimated.

PACS codes: 42.60.Da,07.85.Fv

Keywords: laser-Compton,gamma-ra, storage ring, Fabry-Perot cavity

1. Introduction

A gamma-ray beam, which is produced in the collision between relativistic electrons and a laser beam, has excellent characteristics of tunable energy, narrow energy spread, and high polarization. Using electron beam in the storage ring TERAS in National Institute of Advanced Industrial Science & Technology (AIST), the Laser-Compton back-Scattering (LCS) photon beam[1], whose energy covered from 1 to 40 MeV, were successfully generated and application of the LCS photon beam has been intensively carried out[2]. However, the intensity of LCS photon beam is order of 10^5 and enhancement of the beam intensity is strongly desired. Recycling the laser photons is an effective method to increase the LCS beam intensity[3]. Thus, a 6-m Fabry-Perot (F-P) cavity installed in a straight section of TERAS was designed and a pair of mirror chambers with five-axes manipulators was installed. In this paper we will briefly report on the 6-m F-P cavity installed in TERAS at AIST.

* Corresponding author.
*E-mail address:*ohgaki@iae.kyoto-u.ac.jp

2. Optical Design of 6 m F-P cavity

The electron beam profile is one of key parameters for the design of the F-P cavity. The overlap between the electron beam and the laser beam in the F-P cavity, Form Factor, should be as high as possible to obtain a high intensity gamma-ray beam generated by backward-Compton process. The electron beam profile has been measured and can be expressed as,

$$\sigma_x = \sqrt{\sigma_0^2 + \sigma_a^2}, \sigma_0 = 0.75, \sigma_a = 0.16\sqrt{I},$$

$$\sigma_y = 0.087\sqrt{I}, \tag{1}$$

where σ_x is the horizontal beam size in mm, and σ_y is the vertical one, and I is a electron current (mA). We assumed a constant profile of the electron beam over the laser-electron interaction region where no quadrapoles exist. On the other hand, the laser profile along the longitudinal axis, z, is expressed as $w(z)^2=(z^2+z_r^2)/z_r \ \lambda/\pi$, where z_r is the Rayleigh length of the cavity, λ (=1064 nm), is the wavelength of the laser, and $z=0$ is the centre of the F-P cavity. Thus the overlap between laser and electron beam, Form Factor, can be calculated. Fig 1 shows the Form Factor as a function of Rayleigh length and corresponding laser beam size at the cavity mirror ($z=3.09$ m). It is clear that the Rayleigh length should be above 2 m for our case. However, we have to mention about the curvature, R, of the F-P mirror which is expressed as $R=z(1+z_r^2/z^2)$. It is obvious that a large Rayleigh length requires huge curvature of mirror, which is difficult to obtain. Consequently, the Rayleigh length of 2.43 m, Form Factor = 1.8×10^{-2}, R = 5 m, is chosen.

The reflectivity, which defines the cavity gain and bandwidth, is also important parameter. AIST-LCS facility has already delivered 1-40 MeV gamma-rays by the conventional LCS configuration[1] where Form Factor of 8.3×10^{-3} ($z_r=0.74$ m) and 40W laser power has typically been used. Because of small energy acceptance of the storage ring TERAS, the beam life-time of the ring is reduced to about 6 hours, which is 25% reduction from a normal operation, with the gamma-ray flux of -10^5 γ/s/% at 100 mA stored current. Thus, to keep the beam life-time at the order of hour, the maximum enhancement factor of gamma-ray beam is limited to about 20. Thus the cavity gain should be about 400 when we use 1 W single frequency laser for F-P cavity injector. The

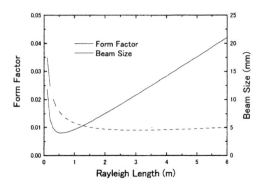

Fig.1 Form Factor and laser beam size at cavity mirror as a function of Rayleigh length of the F-P cavity.

maximum cavity gain is expressed by the reflectivity of the cavity mirror and transmissivity. When we assume the transmissivity of 0.1%, the reflectivity of 99.8% is demanded. In this case the finess of the cavity is 1569 and the bandwidth of the cavity is 15.5 kHz.

3. Conclusion

A 6-m F-P cavity to enhance the gamma-ray flux is designed and installed in a straight section of the storage ring TERAS at AIST. The Rayleigh length of the F-P cavity is determined from Form Factor and mirror curvature. The cavity gain which is limited from the electron beam life-time is 400 and gamma-ray enhancement factor is about 20 with a 1 W laser. The 6-m F-P cavity system consisted high vacuum chambers with five-axis manipulators, an injection laser and matching optics has been installed and the laser storage experiment will be started in near future.

References

[1] H. Ohgaki et al., Nucl. Instr. and Meth. A353 (1994) 384.

[2] e.g. H. Ohgaki et al., Nucl. Phys. A649 (1999) 73c.

[3] K. Imasaki et al., Nucl. Instr. and Meth. A341 (1994) 346; J.P. Jorda et al., Nucl. Instr. and Meth. A412 (1998) 1.

FREE ELECTRON LASERS 2003
E.J. Minehara, R. Hajima and M. Sawamura (Eds.)
Published by Elsevier B.V.

Design of FELICE, the third light source of the free electron laser facility FELIX

B.L. Militsyn[*], P.F.M. Delmee, D. Oepts, A.F.G. van der Meer

FOM Instituut voor Plasma Fysica "Rijnhuizen", P.O. Box 1207, 3430 BE Nieuwegein, The Netherlands

Abstract

As reported last year, a new FEL light source, dedicated to fundamental research of the structure and dynamics of (bio)molecules, clusters, and nano-particles, is presently under development at the FELIX facility. This Free Electron Laser for Intra-Cavity Experiments (FELICE) is designed to generate pulsed infrared radiation, tunable in the region of 3-100 μm. It should allow intra-cavity experiments with optical beam energies in the interaction point of some 10 J in a few μs pulse, which is a factor of 100 higher than currently available for the users. Two permanent intra-cavity set-ups are planned: a high-resolution 7 T FT-ICR mass spectrometer with appropriate ion sources and a molecular/cluster beam set-up. Funding of the project was finally granted in December 2002. R&D and construction stages are estimated to take about three years to be completed.

PACS codes: PACS codes: 41.60.Cr;41.85.-p;42.65Ky

Keywords: Free electron laser; Electron injector; Undulator; Harmonic suppression

1. Introduction

The FELICE [1] will be the third light source of the infrared free electron laser user facility FELIX [2], whose preliminary design was presented in [3]. After the project was finally granted we revised that design. At present FELICE is realised as a FEL light source, which is integrated into FELIX and is driven by the same electron injector. A shunting-yard magnet system, switching the electron beam at a frequency of 20 Hz, will allow simultaneous operation of FELICE and one of the existing two FEL's of FELIX. Thus the optical beam will have a temporal structure consisting of a few-microsecond long, 1 GHz train of ps-micropulses, which is repeated at 10 Hz. The injection line of FELICE, its undulator and a part of the optical cavity are located in the existing FELIX vault. This solution requires the relocation of the existing short wave length FEL.

2. Conceptual design of the FELIX-FELICE facility

General layout of the modified FELIX-FELICE infrared user facility is shown in Fig. 1. The new beam line comprises the magnetic elements between dipole 2B10 behind the second acceleration section, and the beam dump (Dump3) behind the FELICE undulator. The FELIX bending magnets 1B10-1B20 and/or 2B10-2B20 will switch the beam at a frequency of 20 Hz. Their coils will be replaced in order to be able to operate at AC. Matching quadrupoles Q10-Q30 and Q40-Q60 will be also replaced by AC ones. In order to accommodate the beam line in the existing facility vault, a 180° U-turn is used to fold the electron beam. The U-turn is realised as a dispersion-free section, whose chronicity may be adjusted in the range ±0.5 mm/pm in order to compensate possible beam debunching during its transport to undulator.

[*] Corresponding author. Tel.: +31(0)30-609-6999; fax: +31(0)30-603-1204. E-mail address: militsyn@rijnh.nl.

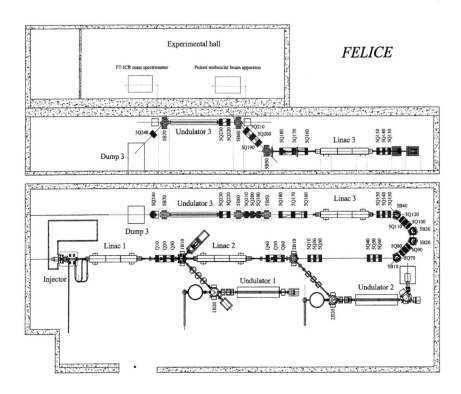

Figure 1. General layout of the modified FELIX-FELICE infrared free electron laser facility (upper panel - side view, lower panel - top view)

Variation of the beam energy in the range 20–65 MeV will be made by an additional acceleration section (Linac3). Injection into the FELICE undulator will be provided by means a 45° S-bend in the vertical plane. A permanent-magnet undulator, with 46 periods of 60 mm and a minimum gap of 22 mm, will be installed just below the ceiling of the vault. To suppress the harmonic content he use of an aperiodic undulator is under consideration. Wavelength tuning of FELICE over a factor of 10 will be achieved by varying the energy of the driving electron beam, while another factor of three will be achieved by changing the undulator gap. A U-shape optical resonator will penetrate the ceiling into a new hall, that will be built on top of the FELIX vault, where the intra-cavity set-ups will be located, thus providing the necessary radiation shielding. At wavelengths longer than about 30 μm, a retractable two-plate system will be used to guide the optical mode in the undulator in order to reduce the diffraction losses. A rectangular bending magnet 5B70, and a quadrupole to suppress the dispersion, which is excited by the bending, magnet will be used to dump the beam. This design allows for transportation of a beam with an energy spread of up to 10% to the dump.

3. Conclusion

This work is part of the research program of the "Stichting voor Fundamenteel Onderzoek der Materie (FOM)", which is financially supported by the "Nederlandse Organisatie voor Wetenschappelijk Onderzoek (NWO)".

4. References

[1] FELICE, a free Electron Laser for Intra-Cavity Experiments. FOM-01.700/D, 2002.

[2] http://www.rijnh.nl/felix.

[3] B.L. Militsyn, G. von Helden, G.J.M. Meijer and A.F.G. van der Meer, FELICE – the free electron laser for intracavity experiments, Nucl. Instr. and Meth. **A507** (2003) 494-497.

FREE ELECTRON LASERS 2003
E.J. Minehara, R. Hajima and M. Sawamura (Eds.)
Published by Elsevier B.V.

Design Considerations for a High-Efficiency High-Gain Free-Electron Laser for Power Beaming

C. Muller and G. Travish *

UCLA Department of Physics & Astronomy, Los Angeles, CA 90095. USA

Abstract

Power beaming from ground-based systems to space-based platforms has been proposed by a number of researchers as a means of delivering energy to orbiting satellites and stations. This paper considers the use of a seeded high-gain high-efficiency Free-Electron Laser (FEL) amplifier based on a conventional linac as the source for power beaming. While the wall-plug efficiency of a single pass FEL is likely to be considerably lower than a recirculating system, electrical efficiency is unlikely to be a serious consideration for first-generation power-beaming systems. Moreover, the simplicity of the proposed scheme scales well from existing and completed experiments.

PACS codes: 41.60.Cr, 42.62.Cf,

Keywords: FEL, Power Beaming, Satellite

1. Introduction

High average-power, high-brightness electron-beams allow for new applications especially in beam-radiation interaction and production. One such application is power beaming from ground-based sources to space-based platforms The concept of power beaming has been around since the 1960's but technical and economic hurdles prevented implementation [1]. The Free Electron Laser has been examined as a possible source for ground-based beaming already [2, 3]. Here we examine a high efficiency, high-gain FEL that can take advantage of recent progress in photoinjectors and high average-power lasers. We consider a system capable of delivering 1 KW of electrical power to a platform in geo-stationary orbit.

2. Assumptions & Efficiencies

An FEL-based system for beaming power from ground to space can be idealized as composed of the following: an accelerator, the radiator (FEL), ground optics, the atmosphere, and the receiving station. An estimate of the energy efficiency (optical or electric) guides the design of the FEL. Simplistic arguments were used; the breakdown of the efficiencies assumed is given in Table 1, and yields an overall optical efficiency of about 0.076% from "wall-plug" to satellite electrical power (at 840 nm).

Table 1: Power beaming assumed efficiencies. The assumptions are based on simplistic arguments, and are meant only to provide an order-of-magnitude estimate of the energy requirements.

Parameter	Efficiency
Geometric (Diffraction)	<64%
Solar Panel Conversion	~50%
Atmospheric Transmission	~80%
Ground Optics Transmission	>50%
Beam to FEL Conversion	≡10%
Wall to Beam Conversion (60% to RF, 10% to Beam)	6%
FEL output to Space Power	*12.7%*
Wall Plug to Space Power	*0.076%*

A more relevant figure of merit is the electron-beam power needed as earth-based electricity is

* gil.travish@physics.ucla.edu

abundant and inexpensive compared to space-based electrical production. The desired figure of merit depends on the FEL efficiency, η_{FEL} (in converting e-beam power to optical power): this study began with $\eta_{FEL} = 10\%$, thus, one might expect an overall electron beam power of about 78 KW. As is discussed in the next section, the FEL efficiency — for the parameters chosen here — is difficult to raise above about 6%. Thus, a 100 KW class accelerator is needed. While the efficiencies listed are reasonable estimates, the strong effect of atmospheric turbulence has not been taken into account [4]. Here we assume that techniques such as adaptive optics can be used to limit the effect of the atmosphere.

The central wavelength of the FEL is chosen to simultaneously maximize the transmission efficiency through the atmosphere and the energy conversion-efficiency of solar panels — 840 nm satisfies these constraints while still being at a wavelength where conventional seed-lasers exist. Setting the electron-beam energy as high as possible maximizes the beam power while lowering the average beam current requirements. However, FEL efficiencies drop and undulator designs become awkward (long periods) with higher beam-energies. On the other hand, peak current is critical to FEL performance as well as to mitigating the effects of diffraction after saturation. Guided by semi-analytic codes, past design and constraining the undulator period to a maximum of about 6 cm and an undulator parameter of no more than 3, a beam energy of 200 MeV was selected initially, and then optimized to 226 MeV.

3. Prototype Design

Unlike conventional FEL sources in which various optical parameters need to be optimized, the primary consideration here is efficiency: the high average-power requirements demand efficient extractions of energy from the electron beam. The initial FEL parameters used for the optimization are listed in Table 2.

A seed laser of 1 KW was used for all simulations — an ambitious but achievable number. The pulse format of the seed laser was assumed to match that of the accelerator. A system scaled from typical UCLA designs was used: an RF system of 4 μs in duration and an uncompressed bunch-length of 5 ps. Here we assume a bunch train of 1000 microbunches, each

with 3.5 nC of charge, and operating at 100 Hz — a selection well-suited to an L-band system. Various other bunch and RF formats have been considered.

Table 2: Initial FEL parameters used in the simulation studies

Parameter	Value
Central wavelength	840 nm
Beam Energy	226 MeV
Beam Current	500 A
Beam Emittance (norm. rms)	5 μm
Beam Energy Spread	0.15%
Undulator Period	6 cm
Undulator Parameter	3.0
Focusing (β-function)	87 cm

3.1 Simulations & Optimization

The parameters that remain to be determined by simulation include the undulator focusing, length, tapering gradient and taper start-point. A series of simulation were performed using the 3D FEL-simulation Genesis 1.3 [5]; keeping the parameters of Table 2 fixed, while varying the remaining parameters. The best efficiencies found for 20 m and 40 m undulators were 2.6% and 6.7%, respectively. These efficiencies were optimized (primarily) by changing the tapering gradient and starting point position. The overall taper for the optimized undulators were 5% and 15% starting at 12.5 m for the 20 m, and 40 m long undulators, respectively. Efficiencies as high as 13% were achieved, but with an unrealistically long (150 m) undulator.

4. Conclusions

Optimization of a high-gain FEL yielded a system capable of producing 1 KW of electric power in space using a 40 m undulator and a $\approx$100 KW electron beam. This design relies on improvements to photoinjectors and lasers that may allow for high repetition-rate, high-brightness beam production and for high-power seeding of the FEL.

5. References

[1] P. Glaser, Science, **162** 3856, pp 857-861.
[2] K.-J. Kim, et al., Proc. FEL Conf. 1997.
[3] M. C. Lampel, et al., Rocketdyne Internal (1993).
[4] G. A. Landis, Acta Astronautica, 25 4 pp. 229-233 (1991).
[5] S. Reiche, NIM A429, 243 (1999).

FREE ELECTRON LASERS 2003
E.J. Minehara, R. Hajima and M. Sawamura (Eds.)
Published by Elsevier B.V.

Proposal and preliminary experiments of high-T_c bulk superconducting wigglers

Hidenori Matsuzawa[a],*, Tatsuya Shibata[a], Takumi Okaya[a], Takumi Kibushi[a],
Eisuke Minehara[b]

[a]*Graduate School of Engineering, University of Yamanashi, Kofu, Yamanashi 400-8511, Japan*
[b]*Free-Electron Laser Laboratory, Advanced Photon Research Center, JAERI, Tokai, Ibaraki 319-1195, Japan*

Abstract

We previously proposed a high-critical temperature (T_c) bulk superconducting wiggler. In this paper, two different microwigglers were made from Bi(2223)-based sintered slabs with a T_c of ~103 K, and were cooled down to a low enough temperature of 32 K. The wigglers had such periodic configurations of surfaces as a quasi-sinusoidal wave with a period of 2 or 4 mm, both with a 100 mm axial length and 20 mm width. The shorter-period wiggler is expected to radiate visible fluorescence when installed in the JAERI linac (20 MeV). To evaluate the utility of the wigglers, a mini-magnet was swept, as an alternative to electron beams, through the gap of the opposing wigglers. A high-speed video camera observed that the magnet approximately described a sinusoidal trajectory.

PACS codes: 41.60.Cr; 41.85.-p; 74.72.Jt; 85.25.-j

Keywords: Superconducting wigglers; Microwigglers; High-T_c bulk superconductors; Mini-magnets; Electron trajectories

1. Introduction

As one of microwigglers, we previously proposed a high-critical temperature (T_c) superconducting wiggler and did a preliminary experiment on it [1]. The principle of the wiggler is simple (Fig. 1). When an electron beam approaches a superconductor tangentially with a low angle of incidence, the beam will be repelled from a point near the superconductor owing to the Meissner effect. Thus, when the beam propagates through the gap of opposing wigglers, the beam would describe a zigzag trajectory. Such an effective, strong repulsive force has already been confirmed using bismuth (Bi)-based superconducting rings [2, 3], for single electron pulses of nanosecond order, in comparison with a

*Corresponding author. Tel.: +81-55-220-8477; fax: +81-55-220-8477.
E-mail address: matuzawa@es.yamanashi.ac.jp (H. Matsuzawa).

normal conducting ring. The wiggler has the advantages of being light in weight, inexpensive, and easily machined into microwigglers.

2. Expererimental results

In this work, two pairs of high-T_c Bi(2223)-based microwigglers ($T_c \doteqdot 103$ K) were fabricated from sintered slabs with a 100 mm length, 20 mm width, and 3.4 mm thickness. After forming lamellar gratings [Fig. 2(a)] with a period of 2 or 4 mm, we cut each corner of the ridges in order for the profile to approximate to a quasi-sinusoidal wave [Fig. 2(b)]. Both the wiggler groove and ridge had the identical basal width, with a 1 mm depth. Therefore, the two wigglers had, respectively, a total number of periods

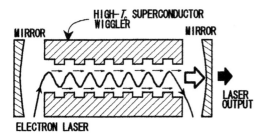

Fig. 1. Principle of high-T_c superconducting wigglers.

Fig. 2. (a) 2- or 4-mm period wigglers machined from Bi(2223)-based slabs, (b) profile of ridge of 2-mm period wiggler, and (c) assembled wiggler in the beam line of JAERI linac.

of 50 and 25. If the wigglers operate successfully for 20 MeV electron beams from the JAERI linac, the shorter-period wiggler would radiate the fluorescence centered at a wavelength of 630 nm. Cooling test of the wigglers installed in the beam line [Fig. 2(c)] confirmed that the lowest temperature, 32 K, was achieved in ~3.5 h.

To simulate utilities of the wigglers, we swept a Nd-based mini-magnet with each edge 0.5 mm long and weight 0.8 mg, through the gap of the opposing wigglers. The magnet gave a 22 G field at a point 1.2 mm apart from the magnet. On the other hand, the 20 MeV electron beams, ~20 A in peak, will induce a 40 G field at a point 1 mm apart from the beam axis. The two fields are thus almost of the same magnitude and configuration. When the magnet, which was vertically magnetized and was suspended by a 90-mm-long silk thread, was pulled horizontally along the axis of the 4-mm-period wigglers at ~77 K, the magnet described a nearly sinusoidal trajectory (Fig. 3). The trajectory was observed using a high-speed video camera (250 f/s) from above the wigglers. The time interval between the consecutive positions is 4 ms. The next step is to inject the 20 MeV electron beams into the wigglers and to observe the radiation by the electron beams.

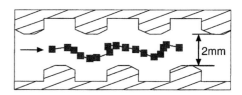

Fig. 3. Example of trajectories of mini-magnet swept through 4-mm period wigglers. The wigglers and magnets are drawn to scale.

References

[1] H. Matsuzawa et al., Appl. Phys. Lett. 59 (1991) 141.
[2] H. Matsuzawa, J. Appl. Phys. 74 (1993) R111.
[3] H. Matsuzawa et al., Appl. Phys. Lett. 78 (2001) 3854.

FREE ELECTRON LASERS 2003
E.J. Minehara, R. Hajima and M. Sawamura (Eds.)
Published by Elsevier B.V.

Design for an Infra-Red Oscillator FEL for the 4GLS Energy Recovery Linac Prototype

N. R. Thompson

ASTeC, CCLRC Daresbury Laboratory, Warrington WA4 4AD, United Kingdom

An infra-red oscillator FEL is planned as a component of the 4th Generation Light Source (4GLS) Energy Recovery Linac Prototype facility at Daresbury Laboratory. In this paper we present the preliminary design and give initial estimates for the expected performance. We also discuss how we plan to use the Prototype IR-FEL facility to inform the design work on 4GLS.

1. Introduction

CCLRC Daresbury Laboratory has funding for the construction of an Energy Recovery Linac Prototype (ERLP) facility [1]. This will facilitate the R&D needed for a design study of 4GLS [2]. Important aims for ERLP are the demonstration of energy recovery from an electron beam with an energy spread such as that induced by a free-electron laser and the development of expertise in the operation of such technology. Other aims are study of the synchronisation of multiple photon sources and the use of bunch compression. It is proposed to pass the beam through both a single undulator (which will soon be de-commissioned from the SRS) for the production of CSR [3], and also an IR oscillator FEL based on a wiggler previously used in the J-Lab IR Demo FEL [4] and supplied on loan as part of a collaboration. This paper presents the preliminary design parameters for the ERLP IR-FEL and discusses how it could be used to inform design work on 4GLS.

2. IR-FEL Parameters

The main parameters of the IR-FEL design are given in Table 1. The injector will be a photocathode design with a high repetition frequency (162.5MHz within macropulse) and the electron bunches will be accelerated in a superconducting linac to at least 30MeV and possibly up to 50MeV. The beam transport arcs will be triple bend achromats (TBAs) with a mechanically ad-

Table 1

ERL Prototype IR-FEL design parameters

PARAMETER	DESIGN
Wiggler Period	0.027 m
Gap	12 mm
Number of periods	40
Length	1.08 m
K	1.0
Beam energy	30-50 MeV
Bunch charge	80 pC
Bunch length (rms)	<0.6 ps
Normalised emittance	5 πmm-mrad
Energy spread (rms)	0.1 %
Optical cavity length	7.3795 m
Mirror radius of curvature	3.8839 m
Mirror reflectivity	99.5 %

justable return arc for path length control. Chicanes will be used for bunch compression.

The optical cavity is near-concentric and the mirror assemblies are likely to be simplified versions of those produced at Daresbury for the EU-FELE project [5], although the required precision for cavity length adjustment and mirror alignment will be less strict due to the longer operating wavelength of the ERLP IR-FEL.

3. IR-FEL Output

The FEL wavelength will vary between 6μm at a beam energy of 30MeV and 2μm at 50MeV. Calculations show that at 30MeV the maximum single pass gain will be approximately 60% with a peak output power of 6MW obtained with an optimised outcoupling of 10%. This leads to an extraction efficiency of approximately 0.4% and a full induced energy spread (6σ) in the electron beam of about 4%. This is therefore a minimum requirement for the energy acceptance of the return arc.

4. Outcoupling

The method of outcoupling has yet to be decided. Options under consideration are hole coupling and scraper mirror coupling (as has been demonstrated successfully at JAERI [6]). Insertable scraper mirror outcoupling allows easier optimisation of the outcoupling fraction as the wavelength range is scanned and the mode size on the mirrors changes. For this reason this technique would appear to be more suitable for the proposed IR-FEL for 4GLS which will be designed to operate over a broad wavelength range of perhaps 3–75μm. It is therefore likely that trials with scraper mirror outcoupling will be undertaken on the ERLP.

5. Short-Pulse Operation

It is also proposed to operate the IR-FEL with *rms* electron bunch lengths <0.4ps and beam energy of around 25MeV. In this regime the bunch length is comparable to the slippage length and this may enable the study of superradiant emission with an enhanced extraction efficiency of possibly 1.5% [7]. The correspondingly increased induced energy spread in the electron beam would allow a more demanding test of the energy acceptance of the return arc.

6. Intracavity backscattering

The high repetition frequency means that to obtain an optical cavity of a practical length the FEL must operate with at least two optical pulses circulating in the cavity simultaneously. Intracavity Thomson backscattering will then occur naturally and it is proposed to take advantage of this by extracting the X-ray photons produced. The effect of the backscattering on the electron beam is expected to be negligible due to the low X-ray photon flux (only about $10^7\mathrm{s}^{-1}$) and the fact that the X-ray photon energy is low (about 20keV) compared to the electron energy (30–50MeV).

7. Seeding studies

The mirror assemblies will be designed to allow easy removal of the mirrors. The relatively high single pass gain (up to 60%) may then enable the wiggler to be used for studies of laser seeding. This will enable research into the synchronisation between seed pulse and electron bunch which will be relevant to the amplifier FEL on 4GLS—simulations with GENESIS indicate that for the 4GLS parameters short femtosecond seed pulses can generate saturated laser pulses with a width of the same order [8] while the energy of the non-saturating electrons in the bunch can be recovered by the ERL.

8. Synchronisation studies

One of the attractions of 4GLS is the synchronisation of multiple sources. It is therefore proposed to use the ERLP to undertake synchronisation and timing studies of the ERLP IR-FEL output in conjunction with the outputs of the other ERLP photon sources. These include CSR from the single undulator and arcs, X-ray photons from the Thomson backscattering and light from the photocathode drive laser.

REFERENCES

1. M. W. Poole *et al*, PAC, 2003.
2. M. W. Poole & B. McNeil, NIM A507 p489, 2003.
3. C. Gerth & B. McNeil, these proceedings.
4. G. R. Neil *et al*, Phys. Rev. Lett. 84(4) p662, 2000
5. M. W. Poole *et al*, NIM A445 p448, 2000.
6. R. Nagai *et al*, NIM A475 p519, 2001.
7. N. Piovella *et al*, Phys. Rev. E, 52(5) p5470, 1995
8. B. McNeil, http://www.astec.ac.uk/conf/ esfri/machine_presentations.htm

FREE ELECTRON LASERS 2003
E.J. Minehara, R. Hajima and M. Sawamura (Eds.)
Published by Elsevier B.V.

XUV Generation at the BNL DUV-FEL Facility

A. Doyuran, S. Hulbert, E. Johnson, H. Loos, W. Li[#], J. B. Murphy,
G. Rakowsky, J. Rose, T. Shaftan, B. Sheehy, Y. Shen, J. Skaritka, A. G. Suits[#],
X.J. Wang [*], Z. Wu, L.H. Yu

BNL, Upton, NY 11973, USA

#Department of Chemistry, Stony Brook University, Stony Brook, NY 11794, USA

Abstract

The performance and future upgrade of the XUV at the Brookhaven DUV-FEL facility are presented in this report. The High Gain Harmonic Generation (HGHG) Free Electron Laser (FEL) at the DUV-FEL now routinely produces fully coherent 266 nm radiation with the 800 nm seed, and the output energy of the HGHG is about 100 μJ with 1 ps pulse duration (FWHM) [1]. The first three harmonics of the HGHG from the nonlinear harmonic generation process were experimented characterized. The third harmonic (89 nm, 14 eV) of the HGHG has been explored for chemical science applications [2].

To further expand the DUV-FEL user base and explore its unique potential, it is proposed that the DUV-FEL to be upgraded to produce XUV (< 100 nm) at the HGHG fundamental. Several options are now under consideration to upgrade the DUV-FEL for XUV radiation generation; which including the electron beam energy upgrade to 300 MeV and installing the shorter period VISA undulator.

PACS codes: 42.72.Bj, 41.60.Cr, 41.60.−m

Keywords: XUV, free electron laser, high gain harmonic generation, coherent radiation

1. The BNL DUV-FEL facility

The Brookhaven National Laboratory (BNL) DUV-FEL is a dedicated platform for FEL and other linac based light source technologies R&D. The main focus at the DUV-FEL is to develop and explore the laser seeded HGHG FEL; we also carry out R&D in electron beam generation, compression and ultra-fast beam instrumentations at the DUV-FEL.

The DUV-FEL is a laser linac facility based on the high-brightness photocathode RF gun injection system. The DUV-FEL facility consists of a 1.6 cell BNL GUN IV, four sections of three-meter long SLAC type linac, a four-magnet chicane bunch compressor, RF synchronized Ti:sapphire laser system and HGHG undulators. The pico-seconds electron beam produced by the RF gun is quickly

accelerated to about 70 MeV by two sections of the linac. The electron beam is compressed down to sub-ps by the chicane before it is further accelerated by the last two sections of the linac. The maximum electron beam energy can be reached at the DUV-FEL is about 200 MeV. The Ti:sapphire laser system is capable of producing 150 fs to several pico-seconds long laser pulses with about 50 mJ output. It is used both to drive the photocathode RF gun and seed the HGHG FEL. One of the unique features of the DUV-FEL laser system is that, one single laser system is capable of producing RF synchronized two laser pulses with both pulse length and relative delay adjustable by employing two grating laser pulse compressors.

[*] Fax:631-344-3029.
E-mail:xwang@bnl.gov

2. XUV by Nonlinear Harmonic Generation

The HGHG FEL reached saturation at the 266 nm with the 800 nm seed in October of 2002[1]. Since then, the harmonic of the HGHG has been experimentally characterized.

As FEL nears the saturation, the strong electron beam micro-bunching at the fundamental drives electron beam micro-bunching at the higher harmonic, and the coherent harmonic radiation by this process is known as nonlinear harmonic generation. A vacuum Acton UV monochromator with 1200 l/mm grating was installed to measure the HGHG harmonic energy. The output of the HGHG is transported to the monochromator by a two-mirror periscope. With the 250 μm slit, the resolution of the monochromator is about 1 nm, which is wider than the HGHG spectrum[1]. The monochromator behaves like a band-pass filter in our measurements. Fig.1 plotted energies of the first three harmonic if the fundamental energy is assumed to be 100 μJ [1]. Significant output was observed at both the second and third harmonic, which confirmed earlier VISA results [3].

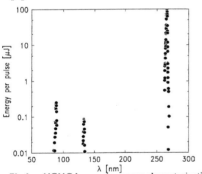

Fig.1. HGHG harmonic energy characterization.

The measured second and third harmonic energies are lower than theoretical predictions (~ 1%) could come following experimental errors:
1. The assumptions on the grating diffraction efficiency for all three harmonics.
2. The alignment error, especially for the second harmonic since it is generated by the off-axis radiation.

The number of the photons/pulse at the third harmonic of the HGHG is about two orders of magnitude higher than third generation light source could generate. This unique XUV source was successfully used for ion pair imaging of methyl fluoride [2]. To open new scientific opportunities in nanotechnology, chemistry, nonlinear UV physics, we propose to upgrade the DUV-FEL to generate more than 100 μJ coherent XUV radiation.

3. Future XUV at the DUV-FEL Facility

It is possible to produce the energy proposed if the HGHG fundamental operating at the XUV. Fig.2 plotted the FEL output for two undulators. For the existing NISUS undulator, electron beam energy upgrade to 300 MeV is required for XUV generation. Since the field gradient of the SLAC structure is limited to about 20 MV/m, a addition linac section and two more klystrons will be installed.

Another option we are considering is to replace last couple meters of the NISUS undualtor with the VISA undulator, and to generate coherent second harmonic radiation with the VISA undulator [4].

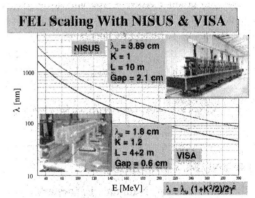

Fig. 2. FEL output as function of the beam energy for NISUS and VISA undualtors.

This work is supported by the US DOE under contact No. DE-AC02-98CH10886

4. References

[1] L.H. Yu et al, Phys. Rev. Lett. 91, 074801 (2003).
[2] W. Li et al, Submitted to Phys. Rev. Lett. for publication.
[3] A. Tremaine et al, Phys. Rev. Lett. 88, 204801 (2002).
[4] R. Bonifacio et al, Nucl. Instr. And Meth. A 296, 787-790(1990).

FREE ELECTRON LASERS 2003
E.J. Minehara, R. Hajima and M. Sawamura (Eds.)
Published by Elsevier B.V.

Experimental verification of velocity bunching at the S-band photoinjector and linac

H. Iijima[a], M. Uesaka[a], L. Serafini[b], T. Ueda[a] and N. Kumagai[c]

[a]Nuclear Engineering Research Laboratory, Graduate School of Engineering, University of Tokyo,
2-22 Shirakata-Shirane, Tokai, Naka, Ibaraki, 319-1188, Japan

[b]Istituto Nazionale di Fisica Nuckeare, Milano, Via, Celoria 16, 20133 Milano, Italy

[c]Japan Synchrotron Radiation Research Institute,
1-1-1 Mikazuki-cho, Sayo-gun, Hyogo, 679-5198, Japan

We report the experimental verification of "velocity bunching" based on the rectilinear compressor. The bunching has been performed by the S-band linac and the Mg photocathode RF injector at Nuclear Engineering Research Laboratory, University of Tokyo. We have achieved compression of 1.0 ps (FWHM) by injection into near zero-cross RF phase. The result is consistent with a numerical analysis by PARMELA.

1. Introduction

The velocity bunching was proposed by Prof. L. Serafini in 2000[1]. The scheme is based on the rectilinear compression, therefore magnetic compressors as such as a chicane-type magnet is not necessary. At Nuclear Engineering Research Laboratory, University of Tokyo, an S-band linac with an Mg photocathode injector (18L) has been utilized for the radiation chemistry. To realize the high time-resolution in order of sub-picosecond, the chicane-type magnetic compressor is normally used[2,3]. Although the chicane-type compressor is able to generate the bunch of less than 1ps, the experimental proof of velocity bunching is interesting.

Normally, two accelerating tubes are used for the complete scheme of velocity bunching as shown in SPARC project[4], because the energy of compressed beam in the first tube is low. Therefore, the electron beam from the photoinjector is compressed in the first tube and accelerated in the second tube. Unfortunately 18L has one accelerating tube, however Helmholtz coils to suppress the emittance growth due to the space-charge effect are located around the tube. Thus, we had demonstrated the velocity bunching with concentrating the verification of compression scheme using one accelerating tube and Helmholtz coils.

2. Experimental setup

The accelerating tube of 18L is the S-band traveling wave structures operating on the $2\pi/3$ mode with 2m long. The beam from RF injector is accelerated up to 22 MeV. In this experiment, the chicane-type compressor is not used, therefore the downstream of accelerating tube including the chicane-type compressor is used as the drift space with approximately 5m long. Eleven Helmholtz coils are located around the accelerating tube. The amplitude of magnetic field is up to 300 Gauss. The Xe gas chamber as a Cherenkov radiator is set at the end of linac. The Cherenkov light is guided to a femtosecond streak camera to measure the electron bunch width.

3. Numerical analysis by PARMELA

Based on numerical analysis used by simulation code of PARMELA, we chose operating parameters in the first experiment. Figure 1 shows rms envelope (beam size), normalized horizontal emittance (rms), bunch width (FWHM) and beam energy along the 18L. Here, we have assumed the initial bunch width of 2.0 ps (rms) and the charge of 1 nC/bunch. Another parameters are

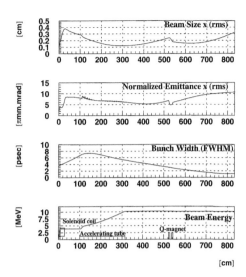

Figure 1. Simulation result of beam size (rms), normalized emittance (rms), bunch width (FWHM) and beam energy along 18L.

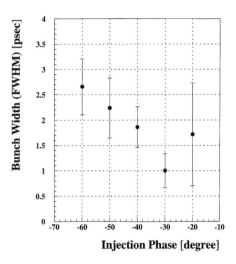

Figure 2. Experimental result of bunch width as a function of injected RF phase.

optimized to obtain minimum bunch width at the end of linac. To shorten the bunch, the injected phase is not 0 degree but -10 degree in this case. In addition, the bunch is also compressed in the drift space. Although this scheme corresponds to "ballistic bunching", there is no solenoid field to suppress the space-charge effect in the drift section. Therefore, one can show the emittance growth in the drift space. Finally, we have obtained the bunch width to be 1.0 ps (FWHM) with optimizing the parameters.

4. Experimental results

We have got streak camera images of the compressed bunches by a single shot for several injected RF phase. Figure 2 shows bunch width as a function of injected RF phase. The each point is the average of 30 shots and the error indicates the root mean square of them. The error without -20 degree is mainly caused by the fluctuation of RF power. The reason why the error at -20 degree is extremely larger than another point is a scarcity of Cherenkov light, because the energy

threshold of Xe gas is 13.6 MeV. Finally, we have achieved in the average bunch width of 1.0 ps for 1nC/bunch at -30 degree.

5. Conclusion

We had demonstrated the velocity bunching experimentally and verified the compression scheme. The bunch width of 1.0 ps is consistent with the numerical analysis and equivalent to the performance of chicane-type magnet. In near future, we are going to evaluate the transverse emittance and the invariant envelope.

REFERENCES

1. L. Serafini and M. Ferrario, Proc. of ICFA Workshop on the Physics of, and Science with, the X-ray Free-Electron Laser, Arcidosso, Italy, Sept. 2000.
2. M. Uesaka, *et al.*, Rad. Phys. Chem. **60**(2001)303.
3. Y. Muroya, *et al.*, Nucl. Instrum. & Meth. A **489**(2002)554.
4. M. Boscolo, *et al.*, Proc. of EPAC, Paris, 2002.

FREE ELECTRON LASERS 2003
E.J. Minehara, R. Hajima and M. Sawamura (Eds.)
Published by Elsevier B.V.

Phase shift induced by Free Electron Laser interaction

N. Nishimori[a]* R. Hajima[a] and R. Nagai[a]

[a]Free Electron Laser Laboratory, Advanced Photon Research Center, Kansai Research Establishment, Japan Atomic Energy Research Institute (JAERI), 2-4 Shirakata-Shirane, Tokai, Naka, Ibaraki 319-1195, Japan

Free-Electron Laser (FEL) interaction with unbunched electrons at resonant energy induces a positive phase shift of the FEL field relative to the input phase. The positive phase shift corresponds to giving electrons positive energy detuning, and results in the FEL amplification. The positive phase shift connects the zero small signal gain in low gain regime for resonant electrons to the exponential growth in high gain regime.

1. Introduction

In a Free-Electron Laser (FEL), a fraction of incident electron beam power is converted into radiation, while a fraction of RF power stored in cavities is given to an incident electron beam in an RF accelerator. If the pulse length of an incident electron beam is much longer than an RF period, energy transfer between the RF and the electrons is canceled out in both systems, while the electrons are bunched every RF period. In FELs, electrons bunched every radiation wavelength of an FEL field are self-injected into the field. In order to realize an energy transfer from electrons to the field, the bunched electrons must be self-injected into the deceleration phase of the field.

In this paper, we show that electrons at resonant energy are bunched every resonant wavelength and self injected into deceleration phase. The self-injection of electrons or the positive phase shift of the field is closely related to the collective instability [1].

2. Positive phase shift

A time domain analysis has shown that an FEL field with uniform phase is formed by longitudinal mixing of shot noises of electrons due to the slippage effect [2]. This corresponds to

the spectrum narrowing in the frequency domain [3]. The interaction between unbunched electrons and the FEL field with uniform amplitude and phase is considered in the time domain. The incident electron is modulated in energy depending on its transversal direction along an undulator field and the sign of an FEL electric field. The transversal velocity of an electron beam is given by $\mathbf{v} = -\sqrt{2}(a_w c/\gamma)\cos(k_w z)\mathbf{x}$ [4] , and the electron trajectory is shown by a dotted line in Fig. 1 (a). The optical electric field $\mathbf{E} = \sqrt{2}E(z,t)\cos[kz - \omega t + \phi(z,t)]\mathbf{x}$ at the entrance to each undulator period is given by a solid line in Fig. 1 (b). The electrons are bunched around $\psi_i(\tau) + \phi(\zeta,\tau) = \pi/2$, as indicated by solid dots in Fig. 1 (b), and are slipped back through the field.

The electron trajectory through an undulator period can be divided into four $\pi/2$ sections shown in Fig. 1 (a). In the section I the electron bunch moves towards the positive potential of the FEL field, resulting in the field absorption. In the section II, III and IV, the field is amplified, absorbed and amplified sequentially, depending on product of the transversal direction of the bunched electrons and the sign of the FEL field. The total gain is zero [4,5], but this process induces a positive phase shift by amount equal to the ratio of the amplitude change to the initial amplitude (see Fig. 1 (b)). The positive phase shift corresponds to giving electrons positive energy detuning, and the field is amplified.

---

*Corresponding author. Tel: +81-29-282-6315; FAX: +81-29-282-6057; E-mail:nisi@milford.tokai.jaeri.go.jp

The previous studies on the collective instability have attributed the physical reason why electrons are bunched around a decelerating phase to the pondermotive phase velocity change based on the dielectric property of an electron beam [6,7]. The phase velocity change is equal to frequency shift given by $(\omega - d\phi/dt)/\omega$. This is different from the present study that the FEL interaction process itself induces the phase shift $d\phi/dt$ and the FEL field is amplified.

Three dimensional effects such as optical guiding appear in the high gain regime as a result of the diffraction parameter change in the transversal direction. The FEL phase shift is related to the diffraction parameter given by $n - 1 = (1 + a_w^2)/(4\pi\gamma^2 N_w)(d\phi/d\tau)$ [5]. The coefficient $(1 + a_w^2)/(4\pi\gamma^2 N_w)$ is several orders of magnitude smaller than 1. The optical guiding effect can be observed with j_0 higher than 10000 [4], whereas the deviation from the antisymmetric shape of the gain curve with respect to detuning energy, which indicates the phase shift or collective instability, is observed with j_0 of order of 1 [8].

3. Conclusion

We have shown that a positive phase shift of an FEL field is induced by FEL interaction with bunched electrons. The positive phase shift corresponds to giving positive detuning to the electron energy, and the FEL field is amplified regardless of the dimensionless current when the initial electron energy is resonant. The present amplification mechanism has been known as the collective instability. The diffraction effect or the frequency shift appears as a result of the phase shift, but is much smaller than the phase shift effect.

REFERENCES

1. R. Bonifacio, C. Pellegrini, and L. M. Narducci, Opt. Comm. 50 (1984) 373.

2. N. Nishimori, R. Hajima, R. Nagai and E.J. Minehara, Nucl. Instr. and Meth. A 507 (2003) 79.

3. Kwang-Je Kim, Phys. Rev. Lett. 57 (1986) 1871; Nucl. Instr. Meth. A 250 (1986) 396.

4. W. B. Colson, in *Laser Handbook*, edited by W. B. Colson, C. Pellegrini, and A. Renieri (North Holland, Amsterdam, 1990), Vol.6, pp. 115 – 193.

5. C. A. Brau, *Free-Electron Lasers* (Academic, San Diego, 1990).

6. R. Bonifacio et al., Riv. Nuovo Cimento 13 (1990) 9.

7. J. B. Murphy and C. Pellegrini, in *Laser Handbook*, edited by W. B. Colson, C. Pellegrini, and A. Renieri (North Holland, Amsterdam, 1990), Vol.6, pp. 47–53.

8. G. Dattoli, H. Fang, L. Giannessi, M. Richetta, A. Torre, and R. Caloi, Nucl. Instr. and Meth. A 285 (1989) 108.

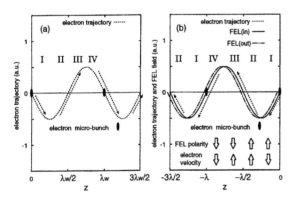

Figure 1. Electron trajectory in an undulator in a laboratory frame (a) and incident FEL field at the entrance to an undulator period together with electron micro-bunches in a frame moving at the vacuum speed of light (b).

FREE ELECTRON LASERS 2003
E.J. Minehara, R. Hajima and M. Sawamura (Eds.)
Published by Elsevier B.V.

Enhance the output power of a high power FEL by using a higher repetition rate

Xiaojian Shu*, Yuanzhang Wang

Institute of Applied Physics and Computational Mathematics, P. O. Box 8009, Beijing 100088, P. R. China

Abstract

Under the condition that the average current of the electron beam is kept a constant with a certain RF power, the influence of peak current or repetition rate of the electron pulse on the output power of a FEL oscillator based on the parameters of Jlab Demo FEL is studied with the help of our FEL codes. It is found that it is more effective and useful to enhance the output power and efficiency of a high power FEL by using a higher repetition rate.

PACS codes: 41.60.Cr, 42.60.Lh, 41.75.Ht

Keywords: Free-electron laser, Repetition rate, Efficiency, Saturation, Numerical simulation

1. Introduction

In a free-electron laser (FEL) oscillator driven by a RF linac, the output power and efficiency can be enhanced by increasing the peak current or repetition rate of the electron pulse. But it is impossible to increase both with a certain RF power and instead the average current is kept a constant. With this condition, the influence of peak current or repetition rate on the output power of a high power FEL based on the parameters of Jlab Demo FEL [1] is studied with the help of our FEL codes [2,3]. It is found that it is more effective to enhance the output power of a high power FEL by using a higher repetition rate than that with a higher peak current. The reason is due to the saturation mechanism of FEL oscillators [4]. The saturation takes place when the optical power in the cavity reaches a certain value, which occurs more early in the case with a higher peak current.

2. Simulation results

In order to reduce the computational time, We firstly use our three-dimensional steady-state code

---

* Corresponding author. Fax: +86-10-62010108. E-mail address:shu_xiaojian@mail.iapcm.ac.cn.

OSIFEL [2,3] to make numerical simulations of the FEL oscillator for various peak current and repetition rate of the electron beam with the parameters based on Jlab Demo FEL as listed in Table 1. In simulations, the distribution functions of transverse position and velocity, energy of the electrons are assumed as Gaussian. The corresponding initial values of the sample electrons are given by Monte Carlo method and the initial phases are loaded according to the "quiet start" scheme to eliminate the numerical noise. The energy spread means FWHM and emittance means RMS. The emittance in y-direction is the same as that in x-direction and the initial size of the electron beam is chosen to obtain a circular cross-section at the centre of the wiggler.

Table 1
The parameters based on Jlab Demo FEL

Electron beam	
energy [MeV]	48
bunch length [ps RMS]	0.4
bunch charge [pC]	60
emittance [mm-mrad]	7.5
energy spread [%]	0.5
Wiggler	
period [cm]	2.7
peak field strength [kG]	5.5
number of periods	40.5
Optical	
cavity length [m]	8.0105
mirror radii of curvature [m]	4.050
mirror radii [cm]	2.54
output coupler refl. [%]	90.5
HR reflectivity [%]	99.85

Table 2 lists the saturated intracavity power P_s and the average output power P_{out} of FEL for various peak current I_P and repetition rate R_r of the electron beam with the average current I_{av} of the electron beam being kept a constant of 4.4 mA. The average output power is calculated from the P_s, the output coupler reflectivity, R_r and the micropulse length of 1ps. The results are in good agreement with the experiments [1]. Although the saturated intracavity power increases along with the peak current, the output power and efficiency decrease due to the reducing of the repetition rate. In other words with a certain RF power and so a constant power of the

electron beam in a linac, the output power and efficiency of the FEL oscillator can be enhanced more effectively by using a higher repetition rate. The reason is due to the saturation mechanism of FEL oscillators [4]. The saturation takes place when the optical power in the cavity reaches a certain value, which occurs more early in the case with a higher peak current.

Table 2
The results from 3D steady-state simulations

I_P (A)	R_r (MHz)	P_P (MW)	P_{out} (kW)
58.82	74.8	310	2.21
117.65	37.4	474	1.68
235.29	18.7	727	1.29

3. Conclusions and discussions

It is found that it is more effective and useful to enhance the output power and efficiency of a FEL oscillator by using a higher repetition rate than a higher peak current with the average current is kept a constant. The reason is due to the saturation mechanism of FEL oscillators. However, with higher peak current, the sidebands will go up, which results in the increase of the output power and efficiency. But this part of power in the sidebands is not desired and must be suppressed since a narrow line-width is usually required for the applications of the FELs. We prefer using a higher repetition rate to lager peak current to obtain a useful higher power FEL.

References

[1] G.R. Neil et al., Nucl. Instr. and Meth. A 445 (2000) 192.

[2] Wang Taichun, Wang Yuanzhang, High Power Laser and Particle Beams, 7 (1995) 25, (in Chinese).

[3] Xiaojian Shu, Yuanzhang Wang, Nucl. Instr. and Meth. A 483 (2002) 205.

[4] N.M. Kroll, P.L. Morton and M.N. Rosenbluth, IEEE J. Quantum Electron. QE-17 (1981) 1436.

FREE ELECTRON LASERS 2003
E.J. Minehara, R. Hajima and M. Sawamura (Eds.)
Published by Elsevier B.V.

Space Charge Effects in a Prebunched Beam Free Electron Maser (FEM)

M. Arbel[*], *A.L. Eichenbaum*[+], *Y. Pinhasi*[+], *Y. Lurie*[+],
H. Kleinman, I.M. Yakover, A. Gover

Dept. of Physical Electronics – Tel –Aviv University (TAU)
[+] Dept. of Electrical & Electronics Engineering – The College of Judea and Samaria, Ariel

An electron beam, prebunched at a frequency near to the synchronous free-electron laser frequency and passing through a magnetic undulator, emits coherent (super-radiant) synchrotron undulator radiation at the bunching frequency. In the collective regime, the spectral properties of the emitted super-radiant power may depend on the electron beam current density (space charge effects). For a low density e-beam (tenuous beam) the monochromatic super-radiant emission power is maximal at a bunching frequency equal to the e-beam synchronism frequency, independent of the beam current. However, for a dense e-beam the space charge effects become dominant and synchronism can take place with either the slow or the fast space-charge (Langmuire) wave, which propagate on the e-beam with different phase velocities. Consequently radiated power maxima can occur at two well-separated frequencies. We report first experimental measurements of super-radiant power emission for a wide range of e-beam currents for which the influences of space charge can be investigated. The results of measurements, carried out on the prebunched e-beam FEM at TAU, are compared to those predicted by an analytical model [1] for a wide range of frequencies and of e-beam currents. Measurement results are in a good agreement with theory.

The analytical model

The prebunched beam radiation measurements were analyzed and compared to the analytical model developed by Schnitzer and Gover [1]. In this model the radiation obtained from a periodically pre-modulated e-beam traversing a waveguiding structure, located in a magnetic undulator has been analyzed and discussed. The analytical model takes into account space charge effects, current density and velocity modulation and characterizes the radiated power in the low and high gain regimes for low and high space charge density e-beam.

The radiated power due to current density modulation ($M_j \neq 0$) or velocity modulation ($M_v \neq 0$) in the e-beam only (without injection of an electromagnetic wave into the interaction region ($P(0)=0$)) is given by

$$P(L_w) = P_B \, F_{PB}(\bar{\theta}, \bar{\theta}_{pr})$$

where P_B is the prebunching power parameter, $F_{PB}(\bar{\theta}, \bar{\theta}_{pr})$ is the prebunched beam detuning function, $\bar{\theta}$ is the detuning parameter and $\bar{\theta}_{pr}$ is the space charge parameter (for more details and definition of parameters see Ref. [1] – [3]). In the space charge dominated regime (dense e-beam, $\bar{\theta}_{pr} >> \pi$) the detuning function F_{PB} reaches maximal values for the slow and the fast space charge waves for $\bar{\theta} \cong -\bar{\theta}_{pr}$ and $\bar{\theta} \cong +\bar{\theta}_{pr}$, respectively, while for a tenuous e-beam $\left(\bar{\theta}_{pr} << \pi\right)$ a single maximum is obtained at exact synchronism $\bar{\theta} \cong 0$.

For a pre-modulated e-beam Schnitzer and Gover calculates the detuning function F_{PB} separately for the case where only a current density modulation is introduced on the e-beam ($M_j \neq 0$) and for the case where only a velocity modulation is introduced ($M_v \neq 0$) in the e-beam. In these cases it was shown that the detuning function F_{PB} is symmetrical around $\bar{\theta} \cong 0$.

Fig. 1 shows the detuning function F_{PB} vs. $\bar{\theta}$ (i.e. vs. frequency) for two values of the space-charge parameter ($\bar{\theta}_{pr} \approx 0$ and $\bar{\theta}_{pr} \approx \pi$) and for $M_j=0.5$, $M_v=0$. For a tenuous e-beam ($\bar{\theta}_{pr} \approx 0$) a single maximum is obtained at exact synchronism $\bar{\theta} \cong 0$ (solid line) while for a dense e-beam $\bar{\theta}_{pr} \approx \pi$ the function reaches two separated maxima at $\bar{\theta} \cong \pm \bar{\theta}_{pr}$ (dashed line). Note that the calculations were carried out assuming $M_v=0$ and that in that case both curves are symmetrical around $\bar{\theta} \cong 0$.

Super-radiance measurements

The prebunched beam radiation measurements were carried out on a unique, table-top, FEM developed at Tel-Aviv University (TAU). A schematic of the experimental system of the prebunched beam TAU FEM is given in Ref. [2], [3]. The very wide range of FEM operating parameters is given in table 1.

The radiated power was measured for a wide range of e-beam modulation frequencies and for a wide range of e-beam currents (i.e. for a wide range values of the space charge parameter $\bar{\theta}_{pr}$). It was found experimentally that for low e-beam currents ($\bar{\theta}_{pr} << \pi$) the super-radiant power indeed reaches a single maximum as a function of frequency as expected theoretically (see Fig. 2). Although there is a small difference in frequency of the maximal radiated power between theory and experiments, qualitatively the results match well. On the other hand, for high e-beam currents ($\bar{\theta}_{pr} \approx \pi$) the experimental results have to be examined carefully, since at first sight they do not match closely those predicted by theory, since the two separated power maxima are of unequal value.

Fig. 3 shows the measured and calculated radiated power P_{PB} vs. frequency (i.e. vs. $\bar{\theta}$) for a high-density e-beam (I=0.85A). The space-charge parameter is $\bar{\theta}_{pr} \approx 3.2$ and the current density modulation parameter is $M_j \approx 0.1$. We assumed a velocity modulation parameter of $M_V=0.05\%$ and a relative phase between the current density and the velocity modulations of $\phi = 1.2\pi$. Note that for this case $\left(\bar{\theta}_{pr} \approx \pi\right)$ two separated power maxima of unequal value are obtained around $\bar{\theta} \approx 0$.

This new result was not discussed in Schnitzer and Gover model, where only the symmetrical cases have been analyzed. A good matching between theory and measurements was possible only if both current density modulation and velocity modulation are introduced in the e-beam with a relative phase between them.

Table 1: *The parameters of the pre-bunched e-beam FEM*

Electron Beam energy	70 keV
e-beam Current	0.1-1.2 Amp
Prebuncher frequency band	3 GHz to 12 GHz
Magnetic Induction	300 Gauss
Length of Period	4.44 cm
Number of Periods	Nw =17
Interaction Length	L_w = 74.8 cm
Rectangular Waveguide	2.21cm X 4.75cm

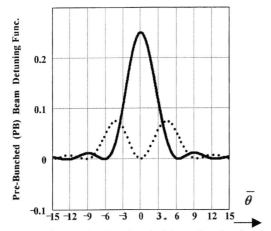

Figure 1: The calculated pre-bunched beam detuning function F_{PB} vs. detuning parameter $\bar{\theta}$ (i.e. vs. frequency) for two values of the space-charge parameter $\bar{\theta}_{pr}$ and for $M_j=0.5$, $M_V=0$. For a tenuous e-beam ($\bar{\theta}_{pr} \approx 0$, solid line) a single maximum is obtained at exact synchronism $\bar{\theta} \cong 0$ while for a dense e-beam ($\bar{\theta}_{pr} \approx \pi$, dashed line) the function reaches two separated maxima at $\bar{\theta} \cong \pm \bar{\theta}_{pr}$.

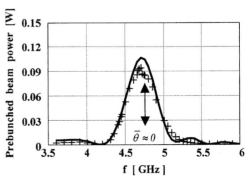

Figure 2: Measured (+ sign) and calculated (solid line) radiated power P_{PB} vs. frequency (i.e. vs. $\bar{\theta}$) for a low-density e-beam (I=0.2A). The space-charge parameter is $\bar{\theta}_{pr} \approx 1.5$, the current density modulation parameter is $M_j \approx 0.05$. For best fit we assumed a velocity modulation parameter of $M_V=0.05\%$ and an e-beam energy of 68 keV (instead of 70 keV used experimentally). Note that for this case, the radiated power is symmetrical around $\bar{\theta} \cong 0$ and a single maximum is obtained at exact synchronism $\bar{\theta} \cong 0$ (although $\bar{\theta}_{pr}$ does not satisfy the condition $\bar{\theta}_{pr} << \pi$).

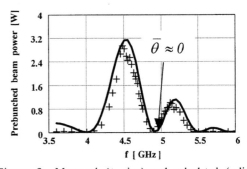

Figure 3: Measured (+ sign) and calculated (solid line) radiated power P_{PB} vs. frequency (i.e. vs. $\bar{\theta}$) for a high-density e-beam (I=0.85A). The space-charge parameter is $\bar{\theta}_{pr} \approx 3.2$, the current density modulation parameter is $M_j \cong 0.1$. We assumed a velocity modulation parameter of $M_V=0.05\%$ and a relative phase between the current density and the velocity modulations of $\phi = 1.2\pi$. Note that for this case $\left(\bar{\theta}_{pr} \approx \pi\right)$ two separated power maxima of unequal value are obtained around $\bar{\theta} \approx 0$.

References

[1]   I. Schnitzer and A. Gover, *Nucl. Instr. and Meth.* A 237 pp. 124-140, 1985.
[2]   M. Arbel et al., *Phys. Rev. Lett.*, vol.86, pp. 2561-4, 2001.
[3]   M. Arbel et al., *Nucl. Inst. and Meth.* A445, 247 (2000)

FREE ELECTRON LASERS 2003
E.J. Minehara, R. Hajima and M. Sawamura (Eds.)
Published by Elsevier B.V.

The Optical Resonator of the IR-FEL at ELBE

W.Seidel[a,*], P.Evtushenko[a], P.Gippner[a], R.Jainsch[a], D.Oepts[b], M.Sobiella[a],
D.Wohlfarth[a], A.Wolf[a], U.Wolf[a], R.Wünsch[a] and B.Wustmann[a]

[a] Forschungszentrum Rossendorf, Postfach 510119, D-01314 Dresden, Germany
[b] FOM Institute for Plasma Physics Rijnhuizen, P.O. Box 1207, 3430 BE Nieuwegein, The Netherlands

Abstract

We outline the actual state of the U27 resonator and its main controlling elements.

PACS codes: 41.60.Cr; 42.60.Da

Keywords: Free-electron laser, Optical resonator, Stabilization

1. Introduction

The radiation source ELBE [1] at Dresden-Rossendorf is centered around a superconducting ELectron accelerator of high Brilliance and low Emittance (ELBE) constructed to produce up to 40 MeV electron beams of 1 mA at 100 % duty cycle. Additionally to the production of secondary radiation such as X-rays, bremsstrahlung, neutrons and positrons, special emphasis will be put on the production of FEL radiation in the IR wavelength range of 5 - 150 µm. Intense infrared radiation in the 5-30 µm range will be produced in the undulator U27 [2]. In the following we outline the actual state of the appropriate resonator and its main controlling elements.

2. Resonator controlling and stabilization

The optical resonator for this ELBE FEL was mounted completely. All mechanical equipment and optical components (alignment system, interferometer for length stabilization, mirror wheels and chambers, temperature stabilization of the mirrors) have been checked thoroughly and tested for long-term stability (partly under vacuum conditions). All components (roughly cleaned), including the undulator and both mirror chambers, have been aligned by means of optical methods. The cavity requirements are summarized in the following table.

Parameter	Requirement
Resonator type	Stable, near concentric, symmetric
Cavity length	11.53 m, stabilized < 0.5 µm
Tilt stability	< 6 µrad
Rayleigh range	1 m
Mirror radii	3.75 cm
Radii of curvature	5.94 m
g^2	0.88
Diameter of outcoupling holes	1.5, 2.0, 3.0 and 4.5 mm

To optimize the extraction ratio in the whole wavelength range we use 5 mirrors with different hole sizes in the upstream mirror chamber. The Au-coated Cu-mirrors are mounted on a revolvable holder (wheel), which is fixed to a high-precision rotational stage. A similar construction with 3 mirrors is used in the downstream chamber.

To ensure the stability of the resonator at wavelengths down to 3 µm we require the mirror angular adjustment to have a resolution and stability in the order of 6 µrad. For the initial alignment of the mirror angles an accuracy in the order of 20 µrad is required. To achieve this accuracy we built up

---

* W.Seidel@fz-rossendorf.de

an alignment system consisting of two collinear He-Ne lasers. It uses two wall markers and 11 moveable adjustment apertures inside the cavity.

A Hewlett-Packard interferometer system is used for monitoring and stabilizing the resonator length (fig. 1). The interferometer beam is split in two beams (70% and 30%). The high intensity beam passes through the same resonator chamber as the

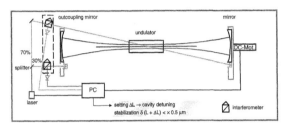

Figure 1. Schematic view of the resonator length control system. Its time constant is one Hz and hence fast in comparison to the thermal time constant of the resonator.

main laser and the electron beam. However, constraints on the width of the vacuum chamber do not leave enough space in the cavity for a separate parallel interferometer beam. Therefore, the latter will pass diagonally from one side of the upstream cavity mirror to a retroreflector on the other side of the downstream mirror. The low intensity beam is directed to one of the five retroreflectors on the front side of the mirror wheel close to the working outcoupling mirror. The control electronics for the two interferometer arms include a servo system to control and stabilize the relative distance between the two cavity mirrors using the motorized micrometer drive on the translation stage of the downstream chamber. There is no active tilt stabilization.

We have estimated the maximum intracavity laser power that can be expected for the ELBE beam when using the smallest outcoupling hole. Despite their high reflectivity of more than 99 % up to 15 W (cw regime) can be absorbed in the mirrors. The entire construction is heated and the resonator adjustment (length and angle) may be disturbed by thermal expansion. Moreover that heat load may affect the precision mechanics. At a movable precision construction in ultra-high vacuum a temperature stabilized system based on water cooling is difficult to be realized. Therefore we introduced special heat isolation between the high-precision rotation stage

and the mirror wheel. The wheel is also made from Cu to reduce mechanical tension between the mirrors and the surrounding material. Furthermore, the heat exchange is improved by a more flexible heat dissipation to the outside of the vacuum chamber (Peltier element or air cooling) rather than by thermal radiation only.

To stabilize the mirror wheel temperature we installed a heater in the center of the wheel. Independently of whether the laser is working or not all components are at the same equilibrium saturation temperature slightly above the saturation temperature.

The temperature behavior of the construction has been studied in vacuum by simulating the laser power by a 15 W heater at one of the mirrors. After 22 hours the saturation temperature of 54°C was achieved at the mirror. The wheel was 4°C colder than the heated mirror. Fig. 2 illustrates the temperature stabilization. The remaining variation of the mirror temperature (3.5°C) changes the resonator length by not more than 1 μm which is much less than one optical wavelength in general. The interferometer system will correct this change.

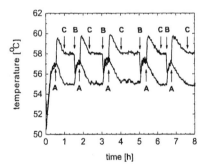

Figure 2. Measured temperature behavior of the mirror (upper curve) and the wheel (lower curve) during a simulation of a working regime of a FEL for the out coupling mirror wheel (see text; arrows A: laser switched on, arrows B: laser switched off, arrows C: heating of the mirror wheel switched on).

References

[1] F. Gabriel, et al., Nucl. Instr. and Meth., B161-163(2000)1143;http://www.fzrossendorf.de/ELBE/en
[2] P. Gippner et al., Contribution to the 23rd Int. Free Electron Laser Conf. Darmstadt, Germany, 2001, Nucl. Instr. and Meth. A483 (2002) II-55

FREE ELECTRON LASERS 2003
E.J. Minehara, R. Hajima and M. Sawamura (Eds.)
Published by Elsevier B.V.

Electron Beam Diagnostics and Undulator Field Adjustment

of the ELBE IR-FEL

P. Evtushenko[* a], P. Gippner[a], P. Michel[a], B. Wustmann[a], W. Seidel[a],
D. Wohlfarth[a], A. Wolf[a], U. Wolf[a], R. Wünsch[a] and C.A.J. van der Geer[b]

[a] Forschungszentrum Rossendorf, Postfach 510119, D-01314 Dresden, Germany
[b] Pulsar Physics, De Bongerd 23, 3762 XA Soest, The Netherlands

Abstract

For the electron beam diagnostics in the undulator of the ELBE FEL two systems are completed. The first one consists of insertable OTR (Optical Transition Radiation) view screens. The same view screens will be used for the FEL optical resonator alignment. This is why there are very strong requirements on the long-term positioning reproducibility of the view screens. The second system is based on a single moveable YAG (Yttrium Aluminum Garnet activated by Cerium) screen, which can travel within the undulator vacuum chamber such that the electron beam position can be measured at any place in the undulator. Both systems are completed and successfully tested. Results of the tests as well as design ideas are presented. After transporting two undulator units to Rossendorf and installing the stainless steel vacuum chamber the magnetic field has been measured again and adjusted applying the pulsed-wire method. Irregularities of the undulator magnetic field could affect gain and wavelength of the FEL. Such effects were analyzed using numerical simulations.

PACS codes: 41.60.Cr, 41.85.Qg,

Keywords: Free electron laser, ELBE, diagnostics, view screen, field adjustment, gain

1. Introduction

It was already reported about design of the electron beam diagnostics for the undulator of the ELBE FEL [1]. This year both diagnostic systems were completed, carefully tested and now are installed at the undulator.

2. Very precise OTR view screens

One diagnostics system is a set of OTR view screens. The view screens are made of Beryllium and have shape of prism with one surface oriented 45° in respect to the beam direction. The view screen has a 1 mm diameter hole in the center that is about the size of the electron beam in the undulator according to the beam transport calculations. The view screens are aligned so that the hole is on the magnetic axis of

the undulator with an accuracy of 10 µm. Once the view screens are aligned they define the axis of the optical resonator. The holes in the view screens are used as apertures during the optical cavity alignment. Long-term reproducibility of the view screen position is required to ensure reproducible beam position measurements and reliable alignment of the optical resonator. The actuator of the view screen, developed at FZR, was tested under vacuum conditions. During the test the actuator was moved in cycle and watched by a stable mounted video camera with zoom lens. The 1 mm hole was giving a scale for the screen displacement. Deviation of the view screen position from its initial position was found to be less than 30 µm after 12500 cycles that is far less than required. Now the system is completed installed at the undulator and is ready for beam measurements.

[*] P.Evtushenko@fz-rossendorf.de

3. Moveable YAG screen

After the first version of the moveable view screen [1] was designed and manufactured considerable amount of time was invested in its tests. First of all it was proved that the mechanical concept of the screen is reliable. Automation software responsible for safe and reproducible motion of the screen was developed during the mechanical tests. For all that motion of the screen in the undulator vacuum chamber was not ideal. This is why the pushcart carrying the screen was redesigned. The new version of the screen was tested again and has shown much better behavior in sense of smoothness of the motion and in sense of particle production what is also very important for a superconducting accelerator. After the first testing phase all pieces of the diagnostic system were cleaned according to our standard procedure. Mounting the very miniature components under clean room conditions was a certain challenge as well but was brilliantly made by our vacuum crew. Since metal parts can adhere to each other when cleaned very well in the last test phase we have tested the screen mechanics again in vacuum under clean conditions. Performances of the mechanics were found to be reliable and reproducible like before.

4. Adjustment of the undulator field

In 2000 the magnetic fields of either section of the U27 undulator were scanned and analyzed at DESY using a hall-probe setup [2]. After transporting the undulator to Rossendorf and installing the stainless steel vacuum chamber the magnetic field has been rechecked by means of the pulsed wire method [3]. We found significant deviations from the field measured at DESY. We suppose a displacement of certain magnets since a low-μ permeability indicator could not detect inhomogeneities in the chamber material.

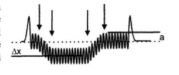

Fig. 1. Electron beam trajectory with the displacement Δx caused by deviations of some magnets from the normal strength.

Deviations of the electron path from the sinusoidal shape affect the lasing process: the gain is reduced and the resonance wavelength is shifted to longer one. To minimize the deviation we have shimmed some of the undulator magnets.

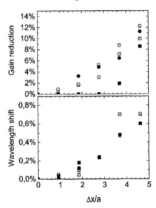

Fig. 2. Gain reduction and wavelength shift as a function of the ratio $\Delta x/a$ (defined in fig.1) calculated for 20 MeV (open symbols) and 40 MeV (full symbols) electron beam. The calculation was performed without (squares) and with taking into account the emittance of the ELBE beam ($\varepsilon^{long} = 50$ ps×keV, $\varepsilon^{trans} = 15$ mm×mrad).

To study the effect of anomalies in the undulator field on the lasing process we have performed numerical calculations using the 3-dimensional particle tracking code GPT [4]. Fig. 1 shows the electron path through one undulator unit for the case that 4 selected undulator magnets (arrows) slightly deviate from the others. As a result the electron path is displaced by Δx from the normal axis.

Fig. 2 shows the gain reduction and the wavelength shift due to this displacement. Independently of the electron energy we can conclude that a significant reduction of the gain or a shift in wavelength is only caused if the electron is displaced by more than twice the amplitude of the regular oscillation in the undulator. Similar results have been obtained for other field anomalies. Anomalies as observed in the field of the U27 undulator can easily be tolerated.

5. References

[1] P. Evtushenko et al.; Proc. of the 23rd Int. FEL Conference, II-57, Darmstadt, Germany,
[2] P. Gippner et al.; Proc. of the 23rd Int. FEL Conference, II-55, Darmstadt,
[3] P. Gippner, A. Schamlott, U. Wolf; http://www.fz-rossendorf.de/ELBE/en/fel/pos.html
[4] M.J. de Loos et al.; EPAC 2002, Paris, France, p. 849; see also http://www.pulsar.nl/

FREE ELECTRON LASERS 2003
E.J. Minehara, R. Hajima and M. Sawamura (Eds.)
Published by Elsevier B.V.

Study of HOM Instability of the JAERI ERL-FEL

M.Sawamura*, R.Hajima, N.Kikuzawa, E.J.Minehara, R.Nagai, N.Nishimori

Free-Electron Laser Laboratory, JAERI, Tokai, Ibaraki 319-1195 Japan

Abstract

The free-electron laser (FEL) linac has been modified to the energy recovery linac (ERL) at the Japan Atomic Energy Research Institute (JAERI) to achieve the higher power FEL of the next stage of 5-10kW. We review transverse instabilities and present experimental data on transverse beam breakup (BBU) obtained at the JAERI ERL-FEL. We compare measurement with simulation.

1. Introduction

JAERI has been developing a high-power FEL with a superconducting linac. After the initial goal of kilowatt FEL lasing was achieved in 2000 [1], the linac has been modified into an ERL. Energy recovery is the process by which the energy invested in accelerating a beam is returned to the rf cavities by decelerating the beam. Energy recovery of an FEL beam driven by a superconducting linac is a possible way of greatly increasing the efficiency of the laser since most of the beam energy remains after lasing occurs. This energy-recovery technology with a superconducting linac is the most promising for the next stage of 10kW FEL lasing owing to increasing the beam current without additional rf power sources.

In a recirculating linac, a feedback system is formed between the beam and the rf cavities, so that instabilities can arise at high currents. These instabilities become important and can potentially limit the average beam current especially for the high-Q superconducting cavities.

In the present paper we will describe transverse HOM instability research by simulation and measurement.

2. Transverse BBU

Transverse beam displacement on successive recirculations can excite HOMs that further deflect the initial beam. The effect is worse in superconducting rf cavities because of higher Q values of HOMs. The threshold current depends on the various parameters of cavity and beam optics such as Q values, frequencies and R/Q of the HOMs, beam energy, beta functions and phase advance in the paths and recirculation path length.

2.1 HOM Instability Simulation

A simulation code, named BBU-R, has been developed to calculate the threshold current at an actual machine configuration. Analytic model for simulation is impulse approximation, where the transverse position of the bunch is treated as one point and the transverse deflection through the cavity as single deflecting force [2].

This simulation code requires the transfer matrices between the adjacent cavities and the HOM parameters such as frequency, R/Q and loaded Q value. The HOM frequencies and R/Q of the JAERI superconducting cavity was calculated with the 2.5-D rf cavity code PISCES II, which can evaluate all the eigenfrequencies and fields for arbitrarily shaped axially symmetric rf cavity [3]. The loaded Q values and frequencies of the HOMs were measured with a network analyzer connected to the HOM coupler, from which reflection power was measured. The transfer matrices were calculated with the code TRANSPORT.

The threshold current is defined as the current where the beam can be transported within 5 times of diameter of the initial. The initial beam has 1mm diameter and 0.5mm offset from the axis. The beam diameter increases with the current. Table 1 lists the threshold currents when one of 10 HOMs is

---
* sawamura@popsvr.tokai.jaeri.go.jp

Table 1: Threshold current

No.	Mode	Threshold Current (A)
#1	TE111 π/5	666.40
#2	TE111 2π/5	487.31
#3	TE111 3π/5	34.36
#4	TE111 4π/5	10.74
#5	TE111 π/5	578.94
#6	TM110 π	32520.32
#7	TM110 4π/5	16.79
#8	TM110 3π/5	5.47
#9	TM110 2π/5	7.03
#10	TM110 π/5	1482.74
#1-#10	All modes	3.42

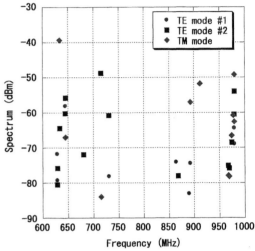

Figure 1: HOM power spectrums

excited and all of 10 HOMs are excited. The threshold current when all of 10 HOMs are excited is 3.42A, which is large enough to increase the beam current for the next stage of our plan.

2.2 HOM Power Spectrums

Each module has five rf couplers such as a main power coupler, a pick-up coupler and three HOM couplers. Two HOM couplers are designed to damp transverse modes and the other to damp longitudinal modes. All HOM couplers are terminated to the dummy loads out of the cryomodules. This makes it possible to measure the excited HOM powers inside the cavity through the HOM couplers. The terminator was exchanged for a real-time spectrum analyzer to measure frequencies and powers of the HOMs. Figure 1 shows the amplitude of the HOMs.

There seems to be four groups of the HOMs or more. The group near 630MHz is considered to be TE111 mode and that near 700MHz to be TM110 mode. The groups over 850MHz have not been identified yet.

The mode of the highest power of TE111 is considered to be 3π/5 mode of 634MHz and that of TM110 to be 4π/5 mode of 715MHz. The calculated threshold current is 34.36A for TE111-3π/5 mode and 16.79A for TM110-4π/5 mode. While these modes of small threshold current show high HOM powers, the modes of the smaller threshold current for such as TM110-3π/5 and TM110-2π/5 modes show fairly low HOM powers.

Measured HOM coupling factors do not vary so much with the difference of the modes. The small spectrums for the low threshold current might be caused by the short macropulse or the low beam current. The further analysis is required to understand the HOM instabilities.

3. Conclusion

The threshold current of the JAERI ERL-FEL limited by the HOM instability is calculated to 3.42A, which is large enough to increase the beam current from 5mA to 40mA of our next stage. The HOM spectrums were measured from three HOM couplers. The mode of the smallest threshold current by calculation is not detected from the HOM couplers. The more precise measurement is required to well-understand the HOM instabilities.

4. References

[1] N.Nishimori et al., Nucl. Instr. Meth. A 475 (2001) 266-269
[2] J.Bisognano, et al. CEBAF-PR-87-007 (1987)
[3] Y.Iwashita, Computational Accelerator Physics, Williamsburg, VA, AIP conference proceedings No.361 Sept. 1996, 119-124

FREE ELECTRON LASERS 2003
E.J. Minehara, R. Hajima and M. Sawamura (Eds.)
Published by Elsevier B.V.

Low Emittance X-band Thermionic Cathode RF Gun for Medical Application

A. Fukasawa[a,*], K. Dobashi[b], T. Imai[a], F. Sakamoto[a], F. Ebina[a], H. Iijima[a], M. Uesaka[a],

J. Urakawa[c], T. Higo[c], M. Akemoto[c], H. Hayano[c], K.. Matsuo[d], H. Sakae[d], M. Yamamoto[e]

[a]*Nuclear Engineering Research Laboratory, Graduated School of Engineering, University of Tokyo, 2-22 Shirakata, Shirane, Tokai, Naka, Ibaraki, 319-1188 Japan*
[b]*National Institute of Radiological Science, 4-9-1, Anagawa, Inage, Chiba, Chiba, 263-8555 Japan*
[c]*High Energy Accelerator Research Organization, 1-1 Oho, Tsukuba, Ibaraki, 305-0801*
[d]*Ishikawajima-Harima Heavy Industries Co., Ltd., 1 Nakahara, Isogo, Yokohama, Kanagawa, 235-8501 Japan*
[e]*Akita National College of Technology, 1-1 Bunkyo, Iijima, Akita, Akita, 011-8511 Japan*

Abstract

Beam loadings in the X-band linac for a compact inverse Compton scattering hard X-ray source is estimated. In the RF gun, 0.14MV beam loading voltage was induced, and in a travelling wave accelerating structure the bunches at the head of the beam gain 2.9MV higher energy than followings. Higher energy part should be compensated since it can be a strong noise source.

PACS codes: 29.27.Bd

Keywords: X-band linac; Thermionic cathode RF gun; Beam loading; Compton scattering

1. Introduction

We are developing a compact hard X-ray source for medical applications, especially the arteriography, and life sciences. It consists of an X-band linac and a YAG laser, and by colliding the relativistic electron beam with the laser Compton scattering hard X-ray is produced. The X-band linac has a thermionic cathode RF gun and a travelling wave accelerating structure. A 50 MW (1μs) klystron drives this linac: 10 MW to the gun and 40MW to the other. Because of the thermionic cathode RF gun, this accelerator produces multibunch beams which cause some undesirable effects on beam transportation. Here we calculate the beam loadings in the gun and the travelling wave accelerating structure.

2. Beam Loading in the RF Gun

2.1 3.5-cell thermionic cathode RF gun

The RF gun has 3.5 cells and coaxial coupler though which RF can be fed with the fine axial symmetry [1][2]. Some parameters of the cavity is calculated by using SUPERFISH (Table 1).

Table 1 Properties of the RF gun cavity

Resonant Frequency	11.424 GHz
Transit Time Factor	0.703
Shunt Impedance	2.46 MΩ
Q Value	9350
Wake Loss Paramete	4.72 V/pC

*   Tel. +81-29-287-8986; Fax. +81-29-287-8488;
    E-mail fukasawa@utnl.jp

2.2 Beam loading in the gun

To analyse the beam dynamics PARMELA code is used. Feeding 6MW of RF into the cavity, the micro-bunch charge at the exit of the gun with the current at the cathode 0.71 A is shown in Figure 1.

Beam Loadings are calculated according to Ref. [3] (Figure 2). In the cavity beam loading cause 0.14 MV decrease of acceleration.

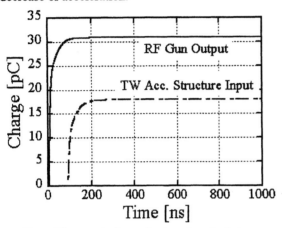

Fig. 1 Time variations of micro-bunch charges

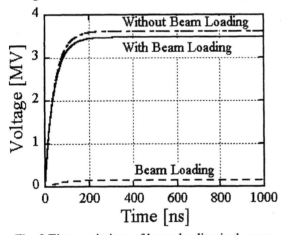

Fig. 2 Time variations of beam loading in the gun

3. Beam Loading in the Travelling Wave Accelerating Structure

Before entering the travelling wave accelerating structure, beams must go through an alpha magnet, which selects the electrons by the energy slit. Assuming the acceptance 3.0 − 3.5 MeV, the micro-bunch charges, which pass through the slit, become

as shown in Figure 1.

Figure 3 shows the beam loadings in the travelling wave accelerating structure. The bunches at the head of beam can obtain higher energy than the followings. Higher energy part of the beam will be lost during the transportation to the colliding point and emit strong noise radiations. Beam energy should be controlled to be the same energy by changing the input RF form.

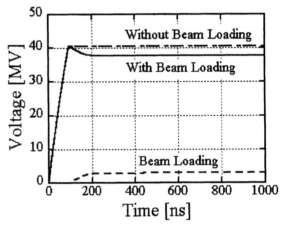

Fig. 3 Time variations of beam loadings in the TW accelerating structure.

4. Summary

Beam loadings in the gun and the travelling wave accelerating structure are estimated. In the gun it decreases the accelerating voltage. In the travelling wave accelerating structure, the bunches at the head of the beam are accelerated higher than following bunches. Since they can be strong noise source, the beam loadings should be compensated by changing the input RF form.

5. References

[1] K. Matsuo, et al., In the Proceedings of the 28th Linear Accelerator Meeting in Japan, 197 (2003)

[2] K. B. Kiewiet, et al., In the Proceedings of EPAC 2000, Vienna, Austria, 1660 (2000)

[3] P. B. Wilson, SLAC Technical Report, SLAC-PUB-2884 (1982)

[4] H. Sakae, In the Proceedings of the 27th Linear Accelerator Meeting in Japan, 175 (2002)

FREE ELECTRON LASERS 2003
E.J. Minehara, R. Hajima and M. Sawamura (Eds.)
Published by Elsevier B.V.

Measurement of the longitudinal phase space at the Photo Injector Test Facility at DESY Zeuthen

J. Bähr[a], I. Bohnet[a], K. Flöttmann[b], J. H. Han[a], M. Krasilnikov[a], D. Lipka[a]*, V. Miltchev[a], A. Oppelt[a], F. Stephan[a]

[a]DESY Zeuthen, Zeuthen, Germany

[b]DESY Hamburg, Hamburg, Germany

A setup for the measurement of the longitudinal phase space at the photo injector test facility at DESY Zeuthen is described. The measurements of the momentum distribution, the length of the electron bunch and of their correlated energy spread are discussed.

1. Introduction

The photo injector test facility at DESY Zeuthen (PITZ) has been developed with the aim to deliver low emittance electron beams and study their characteristics for future applications at free electron lasers and linear accelerators. The energy of the electron beam varies in the range between 4 and 5 MeV. A detailed description of PITZ can be found in [1].

Successful optimisation and improvement of the PITZ performance requires good beam diagnostics for investigation of the electron bunch properties. This contribution is focused on longitudinal emittance measurements.

2. Momentum distribution

A spectrometer dipole and a YAG screen are used for the measurement of the momentum distribution of the electron bunch. The electrons gain different energies depending on the time when the laser hits the cathode. The latter can be expressed in terms of a phase between laser pulse and accelerating field (SP phase). The maximum mean momentum of an electron bunch is (4.72 ± 0.03) MeV/c, the corresponding SP phase is choosen to be $0°$. According to simulation this corresponds to a gun gradient of (41.8 ± 0.3) MV/m and a RF power of (3.15 ± 0.05) MW.

The smallest momentum spread of

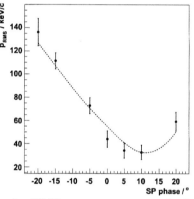

Figure 1. RMS momentum spread of the electron bunch as a function of SP phase for a gradient of 41.8 MV/m compared to simulation results (dashed line) [2].

(33 ± 7) keV/c is observed at a SP phase of $10°$ and an electron bunch charge of 1 nC (see Fig. 1).

3. Bunch length

The electron bunch length is measured using a Cherenkov radiator, an optical transmission line [3] and a streak camera (bunch length measurement system). In order to obtain an adequate photon yield [4] Silica aerogel (SiO_2) with n = 1.03 and thickness of 2 mm is used. The aerogel is placed in vacuum (for details see [5]).

To gain enough intensity multiple photon bunches are collected for the longitudinal distribution measurement. This requires a small jit-

*dlipka@ifh.de

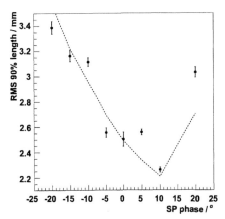

Figure 2. RMS 90 % length as a function of SP phase for electron bunches with a charge of 1 nC compared with a simulation (dashed line).

ter of the whole system (laser, accelerating field, streak camera) compared to the time resolution of the bunch length measurement system. The jitter was investigated by measuring the bunch length as a function of the number of laser pulses in the train. All bunches of the train are overlayed at the streak camera. A significant jitter would lead to an increase of the bunch length with increasing number of laser pulses. The bunch length of $\sigma = 27.2$ ps was measured and appered to be stable within the statistical uncertainties. In addition the standard deviation of the mean of 100 single bunches is calculated to be 1 ps which corresponds to the time jitter.

Dispersion between radiator and streak camera elongates the photon bunch. To suppress the dispersion a bandpass filter with 12 nm bandwidth is used. 10 % of the tails in the longitudinal distribution, originating from the noise were cutted off. The remaining 90 % of the distribution were used for extraction of the bunch length as RMS 90 %. The value of the bunch length as a function of SP phase is shown in Fig. 2 compared to the results (RMS 90 %) from a simulation [2] where corresponding input parameters of the gun are applied. By increasing the phase the accelerating field becomes weaker on the cathode. Therefore the longitudinal space charge becomes more dominant and the bunch is longer at a phase of 20°. The shortest bunch has a length of 2.25 mm, corresponding to a FWHM of 6.3 mm or 21 ps at a phase of 10°.

4. Longitudinal phase space

It is planned to measure the full longitudinal phase space of the electron bunch by using a dipole, a Cherenkov radiator and a streak camera. The produced photon bunch provides the information about the time properties and the momentum spread of the electron bunch. The photon bunch is transported to the streak camera, where its momentum and longitudinal distribution can be measured simultaneously, their correlation will be obtained (for details see [6]).

5. Outlook

The commissioning of the setup to measure the full longitudinal phase space is scheduled for end of 2003. First results are expected in the beginning of 2004.

In 2004 the photo injector will be extended by a booster cavity. The energy of the electrons after the booster will be about 30 MeV. At this energy optical transition radiation can be used in order to transform the electron bunch into a photon bunch instead of Cherenkov radiation mechanism, which provides an even better time resolution than the Cherenkov effect.

REFERENCES

1. F. Stephan et al., Photo Injector Test Facility under Construction at DESY, FEL 2000, Durham, book edition of the proceedings, ISBN 0-444-50939-9.
2. K. Flöttmann, A Space Charge Tracking Algorithm (ASTRA), http://www.desy.de/~mpyflo/, 2003.
3. J. Bähr et al., Optical transmission line for streak camera measurement at PITZ, DIPAC 2003, Mainz.
4. Q. Zhao et al., Design of the Bunch Length Measurement for the Photo Injector Test Facility at DESY Zeuthen, PAC 2001, Chicago.
5. D. Lipka et al., Measurement of the longitudinal phase space at PITZ, EPAC 2002, Paris.
6. D. Lipka et al., Measurement of the longitudinal phase space at PITZ, DIPAC 2003, Mainz.

FREE ELECTRON LASERS 2003
E.J. Minehara, R. Hajima and M. Sawamura (Eds.)
Published by Elsevier B.V.

A Plan of Infrared FEL using PAL Test Linac

H. S. Kang*, D. E. Kim, Y. J. Han, S. H. Nam, W. Namkung, S. G. Baik

Pohang Accelerator Laboratory, San31, Hyoja-dong, Pohang 790-784, Korea

Abstract

An infrared FEL using the PAL Test Linac is proposed as one of new PLS beamline plans for users. For an infrared FEL some parts of Test Linac should be improved such as klystron modulator system and RF-gun. To get a higher intensity of electron beam a magnetic bunch compression system using chicane magnet will be incorporated; which can yield a sub-pico-second beam. Two FEL beam lines with different energy will be provided: one is a low energy FEL (25-35 MeV) and the other one is a high energy FEL (50–70 MeV). The IR FEL beamline will be officially included into the PLS beamline lists. The design FEL wavelength ranges from 1.0 to 20 microns and the repetition rate is up to 30 Hz. The laser macropulse length is more than 5.0 micro-seconds.

PACS codes: 41.60.Cr

Keywords: Linac, RF-gun, Bunch Compression, Infrared FEL

1. Introduction

The PAL Test Linac which consists of a half-cell thermionic RF-gun, an alpha magnet, and two S-band accelerating structures has been in operation since 1999. The Test Linac can provide a bunched electron beam with the beam energy of up to 85MeV, the pulse length of 4μs, the pulse current of 100mA, and the repetition rate of 30Hz [1].

An infrared FEL using the PAL Test Linac is proposed as one of new PLS beamline plans for users. The IR FEL beamline will be officially included into the PLS beamline lists. The design FEL wavelength ranges from 1.0 to 20 microns and the repetition rate is up to 30 Hz. The laser macropulse length is more than 5.0 micro-seconds.

For an infrared FEL some parts of Test Linac should be improved such as klystron modulator system and RF-gun. The RF-gun is a half-cell cavity with a tungsten dispenser cathode of 3-mm diameter on axis. A new RF gun with 1-1/2 cell is being considered as a new gun. The measured micro-bunch length at the end of linac with a streak camera is 4-ps

[2]. A chicane magnet will be installed to get a sub-pico-second beam.

Fig. 1. A photo of Test Linac.

2. Infrared FEL Plan

2.1 FEL layout

Two FEL beam lines with different energy will be provided: one is a low energy FEL (25-35 MeV) and the other one is a high energy FEL (50–70 MeV). Fig. 2 shows the tunnel drawing of Test Linac and the

* Heung-Sik Kang
e-mail address: hskang@postech.ac.kr

FEL layout using Test Linac. The low energy FEL uses an electron beam from AC#1 and the high energy FEL from AC#2. The required electron beam parameters are: the normalized emittance is 10 mm-mrad, the electron macropulse length 7.0 microseconds, the electron micropulse length is smaller than 1 picosecond, and the RF pulse length is more than 8.0 microseconds.

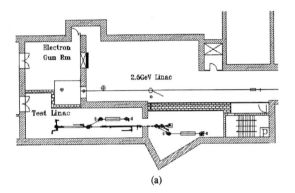

(a)

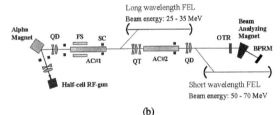

(b)

Fig. 2. (a) The tunnel drawing of Test Linac and (b) the FEL layout using Test Linac. QD: quadrupole doublet, BCT: beam current transformer, FS: focusing solenoid, AC: accelerating column, SC: steering coil, QT: quadrupole doublet, OTR: Optical transition radiation, BPRM: Beam profile monitor.

2.2 FEL Simulation

A simple 1-D simulation was done using the FELBENSON code. The input parameters used in the simulation are as follows: the optical wavelength is 3 μm, the peak micropulse current 100A, the normalized emittance of electron beam 10mm-mrad, the wiggler period length 3cm, the number of wiggler period 50, and the cavity length 4.5meter. Fig. 3 shows the dependence of optical power on cavity detuning (CD). The round-trip frequency of photon beam in a cavity is 33.3 MHz so that the number of pass of 100 in Fig. 3 means 3.0 micro-seconds in time.

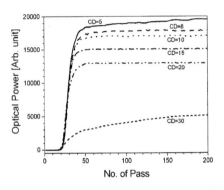

Fig. 3. Dependence of optical power on cavity detuning.

2.3 Bunch compression

To get a higher intensity of electron beam a magnetic bunch compression system using chicane magnet will be incorporated. The location of a chicane magnet is between AC#1 and AC#2 in Fig. 2(b). The PARMELA simulation shows that the bunching system can yield a sub-pico-second beam as shown in Fig. 4.

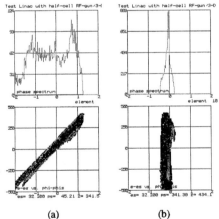

(a)   (b)

Fig. 4. PARMELA simulation results of magnetic bunching system: (a) before chicane and (b) after chicane. The horizontal scales are all ±2 pico-seconds.

3. References

[1] H. S. Kang et al., Proc. of the Linac 2002, Gyeongju, Korea, 2002, p. 58.
[2] H. S. Kang et al., Proc. of the PAC'99, New York, 1999, p. 2501.

FREE ELECTRON LASERS 2003
E.J. Minehara, R. Hajima and M. Sawamura (Eds.)
Published by Elsevier B.V.

A 200keV Gun with NEA-GaAs Photocathode to Produce Low Emittance Electron Beam

M.Yamamoto[a1], K.Wada[a], N.Yamamoto[a], T.Nakanishi[a], S.Okumi[a], F.Furuta[a],
T.Nishitani[a], M.Miyamoto[a], M.Kuwahara[a], K.Naniwa[a],
Y.Takashima[b], H.Kobayakawa[b], M.Kuriki[c], H.Mastumoto[c], M.Yoshioka[c]

[a]*Department of Physics, Nagoya University, Nagoya 464-8602, Japan*
[b]*Faculty of Engineering, Nagoya University, Nagoya 464-8603, Japan*
[c]*High Energy Accelerator Research Organization(KEK), Tsukuba 305-0801, Japan*

Abstract

A high DC gradient gun with a negative electron affinity (NEA) surface GaAs photocathode is expected to produce low emittance (<0.5 πmm•mrad) electron beam which is required by an Energy Recovery Linac. We have developed a 200kV polarized electron gun for a future linear collider and the gun which has the photocathode surface field gradient of 3MV/m could be operated with negligible field emission dark current. We have started the development of the low emittance electron gun by using this system.

PACS codes: 29.25.Bx; 29.27.Ac

Keywords: polarized electron, NEA photocathode, low emittance

1. Introduction

We have been developing a 200keV polarized electron source for feature e^+-e^- linear colliders. This gun aims to produce 0.7ns multi-bunch beam with more than 3A peak current, 90% electron spin polarization and emittance less than 10πmm•mrad, which satisfy the linear collider requirement.[1]

A photoemission from GaAs cathode with negative electron affinity (NEA) enables also to produce a low initial emittance (< 0.1 πmm•mrad) beam.[3] This advantage will make an important role for electron source of a future X-ray source based on energy recovery linac, because it requires very low emittance (< 0.5 πmm•mrad), and high average current (10~100mA) beam for the electron gun.[2]

For the suppression of the emittance growth due to the space charge effect, the high acceleration field gradient is indispensable. We start the study to produce a low emittance beam by using a 200keV gun with an NEA-GaAs photocathode.

2. 200keV Gun[4]

2.1 Gun design

In the case of the NEA GaAs photocathode gun, the maximum field gradient of the gun is limited by the field emission from the cathode electrode, because the field emission dark current more than 1nA is not negligible for the NEA surface lifetime. We employed a load-lock system to prevent the Cs

---

[1] Email: yamamoto@spin.phys.nagoya-u.ac.jp

accumulation on electrodes.

The design of electrodes was done by using a simulation code of POISSON. The maximum field gradient on the cathode electrode is 7.8MV/m, on the center of the photocathode surface is 3.0MV/m.

These electrodes were manufactured from a vacuum re-melted stainless steel, and polished by electro-chemical buffing[5]. As a result, the 200keV gun could be operated with the dark current less than 1nA.

The extremely ultra high vacuum condition is also indispensable for a long NEA surface lifetime. In our system, the vacuum base pressure of the gun reached to $3 \cdot 10^{-9}$ Pa, but NEA surface lifetime was limited <100 hours. We are now trying to improve the vacuum by increasing the non-evaporable getter pumping speed from 850l/s up to 4300l/s.

2.2 Photocathode preparation System

The photocathode for the ERL-gun must offer a high quantum efficiency. In the 200keV gun, the atomic hydrogen cleaning method was employed for the photocathode preparation[6]. After the cleaning, the photocathode was activated by exposure cesium and oxygen, and quantum efficiency of ~13% was obtained for the bulk GaAs at the wave length of 633nm.

2.3 Surface Charge limit (SCL) phenomenon

In order to produce a laser pulse train with high repetition rate (81.25MHz), we prepare a mode locked Ti:sapphire laser pumped by a CW Nd:YVO$_4$ laser.

The highest peak current density extracted from the NEA photocathode is limited by the trapped surface-charge at the band bending region, so called SCL phenomenon. The superlattice photocathode with heavy p-dopant has been overcome this phenomenon at a situation of the current density ~1A/cm^2.[7] For the ERL-gun, the current density of 10~100A/cm^2 in ~20ps pulse is required and the SCL phenomenon should become significant. For the investigation of SLC phenomenon with such high current density, the GaAs photo-cathode with heavy p$^+$ doped (>5×10^{19}/cm^3) surface is scheduled.

3. Emittance simulation

The normalized emittance calculation using a simulation code of general particle tracer (GPT) with POISSON-generated electric field of the 200keV gun. From this result, it seems possible to produce a fairly low emittance (0.5π.mm.mrad) beam at the gun exit for bunch charge of 10pC, the illumination spot size of φ1mm, 20ps pulse width. Although the emittance is strongly enlarged by the space charge effect above bunch charge of 20pC, especially a situation of the small spot size (φ<1mm).

The measurement of the low emittance (~0.1 π.mm.mrad) beam by using a pepper pot method has been prepared in progress. The effort to increase the field gradient of the gun from 3MV/m to ~10MV/m has been also continued.

4. References

[1] T.Nakanishi et al., "Polarized Electron Source" in JLC Design Study, KEK Report 97-01, 36-48 (1997)
[2] S.M.Gruner and M.Tigner, eds., "Study for a proposed Phase I Energy Recovery Linac Synchrotron Light Source at Cornell University, CHESS Technical Memo 02-003, JLAB-ACT-01-04(2001)
[3] S. Pastuszka et al., J. Appl. Phys. 88, 6788 (2000).
[4] K. Wada et al., Proceedings of the PESP2002, Danvers,MA,USA
[5] F. Furuta et al. Proceedings of the LINAC2002, Gyeongju, Korea
[6] C. K. Sinclair, Proceedings of the Particle Accelerator Conference, Vancouver, Canada, 1997
[7] K. Togawa et al., Nucl. Instr. and Meth. A 414(1998) 431

FREE ELECTRON LASERS 2003
E.J. Minehara, R. Hajima and M. Sawamura (Eds.)
Published by Elsevier B.V.

Future plans at the Photo Injector Test Facility at DESY Zeuthen

K.Flöttmann
DESY, 22603 Hamburg, Germany

J.Bähr, I.Bohnet, J.H.Han, M.Krasilnikov, D.Lipka,

V.Miltchev, A.Oppelt*, F.Stephan,
DESY, 15378 Zeuthen, Germany

R.Cee
TU Darmstadt - TEMF, 64289 Darmstadt, Germany

Ph.Piot
FNAL, Batavia IL 60510, USA

V.Paramonov
INR, Moscow, 117312, Russia

Abstract

The Photo Injector Test facility PITZ will be upgraded in 2004. Its main research goals are studies on the production and the conservation of low transverse emittance electron beams. The installation of an additional accelerating cavity is planned, which increases the beam energy from ~ 5 MeV after the gun to about 30 MeV after the booster. The goal is to produce and conserve a normalized transverse emittance of about 1 π mm mrad at 1 nC charge. For the characterization of the low emittance beam at higher beam energy, new diagnostics tools need to be installed. In this article, an overview about the layout of the upgraded facility and the planned physics activities is given.

Keywords: photo injector, booster cavity, emittance conservation principle

1. Introduction

The Photo Injector Test facility at DESY Zeuthen (PITZ) was built in order to test and optimize sources of high brightness electron beams for future free electron lasers and linear colliders. The focus is on the production of intense electron beams with small transverse and longitudinal emittance using the most advanced techniques in combination with key parameters of projects based on TESLA technology like TTF2, TESLA-XFEL and BESSY III.

The first photo electrons were produced at PITZ in January 2002. In the meanwhile, the existing electron source has been fully characterized and beam measurements have been presented at different conferences, see e.g. [1].

The second stage of PITZ (PITZ2) will start in 2004. It is a large extension of the facility and its research program. The concept of PITZ2 is to basically resemble TTF2 up to that critical beam energy where emittance becomes a constant of motion for the rest of acceleration. Thus, the PITZ studies on improvement of electron beam quality can readily be transferred to TTF2 and other facilities. In addition, PITZ will be able to study injector schemes beyond TTF2 demands, e.g. for TESLA-XFEL and BESSY III.

2. Emittance conservation and booster cavity

One of the main objectives of PITZ2 is the proof of the emittance compensation technique and its experimental optimization, since many future FEL proposals rely on this technique [2].

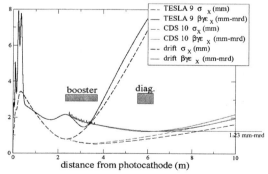

Fig. 1. Emittance conservation at PITZ2 with two different normal conducting booster cavities: TESLA-booster (red) and CDS-booster (green). The drift case (without booster) is shown in blue.

* Corresponding author, email: Anne.Oppelt@desy.de

According to simulations, the minimum normalized transverse beam emittance can be conserved by locating a booster cavity at the position of the beam envelope waist, together with a proper choice of rf-gun and solenoid parameters (Fig. 1).

Since the booster plays such a significant role in the emittance compensation technique, it will be the key technical element of PITZ2. The experimental study will be done in two stages: first, a preliminary TESLA booster will be used. Later, it will be replaced by the CDS booster, being actually under development for PITZ, which will reach the final beam energy of up to 30 MeV.

3. Diagnostics Beamline

In order to provide a most complete characterization of the electron source at higher beam energies, a new diagnostics beamline will be needed. A number of new diagnostics elements have to be developed and installed, including devices that allow efficient and precise measurements of transverse and longitudinal phase space parameters as well as slice parameters for the full range of beam energies, see Fig.2.

4. Further plans

In order to obtain optimum electron beam parameters, a stable and reliable photo cathode laser system with flat-top temporal and transverse laser pulse shape has to be developed and installed. Further optimization of all subsystems requires extensive beam dynamics studies. It is foreseen to design and test a high duty cycle rf-gun, which is important for the operation of high repetition rate FELs and energy recovery linacs. The delivery of improved guns for the VUV FEL at TTF is also envisaged. In addition, studies on the improvement of photocathodes are planned.

PITZ2 will be realized by DESY in collaboration with BESSY Berlin, MBI Berlin and TU Darmstadt. Contributions from INFN Milan (cathodes), INR Troitsk (CDS booster), INRNE Sofia, and LAL Orsay (both diagnostics) are also included. First attempts to extend the collaboration have been started.

5. References

[1] M.Krasilnikov et al., "Characterization of the Electron Source at the Photo Injector Test Facility at DESY Zeuthen", paper to be published at FEL 2003.

[2] M.Ferrario et al. "HOMODYN study for the LCLS rf photo-injector", NLNF-00/004(P) and SLAC-PUB 9400, Contribution to the 2nd ICFA Advanced Accelerator Workshop, Los Angeles, November 1999. M.Ferrario et al., "Conceptual Design of the TESLA XFEL Photoinjector", TESLA-FEL 2001-3. L.Serafini "New perspectives and programs in Italy for advanced applications of high brightness beams", ICFA AABD Workshop, Chia Laguna, Sardenia, July 2002.

Fig. 2. Preliminary layout of PITZ2.

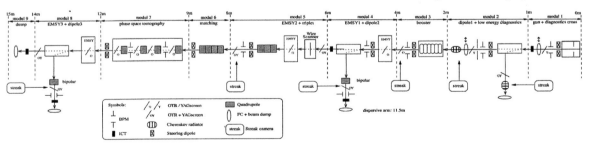

FREE ELECTRON LASERS 2003
E.J. Minehara, R. Hajima and M. Sawamura (Eds.)
Published by Elsevier B.V.

Upgrade of the L-band linac at ISIR, Osaka University for highly stable operation

R. Kato *, G. Isoyama, S. Kashiwagi, T. Yamamoto, S. Suemine, Y. Yoshida,
Y. Honda, T. Kozawa, S. Seki, S. Tagawa

Institute of Scientific and Industrial Research, Osaka University, 8-1 Mihogaoka, Ibaraki, Osaka 567-0047, Japan

Abstract

The L-band linac at the Institute of Scientific and Industrial Research (ISIR), Osaka University, has been renewed in order to realize highly stable operation and good reproducibility. The klystron system was replaced with a new one consisting of a 30 MW klystron and a new pulse modulator, which can provide two different pulse durations, flat top length of 4 or 8 μs. The overall stability of the peak output voltage from the modulator is designed to be less than 0.3 % in peak-to-peak. The RF pulse length for sub-harmonic bunchers is increased to 100 μs for making a flat top in the peak RF voltage. A cooling water system, temperature of which is controlled within ± 0.03 °C was introduced for the accelerating structures, and almost all the magnet power supplies were replaced with new ones. A control system based on personal computers and programmable logic controllers is newly installed.

PACS codes: 29.17.+w; 41.75.Ht; 41.60.Cr; 84.40.Fe

Keywords: linear accelerators; electron beams; Free-electron lasers; klystrons

1. Introduction

The L-band linac at the Institute of Scientific and Industrial Research (ISIR), Osaka University, was constructed in 1978 and has been used mainly for radiation chemistry with pulse radiolysis. In addition to the traditional use, the linac is also used for new research subjects and fields, such as production of the sub-picosecond electron beam, development of a far-infrared free-electron laser (FEL), and basic study of self-amplified spontaneous emission (SASE). We are conducting research and development of FEL and SASE using the linac. The first lasing of the FEL was obtained at wavelengths from 32 to 40 μm in 1994 [1], and the wavelength region has been extended up to 150 μm [2]. The linac is not optimized for development of FEL and accordingly the macropulse length of the electron beam is limited to 2 μs because the filling time of the accelerating structure is 1.8 μs

and the duration of the RF pulse is 4 μs at the maximum. As a result, the FEL could not reach power saturation owing to a small number of amplification times.

We had an opportunity to remodel the linac, so that the linac has been upgraded with the aim of high stability and reproducibility as well as easy operation. Almost all the peripheral components are replaced

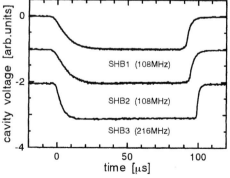

Fig. 1. Temporal profiles of the peak RF voltage in the sub-harmonic buncher cavities.

---

* Corresponding author. Tel.:+81-6-6879-8486; fax: +81-6-6879-8489.
E-mail address: kato@sanken.osaka-u.ac.jp (R. Kato)

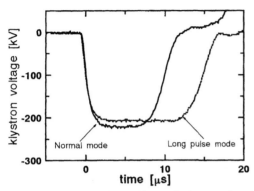

Fig. 2. Voltage pulses for the klystron in the normal mode and the long pulse mode. The pulse modulator is not tuned well to make a good flat top.

with new ones, including the klystron system, magnet power supplies, and the cooling water system, and a computer control system is newly introduced. Taking this opportunity, we modify the linac so that the macropulse can be elongated to a longer duration required for power saturation of the FEL. The modification of the linac has been completed and now commissioning is going on. In this paper, we will report an outline of the upgrade of the L-band linac.

2. New linac system

2.1 Injector System

The electron beam produced with a thermionic gun is injected into the sub-harmonic buncher (SHB) system composed of two 108 MHz and one 216 MHz SHB cavities to produce a high-intensity single-bunch beam. These cavities are separately driven by RF power sources. The single bunch beam is sensitive to changes of the rf power and the phase of the sources and receives harmful influences from them. Since quality factors of the two 108 MHz cavities are about 4400, the peak accelerating voltage in the cavities could not reach flat tops with the previous RF sources with the pulse length of 20 μs. We, therefore, replaced the sources with new ones, which can provide longer RF pulses up to 100 μs.

2.2 RF System

The accelerating system consists of a pre-buncher, a buncher, and an accelerating tube with the accelerating frequency of 1.3 GHz. When the linac was constructed, those three structures were driven by one 20 MW klystron (Thomson CSF, TV-2022A). Three years later, in order to increase bunching force, a 5 MW klystron (Toshiba, E3775A) was installed for providing RF power to both the pre-buncher and the buncher separately. Now we come back to the first configuration that one klystron supplies the RF power to the three structures, but the new klystron (Thales, TV-2022E) has higher output power. This klystron together with a new pulse modulator has two modes for the pulse length. One is called the normal mode, which has 30 MW peak power with a 4 μs long flat top. The other is the long pulse mode for FEL experiments, which has 25 MW peak power with 8 μs duration. The maximum reputation rate is 60 Hz in the normal mode, while it is 30 Hz in the long pulse mode. The pulse-to-pulse fluctuation in the voltage of the modulator is designed to be less than 0.1 % (the peak-to-peak value) and the flatness of the flat top to be below 0.2 % peak-peak.

2.3 Cooling System

The cooling system is an important component for stabilizing operating conditions of the accelerating structures and hence we replace the cooling system with new one. The new system can supply cooling water, temperature of which is stabilized within ± 0.03 °C for the accelerating structures and within ± 0.1 °C for the klystron system.

2.4 Control System

The new control system is a distributed control system, which consists of personal computers (PC) and programmable logic controllers (PLC) connected with a local area network. PLC is a sequence controller widely used in the field of factory automation (FA). The network among the PLCs and PCs we adopted is one called FL-net, which has a standardized protocol for communication with PLCs manufactured by different corporations. The PLCs communicate with common memories on FL-net each other. A data throughput on FL-net is guaranteed to be within 50 ms per 32 nodes.

3. References

[1] S. Okuda, et al., Nucl. Instr. and Meth. A 358 (1995) 244.
[2] R. Kato, et al., Nucl. Instr. and Meth. A445 (2000) 169.

FREE ELECTRON LASERS 2003
E.J. Minehara, R. Hajima and M. Sawamura (Eds.)
Published by Elsevier B.V.

Measurement of the longitudinal wake field in the L-band linac at ISIR

T. Igo[*], R. Kato, S. Kashiwagi, A. Mihara, C. Okamoto, K. Kobayashi,
G. Isoyama

Institute of Scientific and Industrial Research, Osaka University, 8-1 Mihogaoka, Ibaraki, Osaka 567-0047, Japan

Abstract

The wake field is being studied, which is produced in the accelerating structure of the L-band linac by the high-intensity single-bunch beam. The energy spectra of the accelerated beam are measured over the wide range of the phase of the rf field. It is found that the energy spectra change drastically with the phase of the rf field due to the strong wake field produced by the single bunch electron beam with charge per bunch, 32 nC.

PACS codes: 29.17.+w; 41.75.Fr

Keywords: Liner accelerator; Electron beam; Energy spectrum; Wake field

1. Introduction

We are conducting experiments of self-amplified spontaneous emission (SASE) using the L-band electron linac at the Institute of Scientific and Industrial Research (ISIR), Osaka University [1]. A high-intensity single-bunch electron beam, with charge up to 91 nC/bunch, is accelerated with the L-band linac, and used to drive SASE-FEL. When the bunched beam is accelerated through the accelerating structure, it generates the "wake field" after the beam-cavity interaction. The wake field distorts the effective accelerating field and consequently the energy spread of the electron beam is increased. Theoretical and experimental studies have been conducted on effects of the wake field on the energy spectrum of the accelerated electron beam at SLAC for the S-band linac. The high-intensity and high-quality electron beam is required and essential for producing SASE, but these two factors are contradicting requirements. One of the keys for solving the problem is the wake field. We are, therefore, studying the wake field produced in the accelerating structure of the L-band linac, in order to establish bases for SASE in short wavelength region.

In order to experimentally obtain information on the wake field, we measured the energy spectrum of the electron beam as a function of the phase of the rf field for acceleration. In this report, we will report preliminary results of these studies on the wake field in the accelerating structure of the L-band linac.

2. Experiment

The main parameters of the L-band linac are listed in Table 1. The field gradient of the accelerating rf field is relatively low, while charge in a bunch is high compared with that obtained with a typical high-energy linac. It is, therefore, expected that the effective accelerating field is strongly distorted due to the wake field.

The energy spectra of the single bunch beam accelerated to 27 MeV with the L-band linac are measured at various rf phases ranging from −19.5 to +19.5° using a momentum analyser consisting of a bending magnet and a Faraday cup. Figure 1 shows energy spectra measured at various phase angles of

Table 1. Main parameters of the L-band linac and the accelerating tube.

RF frequency	1.3 GHz
Maximum beam energy	38 MeV
Charge in a bunch	< 91 nC/bunch
Accelerating tube	
Type	Traveling wave
	Quasi constant gradient
Shunt impedance	40 MΩ/m
Field gradient	11.6 MV/m

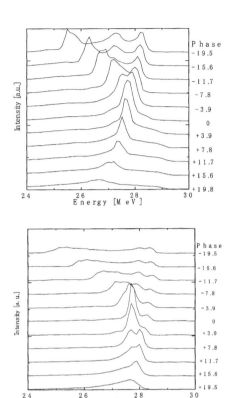

Figure 1. Measured energy spectra: upper is for charge 32 nC, bottom is for charge 2 nC.

the r f field in t he r egion for single bunch electron beams with charges per bunch of Q = 2 and 32 nC. The single bunch beam with charge per bunch of 32 nC is a typical electron beam used in the SASE experiments, while the energy spectra of the beam with Q = 2 nC, for which the effects of the wake field are less significant, are measured for comparison. The energy spectra show complicated features when the phase of the rf field is varied. The energy spread increases, as the phase angle increases from 0 to 19.5°. The peak shifts down to about 26.5 MeV for Q = 32 nC, while such a shift can not be seen for Q = 2 nC. T he energy spread also becomes larger as the phase angle goes down from 0 to −19.8°, but the behavior is more complicated and exaggerated. As the phase angle decreases, the peak splits in two or three for Q = 32 nC. On the other hand, such behaviour does not show up for Q = 2 nC. The difference is due to the stronger wake field produced by the beam with Q = 32 nC, the temporal profile of

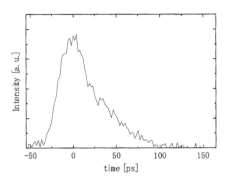

Figure 2. Charge distribution: solid curve is the measured charge distribution. dotted curve is the smoothed it.

which was directly measured with a streak camera and is shown in Fig. 2. It was estimated in a previous study of the wake field produced in the L-band linac that the wake field reached 10 % of the rf field for charge about 30 nC in the single bunch beam [2].

3. Conclusion

The energy spectrum of the single-bunch electron beam was measured as a function of the phase of the rf field in order to study the wake filed induced in the accelerating structure of the L-band linac. The energy spectrum changes significantly with the phase of the rf field, indicating that the wake field is strong. We are now analyzing the measured energy spectra in terms of the Green function wake field.

4. References

[1] R. Kato, et al., "Higher harmonic generation observed in SASE in the far-infrared region", Nucl. Instr. Meth. **A483** (2002) 46.

[2] S. Takeda, et al, "THE WAKE POTENTIAL AND THE ENERGY SPREAD OF A HIGH-CURRENT SINGLE BUNCH", Proceedings of the 5th Symposium on Accelerator Science and Technology, KEK(1984) 80.

FREE ELECTRON LASERS 2003
E.J. Minehara, R. Hajima and M. Sawamura (Eds.)
Published by Elsevier B.V.

II-47

Emittance measurement Using Duo Image Pattern of Cherenkov Radiation on DC-SC photo injector

Jiaer Chen, Anjia Gu, Yuantao Ding, Kui Zhao*, Baocheng Zhang, Shengwen Quan

Institute of Heavy Ion Physics, Peking University, Beijing 100871, P.R. China

Abstract

A novel method to measure the emittance of electron beam is described using Cherenkov radiation. The image patterns of the Cherenkov radiation are formed both in the focal plane and image plane of an achromatic lens with long focus. Both the angular spread and radial distribution of the e-beam are obtained by processing the two patterns and in this way the beam RMS emittance is directly resulted. The emittance of the electron beam produced by the DC-SC photo-injector at Peking University is simulated and a He-Ne laser experiment is presented and discussed in this paper. On the beam line the quadrupole scanning technique is also used to verify this new method.

PACS codes: 41.60.Bq, 42.30.Va

Keywords: Duo Image Patter, Cherenkov radiation, emittance measurement, photoinjector

1. Introduction

A new type of superconducting photo injector, named DC-SC photoinjector, is being developed at Peking University[1]. The major goal is to produce minimum transverse emittance beams at a high average current (~1mA).

According to the parameters of DC-SC photo-injector, the energy of the e-beam is 2-3 MeV. At this level Optical Transition Radiation (OTR) is too weak to be captured by CCD camera. In this paper a new direct way of emittance measurement by using Duo Image Pattern of Cherenkov Radiation (CR) is introduced, which can measure the angular spread and the radial distribution of the electron beam simultaneously.

2. Principle of Duo Image Pattern measurement

As a charged particle travelling at velocity v through a dielectric medium (refractive index n), CR is emitted in a thin cone centered on the trajectory with opening angle θ_c, when $v > c/n$. In Fig. 1, θ_c, θ_i, θ_r, θ_t, and θ_f are angles corresponding to Cherenkov angle, internal incident angle, external refracted angle, tilted angle of the converter, and observation angle, respectively.

Using Snell's law, the observation angle:

$$\theta_f = \arcsin(n\sin(\theta_c - \theta_t)) \approx n(\theta_c - \theta_t) \quad (1)$$

where, $\theta_c = \arccos(1/\beta n)$, the CR angle.

Suppose r'_i is the divergence angle of the ith particle in the beam as it passes through the converter, then:

$$\theta_{f_i} = \arcsin(n\sin(r'_i + \theta_{ci} - \theta_{ti})) = \arcsin(n\sin r'_i) \quad (2)$$

Setting $\theta_t = \theta_c$ approximately $\theta_{f_i} \approx nr'_i$.

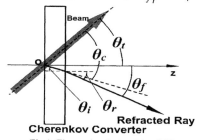

Fig.1. The angle relationship of CR

---

* Corresponding author. Tel.:+86-10-62767160; fax.:+86-10-62758849. E-mail address: kzhao@pku.edu.cn (K.Zhao).

After the one-to-one correspondence between the observation angle of CR and the divergence angle of the e-beam is derived, the information of angular spread can be obtained directly from the pattern on focal plane of the lens[2]. For this purpose, an achromatic lens with long focus is added normally to the direction of observation (Z-axis) (as in Fig.2), then

$$y_f \approx f\,\theta_f = f \cdot n \cdot r' \qquad (3)$$

with small angle approximation,

$$y_f = f \tan \theta_f \qquad (4)$$

where y_f is the vertical ordinate on the focal plane of the lens, f is the focus of the lens.

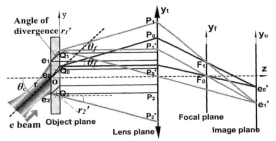

Fig.2. The Optical Path of "Duo Image Pattern of CR"

In principle this process can be called as "Duo Image Pattern": while the radial distribution of the e-beam is attained from the image plane of the lens, the additional information of the angular spread of the e-beam is obtained from the focal plane of the lens at the same time. Consequently, we can work out $<y>$ and $<y'>$ from these two distributions by using statistics. According to the definition, the RMS emittance can be expressed directly as[3]:

$$\varepsilon_{RMS} = 4\,[<(y-<y>)^2><(y'-<y'>)^2> - <(y-<y>)(y'-<y'>)>^2]^{1/2}$$

Fig.3 shows the layout of "Duo Image Pattern" beam diagnostics of DC-SC photo-injector. Passing the combined lens, the CR is split by a light splitter. CCD cameras are placed at each branch of light, where locates the focal plane and image plane of the lens, respectively.

Through simulation experiments, the errors derived from small angle approximation and image processing are rather low (less than 5%); yet the energy divergence of the beam takes a major part. An energy divergence of 2% at 3 MeV results in $\theta_c < 0.04\%$; while 30MeV results in $\theta_c < 0.0007\%$. This method prefers e-beams of higher energy.

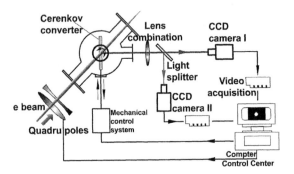

Fig.3. The Schematic view of "Duo Image Pattern" beam diagnostics on PKU-SCAF DC-SC

3. Simulation experiment with He-Ne laser

A He-Ne Laser is used to perform the simulation experiment so as to test the Duo-Image Pattern principles. The images at the focal plane and the image plane are illustrated by Fig.6a, b, respectively. Fig.6 c, d present the 3D plot of the distribution and e, f present the profiles along the center line of a, b.

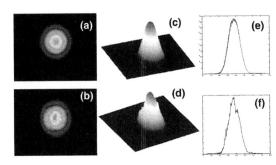

Fig.4. Experiment with He-Ne Laser and the image processing.

The result of Duo image pattern is compared with that of "tri-profile Method", and the difference between these two methods is around 5%. On the beam line the quadrupole scanning technique will be used to verify this new method.

4. References

[1] Zhao Kui, Hao Jiankui, Hu Yanle, et al. Nucl. Instr. & Meth. In Phys. Res., 2001, A475:564-568

[2] GU An-Jia, DING Yuan-Tao, ZHAO Kui, et al. HEP & NP, 2003, 27(2):163-168 (in Chinese)

[3] Stanley Humphries, Jr., Charged Particle Beams. Chinese Edition: Beijing, 1999:57

FREE ELECTRON LASERS 2003
E.J. Minehara, R. Hajima and M. Sawamura (Eds.)
Published by Elsevier B.V.

High temporal resolution, single-shot electron bunch-length measurements

G. Berden[a], B. Redlich[a], A.F.G. van der Meer[a], S.P. Jamison[b] *, A.M. MacLeod[b], and W.A. Gillespie[b].

[a]FOM Institute for Plasma Physics 'Rijnhuizen', Edisonbaan 14, 3439 MN Nieuwegein, The Netherlands

[b]School of Computing and Advanced Technologies, University of Abertay Dundee, Bell Street, Dundee DD1 1HG, United Kingdom

A new technique, combining the electro-optic detection method of the Coulomb field of an electron bunch with the single-shot cross-correlation of optical pulses is used to provide single-shot measurements of the shape and length of sub-picosecond electron bunches. This technique has been applied at the FELIX free electron laser showing bunches of around 600 fs (FWHM) where the resolution is limited primarily by the electro-optic crystal thickness and the relatively low γ of the electrons.

1. INTRODUCTION

X-ray free electron lasers (FELs) require dense, relativistic electron bunches with bunch lengths significantly shorter than a picosecond. For operating and tuning these lasers, advanced electron bunch length monitors with sub-picosecond temporal resolution are essential. Ideally, non-destructive and non-intrusive monitoring of a *single* electron bunch should be available in real-time. A promising candidate for such monitors, and the subject of an on-going research and development project at the Free Electron Laser for Infrared eXperiments (FELIX), is determination of the electron bunch longitudinal-profile via electro-optic (EO) detection of the co-propagating Coulomb field [1,2].

With the EO detection technique, the Coulomb electric field of the bunch induces birefringence in an EO crystal placed adjacent to the beam. The birefringence is determined through ellipsometry using a synchronized ultrafast Ti:sapphire (Ti:S) laser probe pulse. To enable a single-shot measurement, the Ti:S pulse is chirped to a duration exceeding the measurement window, and the induced birefringence is determined as a function of time within the probe pulse. In previously reported experiments, the timing was inferred from the wavelength-time relationship of the chirped probe pulse. However, this (spectrometer) method is intrinsically subject to limitations on the time resolution, and can introduce significant measurement artifacts [3–5].

A recently demonstrated technique [5], which we refer to as the cross-correlator technique, overcomes these limitations. The envelope of the probe pulse is measured directly in the time-domain, using single-shot second harmonic cross-correlation. We report here the results of the first application of this technique to longitudinal-profile electron bunch characterisation.

2. EXPERIMENTAL

The electron bunch induced birefringence in a 0.5 mm thick <110> ZnTe crystal, placed inside the accelerator beam-pipe at the exit of the undulator, is probed by a chirped pulse, essentially as described for previous experiments [2–4]. The crucial improvement of the technique, reported here, is in the method of retrieving the time information encoded in the chirped pulse by the induced birefringence. A single-shot cross-correlation is used, in which the amplitude modulated chirped pulse and a second 30 fs transform limited Ti:S pulse are overlapped spatially, but non-collinearly, in a BBO second harmonic (SH) generation crystal. The non-collinear geometry leads to a (transverse) position-dependent time-

*contact address: Dept. of Physics, University of Strathclyde, Glasgow, UK

delay between the two Ti:S pulses. This position-dependent time-delay then manifests itself in a position dependent emission of SH ($\sim 400\,$nm) radiation. Direct imaging of this radiation emitted from the BBO crystal reproduces the envelope amplitude of the chirped pulse.

3. RESULTS

Fig. 1 shows the intensity of the SH radiation generated with and without a synchronised electron bunch (200 pC, 30 MeV), as a function of transverse position on the CCD array. The position can be converted into time through knowledge of the beam geometry, and confirmed through a calibration procedure (e.g., by varying the time delay between probe laser and electron bunch). Data such as that in Fig. 1, directly representing the electron bunch Coulomb field longitudinal profile, was observed in real-time by displaying a binned trace of the CCD image.

The measured electron bunch of Fig. 1 is asymmetric, as expected from the accelerator operating parameters, with a short leading edge and a longer tail. The FWHM duration was found to be 870 fs under FEL lasing conditions (A), while a shorter duration of 600 fs was obtained by steering the electron beam away from optimum FEL operation (B). The time resolution is limited by the thickness of the EO crystal and the relative low γ of the electrons.

The electron bunch length measured with the cross correlation method is shorter than that previously reported for the spectrometer method (1.7 ps) [2]. While the two measurements were not performed under identical experimental conditions, the observation of shorter bunches is partly attributable to the enhanced time resolution of the cross-correlator technique. The temporal broadening due to the limited resolution of the spectrometer (370 fs [2]) and the broadening as a result of the limited (optical) bandwidth [3–5] are both eliminated in the present measurements.

In summary, using the cross-correlator technique we have overcome some of the time-resolution limitations of previous EO measurements, and measured a 600 fs long electron bunch on the FELIX FEL.

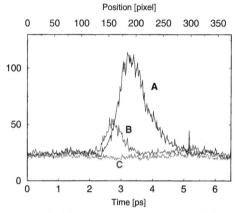

Figure 1. Single-shot measurements of the electric field profile of individual electron bunches under (A) optimal and (B) non-optimal FEL operations. The leading edge is on the left. The background is shown in trace (C).

REFERENCES

1. X. Yan, A.M. MacLeod, W.A. Gillespie, G.M.H. Knippels, D. Oepts, A.F.G. van der Meer, and W. Seidel, Phys. Rev. Lett. 85 (2000) 3404.
2. I. Wilke, A.M. MacLeod, W.A. Gillespie, G. Berden, G.M.H. Knippels, and A.F.G. van der Meer, Phys. Rev. Lett. 88 (2002) 124801.
3. G. Berden, G.M.H. Knippels, D. Oepts, A.F.G. van der Meer, S.P. Jamison, A.M. MacLeod, W.A. Gillespie, J.L. Shen, and I. Wilke, Proceedings PAC 2003, Portland (OR), USA, May 2003.
4. G. Berden, G.M.H. Knippels, D. Oepts, A.F.G. van der Meer, S.P. Jamison, A.M. MacLeod, W.A. Gillespie, J.L. Shen, and I. Wilke, Proceedings DIPAC 2003, Mainz, Germany, May 2003.
5. S.P. Jamison, J.L. Shen, A.M. MacLeod, and W.A. Gillespie, Opt. Lett. (2003) in press.

FREE ELECTRON LASERS 2003
E.J. Minehara, R. Hajima and M. Sawamura (Eds.)
Published by Elsevier B.V.

Emittance measurement for 200keV-electron beam extracted from NEA-GaAs photocathode using a pepper pot method

N.Yamamoto*[A)], M.Yamamoto[A)], T.Nakanishi[A)], S.Okumi[A)], F.Furuta [A)],
T.Nishitani [A)],N.Miyamoto [A)], M.Kuwahara [A)], K.Naniwa [A)], H.Kobayakawa [B)],
Y.Takashima [B)], M.Kuriki[C)], H.Matsumoto[C)], J.Urakawa[C)] and M.Yoshioka [C)]

A) Department of Physics, Nagoya University, Nagoya 464-8602, Japan
B) Department of Materials Processing Engineering, Nagoya University, Nagoya 464-8602, Japan
C) KEK High Energy Accelerator Research Organization, 1-1 Oho, Tsukuba 305-0801, Japan

Abstract

Recently it becomes important to generate an electron beam with very low emittance not only the for future 1TeV linear colliders but also for the next generation accelerators to produce ultra-high brightness X-rays. Especially, the energy recovery linac (ERL) project planned at Cornell, KEK and other labs. demands the emittance less than 0.5pi mm.mrad at exit of the electron gun. To achieve such a low emittance, the GaAs photocathode with NEA (negative electron affinity) surface is considered to have a significant advantage compared with other PEA photocathodes, because it gives the smallest initial emittance such as 0.1pi mm.mrad (1mm radius). However, in order to demonstrate this advantage by experiment, the high field gradient DC-gun must be developed not only to suppress the beam divergence due to space charge effect but also to assure the long lifetime of NEA surface. As the first step to this direction, we plan to measure the emittance of electron beam produced by a 200keV polarized electron gun developed at Nagoya University. An apparatus developed for the emittance measurement of the KEK electron source by pepper-pot method has been modified to match with an ultra-high-vacuum condition ($< 3 \times 10^{-9}$Pa) of the 200keV gun. Present status of preparation of the experiment will be reported at the conference.

PACS codes: 29.25.Bx; 41.85.Qg

Keywords: pepper pot method; NEA-GaAs photocathode; very low emittance; high field gradient DC-gun; energy recovery linac

1. Introduction

An NEA-GaAs photocathode has been studied intensively, because it is indispensable for a source of polarized electron beam, and a 200keV polarized electron gun with NEA-GaAs photocathode has been developed for future linear colliders at Nagoya University [1]. This NEA-surface is so sensitive to a vacuum condition that ultra-high-vacuum ($< 3 \times 10^{-9}$ Pa) is indispensable to keep the good NEA-surface state.

We have started the preparation to measure the emittance of this 200keV gun by the pepper-pot method. This method was first developed by Cutler et al. in 1955 [2][3], and it was applied for emittance measurements over the range 5-160 pi mm mrad for various electron beams. Then, this method was upgraded to measure the emittance less than 1pi mm mrad for the electron beam with 2mm diameter of the KEK gun [4].

2. Advantage of NEA-GaAs photocathode

For an electron source using a photocathode, the

* E-mail: naoto@spin.phys.nagoya-u.ac.jp

initial emittance is affected by an average transverse energy of the electrons before being emitted into vacuum. In order to minimize the initial emittance, it is necessary to decrease the excited energy of electrons above the conduction minimum by decreasing laser photon energy. However the excited electrons are scattered by the phonons in crystal and lose the energy. In case of a photocathode with a PEA (positive electron affinity) surface, the energy-lost electrons below the vacuum level cannot escape into vacuum. Therefore the quantum efficiency decreases with decreased laser photon energy.

On the other hand, in case of NEA-photocathode, the vacuum level is pulled down a little bit lower than the minimum conduction band. Therefore even the band-gap excited electrons could escape into vacuum and both of the high quantum efficiency (several %) and low initial emittance should be obtained.

The initial emittance as function of the excitation energy of electron above the vacuum level can be estimated and it gives the value below 0.1pi mm mrad for the NEA-GaAs surface at room temperature [5][6].

3. Experimental instruments

We employ the same pepper-pot method used to measure the KEK-PF electron source [4]. The vacuum pressure of our 200keV-gun must be kept below 10^{-10}Pa to preserve the good NEA-surface state. However, the vacuum pressure of KEK-PF gun during the emittance measurement was 1.2×10^{-5}Pa. Therefore the following modifications have been made to match the vacuum condition of pepper-pot chamber to that of the 200keV gun.

3.1 Scintillator

A sheet of plastic scintillator is used to measure the size and position of the luminous spots irradiated by beamlets passing through the pinholes. In order to avoid the broadening of the beamlet-size caused by electron scattering in the scintillator sheet, it is required to make the plastic scintillator as thin as possible and the short decay time of scintillation light. Furthermore, the high softening temperature for scintillator is required to enable the baking of the pepper-pot chamber for ultra-high-vacuum. We chose two types of scintillators (Bicron, BC-448 and BC-448M). The characteristics of these scintillators are

as follows: decay time 2.1ns; peak wavelength 425nm; softening point 99°C (BC-448) and 150°C (BC-448M); thickness 0.01mm (BC-448) and 1mm (BC-448M). By using BC-448M, the pepper-pot chamber can be baked up to 150°C and better vacuum condition will be obtained.

1.2 Differential pumping chamber

In order to keep the ultra high vacuum of the gun, the differential pumping chamber is inserted between the gun chamber and the pepper-pot chamber. The pumping units used for the differential pumping chamber consist of 100 L/s ion pump (ULVAC, PST-100-AX), 240 L/s NEG pump (SAES-GETTERS, GP100-MK5). After baked at about 250°C for 5days, a vacuum pressure of 10^{-9} Pa was obtained for this chamber.

4. Scheduled experiments

In order to develop the lowest emittance electron gun, we plan to study the relation of the emittance to following parameters; laser beam size, electron bunch charge, electron energy, and quantum efficiency (the value of NEA).

In parallel, the study to increase the field gradient applied to the NEA-GaAs cathode from the present value of 3 MV/m up to 10 MV/m has been continued.

5. References

[1] M.Yamamoto et al., this proceeding
T.Nakanishi, Linac 2002 proceedings, Gyonju, Korea, Aug.2002
K.Wada et al., PESP2002Proceedings, MIT, Sept., 2002
[2] C.C. Culter and J.A.Saloom, Proc. IRE 43 (1955) 299.
[3] C.C. Culter and M.E. Hines, Proc. IRE 43 (1955) 307.
[4] Y.Yamazaki, et al., Nucl. Instr. and Meth. A 332 (1992) 139-145.
[5] K.Flottmann, DESY-TESLA-FEL-97-01, Feb 1997. 7pp.
[6] J.E.Clendenin and A.Mulhollan, SLAC-PUB-7760 March 1998

FREE ELECTRON LASERS 2003
E.J. Minehara, R. Hajima and M. Sawamura (Eds.)
Published by Elsevier B.V.

Infrared Photoconductivity in Te-doped Ge

H. Nakata[a,*], T. Hatou[b], K. Fujii[b], T. Ohyama[b] and N. Tsubouchi[c]

[a] Osaka-Kyoiku University, 4-698-1 Asahigaoka, Kashihara, Osaka 582-8582, Japan
[b] Osaka University, 1-16 Machikaneyama-cho, Toyonaka, Osaka 560-0043, Japan
[c] Institute of Free Electron Laser, 2-9-5 Tsuda-Yamate, Hirakata, Osaka 573-0128, Japan

Abstract

We investigate infrared photoconductivity by a free electron laser and a CO_2 laser. Ionization energy of Te donor in Ge is estimated to be 93.5 meV from magneto-photoconductivity measurement. Fast photoresponse due to photovoltaic effect was also observed.

PACS codes: 71.55.Cn, 71.70.Di, 72.40.+w

Keywords: Ge, Te, photoconductivity , infrared

1. Introduction

Doped Ge samples such as Ge:Ga etc. have been used for infrared detectors because of high sensitivity and fast response [1]. In the wavelength range of 10 μm, Au-doped Ge is a good photoconductor like mercury telluride which is commercially available. We propose Te-doped Ge (Ge:Te) as an infrared detector with the response time of $\sim \mu$ s at temperatures lower than 100 K.

2. Free Electron Laser Measurements

Infrared photoconductivity was investigated by two kinds of sources, a free electron laser (FEL) and a CO_2 laser. The samples which we used were Ge:Te with donor concentration of 1.1×10^{14} cm^{-3} and their sizes were 3x5x1 mm^3. One of them was placed on a cold finger of a He gas refrigerator at 11K. The FEL beam was focussed by a concave mirror through a ZnSe window. The time-resolved signal was analyzed by a digital oscilloscope.

Two kinds of photoresponses were observed by changing the excitation position as shown in Fig. 1.

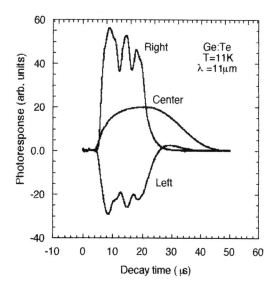

Fig.1 Infrared transient photoresponse of Te-doped Ge for different excitation positions.

---

* Corresponding author. FAX: +81-729-78-3634.
  E-mail address. hynakata@cc.osaka-kyoiku.ac.jp (H. Nakata)

One reveals a slow decay when we shed the beam in the center of the sample. The others reveal a fast decay when the contact region indicated right or left in Fig.1 was illuminated. The slow signal is due to photoconductivity of Te donor., while the fast signal originates in the photovoltaic effect. If we change the excitation position from one side to the other side, the photovoltaic signal changes the sign.

The photoconductive spectrum has a threshold photon energy around 90meV corresponding to the 1s-2p transition or ionization of a donor. The lineshape is influenced by the wavefunction in k space. The peak energy divided ionization energy depends on the detailed wavefunction in k space. The decay time depends on the wavelength and has a minimum value around 11μ m.

3. CO$_2$ Laser Measurements

In order to obtain the accurate ionization energy , we performed magneto-photoconductive measurement by using a CW CO$_2$ laser. The sample was placed in the center of the superconducting magnet up to 5T. It was immersed in liquid He and the chopped CO$_2$ laser beam was guided to the sample in Faraday configuration. The photoconductivity signal was analyzed by a lock-in-amplifier. The mageto-photoconductive signal has small oscillations on the monotonic decrease due to magnetoresistance.

The small oscillations are induced by electronic transitions from the ground state of Te donor to the electron Landau levels in the conduction band. The oscillations in low magnetic fields are fit by a light cyclotron mass (m*=0.082 m$_0$) as shown in Fig. 2. The others may be caused by a heavy- cyclotron-mass electron. We estimate the ionization energy to be 93.5 meV by fitting the low-field data.

The peak positions of the oscillations deviate from the theoretical value neglecting the shift of ground state energy in magnetic fields. The deviation may be induced by increase of ionization energy in magnetic fields and the central cell correction of deep Te donor . Applicaion of magnetic fild induces the shrinkage of the wavefunction of donor electron. Even if the electron is described by effective-mass approximation, the ionization increases with magnetic fields. This tendency is enhanced in the case of deep impurity with central cell correction.

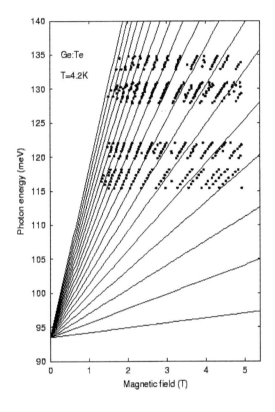

4. Summary

Phtoconductivity and photovoltaic effects were observed in Ge:Te by using an FEL or a CO$_2$ laser. We obtained accurate ionization energy 93.5 meV of Te donor in Ge. It is in a good agreement with previously published value [2]. The peak magnetic fields of magneto-oscillations deviates from theoretical values because of the central cell correction of deep Te donor.

References
[1] P. R. Bratt, Semiconductors and Semimetals, Academic Press, New York **12**, 1977,p39.
[2] H. G. Grimmeis et al. Phys. Rev. B **37** (1988) 6916.

FREE ELECTRON LASERS 2003
E.J. Minehara, R. Hajima and M. Sawamura (Eds.)
Published by Elsevier B.V.

Fragmentation in biological molecules by the irradiation of monochromatized soft X-rays

Yuji Baba, Tetsuhiro Sekiguchi, Iwao Shimoyama, Krishna G. Nath

Synchrotron Radiation Research Center, Japan Atomic Energy Research Institute

Abstract

High-flux photons from FEL can be used not only for the structural analysis but also for the materials modification. In advance of future application of short wavelength light from FEL, we present the results for photo-fragmentation in biological molecules by monochromatized soft X-ray irradiation using synchrotron radiation. For sulfur-containing amino acids, the core-to-valence resonant excitation at the sulfur K-edge induced the desorption of only S^+ ions, which is quite different from the fragmentation patterns by the valence excitation. On the basis of the Auger decay spectra, it is revealed that the resonance excitation from the S 1s to σ^* states is followed by the "spectator-type" Auger decay where the excited electrons remain in the σ^* orbital in the course of Auger decay. We conclude that the specific fragmentation is caused by the fast bond scission due to the highly antibonding character of the σ^* orbital.

PACS codes: 32.80.Hd, 68.43.Tj, 82.50.Kx

Keywords: Core-level excitation, Amino acid, Nucleotide, Synchrotron radiation, Auger decay, Soft X-ray

1. Introduction

High-flux photons from FEL are fascinating tools for materials science, because they can be used not only for the structural analysis but also for the materials modification. The decomposition of materials by the irradiation of FEL has been investigated for organic molecules [1], and it has been elucidated that the multi-photon absorption as well as thermal process plays an important role in the selective bond scission in a molecule. For vacuum ultraviolet light and soft X-rays, experiments using synchrotron radiation have revealed that a single photon absorption induces the bond rapture around a specific atom because a core-level is localized at the atomic site even in multi-element materials. Such kind of specific bond scission by core-level photoexcitation has been found in gas-phase molecules [2] and adsorbed molecules [3] where the effect of secondary electrons on chemical reaction can be ignored. Here we report on the examples for specific bond scissions by soft X-ray irradiation in some of the fundamental biomolecules, in advance of the future application of FEL in short wavelength

region.

2. Experimental

The experiments were performed at the BL-27A station of the Photon Factory in the High Energy Accelerator Research Organization (KEK-PF). The pressed pellets of amino acids (L-cystine, L-cysteine and L-methionine) and nucleotides (ATP) were put into ultra-high vacuum chamber. The desorbed ions were measured in-situ by a quadrupole mass spectrometer. X-ray photoelectron spectra (XPS) and Auger decay spectra were measured by the hemispherical electron energy analyzer.

3. Result and discussion

In this report, we will concentrate on the results for L-cystine. Figure 1 shows the mass spectra of the desorbed ions from L-cystine. The upper spectrum was taken by the irradiation of 14-eV electrons (valence excitation), and the lower one was obtained by the irradiation of 2472.8-eV photons (core-level excitation). This photon energy corresponds to the resonance excitation from the S 1s to the unoccupied σ^* states localized at the S-S bond. The irradiation of 2472.8-eV photons induces only atomic S^+-ion desorption, which is quite different from electron irradiation. The difference suggests that the core-level excitation is fairly localized at the sulfur atoms.

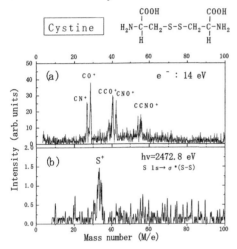

Fig.1. Mass spectra desorbed from cystine excited by, (a) 14-eV electrons, and (b) 2472.8-eV photons. The photon energy of the spectrum (b) is tuned at the resonance maximum from the S 1s to the unoccupied σ^* state localized at the S-S bond.

The photon-energy dependencies of the S^+ yields were compared with the X-ray absorption spectrum, and it was elucidated that the S^+ desorption happens mostly at the resonance excitation from S 1s to the σ^* state localized at the S-S bond (S 1s $\rightarrow \sigma^*_{(S-S)}$). Concerning the chemical bond scission by core-level excitation, it has been established that the positive-ion desorption from adsorbed molecules is well explained on the basis of the localization of the two or more positive valence holes which are created by the sequences of the Auger decays (KF-model) [4]. However, this scenario does not necessarily hold for the biomolecules where the secondary electrons would also contribute to the fragmentation. The time scale of the fragmentation is in the order of picoseconds which is fairly slower than that of core-hole decay.

The Auger decay spectra around the S K-edge excitation are shown in Fig.2. The S 1s ionization at higher photon energy is primarily followed by the S $KL_{2,3}L_{2,3}$ Auger decay (normal Auger decay).

While the S 1s$\rightarrow\sigma^*_{(S-S)}$ resonant excitation is mostly followed by "spectator-type" Auger decay where the excited electrons remain in the $\sigma^*_{(S-S)}$ state during the Auger transition (indicated as broad arrow). The S^+ ions are desorbed only in the photon energy region

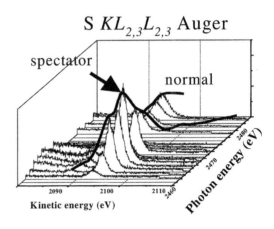

Fig.2. Resonant Auger decay spectra for cystine excited by various energy photons around the sulfur K-edge.

where the "spectator-type" Auger decay happens. Since the σ^* has a highly antibonding character, it was elucidated that the S^+ desorption at the S 1s$\rightarrow \sigma^*_{(S-S)}$ resonance is interpreted by the fast bond scission due to the electrons in the antibonding σ^* orbitals because of the formation of highly repulsive states. The present results show that the time scale of the chemical bond breaking following core-to-valence resonant excitation in biomolecules is comparable to that of the core-hole decay. This proposition will be more clearly confirmed in future by the experiment using short-pulsed, high intensity soft X-rays from FEL.

4. References

[1] T. Yamauchi et al., JAERI-Conf 2002-007 (2002).
[2] W. Eberhardt et al., Phys. Rev. Lett. 50 (1983) 1083.
[3] Y. Baba et al., Surf. Sci. 341 (1995) 190.
[4] M.L. Knotek and P.J. Feibelman, Phys. Rev. Lett. 40 (1978) 964.

FREE ELECTRON LASERS 2003
E.J. Minehara, R. Hajima and M. Sawamura (Eds.)
Published by Elsevier B.V.

Demonstration of material processing using JAERI-FEL

Akihiko Nishimura*, Toshihiko Yamauchi, Eisuke Minehara

Japan Atomic Energy Research Institute, Advanced Photon Research Center,
8-1 Umebidai Kizu-chou Soraku-gun, Kyoto Japan

Abstract

Ablation experiments by the superconducting RF linac FEL in JAERI are planed. The FEL can produce ultrashort laser pulses of 255 fs micro-pulse with the pulse energy of 0.23 mJ. The high operational frequency over 10 MHz is useful for the high repetitive metal processing which is free from thermal damages.

PACS codes: 42.62.

Keywords: free electron laser, laser processing, JAERI

1. Introduction

High power FELs will open the next stage of ultrashort laser material processing. Output laser pulses of the JAERI superconducting RF linac-based FEL consists of 255 fs micro-pulses with energy of 0.23 mJ with an operational frequency of 10.4215 MHz for 1 ms duration of a macro-pulse at a repetition rate of 10 Hz. Because the superconducting RF linac intrinsically has little loss and resultantly has a capability of continuous wave operation, it has inherent potential to extend the FEL duty to CW or 100% and increase the pulse energy. The performance shows a new possibility of industrial applications. Nowadays Q-switched YAG lasers and carbon dioxide lasers are commonly used in many industrial fields of material processing. Drilling, cutting and welding by use of these lasers and O_2 gas assisting are indispensable. For these purposes, the focused intensity on material surface is less than 10^9 W/cm^2 and the laser pulse duration is longer than 10 ns. However, unavoidable side effects of huge debris generation and thermally induced stress around the laser focused area limit further application of the material processing. Since the invention of CPA lasers, the material processing using ultrashort pulse lasers has been widely studied. In particular, the ability of ultrashort pulse laser to fabricate fine microstructures in solid materials is very promising. It is because that the pulse duration is enough short in comparison with a characteristic time of heat conduction in solid materials. However, the repetition rate of solid state CPA lasers is limited in kHz range.

By the success of ultrashort pulse generation by JAERI-FEL, MHz repetitive processing can be possible. Figure 1 shows the view of JAERI-FEL, which has a 22 m long laser beam transport duct which can deliver laser pulses into a user application laboratory.

Fig. 1. View of JAERI-FEL

2. Experiments

2.1 Mechanism of laser ablation

Ultrashort pulse lasers provide the precise

---
* nisimura@apr.jaeri.go.jp

machining generally characterized by the absence of heat diffusion and, consequently, molten layers. This is because the ultrashort interaction is enough short in comparison with the thermal diffusion time. For ultrashort laser pulses, heating process can be separated resulting in the two-temperature model which shows non-equilibrium temperature state in electrons and in lattices. The electron temperature is reached to the maximum value at the end of the ultrashort laser pulse. The lattice temperature is still cold as the absorbed energy has not transferred from the electron to the lattice. By introducing the optical penetration depth δ and the heat penetration depth $d = (D\tau)^{1/2}$, the ablation rate per a single laser pulse can lattice can be shown in the two different regimes which is dependent on the laser fluence F. The gentle ablation regime for lower laser fluence can be characterized by the optical penetration depth, while the strong ablation regime for higher laser fluence can be done by the heat penetration depth d. Especially, the gentle ablation regime shows the highest processing quality which is nearly free of molten layer because the thermal penetration depth is shorter than the optical penetration depth. All metal materials show the same behavior of the two different regimes. In case of copper, the laser fluence of ~ 0.7 J/cm^2 is the transient laser fluence of the two regimes for 150 fs ultrashort pulses [1,2].

When the laser fluence is determined by the optical penetration depth, no molten material can be observed by microscopy. On the other hand, when it is done by the electron penetration depth, ablation has a molten layer and a "corona" of re-solidification around an irradiated surface. Thus, ultrashort pulses cannot give considerable advantages for precise processing in the strong ablation regime. In addition, when ablation is done in air, the ultrashort laser pulse processing might lose the important advantage which is free from plasma generation near the ablation surface. Also, re-deposition of ablated metal and oxidization can occur after the ultrashort pulses which cause unfavorable debris around the surface.

2.2 Materials for laser ablation

Nowadays, JAERI-FEL has a laser pulse sequence which consists of micropulses and macropulses. Time interval of the micropulses of 95 ns is long enough in comparison with the ablation time of 10 ns. Thus, the time interval of the micropulses is convenient for the material processing which is free from heat accumulation. Figure 2 shows an example of laser ablation on SUS316L surface in air by 127 fs pulse duration with 0.46 mJ pulse energy. Accumulation of 1500 laser pulses can make a round hole with 1.01×10^{-3} cm^2 area and 20 μ m depth, which has been demonstrated by a Ti:sapphire CPA laser.

Fig. 2.   Ultrashort laser ablation on SUS316L surface.

Ablation for nuclear grade SUS is very attracting. The in-core structural materials used in BWR are exposed to radiation in high temperature water environment. In Japan, the welded BWR core shrouds have been recently experienced the stress induced cracking [3]. It is very serious because the cracking has occurred in austenitic stainless steels of SUS316L that is more resistant to the thermally induced stress corrosion cracking.

Using JAERI-FEL, the laser beam has 40 mm diameter at the laser beam transport duct and the spot diameter will be 180 μ m with a convex lens of 300 mm focal length. The laser fluence of 0.78 J/cm2 with 250 fs pulse duration will provide the proper ablation condition with high repetition rates. The focused intensity up to 3×10^{12} W/cm^2 will be surely achievable. Thus, the laser energy deposition would be determined by the optical penetration depth, which is very suitable for precise laser machining.

3.   References

[1] C. Momma et al., Optics Comm. Chem. **129**, (1996) 134.

[2] S. Nolte et al., J. Opt. Soc. Am. B. **14**, (1997)2716.

[3]Y. Kaji et a l., J. Nucl. Sci, Technol., **37**,(2000) 949.

FREE ELECTRON LASERS 2003
E.J. Minehara, R. Hajima and M. Sawamura (Eds.)
Published by Elsevier B.V.

Infrared photoinduced alignment change for columnar liquid crystals by using free electron laser

Hirosato Monobe, Kenji Kiyohara, Naohiro Terasawa, Manabu Heya[†],
Kunio Awazu[†], Yo Shimizu[*]

Mesophase Technology Research Group, National Institute of Advanced Industrial Science and Technology, Ikeda, Osaka 563-8577, Japan
[†]*Institute of Free Electron Laser, Graduate School of Engineering, Osaka University, Tsuda-Yamate, Hirakata, Osaka 573-0128, Japan*

Abstract

Infrared photoinduced alignment change of liquid crystal domains was investigated and a uniform and anisotropic change of domains was observed when a polarized infrared light corresponding to the wavelength of the absorption band of triphenylene core was irradiated. The texture observation and polarized microscopic IR spectra show that a change of the molecular alignment occurred and the direction of columns depends on the polarizing angle of an irradiated infrared light. The technique could provide a novel technology to control the columnar alignment of supra-molecular systems.

PACS code: 61.30.Hn   *Keywords:* Alignment, Infrared, Liquid crystals, Orientation, MIR-FEL

1. Introduction

Recent research and developments of organic electronic devices have shown remarkable progress in their fabrication techniques. However, in usage of liquid crystalline semiconductors, it tends to show a certain difficulty in their alignment control, probably due to their viscous state of mesophase derived from relatively strong interaction among the π-electronic conjugation system and the higher order of the molecular orientation. Some trials have been reported to get uniaxial planar and homeotropic alignment of columnar mesophases and a few successes were attained by usage of polytetrafluoroethylene (PTFE) coated by friction transfer technique [1] and the introduction of fluoroalkylated chain into peripheral parts of triphenylene [2] for planar and homeotropic alignment, respectively.

Recently, a linac-driven free electron laser (FEL) system covering a wide range of spectrum from ultraviolet to far-infrared has been developed. Any trials to control the molecular alignment of a columnar mesophase have not been carried out by use of infrared light which enable us to activate a specified local part of the molecule in a mesophase

with the vibrational excitation. We recently reported that infrared laser irradiation of Col_h mesophase shown by a typical triphenylene mesogen causes an alignment change to form a new domain with uniformity of alignment and this phenomena is by way of the excitation of the selected vibrational mode of a chemical bond [3]. Recent studies on this phenomenon have also shown that a relationship was found for the directions of the polarization of incidence and the transition moment of the vibrational excitation [4,5]. These results strongly imply the polarizing infrared irradiation of Col_h mesophase is a possible technique for device fabrication when a liquid crystalline semiconductor is used as a circuit.

2. Experiments

2,3,6,7,10,11-Hexahexyloxytriphenylene (C6OTP) was used for the experiments which is one of a well-known columnar discotics (Cr. 69 °C Col_h 99 °C IL). C6OTP film was prepared between two BaF_2 substrates which have high transparency in infrared region, with 2 μm silica-beads as spacer and mounted on a hot stage which was attached to an optical

---
[*]Corresponding author. Tel.: +81-72-751-9525; fax: +81-72-751-9628. E-mail address: yo-shimizu@aist.go.jp (Y. Shimizu)

polarizing microscope. A homeotropic alignment was spontaneously attained between two BaF_2 plates in the Col_h phase. The incident IR-FEL beam comes normal to the cell through an IR-polarizer. The IR-FEL beam average power was kept on 8 mW in this experiment. The IR-FEL beam was focused to about 0.5 mm in diameter by ZnSe lens. A FT-IR spectrum of 2μm-thick C6OTP cell was shown in Figure 1. A microscopic IR absorption dichroism was measured by a microscopic FT-IR spectrophotometer.

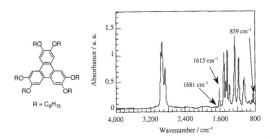

Fig. 1. Chemical structure of C6OTP and IR spectrum at 105 °C.

3. Result and discussions

Figure 2 shows the textures of Col_h phase irradiated by MIR-FEL with wavelength of 1681 and 1615 cm^{-1} at 97 °C. The absorption peak at 1615 cm^{-1} is assigned to an aromatic C-C stretching vibration of the triphenylene core. The bright area appeared only in the latter case, while no change was observed in the former and the bright area appeared only irradiated area. The texture change means a change of the molecular alignment took place. It clearly shows the direction of columnar axis was changed from perpendicular to parallel manner to the substrate and it depends on the polarization direction of incident infrared light. This domain newly formed was quite stable and could be maintained for a few hours to days by keeping the temperature in the mesophase region.

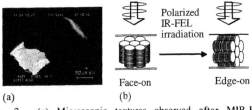

Fig. 2. (a) Microscopic textures observed after MIR-FEL irradiation by 1615 cm^{-1} and (b) a schematic representation of alignment change of C6OTP.

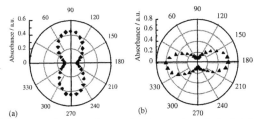

Fig. 3. Angle dependency of microscopic polarized infrared absorption intensity at (a) 1615 cm^{-1} (C-C aromatic stretching vibration) and (b) 839 cm^{-1} (C-H aromatic out-of-plane vibration) of C6OTP film.

Polarized microscopic IR absorption spectra for the domain changed by infrared irradiation was measured (Figure 3) to confirm that the molecular plane aligns normal to the plane formed by electric wave of the incident beam. The angle of the absorption maxim at the peak of 1615 cm^{-1} which is assigned to the C-C aromatic stretching vibration of triphenylene is perpendicular to the polarization direction of irradiated MIR-FEL beam. On the contrary, the angle of the absorption maximum at the peak of 839 cm^{-1} which is assigned to the C-H aromatic out-of-plane vibration is parallel to the polarization direction of incidence. This indicates that the plane of discotic molecules aligns perpendicular to the polarization of incidence. In other words, the results indicate that the relation between the polarizing direction of the incident infrared light and the azimuthal direction of the columnar axis is parallel. These are consistent with the polarizing microscope observation. The in-plane direction of columnar axis is controlled by the angle of polarization plane of infrared incidence.

These results imply that the vibrational excitation could lead to a novel method for controlling the direction of molecular alignment of highly ordered mesophases such as columnar phase.

References

[1] A. M. Craats, N. Stutzmann, O. Bunk, M. M. Nielsen, M. Watson, K. Müllen, H. D. Chanzy, H. Sirringhaus, R. H. Friend, Adv. Mater., **15**, 495, (2003).
[2] N. Terasawa, H. Monobe, K. Kiyohara, Y. Shimizu, Chem. Commun., 1678, (2003).
[3] H. Monobe, A. Awazu, Y. Shimizu, Adv. Mater., **12**, 1495, (2000).
[4] H. Monobe, K. Kiyohara, N. Terasawa, M. Heya, K. Awazu, Y. Shimizu, Thin Solid Films 438 (2003) 418.
[5] H. Monobe, K. Kiyohara, N. Terasawa, M. Heya, K. Awazu, Y. Shimizu, Chem Lett. 32 (2003) 870.

FREE ELECTRON LASERS 2003
E.J. Minehara, R. Hajima and M. Sawamura (Eds.)
Published by Elsevier B.V.

Feasibility of Megagauss-THz Spectrometer combining FEL and Single turn coil

H. Nojiri[a]*

[a]Department of Physics, Faculty of Science, Okayama University, Okayama 700-8530, Japan

A megagauss THz spectroscopy has been made so far by using a farinfrared laser and pulsed magnetic fields. In this work, preliminary results on megagauss-THz spectroscopy by using a single turn coil and a farinfrared laser are presented. On this basis, potential and feasibility of a THz spectrometer combining FEL and single turn coil are discussed.

1. Introduction

Terahertz wave is one of the most developing fields in spectroscopy of condensed matters. The most important reasons are that various elementary excitations are observed in the THz-region and that this is the energy range corresponding to the dynamical response of a crystal lattice. On the other hand, a magnetic field is the most important parameter for magnetic systems. It is because a magnetic field is the directly involved parameter of the Hamiltonian of magnetic compound. Usually, the Zeeman energy of magnetic fields is as low as a few kelvin and an interplay between magnetic excitation and phonon excitation is scarcely observed.

In extremely strong magnetic fields of megagauss range, the Zeeman energy becomes close to the energy of phonon and crystal field splitting. A mixing caused by the proximity of these energy scales may lead to a new complex excitation and a complex quantum state. A magnetic field driven structural phase transition in perovskite oxide is a well known example. An intelligent bi-functional material such as a dielectric-magnetic compound can be designed by understanding the mixing between phonon and magnon. A dynamical control of spin state by structure may be realized by controlling the mixing between the crystal field splitting and the Zeeman energy. A THz spectroscopy in megagauss fields is very useful to study this ad-vancing field of condensed matter physics.

2. Experimental

A megagauss THz spectroscopy has been made so far by using a farinfrared(FIR) laser and a destructive single turn coil. The single turn coil is a method to produce a sinusoidal pulsed field of 100-200 T range. As is well known, a conductor such as Cu cannot withstand the electromagnetic force of megagauss generation. The Maxwell stress of 4 GPa at 1 MG is far beyond the strength of conductors of high magnetic field coils. The principle of a single turn coil is the fast discharge of current. Namely, the coil is kept unbroken in the period of field generation for its finite inertia. In practice, the half width of a pulsed field is about a few micro seconds. The advantage of a single turn coil for spectroscopy is that a sample and a specimen are left unbroken in spite of the hard explosion of the coil. For this special feature, a reproducibility can be checked easily.

On the other hand, a practical THz spectroscopy is very difficult for its very short pulse width. The most crucial point is the response speed of a detector. More than 30 MHz band width is necessary to observe a magnetic field dependence of THz-wave transmission in such sort pulse width. An InSb bolometer cannot be used because its response is only about 1 MHz. A monochromatic pin-diode detector shows a fast response, however, it can be used in very limited

*e-mail:nojiri@cc.okayama-u.ac.jp

frequency range. The response of 30 MHz can be obtained in GeGa detector by adjusting the detector impedance. The typical impedance of GeGa detector at 4.2 K is several MOhm and it is too high to obtain the response of 30 MHz. For the fast response application, the detector impedance is lowered by setting the detector temperature slightly higher than 4.2 K. It should be noted that the sensitivity is also reduced by the decrease of impedance at least by factor of 10. For this low sensitivity, a megagauss THz spectroscopy has been made so far at very limited frequencies.

A new candidate of the detector for a single turn coil experiment is a superconducting bolometer with about 100 MHz response speed. This detector covers from the microwave to the THz range and thus it is the unique broadband detector below 1 THz, where only a monochromatic pin-diode is available. A GeGa detector may be replaced by this superconducting bolometer when a fast response is necessary.

As shown above, the detection of THz wave in the very short period of megagauss generation is still very difficult. Hence, a high intensity THz source such as FEL is necessary to perform high precision tunable THz megagauss spectroscopy.

3. Result and discussion

In this section, we present a preliminary result of THz-electron spin resonance which have been made by using a conventional FIR-laser and a single turn coil. The present results exhibit partly the scientific possibilities which can be fully realized by combining FEL and single turn coils.

Figure 1 shows a spin-phonon coupled mode in the spin-Peierls compound $CuGeO_3$. This compound exhibits a spin-Peierls transition below 14 K and an incommensurate magnetic phase is expected between 12.5 T and 250 T. A conventional paramagnetic like signal is observed at 84 T and a new peak is observed at 107 T. This new peak can be observed only when a radiation has the polarization parallel to the a-axis. This polarization dependence is very similar to that of the spin-phonon coupled mode found by Takehana et al. in low field range. They found that a phonon mode

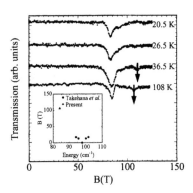

Figure 1. ESR spectra in $CuGeO_3$. The field is tilted 30 degree from the a-axis. In the inset, the present spin phonon mode is plotted together with the result of Takehana et al.

at 98 cm^{-1} splits in the incommensurate phase and that the incommensurate wave vector can be evaluated from this splitting. In the present work, we found a continuous development of the incommensurate wave vector in the very middle of the incommensurate phase as shown in the inset of Fig. 1. The observation of a spin-phonon coupled mode indicates that such complex excitation really exists in THz range and in megagauss fields. The present result exhibit a new possibility of THz-megagauss spectroscopy, which can be realized by combining FEL and single turn coil.

4. Acknowledgement

This work has been performed as the visiting professorship in Institute for Solid State Physics, University of Tokyo. It is partly supported by the Kakanhi(No. 15654048).

REFERENCES

1. M. Hase, I. Terasaki, and K. Uchinokura, Phys. Rev. Lett. 70 (1993) 3651 .
2. K. Takehana, T. Takamasu, M. Hase, G. Kido and K. Uchinokura, J. Phys. Soc. Jpn. 70 (2001) 3391.

FREE ELECTRON LASERS 2003
E.J. Minehara, R. Hajima and M. Sawamura (Eds.)
Published by Elsevier B.V.

Absorption-spectra measurement system synchronized with a free electron laser

T. Kambayashi[a], T. Ookura[a], T. Takahashi[a], H. Kubo[a], N. Mori[a], M. Asakawa[b], and N. Tsubouchi[b]

[a]Department of Electronic Engineering, Osaka University, Suita, Osaka 565-0871, Japan

[b]Institute of Free Electron Laser, Osaka University, Hirakata, Osaka 573-0128, Japan

A measurement system for compound-semiconductor absorption spectra is being implemented in the Institute of Free Electron Laser, Osaka University. Since the FEL operates with ultra-short pulses having a complex pulse structure, it is necessary to generate near-infrared/visible pulses synchronized with the FEL pulses in order to measure absorption spectra of compound semiconductors with high accuracy. We used a mode-locked Nd:YLF laser for generating the synchronous near-infrared/visible beams. The generated beam of the Nd:YLF laser has been confirmed to synchronize with the FEL pluses.

Keywords: free electron laser, compound semiconductors, optical properties
PACS: 78.20.-e, 42.65.-k, 72.20.Ht

1. INTRODUCTION

A free electron laser (FEL) is a coherent optical source using an electron beam in a magnetic field as a gain medium. The FEL system consists of an electron accelerator, an undulator in which the electrons emit the synchrotron radiation, and an optical resonator. By virtue of its simple gain medium, FEL has unique advantages; wide range wavelength tunability, ultrashort pulse operation and intense peak power. A high-power coherent laser beam in terahertz and infrared region reveals novel properties of semiconductors. One of the most interesting issues associated with the high-power laser effects in the mid-infrared (mid-IR) region is the extreme mid-IR nonlinear optics [1]. Although there are many theoretical studies on unusual phenomena in strongly laser driven semiconductors [2–6], relatively few experimental studies of nonlinear optical phenomena in semiconductors have been made using intense light in the mid-IR region [7,8].

We are implementing a measurement system for compound-semiconductor absorption spectra at the Institute of Free Electron Laser (iFEL), Osaka University. As the first step towards developing the absorption-spectra measurement system, in the present study we have generated the second harmonic light of a Nd:YLF-laser beam synchronizing with the mid-IR FEL pluses.

2. GENERATION OF SYNCHRONOUS LASER PULSES

Since the band-gap energy of typical compound semiconductors ranges between near-infrared and visible regions, a near-infrared/visible beam is required as a probe beam to measure the optical properties near band-gap domains. The mid-IR FEL at iFEL operates with a unique pulse structure. The FEL beam has 10 Hz macro-pulse repetition rate and $\approx 15\,\mu s$ macro-pulse duration. Each macro-pulse consists of $\approx 1,300$ micro-pulses. The micro-pulse width is $\approx 5\,ps$ and the micro-pulse separation is 11.2 ns. Because of this unique double-pulse structure, the duty ratio becomes only $\approx 7 \times 10^{-8}$. The use of continuous light to probe the near band-gap properties in compound semiconductors, therefore, will cause a poor signal to noise ratio. In the present study, we have used a mode-locked Nd:YLF laser for generating near-infrared/visible pulses synchronized with the FEL pulses, which enable us to measure the optical properties of compound semiconductors with high accuracy.

The Nd:YLF laser provides picosecond pulses

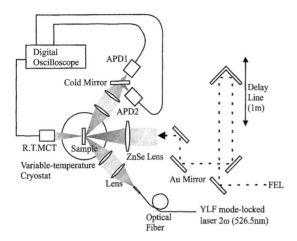

Figure 1. A schematic of the experimental setup.

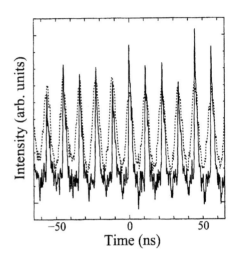

Figure 2. Transmitted mid-IR intensity (solid line) and PL intensity (dotted line).

with wavelength, $\lambda = 1.053 \, \mu m$ and a 89.25 MHz repetition rate. For the Nd:YLF laser oscillating synchronously with the FEL micro-pulses, the Nd:YLF-laser modulator is driven with an electrical signal taken from the FEL master oscillator (178.5 MHz) with a frequency divider. In order to make the macro-pulse structure, we make use of a Pockels-cell optical shutter driven with a 10 Hz FEL macro-pulse trigger signal. The trigger signal is also used for optical amplifiers and a pulse shaper in the Nd:YLF laser system. After optical amplification, the average power reaches ≈ 7 W. The output pulses are then focussed onto a BBO crystal to generate the second ($\lambda = 526.5$ nm) and third ($\lambda = 351$ nm) harmonics.

3. EXPERIMENTS

In order to confirm that the generated Nd:YLF-laser pulses synchronize with the FEL pluses, we have performed simple experiments. A schematic of the experimental setup is shown in Fig. 1. The sample used in the present study is a semi-insulating GaAs crystal, which can prevent the mid-IR induced band-gap luminescence [9]. The sample is mounted on a variable-temperature stage in the vacuum space of a cryostat. The mid-IR FEL beam with $10 \, \mu m$ wavelength was focused with a ZnSe lens onto the surface of the sample. The transmitted mid-IR signal was monitored by a MCT detector. The second harmonics of the Nd:YLF-laser beam was used to excite

electron-hole pairs across the band-gap. The resulting photoluminescence (PL) signal was then detected by an avalanche photo-diode.

An example of the detector signals is given in Fig. 2, which shows the transmitted mid-IR intensity monitored by the MCT detector (solid line) and the PL intensity detected by the avalanche photo-diode (dotted line). We see that the Nd:YLF pulses synchronize with the FEL pulses with the accuracy of the detectors' response time.

REFERENCES

1. J. Kono, Jpn. J. Appl. Phys. **41** (2002) Suppl. 41-1, 76.
2. L.V. Keldysh, Sov. Phys. JETP **20** (1965) 1307.
3. H.H. Nickle, J. Math. Phys. **7** (1966) 1497.
4. H.H. Nickle, Phys. Rev. **160** (1967) 538.
5. Y. Yacoby, Phys. Rev. **169** (1968) 610.
6. H.D. Jones and H.R. Reiss, Phys. Rev. B **16** (1977) 2466.
7. A.H. Chin, J.M. Bakker, and J. Kono, Phys. Rev. Lett. **85** (2000) 3293.
8. A.H. Chin, O.G. Calderón, and J. Kono, Phys. Rev. Lett. **86** (2001) 3292.
9. N. Mori, T. Takahashi, T. Kambayashi, H. Kubo, C. Hamaguchi, L. Eaves, C.T. Foxon, A. Patanè, and M. Henini, Physica B, **314** (2002) 431.

FREE ELECTRON LASERS 2003
E.J. Minehara, R. Hajima and M. Sawamura (Eds.)
Published by Elsevier B.V.

Design of synchronizing system of UV laser pulse and FEL micropulse for MALDI

Sachiko Yoshihashi-Suzuki*, Yasuhide Naito, Katsunori Ishii[1], Kunio Awazu

Institute of Free Electron Laser, Graduate School of Engineering, Osaka University
2-9-5 Tsuda-Yamate, Hirakata, Osaka 573-0128, Japan
[1] Graduate School of Information Science, Nara Institute of Science and Technology
8916-5 Takayama, Ikoma, Nara 630-010, Japan

Abstract

Matrix-assisted laser desorpution ionization (MALDI) is a widely used technique for mass spectrometry (MS) of biological macromolecules. We have been developing a novel MALDI method (UV/FEL MALDI) based on simultaneous irradiation of a nitrogen laser and a free electron laser (FEL). There is a difference the pulse structure between the nitrogen laser and the FEL. To achieve the optimum irradiation condition for UV/FEL-MALDI, the length of the micropulse train and the timing of the optical pulse sequence have to be controlled arbitrarily. To control the length of the micropulse train, we set up the micropulse-picking system using the acousto-optic modulator. This paper reports designs of the opto-electric system which establishes the synchronicity for UV/FEL-MALDI.

PACS codes: 82.80.Ms

Keywords: UV-FEL MALDI, Micropulse-picking, Ge-AOM

1. Introduction

Matrix-assisted laser desorption ionization [1] time-of-flight mass spectrometry (MALDI-TOF MS) is a powerful and robust tool for protein identification. MALDI has been usually carried out by an ultraviolet laser with UV susceptive matrix compound. In the ionization process of UV-MALDI, laser beam is absorbed by matrix, and electronic excitation modes are produced. Following the process toward vibrational mode from electronic excitation modes, the matrix is heated rapidly and analytes are evaporated with matrix molecules. MALDI with an infrared laser (IR-MALDI) has a potential to become an advanced analytical technique for bio-molecules, because those molecules such as protein have rich vibrational absorption bands in an infrared region; however, IR-MALDI has the disadvantage of sensitivity when compared with UV-MALDI.

We have been developing IR-MALDI using a tunable infrared free electron laser (FEL). To address the issue of sensitivity with IR-MALDI, a simultaneous irradiation of UV laser and the FEL (UV-FEL MALDI) is proposed [2]. The UV laser and the FEL can create electronically and vibrationally excited states on a sample in parallel, the result are expected to analyze macromolecules.

The typical pulse width for MALDI was reported to be 100 ps – 200 ns. The FEL has a complex pulse structure. The FEL delivers 15 μm macropulses at a pulse repetition frequency of 10 Hz, and each macropulse consists of a train of 5 ps micropulses spaced 44.8 ns apart. To achieve the optimum irradiation condition, the length of the micropulse train and the timing of the optical pulse sequence have to be controlled arbitrarily. In this paper designs

* Tel.: +81-72-897-6410; FAX: +81-72-897-6419
E-mail address: Suzuki@fel.eng.osaka-u.ac.jp

of the opto-electric system which establishes the synchronicity for UV-FEL MALDI will be discussed.

2. System

A schematic diagram of a synchronizing system of the UV laser pulse and the FEL micropulse for MALDI is shown Fig.1. A micropulse-picking apparatus was designed to switch-out a short micropulse train from an entire macropulse by a deflection of the light path using a germanium acousto-optic modulator (Ge-AOM:AGM-402A1, IntraCation CO.). Fig.2 shows the time chart of the AOM output and the trigger signal. The rise time of AOM depend on the incident beam diameter to AOM. In this system the incident beam diameter and rise time are about 500 μm and 89.2 ns, respectively, thus one FEL micropulse can be selected. The diffracted FEL micropulse are monitored by MCT photodetector (R005, Vigo System), whose output is converted to a TTL level signal by fast discriminator (N-244, SANSHIN Electric CO.), then is used to trigger a nitrogen laser (VSL-337ND, Laser Science, INC.). The trigger signal is controlled by Digital Delay/Pulse generator (DG535, Stanford Research Systems, INC.) because the fast discriminator and the nitrogen laser generate temporal delays about 28.8ns and 700ns, respectively between the trigger and the action.

3. Conclusions

We have designed the opto-electric system which establishes the synchronicity for UV-FEL MALDI. Using the micropulse-picking by Ge-AOM, the length of micropulse train is controlled arbitrarily and the irradiation timimg between the FEL micropulse and N2 laser pulse is adjusted. This system is able to clear on UV-FEL MALDI process.

4. Acknowledgments

This study was supported by a research budget of the Intellectual Cluster Project from Senri Life Science Foundation / the Ministry of Education, Culture, Sports, Science and Technology of Japan.

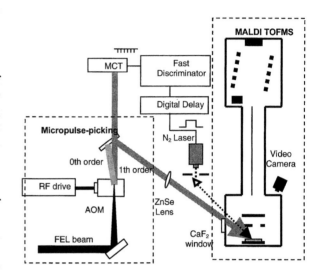

Fig. 1. Schematic diagram of a synchronizing system of the UV laser pulse and the FEL micropulse for MALDI

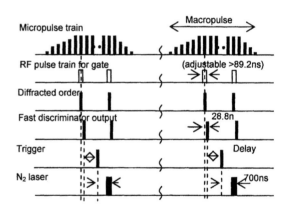

Fig. 2. Time chart of the AOM output and trigger signal.

5. References

[1] K. Tanaka, H. Waki, Y. Ido, S. Akita, Y. Yoshida, and T. Yoshida: Rapid Commun. Mass Spectrum, 2 (1988) 151.

[2] Y. Naito, and K. Awazu: Rev. Laser Engineering, 31 (2003) 16.

[3] C. W. Rella, Ph. D. dissertation, Stanford University (1997)

[4] M. Heya, Y. Fukami, and K. Awazu: J. Appl. Phys. 41 (2002) 143.

FREE ELECTRON LASERS 2003
E.J. Minehara, R. Hajima and M. Sawamura (Eds.)
Published by Elsevier B.V.

Recent status of scanning type synchrotron radiation Mössbauer microscope at the BL11XU (JAERI) of SPring-8

Takaya Mitsui[1] and Makoto Seto[1,2]

[1] Japan Atomic Energy Research Institute, Kouto 1-1-1, Mikazuki-cho, Sayo-gun, Hyogo 679-5198, Japan
[2] Research Reactor Institute, Kyoto University, Noda, Kumatori-cho, Sennan-gun, Osaka 590-0494, Japan

Abstract

A scanning synchrotron radiation Mössbauer microscope was constructed at the BL11XU of SPring-8. The focused Mössbauer probe beam is obtained by a high-resolution monochromator, a multi-layer X-ray focusing mirror and a pinhole slit. For an example experiment, we measured Mössbauer time spectra of an iron foil by a line scanning measurement.

PACS codes: 75.25.+z, 76.

Keywords: Synchrotron Radiation, Mössbauer Spectroscopy, Nuclear Resonant Scattering, X-ray Microscope, X-ray Micro Beam

1. Introduction

The third generation synchrotron radiation (SR) and future's source such as X-ray Free Electron Laser (XFEL) make enable us to perform the Mössbauer spectroscopy using a focusing X-ray. The remarkable beam properties (ultra high brilliance, pulse structure and linear polarization) hold promise to produce new tools for the solid material physics researches not accessible in "classical" Mössbauer spectroscopy with radioactive source. Recently, we have developed a scanning type synchrotron radiation Mössbauer microscope (SSRMM) for ^{57}Fe nucleus using X-ray focusing beam. In this paper, we report on the current status of SSRMM at the BL11XU of SPring-8.

2. Optics of SSRMM

As is shown in Fig.1, SSRMM system is composed of an undulator source, a C(111) pre-monochromator, a high-resolution monochromator (HRM), a multi-layer X-ray focusing mirror (MXFM), a pinhole slit, a precision stage and a fast detector (APD).

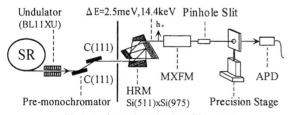

Fig.1. Optics and mechanics of SSRMM.

The incident X-rays are monochromatized at 14.4keV with ΔE=2.5meV by a pre-monochromator [1] and a high-resolution monochromator [2]. The parameters of source and optics are summarized in Table1. After HRM, the beam size is 0.3x1.8mm^2 and the flux is 2.4x10^9 photons/sec at storage ring current of 98mA.

Table 1
Source characteristics and optics parameters

1) Source characteristics		
Type: In-vacuum Undulator		
Beam energy: 8GeV Length:4.5m (Period length:32mm)		
Beam-size / divergence		
σ x: 376 μ m/ σ x': 15.7 μ rad σ y: 32.5 μ m/ σ y': 3.62 μ rad		
2) Beamline position (From Source)		
FE slit (Size:1mmx1mm)	: 28.92m	
Diamond pre-monochromator	: 38.66m	
3) High resolution monochromator for 14.4keV X-ray		
Type : 2-nested channel cuts Si 511xSi 975		
Energy resolution	: ΔE=2.5meV	
Channel-cut crystals	Si(511)	Si(975)
Bragg Angle	24.3°	80.4°
Asymmetric factor	-1/16.8	-4.0
Accept angle of Si(511)	3.2"	

The multi-layer X-ray focusing mirror (MXFM) coated with 50 layers of [W(13Å) /Si(39.5Å)] on SiO$_2$ base is placed after HRM. The parameters are summarized in Table 2. As the result, incident X-rays are focused with a size of 300x35μm^2 at the sample position. A point-like focused Mössbauer probe beam (FMPB) is obtained by using a φ20μm pinhole slit.

Table 2
Parameters of multi-layer focusing mirror

Multilayer focusing mirror for 14.4keV X-ray	
Mirror position : 49.4m To focus point : 600mm	
Focusing type : Bend elliptical mirror (Sagittal focus)	
Energy : 14.4keV Glancing angle : 8.6mrad	
Mirror size : L450mm x H50mm x T10mm	
Bend support length : 400mm	
[Upper steam : 1.164Nm(max) , Down stream : 3.57Nm(max)]	
Mirror material : SiO_2 (Young's modulus:73GPa)	
Coating material : W/Si [W:13Å/Si:39.5Å]	
Layers numbers : 50 Reflectivity : R > 75%	

3. Precision sample stage

Precision sample stage consists of a coarse moving stage and a piezoelectric transducer driven fine moving stage, whose minimum spatial resolution is 10nm. The characteristics are summarized in Table 3.

Table 3
Characteristics of precision sample stage

1) Coarse moving stage	
X : CTS-100X (Sigma Koki)	
Minimum resolution	: < 0.05mm
Travel range	: ±50mm
Z : STM-20ZF (Sigma Koki)	
Minimum resolution	: < 0.05mm
Travel range	: ±10mm
2) Fine moving stage	
x : P-AES-60X(IS)-20 (Sigma Koki)	
z : P-AES-60Z(IS)-20 (Sigma Koki)	
Table dimensions	: 60mm x 60mm
Minimum resolution	: 10nm
Travel range	: ±10mm

4. Beam size evaluation

The beam size of FMPB was evaluated by knife-edge method. As is sown in Fig. 2(a) and 2(b), it was 16μm in vertical direction and 20μm in horizontal direction. Then, the typical photon flux after pinhole slit was about 3.6×10^8 photons/sec at storage ring current of 98mA.

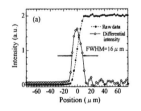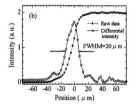

Fig. 2. Beam profiles of FMPB measured by knife-edge method.
(a) Vertical, (b) Horizontal.

5. Example experiment of SSRMM

The advantages of SSRMM are seen with special clarity in the research of the sample whose hyperfine fields don't have a uniform distribution. As an example, in the absence of an external magnetic field, the time spectra of a 2μm thick α-Fe foil enriched to 95% ^{57}Fe were measured with a line scanning of the sample. In H_{ex}=0Oe, α-Fe foil is divided into multi magnetic domains and the typical domain size is below one hundred microns. The irradiation position of FMPB was changed with step of 60μm. As is shown in Fig.3, the measured time spectra revealed some different quantum beat patterns in accordance with the irradiation positions of the sample "clearly".

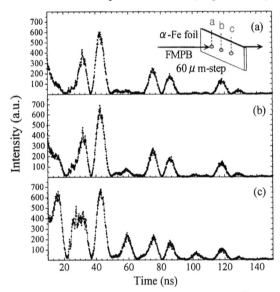

Fig. 3. Measured time spectra at local positions of an ^{57}Fe foil.

6. Summary

We have designed and constructed SSRMM system at the BL11XU of SPring-8. We expect that SSRMM will be a powerful tool for the scientific research fields as follows.
1: Evaluation of Characteristics (Defects etc.)
2: The study of rust process, alloy, polymer science.
3: High pressure physics using diamond anvil cell.
These researches will be started in near future.

References

[1] M.Marushita et al., Nucl. Instr. and Meth. A. **467-468**(2001)392.
[2] T. Mitsui et al., Nucl. Instr. and Meth. A. **467-468**(2001)1105.

FREE ELECTRON LASERS 2003
E.J. Minehara, R. Hajima and M. Sawamura (Eds.)
Published by Elsevier B.V.

Fragmentation pathways caused by soft X-ray irradiation: The detection of desorption products using a rotatable time-of-flight mass-spectrometer combined with pulsed synchrotron radiation

Tetsuhiro Sekiguchi*, Yuji Baba, Iwao Shimoyama, Krishna G. Nath

Synchrotron Radiation Research Center, Japan Atomic Energy Research Institute, Tokai-mura, Naka-gun, Ibaraki, 319-1195, Japan

Abstract

High flux and pulsed photons from FEL are very fascinating for the study of photochemistry. In advance of future application of X-ray-FEL, we have investigated the mechanism of the bond-selective fragmentation occurring at the surface of fluorinated graphite following F 1s core-excitation using time-of-flight mass-spectroscopy (TOF-MS) combined with pulsed synchrotron radiation. Excitation-energy dependence of F^+ fragment yields revealed that C-F bonds are selectively broken by F 1s resonant excitation. Furthermore, translation-energy-resolved excitation spectra of F^+-yields were measured using TOF technique. Slow fragments were found to be involved in the specific dissociation. The result was explained assuming that Auger decay followed by F 1s $\to$ σ^*_{C-F} resonance leaves a long-lived excited electron at σ^*_{C-F} orbital in precursor states, which should be highly antibonding and promote breaking C-F bonds.

PACS codes: 32.80.Hd, 68.43.Tj, 82.50.Kx

Keywords: Desorption induced by electronic transition, Translation energy distribution, Fluorinated graphite, Inner-shell excitation

1. Introduction

Pulsed capability, high photon flux, and high repetition rate of free electron laser (FEL) are powerful advantages for the study of photochemistry. These capabilities enable one to use time-of-flight mass-spectroscopy (TOF-MS), which not only is very sensitive for the detection of photochemical products, but also affords insight into reaction dynamics such as translation energies of products. Although FEL has found wide application, very few examples have been reported on the TOF detection combined with FEL [1]. In advance of the future application of X-ray-FEL, we report on the surface-photochemistry of inorganic materials caused by pulsed synchrotron radiation (SR) using the TOF technique.

Due to intra-atomic nature of inner-shell excitation, chemical bonds can be broken in the vicinity of excited atoms following core-excitation. Although

this phenomenon is recognized, its mechanism has not yet been fully understood. To elucidate the dynamics of such bond-specific dissociation, the present study aims to observe the translation-energy distributions using the TOF-MS technique.

2. Experimental

The experiments were performed at the 13C station in Photon Factory. The dependence of fragment-ion yields on photon incidence angles were measured using a rotatable time-of-flight mass-spectrometer (TOF-MS) that was developed in our group. TOF-MS measurements were done using pulsed soft X-rays during the single-bunched operation of the electron storage ring. Polarized NEXAFS was measured by total electron yield (TEY) method. Fluorinated graphite was prepared by deposition of low energy F^+ ions into highly oriented pyrolitic

---
*Corresponding author. Fax: +81-29-282-5832.
E-mail address: sekiguch@popsvr.tokai..jaeri.go.jp

graphite (HOPG).

3. Result and discussion

Figure 1 compares TOF spectra among three excitation energies: 384 eV and 638 eV photons do not excite F 1s inner-shell level, while 756 eV does. Peaks in the spectra were reasonably assigned to fragments including F^+, F^{2+}, and H^+. No carbon-containing species were detected. To compare F^+ yields with each other, spectra were normalized by H^+ yields. We can find that F^+ yields above the F 1s ionization threshold are roughly ten times larger than that below the threshold. This indicates that C-F bonds are selectively broken by F 1s core-excitation.

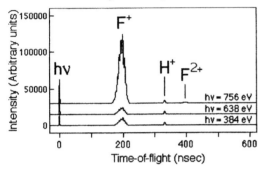

Fig. 1. (a) TOF spectrum of desorbed ions measured for 5×10^{15} F^+-impinged graphite excited by 756 eV photons; (b) TOF spectra obtained for the photon energies of 384, 638, and 756 eV. Symbol "hν" represents photon-scattering signals, which gives the starting time.

Peak shapes of F^+ ions were found to be slightly changed depending on excitation energies near the F 1s threshold. It is well established that flight time of desorbing particles is a function of initial kinetic energies. Thus this observation suggests that the translation-energy distributions of ion products depend on reaction dynamics such as excited states or precursors. To clarify what excited states contribute most to the site-specific fragmentation, we measured the photon-energy dependencies of F^+ yields with various translation energies. As shown in Fig. 2(a), F^+ peak in TOF spectra was separated to six parts. Fig. 2(b) shows the photon-energy dependencies for each component. The slowest component shows the enhanced fragment yields at σ^*_{C-F} resonance, compared with X-ray photoabsorption curve denoted as "TEY". This means that slow components are involved in the selective bond-breaking. Generally, Auger process follows core-excitation and after it

multiply charged states are produced. Translation energy is assumed to be accumulated through Coulombic force between fragment ions in the course of dissociation of such multi-charged precursors. If excited electrons remain after Auger decay, they should shield the Coulombic fields between charged particles. Taking these facts into account, we concluded that resonant electrons at σ^*_{C-F} orbital, which must be strongly antibonding, remain during the bond scission and that they enhance the efficiency of breaking C-F bonds.

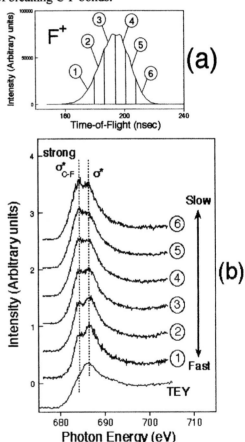

Fig.2 (a) Expanded F^+ peak in TOF spectrum obtained in 756 eV photons. (b) Photon energy dependencies of the yields of F^+ ions with various translation energies. Symbols "①～⑥" represents each component: ① is the fastest component and ⑥ is the slowest.

4. References

[1] J.T. McKinlev et al., AIP Conf. Proc. No.288 (1993) p.70.

FREE ELECTRON LASERS 2003
E.J. Minehara, R. Hajima and M. Sawamura (Eds.)
Published by Elsevier B.V.

A Study on Decomposition of Dioxin Analogues by FEL

Toshihiko Yamauchi*, Eisuke Minehara* and Shinichi Itoh[+]

*Japan Atomic Energy Research Institute, Tokai-mura, Naka-gun, Ibaraki 319-1195, Japan
+Kyouto Biseibutu Kennkyuusyo, Kubo-machi, Kamikazan,Yamashinaku, Kyoto-shi 607-8464, Japan

Abstract

Decomposition of dioxin analogues by infrared (IR) laser irradiation occurs by ways of thermal destruction and multiple-photon dissociation. It is necessary for the decomposition to choose the laser wavelength which is strongly adsorbed.

The thermal decomposition takes place by the irradiation of the low IR laser power. Considering the model of thermal decomposition, it is proposed that the adjacent water molecules assist the decomposition of dioxin analogues.

PACS codes: 33.80.-b
Keywords: dioxin, IR laser, PCDDs, PCDFs, PCB, decomposition, gas-chromatograph, FEL

1. Introduction

The generation of dioxin becomes large problem as a hostile environment material. Measurement and decomposition for controlling the generation of this dioxin become an important research theme nowdays.

The way to detoxify are as follows; 1.incineration, 2.chemical treatment, 3.electron beam irradiation, 4.biological treatment, 5.photolysis, 6.others.

It has been studied to find the condition of compatibility of the decomposition by the wide infrared wavelength range of FEL (free electron laser). At the same time, the possibility of the photolysis using the wavelength 10.6 μm of CO_2 laser was also studied.

In this study, the wavelengths of 7 μm, 10.6 μm, 22 μm and 25 μm were chosen. And, OCDD (octachloro dibenzo-para-dioxin) and OCDF (octachlorodibenzofuran) in which toxicity equivalency quantity (TEQ) was the lowest were mainly chosen, as well as PCB.

2. Decomposition reaction of Dioxins

One of the decomposition of dioxin by IR laser is the dechlorination, which becomes the free radical after removing the chlorine [1]. The reactions are shown in eqs.(1)-(8) below, where the first symbol in eq.(1) is the dioxin, q and q' are the number of photon, qhν (ν is the laser frequency) is the laser energy. The eq.(1) shows the dechlorination by the absorption of laser, and eq.(2) shows the dissociation of

Photophysical Process of Dioxin Decomposition

$$\text{\textcircled{O}Cl} + h\nu \longrightarrow \text{\textcircled{O}}_{\bullet} + {}_{\bullet}Cl \quad (1)$$

$$H_2O + h\nu \longrightarrow {}_{\bullet}H + {}_{\bullet}OH \quad (2)$$

$$\text{\textcircled{O}}_{\bullet} + {}_{\bullet}Cl \longrightarrow \text{\textcircled{O}}Cl \quad (3)$$

$$\text{\textcircled{O}}_{\bullet} + {}_{\bullet}OH \longrightarrow \text{\textcircled{O}}OH \quad (4)$$

$$\text{\textcircled{O}}_{\bullet} + {}_{\bullet}H \longrightarrow \text{\textcircled{O}}H \quad (5)$$

$$_{\bullet}H + {}_{\bullet}Cl \longrightarrow \underline{HCl} \quad (6)$$

$$_{\bullet}Cl + {}_{\bullet}Cl \longrightarrow \underline{Cl_2} \quad (7)$$

$$\text{\textcircled{O}}Cl + {}_{\bullet}H \longrightarrow \text{\textcircled{O}}H + {}_{\bullet}Cl \quad (8)$$

surrounding water on dioxins (dissociation efficiency: 11%). The free radicals of

dioxins combine with the atoms of chlorine, or hydrogen, or hydroxyl radical (eqs.(4)-(6)).

The other strong reaction with high power laser is the combustion reaction, which produces CO_2 gas combined with Oxygen in air, etc.

3. Experimental arrangement

The laser beam was focused on the dioxin sample by the lens and the focusing radius was ~50 μm. The FEL laser of 600 μs pulse duration (macropulse) with the repetition frequency 10 Hz is composed of the pulse train of subpicosecond-20 ps (micropulse) with the repetition rate 10.4 MHz. The laser power of the macropulse was ~0.1 kW, and the power of the micropulse was 1~10 MW. CO_2 laser power was 10 W, and the pulse width was microseconds with the repetition rate of kilo-hertz.

Corresponding author. Tel. +81 292825285, fax +81 292826057, e-mail: ytmauchi@popsvr.tokai.jaeri.go.jp

4. Experimental result

The strong absorption spectra of dioxin OCDD lie at wavenumbers 999 cm^{-1} and 1425 cm^{-1} from the Fourier transform infrared (FTIR) absorption spectrum [2]. Two samples (first one was irradiated and second was not irradiated) were made using the liquid of dioxin dissolved in toluene.

The experimental decomposition efficiency (dioxin quantity of (without-with irradiation) x100/dioxin without irradiation) was 90% as described below. In the first place, the wavelength 10.6 μm of CO_2 laser (pulse width: 80μs, laser power: 4W, 50 minutes' irradiation) was irradiated on the octachloride dioxin, and the high decomposition efficiency 90% was measured by the gaschromatograph. Next, the wavelengths 22 μm and 25 μm of FEL were irradiated on it, and the decomposition could not be clearly observed. This experimental result on the wavelength showed to be connected with the absorption spectrum [3].

The dioxin sample mixed with TCDDs~OCDD (TCDDs, PCDDs, HpCCDs, HxCDDs and OCDD), TCDFs~OCDF (TCDFs, PCDFs, HpCDFs, HxCDFs and OCDF) and PCB sample were used for CO_2 laser irradiation. In our experiment, the decomposition efficiency of TCDD is higher than OCDD, and the decomposition efficiency of polychlorinated dibenzofurans (PCDF) was larger than polychlorinated dibenzo-p-dioxins (PCDD).

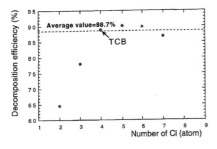

Fig. 1 Decomposition efficiency of PCB by the wavelength 10.6 μm

The decomposition efficiency of PCB is 88.7% in Fig.1 [4].

The strong absorbed wavelength's laser at the wavelength 7.0 μm for TCDDs~OCDD and TCDFs~OCDF was irradiated them. The laser power

was 13 W. The decomposition efficiency

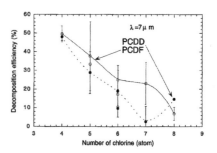

Fig.2 Decomposition efficiency of PCDD and PCDF by the wavelength 7 μm

was ~50 % at TCDDs and TCDFs, and depended on the weight of dioxin molecular as shown in Fig.2. The decomposition efficiency of polychlorinated dibenzofurans was about 24% higher than polychlorinated dibenzo-p-dioxins as same as the above result. The decomposed dioxin did not become the original 4~7 chloride dioxin again, or it decomposed into other chemical substances or the 1~3 chloride dioxin, and lastly the noxious dioxin disappeared.

5. Summary

The decomposition experiment of the dioxin was studied using four infrared wavelengths of FEL and CO_2 laser. The followings were clarified:

1. By the irradiation of the wavelengths 7.0 μm and 10.6 μm which were the strong and weak absorption lines respectively of TCDD~OCDD and TCDF~OCDF, the decomposition efficiency depended on the molecular weight.
2. The resolved dioxin did not become the original 4~7 chloride dioxin again, or it decomposed into other chemical substances or the 1~3 chloride dioxin, and the noxious dioxin disappeared.
3. The hydrogen atom and hydroxyl radical dissociated from water by laser irradiation play an important role for the combination with the free radical of dioxins.

References

[1]T. Yamauchi: Soc. of Environ. Sci., **14** (2001) 567.
[2]S. Sommer et al.: Analy. Chem., **69** (1997) 1113.
[3]T. Yamauchi et al.: Soc. of Environ. Sci., **13** (2000) 383.
[4]T. Yamauchi et al.: Soc. of Environ. Sci., **14** (2001) 73.

FREE ELECTRON LASERS 2003
E.J. Minehara, R. Hajima and M. Sawamura (Eds.)
Published by Elsevier B.V.

Nuclear Isomer Research with FEL-Compton Scattering Photons

T. Shizuma[a]*, T. Hayakawa[a]

[a]Japan Atomic Energy Research Institute, Tokai, Ibaraki, 319-1195 Japan

We describe our research plan for nuclear isomer physics at the Advanced Photon Research Center in the Japan Atomic Energy Research Institute. Photons generated by a superconducting wiggler at SPring-8 and by a laser-Compton scattering with FEL will be used for excitation or deexcitation of the isomers.

1. Introduction

Comparatively long-lived excited states in nuclei are called "nuclear isomers" [1], arbitrarily defined as states having half-lives greater than 1 ns. When an isomer decays, the stored energy is usually released by emission of α, β or γ rays. Gamma-ray transitions, which occur within the same nucleus, usually have transition strength larger than the α or β decay. However, if the γ-ray half-life is long, the α- or β-transition can proceed in competition with the γ decay. This is commonly seen at the lower excitation energy. The spin and parity selection rule governs the decays for this type of isomers (spin isomer). In case that the transitions involve a larger spin difference and/or parity change, the isomer half-life increases, depending on the branching ratios (including the internal conversion) and the energies of transitions decaying out of the isomer. For axially deformed nuclei, an additional selection for the K quantum number, defined as the projection of the total angular momentum onto the nuclear symmetry axis, is applied to a transition which involves a K change. This is known as the K selection rule, and the associated isomer is called K isomer. Another type of isomer known as a fission isomer arises due to a large difference in the shapes of the initial and final states, therefore called a shape isomer. The fission isomer is trapped in the second energy minimum with a large elongated shape in the multi-dimensional nuclear shape coordinates. Due to the small overlap of wave functions between the states at the first and second minima, the corresponding γ-decay strength is weak compared to the fission into two fragments.

In this report, we describe the experiments of the photo-induced excitation and deexcitation of the ^{176}Lu and ^{180}Ta isomers. Figure 1 shows processes of the photo-induced excitation or deexcitation of the isomers through intermediate K-mixing states.

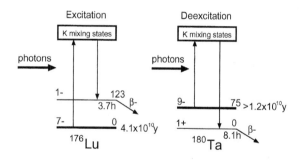

Figure 1. Processes of the excitation [deexcitation] of ^{176}Lu [^{180}Ta] isomers through intermediate K-mixing states.

2. Photon sources

Since the cross sections for the excitation or deexcitation of the isomers in (γ, γ') reactions at incident photon energy of about 1 MeV are expected to be very small, intense photon beams

*E-mail: Shizuma@popsvr.tokai.jaeri.go.jp

are required. One such photon beam can be realized by high-energy synchrotron radiation from a three-pole 10-T superconducting wiggler at SPring-8 [2]. Another possibility is to use a relativistic Compton scattering photons using a free electron laser. In the JAERI-FEL, a stable kW-level lasing has been achieved [3]. This kind of FEL might be used to generate intense γ rays in terms of laser-Compton scattering. The characteristics of FEL; high intensity, tunable wavelength and sharp line width may benefit the nuclear isomer research.

3. Excitation or deexcitation of the isomers

The ^{180}Ta nucleus is famous for two aspects that it is the only naturally occurring isomer and is the nature's rarest isotope. The stellar production of ^{180m}Ta has been a challenging astrophysical problem, since the production process of this isotope is still unknown. As shown in fig. 2, the production of ^{180}Ta is bypassed by the slow neutron-capture (s-)process, and furthermore shielded from the β-decay chains following the rapid neutron-capture (r-)process. Possible

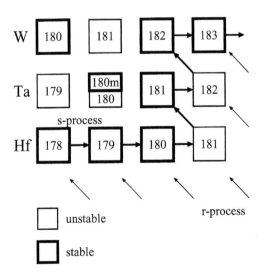

Figure 2. The reaction path of the s- and r-process in the region of the tantalum isotopes.

ways to reach ^{180m}Ta are discussed in term of the γ-process path [4] and the s-process path [5]. In the latter case, ^{180m}Ta may be destroyed in (γ,γ') reactions under the stellar environment such as AGB stars. The cross section measurement for depopulation of ^{180m}Ta with photon beams energies less than 1 MeV could be important.

The nucleus ^{176}Lu has also received much attention since its ground state is long-lived radioactive (see fig. 1) that is comparable to the age of universe. Due to the appropriate half-life of 4.1×10^{10} years, it was long considered to be useful for a chronometer for the s-process [6]. However, the temperature dependence of the effective half-life ruled out the use of ^{176}Lu as an s-process chronometer, but appears to be useful for a potential s-process thermometer [7]. In this case, the detailed knowledge on the coupling between the ground state and the isomeric state in ^{176}Lu (see fig. 1) is necessary. Therefore, the measurement of the excitation cross section of the ^{176}Lu isomer by photons is of great interest.

4. Summary

In summary, we described the photo-induced population and depopulation of the isomers in ^{176}Lu and ^{180}Ta. Since their cross sections are very small, intense photon beams with $E_\gamma \approx 1$ MeV region are required. Photons generated by a superconducting wiggler at SPring-8 and by a laser-Compton scattering with FEL would be used for these isomer experiments.

REFERENCES

1. P.M. Walker et al., Nature 399 (1999) 35.
2. K. Soutome et al., in Proceedings on PAC2003, Portland (2003).
3. E.J. Minehara et al., Nucl. Instr. and Meth. A445 (2000) 183.
4. D.L. Lambert, Astron. Astrophys. Rev. 3 (1992) 201.
5. D. Belic et al., Phys. Rev. Lett. 83 (1999) 5242.
6. J. Audouze et al., Nature 238 (1972) 8.
7. K.T. Lesko et al., Phys. Rev. C 44 (1991) 2850.

FREE ELECTRON LASERS 2003
E.J. Minehara, R. Hajima and M. Sawamura (Eds.)
Published by Elsevier B.V.

Modification of Hydroxyapatite Crystallization Using IR Laser

Weimin Guan*[†], Nobuya Hayashi[†], Satoshi Ihara[†], Saburo Satoh[†], Chobei Yamabe[†]

Masaaki Goto[‡], Yoshimasa Yamaguchi[‡], Atsushi Danjoh[‡]

[†]Department of Electrical and Electronic Engineering, Faculty of Science and Engineering,
Saga University, 1 Honjo-machi, Saga 840-8502, Japan
[‡] Department of Oral and Maxillofacial Surgery, Faculty of Medicine,
Saga University, 5-1-1 Nabeshima , Saga 849-8501, Japan

Abstract

In this study we used laser to remineralize the dentine surface. Experiments have been concentrated to investigate the modification of HAp (Hydroxyapatite) crystal structure using IR laser. The wavelength of incident laser was selected around 10.6μm that matches the characteristic absorption wavelength region of hydroxyapatite. As a preliminary study for FEL (Free Electron Laser) application in biological field, the results we have got indicated that IR laser irradiation (combined with chemical films) offering control over surface modification of human teeth. These results also demonstrate the capability of precisely deliver energy over a wide band of IR wavelength makes FEL as an attractive tool for further ameliorating of biologic HAp crystal.

PACS codes: 42.62.Be

Keywords: Hydroxyapatite; Crystallization; Photoablation; Remineralization; X-ray Diffraction; Tooth dentine and enamel

1. Introduction

To remineralize the dentine surface viz. improves the microhardness and acid-resistance of the surface of teeth, laser application was concerned in this study.

Sound human molar tooth were irradiated with a 10.6μm CO_2 laser under different physical and chemical conditions. Two series of samples covered with different saturated chemical solution coatings during irradiation were investigated. Comparing with the cases that without chemical coating, distinct XRD results were observed.

2. Experiment

The specific experimental conditions are as follows:

(1) Low power high repetition (abundant dosage)
(2) High power low repetition (less dosage).

(3) Irradiation with saturated chemical solution films.

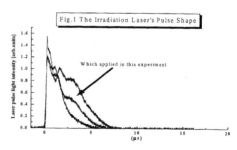

Fig.1 The Irradiation Laser's Pulse Shape

Which applied in this experiment

2.1 The parameter of incident laser

Ablation was performed with 10.6 μm wavelength laser beam, corresponding to the dominant wavelength of a TEA-CO_2 laser. It emits 5 μm duration pulse at a repetition rate of 0.2Hz, with mean energy range from 86mJ to 622mJ. The laser

*Corresponding author. E-mail: 02tj02@edu.cc.saga-u.ac.jp

beam was focused on the target perpendicularly with laser fluence varying from 23.4J/cm^2 to 169J/cm^2. The average power of laser was ranging from 18mW to 130mW. While as the range of power density was from 4.89W/cm^2 to 35.23W/cm^2.

The above figure shows the pulse shape that can be modified and the incident pulse we used in this experiment.

2.2 The XRD analysis

The X-ray diffraction of the profile breadth and relative intensity data was collected from the samples under control on a manual diffractometer using Cu-Kα radiation. The divergence slit was 1° and the receiving slit was 0.3mm. The scan range for relative intensities was from 20° to 120° (2θ), scan rate was 4°/min.

3. Results and Discussion

3.1 Irradiation Without Chemical Solution Film

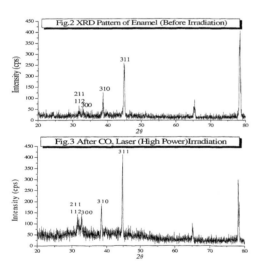

Fig.2 XRD Pattern of Enamel (Before Irradiation)

Fig.3 After CO$_2$ Laser (High Power) Irradiation

After the laser irradiation, the reflection (002) was increased significantly. That indicates the crystal growth in c-axis on the surface of teeth sample. The main peak cluster of HAp (211),(112),(300) was emerged while the reflections (310),(311) were also increased significantly.

The increasing of average reflection magnitude implying that amorphous process occurs due to the thermal effect during laser ablation.

3.2 Irradiation With Chemical Solution Films

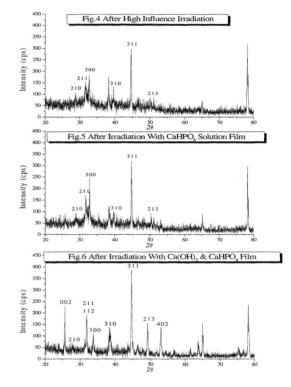

Fig.4 After High Influence Irradiation

Fig.5 After Irradiation With CaHPO$_4$ Solution Film

Fig.6 After Irradiation With Ca(OH)$_2$ & CaHPO$_4$ Film

During the photoablation process, laser power performs superiorly than the energy fluence. Nevertheless, the performance of the power of incidence laser is inferior, comparing with the effect of chemical film.

The irradiation with CaHPO$_4$ saturation solution film increased most of the reflection peaks except for (310), and the pattern of tricalcium phosphate came out. That should be the product of chemical reaction during laser ablation. We believe that the presence of CaHPO$_4$ could be a trace of the original reaction that produces HAp.

For the case of Ca (OH)$_2$ combined CaHPO$_4$ solution, superior remineralization occurred. We inferred that the hydroxyl contained in the solution enhanced the pH value of the surface of sample, so that the surface condition was modified more appropriate for the calcium ion to bond with apatite matrix.

4. Reference: Seiji Ogino, IONICS 1997; 23: 81-87

FREE ELECTRON LASERS 2003
E.J. Minehara, R. Hajima and M. Sawamura (Eds.)
Published by Elsevier B.V.

Linewidth Narrowing and Etalon Fabry-Perot experiment at ELETTRA

D. Garzella[ab*], M. Trovò[c], C. Bruni[b], G. De Ninno[c], B. Diviacco[c], G. L. Orlandi[ab], M. Marsi[c] and M.E. Couprie[ab]

[a]Commissariat à l'Energie Atomique, DSM/DRECAM/SPAM Centre de Saclay, 91191 Gif sur Yvette, FRANCE

[b]LURE, Bât 209D, Centre Universitaire Paris SUD B.P. 34, 91898 Orsay CEDEX, FRANCE

[c]Sincrotrone Trieste S.c.P.A., Basovizza, Trieste, ITALY

The spectral features of a Free Electron Laser oscillator can be easily estimated from the gain spectral distribution and the gain over losses ratio. In Storage Ring Free Electron Lasers, the use of an Optical Klystron and the relatively low gain of the system limit the linewidth narrowing. The use of an intracavity Fabry-Perot Etalon allows to reduce artificially the linewidth by selecting a limited number of longitudinal modes of the cavity, provided the laser systems have enough gain. The main features of an experiment aiming to insert an Etalon in the optical cavity of the European FEL ELETTRA (EUFELE), its state-of-the-art and some preliminary results, are here reported.

1. Introduction

Free Electron Laser (FEL) oscillators are pulsed high coherent and tunable sources, showing quite large linewidths ($10^{-4} - 10^{-5}$). Thus, it turns out that the number of lasing longitudinal modes is quite high, because of the long optical cavity of SRFELs. The longitudinal modes spacing, called the Free Spectral Range (FSR), is given by $FSR = \frac{\lambda^2}{2L}$ where λ is the laser wavelength and L is the optical cavity length. The spectral linewidth can be reduced inserting a Fabry-Perot Etalon in the cavity, allowing the selection of a limited number of longitudinal modes. Such an experiment, already successfully demonstrated in the visible range at VEPP3 in the early 90's [1], is going to be performed at the European FEL ELETTRA [2]. The main purposes of the experiment are: validation of SRFEL theories and codes, characterization of the SRFEL dynamics in a complete and totally controlled Fourier transformed laser pulse and the possible exploitation of the new spectral features for spectroscopy in the UV range, providing a high spectral resolution for FEL users. To achieve these goals, a

new high vacuum vessel, integrating a motorized plate holder, has been manufactured to be put in the optical cavity at ELETTRA. Here we aim to present the state-of-the-art of the experiment and some preliminary results.

2. Fabry-Perot Etalon Theory

At normal incidence, the total transmission of a Fabry-Perot Etalon is given by $T = 1/[1+(4R/1-R^2)\sin^2(4\pi nd/\lambda)]$ where R is the mirrors reflectivity, n is the real part of the refraction index and d is the etalon thickness. The use of slightly tilted Etalons allows to control in the most precise way the wavelength tuning [3]. However, tilted Etalons ask also for accurate compensation of the cavity mismatch and introduce additional losses to the cavity [4]. Thus, for SRFELS, a theory based on supermodes predicts that laser linewidth narrowing without substantial power losses can be achieved, expressed as [1]:

$$\frac{\Delta\lambda}{\lambda} = 0.375 \sqrt{\frac{\lambda^2}{4nd\sigma_b}} \sqrt[4]{\frac{G_0}{R}} \qquad (1)$$

where σ_b is the electron bunch length and G_0 is the initial small signal small gain.

*david.garzella@lure.u-psud.fr

Figure 1. 3D layout of the high-vacuum vessel with the motorized movements and the flipping system of the Etalon plate

3. Experimental Layout

The Etalon plate is placed in a ultrahigh vacuum vessel (see fig. 1), kept in the 10^{-9} $mbar$ range. The vessel includes a plate holder, designed and manufactured to operate with an outer motorized system, aiming to control the two degrees of freedom of plate rotation with respect to the plate center. Preliminary tests indicate that the control accuracy is better than 500 μrad. Moreover, a flipping holder system will allow to put or remove the plate without breaking the vacuum. The plate will be initially a 3 mm thick silica one, with very low absorption losses and a high surface quality (RMS roughness $< 1\mathring{A}$).

The experiment will be held in fall 2003. The vessel will be installed in the back of the cavity (upstream mirror) in order to prevent the etalon plate from optical degradation induced by the incoming higher orders of the spontaneous emission. We will start at a wavelength of $\lambda = 350$ nm (the main characteristics of the EU-FELE experiment are reported elsewhere [2]).The expected value for $\Delta\lambda/\lambda$ from eq. 1 with the Etalon plate characteristics, given above, ranges between 6.5 10^{-6} and 1.6 10^{-5}, following σ_b.

Moreover, a detailed study for developing numerical tools aiming to modelize the insertion of an intra-cavity element are currently taking place within the European EUFELE collaboration. Up to date, a generalized Airy Function has been implemented in a pass-to-pass Storage Ring FEL code [5]. Preliminary tests performed with the operating ELETTRA parameters show an estimated FEL linewidth without Fabry-Perot Etalon insertion $\Delta\lambda/\lambda = 4.7$ 10^{-4}, in a very good agreement with experiment [2]. Further calculations performed by inserting the Etalon plate give $\Delta\lambda/\lambda = 3.3$ 10^{-6}. This latter is quite far from the value given by eq.1, thus calling for further analysis and experimental verification. However, it should be noted that the numerical calculations give also a value for the FSR which is in total agreement (0.014 nm) with the Fabry-Perot theory. This experiment will then allow to validate the theories on the spectral behaviour of SRFELs.

4. Acnowledgments

The support given by EUFELE, a Project funded by the European Commission under FP5 Contract No. HPRI-CT-2001-50025, is acknowledged

REFERENCES

1. M.E. Couprie et al., *Nucl. Inst. Meth.* **A304** (1991) 47-52.

2. G. De Ninno et al.,*Nucl. Inst. Meth.* **A507** (2003) 274-280.

3. M. Hercher *Appl. Optics* **Vol. 8** No. 6 (1969) 1103-1106.

4. W.R. Leeb, *Appl. Physics* **Vol. 6** (1975) 267-272.

5. T. Hara et al.,*Nucl. Inst. Meth.* **A375** (1995) 67-70.

FREE ELECTRON LASERS 2003
E.J. Minehara, R. Hajima and M. Sawamura (Eds.)
Published by Elsevier B.V.

Developments on the EUFELE* project at ELETTRA

B. Diviacco[a,†], M. Trovò[a], G. De Ninno[a], M. Marsi[a], M. Danailov[a],
F. Iazzourene[a], E. Karantzoulis[a], L. Tosi[a], S. Günster[b], A. Gatto[c],
M. E. Couprie[d], D. Garzella[d], F. Sirotti[e], H. Cruguel[e], G. Dattoli[f], L. Giannessi[f],
F. Sarto[f], M. Poole[g], J. Clarke[g], E. Seddon[g]

[a] Sincrotrone Trieste, , Italy
[b] Lazer Zentrum Hannover, Hannover, Germany
[c] Fraunhofer Institut für Angewandte Optik und Feinmechanik, Jena, Germany
[d] , CEA/LURE, Orsay, France
[e] CNRS/LURE, Orsay, France
[f] ENEA, Frascati, Italy
[g] CCLRC Daresbury Laboratory, United Kingdom

Abstract

The European Free-Electron Laser at ELETTRA (EUFELE) is a project funded by the European Community aimed at the development and initial exploitation of a FEL radiation facility in the vacuum ultra-violet spectral region. We present here the status of the project and describe the most recent technical developments.

PACS codes: 29.20 Dh; 41.60 Cr

Keywords: Storage ring, Free-electron laser, Ultra-Violet , Multilayer

1. Introduction

The main features of the ELETTRA Storage Ring FEL have been described in previous papers [1]. It provides intense, monochromatic ($\Delta\lambda/\lambda \sim 10^{-4}$) and tunable radiation in the wavelength range between 190 and 350 nm. The short emitted pulses, the high repetition rate and the natural synchronization with synchrotron radiation make it attractive for different experimental techniques, particularly in the time resolved domain. However, to become truly competitive with conventional lasers, the source properties must still be improved in terms of intensity stability and monochromaticity. Reaching even shorter wavelengths would result in a source with unique characteristics in the important VUV spectral range. The goal of the EUFELE project is to build a FEL user' facility, implementing all the developments necessary to perform selected experiments that will demonstrate the suitability of this radiation source for practical applications. In this context, the most significant development is the recent implementation of the so-called Q-switched operation mode, providing a regularly pulsed output particularly suited to the above mentioned techniques. Given the importance of this topic, it has been presented in detail in a separate paper [2].

2. New instrumentation and diagnostics

Efforts have been made to complement the FEL with suitable diagnostics. A dedicated in-vacuum beamline, including remotely adjustable mirror, fluorescent screen and diaphragm, has been constructed. It deflects the laser beam, outcoupled from the upstream cavity mirror, to a high

* Partly funded under EC contract No. HPRI-CT-2001-50025.

†Corresponding author. Tel. +39-040-3758224; fax. +39-040-3758653. E-mail address: bruno.diviacco@elettra.trieste.it

performance spectrometer providing a resolution of the order of 10^{-5}. A 1 GHz digital oscilloscope has also been purchased, improving the resolution in the measurement of the radiation temporal structure. Moreover, a dedicated vacuum chamber has been designed, constructed and will be soon installed, that will be used for a linewidth narrowing experiment with an intra-cavity Fabry-Perot etalon [3].

3. Storage Ring developments

The FEL performance is strongly influenced by the stability of the electron beam. Until recently, the only available method to obtain sufficient transverse and longitudinal stability consisted in temperature adjustment of the four radio-frequency ring cavities. Proper setting of these temperatures allows a cancellation of the high-order modes driving the multi-bunch instabilities. The availability of a new longitudinal feedback system [4] now provides a fully stable beam with minimum changes to the temperatures used during normal beamtime. The considerable reduction of the time needed to switch from users' to FEL mode of operation and the improved beam stability during the full stored current decay have obvious advantages for the applications. Figure 1 shows the reduction of the longitudinal coupled-bunch modes when the feedback is activated.

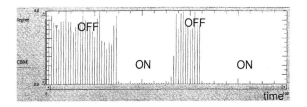

Fig. 1. Measured amplitude of the longitudinal coupled bunch modes with and without the longitudinal feedback.

4. Development of mirrors

Oxide-based (SiO_2, Al_2O_3) multilayer mirrors have been successfully used so far in the optical cavity, showing little or no sign of degradation even after significant use. However, high absorption of these materials in the VUV limits the central wavelenght to about 190 nm. The actual limits are being explored, with a new set of 187 nm mirrors to be tested soon. Below this wavelength the only credible candidate

materials are fluorides (MgF_2, LaF_3, AlF_3). For this reason a systematic study, to exposure of synchrotron radiation, of different single layers and substrates has been performed. The resulting damage level is being used to select the most suitable combination of materials for the realization of actual multilayer mirrors. Irradiation tests performed at increasing beam energies clearly show great sensitivity to the incident power level [5,6]. Despite the minimization of this effect by operating at the lowest possible energy (700 MeV), attempts to lase below 190 nm have failed because of the rapid loss of reflectivity. Work is in progress to improve the mirrors, minimizing the thickness of the most absorbing layers and/or using additional protective coatings.

5. Experimental station

An experimental station based on a Mott detector for spin polarization measurements has been moved from LURE to ELETTRA. It allows the study of the dynamics of magnetization reversal at the ps time scale [7]. Preliminary tests have shown the suitability of the source in terms of obtainable signal level, but also showed that in standard mode the signal-to-noise ratio is degraded by the irregular macrotemporal structure of the FEL beam. The measurements will soon be repeated in the Q-switched mode, from which a significant improvement is expected. In parallel, a new sample manipulator for sub-micron positioning is being developed that will allow to study microscopic specimens. Transportation of a cluster source from the Leicester University is also planned, in order to perform similar measurements on mass-selected Fe clusters [8].

6. References

[1] R. P. Walker et al., Nucl. Instr. and Meth. A 475 (2001) 20; M. Trovò et al., Nucl. Instr. and Meth. A 483 (2002) 157; G. De Ninno et al., Nucl. Instr. and Meth. A 507 (2003) 274.
[2] G. De Ninno et al., these proceedings.
[3] D. Garzella et al., these proceedings.
[4] D. Bulfone et al., PAC 2003 proceedings.
[5] S. Günster et al., SPIE proc. vol 4932, 422.
[6] A. Gatto et al., SPIE proc. vol. 4932, 366.
[7] F. Sirotti et al., Phys. Rev. B 61 (2000) R9221.
[8] C. Binns et al., J. Phys, Condens. Matter 15 (2003) 4287.

FREE ELECTRON LASERS 2003
E.J. Minehara, R. Hajima and M. Sawamura (Eds.)
Published by Elsevier B.V.

Start-to-end simulations for the FERMI project @ ELETTRA

S. Di Mitri*, R.J. Bakker, P. Craievich, G. De Ninno, B. Diviacco, L. Tosi,
V. Verzilov

Sincrotrone Trieste S.C.p.A, Strada Statale 14 - km 163,5 in AREA Science Park
34012 Basovizza, Trieste ITALY

Abstract

The FERMI@ELETTRA project is aiming, in a first stage, for the construction of a single-pass FEL in the wavelength range from 100 nm to 40 nm. As a next step, the energy range of the existing accelerator system will be increased to reach the final wavelength target of 1.2 nm. Theoretical studies have been carried out in order to support the initial phase of the project. In particular, we report here on first integrated tracking simulation from the cathode through the undulators performed using different numerical codes.

PACS codes: 41.60.Ap, 41.60.Cr,, 41.85.Ja, 52.59.Rz

Keywords: Beam quality, Bunch compression , Coherent synchrotron radiation, Saturation length, Power spectrum

1. Introduction

As a first step the FERMI project [1] is based on the existing 1.0 GeV Elettra Linac, provided with a new photoinjector and a bunch compressor to generate high quality beams for the production of 100 ÷ 40 nm radiation. In the simulations an X-band 4th harmonic cavity is used to linearize the bunch compression process in order to permit a uniform charge density in the bunch and to remove excessive current spikes which drive the CSR instability. The electron beam is taken to the surface undulator hall through a dogleg which is achromatic and nearly isochronous (R_{56} = -0.284 mm). A preliminary slice analysis is used for a time independent optimization of the undulators' parameters, in order to achieve the minimum saturation length in the time dependent mode.

2. Injector parameters

For a fixed beam energy and ID gap the FEL process strongly depends on the electron beam parameters. At the same time, the beam quality at the entrance of the undulator cascade is directly related to the injector parameters, where the sections following the gun may only deteriorate the beam quality. For these reasons excellent injector parameters are required, such as:

- Normalized transverse emittance $\approx 2 \times 10^{-6}$ m rad
- Energy spread $\approx 0.1 \div 0.4$ %
- Bunch charge = 1 nC
- Peak current = 100 A (about 10 ps bunch length)

3. Start-to-end simulations

Start-to-end simulations were carried out with such a beam using three numerical codes: *Astra* [2], *Elegant* [3] and *Genesis* [4]. *Astra* was used for the generation of the electron beam from a LCLS gun model, taking into account space charge forces and the consequent degradation of the transverse emittances [5]. The output of Astra is read by *Elegant*, which transports the bunch to the magnetic chicane with the appropriate correlation between the longitudinal phase space coordinates. Elegant takes into account CSR effects and allows to vary the

---

* E-mail: simone.dimitri@elettra.trieste.it

parameters of the chicane and those of the accelerating sections in order to optimize the energy spread and the bunch current distributions at the end of the linac. The bunch is then transported to the end of the dogleg at the entrance of the undulator cascade, where the input file for Genesis is generated.

In the Elegant simulations the longitudinal wakefields have been included for the X-band cavity and all the linac sections. The wakefields modify the phase space portrait and the energy spread distribution, leading to a 40% loss in the compression factor with respect to the unperturbed case, after a new re-optimization of the whole structure.

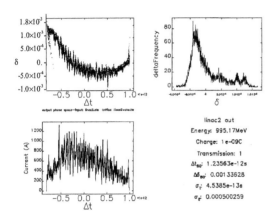

Fig. 1. Longitudinal phase space (up left), relative energy spread distribution (up right) and bunch current distribution (down left) at the end of the linac. Wakefields are included.

The beam slice analysis performed by Elegant was used by *Genesis* for a preliminary optimization of the saturation in the time independent mode. Genesis permits then a study of the radiation power and spectrum along the undulators in order to reach the maximum lasing power and/or the minimum saturation length in the time dependent mode. The simulated undulator chain is composed of APPLE-II type modules with period length of 62 mm and 40 periods each, providing a horizontal polarization with a maximum on axis field of 0.82 T and roll-off coefficients $K_x = 38$ m^{-1} and $K_y = 84$ m^{-1}. The optics is defined by a FODO type layout, giving an average beta function required for the saturation of 7.6 m.

Figure 3 shows the behaviour of the radiation power along the undulator cascade in the time independent mode without wakefields. Figure 4 depicts the radiation power in the time dependent mode close to the saturation length when the wakefields are included.

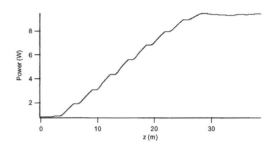

Fig. 2. Radiation power as a function of z. Wavelength = 98.263 nm, saturation length = 27.5 m, power at saturation = 2.2×10^9 W.

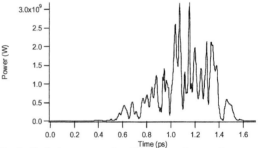

Fig. 3. Radiation power after 28 m (i.e. close to the saturation length). Wakefields are included.

According to these calculations, the time dependent simulations in absence and in presence of wakefields give essentially the same result.

4. References

1. R.J. Bakker, Fermi@Elettra: Project Update, this conference.
2. K.Flottmann, Astra User's Manual 3 1.0.0. *http://www.desy.de/~mpyflo/Astra_dokumentation/Manual.pdf.*
3. M.Borland, User's Manual for Elegant 14.8. *http://www.aps.anl.gov/asd/oag*
4. S. Reiche, Genesis 1.3 Operation Manual. *corona.physics.ucla.edu/~reiche/*
5. V.A.Verzilov, Photo-injector study for the ELETTRA linac FEL, this conference.

FREE ELECTRON LASERS 2003
E.J. Minehara, R. Hajima and M. Sawamura (Eds.)
Published by Elsevier B.V.

II-83

Upgrading Optical klystron FOR HEFEI SRFEL[*]

LI Ge, ZHANG Pengfei, CHEN Nian, HE Duohui, ZHANG Shancai[*]

National Synchrotron Radiation Laboratory of USTC, Anhui Hefei 230029, P.R. China

Abstract

A symmetry Optical Klystron (OK) is being upgraded to asymmetry structure for harmonic generation (HG) FEL. The period of independently adjusted OK modulator is extended from 7.2cm to 9.2cm for higher OK operation energy that can be easily matched by HLS storage ring. The upgraded OK parameters are listed. The magnetic gaps of the OK three sections can be independently controlled with fully closed loops. The Energy range of the upgraded OK in Hefei Light Source (HLS) for HG FEL Experiments is given and then, several experimental schemes are sorted out.

PACS codes: 52.59.Rz, 41.60.Cr

Keywords: Asymmetry Optical Klystron, Harmonic generation FEL

1. Basic Computation

One way of FEL is coherent harmonic generation by firing the electron beam passing through an OK with an external laser, which induces the energy modulation of the electron beam in the modulator by interaction with the optical field. The wavelength of seeding laser is given as $\lambda_s = [\lambda_m /(2\gamma^2)] \cdot (1 + k_m^2/2)$ (1). Here, λ_m is the modulator period, γ is Lorentz factor of the beam, $k_m = 0.934 \cdot (\hat{B}_y /T) \cdot (\lambda_m/cm)$ is the undulator deflection factor. It is transformed into density modulation (bunching) after passing the buncher. Then, the fundamental radiation is produced in the tuned radiator to a harmonic of the seed laser, which is given as [1-2] $\lambda_s /i = [\lambda_r /(2\gamma^2)] \cdot (1 + k_r^2/2)$, i=1,2,3,4,5. (2). Here, i is harmonic number, k_r is the radiator deflection factor.

The emitted peak intensity *of* nth harmonic on axis from the radiator in practical units is given by [3]:

$$\frac{dn_0 /dt}{d\Omega d\omega_n /\omega_n} = 1.744 \times 10^{14} N_r^2 (\frac{E}{GeV})^2 \frac{I}{A} F_n(K_r)/(s \cdot mrad^2 \cdot 0.1\% bw)$$

$$n=1,3,5,7\ldots\ldots \quad (3)$$

Where $F_n(K_r) = \frac{n^2 K_r^2}{(1 + K_r^2/2)^2} \{J_{\frac{n-1}{2}}[\frac{n \cdot K_r^2}{4(1 + K_r^2/2)}] - J_{\frac{n+1}{2}}[\frac{n \cdot K_r^2}{4(1 + K_r^2/2)}]\}^2$;

E, I, J_n are respectively the electron beam energy, beam current and nth Bessel functions.

For nth harmonic radiation of the electron bunch in Gaussian energy distribution, the spectral ratio of the coherent (with the laser on) and incoherent (without laser) intensities is given by[4] $R_n = N_e J_n^2(\Delta\xi) f_\gamma^2 /3$ (4). Here N_e is the total number of electrons in the bunch., $f_\gamma = \exp\{-[4\pi n(N + N_d)\sigma_\gamma /\gamma]^2/2\}$ is the energy spread factor, $\Delta\xi = 4\pi(N + N_d) \cdot \Delta\gamma_m /\gamma$ is the phase difference of electrons induced by energy modulation, $\Delta\gamma_m /\gamma = 2\pi N K_m a_s /(1 + K_m^2/2)$ is the induced maximum energy modulation of the electron beam with the optical field in the modulator, $a_s = eE_s /(mc^2 k_s)$ is the demensionless vector of the seeding optical field. N_d is the dimensionless parameter of dispersive section, given as [5]:

$$N_d = \frac{d}{2\lambda\gamma^2} \cdot \{1 + \frac{e^2}{dm^2 c^2} \int_0^d [\int_0^u B(z)dz]^2 du + \gamma^2 \vartheta^2\} \quad (5).$$

---

[*]Revised by LI Ge, Aug. 20, 2003.
lige@ustc.edu.cn, Projects of foundation of National Key Program for Basic Research of China (2001CCB01000), 211 Engineering of Chinese Key universities, Chinese high-tech 863 and 973.

1.1 Computation to the Upgraded Asymmetry OK

The technical parameters of the upgraded OK are computed and listed in table 1.

Table 1: the main parameters of the upgraded optical klystron

	Modulator	Buncher	Radiator
B_r/T	1.25	1.2	1.2
Gap/mm	36-140	39-140	36-140
Period/mm	92	216	72
period number	10	1	12
Peak Field/T	0.352-0.014	0.707-0.21	0.338-0.0036
Deflection factor	3.025-0.122	15.2-4.281	2.273-0.024
Gap resolution/μm	2.5	2.5	2.5
Field resolution /T	1.38×10^{-3}	4.28×10^{-3}	1.33×10^{-3}

1.2 Experimental Analysis to Upgrade schemes

Experimental Scheme of OK with the injected energy of HLS is first computed by analysis equation 1-5 for the case of $i=2$ and $n=1$[6-7]. The results are listed in table 2. Four experimental Schemes of OK with maximum beam energy are also sorted out using analysis equation 1-5, which are listed in table 3.

Table 2: FEL experiment in the injected energy condition of HLS

λ_s / nm	532	N_m	10
λ_s / i / nm	266	N_r	12
E/MeV	200	g_r/mm	70
g_m/mm	72.061	K_r	0.51368
K_m	1.24	P_s/MW	40
g_d/mm	71.6	$\Delta\gamma_m / \gamma$	0.002
B_d/T	0.47	$E^2 F_l / 10^3$	7.8332
N_d	62	R_l / N_e	0.34

Table 3: OK FEL experiments with maximum beam energy

No.	1	2	3	4
E/MeV	250.88	354.8	250.88	250.88
k_m	3.025	3.025	3.025	3.025
g_m/mm	36	36	36	36
k_d	18.6-4.281	18.6-4.281	18.6-4.281	18.6-4.281
g_d/mm	39-140	39-140	39-140	39-140
N_d	43.3-4.28	43.3-4.28	43.3-4.28	43.3-4.28
k_r	2.2636	2.273	1.2498	0.9217
g_r/mm	36.092	36.092	49.7	56.68
λ_s / nm	1064	532	1064	1064
I	2	2	4	5
λ_s / i / nm	532	266	266	212.8

2. Conclusions

2.1 For HG FEL experiments, the upgraded OK can be operated at any energy in HLS between 200MeV and 354.8MeV, which is resonant with both modulator and radiator of the OK. Selecting N_d matched with the seeding laser power could optimise maximum FEL photos.

2.2 Matching between the OK and the storage ring with electron beams is the key issue for successful FEL Experiments. Experiments show that HLS storage ring can be normally operated with the old OK above 530MeV without compensation. The present scheme, which use 9.2cm period PPM undulator as modulator of FEL experiments, still needs matching exploration between HLS storage ring and the upgraded OK because the highest energy of electron beam in this scheme is 354.8MeV even with extremism condition by closing the modulator gap to 36mm. Tune shifts caused by the inserted OK on the HLS storage ring still needs to be compensated while implementing the on going FEL experiments with the present schemes.

3. References

[1] S. V. Milton, E. Gluskin and N. D. Arnold etc. 'Exponential Gain and Saturation of a Self-Amplified Spontaneous Emission FEL', Science, Vol. 292, 15 JUNE 2001, PP2037-2041.

[2] L.-H. Yu, M. Babzien, I.Ben-Zvi etc. 'High Gain Harmonic Generation FEL', Science, Vol. 289, Aug. 11, 2000, PP932-934.

[3] X-Ray Data Booklet, BNL Jan. 2001.

[4] R. Prazeres, P. Guyot-Sionnest and J.M. Ortega et al. Production of VUV Coherent Light by HG with OK of Super-ACO, IEEE Journal of Quantum Electronics. Vol. 27. No 4, 1991.

[5] P. Elleaume, Optical Klystron, Journal de Physique, Colloque C1-333-352,supplement au1,Tome 45, janvier1984.

[6] CHEN, Nian et al. THE WORKING CONDITION ANALYSIS OF RECONSTRUCTED OK OF CHG-SRFEL, High Power Laser and Particle Beams v 13 n 2 Mar. 2001. p164-168.

[7] Walker R.P., Undulators and Wigglers, Proc. CERN Accelerator School, CERN 98-04 p.129.

FREE ELECTRON LASERS 2003
E.J. Minehara, R. Hajima and M. Sawamura (Eds.)
Published by Elsevier B.V.

Study of free electron laser and ultra-short pulse x-ray generation on NewSUBARU

S.Amano*, S.Miyamoto, T.Inoue, Y.Shoji, T.Mochizuki

Laboratory of Advanced Science and Technology for Industry, Himeji Institute of Technology
3-1-2 Kouto, Kamigori, Ako, Hyogo 678-1205, Japan

Abstract

A FEL experiment, using a NewSUBARU storage ring and an optical klystron (OK), is being carried out to achieve laser oscillation at around 488nm. The storage ring has a maximum energy of 1.5GeV and the OK is made of electromagnets with a variable period of 160 to 320mm. The FEL wavelength was designed to be from 10μm to 200nm with an electron energy of 0.5-1GeV. Design calculations for ultra-short pulse x-ray SR generation were also carried out. It was found that an x-ray pulse would be obtained with a critical photon energy of 0.69keV, a pulse width of 250fs and an average brightness of 3×10^6 photons/s/ mm^2/mrad^2/ 0.1%BW.

PACS codes: 41.60.Cr, 41.60.Ap, 29.20.Dh, 07.85.Qe

Keywords: free electron laser, storage ring, optical klystron, femtosecond x-ray pulse

1. Introduction

NewSUBARU is a 1.5GeV electron storage ring constructed in the Spring-8 site, Harima Science City, Japan. In NewAUBARU, the linac for Spring-8 is also used as an injector; NewSUBARU covers a region of photon energy ranging from infrared, visible, to soft x-ray, which is complementarily used with Spring-8. The aims of NewSUBARU are to promote industrial applications and R&D towards new light sources [1]. The NewSUBARU is a racetrack-type ring with a circumference of 118.7m; it has two 14-m and four 4-m straight sections. These 14-m sections are unique to this ring. An optical klystron (OK) was installed at one for the R&D of new light sources [2].

In this paper, we report free electron laser and ultra-short pulse x-ray generation using the OK.

2. Free electron laser

The main parameters of the NewSUBSRU ring and the OK are summarized in Table 1 and 2,

respectively. The OK, which is 10.9m, was developed for the free electron laser (FEL) with a wide photon energy range from the mid-infrared to the ultra-violet region. This OK is unique because its undulators are made of electromagnets with a variable period of 160 to 320mm by changing the current connection. In each period, the value of the K-parameter can be taken to be 0.3-4.4 and 0.3-11, respectively. Using this OK, the FEL wavelength was designed to be from 10μm to 200nm with an electron energy of 0.5-1GeV. In future, we will also challenge developments of a shorter wavelength FEL using a higher harmonic generation and a tunable g-ray source in the FEL-Compton backscattering process.

The FEL cavity is 29.7m-lomg, or one-quarter of the ring circumference. The cavity is a sable resonator, in which a laser beam waist is positioned at the center of the OK and its size is designed to be 0.6mm so that it overlaps with the electron beam efficiently. It was decided to first try FEL oscillation at a wavelength of 488nm with an electron energy of 1GeV. Two laser mirrors were then installed with

* Corresponding author, E-mail: sho@lasti.himeji-tech.ac.jp

reflectivities of 99% and 99.7% at 488nm, respectively.

Before FEL experiments, the performance of the OK was tested. Spontaneous emissions were observed for the varying values of the K-parameter by changing the OK magnet current. The measured spectral peaks were found to be in good agreement with the calculated results. From the spectra, we could estimate the relative energy spread of the electron beam, which was 6×10^{-4} at a storage current of 2.6mA. The bunch length of the electron beam was also measured with a streak-camera. Figure 1 shows the measured bunch length and peak current in a single-bunch mode, dependent on the NewSUBARU operation current. From these results, it was confirmed that the peak current of 30A could be obtained. A simple calculation gave the FEL gains of a few percent at least. We tuned the FEL cavity length precisely to synchronize the round-trip laser pulse with the electron bunches using a streak-camera [3]. The FEL experiment is now in progress to achieve laser oscillation at 488nm.

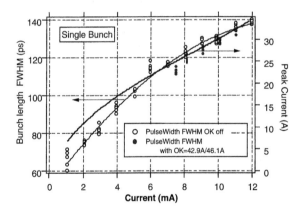

Fig. 1. Bunch length and peak current as a function of storage beam current.

Table 1
Parameters of the storage ring, NewSUBARU

Energy	$0.5 \sim 1.5$GeV
Critical Energy (@1.5GeV)	2.33keV
Circumference	118.716 m
RF frequency	500MHz
Harmonic number	198
Maximum stored current	500mA
Relative energy spread	$0.04 \sim 0.072\%$
Momentum compaction factor	$-0.001 \sim +0.001$

Table 2
Parameters of the optical klystron

Magnet		Electromagnet
Total length		10.92m
Undulator	Period (changeable)	160mm / 320mm
	Number of periods	32.5x2 / 16.5x2
	K-parameter	$0.3 \sim 4.4 / 0.3 \sim 11$
Dispersion region Nd		$6 \sim 136$
Resonant wavelength		200nm $\sim 10\mu$m

3. Ultra-short pulse x-ray generation

The OK can also work as an interactor between the electron beam and an external laser. By using a femtosecond laser of Ti:sapphire as an external laser, the energy at a small part of the electron bunch is modulated in the OK. This energy-modulated part is sliced from the other part of the electron bunch by a bending magnet and the sliced parts emit an ultra-short pulse of an x-ray SR. This laser-slicing technique has already been successful on ALS [4] and will be possible on the NewSUBARU. To evaluate the performance of this technique on the NewSUBARU, numerical designs were carried out [5]. As a result, an x-ray pulse will be obtained with a critical photon energy of 0.69keV and a pulse width of 250fs, and its average brightness is 3×10^6 photons/s/mm^2/mrad^2/0.1%BW and its average photon flux is 10^5 photons/s/0.1%BW when a Ti:sapphire laser pulse with a pulse energy 100μJ, a pulse width of 150fs, and a repetition rate of 20kHz is used with an 800μm-diaameter collimator. At this calculation, the electron beam energy was 1GeV.

References

[1] S. Hashimoto et al., Trans. Materials Research Soc. Japan 26 (2001) 783.

[2] S. Miyamoto et al., Proc. 13th International Conf. on High-Power Particle Beams, Nagaoka, Japan, June25-30, O-13-1 (2000).

[3] K. Yamada et al., Nucl. Instr. and Meth. A 304 (1991) 86.

[4] A. A. Zholents and M. S. Zolotorev, Phys. Rev. Lett. 76 (1996) 912; R.W. Schoenlein et al., Science 287(2000) 2237.

[5] T. Inoue, S. Miyamoto, S. Amano, T. Mochizuki, and M. Yatsuzuka, Rev. Laser Eng. 30 (2002) 531.

FREE ELECTRON LASERS 2003
E.J. Minehara, R. Hajima and M. Sawamura (Eds.)
Published by Elsevier B.V.

Development of the far-infrared FEL and SASE at ISIR, Osaka University

Goro Isoyama*, Ryukou Kato, Shigeru Kashiwagi, Shoji Suemine, Tetusya Igo,
Akihito Mihara, Chikara Okamoto,Takanori Noda, Kenji Kobayashi

Institute of Scientific and Industrial Research, Osaka University, 8-1 Mihogaoka, Ibaraki, Osaka 567-0047, Japan

Abstract

Reviewed briefly is development of the FEL and SASE in the far-infrared region using the L-band linac at the Institute of Scientific and Industrial Research (ISIR), Osaka University. A future plan is also presented.

PACS codes: 41.60.Cr; 29.17.+w; 07.57.Hm

Keywords: FEL, SASE, Far-Infrared, RF Linac

1. Introduction

We have been developing a far-infrared free electron laser (FEL) since the end of 1980s and also conducting basic study on SASE in the far-infrared region, using the 38-MeV, L-band (1.3 GHz) linac at the Institute of Scientific and Industrial Research (ISIR), Osaka University. The linac, which was built in 1978 mainly for radiation chemistry using pulse radiolysis, is equipped with a three-stage sub-harmonic buncher system, which makes it possible to accelerate the high-intensity single-bunch beam with charge up to 91 nC/bunch an well as a long pulse beam with bunch intervals as long as 9.2 ns. The single bunch beam is used for SASE and the long pulse beam for the FEL.

The 25 year old linac was not necessarily suitable for study to develop FEL and SASE, to say nothing of a user facility for FEL and SASE in the far-infrared region. The linac was one of the main facilities of the Radiation Laboratory attached to the Nanosicence and Nanotechnology Center newly found in ISIR. The Radiation Laboratory was formerly a laboratory directly attached to ISIR, but it was reorganized recently to be a part of the Nanoscience and Nanotechnology Center. To accomplish its missions, we had a budget to modify the linac for the new objective. The linac was completely remodelled and renewed, though superficial specifications are not much different or even worse, detailed of which are reported at this conference. We plan to establish a user facility of FEL in the far-infrared region in the near future using the renewed linac.

In this paper, we will briefly review our activities to develop FEL and to conduct basic studies on SASE as well as the modifications of the linac.

2. L-band linac

Since the linac is optimized for acceleration of the high-intensity single-bunch beam, the filling time of the RF power in the accelerating structure is as long as 2 µs and the duration of the macropulse for FEL is limited to approximately 2 µs on the condition that the pulse duration of the RF power is 4 µs long. Due to this limitation, we have not yet been able to obtain saturation of the FEL power. The linac is now renewed drastically and its commissioning is going on. The main objectives of the modification are to realize high stability of the accelerator or the electron beam, and an easy change of operation modes for various kinds of experiments, including FEL. Taking this opportunity, we replaced the klystron and the pulse modulator for it to provide the RF power with the pulse duration up to 8 µs to realize power

---

* Corresponding author, isoyama@sanken.osaka-u.ac.jp

saturation of the FEL. The modifications of the linac include replacement of almost all the power supplies in pursuit of higher stability and reproducibility, and introduction of a computer control system for easy operation of the system.

3. Basic study on SASE

We observed non-linear increase of infrared light intensities at 20 and 40 μm with charge in the single bunch beam passing through a 1.92 m long, 32 period wiggler with the period length of 6 cm used for FEL in 1992 [1], which was, to our knowledge, the first indication of SASE in the Compton Regime. After a break over some years, we began basic study of SASE again in the far-infrared region in 1999. First, we measured characteristics of the electron beam to be used for SASE, such as charge in a pulse, emittance, and the energy spread, and estimated expected performance of SASE using one-dimensional theory (1-D theory). The FEL parameter is 0.02, which is an order of magnitude larger than most of the on-going experiments and proposed SASE facilities, the gain length is 15 cm, and the power gain is ~10^6. We observed SASE in the wavelength region 100~200 μm and measured its intensity as a function of the wiggler K-value. The measured intensity variations agree well with the prediction of the 1-D theory. The wavelength spectrum of SASE was measured over the wide spectral range and we observed the second and the third harmonic peaks due to non-linear harmonic generation of SASE [2]. We measured angular distributions of SASE at the fundamental, the second harmonic and the third harmonic peaks, together with their spectral widths [3]. We are conducting SASE experiments to understand mechanism and physics involved in SASE. The wiggler we use for SASE experiments is only 2 m long and consequently we cannot obtain power saturation of SASE, though the power gain is extremely high. We plan to make a new wiggler for SASE and basic study for a new type of wiggler called the edge-focus wiggler is in progress.

4. Development of the far-infrared FEL

We obtained the first lasing of the FEL at wavelengths between 32 and 40 μm in 1994 and measured its characteristics [4]. The FEL gain was 68 %, the loss of the optical cavity was 6 %, and the energy in the macropluse is 0.8 mJ at 40 μm. The FEL is suitable for the long wavelength regions since the micropulse duration of the electron beam from the L-band linac, typically 20-30 ps, is significantly longer than that from the S-band linac, a few ps. We, therefore, began modifying the FEL system to expand the wavelength region towards the longer wavelength side beyond 100 μm wavelength. At that time, FELIX was the only one working beyond the wavelength among FELs based on RF linacs. The remodelling covers almost all the components, including the wiggler, the optical cavity, and the detection system. The system was modified progressively and experiments to expand the wavelength region were conducted in parallel. We obtained lasing at 126 μm in 1997 and at 150 μm the next year, which were, at that time, the longest wavelengths obtained with the FEL based on the RF linac [5-6].

As it was mentioned previously, the macropule of the electron beam was 2 μs long and it was not long enough to obtain power saturation of FEL, though the gain was relatively high. There are four optical pulses bouncing in the optical cavity and hence the number of amplification times is approximately 50. With the renewed linac, the macropluse will be as long as 6 μs in the long pulse mode, or the FEL mode, and we expect the FEL power will reach saturation. We plan to establish a user facility of FEL in the far-infrared region. The wavelength range will be 40~150 μm, which has been already achieved, and plan to conduct research and development to expand the wavelength region on the both sides. The average power is expected to be ~0.3 W at 10 Hz operation of the linac. The peak power will be 15 kW in the macropulse and 67 MW in the micropulse in a few ps duration.

5. References

[1] S. Okuda et al., Nucl Instr. and Meth. A331 (1993) 76.
[2] R. Kato et. al., Nucl Instr. and Meth. A475 (2001) 334.
[3] R. Kato et. al., Nucl Instr. and Meth. A507 (2003) 409.
[4] S. Okuda et. al., Nucl Instr. and Meth. A358 (1995) 224.
[5] R. Kato et. al., Nucl Instr. and Meth. A407 (1998) 157.
[6] R. Kato et. al., Nucl Instr. and Meth. A445 (2000) 169.

FREE ELECTRON LASERS 2003
E.J. Minehara, R. Hajima and M. Sawamura (Eds.)
Published by Elsevier B.V.

FERMI@ELETTRA: Project Update

R.J. Bakker*, C.J. Bocchetta, P. Craievich, M.B. Danailov, G. D'Auria,
B. Diviacco, S. Di Mitri, G. De Ninno, L. Tosi, V. Verzilov

Sincrotrone Trieste, S.S. 14 Km. 163.5 – in Area Science Park, 34012 Basovizza (Trieste), Italy

Abstract

The FERMI@ELETTRA project is an initiative from ELETTRA, INFM and other Italian institutes, to construct a single-pass FEL user-facility for the wavelength range from 100 nm (12 eV) to 1.2 nm (1 keV), to be located next to the third-generation synchrotron radiation facility ELETTRA in Trieste, Italy. The initial stage of the project is concentrated around the existing 1.2-GeV linac and aims for lasing and initial user experiments in the wavelength range from 100 nm to 10 nm.

PACS codes: 41.60.Cr, 42.72.Bj, 07.85.Fv, 29.17.+w

Keywords: Free-Electron Laser, User facility, VUV, Soft X-ray, Proposal

1. Introduction

FERMI@ELETTRA [1] is a research and proposal for the construction of a single-pass FEL user-facility for the Vacuum Ultra-Violet (VUV) to the X-ray spectral region, i.e., from 100 nm (12 eV) to 1.24 nm (1 keV). It is foreseen to build-up this machine in three successive stages, aiming for lasing from 100 nm to 40 nm, from 40 nm to 10 nm, and from 10 nm to 1.24 nm, respectively. The first two stages are based on use of the existing 1.2-GeV linac. For the latter stage, the energy of the linac will be extended up to 3.0 GeV. An overview is presented in Fig. 1.

The FEL is intended to operate as a user-facility. Hence, the target capabilities and specifications have been defined in close collaboration with potential users [2]. The most important machine properties are:

- An S-band accelerator with advanced feedback and feed-forward systems to improve the stability
- High-power short optical pulses (~100 fs) with a high pulse-to-pulse reproducibility
- APPLE II type undulators to enable flexible tuning of both the wavelength and the polarization
- Implementation of seeding schemes for further stabi-lization of the FEL process.

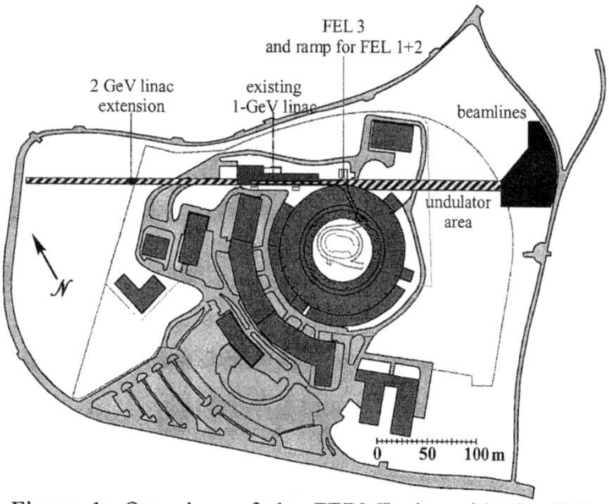

Figure 1: Overview of the FERMI site with the FEL positioned at the top from left to right and the ELETTRA storage ring below. An additional 350 m right of the user hall is reserved for future extensions.

The key-parameters are summarized in Tab. 1. To en-sure effective operation, the machine will be built using a progressive approach where users will gain access to the machine in an as early as possible stage. Feedback from this approach is then used to refine

*R.J. Bakker (rene.bakker@elettra.trieste.it), Sincrotrone
Trieste, S.S. 14 Km. 163.5 – in Area Science Park, 34012
Basovizza (Trieste), Italy, tel/fax: +39 040 375 8551/8028

the design for the successive stages of the project.

Presently, the project focuses on: (1) a detailed planning of the project, (2) a build-up of expertise in areas beneficial to FEL development and (3) steps for implementation of the FEL.

Table 1: FERMI@ELETTRA machine parameters

	Stage 1+2	Stage 3
Wavelength (nm)	100/40/10	1.2
Undulator period (mm)	62/51/51	36.6
Peak brightness*	0.5/1.7/6.0	170
Peak power (GW)	3	10
Beam Energy (GeV)	1.0	3.0
Norm. emittance (mm-mrad)	2.0	1.5
Peak current (kA)	0.6/1.0/2.5	3.5
Charge per pulse (nC)	1	1
Energy spread (MeV)	0.5/0.7/1.0	1.5
Repetition frequency (Hz)	10	10 – 50
FEL parameter r (10^{-3})	3.2/2.6/1.4	1.0
Gain length (m)	0.9/0.9/1.7	1.8

* (ph/s/mm^2/mrad^2/0.1%BW)($\times 10^{30}$)

2. Upgrade of the accelerator

To transform the present infrastructure into an FEL driver requires some modifications of which the essential ones are:

- Replacement of the existing electron beam source with a low-emittance.
- An upgrade of the RF system to improve stability. In addition, we intend to install an harmonic cavity to improve the bunching process.
- Modification of the electron beam optics, including bunch compressors, to increase the peak current.
- Installation of an undulator line, possibly added with an FEL seeding option.
- Installation of additional electron-beam and photon-beam diagnostics.

For ELETTRA, it is also important that the linac can serve as an injector for the storage ring until a new full-energy injector will take over this task in 2005.

A possible implementation of the 1-GeV accelerator section for a 10-nm FEL is sketched in Fig. 2. It illustrates the necessary:

- Installation of a high brightness electron source in parallel with the present injector (1,2). The drive laser for the new injector will be positioned in an adjacent shielded area (3).
- Moving of two accelerator sections downstream to create sufficient space for a harmonic cavity (4) and a magnetic bunch compressor (5). For phase 2 of the project, i.e., construction of a 40 nm to 10 nm FEL, a second magnetic bunch compressor (7) is needed to boost the beam current into the kA regime.
- Minor adjustments to the electron beam optics in the accelerating section as well as the installation of additional diagnostics to monitor electron beam and the bunching process.

Note that the scenario is intended for a 10-nm FEL user facility. A 1.24 nm FEL requires an extension of the electron beam energy and linac section.

The expected performance, based on analytical formulas, of the machine is summarized in Tab. 1. Details and more elaborate start-to-end simulations are presented in Ref. [4].

References

[1] C. J. Bocchetta et al, Nucl. Instr. Meth. in Phys. Research, **A507**, p. 484 (2003)

[2] FERMI@ELETTRA, Conceptual Design Report, Sincrotrone Trieste, Feb. 2002

[3] A Photo-Injector Study for the ELETTRA Linac FEL, V. Verzilov et al., this conference, Tu-P-70

[4] S2E Simulations for the FERMI Project, S. Di Mitri et al., this conference, We-P-28.

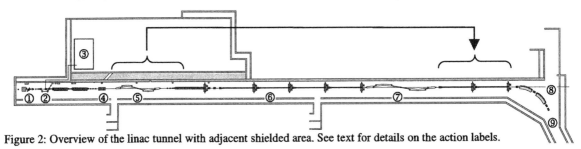

Figure 2: Overview of the linac tunnel with adjacent shielded area. See text for details on the action labels.

FREE ELECTRON LASERS 2003
E.J. Minehara, R. Hajima and M. Sawamura (Eds.)
Published by Elsevier B.V.

Start-End Simulations for the LCLS X-Ray Free-Electron Laser

S. Reiche[a], P. Emma[b], H.-D. Nuhn[b], C. Pellegrini[a]

[a]University of California, Los Angeles, CA 90095-1547, USA

[b]Stanford Linear Accelerator Laboratory, Stanford, CA 94309, USA

In this paper we evaluate the performance of the LCLS FEL, including all known physical effects -from the electron source to the undulator exit- which determine the X-ray pulse characteristics. The wavelength range considered is from 15 to 1.5 Ångstrom, with an extension to 0.5 Ångstrom using the third harmonic. The results of this work have been useful to identify areas in the LCLS design where improvements could be made, thus optimizing the system performance.

1. Introduction

The successful operation of Self-Amplifying Spontaneous Radiation Free-Electron Lasers (SASE FEL) down to a wavelength of 80 nm [1] are essential experimental results for a better understanding of the beam dynamics of the injector, linear accelerator and FEL and to the check of numerical codes with experimental data. The obtained experience on start-end simulations can be applied for LCLS [2], operating in the Ångstrom level.

We discuss the results of the start-end simulation results for the LCLS X-ray FEL. For the simulation we use the codes Parmela [3] for the injector, Elegant [4] for the LCLS beam line and Genesis 1.3 [5] for the FEL with macro particle distributions of $10^5 - 10^6$ particles.

2. The LCLS Design Case

The operation point of LCLS is limited by charge and energy. Beam diagnostics loose in resolution for less than 200 pC while bunches with more than 1 nC have stronger wakefield and CSR effects, resulting in an insufficient energy spread and emittance. The maximum beam energy is 14.5 GeV while the lower limit of 4.5 GeV is given by the final bunch compressor, which operates at this energy, and the stability of the quadrupole lattice along the undulator.

The results for the four corner points are shown in Fig. 1. Although the emittance scaling suggests

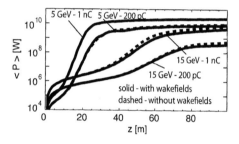

Figure 1. Radiation power for low and high charge case at 15 and 1.5 Å and with and without wakefields.

a better performance the performance is worse at lower charge. The reason is that for the simulation for 200 pC the input deck was modeled after the results of GTF [6], which measured a larger slice emittance than theoretically expected. Compared to the 1 nC, which use the theoretically value for the slice emittance, the beam brightness is reduced.

The radiation profile of each case does not reflect the current profile, because CSR [7] effects and undulator wakefields [8] spoil the beam quality and the amplification process is inhibited locally. The result for the 1 nC case at 1.5 Å is shown in Fig. 2, where gaps are visible at the beginning and end of the time-window, which covers

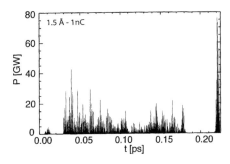

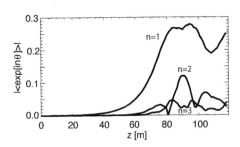

Figure 2. Radiation pulse profile at saturation at 1.5 Å and 1nC.

Figure 3. Evolution of the bunching factor at the fundamental and next two higher harmonics along the LCLS undulator.

the entire electron bunch. The impact by wakefields and CSR is much reduced at 15 Å due to the shorter saturation length and wider FEL bandwidth. The spectra are slightly shifted towards larger wavelength than 1.5 Å, which can be explained by the net loss due wakefields.

3. Emission at Higher Harmonics

The non-linear dynamics of the FEL process drive also the microbunching on harmonics of the resonant wavelength. This process dominates over the bunching of a higher harmonics due to the interaction with the harmonics in the radiation field [9]. For LCLS the modulation at the harmonics is about 10% for the second harmonic and 5% for the third (see Fig. 3). This allows for a succeeding undulator, tuned to the higher harmonics. The coherent emission is accumulated and a radiation power level at 0.5 Å in the GW range can be expected.

4. Conclusion

Start-end simulations show a successful operation of the LCLS X-ray Free-Electron Laser within the design operation space (4.5 – 14.5 GeV and 0.2 – 1.0 nC). While theory suggest lower values for slice emittance, the simulations at 200 pC were match to the experimental results of GTF with larger slice emittances. The satura-

tion length lies well within the length of the LCLS undulator. Finally, the harmonic content in the current modulation suggest a successful operation of a second undulator, tuned to one of the harmonics.

REFERENCES

1. V. Ayvazyan *et al.*, Phys. Ref. Lett. **88** (2002) 104802
2. *Linac Coherent Light Source (LCLS)*, SLAC-R-521, UC-414 (1998)
3. L. M. Young, J. H. Billen, Parmela, LA-UR-96-1835, Rev. Jan. 2000.
4. M. Borland, Elegant, APS LS-287, Sep. 2000.
5. S. Reiche, Nucl. Inst. & Meth. **A429** (1999) 243
6. D.H. Dowell *et al..*, Proc. of the 24th International Free Electron Laser Conference, Argone, 2002, USA
7. Y.S. Derbenev *et al.* TESLA-FEL 95-05, DESY, Hamburg, Germany (1995)
8. S. Reiche *et al.* Nucl. Inst. & Meth. **A475** (2001) 328
9. Z. Huang and K.-J. Kim, Phys. Rev. **E62** (2000) 7295

FREE ELECTRON LASERS 2003
E.J. Minehara, R. Hajima and M. Sawamura (Eds.)
Published by Elsevier B.V.

Start-End Simulations for the LCLS X-Ray Free-Electron Laser

S. Reiche[a], M. Borland[b], P. Emma[c], W. Fawley[d], C. Limborg[c], H.-D. Nuhn[c], C. Pellegrini[a]

[a]University of California, Los Angeles, CA 90095-1547, USA

[b]Argonne National Laboratory, Argonne, IL 60439, USA

[c]Stanford Linear Accelerator Laboratory, Stanford, CA 94309, USA

[d]Lawrence Berkeley National Laboratory, Berkeley, CA 94720, USA

The LCLS Free-Electron Lasers operates in the wavelength range of 1.5 - 15 Angstrom, using an electron beam with an energy between 4.5 and 14.5 GeV. The generation of the electron beam, the preservation of its brightness during acceleration and compression, and the amplification of the spontaneous radiation within the FEL can only be described by a consistent set of simulation codes.

We present the change in the FEL performance with respect to the LCLS design case, when various effects are included, altering the electron beam distribution and motion (e.g. wake fields, CSR, magnet misalignment or field errors of the undulator field). To distinguish the individual contribution of each effect, multiple start-end simulations are performed, including step by step additional effects and, thus, approaching a more and more realistic model of the LCLS FEL.

1. Introduction

Improvements of high brightness electron beam sources and the successful operation of Self-Amplifying Spontaneous Radiation Free-Electron Lasers (SASE FEL) down to a wavelength of 80 nm [1] are essential experimental results for a successful operation of proposed X-ray Free-Electron Lasers such as LCLS [2] and TESLA X-FEL [3]. These experiments have been sucessfully modeled with start-end simulations. In the LCLS case, discussed here, we are using Parmela for the injector, Elegant for the main linac and Ginger and Genesis for the FEL. We study the influence of various effects on the FEL performance: coherent synchrotron radiation, wakefields in the linac and undulator, misalignment of linear accelerator and undulator components, and tolerance studies on linac jitters and undulator field errors.

2. CSR and Undulator Wakefields

The emission of coherent synchrotron radiation [4] in the bunch compressor and the energy change during the FEL process due to undulator wakefields [5] have the strongest impact on the FEL performance, reducing the FEL output power by up 40%.

Simulations were done for a 50 fs subsection in the central and head region of the bunch, deriving the particle distribution from the Elegant output. The results are shown in Fig. 1. The Genesis 1.3 results have a lower saturation power, because undulator wakefields were included. The calculated radiation spectra were in good agreement

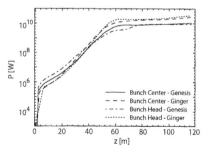

Figure 1. Radiation power along the undulator for a core and head region of the electron bunch. Undulator wakefields were included only in the Genesis simulations.

3. Beam Line Misalignment

For this study, transverse misalignments are added to all quadrupole magnets, accelerator structures, and beam position monitors (BPMs) with a Gaussian uncorrelated level of 300 microns rms. The emittance growth is effectively a head-to-tail varying centroid kick along the bunch length. Correction is accomplished by empirically varying the linac trajectory, using two pairs of steerers at the beginning of the linac, while minimizing the measured emittance at the end of the linac [6]. Final emittance levels are typically correctable to within $\approx 10\%$ of their initial values.

4. Jitter Tolerances

We estimate the jitter tolerance by 200 independent runs with Parmela [7], Elegant and Genesis, where various beam line parameters such as rf phases and voltages has been varied. The resulting tolerances has been reported elsewhere [8]. The strongest impact on the FEL performance is the resulting jitter in the beam centroid position at the entrance of the undulator (see Fig. 2). It results in a strong fluctuation of the FEL output power of about 25% rms.

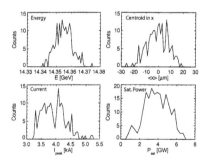

Figure 2. Jitter in the beam energy, beam centroid in x, current and saturation power for 200 random variations in beam line parameters.

5. Undulator Errors

For our study we include three sources of undulator errors: quadrupole misalignment, variation in the field strength of the undulator poles and misalignment of undulator modules. The tolerance of the BPM resolution for the beam-based alignment procedure[9]is 3 μm while the module alignment tolerance is 100 μm. With the achievable precision in the magnetic field measurement the resulting degradation is less than 15%.

6. Conclusion

The simulation codes Parmela, Elegant, Ginger and Genesis have been successfully used for start-end simulations of the LCLS X-ray Free-Electron Laser, including effects such as misalignment and beam jitter, the results yield tolerances of the machine alignment and stability. However a degradation of up to 60% and a fluctuation in the FEL output power of more than 25% cannot be removed.

REFERENCES

1. V. Ayvazyan *et al.*, Phys. Ref. Lett. **88** (2002) 104802
2. *Linac Coherent Light Source (LCLS)*, SLAC-R-521, UC-414 (1998)
3. TESLA-FEL 2001-05, Deutsches Elektronen Synchrotron, Hamburg, Germany (2001)
4. Y.S. Derbenev *et al.* TESLA-FEL 95-05, DESY, Hamburg, Germany (1995)
5. S. Reiche and H. Schlarb, Nucl. Inst. & Meth. **A445** (2000) 155
6. J. T. Seeman et al., 15th International Conference on High-Energy Accelerators, Hamburg, Germany, July 1992.
7. C. Limborg *et al.*, Proc. of the 25th Free-Electron Laser Conference, Tsukuba, Japan, 2003
8. P. Emma, Proc. of the Start-end Workshop, Desy-Zeuthen, Germany, 2003
9. P. Emma, R. Carr, H.-D. Nuhn, Nucl. Inst. & Meth. **A 429** (1999) 407

FREE ELECTRON LASERS 2003
E.J. Minehara, R. Hajima and M. Sawamura (Eds.)
Published by Elsevier B.V.

Short Wavelength Free Electron Lasers in 2003

W. B. Colson*

Physics Department, Naval Postgraduate School, Monterey CA 93943 USA

Abstract

Twenty-six years after the first operation of the short wavelength free electron laser (FEL) at Stanford University, there continue to be many important experiments, proposed experiments, and user facilities around the world. Properties of FELs operating in the infrared, visible, UV, and x-ray wavelength regimes are listed and discussed.

PACS codes: 41.60Cr

Keywords: free electron lasers

1. Introduction

The following tables list existing (Table 1) and proposed (Table 2) relativistic free electron lasers (FELs) in 2003. Each FEL is identified by a location or institution, followed by the FEL's name in parentheses.

The first column of the table lists the operating wavelength λ, or wavelength range. The large range of operating wavelengths, six orders of magnitude, indicates the flexible design characteristics of the FEL mechanism. In the second column, σ_z is the electron pulse length divided by the speed of light c, and ranges from almost CW to short sub-picosecond pulse time scales. The expected optical pulse length can be 3 to 5 times shorter or longer than the electron pulse depending on the optical cavity Q, the FEL desynchronism, and the FEL gain. If the FEL is in an electron storage-ring, the optical pulse is typically much shorter than the electron pulse. Most FEL oscillators produce an optical spectrum that is Fourier transform limited by the optical pulse length.

The electron beam energy E and peak current I provided by the accelerator are listed in the third and fourth columns. The next three columns list the number of undulator periods N, the undulator wavelength λ_0, and the undulator parameter $K=eB\lambda_0/2\pi mc^2$ where e is the electron charge magnitude, B is the rms undulator field strength, and m is the electron mass. For an FEL klystron undulator, there are multiple undulator sections as listed in the N-column. Note that the range of values

---

* Corresponding Author: 831-656-2765, Colson@nps.navy.mil

for N, λ_0, and K are much smaller than for the other parameters, indicating that most undulators are similar. Only a few of the FELs use the klystron undulator at present, and the rest use the conventional periodic undulator. The FEL resonance condition, $\lambda=\lambda_0(1+K^2)/2\gamma^2$ where γ is the relativistic Lorentz factor, provides a relationship that can be used to derive K from λ, E, and λ_0. The middle entry of the last column lists the accelerator type (RF for Radio Frequency Linear Accelerator, MA for Microtron Accelerator, SR for Storage Ring, EA for Electrostatic Accelerator), and the FEL type (A for FEL Amplifier, O for FEL Oscillator, S for SASE FEL, H for a high-gain high harmonic HGHG FEL). Most of the FELs are oscillators, but recent progress has resulted in short wavelength FELs using SASE (Stimulated Amplification of Spontaneous Emission). A reference [..] describing the FEL is provided at the end of each line entry.

For the conventional undulator, the peak optical power can be estimated by the fraction of the electron beam peak power that spans the undulator spectral bandwidth, $1/4N$, or $P\approx EI/4eN$. For the FEL using a storage ring, the optical power causing saturation is substantially less than this estimate and depends on ring properties. For the high-gain FEL amplifier, the optical power at saturation can be substantially greater. The average FEL power is determined by the duty cycle, or spacing between electron pulses, and is typically many orders of magnitude lower than the peak power. The TJNAF IRFEL has now reached an average power of 2 kW with the recovery of the electron beam energy in superconducting accelerator cavities.

In the FEL oscillator, the optical mode that best couples to the electron beam in an undulator of length $L=N\lambda_0$ has Rayleigh length $z_0\approx L/12^{1/2}$ and has a mode waist radius of $w_0\approx N^{1/2}\gamma\lambda/\pi$. The FEL optical mode typically has more than 90% of the power in the fundamental mode described by these parameters.

2. Acknowledgments

The author is grateful for support from ONR, NAVSEA, and the JTO.

3. References

[1] G. Ramian, Nucl. Inst. and Meth. **A318**, 225-229 (1992).

[2] E.A. Antokhim et. al, Nucl. Inst. and Meth. **AXXX**, xx (2004); Twenty-Fifth International Free Electron Conference, Tsukuba, Japan (Mo-O-03, Sept 2003).

[3] T. Mochizuki et. al., Nucl. Inst. and Meth. **A393**, II-47 (1997); S. Miyamoto and T. Mochizuki, J. Japan Society of Infrared Science and Technology, **7**, 73-78 (1997).

[4] N. Ohigashi et. al., Nucl. Inst. and Meth. **A375**, 469 (1996).

[5] S. Okuda et. al., Nucl. Inst. and Meth. **A341**, 59 (1994).

[6] R. Hajima et. al., Nucl. Inst. and Meth. **A507**, 115 (2003).

[7] P. Guimbal et. al., Nucl. Inst. and Meth. **A341**, 43 (1994).

[8] T. Takii et. al., Nucl. Inst. and Meth. **A407**, 21-25 (1998).

[9] M. Hogan et. al., Physical Review Letters **80**, 289 (1998).

[10] D. C. Nguyen et. al., Nucl. Inst. and Meth. **A429**, 125-130 (1999).

[11] K. W. Berryman and T. I. Smith, Nucl. Inst. and Meth. **A375**, 6 (1996).

[12] M. Hogan et. al., Physical Review Letters **81**, 4867 (1998).

[13] I. S. Lehrman et. al., Nucl. Inst. and Meth. **A393**, 178 (1997).

[14] J. Xie et. al., Nucl. Inst. and Meth. **A341**, 34 (1994)

[15] J. Auerhammer et. al., Nucl. Inst. and Meth. **A341**, 63 (1994).

[16] A. Doyuran et. al., Nucl. Inst. and Meth. **A475**, 260 (2001).

[17] A. Kobayashi, et. al., Nucl. Inst. and Meth. **A375**, 317, (1996).

[18] D. Oepts et. al., Infrared Phys. Technol. **36**, 297 (1995).

[19] S.V. Benson et. al., Nucl. Inst. and Meth. **A250**, 39 (1986).

[20] S. Benson et. al., Nucl. Inst. and Meth. **A429**, 27-32 (1999).

[21] H. A. Schwettman et. al., Nucl. Inst. and Meth. **A375**, 662 (1996);

[22] J. M. Ortega et. al., Nucl. Inst. and Meth. **A375**, 618 (1996).

[23] C. Brau, Nucl. Inst. and Meth. **A318**, 38 (1992).

[24] K. Batchelor et. al., Nucl. Inst. and Meth. **A318**, 159 (1992).

[25] D. Nolle et. al., Nucl. Inst. and Meth. **A341**, ABS7 (1994); Schmidt et. al., Nucl. Inst. and Meth. **A341**, ABS9 (1994).

[26] M. E. Couprie et. al., Nucl. Inst. and Meth. **A407**, 215-220 (1998).

[27] T. Tomimasu et. al., Nucl. Inst. and Meth. **A393**, 188-192 (1997).

[28] H. Hama et. al., Nucl. Inst. and Meth. **A341**, 12 (1994).

[29] K.Yamada et. al., Nucl. Inst. and Meth. **A475**, 205 (2001).

[30] M. Shane Hutson et. al., Nucl. Inst. and Meth. **A483**, 560 (2003).

[31] T. Shintake et. al., Nucl. Inst. and Meth. **A507**, 382 (2003); Tsukuba We-P-59 (Sept 2003).

[32] J. I. M. Botman et. al., Nucl. Inst. and Meth. **A341**, 402 (1994).

[33] E. D. Shaw et. al., Nucl. Inst. and Meth. **A318**, 47 (1992).

[34] G. R. Neil et. al., Phys. Rev.Lett. **84**, 662 (2000); Nucl. Inst. and Meth. **A507**, II-5 (2003).

[35] N. G. Gavrilov et. al., Status of Novosibirsk High Power FEL Project, SPIE Proceedings, vol. **2988**, 23 (1997); N. A. Vinokurov et. al., Nucl. Inst. and Meth. **A331**, 3 (1993).

[36] R. J. Burke et al, Proc. SPIE: Laser Power Beaming, Los Angeles, Jan. 27-28, 1994, Vol **2121**.

[37] S. Miyamoto et. al., Report of the Spring-8 International Workshop on 30 m Long Straight Sections, Kobe, Japan (August 9, 1997).

[38] R. P. Walker et. al., Nucl. Inst. and Meth. **A429**, 179-184 (1999).

[39] E. D. Johnson, Nucl. Inst. and Meth. **A393**, II-12 (1997).

[40] J. W. Lewellen et. al., Nucl. Inst. and Meth. **A483**, 40 (2002).

[41] F. Ciocci et. al., A. Torre, IEEE J.Q.E. **31**, 1242 (1995).

[42] J. Andruszkow et. al., Physical Review Letters **85**, 3825 (2000).

[43] V. N. Litvinenko et. al., Nucl. Inst. and Meth. **A358**, 369 (1995).

[44] W. Brefeld et. al., Nucl. Inst. and Meth. **A375**, 295 (1996).

[45] M. Cornacchia, Proc. SPIE 2998, 2-14 (1997); LCLS Design Study Report, SLAC R-521 (1998).

[46] R. Brinkmann et. al., Nucl. Inst. and Meth. **A393**, 86 (1997); TESLA Technical Design Report.

[47] Young Uk Jeong et. al., Nucl. Inst. and Meth. **A475**, 47 (2001).

[48] M. Yokoyama et. al., Nucl. Inst. and Meth. **A475**, 38 (2001).

[49] Y. Hayakawa et. al., Nucl. Inst. and Meth. **A483**, 29 (2002).

[50] A. Tremaine et. al., Nucl. Inst. and Meth. **A483**, 24 (2002).

[51] M. Abo-Bakr et. al., Nucl. Inst. and Meth. **A483**, 470 (2002); Tsukuba Mo-P-07,Mo-P-08,We-P-51 (Sept 2003).

[52] F. Gabriel et. al., Nucl. Inst. and Meth. **B161**, 1143 (2000).

[53] V. P. Bolotin et. al., Nucl. Inst. and Meth. **A475**, II-37 (2001).

[54] L. DiMauro et. al., Nucl. Inst. and Meth. **A507**, 15 (2003).

[55] M. W. Poole and B. W. J. McNeil, Nucl. Inst. and Meth. **A507**, 489 (2003).

[56] H. Koike et. al., Nucl. Inst. and Meth. **A483**, II-15 (2002).

[57] C. J. Bocchetta et. al., Nucl. Inst. and Meth. **A507**, 484 (2003); Tsukuba We-P-53 (Sept 2003).

[58] A. Renieri et. al., Nucl. Inst. and Meth. **A507**, 507 (2003).

[59] A. Doria et. al, Phys. Rev. Lett. **80**, 2841 (1998); in these proceedings.

[60] B.C. Lee et. al., in these proceedings.

[61] Z. T. Zhao et. al, in these proceedings.

[62] W.S. Graves et. al., in these proceedings.

Table 1: Short Wavelength Free Electron Lasers (2003)

EXISTING FELs	$\lambda(\mu m)$	σ_z(ps)	E(MeV)	I(A)	N	λ_0 (cm)	K(rms)	Acc.,Type
Italy(FEL-CAT)	760	15-20	1.8	5	16	2.5	0.75	RF,O[59]
UCSB(mm FEL)	340	25000	6	2	42	7.1	0.7	EA,O[1]
Novosibirsk(RTM)	120-180	70	12	10	6	12	0.71	RF,O[2]
Korea(KAERI-FEL)	97-300	25	4.3-6.5	0.5	80	2.5	1.0-1.6	MA,O[47]
Himeji(LEENA)	65-75	10	5.4	10	50	1.6	0.5	RF,O[3]
UCSB(FIR FEL)	60	25000	6	2	150	2	0.1	EA,O[1]
Osaka(ILE/ILT)	47	3	8	50	50	2	0.5	RF,O[4]
Osaka(ISIR)	40	30	17	50	32	6	1	RF,O[5]
Tokai(JAERI-FEL)	22	15	17	10	52	3.3	0.7	RF,O[6]
Bruyeres(ELSA)	20	30	18	100	30	3	0.8	RF,O[7]
Osaka(FELI4)	18-40	10	33	40	30	8	1.3-1.7	RF,O[8]
UCLA-Kurchatov	16	3	13.5	80	40	1.5	1	RF,A[9]
LANL(RAFEL)	15.5	15	17	300	200	2	0.9	RF,O[10]
Stanford(FIREFLY)	15-80	1-5	15-32	14	25	6	1	RF,O[11]
UCLA-Kurchatov-LANL	12	5	18	170	100	2	0.7	RF,A[12]
Maryland(MIRFEL)	12-21	5	9-14	100	73	1.4	0.2	RF,O[13]
Beijing(BFEL)	5-20	4	30	15-20	50	3	1	RF,O[14]
Korea(KAERI HP FEL)	3-20	10-20	20-40	30	30x2	3.5	0.5-0.8	RF,O[60]
Newport News(IRdemo)	6, 10	0.2	80	270	25	20	4.5	RF,O[34]
Darmstadt(FEL)	6-8	2	25-50	2.7	80	3.2	1	RF,O[15]
BNL(HGHG)	5.3	6	40	120	60	3.3	1.44	RF,A[16]
Osaka(iFEL1)	5.5	10	33.2	42	58	3.4	1	RF,O[17]
Tokyo(KHI-FEL)	4-16	2	32-40	30	43	3.2	0.7-1.8	RF,O[48]
Nieuwegein(FELIX)	3-250	1	50	50	38	6.5	1.8	RF,O[18]
Duke(MARKIII)	2.7-6.5	3	31-41.5	20	47	2.3	1	RF,O[19]
Stanford(SCAFEL)	3-13	0.5-12	22-45	10	72	3.1	0.8	RF,O[21]
Orsay(CLIO)	3-53	0.1-3	21-50	80	38	5	1.4	RF,O[22]
Vanderbilt(FELI)	2.0-9.8	0.7	43	50	52	2.3	1.3	RF,O[23]
Osaka(iFEL2)	1.88	10	68	42	78	3.8	1	RF,O[17]
Nihon(LEBRA)	1.5	1-10	86.8	10-20	50	4.8	0.92	RF,O[49]
BNL(VISA)	0.8	2	70.9	250	220	1.8	1.2	RF,S[50]
BNL(ATF)	0.6	6	50	100	70	0.88	0.4	RF,O[24]
Dortmund(FELICITAI)	0.42	50	450	90	17	25	2	SR,O[25]
BNL NSLS(DUVFEL)	0.266	0.7	172	500	256	3.9	0.7	RF,SH[54]
Orsay(Super-ACO)	0.3-0.6	15	800	0.1	2x10	13	4.5	SR,O[26]
Osaka(iFEL3)	0.3-0.7	5	155	60	67	4	1.4	RF,O[27]
Okazaki(UVSOR)	0.2-0.6	6	607	10	2x9	11	2	SR,O[28]
Tsukuba(NIJI-IV)	0.2-0.6	14	310	10	2x42	7.2	2	SR,O[29]
Italy(ELETTRA)	0.2-0.4	28	1000	150	2x19	10	4.2	SR,O[38]
Duke(OK-4)	0.193-2.1	0.1-10	1200	35	2x33	10	0-4.75	SR,O[30]
ANL(APSFEL)	0.13	0.3	399	400	648	3.3	2.2	RF,S[40]
DESY(TTF1)	0.109	0.5	233	300	492	2.73	0.81	RF,S[42]

Table 2: Proposed Short Wavelength Free Electron Lasers (2003)

PROPOSED FELs	$\lambda(\mu m)$	σ_z (ps)	E(MeV)	I(A)	N	$\lambda_0(cm)$	K(rms)	Acc.,Type
Tokyo(FIR-FEL)	300-1000	5	10	30	25	7	1.5-3.4	RF,O[56]
Netherlands(TEUFEL)	180	20	6	350	50	2.5	1	RF,O[32]
Rutgers(IRFEL)	140	25	38	1.4	50	20	1	MA,O[33]
Novosibirsk(RTM1)	100-200	50	14	20-100	2x36	12	1.2	RF,O[53]
Dresden(ELBE)	30-750	1-5	10-40	30	45	5	0.4-1.6	RF,O[52]
Daresbury(4GLS)	5-100	0.2-1	50	100	100	4	2	RF,O[55]
Dresden(ELBE1)	3-30	0.5-1	10-40	30	2x34	2.73	0.3-0.8	RF,O[52]
Novosibirsk(RTM)	2-11	20	98	100	4x36	9	1.6	RF,O[35]
Frascati(SPARC)	0.533	0.1	142	500	6x71	3	1.3	RF,S[58]
TJNAF(UVFEL)	0.25-1	0.2	160	270	60	3.3	1.3	RF,O[20]
Rocketdyne/Hawaii(FEL)	0.3-3	2	100	500	84	2.4	1.2	RF,O[36]
Harima(SUBARU)	0.2-10	26	1500	50	33,65	16,32	8	SR,O[37]
Shanghai(SDUV-FEL)	0.5-0.088	1	300	400	400	2.5	1.025	RF,O[61]
Frascati(COSA)	0.08	10	215	200	400	1.4	1	RF,O[41]
Daresbury(4GLS)	0.1	0.1-1	600	300	150	5	2	RF,O[55]
Daresbury(4GLS)	0.01	0.1-1	600	2000	1000	2	1	RF,S[55]
Duke(OK-5,VUV)	0.03-1	0.1-10	1200	50	4x32	12	3	SR,O[43]
DESY(TTF2)	0.006	0.17	1000	2500	981	2.73	0.9	RF,S[44]
Italy(SPARX)	0.0015	0.1	2500	2500	1000	3	1.2	RF,S[58]
BESSY(Soft X-ray)	0.0012	0.08	2300	3500	1450	2.75	0.9	RF,S[51]
Trieste(FERMI)	0.001-0.1	0.1	3000	2500	570-1140	3.5	1.2	RF,S[57]
RIKEN(SPring8 SCSS)	0.00036	0.5	1000	2000	1500	1.5	1.3	RF,S[31]
MIT(Bates X-Ray FEL)	0.0003	0.05	4000	1000	1500	1.8	2	RF,S[62]
SLAC(LCLS)	0.00015	0.07	14350	3400	3328	3	3.7	RF,S[45]
DESY(TESLA)	0.0001	0.08	30000	5000	4500	6	3.2	RF,S[46]

FREE ELECTRON LASERS 2003
E.J. Minehara, R. Hajima and M. Sawamura (Eds.)
Published by Elsevier B.V.

Multimode Simulations of a Short-Rayleigh Length FEL

J. Blau, G. Allgaier, S. Miller, T. Fontana, E. Mitchell, B. Williams,
P. P. Crooker and W. B. Colson[*]

Physics Department, Naval Postgraduate School, Monterey, California 93943 USA

Abstract

A high-power free electron laser (FEL) is being designed in collaboration with Jefferson Laboratory, University of Maryland and Advanced Energy Systems. The proposed system consists of a superconducting accelerator with energy recovery, a relatively short undulator, and a short-Rayleigh length resonator with cryogenic mirrors. Multi-mode simulations are used to study the effects of varying the electron bunch charge, beam radius, number of undulator periods, Rayleigh length, and cavity losses. New design parameters are proposed to reduce the system size and enhance the output power while minimizing the potential for mirror damage.

PACS codes: 41.60 Cr

Keywords: Free electron laser, Short Rayleigh length

1. Introduction

At the Naval Postgraduate School, we are designing a high-power free electron laser (FEL) in collaboration with Jefferson Laboratory, University of Maryland and Advanced Energy Systems [1]. The design is similar to the Jefferson Lab IR FEL, utilizing a superconducting accelerator with energy recovery [2]. To obtain high average power without damaging the mirrors in a compact system, we have proposed a short-Rayleigh length resonator and a short undulator. The electron beam is focused to a small waist to overlap the intense optical fields in the center of the undulator. We studied the effects of varying electron beam, undulator, and resonator parameters, using a three-dimensional, multi-mode simulation [3]. For each set of parameters, the FEL is started in weak fields and allowed to evolve over many passes to steady-state, and the single-pass extraction is calculated (the ratio of output optical power to initial electron beam power).

[*] Corresponding author. Tel: 831-656-2765;
Email: colson@nps.navy.mil.

2. Simulation results

Fig. 1 shows the simulation results for extraction η as a function of number of undulator periods N. For N<10, the FEL gain is below threshold. The optimal extraction $\eta \approx 3\%$ is achieved with N=15 periods. Further increasing the number of periods causes the extraction to slowly decline, due to lower optical power saturation. For N<20, there is very little optical power scraping at the ends of the undulator.

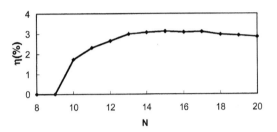

Fig. 1. Simulation results for extraction η versus number of undulator periods N.

Fig. 2 shows the simulation results for extraction η as a function of electron beam waist radius σ, normalized to $(\pi/L\lambda)^{1/2}$, where L is the undulator length and λ is the optical wavelength. The transverse emittance was held constant. The results show that for a large beam radius, the extraction drops due to reduced overlap with the intense optical fields in the center of the undulator, whereas for a small beam radius the extraction drops due to the corresponding large angular spread of the electron beam. Typically there is an optimum radius between these extremes; in this case it was $\sigma=0.12$, with an extraction of $\eta=3.8\%$.

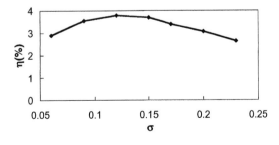

Fig. 2. Simulation results for extraction η versus normalized electron beam radius σ.

Fig. 3 shows the simulation results for extraction η and mirror intensity I as a function of the Rayleigh length z_o, normalized to L. The units of I are arbitrary, depending on the output power, mirror transmission and cavity length. As z_o is reduced, the extraction (solid line) slowly drops, but the mirror intensity (dotted line) falls rapidly, since the mode spot size at the mirrors increases with z_o. The horizontal dashed line indicates the maximum intensity that the mirrors are expected to withstand in our design, implying that the normalized Rayleigh length should be less than 0.06.

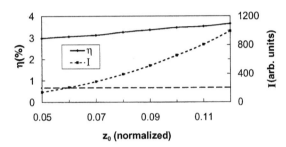

Fig. 3. Simulation results for extraction η and mirror intensity I versus normalized Rayleigh length z_o.

We also looked at varying the electron bunch charge, assuming constant emittance. The extraction increased linearly with bunch charge, but in a real experiment, emittance would typically increase with bunch charge, reducing the extraction.

Finally, we considered the mirror transmission losses. Our results show that reducing the losses improves the extraction, but at the expense of a rapid increase in intra-cavity power, so the losses must be > 50% to stay below the mirror damage threshold.

The authors are grateful for the support of NAVSEA and the Joint Technology Office.

3. References

[1] W. B. Colson, A. Todd, and G. R. Neil, Nucl. Inst. Meth. A483 (2002), II-9.
[2] G. R. Neil, et al., Phys. Rev. Lett. 84 (2000) 662.
[3] J. Blau, Ph.D. Dissertation, Naval Postgraduate School, http://handle.dtic.mil/100.2/ADA401739 (2002).

Author index

Printed and bound by CPI Group (UK) Ltd, Croydon, CR0 4YY

08/05/2025

01864932-0003